CHAPMAN & HALL/CRC COMPUTER and INFORMATION SCIENCE SERIES

Handbook of Algorithms for Wireless Networking and Mobile Computing

CHAPMAN & HALL/CRC
COMPUTER and INFORMATION SCIENCE SERIES

Series Editor: Sartaj Sahni

PUBLISHED TITLES

HANDBOOK OF SCHEDULING: ALGORITHMS, MODELS, AND PERFORMANCE ANALYSIS
Joseph Y.-T. Leung

THE PRACTICAL HANDBOOK OF INTERNET COMPUTING
Munindar P. Singh

HANDBOOK OF DATA STRUCTURES AND APPLICATIONS
Dinesh P. Mehta and Sartaj Sahni

DISTRIBUTED SENSOR NETWORKS
S. Sitharama Iyengar and Richard R. Brooks

SPECULATIVE EXECUTION IN HIGH PERFORMANCE COMPUTER ARCHITECTURES
David Kaeli and Pen-Chung Yew

SCALABLE AND SECURE INTERNET SERVICES AND ARCHITECTURE
Cheng-Zhong Xu

HANDBOOK OF BIOINSPIRED ALGORITHMS AND APPLICATIONS
Stephan Olariu and Albert Y. Zomaya

HANDBOOK OF ALGORITHMS FOR WIRELESS NETWORKING AND MOBILE COMPUTING
Azzedine Boukerche

CHAPMAN & HALL/CRC COMPUTER and INFORMATION SCIENCE SERIES

Handbook of Algorithms for Wireless Networking and Mobile Computing

Edited by

Azzedine Boukerche

University of Ottawa
Canada

Chapman & Hall/CRC
Taylor & Francis Group
Boca Raton   London   New York

Published in 2006 by
Chapman & Hall/CRC
Taylor & Francis Group
6000 Broken Sound Parkway NW, Suite 300
Boca Raton, FL 33487-2742

© 2006 by Taylor & Francis Group, LLC
Chapman & Hall/CRC is an imprint of Taylor & Francis Group

No claim to original U.S. Government works
Printed in the United States of America on acid-free paper
10 9 8 7 6 5 4 3 2 1

International Standard Book Number-10: 1-58488-465-7 (Hardcover)
International Standard Book Number-13: 978-1-58488-465-1 (Hardcover)
Library of Congress Card Number 2005049375

Library of Congress Cataloging-in-Publication Data

Handbook of algorithms for wireless networking and mobile computing / edited by Azzedine Boukerche.
    p. cm. -- (Chapman & Hall/CRC computer and information science series)
  Includes bibliographical references and index.
  ISBN 1-58488-465-7
    1. Wireless communication systems--Design and construction--Mathematics. 2. Mobile communication systems--Design and construction--Mathematics. 3. Mobile computing--Design and construction--Mathematics. I. Boukerche, Azzedine. II. Series.

TK5103.2.H335 2005
621.384--dc22
                                                        2005049375

Taylor & Francis Group
is the Academic Division of Informa plc.

Visit the Taylor & Francis Web site at
http://www.taylorandfrancis.com

and the CRC Press Web site at
http://www.crcpress.com

This book is dedicated to my parents, brothers and sisters with love

Editor

Prof. Azzedine Boukerche is a full professor at the University of Ottawa. He is the founding director of PARADISE Research Laboratory at the University of Ottawa and holds a Canada Research Chair in Large-Scale Distributed Interactive Simulation and Wireless Networking and Mobile Computing. Previously, he was a faculty member in the Department of Computer Sciences, University of North Texas, USA. Dr. Boukerche's current research interests include: wireless and mobile networking, wireless sensors, mobile ad hoc networks, wireless multimedia, distributed management and security systems for wireless and mobile networks, large-scale distributed interactive simulation, distributed and collaborative environment, distributed systems and parallel simulation. Dr. Boukerche has published several research papers in these areas. He was nominated for the best research paper award at the IEEE/ACM PADS '97, IEEE/ACM PADS '99, and ACM MSWiM 2001. In 1999 he was the recipient of the 3rd National Award for Telecommunication Software for his work on a distributed security system for mobile phone operations.

Dr. Boukerche has served as the general chair and program chair for several IEEE/ACM International conferences, and he is the co-founder of the 1st International Conference on QoS in Heterogenous Wired/Wireless Networks (QShine 2004). He is a Steering Committee Chair for ACM MSWiM Symposium, IEEE DS-RT Symposium, ACM Workshop on Mobility Management and Wireless Access, ACM Workshop on Performance Evaluation of Wireless Ad Hoc, Sensors and Ubiquitous Networks; and has been selected as an IEEE Computer Society Distinguished Visitor. He also serves as an Associate Editor for the *International Journal of Parallel and Distributed Computing (JPDC)*, *ACM Wireless Networks*, the *International* (Wiley) *Journal on Wireless Communication and Mobile Computing*, and the *SCS Transactions on Simulation*.

Contributors

Dharma P. Agrawal
Department of Electrical and
 Computer Engineering and
 Computer Science
University of Cincinnati
Cincinnati, Ohio

Mohammed Al-Kahtani
School of Information Technology
 and Engineering (SITE)
University of Ottawa
Ottawa, Ontario, Canada

G. Anastasi
Department of Information
 Engineering
University of Pisa
Pisa, Italy

Srinivas R. Avasarala
Department of Computer
 Sciences
Purdue University
West Lafayette, Indiana

Lichun Bao
Computer Science Department
School of Information and
 Computer Sciences
University of California
Irvine, California

Bharat Bhargava
Department of Computer
 Sciences
Purdue University
West Lafayette, Indiana

Luciano Bononi
Department of Computer Science
University of Bologna
Bologna, Italy

Azzedine Boukerche
University of Ottawa
Ottawa, Ontario, Canada

Tracy Camp
Department of Mathematics and
 Computer Sciences
Colorado School of Mines
Golden, Colorado

Xiang Chen
Wireless Networks Laboratory
Department of Electrical and
 Computer Engineering
University of Florida
Gainesville, Florida

Whai-En Chen
Department of Computer Science
 and Information Engineering
National Chiao Tung University
Hsinchu, Taiwan, R.O.C.

Imrich Chlamtac
Erik Jonsson School of
 Engineering and Computer
 Science
The University of Texas
Richardson, Texas

M. Conti
CNR-IIT Institute
Pisa, Italy

Fei Dai
Department of Computer Science
 and Engineering
Florida Atlantic University
Boca Raton, Florida

Samir R. Das
Department of Computer Science
State University of New York
Stony Brook, New York

Sajal K. Das
Center for Research in Wireless
 Mobility and Networking
Department of Computer Science
 and Engineering
The University of Texas at
 Arlington
Arlington, Texas

Hongmei Deng
Department of Electrical and
 Computer Engineering and
 Computer Science
University of Cincinnati
Cincinnati, Ohio

G. Ding
Department of Computer
 Sciences,
Purdue University
West Lafayette, Indiana

Lorenzo Donatiello
Department of Computer Science
University of Bologna
Bologna, Italy

Sonia Fahmy
Department of Computer
 Sciences
Purdue University
West Lafayette, Indiana

Kai-Wei Fan
Computer Science and
 Engineering
Ohio State University
Columbus, Ohio

Yuguang Fang
Wireless Networks Laboratory
Department of Electrical and
 Computer Engineering
University of Florida
Gainesville, Florida

Carlos M.S. Figueiredo
Federal University of Minas Gerais
Belo Horizonte, MG, Brazil
and
Research and Technological
    Innovation Center
FUCAPI, Brazil

J.J. Garcia-Luna-Aceves
Computer Engineering
    Department
Baskin School of Engineering
University of California at
    Santa Cruz
Santa Cruz, California

Nicolas D. Georganas
School of Information Technology
    and Engineering
University of Ottawa
Ottawa, Ontario, Canada

Andreas Görlach
Department of Computer Science
Technical University of Darmstadt
Darmstadt, Germany

Hossam Hassanein
Telecommunications Research Lab
School of Computing
Queen's University
Kingston, Ontario, Canada

Andreas Heinemann
Department of Computer Science
Technical University of Darmstadt
Darmstadt, Germany

Wendi B. Heinzelman
Department of Electrical and
    Computer Engineering
University of Rochester
Rochester, New York

Ahmed Helmy
Department of Electrical
    Engineering
University of Southern California
Los Angeles, California

Tingxue Huang
PARADISE Research Laboratory
School of Information Technology
    and Engineering (SITE)
University of Ottawa
Ottawa, Ontario, Canada

Ray Hunt
Department of Computer Science
    and Software Engineering
College of Engineering
University of Canterbury
Christchurch, New Zealand

Tom Jacob
Department of Computer Science
    and Engineering
University of North Texas
Denten, Texas

Evangelos Kranakis
School of Computer Sciences
Carleton University
Ottawa, Ontario, Canada

Victor C.M. Leung
Department of Electrical and
    Computer Engineering
The University of British
    Columbia
Vancouver, BC, Canada

Bo Li
Department of Computer Science
Hong Kong University of Science
    and Technology
Kowloon, Hong Kong

Xiang-Yang Li
Department of Computer Science
Illinois Institute of Technology
Chicago, Illinois

Yi-Bing Lin
Department of Computer Science
    and Information Management
Providence University
Taichung, Taiwan, R.O.C.

Sha Liu
Computer Science and
    Engineering
Ohio State University
Columbus, Ohio

Wei Liu
Wireless Networks Laboratory
Department of Electrical and
    Computer Engineering
University of Florida
Gainesville, Florida

Antonio A.F. Loureiro
Federal University of Minas Gerais
Belo Horizonte, MG, Brazil

Lewis M. MacKenzie
Department of Computing
    Science
University of Glasgow
Glasgow, Scotland

Marco Ajmone Marsan
Dipartimento di Elettronica
Politecnico di Torino
Torino, Italy

Geraldo R. Mateus
Federal University of Minas Gerais
Belo Horizonte, MG, Brazil

Michela Meo
Dipartimento di Elettronica
Politecnico di Torino
Torino, Italy

Jelena Mišić
Department of Computer Science
University of Manitoba
Winnipeg, Manitoba, Canada

Vojislav Mišić
Department of Computer Science
University of Manitoba
Winnipeg, Manitoba, Canada

Marco Mamei
Dipartimento di Scienze e Metodi
  dell'Ingegneria
Università di Modena e Reggio
  Emilia
Reggio Emilia, Italy

Hussein T. Mouftah
School of Information Technology
  and Engineering (SITE)
University of Ottawa
Ottawa, Ontario, Canada

Max Mühlhäuser
Department of Computer Science
Technical University of Darmstadt
Darmstadt, Germany

Eduardo F. Nakamura
Federal University of Minas Gerais
Belo Horizonte, MG, Brazil
and
Research and Technological
  Innovation Center
FUCAPI, Brazil

Asis Nasipuri
Department of Electrical and
  Computer Engineering
The University of North Carolina
  at Charlotte
Charlotte, North Carolina

Nidal Nasser
Department of Computing and
  Information Science
University of Guelph
Guelph, Ontario, Canada

Sotiris Nikoletseas
Computer Engineering and
  Informatics Department
Patras University, Greece

José Marcos S. Nogueira
Federal University of Minas Gerais
Belo Horizonte, MG, Brazil

Stephan Olariu
Department of Computer Science
Old Dominion University
Norfolk, Virginia

Mohamed Ould-Khaoua
Department of Computing
  Science
University of Glasgow
Glasgow, Scotland

Ai-Chun Pang
Graduate Institute of Networking
  and Multimedia
Department of Computer Science
  and Information Engineering
National Taiwan University
Taipei, Taiwan, R.O.C.

Stylianos Papanastasiou
Department of Computing
  Science
University of Glasgow
Glasgow, Scotland

A. Passarella
Department of Information
  Engineering
University of Pisa
Pisa, Italy

Mark A. Perillo
Department of Electrical and
  Computer Engineering
University of Rochester
Rochester, New York

Venkatesh Prabhakar
Department of Computer
  Sciences
Purdue University
West Lafayette, Indiana

Viktor K. Prasanna
University of Southern California
Los Angeles, California

Aniruddha Rangnekar
Department of CSEE
University of Maryland
Baltimore, Maryland

Mona El-Kadi Rizvi
Department of Physics, Computer
  Science and Engineering
Christopher Newport University
Newport News, Virginia

Linnyer B. Ruiz
Federal University of Minas Gerais
Belo Horizonte, MG, Brazil

Karim Seada
Department of Electrical
  Engineering
University of Southern California
Los Angeles, California

Tarek Sheltami
School of Information Technology
  and Engineering (SITE)
University of Ottawa
Ottawa, Ontario, Canada

Mitali Singh
Department of Computer Science
University of Southern California
Los Angeles, California

Prasun Sinha
Computer Science and
  Engineering
Ohio State University
Columbus, Ohio

Krishna M. Sivalingam
Department of CSEE
University of Maryland
Baltimore, Maryland

Paul Spirakis
Computer Engineering and
  Informatics Department
Patras University
Patras, Greece

Ladislav Stacho
Department of Mathematics
Simon Fraser University
Burnaby, British Columbia
Canada

Ivan Stojmenovic
School of Information Technology
    and Engineering (SITE)
University of Ottawa
Ottawa, Ontario, Canada

Riky Subrata
Parallel Computing Research
    Laboratory
School of Electrical, Electronic
    and Computer Engineering
The University of Western
    Australia
Perth, Australia

Violet R. Syrotiuk
Department of Computer Science
    and Engineering
Arizona State University
Tempe, Arizona

Glen Takahara
Department of Mathematics and
    Statistics
Queen's University
Kingston, Ontario, Canada

Wesley W. Terpstra
Department of Computer Science
Technical University of Darmstadt
Darmstadt, Germany

Di Tian
Distributed and Collaborative
    Virtual Environments Research
    Laboratory (DISCOVER)
School of Information Technology
    and Engineering
University of Ottawa
Ottawa, Ontario, Canada

Terence D. Todd
Electrical and Computer
    Engineering
McMaster University
Hamilton, Ontario, Canada

Upkar Varshney
Department of Computer
    Information Systems
Georgia State University
Atlanta, Georgia

Chonggang Wang
Department of Computer Science
Hong Kong University of Science
    and Technology
Kowloon, Hong Kong

Vincent W.S. Wong
Department of Electrical and
    Computer Engineering
The University of British
    Columbia
Vancouver, BC, Canada

Jenne Wong
Department of Computer Science
    and Software Engineering
College of Engineering
University of Canterbury
Christchurch, New Zealand

Jie Wu
Department of Computer Science
    and Engineering
Florida Atlantic University
Boca Raton, Florida

Yun Wu
Electrical and Computer
    Engineering
McMaster University
Hamilton, Ontario, Canada

Quincy Wu
Institute of Communication
    Engineering
National Chi Nan University
Puli, Taiwan, R.O.C.

X. Wu
Department of Computer
    Sciences
Purdue University
West Lafayette, Indiana

Kenan Xu
Department of Electrical and
    Computer Engineering
Queen's University
Kingston, Ontario, Canada

Ossama Younis
Department of Computer
    Sciences
Purdue University
West Lafayette, Indiana

Yang Yu
University of Southern California
Los Angeles, California

Franco Zambonelli
Dipartimento di Scienze e Metodi
    dell'Ingegneria
Università di Modena e Reggio
    Emilia
Reggio Emilia, Italy

Chu Zhang
Department of Electrical and
    Computer Engineering
The University of British
    Columbia
Vancouver, BC, Canada

Albert Y. Zomaya
School of Information
    Technologies
The University of Sydney
Sydney, NSW, Australia

Contents

1 Fundamental Algorithms and Protocols for Wireless Networking
   and Mobile Computing
   Azzedine Boukerche **1**-1

2 Wireless Communication and Mobile Computing
   Azzedine Boukerche, Antonio A.F. Loureiro, Geraldo R. Mateus,
   José Marcos S. Nogueira, and Linnyer B. Ruiz **2**-5

SECTION I MAC Protocols and Scheduling Strategies in Wireless Networks

3 Medium Access Control (MAC) Protocols for Wireless Networks
   Violet R. Syrotiuk **3**-23

4 Distributed Channel Access Scheduling for Ad Hoc Networks
   Lichun Bao and J.J. Garcia-Luna-Aceves **4**-45

5 Multiple Access Protocols and Scheduling Algorithms for Multiple
   Channel Wireless Networks   *Aniruddha Rangnekar,*
   Chonggang Wang, Krishna M. Sivalingam, and Bo Li **5**-77

6 Multichannel MAC Protocols for Mobile Ad Hoc Networks
   Asis Nasipuri and Samir R. Das **6**-99

SECTION II Routing Protocols and Location Awareness Strategies in Wireless Networks

7 Distributed Algorithms for Some Fundamental Problems in Ad Hoc
   Mobile Environments   *Sotiris Nikoletseas and Paul Spirakis* .   **7**-123

8 Routing and Traversal via Location Awareness in Ad Hoc Networks
   Evangelos Kranakis and Ladislav Stacho **8**-165

9 Ad Hoc Routing Protocols   *Kai-Wei Fan, Sha Liu, and*
   Prasun Sinha **9**-183

10 Cluster-Based and Power-Aware Routing in Wireless Mobile Ad Hoc
    Networks   *Tarek Sheltami and Hussein T. Mouftah* **10**-217

11 Broadcasting and Topology Control in Wireless Ad Hoc Networks
    Xiang-Yang Li and Ivan Stojmenovic **11**-239

12 Cross-Layer Optimization for Energy-Efficient Information
    Processing and Routing   *Yang Yu and Viktor K. Prasanna*   . .   **12**-265

13 Modeling Cellular Networks for Mobile Telephony Services
    Marco Ajmone Marsan and Michela Meo **13**-287

14 Location Information Services in Mobile Ad Hoc Networks
    Tracy Camp **14**-319

15 Geographic Services for Wireless Networks   *Karim Seada and
    Ahmed Helmy* **15**-343

16 Location Management Algorithms Based on Biologically Inspired
    Techniques   *Riky Subrata and Albert Y. Zomaya* **16**-365

17 Location Privacy   *Andreas Görlach, Andreas Heinemann,
    Wesley W. Terpstra, and Max Mühlhäuser* **17**-393

SECTION III   Resource Allocation and Management in Wireless Networks

18 Radio Resource Management Algorithms in Wireless Cellular
    Networks   *Nidal Nasser and Hossam Hassanein* **18**-415

19 Resource Allocation and Call Admission Control in Mobile
    Wireless Networks   *Xiang Chen, Wei Liu, and Yuguang Fang*   **19**-439

20 Distributed Channel Allocation Protocols for Wireless and Mobile
    Networks   *Azzedine Boukerche, Tingxue Huang, and
    Tom Jacob* **20**-459

SECTION IV   Mobility and Location Management in Wireless Networks

21 Adaptive Mobility Management in Wireless Cellular Networks
    Sajal K. Das **21**-477

22 Mobility Control and Its Applications in Mobile Ad Hoc Networks
    Jie Wu and Fei Dai **22**-501

23   Self-Organization Algorithms for Wireless Networks
     Carlos M.S. Figueiredo, Eduardo F. Nakamura, and
     Antonio A.F. Loureiro **23**-519

24   Power Management in Mobile and Pervasive Computing Systems
     G. Anastasi, M. Conti, and A. Passarella **24**-535

25   Collision Avoidance, Contention Control, and Power Saving
     in IEEE 802.11 Wireless LANs   *Luciano Bononi and*
     Lorenzo Donatiello **25**-577

SECTION V   QoS in Wireless Networks

26   QoS Enhancements of the Distributed IEEE 802.11 Medium Access
     Control Protocol   *Kenan Xu, Hossam Hassanein, and*
     Glen Takahara **26**-613

27   QoS Support in Mobile Ad Hoc Networks
     Mohammed Al-Kahtani and Hussein T. Mouftah **27**-633

28   QoS Provisioning in Multi-Class Multimedia Wireless Networks
     Mona El-Kadi Rizvi and Stephan Olariu **28**-647

SECTION VI   TCP Studies in Wireless Networks

29   TCP over Wireless Networks   *Sonia Fahmy, Ossama Younis,*
     Venkatesh Prabhakar, and Srinivas R. Avasarala **29**-679

30   TCP Developments in Mobile Ad Hoc Networks
     Stylianos Papanastasiou, Mohamed Ould-Khaoua, and
     Lewis M. MacKenzie **30**-699

SECTION VII   Algorithms and Protocols for Bluetooth Wireless PAN

31   Intra-Piconet Polling Algorithms in Bluetooth   *Jelena Mišić and*
     Vojislav B. Mišić **31**-725

32   A Survey of Scatternet Formation Algorithms for
     Bluetooth Wireless Personal Area Networks   *Vincent W.S. Wong,*
     Chu Zhang, and Victor C.M. Leung **32**-735

33   Bluetooth Voice Access Networks   *Yun Wu and*
     Terence D. Todd **33**-755

34 NTP VoIP Test Bed: A SIP-Based Wireless VoIP Platform
    *Quincy Wu, Whai-En Chen, Ai-Chun Pang, Yi-Bing Lin, and
    Imrich Chlamtac* **34**-773

35 Cross-Layer Algorithms for Video Transmission over Wireless
    Networks *G. Ding, Bharat Bhargava, and X. Wu* **35**-789

SECTION VIII Wireless Sensor Networks

36 Wireless Sensor Network Protocols *Mark A. Perillo and
    Wendi B. Heinzelman* **36**-813

37 Information Fusion Algorithms for Wireless Sensor Networks
    *Eduardo F. Nakamura, Carlos M.S. Figueiredo, and
    Antonio A.F. Loureiro* **37**-843

38 Node Activity Scheduling Algorithms in Wireless Sensor Networks
    Di Tian and Nicolas D. Georganas **38**-867

39 Distributed Collaborative Computation in Wireless Sensor Systems
    Mitali Singh and Viktor K. Prasanna **39**-885

SECTION IX Security Issues in Wireless Networks

40 Security Architectures in Wireless LANs *Ray Hunt and
    Jenne Wong* **40**-905

41 Security in Wireless Ad Hoc Networks *Hongmei Deng and
    Dharma P. Agrawal* **41**-937

42 Field-Based Motion Coordination in Pervasive Computing Scenarios
    Marco Mamei and Franco Zambonelli **42**-959

43 Technical Solutions for Mobile Commerce *Upkar Varshney* . **43**-983

Index **I**-997

1

Fundamental Algorithms and Protocols for Wireless Networking and Mobile Computing

Azzedine Boukerche

Recent advances in wireless networking and mobile computing technology coupled with the increasing proliferation of portable computers have led development efforts for mobile systems toward *distributed* and *mobile computing* — a new dimension for future generation of mobile computing networks. Wireless and mobile technology is being used to link portable computer equipment to corporate distributed computing and other sources of necessary information. Many researchers and scientists from both academia and industry are undertaking efforts to explore new technology for mobile computing and wireless communication, in a heterogeneous wireless and wired network environment, with mobile computing applications.

Most of the available literature in this technology concentrates on the physical aspects of the subject, such as spectrum management and cell reuse, just to mention a few. However, in most of the literature, a description of *fundamental distributed algorithms* that support mobile host in a wireless environment is either not included or briefly discussed. As such, there is a need to emphasize other aspects of wireless networking and mobile computing systems. Another important requirement for successful deployment of mobile and wireless applications is the careful evaluation of performance and investigation of alternatives algorithms, prior to implementation. Consequently, research in the emerging wireless communication and mobile computing field needs to cover a wide area of topics and scenarios.

The primary goal of this book is to focus on several aspects of mobile computing and in particular algorithmic methods and distributed computing with mobile communication capability. We believe that these topics are necessary to set the background and build the foundation for the design and construction of future generation of mobile and wireless networks, including cellular networks and wireless ad hoc, sensor, and ubiquitous networks. This book comprises 43 chapters written by a group of well-known experts in their respective fields. All chapters cover a variety of topics, and provide not only a good background but also an in-depth analysis of fundamental algorithms for the design and development of wireless and mobile networks, in general.

The rest of this book is organized as follows. In Chapter 2, we discuss the convergence between communication and computation, followed by the basic background in wireless networks and a brief overview of different wireless technologies. Chapter 3 introduces the basic techniques of medium access. Medium access control (MAC) protocols for centralized network architectures, such as cellular networks, paging systems, and satellite networks, are overviewed. A discussion of MAC protocols in distributed network architectures such as mobile ad hoc networks (MANETs) and sensor networks is provided. The protocols are largely classified according to whether time is viewed as continuous or discrete, giving rise to contention (or random access) and allocation (or scheduled) protocols, respectively. Research issues for MANETs are explored since these networks are among the least mature. Chapter 4 takes an in-depth look into distributed scheduling (also called topology-dependent, or spatial reuse TDMA) protocols for MANETs, providing an example of node-activation, link-activation, as well as a hybrid. In addition, a throughput analysis is provided for each protocol, as well as its channel access probability compared to basic random access protocols. A throughput comparison obtained through simulation is also presented. While Chapters 3 and 4 focus on a single broadcast channel shared by all the nodes in the network, Chapters 5 and 6 extend the protocols for the case when multiple channels are available; this may arise response to applications demanding higher bandwidth. Specifically, Chapter 5 summarizes extensions of several contention and allocation MAC protocols proposed for wireless networks to the multichannel case. Chapter 6 focuses more narrowly on multichannel MACs for MANETs, and primarily on dynamic channel selection in multichannel MAC protocols.

Chapter 7 addresses several important algorithmic issues arising in mobile ad hoc computing environment, and highlights the main differences to classical distributed computing. Unlike cellular and satellite networks wireless and mobile ad hoc networks are rapidly deployable networks that do not rely on a pre-existing infrastructure. MANET finds applications to search and rescue operation, law enforcement efforts, and on-the-fly collaborative environments such as, multimedia classrooms where the participating students, equipped with laptops, self-organize into a short-lives MANET for the purpose of exchanging class notes or other teaching related material. However, before MANETs become a common place, several technical challenges must be overcome. Chapter 8 presents an in-depth analysis of the basic information dissemination protocols via location awareness in wireless ad hoc networks. Chapters 9 and 10 present several key routing protocols that have been proposed for one-to-one (unicast), one-to-many (multicast), and one-to-all (broadcast) communication in wireless ad hoc networks. A more detailed look at some of the recent work on cluster-based, power-aware, and hierarchical unicast routing protocols is presented in Chapter 10, and multicast and broadcast routing protocols are discussed in Chapter 11. Chapter 12 present cross layers optimization mechanisms for energy efficient information processing a and routing.

In cellular networks, the presence of the infrastructure has a twofold effect. On the one hand, it allows the effective exploitation of the still limited resources at the wireless access. On the other hand, the system cannot rely on protocols dynamically adapting to the network and traffic conditions, but, rather, careful planning and design of mobility management, location management, and resource allocation strategies are needed. Chapter 13 reports on a number of analytical models for the design and planning of cellular systems. The models allow the prediction of the system performance subject to given input traffic specifications and user mobility profile. By means of performance predictions, the comparisons between system configurations can be easily done, as well as the computation of the number of radio resources required to meet some design objectives. Thanks to theirs simplicity, the models can be tailored to describe different scenarios, which include users with different degree of mobility, system configurations with channel reservation to handovers, systems providing services with different bandwidth requirements. In these scenarios, a proper exploitation of the knowledge of the hosts' geographic location may partially compensate the lack of infrastructure and the variable topology in wireless ad hoc networks.

Basic geographic services and protocols are described in Chapters 14–15, which focuses on network layer services and presents three families of geographic protocols. The first family of protocols are about geographic routing. These protocols consist in using location information to efficiently discover the path from a source node to a destination; they require nodes to maintain knowledge of the location of both their neighboring nodes and the destination node. As a second kind of service, geocasting protocol is described. Geocasting is used to deliver to all nodes in a given area some information relevant for the area itself. Finally, geographic rendezvous mechanisms are reviewed. In this case, resource and service discovery, as well as efficient data dissemination, are performed by means of rendezvous based on geographic location. Chapter 14 deals also with the location-based services, where knowledge of the position of the nodes is exploited in routing protocols so that a number of benefits can be achieved. Location information services are classified in two categories: proactive and reactive location services. According to proactive schemes, nodes periodically exchange location information; on the contrary, location information is exchanged only when needed in reactive services. Two families of proactive schemes can be further identified. Proactive location database systems, in which some nodes act as servers of a distributed database system, and proactive location dissemination systems, in which all nodes periodically receive updates on the nodes' location. Chapter 16 discusses the use of biological inspired techniques to solve the location management (or cell planning) problem. The goal mobility tracking or location management is to balance the registration and search operation so as to minimize the cost of mobile terminal location tracking. Chapter 17 presents an in-depth analysis of location privacy.

Next generations of Wireless Cellular Networks (WCNs), including 3G and 4G technologies are envisaged to support more mobile users and variety of Wireless Multimedia Services (WMSs). WMS enables the simultaneous transmission of voice, data, texts, and images through radio links by means of the new wireless technologies. With an increasing demand for WMSs, better Radio Resource Management (RRM) is needed to improve system performance by maximizing the overall system capacity and maintaining the Quality of Service (QoS) of multimedia traffic. RRM in present cellular systems has essentially been optimized for voice services. However, this is not valid if WMSs are to be supported, since these have different traffic characteristics (voice, data, or video) that need to meet their QoS requirements. These QoS can be parameterized as throughput, delay, delay variation (jitter), loss and error rates, security guarantees, etc. Therefore, the need for new RRM strategies to satisfy diverse QoS requirements of WMSs becomes more important. Chapters 18 and 19 address the problem of developing efficient resource management techniques for next generation WCNs. These chapters discuss some representative research works in this area. Chapter 20 presents several well-known channel and resource allocation protocols for wireless and mobile networks.

The next three chapters investigate issues related to mobility with the goal to devise techniques to keep associated uncertainties under control. Chapter 22 identifies the uncertainty due to user mobility as the complexity of the location management problem in cellular environment and presents a set of optimal location tracking protocols based on the information theory paradigm. Chapter 21 addresses issues related to mobility control in MANETs. It focuses on two localized protocols, one form connected dominant set and the other for topology control. Chapter 23 presents a set of self-organization algorithms for wireless networks.

The limited energy mobile devices constitute another challenging issue that needs to be taken seriously before wireless and mobile network become a common place. Chapter 24 provides an up-to-date state of the art of power management algorithms for mobile and pervasive computing environments, and highlights some open issues. Whilst, Chapter 25 presents well-known collision avoidance, contention control and power saving in IEEE 802.11 wireless LAN.

The next three chapters in this book, address other important aspects of QoS provisioning in wireless networks. Specifically, Chapter 26 discusses enhancements to the IEEE standard 802.11 MAC protocol to enable it to provide some elements of QoS at the MAC layer. The original version of IEEE 802.11 was notorious for lacking in the areas of both fairness and QoS provisioning. Later enhancements have addressed this deficiency to various degrees. Chapter 27 addresses a companion problem — namely that of providing QoS support in MANET. Chapter 28 presents an in-depth analysis of QoS provisioning in multiclass multimedia wireless networks.

The next two chapters compare and contrast a number of wireless TCP proposals. Infrastructure and ad hoc networks are environments with varying characteristics unlike those of wired LANs, which TCP was originally designed for, and as a result TCPs wireless performance is less than optimal. Several alterations and additions to TCP have been proposed in literature to allow its smoother and more efficient operation over wireless media and Chapters 29 and 30 provide a concise and up-to-date overview of wireless TCP. Specifically, Chapter 28 issues and proposed solutions in wireless environments that presuppose existing infrastructure to some degree, whilst Chapter 30 deals with the challenges and optimizations for TCP in MANETs.

The two chapters are complementary, although each may be read and understood on its own.

Chapters 31 and 32 review fundamental algorithms for bluetooth wireless personal area networks (WPAN). Bluetooth is an emerging standard for WPAN: short range, ad hoc wireless networks. However, the number of possible uses of bluetooth have increased, which includes several networking tasks as illustrated in Chapter 31. Fundamental intra-piconet polling algorithms in bluetooth are presented in the same chapter. Chapter 32 describes both the asymmetric and symmetric link establishment procedures for piconet formation, and present a set of well-known scatternet formation algorithms for bluetooth wireless personal area networks.

The next three chapters investigate future voice and video access networks. Chapter 33 provides an overview of recent work in the area of real-time voice over blue tooth, whilst Chapter 34 discusses a novel approach to support real-time VoIP through the General Packet Radio Service (GPRS) that interworks with the NTP VoIP testbed. Chapter 34 reviews current cross layer algorithms for video transmission over wireless networks, including application-based, network-based, and cross layer approaches.

Chapter 36 describes protocols and algorithms that are used to provide a variety of services in wireless sensor networks. Wireless sensors have received a great deal of attention in recent years, mainly due to their diverse applications and their ability that connects the physical world to the virtual world. Chapter 37 reviews information fusion algorithms for wireless sensor networks, while Chapter 38 presents fundamental node activity scheduling algorithms that have been proposed in the literature. In Chapter 39, distributed collaborative computation in wireless sensor systems is discussed.

It is well-known that the future of wireless networks depends greatly on how the security and privacy will be overcome. Chapter 40 discusses the existing and proposed WLAN security technologies designed to improve the 802.11 standard including some of the current proposals coming from IEEE 802.11 security working group. Whilst, Chapter 41 provides a general understanding on the security issues in wireless ad hoc networks.

The last two chapters investigate two potential applications in pervasive computing and mobile e-commerce. Chapter 42 describes the problem of motion coordination in pervasive computing scenarios, while Chapter 43 addresses several technical challenges in mobile commerce including infrastructure issues, location management, and transactions' support.

The Editor believes that this book covers the basic and fundamental algorithms and protocols for wireless and mobile networks that can be used as a textbook for a number of computer sciences, computer engineering, and electrical engineering undergraduate and graduate courses in wireless networking and mobile computing.

Special thanks are due to all authors for their thorough chapters and the reviewers for their hard work and timely report, which made this book truly special. Last but certainly not the least, our thanks to Prof. S. Sahni, for his encouragement, support, and guidance throughout this project.

2

Wireless Communication and Mobile Computing

2.1	The Convergence between Communication and Computation	2-6
2.2	Cellular Networks	2-6
	The 2G Cellular Mobile Networks • 2.5G Cellular Mobile Networks • 3G Cellular Mobile Networks • 4G Cellular Mobile Networks	
2.3	Wireless Local Area Networks	2-11
	IEEE 802.11 • HiperLAN • Digital Enhanced Cordless Telecommunications	
2.4	Short-Range Wireless Communication	2-13
	Wireless Personal Area Networks • ZigBee and IEEE 802.15 • Ultra Wideband • Infrared Data Association • HomeRF	
2.5	Wireless Metropolitan Area Networks (WiMAX)	2-16
2.6	Ad Hoc Networks	2-17
	Mobile Ad Hoc Networks • Wireless Sensor Networks (WSNs)	
2.7	Satellite Communication	2-18
2.8	Conclusion	2-18
	References	2-19

Azzedine Boukerche
Antonio A.F. Loureiro
Geraldo R. Mateus
José Marcos S. Nogueira
Linnyer B. Ruiz

In 1860s, the Scottish physicist James Clerk Maxwell predicted the existence of propagating electromagnetic waves, which were first demonstrated by the German physicist Heinrich Rudolf Hertz in 1888. This was just the beginning of a long list of achievements by other scientists and engineers in more than 100 years of wireless communication.

While wired communication networks have been the dominant technology used since the beginning of the Arpanet, during the last decade, we have seen a fast convergence between wired and wireless data communications. The field of wireless and mobile computing emerges from the integration among personal computing, distributed computing, wireless networks, and Internet applications. This convergence is supported by a plethora of devices and wireless networks, which is based on a continuously and increasing interaction between communication and computing. The success of mobile data communication lies in the vision to provide different services to users anytime and anywhere.

Currently, there are different kinds of wireless networks spanning from the well-known infrastructured cellular networks to infrastructureless wireless networks. Typical examples of the former class are 2G, 2.5G, 3G cellular networks and wireless local area networks, and of the latter class are mobile ad hoc networks and sensor networks. This scenario raises a number of interesting and difficult algorithmic issues in different areas such as routing, location management, resource allocation, information retrieval, network connectivity, reliability and security, and energy consumption. This book is intended to cover contributions of different algorithms in the context of mobile, wireless, ad hoc, and sensor networks.

2.1 The Convergence between Communication and Computation

The current infrastructured wireless technology extends the reach of the Internet but provides a limited mobility to users because of the limited coverage, that is, users are still connected to a fixed location. This may be acceptable in many situations but there will be scenarios where users will desire to stay connected to the Internet even as they move around. It is the user's demand for mobility and computation that will forge this convergence. In this scenario, the challenge is to design new applications and services for mobile users that will enable broadband wireless communications on a wide scale. This is an important point to be considered since the current 1 billion Internet users are increasingly demanding wireless Internet access.

Probably there will be no single technology that will dominate the world of wireless communications. We can expect to have coexisting and overlapping technologies that will create a wireless high-speed communication infrastructure to Internet access worldwide, where each technology is important in a given segment. For instance, Wi-Fi is strengthening its presence in hotspots and will likely become a standard interface while using laptops and PDAs.

Handset manufactures will likely integrate multiple wireless technologies into their mobile devices to maximize user ability to stay connected. Thus, for instance, we can expect the emergence of dual-mode cellular and Wi-Fi phones, given the support from chip vendors and handset manufacturers and the interest from telecom operators in these hybrid type of devices. Users will probably mix and match mobile devices and wireless technologies to meet their own requirements, thereby, enabling them to stay connected virtually anytime and anywhere.

Mobility and convenience are basically the main benefits of wireless communication provided to users [?]. Handheld devices and laptop computers are currently available with wireless network cards that allow them to stay connected at the office, home, and other places without needing to be wired-up to a traditional network. On the other hand, data throughput in wireless networks is much slower when compared to a wired network such as a Gigabit Ethernet (1 Gbps).

All of these wireless technologies are useless without the availability of interesting services and applications for users, which in turn are based on different kinds of algorithms. The primary goal of this book is to focus on several aspects of wireless networking and mobile computing and in particular algorithmic methods and distributed computing with mobile communication capability.

Before, we proceed further, we wish to present a brief overview of some of the wireless technologies currently available. The goal is to provide a brief description and present some of the main characteristics that may be important in the design of algorithms for a particular technology.

2.2 Cellular Networks

Probably, the most important 2G cellular networks that exist today are global system for mobile communications (GSM), time division multiple access (TDMA — IS-136), and code division multiple access (CDMA — IS-95). GSM has been largely used all over the world and it is the most successful standard in terms of coverage. These networks have different features and capabilities. Both GSM and TDMA networks use time division multiplexing on the air interfaces, but have different channel sizes, structures, and core networks. CDMA has an entirely different air interface.

Standards like general packet radio service (GPRS), high speed circuit switched data (HSCSD), and enhanced data rates for GSM evolution (EDGE) were designed for packet data service and increased data rates in the existing GSM/TDMA networks. They are collectively called 2.xG cellular networks.

2.2.1   The 2G Cellular Mobile Networks

In mobile telephony, second generation protocols use digital encoding and include TDMA, GSM, and CDMA. Second generation networks convert audio signals into a stream of binary information that is transmitted according to the rules of the particular technology. These protocols support high bit rate voice and limited data communications. They offer auxiliary services such as data, fax, and short message service (SMS), defining different levels of encryption. These networks are in current use around the world.

2.2.1.1   Time Division Multiple Access

The TDMA standard was first specified in the Interim Standard 54 (IS-54), and the United States uses IS-136 for TDMA for both the cellular (850 MHz) and PCS (1900 MHz) spectrums [7]. It is also the name of the digital technology based on the IS-136 standard. TDMA is the current designation for what was formerly known as digital advanced mobile phone service or simply D-AMPS. It uses time division multiplexing to divide each channel into a number of time slots to increase the amount of data transmitted. TDMA divides a radio frequency into time slots and then allocates slots to multiple calls. In this way, a single frequency can support multiple, simultaneous data channels. Users access sequentially a single radio frequency channel without interference. The communication is performed in data bursts that are reassembled at the receiving end.

The TDMA scheme is present in different wireless networks such as D-AMPS, GSM, digital enhanced cordless telecommunications (DECT), and personal digital cellular (PDC) — a Japanese cellular standard. However, TDMA is implemented within each of these systems differently.

2.2.1.2   Global System for Mobile Communications

The GSM is a digital cellular phone technology based on TDMA that is currently available worldwide [11]. It operates in the 900 MHz and 1.8 GHz bands in Europe and the 1.9 GHz PCS band in the United States. GSM defines the entire cellular system, not just the air interface (e.g., TDMA and CDMA). GSM cellular handsets use a subscriber identity module (SIM) smart card, which is a vital component in GSM operation, that contains information about the user account. A very interesting aspect is that a GSM phone becomes immediately programmed after plugging in the SIM card. Furthermore, SIM cards can be programmed to display custom menus for personalized services. Once the card is plugged into any GSM-compatible phone, the cellular handset is instantly personalized to the user — a feature that allows GSM phones to be easily rented or borrowed.

The GSM air interface is based on narrowband TDMA technology with available frequency bands being divided into time slots. Each user has access to one time slot at regular intervals. Narrowband TDMA allows eight simultaneous communications on a single radio multiplexor and is designed to support 16 half-rate channels. GSM provides incoming and outgoing data services, such as email, fax, and Internet access.

2.2.1.3   Code Division Multiple Access

The CDMA is a spread spectrum cellular technology for the digital transmission of data and voice over radio frequencies [13]. The audio media is digitized and the data is split into data packets that are assigned unique codes to each communication to distinguish them in the same spectrum. In CDMA, signals are encoded using a pseudo-random sequence that corresponds to a different communication channel, also known by the receiver, and used to decode the received signal. CDMA allows several users to share the same time and frequency allocations in a given band/space, generating a controlled level of interference into other users. The receiver recovers the desired information and rejects unwanted information by means of this unique code.

The CDMA air interface is used in both 2G and 3G networks. The second generation CDMA standard is currently known as cdmaOne and includes IS-95A and IS-95B. The IS-95A protocol employs a 1.25 MHz carrier, operates at 800 MHz or 1.9 GHz, and supports data speeds of up to 14.4 kbps. IS-95B supports data speeds of up to 115 kbps by grouping up to eight channels. The cdmaOne describes a wireless system based on the TIA/EIA IS-95 CDMA standard. It encompasses the end-to-end wireless system and all the specifications that control its operation. CDMA is also the basis for 3G cellular networks that include the two dominant IMT-2000 standards, CDMA2000, and WCDMA. It is worthwhile to mention that an important advantage of CDMA over the other approaches, is that it allows more users to share the frequency spectrum — a very limited resource — when compared to alternative technologies either analog or digital. The origins of the CDMA scheme date back to the 1940s, during World War II.

2.2.2 2.5G Cellular Mobile Networks

A serious limitation of the 2G cellular systems is that data can only be transferred once a connection has been set up. This is not efficient whenever a user wants to transmit just a small amount of data or data is transmitted in bursts. The so-called 2.5G cellular systems allow a cellular handset to be always "connected." It means that at any moment it is possible to send and receive a packet data. In this way, the user does not incur in the cost of establishing a connection for sending/receiving small amounts of data or transmitting bursty data. This allows efficient transfer of small amounts of data, without the overhead of establishing a connection for each transfer. Currently, there are two major 2.5G enhancements to second generation cellular systems: GPRS and EDGE.

2.2.2.1 General Packet Radio Service

GPRS is the first implementation of packet switching within GSM, that is, it only utilizes the network whenever there is data to be sent [2,4]. It will complement existing services based on circuit switching and the SMS. The upgrade of a GSM network to GPRS will allow a GSM operator to eventually migrate to a third generation network, as universal mobile telecommunications system (UMTS) is based on packet switching.

In GPRS, users can send and receive data at speeds up to 115 kbps. The deployment of GPRS networks provides several advantages to GSM network operators. It allows to integrate the IP protocol to the GSM network, and, consequently, the TCP/IP protocol suite. Furthermore, GPRS is efficient in its use of the spectrum and allows GSM operators to offer new data services, mainly those based on bursty data communication such as e-mail, Web access, and similar applications using mobile handheld devices and laptop computers.

In practice, GPRS enables the connection to different public and private data networks. Service and application development and deployment should be easier to make since the faster data rate means that a middleware needed to adapt applications to the slower speed of wireless systems will be no longer needed.

2.2.2.2 Enhanced Data Rates for GSM Evolution

As its name suggests, EDGE is a faster version of GSM networks designed to provide data rates of up to 384 kbps [12]. The goal is to enable and deliver multimedia and other broadband applications to cellular phones and mobile handheld devices. The EDGE standard reuses the carrier bandwidth and time slot structure of GSM, but it is not limited to be used in GSM cellular systems only. In fact, EDGE provides a generic air interface for higher data rates.

EDGE allows an evolutionary migration path from GPRS to UMTS by defining a new modulation scheme that will be necessary for implementing UMTS later. GPSR uses the Gaussian minimum-shift keying (GMSK) modulation whereas EDGE is based on the eight-phase-shift keying (8PSK), a new modulation scheme that allows a much higher bit rate across the air interface. The change in the modulation scheme from GMSK to 8PSK is the main point to the EDGE standard to lead the GSM world (and TDMA in general) for UMTS.

2.2.3 3G Cellular Mobile Networks

Third generation wireless systems, or simply 3G, is a generic term used for a new generation of mobile communications systems aimed to provide enhanced voice, text and data services, with greater capacity, quality and data rates than the current cellular networks. The World Administrative Conference assigned 230 MHz of spectrum at 2 GHz for multimedia 3G networks. Third generation systems are defined by the International Telecommunications Union, under initiative called IMT-2000, as being capable of supporting high-speed data rates, depending on the conditions and mobile speed. These networks must be able to transmit wireless data at 144 kbps at mobile user speeds, 384 kbps at pedestrian user speeds, and 2 Mbps in fixed locations.

Third generation cellular mobile networks will provide advanced services transparently to the end user independent of the underlying wireless technology. One of the goals of 3G systems is to enable the convergence between the wireless world and the computing/Internet world, making inter-operation apparently seamless.

2.2.3.1 Migration Paths to 3G

There is no clear understanding what exactly constitutes a 3G cellular system and a 2.5G system. Technically, many CDMA2000 systems are modifications to existing 2G architectures, but they are announced as 3G systems. The GSM variant on the other hand, while theoretically capable of achieving the performance of 3G networks, is often classified as a 2.5G system. The main reason for that is because GSM manufacturers and telecom operators developed the wideband-CDMA (W-CDMA) standard that is expected to be deployed by existing GSM network operators.

There are two migration paths to the 3G cellular systems based on a number of different technologies (depicted in Figure 2.1 [19]), mainly based on a variation of 2G systems, which fulfill the requirements

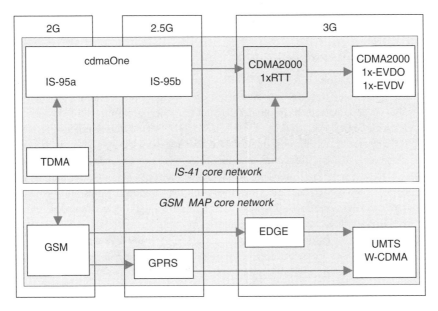

FIGURE 2.1   Migration paths to 3G systems.

of the ITU: the first one based on CDMA 1xRTT standard [5], with data rates of up to 144 kbps, and the second one based on GPRS [27].

2.2.3.2 CDMA2000

The CDMA2000 is a general name for a group of standards based on enhancements introduced to the 2G cdmaOne technology [5]. The goal is to provide telecom operators 3G data rates using their existing networks, with minimal changes in the infrastructure. CDMA2000 standards are defined and controlled by the 3GPP2 and some of them are also part of the W-CDMA specification group 3GPP.

The standards that constitute the CDMA2000 system are:

- 1xRTT: cdmaOne system enhancement designed to double voice capacity and support always-on data transmission. 1x stands for a single carrier, while RTT stands for radio transmission technology.
- 1xEV-DO: Enhancement to 1xRTT system that can support asymmetric non-real time data services. Telecom operators can use it to deploy a data-only broadband system or use it to add or increase the data capacity of their existing cdmaOne or 1xRTT networks.
- 1xEV-DV: Enhancement to 1xRTT system that supports voice and packet services on the same spectrum of a single RF carrier. It supports integrated real-time voice and video with nonreal time packet data services (e.g., ftp and http), while also supporting simultaneous voice and data services for a single subscriber.
- 3xRTT: Uses three 1x CDMA channels grouped together to provide higher data rates.

2.2.3.3 Wideband Code Division Multiple Access

W-CDMA is known as IMT-2000 direct spread spectrum. W-CDMA was selected as the air interface for UMTS, the 3G successor to GSM [12,27]. It is a 3G mobile wireless technology that supports new mobile multimedia services (voice, image, data, and video communications) at data rates of up to 2 Mbps. The input data signals are digitized and transmitted in coded, spread-spectrum mode over a wide range of frequencies. A 5 MHz-wide carrier is used, compared with 200 kHz-wide carrier for narrowband CDMA.

The W-CDMA technology borrows some ideas from CDMA, as the name suggests, but it is in fact very different and incompatible with phones and networks using the CDMA family of standards (including cdmaOne and CDMA2000). W-CDMA is more than a multiplexing technique. It describes a complete set of specifications, and a detailed protocol that defines how a mobile phone communicates with the base station, signals are modulated, and datagrams are structured.

2.2.4 4G Cellular Mobile Networks

In the 1980s, the first generation (1G) mobile communication systems was introduced. These systems were intended primarily for voice transmission and based on analog amplitude and frequency modulation. However, during the last decade, technological advance in the development of powerful processors, multiple access technologies (e.g., FDMA, TDMA, and CDMA), and signal processing techniques has allowed a better coverage, capacity, and quality of voice and data services over the 1G systems. As a consequence, 2G systems, which use digital encoding to convert voice signals into a stream of bits, have been introduced, followed by the third generation (3G) of wireless systems. 3G mobile systems offer broader connectivity and higher data rates varying from 384 kbps to 2 Mbps and provide significant improvements over the 2G systems. In this way, 3G systems are a natural way to enable the convergence between the worlds of wireless communication and the Internet.

The fourth generation (4G) mobile systems are intended to promote digital mobile multimedia (voice, streaming video, and data) services with data rates of up to 20 Mbps or more. The so-called 4G technologies may start appearing even before the full deployment of 3G systems. It is not very clear as to how these developments will influence an already complex scenario of mobile cellular systems. The point is that researchers are investigating the technologies that may constitute a 4G system. For instance, different modulation technologies, such as orthogonal frequency division modulation (OFDM) that is used in other wireless systems, such as wireless metropolitan area networks (MAN) or WiMAX (IEEE802.16) and digital video broadcast (DVB).

It is our belief that IP traffic and 4G systems will continue to grow in telecommunications networks, demanding efficient IP end-to-end transport service. Next generation IP-based access networks will have to support Quality of service (QoS), bandwidth control and management, multicast capability, and mainly security. QoS mechanisms have to be proposed and adopted to support packet-based multimedia services. With such features, IP-based 4G mobile solutions can enable seamless integration of wireless access networks and an IP core-based network infrastructure. IP packet-based 4G mobility systems will carry voice, data, and multimedia traffic with QoS based on various service types.

Fourth generation systems must be able to seamlessly switch between different wireless broadband access standards. The system must also provide global roaming facilitating coverage, service portability, and scalability to mobile users. There are some challenges ahead for 4G systems since they are expected to operate with other access interfaces and standards such as convergence with other wireless systems like Bluetooth, HiperLAN, and Wireless LAN (802.11); adaptation to channel conditions and support end-to-end IP QoS using a unified set of networking protocols; and issues related to timing, synchronization, and hand-off associated with multiple access interfaces and standards.

2.3   Wireless Local Area Networks

A wireless local area network (WLAN) is a generic term to refer to different technologies providing local area networking via a radio link. Examples of WLAN technologies include IEEE 802.11, Hiper-LAN, and DECT. A WLAN connects wireless devices such as laptops and PDAs to an access point (base station) using radio propagation as the transmission medium. A WLAN can work completely independent of a traditional network, but it is usually connected to an existing wired LAN such as an Ethernet.

A WLAN can be set up in both indoor and outdoor environments, but is especially suitable for indoor locations such as hotspots — a public access WLAN installed in places such as coffee shops, airports, and railway stations, offices, buildings, hospitals, and universities. In all these places, a WLAN allows mobile users to have a high-speed access to the Internet using a wireless device. Furthermore, these users will probably want to have access to the applications they run over a traditional wired network, mainly those clients with a laptop computer. However, the network throughput depends on different factors, such as the building architecture, objects and material present in the radio propagation environment, interference, and the distance between the wireless device and the access point.

2.3.1   IEEE 802.11

The term 802.11 refers to a family of standards developed by the IEEE for WLAN [15]. It defines an over-the-air interface between a wireless device and an access point or between a pair of clients. It is the first wireless standard that spans home and business environments.

The 802.11 protocol supports two communication modes for wireless devices: distributed coordination function (DCF) and point coordination function (PCF). The DCF mode is based on carrier sense multiple access with collision avoidance (CSMA/CA), also known as "listen before talk" or Wireless Ethernet. A wireless device waits for a quiet period on the network and begins to transmit data and detect collisions. DCF provides coordination among stations, but it does not define any type of priority access of the wireless medium for a station.

On the other hand, the PCF mode supports time-sensitive traffic flows. An access point periodically broadcasts beacon frames with network identification and management parameters related to the WLAN. Intervals between the transmission of beacon frames are divided into a contention-free period and a contention period. When PCF is enabled, a station can transmit data during contention-free polling periods.

There are several specifications in the 802.11 family. In the following, we briefly discuss each member of the 802.11 specification:

- 802.11a operates in the unlicensed 5-GHz band and allows data transmission rates of up to 54 Mbps. This standard uses OFDM as opposed to frequency hopping spread spectrum (FHSS) or direct sequence spread Spectrum (DSSS).
- 802.11b operates in the unlicensed 2.4-GHz band, allowing data transmission rates of up to 11 Mbps with a fallback to 5.5, 2, and 1 Mbps, depending on communication range and signal strength, and it supports only DSSS signaling. Its maximum range at full speed is 100 m (300 feet) in optimal conditions. It supports encryption using 40- and 128-bit keys (actual key length 112-bits). The 802.11b standard is a 1999 IEEE ratification to the original 802.11 protocol, allowing wireless functionality comparable to Ethernet, giving origin to the so-called "Wireless Ethernet". 802.11b was the first WLAN standard offered to consumers and enabled the proliferation of WLANs in different places such as hotspots, offices, and homes.
- 802.11e defines QoS features and multimedia support to the existing 802.11a and 802.11b standards. The support of QoS for multimedia applications is a fundamental aspect to a broad deployment of WLANs, in particular wireless home networks where multimedia data (voice, video, and audio) will be delivered. Some of the applications foreseen to be delivered in these networks are video on demand (VoD), audio on demand (AoD), voice over IP (VoIP) and high-speed Internet access.

  The DCF and PCF communication modes described earlier do not distinguish traffic types or sources. The IEEE 802.11e proposes enhancements to both modes to support QoS. The goal is to deliver critical service requirements but keeping backward compatibility with current 802.11 standards.
- 802.11g is a successor to the 802.11b standard, with data rates of up to 54 Mbps over the same 2.4 GHz band used by 802.11b. The standard 802.11g uses the OFDM signaling mechanism, which allows to achieve this higher data rate. An 802.11g access point can provide simultaneous WLAN connectivity for both 802.11b and 802.11g mobile clients, that is, 802.11g can be viewed as an extension of the 802.11b standard.
- 802.11h is an addition to the 802.11 standards to provide a solution to interference issues introduced by the use of 802.11a (5-GHz band) in some applications (e.g., military radar systems and medical devices) especially in Europe.
- 802.11i provides improved encryption for the 802.11a, 802.11b, and 802.11g standards. The new encryption key schemes are the Temporal Key Integrity Protocol (TKIP) and Advanced Encryption Standard (AES), which requires a dedicated chip. In this case, it may be necessary to have hardware upgrades for most existing Wi-Fi networks. The 802.11i standard also defines other features such as key caching to facilitate fast reconnection to the server when users disconnect temporarily, and preauthentication to allow fast roaming — an important support for applications that demand QoS such as VoIP.
- 802.11j is an addition to the 802.11 standards to incorporate Japanese regulatory extensions to 802.11a. The main goal of 802.11j is to add channels in the band of 4.9 to 5.0 GHz. It also deals with certain changes proposed to satisfy Japanese legal requirements concerning wireless transmitter output power, operational modes, channel arrangements, and spurious emission levels. The 802.11j standard provides data rates of up to 54 Mbps and employs the OFDM signaling mechanism. It also specifies how Japanese 802.11 WLANs and other wireless networks, in particular HiperLAN2, can operate in geographic proximity without mutual interference.
- 802.11n is being proposed to double the speed of 802.11a/g WLANs (expected for 2006).

As mentioned earlier, the 802.11 specification defines a family of standards. The different members with their goals lead naturally to some issues such as the frequency band used, interoperability, conformance test, and quality of service, which are discussed below:

The ISM band: The radio bands of 2.4 and 5 GHz, the base of the 802.11 standards, are referred to as industrial, scientific, and medical (ISM) frequency bands. These are unlicensed bands available for industrial, scientific, and medical applications and are subject to interferences from other devices such as handhelds or even microwave ovens operating at the same frequency. This is the case of Bluetooth devices that operate at 2.4 GHz.

Interoperability: An 802.11a client cannot interoperate with 802.11b and 802.11g wireless devices. A possible solution to overcome this problem is to have multiprotocol network interface cards (NICs). Due to the higher frequency of 802.11a, its range is shorter than lower frequency systems such as 802.11b and 802.11g. To have the same coverage of the b and g networks it may be required to have a greater number of 802.11a access points.

A clear advantage of 802.11a over 802.11g (recall that both standards can achieve the same data rate) is that in the United States the 5-GHz band defines 12 channels versus 3 channels in 2.4-GHz band. This larger value allows a higher density of mobile users per access point in a given area, which may be valuable for large-scale enterprise.

Wi-Fi: This term refers to a certification program established by the Wi-Fi Alliance [28] to perform conformance tests to wireless devices and, thus, ensure their interoperability. Initially, the term Wi-Fi was associated with the 802.11b standard, but currently it applies to any 802.11 network. Therefore, devices that are in conformance with the IEEE 802.11 standard are Wi-Fi devices and bear the official Wi-Fi logo, but the reverse is not always true.

QoS: Quality of service (QoS) plays an important role in WLANs in general and in 802.11 networks in particular for two reasons. First, the available bandwidth in WLANs is very restricted when compared to a wired medium such as a fiber optic. Second, mobile users will demand adequate support for the same set of applications they run in a wired network. The IEEE 802.11e standard proposes mechanisms to support QoS-sensitive applications (e.g., voice and video).

2.3.2 HiperLAN

HiperLAN refers to a set of WLANs primarily used in Europe [10]. It defines the specifications HiperLAN/1 and HiperLAN/2 — both adopted by the European Telecommunications Standards Institute (ETSI). The HiperLAN standards define features and capabilities comparable to the IEEE 802.11 WLAN standards. HiperLAN/1 provides data rates of up to 20 Mbps and HiperLAN/2 data rates of up to 54 Mbps, both in the 5-GHz band. HiperLAN/2 is compatible with 3G WLAN systems for sending and receiving multimedia data and is intended for being deployed worldwide.

2.3.3 Digital Enhanced Cordless Telecommunications

The DECT is a digital wireless telephone technology with a more ambitious goal than just telephony [8]. DECT was designed especially to operate on a small area with a large number of users such as a corporation or even a city, using TDMA to transmit radio signals to phones. A user equipped with a DECT/GSM dual-mode telephone can operate seamlessly in both networks. In this case, a dual-mode phone would automatically search first for a DECT network, then for a GSM network if DECT is not available.

2.4   Short-Range Wireless Communication

As the name suggests, short-range wireless communication allows communication typically around a few tens of meters. Short-range communication is the basis of wireless personal area networks (WPAN) and other applications for monitoring and controlling systems. A basic goal of short-range wireless communication is to replace cables.

Short-range technologies do not compete with other wireless solutions such as 3G, Wi-Fi, and WiMAX, but complement them. In fact, this technology allows, for instance, to turn 3G cellular phones into modems for laptops. Examples of short-range wireless technologies include Bluetooth, ZigBee, Ultra Wideband, and IrDA.

2.4.1 Wireless Personal Area Networks

A WPAN typically covers few meters around a user's location and provides the capacity to communicate and synchronize a wireless device to other computing equipments, peripherals, and a range of pocket hardware (e.g., dedicated media devices such as digital cameras and MP3 players). The goal is to allow seamless operation among home or business devices and systems. It is based on a technology that allows communication within a short range, typically around 10 m.

A WPAN exists where the user/device is, that is, it does not depend on an access point to establish a communication among devices in the network. This leads to a concept known as plugging in. It means that when any two WPAN devices are close enough (within radio communication of each other), they can communicate directly as if they were connected by a cable. If there are many communications among devices, each equipment may stop communicating with specific devices, preventing needless interference or unauthorized access to information.

WPAN is a generic term that refers to different technologies providing personal area networking. Examples of WPAN technologies include Bluetooth and IEEE 802.15, which is based on Bluetooth.

2.4.1.1 Bluetooth

Bluetooth is a short-range wireless technology that has been widely used for interconnecting computers, mobile phones, PDAs, headsets, and automotive hands-free systems [3]. Bluetooth was initially designed for WPAN applications but it is being applied in new areas such as sensor networks, streaming applications, and multiplayer gaming. The core specification consists of an RF transceiver, baseband, and protocol stack that includes both link layer and application layer protocols, unlike other wireless standards. Bluetooth offers services that allow the communication among wireless devices using different classes of data, distant from 10 to 100 m.

In a Bluetooth WPAN, up to eight devices can be connected to form what is called a piconet — a fundamental form of communication in this network. In a piconet, one of the devices becomes a master, which is responsible for controlling the traffic on the data communication channel, while all other devices become slaves. In this scenario, the master defines the synchronization reference — based on a common clock — and a frequency hopping pattern to the slaves. The roles of master and slave can be exchanged along the time.

Several piconets can be linked through a bridge to create a scatternet, which demands from a Bluetooth device the capability to establish a point-to-multipoint communication. A bridge is a Bluetooth device that works as a slave in both piconets or as a slave in one and master in the other, jumping between them.

Bluetooth provides data rates of up to 1 Mbps (basic rate) or a gross data rate of 2 or 3 Mbps with Enhanced Data Rate. It transmits in the 2.4-GHz ISM band and uses the FHSS signaling technique that changes its signal 1600 times per sec. In case there is an interference from other Bluetooth devices, the transmission does not stop, but its data rate is downgraded.

For a Bluetooth to achieve its potential as a useful wireless technology, it is fundamental that different devices interoperate. In the Bluetooth standard, this is accomplished by profiles, which describe how to use a specification to implement a given end-user function. Profiles present minimum implementations of the Bluetooth protocol stack for some typical applications. If a Bluetooth device implements an end-user function present in a profile, it guarantees interoperability. It can also implement a proprietary method providing flexibility.

The name Bluetooth comes from the 10th century Danish King Harald Blåtand (Harold Bluetooth in English) who was responsible for uniting parts of what is now Norway, Sweden, and Denmark. The

Bluetooth technology has a "similar uniting principle" since it was proposed to allow collaboration between differing industries such as the computing and wireless communication.

2.4.2 ZigBee and IEEE 802.15

The ZigBee protocol aims to provide longer battery life (months or even years on a single battery charge) [30]. It defines a wireless network used for monitoring and controlling systems that demand low throughput. A ZigBee network provides data rates of 20, 40, and 250 kbps covering a distance of 10 to 100 m. ZigBee takes its name from the zig-zag movement of bees that can be represented by mesh networks.

The low power demand and limited data rates of ZigBee makes it appropriate for applications with static or nearly static nodes. Some of the applications foreseen for ZigBee are industrial automation, building automation, home control, consumer electronics, computer peripherals, medical applications, and toys. These applications typically need simple control networks that periodically send a small amount of data from sensors and switches to regulate and turn devices on and off.

Both ZigBee and IEEE 802.15.4 specifications are open standards that consider networks with low-power operation and consisting of thousands of devices [16]. The IEEE 802.15.3a (high data rate) and IEEE 802.15.4a (very low data rate) standards define both the physical and MAC layers of the architecture whereas the ZigBee Alliance defines the networking layer, security protocol, and application profiles.

The ZigBee allows to configure a network using different topologies. Its base mode uses a master-and-slave model, but ZigBee can also do peer-to-peer networking. The basic topologies are star, cluster tree, and mesh. The simplest configuration is a star topology with a single network coordinator connected to a number of slaves. A cluster tree or a connected star network may be used to extend the range of a star network or to link two networks. A mesh network may spread over a large area and contain thousands of nodes. The ZigBee routing protocol chooses the shortest and most reliable path through the network. Based on network conditions, this path can dynamically change, leading to a more reliable network.

2.4.3 Ultra Wideband

The term ultra wideband (UWB) refers to a technology that is known since the early 1960s as carrier-free, baseband, or impulse technology [25]. The UWB technology transmits an extremely short duration pulse of radio frequency in the order of 10^{-12} to 10^{-9} sec, not using power during the periods between pulses. These bursts represent only a few cycles of an RF carrier wave. The resultant waveforms are extremely broadband, which make them often difficult to identify an actual center frequency.

The very short time period of UWB pulses exhibits some unique physical properties of these waveforms. For instance, UWB pulses can be used to provide an extremely high data rate for network applications and are relatively immune to multipath cancellation effects present in mobile and in-building environments. Multipath cancellation occurs when a strong wave arrives partially or totally out of phase with the direct path signal, causing a reduced amplitude response in the receiver.

An important aspect of the UWB technology is its low system complexity and low cost. UWB systems can be produced almost completely in digital circuitry, with minimal RF or microwave electronics. The UWB technology is frequency adaptive because of the RF simplicity of its design. This characteristic avoids interference to existing systems and makes use of the available spectrum.

Currently, there are two competing transmission schemes in UWB. One scheme transmits the pulses in continuous varying time slots using a pseudo-random number sequence similar to CDMA. The second scheme divides the spectrum into smaller frequency bands that can be added and dropped as necessary.

The UWB demands less power and provides higher data rates than the IEEE 802.11 and Bluetooth protocols. This technology is expected to be used in applications such as wireless multimedia transmission, auto safety, localization and navigation, and security surveillance.

2.4.4 Infrared Data Association

Infrared data association (IrDA) is a wireless communication technology developed by the infrared data association for interoperable and low-cost infrared data communication [20]. IrDA defines a set of protocols from lower to upper layers covering data transfer and network management. The IrDA transfer message comprises of an IrDA control part and an IrDA data part, which specifies a standard for an interoperable two way cordless infrared communication.

In IrDA, the physical communication depends on the objects present along the path between source and destination, and their alignment. IrDA communicates using the nonvisible infrared light spectrum and, thus, cannot pass through obstacles that block light such as walls and people. The physical transmission between transmitter and receiver requires that the two devices be aligned with each other with a certain precision. Furthermore, only two IrDA devices may communicate at a given time.

IrDA was designed for high data rate, low power consumption, and low cost. The 2001 IrDA specification defines data rates of up to 115.2 kbps, 576 kbps, 1.152 Mbps, 4 Mbps, and 16 Mbps. IrDA consumes low power because infrared transceivers require far less power than for RF transceivers. Its hardware is less expensive mainly due to the maturity and wide deployment of the IrDA standard.

2.4.5 HomeRF

The HomeRF standard provides a low-power wireless data communication solution for in-home networks enabling both voice and data traffic, developed by the former HomeRF working group [14]. HomeRF is a relatively short-range wireless communication scheme that operates in the 2.4 GHz ISM band. HomeRF version 2.0 defines data rates of up to 10 Mbps and communication range about 50 m to cover a typical home. It communicates using frequency hopping spread spectrum at 50 to 100 hops per sec. It supports both a TDMA service to provide delivery of interactive voice and CSMA/CA service for delivery of high speed data packets.

The HomeRF Working Group has also proposed the shared wireless access protocol (SWAP). It is a specification that allows personal computers, peripherals, cordless phones, and other mobile devices to transmit voice and data. SWAP is a protocol similar to the CSMA/CA incremented with features to support voice traffic. A SWAP network can operate either in ad hoc mode or infrastructured mode. In the former case, all devices play the same role and the network works in a distributed way. In the latter case, the network operates under the control of a connection point that provides a gateway to the public switched telephone network (PSTN). HomeRF provides a security mechanism based on unique network IDs. The HomeRF standard assigns a different QoS for each data traffic. It supports simultaneously up to eight voice connections and eight prioritized streaming multimedia connections. In case there are no voice connections and no multimedia traffic, all bandwidth can be allocated for data transmission. If multimedia or voice connections are initiated, the bandwidth is automatically re-allocated in such a way that voice connections receive half and data connections the other half with streaming media content receiving automatically a higher priority.

2.5 Wireless Metropolitan Area Networks (WiMAX)

WiMAX [29] or IEEE 802.16 [17] is a standard for air interface for fixed broadband wireless access systems. The goal of this technology is to provide wireless access in a metropolitan area network, that is, it works as a wireless last-mile broadband access in a MAN. It is a low-cost alternative to other last mile wired solutions such as digital subscriber line (DSL) and cable broadband access.

The WiMAX provides data rates of up to 75 Mbps per cell, which has a size from 2 to 10 km. This is equivalent to support simultaneously, using a single base station, more than 60 T1/E1 channels and hundreds of DSL-type connections. It is worthwhile to mention that an important merit of the WiMAX Forum and IEEE 802.16 is to standardize broadband wireless services, already available all over the world. WiMAX is proposing a set of baseline features, grouped in what is referred to as system profiles, that manufactures of wireless broadband equipments must comply. The goal is to establish a baseline protocol that allows equipment from multiple vendors to interoperate. System profiles are responsible for dealing with regulatory spectrum constraints faced by operators in different regions of the planet.

2.6 Ad Hoc Networks

The term "ad hoc" is a Latin expression that means *for this* or more specifically *for this purpose only*. This term has been applied especially to wireless networks comprised of communicating entities that belong to a network only during a communication session and are within radio range with other entities. In an ad hoc network, connections among entities are usually temporary since nodes may be added or removed either logically or physically. In general, ad hoc networks operate in a standalone way, but may be connected to another network such as the Internet.

Ad hoc networks pose new research challenges and, in general, traditional distributed algorithms need to be revised before being applied to this network [?]. This is the case, for instance, with routing protocols that have a fundamental role in ad hoc networks. In fact, without an appropriate ad hoc routing protocol it is very difficult to offer any other service in this network. It is also important to have medium access control protocols that support an ad hoc communication such as Bluetooth [3] and IEEE 802.11 [15]. Bluetooth was designed primarily for ad hoc networks whereas IEEE 802.11 also supports an ad hoc networking system when there are no wireless access points.

In the following, we present two kinds of ad hoc networks: mobile ad hoc networks and wireless sensor networks.

2.6.1 Mobile Ad Hoc Networks

A mobile ad hoc network (MANET) is comprised of mobile hosts that communicate with each other using wireless links and based on the peer-to-peer paradigm [18]. A MANET is a self-configuring network that can have an arbitrary topology along the time. Each mobile host works as a router and it is free to move randomly and connect to other hosts arbitrarily. Thus, the network topology can change quickly and unpredictably since there may exist a large number of independent ad hoc connections. In fact, it is possible to have different applications running on the same MANET.

MANETS tend to play an important role in new distributed applications such as distributed collaborative computing, distributed sensing applications, next generation wireless systems, and response to incidents without a communication infrastructure. Some scenarios where an ad hoc network could be used are business associates sharing information during a meeting, military personnel relaying tactical and other types of information in a battlefield, and emergency disaster relief personnel coordinating efforts after a natural disaster such as a hurricane, earthquake, or flooding.

In a MANET a route between two hosts may consist of hops through one or more nodes. An important problem in a MANET is finding and maintaining routes since host mobility can cause topology changes. Several routing algorithms for MANETs have been proposed in the literature such as ad hoc on-demand distance vector routing (AODV) [26], dynamic source routing protocol (DSR) [21], optimized link state routing protocol (OLSR) [6], and topology dissemination based on reverse-path forwarding (TBRPF) [24]. These algorithms differ in the way new routes are found and existing ones are modified.

2.6.2 Wireless Sensor Networks (WSNs)

Wireless sensor networks are those in which nodes are low-cost sensors that can communicate with each other in a wireless manner, have limited computing capability, and memory and operate with limited battery power [1,9]. These sensors can produce a measurable response to changes in physical conditions, such as temperature or magnetic field. The main goal of such networks is to perform distributed sensing tasks, particularly for applications like environmental monitoring, smart spaces and medical systems, that is, in this network, all nodes are often involved in the same sensing task. WSNs form a new kind of ad hoc networks with a new set of characteristics and challenges.

Unlike conventional WSN, a WSN potentially has hundreds to thousands of nodes. Sensors have to operate in noisy environments and higher densities are required to achieve a good sensing resolution. Therefore, in a sensor network, scalability is a crucial factor. Different from nodes of a customary ad hoc network, sensors are generally stationary after deployment. Although nodes are static, these networks still have dynamic network topology. During periods of low activity, the network may enter a dormant state in which many nodes go to sleep to conserve energy. Also, nodes go out of service when the energy of the battery runs out or when a destructive event takes place. Another characteristic of these networks is that sensors have limited resources, such as limited computing capability, memory and energy supplies, and they must balance these restricted resources to increase the lifetime of the network. In addition, sensors will be battery powered and it is often very difficult to change or recharge batteries for these nodes. Therefore, in sensor networks, we are interested in prolonging the lifetime of the network and thus the energy conservation is one of the most important aspects to be considered in the design of these networks.

2.7   Satellite Communication

Satellite communications are becoming increasingly important for connecting remote regions, regions underserved, and a large number of users [22]. A satellite is capable of establishing a communication between any two communicating entities that are within its view using a space vehicle in orbit over the Earth. Basically, a satellite works as a repeater that receives a signal from a transmitter and restores it before sending it to a receiver.

The number of transponders of a satellite limits the amount of channels it can receive and retransmit. However, an important advantage of a satellite is that it can transmit a given data to an unlimited number of communication devices on the ground. This makes satellites an ideal solution for broadcast communication.

Currently, satellites are digital, offer more capacity and less cost, and are designed to handle different altitudes. A geosynchronous-orbit satellite (GEO) orbits the earth over the Equator line. This satellite appears stationary above the surface and it is primarily used for television broadcast services. Medium-earth-orbit satellites (MEO) and low-earth-orbit satellites (LEO) are used for mobile communications as they are located much closer to the earth and, thus, provide lower latency to applications. The Mission and Spacecraft Library of NASA [23] defines the orbits of GEO, MEO, and LEO satellites as follows: a GEO has a circular orbit with apogees and perigees between 30,000 and 40,000 km, respectively; a MEO has apogees greater than 3,000 km but less that 30,000 km; and a LEO has apogees and perigees below 3,000 km.

LEO and MEO satellites are typically deployed in constellations — a set of satellites designed to provide some service. Each one of these satellites covers just a small area on the Earth surface and completes an entire revolution around the globe in a few hours. A constellation provides continuous communications coverage on a given region of interest, that is, along the time, inside this coverage area, there is always a satellite of the constellation providing a service for a mobile user.

2.8   Conclusion

In the past few years, there has been a growing interest about the convergence of mobility, wireless data communications, and computation. This convergence extends the reach of the Internet to mobile users allowing them to have instant access, anytime and anywhere, to services and applications spread over wired networks. Users will clearly benefit from this convergence once it is widely available.

Given this context, there are new algorithms that need to be designed considering different types of wireless networks. These algorithms are the basis for new applications and services in mobile environments. Furthermore, each wireless technology is designed to attend a specific set of requirements, which

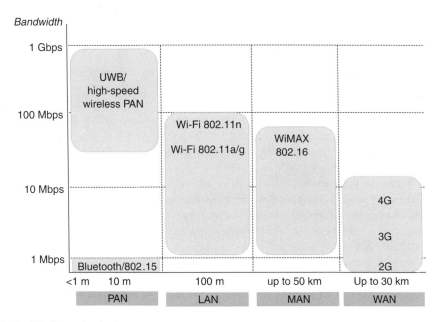

FIGURE 2.2 Wireless technologies.

in turn lead to different algorithmic solutions. For instance, we have to design different routing protocols for ad hoc networks and infrastructured networks.

This chapter provided a brief overview of different wireless technologies considering different segments: personal area networks, local area networks (LANs), metropolitan area networks (MANs), and wide area networks as shown in Figure 2.2. The requirements for each segment are based on a variety of parameters, including bandwidth, communication distance and user location, and protocols, which drive the design of algorithms for wireless networks.

Each network segment may have different potential wireless standards, but there are segment overlaps and complementary technologies. For instance, UWB supports faster file transfers than Wi-Fi or WiMAX, but distance limitations prevent it from being commonly used in LANs or MANs. The same problem happens to Bluetooth and other short-range protocols that cannot communicate directly with an entity distant more than 100 m. These wireless standards possibly will need other complementary technologies. On the other hand, technologies such as ZigBee, IrDA, and HomeRF provide similar or related services to Bluetooth and they compete among themselves.

However, depending on the acceptance of a given technology, when supposedly another one should prevail, it is natural to question whether the two technologies are in competition. This is possibly the case with 3G and WLAN technologies. Most experts say no — at least for now. Both technologies are seen as complementary, because they service different usage scenarios and business models. Yet the two great market domains of telecom and datacom tend to converge. VoIP and other multimedia systems are expected to be common applications in all wireless networks.

From the point of view of a user, services and applications should be deployed considering a seamless integration of different wireless technologies. It is difficult to anticipate the level of such an integration in the future but algorithms will play, for sure, a very important role in this process.

References

[1] I.F. Akyildiz, W. Su, Y. Sankarasubramaniam, and E. Cayirci. Wireless sensor networks: a survey. *Computer Networks*, 38: pp. 393–422, 2002.

[2] Regis J. Bates and Regis J.B. Bates. *GPRS: General Packet Radio Service*. McGraw-Hill Professional, New York, 2001.

[3] Bluetooth Special Interest Group. http://www.bluetooth.org/.

[4] Simon Buckingham. What is general packet radio service? http://www.gsmworld.com/technology/gprs/intro.shtml, January 2000.

[5] CDMA Development Group. http://www.cdg.org/.

[6] T. Clausen and P. Jacquet. Optimized link state routing protocol (OLSR). Request for comments 3626, October 2003. http://www.ietf.org/rfc/rfc3626. txt.

[7] Cameron Kelly Coursey. *Understanding Digital PCS: The TDMA Standard*. Artech House Publishers, 1998.

[8] DECT Forum. http://www.dect.ch/.

[9] Deborah Estrin, Ramesh Govindan, John S. Heidemann, and Satish Kumar. Next century challenges: scalable coordination in sensor networks. In *Proceedings of the 5th Annual ACM/IEEE International Conference on Mobile Computing and Networking*, pp. 263–270, 1999.

[10] ETSI HIPERLAN/2 Standard. http://portal.etsi.org/bran/kta/Hiperlan/hiperlan2.asp.

[11] GSM World. http://www.gsmworld.com/.

[12] Timo Halonen, Javier Romero, and Juan Melero, editors. *GSM, GPRS and EDGE Performance: Evolution Towards 3G/UMTS*, 2nd ed. John Wiley & Sons, New York, 2003.

[13] Lawrence Harte, Morris Hoenig, Daniel McLaughlin, and Roman Kikta. *CDMA IS-95 for Cellular and PCS: Technology, Applications, and Resource Guide*. McGraw-Hill Professional, New York, 1999.

[14] HomeRF Working Group. http://www.consortiuminfo.org/links/homerf.php.

[15] IEEE 802.11. http://grouper.ieee.org/groups/802/11/.

[16] IEEE 802.15. http://grouper.ieee.org/groups/802/15/.

[17] IEEE Working Group 802.16 on Broadband Wireless Access Standards. http://grouper.ieee.org/groups/802/16/.

[18] IETF. Manet group. http://www.ietf.org/html.charters/manet-charter.html.

[19] International Telecommunication Union. The internet for a mobile generation. ITU Internet Reports, 2002. http://www.itu.int/osg/spu/publications/sales/mobileinternet/exec_summary.html.

[20] IrDA — The Infrared Data Association. http://www.irda.org/.

[21] David B. Johnson, David A. Maltz, and Yih-Chun Hu. The dynamic source routing protocol for mobile ad hoc networks (DSR). IETF MANET working group, Internet-draft, July 2004. http://www.ietf.org/internet-drafts/draft-ietf-manet-dsr-10.txt.

[22] Grard Maral and Michel Bousquet. *Satellite Communications Systems: Systems, Techniques and Technology*, 4th ed. John Wiley & Sons, New York, 2002.

[23] NASA. Mission and spacecraft library. http://msl.jpl.nasa.gov/glossary.html.

[24] R. Ogier, F. Templin, and M. Lewis. Topology dissemination based on reverse-path forwarding (TBRPF). Request for Comments 3684, February 2004. http://www.ietf.org/rfc/rfc3684.txt.

[25] Ian Oppermann, Matti Hämäläinen, and Jari Iinatti, editors. *UWB: Theory and Applications*. John Wiley & Sons, New York, 2004.

[26] C. Perkins, E. Belding-Royer, and S. Das. Ad hoc on-demand distance vector (AODV) routing. Request for Comments 3561, July 2003. http://www.ietf.org/rfc/rfc3561.txt.

[27] UMTS Forum. http://www.umts-forum.org/.

[28] Wi-Fi Alliance. http://www.wi-fi.org/.

[29] WiMAX Forum. http://www.wimaxforum.org/.

[30] Zigbee Alliance. http://www.zigbee.org/.

I

MAC Protocols and Scheduling Strategies in Wireless Networks

3

Medium Access Control (MAC) Protocols for Wireless Networks

3.1	Introduction..	**3**-23
3.2	Basic Multiple Access Techniques......................	**3**-24
	Frequency Division Multiple Access • Time Division Multiple Access • Spread Spectrum Multiple Access	
3.3	Cellular Networks and Paging Systems.................	**3**-26
	Cellular Networks • Paging Systems	
3.4	Satellite Networks......................................	**3**-28
	Demand Access Multiple Access (DAMA)	
3.5	Mobile Ad Hoc Networks	**3**-30
	Contention MAC Protocols for MANETs • Allocation MAC Protocols for MANETs • Hybrid MAC Protocols for MANETs	
3.6	Sensor Networks	**3**-37
3.7	Research Directions in MAC Protocols................	**3**-38
	Quality of Service Support • Fairness • Power Management • Smart Antennas	
3.8	Summary	**3**-41
	Acknowledgment..	**3**-41
	References	**3**-41

Violet R. Syrotiuk

3.1 Introduction

Wireless networks include a diverse range of networks such as cellular networks and paging systems, mobile ad hoc networks, sensor networks, satellite systems, and wireless local area networks. Despite the diversity of these systems, a commonality of all wireless networks is a *broadcast channel* as the basis of their communication. In contrast to point-to-point channels, a broadcast channel is a single communication channel that is shared by all the nodes in the network. In general, this means that a packet transmitted by a node is received by all other nodes. This is an advantage if the packet is addressed to all destinations; a *broadcast* may be achieved by a single transmission. Other primitives typically provided in support of higher level network functions include *multicast* (transmission to a subset, or group, of the nodes) and *unicast* (transmission to an individual node).

This chapter discusses the *medium access control* (MAC) protocol — the protocol used to determine which node transmits next on a broadcast channel being accessed concurrently by multiple users. The role of the MAC protocol is critical to the overall performance of the network since it directly controls the sharing, and therefore the utilization, of the communication resources. Since the technology underlying different wireless devices is quite vast and the optimization criteria of the objective function impacts the protocol design, a large number of MAC protocols have been proposed. Consequently, the goal of this chapter is to broadly overview fundamental wireless MAC protocols.

The architecture of a network describes the components of the network infrastructure and their interaction. From the perspective of the MAC protocol, the network architecture affects where the access decisions are made. In general, decisions may be made either in a centralized or in a distributed manner. In a *centralized* architecture, one of the nodes is distinguished (e.g., the base station of a cellular network) and it is responsible to make channel access decisions for the nodes it serves (e.g., the mobile stations in its cell). Typically, the distinguished node has global information of its service area, which facilitates the support of quality of service (QoS), fair access, and other decisions made by the MAC protocol. In a (flat) *distributed* architecture, each node makes channel access decisions individually, usually only making use of information local to its vicinity.

Examples of networks with centralized architectures include cellular networks, paging systems, and satellite systems. Examples of networks with distributed architectures include mobile ad hoc networks and sensor networks. For each class within a network architecture we identify distinguishing characteristics of the network along with the resulting challenges to MAC protocol design. Mobility of the nodes is only one issue from which the challenges arise.

A number of representative MAC protocols for each class in a network architecture are highlighted giving an algorithmic emphasis. The protocols are broadly classified according to whether time is continuous or discrete. In *continuous* time, transmission of a packet may begin at any time. There is no need for a global clock in the system. When time is divided into *discrete* intervals called slots, transmission of a packet usually begins at the start of a slot. That is, nodes are synchronized in time on frame or slot boundaries.

Finally, research directions in MAC protocols are explored for mobile ad hoc and sensor networks since these wireless networks are less mature than the other wireless networks considered. Topics of active research include energy efficiency, how to exploit the use of directional (or smart) antennas, QoS, and fair access to the medium, among others. Chapter 4 provides an in-depth look at distributed scheduling protocols, while Chapters 5 and 6 examine extensions to allocation and random access protocols for the multichannel case.

For an overview of wireless MAC protocols that also includes wireless ATM see Myers and Basagni [36]. Chandra et al. [9] provide additional details on centralized random access protocols and hybrid protocols. A survey of MAC protocols that also includes wired networks can be found in Reference 22.

3.2 Basic Multiple Access Techniques

Full-duplex communication, where simultaneous transmission and reception between a pair of nodes is possible, may be achieved in either the frequency or the time domain. *Frequency division duplexing* (FDD) provides two distinct frequency bands for every user. For a channel pair, a *duplexer* allows transmission while simultaneously receiving signals at a node. *Time division duplexing* (TDD) shares a single radio channel by taking turns in time. In TDD, each full-duplex channel requires two time slots to facilitate communication in each direction.

Frequency, time, and spread spectrum multiple access are the three major access techniques used to share the available bandwidth in a wireless communication system. The duplexing technique together with the access technique fully defines the access scheme of the system.

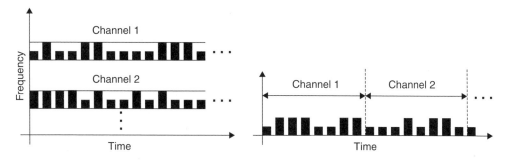

FIGURE 3.1   Digital data transmitted using FDMA and TDMA.

3.2.1   Frequency Division Multiple Access

Frequency division multiple access (FDMA) assigns individual channels to individual users. Figure 3.1 shows that in FDMA, each user is allocated a unique frequency band or channel. These channels are assigned on demand to users who request service. During the period of the call, no other user can share the same channel. In FDD systems, the users are assigned a channel as a pair of frequencies; one frequency is used for the forward channel, while the other frequency is used for the reverse channel (more on this in Section 3.3.1). In FDMA, a user is assigned a particular channel that is not shared by other users in the vicinity, and if FDD is used then the system is FDMA/FDD.

FDMA is a continuous transmission scheme and is therefore suited for both analog and digital systems. It requires the use of a duplexer in each node since both the transmitter and receiver operate at the same time.

3.2.2   Time Division Multiple Access

In *time division multiple access* (TDMA) systems, the radio spectrum is divided into nonoverlapping time slots. Figure 3.1 also shows TDMA, where only one user is allowed to transmit in each slot. Since the *frame* (which is an assignment of slots to users) cyclically repeats, a channel for a user may be thought of as a particular time slot that recurs in every frame. As a result, the transmission for any user is not continuous. This implies that, unlike in FDMA systems, which accommodate analog frequency modulation, digital data and digital modulation must be used with TDMA.

In TDMA/TDD systems, half of the time slots in the frame are used for the forward channel and half are used for the reverse channel. In TDMA/FDD systems, a similar frame structure is used for the forward and the reverse channel, but on different carrier frequencies. Since TDMA uses different time slots for transmission and reception, duplexers are not required. Even if FDD is used, a switch rather than a duplexer is all that is required to switch between a transmitter and a receiver using TDMA.

The TDMA has an advantage in that it is possible to dynamically allocate different numbers of time slots per frame to different users. Thus, bandwidth can be supplied on demand to different users by assigning time slots based on priority or by request.

3.2.3   Spread Spectrum Multiple Access

Spread spectrum multiple access uses signals that have a bandwidth that is several orders of magnitude greater than the minimum required bandwidth. Such wideband systems are efficient when many users share the bandwidth without interfering with one another.

There are two forms of spread spectrum multiple access techniques: frequency hopped multiple access spreads the signal over time and narrowband frequency channels, and direct sequence multiple access (also called code division multiple access [CDMA]), where the signal energy is spread over the wideband channel.

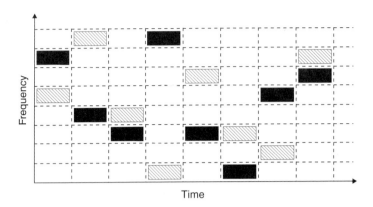

FIGURE 3.2   Frequency hopped multiple access (FHMA).

3.2.3.1   Frequency Hopped Multiple Access

In *frequency hopped multiple access* (FHMA), the digital data of each user are transmitted in uniform sized bursts on different channels within the allocated spectrum band. The carrier frequency of a user is varied according to a hopping sequence. The difference between an FHMA and an FDMA system is that the frequency hopped signal changes channels at rapid intervals. Figure 3.2 shows how two users may simultaneously occupy the spectrum at the same time using FHMA.

In FHMA, the receiver uses a locally generated hopping sequence to synchronize the frequency of the receiver with that of the transmitter. Without the proper sequence only noise is received, since the signal is hidden in the wideband channel in which the overall noise may be stronger than the narrowband burst. An FHMA system provides a level of security, especially when a large number of narrowband channels are used, since the signal is difficult to detect and intercept without the hopping sequence.

3.2.3.2   Code Division Multiple Access

In *direct sequence* or CDMA, the narrowband signal is multiplied by a large bandwidth signal called the spreading signal. All users in a CDMA system use the same carrier frequency and may transmit simultaneously. Either TDD or FDD may be used.

Each user has its own codeword that is orthogonal to all other codewords. To detect the codeword of a specific user, the receiver needs to know the codeword used by the transmitter. The receiver performs a time correlation operation of the signal with the codeword of the transmitter. Since the codewords are pairwise orthogonal, if it is the same codeword then the correlation is exact, otherwise it is zero (or, in the case of approximately orthogonal codewords, the correlation is high if it is the same codeword, and low otherwise).

If the power of each user is not controlled such that they do not appear equal at the receiver, then the near–far problem occurs. The *near–far* problem occurs when many users share the same channel. In general, the strongest signal received is captured. To combat the near–far problem, power control is used in most implementations of CDMA to assure that each node provides the same signal level to the receiver. This solves the problem of a nearby node overpowering nodes farther away from the receiver.

3.3   Cellular Networks and Paging Systems

Cellular networks and paging systems have a centralized architecture, relying on the existence of an underlying fixed infrastructure. The primary difference between them is that the channels in cellular networks are *full-duplex* (simultaneous transmission and reception between a pair of nodes is possible), while in most paging systems they are *simplex* (communication is possible in one direction only). Paging systems are primarily broadcast while cellular is unicast.

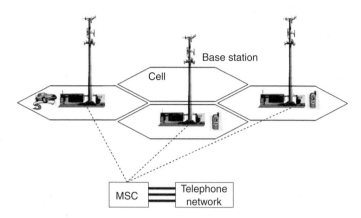

FIGURE 3.3 Cellular network architecture.

3.3.1 Cellular Networks

A cellular network provides a wireless connection to the wired telephone network for any user within the radio range of the system. Figure 3.3 shows a basic cellular network architecture. It consists of *mobile stations* (MSs), *base stations* (BSs), and a *mobile switching center* (MSC).

A large number of users is supported in a cellular system by limiting the coverage of each BS transmitter to a *cell*, a small geographic area. This allows *frequency reuse*, that is, the same radio channels may be reused by another BS located some distance away. A BS generally consists of several transmitters and receivers that simultaneously handle full-duplex communications. The BS serves as a bridge between all MSs in the cell to the MSC, connecting the calls through telephone lines or microwave links. Each MS communicates by radio with one of the BSs. If the MS moves from one cell to another, a *handoff* enables a call in progress to proceed uninterrupted. Throughout the duration of a call, it may experience several handoffs. The MSC coordinates the activities of all the BSs and is responsible for connecting all MSs in the entire cellular system to the telephone network.

Communication between the BS and the mobiles is defined by a *common air interface* that uses two full-duplex channels, each consisting of two simplex channels. One full-duplex channel is used for voice (and data) transmission, while the other is used for control purposes. The simplex channel from the BS to a mobile is the *forward* channel, while the channel from a mobile to the BS is the *reverse* channel. The control channel is used to initiate mobile calls and service requests, such as facilitating automatic changes of channel and handoff instructions for the mobiles before and during a call. The control channel is monitored by mobiles when they do not have a call in progress to detect incoming calls.

3.3.1.1 Cellular Systems and Standards

The AT&T Bell Laboratories developed the *Advanced Mobile Phone Service* (AMPS), the first US cellular telephone system; it was deployed in 1983 [60]. Like all first generation analog cellular systems, AMPS used FDMA and FDD for radio transmission.

The *United States Digital Cellular* (USDC) system was designed to support improved capacity and performance over AMPS [43]. USDC is a TDMA system that uses the same FDD scheme as AMPS. The dual mode USDC/AMPS system was standardized by the Electronic Industries Association and Telecommunication Industry Association as Interim Standard 54 (IS-54) and was later upgraded to IS-136. The USDC system was designed to be compatible with AMPS, so that the base- and mobile-stations could be equipped with both AMPS and USDC channels using the same equipment. A USDC system based on CDMA was standardized as Interim Standard 95 (IS-95). It too was designed to be compatible with AMPS.

The *global system for mobiles* (GSM) is a second generation (2G) cellular system, developed to unify the first European cellular systems. Introduced in 1991, it has been very successful in this goal and is the most popular 2G technology in the world. The organization that standardizes GSM is the European Technical

TABLE 3.1   Access and Duplexing Techniques Used in Different Cellular Systems

Cellular network	Access scheme
Advanced mobile phone system (AMPS)	FDMA/FDD
United States Digital Cellular (USDC) (IS-54, IS-136)	TDMA/FDD
US narrowband spread spectrum (IS-95)	CDMA/FDD
Global system for mobile (GSM)	TDMA/FDD
Pacific digital cellular (PDC)	TDMA/FDD
W-CDMA (3GPP)	CDMA/(FDD and TDD)
cdma2000 (3GPP2)	CDMA/(FDD and TDD)

Standards Institute (ETSI) [28]. GSM uses FDD and a combination of TDMA and FHMA to provide multiple access to mobile users [29,45].

The *Pacific Digital Cellular* (PDC) standard, also known as *Japanese Digital Cellular* (JDC), was developed in 1991. PDC has similarities to the IS-54 and the IS-136 standards, using FDD and TDMA for channel access.

Since the mid 1990s, the 2G digital standards have been widely deployed by wireless carriers. In an effort to upgrade the 2G standards for increased data rates for support of new data applications, interim 2.5G standards have been developed on the way to third generation (3G) technology. Three TDMA upgrade paths have been developed for GSM: *high speed circuit switched data* (HSCSD), *general packet radio service* (GPRS), and *enhanced data rates for GSM evolution* (EDGE). Both GPRS and EDGE support IS-136. The eventual evolution for GSM, IS-136, and PDC systems lead to *wideband CDMA* (W-CDMA), also called *universal mobile telecommunications service* (UMTS). There is only a single upgrade path for CDMA systems for 3G operation. The interim data solution for CDMA is IS-95B, leading to *cdma2000*. There are two major 3G technologies, with the standards organizations correspondingly fragmented: 3GPP, the 3G Partnership Project for CDMA standards based on backward compatibility with GSM and IS-136/PDC, and 3GPP2, the 3G Partnership Project for cdma2000 standards based on backward compatibility with IS-95.

Table 3.1 summarizes the cellular standards and the access schemes used.

3.3.2 Paging Systems

Most paging systems are examples of simplex communication systems in a centralized network architecture, where a brief message is received but not acknowledged by the subscriber. Depending on the type of service, the message may be numeric, alphanumeric, or a voice message. Paging systems are typically used to notify a subscriber of the need to retrieve further instructions. More recently, paging systems have also been used to provide extended services such as stock quotes and news headlines.

Though paging receivers are simple and inexpensive, the transmission system required is quite sophisticated since the service area may extend worldwide. Wide area paging systems consist of a network of telephone lines and many BS transmitters. A message is sent to a paging subscriber through the paging system access number. The issued message is called a *page*. The paging system then simultaneously broadcasts the page from each BS throughout the service area.

Paging systems are designed to provide reliable communication to subscribers whether inside a building, driving on a highway, or flying in an airplane. This requires large transmitter powers and low data rates for maximum coverage from each BS.

Table 3.2 summarizes the medium access techniques used in some paging system standards.

3.4 Satellite Networks

Satellite communication systems were originally developed to provide long-distance telephone service. At the end of the 1960s an era of expansion began for telecommunication satellites when it became possible

TABLE 3.2 Multiple Access
Techniques Used in Different
Paging Systems

Paging system	MAC technique
ERMES	FDMA
NTT	FDMA
NEC	FDMA

to launch a satellite in *geostationary earth orbit* (GEO) [42]. A satellite in GEO appears stationary over a fixed point on the earth. GEO satellites are preferred for high capacity communication satellite systems, since each satellite can establish links to one-third of the surface of the earth using fixed antennas at the earth stations. This is particularly valuable for broadcasting, since a single satellite can serve an entire continent. Although television accounts for much of the traffic carried by the GEO satellites, they also support international and regional telephony, data transmission, and Internet access.

A satellite *transponder* is essentially a repeater in GEO — a *repeater* is a receiver linked to a transmitter that can receive a signal from one earth station, amplify it, and retransmit it to another earth station on a different frequency (so that the outgoing channel does not interfere with the incoming one).

Satellites are built with the intention that many users will share the bandwidth allocated to the satellite, allowing many separate communication links to be established through the transponders of the satellite.

The three basic multiple access techniques of FDMA, TDMA, and CDMA (described in Section 3.2) are also used by all communications satellites. FDMA was the first multiple access technique used in satellite systems. In FDMA all users share the satellite at the same time, but each user transmits on a unique allocated frequency.

In TDMA each user is allocated a unique time slot at the satellite so that signals pass through the transponder sequentially. TDMA systems are easily reconfigured for changing traffic demands and traffic mixes, and are resistant to noise and interference. However, using all of the transponder bandwidth requires every earth station to transmit at a high bit rate; this requires high transmitter power.

In CDMA all users transmit to the satellite on the same frequency and at the same time. The earth stations transmit orthogonal spread spectrum signals that can be separated at the receiving earth station by correlation with the transmitted code. While CDMA is popular in cellular systems, it has not been widely adopted in satellite communication systems because it usually proves to be less efficient in terms of capacity than FDMA and TDMA. The Globalstar *low earth orbit* (LEO) satellite system was designed to use CDMA for multiple access by satellite telephones [25].

The *global positioning system* (GPS) is a worldwide radio-navigation system formed from a constellation of twenty-four satellites and their earth stations. It uses CDMA for the transmission of signals that permit precise location of a receiver in three dimensions. CDMA is used to share a single channel at the receiver between all of the GPS satellite transmissions. Chapter 12 of Reference 42 details the GPS signal structure and the process of data recovery from the CDMA satellite signals.

3.4.1 Demand Access Multiple Access (DAMA)

Demand access can be used in any satellite communication link where traffic from an earth station is intermittent. Cellular systems use demand access techniques similar to those used by satellite systems in the allocation of channels to users. The major difference between a cellular system and a satellite system is that in a cellular system the user is connected to the BS by a single hop radio link. In a satellite system, there is always a two hop link through the satellite to the earth station, where the decisions are made. Controllers are not placed on the satellites to simplify the overall design.

Demand access systems require two different types of channels: a control channel and a data channel. A user wishing to enter the network first calls the controlling earth station using the control channel. The controller then allocates a pair of channels to that user. The control channel usually operates in a random

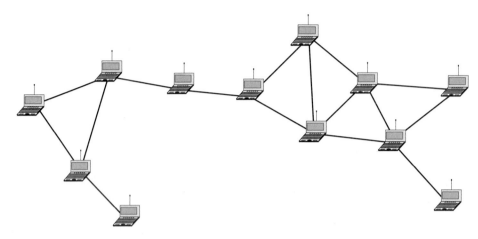

FIGURE 3.4   Each edge indicates nodes in the transmission range of each other.

access mode because the demand for use of the control channel is relatively low, messages are short, and the channel is therefore lightly loaded. Random access is a widely used satellite multiple access technique where the traffic density from individual users is low (see Section 3.5.1 for a discussion of MAC protocols based on random access techniques).

A thorough survey of MAC protocols for satellite systems, including demand access protocols and protocols for multimedia, can be found in References 41 and 38, respectively.

3.5   Mobile Ad Hoc Networks

A *mobile ad hoc network* (MANET) is a collection of mobile wireless nodes that self-organizes without the aid of any centralized control or any fixed infrastructure. Since the radio transmission range of each node is limited, it may be necessary to forward a packet over multiple hops in order for it to reach its destination. As such, MANETs have also been called mobile multi-hop and packet radio networks (see Figure 3.4). While the transmission range is limited, this also offers the opportunity for *spatial reuse*, that is, concurrent transmissions *on the same channel* when nodes are separated sufficiently far apart (in contrast to frequency reuse in cellular networks).

The challenge for MAC protocols for MANETs is to find a satisfactory trade-off between the two primary objectives of minimizing delay and maximizing throughput.

MANETs and sensor networks (see Section 3.6) both use half-duplex communication to reduce cost and complexity of the transceiver. In *half-duplex* radio systems, the communication is two-way, but the same channel is used for both transmission and reception. At any point in time, a node can only transmit or receive.

New problems such as the hidden and exposed terminal problems arise in half-duplex systems. The *hidden terminal* problem occurs when the destination of a transmitting node suffers a collision as a result of an interfering transmission from another node that is out of range of the transmitter. In Figure 3.5, if both nodes A and C transmit concurrently to node B their packets *collide* at B with the result that neither packet is successfully received. A and C are not in the transmission range of each other and are therefore hidden from each other. A goal of the MAC protocol is to reduce the hidden terminal problem to increase utilization of the channel.

If a node is transmitting, another node in its transmission range may transmit as long as their destinations are not in the union of the range of both transmitters. Since the transmitters cannot receive at the same time as when they transmit, their transmissions do not interfere with each other. For example, in Figure 3.5 if node B is transmitting to A, C can transmit concurrently to D since neither A nor D is in

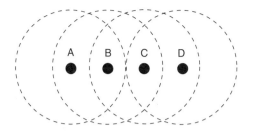

FIGURE 3.5   Each circle indicates the transmission range of the node at its center.

the union of the transmission range of both B and C. This *exposed terminal* problem results in bandwidth being underutilized; satisfactory solutions to this problem still do not exist.

In the following, we consider contention based MAC protocols and allocation based protocols, which correspond to time being viewed as continuous and discrete, respectively. We also overview some hybrid approaches that combine contention and allocation to exploit the best features from each class.

3.5.1   Contention MAC Protocols for MANETs

Contention based MAC protocols can be classified into four categories:

1. *No coordination*: Nodes transmit at will when they have data to send (e.g., ALOHA).
2. *Carrier sensing*: Nodes listen to the channel before transmitting a data packet (e.g., CSMA).
3. *Carrier sensing and collision detection*: Nodes listen before and during transmission, and stop if a collision, that is, noise, is heard when transmitting (e.g., CSMA/CD).
4. *Collision avoidance*: A handshake is typically used to determine the node that can send a data packet (e.g., MACA, FAMA, IEEE 802.11, CSMA/CA, RIMA, and many others).

Collision resolution combined with a backoff strategy may be used in each of these categories to determine when a node should try to retransmit after a collision.

3.5.1.1   The ALOHA Protocol

Historically, ALOHA was the first contention protocol for multiple access channels [2]; Abramson also provided the first analysis of such protocols. The protocol was originally designed for systems with a central BS or a satellite transponder, but it was also used in packet radio networks.

In ALOHA, when a node has a packet to transmit, it transmits immediately. An integral part of the ALOHA protocol is feedback from the receiver. Feedback occurs after a packet is transmitted and consists of an acknowledgment sent or piggybacked onto a packet from the receiver. On receipt of an acknowledgment for the packet, the node advances to the next enqueued packet. Otherwise, after a random delay, the node retransmits the packet. Of course, it is possible that successive transmission attempts are not successful, but this has lower probability, assuming that the network is not overloaded. After a number of retransmissions the packet is likely to be successful.

The throughput of ALOHA can be improved by reducing the time that a packet is vulnerable to interference from other packets. In slotted ALOHA, the only difference in the protocol is that time is discrete meaning that packet transmission may begin only at slot boundaries. This reduces packet vulnerability by half and doubles throughput. The operation of ALOHA and slotted ALOHA is described by the flow charts in Figure 3.6.

3.5.1.2   Carrier Sense Multiple Access (CSMA)

The capacity of pure and slotted ALOHA is limited by the large period of vulnerability of a packet. By *carrier sensing*, that is, listening to the medium before transmitting, nodes try to reduce the vulnerability

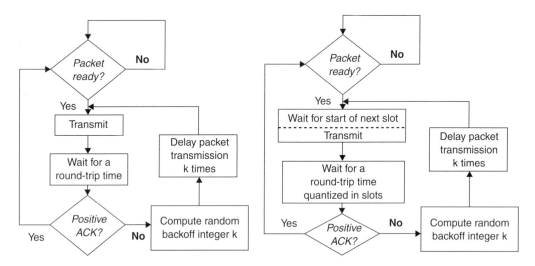

FIGURE 3.6   Flow chart for the ALOHA and the slotted ALOHA protocol.

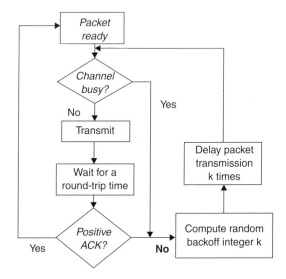

FIGURE 3.7   Flow chart for the CSMA protocol.

period to the propagation delay of one packet. This is the basis of the CSMA protocol [56], which otherwise operates as ALOHA (see Figure 3.7).

If the propagation delay of the network was zero, then a collision could only occur if two nodes start transmission of a packet at the same time. The probability of this happening when time is continuous is essentially zero. With finite propagation delay, however, a node may sense the channel is idle even if another node has already started transmitting, but due to the propagation delay the signal has not yet reached the sensing node.

3.5.1.3 Carrier Sense Multiple Access with Collision Detection (CSMA/CD)

CSMA improves on the performance of ALOHA tremendously. The remaining limitation is that, once a packet is sent, feedback only occurs a roundtrip time after the packet is transmitted. The solution to improve on the performance of CSMA is to use *collision detection*, that is, to listen to the channel while a

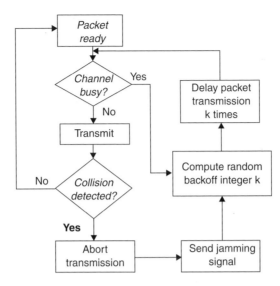

FIGURE 3.8   Flow chart for the CSMA/CD protocol.

packet is being transmitted to determine if the signals sent and heard are different. If a collision is detected, the packet transmission is aborted and a jamming signal is sent to ensure all nodes know of the collision. The detection of the collision serves as a negative acknowledgment. The basic protocol, given in Figure 3.8, is the basis of Ethernet protocol (IEEE 802.3) in widespread use in *local area networks* (LANs) [34].

The problem is that collision detection is not usually available in wireless nodes. Collision detection requires that a node both transmit and receive simultaneously and so it is not achievable with a single half-duplex transceiver.

3.5.1.4 Carrier Sense Multiple Access with Collision Avoidance (CSMA/CA)

Collision avoidance (CA) emulates collision detection in networks where nodes are half-duplex. Another objective of a collision avoidance protocol is to reduce the hidden-terminal problem of CSMA. The first protocol proposed to use collision avoidance was split reservation multiple access [57]. Since then, many protocols have been proposed including MACA, MACAW, FAMA, and RIMA [31, 7, 23, 24].

In the most common implementation of collision avoidance, nodes exchange a control packet handshake to determine which sender may transmit to a receiver. The sender initiates the handshake by transmitting a *request-to-send* (RTS) control packet addressed to the receiver. On receipt of the RTS, the receiver responds with a *clear-to-send* (CTS) control packet. The idea is to force hidden terminals of the sender to hear the feedback from the receiver, making them aware of the impending data transmission. The collision avoidance handshake is usually initiated by the sender, but can be initiated by a receiver (see, e.g., the *receiver initiated multiple access* (RIMA) protocol [24]).

The throughput of all collision avoidance protocols is always below that of CSMA/CD. The collision interval in CA in much longer than in CD because detecting collisions is done using small packets rather than listening to self-transmission.

The most prevalent contention MAC protocol for MANETs is the *distributed coordination function* (DCF) of the IEEE 802.11b wireless LAN standard [1]. It is a CSMA/CA protocol, and in addition to carrier sensing provided by the physical layer, it also uses virtual carrier sensing (Figure 3.9).

A *virtual carrier sense* mechanism is provided by the *network allocation vector* (NAV). The RTS and CTS packets contain a duration field that defines the period of time that the medium is to be reserved to transmit the data packet and acknowledgment. All nodes within the reception range of either the source node (that transmits the RTS) or the destination node (that transmits the CTS) learn of the medium

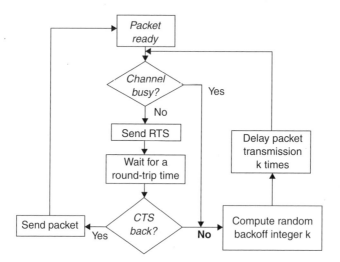

FIGURE 3.9 Flow chart for a CSMA/CA protocol.

reservation. The NAV maintains a prediction of future traffic on the medium based on duration information overheard in the handshake. The NAV may be thought of as a counter, that counts down to zero at
a uniform rate. When the counter is zero, the virtual carrier sense indication is that the medium is idle;
when nonzero, that it is busy.

Another solution proposed for the hidden-terminal problem is to make use of a separate control
channel. In the *busy tone multiple access* (BTMA) protocol [56], a node transmits a "busy tone" on the
control channel while receiving a data packet. This informs all nodes in the range of the receiver that they
should refrain from transmission, preventing collision at the receiver. In *dual busy tone multiple access*
(DBTMA) [20] two control channels are used to further distinguish between the transmission of the
sender and the receiver.

A difficulty that arises with the use of such control channels stems from the radio propagation characteristics. Since propagation is frequency dependent, the range of a data transmission and a transmission
on the control channel may not coincide.

3.5.1.5 Backoff Strategies for Collision Resolution

An important issue in all contention protocols is the backoff strategy used for collision resolution [6].
When a collision occurs, additional attempts may be required to transmit the packet before advancing to
the next enqueued packet. The collision is resolved through a backoff strategy that determines the time
between retransmissions.

One of the most common strategies is *binary exponential backoff*, where the next transmission slot is
drawn uniformly at random from an interval (a contention window) and after each unsuccessful trial the
size of the interval is doubled. In protocols such as IEEE 802.11, if the contention window reaches its
maximum size of 1024, it remains fixed at that size. If the retransmission is successful, the interval returns
to its minimum size.

Despite its widespread usage, under certain modeling assumptions, analysis shows that binary exponential backoff results in a protocol that is provably unstable (meaning that the queue of packets to transmit
grows infinitely) [4]. The existence of stable protocols depends on the type of feedback available from the
channel and on how the user population is modeled. For acknowledgment based protocols, it is known
that a large class of backoff strategies, including polynomial backoff, is unstable in the infinite user population model [32]. However, for a finite user population, any protocol using a super-linear polynomial
backoff strategy has been proven stable, while binary exponential backoff still remains unstable above a
certain arrival rate [26].

3.5.2 Allocation MAC Protocols for MANETs

While contention-based protocols achieve high throughput with a reasonable expected delay, in the worst-case the delay is very poor. As the interest in delay sensitive (e.g., voice, video, multimedia) applications grows it is clear that the MAC protocol must be QoS aware, that is, it must provide a delay guarantee to the real-time traffic. As a result, a number of ideas to support real-time and best-effort packets in IEEE 802.11 have emerged (see Section 3.7 for some details), however, the delay guarantee remains probabilistic. The fact is, to obtain a deterministic delay guarantee, an allocation approach is necessary.

While simple TDMA (one slot per user) is certainly a solution providing a delay bound, it does not meet a reasonable throughput objective because it does not take advantage of spatial reuse in MANETs. There are two approaches to take advantage of spatial reuse: topology-dependent protocols (also called *spatial reuse* TDMA) and topology-transparent protocols.

A *topology-dependent* protocol is a hybrid, combining contention with allocation. One way to combine approaches is to alternate between a contention phase in which neighbor information is collected, and an allocation phase in which nodes transmit according to a schedule constructed using the neighbor information (see, e.g., References 14 and 61). The task can be mathematically modeled as a graph coloring problem where nodes of the same color may transmit concurrently. This problem is NP-complete even for the restricted graphs topologies that correspond to MANETs [48]. While acceptable heuristic solutions can be found, these topology-dependent MAC protocols may be unstable under high load or mobility conditions, becoming unable to converge on a new schedule, and as a result lose their ability to deliver packets with a delay guarantee. Chapter 4 provides an in-depth look at distributed channel access scheduling for MANETs.

Alternatively, *topology-transparent* protocols do not use any neighbor information. The existing topology-transparent access protocols [10,30] depend on two design parameters: N, the number of nodes in the network, and D, the maximum (active) node degree. This creates complex trade-offs between the design parameters and the delay and throughput characteristics of the resulting schedules. While it is often possible to construct schedules that are significantly shorter than simple TDMA, if the number of active neighbors exceeds D, contention among the assigned slots results in the loss of the delay guarantee (more precisely, it becomes probabilistic rather than deterministic).

Recently, several new results relate topology-transparent scheduling to some combinatorial objects. Specifically, the known topology-transparent MAC protocols are generalized by observing that their transmission schedule corresponds to an orthogonal array. A Steiner system supports the largest number of nodes for a given frame length. Both orthogonal arrays and Steiner systems are specific types of cover-free families [17,18,53].

To be specific, let us construct a schedule corresponding to a codeword from an orthogonal array. Let V be a set of v symbols, usually denoted by $0, 1, \ldots, v - 1$. A $k \times v^t$ array A with entries from V is an *orthogonal array* with v levels and strength t (for some t in the range $0 \le t \le k$) if every $t \times v^t$ subarray of A contains each t-tuple based on V exactly once[1] as a column [16]. We denote such an array by $OA(t, k, v)$.

Table 3.3 shows an example from Reference 27 of an orthogonal array $OA(2, 4, 4)$ of strength two with $v = 4$ levels, that is, $V = \{0, 1, 2, 3\}$. Select any two rows, say the first and the fourth. Each of the sixteen ordered pairs $(x, y), x, y \in V$ appears the same number of times, once in this case.

Each column, called a *codeword*, gives rise to a transmission schedule. In our application, that of assigning transmission schedules to nodes, the number of columns in the orthogonal array is an upper bound on the number of nodes in the network. Each codeword intersects every other in fewer than t positions. For example, the first and the eighth column intersect in no positions, while the first and the second column intersect in a zero in the first position (i.e., the first row).

The importance of this intersection property is as follows. Select any column. Since any of the other columns can intersect it in at most $t - 1$ positions, any collection of D other columns has the property that our given column differs from all of these D in at least $k - D(t - 1)$ positions. Provided this difference is

---

[1] Here, we assume the index of the orthogonal array $\lambda = 1$.

positive, the column therefore contains at least one symbol appearing in that position, not occurring in any of the D columns in the same position. For scheduling this means that at least one collision-free slot to each neighbor exists when a node has at most D neighbors. Thus, as long as the number of neighbors is bounded by D, the delay to reach each neighbor is bounded, even when each neighbor is transmitting. (The orthogonal array gives a D cover-free family.)

A large number of techniques are known for constructing orthogonal arrays, usually classified by the essential ideas that underlie them. There is a classic construction based on Galois fields and finite geometries; both Chlamtac and Faragó [10] and Ju and Li [30] use this construction implicitly.

Given a codeword W, that is, a column of length k from the orthogonal array, the corresponding schedule is constructed as follows. Each symbol w_i in $W = w_0 w_1 \ldots w_{k-1}$ gives rise to a binary *subframe* F_j of length v (since each symbol is from the set V), $j = 0, \ldots, k-1$. Specifically, $F_j[i] = 1$ if and only if $w_i = i$, for $i = 0, \ldots, v-1$ and $F_j[i] = 0$ otherwise. The schedule S is a $k \times v$ binary vector made up of the concatenated subframes, that is, $S = F_0 F_1 \ldots F_{k-1}$. Figure 3.10 shows an example of a schedule constructed from the codeword 1230 (column seven of Table 3.3).

An important point is that the resulting transmission schedules *do not* depend on the construction of the orthogonal array. Only the combinatorial properties of the orthogonal array itself are exploited.

3.5.3 Hybrid MAC Protocols for MANETs

A number of MAC protocols for MANETs combine the advantages of allocation and contention. Already mentioned (in Section 3.5.2, and in more detail in Chapter 4) are the spatial reuse TDMA protocols that alternate periods of contention and allocation.

Alternatively, some protocols use contention *within* allocation. In the *ADAPT* protocol [11], there is an underlying simple TDMA schedule. If a node does not use its assigned slot, then other nodes sense this and can contend for the slot using CSMA. The CATA protocol [55] also incorporates contention within a slot, and provides explicit support for unicast, multicast, and broadcast packet transmissions. However, it is subject to instability at high traffic load due to the lack of a fixed frame length.

A different type of combination is implemented in the *meta-MAC* protocol [21]. Here a higher layer, or meta, protocol combines the decision of any set of MAC protocols that run independently using a technique from computational learning theory. The combination is based on a weighed majority decision with randomized rounding, using continuously updated weights depending on the feedback obtained from the channel. For protocols using discrete time, the meta-MAC protocol has been shown to automatically and adaptively select the best protocol for unknown or unpredictable channel conditions from a given set of protocols. While this set may contain different MAC protocols, it may instead consist of a single MAC protocol with different parameter settings. Hence, the meta-MAC protocol also provides a way to optimize the parameters of a given MAC protocol.

TABLE 3.3   Orthogonal Array $OA(2, 4, 4)$

0	0	0	0	1	1	1	1	2	2	2	2	3	3	3	3
0	1	2	3	0	1	2	3	0	1	2	3	0	1	2	3
0	1	2	3	1	0	3	2	2	3	0	1	3	2	1	0
0	1	3	2	3	2	0	1	2	3	1	0	1	0	2	3

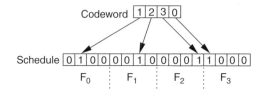

FIGURE 3.10   Construction of schedule from codeword 1230 from OA (2,4,4) in Table 3.3.

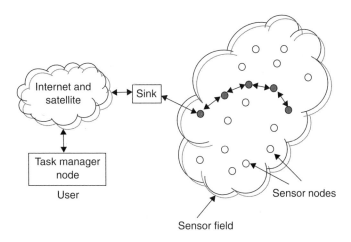

FIGURE 3.11   Sensor network architecture.

3.6   Sensor Networks

Advances in wireless communications, micro-electro-mechanical systems (MEMS) technology, and digital electronics have contributed to the development of small and inexpensive low-power sensor nodes. A sensor node consists of some type of sensor, with data processing and communications capabilities. A sensor network is a collection of sensor nodes that communicate what is sensed continuously, or to detect specific events. Since the position of the sensor nodes may not be known in advance, there is a need for the network to coordinate in a distributed manner, similar to the self-organizing capabilities of a MANET (refer to Section 3.5). Figure 3.11 shows the architecture of a typical sensor network.

Sensors come in many types (e.g., thermal, infrared, acoustic) and can monitor a wide variety of conditions (e.g., temperature, humidity, pressure). The variety in types of sensors and in their usage has precipitated applications for sensor networks that span a range of personal, corporate, and national interests [3]. As a result, research on sensor networks has grown rapidly in order to support the implementation of these emerging applications.

There are several differences between sensor networks and MANETs [40]. In particular, the number of sensor nodes deployed in a sensor network is expected to be several orders of magnitude higher than the number of nodes in a MANET. While sensor nodes may be more densely deployed, they are also limited in power, computational capability, and memory capacity. Since sensor nodes are more prone to failure, destruction, and energy depletion, the topology of the network changes frequently. Sensor nodes may not have global identifiers because of the amount of overhead assigning such identifiers for a large numbers of sensors. Furthermore, the network tends to operate as a collective structure, rather than supporting many independent point-to-point flows. Traffic tends to be variable and highly correlated. The protocols designed for MANETs are not directly useable in sensor networks because of these differences. For the MAC protocol for sensor networks, not only is it important that the channel be shared efficiently, but it should also be shared fairly and conserve energy.

In general, contention-based protocols are less satisfactory when the traffic is not independent. Furthermore, the overhead of the control packets in the handshake can be considerable (as high as 40% [59]) in sensor networks where the data packets are typically not very large.

The SMAC protocol [51] enables nodes to discover their neighbors and establish transmission and reception schedules for communication in a distributed manner. A communication link consists of a pair of time slots operating at a randomly chosen, but fixed, frequency. Power conservation is achieved using a random wake-up schedule during the establishment phase and by turning the radio off during idle slots.

A CSMA based MAC protocol is proposed for sensor networks in Reference 59. Constant periods of listening, together with random delays, provide robustness against repeated collisions. A fixed contention window and a linear increase multiplicative decrease backoff strategy is recommended for proportional fairness in the system.

3.7   Research Directions in MAC Protocols

Research directions in MANETs and sensor networks are explored since these networks are among the least mature of existing wireless networks. Research issues in MAC include, among others:

- The support for QoS
- Providing fair access to the channel
- Energy-aware protocols since the nodes are energy limited devices
- Effective utilization of directional (or smart) antennas

Many of these issues have been integrated into, or are a modification of, the IEEE 802.11 DCF protocol. The reason for the interest in this protocol is that it is widely deployed in real wireless local area networks and in emerging MANET testbeds.

3.7.1   Quality of Service Support

A number of ideas to support real-time and best-effort packets in IEEE 802.11 have been proposed. The *enhanced DCF* (EDCF) standard [5] is an extension of IEEE 802.11 for this purpose. It introduces separate queues for different traffic classes and assigns contention window sizes based on the priority of traffic flow. However, since it is a pure contention protocol, the collision rate increases very quickly as the contention for the medium increases. This greatly affects the delay and thereby significantly degrades the performance of delay-bounded applications.

The *adaptive EDCF* (AEDCF) protocol [47] tries to remedy EDCF by addressing how the contention window is reset. Rather than using a static reset, AEDCF uses a dynamic procedure to alter the contention window size after each successful transmission and also after each collision. As a result, AEDCF improves throughput when compared to EDCF especially under high load conditions. In both these protocols, the delay guarantee remains probabilistic since underlying the protocol is a pure contention-based access scheme.

Another approach to support QoS is given by Sobrinho and Krishnakumar [50], where "black-bursts" are used to identify high priority nodes and allocate the channel in priority order. The duration of a black-burst of a node is a function of the delay it incurred until the channel became idle. Hence the node whose wait is longer sends the black-burst for a longer duration, thus having higher priority to access the channel.

In general, contention-based approaches to support QoS such as References 5, 47, and 50 are able to meet delay requirements when the network load is low; however the delay guarantee remains probabilistic.

Topology-dependent and topology-transparent protocols (see Section 3.5.2) can provide bounded delay. The open issues with topology-dependent schemes are related to how to set parameters of the protocol for the given network conditions. One issue is to determine what is the appropriate length of the contention period to resolve the contention. In addition, the contention and allocation phases need not strictly alternate. Hence a second issue is to determine at what frequency the contention should be run.

For topology-transparent schemes, an open issue is how to adapt when the design parameter D is exceeded. The analysis in Reference 17 focusses on a single frame; while this is a reasonable view for studying minimum throughput, it ignores what happens when the neighborhood limitation is exceeded. Ju and Li [30] argue that by selecting a larger v, their choice is (relatively) insensitive to the degree limitation; essentially they "over-engineer" the system to permit more neighbors than the design criteria stipulate. Nevertheless, their scheme can fail totally when the number of neighbors exceeds the capacity of their chosen scheme, despite its continued operation for some numbers of neighbors larger than the stipulated number.

Chlamtac et al. [13] propose a different solution, interleaving or "threading" different schemes each supporting a different degree limitation; however, this incurs a dramatic slowdown in the initial scheme and hence a substantial penalty is paid whether or not the degree limit is exceeded. What is needed is a scheme that, if the degree limitation might be exceeded, degrades gracefully rather than failing completely (as in Reference 30) or imposing a large slowdown factor (as in Reference 10). Ju and Li's approach fails to provide a guarantee when the number of neighbors is too large. This in itself is reasonable. However, their scheme repeatedly uses the same frame schedule in each subsequent frame. Thus loss of transmission opportunity within one frame is extended to all subsequent frames unless and until transmitting nodes move or cease transmission. Of course, this does not impact the *expected* throughput. But when the degree limit is exceeded, it can result in a denial of service to some nodes.

The results in Colbourn et al. [17] demonstrate, however, that *expected* throughput can remain quite acceptable when the degree limitation is exceeded. The task, then, is to ensure that a situation resulting in catastrophic collisions in one frame is not automatically repeated in the next frame by simply repeating the schedule. One approach is to allow nodes to remain silent throughout a frame despite having a pending transmission, that is, "backing off" when it is detected that the degree limit is exceeded. When the degree limit is satisfied, such a scheme ensures nonzero minimum throughput; when exceeded, the scheme employs contention.

In general, hybrid protocols that take advantage of the strengths of both contention and allocation approaches to access control, perhaps using measurements of the actual network conditions for adaptation, that is, to effectively transition between approaches given the current network conditions, may provide robust protocols in support of QoS.

3.7.2 Fairness

Fairness is an important issue when accessing a shared wireless channel. In Vaidya et al. [58], a distributed algorithm for fair scheduling in a wireless local area network is derived from the IEEE 802.11 DCF. The protocol described in the standard is not fair; once a node has transmitted a packet successfully, the contention window is reset to the minimum size. This effectively gives the successful node a higher probability of accessing the channel with its next attempt since a delay is drawn at random from the minimum size interval. In *fair scheduling*, different flows sharing a wireless channel should be allocated bandwidth in proportion to their weights. The essential idea used is to choose a backoff interval that is proportional to the finish tag of the packet to be transmitted, and transmitting packets in increasing order of their finish tag.

Ozugur et al. [39] also tackled the problem of fairness. In their solution, each active user broadcasts information about its average access time, or number of connections, to its neighbors. Based on this information exchange, each node computes a link access probability. Then a CSMA based algorithm is proposed in which fair access is accomplished using this link access probability.

Fairness may be achieved in allocation approaches through appropriate scheduling [12]. TDMA and the existing topology-transparent protocols are examples of constant-weight codes, that is, each node has the same (a constant) number of transmission opportunities in each frame. This need not be the case, as the allocation can proportional to the traffic requirements of flows. However, this requires knowledge of the traffic in the network, which is a challenge to obtain in a distributed network architecture.

Research in fairness may benefit from a cross-layer design, in which the MAC layer is aware of information about flows from the network or transport layer protocol. Such knowledge may improve both contention and allocation based approaches to fairness.

3.7.3 Power Management

Recent measurements of power consumption of devices with wireless network interfaces shows that the transmit power is typically a constant multiple higher than the receive power, which is an order of magnitude higher than the idle state [52]. In light of these measurements, one way to reduce the power consumption of a transmission is to reduce the number of nodes that "overhear" the transmission.

One way to reduce packet overhearing is for nodes to power off when not transmitting or receiving a packet. This approach introduces many new problems such as where to buffer packets for the node that is off, how long a node should remain off, etc., not to mention to decide which nodes to power off. Another approach to reduce packet overhearing is through power-control, that is, by controlling the transmit power level of a node. The *power aware multiple access* (PAMAS) protocol [49] is an example that takes the former approach while the *power controlled multiple access* (PCMA) protocol [35] is an example that takes the latter approach.

Power management is especially important in sensor networks since it may not be feasible to replace or recharge the batteries the nodes use as their power source. In addition, how to transition from a low power state to respond to an event detected by a group of sensors is a challenging issue.

3.7.4 Smart Antennas

An assumption underlying the discussion on MANETs has been the use of an omnidirectional antenna system. Smart antennas are directional antennas that can change the direction, shape, and power of the antenna beam. The use of directional antennas in cellular networks has been well demonstrated [46], however, the flexibility they offer has not been well explored in MANETs.

The D-MAC protocol, introduced by Ko et al. [33], has two schemes similar to IEEE 802.11. It is assumed that nodes can transmit both omnidirectionally and directionally. In the first scheme, the RTS packet is sent directionally (DRTS) whereas the CTS reply is transmitted omnidirectionally (OCTS). If this exchange is successful, the data and ACK are also transmitted directionally. The difference from IEEE 802.11 for omnidirectional antennas only is that the neighbor of the destination can initiate a concurrent transmission assuming its destination is in a different direction. Since using DRTS can increase the probability of collisions, the second scheme uses either DRTS or ORTS depending on whether the channel is blocked in all directions. This protocol assumes that the location of nodes is known.

Takai et al. [54] present a directional virtual carrier sensing (DVCS) mechanism for directional antennas. DVCS determines the availability of a channel in a specific direction. To implement DVCS, three capabilities are added to IEEE 802.11. First, the incoming angle of arrival information is cached. This is used to make decisions about whether to transmit directionally or omnidirectionally. Second, the use of beam pattern locking during transmission helps maximize signal power at the receiver. Third, the protocol makes use of a directional network allocation vector (DNAV). A direction, beam width, and timer is associated with each entry, indicating how long the channel is unavailable in a particular direction.

A DNAV was proposed independently by Choudhury et al. [15], however, it is less general than that described in Reference 54. The proposed protocol attempts to take advantage of the higher gain of directional antennas. Though a destination may not be reachable through an omnidirectional transmission, the RTS packet is routed to the destination across multiple hops. On successful receipt of the routed RTS, the destination responds with a directional CTS (DCTS). On receipt of the DCTS, the data and ACK are also transmitted directionally.

Nasipuri et al. [37] propose a protocol similar to IEEE 802.11 but do not assume the location of neighbors is known. In this protocol, the RTS and CTS are both sent omnidirectionally. The received signal strength is used to set the direction of the sender and receiver, with data and ACK transmitted directionally.

Ramanathan [44] extends the CSMA/CA protocol with directional antennas using an aggressive and a conservative collision avoidance model. In the aggressive model, the RTS/CTS handshake is used only to ensure that the receiver is not already busy. In the conservative model, a node always blocks on receipt of a RTS or CTS. Experimental results show that the aggressive approach consistently outperforms the conservative approach.

In general, how to exploit the higher gain and directionality offered by smart antennas is not well understood.

An alternate method for improving throughput in MANETs is the use of multiple orthogonal channels that can be shared by all the nodes in the network. Chapter 5 emphasizes allocation approaches to access control in multichannel architectures, while Chapter 6 emphasizes contention approaches in this setting.

3.8 Summary

The problem of medium access control MAC is fundamental in all wireless networks since a broadcast channel underlies each one. We first introduced the three basic techniques of medium access: FDMA, TDMA, and spread spectrum. MAC protocols for centralized network architectures were overviewed. This included cellular networks and a discussion of the progression of the standards path from 2G to 3G. Paging systems, which are among the few simplex systems, were also discussed. Satellite networks are unique because of their coverage, and also the transmission delays involved. A discussion of MAC protocols for distributed network architectures then follows. In MANETs, protocols are roughly categorized as contention based or allocation based, according to whether time is viewed as continuous or discrete, respectively. Some hybrids, that combine contention and allocation were highlighted. MAC protocols under development for sensor networks, a form of energy limited ad hoc network, are summarized. Research issues for MANETs and sensor networks are explored since these networks are among the least mature of existing wireless networks. These research issues include support for QoS, fairness, improving power management, and exploiting smart antennas and cross-layer information. The design of MAC protocols critically depends on the technological constraints. Thus, the emergence of new wireless technologies and systems will continue to pose new challenges to MAC protocol design.

Acknowledgment

This work was supported in part by National Science Foundation grant NSF ANI-0105985.

References

[1] IEEE Standard 802.11: Wireless LAN Medium Access Control and Physical Layer Specifications, December 1999.

[2] N. Abramson, "The ALOHA system — another alternative for computer communications," in *Proceedings of the AFIPS Fall Joint Computer Conference*, 1970, pp. 281–285.

[3] I.F. Akyildiz, W. Su, Y. Sankarasubramaniam, and E. Cayirci, "wireless sensor networks: a survey," *Computer Networks*, Vol. 38, 2002, pp. 393–422.

[4] D. Aldous, "Ultimate stability of exponential backoff protocol for acknowledgement based transmission control of random access communication channels," *IEEE Transactions on Information Theory*, Vol. 33, 1987, pp. 219–223.

[5] M. Benveniste, G. Chesson, M. Hoeben, A. Singla, H. Teunissen, and M. Wentink, "Enhanced distributed coordination function (EDCF) proposed draft text," IEEE working document 802.11-01/131r1, March 2001.

[6] D. Bertsekas and R. Gallager, *Data Networks*, Prentice Hall, Inc., New York, 1992.

[7] V. Bharghavan, A. Demers, S. Shenker, and L. Zhang, "MACAW: a media access protocol for wireless LANs," in *Proceedings of the SIGCOMM'94*, 1994, pp. 212–225.

[8] A. Boukerche, "Special issue on wireless and mobile ad hoc networking and computing," *International Journal of Parallel and Distributed Computing*, Vol. 63, 2003.

[9] A. Chandra, V. Gummalla, and J.O. Limb, "Wireless medium access control protocols," *IEEE Communications Surveys*, Second Quarter, 2000, pp. 2–15.

[10] I. Chlamtac and A. Faragó, "Making transmission schedules immune to topology changes in multi-hop packet radio networks," *IEEE/ACM Transactions on Networking*, Vol. 2, 1994, pp. 23–29.

[11] I. Chlamtac, A. Faragó, A.D. Myers, V.R. Syrotiuk, and G. Záruba, "ADAPT: a dynamically self-adjusting media access control protocol for ad hoc networks," in *Proceedings of the IEEE Global Telecommunications Conference (Globecom'99)*, December 1999, pp. 11–15.

[12] I. Chlamtac, A. Faragó, and H. Zhang, "A fundamental relationship between fairness and optimum throughput in TDMA protocols," in *Proceedings of the IEEE International Conference on Universal Personal Communications (ICUPC'96)*, September 1996, pp. 671–675.

[13] I. Chlamtac, A. Faragó, and H. Zhang, "Time-spread multiple-access (TSMA) protocols for multihop mobile radio networks," *IEEE/ACM Transactions on Networking*, Vol. 5, 1997, pp. 804–812.

[14] I. Chlamtac and S. Pinter, "Distributed node organization algorithm for channel access in a multi-hop packet radio network," *IEEE Transactions on Computers*, Vol. 36, 1987, pp. 728–737.

[15] R.R. Choudhury, X. Yang, R. Ramanathan, and N.H. Vaidya, "Using directional antennas for medium access control in ad hoc networks," in *Proceedings of the ACM International Conference on Mobile Networking and Computing (MobiCom'02)*, September 2002, pp. 59–70.

[16] C.J. Colbourn and J.H. Dinitz (eds.), *The CRC Handbook of Combinatorial Designs*, CRC Press, Inc., Boca Raton, FL, 1996.

[17] C.J. Colbourn, A.C.H. Ling, and V.R. Syrotiuk, "Cover-free families and topology-transparent scheduling for MANET," *Designs, Codes, and Cryptography*, Vol. 32, 2004, pp. 65–96.

[18] C.J. Colbourn, V.R. Syrotiuk, and A.C.H. Ling, "Steiner systems for topology-transparent access control in MANETs," in *Proceedings of the 2nd International Conference on Ad Hoc Networks and Wireless (AdHoc Now'03)*, October 2003, pp. 247–258.

[19] W. Crowther, R. Rettberg, D. Walden, S. Orenstein, and F. Heart, "A system for broadcast communication: reservation-ALOHA," in *Proceedings of the 6th Hawaii International System Science Conference*, January 1973, pp. 596–603.

[20] J. Deng and Z.J. Haas, "Dual busy tone multiple access (DBTMA): a new medium access control for packet radio networks," in *Proceedings of the IEEE ICUPC'98*, Vol. 2, October 1998, pp. 973–977.

[21] A. Faragó, A.D. Myers, V.R. Syrotiuk, and G. Záruba, "Meta-MAC protocols: automatic combination of MAC protocols to optimize performance for unknown conditions," *IEEE Journal on Selected Areas in Communications*, Vol. 18, 2000, pp. 1670–1681.

[22] A. Faragó and V.R. Syrotiuk, "Medium Access Control (MAC) protocols," in *Encyclopedia of Telecommunications*, J. Proakis (ed.), John Wiley & Sons, New York, 2002.

[23] J.J. Garcia-Luna-Aceves and C. Fullmer, "Floor acquisition multiple access (FAMA) in single-channel wireless networks", *ACM Mobile Networks and Applications Journal*, Vol. 4, 1999, pp. 157–174.

[24] J.J. Garcia-Luna-Aceves and A. Tzamaloukas, "Reversing the collision-avoidance handshake in wireless networks," in *Proceedings of the ACM/IEEE International Conference on Mobile Networking and Computing (Mobicom'99)*, August 1999, pp. 120–131.

[25] Globalstar, San Jose, CA, http://www.globalstar.com

[26] J. Hastad, F.T. Leighton, and B. Rogoff, "Analysis of backoff protocols for multiple access channels," in *Proceedings of the ACM Symposium on the Theory of Computing (STOC'87)*, May 1987, pp. 241–253.

[27] A.S. Hedayat, N.J.A. Sloane, and J. Stufken, *Orthogonal Arrays, Theory and Applications*, Springer-Verlag, New York, Inc., 1999.

[28] European Telecommunications Standards Institute (ETSI), http://www.etsi.org

[29] M.R.L. Hodges, "The GSM radio interface," *British Telecom Technological Journal*, Vol. 8, 1990, pp. 31–43.

[30] J.-H. Ju and V.O.K. Li, "An optimal topology-transparent scheduling method in multihop packet radio networks," *IEEE/ACM Transactions on Networking*, Vol. 6, 1998, pp. 298–306.

[31] P. Karn, "MACA – a new channel access protocol for packet radio," in *Proceedings of the ARRL/CRRL Amateur Radio 9th Computer Networking Conference*, 1990, pp. 134–140.

[32] F.P. Kelly, "Stochastic models of computer communication systems," *Journal of Royal Statistical Society (B)*, Vol. 47, 1985, pp. 379–395.

[33] Y.-B. Ko, V. Shankarkumar, and N. Vaidya, "Medium access control protocols using directional antennas in ad hoc networks", in *Proceedings of the IEEE Infocom'00*, March 2000, pp. 13–21.

[34] R.M. Metcalfe and D.R., Boggs, "Ethernet: distributed packet switching for local computer networks," *Communications of the ACM*, Vol. 19, 1976, pp. 395–404.

[35] J.P. Monks, V. Bharghavan, and W.-M.W. Hwu, "A power controlled multiple access protocol for wireless packet networks," in *Proceedings of the IEEE Infocom'01*, April 2001, pp. 219–228.

[36] A.D. Myers and S. Basagni, "Wireless media access control," in *Handbook of Wireless Networks and Mobile Computing*, Chapter 6, I. Stojmenovic (ed.), John Wiley & Sons, Inc., New York, 2002, pp. 119–143.

[37] A. Nasipuri, S. Ye, J. You, and R.E. Hiromoto, "A MAC protocol for mobile ad hoc networks using directional antennas," in *Proceedings of the IEEE Wireless Communications and Networking Conference (WCNC'00)*, September 2000, pp. 23–28.

[38] T. Nguyen and T. Suda, "Survey and evaluation of multiple access protocols in multimedia satellite networks," in *Proceedings of the IEEE Southeastcon'90*, 1990, pp. 408–413.

[39] T. Ozugur, M. Naghshineh, P. Kermani, and J.A. Copeland, "Fair media access for wireless LANs," in *Proceedings of the Globecom'99*, Vol. 1b, December 1999, pp. 570–579.

[40] C.E. Perkins (ed), *Ad Hoc Networking*, Addison-Wesley, Inc., New York, 2001.

[41] H. Peyravi, "Medium access control protocols for space and satellite communications: a survey and assessment," manuscript, August 1997. http://www.cosy.sbg.ac.at/ pkubesch/studium/mac_protocols/mac_protocols.html

[42] T. Pratt, C. Bostian, and J. Allnutt, *Satellite Communications*, John Wiley & Sons, Inc., New York, 2003.

[43] K. Raith and J. Uddenfeldt, "Capacity of digital cellular TDMA systems," *IEEE Transactions on Vehicular Technology*, Vol. 40, 1991, pp. 323–331.

[44] R. Ramanathan, "On the performance of ad hoc networks with beamforming antennas," in *Proceedings of the 2nd ACM International Symposium on Mobile Ad Hoc Networking and Computing (MobiHoc'01)*, 2001, pp. 95–105.

[45] M. Ranhema, "Overview of the GSM system and protocol architecture," *IEEE Communications Magazine*, April 1993, pp. 92–100.

[46] T.S. Rappaport, *Wireless Communications, Principles and Practice*, 2nd Ed., Prentice-Hall, Inc., New York, 2002.

[47] L. Romdhani, Q. Ni, and T. Turletti, "AEDCF: enhanced service differentiation for IEEE 802.11 wireless ad-hoc networks," *INRIA Research Report*, 2002, No. 4544.

[48] A. Sen and M.L. Huson, "A new model for scheduling packet radio networks," in *Proceedings of the IEEE INFOCOM'96*, 1996, pp. 1116–1124.

[49] S. Singh, C.S. Raghavendra, and J. Stepanek, "Power-aware broadcasting in mobile ad hoc networks," in *Proceedings of the 10th IEEE International Symposium on Personal, Indoor and Mobile Radio Communication (PIMRC'99)*, September 1999, pp. 22–31.

[50] J.L. Sobrinho and A.S. Krishnakumar, "Quality-of-service in ad hoc carrier sense multiple access wireless networks," *IEEE Journal on Selected Areas in Communications*, Vol. 17, 1999, pp. 1352–1368.

[51] K. Sohrabi, J. Gao, V. Ailawadhi, and G.J. Pottie, "Protocols for self-organization of a wireless sensor network," *IEEE Personal Communications*, Vol. 7, 2000, pp. 16–27.

[52] M. Stemm, P. Gauthier, and D. Harada, "Reducing power consumption of network interfaces in hand-held devices," in *Proceedings of the 3rd International Workshop on Mobile Multimedia Communications (MOMUC'96)*, September 1996.

[53] V.R. Syrotiuk, C.J. Colbourn, and A.C.H. Ling, "Topology-transparent scheduling in MANETs using orthogonal arrays," in *Proceedings of the DIALM-POMC Joint Workshop on Foundations of Mobile Computing*, September 2003, pp. 43–49.

[54] M. Takai, J. Martin, A. Ren, and R. Bagrodia, "Directional virtual carrier sensing for directional antennas in mobile ad hoc networks," in *Proceedings of the 3rd ACM International Symposium on Mobile Ad Hoc Networking and Computing (MobiHoc'02)*, 2002, pp. 183–193.

[55] Z. Tang and J.J. Garcia-Luna-Aceves, "A protocol for topology-dependent transmission scheduling in wireless networks," in *Proceedings of the IEEE WCNC'99*, Vol. 3, September 1999, pp. 1333–1337.

[56] A. Tobagi and L. Kleinrock, "Packet switching in radio channels, part II: the hidden terminal problem in carrier sense multiple access and the busy-tone solution," *IEEE Transactions of Communications*, Vol. 23, 1975, pp. 1417–1433.

[57] F. Tobagi and L. Kleinrock, "Packet switching in radio channels: part III — Polling and (Dynamic) split-channel reservation multiple access," *IEEE Transactions of Communications*, Vol. 24, 1976, pp. 832–845.

[58] N.H. Vaidya, P. Bahl, and S. Gupta, "Distributed fair scheduling in a wireless LAN," in *Proceedings of the MobiCom'00*, 2000, pp. 167–178.

[59] A. Woo and D.E. Culler, "A transmission control scheme for media access in sensor networks," in *Proceedings of the Mobicom'01*, July 2001, pp. 221–235.

[60] W.R. Young, "Advanced mobile phone service: introduction, background, and objectives," *Bell Systems Technical Journal*, Vol. 58, 1979, pp. 1–14.

[61] C. Zhu and S. Corson, "A five-phase reservation protocol (FPRP) for mobile ad hoc networks," in *Proceedings of the IEEE INFOCOM'98*, March/April 1998, pp. 322–331.

4

Distributed Channel Access Scheduling for Ad Hoc Networks

4.1 Introduction... 4-45
4.2 Background Review.. 4-46
   Dynamic Reservation TDMA • Graph Theoretic Approaches • Topology-Transparent Scheduling
4.3 Neighbor-Aware Contention Resolution................ 4-51
   Specification • Dynamic Resource Allocation • Performance
4.4 Channel Access Scheduling for Ad Hoc Networks..... 4-55
   Modeling of Network and Contention • Code Assignment • Node Activation Multiple Access (NAMA) • Link Activation Multiple Access (LAMA) • Pairwise-Link Activation Multiple Access (PAMA) • Hybrid Activation Multiple Access (HAMA)
4.5 Throughput Analysis in Ad Hoc Networks 4-62
   Geometric Modeling • Throughput Analysis for NAMA • Throughput Analysis for HAMA • Throughput Analysis for PAMA • Throughput Analysis for LAMA • Comparisons among NAMA, HAMA, PAMA, and LAMA • Comparisons with CSMA and CSMA/CA
4.6 Simulations... 4-69
4.7 Research Challenges 4-72
References 4-74

Lichun Bao
J.J. Garcia-Luna-Aceves

4.1 Introduction

In this chapter, we present and analyze protocols for time-division multiple access (TDMA) scheduling in ad hoc networks with omnidirectional antennas. These protocols use a neighbor-aware contention resolution (NCR) algorithm to elect deterministically one or multiple winners in a given contention context based on the topology information within two hops. In NCR, the identifiers and the current contention context number are used to derive a randomized priority for each contender in a given contention context. A contention context corresponds to a time slot in TDMA schemes, and the contenders of a given node during a contention context are those nodes one and two hops away from the node. Each contender runs

NCR to determine locally its eligibility to access the resource in the contention context by comparing its priority with the priorities of the rest of the contenders.

Based on NCR, the node activation multiple access protocol (NAMA) elects nodes for collision-free broadcast transmissions over a single channel. The link activation multiple access (LAMA), the pairwise link activation multiple access (PAMA), and the hybrid activation multiple access (HAMA) protocols operate over multiple channels that are orthogonal by codes or frequencies to elect either links or nodes for collision-free unicast transmissions, or a mix of broadcast and unicast transmissions. The throughput and delay characteristics of these protocols are studied by analysis and simulation in multihop wireless networks with randomly generated topologies. The performance of the new protocols is also compared against a well-known static scheduling algorithm based on complete topology information, and the ideal carrier sensing multiple access (CSMA) and carrier sensing multiple access with collision avoidance (CSMA/CA) protocols.

After Section 4.2 reviewing the prior work on schedules channel access in wireless ad hoc networks, Section 4.3 presents the NCR algorithm and analyzes the packet delay encountered in a general queuing model under certain contention level. Section 4.4 describes four scheduling protocols based on NCR, including the packet transmission scheduling based on NAMA suitable for broadcast transmission, LAMA, and PAMA for unicast packet transmission, and HAMA for both unicast and broadcast transmissions. Section 4.5 derives the channel access probabilities of the four protocols in randomly generated ad hoc networks, and compares the throughput of the protocols with that of the ideal CSMA and CSMA/CA schemes. Section 4.6 presents the results of simulations that provide further insights to the performance differences among the four scheduling protocols and the corresponding static scheduling approaches based on a unified multiple access scheduling framework Unified T/O/C Division Multiple Access (UxDMA) [22]. Section 4.7 summarizes the research and proposes open problems in the area.

4.2 Background Review

There has been considerable work on cellular networks using the TDMA for data communication. These multiple access protocols allocate reservation slots to resolve contentions and data information slots to transmit data packets. Examples of this type of approach are D-TDMA (Dynamic TDMA) [12], PRMA (Packet Reservation Multiple Access) [19], RAMA (Resource Auction Multiple Access) [18], and DRMA (Dynamic Reservation Multiple Access) [21]. Despite there are many similarities between the medium access control protocols for cellular and ad hoc wireless networks, ad hoc networks present a distinct multihop and distributed scenario in nature, and require special considerations in protocol designs. We provide a brief survey of related time division multiplexing approaches that handles the characteristics of ad hoc networks.

Channel access protocols for ad hoc networks can be nondeterministic or deterministic. The nondeterministic approach started with ALOHA and CSMA [15] and continued with several collision avoidance schemes, of which the IEEE 801.11 series standards for wireless LANs [6] being the most popular examples to date. However, as the network load increases, network throughput drastically degrades in these nondeterministic approaches because the probability of collisions rises, preventing stations from acquiring the channel.

On the other hand, deterministic access schemes set up timetables for individual nodes or links, such that the transmissions from the nodes or over the links are conflict-free in the code, time, frequency, or space divisions of the channel. The schedules for conflict-free channel access can be established based on the topology of the network, or it can be topology independent.

Topology-dependent channel access control algorithms can establish transmission schedules by either dynamically exchanging and resolving time slot requests [5,28], or pre-arrange a timetable for each node based on the network topologies. Setting up a conflict-free channel access time-table is typically treated as node- or link-coloring problems on graphs representing the network topologies. The problem of optimally scheduling access to a common channel is one of the classic NP-hard problems in graph theory

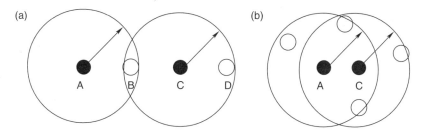

FIGURE 4.1   The hidden terminal and exposed terminal problems.

(k-colorability on nodes or edges) [9,10,23]. Polynomial algorithms are known to achieve suboptimal solutions using randomized approaches or heuristics based on such graph attributes as the degree of the nodes.

A unified framework for TDMA/FDMA/CDMA channel assignments, called UxDMA algorithm, was described by Ramanathan [22]. UxDMA summarizes the patterns of many other channel access scheduling algorithms in a single framework. These algorithms are represented by UxDMA with different parameters. The parameters in UxDMA are the constraints put on the graph entities (nodes or links) such that entities related by the constraints are colored differently. Based on the global topology, UxDMA computes the node or edge coloring, which correspond to channel assignments to these nodes or links in the time, frequency or code domain.

A number of topology-transparent scheduling methods have been proposed [4,13,16] to provide conflict-free channel access that is independent of the radio connectivity around any given node. The basic idea of the topology-transparent scheduling approach is for a node to transmit in a number of time slots in each time frame. The times when node i transmits in a frame corresponds to a unique code such that, for any given neighbor k of i, node i has at least one transmission slot during which node k and none of k's own neighbors are transmitting. Therefore, within any given time frame, any neighbor of node i can receive at least one packet from node i conflict-free. An enhanced topology-transparent scheduling protocol, time spread multiple access (TSMA), was proposed by Krishnan and Sterbenz [16] to reliably transmit control messages with acknowledgments. However, TSMA performs worse than CSMA in terms of delay and throughput [16].

In general, the medium access control protocols in ad hoc networks have to combat two problems, namely, "hidden terminal problem" [25] and "exposed terminal problem." The "exposed terminal problem" is exacerbated in providing reliable broadcast services. The two problems are illustrated in Figure 4.1.

4.2.1 Dynamic Reservation TDMA

The TDMA schemes based on dynamic reservations constitute a commonly adopted approach that adjusts the time slot assignments according to the traffic requirements of individual nodes. The reservation can be made by either in-band signaling or out-band signal, where in-band signals are exchanged in the same channel as data traffic, and out-band signaling uses a separate channel for control information exchanges. However, these two methods are equivalent.

Five-Phase Reservation Protocol (FPRP) was designed to produce TDMA broadcast schedule in mobile ad hoc networks [28].

As shown in Figure 4.2, FPRP uses a similar scheme as D-TDMA in that the time is separated into two alternating activities — a reservation period and multiple information transmission periods. A successful time-slot reservation made during the reservation period lasts for a series of information periods, called information frames (IFs). Each information frame contains multiple time slots, called information slots (ISs), which are assigned to individual nodes according to the result of the reservation. Correspondingly, FPRP sets aside a reservation slot (RS) for each IS during the reservation period for nodes to contend using five-phase handshakes for m times, as indicated by 1, 2, 3, 4, 5 in Figure 4.2. Using multiple five-phase handshakes, an information slot can be allocated with high probability.

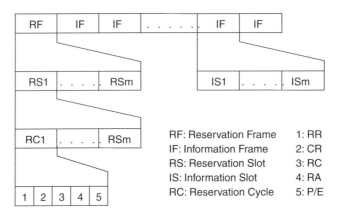

FIGURE 4.2   The five-phase reservation protocol.

The five phases are designated according to senders or receivers. The first phase is called "Reservation Request (RR)," where senders send out probes to the receivers of the communication. If the RR reaches the receiver successfully, the receiver keeps silent in the second phase to indicate a clear channel to the sender. Otherwise, the receiver replies with a "collision report (CR)." If the sender does not hear anything in the channel during the second phase, it is assured of the clear channel, and sends out "reservation confirm (RC)" in the third phase so that all the one-hop neighbors know about the reservation. Then the receiver responds with a "reservation acknowledgment (RA)" in the fourth phase, which establishes the connection with the sender. Note that an isolated node may get into phase three without any channel response, and only phase four confirms the existence of the receiver. In phase five, called "packing and elimination (P/E)," the neighbors of the receivers send "Packing Packet" to notify their one-hop neighbors that less nodes will compete in the next contention phases. In phase five, senders may also probabilistically send "Elimination Packet" in order to break the tie where two senders are adjacent to each other.

Because of the relative simple meaning of each phase, the handshake packets can be made small enough to indicate the channel busy state, and the five-phase reservation cycle becomes compact.

The FPRP resembles IEEE 802.11 DCF [20] in collision avoidance and HiperLAN [8] in sender deadlock elimination. The only difference is that FPRP is a synchronous protocol, requiring tight timing between phases and short propagation delay between nodes, which are stringent in mobile ad hoc networks. In addition to its complexity, the fixed numbers of information frames and reservation cycles in each reservation slot are not suitable in mobile and ad hoc environments where the density and topology of the network change frequently. Although FPRP successfully avoids the hidden terminal problem, it still suffers from the "exposed terminal problem" with certain "deadlock" probability as shown in the fifth phase in FPRP. Therefore, FPRP can only provide reliable broadcast schedules with high probability.

Other approaches in the same reservation-based vein differ in the handshaking procedures and the arrangement of reservation phases with regard to the data transmission phases, such as [5,24].

4.2.2   Graph Theoretic Approaches

Besides the various contention based resource reservation protocols, the channel access problem is often reduced to graph coloring problems, where the edge- and node-coloring solutions are directly applied to the link- and node-activation schemes of the channel access problems in ad hoc networks.

In Reference 9, Ephremides and Truong proved that the optimal broadcast scheduling problem (BSP) in ad hoc networks is NP-complete by showing that the well-known NP-complete problem of finding the maximum independent set (MIS) in a graph reduces to the BSP. We briefly describe the reduction steps.

The basic approach in reducing the MIS of a graph is to add a vertex for each edge in the graph, and connect the vertex with the end-points of the edge. In addition, the newly added vertexes are fully connected.

For instance in Figure 4.3(a), the original graph G is place in plane 1, and the added vertexes are in plane 2. When the original edges are removed, it gives a reduced graph G' as shown in Figure 4.3(b). Because a broadcast schedule requires the set of broadcasting nodes be separated by at least two hops to avoid hidden terminal problem, it is easy to see that the maximum broadcast set in G' consists of vertexes only from the original graph G. Otherwise there can be only one broadcast vertex if it is selected from plane 2. Then, finding the maximum broadcast set in G' is identical to finding the maximum independent set of graph G and vice versa.

A distributed implementation of finding the maximal-throughput broadcast schedule was provided in Reference 9, which starts from a simple and inefficient time slot assignment and iteratively assigns more spatially reusable time slots. The drawback of the approach is that the convergence of the algorithm is proportional to the diameter of the network topology. Because the distributed protocol is sensitive to topological change, a topological change may be required for computing the broadcast schedule from scratch.

Channel assignment problems in the time, frequency, and code domains are studied in other works as well. Ramanathan explored a set of eleven atomic "constraints" that characterize the assignment problems within and across these domains, and introduced a unified framework, called UxDMA, for the study of assignment problems using graph coloring algorithms. Most assignment problems can be represented as various combinations of these atomic constraints. We describe UxDMA as follows.

In UxDMA, a wireless network is represented as a directed graph $G = (V, A)$, where V is a set of vertexes denoting the nodes in the network, and A is a set of directed edges between vertexes representing wireless links in the network. A directional edge $(i, j) \in A$ means vertex v can receive packets from vertex u. Communication constraints over the shared wireless media in the frequency, time, and code division domains are shown in Figure 4.4, which illustrates the eleven atomic relations between vertexes and directional edges. The solid dots are transmitters, and the circles are receivers. Wide lines indicate that the lines are activated, and thin lines are interferences.

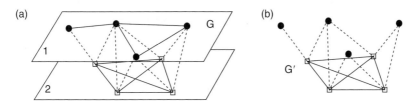

FIGURE 4.3 The reduction of maximum independent set to broadcast schedule.

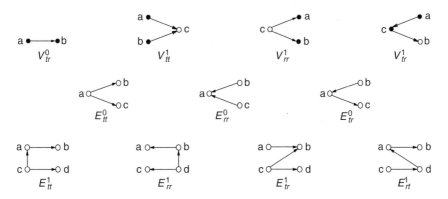

FIGURE 4.4 Constraints used by UxDMA for channel access scheduling.

A constraint is represented using syntax X_z^y, where X indicates whether the constrained entities are vertexes or edges, y indicates separation distance between the constrained entities, and z indicates the transmission or reception relation between the constrained entities. For instance, constraints V_{tt}^1, E_{rr}^0, and E_{tr}^1 forbid hidden terminal problem in the channel access schedules, while V_{tr}^1 and E_{tt}^1 forbid exposed terminal problem in the channel access schedules, as shown in Figure 4.1.

Assignment problems are characterized by a set of constraints. For instance, $C = \{V_{tr}^0, V_{tt}^1, V_{tr}^1\}$ is a constraint set for broadcast scheduling. An assignment problem specified by a constraint set requires an assignment that eliminates all the possible scenarios in the constraint set. For instance, a broadcast schedule under constraint set $C = \{V_{tr}^0, V_{tt}^1, V_{tr}^1\}$ forbids that two vertexes can have the same color if they are either adjacent (V_{tr}^0, V_{tr}^1) or have a common neighbor (V_{tt}^1).

Given the constraint set of a channel access scheduling problem, the coloring of the graph becomes an orderly procedure on the vertexes or edges based on certain heuristics, such as "maximum-degree vertex first" approach. Each step walks through uncolored vertexes in the ordered list of vertexes, and the minimum unused color is found and assigned to a vertex or an edge, subject to the constraint set with regard to the already colored vertexes or edges. Such steps repeat until all the entities in the graph are colored. As we can see, the complexity of the UxDMA for any channel access scheduling problem is polynomial based on the heuristics.

The UxDMA can represent a large class of channel access problems, and provides a theoretic bound for such problems. However, just like other approaches that study the channel access problems from a centralized point of view, the scalability and adaptiveness of such algorithms in mobile ad hoc network are the main issues.

4.2.3 Topology-Transparent Scheduling

Schedules based on TDM depend critically on the actual network topology. So in a mobile environment, due to nodal mobility and potentially rapid change of topology, TDM schemes may require prohibitive overhead associated with constant updating of schedules.

Several approaches based on the topology-transparent scheduling schemes that are based on different theories have been proposed [3,4,13,14] to reduce the overheads incurred from the conflict-free schedule maintenance. However, they require that the maximum number of nodes and the maximum degree of nodes in the network is known beforehand to guarantee successful transmission of each data frame to all one-hop neighbors.

The TSMA proposed by Chlamtac et al. [3,4], and further refined by Ju and Li [13] uses polynomials over Galois fields to produce appropriate TSMA codes, and assigns them to the network nodes. In addition, Ju and Li proposed another topology-transparent channel access scheduling approach based on Latin Squares [7] for ad hoc networks with multiple channels [14].

A Latin square of order n is defined as an $n \times n$ matrix composed of n symbols $\{1, 2, \ldots, n\}$ such that symbols in each row and column are distinct. Two Latin squares $A = (a_{i,j})$ and $B = (b_{i,j})$ are said to be orthogonal if the n^2 ordered pairs $(a_{i,j}, b_{i,j})$ are all different. For example, the following two square matrices A and B using symbols $\{1, 2, 3, 4\}$ are both Latin squares, and mutually orthogonal. Generally, latex squares $A^{(1)}, A^{(2)}, \ldots, A^{(r)}$ form an orthogonal family if every pair of them is orthogonal.

$$A = \begin{bmatrix} 1 & 2 & 3 & 4 \\ 2 & 3 & 4 & 1 \\ 3 & 4 & 1 & 2 \\ 4 & 1 & 2 & 3 \end{bmatrix} \qquad B = \begin{bmatrix} 4 & 1 & 2 & 3 \\ 3 & 2 & 1 & 4 \\ 1 & 4 & 3 & 2 \\ 2 & 3 & 4 & 1 \end{bmatrix}$$

The TDMA scheduling algorithm in Reference 14 maps a $p \times p$ Latin square onto an $M \times p$ time division multiple channels, where the number of channels is M, and the number of time slots in each frame is p. Then it assigns a unique symbol from the Latin square to each node, and the positions of the symbol in the Latin square determines the time slots assigned to the node.

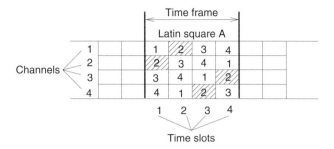

FIGURE 4.5   Multi-channel TDMA time frame structure.

For example, in Figure 4.5, a network is given four channels, and each frame contains four time slots. The assignments of time slots are given in the figure according to the symbols 1, 2, 3, 4 in the Latin square A given above. Therefore, if a node is assigned symbol 2, it can access the first channel in time slot 2, the second channel in time slot 1, the third channel in time slot 4, and the fourth channel in time slot 3.

In addition, it is assumed that each node is equipped with p receivers for all the channels, and one transmitter to send data frames in the assigned time slots. Each data frame is transmitted for p times in each time frame.

However, the available symbols would have been scarce if only p symbols are available to networks with more than p nodes. This is where the orthogonal Latin matrices come into the solution in Reference 14 by assigning the same symbol from different orthogonal Latin matrices to nodes. This is explained as follows.

From the definition of orthogonal matrix, we know that any pair of order symbol (a, b) appears at most once in all positions of two orthogonal matrices. Therefore, the symbol/time slot assignments to two nodes conflict in one position at most. Given that a data frame is transmitted multiple times in multiple channels, the successful transmission chances between two nodes are guaranteed. By requiring the minimum order of the orthogonal matrices, which is a function of the nodal degree, the number of collisions at a node from other neighbors with different symbol assignments is kept below p. Consequently, every data frame from a node can be successfully received by its neighbors in one of the p transmissions. Given an orthogonal Latin matrix family of r members, a total of $p \times r$ symbols can be assigned in a network.

Overall, the topology-transparent TDMA scheme based on Galois Field needs only a single channel and multiple frames for each transmission session, while the scheme based on Latin square specifies a multiple-channel TDMA scheduling protocol. It remains to be proved that these approaches based on Galois field and Latin square are based on the same theory because of their many similarities.

4.3   Neighbor-Aware Contention Resolution

4.3.1   Specification

The NCR solves a special election problem where an entity to locally decide its leadership among a known set of contenders in any given contention context. We assume that the knowledge of the contenders for each entity is acquired by some means. For example, in the ad hoc networks of our interest, the contenders of each node are the neighbors within two hops, which can be obtained by each node periodically broadcasting the identifiers of its one-hop neighbors [1]. Furthermore, NCR requires that each contention context be identifiable, such as the time slot number in networks based on a TDMA scheme.

We denote the set of contenders of an entity i as M_i, and the contention context as t. To decide the leadership of an entity without incurring communication overhead among the contenders, we assign each entity a priority that depends on the identifier of the entity and the current contention context.

```
NCR(i, t)
{
        /* Initialize. */
1       for (k ∈ M_i ∪ {i})
2           k.prio = Hash(k ⊕ t) ⊕k;

        /* Resolve leadership. */
3       if (∀k ∈ M_i, i.prio > k.prio)
4           i is the leader;
} /* End of NCR. */
```

FIGURE 4.6   NCR specification.

Equation (4.1) provides a formula to derive the priority, denoted by i.prio, for entity i in contention context t.

$$i.prio = \text{Hash}(i \oplus t) \oplus i, \tag{4.1}$$

where the function $\text{Hash}(\cdot)$ is a fast message digest generator that returns a random integer in a predefined range, and the sign '$\oplus$' is designated to carry out the concatenation operation on its two operands. Note that, although the $\text{Hash}(\cdot)$ function can generate the same number on different inputs, each priority number is unique because the priority is appended with identifier of the entity.

Figure 4.6 shows the NCR algorithm. Basically, NCR generates a *permutation* of the contending members, the order of which is decided by the priorities of all participants. Since the priority is a pseudo-random number generated from the contention context that changes from time to time, the permutation also becomes random such that each entity has certain probability to be elected in each contention context, which is inversely proportional to the contention level as shown in Equation (4.2).

$$q_i = \frac{1}{|M_i \cup \{i\}|}. \tag{4.2}$$

Conflicts are avoided because it is assumed that contenders have mutual knowledge and t is synchronized, the order of contenders based on the priority numbers is consistent at every participant.

4.3.2 Dynamic Resource Allocation

The description of NCR provided so far evenly divides the shared resource among the contenders. In practice, the demands from different entities may vary, which requires allocating different portion of the shared resource. There are several approaches for allocating variable portion of the resource according to individual demands. In any approach, an entity, say i, needs to specify its demand by an integer value chosen from a given integer set, denoted by p_i. Because the demands need to be propagated to the contenders before the contention resolution process, the integer set should be small and allow enough granularity to accommodate the demand variations while avoiding the excess control overhead caused by the demand fluctuations.

Suppose the integer set is $\{0, 1, \ldots, P\}$, the following three approaches provide resource allocation schemes, differing in the portion of the resource allocated on a given integer value. If the resource demand is 0, the entity has no access to the shared resource.

4.3.2.1 Pseudo-Identities

An entity assumes p pseudo-identities, each defined by the concatenation of the entity identifier and a number from 1 to p. For instance, entity i with resource demand p_i is assigned with the following pseudo-identities: $i \oplus 1, i \oplus 2, \ldots, i \oplus p_i$. Each identity works for the entity as a contender to the shared resource.

```
NCR-PI(i, t)
{
        /* Initialize each entity k with demand pₖ. */
1       for (k ∈ Mᵢ ∪ {i} and 1 ≤ l ≤ pₖ)
2           (k ⊕ l).prio = Hash(k ⊕ l ⊕ t) ⊕ k ⊕ l;

        /* Resolve leadership. */
3       if (∃k, l : k ∈ Mᵢ, 1 ≤ l ≤ pₖ and
4           ∀m : 1 ≤ m ≤ pᵢ, (k ⊕ l).prio > (i ⊕ m).prio)
5           i is not the leader;
6       else
7           i is the leader;
} /* End of NCR-PI. */
```

FIGURE 4.7   NCR-PI specification.

Figure 4.7 specifies NCR with pseudo-identities (NCR-PI) for resolving contentions among contenders with different resource demands.

The portion of the resource available to an entity i in NCR-PI according to its resource demand is as follows:

$$q_i = \frac{p_i}{\sum_{k \in M_i \cup \{i\}} p_k}. \tag{4.3}$$

4.3.2.2 Root Operation

Assuming enough computing power for floating point operations at each node, we can use the root operator to achieve the same proportional allocation of the resource among the contenders as in NCR-PI.

Given that the upper bound of function Hash in Equation (4.1) is M, substituting line 2 in Figure 4.6 with the following formula generates a new algorithm, which provides the same resource allocation characteristic as shown in Equation (4.3).

$$k.\text{prio} = \left(\frac{\text{Hash}(k \oplus t)}{M}\right)^{\frac{1}{p_k}}. \tag{4.4}$$

4.3.2.3 Multiplication

Simpler operations, such as multiplication in the priority computation, can provide non-linear resource allocation according to the resource demands. Substituting line 2 in Figure 4.6, Equation (4.5) offers another way of computing the priorities for entities.

$$k.\text{prio} = (\text{Hash}(k \oplus t) \cdot p_k) \oplus k. \tag{4.5}$$

According to Equation (4.5), the priorities corresponding to different demands are mapped onto different ranges, and entities with smaller demand values are less competitive against those with larger demand values in the contentions, thus creating greater difference in resource allocations than the linear allocation schemes provided by Equation (4.3) and Equation (4.4). For example, among a group of entities, a, b, and c, suppose $p_a = 1$, $p_b = 2$, $p_c = 3$, and $P = 3$. Then the resource allocations to a, b, and c are statistically $\frac{1}{3} \cdot \frac{1}{3} = 0.11$, $\frac{1}{3} \cdot \frac{1}{3} + \frac{1}{3} \cdot \frac{1}{2} = 0.28$, and $\frac{1}{3} \cdot \frac{1}{3} + \frac{1}{3} \cdot \frac{1}{2} + \frac{1}{3} \cdot \frac{1}{1} = 0.61$, respectively.

For simplicity, the rest of this chapter addresses NCR without dynamic resource allocation.

4.3.3 Performance

4.3.3.1 System Delay

We assume NCR as an access mechanism to a shared resource at a server (an entity), and analyze the average delay experienced by each client in the system according to the M/G/1 queuing model, where clients arrive at the server according to a Poisson process with rate λ and are served according to the first-come-first-serve (FIFO) discipline. Specifically, we consider the time-division scheme in which the server computes the access schedules by the time-slot boundaries, and the contention context is the time slot. Therefore, the queuing system with NCR as the access mechanism is an M/G/1 queuing system with server vacations, where the server takes a fixed vacation of one time slot when there is no client in the queue at the beginning of each time slot.

The system delay of a client using NCR scheduling algorithm can be easily derived from the extended Pollaczek–Kinchin formula, which computes the service waiting time in an M/G/1 queuing system with server vacations [2]

$$W = \frac{\lambda \overline{X^2}}{2(1 - \lambda \overline{X})} + \frac{\overline{V^2}}{2\overline{V}},$$

where X is the service time, and V is the vacation period of the server.

According to the NCR algorithm, the service time X of a head-of-line client is a discrete random variable, governed by a geometric distribution with parameter q, where q is the probability of the server accessing the shared resource in a time slot, as given by Equation (4.2). Therefore, the probability distribution function of service time X is

$$P\{X = k\} = (1 - q)^{k-1}q,$$

where $k \geq 1$. Therefore, the mean and second moments of random variable X are:

$$\overline{X} = \frac{1}{q}, \qquad \overline{X^2} = \frac{2 - q}{q^2}.$$

Because V is a fixed parameter, it is obvious that $\overline{V} = \overline{V^2} = 1$. Therefore, the average waiting period in the queue is:

$$W = \frac{\lambda(2 - q)}{2q(q - \lambda)} + \frac{1}{2}.$$

Adding the average service time to the queuing delay, we get the overall delay in the system:

$$T = W + \overline{X} = \frac{2 + q - 2\lambda}{2(q - \lambda)}. \tag{4.6}$$

The probabilities of the server winning a contention context are different, and so are the delays of clients going through the server. Figure 4.8 shows the relation between the arrival rate and the system delay of clients in the queuing system, given different resource access probabilities. To keep the queuing system in a steady state, it is necessary that $\lambda < q$ as implied by Equation (4.6).

4.3.3.2 System Throughput

Because of the collision freedom, NCR guarantees successful service to the clients. Therefore, the throughput of the server (the entity) over the shared resource is the minimum of the client arrival rate and the resource access probability. Considering all contenders for the shared resource, the overall system throughput is the summary of the throughput at individual entities. We have the following system throughput S combined from each and every entity k that competes for the shared resource:

$$S = \sum_k \min(\lambda_k, q_k), \tag{4.7}$$

where q_k is the probability that k may access the resource, and λ_k is the client arrival rate at k.

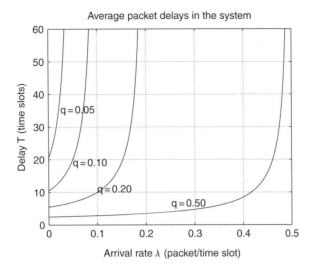

FIGURE 4.8 Average system delay of packets.

4.4 Channel Access Scheduling for Ad Hoc Networks

In this section, we apply the NCR algorithm to derive four channel access protocols in ad hoc networks with omnidirectional antennas.

4.4.1 Modeling of Network and Contention

We assume that each node is assigned a unique identifier, and is mounted with an omnidirectional radio transceiver that is capable of communicating using direct sequence spread spectrum (DSSS) on a pool of well-chosen spreading codes. The radio of each node only works in half-duplex mode, that is, either transmit or receive data packet at a time, but not both.

In multihop wireless networks, signal collisions may be avoided if the received radio signals are spread over different codes or scattered onto different frequency bands. Because the same codes on certain different frequency bands can be equivalently considered to be on different codes, we only consider channel access based on a code division multiple access scheme.

Time is synchronized at each node, and nodes access the channel based on slotted time boundaries. Each time slot is long enough to transmit a complete data packet, and is numbered relative to a consensus starting point. Although global time synchronization is desirable, only limited-scope synchronization is necessary for scheduling conflict-free channel access in multihop ad hoc networks, as long as the consecutive transmissions in any part of the network do not overlap across time slot boundaries. Time synchronization has to depend on physical layer timing and labeling for accuracy, and is outside the scope of this chapter.

The topology of a packet radio network is represented by a graph $G = (V, E)$, where V is the set of network nodes, and E is the set of links between nodes. The existence of a link $(u, v) \in E$ implies that $(v, u) \in E$, and that node u and v are within the transmission range of each other, so that they can exchange packets via the wireless channel. In this case, node u and v are called *one-hop neighbors* to each other. The set of one-hop neighbors of a node i is denoted as N_i^1. Two nodes are called *two-hop neighbors* to each other if they are not adjacent, but have at least one common one-hop neighbor. The neighbor information of node i refers to the union of the one-hop neighbors of node i itself and the one-hop neighbors of i's one-hop neighbors, which equals

$$N_i^1 \cup \left(\bigcup_{j \in N_i^1} N_j^1 \right).$$

In multihop wireless networks, a single radio channel is spatially reused at different parts of the network. Hidden-terminal problem is the main cause of interference and collision in ad hoc networks, and involves nodes within at most two hops. To ensure conflict-free transmissions, it is sufficient for nodes within *two hops* to not transmit on the same time, code, and frequency coordinates. Therefore, the topology information within two hops provides the contender information required by the NCR algorithm. When describing the operation of the channel access protocols, we assume that each node already knows its neighbor information within two hops. Bao and Garcia-Luna-Aceves described a neighbor protocol for acquiring this information in mobile ad hoc networks [1].

4.4.2 Code Assignment

We assume that the physical layer is capable of DSSS transmission technique. In DSSS, the code assignments are categorized into transmitter-oriented, receiver-oriented, or per-link-oriented schemes, which are also referred to as TOCA, ROCA, and POCA, respectively (e.g., [11,17]). The four channel access protocols described in this chapter adopt different code assignment schemes, thus providing different features.

We assume that a pool of well-chosen orthogonal pseudo-noise codes, $C_{pn} = \{c_k \mid k = 0, 1, \ldots\}$, is available in the signal spreading function. The spreading code assigned to node i is denoted by i.code. During each time slot t, a new spreading code is assigned to node i derived from the priority of node i, using Equation (4.8).

$$i.\text{code} = c_k, \quad k = i.\text{prio} \bmod |C_{pn}|. \tag{4.8}$$

Table 4.1 summarizes the notation used in the chapter to describe the channel access protocols.

4.4.3 Node Activation Multiple Access (NAMA)

The NAMA protocol requires that the transmission from a node is received by the one-hop neighbors of the node without collisions. That is, when a node is activated for channel access, the neighbors within two hops of the node should not transmit. Therefore, the contender set M_i of node i is the one-hop and two-hop neighbors of node i, which is $N_i^1 \cup (\bigcup_{j \in N_i^1} N_j^1) - \{i\}$.

Figure 4.9 specifies NAMA. Because only node i is able to transmit within its two-hop neighborhood when node i is activated, data transmissions from node i can be successfully received by all of its one-hop neighbor. Therefore, NAMA is capable of collision-free broadcast, and does not necessarily require code-division channelization for data transmissions.

Figure 4.10 provides an example of how NAMA operates in a multihop network. In the figure, the lines between nodes indicate the one-hop relationship, the dotted circles indicate the effective transmission ranges from nodes, and the node priorities in the current time slot are given beside each node. According

TABLE 4.1 Notation

i.prio	The priority of node i
(u, v).prio	The priority of link (u, v)
i.code	The code assigned to node i for either reception or transmission
i.state	The activation state of node i for either reception or transmission
Tx	Transmission state
Rx	Reception state
i.in	The sender to node i
i.out	The receiver set of node i
i.Q(i.out)	The packet queues for the eligible receivers in i.out
N_i^c	The set of one-hop neighbors assigned with code c at node i
[statement]	A more complex and yet easy-to-implement operation than an atomic statement, such as a function call

NAMA(i, t)

{

    /* Initialize. */

1    $M_i = N_i^1 \cup \left(\bigcup_{j \in N_i^1} N_j^1 \right) - \{i\}$;

2   **for** $(k \in M_i \cup \{i\})$

3        $k.\texttt{prio} = \texttt{Hash}(k \oplus t) \oplus k$;

    /* Resolve nodal state. */

4   **if** $(\forall k \in M_i, i.\texttt{prio} > k.\texttt{prio})$ {

5        $i.\texttt{state} = \texttt{Tx}$;

6        $i.\texttt{out} = N_i^1$;

7   [Transmit the earliest packet in $i.\texttt{Q}$ ($i.\texttt{out}$)];

8   }

9   **else** {

10      $i.\texttt{state} = \texttt{Rx}$;

11   [Listen to the channel];

12   }

} /* End of **NAMA**. */

FIGURE 4.9   NAMA specification.

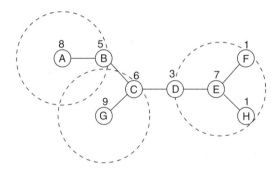

FIGURE 4.10   An example of NAMA operation.

to NAMA, there are three nodes A, G, and E able to transmit because their priorities are the highest in their respective two-hop neighborhood.

4.4.4 Link Activation Multiple Access (LAMA)

In LAMA, the code assignment for data transmission is receiver-oriented, which is suitable for unicasting using a link-activation scheme. The purpose of LAMA is to determine which node is eligible to transmit, and find out which outgoing link from the node can be activated in the current time slot.

Figure 4.11 specifies LAMA for activating a link from node i in time slot t. Node i first initializes the priorities and code assignments of nodes within two hops (lines 1–5), and determines its eligibility to transmit (line 6). If eligible, node i examines each reception code c assigned to its one-hop neighbors, and decides whether node i can activate links to the one-hop neighbor subset N_i^c, in which all nodes are assigned code c (lines 9–12). Here, the set of contenders to node i is N_i^c and one-hop neighbors of nodes in N_i^c, excluding node i (line 10). Then node i selects and transmits the earliest packet to one of the receivers in $i.\texttt{out}$ (lines 14–15 according to FIFO). If node i is not able to transmit, it listens on the code assigned to itself (lines 17–20).

Figure 4.12 illustrates a contention situation at node i in a time slot. The topology is represented by a undirected graph. The number beside each node represents the current priority of the node. Node j

LAMA(i, t)
{
    /* Initialize. */

1   **for** $(k \in N_i^1 \cup \left(\bigcup_{j \in N_i^1} N_j^1 \right))$

2      $k.\mathtt{prio} = \mathtt{Hash}(k \oplus t) \oplus k;$

3      $n = k.\mathtt{prio} \bmod |C_{pn}|;$

4      $k.\mathtt{code} = c_n;$

5   }

    /* Resolve nodal state. */

6   **if** $(\forall k \in N_i^1,\ i.\mathtt{prio} > k.\mathtt{prio})$ {

7      $i.\mathtt{state} = \mathtt{Tx};$

8      $i.\mathtt{out} = \emptyset;$

9   **for** $(c : \exists k \in N_i^1,\ c \equiv k.\mathtt{code})$ {

10        $M_i = N_i^1 \cup \left(\bigcup_{j \in N_i^c} N_j^1 \right) - \{i\};$

11   **if** $(\forall j \in M_i,\ i.\mathtt{prio} > j.\mathtt{prio})$

12          $i.\mathtt{out} = i.\mathtt{out} \cup N_i^c;$

13   }

14   **if** $(\exists k : k \in i.\mathtt{out}$ **and**
        [k has the earliest packet in $i.\mathtt{Q}\,(i.\mathtt{out})$])

15   [Transmit the packet in $i.\mathtt{Q}\,(\{k\})$ on $k.\mathtt{code}$];

16   }

17   **else** {

18      $i.\mathtt{state} = \mathtt{Rx};$

19   [Listen to transmissions on $i.\mathtt{code}$];

20   }
} /* End of **LAMA**. */

FIGURE 4.11   LAMA specification.

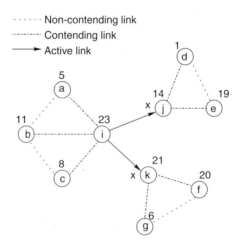

FIGURE 4.12   An example of LAMA operation.

and k happen to have the same code x. To determine if node i can activate links on code x, we compare priorities of nodes according to LAMA. Node i has the highest priority within one-hop neighbors, and higher priority than node j and k as well as their one-hop neighbors. Therefore, node i can activate either (i, j) or (i, k) in the current time slot t depending on the back-logged data flows at node i. In addition, node e may activate link (e, d) if node d is assigned a code other than code x.

4.4.5 Pairwise-Link Activation Multiple Access (PAMA)

The PAMA, is different from NAMA and LAMA in that the link priorities are used in the contention resolution for channel access, instead of the node priorities. The priority of link (u, v) is computed according to Equation (4.9), which is an adaptation of Equation (4.1).

$$(u, v).\texttt{prio} = \texttt{Hash}(u \oplus v \oplus t) \oplus u \oplus v. \tag{4.9}$$

Based on the priorities of the incident links to a node, PAMA chooses the link with the highest priority for reception or transmission at the node. Hence, the set of contenders of a link includes all other links incident to the endpoints of the link.

PAMA uses POCA code assignment scheme, in which a code is assigned per sender–receiver pair. However, because a node can activate only one incident link for either transmission or reception in each time slot, the POCA is equivalent to the transmitter-oriented (TOCA) scheme.

Figure 4.13 specifies PAMA. Lines 1–5 assign codes to the nodes in the two-hop neighborhood of node i. Then the priorities of the incident links at node i and its one-hop neighbors are computed (lines 7–10). The link with the highest priority at each node is marked for active incoming link (lines 13–16) or active outgoing link (lines 17–20). If node i has an active outgoing link, which is also an active incoming link at the receiver (line 21), node i further examines the hidden terminal problem at other nodes (lines 23–26). If node i can still transmit, it selects the packet for the active outgoing link and transmits on $i.\texttt{code}$ (lines 28–29). Otherwise, node i listens on the code assigned to the active incoming link (lines 31–34).

Figure 4.14 illustrates a simple example network, in which a collision happens at node b when link (a, b) and (c, d) are activated using the same code k. However, node c is able to know the possible collision and deactivate link (c, d) for the current time slot using PAMA lines 23–26 in Figure 4.13.

PAMA(i, t)

```
{
        /* Initialize. */
1       for (k ∈ N_i^1 ∪ (⋃_{j∈N_i^1} N_j^1)) {
2           k.prio = Hash(k ⊕ t) ⊕ k;
3           n = k.prio mod |C_{pn}|;
4           k.code = c_n;
5       }

6       for (k ∈ N_i^1 ∪ {i}) {
        /* Link priorities. */
7           for (j ∈ N_k^1) {
8               (k, j).prio = Hash(k ⊕ j ⊕ t) ⊕ k ⊕ j;
9               (j, k).prio = Hash(j ⊕ k ⊕ t) ⊕ j ⊕ k;
10          }

11          k.in = -1;
12          k.out = ∅;
        /* Active incoming or outgoing link. */
13          if (∃j ∈ N_k^1, ∀u ∈ N_k^1,
14              ((j, k).prio > (u, k).prio | u ≠ j) and
15              (j, k).prio > (k, u).prio)
16              k.in = j;

17          else if (∃j ∈ N_k^1, ∀u ∈ N_k^1,
18              ((k, j).prio > (k, u).prio | u ≠ j) and
19              (k, j).prio > (u, k).prio)
20              k.out = {j};

        /* Nodal states. */
21      if (i.out ≡ {k} and k.in ≡ i) {
22          i.state = Tx;
        /* Hidden terminal avoidance. */
23          if (∃u ∈ N_i^1 − {k}, u.in ≡ v, v ≠ i and
24              i.code ≡ v.code and
25              ((v ∈ N_i^1 and u ∈ v.out) or (v ∉ N_i^1)))
26              i.out = ∅;

27          if ([ There is a packet in i.Q (i.out) ])
28              [ Transmit the packet on i.code ];
29      }
30      else if (i.in ≡ k) {
31          i.state = Rx;
32          [ Listen to transmissions on k.code ];
33      }
} /* End of PAMA. */
```

FIGURE 4.13 PAMA specification.

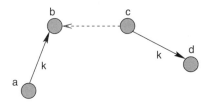

FIGURE 4.14   An example of hidden terminal problem in PAMA.

4.4.6 Hybrid Activation Multiple Access (HAMA)

Unlike previous channel access scheduling protocols that activate either nodes or links only, HAMA, is a node-activation channel access protocol that is capable of broadcast transmissions, while also maximizing the chance of link activations for unicast transmissions. The code assignment in HAMA is the TOCA scheme.

In each time slot, a node derives its state by comparing its own priority with the priorities of its neighbors. We require that only nodes with higher priorities transmit to those with lower priorities. Accordingly, HAMA defines the following node states:

- (R) *Receiver*: The node has an intermediate priority among its one-hop neighbors.
- (D) *Drain*: The node has the lowest priority among its one-hop neighbors, and can only receive a packet in the time slot.
- (BT) *Broadcast Transmitter*: The node has the highest priority within its two-hop neighborhood, and can broadcast to its one-hop neighbors.
- (UT) *Unicast Transmitter*: The node has the highest priority among its one-hop neighbors, instead of two-hop. Therefore, the node can only transmit to a selected subset of its one-hop neighbors.
- (DT) *Drain Transmitter*: The node has the highest priority among the one-hop neighbors of a *Drain* neighbor.
- (Y) *Yield*: The node could have been in either UT- or DT-state, but chooses to abandon channel access because its transmission may incur unwanted collisions due to potential hidden sources from its two-hop neighbors.

Figure 4.15 specifies HAMA. Lines 1–8 compute the priorities and code assignments of the nodes within the two-hop neighborhood of node i using Equation (4.1) and Equation (4.8), respectively. Depending on the one-hop neighbor information of node i and node $j \in N_i^1$, node i classifies the status of node j and itself into receiver (R or D) or transmitter (UT) state (lines 9–14).

If node i happens to be a UT, then i further checks whether it can broadcast by comparing its priority with those of its two-hop neighbors (lines 15–17). If node i is a *Receiver* (R), it checks whether it has a neighbor j in *Drain* state (D) to which it can transmit, instead (lines 18–21). If yes, before node i becomes the *drain* transmitter (DT), it needs to make sure that it is not receiving from any one-hop neighbor (lines 22–25).

After that, node i decides its receiver set if it is in transmitter state (BT, UT or DT), or its sources if in receiver state (R or D). A receiver i always listens to its one-hop neighbor with the highest priority by tuning its reception code into that neighbor's transmission code (lines 26–42).

If a transmitter i unicasts (UT or DT), the hidden terminal problem should be avoided, in which case node i's one-hop receiver may be receiving from two transmitters on the same code (lines 43–45).

Finally, node i in transmission state may send the earliest arrived packet (FIFO) to its receiver set i.out, or listens if it is a receiver (lines 46–58). In case of the broadcast state (BT), i may choose to send a unicast packet if broadcast buffer is empty.

Figure 4.16 provides an example of how HAMA operates in a multihop network during a time slot. In the figure, the priorities are noted beside each node. Node A has the highest priority among its two-hop neighbors, and becomes a broadcast BT. Nodes F, G and H are receivers in the *drain* state, because they

```
HAMA(i, t)
{
      /* Every node is initialized in Receiver state. */
1     i.state = R;
2     i.in = -1;
3     i.out = ∅;

      /* Priority and code assignments. */
4     for (k ∈ N_i^1 ∪ (∪_{j∈N_i^1} N_j^1)) {
5         k.prio = Hash(t ⊕ k);
6         n = k.prio mod |C_pn|;
7         k.code = c_n;
8     }

      /* Find UT and Drain. */
9     for (∀j ∈ N_i^1 ∪ {i}) {
10        if (∀k ∈ N_j^1, j.prio > k.prio)
11            j.state = UT; /* May unicast. */
12        elseif (∀k ∈ N_j^1, j.prio < k.prio)
13            j.state = D; /* A Drain. */
14    }

      /* If i is UT, see further if i can become BT */
15    if (i.state ≡ UT and
16        ∀k ∈ ∪_{j∈N_i^1} N_j^1, k ≠ i, i.prio > k.prio)
17        i.state = BT;

      /* If i is Receiver, i may become DT. */
18    if (i.state ≡ R and
19        ∃j ∈ N_i^1, j.state ≡ D and
20        ∀k ∈ N_j^1, k ≠ i, i.prio > k.prio) {
21        i.state = DT;

      /* Check if i should listen instead. */
22    if (∃j ∈ N_i^1, j.state ≡ UT and
23        ∀k ∈ N_i^1, k ≠ j, j.prio > k.prio)
24        i.state = R; /* i has a UT neighbor j. */
25    }

      /* Find dests for Txs, and srcs for Rxs. */
26    switch (i.state) {
27        case BT:
28            i.out = {-1}; /* Broadcast. */
29        case UT:
30            for (j ∈ N_i^1)
31                if (∀k ∈ N_j^1, k ≠ i, i.prio > k.prio)
32                    i.out = i.out ∪{j};
33        case DT:
34            for (j ∈ N_i^1)
35                if (j.state ≡ D and ∀k ∈ N_j^1, k ≠ i, i.prio > k.prio)
36                    i.out = i.out ∪{j};
37        case D, R:
38            if (∃j ∈ N_i^1 and ∀k ∈ N_i^1, k ≠ j, j.prio > k.prio) {
39                i.in = j;
40                i.code = j.code;
41            }
42    }

      /* Hidden Terminal Avoidance. */
43    if (i.state ∈ { UT, DT } and ∃j ∈ N_i^1, j.state ≠ UT and
44        ∃k ∈ N_j^1, k.prio > i.prio and k.code ≡ i.code)
45        i.state = Y;

      /* Ready to communicate. */
46    switch (i.state) { /* FIFO */
47        case BT:
48            if (i.Q(i.out) ≠ ∅)
49                pkt = The earliest packet in i.Q(i.out);
50            else
51                pkt = The earliest packet in i.Q(N_i^1);
52            Transmit pkt on i.code;
53        case UT, DT:
54            pkt = The earliest packet in i.Q(i.out);
55            Transmit pkt on i.code;
56        case D, R:
57            Receive pkt on i.code;
58    }
} /* End of HAMA. */
```

FIGURE 4.15   HAMA specification.

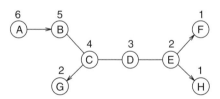

FIGURE 4.16   An example of HAMA operation.

have the lowest priorities among their one-hop neighbors. Nodes C and E become transmitters to *drain*s, because they have the highest priorities around their respective *drain*s. Nodes B and D stay in *receiver* state because of their low priorities. Notice that in this example, only node A would be activated in NAMA, because node C would defer to node A, and node E would defer to node C. This illustrates that HAMA can provide better channel access opportunities over NAMA, although NAMA does not requires code-division channelization.

In contrast to NAMA, HAMA provides similar broadcasting capability, in addition to the extra opportunities for sending unicast traffic with only a little more processing required on the neighbor information.

4.5   Throughput Analysis in Ad Hoc Networks

In a fully connected network, it comes natural that the channel bandwidth is evenly shared among all nodes using any of the above channel access protocols, because the priorities of nodes or links are uniformly distributed. However, in an ad hoc network model where nodes are randomly placed over an infinite plane, bandwidth allocation to a node is more generic, and much more complex. We first analyze the accurate channel access probabilities of HAMA and NAMA, then the upper bound of the channel access probability of PAMA and LAMA in this model. Using the results in References 27 and 26, the throughput of NAMA and HAMA is compared with that of ideal CSMA and CSMA/CA.

For simplicity, we assumed that infinitely many codes are available such that hidden terminal collision on the same code was not considered.

4.5.1   Geometric Modeling

Similar to the network modeling in References 27 and 26, the network topology is generated by randomly placing many nodes on an infinitely large two-dimensional area independently and uniformly, where the node density is denoted by ρ. The probability of having k nodes in an area of size S follows a Poisson distribution:

$$p(k, S) = \frac{(\rho S)^k}{k!} e^{-\rho S}.$$

The mean of the number of nodes in the area of size S is ρS.

Based on this modeling, the channel access contention of each node, is related with node density ρ and node transmission range r. Let N_1 be the average number of one-hop neighbors covered by the circular area under the radio transmission range of a node, we have $N_1 = \rho \pi r^2$.

Let N_2 be the average number of neighbors within two hops. As shown in Figure 4.17, two nodes become two-hop neighbors only if there is at least one common neighbor in the shaded area. The average number of nodes in the shaded area is:

$$B(t) = 2\rho r^2 a(t),$$

where

$$a(t) = \arccos \frac{t}{2} - \frac{t}{2}\sqrt{1 - \left(\frac{t}{2}\right)^2}. \tag{4.10}$$

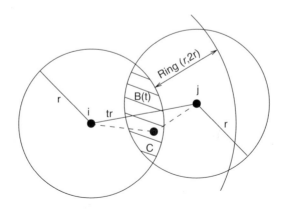

FIGURE 4.17   Becoming two-hop neighbors.

Thus, the probability of having at least one node in the shaded area is $1 - e^{-B(t)}$. Adding up all nodes covered by the ring $(r, 2r)$ around the node, multiplied by the corresponding probability of becoming two-hop neighbors, the average number of two-hop neighbors of a node is:

$$n_2 = \rho \pi r^2 \int_1^2 2t(1 - e^{-B(t)})dt.$$

Because the number of one-hop neighbors is $N_1 = \rho \pi r^2$, adding the average number of one-hop and two-hop neighbors, we obtain the number of neighbors within two hops as:

$$N_2 = N_1 + n_2 = N_1 \left(1 + \int_1^2 2t(1 - e^{-B(t)})dt\right).$$

For convenience, symbols $T(N)$, $U(N)$, and $W(N)$ are introduced to denote three probabilities when the average number of contenders is N.

$T(N)$ denotes the probability of a node winning among its contenders. Because the number of contenders follows Poisson distribution with mean N, and that all nodes have equal chances of winning, the probability $T(N)$ is the average over all possible numbers of the contenders using Equation (4.2):

$$T(N) = \sum_{k=1}^{\infty} \frac{1}{k+1} \frac{N^k}{k!} e^{-N} = \frac{e^N - 1 - N}{Ne^N}.$$

Note that k starts from 1 in the expression for $T(N)$, because a node with no contenders does not win at all.

$U(N)$ is the probability that a node has at least one contender, which is simply

$$U(N) = 1 - e^{-N}.$$

$W(N)$ is introduced to denote

$$W(N) = U(N) - T(N) = 1 - \frac{1}{N}(1 - e^{-N}).$$

4.5.2 Throughput Analysis for NAMA

Because N_2 denotes the average number of two-hop neighbors, which is the number of contenders for each node in NAMA, it follows that the probability that the node broadcasts is $T(N_2)$. Therefore, the channel access probability of a node in NAMA is

$$q_{\text{NAMA}} = T(N_2). \tag{4.11}$$

4.5.3 Throughput Analysis for HAMA

The HAMA includes the node activation cases in NAMA in the BT. In addition, HAMA provides two more states for a node to transmit in the unicast mode (UT and DT). Overall, if node i transmits in the unicast state (UT and DT), node i must have at least one neighbor j, of which the probability is

$$p_u = U(N_1).$$

In addition, the chances of unicast transmissions in either the UT or the DT states depend on three factors: (a) the number of one-hop neighbors of the source, (b) the number of one-hop neighbors of the destination, and (c) the distance between the source and destination.

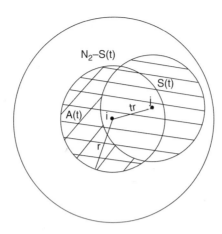

FIGURE 4.18   The unicast between two nodes.

First, we consider the probability of unicast transmissions from node i to node j in the UT state, in which case, node i contend with nodes residing in the combined one-hop coverage of nodes i and j, as illustrated in Figure 4.18. Given that the transmission range is r and the distance between nodes i and j is tr ($0 < t < 1$), we denote the number of nodes within the combined coverage by k_1 excluding nodes i and j, of which the average is

$$S(t) = 2\rho r^2 [\pi - a(t)].$$

$a(t)$ is defined in Equation (4.10). Therefore, the probability of node i winning in the combined one-hop coverage is:

$$p_1 = \sum_{k_1=0}^{\infty} \frac{1}{k_1 + 2} \frac{S(t)^{k_1}}{k_1!} e^{-S(t)} = \frac{W(S(t))}{S(t)}.$$

Furthermore, because node i cannot broadcast when it enters the UT state, there has to be at least one two-hop neighbor with higher priority than node i outside the combined one-hop coverage in Figure 4.18. Denote the number of nodes outside the coverage by k_2, of which the average is $N_2 - S(t)$. The probability of node i losing outside the combined coverage is thus:

$$p_2 = \sum_{k_2=1}^{\infty} \frac{[N_2 - S(t)]^{k_2}}{k_2!} e^{-(N_2 - S(t))} \frac{k_2}{k_2 + 1} = W(N_2 - S(t)).$$

In all, the probability of node i transmitting in the UT state is:

$$p_3 = p_1 \cdot p_2 = \frac{W(N_2 - S(t))\, W(S(t))}{S(t)}.$$

The probability density function (PDF) of node j at position t is $p(t) = 2t$. Therefore, integrating p_3 on t over the range $(0, 1)$ with PDF $p(t) = 2t$ gives the average probability of node i becoming a transmitter in the UT state:

$$p_{\text{UT}} = \int_0^1 p_3 2t \, dt = \int_0^1 2t \, \frac{W(N_2 - S(t))\, W(S(t))}{S(t)} \, dt.$$

Second, we consider the probability of unicast transmissions from node i to node j in the DT state. We denote the number of one-hop neighbors of node j by k_3, excluding nodes i and j, of which the average is N_1. Then, node j requires the lowest priority among its k_3 neighbors to be a *drain*, and node i requires the highest priority to transmit to node j, of which the average probability over all possible values of k_3 is:

$$p_4 = \sum_{k_3=0}^{\infty} \frac{N_1^{k_3}}{k_3!} e^{-N_1} \frac{1}{k_3 + 2} \frac{1}{k_3 + 1} = \frac{T(N_1)}{N_1}.$$

In addition, node i has to lose to nodes residing in the side lobe, marked by $A(t)$ in Figure 4.18. Otherwise, node i would enter the UT state. Denote the number of nodes in the side lobe by k_4, of which the average is

$$A(t) = 2\rho r^2 \left[\frac{\pi}{2} - a(t) \right].$$

The probability of node i losing in the side lobe is thus

$$p_5 = \sum_{k_4=1}^{\infty} \frac{A(t)^{k_4}}{k_4!} e^{-A(t)} \frac{k_4}{k_4 + 1} = W(A(t)).$$

In all, the probability of node i entering the DT state for transmission to node j is the product of p_4 and p_5:

$$p_6 = p_4 \cdot p_5 = \frac{T(N_1)}{N_1} W(A(t)).$$

Using the PDF $p(t) = 2t$ for node j at position t, the integration of the above result over range $(0, 1)$ gives the average probability of node i entering the DT state, denoted by p_{DT}:

$$p_{DT} = \int_0^1 p_6 2t \, dt = \frac{T(N_1)}{N_1} \int_0^1 2t W(A(t)) dt.$$

In summary, the average channel access probability of a node in the network is the chance of becoming a transmitter in the three mutually exclusive broadcast or unicast states (BT, UT or DT), which is given by

$$q_{HAMA} = q_{NAMA} + p_u(p_{UT} + p_{DT})$$

$$= T(N_2) + U(N_1) \cdot \left(\frac{T(N_1)}{N_1} \int_0^1 2t \, W(A(t)) \, dt + \int_0^1 2t \, \frac{W(N_2 - S(t)) \, W(S(t))}{S(t)} \, dt \right).$$

$$(4.12)$$

The above analysis for HAMA have made four simplifications. First, we assumed that the number of two-hop neighbors also follow Poisson distribution, just like that of one-hop neighbors. Second, we let $N_2 - S(t) \geq 0$ even though N_2 may be smaller than $S(t)$ when the transmission range r is small. Third, only one neighbor j is considered when making node i to become a unicast transmitter in the DT or the UT state, although node i may have multiple chances to do so owning to other one-hop neighbors. The results of the simulation experiments reported in Section 4.6 validate these approximations.

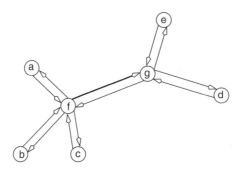

FIGURE 4.19   Link activation in PAMA.

4.5.4 Throughput Analysis for PAMA

In PAMA, a link is activated only if the link has the highest priority among the incident links of the head and the tail of the link. For example, in Figure. 4.19, link (f, g) is activated only if it has the highest priority among the links with f and g as the heads or tails.

To analyze the channel access probability of a node in PAMA, we simplify the problem by assuming that the one-hop neighbor sets of the one-hop neighbors of a given node are disjoint (i.e., any two-hop neighbor of a node is reachable through a single one-hop neighbor only). Using the simplification, the sizes of the two-hop neighbor sets become identical independent random variables following Poisson distribution with mean N_1, so as to avoid handling the correlation between the sizes of the two-hop neighbor sets.

Suppose that a node i has $k_1 \geq 1$ one-hop neighbors. The probability that the node is eligible for transmission is $k_1/2k_1 = 1/2$ because the node has $2k_1$ incident links, and k_1 of them are outgoing. Further suppose that link (i, j) out of the k_1 outgoing links has the highest priority, then node i is able to activate link (i, j) if link (i, j) also has the highest priority among the links incident to node j. Denote the number of one-hop neighbors of node j by k_2. Then the probability of link (i, j) having the highest priority among the incident links of node j is a conditional probability, based on the fact that link (i, j) already has the highest priority among the incident links of node i.

We denote the conditional probability of link (i, j) having the highest priority among the incident links of node j as $P\{A \mid B\}$, where A is the event that link (i, j) wins among the $2k_2$ incident links of node j, and B is the event that link (i, j) wins among the $2k_1$ incident links of node i. We have:

$$P\{B\} = \frac{1}{2k_1}, \qquad P\{A \cap B\} = \frac{1}{2k_1 + 2k_2},$$

$$P\{A \mid B\} = \frac{P\{A \cap B\}}{P\{B\}} = \frac{k_1}{k_1 + k_2}.$$

Therefore, the condition of node i being able to transmit is that node i has an outgoing link (i, j) with the highest priority, of which the probability is $\frac{1}{2}$, and that link (i, j) has the highest priority among the incident links of node j, of which the probability is $(k_1/k_1 + k_2)$. Considering all possible values of random variables k_1 and k_2, which follow the Poisson distribution, we have:

$$q_{\text{PAMA}} = \sum_{k_1=1}^{\infty} \frac{N_1^{k_1}}{k_1!} e^{-N_1} \frac{1}{2} \sum_{k_2=0}^{\infty} \frac{N_1^{k_2}}{k_2!} e^{-N_1} \frac{k_1}{k_1 + k_2} = \frac{N_1}{2} (e^{-2N_1} + T(2N_1)). \qquad (4.13)$$

The q_{PAMA} is the upper bound of the channel access probability of a node in PAMA, because if we have not assumed that the one-hop neighbor sets of the head and tail of a link are disjoint, the number of one-hop neighbors of the tail of the activated link, k_2, could have started from a larger number than zero in the expressions earlier, and the actual channel access probability in PAMA would be less than q_{PAMA}.

4.5.5 Throughput Analysis for LAMA

In LAMA, a node can activate an outgoing link only if the node has the highest priority among its one-hop neighbors, as well as among its two-hop neighbors reachable through the tail of the outgoing link. For convenience, we make the same assumption as in the analysis of PAMA that the one-hop neighbor sets of the one-hop neighbors of a given node are disjoint.

Similarly, suppose a node i has k_1 one-hop neighbors, and the number of the two-hop neighbors reachable through a one-hop neighbor j is k_2. The probability of node i winning in its one-hop neighbor set N_i^1 is $1/(k_1 + 1)$. The probability of node i winning in the one-hop neighbor set of node j is $(k_1 + 1)/(k_1 + k_2 + 1)$, which is conditional upon the fact that node i already wins in N_i^1, and is derived in the same way as in the PAMA analysis. Because k_2 is a random variable following the Poisson distribution,

$$p_7 = \sum_{k_2=0}^{\infty} \frac{N_1^{k_2}}{k_2!} e^{-N_1} \frac{k_1 + 1}{k_1 + k_2 + 1}$$

is the average conditional probability of node i activating link (i, j). Besides node j, node i has other one-hop neighbors. If node i has the highest priority in any one-hop neighbor set of its one-hop neighbors, node i is able to transmit. Therefore, the probability of node i being able to transmit is

$$p_8 = 1 - (1 - p_7)^{k_1}.$$

Because k_1 is also a random variable following the Poisson distribution, the channel access probability of node i in LAMA is:

$$p_9 = \sum_{k_1=1}^{\infty} \frac{N_1^{k_1}}{k_1!} e^{-N_1} \frac{1}{k_1 + 1} p_8.$$

When k_1 increases, p_8 edges quickly toward the probability limit 1. Since we are only interested in the upper bound of channel access probability in LAMA, assuming $p_8 = 1$ simplifies the calculation of p_9 and provides a less tight upper bound. Let $p_8 = 1$, the upper bound of channel access probability in LAMA is thus:

$$q_{LAMA} = \sum_{k_1=1}^{\infty} \frac{N_1^{k_1}}{k_1!} e^{-N_1} \frac{1}{k_1 + 1} = T(N_1). \tag{4.14}$$

4.5.6 Comparisons among NAMA, HAMA, PAMA, and LAMA

Assuming a network density of $\rho = 0.0001$, equivalent to placing 100 nodes on a 1000×1000 square plane, the relation between transmission range and the channel access probability of a node in node activation based NAMA, hybrid activation based HAMA, pairwise link activation based PAMA and link activation based LAMA is shown in Figure 4.20, based on Equation (4.11) to Equation (4.14), respectively.

Because a node barely has any neighbor in a multihop network when the node transmission range is too short, Figure 4.20 shows that the system throughput is close to none at around zero transmission range, but it increases quickly to the peak when the transmission range covers around one neighbor on the average, except for that of PAMA, which is an upper bound. Then network throughput drops when more and more neighbors are contacted and the contention level increases.

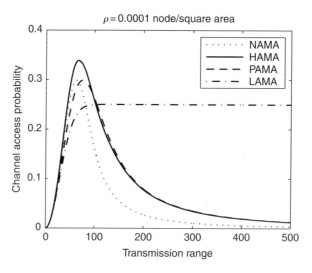

FIGURE 4.20   Channel access probability of NAMA, HAMA, PAMA, and LAMA.

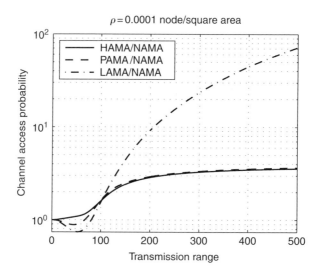

FIGURE 4.21   Channel access probability ratio of HAMA, PAMA, and LAMA to NAMA.

Figure 4.21 shows the performance ratio of the channel access probabilities of HAMA, PAMA, and LAMA to that of NAMA. At shorter transmission ranges, HAMA, PAMA, and LAMA perform very similar to NAMA, because nodes are sparsely connected, and node or link activations are similar to broadcasting. When transmission range increases, HAMA, LAMA, and PAMA obtains more and more opportunities to leverage their unicast capability and the relative throughput also increases more than three times that of NAMA. HAMA and LAMA perform very similarly.

4.5.7   Comparisons with CSMA and CSMA/CA

Because the analysis about NAMA and PAMA is more accurate than the analysis of PAMA and LAMA, which simply derives the upper bounds, we only compare the throughput of HAMA and NAMA that of the idealized CSMA and CSMA/CA protocols, which are analyzed in References 27 and 26. We consider only unicast transmissions, because CSMA/CA does not support collision-free broadcast.

Scheduled access protocols are modeled differently from CSMA and CSMA/CA. In time-division scheduled channel access, a time slot can carry a complete data packet, while the time slot for CSMA and CSMA/CA only lasts for the duration of a channel round-trip propagation delay, and multiple time slots are used to transmit a data packet once the channel is successfully acquired. In addition, Wang and Garcia-Luna-Aceves [26] and Wu and Varshney [27] assumed a heavily loaded scenario in which a node always has a data packet during the channel access, which is not true for the throughput analysis of HAMA and NAMA, because using the heavy load approximation would always result in the maximum network capacity according to Equation (4.7).

The probability of channel access at each time slot in CSMA and CSMA/CA is parameterized by the symbol p'. For comparison purposes, we assume that *every attempt* to access the channel in CSMA or CSMA/CA is an *indication* of a packet arrival at the node. Though the attempt may not succeed in CSMA and CSMA/CA due to packet or RTS/CTS signal collisions in the common channel, and end up dropping the packet, conflict-free scheduling protocols can always deliver the packet if it is offered to the channel. In addition, we assume that no packet arrives during the packet transmission. Accordingly, the traffic load for a node is equivalent to the portion of time for transmissions at the node. Denote the average packet size as l_{data}, the traffic load for a node is given by

$$\lambda = \frac{l_{data}}{1/p' + l_{data}} = \frac{p' l_{data}}{1 + p' l_{data}}$$

because the average interval between successive transmissions follows Geometric distribution with parameter p'.

The network throughput is measured by the successful data packet transmission rate within the one-hop neighborhood of a node in References 26 and 27, instead of the whole network. Therefore, the comparable network throughput in HAMA and NAMA is the sum of the packet transmissions by each node and all of its one-hop neighbors. We reuse the symbol N in this section to represent the number of one-hop neighbors of a node, which is the same as N_1 defined in Section 4.5.1. Because every node is assigned the same load λ, and has the same channel access probability (q_{HAMA}, q_{NAMA}), the throughput of HAMA and NAMA becomes

$$S_{HAMA} = N \cdot \min(\lambda, q_{HAMA}).$$

$$S_{NAMA} = N \cdot \min(\lambda, q_{NAMA}).$$

Figure 4.22 compares the throughput attributes of HAMA, NAMA, the idealized CSMA [27], and CSMA/CA [26] with different numbers of one-hop neighbors in two scenarios. The first scenario assumes that data packets last for $l_{data} = 100$ time slots in CSMA and CSMA/CA, and the second assumes a 10-time-slot packet size average.

The network throughput decreases when a node has more contenders in NAMA, CSMA, and CSMA/CA, which is not true for HAMA. In addition, HAMA and NAMA provide higher throughput than CSMA and CSMA/CA, because all transmissions are collision-free even when the network is heavily loaded. In contrast to the critical role of packet size in the throughput of CSMA and CSMA/CA, it is almost irrelevant in that of scheduled approaches, except for shifting the points of reaching the network capacity.

4.6 Simulations

The delay and throughput attributes of NAMA, LAMA, PAMA, and HAMA are studied by comparing their performance with UxDMA [22] in two simulation scenarios: fully connected networks with different number of nodes, and multihop networks with different radio transmission ranges.

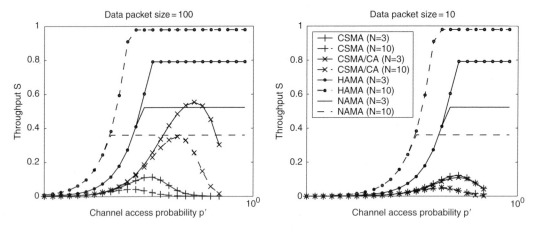

FIGURE 4.22　Comparison between HAMA, NAMA and CSMA, CSMA/CA.

In the simulations, we use the normalized *packets per time slot* for both arrival rates and throughput. This metric can be translated into concrete throughput metrics, such as megabits per second (Mbps), if the time slot sizes and the channel bandwidth are instantiated.

Because the channel access protocols based on NCR have different capabilities regarding broadcast and unicast, we only simulate unicast traffic at each node in all protocols. All nodes have the same load, and the destinations of the unicast packets at each node are evenly distributed over all one-hop neighbors.

In addition, the simulations are guided by the following parameters and behavior:

- The network topologies remain static during the simulations to examine the performance of the scheduling algorithms only.
- Signal propagation in the channel follows the free-space model and the effective range of the radio is determined by the power level of the radio. Radiation energy outside the effective transmission range of the radio is considered as a negligible interference to other communications. All radios have the same transmission range.
- Each node has an unlimited buffer for data packets.
- Thirty pseudo-noise codes are available for code assignments, that is, $|C_{pn}| = 30$.
- Packet arrivals are modeled as Poisson arrivals. Only one packet can be transmitted in a time slot.
- The duration of the simulation is 100,000 time slots, long enough to collect the metrics of interests.

We note that assuming static topologies does not favor NCR-based channel access protocols or UxDMA, because the same network topologies are used. Nonetheless, exchanging the full topology information required by UxDMA in a dynamic network would be far more challenging that exchanging the identifiers of nodes within two hops of each node.

Except for HAMA, which schedules both node- and link activations, UxDMA has respective constraint sets for NAMA, LAMA, and PAMA. Table 4.2 gives the corresponding constraint sets for NAMA, LAMA, and PAMA. The meaning of each symbol is illustrated by Figure 4.4.

Simulations were carried out in four configurations in the fully connected scenario: 2-, 5-, 10-, 20-node networks, to manifest the effects of different contention levels. Figure 4.23 shows the maximum throughput of each protocol in fully connected networks. Except for PAMA and UxDMA-PAMA, the maximum throughput of every other protocol is one because their contention resolutions are based on the node priorities, and only one node is activated in each time slot. Because PAMA schedules link activations

TABLE 4.2 Constraint Sets for
NCR-Based Protocols

Protocol	Entity	Constraint set
UxDMA-NAMA	Node	$\{V_{tr}^0, V_{tt}^1\}$
UxDMA-LAMA	Link	$\{E_{rr}^0, E_{tr}^0\}$
UxDMA-PAMA	Link	$\{E_{rr}^0, E_{tt}^0, E_{tr}^0, E_{tr}^1\}$

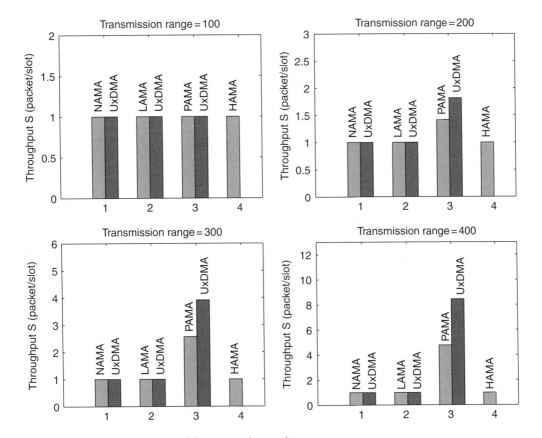

FIGURE 4.23 Packet throughput in fully connected networks.

based on link priorities, multiple links can be activated on different codes in the fully connected networks, and the channel capacity is greater in PAMA than in the other protocols.

Figure 4.24 shows the average delay of data packets in NAMA, LAMA, and PAMA with their corresponding UxDMA counterparts, and HAMA with regard to different loads on each node in fully connected networks. NAMA, UxDMA-NAMA, LAMA, UxDMA-LAMA, and HAMA have the same delay characteristic, because of the same throughput is achieved in these protocols. PAMA and UxDMA-PAMA can sustain higher loads and have longer "tails" in the delay curves. However, because the number of contenders for each link is more than the number of nodes, the contention level is higher for each link than for each node. Therefore, packets have higher starting delay in PAMA than other NCR-based protocols.

Figure 4.25 and Figure 4.26 show the throughput and the average packet delay of NAMA, LAMA, PAMA, HAMA, and UxDMA variations.

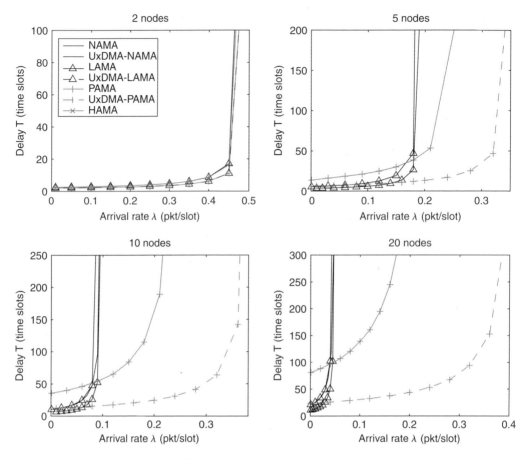

FIGURE 4.24   Average packet delays in fully connected networks.

Except for the ad hoc network generated using transmission range one hundred meters in Figure 4.25, UxDMA always outperforms its NCR-based counterparts — NAMA, LAMA, and PAMA at various levels. For example, UxDMA-NAMA is only slightly better than NAMA in all cases, and UxDMA-PAMA is 10% to 30% better than PAMA. LAMA is comparatively the worst, with much lower throughput than its counterpart UxDMA-LAMA. One interesting point is the similarity between the throughput of LAMA and HAMA, which has been shown by Figure 4.22 as well, even though they have different code assignment schemes and transmission schedules. Especially, the network throughput of NAMA, LAMA, PAMA, and HAMA based on Equation (4.7) and the analysis in Section 4.5 is compared with the corresponding protocols in the simulations. The analytical results fits well with the simulations results. Note that the analysis bars with regard to PAMA and LAMA are the upper bounds, although the analysis of LAMA is very close to the simulation results.

In Figure 4.26, PAMA still gives higher starting point to delays than the other two even when network load is low due to similar reasons as in fully connected scenario. However, PAMA appears to have slower increases when the network load goes larger, which explains the higher spectrum and spatial reuse of the common channel by pure link-oriented scheduling.

4.7 Research Challenges

We have introduced a new approach to the contention resolution problem by using only two-hop neighborhood information to dynamically resolve channel access contentions, therefore eliminating

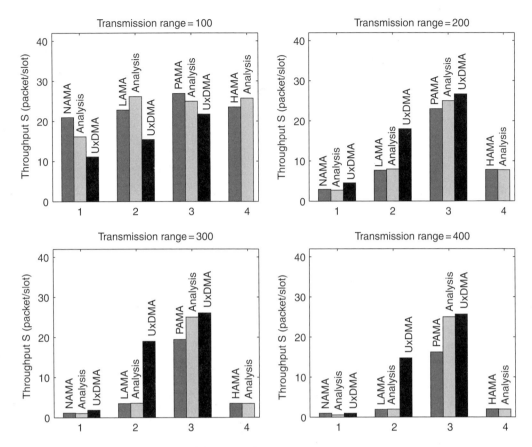

FIGURE 4.25 Packet throughput in multihop networks.

much of the schedule exchanging overhead in prior collision-free scheduling approaches. Based on this approach and time-division channel access, four protocols were introduced for both node-activation and link-activation channel access scheduling in ad hoc networks. Nonetheless, the several problems remain open for further discussions and improvements:

1. Traffic-adaptive resource allocation: Section 4.3.2 discussed the dynamic resource allocation provisioning in the contention resolution algorithm. However, the protocol that can quickly lead to a resource allocation equilibrium remains an open issue.
2. The priority computation is based on a hash function using the node identity and the time slot number as the randomizing seed. Therefore, the interval of successive node or link activations is a random number, which cannot provide delay guarantee required by some real-time applications. Prolonged delay in channel access may even impact regular best-effort traffic, such as TCP connection management and congestion control. Therefore, the choice of an optimum hashing function remain a challenge.
3. The duration of a time slot is based on the transmission time of a whole packet in TDMA schemes. When a network node does not have traffic to send in its assigned slot, the time slot is wasted in TDMA scheme. A mechanism that conserves the channel bandwidth is needed in such scenarios.
4. The hash function computing the priorities was originally proposed to use a pseudo-random number generator, which is computationally expensive, especially for mobile devices with limited power supply. A more simplified hash function with sufficient random attribute is needed.

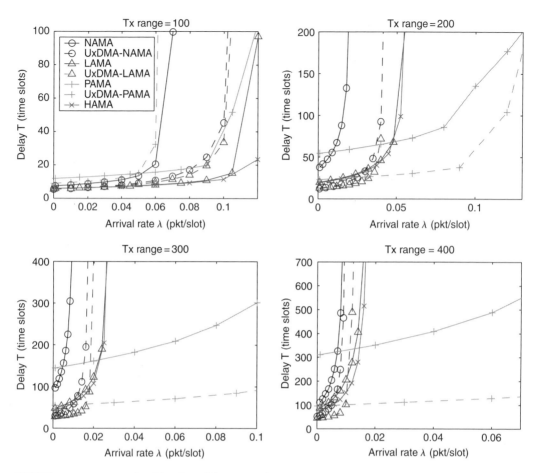

FIGURE 4.26 Average packet delays in multihop networks.

References

[1] L. Bao and J.J. Garcia-Luna-Aceves. Transmission scheduling in ad hoc networks with directional antennas. In *Proceedings of the ACM Eighth Annual International Conference on Mobile Computing and Networking*, Atlanta, Georgia, USA, September 23–28, 2002.

[2] D. Bertsekas and R. Gallager. *Data Networks, 2nd ed.* Prentice Hall, Englewood Cliffs, NJ, 1992.

[3] I. Chlamtac and A. Farago. Making transmission schedules immune to topology changes in multihop packet radio networks. *IEEE/ACM Transactions on Networking*, 2: 23–9, 1994.

[4] I. Chlamtac, A. Farago, and H. Zhang. Time-spread multiple-access (TSMA) protocols for multihop mobile radio networks. *IEEE/ACM Transactions on Networking*, 6: 804–12, 1997.

[5] I. Cidon and M. Sidi. Distributed assignment algorithms for multihop packet radio networks. *IEEE Transactions on Computers*, 38: 1353–61, 1989.

[6] B.P. Crow, I. Widjaja, L.G. Kim, and P.T. Sakai. IEEE 802.11 Wireless local area networks. *IEEE Communications Magazine*, 35: 116–26, 1997.

[7] J. Denes and A.D. Keedwell. *Latin Squares and Their Applications*. Academic Press, New York, 1974.

[8] draft pr ETS 300 652. Radio equipment and systems (RES): high performance radio local area network (HIPERLAN), Type 1 Functional Specification. Technical report, European Telecommunications Standards Institute, Dec. 1995.

[9] A. Ephremides and T.V. Truong. Scheduling broadcasts in multihop radio networks. *IEEE Transactions on Communications*, 38: 456–60, 1990.

[10] S. Even, O. Goldreich, S. Moran, and P. Tong. On the NP-completeness of certain network testing problems. *Networks*, 14: 1–24, 1984.

[11] M. Joa-Ng and I.T. Lu. Spread spectrum medicum access protocol with collision avoidance in mobile ad hoc wireless network. In *Proceedings of the IEEE Conference on Computer Communications (INFOCOM)*, pp. 776–83, New York, USA, Mar. 21–25, 1999.

[12] K. Joseph, N.D. Wilson, R. Ganesh, and D. Raychaudh. Packet CDMA versus dynamic TDMA for multiple access in an integrated voice/data PCN. *IEEE Journal of Selected Areas in Communications*, 11: 870–84, 1993.

[13] J.H. Ju and V.O.K. Li. An optimal topology-transparent scheduling method in multihop packet radio networks. *IEEE/ACM Transactions on Networking*, 6: 298–306, 1998.

[14] J.H. Ju and V.O.K. Li. TDMA scheduling design of multihop packet radio networks based on latin squares. *IEEE Journal on Selected Areas in Communications*, 17: 1345–52, 1999.

[15] L. Kleinrock and F.A. Tobagi. Packet switching in radio channels. i. carrier sense multiple-access modes and their throughput-delay characteristics. *IEEE Transactions on Communications*, 23: 1400–16, 1975.

[16] R. Krishnan and J.P.G. Sterbenz. An evaluation of the TSMA protocol as a control channel mechanism in MMWN. Technical report, BBN Technical Memorandum No. 1279, Apr. 26, 2000.

[17] T. Makansi. Trasmitter-oriented code assignment for multihop radio net-works. *IEEE Transactions on Communications*, 35: 1379–82, 1987.

[18] N. Amitay. Resource auction multiple access (RAMA): efficient method for fast resource assignment in decentralised wireless PCS. *Electronics Letters*, 28: 799–801, 1992.

[19] S. Nanda, D.J. Goodman, and U. Timor. Performance of PRMA: a packet voice protocol for cellular systems. *IEEE Transactions on Vehicular Technology*, 40: 584–98, 1991.

[20] Draft p802.11d5.0. IEEE P802.11 Draftstandard for wireless LAN: medium access control (MAC) and physical layer (PHY) specification. Technical report, Institute of Electrical and Electronics Engineers (IEEE), 1996.

[21] X. Qiu and V.O.K. Li. Dynamic reservation multiple access (DRMA): a new multiple access scheme for personal communication systems (PCS). *Wireless Networks*, 2: 117–28, 1996.

[22] R. Ramanathan. A unified framework and algorithm for channel assignment in wireless networks. *Wirelss Networks*, 5: 81–94, 1999.

[23] R. Ramaswami and K.K. Parhi. Distributed scheduling of broadcasts in a radio network. In *Proceedings of the IEEE Conference on Computer Communications (INFOCOM)*, Vol. 2, pp. 497–504, Ottawa, Ont., Canada, Apr. 23–27, 1989. IEEE Computer Society Press, Washington, DC.

[24] Z. Tang and J.J. Garcia-Luna-Aceves. A protocol for topology-dependent transmission scheduling. In *Proceedings of the IEEE Wireless Communications and Networking Conference 1999 (WCNC 99)*, New Orleans, Louisiana, Sep. 21–24, 1999.

[25] F. A. Tobagi and L. Kleinrock. Packet switching in radio channels: part II — the hidden terminal problem in carrier sense multiple-access modes and the busy-tone solution. *IEEE Transactions on Communications*, 23: 1417–33, 1975.

[26] Y. Wang and J.J. Garcia-Luna-Aceves. Performance of collision avoidance protocols in single-channel ad hoc networks. In *Proceedings of the IEEE International Conference on Network Protocols (ICNP)*, Paris, France, Nov. 12–15, 2002.

[27] L. Wu and P.K. Varshney. Performance analysis of CSMA and BTMA protocols in multihop networks (I), Single Shannel Case. *Information Sciences*, 120: 159–77, 1999.

[28] C. Zhu and M.S. Corson. A five-phase reservation protocol (FPRP) for mobile ad hoc networks. In *Proceedings of the IEEE Conference on Computer Communications (INFOCOM)*, Vol. 1, pp. 322–31, San Francisco, CA, USA, Mar. 29–Apr. 2, 1998.

5

Multiple Access Protocols and Scheduling Algorithms for Multiple Channel Wireless Networks

5.1 Introduction... **5**-77
5.2 Background **5**-78
5.3 Contention-Based Protocols **5**-79
Group Allocation Multihop Multiple Access • Multichannel CSMA • Collision Avoidance and Resolution Multiple Access for Multichannel Wireless Networks • Multichannel MAC (MMAC) Protocol
5.4 Scheduling Algorithms for Contention-Free Protocols **5**-84
Algorithm Notations and Basics • Contiguous Allocation Algorithm for Single Priorities • Non-Contiguous Allocation Algorithm for Single Priorities • Contiguous Sorted Sequential Allocation • Non-Contiguous Round Robin Allocation • Non-Contiguous Sequential Round Robin Allocation • Contiguous Allocation Algorithm for Multiple Priorities • Non-Contiguous Allocation Algorithm for Multiple Priorities
5.5 Conclusions **5**-94
Acknowledgments... **5**-96
References **5**-96

Aniruddha Rangnekar
Chonggang Wang
Krishna M. Sivalingam
Bo Li

5.1 Introduction

Wireless Local Area Networks (WLAN) and Personal Area Networks (WPAN) have become increasingly important in today's computer and communications industry. In addition to traditional data services, there is also an increasing demand for support of multimedia services such as voice and video. This demand translates to an ever growing demand for channel bandwidth. The various standard groups are addressing these issues in different ways, for example, using OFDM techniques to achieve 54 Mbps as in IEEE 802.11a [1] and using UWB techniques to achieve hundreds of Mbps as in IEEE 802.15.3a [2].

In current generation wireless networks, the available wireless spectrum is divided into several channels. The number of channels and capacity of each channel depends on the particular technology. For example, the IEEE 802.11b standard specifies 11 channels operating in the 2.4-GHz ISM band with 80 MHz of reusable spectrum. However, each basic wireless network (consisting of an access point and mobile devices communicating through it) uses only a single channel for communication. The multiple channels are used more for spatial coexistence of multiple wireless networks in the same area.

Even though it may be possible to create many channels from the wireless spectrum, the actual number of available channels is quite small. To prevent interference of signals from communication between adjacent channels (cochannel interference), adjacent channels need to be separated by guard bands. To increase the number of channels, this criterion may be relaxed. Even though IEEE 802.11b communication has eleven channels for communication, at any instant a maximum of three channels can be used simultaneously as the channels overlap each other. Due to the small number of channels that are available, each channel has to be shared by many nodes using either a random access method such as Slotted Aloha or by scheduling the channel to the nodes based on user requests. As the number of nodes that access the same wireless medium increases, the amount of bandwidth available to each node drops. This effect is acerbated by the fact that the available bandwidth is scarce to begin with.

In order to improve the per-network capacity, especially as we move to higher operating bandwidths, it is possible to provide multiple transmission channels in a given region. For such a system, the access point and the mobile device should be able to access the different transmission channels. For such a network, an appropriate medium access control (MAC) protocol needs to be specified to share the multiple channels among the network nodes. The objective of this chapter is to present a summary of several MAC protocols proposed for wireless networks with multiple transmission channels [3–17].

5.2 Background

There are two major classes of packet switched wireless networks. In the *infrastructure* or centralized architecture, there is a specialized node called the *base station* (BS) or *access point* (AP) that is responsible for maintaining the network. If the network uses a reservation-scheduling approach or a polling-based approach, the BS performs the scheduling operations. The BS needs to consider several factors like node demand, priorities, and tuning latencies while computing the schedule.

In the distributed architecture, nodes are deployed in an *ad hoc* manner and a BS cannot be deployed as the nodes move in and out of the network. In such networks, the medium access is resolved in a contention-based manner. One common problem with such networks is the *hidden terminal* problem where two nodes that are not within range of each other may communicate simultaneously with a third intermediate node resulting in a collision. Several protocols have been used to solve the hidden terminal problem [18–20].

The MAC protocol is a critical component of the network protocol stack and much work has been done in standardizing the MAC protocols for wireless LANs.

In designing MAC protocols for multiple channels, several design choices have to be considered. One of them is the node transceiver architecture and whether tunable or fixed transmitters (or receivers) have to be used. In some protocols, the channel on which a node can receive data is preassigned [6,7]. In such cases it may be sufficient to equip the node with a fixed receiver. But since the node will now have to transmit on the receiving channel of different nodes, it will need a tunable transmitter. This architecture is classified as a *tunable transmitter – fixed receiver* (TT–FR) architecture. Similarly in a *fixed transmitter–tunable receiver* (FT–TR) architecture, a node transmits on a preassigned channel but tunes its receiver to the transmitting channel of the sender. Other protocols assign channels to nodes dynamically, based on the traffic load and channel availability [3,4,9]. This approach, *tunable transmitter–tunable receiver*

(TT–TR) architecture, requires that both the transmitter and the receiver be capable of tuning to any of the available channels. The design choice usually depends on the tuning latency of the components and cost associated with adding tuning capability.

In addition, the protocols vary in the number of transceivers per node. Some protocols require multiple transceivers per node so that they can operate on multiple channels simultaneously [3,4]. On the contrary, other protocols reduce the transceiver requirement by increasing the protocol complexity [9]. Further, it is possible to designate one or more channels as control channels, that are exclusively used for control traffic. Other protocols do not make this differentiation as they allow both control and data traffic on the same channels [9].

Another important design choice is whether medium access is based on contention-free, contention-based, or a combination of both mechanisms. The IEEE 802.1s standard [1] for wireless LANs defines multiple access using both contention-based and contention-free modes. The contention-based mode is based on the Carrier Sense Multiple Access/Collision Avoidance (CSMA/CA) mechanism originally defined in [18] and enhanced in Reference 19. The contention-free mode is based on polling by the BS. In Reference 13, an energy-conserving medium access control (EC-MAC) protocol that is based on reservation and scheduling with energy efficiency design considerations is presented.

In the following sections, we first present different contention-based MAC protocols that have been proposed for multichannel wireless networks. We then present scheduling algorithms for contention-free MAC protocols.

5.3 Contention-Based Protocols

In contention-based protocols, nodes compete to gain control of the transmission medium for transmission of every packet. This strategy is efficient for lightly loaded systems. But as the system load increases, competing nodes prevent each other from gaining control of the channel and hence data transmission is impaired and packet delay increases.

5.3.1 Group Allocation Multihop Multiple Access

The Group Allocation Multihop Multiple Access (GAMMA) [6] is a MAC protocol designed for ad hoc networks with multiple transmission channels. Each node in the network is assigned a unique dedicated channel. The channel may be a unique code, a dedicated frequency, or a time-hopping sequence. Each node's receiver is permanently tuned to this channel and all transmissions to this node must be sent on this channel. The access to this channel is arbitrated by this node. Each node maintains a list of nodes called the "transmission group" and only nodes in the transmission group can send data to this node. Before a node starts transmission, it has to join the transmission group of the receiver. Once it has joined the transmission group, it can send data collision-free as long as it has data to send. This is done by dividing the transmission channels into cycles and further partitioning each cycle into variable number of slots. There are two types of slots: *contention* and *data* slots. Contention slots are used to transmit control packets like request-to-send (RTS)/clear-to-send (CTS) to gain admission into the transmission group. Once the node gains membership of a group, it is assigned a data slot for transmission.

The number of data slots in a cycle is variable since it depends on the number of stations transmitting to this node at any given instant. Due to this, two nodes may have different data cycle lengths. For these two nodes to exchange RTS/CTS, their contention slots must be aligned. To satisfy this condition, the number of data slots between two consecutive contention slots is fixed at some constant (q). The value of q is constant throughout the system and is independent of the cycle length. Figure 5.1 shows the structure of a cycle where the cycle length is four slots and there are five data slots between consecutive contention slots ($q = 4$).

Once the source has a packet to send, it needs to join the transmission group of the destination. It does so by sending a RTS during the next contention slot. The destination responds with a CTS that implies that the source has been added to the transmission group. The RTS and CTS include information about

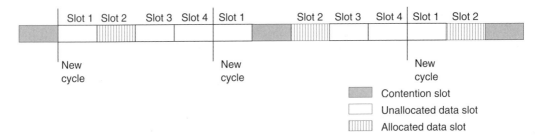

FIGURE 5.1 Cycle structure of GAMMA; data slots per cycle = 4; data slots between consecutive contention slots $(q) = 5$.

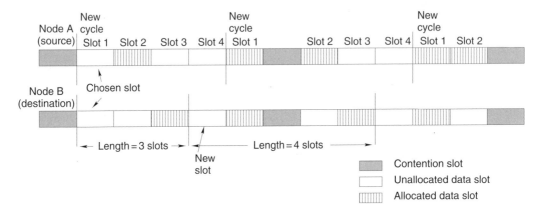

FIGURE 5.2 Slot selection for GAMMA with unequal cycle lengths.

the respective node's cycle, for example, number of data slots, unallocated data slots. The source and the destination nodes compare this information to determine which slot to use for data transfer. To maintain synchronization of the assigned data slot over multiple cycles, the two communicating nodes must have the same cycle length. If the cycle length of the two nodes is not same, the node with the shorter cycle length adds additional data slots to its cycle. Before doing so, it also needs to inform its member nodes about the increase in cycle length. The member nodes in turn increase their cycle lengths to stay aligned. Thus a ripple effect is observed where an expanding circle of nodes increase their cycle length.

The allocation of the data slot to a connection depends on the length of the cycle and the position of the unallocated data slots at each node. If the cycle lengths are the same and both nodes have an unallocated data slot that is aligned to each other, the aligned slot is chosen. If the two nodes have different cycle lengths but at least one unallocated slot on the longer cycle is aligned with an unallocated slot on the shorter cycle, then the length of the shorter cycle is increased and the matching slot is chosen for data transfer. This case can be observed in Figure 5.2 where the cycle length of node B is increased from 3 to 4. If none of the unallocated slots on the longer cycle align with an unallocated slot on the shorter cycle, then an additional data slot is added to the longer cycle and the length of the shorter cycle is increased to match that of the longer cycle. The new slot on the longer cycle is chosen for the connection as it is guaranteed to align with one of the new slots in the shorter cycle. Once the source finishes its data transfer, it informs the destination of its intent to leave the group. The destination then removes the source from the transmission group by marking its data slot as being unallocated.

The GAMMA performs similar to CSMA when the network is lightly loaded. But as the load in the network increases, GAMMA behaves like TDMA as it assigns one data slot in each cycle to a requesting node. This work has been extended in References 7 and 8 to describe new receiver-initiated multiple access protocols.

5.3.2 Multichannel CSMA

Nasipuri and Das [3,4] describe a multichannel variation of the basic CSMA/CA protocol described in the IEEE 802.11 standard. Here the channels are not preassigned to each node but are selected dynamically after sensing the carrier. The available bandwidth is divided into N nonoverlapping channels. The channels may be created in the frequency domain (FDMA) or code domain (CDMA). Each node monitors the N channels continuously and marks them as IDLE depending on the received signal strength and its sensing threshold. When a node has data to send, it chooses one of the IDLE channels based on its characteristics. "Soft" channel reservation is employed to reduce the impact of tuning the transmitter by remembering the last used channel for transmission. If the "last_channel" is IDLE, it is used for the current transmission. Else a channel is randomly chosen from the other IDLE channels and used for transmission. If none of the N channels are IDLE, the node waits for the first channel that becomes IDLE. The node then waits for a random backoff period before transmitting on the chosen channel. An implicit assumption made in this protocol is that the receiver can hear data sent on any of the N channels simultaneously. Hence, for this protocol to function correctly, each node would need N receivers and one tunable transmitter. An optional criteria for channel selection is the power sensed on each channel. The channel with the lowest sensed power is chosen based on the assumption that the channel with the lowest signal is used at a point farthest away from the transmitter. This in turn lowers interference by distributing the radio interference equally over all channels.

This protocol also has a newer version [5] where the channel selection is receiver-based. Here, the available bandwidth is divided into one control channel and N data channels. The control channel is used by the nodes for exchanging RTS/CTS control packets to select the channel for data transmission. The channel selection is based on the interference sensed at the receiver. When a node has a packet to transmit, it senses the carriers on all the data channels and compiles a list of channels that are IDLE. It then sends a RTS packet, containing the list of IDLE channels, to the destination on the control channel. If there are no IDLE channels, the node backsoff and retransmits the RTS after the backoff interval. Upon reception of the RTS, the destination senses its channels and creates its list of IDLE channels. It compares the two lists, chooses any one of the common channels for data transmission and conveys this information to the source by sending a CTS packet on the control channel. If no common idle channels are available, the destination refrains from sending a CTS. This causes the sender to timeout and retransmit RTS. After receiving the CTS, the sender transmits data packets on the channel indicated in the CTS. All other nodes in the vicinity of the source and destination overhear either RTS or CTS and refrain from sending packets on this channel for the duration of the data transfer.

5.3.3 Collision Avoidance and Resolution Multiple Access for Multichannel Wireless Networks

Collision Avoidance and Resolution Multiple Access for Multichannel Wireless Networks (CARMA-MC) [7] is a receiver initiated multiple access protocol with collision resolution. This protocol also assumes that each node, identified by a unique ID, is assigned a channel that is unique at least within a two-hop neighborhood. A node is permitted to transmit packets on the unique receiving channels of any of its one-hop neighbors. Each node divides its channel into receiving and transmitting periods. The transmission period has a fixed maximum length and the receiving period, also called contention interval, consists of collision resolution steps.

Basic Operation: If a node does not have a data packet to send, it enters the contention interval on its designated channel. If an end of the current contention interval it has data packets to be sent, it then enters the transmitting period and switches to the channel of the intended receiver. If an ready-to-receive (RTR) is heard on the intended receiver's channel, the node contends for the intended receiver's channel as a sender. If, on the other hand, the node does not have any data packets, it remains as a receiver and initiates a new contention interval.

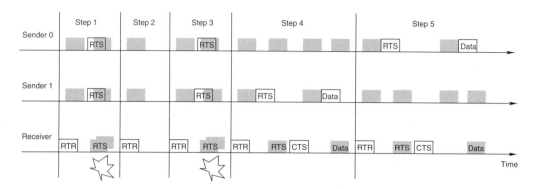

FIGURE 5.3   Operation of CARMA-MC protocol.

The contention interval is initiated by the receiver sending a RTR message on its channel. The RTR message contains an allowed ID-number interval. All nodes, which wish to transmit to this receiver and whose ID falls between the advertised ID interval, contend for the channel by sending an RTS. If the RTS is acceptable, the receiver sends a CTS to the chosen sender. The sender of the RTS waits for one maximum round-trip-time delay plus the time required for the CTS. If the CTS arrives on time, the sender of the RTS acquires the channel and starts data transmission. If the CTS is not received within a specified time interval, a deterministic tree-splitting algorithm is invoked and collisions among RTSs in the same receiving period are resolved. All control packets are exchanged on the receiver's designated channel. Thus the sender waits on the intended receiver's channel for the RTRs and CTSs, and transmits RTS and the data packets on the receiver's channel.

The tree-splitting algorithm [21], used to resolve collisions, is initiated with the first round of RTS collisions. On detecting a collision, the receiver splits the allowed ID interval into two equal components. An RTR message with one of the components is sent out while a backoff counter is set for the second component. This operation is repeated each time there is a collision. The result of the contention resolution can be one of the following three cases:

Case 1 — *Idle*:   None of the nodes in the ID interval have to send a RTS. The channel remains idle for one propagation delay and the receiver sends another RTR with the ID interval updated from the second component.

Case 2 — *Success*:   Only one node in the ID interval has an RTS to send. The receiver responds with a CTS and data transfer takes place.

Case 3 — *Collision*:   There are more than one nodes in the ID interval that have pending RTSs. In this case, the ID interval is split into two new ID intervals and an RTR message with the updated values of the ID interval is sent.

Figure 5.3 shows the operation of the contention resolution algorithm in CARMA-MC. Here "Sender 0" and "Sender 1" want to transmit to the "Receiver." On receiving an RTR from the receiver, both senders transmit RTS in step 1. These packets collide at the receiver and the collision resolution algorithm is initiated. The allowed ID interval is split and an RTR is sent with the new interval in step 2. None of the nodes fall within the allowed interval (case 1) and hence the channel remains idle. In step 3, an RTR is sent with the second component of the interval list. Both the senders fall within the allowed interval and hence both respond with an RTS (case 3) resulting in a collision. The ID interval is split further and this time only one node responds with RTS (case 2). Thus sender 1 manages to gain control of the channel in step 4 and transmits data to the receiver. Similarly sender 0 transmits in step 5.

5.3.4   Multichannel MAC (MMAC) Protocol

Current IEEE 802.11 devices are generally equipped with a single half-duplex transceivers. Although these transceivers can tune to different channels, they can only listen and transmit on one channel at a given

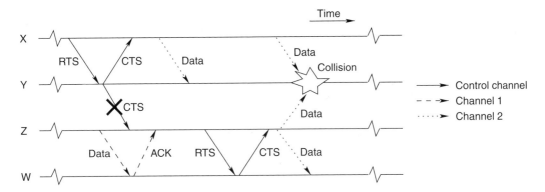

FIGURE 5.4   Multichannel hidden terminal problem.

time. Thus when a host is listening on a particular channel, it cannot listen to communications on any other channel. This problem is described as the *multichannel hidden terminal problem* [9]. This problem can be further explained by the following example. Consider a MMAC protocol that does not consider this problem. For sake of explanation, consider that the protocol uses one channel for control messages and the remaining channels for data traffic. When a node is not transmitting or receiving data, it tunes its receiver to the control channel. When node X has a packet for node Y, it sends a RTS on the control channel. Node Y decides which channel to choose and conveys this information to node X by sending a CTS on the control channel. Once this CTS reaches node X, both the nodes tune their receivers to the chosen channel for further communication.

Consider the scenario in Figure 5.4 where nodes X and Y have exchanged RTS/CTS and have chosen channel 2 for data communication. Consider a node Z in the vicinity of node Y that was communicating with some other node during this RTS/CTS exchange. Since this node was listening to some other channel, it does not hear the CTS that node Y sent to node X on the control channel. Unaware that node Y is communicating on channel 2, node Z may choose the same channel for initiating communication with some other node. This will result in a collision at node Y.

A common solution to avoid this problem is the use of multiple transceivers in each host. MMAC protocol [9] solves the multiple-channel hidden terminal problem by using just a single transceiver. MMAC borrows the idea of Ad hoc Traffic Indication Messages (ATIM) from the power saving mechanism (PSM) of IEEE 802.11. ATIM windows are used by IEEE 802.11 to put nodes into *doze* mode, where the node consumes much less energy by not sending or receiving packets. MMAC adopts this concept by periodically sending beacons to divide time into beacon intervals. The nodes are synchronized so that their beacon intervals begin at the same time. Each beacon interval starts with an "ATIM window" which is used by the communicating nodes to exchange control information. One of the N data channels is chosen as the default channel and all nodes listen to this channel during the ATIM window of each beacon interval. The control packets are sent during the ATIM window on the default channel only. Thus instead of having a separate channel just for control traffic, MMAC uses one of the data channels for a fraction of the time. This technique is especially useful when the number of available channels is low and allocation of a separate control channel would be wasteful.

In MMAC, each node maintains a data structure called *Preferable Channel List* (PCL) to record the usage of different channels in its neighborhood. Each channel is categorized based on its preference as:

(1) *HIGH*: This channel is being used by the node in the current beacon interval. Only one channel can be in this state at a time.
(2) *MID*: This channel is not being used by any of the node's neighbors.
(3) *LOW*: This channels is already being used by one or more nodes in this node's vicinity.

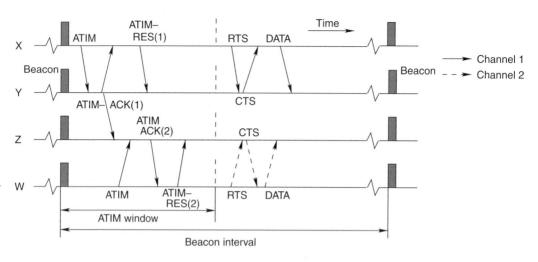

FIGURE 5.5   Channel negotiation in MMAC protocol.

The state of the channel, in the PCL, is changed as follows: All the channels in the PCL are in MID state at the start of each beacon interval. If two nodes choose a channel for communication, that channel is moved to HIGH state. If a node overhears control messages that specify that a particular channel is used by some other node in this node's vicinity, then this node moves the chosen channel to the LOW state.

When node X has packets for node Y, it will send an ATIM packet to Y that contains the PCL of X. On receiving this information, Y chooses a channel for communication based on the received PCL and its own PCL. The chosen channel information is included in an ATIM–ACK and sent to X. If the chosen channel is acceptable to X, it sends an ATIM–RES packet to Y to reserve the channel and also to let other nodes in its neighborhood know that the particular channel has been reserved. This information is used by its neighbors to update their PCLs. After the ATIM window, nodes X and Y switch to the chosen channel and start data transfer. On the other hand if the chosen channel is not acceptable to X, it will have to wait until the next ATIM window and renegotiate (Figure 5.5). Since node Z also tunes to the default channel during the ATIM window, it hears the control exchange between X and Y. So if it receives an ATIM packet from some other node during the ATIM window, it chooses another channel and avoids collision. Random backoff is used to resolve collision of ATIM packets when multiple nodes contend for the control channel.

5.4   Scheduling Algorithms for Contention-Free Protocols

In the previous section, contention-based schemes were presented for multiple channel wireless LANs. In this section, we will consider contention-free schemes. There are several contention-free MAC protocols derived from polling and reservation-based mechanisms. In this section, we will use the EC-MAC protocol presented in Reference 13 as the basis for the presented scheduling algorithms [14,15].

The EC-MAC access protocol is defined for an infrastructure network with a single BS serving nodes in its coverage area. The BS organizes the transmission into frames. The frame is divided into multiple phases and contains a fixed number of slots for data transmission. The frame structure for the EC-MAC protocol is presented in Figure 5.6. At the start of each frame, the BS transmits the frame synchronization message (FSM) which contains synchronization information and the uplink transmission order for the subsequent reservation phase. During the request/update phase, each registered node transmits new connection requests and status of established queues according to the transmission order received in the FSM. New nodes that have entered the BSs area register with the BS during the new-user phase. The BS then broadcasts a schedule message that contains the slot permissions for the subsequent data phase.

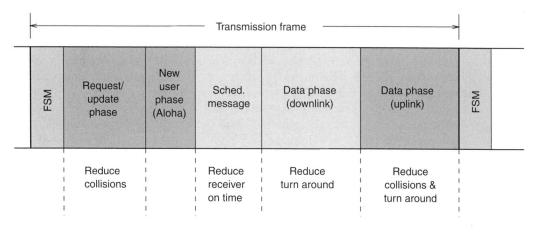

FIGURE 5.6 EC-MAC protocol frame structure.

TABLE 5.1 Notations Used in Algorithm Description

M	Number of nodes
C	Number of channels
K	Maximum value, over all entries, in traffic vector
P	Number of priority queues within a node
T	Tuning time, in slots
Φ	Optimal schedule length, in slots
$\mathcal{L}$	Schedule length, in slots
L	Length of each packet, in bits
S	Transmission rate, in Mbps
Ω	Slot wastage, in slots

Downlink transmission from the BS to the node is scheduled considering the QoS requirements. Likewise, the uplink slots are allocated using a suitable scheduling algorithm. The original EC-MAC protocol did not consider multiple channels but can be easily adapted by simply modifying the scheduling algorithm. With multiple channels available per cell, the frame synchronization, reservation, and schedule messages are sent over a fixed channel, referred to as the control channel. All nodes tune into this channel during the appropriate phases. During the data phase, the nodes tune their transmitters and receivers to the corresponding channels when it is their "turn" to transmit or receive data. The protocol assumes a time division duplexed (TDD) approach where all the transmission channels are available for either uplink or downlink transmissions at one time.

The goal of the traffic scheduling algorithm is to efficiently share the limited number of available channels amongst the nodes such that the network utilization is maximized, packet delay is minimized, and the available bandwidth is efficiently utilized.

The scheduling algorithms presented include *Contiguous Allocation algorithm for Single Priority* (CASP) and *Non-Contiguous Allocation algorithm for Single Priority* (NCASP) described in References 14 and 15; and *Contiguous Sorted Sequential Allocation* (CSSA), *Non-Contiguous Round Robin Allocation* (NCRRA), and *Non-Contiguous Sequential Round Robin Allocation* (NCSRRA) described in Reference 16. These algorithms consider single-priority requests and have been extended to handle multiple priorities as described in the *Contiguous Allocation algorithm for Multiple Priorities* (CAMP) and *Non-Contiguous Allocation algorithm for Multiple Priorities* (NCAMP) algorithms [15].

5.4.1 Algorithm Notations and Basics

The notations used in the following sections are summarized in Table 5.1. Let the number of nodes in a given cell be denoted by M and the number of channels by C. The uplink requests made by the M nodes to the BS and the downlink traffic from the BS to the M nodes over the C channels are represented by a $M \times 2$ matrix. This matrix is also referred to as the *traffic demand* matrix (or the *demand* matrix). The basic unit of transmission is a slot. Each request can be no larger than K slots. Column 1 in the demand matrix indicates the slot requests from the M nodes to the BS (uplink traffic). Column 2 in the demand matrix indicates the slot requests for the M nodes from the BS. We now need to find a conflict-free assignment of the requests on the C channels such that the frame length is minimized.

This scheduling problem is similar to one of the basic, well-studied problems of scheduling theory, that of nonpreemptively scheduling "M" independent tasks on "C" identical, parallel processors. The objective is to minimize the total time required to complete all the tasks. This problem is known to be NP-complete [22] and approximations to this problem such as MULTI-FIT [23] for finding near-optimal schedules have been studied earlier.

Since uplink transmissions and downlink transmissions are separated in time domain, they can be scheduled independent of each other. The rest of this section considers only uplink transmissions with similar results applicable to the downlink.

The data structures used are:

1. Let $D = (d_i)$ denote the demand vector, and let d_i denote the request made by node i where $0 \le d_i \le K, i = 1, \ldots, M$.
2. Let $S = (S_i)$ denote the allocation vector, and let S_i denote the total amount of slots currently allocated to channel i where $0 \le S_i, i = 1, \ldots, C$.

The following are some of the key definitions used.

Definition 5.1 *The Optimal Schedule Length,* $\Phi = \max(K, (\sum_i d_i / C))$, *where K is the largest entry in the demand vector.*

Definition 5.2 *A channel C_k is said to overflow if $S_k > \Phi, k = 1, 2, \ldots, C$.*

Definition 5.3 *The channel capacity,* $\bar{\Phi} = (\sum_i d_i / C) + \min_j d_j$.

5.4.2 Contiguous Allocation Algorithm for Single Priorities

The CASP systems maintain an allocation vector S, where S_i ($i = 1, 2, \ldots, C$), represents the partial sum of slots currently allocated to channel C_i. For a given request, the scheduling algorithm allocates the requests contiguously (without any splitting of the requests) on the channel, which currently has the least partial sum of allocation. If more than one channel has the same least partial sum, one channel is chosen at random for the next allocation. The algorithm is described as pseudo-code in Figure 5.7.

Consider a system with $M = 5$ and $C = 3$. Let the given uplink vector be $D = [5\ \ 7\ \ 4\ \ 9\ \ 2]$. The channel assignments are computed as shown in Figure 5.8. Following the algorithm, we calculate the schedule length, $\mathcal{L} = 13$ slots; slot wastage, $\Omega = 12$ slots. The allocation vector S, at the end of the algorithm contains: $S_1 = 7, S_2 = 7, S_3 = 13$.

The following are some of the properties of the CASP algorithm:

1. The time complexity of the CASP algorithm is $O(M \lg C)$.
2. Each channel overflows at most once. That is, $\forall k, S_k > \Phi$ at most once.
3. An upper bound for the schedule length computed by CASP algorithm is $\Phi + K$.

The reader is referred to Reference 14 for the proofs. From the properties, it follows that the schedule length will be greater than the optimal value, by a factor of at most K/Φ. If the standard deviation between the requests is small (all nodes in the system approximately ask for the same number of slot requests — say x), then the percentage overflow will be $(x/(Mx/C))\% = C/M$. This implies that if the number of

Algorithm CASP

> **For** $(j = 1, 2, \ldots, C)$
> $\quad S_j = 0$
> **Endfor**
> **For** $(i = 1, 2, \ldots, M)$
> $\quad$ find k, such that $S_k \leq S_l, \forall l = 1, 2 \ldots, C$.
> $\qquad$ In the event of a tie, choose k randomly.
> $\quad$ Allocate request d_i on channel C_k
> $\quad$ Update $S_k = S_k + d_i$
> $\quad$ **Endfor**
> **Endfor**
> **Compute** $\mathcal{L} = \max_{l=1}^{C} S_l$
> **Compute** $\Omega = C * \mathcal{L} - (\sum_i d_i)$.

End.

FIGURE 5.7 Contiguous allocation single priority (CASP) algorithm.

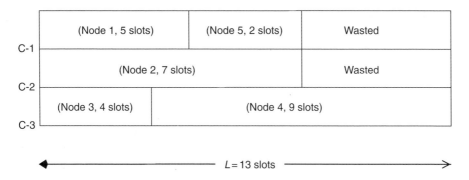

FIGURE 5.8 Contiguous allocation of the requests based on CASP algorithm.

channels is around 4 or 8 and the number of nodes in the system is around 100, the percentage overflow will be around 10%.

5.4.3 Non-Contiguous Allocation Algorithm for Single Priorities

The NCASP system is very similar to the CASP algorithm except that the CASP algorithm allocates a request in contiguous slots, while NCASP algorithm uses non-contiguous allocations. When allocating a request to a channel, C_k, we compare it against Φ. If allocating this request will cause S_k to exceed Φ by a certain amount (the overflow), we try to avoid the situation by:

1. Allocating the current request in its entirety.
2. Splitting an earlier assigned request (if it is big enough to yield as much slots as the overflow) in the same channel so that the overflow in the current channel is reduced to zero.
3. If the above is not possible, that is, a single request is not big enough to yield as much slots as the overflow, we need to walk back through the channel allocation and pick multiple relocatable requests, until we get enough slots to eliminate the overflow.
4. Relocating the selected requests to a next channel which will not overflow.

Algorithm NCASP

    Calculate $\Phi = max(K, \sum_i d_i/C)$
    For $(j = 1, 2, \ldots, C)$
        $S_j = 0$
    Endfor
    For $(i = 1, 2, \ldots, M)$
        find k, such that $S_k \leq S_l, \forall l = 1, 2, \ldots, C.$
            In the event of a tie, choose k randomly.
        If $(S_k + d_i > \Phi)$
            Compute the split s, by breaking an earlier assigned request
            in channel C_k to get the difference $S_k + d_i - \Phi$.
            Find l such that, $S_l \leq S_t, \forall_t = 1, 2, \ldots, C$ and $(t\,! = k)$
            Relocate the spilt request(s), to the next channel C_l
            Update $S_k = \Phi$
    Endfor
    Compute $\mathcal{L} = max_{l=1}^{C} S_l$
    Compute $\Omega = C * \mathcal{L} - (\sum_i d_i).$

End.

FIGURE 5.9   Non-contiguous allocation single priority (NCASP) algorithm.

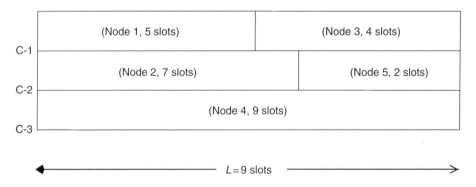

FIGURE 5.10   Non-contiguous allocation of the requests based on NCASP algorithm.

The algorithm is described as pseudo-code in Figure 5.9.

Consider a system with $M = 5$ and $C = 3$. Let the given uplink vector be $D = [5 \quad 7 \quad 4 \quad 9 \quad 2]$. The channel assignments are computed as shown in Figure 5.10. Following the algorithm, we calculate the schedule length, $\mathcal{L} = 9$ slots, which is equal to the optimal value, $\Phi = 9$; and slot wastage, $\Omega = 0$ slots. The allocation vector S, at the end of the algorithm contains: $S_1 = 9, S_2 = 9, S_3 = 9$.

Compare this to the CASP algorithm allocation where the schedule length, $\mathcal{L} = 13$ slots, and slot wastage, $\Omega = 12$ slots.

One of the key requirements of the NCASP algorithm is that subrequests assigned to two different channels do not overlap in time. This is needed because a node cannot tune to two different frequencies at the same time when transmitting or receiving data from the BS. This is the reason for the algorithm to

```
Algorithm CSSA
    For (j = 1, 2, ..., C)
        S_j = 0
    Endfor
    HeapSort the demand vector D in decreasing order
    For (i = 1, 2, ..., M)
        find k, such that S_k ≤ S_l, ∀l = 1, 2, ..., C.
            In the event of a tie, choose k randomly.
        Allocate request d_i to channel C_k
        Update S_k = S_k + d_i
    Endfor
    Compute L = max_{l=1}^{C} S_l
    Compute Ω = C * L − (∑_i d_i).

End.
```

FIGURE 5.11   Contiguous sorted sequential allocation (CSSA) algorithm.

relocate an earlier assigned request, instead of splitting the current request and assigning it to a different channel.

The following are some of the properties of the NCASP algorithm:

1. The time complexity of the NCASP algorithm lies between $O(M \lg C)$ and $O(M^2 \lg C)$.
2. For a request that has been split across two different channels, the split subrequests do not overlap with time.

The reader is referred to Reference 14 for the proofs.

5.4.4 Contiguous Sorted Sequential Allocation

The CSSA algorithm first sorts the demand vector in decreasing order and then sequentially schedules each demand in order. While scheduling each item in the demand vector, the algorithm will select a channel, C_k, which has the least partial sum of allocation. CSSA has smaller schedule length due to the fact that the demands are considered in decreasing order of length. The algorithm is described as pseudo-code in Figure 5.11.

Consider a system with $M = 5$ and $C = 3$. Let the given uplink vector be $D = [5 \quad 7 \quad 4 \quad 9 \quad 2]$. After sorting the requests, the algorithm schedules the requests in the node ID order: 4, 2, 1, 3, 5. While scheduling station 3, after stations 4, 2, and 1 have been scheduled, the algorithm selects channel 3 because channel 3 is occupied by the least number of slots. The algorithm will similarly schedule station 5 to channel 2. The final channel assignments are computed as shown in Figure 5.12.

The schedule length, $L = 9$ slots is better than that of CASP algorithm. The slot wastage, $\Omega = 0$ slots. The allocation vector S, at the end of the algorithm contains: $S_1 = 9, S_2 = 9, S_3 = 9$.

5.4.5 Non-Contiguous Round Robin Allocation

The NCRRA is a non-contiguous allocation algorithm where the demand vector is not sorted (Figure 5.13). Similar to NCASP, it computes the optimal channel schedule length, $\Phi = \max(\max_i(d_i), \lceil (\sum_i d_i)/C \rceil)$, for arbitrating scheduling. Each demand is then scheduled in order and sequentially. While scheduling a demand, the channel is chosen in a round robin manner. If the scheduling of a demand causes the channel

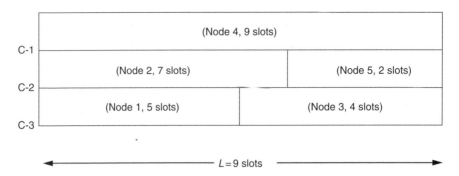

FIGURE 5.12 Contiguous sorted sequential allocation (CSSA) of the requests on the channels.

Algorithm NCRRA

> **BeginSearchChannel** $BSC = 1$
> **For** $(j = 1, 2, \ldots, C)$
> > $S_j = 0$
> **Endfor**
> **Calculate** $\Phi = max(max_i(d_i), \lceil (\sum_i d_i)/C \rceil)$
> **For** $(i = 1, 2, \ldots, M)$
> > **For** $(j = BSC, BSC + 1, \ldots, C)$
> > > If $d_i \leq \Phi - S_j$
> > > > ```
> > > > Allocate request  d_i to channel  C_j
> > > > Update  S_j = S_j + d_i
> > > > ```
> > > > **Break**
> > > If $d_i > \Phi - S_j$
> > > > ```
> > > > Allocate request  Φ − S_j to channel  C_j
> > > > d_i = d_i − Φ − S_j
> > > > Update  S_j = S_j + Φ − S_j = Φ
> > > > BSC = BSC + 1
> > > > ```
> > **Endfor**
> **Endfor**
> **Compute** $\mathcal{L} = max_{l=1}^{C} S_l$
> **Compute** $\Omega = C * \mathcal{L} - (\sum_i d_i)$.

End .

FIGURE 5.13 Non-contiguous round robin allocation (NCRRA) algorithm.

to exceed its optimal allocation, the excess demand is allocated to the next channel. Thus some demands may be split and allocated to separate channels. The complexity of NCRRA is $O(M)$.

Consider a system with $M = 5$ and $C = 3$. Let the given uplink vector be $D = [\ 5\quad 7\quad 4\quad 9\quad 2\ ]$. The optimal channel schedule length $\Phi = \max(9, \lceil (27)/3 \rceil) = 9$. This algorithm allocates the demand of station 1 to channel 1. The demand of station 2 is split such that channel 1 is allocated 4 units and channel 2 is allocated the remaining 3 units. The final channel assignments are computed as shown in Figure 5.14.

The following is an important property of the NCRRA algorithm:

1. For a request that has been split across two different channels, the split subrequests do not overlap in time.

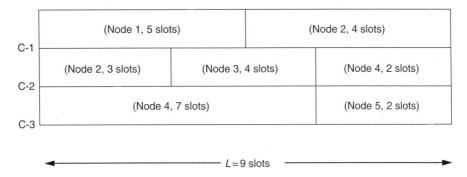

FIGURE 5.14 Non-contiguous round robin allocation (NCRRA) of the requests on the channels.

The proof is available in Reference 16. Note that some channels may not be scheduled completely if $\max(d_i) > \lceil(\sum_i d_i)/C\rceil$. However if $\max(d_i) \leq \lceil(\sum_i d_i)/C\rceil$, the slot wastage $\Omega = 0$ slots and NCRRA will achieve complete load balancing as all channels are fully scheduled.

5.4.6 Non-Contiguous Sequential Round Robin Allocation

NCSRRA is quite similar to NCRRA except that it sorts the demand vector in a decreasing order and re-computes the optimal channel schedule length after scheduling each demand (Figure 5.15). Once sorted, the algorithm sequentially schedules each demand in order and operates similar to NCRRA. Due to the additional sorting performed by NCSRRA, its complexity is increased to $O(M \operatorname{Log} M + M)$ where the $M \operatorname{Log} M$ component is due to the sorting and M component is due to the scheduling. NCSRRA offsets the additional complexity by performing better load balancing when $\max(d_i) > \lceil(\sum_i d_i)/C\rceil$.

Consider a system with $M = 5$ and $C = 3$. Let the given uplink vector be $D = [5 \quad 7 \quad 4 \quad 9 \quad 2]$. The optimal channel schedule length $\Phi = \max(9, \lceil(27)/3\rceil) = 9$. This algorithm sorts the demands and allocates the demand of station 4 to channel 1 and of stations 2 to channel 2. The demands of station 1 are split over channel 2 and 3. The final channel assignments are computed as shown in Figure 5.16. The schedule length, $\mathcal{L} = 9$ slots and the slot wastage, $\Omega = 0$ slots are comparable to those of NCRRA. The allocation vector S, at the end of the algorithm contains: $S_1 = 9, S_2 = 9, S_3 = 9$.

The non-contiguous algorithms have the best scheduling efficiency due to the "splitting mechanism." NCSRRA has lower complexity than NCASP and NCRRA has the least complexity. But NCRRA does not perform load balancing as good as NCSRRA and NCASP when $\max(d_i) > \lceil(\sum_i d_i)/C\rceil$. There can be at most $C - 1$ demands scheduled non-contiguously in NCRRA and NCSRRA. But NCASP may have M such demands in the worst case.

The protocols presented in this section so far are summarized in Table 5.2 with respect to their complexity and worst-case non-contiguous splits for a given request. The extension to multiple priorities is presented in the following sections.

5.4.7 Contiguous Allocation Algorithm for Multiple Priorities

The following two protocols introduce the concept of priority amongst packets, which compete for slots in the C channels. A node may have multiple communication sessions from applications like telnet, ftp, email, Internet Phone, Web Browsing, and Video Conferencing. Each of the applications may have different need and service characteristics. The following two protocols extend the previous scheduling algorithms to accommodate multiple priority sessions of a node sharing the multiple channels. Each session traffic within a node is organized in the order of priority.

The uplink requests made by the M nodes to the BS and the downlink traffic from the BS to the M nodes over the C channels are represented by a $M \times 2P$ matrix. As before, K represents the largest value of a request in slots. Columns $1, \ldots, P$ in the traffic vector indicate the slot requests from the M nodes

Algorithm NCSRRA

> **BeginSearchChannel** $BSC = 1$
> **For** $(j = 1, 2, \ldots, C)$
> $S_j = 0$
> **Endfor**
> **HeapSort** the demand vector D in decreasing order
> **Calculate** $\Phi = max(max_i(d_i), \lceil(\sum_i d_i)/C\rceil)$
> **For** $(i = 1, 2, \ldots, M)$
>   **For** $(j = BSC, BSC + 1, \ldots, C)$
>   If $d_i \leq \Phi - S_j$
>   Allocate request d_i to channel C_j
>   Update $S_j = S_j + d_i$
>   **Break**
>   If $d_i > \Phi - S_j$
>   Allocate request $\Phi - S_j$ to channel C_j
> $d_i = d_i - \Phi - S_j$
>   Update $S_j = S_j + \Phi - S_j = \Phi$
> $BSC = BSC + 1$
>   **Endfor**
>   **If** $(\Phi = d_i)$
>   /* Dynamically re-compute the optimal schedule length
>   to guarantee load balancing */
> $\Phi = max(d_{i+1}, \lceil(\sum_{k=i+1}^{M} d_k)/C\rceil)$
> **Endfor**
> **Compute** $\mathcal{L} = max_{l=1}^{C} S_l$
> **Compute** $\Omega = C * \mathcal{L} - (\sum_i d_i)$.

End.

FIGURE 5.15   Non-contiguous sequential round robin allocation (NCSRRA) algorithm.

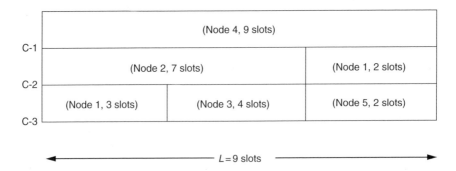

FIGURE 5.16   Non-contiguous sequential round robin (NCSRRA) allocation of the requests on the channels.

to the BS (uplink traffic), where each column represents a priority queue. Columns $P + 1, \ldots, 2P$ in the traffic vector indicates the slot requests for the M nodes from the BS. Thus, we only consider the first P columns of the matrix. We now need to find a conflict-free assignment of the C channels such that the frame length is minimized.

TABLE 5.2   Comparison of the Different Single-Priority Algorithms

Algorithm	Complexity	NNC
CASP	$O(M \log C)$	0
NCASP	$O(M \log C)$ $O(M^2 \log C)$	M
CSSA	$O(M \log M + M \log C)$	0
NCRRA	$O(M)$	$C - 1$
NCSRRA	$O(M \log M + M)$	$C - 1$

Here, NNC refers to the number of demands scheduled non-contiguously in the worst case.

This problem is an extension of the problem studied in the previous sections. The notations used in the following sections are explained in Table 5.1.

The data structures used are:

1. Let $D = (d_{ij})$ denote the demand matrix, and let d_{ij} denote the request made by node i for priority j traffic, $0 \leq d_{ij} \leq K, i = 1, \ldots, M$ and $j = 1, \ldots, P, P$ represents the number of priorities. The actual value of P depends upon implementation of the priority scheme chosen.
2. Let $S = (S_i)$ denote the allocation vector and let S_i denote the total amount of slots allocated on channel i, where $0 \leq S_i, i = 1, \ldots, C$.

The idea behind the scheduling algorithm is to schedule traffic of priority 1 first, followed by that of priority 2, and so on.

The contiguous allocation algorithm for multiple priority systems (CAMP) maintains an allocation vector S, where S_i $(i = 1, 2, \ldots, C)$, represents the partial sum of slots currently allocated to channel C_i. For a given request, the scheduling algorithm allocates the requests d_{i1} $(i = 1, 2, \ldots, M)$, contiguously (without any splitting of the requests) on the channel which currently has the least partial sum of allocation. If more than one channel has the same least partial sum, one channel is chosen at random for the next allocation. Once this is done, it computes the schedule length, $\mathcal{L}$, of the new schedule, initializes $S_i = \mathcal{L}$ $(i = 1, 2, \ldots, C)$, and then continues by allocating the requests in the next priority queue. This algorithm essentially schedules the requests in different priority queues in separate groups. The algorithm is described as pseudo-code in Figure 5.17. The time complexity of the CAMP algorithm is $O(PM \log C)$ as shown in Reference 14.

Consider a system with $M = 5, C = 3, P = 3$.

Let the given uplink vector be $D = \begin{bmatrix} 5 & 3 & 2 \\ 7 & 8 & 3 \\ 4 & 6 & 2 \\ 9 & 5 & 7 \\ 2 & 5 & 4 \end{bmatrix}$.

The channel assignments are computed as shown in Figure 5.18. Following the algorithm, we calculate the schedule length, $\mathcal{L} = 33$ slots; the slot wastage, $\Omega = 27$ slots. The allocation vector, S, at the end of the algorithm contains: $S_1 = 33, S_2 = 27, S_3 = 30$.

5.4.8   Non-Contiguous Allocation Algorithm for Multiple Priorities

The NCAMP system has the NCASP algorithm as the basis, and is extended to handle multiple priority queues within a node. Once the requests in a priority queue have been scheduled, the schedule length, $\mathcal{L}$, is computed and S_i is initialized to $\mathcal{L}$ $(i = 1, 2, \ldots, C)$. We then continue by allocating the requests in the

Algorithm CAMP

> **For** $(j = 1, 2, \ldots, C)$
>> $S_j = 0$
>
> **Endfor**
> **For** $(j = 1, 2, \ldots, P)$
>> **For** $(i = 1, 2, \ldots, M)$
>>> find k, such that $S_k \leq S_l, \forall l = 1, 2, \ldots, C$.
>>>> In the event of a tie, choose k randomly.
>>>
>>> Allocate request d_{ij} on channel C_k
>>> Update $S_k = S_k + d_{ij}$
>>
>> **Endfor**
>> **Compute** $S_{max} = \max_{l=1}^{C} S_l$
>> **For** $(j = 1, 2, \ldots, C)$
>>> $S_j = S_k$
>>
>> **Endfor**
>
> **Endfor**

End.

FIGURE 5.17   Contiguous allocation algorithm for multiple priorities (CAMP).

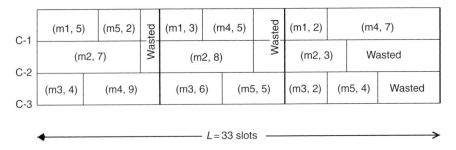

FIGURE 5.18   Contiguous allocation of the requests on the channels.

next priority queue. The algorithm is described as pseudo-code in Figure 5.19. The time complexity of the NCAMP algorithm is between $O(PM \lg C)$ and $O(PM^2 \lg C)$ as shown in Reference 14.

Consider a system with $M = 5, C = 3, P = 3$. Let the given uplink vector be the same as in the previous example. The channel assignments are computed as shown in Figure 5.20. Following the algorithm, we calculate the schedule length, $\mathcal{L} = 25$ slots. The slot wastage, $\Omega = 3$ slots. The allocation vector, S, at the end of the algorithm contains: $S_1 = 25, S_2 = 25, S_3 = 22$.

Compare this to the CAMP algorithm allocation where the schedule length, $\mathcal{L} = 33$ slots, and slot wastage, $\Omega = 27$ slots.

In other related work, algorithms that maximize the system throughput subject to fairness and resource constraints [10–12] are presented. We do not discuss these here due to lack of space.

5.5 Conclusions

This chapter presented an overview of multiple access control protocols for multiple channel wireless networks. Simultaneous use of multiple channels increases the available bandwidth of the network and

Algorithm NCAMP

> **For** $(j = 1, 2, \ldots, C)$
>> $S_j = 0$
>
> **Endfor**
> **Let** $\mathcal{L} = 0$
> **For** $(j = 1, 2, \ldots, P)$
>> **For** $(i = 1, 2, \ldots, C)$
>>> $S_i = \mathcal{L}$
>>
>> **Endfor**
>> **Calculate** $\Phi = \mathcal{L} + max(K, \sum_i d_{ij}/C)$
>> **For** $(i = 1, 2, \ldots, M)$
>>> find k, such that $S_k \leq S_l, \forall l = 1, 2, \ldots, C$.
>>> In the event of a tie, choose k randomly.
>>> If $(S_k + d_{ij} > \Phi)$
>>>> **Compute** the split s, by breaking an earlier assigned request in channel C_k to get the difference $S_k + d_{ij} - \Phi$.
>>>> Find l such that, $S_l \leq S_t, \forall t = 1, 2, \ldots, C$ and $(t \mathrel{!=} k)$
>>>> **Relocate** the split request(s), to the next channel C_l
>>>> Update $S_k = \Phi$
>>>
>> **Endfor**
>> find k, such that $S_k \geq S_i, \forall i = 1, 2, \ldots, C$.
>> **Update** $\mathcal{L} = S_k$
> **Endfor**

End.

FIGURE 5.19   Non-contiguous allocation algorithm for multiple priorities (NCAMP).

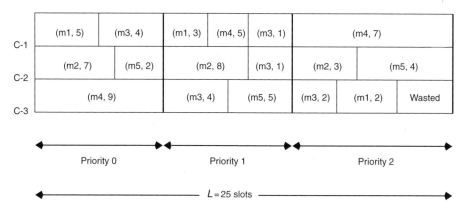

FIGURE 5.20   Non-contiguous allocation of the requests on the channels.

addresses the ever-increasing demand for higher bandwidth. This chapter surveyed some contemporary protocols studied for such systems. The protocols are classified into two categories: (i) contention-based protocols and (ii) contention-free protocols. Contention-based protocols rely on random access carrier sensing techniques whereas the contention-free protocols are based on reservation and scheduling techniques. This chapter also presented various scheduling algorithms for contention-free MAC protocols.

Acknowledgments

Part of the research was supported by Air Force Office of Scientific Research grant F-49620-99-1-0125, NSF grant No. CCR-0209211 and UMBC SRIS award.

References

[1] IEEE 802.11 Working Group for WLAN, "Part 11: Wireless LAN Medium Access Control (MAC) and Physical Layer (PHY) Spec: High Speed Physical Layer in the 5 GHZ Band," http://grouper.ieee.org/groups/, 1999.

[2] IEEE 802.15 Working Group for WPAN, "Part 15.3: Wireless Medium Access Control (MAC) and Physical Layer (PHY) Specifications for High Rate Wireless Personal Area Networks (WAPN)," Draft P802.15.3/D17-pre, 2003.

[3] Asis Nasipuri and Samir R. Das, "A Multichannel CSMA MAC Protocol for Mobile Multihop Networks," in *Proceedings of the IEEE Wireless Communications and Networking Conference (WCNC'99)*, New Orleans, LA, September 1999.

[4] Asis Nasipuri and Samir R. Das, "Multichannel CSMA with Signal Power-Based Channel Selection for Multihop Wireless Networks," in *Proceedings of the IEEE Fall Vehicular Technology Conference*, Boston, MA, September 2000.

[5] Nitin Jain, Asis Nasipuri, and Samir R. Das, "A Multichannel MAC Protocol with Receiver-Based Channel Selection for Multihop Wireless Networks," in *Proceedings of the IEEE International Conference on Computer Communication and Networks*, Phoenix, AZ, October 2001.

[6] Andrew Muir and J.J. Garcia-Luna-Aceves, "A Channel Access Protocol for Multihop Wireless Networks with Multiple Channels," in *Proceedings of the International Conference on Communications (ICC)*, Atlanta, GA, July 1998.

[7] R. Garces and J.J. Garcia-Luna-Aceves, "Collision Avoidance Resolution Multiple Access for Multichannel Wireless Networks," in *Proceedings of the IEEE INFOCOM*, Tel-Aviv, Israel, March 2000.

[8] Asimakis Tzamaloukas and J.J. Garcia-Luna-Aceves, "A Receiver-Initiated Collision-Avoidance Protocol for Multi-Channel Networks," in *Proceedings of the IEEE INFOCOM*, Anchorage, AK, April 2001.

[9] Jungmin So and Nitin Vaidya, "A Multi-Channel MAC Protocol for Ad Hoc Wireless Networks," Technical Report, CSL/UIUC Technical Report, January 2003.

[10] V. Kanodia, C. Li, A. Sabharwal, B. Sadeghi, and E. Knightly, "Distributed Multi-Hop Scheduling and Medium Access with Delay and Throughput Constraints," in *Proceedings of the ACM MobiCom*, Rome, Italy, July 2001.

[11] V. Kanodia, C. Li, A. Sabharwal, B. Sadeghi, and E. Knightly, "Ordered Packet Scheduling in Wireless Ad Hoc Networks: Mechanisms and Performance Analysis," in *Proceedings of the ACM MobiHoc*, Lausanne, Switzerland, June 2002.

[12] Y. Liu and R. Knightly, "Opportunistic Fair Scheduling over Multiple Wireless Channels," in *Proceedings of the IEEE INFOCOM*, San Francisco, CA, April 2003.

[13] Krishna M. Sivalingam, Jyh-Cheng Chen, Prathima Agrawal, and Mani Srivastava, "Design and Analysis of Low-Power Access Protocols for Wireless and Mobile ATM Networks," *ACM/Baltzer Wireless Networks*, vol. 6, pp. 73–87, 2000.

[14] Satish Damodaran and Krishna M. Sivalingam, "Scheduling Algorithms for Multiple Channel Wireless Local Area Networks," *Computer Communications*, vol. 25, pp. 1305–1314, 2002.

[15] Satish Damodaran and Krishna M. Sivalingam, "Scheduling in wireless networks with multiple transmission channels," in *Proceedings of the International Conference on Network Protocols*, Toronto, Canada, October 1999, pp. 262–269.

[16] Chonggang Wang, Bo Li, and Krishna M. Sivalingam, "Scalable Multiple Channel Scheduling with Optimal Utility in Wireless Local Area Networks," *ACM/Kluwer Wireless Networks Journal*, July 2004, (Accepted for).

[17] Aniruddha Rangnekar and Krishna M. Sivalingam, "Multiple Channel Scheduling in UWB Based IEEE 802.15.3 Networks," in *Proceedings of the First International Conference on Broadband Networks — Wireless Networking Symposium*, San Jose, CA, October 2004.

[18] P. Karn, "MACA — A New Channel Access Method for Packet Radio," in *Proceedings of the ARRL/CRRL Amateur Radio 9th Computer Networking Conference*, Ontario, Canada, September 1990, pp. 134–140.

[19] V. Bhargavan, A. Demers, S. Shenker, and L. Zhang, "MACAW: A Media Access Protocol for Wireless LANs," in *Proceedings of the ACM SIGCOMM*, London, UK, August 1994, pp. 212–225.

[20] F. Talucci and M. Gerla, "MACA-BI (MACA by Invitation): A Wireless MAC Protocol for High Speed Ad Hoc Networking," in *Proceedings of the IEEE International Conference on Universal Personal Communications (ICUPC)*, San Diego, CA, October 1997, pp. 913–917.

[21] R. Garces and J.J. Garcia-Luna-Aceves, "Collision Avoidance and Resolution Multiple Access (CARMA)," *Cluster Computing*, vol. 1, pp. 197–212, 1998.

[22] J.D. Ullman, "Complexity of Sequencing Problems," in *Computer and Job/Shop Scheduling Theory*, E.G. Coffman, Ed., Chapter 4. John Wiley, New York, 1976.

[23] E. Coffman, M.R. Garey, and D.S. Johnson, "An Application of Bin-Packing to Multiprocessor Scheduling," *SIAM Journal of Computing*, vol. 7, pp. 1–17, 1978.

6

Multichannel MAC Protocols for Mobile Ad Hoc Networks

6.1   Introduction **6**-99
6.2   The Medium Access Problem in Wireless **6**-100
     Basic Concepts • Using CSMA in Wireless • The IEEE
     802.11 MAC • Concerns with 802.11 MAC • The
     Multichannel MAC Solution
6.3   Multichannel Support in Ad Hoc Standards **6**-106
6.4   Multichannel MAC Protocols with Dynamic
     Channel Selection **6**-107
     The MMAC with Soft Reservation (MMAC-SR) • The
     MMAC Using Clearest Channel at Transmitter (MMAC-CT)
     • The MMAC Using Clearest Channel at Receiver
     (MMAC-CR) • The MMAC with Cooperative Channel
     Selection (MMAC-CC) • Performance of MMAC Protocols
6.5   Related Work on Multichannel MAC Protocols **6**-114
     Networks Using FH-SSMA Radio • Separate Control and
     Data Channels • Multiple Data Channels with Channel
     Selection
6.6   Conclusion **6**-118
References **6**-118

Asis Nasipuri
Samir R. Das

6.1   Introduction

A mobile ad hoc network (MANET) [6] is a multihop wireless network of mobile terminals (nodes) that can be established instantly, without requiring any pre-existing fixed network infrastructure. All nodes possess traffic routing and relaying capabilities so that any node can communicate with any other node in the network using multihop packet transmissions over other intermediate nodes. The nodes can be highly mobile and may leave or join the network dynamically. A MANET is therefore designed to dynamically adapt to frequent changes in network topology.

MANETs are typically designed to operate using a random channel access scheme over a common channel. There are many challenging design issues in MANETs that arise due to the characteristics of the wireless medium and the dynamic nature of the network. Medium access is one of the key challenges, which is distinctively different from that in a wired LAN due to the differences in the propagation characteristics, channel capacity, and receiver performance in the two mediums. The medium access control (MAC) protocol plays the key role in determining the efficiency of channel usage, and therefore, is the primary factor determining the network capacity. Despite a tremendous amount of research on MAC protocols for multihop wireless networks, the network capacity continues to be the primary limitation of these networks. Studies have shown that the end-to-end throughput degrades rapidly as the number of hops increases, which is a major concern in large and dense ad hoc networks.

The IEEE 802.11 MAC is the most popular established standard for wireless LANs that is applied to MANETs [5]. Traditional 802.11 compliant network interfaces use a single DS/SSMA channel and one radio per node. Contention in the channel is avoided using a combination of carrier sensing, random backoffs, and an optional channel reservation scheme using an exchange of control packets. Numerous studies have shown that this design has significant disadvantages under heavy traffic, which is exacerbated in dense networks or under situations requiring a number of concurrent transmissions in the same region. The main factors responsible for such performance degradation are the isotropic nature of signal propagation and the exponential path loss characteristics in the wireless medium. Some of the well-known problems arising out of these are the hidden terminal and exposed terminal problems, which have been investigated thoroughly.

Most of the techniques used to improve the throughput in MANETs incorporate features in the MAC protocol that reduce the hidden and exposed terminal problems. A broader objective is to reduce the effect of multiuser interference, which is the key factor that determines the probability of successful transmission of packets. The use of directional antennas is regarded as one of the viable solutions to achieve this objective, which led to several directional-antenna based MAC protocols for ad hoc networks. Directional antennas can provide benefits by reducing the interference in unwanted directions while providing additional gain in the desired direction of transmission. Hence, it improves the throughput in the network by increasing the *spatial reuse* of the wireless channel.

An alternate method for improving the throughput in MANETs, which is the focus of this chapter, is the use of multiple orthogonal channels that can be shared by all the nodes in the network. Although MANETs do not have centralized control or base stations that can perform channel allocation, appropriately designed MAC protocols can allow the transmitting nodes to *dynamically* select distinct channels in the same neighborhood to reduce the possibility of destructive interference or packet collisions in the network. In recent times several multichannel MAC protocols have been proposed with this objective [8,13,14,19,27,30]. These protocols provide significant improvement in the average network throughput in comparison to single channel protocols, even if the same total channel bandwidth is used. In addition, use of multiple channels can provide a simpler way to support quality of service (QoS) in ad hoc networks. The study of multichannel MAC protocols is also important from the perspective of resource utilization in wireless LANs. Although not used, the physical layer specifications of IEEE 802.11b as well as 802.11a define multiple channels, of which several can be used for orthogonal transmissions. This additional bandwidth is wasted when all transmissions are performed on the same channel. Hence, efficient mechanisms for utilizing multiple channels for improving the throughput in MANETs is an important issue.

In this chapter, we first explain the possible advantages of using multiple channels in random access protocols for MANETs. We also discuss the theoretical issues governing efficient medium access and the shortcomings of the IEEE 802.11 MAC using a single channel. Details of the physical layer specifications that allow the use of multiple orthogonal channels are explained. Finally, a discussion on design aspects of multichannel MAC protocols with dynamic channel selection are presented. We also present some comparative performance results of various existing multichannel MAC protocols vis-à-vis those of a single-channel 802.11 MAC in the same network.

6.2 The Medium Access Problem in Wireless

A key difference between the MAC protocols in wired and wireless LANs is that protocols for wireless networks must take into account the path loss characteristics of the propagating signal. Whereas a transmitted signal in the wired medium is received at all nodes connected to the medium with almost equal signal strength (albeit after a propagation delay), the same is not true in wireless. In wireless, the signal strength drops exponentially with distance. Hence, for successful packet transmission, transmitters in wireless networks must employ special techniques to ascertain that the strength of the signal as well as the ambient interference level at the receiver is adequate for it to receive the transmitted packet correctly.

6.2.1 Basic Concepts

Some of the key concepts for understanding the design issues of MAC protocols in MANETs are briefly described below:

Frequency reuse: The average power of a wireless signal varies inversely to the nth power of the distance, where n ranges between 2 and 4 [17]. Hence, the same channel can be simultaneously used at two different locations that are sufficiently far apart so that the interference from one to the other is within an acceptable level, a principle known as "frequency reuse". The acceptable level of interference depends on the receiver properties, such as the minimum signal-to-interference-plus-noise ratio (SINR) required by the receiver to detect the packet correctly. This threshold, which we denote by $SINR_{min}$, depends on the modulation used and the receiver hardware.

Radio transmission range: The transmission range is determined by the maximum distance from a transmitter at which the SINR in the absence of any interference is above the minimum receiver threshold $SINR_{min}$. Within the transmission range, a receiver can detect (receive) a transmitted packet correctly as long as the ratio of the received signal power of the packet to the total noise and interference from all other transmissions on the channel (SINR) exceeds $SINR_{min}$. Hence, a receiver located within the radio range of a transmitter node will be unable to receive a packet when the interference level on the channel is high, which is often described by a "busy" carrier state.

Carrier sensing: The process of determining whether the channel is busy or idle is termed *Clear Channel Assessment* (CCA) or carrier sensing. Carrier sensing is performed by comparing the power of the received signal on the channel to a threshold T_{CS}. Typically T_{CS} is chosen such that a receiver can detect a transmitted signal from a distance larger than its radio range. The illustration in Figure 6.1 explains these concepts.

Packet collision: If the transmission times of two or more packets transmitted from nodes located within the radio range of the same receiver overlap such that none of the packets can be received correctly, then the packets are termed as *collided*. Technically, a packet suffers a collision whenever its SINR falls below the $SINR_{min}$ threshold of the receiver. This may happen even if the combined interference from transmissions *outside* the radio range of the receiver reduces the SINR of the packet to fall below $SINR_{min}$.

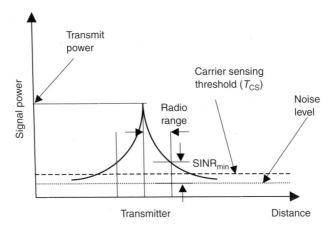

FIGURE 6.1 Illustration of wireless signal propagation and receiver thresholds.

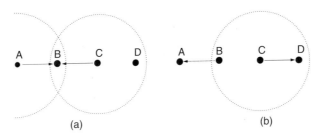

(a)   (b)

FIGURE 6.2   Illustrations of the hidden and exposed terminal problems.

Capture: If two different transmissions are received at the same time and both signals are strong enough to be independently received in the absence of the other, the receiver may still be able to receive one of the signals correctly, typically known as capture. The actual principle of capture depends on the physical layer design issues, such as modulation and receiver structure. A commonly used capture model is known as power-capture, which is the mechanism by which the receiver detects the strongest of several overlapping signals as long as its SINR is greater than a threshold called the capture threshold.

6.2.2 Using CSMA in Wireless

Carrier sense multiple access or CSMA is an effective technique for random channel access in mediums where the propagation characteristics such as path loss and delay are small. Hence, CSMA based protocols in wired networks can achieve high probabilities of successful packet transmission by using a "listen before transmit" principle. In the wired medium, CSMA protocols can also benefit from *collision detection*, where a sender monitors the channel while it is transmitting and aborts its transmission when it detects a collision, thereby minimizing unnecessary wastage of bandwidth during packet collisions.

    In a wireless network, a transmitter cannot truly detect the state of the channel at the receiver by performing CCA, because propagation path losses in wireless make the signal strengths quite different at two different locations. As a result, CSMA-based protocols in wireless networks suffer from the *hidden* and *exposed terminal* problems that are illustrated in Figure 6.2.

- *The hidden terminal problem*: This arises when two nodes that are not within the carrier sensing range of each other (i.e., hidden) try to transmit to the same node, which is within range of both. An illustration is shown in Figure 6.2(a), where nodes A and C are hidden from each other while both are within range of B. Hence, while either A or C is transmitting to B, the other will not be able to detect the transmission and might transmit at the same time, causing a packet collision at B. Because of the isotropic nature of signal propagation in wireless, even if C is transmitting a packet to D, A faces the same problem while transmitting to B.
- *The exposed terminal problem*: This describes the complimentary situation where a node that is within the range of a transmitting node finds the channel busy and defers its transmission until the channel is perceived to be free, where its concurrent transmission would not have affected either of the receivers. This is illustrated in Figure 6.2(b), where B wanting to transmit to A waits until the detected transmission from C to D is over. However, this waiting period is unnecessary as both transmissions can take place at the same time. This is true when A(D) is sufficiently far from C(D) so that A(D) can correctly receive the packet from B(C) despite the transmission from C(B). The unnecessary waiting caused by the exposed terminal problem leads to inefficient channel utilization.

The above problems are typical examples of situations arising in wireless networks where carrier sensing fails to be effective. An additional issue is that spatial variations of signal strengths makes it impossible for a transmitting node to detect a collision at the corresponding receiver. Usually, wireless networks rely on the use of a MAC layer acknowledgment packet (ACK) returned from a receiver for confirmation of successful transmissions. This introduces an additional transmission overhead over that required in wired networks. In addition, a successful packet transfer depends on correct packet reception at both ends, that is, the success of the data packet as well as the ACK. These are additional factors that reduce the efficiency of random access protocols in wireless.

6.2.3 The IEEE 802.11 MAC

The IEEE 802.11 [5] is an international standard of physical and MAC layer specifications for wireless LANs that addresses the issues presented in the previous section. It defines three physical layers of which two are based on radio transmission on the 2.4 GHz unlicensed ISM radio band: one using direct sequence (DS), and the other using frequency hopped (FH) spread spectrum. The third physical layer uses infrared signals. 802.11 supports mandatory support for 1 Mb/sec data rate with optional support for 2 Mb/sec in wireless LANs. Subsequent extensions of this standard, 802.11b and 802.11a, can support data rates of 11 and 54 Mb/sec, respectively. While 802.11b uses the 2.4 GHz band, 802.11a uses the 5 GHz ISM band. In this discussion, we are not particularly concerned with any specific physical layer as the fundamental aspects of the MAC layer performance is the same for all. For some numerical performance examples, we use parameters that relate to 802.11 with 1 Mb/sec data rate.

The MAC layer specification in 802.11 for noninfrastructured wireless LANs uses a scheme that is known as *carrier sense multiple access with collision avoidance* (CSMA/CA) with an optional channel reservation scheme that uses a four-way handshape between sending and receiving nodes. These two methods constitute the *Distributed Coordination Function* (DCF), which is designed to provide distributed coordination amongst asynchronous nodes for sharing a common channel for communication.

Basic DCF using CSMA/CA: The basic CSMA/CA scheme relies on two techniques to avoid collisions amongst contending nodes: *carrier sensing* and *random backoffs*. Before transmitting a data packet, the sender performs CCA to determine if the channel is free from other transmissions in its neighborhood. The data packet is transmitted if the channel is observed to be free for a period of time exceeding a *distributed interframe space* (DIFS). If the channel is found busy, it retransmits the packet after a slotted random backoff period that is counted over the subsequent period of time in which it senses the channel to be free. The backoff period is chosen to be B times a backoff slot duration, where B is an integer that is chosen uniformly in $[0, \mathrm{CW} - 1]$ and CW is the length of the contention window. The initial (default) value of the contention window is a parameter $\mathrm{CW_{min}}$, which is chosen as 8 in 802.11. If a data packet is received correctly, the receiving node transmits an ACK packet after a *short interframe space* (SIFS). If an ACK is not received within the stipulated time period, the sender attempts to transmit the data packet again following the same procedure. With every subsequent retry attempt, the value of CW is doubled until it reaches the maximum value of 256. The adaptive backoff period is designed to account for variations of the congestion in the network.

To minimize packet collisions due to the hidden terminal problem, it is recommended to use a T_{CS} value for which the carrier sensing range is approximately twice as much as its radio range. However, increasing the carrier sensing range also increases the effect of the exposed terminal problem and essentially reduces spatial reuse of the wireless channel. An alternative technique for reducing the hidden terminal problem without having to use a high carrier sensing range is the use of the channel reservation option.

The CSMA/CA with RTS/CTS extension: With this option, the sending and receiving nodes exchange two short control packets, known as the *request-to-send* (RTS) and the *clear-to-send* (CTS) packets, respectively, before initiating a data packet transmission. If the receiver correctly receives the RTS packet, it replies by sending a CTS, thereby confirming that adequate conditions exist for data transmission. This serves as a "virtual" carrier sensing at the location of the receiver from that at the sender. Neighboring nodes that receive either the RTS or the CTS or both are required to delay their transmissions for a period of time specified in the corresponding packet. The duration of silencing period specified by the RTS is equal to the time by which the sender expects to receive the ACK packet, and the corresponding duration specified by the CTS packet is the time until which the receiver expects the data packet transmission to be over. This results in the sending and receiving nodes to reserve the channel until the end of data packet transmission cycle.

FIGURE 6.3   Illustration of scenario where RTS–CTS may not silence transmissions from a hidden terminal.

6.2.4   Concerns with 802.11 MAC

The basic DCF does not solve the hidden terminal problem, although it still is fairly efficient in resolving contention amongst neighboring nodes that are contending to transmit packets. The RTS–CTS option does reduce the hidden terminal problem, but is not successful in dealing with the inefficiency caused by the exposed terminal problem. In addition, several other issues add to the concerns related to the efficiency of the 802.11 DCF. We describe some of these issues below, which motivate alternative design efforts toward more efficient MAC protocols for ad hoc networks:

Additional overhead due to RTS and CTS transmissions: The RTS–CTS option requires additional transmissions on the shared channel. These packets must rely on the basic CSMA/CA mechanism for transmission and suffer from inefficiencies caused by the hidden and exposed terminal problems. Although these packets are much smaller than data packets and they waste a smaller amount bandwidth during collisions in comparison to data packets, they still introduce additional overhead on the channel. This concern is particularly serious under heavy traffic when there are a large number of RTS transmissions trying to establish channel reservation.

Channel reservation with RTS/CTS can fail: All nodes located within range of the sending the receiving nodes must correctly set their NAVs in order to make channel reservation complete. However this might not happen due to a number of reasons. For instance, the illustration in Figure 6.3 shows how node C fails to receive the CTS packet from B due to a collision with another RTS packet from D, and becomes a potential source of destructive interference for B.

Interference from out-of-range transmitters: Nodes that are beyond the range of the sender and the receiver do not learn about the ongoing data packet transmission. However, when they transmit, the resulting signal reaching the receiver adds to its total interference power. With sufficient amount of interference from out-of-range nodes, the SINR at a receiving node can drop below the $SINR_{min}$ threshold, causing packet loss. This is illustrated in Figure 6.4, where C cannot correctly receive transmissions from A or B as it is outside their transmission ranges. The SNR of the packet received by B from A (P_{AB}/N), is greater than the minimum SINR threshold of the receiver $SINR_{min}$ in the absence of any other transmission, and hence B can correctly receive the packet from A in that situation. However, if C starts transmitting at the same time, the SINR of the same packet $(P_{AB}/(N + P_{CB}))$ may fall below $SINR_{min}$ causing the packet to be decoded incorrectly. Such packet loss due to interference can occur frequently due to the combined interference from a large number of out-of-range transmissions at a receiver.

6.2.5   The Multichannel MAC Solution

The advantages of using multiple channels in random access protocols was first explored for wired LANs by Marsan and Roffinella [2]. There, it was shown that the use of multiple channels with the *same aggregate channel capacity* improves the throughput of CSMA [9] based MAC protocols in wired LANs. This is attributed to the fact that the normalized propagation delay in a wired medium, defined as the ratio of the propagation delay and the packet transmission time, decreases with increasing number of channels when the total bandwidth is conserved. Since the probability of packet collisions[1] increases with increasing delay

---

[1]A packet collision in a wired medium occurs whenever two packet transmissions overlap, making it impossible to recover either of the overlapping packets. The same term when applied to wireless LANs, has a slightly different physical interpretation.

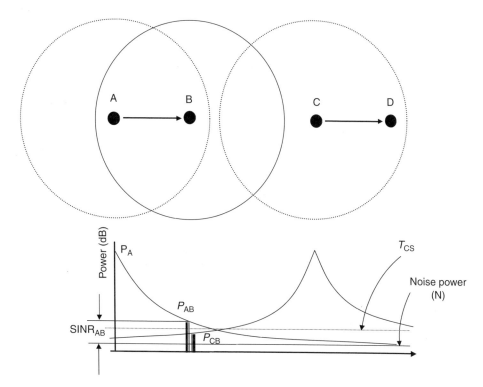

FIGURE 6.4   Illustrations of packet loss due to out-of-range transmissions.

between the start of transmissions and when it can be sensed by another node, use of multiple channels leads to a lower probability of packet collisions.

As described in the earlier sections, the primary causes of packet loss in wireless networks are due to the hidden and exposed terminal problems and other effects of radio interference. The relatively small propagation times do not pose a serious problem to the carrier sense based MAC protocols. However, use of multiple channels provide some advantages in wireless networks as well, although for different reasons. We explain these issues here.

The primary advantage of using multiple channels in a wireless network is that it provides a mechanism for multiple concurrent transmissions amongst independent source–destination pairs in a given region. Typically, a random medium access protocol uses techniques that enable different transmitting nodes to space out their packet transmissions over mutually nonoverlapping *time intervals* to avoid interfering with one another. The use of multiple channels provides a mechanism by which collisions may be reduced by spacing out transmissions *over channels* as well as over time, thereby introducing an additional dimension for controlling interference.

We provide an explanation of this fundamental advantage using the illustration in Figure 6.5. The example shows three pairs of communicating nodes with independent transmission times. Consider that the packet arrival times at nodes A, C, and E are T_A, T_C, and T_D, respectively. If all transmissions are performed on a single channel, nodes C and E will find the channel busy at the times of their packet transmissions and have to backoff for random time intervals before they are eventually transmitted. On the other hand, if three channels are available with each node capable of finding a free channel for its transmission, there would be no need for backoffs. Hence, the average channel utilization and throughput in this scenario is better with three channels instead of one, even if the same aggregate channel capacity is used in both cases.

Although this simple illustration explains the advantages of using three channels in place of one in this scenario, several points are worth mentioning. First, when the number of transmitting nodes is higher than

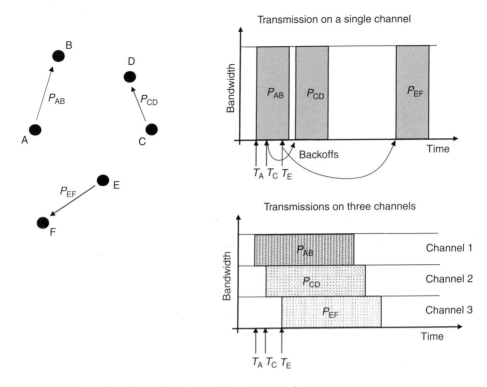

FIGURE 6.5　Example showing the benefit of using multiple channels.

the number of available channels, contention and backoffs cannot be avoided. The performance of using multiple channels in that case will depend on how channels are selected. An efficient channel selection scheme can distribute the transmissions over different channels and times to provide a comparatively higher throughput than a scheme using a single channel with the same overall bandwidth. Second, when multiple channels are used, the *quality* or clarity of transmissions will vary over channels. This is particularly important in wireless where the quality of a transmission depends on the SINR at the receiver. Hence, multichannel MAC protocols can use channel selection schemes for achieving desired levels of quality (such as specific QoS constraints) and simultaneously maximize the utilization of the available bandwidth. Lastly, it should be noted that if the same total aggregate bandwidth is to be used for a multichannel protocol, the bandwidth gets reduced. This leads to higher packet transmission times, which can cause a negative effect on the throughput if there are too many backoffs in the multichannel scheme. The above factors generate challenging issues for designing efficient multichannel MAC protocols for ad hoc networks.

6.3　Multichannel Support in Ad Hoc Standards

Technically, multiple independent channels can be implemented by dividing the available bandwidth into nonoverlapping frequency bands (FDMA), time-slots (TDMA), or codes (CDMA). The IEEE 802.11 physical layer standard already has multiple channels defined. There are currently four specifications in the family: 802.11, 802.11a, 802.11b, and 802.11g. The differences in these physical layers are in the modulation types used. 802.11 uses phase-shift keying (PSK) with CDMA channels in the 2.4 GHz frequency band. The modulation in 802.11b is complimentary code keying (CCK), which allows higher data rates and is less susceptible to multipath propagation. The 802.11a uses a modulation scheme called orthogonal frequency-division multiplexing (OFDM) in the 5-GHz frequency band that makes it possible to achieve

TABLE 6.1 Channels in the IEEE 802.11 Standards Operated in North America

Standard	Frequency band (GHz)	Modulation	Data rate (Mb/sec)	No. of Channels
802.11	2.4	PSK	2	3
802.11b	2.4	CCK	11	3
802.11a	5	OFDM	54	8
802.11g	2.4	OFDM	54	8

data rates as high as 54 Mb/sec. All the physical layer specifications allow multiple channels, which are summarized in Table 6.1. However, due to problems associated with radio interference and frequency coordination in the unlicensed ISM band, the number of usable orthogonal channels differs from the actual number of available channels. In North America, 3 of the 11 802.11b channels, channels 1, 6, and 11, are found to be compatible for independent use without interference problems. Similarly, 802.11a offers eight independently usable channels.

Traditionally, standard-compliant 802.11 wireless LAN radios operate on only one channel at any given time. The additional capacity is intended for infrastructured wireless LANs, where multiple nonoverlapping channels can be used by adjacent access points. However, needs for higher end-to-end throughputs in multihop ad hoc networks have led to the interest in using multiple channels within the noninfrastructured or ad hoc networking mode as well. In such cases, the MAC needs to be designed for interacting with the physical layer for optimum channel selection.

6.4 Multichannel MAC Protocols with Dynamic Channel Selection

Based on the discussion presented in Section 2.5, we note that in addition to the obvious benefit of utilizing more communication resources (channels), multichannel MAC protocols can also improve the throughput performance due to the following reasons:

1. *Fewer backoffs*: If contending transmitters in a given neighborhood distribute their transmissions over different channels so as to gain better access to free channels independently, the need for backoffs is reduced. This automatically improves the channel utilization due to smaller idle periods in the channel.
2. *Reduced interference*: Because of the spatial distribution of interference and the option of transmitting on one of several available channels, transmissions can be distributed on appropriate channels so that each transmission experiences the lowest possible interference. This can lead to lower probabilities of packet collisions and better quality of transmissions on the average.

These issues pose a complex set of design criteria that motivate exploration of newer multichannel MAC protocols for MANETs. These protocols try to maximize the utilization of multiple orthogonal channels for achieving higher throughputs. The assumptions that differentiate the conditions under which multichannel MAC protocols can be implemented over that used by the traditional 802.11 standard may be stated as follows:

- N nonoverlapping data channels are available, all having identical bandwidths and propagation characteristics. For the sake of discussion N is assumed to be arbitrary, but is much smaller than the number of nodes in the network.
- Each node has a single half-duplex transceiver that can operate on any of the N channels. It is assumed that a node can transmit or receive on only one channel at any given time. The MAC can dynamically select the channel that is to be used by the network interface card.

- Each node is capable of sensing the carriers on all channels. The CCA is performed sequentially over all channels to identify the free and busy channels. The channel switching time is usually less than 1 μsec [27], which is negligible compared to packet transmission times.

These assumptions create a framework where a sender first determines a *free-channel list*, comprising of the channels in which the carrier strength is found to be smaller than T_{CS}. The sender can choose *any one* of these free channels for transmitting its data using a 802.11-like mechanism to take advantage of the combined bandwidth available on multiple orthogonal channels. In the absence of any channel selection criterion, the sender can choose a channel *randomly* from the *free-channel* list. However, higher channel utilization can be achieved if each packet is transmitted on the channel which would have the lowest collision probability. Determination of the best channel would typically require additional information. Various different schemes for channel selection have been proposed that are based on the past and current channel usage in the vicinity of the sender and the receiver.

A related issue is the mechanism for acquiring relevant information for channel selection. Since wireless signals are location-dependent, the information available at the sender is not sufficient to determine the most appropriate channel for transmission. The usage of MAC-layer control packets, such as RTS and CTS, is a typical solution for exchanging relevant information between a sender and the intended receiver. However, additional mechanisms may be necessary if it is deemed necessary to get information from other nodes (such as neighbors of the receiver or the transmitter) as well.

In the following, we describe some multichannel MAC protocols that use dynamic channel selection, to provide a better understanding of the issues discussed earlier. These do not represent the exhaustive list of multichannel protocols that have been presented in literature. However, some of the main aspects of using multiple channels in ad hoc networks are captured in these protocols.

6.4.1 The MMAC with Soft Reservation (MMAC-SR)

This protocol tries to reduce contention on the channels by confining the transmissions from a particular node on the same channel whenever possible [15]. Each node determines the set of free channels whenever it needs to transmit a data packet. It then selects the channel that it has used most recently without experiencing a collision, if that channel is free. If the last used channel is not free, the node chooses another free channel randomly.

If there are as many channels as the number of nodes that are active in transmission in a given neighborhood, then this procedure leads to a situation where each active node continuously uses the same channel for data transmission, thereby eliminating contention and the need for backoffs. This results in a "soft" reservation of a channel for every node, much like the distribution of nonoverlapping channels among users in a cellular network. If the number of transmitting nodes exceeds the number of free channels, then some nodes will occasionally seek alternative channels, which will result in the access and backoff phenomenon similar to that in 802.11. However, due to the fact that only unsuccessful transmissions result in channel switching, the amount of contention is still expected to be lower than that in the traditional single channel protocol as long as the network is operating within acceptable limits of traffic (i.e., when the congestion is not too high).

One of the characteristics of this protocol is that it can be implemented without using RTS and CTS packets. A positive acknowledgment using an ACK packet is required to allow the transmitter to know when a transmitted packet has been successfully received. Although eliminating the RTS/CTS handshake reduces the channel overhead, it has the drawback that there is no mechanism for the transmitter to inform the receiver about the chosen channel. Hence, this protocol requires that nodes simultaneously monitor all data channels for incoming data packets. This introduces additional hardware constraints. To eliminate this problem, the protocol may be extended to include an exchange of RTS and CTS packets for conveying channel selection information. This requires that one of the N channels be designated as a *control channel*, to be used for RTS and CTS packets only. All nodes monitor this channel when they are idle. The concept of a common control channel has been used in a number of multichannel MAC protocols [13,19,27]. With this extension, when a node wants to transmit, it performs carrier sensing on all the $N - 1$ data channels to decide the channel that it wants to use based on the "soft reservation" mechanism. It includes this information in the RTS packet that it sends over the control channel to the receiver. If the receiver receives the RTS correctly, it replies with a CTS, after which both nodes tune to the chosen data channel for the data packet transmission. The ACK packet is also transmitted on the same data channel, after which both nodes return to the control channel. In this framework, the NAV is to be maintained only over the control channel. Furthermore, since control and data channels are separate, it is not required for nodes receiving an RTS or CTS to wait until the end of the ACK packet transmission. This reduces waiting time without affecting the throughput.

6.4.2 The MMAC Using Clearest Channel at Transmitter (MMAC-CT)

An alternative mechanism for data channel selection that has been used in References 8, 13, and 14 is based on carrier signal strength measurements on free channels. These protocols try to determine the "clearest" channel for sending data with the intention of maximizing the probability of success for every data packet transmitted. The clearest channel is the channel that has the lowest detected carrier signal strength. This requires that while performing CCA, nodes also estimate the strengths of carrier signals on all the free channels.

The detection of the clearest channel can be implemented in a number of ways. In Reference 14, a multichannel MAC protocol was introduced in which each transmitter selects the channel in which it detects the lowest carrier signal power before transmitting a data packet. The motivation for such a choice is based on the fact that ideally the strength of a wireless signal degrades with distance according to the same path loss function in all directions. Hence, the channel that has the lowest carrier signal is the channel which is least used in that location, that is, the nearest co-channel user on this channel is located farthest. This implies that the probability of packet collisions due to the hidden terminal problem is likely to be the smallest on this channel.

To gain a better understanding of the possible benefits of this scheme we refer the reader to Figure 6.4 and Figure 6.6. Figure 6.4 illustrates that any transmission on the channel in which a receiver is currently

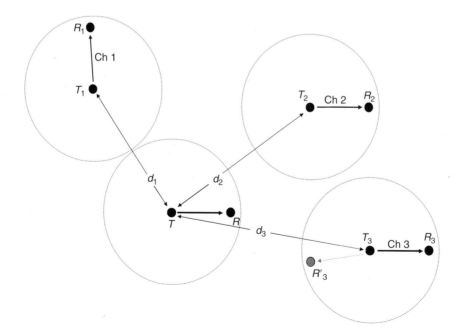

FIGURE 6.6 Example illustrating channel selection based on the clearest channel at the transmitter.

receiving a packet causes interference resulting in its SINR to drop. Now, assume that there are only three data channels, which are being used by the node pairs T_1–R_1, T_2–R_2, and T_3–R_3, respectively, as shown in Figure 6.6, when a new packet is to be transmitted from T to R. Since $d_1 < d_2 < d_3$, T would detect the lowest carrier signal strength on channel 3. Hence, it would select channel 3 for transmitting its packet to R. This is intended to achieve two objectives: (a) the resulting interference caused by T to R_3 (which is the only other receiver that is receiving in channel 3) is expected to be smaller than what it would cause to either R_1 or R_2 if channel 1 or channel 2 were chosen, respectively; (b) the SINR of the received packet at R is expected to be highest on channel 3.

We make the observation that although lowest carrier power indicates highest distance from the corresponding transmitter, the above two objectives may not always be achieved with this channel selection scheme. For instance, if R_3 was located at the position indicated by R_3' in Figure 6.6, T would still select channel 3 for transmission, but the resulting interference to R_3 would possibly be greater than what it would have caused to R_1 or R_2 had it chosen channel 1 or 2, respectively. This is because of the fact that the transmitter-based clearest channel selection scheme estimates the distances from active *transmitters*, but its transmission can affect the active *receivers* the network. The transmitter-based channel selection scheme may also fail to select the channel that can provide the maximum SINR at the intended receiver. For instance, the clearest channel at R in Figure 6.6 is likely to be channel 1 as T_1 is farthest from R.

The multichannel MAC protocol with transmitter-based clearest channel selection presented in Reference 14 still provides significant benefits over one using random channel selection as most of the time the clearest channel at the transmitter is better than a random choice. This protocol also works without the RTS–CTS option, and is based on the assumption that the receiver is able to monitor all the channels for incoming data packets. To avoid the costly implementation of multiple receivers monitoring all the channels simultaneously, the protocol can be modified to include RTS and CTS transmissions over a common control channel. As in the MMAC-SR, the transmitter can inform the receiver about its channel selection on the RTS packet and the receiver can tune to this channel for receiving the data packet after sending the CTS packet over the control channel.

6.4.3 The MMAC Using Clearest Channel at Receiver (MMAC-CR)

One of the ways to achieve the best SINR at the receiver is to select the clearest channel that is detected by the receiver. In order for the transmitter to learn about this, an exchange of RTS and CTS packets can be performed over a control channel. This scheme was proposed in Reference 8 and variants of the same has been studied by other researchers as well. The protocol presented in Reference 8 uses an RTS packet to send its list of free channels to the intended receiver. If the receiver receives the RTS, it also performs CCA over all channels and selects the channel included in the sender's free-channel list that has the lowest carrier signal strength at its own location. The receiver includes this channel number in the CTS packet that it sends to the receiver and simultaneously tunes to that channel awaiting the data packet. If a common channel is not found between the transmitter and receiver, the transmitter retries the transaction by sending another RTS packet after a backoff delay.

This protocol allows each transmitter-receiver pair to select the channel that has the smallest interference at the receiver. The idea is to maximize the probability of success by using the channel that has the highest margin between the received signal power of the transmitted packet and the existing interference on the channel. Hence the probability of collisions or failure of correct reception due to co-channel interference is expected to be minimized using this channel selection process. However, since this protocol relies on the success of the RTS–CTS handshake to initiate data packet transmission, the throughput is also dependent on the success of the RTS and CTS transmissions on the control channel, which use the basic CSMA/CA scheme. If the control channel has the same bandwidth as all the data channels, the contention on the control channel becomes a minor issue as the RTS and CTS packets are much smaller than data packets. On the other hand, the control channel may be somewhat under-utilized in that case. An alternative is to design a "tailored" control channel that has just as much bandwidth as required corresponding to the data channels available for use. We consider this issue when the total bandwidth occupied by all channels

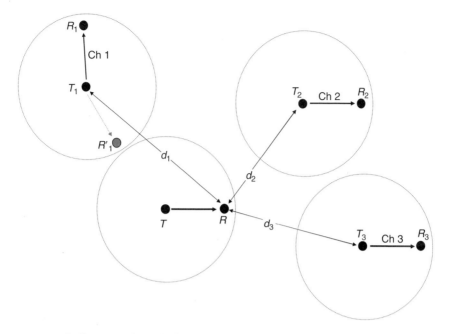

FIGURE 6.7 Example illustrating channel selection based on the clearest channel at the receiver and the associated problem.

(control as well as data) is considered to be fixed, and it is possible to assume some optimum bandwidth allocation for the control channel with respect to data.

6.4.4 The MMAC with Cooperative Channel Selection (MMAC-CC)

Of the two objectives targeted by the signal power based channel selection schemes mentioned in Section 6.4.2, the MMAC-CR protocol definitely achieves the second. However, it does not always meet the first objective. In other words, the channel selection scheme in MMAC-CR does not consider the interference that would be caused to active receivers on the chosen channel. An example to illustrate this event and the possible effect is shown in Figure 6.7. This is the same scenario as in Figure 6.6 but here we depict the distances from all active transmitters to R instead of T. Since here $d_1 > d_3 > d_2$, it is expected that under isotropic propagation conditions the carrier signal strength at R will be smallest on channel 1. Hence, MMAC-CR would select channel 1. This does not seem to pose any problem under the described scenario, where the closest active receiver on channel 1, R_1, is located sufficiently far from T. However, the transmission from T might generate significant interference to R_1 if it was located at R_1' instead. Note that MMAC-CR chooses channel 1 irrespective of the location of R_1 as its channel selection is based on the distances from the active transmitters to R. However, when R_1 is in the latter position, the transmission from T on channel 1 could cause significant interference to it, possibly resulting in the loss of the packet that it is receiving.

We now describe MMAC-CC, a multichannel MAC protocol that attempts to meet both the objectives described above by using a "cooperative" channel selection scheme [14]. This channel selection scheme aims to minimize the amount of interference from new transmitters to currently *active receivers* while selecting the best channel for their own transmissions. The main hurdle for achieving this is to implement a mechanism by which a transmitter can estimate the amount of interference that it would cause to the closest active receiver on that channel. The protocol presented in Reference 14 proposes the use of *receiver-initiated busy-tones* to serve this purpose. Busy-tones are narrow out-of-band tones [21,4] that serve as a mechanism for neighbors or active receivers to know about an ongoing transmission. Here, we consider the use of busy-tones on each data channel, so that an active receiver can allow other nodes intending to transmit data packets to estimate their distances from it by detecting the strength of busy-tone signal. The idea is for the transmitter to check the strengths of busy-tones from active receivers to estimate the amount of interference it would cause to them, and the receiver to use carrier sensing on the data channels to determine the clearest data channel for receiving. This information can be exchanged over RTS and CTS packets to select the optimum channel from the point of view of all active nodes in the network as well as the new transmission.

6.4.4.1   Details of the MMAC-CC Protocol

Because of the fact that MMAC-CC takes into account all the relevant issues for optimum distributed channel selection that were discussed earlier, we present the details of a possible implementation of MMAC-CC:

1. When a node has a data packet to send, it first transmits an RTS packet to the receiving node in the control channel.
    a. Before transmitting the RTS, the sender builds a list of free data channels available for transmission. Free channels are those for which the *busy tone signals* lie below the carrier-sensing threshold. This list is sorted in ascending order of the signal powers of the busy tones, and embedded in the RTS packet.
    b. If the free-channel list is empty, that is, the free channel count is zero, the node initiates a backoff and re-attempts transmission of the RTS packet later.
    c. Unlike 802.11, other nodes receiving the RTS on the control channel defer their transmissions only until the duration of CTS and not until the duration of ACK. This is because data and ACK are transmitted on the data channel and cannot interfere with other RTS–CTS transmissions.
2. Upon successful reception of the RTS packet, the receiver node creates its own free-channel list by sensing the *carrier* signal on the data channels. This list is also sorted in ascending order of signal strengths.
    a. If there are channels that are included in the free-channel list sent by the sender as well as that obtained at the receiver, the receiver selects the *best common channel* by going down the lists and selecting the first channel that is common to both. In case of a tie, the best channel from the receiver's list gets preference. The receiver then sends this channel information in the CTS packet and switches to the chosen data channel to receive the data. If it does not receive the data within a certain period, it reverts to idle mode and continues monitoring the control channel.
    b. If no common free channel is available, the no CTS packet is sent.
3. If the sender receives a CTS packet, it verifies that the selected channel is still free and transmits the data packet on it. Thus, the data is transmitted on a channel that is the clearest at the receiver as well as one that causes least possible interference to active receivers near the sender. The receiver node transmits the busy tone signal corresponding to the selected channel as long as it is receiving the data packet.

    If the transmitter does not receive a CTS packet within a certain period of time, it enters a "no-CTS backoff" and re-attempts transmission of the RTS packet later.
4. There is no "off the air" wait period for the nodes that receive a CTS packet.

6.4.5   Performance of MMAC Protocols

In order to quantitatively determine the benefits obtained from each of the features discussed in multichannel MAC protocols in the previous section, we present some comparitive performance evaluations of some if these protocols obtained from simulations. These simulations are performed using the RFMACSIM simulator, which is a discrete event simulator that has accurate wireless physical layer and the IEEE 802.11 MAC implementations [16]. Additional modifications were incorporated into RFMACSIM to simulate the following multichannel MAC protocols that use carrier signal strengths for channel selection: MMAC-CT (clearest channel at the transmitter), MMAC-CR (clearest channel at the receiver), and MMAC-CC (clearest channel based on cooperative channel selection). In addition, a multichannel MAC that uses

TABLE 6.2 Parameter Values
Used in Simulations

Parameter	Values used
Transmit power	25 dBm
Carrier sense threshold	−90 dBm
Noise floor	−110 dBm
Minimum SIR	10 dB
Data packet size	1000 Bytes
RTS packet size	10 Bytes
CTS packet size	10 Bytes
SIFS	10 μsec
DIFS	60 μsec
Total bandwidth	1 Mb/sec

random channel selection (MMAC-random) is also implemented, that is, one in which the sender selects a channel randomly from the set of free channels detected at the time of transmission. The performances of these multichannel MAC protocols are compared to that of a single channel 802.11 MAC under identical conditions.

6.4.5.1 Network Model

The simulations were performed assuming a grid network topology, where static nodes are placed in a uniform grid in an area of 2100 m × 2100 m. Table 6.2 lists the parameters used in all MAC implementations. Traffic is considered to be uniformly distributed over the whole network, with each node generating data packets for transmission according to a Poisson distribution. The destination of a generated packet is selected randomly from the set of neighbors of the node in which the packet is generated. This network and traffic model is aimed to simulate a uniform density of nodes as well as traffic. On an average, this represents the MAC layer packet transmissions in a network, which is experiencing a large number of multihop communications. In order to make a fair comparison amongst the various MAC protocols, we use the same aggregate channel bandwidth and packet sizes for all simulations. These are also shown in Table 6.2.

6.4.5.2 Optimizations

Except for the addition of channel selection and transmission of packets over chosen channels, the basic framework of the multichannel MAC protocols described earlier follow the same principles as the IEEE 802.11 MAC. This includes the mechanisms followed for carrier sensing, backoffs, and access retries. However, there are several aspects in the implementation of the multichannel MAC protocols that warrant reviewing these mechanisms and the corresponding parameters used.

Optimizing the control channel bandwidth: The first aspect is that of having a separate control channel that carries only the RTS and CTS packets. In a single-channel MAC, RTS, and CTS packets are transmitted using carrier sensing and random backoffs on a *common channel* carrying RTS, CTS, as well as data. Having a separate control channel brings up the issue of determining its optimum bandwidth to maximize bandwidth utilization. The optimum solution would provide sufficient control channel bandwidth to minimize the contention amongst RTS and CTS packets while leaving enough channel capacity for the data channels. This depends on a number of parameters that include the data and control packet lengths, the node density, traffic, and the total channel bandwidth. The optimum solution under heavy traffic conditions for the grid network topology has been found to be 0.1 times the total bandwidth, approximately, for a wide range of parameters from computer simulations.

Optimizing NAVs: A related design issue when using a separate control channel is the modification of the NAV parameter, which is used to keep neighboring nodes from transmitting when a data transmission is in progress. In single-channel 802.11 MAC, when a node receives an RTS packet that has a different destination, it sets its NAV to $NAV_{RTS} = $ CTS duration + data frame duration + ACK duration + 3 × SIFS

TABLE 6.3   Optimized Backoff Parameters for the Grid Network

No. of data channels	Slot time (μsec)	Minimum window size (slots)	Maximum window size (slots)
2	25	8	1024
4	15	128	1024
9	10	256	1024

$+ 3 \times$ propagation time after the RTS is received. Similarly, on receiving a CTS packet, a node sets its NAV to $\text{NAV}_{\text{CTS}} =$ data frame duration $+$ ACK duration $+ 2 \times$ SIFS $+ 2 \times$ propagation time. Since RTS and CTS transmissions cannot interfere with data, these parameters need to be modified to:

$$\text{NAV}_{\text{RTS}} = \text{CTS duration} + \text{SIFS} + \text{propagation time}$$

$$\text{NAV}_{\text{CTS}} = 0$$

Optimizing the backoff parameters: Another issue concerns the backoff parameters, that is, the minimum and maximum backoff window sizes and the backoff slot period, used for calculating the random backoff periods during access retries. Backoffs are required when the carrier on the control channel is found to be high (implying that the channel is in use) during the transmission of a control packet. Since data packet transmission times are higher for multichannel MAC protocols (due to the reduction of the bandwidth per data channel, under the same aggregate bandwidth assumption), these parameters also depend on N. The optimum values of these parameters have also been found experimentally from computer simulations to maximize the throughput in the grid network topology. The results are shown in Table 6.3.

6.4.5.3   Throughput Performance

The variation of the average network throughput versus the offered load in the network for the selected MAC protocols in a 225 node network are shown in Figure 6.8. These results indicate that the average throughput performance in all multichannel protocols employing dynamic channel selection is better than that of the single channel 802.11 MAC. The multichannel MAC using random channel selection actually has a lower throughput in comparison to the single-channel 802.11 MAC. The performance improves with MMAC-CT, which is further improved by the MMAC-CR protocol. The MMAC-CC protocol with $N = 4$ generates a higher throughput than MMAC-CR with $N = 9$. The performance of MMAC-CC is even better with $N = 9$, but we observe that higher values of N does not improve the throughput proportionately. This is because of the fact that a larger value of N increases the packet transmission times due to a smaller data channel bandwidth and hence the number of retransmission attempts increase due to a higher number of "destination busy" events.

The variation of the throughput per node with increasing node density at an offered load of 250 KBytes/sec in the network is shown in Figure 6.9. These results are obtained assuming the same network area. Hence, they indicate that the achievable throughput per node decreases logarithmically with the number of nodes in the network for all MAC protocols. A theoretical treatment of the capacity of multichannel MAC protocols may be found in References 10 and 11.

6.5   Related Work on Multichannel MAC Protocols

A large volume of work have been reported on the utilization of multiple channels in ad hoc networks. In order to review the existing work on multichannel MAC protocols, it is useful to classify them into three broad classes: (a) protocols employing channel hopping using frequency-hopped spread spectrum (FH-SSMA) radio, where the primary emphasis is on the determination of appropriate frequency slots for data transmission, (b) protocols that use a separate signaling or control channel but only one data

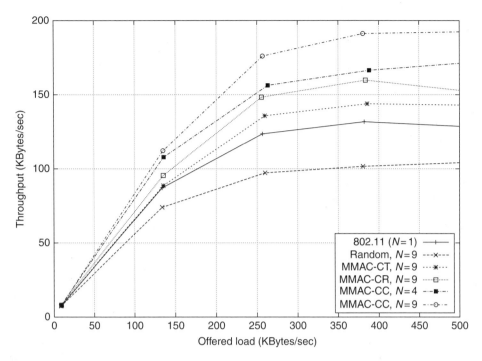

FIGURE 6.8 Comparison of average MAC layer throughput obtained from using various MAC protocols in an ad hoc network of 225 nodes placed in a uniform grid. The total aggregate channel capacity is assumed to be 1 Mb/sec for all cases.

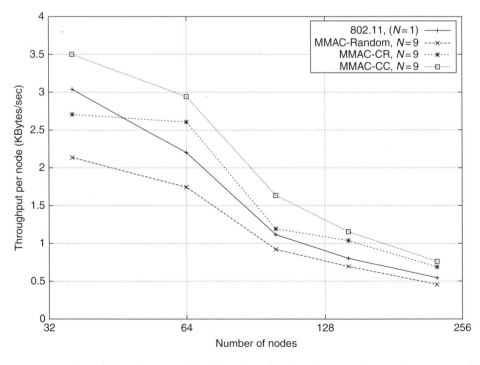

FIGURE 6.9 Variation of throughput per node with number of nodes in the network, using the same network area.

channel, and (c) protocols using a certain fixed number of channels with the idea of distributing packet transmissions over channels selectively to minimize errors. Of these, the second and third categories can potentially be applied to systems having multiple orthogonal channels constructed in any fashion, that is, by frequency division, time division, or code division. Since creating channels by time division requires accurate synchronization, it is not likely to be used in an ad hoc network. The availability of multiple channels in existing wireless standards has been described in Section 6.3.

6.5.1 Networks Using FH-SSMA Radio

The 2.4 GHz ISM band in the United States supports FH-SSMA by providing 79 channel frequencies in the hopping set, with a channel spacing of 1 MHz. There are three different hopping sequence sets in the United States, with 26 hopping sequences in each set. All the terminals in any given ad hoc network must use the same hopping sequence. However, the availability of multiple sets allow multiple systems or networks to coexist in the same location.

Several MAC protocols have been proposed based on slow FH where the hopping rate is assumed to be slower than the transmission times of control packets [20,23–25]. The general principle used in these protocols is that all nodes listen using a common FH sequence. To send data, nodes exchange control packets on the current FH. If this handshake is successful they use the same frequency slot for communication. Other hosts continue hopping and the communicating pair synchronizes with the common hopping sequence after the data packet transmission is over. Advantages of these protocols include the elimination of the needs for carrier sensing as well as distributed code assignment. The Hop Reservation Multiple Access (HRMA) protocol proposed in Reference 20 uses the traditional RTS and CTS based handshake for initiating data packet exchange. The Receiver Initiated Collision Avoidance (RICH) [23] and Receiver Initiated Collision Avoidance with dual polling (RICH-DP) [24] protocols use receiver-initiated polling with ready-to-receive (RTR) packets. The Channel Hopping Access with Trains (CHAT) protocol [25] enhances the control handshake to allow collision-free transmissions of packet trains, multicast packets, and broadcast packets. These protocols can be implemented with one FH radio receiver per node.

6.5.2 Separate Control and Data Channels

A markedly different mechanism for using multiple channels is the separation of transmissions of control signals from data. The concept of using a out-of-band busy tone, introduced by Tobagi and Kleinrock [21], is an example of sending control signals without interfering with data. This protocol, called Busy Tone Multiple Access (BTMA), was designed for packet radio networks with base stations, but can potentially be used in ad hoc networks as well. When a base station senses the transmission from any terminal, it broadcasts a busy tone signal to prevent all other terminals from accessing the channel. A subsequently developed protocol, called Split-channel Reservation Multiple Access (SRMA) [22] considers splitting the available bandwidth into either two subchannels for data and control packets (RAM mode), or three subchannels for data, request transmission, and answer-to-request transmission, respectively (RA mode). This protocol is also designed for packet radio networks working in the infrastructure mode, that is, having base stations. The Receiver-initiated Busy-Tone Multiple access scheme (RI-BTMA) [26] uses a similar principle as BTMA. Here, a transmitter sends a packet preamble to the intended receiver before initiating message transmission. If the preamble is received correctly, the receiver sends an out-of-band busy tone signal that serves the purpose of acknowledging the transmission request from the transmitter as well as preventing other nodes from transmitting at the same time. The Dual Busy Tone Multiple Access Protocol (DBTMA) [4] proposes to use two busy tones, one from the transmitter and the other from the receiver. An RTS packet is sent by the transmitter to initiate the data transmission cycle. An out-of-band transmit busy tone BT_t is transmitted simultaneously to protect the RTS packet by preventing those nodes that hear the tone from transmitting. The receiver busy tone BT_r is used to acknowledge the RTS packet as well as silence neighbors of the receiver while the data packet is being received. The BT_r is transmitted as long as the receiver is receiving the data packet. The usage of dual tones allow exposed terminals to initiate data transmissions and also hidden terminals to respond to RTS packets without affecting the ongoing data packet exchange. The Power Aware Multi-Access protocol presented in Reference 18 also relies on a separate signaling channel for exchanging control packets. The main objective of this protocol is to use the exchange of control packets to implement a power conservation mechanism by allowing nodes to power off while not involved in communication. The control packets carry information on the times and durations of periods when the nodes are to be powered off. An 802.11-like protocol with a separate control channel was also considered in Reference 29.

All of these schemes use only one data channel. As discussed by Tobagi and Kleinrock in Reference 21 the transmission of out-of-band busy tones require about 0.1 to 10 KHz bandwidth when the main channel has a bandwidth of 100 KHz. Little additional information is available about the implementation of busy tones.

6.5.3 Multiple Data Channels with Channel Selection

The largest category of work on multichannel MAC protocols is on protocols that assume the availability of an arbitrary number (>2) of channels where nodes can select any channel for transmitting its data frames. The main objective of these protocols is to improve the channel utilization by partitioning the available bandwidth into multiple orthogonal channels to allow concurrent nonoverlapping transmissions. A key issue in such protocols is the mechanism employed for channel selection. While similar ideas have been applied successfully for improving the frequency reuse factor in infrastructured networks such as cellular, designing the channel selection scheme in wireless ad hoc networks poses additional challenges due to the absence of central control. The details presented in Section 6.4 are related to this domain of work.

Early work on this concept include the multichannel MAC protocol proposed in Reference 12, where the nodes identify channel usage in their neighborhood by carrier sensing and exchange this information during data transmission. An appropriate free channel (the first free channel) is selected for data transmission. A multichannel MAC protocol that uses a polling mechanism to schedule data transmissions based on the exchange of usage tables was presented in Reference 7. The multichannel MAC with "soft reservation" presented in Reference 15 introduced the idea of using a specific criterion for choosing the best channel among a set of free channels. This protocol tries to restrict channel usage for each node on the same channel whenever possible, thereby reducing contention during channel access. The idea of using signal power measurements on free channels for dynamic channel selection is proposed in Reference 14. Details of these protocols have been described in Section 6.4.1. In [28] proposed a multichannel Dynamic Channel Assignment (DCA) MAC. They consider the use of a dedicated control channel and two transceivers per node. Nodes can monitor the control channel for RTS and CTS packets simultaneously while exchanging data on any one of n data channels. RTS and CTS packets are used to exchange free channel information amongst a pair of communicating nodes and a randomly selected free channel is used for data transmission. In the sequel [27], the authors include power control to the DCA protocol. The multichannel protocols presented in References 8 and 14 also consider the exchange of free channel lists over RTS and CTS packets. However they perform channel selection using signal power measurements at the receiver or at both the transmitter and receiver nodes. In Reference 30, the authors propose an Opportunistic multichannel MAC that uses the RTS packet to also act as a "pilot signal" for channel estimation. The RTS packet is transmitted on all channels to allow the receiver to estimate the best channel for data transmission. This protocol assumes that the physical layer is capable of multirate data transmission. The receiver determines the optimum transmission rate from channel estimation and sends this information back to the transmitter over the CTS packet on the chosen channel. The authors suggest that this protocol can be implemented using only two transceivers per node, one for data transmission on any chosen channel, and the other for listening on all other channels. A multichannel MAC protocol that provides differentiated services over multiple channels is proposed in Reference 3. This protocol also considers negotiation over a common control channel but assumes that only one transceiver is used in each node. Each node maintains channel usage tables that includes channel-related information for all data channels. Channel selection is based on data packet lengths (considered variable), priorities, and data channel availability. The multichannel MAC proposed in Reference 19 uses only one transceiver per node. It proposes to use time synchronization using beacon signals and Ad Hoc Traffic Indication Messages (ATIM) messages at the start of all beacon intervals to exchange Preferable Channel Lists (PCL). Channel selection is aimed to balance the channel load as much as possible.

Along with the development of multichannel MAC protocols, some work has also been reported on evaluation of the capacity of CSMA based MAC protocols using multiple data channels and dynamic channel selection [10,11]. The theoretical guaranteed throughput depends on a number of parameters that include the topology, the packet loss rate, and the aggregate bandwidth. However, the authors show that for medium network sizes (100 to 300 nodes) with a 2 Mbps channel rate, multichannel MAC protocols can theoretically achieve more than nine times the throughput capacity as can be achieved by using a single channel protocol.

6.6 Conclusion

The usage of multiple channels in an ad hoc wireless network provides many advantages. An ad hoc network experiences multiple simultaneous transmission attempts caused by multihop communications and the existence of multiple conversations. The usage of mutliple channels provides a natural framework for concurrent wireless transmissions in the same region. In addition, it provides a mechanism for reducing contention in random channel access by allowing contending transmitters to distribute their transmissions over channels as opposed to using random backoffs over a single shared channel. Multiple channels can also be used to create a framework for providing priorities and differentiated service types.

However, the requirement of implementing a distributed (i.e., noncentralized) channel selection scheme in a dynamically changing network poses challenging design issues. In this chapter, we present a review of the extensive body of work that exists on the development of MAC protocols using multiple channels. These protocols differ in hardware requirements, complexity, and performance. While there is still significant scope of designing more efficient multichannel MAC protocols, the literature on these protocol provide valuable insights into the applicable methods for improving the throughput by utilizing multiple channels.

Future directions in the usage of multiple channels in wireless LANs include investigations on multiple network interface cards operating on multiple channels [1]. While this scheme has the same objective of exploiting the available spectrum for deriving higher bandwidths, usage of multiple interface cards would require modification of the link layer without affecting the MAC. This would allow the use of existing hardware built, for instance, over the 802.11 standard.

References

[1] A. Adya, P. Bahl, J. Padhye, A. Wolman, and L. Zhou, "A multi-radio unification protocol for IEEE 802.11 wireless networks," *Microsoft Technical Report, Microsoft Research-TR-2003-41*, June 2003.

[2] M. Ajmone-Marsan and D. Roffinella, "Multichannel local area network protocols," *IEEE Journal on Selected Areas of Communication*, 1: 885–897, 1983.

[3] N. Choi, Y. Seok, and Y. Choi, "Multichannel mac for mobile ad hoc networks," *Proceedings of the IEEE VTC*, Fall 2003.

[4] Jing Deng and Zygmunt J. Haas, "Dual busy tone multiple access (DBTMA): A new medium access control for packet radio networks," *Proceedings of the IEEE ICUPS'98*, October 1998.

[5] IEEE Standards Department, "Wireless LAN medium access control (MAC) and physical layer (PHY) specifications," IEEE Standard 802.11–1997, 1997.

[6] IETF MANET Working Group, http://www.ietf.org/html.charters/manet-charter.html.

[7] Z.J. Haas, "On the performance of a medium access control scheme for the reconfigurable wireless networks," *Proceedings of the IEEE MILCOM'97*, November 1997.

[8] N. Jain, S.R. Das, and A. Nasipuri, "A multichannel MAC protocol with receiver based channel selection for multihop wireless networks," *Proceedings of the IEEE International Conference of Computer Communication and Networks (ICCCN 2001)*, October 2001.

[9] L. Kleinrock and F.A. Tobagi, "Packet switching in radio channels: Part-I — carrier sense multiple access modes and their throughput-delay characteristics," *IEEE Transactions in Communications*, COM-23(12): 1400–1416, 1975.

[10] J. Li, Z.J. Haas, and M. Sheng, "Capacity evaluation of multi-channel wireless ad hoc networks," *Journal of Electronics (China)*, 20(5): 344–352, 2003.

[11] J. Li, Z.J. Haas, M. Sheng, and Y. Chen, "Performance evaluation of modified IEEE 802.11 MAC for multi-channel multi-hop ad hoc networks," *Journal of Interconnection Networks*, 4(3), pp. 345–359, 2003.

[12] D. Lopez-Rodriquez and R. Perez-Jimenez, "Distributed method for channel assignment in CDMA based "ad-hoc" wireless local area networks," *Proceedings of the IEEE MTT-S Symposium on Technologies for Wireless Applications*, pp. 11–16, 1999.

[13] A. Nasipuri and S.R. Das, "Multichannel CSMA with signal power-based channel selection for multihop wireless networks," *Proceedings of the IEEE Fall Vehicular Technology Conference (VTC 2000)*, September 2000.

[14] A. Nasipuri and J. Mondhe, "Multichannel mac with dynamic channel selection for ad hoc networks," *Technial Report, University of North Carolina at Charlotte*, January 2004.

[15] A. Nasipuri, J. Zhuang, and S.R. Das, "A multichannel CSMA MAC protocol for multihop wireless networks," *Proceedings of the IEEE Wireless Communications and Networking Conference (WCNC'99)*, September 1999.

[16] C.M. Puig, "RF MAC simulator documentation: RFMACSIM version 0.32," *Apple Computer, Inc.*, July 1993.

[17] T.S. Rappaport, *Wireless Communications Principles and Practice*, Prentice Hall, Englewood Cliffs, NJ, 2002, pp. 105–176.

[18] S. Singh and C.S. Raghavendra, "PAMAS: Power aware multi-access protocol with signalling for ad hoc networks," *ACM Computer Communications Review*, July 1998.

[19] J. So and N.H. Vaidya, "Multi-channel MAC for ad hoc networks: Handling multi-channel hidden terminals using a single transceiver," *ACM/IEEE International Conference on Mobile Computing and Networking (MOBICOM)*, pp. 66–75, November 1998.

[20] Z. Tang and J.J. Garcia-Luna-Aceves, "Hop-reservation multiple access (HRMA) for ad hoc networks," *Proceedings of the IEEE INFOCOM'99*, March 1999.

[21] F.A. Tobagi and L. Kleinrock, "Packet switching in radio channels: Part-II — the hidden terminal problem in carrier sense multiple-access and the busy-tone solution," *IEEE Transactions in Communications*, COM-23(12): 1417–1433, 1975.

[22] F.A. Tobagi and L. Kleinrock, "Packet switching in radio channels: Part-III — polling and (dynamic) split-channel reservation multiple access," *IEEE Transactions in Communications*, COM-24(12): 336–342, August 1976.

[23] A. Tzamaloukas and J.J. Garcia-Luna-Aceves, "Receiver-initiated channel hopping for ad hoc networks," *Proceedings of the IEEE Wireless Communications and Networking Conference (WCNC)*, September 2000.

[24] A. Tzamaloukas and J.J. Garcia-Luna-Aceves, "A receiver-initiated collision-avoidance protocol for multi-channel networks," *Proceedings of the IEEE Mobile Multimedia Communications (MoMuC)*, November 2000.

[25] A. Tzamaloukas and J.J. Garcia-Luna-Aceves, "A channel hopping protocol for ad hoc networks," *Proceedings of the IEEE IC3N'00*, October 2000.

[26] C. Wu and V.O.K. Li, "Receiver-initiated busy-tone multiple access in packet radio networks," *Proceedings of ACM SIGCOMM'87*, pp. 336–342, 1987.

[27] S.-L. Wu, C.-Y. Lin, J.-P. Shen, and Y.-C. Tseng, "A multi-channel MAC protocol with power control for multi-hop mobile ad hoc networks," *The Computer Journal*, 45, pp. 101–110, 2002.

[28] S.-L. Wu, C.-Y. Lin, Y.-C. Tseng, and J.-P. Sheu, "A new multi-channel MAC protocol with on-demand channel assignment for multi-hop mobile ad hoc networks," *Proceedings of the IEEE Wireless Communications and Networking Conference (WCNC 2000)*, September 2000.

[29] S.-L. Wu, Y.-C. Tseng, and J.-P. Sheu, "Intelligent medium access for mobile ad hoc networks with busy tones and power control," *IEEE Journal on Selected Areas in Communications*, 18: 1647–1657, 2000.

[30] D. Zheng and J. Zhang, "Protocol design and performance analysis of opportunistic multi-channel medium access control," *Technical Report, Arizona State University*, 2003.

II

Routing Protocols and Location Awareness Strategies in Wireless Networks

7

Distributed Algorithms for Some Fundamental Problems in Ad Hoc Mobile Environments

7.1　Introduction.. 7-123

7.2　Mobile versus Distributed Computing................. 7-124
Models for Mobile Algorithms • Some Basic Problems in
Mobile Computing • Mathematical Tools for the Basic
Problems • The Behavior of Mobile Computing Protocols •
Some More Algorithmic Problems in Mobile Environments

7.3　An Algorithm for Leader Election 7-129
The Problem • The Model • The Algorithm • Analytic
Considerations

7.4　A Communication Algorithm for High Mobility
Networks 7-135
The Case of High Mobility Rate • The Basic Idea • The
Model • The Problem • A Set of Algorithms • Analysis of
the SNAKE Algorithm • Experimental Evaluation

7.5　An Algorithm for Reconfiguration of Robot Chains .. 7-154
Introduction • The Model

7.6　Conclusions and Future Research 7-160

Acknowledgment... 7-161

References 7-161

Sotiris Nikoletseas
Paul Spirakis

7.1　Introduction

We address several important modeling and algorithmic issues arising in ad hoc *mobile* computing environments, with an emphasis on the case where all devices can move. Our main goal is to highlight *essential differences* (in models, problems and algorithms' correctness, design and efficiency) to classical *distributed* computing. We concentrate on three basic problems: leader election, point-to-point communication, and reconfiguration of positions of robotic devices. We discuss some characteristic solutions to these problems that have appeared in the recent state-of-the-art literature. In addition, we informally present and criticize some general algorithmic and modeling notions and methodologies.

The chapter is organized as follows. In Section 7.2 we first discuss some important modeling differences between mobile and distributed computing. Then we describe the different meaning of several fundamental problems in mobile computing. In Section 7.3 we deal with the Leader Election Problem, by presenting and discussing in detail a basic algorithm and also briefly discussing a few others. In Section 7.4 we study the problem of pairwise communication in ad hoc mobile networks, by introducing appropriate motion and network area models and providing and analyzing rigorously a set of algorithms. In Section 7.5 we study motion coordination in systems of robots, giving an algorithm for system reconfiguration. We conclude with a section discussing open problems and future research.

7.2   Mobile versus Distributed Computing

Major technological advances in mobile networks have recently motivated the introduction of a *new* computing environment, called *mobile computing*. Mobile computing is a new kind of distributed computing; its several important features introduce essential modeling and abstraction differences. Among such features the most important is that some (or all) of the computing devices in the environment may move (in general, in arbitrary ways). In addition, usually the mobile computing elements, suffer from severe limitations with respect to:

- a. *Communication*: They can communicate up to a certain range.
- b. *Power (energy)*: They rely on a finite, quite limited, energy source (battery).
- c. *Reliability*: They are exposed to unexpected conditions while moving.

These limitations, when put together with the feature of motions, make the design of mobile information systems a quite hard task; communication among two devices may be broken due to distance while they move; the battery may fail in the middle of a global computing task; very fast motion may even introduce clock relativistic effects, thus making synchronism a hard task. Hence, many distributed computing tasks, even the most elementary (leader election, counting, pair-wise communication, etc.) have to be reconsidered in this new setting. As a consequence, new approaches are usually required.

An important special case of mobile computing environments is that in which all the devices are mobile (no fixed network exists). Such an environment is called an *ad hoc* mobile computing environment. In contrast, a *hybrid* mobile computing environment allows for the coexistence of a fixed (backbone) network of stations that do not move and at the same time the existence of (usually many) moving devices.

7.2.1   Models for Mobile Algorithms

The design of high-level protocols or algorithms for mobile computing environments, requires the adoption of a suitable abstraction-model that precisely captures the new features. Any such model has two sides:

- *Local*, that is, the modeling of each computing device
- *Global*, that is, the modeling of motions and motion/communication territory and constraints

A local model of a mobile computing device, m, usually includes: (a) A transmission radius, $R(t)$, at which m can send (or from which it can receive) messages. $R(t)$ may vary with time if the mobile device can adjust its power supply. (b) A direction (or an angle) for communication at any position of the device; usually this is 360°, that is, a broadcast is assumed. (c) The amount $e(t)$ of energy still available at time t. (d) The computing model of the device (i.e., an automaton, a machine with memory, a Turing Machine, etc.)

Although the motion pattern of each device is a local feature, many models prefer to describe this in their global part, via a possible set of, for example, stochastic motions or in other summarizing ways. Thus, the *global* model part of a mobile computing environment must characterize: (a) The *motion space*, that is, the allowable motions (due to, for example, existence of obstacles or variable geometry), (b) The form of motions allowed in that space (ideally, the instant velocity vector as a function of time), and (c) How energy is dissipated while moving. Global models are either *explicit* (i.e., the motion space and patterns are given) or *implicit*.

Definition 7.1 *An* implicit *global model for a mobile computing environment is usually a graph $G_t(V, E_t)$ where nodes correspond to devices and links* (time dependent) *represent direct communication capability due to instant proximity.*

Usually, the form of G_t is assumed *not to be known* to the nodes (an *unknown* graph of communication).

Many research works assume G_t to be *quasi-static*, that is, they assume the existence of a time period Δ during which at most a certain number κ of links may appear or disappear. Indeed, in some cases it is assumed that $\kappa = 1$.

Definition 7.2 *The rate of change, h, of G_t is defined as $h = \kappa/\Delta$.*

Notice that implicit representations *avoid* the geometry issues of the motion space; thus they are in essence combinatorial abstractions. *Intermediate* global models may represent the motion-allowable paths as a graph (the motion graph M). Notice that M does not (usually) change with time. Then, moving devices are seen as distinct particles, that occupy a node of M at each time and that move according to some law (e.g., from a vertex to some neighbor of it). None of these abstractions should exclude *unnatural motions* (e.g., the instant jump from a point in space to another, not nearby. Examples here are quantum phenomena and telepresense). So, we have:

Definition 7.3 *A mobile computing model is a tuple $T = (\Gamma, E)$ where Γ is a global model and E a set of local models.*

Definition 7.4 *A problem specification P_i for the model T is a set of properties that T must satisfy.*

Definition 7.5 *A mobile computing protocol P, on T, for the problem Π, is a set of programs, P_I, for each device I, where each P_I has computing, communication, or motion commands, that satisfy the properties of Π when executed.*

Note that we allow mobile protocols P to specify the motion of *some* of the devices to the benefit of the protocol. Such protocols are called semicompulsory [19]. (Compulsory when the motions of *all* devices *obey* the protocol.) Else, the protocol P is noncompulsory. The usefulness scenarios of semicompulsory protocols include, for example, the case of a small set of moving devices that their purpose is to help in managing the mobile environment (this set is called the *support*, see Reference 19) or the case of some robots that move according to their program in order to carry out a global task (e.g., to coordinate, to rearrange, to achieve a formation, to help each other).

We should stress here that a model should allow (as a trivial case) some devices to stay still (*fixed stations*). Fixed stations may be connected via permanent channels (*fixed* or *backbone* network) and may communicate in very different ways than the mobile devices (e.g., two fixed stations may be connected via a high-speed optical link while mobile devices usually are equipped only with digital radio capabilities).

We should also stress that explicit representation of geometry, in addition to the way local communication is modeled, introduces nice and deep problems; the graph G_t becomes, for example, an instant *geometric graph* (edges are where cycles of radius $R(t)$, centered at devices, intersect). Even *random* geometric graphs are extremely interesting (with respect to connectivity, coloring, diameter, etc., see Reference 30). Failure patterns of individual devices may be modeled, for example, by a failure probability (or rate) per device and for a given (short) time, or for the whole computation.

The above modeling considerations necessarily introduce some important *parameters* for a mobile computing environment. Such parameters are, for example:

- The communication radius $R(t)$
- The number of (mobile) devices
- The number of compulsory devices (support size)
- The rate of motion graph links updates h
- The motion space size (volume) $V(M)$ (e.g., the numbers of nodes of the graph representing M, or the actual volume)
- The total available energy $e(t)$

7.2.2 Some Basic Problems in Mobile Computing

While a mobile computing environment may execute complicated protocols at a higher layer for example, information system applications; however, (as in the case of distributed computing) some *basic problems* are usually required to be faced by any protocol. Such problems, in distributed computing, are for example:

- Leader Election
- Routing
- Communication of two stations a, b (pairwise communication)
- Counting the number of processes
- Critical regions/mutual exclusion, etc.

Note that (a) even the meaning of the problem may change in a mobile environment setting (b) new basic problems arise.

As an example, let us examine carefully the problem of Leader Election. Clearly, if two devices never arrive within communication range, the possibility of electing two "leaders" is apparent. Even if one requires a recognized leader device for a "connected" part of the communication graph, one then falls into the problem of specifying the duration and form of those "connected" parts.

There are at least two ways to make the problem meaningful here:

- *Way 1*: Given that a subset of devices $P \subseteq V$ of G_t is a *connected component* $\forall t \in I$ (where I is some time interval), the protocol should elect a unique leader in P, for times in I.
- *Way 2*: Let us assume that, after a finite number of topology changes (link updates) of G_t, G_t becomes a fixed G (*quiescent*). Then, it should eventually hold that: For each connected component, C, of G, there is a node $l \in V$ (the leader) whose identity is known to all nodes $i \in C$.

In many applications, it helps to ask for a second requirement in Way 2 (a *routing information*): eventually, each edge of G has a direction (imposed to it by its endpoint nodes) so that each connected component C of G becomes a *directed acyclic graph* (DAG) with the leader as *the single* sink (such graphs are called leader-oriented graphs).

Note that the second way of specifying the problem is (perhaps) easier to handle mathematically. Note also that the requirement of quiescence after a "finite number" of link changes is semantically equivalent to the assumption that link changes *cease to happen* after some (long maybe) time. It is very tempting to us to stress here the similarity to on-line distributed algorithm. Indeed, any protocol that tries to *maintain* some global properties of G_t while links come and go, is *exactly* a distributed on-line algorithm. This remark opens the ground for efficiency measures of such protocols other than the usual, such as the competitive ratio, where the protocol P competes (continuously) against some (oblivious to P or not) "adversary" A that selects, at each time, which link to change.

Let us now carefully examine another problem: to implement the sending of a message from node a to node b (pairwise communication). This is very easy in distributed computing (e.g., solved by flooding the network with the message from node a). However, the flooding way is not only energy nonefficient in a mobile setting, but also it may *fail* to solve the problem, simply because two parts of the network remain always disconnected. Here, an assumption about motions is necessary for the problem's *solvability*: it is necessary to assume that, given enough time, moving stations meet pairwise in such a way that a temporal path Π from a to b is formed. (A temporal path $\Pi = (a = i_0, i_1, i_2, \ldots, i_k = b)$ is a sequence of triples $(i_\lambda, i_{\lambda+1}, t_\lambda)$ meaning that i_λ and $i_{\lambda+1}$ communicate directly at time t_λ and $\forall \lambda, t_\lambda > t_{\lambda-1}$.) Indeed, nodes a, b may *never meet*, however, the message may go via intermediate passes between pairs of nodes that meet.

Yet a problem of a different taste (that *does not have* a counterpart in classical distributed computing) is that of reconfiguring a "configuration" (i.e., some initial placement of devices in M). This problem comes from the areas of Robotics (see Reference 27) or puzzles (e.g., the famous 16-puzzle or the Rubic cube) and it might be very hard even in its centralized version. Any way to solve a reconfiguration problem includes the subproblem of moving from some word of a finite group or semigroup to another, and the number of elementary motions (e.g., nearby station movements one at a time) is lower-bounded by the group diameter.

Distributed ways to solve such problems must specify carefully how the devices move and in which sequence. Not all motions may be possible due to, for example, geometry or robot shape restrictions; also care is required so that, for example, the devices do not become disconnected and "lose their way" when they move.

In the sequel of the chapter we have selected three mobile computing algorithms (protocols) each solving one of the problems discussed earlier. We did that in order to provide the reader with an as wide as possible set of problems, solutions, and modeling assumptions. One of the protocols is due to our team, while the other two (and their analyses) are adapted from References 42 and 52. We do not insist on this being a complete set but rather a biased choice with the criteria of clarity, variability of model and approach, and the fundamental nature of the problems.

7.2.3 Mathematical Tools for the Basic Problems

Each of the problem discussed earlier hints a different area of analytical tools, mainly due to the modeling assumptions about motion. Thus, the implicit representation (via link changes) of motion in the leader-election problem recalls the use of combinatorics (graph theory) and some use of temporal logic and inductive inference.

On the contrary, the explicit representation of motion in the basic communication problem, via independent random *walks* (stochastic) that some devices do, leads to tools from the rich and recent theory of concurrent random walks and stochastic multiparticle interaction. Finally, the reconfiguration problems hint either to group theory or to lattice distances; a flavor of algebraic combinatorics. We will see these in detail in the next chapters.

Let us note here that, in the problems of maintenance of properties of G_t (e.g., leaders, connectivity, etc.) while links change, we believe that all the techniques of on-line algorithms can be re-used here (even statements of system "potential" changes).

To this end there is a relation with the approach of (stochastic) multiparticle interactions theory. That theory (originated in Statistical Physics) also examines the duration (or the eventual satisfaction) of global properties, and sometimes uses summary measures of the whole mobile system (e.g., "heat"). There is a rich theory of phase transitions in multiparticle systems and quite similar phenomena are expected to appear in the system of many moving and locally interacting devices.

7.2.4 The Behavior of Mobile Computing Protocols

Many performance issues of mobile computing protocols (algorithms) are motivated by the Theory of Networks. We informally discuss some of them here:

a. *Throughput.* It can be understood as the number of messages delivered to final destinations per unit time.

b. *Scalability*. A protocol P is scalable when (1) its correctness is not affected when the network size changes and (2) its "performance" admits a smooth change when the size of the network changes. Here, the notion of performance can be quantified in various ways, for example, energy loss, time for a message to be finally delivered to its destination, number of compulsory motions, etc.

c. *Stability*. The networks-motivated definition of stability is that the message population (number of messages originated but not yet delivered) in the system remains bounded as time grows. Although this definition still holds in mobile networks of devices, it does not capture entirely our intuition about a stable system. For example, an abnormal behavior in link updates may cause certain protocol states to be repeated infinitely often, disallowing termination. The issue is quite open for research.

d. *Fault-tolerance*. A mobile computing algorithm *tolerates k faults* when its correctness is not violated by up to k faults of devices during execution. A similar statement can be made for a protocol tolerating up to a certain *faults rate r* without correctness violation.

The semantics of faults are identical to those in Distributed Computing (random delays, permanent device failures, adversarial behavior of some mobile stations, i.e., not following the protocol code, etc.).

A quite interesting research aspect of mobile computing (with applications to mobile viruses in, for example, the Internet) is that of protocols of *pursuit-evasion*: the mobile devices (or entities) here are separated into two groups with different objectives. The *fugitives* group wants to remain in the motion space as long as possible. The *hunters* group of devices has as its goal to "capture" or "find" the fugitives. The semantics of capturing a fugitive may vary from a simple meeting (or coexistence within some range) of the fugitive with one or more hunters, to a blocking of all possible moves of the fugitive by a set of hunters. Such protocols are motivated by the Graph Searching problem and have been proposed and analyzed in Reference 1. It should be clear that in pursuit–evasion scenarios, the hunter devices are compulsory, since their motions are specified by the protocol. The motion *strategy* of fugitive devices must be assumed to be a *best response* to the hunters protocol (i.e., to be as adversarial as possible so that the task of the hunters is nontrivial). Clearly, game-theoretic notions can be applied in such settings.

7.2.5 Some More Algorithmic Problems in Mobile Environments

Some of the distinctive features of mobile computing are also the source of new problems. As an example, let us consider power-controlled networks, that is, mobile networks where devices are able to change their transmission power (by controlling their battery according to the protocol). In such mobile environments, if a device u attempts to send a message with transmission power P_u, then all devices within range but with power less than αP_u ($\alpha > 0$ is a constant) cannot receive the message. Such mobile computing environments were considered in Reference 2. The authors of Reference 2 discuss the complexity of routing in such environments, and propose near-optimal routing protocols for them.

Another algorithmic problem worthy of mentioning is that of devising protocols for ad hoc mobile networks that manage to implement group communication while ensuring a total order in the delivery of messages. There, motion makes preservation of delivery order hard (see, e.g., Reference 53). A possible solution is to provide a subprotocol that causes a token to continually circulate through all the mobile devices. Such an idea is presented in Reference 41, with a nice discussion about the performance of the protocol.

We remark here that the solvability and performance of many algorithmic problems in mobile computing depends (sometimes crucially) on the assumed properties of the mobile devices (that are abstracted from technology). For example, the assumption of existence of GPS technology [8,10,32,55,56] makes the mobile devices position-aware with respect to a global system of coordinates. Position-awareness may trivialize some problems of reconfiguration.

For a further discussion of some principles of algorithmic design in mobile ad hoc networks, see Reference 9, the book chapter by Boukerche and Nikoletseas [14] and the book chapter by Chatzigiannakis, Nikoletseas, and Spirakis [18].

7.3 An Algorithm for Leader Election

7.3.1 The Problem

The leader election problem is one of the most fundamental problems in distributed computing. Leader election is in fact a building block in the design of distributed protocols. For example, electing a unique node that generates a single token controlling the entrance to the critical section is essential to mutual exclusion algorithms. Further important uses of the leader election problem are the coordination in resource sharing among a set of nodes toward achieving efficiency and fault-tolerance, and symmetry breaking (such as in a deadlock situation where a single node elected as a leader may undertake deadlock repairing actions). (See References 43 and 7 for a detailed study of leader election algorithms.) The leader election problem has many variants. For static networks, its most general version is, informally, the following: "Eventually, exactly one node in the network is designated a leader, while there should never be more than one leader during the execution of the leader election algorithm."

Leader election becomes especially important in a mobile setting, due to the frequent changes in the network topology because of the movement of hosts and other special reasons (permanent or temporary disconnections from the network, failures, etc.). As an example, in a mutual exclusion algorithm, link/node failures may frequently result in token loss, thus it is crucial to elect a new leader for creating a new token. As an another example, consider a group communication algorithm for a mobile network (see, e.g., Reference 41). In such a setting, mobility of hosts results in frequent group membership changes and a new leader must be elected among the new members of the group.

However, mainly because of the dynamically changing network topology, solving the leader election problem in a mobile setting becomes more difficult and challenging compared to the static case. A usual complication is the possibility of disconnection of the ad hoc network into connected components, or the merging of two or more connected components into one. In the first case it is essential to eventually elect a unique node in each connected component (note that after the disconnection and for a certain period, there might be components with no leader node). Similarly, in the case of two or more components that merge, there may be more than one leaders. In such a case, the algorithm must resolve the presence of multiple leaders by guaranteeing that only one leader survives.

In the light of the above, Malpani *et al.* [42] propose the following modified informal definition of the leader election problem in an ad hoc mobile network: "Any connected component whose topology is static for sufficiently long time will eventually have exactly one leader." More formally, Malpani *et al.* [42] provide the following problem statement for leader election in ad hoc mobile networks.

Definition 7.6 [42] *Let in each node i a variable* lid_i *that holds the identifier of the node currently considered the leader in the component of i. In every execution with a finite number of topology changes, eventually we must have that, for every connected component C of the topology graph, there is a node l in C such that* $\text{lid}_i = l$ *for all nodes i in C.*

Note that the assumption in the above definition of "a finite number of topology changes" suggests (it is in fact equivalent) that the topology remains "for a sufficiently long time" unchanged.

7.3.2 The Model

The ad hoc mobile networks considered in Reference 42 are modeled as follows: the network is comprised of a set of n independent mobile nodes, communicating in a wireless manner by message passing. The network is modeled as a dynamically changing, not necessarily connected, undirected graph, with vertices representing nodes and edges between vertices corresponding to nodes that are within each other's transmission range and can communicate.

The following assumptions on the mobile nodes and the network are made:

1. The nodes have unique node identities.
2. Communication links are bidirectional, reliable, and FIFO. Unidirectional links, if any, are not used and ignored.
3. A protocol at the link layer ensures that each node knows the set of nodes with which it can currently directly (in one "hop") communicate (by providing indications of link formations and failures).

As far as the rate of topology changes are concerned, Malpani *et al.* [42] distinguish two cases:

a. The case of a single topology change: In this case it is assumed that only one change (either a link failure or a link formation) can occur at a time, in the sense that the next change occurs only after the network has been updated after the previous change.
b. The case of multiple concurrent topology changes.

Note that the latter case is much more difficult to cope with. Malpani *et al.* [42] indeed proposes two different algorithms (one for each case). We discuss in the following section only the first one.

7.3.3 The Algorithm

The algorithm in Reference 42 is in fact a modification of the Temporally Ordered Routing Algorithm (TORA) [46] routing algorithm, which in turn is based on a routing algorithm by Gafni and Bertsekas (GB, [34]). Both GB and TORA follow the so-called "link reversal routing" (LRR) paradigm (see Reference 47 for an in-depth discussion). The common characteristic of all LRR techniques is their goal to reduce (as much as possible) the routing overhead communication complexity due to topology changes. In LRR, the reaction to topological changes is local, limited in frequency and scope of application. In contrast to trying to maintain at each node very powerful information (e.g., shortest paths) toward a destination node, LRR just maintains information necessary to support a DAG, with the only sink being the destination node.

Maintaining a DAG guarantees the important property of loop freedom in the routing tables. Moreover, a DAG provides multiple, redundant paths to the destination node. However, a node in the DAG structure knows only its distance to its one-hop neighbors (i.e., those nodes within its transmission range) and is not aware of its distance to the destination node, that is, no additive distance metric or topology composition information for a multihop propagation is supported.

Thus, LRR techniques may trade-off routing optimality for reduced topology adaptation communication overhead. Thus, LRR tends to be very adaptive and scalable. These properties are further supported by minimizing the frequency and scope of reactions to topology changes. In LRR, route maintenance (triggered by the failure of all outgoing links of a node) affects only the set of nodes whose directed paths (prior to failure) to the destination include the failed links. This is in sharp contrast to other approaches: in link-state approaches topological changes must be reported to all the nodes in the network, while in path finding techniques topology changes must be propagated to all nodes whose shortest path spanning trees are affected by the change.

Because of the above, it seems that LRR techniques are best suitable for networks of rather "medium" rate of topology changes, that is, when the mobility rate is not very slow to allow path maintenance or link-state techniques to be applicable, and not so fast to make any techniques trying to maintain some topological information not applicable, inefficient or even erroneous. For a rigorous analysis of the performance of link reversal techniques see Reference 15.

In the next section we present the first ever LRR algorithm. Other efficient routing protocols for ad-hoc mobile environments can be found in references 12 and 13.

7.3.3.1 A Brief Description of the GB Algorithm

Gafni and Bertsekas introduce in Reference 34 for the first time a LRR approach, by providing two algorithms for maintaining a destination-oriented DAG in a network undergoing link failures. The algorithms are called the "Partial Reversal Algorithm" and the "Full Reversal Algorithm." Both algorithms assign and dynamically update a unique "height" to each node, which is drawn from a totally ordered set of possible values. Each link between two nodes is directed from the node with the higher height to that with the lower height. The goal of the algorithm is to appropriately update the link directions (after the link failures) for maintaining a DAG in which the destination is the only sink. For the algorithms to achieve this goal, whenever a node (other than the sink) loses all its outgoing links (i.e., when it loses its *last* outgoing or "downstream" link), either because of a link failure or because of a change in a neighbor's height, it calculates a new height for itself.

The two algorithms differ in the rule for calculating a new height. We here outline the Partial Reversal Algorithm, because it is this algorithm TORA and the algorithm of Reference 42 are based upon. Furthermore, the Partial Reversal Algorithm tends to be the most efficient of the two, because the Full Reversal Algorithm (when a node has no outgoing links) reverses the direction of *all* of its links.

The height of a node i is a triple (α_i, β_i, i) of integers; the last component in the triple is the node's identity and is included in order to assure uniqueness. Indeed, assuming unique node identities the triples form a total order. Triples are compared lexicographically. In an abstract way, the triple of each node can be seen as its "height," where links between neighbor nodes are directed according to their relevant heights, that is, from the higher to the lower. Thus, messages flow downstream from higher to lower height nodes. In fact, parameter α_i can be seen as a "base" or "reference" height level, with the other two parameters in the triple distinguishing heights of nodes with the same reference level.

A new iteration of the distributed algorithm is triggered whenever one or more nodes have no downstream links anymore. If node i loses all its outgoing (i.e., when it loses its last outgoing) links, it must compute a new height for itself. It chooses it to be (α_i', β_i', i), where α_i' is by one larger than the smallest α value among all its neighbors' heights. If i has a neighbor whose α height component is equal to α_i', then β_i' is set to be by one less than the smallest β value among all neighbors of i whose α height component equals α_i'. Otherwise the β component of i's height remains unchanged.

The above rule for setting α_i' ensures that node i will have at least one outgoing link, that is, that (α_i', β_i', i) will be larger than the height of at least one neighbor, the one with the smallest height. The rule for setting β_i' aims at limiting the number of links incident on i that will reverse their direction, by keeping i's height smaller than that of any neighbors whose α height component is not smaller than α_i'. Reducing the number of link reversals contributes to the efficiency of the algorithm toward limiting the propagation of height changes (in contract to the Full Reversal Algorithm that reverses all link directions).

The GB algorithm is shown to be deadlock-free and loop-free. Furthermore, the maintenance of a DAG exhibits redundancy features by providing multiple routes to the destination. Assuming that the network remains connected, the GB algorithm converges in finite time (i.e., after a finite number of iterations). When, however, the topological changes result in the disconnection of the network, the algorithm never converges by trapping in an infinite number of loops.

7.3.3.2 A Brief Description of the TORA Algorithm

Park and Corson [46] adapted the GB algorithm for routing in ad hoc mobile networks, proposing TORA. Their main contribution was a mechanism (maintaining a temporal marker) for distributed detection of network disconnections that result in the destination being no longer reachable. As mentioned in the previous section, the GB algorithms would cause an infinite cycle of messages in that case. TORA seems to be adaptive and scalable, thus it is possibly suitable for large ad hoc mobile networks.

In TORA, the height of node i is a 5-tuple, $(\tau_i, \text{oid}_i, r_i, \delta_i, i)$. As in the GB algorithm, the last component is the node's identity, to ensure uniqueness. The first three components form a "reference" (or "base") level with the rest two differentiating node heights of the same reference level. A new reference level is started by node i if it loses all (i.e., its last) outgoing links due to a link failure. Then τ_i is set to the time when this occurs and oid_i is set to i, the originator node of this reference level. Component r_i in turn is used to modify the reference level. Initially, it is set to 0, the so-called "unreflected" reference level. It can be changed at some times to 1, indicating a "reflected" reference level, which is used in detecting network disconnections.

If the node i loses all its outgoing links due to a link failure, then the following route maintenance action is taken:

Case 1 (Generate new reference level): The δ_i, i components of the triple decide the link directions among all nodes with the same reference level so as to form a destination-oriented DAG. The originator of a new reference level assigns 0 to its δ component.

When a new reference level is created, say by node i, it is larger than any pre-existing reference level, since it is based on the current time. The originator notifies its neighbors of its new height. As shown in Reference 42, this change eventually propagates among all nodes for whom i was on their only path to the destination. These are the nodes that must either form new paths to the destination or discover that, due to partitioning, there is no such path anymore. Note that a node i can lose its last outgoing link due to a neighbor's height change under a variety of circumstances, which are explained in the sequel.

Case 2 (Propagate the highest neighbor's reference level): The first of these cases is when *not all* the neighbors of i have the same reference level. Then i sets its reference level to the largest among all its neighbors and sets its δ component to one less than the minimum δ value among all neighbors with the largest reference level (this partial reversal is similar to that in the GB algorithm).

Case 3 (Reflect back a higher sublevel): If all of i's neighbors do have the same reference level and it is an unreflected one, then i starts a reflection of this reference level by setting its reference level to the reflected version of its neighbors' (with $r_i = 1$) and its δ to zero.

Case 4 (Partition detected, deletion of invalid routes): If all of i's neighbors have the same reflected reference level with i as the originator, then i has detected a partition and takes appropriate action by deleting invalid routes.

Case 5 (Generate a new reference level): If all of i's neighbors have the same reference level with an originator other than i, then i starts a new reference level. This situation only happens if a link fails while the system is recovering from an earlier link failure.

7.3.3.3 Synopsis of the Leader Election Algorithm

As the authors state in Reference 42, they made the following changes to TORA. The height of each node i in their algorithm is now a 6-tuple $(lid_i, \tau_i, oid_i, r_i, \delta_i, i)$. The first component is the id of a node believed to be the leader of i's component. The remaining five components are the same as in TORA. The reference level $(-1, -1, -1)$ is used by the leader of a component to ensure that it is a sink.

Remark that in TORA, once a network disconnection has been detected, the node that first detected the partition notifies the other nodes in its component so that they stop updating heights and sending (useless) messages. In the leader election algorithm, the node that detected the partition elects itself as the leader of the newly created component. It then transmits this information to its neighbors, who in turn propagate this information further to their neighbors and so on. Eventually all the nodes in the new component are notified of the leader change. In the other case, that is, when two or more components merge due to the activation of new links, the leader of the component whose identity is the smallest eventually becomes the only leader of the entire newly created component.

7.3.3.4 A Pseudo-Code Description of the Algorithm

Malpani *et al.* [42] describes the pseudo-code executed by node i, which we present below. Recall that each step is triggered either by the notification of the failure or formation of an incident link or by the receipt of a message from a neighbor. Node i stores its neighbors' identities in a local variable N_i. When an incident link fails, i updates N_i. When an incident link is formed, i updates N_i and sends an Update message over that link with its current height. Upon receipt of an Update message, i updates a local data structure that keeps the current height reported for each of its neighbors. Node i uses this information to decide the direction of its incident links.

References in the pseudo-code below to variables $lid_j, \tau_j, oid_j, r_j, \delta_j$ for a neighbor node j of node i actually refer to the information that i has stored about j's height, in variable $height_i[j]$. At the end of each step, if i's height has changed, then it sends an Update message with the new height to all its neighbors. The pseudo-code below explains how and when node i's height is changed. Parts B through D are executed only if the leader identity in the received Update message is the same as lid_i.

A. When node i has no outgoing links due to a link failure, then:
   (1) if node i has no incoming links as well then
   (2) $lid_i := i$
   (3) $(\tau_i, oid_i, r_i) := (-1, -1, -1)$
   (4) $\delta_i := 0$
   (5) else
   (6) $(\tau_i, oid_i, r_i) := (t, i, 0)$ (t is the current time)
   (7) $\delta_i := 0$

B. When node i has no outgoing links due to a link reversal following reception of an Update message and the reference levels (τ_j, oid_j, r_j) are not equal for all $j \in N_i$, then:
   (1) $(\tau_i, oid_i, r_i) := \max\{(\tau_j, oid_j, r_j) \mid j \in N_i\}$
   (2) $\delta_i := \min\{\delta_j \mid j \in N_i \text{ and } (\tau_j, oid_j, r_j) = (\tau_i, oid_i, r_i)\} - 1$

C. When node i has no outgoing links due to a link reversal following reception of an Update message and the reference levels (τ_j, oid_j, r_j) are equal with $r_j = 0$ for all $j \in N_i$, then:
   (1) $(\tau_i, oid_i, r_i) := (\tau_j, oid_j, 1)$ for any $j \in N_i$
   (2) $\delta_i := 0$

D. When node i has no outgoing links due to a link reversal following reception of an Update message and the reference levels (τ_j, oid_j, r_j) are equal with $r_j = 1$ for all $j \in N_i$ and $oid_j = i$, then:
   (1) $lid_i := i$
   (2) $(\tau_i, oid_i, r_i) := (-1, -1, -1)$
   (3) $\delta_i := 0$

E. When node i receives an Update message from neighboring node j such that $lid_j \neq lid_i$, then:
   (1) if $lid_j > lid_i$ or ($oid_i = lid_j$ and $r_i = 1$) then
   (2) $lid_i := lid_j$
   (3) $(\tau_i, oid_i, r_i) := (0, 0, 0)$
   (4) $\delta_i := \delta_j + 1$

In part E, if the new identity is smaller than yours, then adopt it. If the new id is larger than yours, then adopt it, but only if it is the case that the originator of a new reference level has detected a partition and elected itself.

7.3.4 Analytic Considerations

We now present part of the correctness proof of Reference 46 for their leader election algorithm. The authors of Reference 42 assume that each connected component is a leader-oriented DAG initially and that only one change (either a link failure or a link formation) can occur at a time. The next change occurs only after the entire network has recovered from the previous change (i.e., this is the case of single topology changes). They also assume that the system is synchronous, that is, the execution occurs in rounds. Messages are sent at the beginning of each round and are received by the nodes to whom they were sent before the end of each round.

Theorem 7.1 [42] *The algorithm ensures that each component eventually has exactly one unique leader.*

Proof (sketch following [42]). We consider the following three cases (the only ones causing changes):

Case 1: A link disappears at time t, causing node i to lose its last outgoing link but not disconnecting the component.

Case 2: A link appears at time t, joining two connected components.

Case 3: A link disappears at time t, causing node i to lose its last outgoing link and disconnecting the component.

In each case, Malpani *et al.* [42] show that eventually each component becomes a leader-oriented DAG:

Case 1: A link disappears at time t, causing node i to lose its last outgoing link but not disconnecting the component.

Let G be the directed graph representing the resulting component topology. Let l be the leader of the component. Then the component was an l-oriented DAG before the link was lost. Let V_l be the set of nodes that still have a path to l. At time t, the remaining nodes have a path to i; let this set be V_i. Let G_l be the graph induced by V_l, and G_i be the graph induced by V_i. ∎

Definition 7.7 *The frontier nodes of V_i are nodes that are adjacent to nodes in V_l. The edges between V_i and V_l are the frontier edges.*

Let k be any node in V_i.

Definition 7.8 *Level(k) is the length of the longest path in G_i from k to i.*

Note that level is defined with respect to the fixed G_i. Thus, even though the direction of edges change as the algorithm proceeds, the levels remain the same.

Lemma 7.1 *If k is on a path in G_i from a frontier node to i, then k's final height is $(l, t, i, 0, -\text{level}(k), k)$. Otherwise, k's final height is $(l, t, i, 1, -\text{diff}(k), k)$, where $\text{diff}(k) = \max\{\text{level}(h) \mid h \in V_i$ and k is reachable from h in $G_i\} - \text{level}(k)$.*

Proof (sketch). By induction on the number of rounds r after t, Malpani *et al.* [42] show that at the end of round r:

    a. If $r < \text{level}(k)$, then k's height is the same as it was at time t.
    b. If k is on a path from a frontier node to i and $r \geq \text{level}(k)$, then k's height is $(l, t, i, 0, -\text{level}(k), k)$.
    c. If k is not on a path from a frontier node to i and $\text{level}(k) \leq r < \text{level}(k) + 2 \cdot \text{diff}(k)$, then k's height is $(l, t, i, 0, -\text{level}(k), k)$.
    d. If k is not on a path from a frontier node to i and $r \geq \text{level}(k) + 2 \cdot \text{diff}(k)$, then k's height is $(l, t, i, 1, -\text{diff}(k), k)$. ∎

Thus, Lemma 7.1 implies that the resulting graph is an l-oriented DAG, since all nodes in G_i now have paths to frontier nodes. The frontier edges are now directed from V_i to V_l because the τ-component in the heights of nodes in V_i is larger than for V_l (since the algorithm has access to synchronized or at least logical clocks).

Case 2: A link appears at time t, joining two formerly separate components C_1 and C_2 into a new component C.

Let l_1 be the leader of C_1 and l_2 the leader of C_2. Assume, without loss of generality, that $l_1 < l_2$. Suppose a link appears at time t between k_1 (a node in C_1) and k_2 (a node in C_2).

Lemma 7.2 *Eventually l_1 becomes the leader of component C and C is an l_1-oriented DAG.*

Proof (sketch). Let r be the number of rounds after t. At $r = 0$, k_1 and k_2 send Update messages to each other. Since k_1's leader l_1 is smaller than k_2's leader, l_2, k_2 updates its height to $(l_1, 0, 0, 0, \delta_{k_1+1}, k_2)$ and gets an outgoing link to k_1. Let the value of dist(k) for any node k in partition$_2$ be the shortest path distance from that node to node k_2 (the path distance is in terms of number of links). When $r < $ dist(k), the height of k remains unaffected since it has not yet received the Update message regarding the change in leadership. When $r = $ dist(k), k (including l_2) changes its height to $(l_1, 0, 0, 0, \delta_{k_1} + $ dist$(k) + 1, k)$. Thus k now has a path to k_1 and its leader id has also changed to indicate a change in leadership. When $r > $ dist(k), the height of k remains unchanged.

Thus we see that when $r = $ dist(k), where k is the farthest node from k_2, all the nodes in partition$_2$ have updated their heights and have a route to k_1. The resulting graph (for the merged component) will be an l_1-oriented DAG, since k_1 is a node in partition$_1$, which is an l_1-oriented DAG. ∎

Case 3: A link disappears at time t, causing node i to lose its last outgoing link and disconnecting the component.

The proof for case 3 is very similar to case 1, except that there will be no path from node i to a frontier node. The following condition will arise, which is different from the conditions in case 1: Let $r_1 = \max\{$level$(k) + 2 \cdot $diff$(k)\}$ for all k adjacent to i. At round r_1, the heights of all the adjacent nodes k will be $(l, t, i, 1, -$diff$(k), k)$ and node i will detect that a partition has occurred and will elect itself as the leader.

Lemma 7.3 *At round r_1 a DAG with node i as the sink has already been formed.*

Proof. We know from the proof of case 1 that, when $r > $ level$(k) + 2 \cdot $diff$(k)$ for any node k other than i, node k has changed its height to $(l, t, i, 1, -$diff$(k), k)$ and has an outgoing link toward node i. This height of k will not change when $r > $ level$(k) + 2 \cdot $diff$(k)$ and $r < r_1$. Also when $r = r_1 - 1$, one of the nodes k, which is adjacent to i will change its height to $(l, t, i, 1, -$diff$(k), k)$ and have an outgoing link to node i. This node k will also be the last adjacent node of i to do so. ∎

Thus at r_1, when node i detects the partition, it changes its height to $(l, t, i, 1, -$diff$(k), k)$ and sends an Update message to its neighbors. This message is propagated throughout the new component. The resulting graph is an i-oriented DAG. The proof for this is the same as the proof for Lemma 7.2.

Thus we see from all the three cases that the algorithm will eventually ensure that each component has exactly one unique leader. Thus Theorem 7.1 has been proved.

7.4 A Communication Algorithm for High Mobility Networks

In this section, we focus on networks with *high rate of mobility changes*. For such dynamically changing networks we propose protocols, which exploit the *coordinated* (by the protocol) motion of a *small part* of the network. We show that such protocols *can be designed to* work correctly and efficiently even in the case of *arbitrary* (but not malicious) movements of the hosts not affected by the protocol.

We also propose a methodology for the *analysis of the expected* behavior of protocols for such networks, based on the *assumption* that mobile hosts (those whose motion is not guided by the protocol) conduct concurrent *random walks* in their motion space.

In particular, our work examines the fundamental problem of communication and proposes distributed algorithms for it. We provide *rigorous proofs* of their correctness, and also give performance analyses by combinatorial tools. Finally, we have evaluated these protocols by large scale simulation experiments.

7.4.1 The Case of High Mobility Rate

Ad hoc mobile networks have been modeled by most researchers by a set of independent mobile nodes communicating by message passing over a wireless network. The network is modeled as a dynamically changing, not necessarily connected, undirected graph, with nodes as vertices and edges (virtual links) between vertices corresponding to nodes that can currently communicate. Such a model is used, for example, in References 42, 39, and 47. See also Section 7.3.

However, the proof of correctness of algorithms presented under such settings requires *a bound on the virtual link changes* that can occur at a time. As Reference 47 also states, in ad hoc mobile networks the topological connectivity is subject to frequent, unpredictable changes (due to the motion of hosts). If the *rate* of topological change is very high, then *structured* algorithms (i.e., algorithms that try to maintain connectivity related data structures, such as shortest paths, DAGs, etc.) *fail to react fast enough* and Reference 47 suggests that, in such cases, the only viable alternative is flooding. If the rate is low ("quasi-static" networks) or medium, then adaptive algorithmic techniques can apply. In such settings many nice works have appeared like, for example, the work of Reference 42 on leader election, the work of Reference 2 on communication (which, however, examines only *static* transmission graphs), etc. See the discussion on LRR in Section 7.3.

The most common way to establish communication in ad hoc networks is to form paths of intermediate nodes that lie within one another's transmission range and can directly communicate with each other [38,39,48]. The mobile nodes act as hosts and routers at the same time in order to propagate packets along these paths. Indeed, this approach of exploiting pairwise communication is common in ad hoc mobile networks that cover a relatively small space (i.e., with diameter that is small with respect to transmission range) or are dense (i.e., thousands of wireless nodes) where all locations are occupied by some hosts; broadcasting can be efficiently accomplished.

In wider area ad hoc networks with less users, however, broadcasting is impractical: two distant peers will not be reached by any broadcast as users may not occupy all intermediate locations (i.e., the formation of a path is not feasible). Even if a valid path is established, single link "failures" happening when a small number of users that were part of the communication path move in a way such that they are no longer within transmission range of each other, will make this path invalid. Note also that the path established in this way may be very long, even in the case of connecting nearby hosts.

We conjecture that, in cases of high mobility rate and of low density of user spreading, one can even state an *impossibility result*: any algorithm that tries to maintain a global structure with respect to the temporary network will be erroneous if the mobility rate is faster than the rate of updates of the algorithm. An experimental validation of conjecture is partially achieved in Reference 16, where a detailed study of the impact of user mobility and user density on the performance and correctness of communication protocols is performed.

In contrast to all such methods, we try to avoid ideas based on paths finding and their maintenance. We envision networks with *highly dynamic movement* of the mobile users, where the idea of "maintenance" of a valid path is inconceivable (paths can become invalid immediately after they have been added to the routing tables). Our approach is to take advantage of the mobile hosts natural movement by exchanging information whenever mobile hosts meet incidentally. It is evident, however, that if the users are spread in remote areas and they do not move beyond these areas, there is no way for information to reach them, unless the protocol takes special care of such situations.

7.4.2 The Basic Idea

In the light of the above, we propose the idea of forcing a *small* subset of the deployed hosts *to move as per the needs of the protocol*. We call this set the *"support"* of the network. Assuming the availability of such hosts, the designer can use them suitably *by specifying their motion* in certain times that the algorithm dictates.

We wish to admit that the assumption of availability of such hosts for algorithms deviates from the "pure" definition of ad hoc mobile networks. However, we feel that our approach opens up an area for distributed algorithms design for ad hoc mobile networks of high mobility rate.

Our approach is motivated by two research directions of the past:

     a. The *"two tier principle"* [38], stated for mobile networks with a fixed subnetwork, however, which says that any protocol should try to move communication and computation to the fixed part of the network. Our assumed set of hosts that are coordinated by the protocol *simulates* such a (skeleton) network; the difference is that the simulation actually constructs a coordinated *moving* set.

b. A usual scenario that fits to the ad hoc mobile model is the particular case of *rapid deployment* of mobile hosts, in an area where there is no underlying fixed infrastructure (either because it is impossible or very expensive to create such an infrastructure, or because it is not established yet, or because it has become temporarily unavailable, that is, destroyed or down).

In such a case of rapid deployment of a number of mobile hosts, it is possible to have a small team of fast moving and versatile vehicles, to implement the support. These vehicles can be cars, jeeps, motorcycles, or helicopters. We interestingly note that this small team of fast moving vehicles can also be a collection of independently controlled mobile modules, that is, robots. This specific approach is inspired by the recent paper of Walter, Welch, and Amato. In their paper "Distributed Reconfiguration of Metamorphic Robot Chains" [52] the authors study the problem of motion coordination in distributed systems consisting of such robots, which can connect, disconnect, and move around. The paper deals with metamorphic systems where (as is also the case in our approach) all modules are identical. See Section 7.5 for a detailed discussion of the paper [52]. Note in particular that the approach of having the support moving in a coordinated way, that is, *as a chain of nodes*, has some similarities to Reference 52.

Related Work. In a recent paper [40], Li and Rus present a model, which has some similarities to ours. The authors give an interesting, yet different, protocol to send messages, which forces *all the mobile hosts to slightly deviate (for a short period of time) from their predefined, deterministic routes, in order to propagate the messages.* Their protocol is, thus, *compulsory* for any host, and it works only for deterministic host routes. Moreover, their protocol considers the propagation of only one message (end to end) each time, in order to be correct. In contrast, our support scheme allows for simultaneous processing of many communication pairs. In their setting, they show optimality of message transmission times.

Adler and Scheideler [2] deal only with *static* transmission graphs, that is, the situation where the positions of the mobile hosts and the environment do not change. In Reference 2 the authors pointed out that static graphs provide a starting point for the dynamic case. In our work, we consider the *dynamic case* (i.e., mobile hosts move *arbitrarily*) and in this sense we extend their work. As far as performance is concerned, their work provides time bounds for communication that are proportional to the diameter of the graph defined by random uniform spreading of the hosts.

In Reference 31, Dimitriou et al. study the "infection time" of graphs, that is, the expected time needed for red and white particles moving randomly on the vertices of a graph and "infecting" each other with their color when they meet, to become all red. By using properties of stochastic interaction of particles, the authors provide tight bounds on the infection time. We note that the infection time may capture the communication time of mobile hosts in an ad hoc network. Other communication algorithms extending the support approach to other important network types (such as highly changing and hierarchical networks), can be found in references 20, 22, 25, 26 and 29.

7.4.3 The Model

7.4.3.1 The Motion Space

The set of previous research that follows the approach of slowly changing communication graphs, models hosts motions only implicitly, that is, via the pre-assumed upper bound on the rate of virtual link changes. In contrast, we propose an *explicit* model of motions because it is apparent that the motions of the hosts are the cause of the fragility of the virtual links. Thus we distinguish explicitly between (a) the *fixed* (for any algorithm) *space* of possible motions of the mobile hosts and (b) the kind of motions that the hosts *actually perform* inside this space. In the sequel we have decided to model the space of motions only combinatorially, that is, as a graph. We however believe that future research will complement this effort by introducing geometry details into the model.

In particular, we abstract the environment where the stations move (in three-dimensional space with possible obstacles) by a *motion-graph* (i.e., we neglect the detailed geometric characteristics of the motion).

In particular, we first assume that each mobile host has a transmission range represented by a sphere tr centered by itself. This means that any other host inside tr can receive any message broadcast by this host. We approximate this sphere by a cube tc with volume $\mathcal{V}(tc)$, where $\mathcal{V}(tc) \simeq \mathcal{V}(tr)$. The size of tc can be chosen in such a way that its volume $\mathcal{V}(tc)$ is the maximum that preserves $\mathcal{V}(tc) < \mathcal{V}(tr)$, and if a mobile host inside tc broadcasts a message, this message is received by any other host in tc. Given that the mobile hosts are moving in the space $\mathcal{S}$, $\mathcal{S}$ is divided into consecutive cubes of volume $\mathcal{V}(tc)$.

Definition 7.9 *The motion graph $G(V, E)$, $(|V| = n, |E| = m)$, which corresponds to a quantization of $\mathcal{S}$ is constructed in the following way: a vertex $u \in G$ represents a cube of volume $\mathcal{V}(tc)$. An edge $(u, v) \in G$ if the corresponding cubes are adjacent.*

The number of vertices n, actually approximates the ratio between the volume $\mathcal{V}(\mathcal{S})$ of space $\mathcal{S}$, and the space occupied by the transmission range of a mobile host $\mathcal{V}(tr)$. In the case where $\mathcal{V}(\mathcal{S}) \simeq \mathcal{V}(tr)$ (the transmission range of the hosts approximates the space where they are moving), then $n = 1$. Given the transmission range tr, n depends linearly on the volume of space $\mathcal{S}$ regardless of the choice of tc, and $n = O(V(\mathcal{S})/V(tr))$. Let us call the ratio $V(\mathcal{S})/V(tr)$ by the term *relative motion space size* and denote it by ρ. Since the edges of G represent neighboring polyhedra each node is connected with a constant number of neighbors, which yields that $m = \Theta(n)$. In our example where tc is a cube, G has maximum degree of six and $m \leq 6n$. Thus *motion graph G* is (usually) a *bounded degree graph* as it is derived from a regular graph of small degree by deleting parts of it corresponding to motion or communication obstacles. Let Δ be the maximum vertex degree of G.

7.4.3.2   The Motion of the Hosts — Adversaries

The motion that the hosts perform (we mean here the hosts that are not part of the support, that is, those whose motion is *not specified* by the distributed algorithm) is *an input* to any distributed algorithm.

     a. In the general case, we assume that the motions of such hosts are decided by an *oblivious adversary*. The adversary determines motion patterns in any possible way but independently of the part of the distributed algorithm that specifies motions of the support. In other words, we exclude the case where some of the hosts not in the support are deliberately trying to *maliciously affect* the protocol (e.g., to avoid the hosts in the support). This is a pragmatic assumption usually followed by applications. We call such motion adversaries the *restricted motion adversaries*.

     b. For the purpose of studying efficiency of distributed algorithms for ad hoc networks *on an average*, we examine the case where the motions of any host not affected by the algorithm are modeled by *concurrent and independent random walks*. In fact, the assumption that the mobile users move randomly, either according to uniformly distributed changes in their directions and velocities or according to the random waypoint mobility model by picking random destinations, has been used by other research (see, e.g., References 35 and 37).

We interestingly note here a fundamental result of graph theory, according to which dense graphs look like random graphs in many of their properties. This has been noticed for expander graphs at least by References 3 and 6 and is captured by the famous Szemeredi's Regularity Lemma [50], which says that in some sense most graphs can be approximated by random-looking graphs.

In analogy, we conjecture that any set of *dense (in number)* but arbitrary otherwise motions of many hosts in the motion space can be approximated (at least with respect to their meeting and hitting times statistics) by a set of concurrent dense *random walks*. But the meeting times statistics of the hosts essentially determine the virtual fragile links of any ad hoc network. We thus believe that our suggestion for adoption of concurrent random walks as a model for input motions, is not only a tool for average case performance analysis but it might in fact approximate well any ad hoc network of dense motions.

7.4.4 The Problem

A *basic communication problem*, in ad hoc mobile networks, is to send information from some *sender* user, MH_S, to another designated *receiver* user, MH_R.

One way to solve this problem is the protocol of notifying every user that the sender MH_S meets (and providing *all the information to it*) hoping that some of them will eventually meet the receiver MH_R.

> Is there a more efficient technique (other than notifying every user that the sender meets, in hope that some of them will then eventually meet the receiver) that will effectively solve the basic communication problem without flooding the network and exhausting the battery and computational power of the hosts?

A protocol solving this important communication problem is *reliable* if it allows the sender to be notified about delivery of the information to the receiver.

Note that this problem is very important and in fact fundamental, in the sense that no distributed computing protocol can be implemented in ad hoc mobile networks without solving this basic communication problem.

7.4.5 A Set of Algorithms

7.4.5.1 The Basic Algorithm

In simple terms, the protocol works as follows: the nodes of the support move fast enough so that they visit (in a sufficiently short time) the entire motion graph. Their motion is accomplished in a distributed way via a *support motion subprotocol* P_1. When some node of the support is within communication range of a sender, an underlying *sensor subprotocol* P_2 notifies the sender that it may send its message(s). The messages are then stored "somewhere within the support structure." When a receiver comes within the communication range of a node of the support, the receiver is notified that a message is "waiting" for him and the message is then forwarded to the receiver. The messages received by the support are propagated within the structure when two or more members of the support meet on the same site (or are within the communication range). A synchronization subprotocol P_3 is used to dictate the way that the members of the support exchange information.

In a way, the support Σ plays the role of a (moving) skeleton subnetwork (whose structure is defined by the motion subprotocol P_1), through which all communication is routed. From the above description, it is clear that the size, k, and the shape of the support may affect performance.

Note that the proposed scheme does not require the propagation of messages through hosts that are not part of Σ, thus its security relies on the support's security and is not compromised by the participation in message communication of other mobile users. For a discussion of intrusion detection mechanisms for ad hoc mobile networks see Reference 54.

7.4.5.2 The SNAKE Algorithm

The main idea of the protocol proposed in References 19 and 21 is as follows. There is a set-up phase of the ad hoc network, during which a predefined set, k, of hosts, become the nodes of the support. The members of the support perform a leader election by running a randomized breaking symmetry protocol in anonymous networks (see, e.g., Reference 36). This is run once and imposes only an initial communication cost. The elected leader, denoted by MS_0, is used to coordinate the support topology and movement. Additionally, the leader assigns local names to the rest of the support members MS_1, $MS_2, \ldots, MS_{k-1}$.

The nodes of the support move in a coordinated way, always remaining pairwise adjacent (i.e., forming a chain of k nodes), so that they sweep (given some time) the entire motion graph. This encapsulates the *support motion subprotocol* P_1^S. Essentially the motion subprotocol P_1^S enforces the support to move as a "snake," with the head (the elected leader MS_0) doing a random walk on the motion graph and each of the other nodes MS_i executing the simple protocol "move where MS_{i-1} was before." More formally, the movement of Σ is defined as follows:

Initially, MS_i, $\forall i \in \{0, 1, \ldots, k - 1\}$, start from the same area-node of the motion graph. The direction of movement of the leader MS_0 is given by a memoryless operation that chooses *randomly the direction* of the next move. Before leaving the current area-node, MS_0 sends a message to MS_1 that states the new direction of movement. MS_1 will change its direction as per instructions of MS_0 and will propagate the message to MS_2. In analogy, MS_i will follow the orders of MS_{i-1} after transmitting the new directions to MS_{i+1}. Movement orders received by MS_i are positioned in a queue Q_i for sequential processing. The very first move of MS_i, $\forall i \in \{1, 2, \ldots, k - 1\}$ is delayed by a δ period of time.

We assume that the mobile support hosts move with a common speed. Note that the above described motion subprotocol P_1^S enforces the support to move as a "snake," with the head (the elected leader MS_0) *doing a random walk on the motion graph G* and each of the other nodes MS_i executing the simple protocol "move where MS_{i-1} was before." This can be easily implemented because MS_i will move following the edge from which it received the message from MS_{i-1} and therefore our protocol does not require common sense of orientation.

The purpose of the random walk of the head is to ensure a *cover* (within some finite time) of the whole motion graph, without memory (other than local) of topology details. Note that this memoryless motion also ensures fairness. The value of the *random walk principle* of the motion of the support will be further justified in the correctness and the efficiency parts of the paper, where we wish our basic communication protocol to meet some performance requirements *regardless of the motion of the hosts not in Σ*. A modification of M_Σ^S is that the head does a random walk on *a spanning subgraph* of G (e.g., a spanning tree). This modified M_Σ (call it M_Σ') is more efficient in our setting since "edges" of G just represent adjacent locations and "nodes" are really possible host places.

A simple approach to implement the support *synchronization subprotocol P_3^S* is to use a forward mechanism that transmits incoming messages (in-transit) to neighboring members of the support (due to P_1^S at least one support host will be close enough to communicate). In this way, all incoming messages are eventually copied and stored in every node of the support. This is not the most efficient storage scheme and can be refined in various ways in order to reduce memory requirements.

7.4.5.3 The RUNNERS Algorithm

A different approach to implement M_Σ is to allow each member of Σ not to move in a snake-like fashion, but to perform an *independent* random walk on the motion graph G, that is, the members of Σ can be viewed as "runners" running on G. In other words, instead of maintaining at all times pairwise adjacency between members of Σ, all hosts sweep the area by moving independently of each other. When two runners meet, they exchange any information given to them by senders encountered using a new synchronization subprotocol P_3^R. As in the snake case, the same underlying sensor subprotocol P_2 is used to notify the sender that it may send its message(s) when being within communication range of a node of the support.

As described in Reference 19, the runners protocol does not use the idea of a (moving) backbone subnetwork as no motion subprotocol P_1^R is used. However, all communication is still routed through the support Σ and we expect that the size k of the support (i.e., the number of runners) will affect performance in a more efficient way than that of the snake approach. This expectation stems from the fact that each host will meet each other in parallel, accelerating the spread of information (i.e., messages to be delivered).

A member of the support needs to store all undelivered messages, and maintain a list of receipts to be given to the originating senders. For simplicity, we can assume a generic storage scheme where all undelivered messages are members of a set S_1 and the list of receipts is stored on another set S_2. In fact, the unique ID of a message and its sender ID is all that is needed to be stored in S_2. When two runners meet at the same vertex of the motion graph G, the synchronization subprotocol P_3^R is activated. This subprotocol imposes that when runners meet on the same vertex, their sets S_1 and S_2 are synchronized. In this way, a message delivered by some runner will be removed from the set S_1 of the rest of runners encountered, and similarly delivery receipts already given will be discarded from the set S_2 of the rest of runners.

The synchronization subprotocol P_3^R is partially based on the *two-phase commit* algorithm as presented in Reference 43 and works as follows:

Let the members of Σ residing on the same site (i.e., vertex) u of G be $MS_1^u, \ldots, MS_j^u$. Let also $S_1(i)$ (resp. $S_2(i)$) denote the S_1 (resp. S_2) set of runner MS_i^u, $1 \leq i \leq j$. The algorithm assumes that the underlying sensor subprotocol P_2 informs all hosts about the runner with the lowest ID, i.e., the runner MS_1^u. P_3^R consists of two rounds:

Round 1: All $MS_1^u, \ldots, MS_j^u$ residing on vertex u of G, send their S_1 and S_2 to runner MS_1^u. Runner MS_1^u collects all the sets and combines them with its own to compute its new sets S_1 and S_2: $S_2(1) = \bigcup_{1 \leq l \leq j} S_2(l)$ and $S_1(1) = \bigcup_{1 \leq l \leq j} S_1(l) - S_2(1)$.

Round 2: Runner MS_1^u broadcasts its decision to all the other support member hosts. All hosts that received the broadcast apply the same rules, as MS_1^u did, to join their S_1 and S_2 with the values received. Any host that receives a message at Round 2 and which has not participated in Round 1, accepts the value received in that message as if it had participated in Round 1.

This simple algorithm guarantees that mobile hosts which remain connected (i.e., are able to exchange messages) for two continuous rounds, will manage to synchronize their S_1 and S_2. Furthermore, on the event that a new host arrives or another disconnects during Round 2, the execution of the protocol will not be affected. In the case where runner MS_1^u fails to broadcast in Round 2 (either because of an internal failure or because it has left the site), then the protocol is simply re-executed among the remaining runners.

Remark that the algorithm described above does not offer a mechanism to remove message receipts from S_2; eventually the memory of the hosts will be exhausted. A simple approach to solve this problem and effectively reduce the memory usage is to construct an order list of IDs contained in S_2, for each receiver. This ordered sequence of IDs will have gaps — some messages will still have to be confirmed, and thus are not in S_2. In this list, we can identify the maximum ID before the first gap. We are now able to remove from S_2 all message receipts with smaller ID than this ID.

7.4.6 Analysis of the SNAKE Algorithm

7.4.6.1 Preliminaries

In the sequel, we assume that the head of the snake does a *continuous time random walk* on $G(V, E)$, without loss of generality (if it is a discrete time random walk, all results transfer easily, see Reference 5). We define the random walk of a host on G that induces a continuous time Markov chain M_G as follows: the states of M_G are the vertices of G and they are finite. Let s_t denote the state of M_G at time t. Given that $s_t = u$, $u \in V$, the probability that $s_{t+dt} = v$, $v \in V$, is $p(u, v) \cdot dt$, where

$$p(u, v) = \begin{cases} \dfrac{1}{d(u)}, & \text{if } (u, v) \in E \\ 0, & \text{otherwise} \end{cases}$$

and $d(u)$ is the degree of vertex u. We assume that all random walks are *concurrent* and that there is a global time t, not necessarily known to the hosts.

Note 7.1 Since the motion graph G is finite and connected, the continuous Markov chain abstracting the random walk on it is automatically time-reversible.

Definition 7.10 $P_i(E)$ *is the probability that the random walk satisfies an event E given it started at vertex i.*

Definition 7.11 *For a vertex j, let $T_j = \min\{t \geq 0 : s_t = j\}$ be the first hitting time of the walk onto that vertex and let $E_i T_j$ be its expected value, given that the walk started at vertex i of G.*

Definition 7.12 *For the random walk of any particular host, let $\pi(\ )$ be the stationary distribution of its position after a sufficiently long time.*

We denote $E_\mu[\cdot]$ the expectation for the chain started at time 0 from any vertex with distribution μ (e.g., the initial distribution of the Markov chain). We know (see Reference 5) that for every vertex σ, $\pi(\sigma) = d(\sigma)/2m$ where $d(\sigma)$ is the degree of σ in G and $m = |E|$.

Definition 7.13 *Let $p_{j,k}$ be the transition probability of the random walk from vertex j to vertex k. Let $p_{j,k}(t)$ be the probability that the random walk started at j will be at $k \in V$ in time t.*

Definition 7.14 *Let $X(t)$ be the position of the random walk at time t.*

7.4.6.2 Correctness Guarantees under the Restricted Motion Adversary

Let us now consider the case where any sender or receiver is allowed a general, unknown motion strategy, but its strategy is provided by a restricted motion adversary. This means that each host not in the support either executes a deterministic motion (which either stops at a node or cycles forever after some initial part) or it executes a stochastic strategy, which however is *independent* of the motion of the support.

Theorem 7.2 *The support Σ and the management subprotocol M_Σ guarantee reliable communication between any sender–receiver $(\text{MH}_S, \text{MH}_R)$ pair in finite time, whose expected value is bounded only by a function of the relative motion space size ρ and does not depend on the number of hosts, and is also independent of how MH_S, MH_R move, provided that the mobile hosts not in the support do not deliberately try to avoid the support.*

Proof. For the proof purpose, it is enough to show that a node of Σ will meet MH_S and MH_R infinitely often, with probability 1 (in fact our argument is a consequence of the Borel–Cantelli Lemmas for infinite sequences of trials). We will furthermore show that the first meeting time M (with MH_S or MH_R) has an expected value (where expectation is taken over the walk of Σ and any strategy of MH_S [or MH_R] and any starting position of MH_S [or MH_R] and Σ), which is bounded by a function of the size of the motion graph G only. This then proves the theorem since it shows that MH_S (and MH_R) meet with the head of Σ infinitely often, each time within a bounded expected duration. So, let EM be the expected time of the (first) meeting and $m^* = \sup EM$, where the supremum is taken over all starting positions of both Σ and MH_S (or MH_R) and all strategies of MH_S (one can repeat the argument with MH_R).

We proceed to show that we can construct for the walk of Σs head a *strong stationary time sequence* V_i such that for all $\sigma \in V$ and for all times t

$$P_i(X(V_i) = \sigma \mid V_i = t) = \pi(\sigma)$$

Remark that strong stationary times were introduced by Aldous and Diaconis in Reference 4. Notice that at times V_i, MH_S (or MH_R) will necessarily be at some vertex σ of V, either still moving or stopped. Let u be a time such that for X,

$$p_{j,k}(u) \geq \left(1 - \frac{1}{e}\right)\pi(k)$$

for all j, k. Such a u always exists because $p_{j,k}(t)$ converges to $\pi(k)$ from basic Markov Chain Theory.

Note that u depends only on the structure of the walk's graph, G. In fact, if one defines separation from stationarity to be $s(t) = \max_j s_j(t)$ where $s_j(t) = \sup\{s : p_{ij}(t) \geq (1 - s)\pi(j)\}$ then

$$\tau_1^{(1)} = \min\{t : s(t) \leq e^{-1}\}$$

is called the *separation threshold time*. For general graphs G of n vertices this quantity is known to be $\mathcal{O}(n^3)$ [11]. Now consider a sequence of stopping times $U_i \in \{u, 2u, 3u, \dots\}$ such that

$$P_i(X(U_i) = \sigma \mid U_i = u) = \frac{1}{e}\pi(\sigma) \tag{7.1}$$

for any $\sigma \in V$. By induction on $\lambda \geq 1$ then

$$P_i(X(U_i) = \sigma \mid U_i = \lambda u) = e^{-(\lambda-1)}\frac{1}{e}\pi(\sigma)$$

This is because of the following: first remark that for $\lambda = 1$ we get the definition of U_i. Assume that the relation holds for $(\lambda - 1)$, that is,

$$P_i(X(U_i) = \sigma \mid U_i = (\lambda - 1)u) = e^{-(\lambda-2)}\frac{1}{e}\pi(\sigma)$$

for any $\sigma \in V$. Then $\forall \sigma \in V$

$$P_i(X(U_i) = \sigma \mid U_i = \lambda u) = \sum_{\alpha \in V} P_i(X(U_i) = \alpha \mid U_i = (\lambda - 1)u)$$

$$\times P(X(U_i + u) = \sigma \mid X(U_i) = \alpha)$$

$$= e^{-(\lambda-2)}\frac{1}{e}\sum_{\alpha \in V}\pi(\alpha)\frac{1}{e}\pi(\sigma) \qquad \text{from (7.1)}$$

$$= e^{-(\lambda-1)}\frac{1}{e}\pi(\sigma)$$

which ends the induction step. Then, for all σ

$$P_i(X(U_i) = \sigma) = \pi(\sigma) \tag{7.2}$$

and, from the geometric distribution with $q = 1/e$,

$$E_i U_i = u\left(1 - \frac{1}{e}\right)^{-1}$$

Now let $c = u(e/(e-1))$. So, we have constructed (by (7.2)) a *strong stationary time sequence* U_i with $EU_i = c$. Consider the sequence $0 = U_0 < U_1 < U_2 < \cdots$ such that for $i \geq 0$

$$E(U_{i+1} - U_i \mid U_j), \qquad j \leq i) \leq c$$

But, from our construction, the positions $X(U_i)$ are *independent* (of the distribution $\pi()$), and, in particular, $X(U_i)$ *are independent* of U_i. Therefore, regardless of the strategy of MH$_S$ (or MH$_R$) and because of the independence assumption, the support's head has chance at least $\min_\sigma \pi(\sigma)$ to meet MH$_S$ (or MH$_R$) at time U_i, independently of how i varies. So, the meeting time M satisfies $M \leq U_T$, where T is a stopping time with mean

$$ET \leq \frac{1}{\min_\sigma \pi(\sigma)}$$

Note that the idea of a stopping time T such that $X(T)$ has distribution π and is independent of the starting position is central to the standard modern Theory of Harris — recurrent Markov Chains (see, e.g., Reference 33). From Wald's equation for the expectation of a sum of a random number of independent random variables (see, e.g., Reference 5, 49) then

$$EU_T \leq c \cdot ET \quad \Rightarrow \quad m^* \leq c \frac{1}{\min_\sigma \pi(\sigma)}$$

Note that since G is produced as a subgraph of a regular graph of fixed degree Δ we have

$$\frac{1}{2m} \leq \pi(\sigma) \leq \frac{1}{n}$$

for all σ ($n = |V|$, $m = |E|$); thus

$$ET \leq 2m$$

Hence

$$m^* \leq 2mc = \frac{e}{e-1} 2mu$$

Since m, u only depend on G, this proves the theorem. ∎

Corollary 7.1 *If Σs head walks randomly in a* regular *spanning subgraph of G, then $m^* \leq 2cn$.*

7.4.6.3 Worst Case Time Efficiency for the Restricted Motion Adversary

Clearly, one intuitively expects that if $k = |\Sigma|$ is the size of the support then the higher k is (with respect to n), the best the performance of Σ gets. By working as in the construction of the proof of Theorem 7.2, we can create a sequence of strong stationary times U_i such that $X(U_i) \in F$, where $F = \{\sigma : \sigma$ is a position of a host in the support$\}$. Then $\pi(\sigma)$ is replaced by $\pi(F)$, which is just $\pi(F) = \sum \pi(\sigma)$ over all $\sigma \in F$. So now m^* is bounded as follows:

$$m^* \leq c \frac{1}{\min_{\sigma \in J} (\sum \pi(\sigma))},$$

where J is any induced subgraph of the graph of the walk of Σs head such that J is the neighborhood of a vertex σ of radius (maximum distance simple path) at most k. The quantity

$$\min_J \left(\sum_{\sigma \in J} \pi(\sigma) \right)$$

is then at least $k/2m$ (since each $\pi(\sigma)$ is at least $1/2m$ and there are at least k vertices σ in the induced subgraph), hence, $m^* \leq c(2m/k)$. The overall communication is given by:

$$T_{\text{total}} = X + T_\Sigma + Y, \tag{7.3}$$

where

- X is the time for the sender node to reach a node of the support.
- T_Σ is the time for the message to propagate inside the support. Clearly, $T_\Sigma = \mathcal{O}(k)$, that is, linear in the support size.
- Y is the time for the receiver to meet with a support node, after the propagation of the message inside the support.

We have for all MH$_S$, MH$_R$:

$$E(T_{\text{total}}) \leq \frac{2mc}{k} + \Theta(k) + \frac{2mc}{k}$$

The upper bound achieves a minimum when $k = \sqrt{2mc}$.

Lemma 7.4 *For the walk of Σs head on the entire motion graph G, the communication time's expected value is bounded above by $\Theta(\sqrt{mc})$ when the (optimal) support size $|\Sigma|$ is $\sqrt{2mc}$ and c is $(e/(e-1)u)$, u being the "separation threshold time" of the random walk on G.*

Remark that other authors use for u the symbol $\tau_1^{(1)}$ for a symmetric Markov Chain of continuous time on n states.

7.4.6.4 Tighter Bounds for the Worst Case of Motion

To make our protocol more efficient, we now force the head of Σ to perform a random walk on a *regular spanning graph* of G. Let $G_R(V, E')$ be such a subgraph. Our improved protocol versions assume that (a) such a subgraph exists in G and (b) is given in the beginning to all the stations of the support. Then, for any $\sigma \in V$, and for this new walk X', we have for the steady state probabilities $\pi_{X'}(\sigma) = 1/n$ for all σ (since in this case $d(\sigma) = r$ and $2m = rn$, for all σ, where r is the degree of any σ).

Let now M be again the first meeting time of MH$_S$ (or MH$_R$) and Σs head. By the stationarity of X' and since all $\pi_{X'}(\sigma)$ are equal,

$$\int_0^t P(\text{MH}_S, X' \text{ are together at time s })ds = \frac{t}{n} \qquad (7.4)$$

Now let

$$p^*(t) = \max_{x,v} p_{x,v}(t)$$

Regardless of initial positions, the chance that MH$_S$ (or MH$_R$), Σs head are together (i.e., at the same vertex) at time u is *at most $p^*(u)$*. So, by assuming first that Σs head starts with the stationary distribution, we get

$$P(\text{MH}_S, X' \text{are together at time s}) = \int_0^s f(u)P(\text{together at time s} \mid M = u)du$$

where $f(u)$ is the (unknown) probability density function of the meeting time M. So

$$P(\text{MH}_S, X' \text{together at time s}) \leq \int_0^s f(u)p^*(s-u)du$$

But we know (see Reference 5) the following:

Lemma 7.5 [5] *There exists an absolute constant K such that*

$$p^*(t) \leq \frac{1}{n} + Kt^{-1/2} \quad \text{where } 0 \leq t < \infty$$

Thus

$$P(\text{MH}_S, X' \text{ are together at time s}) \leq \frac{1}{n}P(M \leq s) + K\int_0^s f(u)(s-u)^{-1/2}du$$

So, from (7.4), we get

$$
\frac{t}{n} \leq \frac{1}{n} \int_0^t P(M \leq s)\,ds + K \int_0^t f(u)\,du \int_0^t (s-u)^{-1/2}\,ds
$$

$$
= \frac{t}{n} - \frac{1}{n} \int_0^t P(M > s)\,ds + 2K \int_0^t f(u)(t-u)^{-1/2}\,du
$$

$$
\leq \frac{t}{n} - \frac{1}{n} E\min(M,t) + 2Kt^{-1/2}
$$

So, the expected of the minimum of M and t is

$$
E\min(M,t) \leq 2Knt^{-1/2}
$$

Taking $t_0 = (4Kn)^2$, from Markov's inequality we get $P(M \leq t_0) \geq \frac{1}{2}$. When Σ starts at some arbitrary vertex, we can use the notion of *separation time $s(u)$* (a time to approach stationarity by at least a fraction of $(1 - (1/e))^{-1}$) to get

$$
P(M \leq u + t_0) \geq \frac{1 - s(u)}{2}
$$

that is, by iteration,

$$
EM \leq \frac{2(u + t_0)}{1 - s(u)},
$$

where u is as in the construction of Theorem 7.2. Thus,

$$
m^* \leq \frac{2}{1 - \left(\frac{1}{e}\right)}(t_0 + u)
$$

but for regular graphs it is known (see, e.g., Reference 5) that $u = \mathcal{O}(n^2)$ implying (together with our remarks about Σs size and set of positions) the following theorem:

Theorem 7.3 *By having Σs head to move on a regular spanning subgraph of G, there is an absolute constant $\gamma > 0$ such that the expected meeting time of $\mathrm{MH_S}$ (or $\mathrm{MH_R}$) and Σ is bounded above by $\gamma(n^2/k)$.*

Again, the total expected communication time is bounded above by $2\gamma(n^2/k) + \Theta(k)$ and by choosing $k = \sqrt{2\gamma n^2}$ we can get a best bound of $\Theta(n)$ for a support size of $\Theta(n)$. Recall that $n = \mathcal{O}(V(\mathcal{S})/V(tr))$, that is, n is linear to the ratio ρ of the volumes of the space of motions and the transmission range of each mobile host. Thus,

Corollary 7.2 *By forcing the support's head to move on a regular spanning subgraph of the motion graph, our protocol guarantees a total expected communication time of $\Theta(\rho)$, where ρ is the relative motion space size, and this time is independent of the total number of mobile hosts, and their movement.*

Note 7.2 Our analysis assumed that the head of Σ moves according to a continuous time random walk of total rate 1 (rate of exit out of a node of G). If we select the support's hosts to be ψ *times faster* than the rest of the hosts, all the estimated times, except of the inter-support time, will be divided by ψ. Thus,

Corollary 7.3 *Our modified protocol where the support is ψ times faster than the rest of the mobile hosts guarantees an expected total communication time, which can be made to be as small as $\Theta(\gamma(\rho/\sqrt{\psi}))$ where γ is an absolute constant.*

7.4.6.5 Time Efficiency — A Lower Bound

Lemma 7.6

$$m^* \geq \max_{i,j} E_i T_j$$

Proof. Consider the case where MH_S (or MH_R) just stands still on some vertex j and Σs head starts at i. ∎

Corollary 7.4 *When Σ starts at positions according to the stationary distribution of its head's walk then*

$$m^* \geq \max_j E_\pi T_j$$

for $j \in V$, where π is the stationary distribution.

From a lemma of [5, ch. 4, pp. 21], we know that for all i

$$E_\pi T_i \geq \frac{(1 - \pi(i))^2}{q_i \pi(i)}$$

where $q_i = d_i$ is the degree of i in G, that is,

$$E_\pi T_i \geq \min_i \frac{(1 - (d_i/2m))^2}{d_i \frac{d_i}{2m}} \geq \min_i \frac{1}{2m} \frac{(2m - d_i)^2}{d_i^2}$$

For regular spanning subgraphs of G of degree Δ we have $m = \Delta n/2$, where $d_i = \Delta$ for all i. Thus,

Theorem 7.4 *When Σs head moves on a regular spanning subgraph of G, of m edges, we have that the expected meeting time of MH_S (or MH_R) and Σ cannot be less than $(n - 1)^2/2m$.*

Corollary 7.5 *Since $m = \Theta(n)$ we get a $\Theta(n)$ lower bound for the expected communication time. In that sense, our protocol's expected communication time is* optimal, *for a support size, which is $\Theta(n)$.*

7.4.6.6 Time Efficiency on an Average

Time-efficiency of semicompulsory protocols for ad hoc networks is not possible to estimate without a scenario for the motion of the mobile users not in the support (i.e., the noncompulsory part). In a way similar to References 23 and 36, we propose an "on-an-average" analysis by assuming that the movement of each mobile user *is a random walk on the corresponding motion graph G.* We propose this kind of analysis as a necessary and interesting first step in the analysis of efficiency of any semicompulsory or even noncompulsory protocol for ad hoc mobile networks. In fact, the assumption that the mobile users are moving randomly (according to uniformly distributed changes in their directions and velocities, or according to the random waypoint mobility model, by picking random destinations) has been used in References 37 and 35. In the light of the above, we assume that any host not belonging in the support, conducts a random walk on the motion graph, independently of the other hosts.

The communication time of M_Σ^S is the total time needed for a message to arrive from a sending node u to a receiving node v. Since the communication time, T_{total}, between MH_S, MH_R is bounded above by $X + Y + Z$, where X is the time for MH_S to meet Σ, Y is the time for MH_R to meet Σ (after X) and Z is the message propagation time within Σ, in order to estimate the expected values of X and Y (these are random variables), we work as follows:

a. Note first that X, Y are, statistically, of the same distribution, under the assumption that u, v are randomly located (at the start) in G. Thus $E(X) = E(Y)$.

b. We now replace the meeting time of u and Σ by a hitting time, using the following thought experiment:

(b1) We fix the support Σ in an "average" place inside G.

(b2) We then collapse Σ to a single node (by collapsing its nodes to one but keeping the incident edges). Let H be the resulting graph, σ the resulting node and $d(\sigma)$ its degree.

(b3) We then estimate the hitting time of u to σ assuming u is somewhere in G, according to the *stationary distribution*, π of its walk on H. We denote the expected value of this hitting time by $E_\pi T_\sigma^H$.

Thus, now $E(T_{\text{total}}) = 2E_\pi T_o^H + \mathcal{O}(k)$.

Note 7.3 The equation above amortizes over meeting times of senders (or receivers) because it uses the stationary distribution of their walks for their position when they decide to send a message.

Now, proceeding as in Reference 5, we get the following lemma.

Lemma 7.7 *For any node σ of any graph H in a continuous-time random walk*

$$E_\pi T_\sigma^H \leq \frac{\tau_2(1 - \pi(\sigma))}{\pi(\sigma)}$$

where $\pi(\sigma)$ is the (stationary) probability of the walk at node (state) σ and τ_2 is the relaxation time of the walk.

Note 7.4 In the above bound, $\tau_2 = 1/\lambda_2$ where λ_2 is the second eigenvalue of the (symmetric) matrix $S = \{s_{i,j}\}$ where $s_{i,j} = \sqrt{\pi(i)}\, p_{i,j}\, (\sqrt{\pi(i)})^{-1}$ and $P = \{p_{i,j}\}$ is the transition matrix of the walk. Since S is symmetric, it is diagonalizable and the spectral theorem gives the following representation: $S = U \Lambda U^{-1}$ where U is orthonormal and Λ is a diagonal real matrix of diagonal entries $0 = \lambda_1 < \lambda_2 \leq \cdots \leq \lambda_{n'}$, $n' = n - k + 1$ (where $n' = |V_H|$). These λs are the eigenvalues of both P and S. In fact, λ_2 is an indication of the *expansion* of H and of the asymptotic rate of convergence to the stationary distribution, while relaxation time τ_2 is the corresponding interpretation measuring time.

It is a well-known fact (see, e.g., Reference 45) that $\forall v \in V_H$, $\pi(v) = d(v)/2m'$ where $m' = |E_H|$ is the number of the edges of H and $d(v)$ is the degree of v in H. Thus $\pi(\sigma) = d(\sigma)/2m'$. By estimating $d(\sigma)$ and m' and remarking that the operation of locally collapsing a graph does not reduce its expansion capability and hence $\lambda_2^H \geq \lambda_2^G$ (see lemma 1.15 [28, ch. 1, p. 13]).

Theorem 7.5

$$E(X) = E(Y) \leq \frac{1}{\lambda_2(G)} \Theta\left(\frac{n}{k}\right)$$

Theorem 7.6 *The expected communication time of our scheme is bounded above by the formula*

$$E(T_{\text{total}}) \leq \frac{2}{\lambda_2(G)} \Theta\left(\frac{n}{k}\right) + \Theta(k)$$

Note 7.5 The above upper bound is minimized when $k = \sqrt{2n/\lambda_2(G)}$, a fact also verified by our experiments (see Reference 23 and also Section 7.4.7).

We remark that this upper bound is tight in the sense that by Reference 5, Chapter 3,

$$E_\pi T_\sigma^H \geq \frac{(1 - \pi(\sigma))^2}{q_\sigma\, \pi(\sigma)},$$

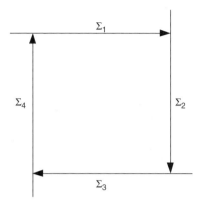

FIGURE 7.1   Deadlock situation when four "snakes" are about to be merged.

where $q_\sigma = \sum_{j \neq \sigma} p_{\sigma j}$ in H_j is the total exit rate from σ in H. By using martingale type arguments we can show sharp concentration around the mean degree of σ.

7.4.6.7 Robustness

Now, we examine the robustness of the support management subprotocol M_Σ^S and M_Σ^R under single stop-faults.

Theorem 7.7 *The support management subprotocol M_Σ^S is 1-fault tolerant.*

Proof. If a single host of Σ fails, then the next host in the support becomes the head of the rest of the snake. We thus have two, independent, random walks in G (of the two snakes), which, however, will meet in expected time at most m^* (as in Theorem 7.2) and re-organize as a single snake via a very simple re-organization protocol, which is the following: when the head of the second snake Σ_2 meets a host h of the first snake Σ_1 then the head of Σ_2 follows the host, which is "in front" of h in Σ_1, and all the part of Σ_1 after and including h waits, to follow Σ_2s tail.   ∎

   Note that in the case that more than one faults occur, the procedure for merging snakes described above may lead to deadlock (see Figure 7.1).

Theorem 7.8 *The support management subprotocol M_Σ^R is resilient to t faults, for any $0 \leq t < k$.*

Proof. This is achieved using redundancy: whenever two runners meet, they create copies of all messages in transit. In the worst-case, there are at most $k - t$ copies of each message. Note, however, that messages may have to be re-transmitted in the case that only one copy of them exists when some fault occurs. To overcome this limitation, the sender will continue to transmit a message for which delivery has not been confirmed, to any other member of the support encountered. This guarantees that more than one copy of the message will be present within the support structure.   ∎

7.4.7 Experimental Evaluation

To experimentally validate, fine-tune, and further investigate the proposed protocols, we performed a series of algorithmic engineering experiments. All of our implementations follow closely the support motion subprotocols described earlier. They have been implemented as C++ classes using several advanced data types of LEDA [44]. Each class is installed in an environment that allows to read graphs from files, and to perform a network simulation for a given number of rounds, a fixed number of mobile users and certain communication and mobility behaviors. After the execution of the simulation, the environment stores the results again on files so that the measurements can be represented in a graphical way.

7.4.7.1 Experimental Set-Up

A number of experiments were carried out modeling the different possible situations regarding the geographical area covered by an ad hoc mobile network. We considered three kinds of inputs, unstructured (random) and more structured ones. Each kind of input corresponds to a different type of motion graph. For all the cases that we investigated, each $h \notin \Sigma$, generates a new message with probability $p_h = 0.01$ (i.e., a new message is generated roughly every 100 rounds) by picking a random destination host. The selection of this value for p_h is based on the assumption that the mobile users do not execute a real-time application that requires continuous exchange of messages. Based on this choice of p_h, each experiment takes several thousands of rounds in order to get an acceptable average message delay. For each experiment, a total of 100,000 messages were transmitted. We carried out each experiment until the 100,000 messages were delivered to the designated receivers. Our test inputs are as follows:

Random Graphs. This class of graphs is a natural starting point in order to experiment on areas with obstacles. We used the $G_{n,p}$ model obtained by sampling the edges of a complete graph of n nodes independently with probability p. We used $p = 1.05 \log n/n$, which is marginally above the connectivity threshold for $G_{n,p}$. The test cases included random graphs with $n \in \{1,600, 3,200, 6,400, 40,000, 320,000\}$ over different values of $k \in [3, 45]$.

2D Grid Graphs. This class of graphs is the simplest model of motion one can consider (e.g., mobile hosts that move on a plane surface). We used two different $\sqrt{n} \times \sqrt{n}$ grid graphs with $n \in \{400, 1600\}$ over different values of k. For example, if $\mathcal{V}(tc) = 50 \text{ m}^2$ then the area covered by the ad hoc network will be 20,000 m^2 in the case where $n = 400$.

Bipartite multistage graphs. A bipartite multistage graph is a graph consisting of a number ξ of stages (or levels). Each stage contains n/ξ vertices and there are edges between vertices of consecutive stages. These edges are chosen randomly with some probability p among all possible edges between the two stages. This type of graphs is interesting as such graphs may model movements of hosts that have to pass through certain places or regions, and have a different second eigenvalue than grid and $G_{n,p}$ graphs (their second eigenvalue lies between that of grid and $G_{n,p}$ graphs). In our experiments, we considered $\xi = \log n$ stages and chose $p = (n/\xi) \log(n/\xi)$, which is the threshold value for bipartite connectivity (i.e., connectivity between each pair of stages). The test cases included multistage graphs with 7, 8, 9, and 10 stages, number of vertices $n \in \{1,600, 3,200, 6,400, 40,000\}$, and different values of $k \in [3, 45]$.

7.4.7.2 Validation of the SNAKE Protocol's Performance

In References 19 and 23 we performed extensive experiments to evaluate the performance of the "snake" protocol. A number of experiments were carried out modeling the different possible situations regarding the geographical area covered by an ad hoc mobile network. We only considered the *average case where all users (even those not in the support) perform independent and concurrent random walks.*

In the first set of experiments, we focused on the validation of the theoretical bounds provided for the performance of the "Snake" protocol. For each message generated, we calculated the delay X (in terms of rounds) until the message has been transmitted to a member of Σ. Additionally, for each message received by the support, we logged the delay Y (in terms of rounds) until the message was transmitted to the receiver. We assumed that no extra delay was imposed by the mobile support hosts (i.e., $T_{\Sigma} = 0$). We then used the average delay over all measured fractions of message delays for each experiment in order to compare the results and further discuss them.

The experimental results have been used to compare the average message delays ($E(T_{\text{total}})$) with the upper bound provided by the theoretical analysis of the protocol. We observe that as the total number n of motion-graph nodes remains constant, as we increase the size k of Σ, the total message delay (i.e., $E(T_{\text{total}})$) is decreased. Actually, $E(T_{\text{total}})$ initially decreases very fast with k, while having a limiting behavior of no further significant improvement when k crosses a certain threshold value (see, e.g., Figure 7.2). Therefore,

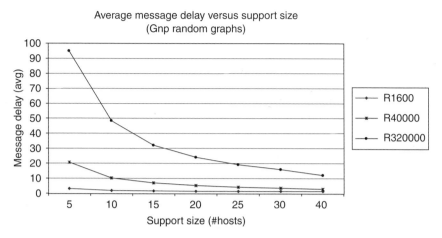

FIGURE 7.2 Average message delay for the "snake" approach over $G_{n,p}$ types of motion graphs with different sizes n and varied support size k.

taking into account a possible amount of statistical error, the following has been experimentally validated:

$$\text{if } k_1 > k_2 \;\Rightarrow\; E_1(T_{\text{total}}) < E_2(T_{\text{total}})$$

Throughout the experiments we used small and medium sized graphs, regarding the total number of nodes in the motion graph. Over the same type of graph (i.e., grid graphs) we observed that the average message delay $E(T_{\text{total}})$ increases as the total number of nodes increases. This can be clearly seen from Figure 7.3 where the curves of the four different graph-sizes are displayed. It is safe to assume that for the same graph type and for a fixed-size Σ the overall message delay increases with the size of the motion graph. We further observe that $E(T_{\text{total}})$ does not increase linearly but is affected by the expansion rate of the graph, or otherwise, as stated in the theoretical analysis, by the second eigenvalue of the matrix of the graph. If we take into account possible amount of statistical error that our experiments are inherently prone to, we can clearly conclude the following:

$$\text{if } n_1 > n_2 \;\Rightarrow\; E_1(T_{\text{total}}) > E_2(T_{\text{total}})$$

Furthermore, we remark that $E(T_{\text{total}})$ does not only depend on the actual size of the graph and the size of the support Σ, but it is also directly affected by the type of the graph. This is expressed throughout the theoretical analysis by the effect of the second eigenvalue of the matrix of the graph on communication times. If we combine the experimental results of Figures 7.2–7.4 for fixed number of motion-graph nodes (e.g., $n = 400$) and fixed number of Σ size (e.g., $k = 40$) we observe the following:

$$E_{\text{grid}}(T_{\text{total}}) > E_{\text{multistage}}(T_{\text{total}}) > E_{G_{n,p}}(T_{\text{total}})$$

which validates the corresponding results due to the analysis.

In Section 7.4.6.6, we note that the upper bound on communication times $E(T_{\text{total}}) \leq (2/\lambda_2(G))\Theta(n/k) + \Theta(k)$ is minimized when $k = \sqrt{2n/\lambda_2(G)}$. Although our experiments disregard the delay imposed by Σ (i.e., we take $\Theta(k) \to 0$) this does not significantly affect our measurements. Interestingly, the values of Figures 7.2–7.4 imply that the analytical upper bound is not very tight. In particular, for the $G_{n,p}$ graphs we clearly observe that the average message delay drops below 3 rounds (which is a very low, almost negligible, delay) if Σ size is equal to 10; however, the above formula implies that this size should be higher, especially for the medium sized graphs. Actually, it is further observed that there exists a certain threshold value (as also implied by the theoretical analysis) for the size of Σ above,

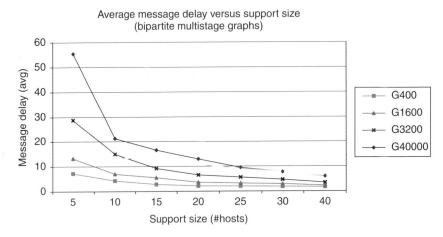

FIGURE 7.3   Average message delay of the "Snake" approach over bipartite multistage types of motion graphs with different sizes (n) and varied support size k.

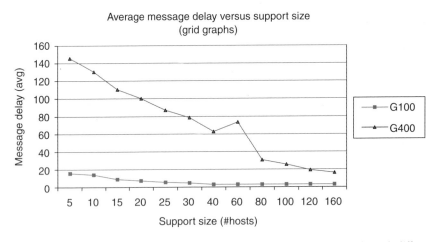

FIGURE 7.4   Average message delay of "snake" approach over grid types of motion graphs with different sizes (n) and varied support size k.

which the overall message delay does not further significantly improve. The threshold size for the support validated experimentally is smaller than the threshold implied by the analytical bound.

7.4.7.3   Experimental Comparison of the SNAKE and the RUNNERS Protocols

In Reference 24 we performed a comparative experimental study of the "snake" and the "runners" protocols. Our test inputs were similar to that used in Section 7.4.7.2. We measured the total delay of messages exchanged between pairs of sender–receiver users. For each motion graph constructed, we injected 1,000 users (mobile hosts) at random positions that generated 100 transaction message exchanges of one packet each by randomly picking different destinations. For each message generated, we calculated the overall delay (in terms of simulation rounds) until the message was finally transmitted to the receiver. We used these measurements to experimentally evaluate the performance of the two different support management subprotocols.

The reported experiments for the three different test inputs we considered are illustrated in Figures 7.5–7.7. In these figures, we have included the results of two instances of the input graph w.r.t. n, namely, a smaller and a larger value of n (similar results hold for other values of n). Each curve in the

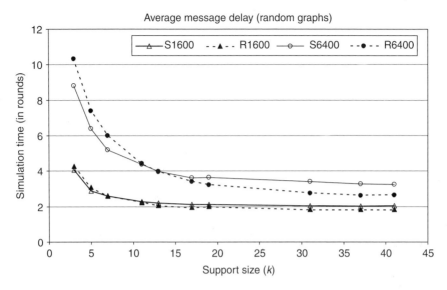

FIGURE 7.5 Average message delay over random graphs with different sizes (n) and varied support size k, for the snake (S) protocol and runners (R) protocol.

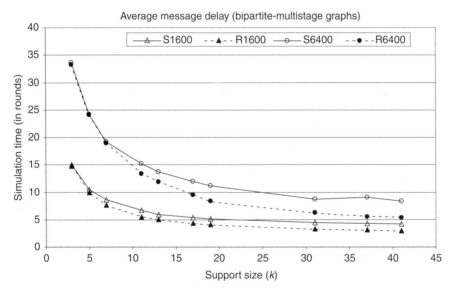

FIGURE 7.6 Average message delay over bipartite multistage motion graphs with different sizes (n) and varied support size k, for the snake (S) protocol and runners (R) protocol.

figures is characterized by the name "Px," where P refers to the protocol used (S for snake and R for runners) and x is a 3 or 4 digit number denoting the value of n.

The curves reported for the snake protocol confirm the theoretical analysis in Section 7.4.6.6. That is, the average message delay drops rather quickly when k is small, but after some threshold value stabilizes and becomes almost independent of the value of k. This observation applies to all test inputs we considered. We also note that this behavior of the snake protocol is similar to the one reported in Section 7.4.7.2, although we count here the extra delay time imposed by the synchronization subprotocol P_3.

Regarding the "runners" protocol, we first observe that its curve follows the same pattern with that of the snake protocol. Unfortunately, we do not have yet a theoretical analysis for the new protocol to see

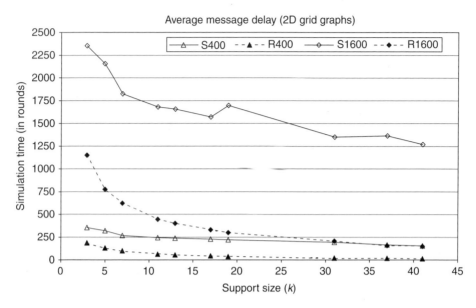

FIGURE 7.7 Average message delay over grid motion graphs with different sizes (n) and varied support size k, for the snake (S) protocol and runners (R) protocol.

whether it behaves as should expected, but from the experimental evidence we suspect that it obeys a similar but tighter bound than that of the snake protocol. Our second observation is that the performance of the "runners" protocol is slightly better than that of the "snake" protocol in random graphs (except for the case of random graphs with small support size; see Figure 7.5) and bipartite multistage graphs (Figure 7.6), but is substantially better in the more structured case (grid graphs; see Figure 7.7). This could be partly explained by the fact that the structured graphs have a smaller second eigenvalue than that of bipartite multistage and random graphs. A small value for this eigenvalue makes the average message delay to be more dependent on the hidden constant in the asymptotic analysis of the "snake" protocol (see Section 7.4.6.6), which is apparently larger than the corresponding constant of the runners protocol.

7.5   An Algorithm for Reconfiguration of Robot Chains

7.5.1   Introduction

The aspect of motion planning in robotic systems has become recently a topic of significant research and development activity. We study here a particular motion planning problem, that of reconfiguration. Roughly speaking, the reconfiguration problem for a robotic system comprised of several modules (robots) consists in reshaping the system by motions of the robots that change their relative positions and thus allow the system to go from an initial configuration to a desired goal configuration.

A reason for the increasing interest in the reconfiguration problem of robotic systems is its wide range of important applications. To name a few characteristic examples, reshaping of robotic systems may allow such systems to perform specific shape-dependent tasks (e.g., "bridge-like" shapes may be useful in crossing rivers, adjustable buttress shapes may be used in supporting collapsing buildings, etc.). Also, the operation of systems of robots in areas where human presence is either impossible or dangerous (such as in disaster places, or deep undersea areas and space missions) necessitates the ability of such systems to self-organize themselves into (possibly changing) formations and shapes. Furthermore, the reshaping feature significantly increases the fault-tolerance of robotic systems since it allows them to better adapt to external (possibly adversarial) conditions or to dynamically coordinate responsibilities in case of failure of certain robots in the system.

Such robotic systems, able to change their shape by individual robot motions, are called self-configurable or metamorphic. More specifically, a self-configurable robotic system is a collection of independently controlled, mobile modules (robots), each of which can connect, disconnect, and move around adjacent modules. A metamorphic robotic system is a special case of self-configurable systems, where additionally all modules (robots) are identical (in structure, motion, and computing capabilities). For achieving shape formation, the modules should be able to pack together as densely as possible (i.e., by minimizing gaps between modules). A module shape allowing their dense packing with each other, is one having both regularity and symmetry. A typical example of such module structure (studied here as well) is that of regular hexagons of equal size.

Two major, different approaches in the design and operation of metamorphic systems use (a) centralized motion planning algorithms and (b) distributed algorithms. Most of the existing strategies are centralized. The fundamental work of Reference 52 that we present in this section is in fact the first distributed approach to reconfiguration in metamorphic robotic systems. We wish to note the following three basic advantages of a distributed approach over a centralized one: (a) the motion decisions are local, thus such algorithms can operate under lack of knowledge of the global system structure and are potentially more efficient, (b) the algorithm's fault-tolerance is greater compared to the centralized case, where failure of the central module is fatal, (c) distributed algorithms can operate under weaker modeling assumptions, such as that of lack of communication between modules.

For another problem of motion coordination (in particular, cycle formation) of systems of robots see Reference 17. We study there a system of anonymous, oblivious robots, which running an efficient distributed algorithm, place themselves in equal distances on the boundary of a cycle.

7.5.2 The Model

The modeling assumptions in Reference 52 include basically the coordination system as well as the hardware and motion characteristics of the robots.

7.5.2.1 The Coordinate System

The two-dimensional (plane) area is partitioned into regular hexagonal cells of equal size. The coordinate system exemplified in Figure 7.8 is used for labeling the cells.

A key concept for motion planning by the reconfiguration algorithm, is that of the "lattice distance" in the cell structure. Given the coordinates of two cells, $c_1 = (x_1, y_1)$ and $c_2 = (x_2, y_2)$, the lattice distance, LD, between them is defined as follows: Let $\Delta x = x_1 - x_2$ and $\Delta y = y_1 - y_2$. If $\Delta x \cdot \Delta y < 0$, then $\text{LD}(c_1, c_2) = \max\{|\Delta x|, |\Delta y|\}$. Otherwise, $\text{LD}(c_1, c_2) = |\Delta x| + |\Delta y|$. The lattice distance of cells in the plane area is very important, since it is equal to the minimum number of cells a robot must move through to go from the one cell to the other. As an example, consider cells $c_1 = (0, 2)$ and $c_2 = (-1, 1)$. It is $\Delta x = 1, \Delta y = 1$, so $\text{LD}(c_1, c_2) = |\Delta x| + |\Delta y| = 2$. Consider now c_1 and a third cell $c_3 = (1, -1)$. It is $\Delta x = -1, \Delta y = 3$, so $\text{LD}(c_1, c_3) = \max\{|\Delta x|, |\Delta y|\} = 3$.

7.5.2.2 The Modules (Robots)

The model in Reference 52 includes the following assumptions about the modules (imposed by hardware constraints and application needs):

- Modules are identical in terms of computing capability.
- Each module runs the same program.
- Each module is a regular hexagon of the same size as the cells of the plane area. At any point in time each module occupies exactly one of the cells.
- Each module knows at all times the following: (a) its location (i.e., the x, y coordinates of the cell that it currently resides on), (b) its orientation (which edge it is facing in which direction), and

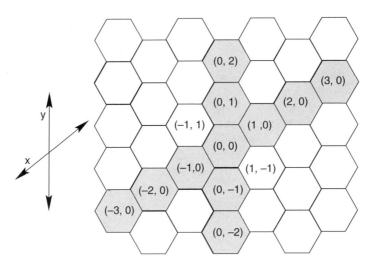

FIGURE 7.8 The coordinate system.

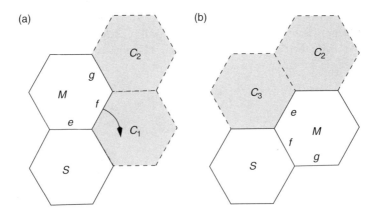

FIGURE 7.9 Example of movement.

(c) which of its neighboring cells is occupied by another module (the latter piece of information is assumed to be provided by an appropriate link layer protocol).

Furthermore, Walter *et al.* [52] makes the following assumptions about the possible motion of the robots:

- Modules move in synchronous rounds.
- A module M is capable of moving to an adjacent cell, C_1, if and only if the following conditions hold: (a) cell C_1 is empty, (b) module M has a neighbor S that does not move in the current round (called the substrate) and S is also adjacent to cell C_1, and (c) the neighboring cell to M on the other side of C_1 from S, C_2, is empty. See Figure 7.9 for an example. Note that the algorithm must ensure that each module trying to move has a substrate that is not moving, otherwise the algorithm may engage in unpredictable behavior.
- Only one module tries to move into a particular cell in each round. Note that if the algorithm does not ensure collision freedom, then unpredictable behavior may emerge.

7.5.2.3   The Problem

The problem consists in designing a distributed algorithm running locally at each module, putting the modules to move from an initial configuration, I, to a (known to all modules) goal configuration, G. We here in fact deal with a special case of the problem where the initial and the goal configurations are straight-line sets of modules that overlap in exactly one cell. See Figure 7.11 for an example of a reconfiguration.

7.5.2.4   The Algorithm

In this section we present the algorithm of Walter et al. [52] only for the case where the initial configuration and the goal configuration are collinear. The interested reader can find in their paper an extension of this algorithm to the noncollinear case (i.e., when the goal configuration is allowed to intersect the initial one in any orientation).

Walter *et al.* [52] categorizes modules based on their possible types of edge contacts during an execution of the algorithm. In particular, modules are classified into three categories (trapped, free, and other) according to the number and orientation of their contact and noncontact edges. Noncontact edges are those edges where the module is adjacent to an empty cell. Contact edges connect a module to an occupied adjacent cell. In the graphical presentation in Figure 7.10, there are vertical line segments toward contact edges. We note that the classification above applies to any rotation of a module.

Trapped modules cannot move because of the number and orientation of their contact edges (this represents hardware constraints on movement, explained in Section 7.5.2.2). Free modules are required to move in the algorithm. The other category includes modules whose movement is restricted by the algorithm, even though their movement is not constrained by the number and orientation of contact edges.

The algorithm in Reference 52 operates under the following assumptions:

a. Each module knows the total number n of modules in the system. Modules also know the goal configuration G.
b. Initially, exactly one module occupies each cell of the initial configuration I.
c. Configurations I and G are collinear, while they overlap in exactly one cell.

The algorithm works in synchronous rounds. Each module moves (throughout the execution) either clockwise (CW) or anticlockwise (CCW). This direction of movement for each module is decided by the algorithm in a way that creates alternate direction for adjacent modules. This is achieved by using the parity of each module's distance from the (single) overlap cell.

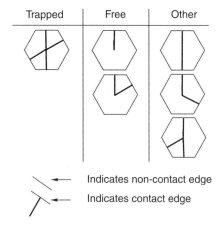

FIGURE 7.10   Module types.

In each round, each module calculates whether it is free, and if free it moves in the direction (CW or CCW) calculated initially.

We below provide the pseudo-code description (as given in Reference 52) of the reconfiguration algorithm. Before, we present some data structures kept at each module and used in the algorithm:

- The boolean array *contacts* show which edges have neighboring modules. (It is assumed that this information is accordingly updated at each round by some underlying low layer protocol.)
- The coordinates of a module according to the coordinate system explained earlier are kept in data structure *coord*.
- The array *goalCell* contains all coordinates of cells in the goal configuration in decreasing order of the y coordinate.
- Variable d keeps the direction of movement, CW or CCW.
- *flips* is a counter used to represent whether a module is free or not.

The pseudo-code of the reconfiguration algorithm follows:
Each module $\notin G$ executes:

Initially:
1. if $((n - LD(myCoord, goalCell[1]))$ is even)
2. then $d = CCW$
3. else $d = CW$

In round $r = 1, 2, \ldots$:
4. if $(isFree())$
5. then move d

Procedure *isFree*():
6. *flips* $= 0$
7. for $(i = 0$ to $5)$ do
8. if $(contacts[i] \neq contacts[(i + 1)\%6])$
9. then *flips* $+ +$
10. return $(flips == 2)$

Note that modules in goal cells are not allowed to move. In this sense the above algorithm is called a "goal-stopping" algorithm.

Figure 7.11 presents an execution of the above reconfiguration algorithm when the number of robots in the system is $n = 4$. In this figure, occupied cells have solid borders and goal cells are shaded. For purposes of analysis, modules are labeled 1 through 4. Remark that nine rounds are required for this reconfiguration.

7.5.2.5 Analysis

We here present sketches of the correctness and performance proofs of Reference 52. Assume, without loss of generality, that configurations I and G are directed from north to south and also that I is north of G. For reference clarity in the following proofs, number the cells in I and G from 1 through $2n - 1$ from north to south. We will refer to the module occupying at the beginning of the algorithm cell i as module i, where $1 \leq i \leq n$. We will refer to a cell's neighboring cells as north (N), northeast (NE), southeast (SE), south (S), southwest (SW), and northwest (NW).

Theorem 7.9 [52] *The reconfiguration algorithm is correct.*

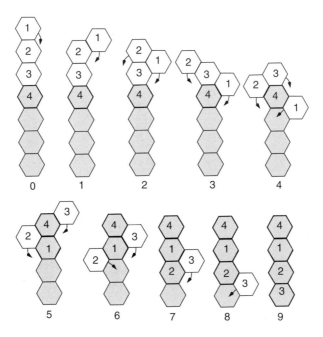

FIGURE 7.11 Reconfiguration example.

Proof (sketch following [52]). We show that the following properties $P1$–$P3$ are invariant throughout the execution. For each i, $1 \leq i \leq n$, and each round $S \geq 1$, at the end of round S, it is:

- $P1$: If $S < 2i - 1$, then module i is in cell i.
- $P2$: If $S = 2i - 1 + j$, $0 \leq j \leq n - 1$, then:
    a. If i is odd, then module i is SE of cell $i + j$,
    b. If i is even, then module i is SW of cell $i + j$,
- $P3$: If $S \geq 2i - 1 + n$, then module i is in cell $i + n$.

The proof is by induction on the number of rounds, S, in the execution. The basis ($S = 0$) is easy since initially $P2$ and $P3$ are not applicable, and $P1$ is clearly true by assumption. For the inductive hypothesis, assume that the invariants hold for round $S - 1$, $S > 0$. Figure 7.12 shows the configuration at the end of round $2k - 1$. In this figure, occupied cells have solid borders and goal cells are shaded. Without loss of generality, we take i to be even.

Case 1: $S < 2i - 1$. Thus, $S - 1 < 2i - 2$. By the inductive hypothesis and $P1$, module $i - 1$ occupies cell $i - 1$ and module $i + 1$ is in cell $i + 1$ at the end of round $S - 1$. Therefore, module i does not move in round S because it will not be free in this round. Referring to Figure 7.12, module $k + 2$ has contacts that correspond to those described for module i. By property $P1$, module i is in cell i at the end of round $S - 1$, thus (since it does not move in next round) it remains in cell i at the end of round S.

Case 2: $S = 2i - 1 + j$ for some j, $0 \leq j \leq n - 1$. Thus, $S - 1 = 2i - 1 + (j - 1) = 2(i - 1) + j + 1$. If $j = 0$, then by the inductive hypothesis $P1$ implies that module i is in cell i at the end of round $S - 1$ and because of $P2$ module $i - 1$ is SE of cell i at the end of round $S - 1$, since $S - 1 = 2(i - 1) + 0 + 1$. Thus module i is free at the end of round $S - 1$. So module i moves anticlockwise, and gets SW of cell i in round S. In Figure 7.12, module k corresponds to the position described for module i and module $k - 1$ corresponds to the position described for module $i - 1$ in round S. If $j > 0$, then property $P1$ implies that module i is in the cell SW of cell $i + (j - 1)$ at the end of round $S - 1$. Module i will be free at the end of round $S - 1$ because i will have contacts only on its NE and SE edges. Therefore, module i will

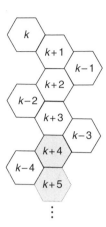

FIGURE 7.12 Configuration at the end of round $2k - 1$, where k is even.

move CCW in round S to be SW of cell j. In Figure 7.12, module $k - 2$ is like module i at the end of round $S - 1$.

Case 3: $S \geq 2i - 1 + n$. Then for the equality case it is $S - 1 = 2i - 2 + n$, thus by $P3$ module $i - 1$ is in cell $i + n - 1$ in round $S - 1$, so module $i - 1$ will not move in round $S - 1$ or later. By $P2$, module i is in the cell SW of cell $i + n - 1$ at the end of round $S - 1$. Therefore, module i is free at the end of round $S - 1$ because it has only one contact. Module i has no other neighbors in round $S - 1$. So in round S, module i moves to cell $i + n$ and stops moving in this round by the code. In Figure 7.12, module $k - 4$ is in a position like that described for module i, and module $k + 4$ is like that $i - 1$ at the end of round $S - 1$. If $S > 2i - 1 + n$, then, since at the end of round S module i is already in cell $i + n$ by $P3$, then it is still in cell $i + n$ in every round following S. By the code, once a module is in a goal position, it does not move further. ∎

The above invariants imply that the modules only use three columns of the cell structure. At the initial configuration, modules are all in the center column and, during the execution, modules in outer columns are spaced or staggered such that there is an empty cell between any two modules. Invariant $P3$ implies that, after round $3(n - 1)$, all modules are in goal positions.

As far as the performance of the algorithms is concerned, Walter *et al.* [52] prove the following:

Theorem 7.10 [52] *The reconfiguration algorithm takes $3(n - 1)$ rounds and makes $n^2 - 1$ module movements.*

Proof. Clearly, invariants $P1$ through $P3$, imply that the algorithm takes $3(n - 1)$ rounds. Invariants $P2$ and $P3$ show that $n - 1$ (all but the overlapping one) of the modules in the initial configuration make $n + 1$ moves each, thus we get a total of $n^2 - 1$ module movements. ∎

By estimating lower bounds, the authors in Reference 56 prove that their algorithm is asymptotically optimal, both with respect to number of moves and rounds (time).

7.6 Conclusions and Future Research

In this chapter we focused on some characteristic algorithms for three fundamental problems in ad hoc mobile networks. The material presented is far from being exhaustive, and also there is a rich research activity going on.

In particular, we intend to work on validating the "high mobility rate impossibility conjecture" by also using analytic means. Also, to study hybrid words of ad hoc mobile networks and other important network environments (such as peer-to-peer [51], wireless sensor networks, and fixed network topologies). Such hybrid combinations may be very helpful in achieving goals like fault-tolerance, scalability, etc.

Acknowledgment

This work has been partially supported by the IST Programme of the European Union under contract number IST-2001-33116 (FLAGS) and by the 6th Framework Programme under contract number 001907 (DELIS).

References

[1] M. Adler, H. Racke, C. Sohler, N. Sivadasan, and B. Vocking: Randomized Pursuit-Evasion in Graphs. In *Proceedings of the 29th ICALP*, pp. 901–912, 2002.

[2] M. Adler and C. Scheideler: Efficient Communication Strategies for Ad Hoc Wireless Networks. In *Proceedings of the 10th Annual Symposium on Parallel Algorithms and Architectures* — SPAA'98, 1998.

[3] M. Ajtai, J. Komlos, and E. Szemeredi: Deterministic Simulation in LOGSPACE. In *Proceedings of the 19th Annual ACM Symposium on Theory of Computing* — STOC'87, 1987.

[4] D. Aldous and P. Diaconis: Strong Stationary Times and Finite Random Walks. *Adv. Appl. Math.*, 8: 69–97, 1987.

[5] D. Aldous and J. Fill: Reversible Markov Chains and Random Walks on Graphs. Unpublished manuscript. http://stat-www.berkeley.edu/users/aldous/book.html, 1999.

[6] N. Alon and F.R.K. Chung: Explicit Construction of Linear Sized Tolerant Networks. *Discrete Math.*, 72: 15–19, 1998.

[7] H. Attiya and J. Welch: *Distributed Computing: Fundamentals, Simulations and Advanced Topics*. McGraw-Hill, New York, 1998.

[8] A. Boukerche and S. Rogers: GPS Query Optimization in Mobile and Wireless Ad Hoc Networks. In *Proceedings of the 6th IEEE Symposium on Computers and Communications*, pp. 198–203, July 2001.

[9] A. Boukerche: Special Issue on Wireless and Mobile Ad Hoc Networking and Computing. *Int. J. Parallel and Distributed Comput.*, 60: 2003.

[10] A. Boukerche and S. Vaidya: A Performance Evaluation of a Dynamic Source Routing Discovery Optimization Using GPS Systems. *Telecommun. Syst.*, 22: 337–354, 2003.

[11] G. Brightwell and P. Winkler: Maximum Hitting Times for Random Walks on Graphs. *J. Random Struct. Algorithms*, 1: 263–276, 1990.

[12] J. Broch, D.B. Johnson, and D.A. Maltz: The Dynamic Source Routing Protocol for Mobile Ad Hoc Networks. IETF, Internet Draft, draft-ietf-manet-dsr-01.txt, December 1998.

[13] A. Boukerche: A Source Route Discovery Optimization Scheme in Mobile Ad Hoc Networks. In *Proceedings of the 8th International Conference on Parallel and Distributed Computing* — EUROPAR'02, 2002.

[14] A. Boukerche and S. Nikoletseas: Algorithmic Design for Communication in Mobile Adhoc Networks. In *Performance Tools and Applications to Networked Systems*, M.C. Calzarossa and Erol Gelenbe (Eds), Lecture Notes in Computer Science (LNCS), Vol. 2965, Spring, Berlin, pp. 235–254, 2004.

[15] C. Busch, S. Surapaneni, and S. Tirthapura: Analysis of Link Reversal Routing Algorithms for Mobile Ad Hoc Networks. In *Proceedings of the 15th ACM Symposium on Parallelism in Algorithms and Architectures (SPAA 2003)*, pp. 210–219, 2003.

[16] I. Chatzigiannakis, E. Kaltsa, and S. Nikoletseas: On the Effect of User Mobility and Density on the Performance of Routing Protocols for Ad Hoc Mobile Networks. In *Wireless Commun. Mobile Comput. J.*, 4: 609–621, 2004. Also, in *Proceedings of the 12th IEEE International Conference on Networks (ICC)*, pp. 336–341, 2004.

[17] I. Chatzigiannakis, M. Markou, and S. Nikoletseas: Distributed Circle Formation for Anonymous Oblivious Robots. In *Proceedings of the 3rd Workshop on Efficient and Experimental Algorithms (WEA 2003)*, pp. 159–174, Springer Verlag, LNCS. Also, to appear in *Int. J. Wireless Mobile Comput.*, Special Issue on the 2nd International Technology Symposium (I2TS), Inderscience Publishers, 2005.

[18] I. Chatzigiannakis, S. Nikoletseas, and P. Spirakis: *Distributed Communication Algorithms for Ad-Hoc Mobile Networks*, in "Current Trends in Theoretical Computer Science, The Challenge of the New Century, Vol 1: Algorithms and Complexity," G. Paun and Rovira I. Virgili (Eds), World Scientific, pp. 337–372, 2004, ISBN 981-238-966-0.

[19] I. Chatzigiannakis, S. Nikoletseas, and P. Spirakis: Distributed Communication and Control Algorithms for Adhoc Mobile Networks. *J. Parallel Distributed Comput.* (Special Issue on Mobile Ad hoc Networking and Computing) 63: 58–74, 2003.

[20] I. Chatzigiannakis and S. Nikoletseas: Design and Analysis of an Efficient Communication Strategy for Hierarchical and Highly Changing Ad Hoc Mobile Networks. *Mobile Networks Appl. J.*, 9: 319–332, 2004.

[21] I. Chatzigiannakis, S. Nikoletseas, and P. Spirakis: An Efficient Communication Strategy for Ad Hoc Mobile Networks. In *Proceedings of the 15th International Symposium on Distributed Computing —* DISC'01. Lecture Notes in Computer Science, Vol. 2180, Springer-Verlag, Berlin, pp. 285–299, 2001. Also Brief Announcement in *Proceedings of the 20th Annual Symposium on Principles of Distributed Computing — PODC'01*, pp. 320–322, ACM Press, 2001.

[22] I. Chatzigiannakis, S. Nikoletseas, and P. Spirakis: On the Average and Worst-case Efficiency of Some New Distributed Communication and Control Algorithms for Ad Hoc Mobile Networks. Invited Paper in *Proceedings of the 1st ACM International Workshop on Principles of Mobile Computing (POMC'01)*, pp. 1–16, 2001.

[23] I. Chatzigiannakis, S. Nikoletseas, and P. Spirakis: Analysis and Experimental Evaluation of an Innovative and Efficient Routing Approach for Ad Hoc Mobile Networks. In *Proceedings of the 4th Annual Workshop on Algorithmic Engineering — WAE'00*, 2000. Lecture Notes in Computer Science, Volume 1982, Springer-Verlag, Berlin, 2000.

[24] I. Chatzigiannakis, S. Nikoletseas, and P. Spirakis: Experimental Evaluation of Basic Communication for Ad Hoc Mobile Networks. In *Proceedings of the 5th Annual Workshop on Algorithmic Engineering — WAE'01*, pp. 159–171, 2001. Lecture Notes in Computer Science, Vol. 2141, Springer-Verlag, Berlin, 2001.

[25] I. Chatzigiannakis, S. Nikoletseas, and P. Spirakis: An Efficient Routing Protocol for Hierarchical Ad Hoc Mobile Networks. In *Proceedings of the 1st International Workshop on Parallel and Distributed Computing Issues in Wireless Networks and Mobile Computing*, Satellite Workshop of *15th Annual International Parallel & Distributed Processing Symposium — IPDPS'01*, 2001.

[26] I. Chatzigiannakis and S. Nikoletseas: An Adaptive Compulsory Protocol for Basic Communication in Highly Changing Ad Hoc Mobile Networks. In *Proceedings of the 2nd International Workshop on Parallel and Distributed Computing Issues in Wireless Networks and Mobile Computing*, Satellite Workshop of *16th Annual International Parallel & Distributed Processing Symposium — IPDPS'02*, 2002.

[27] G. Chirikjian: Kinematics of a metamorphic robotic system. In *Proceedings of the IEEE International Conference on Robotics and Automation*, pp. 449–455, 1994.

[28] F.R.K. Chung: Spectral Graph Theory. Regional Conference Series in Mathematics, ISSN 0160-7642; no. 92. CBMS Conference on Recent Advances in Spectral Graph Theory, California State University at Fesno, 1994, ISBN 0-8218-0315-8.

[29] D. Coppersmith, P. Tetali, and P. Winkler: Collisions among Random Walks on a Graph. *SIAM J. Disc. Math.*, 6: 363–374, 1993.

[30] J. Diaz, J. Petit, and M. Serna: A Random Graph Model for Optical Networks of Sensors. *IEEE Trans. Mobile Comput.*, 2: 186–196, 2003.

[31] T. Dimitriou, S. Nikoletseas, and P. Spirakis: *Analysis of the Information Propagation Time Among Mobile Hosts*. In *Proceedings of the 3rd International Conference on AD HOC Networks & Wireless Networks (ADHOC-NOW)*, Lecture Notes in Computer Science (LNCS), Springer-Verlag, Berlin, pp. 122–134, 2004.

[32] G. Dommety and R. Jain: Potential Networking Applications of Global Positioning Systems (GPS). Technical Report TR-24, CS Dept., The Ohio State University, April 1996.

[33] R. Durret: *Probability: Theory and Examples*. Wadsworth, Belmont, CA, 1991.

[34] E. Gafni and D. Bertsekas: Distributed Algorithms for Generating Loop-Free Routes in Networks with Frequently Changing Topology. *IEEE Trans. Commun.*, C-29: 11–18, 1981.

[35] Z.J. Haas and M.R. Pearlman: The Performance of a New Routing Protocol for the Reconfigurable Wireless Networks. In *Proceedings of the IEEE International Conference on Communications — ICC'98*, 1998.

[36] K.P. Hatzis, G.P. Pentaris, P.G. Spirakis, V.T. Tampakas, and R.B. Tan: Fundamental Control Algorithms in Mobile Networks. In *Proceedings of the 11th Annual ACM Symposium on Parallel Algorithms and Architectures — SPAA'99*, 1999.

[37] G. Holland and N. Vaidya: Analysis of TCP Performance over Mobile Ad Hoc Networks. In *Proceedings of the 5th Annual ACM/IEEE International Conference on Mobile Computing — MOBICOM'99*, 1999.

[38] T. Imielinski and H.F. Korth: *Mobile Computing*. Kluwer Academic Publishers, Dordrecht, 1996.

[39] Y. Ko and N.H. Vaidya: Location-Aided Routing (LAR) in Mobile Ad Hoc Networks. In *Proceedings of the 4th Annual ACM/IEEE International Conference on Mobile Computing — MOBICOM'98*, 1998.

[40] Q. Li and D. Rus: Sending Messages to Mobile Users in Disconnected Ad Hoc Wireless Networks. In *Proceedings of the 6th Annual ACM/IEEE International Conference on Mobile Computing — MOBICOM'00*, 2000.

[41] N. Malpani, N. Vaidya, and J.L. Welch: Distributed Token Circulation on Mobile Ad Hoc Networks. In *Proceedings of the 9th International Conference on Network Protocols (ICNP)*, 2001.

[42] N. Malpani, J.L. Welch, and N. Vaidya: Leader Election Algorithms for Mobile Ad Hoc Networks. In *Proceedings of the 4th International Workshop on Discrete Algorithms and Methods for Mobile Computing and Communications (DIAL M for Mobility)*, pp. 96–103, 2000.

[43] N.A. Lynch: *Distributed Algorithms*. Morgan Kaufmann, San Francisco, CA, 1996.

[44] K. Mehlhorn and S. Näher: *LEDA: A Platform for Combinatorial and Geometric Computing*. Cambridge University Press, Cambridge, UK, 1999.

[45] R. Motwani and P. Raghavan: *Randomized Algorithms*. Cambridge University Press, Cambridge, UK, 1995.

[46] V.D. Park and M.S. Corson: Temporally-Ordered Routing Algorithms (TORA) Version 1 Functional Specification. IETF, Internet Draft, draft-ietf-manet-tora-spec-02.txt, October, 1999.

[47] C.E. Perkins: *Ad Hoc Networking*. Addison-Wesley, Boston, USA, January 2001.

[48] C.E. Perkins and E.M. Royer: Ad Hoc on Demand Distance Vector (AODV) Routing. IETF, Internet Draft, draft-ietf-manet-aodv-04.txt, 1999.

[49] S. Ross: *Stochastic Processes*. John Wiley & Sons, New York, 1996.

[50] E. Szemeredi: *Regular Partitions of Graphs*. Colloques Internationaux C.N.R.S., Nr 260, Problems Combinatoires et Theorie des Graphes, Orsay, pp. 399–401, 1976.

[51] P. Triantafillou, N. Ntarmos, S. Nikoletseas, and P. Spirakis: NanoPeer Networks and P2P Worlds. In *Proceedings of the 3rd IEEE International Conference on Peer-to-Peer Computing — P2P 2003*, pp. 40–46, 2003.

[52] J.E. Walter, J.L. Welch, and N.M. Amato: Distributed Reconfiguration of Metamorphic Robot Chains. In *Proceedings of the 19th ACM Annual Symposium on Principles of Distributed Computing — PODC'00*, 2000.

[53] J. Walter, J. Welch, and N. Vaidya: A Mutual Exclusion Algorithm for Ad Hoc Mobile Networks. Accepted to *Wireless Networks*. Also in *Proceedings of the 2nd International Workshop on Discrete Algorithms and Methods for Mobile Computing and Communications (DIAL M for Mobility)*, 1998.

[54] Y. Zhang and W. Lee: Intrusion Detection in Wireless Ad Hoc Networks. In *Proceedings of the 6th Annual ACM/IEEE International Conference on Mobile Computing — MOBICOM'00*, 2000.

[55] Iowa State University GPS page. http://www.cnde.iastate.edu/ gps.html.

[56] NAVSTAR GPS operations. http://tycho.usno.navy.mil/gpsinfo.html.

8

Routing and Traversal via Location Awareness in Ad Hoc Networks

8.1 Challenges in Ad Hoc Networking **8**-165
8.2 Modeling Ad Hoc Networks **8**-166
    Communication and Information Dissemination
8.3 Ad Hoc Communication Infrastructure **8**-167
    Geometric Spanners • Tests for Preprocessing the Ad Hoc
    Network • Hop Spanners
8.4 Information Dissemination **8**-173
    Routing in Undirected Planar Graphs • Routing in Directed
    Planar Graphs • Traversal of Nonplanar Graphs
8.5 Conclusion **8**-181
Acknowledgments... **8**-181
References **8**-181

Evangelos Kranakis
Ladislav Stacho

8.1 Challenges in Ad Hoc Networking

The current rapid growth in the spread of wireless devices of ever increasing miniaturization and computing power has greatly influenced the development of ad hoc networking. Ad hoc networks are wireless, self-organizing systems formed by cooperating nodes within communication range of each other that form temporary networks with a dynamic decentralized topology. It is desired to make a variety of services available (e.g., Internet, GPS, service discovery) in such environments and our expectation is for a seamless and ubiquitous integration of the new wireless devices with the existing wired communication infrastructure. At the same time we anticipate the development of new wireless services that will provide solutions to a variety of communication needs (e.g., sensing and reporting an event, integrating hosts in a temporary network) among hosts without the interference of an existing infrastructure.

To comprehend the magnitude of our task and the challenges that need to be addressed it is important to understand that currently hosts of an ad hoc network are energy limited and must operate in a bandwidth limited communication environment. Providing for efficient information dissemination in such an ad hoc network is a significant challenge. The present chapter is a survey with a threefold goal. First, to understand and propose a reasonable abstract model of communication that integrates well with the expected "location awareness" of the hosts, second to explain how by preprocessing the underlying wireless network we can simplify the communication infrastructure, and third propose solutions for achieving efficient information dissemination.

A brief outline of the chapter is as follows. After discussing various abstract models in Section 8.2 we address how neighborhood proximity can be used to produce geometric and hop spanners in Section 8.3. Finally, in Section 8.4 we outline several algorithms for route discovery and traversal in undirected and directed ad hoc networks.

8.2 Modeling Ad Hoc Networks

An ad hoc network can be modeled as a set of points, representing network nodes (hosts), in two- or three-dimensional Euclidean space $\mathbb{R}^2$ or $\mathbb{R}^3$, with a relation among the nodes representing communication links (or wireless radio channels). One node is linked with another if the signal of the former can reach the latter. Hosts have limited computational and communication power. The computational power corresponds to the level of coding the host can accomplish as well as the amount of local memory accessible to the host. The communication power reflects how far a signal can reach and is determined by propagation characteristics of the radio channel in the environment as well as the power control capabilities of the host.

Modeling a radio channel is a nontrivial task due to possible signal loss or signal degeneracy caused by interference with other channels and blockages by physical obstructions. Radio propagation and interference models based on the physical layer have been developed in [12,13]. These models provide accurate information on the capacity and limitations of an ad hoc network. However, they are too general for the design and analysis routing protocols in higher layers. Thus, a simpler model that abstracts the wireless ad hoc nature of the network is being used. In this model, the wireless network is modeled as a graph $G = (V, E)$ where V is the set of all hosts/nodes and E contains an edge joining u to v if the node u can directly transmit to v. Adjacency between nodes can be determined, for example, by a pass loss equation and a signal to noise formula [19]. If we assume that the transmission power of all the nodes is the same and fixed, then the graph G will be an undirected graph, called a *unit disk graph*. This is the standard theoretical model of wireless ad hoc network. On the other hand, if hosts can adjust their transmission power and the transmission power is not the same for all hosts, then the resulting graph G will be a directed graph. The directed graph also models the signal loss in the wireless radio channels and as such can be viewed as a more realistic model of a wireless ad hoc network.

8.2.1 Communication and Information Dissemination

An essential requirement of information dissemination in communication networks is that the nodes be endowed with a consistent addressing scheme. IP-addresses, so typical of traditional Internet networking, can be either setup at initialization or else acquired by means of an address configuration protocol. Unfortunately none of these techniques are suitable for wireless networking where addresses have to be built "on the fly" and are usually based on the geographic coordinates of the hosts. The latter can be acquired either directly from a participating satellite of an available GPS or via radio-location by exchanging signals with neighboring nodes in the wireless system. Thus, the coordinates of the hosts form the foundation of any addressing scheme in such infrastructureless networks. The corresponding model is then a unit disk graph together with an embedding (not necessarily planar) into $\mathbb{R}^2$. We refer to the graph with an embedding as *geometric graph* if we want to emphasize the fact that nodes are aware of their coordinates.

In general, ad hoc networks are infrastructureless, have no central components and the nodes may have unpredictable mobility patterns. In order to produce dynamic solutions that take into account the changing topology of the backbone the hosts should base their decision only on factors whose optimality depends on geographically local conditions.

A common approach to ad hoc wireless network design is to separate the infrastructure design and the routing scheme design. However, these two parts are closely interlaced as the choice of a particular infrastructure control algorithm may have a considerable impact on the choice of the routing scheme.

In Section 8.3, we discuss the infrastructure control techniques, and in Section 8.4 we survey some well-known routing techniques in wireless ad hoc networks. Since hosts of a wireless ad hoc network are embedded in two-dimensional (or three-dimensional) Euclidean space, many of these techniques borrow tools from combinatorial geometry.

8.3   Ad Hoc Communication Infrastructure

Due to the host's limited resources in a wireless ad hoc network, an important task is to compute and maintain a suitable topology over which high level routing protocols are implemented. One possible approach is to retain only a linear number of radio channels using some local topology control algorithm. A fundamental requirement on the resulting sparse network topology is that it has similar communication capabilities and power consumption as the original network. In the unit disk graph model, this directly leads to the problem of finding a sparse subgraph so that for each pair of nodes there exists an efficient path in the subgraph. A path is *efficient* if its Euclidean length or number of edges/hops is no more than some small constant factor (*length-stretch factor* in the former case and *hop-stretch factor* in the latter case) of the minimum needed to connect the nodes in the unit disk graph.

If our objective is to optimize the power consumption and the ad hoc network has hosts with variable transmission ranges, then the resulting communication infrastructure should have good (small) length-stretch factor. Constructions of such topologies are based on various types of proximity graphs, and we review some of them in Section 8.3.1.

If the hosts of the wireless ad hoc network have fixed transmission power, then the power consumption is minimized when the routes in the communication infrastructure minimize hops. We present a technique for maintaining communication infrastructure with optimal (up to a small constant) hop-stretch factor in Section 8.3.3.

Fundamental criteria for evaluation of the communication infrastructure include the following:

- *Bounded Degree*: Each node has only a bounded number of neighbors in the subgraph. This is crucial since nodes have limited memory and power resources.
- *Localized Construction*: This is one of the most important requirements on the infrastructure control algorithm since nodes have only access to the information stored in nodes within a constant number of hops away. Thus, distributed local algorithms for maintenance of the topology are desirable.
- *Planarity*: This strong geometric requirement forces edges not to cross each other in the embedding. Many routing algorithms rely on planarity of the communication topology.
- *Sparseness*: The topology should be a sparse subgraph, that is, the number of links should be linear in the number of nodes of the network. Sparseness enables most routing algorithms to run more efficiently. Note that planarity as previously mentioned implies sparseness.
- *Spanner*: Nodes in the substructure should have distance "comparable" to their distance in the original network. In particular, a subgraph $S \subseteq G$ is a spanner of G if there is a positive constant t so that for every pair of nodes u and v, $d_S(u, v) \leq t \cdot d_G(u, v)$. This general definition gives rise to various types of spanners: if the distance function d measures the Euclidean length, then the subgraph is *geometric spanner* and the constant t is the *length-stretch factor*. On the other hand, if d is the number of edges/hops, then the subgraph is *hop spanner* and the constant t is the *hop-stretch factor*.

8.3.1   Geometric Spanners

In this section, we survey several results on constructing sparse communication topology (spanner) for ad hoc networks with hosts of variable transmission ranges. Consequently the obtained spanners will be geometric spanners.

Suppose that P is a set of n points in the plane. Every pair (p, q) of points is associated with a planar region $S_{p,q}$ that specifies "the neighborhood association" of the two points. Let S be the set of neighborhood associations in the pointset. Formally, the neighborhood graph $G_{\mathcal{S},\mathcal{P}}$ of the planar pointset P is determined by a property $\mathcal{P}$ on the neighborhood associations:

1. $(p, q) \rightarrow S_{p,q} \subseteq \mathbb{R}^2$, for $p, q \in P$.
2. $\mathcal{P}$ is a property on $\mathcal{S} := \{S_{p,q} : p, q \in P\}$.

The graph $G_{\mathcal{S},\mathcal{P}} = (P, E)$ has the pointset P as its set of nodes and the set E of edges is defined by $(p, q) \in E \Leftrightarrow S_{p,q}$ has property $\mathcal{P}$. There are several variants on this model including Nearest Neighbor Graph, Relative Neighbor Graph, and Gabriel Graph which we define in the sequel.

8.3.1.1 Nearest Neighbor Graph

The edges of a nearest neighbor graph (NNG) are determined by minimum distance. More precisely, for two points p and q $(p, q) \in E \Leftrightarrow p$ is nearest neighbor of q. If $S_{p,q}$ is the disk with diameter determined by the points p and q then it is easy to see that the NNG is the same as the graph $G_{\mathcal{S},\mathcal{P}}$, where the property $\mathcal{P}$ is defined by $S_{p,q} = \emptyset$. Although the NNG incorporates a useful notion it has the disadvantage that the resulting graph may be disconnected. A generalization of the NNG is the k-nearest neighbor graph (k-NNG), for some $k \geq 1$. In this case, $(p, q) \in E \Leftrightarrow p$ is kth nearest neighbor of q or q is kth nearest neighbor of p (Figure 8.1).

8.3.1.2 Relative Neighbor Graph

The neighborhood association in the relative neighbor graph (RNG) is determined by the lune. Formally, the *lune* $L_{p,q}$ of p and q is the intersection of the open disks with radius the distance $d(p, q)$ between p and q and centered at p and q, respectively. The set of edges of the graph is defined by $(p, q) \in E \Leftrightarrow$ the lune $L_{p,q}$ does not contain any point in the pointset P (Figure 8.2).

8.3.1.3 Gabriel Graph

Gabriel graphs (GG; also known as Least squares adjacency graphs) where introduced by Gabriel and Sokal [11] and have been used for geographic analysis and pattern recognition. The region of association between two points p, q is specified by the disk with diameter determined by p and q. Formally, the set of

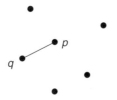

FIGURE 8.1 q is the nearest neighbor of p.

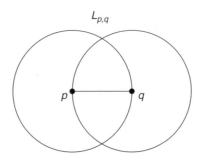

FIGURE 8.2 The lune defined by points p and q.

edges of the GG is defined by $(p, q) \in E \Leftrightarrow$ the disk centered at $(p + q)/2$ and radius $d(p \cdot q)/2$ does not contain any point in the pointset P (Figure 8.3).

8.3.1.4 Relation between the Neighborhood Graphs

We mention without proof the following theorem.

Theorem 8.1 (O'Rourke and Toussaint [16,20]).   *Let P be a planar pointset. Then the following inclusions are satisfied for the previously mentioned neighborhood graphs*

$$NNG \subset MST \subset RNG \subset GG \subset DT,$$

where MST denotes the Minimum spanning tree, and DT the Delaunay triangulation of the pointset P.

8.3.2 Tests for Preprocessing the Ad Hoc Network

In this section, we describe two procedures for constructing a geometric spanner in a given ad hoc network: the Gabriel test is due to Bose et al. [5] and the Morelia test is due to Boone et al. [3].

8.3.2.1 Gabriel Test

One of the most important tests for eliminating crossings in a wireless network is called Gabriel test, which is applied to every link of the network. Assume that all nodes have the same transmission range R. Let A, B be two nodes whose distance is less than the transmission range R of the network. In the Gabriel test, if there is no node in the circle with diameter AB then the link between A and B is kept. If however there is a node C in the circle with diameter AB, as depicted in Figure 8.4, then nodes A and B remove

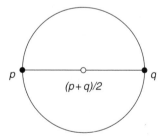

FIGURE 8.3   The circle with diameter the line segment determined by p, q and centered at the point $(p + q)/2$.

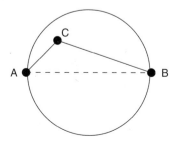

FIGURE 8.4   Eliminating an unnecessary link (dashed line AB) with the Gabriel test.

their direct link. Formally, the Gabriel test is as follows:

Gabriel Test
Input: Pointset
Output: Gabriel Graph of Pointset

  1: **for** for each node v **do**
  2:     **for** for each u neighbor of v **do**
  3:        **if** circle with diameter uv contains no other points in P **then**
  4:           remove the edge uv
  5:        **end if**
  6:     **end for**
  7: **end for**

In particular, when A (respectively, B) is queried on routing data to B (respectively, A) the routing table at A (respectively, B) forwards the data through C (or some other similar node if more than one node is in the circle with diameter AB. The advantage of doing this re-routing of data is that the resulting graph is a planar spanner on which we can apply the face routing algorithm (to be described later) for discovering a route from source to destination.

Theorem 8.2 (Bose et al. [5]). *If the original network is connected then the Gabriel test produces a connected planar network.*

8.3.2.2   Morelia Test

The Gabriel test suffers from the multiple hop effect. Consider a set of pairwise mutually reachable nodes as depicted on the left-hand side of Figure 8.5. When we apply the Gabriel test the configuration on the right-hand side of Figure 8.5 results. We can see that although nodes A and B could have reached each other directly in a single hop instead they must direct their data through a sequence of hops.

The Morelia test takes into account the "distance one" neighborhood of the nodes prior to deciding on whether it should delete an edge. It is similar to the Gabriel test in that given two nodes A and B it eliminates links based on the inspection of the circle with diameter AB. Unlike the Gabriel test it does not necessarily eliminate the direct link AB when it finds another node inside the circle with diameter AB. Instead, it verifies whether the nodes inside the circle create any crossing of the line AB. If no crossing is created the line AB is kept, otherwise it is removed. The verification of the existence of crossing is done in most cases by inspecting only the neighborhood of nodes A and B at the transmission distance R. In a few cases, the neighborhood of some of the nodes in the circle around AB is inspected. We subdivide the area of the circle with diameter AB into four regions X_1, X_2, Y_1, and Y_2 (see Figure 8.6) as determined by an arc of radius R.

Figure 8.7 depicts several scenarios being considered prior to determining whether or not an edge should be deleted. Details of the precise specification of the Morelia test applied to a link AB can be found in Boone et al. [3].

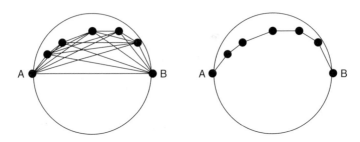

FIGURE 8.5   Multiple hop effect when eliminating a link (line segment AB) via the Gabriel test.

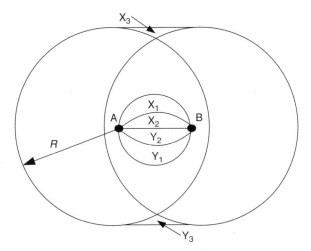

FIGURE 8.6 Morelia test.

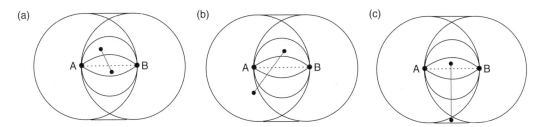

FIGURE 8.7 Examples of the Morelia test scenarios: (a) *AB* deleted by rule 1, (b) *AB* deleted by rule 2, and (c) *AB* deleted by rule 4.

We can prove the following theorem:

Theorem 8.3 (Boone et al. [3]). *If the original network is connected then application of the Morelia test produces a connected planar spanner of the original network, which contains the GG of the network as a subgraph.*

8.3.3 Hop Spanners

In all constructions described in the previous sections, the computed topology is a sparse geometric spanner. If the hosts of the wireless network can adjust their transmission power, then the topology will guarantee optimal power consumption. However, the communication delay is not bounded.

To bound the communication delay, one must construct a topology with a small hop-stretch factor. Hop spanners were introduced by Peleg and Ullman [18] and were used as network synchronizers. A survey of results on (sparse) hop spanners is given in Reference 17. The problem of finding a sparsest hop spanner with a given hop-stretch factor t was shown to be NP-complete for most values of t in general graphs, see Reference 6. In the case $t = 2$ (i.e., the hop length of a shortest path in a spanner is at most twice the hop length of a shortest path in the original graph) an approximation algorithm to construct a hop spanner with hop-stretch factor two is given in Reference 14. However, such a spanner cannot have a linear number of edges in general networks [14]. Recently a distributed nonconstant hop-stretch factor $O(\log n)$ spanner algorithm appeared in Reference 10.

For unit disk graphs, various hierarchical structures, based on the concepts of dominating sets and coverings, have been proposed. Although hop spanners in unit disk graphs can be made sparse, they cannot have every node of constant degree. This follows from a more general result in Reference 2 where the following generalization of a hop spanner has been introduced.

Let S be a subgraph of G. We say that u and v are *quasi-connected* (in S) if there exists a path $(u, u_1, \ldots, u_k, v)$ in G with all but the last edge in $E(S)$. Such a path will be referred to as a *quasi-path* in S. Given two quasi-connected nodes u and v, their *quasi-distance* $qd_S(u, v)$ is the number of edges of a shortest $u - v$ quasi-path in S. A subgraph $S \subseteq G$ is a hop quasi-spanner of G if there is a positive constant t so that for every pair of nodes u and v, $qd_S(u, v) \leq t(d_G(u, v) - 1) + 1$. Notice that according to this definition every hop spanner is also a hop quasi-spanner with the same hop-stretch factor. Thus, the notion of quasi-spanner generalizes the notion of hop spanner. Furthermore, in a wireless ad hoc network it completely suffices to route a message to a host that is in the transmission range of the destination host. Indeed, if a host receives a message that is destined for it, it can accept the message, disregarding the fact that the sender may not be in its routing table based on the communication topology.

Now the following result shows that hop spanners must contain some nodes of large degree.

Theorem 8.4 (Berenbrink et al. [2]).   *Let u be a node of a unit disk graph G. Let d and D be the number of nodes at distance 1 and 2 from u in G, respectively. Let H be a hop quasi-spanner of G with hop-stretch factor two. Then some node in G has degree at least $\sqrt{D/d}$.*

In the remainder of this section, we survey some known results on computing hop spanners with small hop-stretch factor. The most common approach is to first compute a dominating set of the wireless network G, then to add some connector nodes to transform the dominating set into a connected subgraph CDS of G, and finally to connect all remaining nodes to the nodes of CDS. The best result using this technique is the following.

Theorem 8.5 (Alzoubi et al. [1]).   *There exists a local distributed algorithm that will compute a hop spanner of a unit disk graph with the hop-stretch factor 5. Moreover, the hop spanner is planar.*

From the practical point of view, the hop-stretch factor of the spanner from Theorem 8.5 is large. Another approach was taken in Reference 2 where the hop quasi-spanner was proposed as the communication topology for wireless ad hoc networks. The quasi-spanner is build on the concept of coverings: every node selects a constant number of nodes covering all other nodes in the circle with radius two centered at the node. Since the algorithm is simple, we present it in detail.

The *k-neighborhood* of a node v, denoted as $N_G^k(v)$, is the set of all nodes at distance k from v. Similarly, we denote the set of all nodes at distance at most k from v as $N_G^{\leq k}(v)$. A subset C of a set S is called a *covering* of S if for every point $s \in S$ there is a point $c \in C$ such that the distance between s and c is at most 1. Given a unit disk graph G and a node $u \in V$, a covering of $N_G^2(u)$ will be denoted as $T(u)$. A covering $T(u)$ is *minimal* if no other covering of $N_G^2(u)$ is a subset of $T(u)$.

The following is a simple greedy procedure constructing a minimal covering.

Min-Covering(u, G)
Output: minimal covering $T(u)$ of $N_G^2(u)$

1: $N := N_G^2(u)$
2: $T(u) = \emptyset$
3: **while** $N \neq \emptyset$ **do**
4:     choose $v \in N$
5:     $T(u) := T(u) \cup \{v\}$
6:     **for** every $x \in N_G^1(v) \cap N$ **do**
7:         $N := N - \{x\}$
8:     **end for**
9: **end while**

The algorithm for constructing a hop quasi-spanner follows. The algorithm is very simple — every node will build a minimal covering around it.

Hop Quasi-Spanner
Input: Unit disk graph $G = (V, E(G))$
Output: Spanning subgraph $K = (V, E(K))$ of G

1: **for** each node u **do**
2:    $E(u) := \emptyset$
3:    $T(u) := \text{Min-Covering}(u, G)$
4:    **for** every $v \in T(u)$ **do**
5:       find a $w \in N_G^1(u) \cap N_G^1(v)$
6:       $E(u) := E(u) \cup \{(u, w), (w, v)\}$
7:    **end for**
8: **end for**
9: $E(H) := \cup_{u \in V} E(u)$

The hop quasi-spanner algorithm produces a sparse hop quasi-spanner. This is stated in the following theorem.

Theorem 8.6 (Berenbrink et al. [2]). *The graph K produced by the hop quasi-spanner algorithm is a hop quasi-spanner with the hop-stretch factor two and has at most $36|V|$ edges, that is, the average degree of K is at most 72.*

8.4 Information Dissemination

After producing an underlying communication topology, an important task is how to disseminate information efficiently. For example, in routing we want to guarantee real-time message delivery and at the same time: (1) avoid flooding the network, (2) discover an optimal path (the existence of such a path strongly depends on our choice of communication topology), (3) use only geographically local information. It is inevitable that we may not be able to satisfy all three goals at the same time. Traditional methods discover routes using a greedy approach in which the "next-hop" is determined by iteratively (hop-to-hop) reducing the value of a given function of a distance and angle determined by the source and destination hosts. Unfortunately this does not always work either because of loops (return to the source host without finding the destination) or voids (hosts with no neighbors closer to the destination). In this section, we look at the problem of determining in which networks it is possible to attain geographically local routing with guaranteed delivery.

8.4.1 Routing in Undirected Planar Graphs

In this section, we will consider planar geometric graphs as the communication topology and will describe several routing algorithms for them. Let G be a planar geometric graph. Recall that V denotes the set of nodes, and E the set of edges in G. We denote the set of faces of G by F. We will assume that nodes of G are in general position, that is, no three are collinear.

8.4.1.1 Compass Routing

The first routing algorithm we consider is greedy in the sense that it always selects its next edge/hop on the basis of the smallest angle between the "current edge" and the "straight line" formed by the current node and the target node. More precisely the compass routing is achieved by the following algorithm.

Compass Routing Algorithm

Input: connected planar graph $G = (V, E)$
Source node: s
Destination node: t

1: Start at source node $c := s$.
2: **repeat**

3:     Choose an edge incident to current node c forming the smallest angle with straight line c–t connecting c to t.

4:     Traverse the chosen edge.

5:     Go back to 3

6: **until** target t is found

In a deterministic setting, compass routing guarantees delivery if the planar graph is, for example, a Delaunay Triangulation (see Kranakis et al. [15]). It is easy to show that in a random setting, whereby the points of the set P are uniformly distributed over a given convex region, compass routing can reach the destination with high probability. In general, however, compass routing cannot guarantee delivery. This is not only due to possible voids (i.e., hosts with no neighbors closer to the destination) along the way but also inherent loops in the network (see Figure 8.8).

In the next section, we describe a routing algorithm with guaranteed delivery on geometric planar graphs.

8.4.1.2   Face Routing Algorithm

Failure of compass routing to guarantee delivery is due to its "greedy" nature in order to minimize memory storage. Face routing overcomes this problem by remembering the straight line connecting the source s to the target node t.

Face Routing Algorithm
Input: Connected Geometric Planar Network $G = (V, E)$
Source node: s
Destination node: t

1: Start at source node $c := s$.

2: **repeat**

3:     **repeat**

4:        Determine face f incident to current node c and intersected by the straight line s–t connecting s to t.

5:        Select any of the two edges of f incident to c and traverse the edges of f in the chosen direction.

6:     **until** we find the second edge, say xy, of f intersected by s–t.

7:     Update the face f to the new face of the graph G that is "on the other side" of the edge xy.

8:     Update current node c to either of the nodes x or y.

9: **until** target t is found.

An example of the application of face routing is depicted in Figure 8.9. Face routing always advances to a new face. We never traverse the same face twice. Since each link is traversed a constant number of times we can prove the following theorem.

Theorem 8.7 (Kranakis et al. [15]).   *Face routing in a geometric planar graph always guarantees delivery and traverses at most $O(n)$ edges.*

The reader may wish to consult [5] for additional variants of the face routing algorithm.

8.4.2   Routing in Directed Planar Graphs

As discussed earlier the unit disk graph is the graph of choice used in modeling ad hoc networks when the communication links are bidirectional. However, bidirectionality cannot be assured in a system consisting of diverse hosts having variable power ranges, and nodes facing obstructions that may attenuate the signals (Figure 8.9). This gives rise to the "directed link model," which provides a natural direction between links

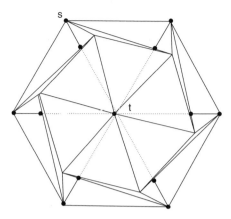

FIGURE 8.8 An example of a planar graph over which compass routing fails to find the destination.

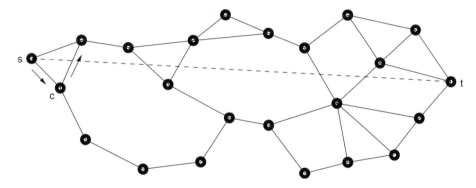

FIGURE 8.9 Face routing on a planar graph.

as follows. If a node A can reach a node B (but it is not necessarily the case that B can also reach A) then we say that there is a directed link from A to B in the graph. This gives rise to an orientation in the network.

An *orientation* $\rightarrow$ of a graph G is an assignment of a direction to every edge e of G. For an edge e with endpoints u and v, we write $e = (u, v)$ if its direction is from u to v. A graph G together with its orientation is denoted by $\vec{G}$. In this section, we will consider two classes of directed planar geometric graphs: Eulerian and Outerplanar. For both classes, we present a simple routing algorithm.

8.4.2.1 Eulerian Networks

Consider a connected planar geometric graph $G = (V, E)$. If G is oriented, then the face routing algorithm will not necessarily work since some edges may be directed in an opposite direction while traversing a face. We describe a simple method from Reference 7 for routing a message to the other end of an oppositely directed edge in Eulerian geometric networks. The idea is simple: imitate the face routing algorithm but when an edge that needs to be traversed is misdirected find a way to circumvent it using only "constant" memory and geographically local information.

We say that $\vec{G}$ is *Eulerian* if for every node $u \in V$, the number of edges outgoing from u equals the number of edges ingoing into u, that is, the size of $N^+(u) = \{x, (u, x) \in E\}$ equals the size of $N^-(u) = \{y, (y, u) \in E\}$. Now suppose that $\vec{G}$ is an Eulerian planar geometric graph. For a given node u of $\vec{G}$, we order edges (u, x) where $x \in N^+(u)$ clockwise around u starting with the edge closest to the vertical line passing through u. Similarly we order edges (y, u) where $y \in N^-(u)$ clockwise around u (see Figure 8.10).

Let $e = (y, u)$ be the ith ingoing edge to u in $\vec{G}$. The function **succ(e)** will return a pointer to the edge (u, x) so that (u, x) is the ith outgoing edge from u. For an illustration of the function see Figure 8.11.

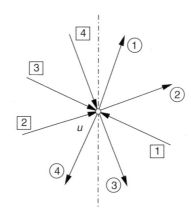

FIGURE 8.10　Circled numbers represent the ordering on outgoing edges, squared numbers on ingoing ones.

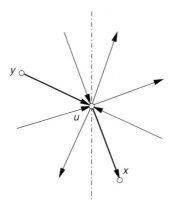

FIGURE 8.11　In this example the ingoing edge (y, u) is third, so the chosen outgoing edge (u, x) is also third. Both these edges are depicted bold.

Again, this function is easy to implement using only local information. Obviously, the function **succ**() is injective, and thus, for every edge $e = (u, v)$ of $\vec{G}$, we can define a closed walk by starting from $e = (u, v)$ and then repeatedly applying the function **succ**() until we arrive at the same edge $e = (u, v)$. Since $\vec{G}$ is Eulerian, the walk is well-defined and finite. We call such a walk a *quasi-face* of $\vec{G}$.

We modify the face routing algorithm from Reference 15 so that it will work on Eulerian planar geometric graphs. This requires extending the face traversal routine so that whenever the face traversal routine wants to traverse an edge $e = (u, v)$ that is oppositely directed, we traverse the following edges in this order: $\mathbf{succ(e), succ(e)^2, \ldots, succ(e)^k}$, so that $\mathbf{succ(e)^{k+1}} = (\mathbf{u}, \mathbf{v})$. After traversing $\mathbf{succ(e)^k}$, the routine resumes to the original traversal of the face. Formally the algorithm is as follows:

Eulerian Directed Graph Algorithm
Input: Connected Eulerian geometric graph $\vec{G} = (V, E)$
Source node: s
Destination node: t

1: $v \leftarrow s$ {Current node = source node.}
2: **repeat**
3:　　Let f be a face of G with v on its boundary that intersects the line v-t at a point (not necessarily a node) closest to t.
4:　　**for all** edges xy of f **do**
5:　　　　**if** $xy \cap v$-$t = p$ and $|pt| < |vt|$ **then**

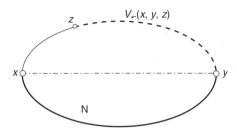

FIGURE 8.12 The dashed part of the outer face represents the nodes in $V_\frown(x, y, z)$ and the bold solid part represents nodes in $V_\frown(x, y, z)$, respectively. Note that y belongs to both these sets and is in fact the first element of those sets.

6:        $v \leftarrow p$

7:     **end if**

8:     **end for**

9:     Traverse f until reaching the edge xy containing the point p.

10: **until** $v = t$

We can prove the following theorem:

Theorem 8.8 (Chavez et al. [8]). *The Eulerian Directed Graph Algorithm will reach t from s in at most $O(n^2)$ steps.*

8.4.2.2 Outerplanar Networks

If $\vec{G}$ is not Eulerian, then the routing algorithm above will fail. It is an important open problem to design routing algorithms for general directed geometric graphs. A fundamental question is whether such an algorithm exists for directed planar geometric graphs. In all cases, the directed graphs must be strongly connected so that we have a directed path guaranteed.

In this section, we describe a routing algorithm from Reference 8 that works on outerplanar graphs. A planar geometric graph G is *outerplanar* if a single face (called outer-face) contains all the nodes. We will assume that this face is a convex polygon in $\mathbb{R}^2$. For a given triple of nodes x, y, and z, let $V_\frown(x, y, z)$ (respectively, $V_\frown(x, y, z)$) denote the ordered set of nodes distinct from x and z that are encountered while moving from y counterclockwise (respectively, clockwise) around the outer-face of G until either x or z is reached; see Figure 8.12. Now consider an orientation $\vec{G}$ of the geometric graph G and let

$$N_\frown(x, y, z) = V_\frown(x, y, z) \cap N^+(x) \text{ and } N_\frown(x, y, z) = V_\frown(x, y, z) \cap N^+(x).$$

If $N_\frown(x, y, z) \neq \emptyset$, let $v_\frown(x, y, z)$ denote the first node in $N_\frown(x, y, z)$. Similarly we define $v_\frown(x, y, z)$ as the first node in $N_\frown(x, y, z)$, if it exists. A geometric network with fixed orientation is *strongly connected* if for every ordered pair of its nodes, there is a (directed) path joining them.

The intuitive idea of the algorithm is to start at the source node s. It specifies the two nodes v_1, v_2 adjacent to the current node c such that the straight line determined by nodes c and t lies within the angle $v_1 c v_2$. It traverses one of the nodes and backtracks if necessary in order to update its current position. Formally, the algorithm is as follows.

Outer Planar Directed Graph Algorithm

Input: Strongly connected outerplanar geometric graph $\vec{G} = (V, E)$

Source node: s

Destination node: t

1: $v \leftarrow s$ {Current node = source node.}

2: $v_\frown, v_\frown \leftarrow s$ {counterclockwise and clockwise bound = starting node.}

3: **while** $v \neq t$ **do**

4:   **if** $(v, t) \in E$ **then**

5:     $v, v_\curvearrowleft, v_\curvearrowright \leftarrow t$ {Move to t.}

6:   **else if** $N_\curvearrowleft(v, t, v_\curvearrowright) \neq \emptyset$ and $N_\curvearrowright(v, t, v_\curvearrowright) = \emptyset$ **then** {No-choice node; greedily move to the only possible counterclockwise direction toward t.}

7:     $v, v_\curvearrowleft \leftarrow v_\curvearrowleft(v, t, v_\curvearrowright)$

8:   **else if** $N_\curvearrowleft(v, t, v_\curvearrowright) = \emptyset$ and $N_\curvearrowright(v, t, v_\curvearrowright) \neq \emptyset$ **then** {No-choice node; greedily move to the only possible clockwise direction toward t.}

9:     $v, v_\curvearrowright \leftarrow v_\curvearrowright(v, t, v_\curvearrowright)$

10:   **else if** $N_\curvearrowleft(v, t, v_\curvearrowright) \neq \emptyset$ and $N_\curvearrowright(v, t, v_\curvearrowright) \neq \emptyset$ **then** {Decision node; first take the "counterclockwise" branch but remember the node for the backtrack purpose.}

11:     $b \leftarrow v;\ v, v_\curvearrowleft \leftarrow v_\curvearrowleft(v, t, v_\curvearrowright)$

12:   **else if** $N_\curvearrowleft(v, t, v_\curvearrowright) = \emptyset$ and $N_\curvearrowright(v, t, v_\curvearrowright) = \emptyset$ **then** {Dead-end node; backtrack to the last node where a decision has been made. No updates to $v_\curvearrowleft$ and $v_\curvearrowright$ are necessary.}

13:     **if** $v \in V_\curvearrowleft(t, b, t)$ **then**

14:       **while** $v \neq b$ **do**

15:         $v \leftarrow v_\curvearrowleft(v, b, v)$

16:       **end while**

17:     **end if**

18:     **if** $v \in V_\curvearrowright(t, b, t)$ **then**

19:       **while** $v \neq b$ **do**

20:         $v \leftarrow v_\curvearrowright(v, b, v)$

21:       **end while**

22:     **end if**

23:     $v, v_\curvearrowright \leftarrow v_\curvearrowright(v, t, v_\curvearrowright)$ {Take the "clockwise" branch toward t.}

24:   **end if**

25: **end while**

We have the following theorem.

Theorem 8.9 (Chavez et al. [8]). *The Outer Planar Directed Graph Algorithm will reach t from s in at most $2n - 1$ steps.*

8.4.3 Traversal of Nonplanar Graphs

Network traversal is a technique widely used in networking for processing the nodes, edges, etc., of a network in some order. For example, it may involve reporting each node, edge, and face of a planar graph exactly once, in order to apply some operation to each. As such it can be used to discover network resources, implement security policies, and report network conditions. Although traversal can be used to discover routes between two hosts, in general it will be less efficient than routing since it cannot guarantee that its "discovery process" will be restricted to employing only information relevant to routing.

The Depth First Search (DFS) of the primal nodes and edges or dual faces and edges of the graph is the usual approach followed for implementing traversal but it cannot be implemented without using mark bits on the nodes, edges, or faces, and a stack or queue. In this section we discuss a traversal technique from Reference 7 that is applicable to a class of nonplanar networks (to be defined further) and which is an improvement of a technique introduced by de Berg et al. [9] and further elaborated by Bose and Morin [4].

A *quasi-planar graph* $G = (V, E)$ has nodes embedded in the plane and partitioned into $V_p \cup V_c = V$ so that

- Nodes in V_p induce a connected planar graph P.
- The outer-face of P does not contain any node from V_c or edge of $G - P$.
- No edge of P is crossed by any other edge of G.

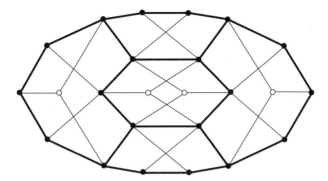

FIGURE 8.13 An example of a quasi-planar graph that satisfies the Left-Neighbor Rule. The filled nodes are in V_p and bold edges are edges of the underlying planar subgraph P.

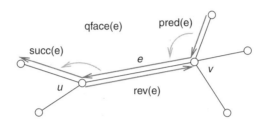

FIGURE 8.14 Illustration of basic functions on quasi-planar graphs.

An example of a quasi-planar graph is depicted in Figure 8.13.

We will refer to the graph P as an *underlying planar subgraph* and to its faces as *underlying faces.* The notion of nodes and edges is explicit in the definition of quasi-planar graph, however, the notion of faces is not. To define the notion of a face, we need to introduce some basic functions on quasi-planar graphs.

A node u is uniquely determined by the pair $[x, y]$, of its horizontal and vertical coordinates. Every edge $e = uv$ is stored as two oppositely directed edges (u, v) and (v, u). The functions **xcor(v)** (respectively, **ycor(v)**) will return the horizontal (respectively, vertical) coordinate of the node v and **rev(e)** will return a pointer to the edge (v, u). Similarly the function **succ(e)** will return a pointer to the edge (v, x) so that (v, x) is the first edge counter-clockwise around v starting from the edge (v, u), and the function **pred(e)** will return a pointer to the edge (y, u) so that (u, y) is the first edge clockwise around u starting from the edge (u, v). For an illustration of these functions, see Figure 8.14.

For every edge $e = (u, v)$ of G, we can define a closed walk by starting from $e = (u, v)$ and then repeatedly applying the function **succ()** until we arrive at the same edge $e = (u, v)$. Such a walk is called a *quasi-face* of G and the set of all quasi-faces of G is denoted by F. The function **qface(e)** will return a pointer to the quasi-face determined by edge $e = (u, v)$.

A quasi-planar graph G satisfies the *Left-Neighbor Rule* if every node $v \in V_c$ has a neighbor u so that **xcor(u)** < **xcor(v)**. For an example of G that satisfies the Left-Neighbor Rule, see Figure 8.13.

8.4.3.1 Quasi-Planar Graph Traversal Algorithm

As in de Berg et al. [9], the general idea of the algorithm is to define a total order $\preceq$ on all edges in E. This gives rise to a unique predecessor for every quasi-face in F. The predecessor relationship imposes a virtual directed tree $G(F)$. The algorithm (due to Chavez et al. [7]) will search for the root of $G(F)$ and then will report quasi-faces of G in DFS order on the tree $G(F)$. For this, a well-known tree-traversal technique is used in order to traverse $G(F)$ using $O(1)$ additional memory (Figure 8.15).

In order to define the virtual tree $G(F)$, we determine a unique edge, called an *entry edge*, in each quasi-face. We first define a total order on all edges in E. We write $u \ll v$ if

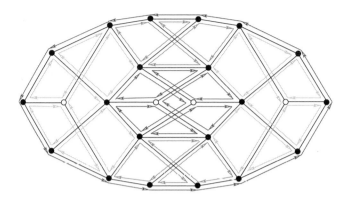

FIGURE 8.15 A quasi-planar graph and its six quasi-faces.

$(\mathbf{xcor(u), ycor(u))} \leq (\mathbf{xcor(v), ycor(v))}$ by lexicographic comparison of the numeric values using $\leq$. For an edge $e = (u, v)$, let

$$\mathbf{left(e)} = \begin{cases} u, & \text{if } u \ll v \\ v, & \text{otherwise} \end{cases} \qquad \mathbf{right(e)} = \begin{cases} v, & \text{if } u \ll v \\ u, & \text{otherwise} \end{cases}$$

and $\check{u} = [\mathbf{xcor(u), ycor(u)} - 1]$. Now let $\mathbf{key(e)}$ be the 5-tuple

$$\mathbf{key(e)} = (\mathbf{xcor(left(e)), ycor(left(e))},$$

$$\angle \check{\mathbf{left}}(\mathbf{e})\mathbf{left(e)right(e), xcor(u), ycor(u)}).$$

By $\angle abc$ we always refer to the counter-clockwise angle between rays ba and bc with b being the apex of the angle. It follows by our assumption that edges cannot cross vertices that if two edges $e \neq e'$ have the same first three values in their $\mathbf{key}()$, then $e' = e^-$ and hence their last two values in $\mathbf{key}()$ cannot both be the same. Hence it follows that $e = e'$ if and only if $\mathbf{key(e)} = \mathbf{key(e')}$. We define the total order $\preceq$ by lexicographic comparison of the numeric $\mathbf{key}()$ values using $\leq$. For a quasi-face $f \in F$, we define

$$\mathbf{entry(f)} = \mathbf{e} \in \mathbf{f} : \mathbf{e} \preceq \mathbf{e'} \quad \text{for all } \mathbf{e'} \neq \mathbf{e} \in \mathbf{f},$$

that is, $\mathbf{entry(f)}$ is the minimum edge (with respect to the order $\preceq$) on the quasi-face f. Such an edge e will be called the *entry edge* of f. Note that this function is easy to implement using the function $\mathbf{succ}()$, and the total order $\preceq$ using only $O(1)$ memory.

The main algorithm for traversal from Reference 7 is as follows:

Quasi-Planar Traversal Algorithm
Input: $e = (u, v)$ of $G(V, E)$;
Output: List of nodes, edges, and quasi-faces of G.

```
 1: repeat {find the minimum edge e₀}
 2:     e ← rev(e)
 3:     while e ≠ entry(qface(e)) do
 4:         e ← succ(e)
 5:     end while
 6: until e = e₀
 7: p ← ľeft(e)
 8: repeat {start the traversal}
 9:     e ← succ(e)
```

```
10:    let e = (u, v) and let succ(e) = (v, w)
11:    if p is contained in cone(u, v, w) then {report u if necessary}
12:        report u
13:    end if
14:    if |up| < |vp| or (|up| = |vp| and →up < →vp) then {report e if necessary}
15:        report e
16:    end if
17:    if e = entry(qface(e)) then {report e and return to parent of qface(e)}
18:        report qface(e)
19:        e ← rev(e)
20:    else {descend to children of qface(e) if necessary}
21:        if rev(e) = entry(qface(rev(e))) then
22:            e ← rev(e)
23:        end if
24:    end if
25: until e = e₀
26: report qface(e₀)
```

We have the following theorem:

Theorem 8.10 (Chavez et al. [7]).   *The Quasi-Planar Traversal Algorithm reports each node, (undirected) edge, and quasi-face of a quasi-planar graph G that satisfies the Left-Neighbor Rule exactly once in $O(|E| \log |E|)$ time.*

8.5 Conclusion

In this chapter we provided a survey of recent results on information dissemination via location awareness in ad hoc networks. Although routing and traversal may be wrongfully considered as overworked research topics we believe that information dissemination in ad hoc networks provides new challenges that go beyond trivial extensions of existing or old results. We expect that exploring tradeoffs between location awareness and reducing preprocessing time for simplifying a wireless system is of vital importance to our understanding of the efficiency of information dissemination. Future studies in a variety of graph models (oriented and otherwise) should provide better clues to maintaining a seamless, ubiquitous, and well-integrated communication infrastructure.

Acknowledgments

Many thanks to the Morelia group (Edgar Chavez, Stefan Dobrev, Jarda Opartny, and Jorge Urrutia) for many hours of enjoyable discussions on the topics reviewed in this chapter and to Danny Krizanc for his comments. Research of both authors is supported in part by NSERC (Natural Sciences and Engineering Research Council of Canada) and MITACS (Mathematics of Information Technology and Complex Systems) grants.

References

[1] K. Alzoubi, X.Y. Li, Y. Wang, P.J. Wan, and O. Frieder. Geometric spanners for wireless ad hoc networks. *IEEE-Transactions on Parallel and Distributed Systems*, 14: 1–14, 2003.

[2] P. Berenbrink, T. Friedetzky, J. Manuch, and L. Stacho. (Quasi) spanner in mobile ad hoc networks. *Journal of Interconnection Networks*.

[3] P. Boone, E. Chavez, L. Gleitzky, E. Kranakis, J. Opartny, G. Salazar, and J. Urrutia. Morelia test: Improving the efficiency of the gabriel test and face routing in ad hoc networks. In R. Kralovic and O. Sykora, ed, *Proceedings of SIROCCO 2004, Springer-Verlag, LNCS*, Vol. 3104, 2004.

[4] P. Bose and P. Morin. An improved algorithm for subdivision traversal without extra storage. *International Journal of Computational Geometry and Applications*, 12: 297–308, 2002.

[5] P. Bose, P. Morin, I. Stojmenovic, and J. Urrutia. Routing with guaranteed delivery in ad hoc wireless networks. In *Proceedings of the Discrete Algorithms and Methods for Mobility (DIALM'99)*, pp. 48–55, 1999.

[6] L. Cai. NP completeness of minimum spanner problems. *Discrete Applied Mathematics*, 48: 187–194, 1994.

[7] E. Chavez, S. Dobrev, E. Kranakis, J. Opartny, L. Stacho, and J. Urrutia. Route discovery with constant memory in oriented planar geometric networks. In S. Nikoletseas and J. Rolim, eds, *Proceedings of Algosensors 2004, Springer Verlag, LNCS*, Vol. 3121, pp. 147–156, 2004.

[8] E. Chavez, S. Dobrev, E. Kranakis, J. Opartny, L. Stacho, and J. Urrutia. Traversal of a quasi-planar subdivision without using mark bits. In *WMAN (Workshop on Wireless Mobile Ad hoc Networks), IPDPS*, Santa Fe, New Mexico, April 26–30, 2004.

[9] M. de Berg, M. van Kreveld, R. van Oostrum, and M. Overmars. Simple traversal of a subdivision without extra storage. *International Journal of Geographic Information Systems*, 11: 359–373, 1997.

[10] D. Dubhashi, A. Mei, A. Panconesi, J. Radhakrishnan, and A. Srinivasan. Fast distributed algorithms for (weakly) connected dominating sets and linear-size skeletons. In *Proceedings of the 14th ACM-SIAM Symposium on Discrete Algorithms (SODA 03)*, 2003.

[11] K.R. Gabriel and R.R. Sokal. A new statistical approach to geographic variation analysis. *Systemic Zoology*, 18: 259–278, 1969.

[12] M. Grossglauser and D. Tse. Mobility increases the capacity of wireless adhoc networks. In *Proceedings of the IEEE Infocom*, 2001.

[13] P. Gupta and P. Kumar. Capacity of wireless networks. *IEEE Transactions on Information Technology*, 46: 388–404, 2000.

[14] G. Kortsarz and D. Peleg. Generating sparse 2-spanners. *Journal of Algorithms*, 17: 222–236, 1994.

[15] E. Kranakis, H. Singh, and J. Urrutia. Compass routing on geometric networks. In *Proceedings of the 11th Canadian Conference on Computational Geometry*, pp. 51–54, August 1999.

[16] J. O'Rourke and G.T. Toussaint. Pattern recognition. In J.E. Goodman and J. O'Rourke, eds, *Handbook of Discrete and Computational Geometry*, CRC Press, New York, pp. 797–813, 1997.

[17] D. Peleg and A.A. Schaffer. Graph spanners. *Journal of Graph Theory*, 13: 99–116, 1989.

[18] D. Peleg and J.D. Ullman. An optimal synchronizer for the hypercube. *SIAM Journal of Computing*, 18: 740–747, 1989.

[19] T. Rappaport. *Wireless Communications: Principles and Practices*, Prentice Hall, New York, 1996.

[20] G.T. Toussaint. The relative neighborhood graph of a finite planar set. *Pattern Recognition*, 12: 261–266, 1980.

9

Ad Hoc Routing Protocols

9.1 Introduction... **9**-183
9.2 Unicast Routing Protocols **9**-184
   ID-Based Flat Routing Protocols • ID-Based Hierarchical
   Routing Protocols • Location-Based Routing Protocols
9.3 Broadcast Routing Protocols........................... **9**-201
9.4 Multicast Routing Protocols **9**-205
   Forwarding Group Multicast Protocol • On-Demand
   Multicast Routing Protocol • Ad Hoc Multicast Routing
   Protocol • Multicast Operation of AODV Routing Protocol
   • Core-Assisted Mesh Protocol • Ad Hoc Multicast Routing
   Protocol Utilizing Increasing id-NumberS • Multicast
   Core-Extraction Distributed Ad Hoc Routing • Adaptive
   Backbone-Based Multicast • Differential Destination
   Multicast • Location-Guided Tree Construction Algorithms
   • Route-Driven Gossip
9.5 Summary **9**-213
References **9**-213

Kai-Wei Fan
Sha Liu
Prasun Sinha

9.1 Introduction

Ad hoc routing protocols enable communication between nodes in ad hoc networks. Protocols for routing in ad hoc networks has been an active area of research since the early 1990s. Early research in this area focused on adapting Internet routing protocols for ad hoc networks. Frequent disconnections due to node mobility, transient characteristics of the wireless channel, and channel interference, posed new challenges, which were not considered in the design of routing protocols for the wired Internet. This chapter presents various key routing protocols that have been proposed for one-to-one (unicast), one-to-many (multicast), and one-to-all (broadcast) communication in ad hoc networks. Figure 9.1 shows the classification of routing protocols followed in this chapter.

Unicast routing protocols make it possible for any node in the network to communicate with any other node in the network. Protocols for flat as well as hierarchical network organizations have been explored for ad hoc networks. With decreasing price of GPS devices, it is not hard to imagine that nodes will know their geographic coordinates. The location information of nodes can be smartly used to design efficient routing protocols with the help of location-based unicast routing protocols that make use of location

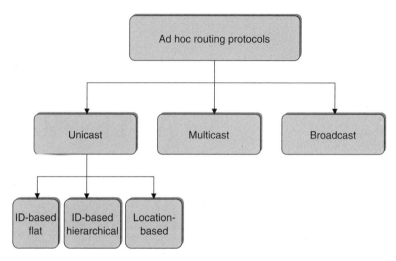

FIGURE 9.1   Classification of routing protocols.

services. In this chapter, we explore three categories of unicast routing protocols, namely, ID-based routing protocols, hierarchical routing protocols, and location-based routing protocols.

Reliable dissemination of code and data to all the network nodes is critical for several applications. Applications and services that will benefit from a network-wide broadcast protocol include network reprogramming, health monitoring, network-wide queries, and critical information dissemination. A sudden flow of packets from a single node can easily cause network-wide congestion leading to low packet delivery rate for broadcast routing protocols. Various broadcast routing protocols use combinations of redundant transmission suppression and transmission pacing to achieve high packet delivery rate and low delivery latency.

In various ad hoc networking scenarios such as battlefield and disaster recovery, people typically work in groups. Multicast routing protocols enable efficient communication among group members. Such protocols typically require computation and maintenance of a substructure (often a tree or a mesh) to facilitate efficient data dissemination to all group members. Communication for maintenance of such substructures can be demanding on the battery-life and the available bandwidth, which are both scarce resources in ad hoc networks.

The rest of this chapter is organized as follows: Section 9.2 presents various unicast routing protocols. Section 9.4 presents multicast routing protocols. Section 9.3 presents various broadcast routing protocols. Finally, Section 9.5 summarizes the chapter with pointers to some future research directions on ad hoc routing protocols.

9.2   Unicast Routing Protocols

Routing protocols can be primarily categorized as proactive (table-driven) and reactive (on-demand) protocols. Proactive routing protocols maintain routes between all pairs of nodes at all times. Reactive routing protocols, on the other hand, do not maintain routes. When a route is needed, they will initiate route discovery process to find a route (or possibly multiple routes). Only after the route is found, the node can start the communication.

Another way is to organize them as flat and hierarchical routing protocols. Flat routing protocols regard the whole network as uniform where each node in the network has same functions and responsibilities. Each node has uniform responsibility for constructing routes. Hierarchical routing protocols organize the network as a tree of clusters, where the roles and functions of nodes are different at various levels of the hierarchy. Routes are constructed according to the node's position in the virtual hierarchy.

Yet another way is to classify them as ID-based or location-based protocols. ID-based routing protocols do not make use of the geographical location of the destination node, they route traffic based on the destination node's ID or address. Location-based protocols route packets according to node's geographical location.

This section presents the key principles of some well-known unicast routing protocols. We organize the unicast routing protocols as ID-based flat routing protocols, ID-based hierarchical routing protocols, and location-based routing protocols.

9.2.1 ID-Based Flat Routing Protocols

ID-based routing protocols route packets according to node's ID or address. This requires nodes to disseminate their ID or address through their neighbors. This information dissemination can be periodical (proactive) or on-demand (reactive). In the rest of the section, we present the key ideas and the principles behind the design of the following routing protocols: DSDV, DSR, AODV, TORA, ABR, R-DSDV, and P_rAODV.

Destination Sequenced Distance Vector (DSDV) (Perkins and Bhagwat, 1994) is a proactive, distance-vector routing protocol based on the classical Bellman–Ford routing algorithm. It is a distributed, self-organized, and loop-free routing protocol suitable for dynamic networks. Each node maintains a routing table that contains routing entries for all nodes in the network. Nodes periodically advertise their routing information to their neighbors. Each entry in the routing table contains the destination node's address, next-hop node's address, the number of hops to reach, and the sequence number originated by the destination node. Nodes can forward packets to next-hops, and so on to the destination according to their routing tables.

In DSDV, nodes maintain routes by exchanging routing information with their neighbors, which propagate it to other nodes. Each node periodically broadcasts routing information containing the following data for each routing entry:

- The address of the destination
- The number of hops to reach the destination
- The sequence number originated by the destination

Each node maintains an even sequence number, which is increased whenever new routing information is advertised. It is used to distinguish which routing information is fresher for that node. Nodes that receive the routing information will update the entries in their routing table based on the following rules:

1. It is a new route entry.
2. The sequence number in the received routing information is higher than it is in the routing table.
3. The sequence numbers are equal, but the metric in the received routing information is better.

The updated routing information will be advertised and propagated to its neighbors, and so on. Finally, every node will have routing information for all other nodes in the network. Table 9.1 shows a possible routing table of node 5 for an ad hoc network topology as shown in Figure 9.2.

When a node moves away or is down, the link to that node as next-hop will be broken. The node that detects the broken link will generate a routing update of odd sequence numbers with metric set to ∞ and advertise it. Any route through that node will be updated by setting the metric to ∞, indicating that the route is broken. Any later routing update of that node with higher sequence number will supersede the ∞ metric and the new route will be created again.

Each node in DSDV also maintains a "settling time table" to prevent fluctuations of the routing update. A node may receive a new routing update with a worse metric, and later receive a routing update with the same sequence number but with a better metric. This may happen due to network congestion or different advertisement schedules of nodes. It will be a redundant advertisement if a node receives the routing update with a worse metric and propagates it immediately, if it will be overwritten with a better metric

TABLE 9.1   Routing Table of Node 5

Destination	Next hop	Hop count	Sequence number
N1	N4	2	S1_105
N2	N4	2	S2_092
N3	N3	1	S3_101
N4	N4	1	S4_118
N5	N5	0	S5_093
N6	N6	1	S6_128
N7	N6	2	S7_089
N8	N6	3	S8_111

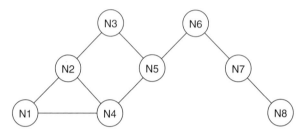

FIGURE 9.2   Ad hoc network topology.

later. "Settling time table" stores the history of past update time for each destination, and nodes will calculate the next time to advertise according to the past update history if the routing update is not a significant one that needs to be disclosed immediately.

In DSDV, routing updates have to be propagated throughout the whole network. As the network grows, the routing update becomes inefficient and consumes more resources such as network bandwidth and power. In a highly dynamic network, most of the routing updates may be useless because it will become invalid before the updated routes will be used, which cause more unnecessary wastes. But DSDV provides a simple and distributed routing mechanism that can be easily implemented and deployed, and still works efficiently in general network topology.

Dynamic Source Routing (DSR) (Johnson and Maltz, 1996) is a reactive, source-routed routing protocol designed for highly dynamic networks. In DSR, each node maintains a route cache, which stores routing paths from itself to other nodes. A source node initiates the "Route Discovery" process to find a routing path to a destination node when it cannot find the route in its route cache. DSR uses "Route Maintenance" mechanisms to maintain the routes when there are link breakages that result in an invalid route, and makes the topology changes converge to stable routes quickly. When a route is found, the source sends packets with headers containing complete routing path to the destination. Each node that receives the packets can forward them to the next node according to the routing path in the header.

"Route discovery" is a process used to construct a routing path from a source to destination dynamically. The source node initiates a route discovery by broadcasting a route request (RREQ) packet to its neighbors. The RREQ packet contains the following information: source address, destination address, request-ID, and routing records. Each source assigns a unique identifier (Request-ID) to each new RREQ packet that it generates. This enables other nodes to eliminate duplicate RREQ packets. Routing records is the nodes sequence in which the RREQ is propagated. A node that receives the RREQ packet processes it as follows:

1. If the ⟨source address, request-ID⟩ is found in its recently processed requests list, the RREQ is discarded.
2. If its own address is found in the recorded route, the RREQ is discarded.
3. If it is the destination, it replies back to the source using a route reply packet (RREP).

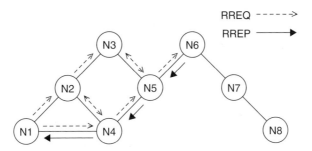

FIGURE 9.3   Route discovery. N1 is the source and N6 is the destination. The route caches are assumed to be empty in the beginning.

4. If the route to the destination is found in its route cache, it appends the route in cache to the recorded route and replies to the source using a source routed RREP packet.
5. Otherwise, it appends its address to the recorded route and rebroadcasts the RREQ to all its neighbors.

In this route discovery process, the RREQ message is flooded throughout the network; eventually the destination receives the request, and replies with a RREP packet. If the links in the network are bidirectional, the RREP can be propagated to the source using the recorded route. If the links are unidirectional, the node replying the RREP can broadcast a RREQ packet piggybacking the recorded route to find the source. After the source receives the RREP message, the source learns the full route and starts source routed transmission to send packets to the destination. Figure 9.3 shows the route discovery process in a network using symmetric links.

When topology changes due to movement or failure of nodes, "route maintenance" is triggered. Topology changes will not cause any reaction if they do not interfere with established routing paths. When a node detects a link failure to its neighbor, which is its next hop in its route cache for some destinations, it will send a route error (RERR) packet. RERR contains the address of nodes at both end of the broken link. The RERR packet will be forwarded to the source, and the source can initiate another route discovery after receiving the RERR packet.

Unlike DSDV, DSR requires no periodic advertisements about its connectivity to other nodes, and maintains only routes for destinations that are required. Nodes in DSR can increase its cache entries by listening to RREPs broadcasts, which can increase the amount of routing information without incurring any extra overhead. Nodes can also maintain multiple routes for each destination to reduce the overhead of route maintenance. However, the nature of source routing in which route control messages have to propagate full source routing information, it is inevitable that these routing control messages will become larger as the number of nodes grow and make the overhead of routing protocol become more dominant. Hence is not scalable in large networks.

Ad hoc On-demand Distance Vector (AODV) (Perkins and Royer, 1999; Perkins, 1999) is a reactive, distance-vector routing protocol suitable for highly dynamic networks. Like in DSDV, each node in AODV maintains a routing table, but the routing table only contains active routing entries. Its route construction process and maintenance mechanisms are similar to those in DSR. A source node initiates the "path discovery" process to construct a routing path to the destination node. Nodes use "path maintenance" mechanism to maintain the routes when there are link breakages. In AODV, no periodic advertisement is required.

The AODV uses "path discovery" process to construct a routing path from a source to the destination when needed. Each node maintains two counters, a sequence number and a broadcast-ID. The source node initiates path discovery by broadcasting RREQ, which contains source address (src_addr), source sequence number (src_seq_no), broadcast id (broadcast_id), destination address (dst_addr), the latest

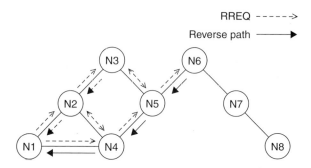

FIGURE 9.4   RREQ and reverse path.

known sequence number of the destination (dst_seq_no), and number of hops to reach (hop_cnt), to its neighbors. When a node receives the RREQ, it processes the RREQ as follows:

1. If it has already received an RREQ packet with same ⟨src_addr, broadcast_id⟩, the RREQ is dropped.
2. If it is the destination node, or it has a routing entry for that destination and the destination nodes' sequence number in its routing entry is greater or equal to the one in the RREQ packet, it will reply with a RREP packet back to the source, indicating that a routing path has been found.
3. Else, it rebroadcasts the RREQ after increasing the hop_cnt, and sets up a "reverse path" pointed to the node from which the RREQ came.

The "reverse path" is set up when the RREQ travels from the source to the destination. It is used to send the RREP packet from the destination back to the source. Figure 9.4 illustrates reverse path setup in a path discovery process as the RREQ propagates from N1 to N6.

When the RREQ reaches the destination or the node that has a fresher routing entry for that destination, a RREP is sent back to the source. RREP contains the following information: src_addr, dst_addr, dst_seq_no, hop_cnt, and lifetime. The lifetime is used to specify how long the routing information should be cached before it is removed if the route is not used. The node that receives the RREP packet will update its routing table and forward RREP back to the source through the reverse path in the following conditions:

- There is no routing entry for the dst_addr
- The dst_seq_no in RREP is greater than it in the routing table
- The dst_seq_no are the same, but the metric is better (hop_cnt is less)

Otherwise, the RREP is discarded. After the source node receives the RREP message, it can send data to the destination node, and the data will be forwarded hop by hop according to each node's routing table.

When a link breaks along an active routing path, the "path maintenance" is triggered. The node that detects the broken link will issue a special unsolicited RREP with a dst_seq_no, one greater than the previously known sequence number of the destination, and with hop_cnt set to ∞, to its active neighbors of the affected routes. An active neighbor of a route is a node from which packets are received for that route. In this way, all nodes can be notified of the broken link, and may initiate another path discovery process to find another route to the destination.

The AODV takes the merit of DSDV by maintaining only local connectivity to its neighbors. It also takes the advantages of DSR, such as, requires no periodic advertisement, caches only required routing entries, and avoids redundant routes maintenance when topology changes.

Temporary Ordered Routing Algorithm (TORA) (Park and Corson, 1997a) is a reactive, distributed routing protocol based on "link reversal" algorithm. It uses a directed acyclic graph (DAG) to construct multiple

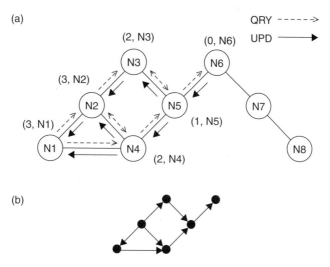

FIGURE 9.5   Route creation in TORA. (a) QRY, UPD, and height of nodes; (b) constructed routing paths (DAG).

routing paths from source to destination on-demand. When the network topology changes, the reaction is localized only to the vicinity of changed links, preventing the routing control messages from propagating throughout the network.

The basic idea of TORA is to assign a "height" to each node when a route is constructed. Height is assigned in a manner such that nodes in the upstream of a route always have higher height than nodes in the downstream of the route, that is, the route is constructed by assigning directed links from high nodes to low nodes from source to destination. When a link is broken, the node that detects the broken link adjusts its height if it does not have downstream links anymore. The node changes its height to be higher than all of its neighbors, and new downstream links can be assigned from the node to all its neighbors.

The TORA can be categorized into three basic functions: route creation, route maintenance, and route erasure, which stand for creating route from source to destination, maintaining routing paths with topology changes, and deleting route when topology changes result in a partitioned network, respectively.

When a source wants to create a route to a destination, it broadcasts a query (QRY) packet, to find the destination. The QRY packet will be propagated throughout the network and will eventually reach the destination. The destination will send a reply packet, UPD, which will contain its height. The UPD packet will be propagated back to the source. Nodes that receive the first UPD packet will set its height so that it is always higher than the node from which the UPD came. When the UPD reaches the source, a directed acyclic graph will be constructed, forming multiple routing paths from the source to the destination. Figure 9.5(a) illustrates the route creation process where source is node N1 and destination is node N5. Figure 9.5(b) is the constructed routing paths in Figure 9.5(a). Here we use $(t, \text{node\_id})$ as the height of nodes, and we define $(t_i, \text{node\_id}_i) > (t_j, \text{node\_id}_j)$ if $t_i > t_j$, or $t_i = t_j$ and $\text{node\_id}_i > \text{node\_id}_j$.

When a link of an active route breaks, if the upstream node of the broken link has other downstream links to other neighbors, no reaction is required (as in Figure 9.6[a]). When the last downstream link of a node breaks, the node initiates a process termed "link reversal" (Figure 9.6[b]). The node changes its height to be higher than all its neighbors such that all links of the node are reversed from upstream link to downstream link. After the node changes its height, it broadcasts a UPD packet to notify its neighbors such that all its neighbors know the links are reversed.

However, when a broken link causes network to the partition, the "link reversal" route maintenance process fails to work. The node that broadcasts the UPD message finds that the same UPD message is reflected back from all its neighbors. This is used as an indication of network partitioning. In this situation,

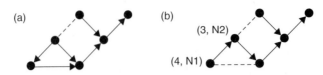

FIGURE 9.6 Route maintenance: (a) Link between N2 and N3 is broken in Figure 9.5(b); (b) Link between N1 and N4 is broken in Figure 9.6(a).

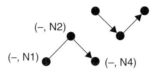

FIGURE 9.7 Route erasure. Link between N4 and N5 is broken in Figure 9.6(b).

it will broadcast CLR packet to notify all nodes in the partitioned network, set all nodes' height to NULL, and remove all invalid routes (Figure 9.7).

The TORA provides multiple routing paths from source to destination such that link breakages some-time require no reaction to reconstruct a route. Even if link breakages result in route failures, the route reconstruction process is limited to the vicinity of the broken link, which reduces the number of routing control messages and decreases the overhead of route maintenance. But over a sufficiently long period, because of the way TORA maintains routes, the optimality of routes may no longer hold.

Associativity-Based Routing (ABR) (Toh, 1996) is a reactive, associativity-based routing protocol that constructs routes based on the stability of links. Associativity is a measure of the node's association with its neighbors, that is, the mobility of nodes, link delay, signal strength, power life, and relaying load. Each node maintains the association of its neighbors, called associativity ticks, by periodically broadcasting and receiving beacons. ABR constructs a routing path by selecting nodes that have the higher associativity ticks, which means more stable links, as routers to forward packets. This increases the lifetime of a route, decreases the number of required route reconstruction, and reduces the overhead of route maintenance.

When a source node needs to find a path to a destination node, it broadcasts a broadcast query (BQ) message including the associativity ticks with its neighbors. The node that receives the BQ message will discard it if the BQ message has been previously processed. If the node is not the destination, it will append its address to the BQ message, update the associativity ticks with its own associativity ticks, and rebroadcast it. Eventually the BQ message will reach the destination and contain the routing nodes, their associativity ticks, and other routing information. Because the destination may receive multiple BQ messages from different routes, it can then select the best route and send the route in a REPLY packet back to the source. Various metrics such as higher stability and lower relaying load can be used for selecting the best route. The intermediate nodes that receive the REPLY packet can save its route to the destination.

When a link from a source node breaks, it may reconstruct the route by reinitiating route discovery process. If it is the intermediate node that moves, the upstream node of the broken link will initiate the route reconstruction (RRC) process. It will start a localized query (LQ) process to determine if the destination is still reachable by broadcasting a LQ{H} packet, where H indicates the hop count between the node and the destination. If the destination node can receive the LQ{H} packet, it will find a best partial path and send a REPLY back to the node. If the node does not receive the REPLY in a period, the LQ process will be time-out and the node will backtrack to the next upstream node to start another LQ process by sending a FQ{0} packet. If the destination cannot be found when the LQ process has been

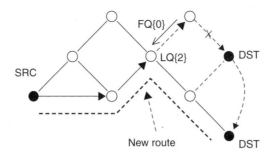

FIGURE 9.8   RRC process.

backtracked halfway to the source, the LQ process will stop and a FQ{1} packet will be sent directly back to the source to start a new route discovery process. Figure 9.8 illustrates the RRC process.

In ABR, because routes are determined by the stability of links, called associativity ticks, the routes are more stable, that is, have longer lifetime. This property improves the routing performance because the number of RRC will be less. When node movement results in an invalid routing path, the RRC process may find a partial path to reconstruct the route, localize the reaction to the node movement, thereby reducing the overhead of RRC.

Randomized Destination Sequenced Distance Vector (R-DSDV) (Boukerche et al., 2001) is a modification to the DSDV protocol by including a randomized scheme. In DSDV, the routing updates have to be advertised as soon as possible whenever the routing information changes. This ensures the correctness and freshness of the routing information. In R-DSDV, however, nodes do not always advertise or propagate the routing updates when the routing information changes. Nodes advertise the routing updates probabilistically.

By including the randomized scheme, the overhead of messages for route control can be reduced since the number of updates is reduced. Furthermore, because the routing updates are randomized, the routes may change for the same source–destination even if the network is static. This is because the routing updates may be propagated on different paths, and therefore the updates to the routing table may be different.

With the randomized routing update, the reaction to the topology changes may be slower. However, the traffic can be distributed to different routes because of the randomization. Furthermore, when some nodes experience congestion, the network can adjust to it by decreasing the probability of route updates of those nodes, then directing the traffic to other routes.

Preemptive Ad Hoc On-Demand Distance Vector ($P_r AODV$) (Boukerche and Zhang, 2003) is a reactive routing protocol based on the AODV. Whenever a route is required, the source node initiates "path discovery" process to find a routing path as it does in AODV, and when a link breaks, it uses "path maintenance" mechanism to reconstruct the path.

The difference is that $P_r AODV$ adopts preemptive mechanism to find another routing path before the existing path breaks. $P_r AODV$ uses the following two approaches:

- *Schedule a rediscovery in advance:* In "path discovery" phase, the links information along the discovered route are collected and forwarded back to the source. Therefore, the source can learn the condition of all links, including how long the lifetime of a link will be. Hence the source can schedule another "path discovery" process before the path breaks.
- *Warn the source before the path breaks:* Nodes monitor the signal power of packets to check if the link is going to break. When the signal power is below a threshold, it begins the *ping–pong process*. The node that detects the low power signal sends a *ping* packet to the node from which the packet was sent. The node that receives the *ping* packet replies with a *pong* packet. If the *pong* packet cannot be received successfully, a warning message will be sent back to the source. Upon receiving the warning message, the source will initiate "path discovery" to find another feasible route.

Using these two mechanisms, P_rAODV can find new paths before current in-used routes break. By switching to new routing path before current in-used route breaks, P_rAODV can increase number of packets delivered and decrease the average packet delay.

9.2.2 ID-Based Hierarchical Routing Protocols

Unlike flat routing protocols that regard the network as uniform, hierarchical routing protocols divide network nodes into different groups, or clusters. Routes are constructed according to the hierarchy created by these groups. In the rest of the section, we present the key ideas and the principles behind the design of the following routing protocols: ZRP, HSR, and CEDAR.

Zone Routing Protocol (ZRP) (Zygmunt J. Haas et al., 2002d) is a hybrid routing protocol framework that takes the merits of proactive and reactive routing protocols. Each node in ZRP maintains the local routing information within its vicinity, called the routing zone. ZRP uses proactive routing protocol, intrazone routing protocol (IARP) [Zygmunt J. Haas et al., 2002c], to maintain routes. Routes across different routing zones use the reactive routing protocol interzone routing protocol (IERP) (Zygmunt J. Haas et al., 2002b). IERP uses bordercast resolution protocol (BRP) (Zygmunt J. Haas et al., 2002a) to deliver route query messages across routing zones.

The IARP is a family of proactive routing protocols that offers routing services based on local connectivity. IERP is a family of reactive routing protocols that provides on-demand routing service. All routing protocols conforming to the properties of IARP or IERP can be adapted in the ZRP protocol.

In ZRP, each node maintains its own routing zone. Routing zone is defined by the nodes that are reachable within a certain number of hops. Routes within a routing zone can be constructed quickly and locally because routing zone uses proactive routing protocol. Routes across different routing zones have to be constructed using route discovery process just like all other on-demand routing protocols. But in ZRP, route queries are not flooded throughout the network. Routes across different routing zones are constructed using IERP, which uses BRP protocol to deliver route queries.

The BRP is a bordercast protocol, which is a multicast-like protocol, which sends packets only to some specific nodes. BRP transfers route queries to peripheral nodes of the routing zone. Peripheral nodes are nodes at the routing zone border. Figure 9.9 illustrates the routing zone and peripheral nodes of the routing zone. N2, N4, N5, N6, N8, and N10 are within N6's routing zone with radius equals to two hops. N2 and N10 are the peripheral nodes of N6's routing zone. BRP transfers route query only to peripheral nodes instead of flooding it, which reduces a lot of overhead. When a node receives the route query, if the destination is in its routing zone, the node can send route reply back to the source; otherwise, it will bordercast the route query to its peripheral nodes, and so on, until the destination is found.

ZRP limits the routing information exchange in routing zone, and reduces global route query flooding by bordercasting the query directly to the farthest nodes from the source in a routing zone. This results in

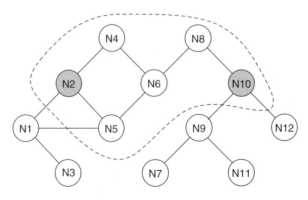

FIGURE 9.9 Routing zone and peripheral nodes with routing radius is 2.

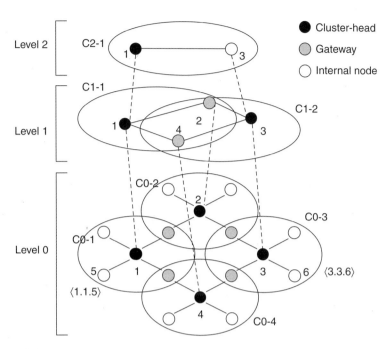

FIGURE 9.10 Multilevel clusters hierarchy.

the query flowing outward while reducing routing control message overhead. Because of the properties of the proactive routing, node movements cause nearly no extra overhead to maintain. The information of a broken link is only propagated within the zone-radius.

Hierarchical State Routing (HSR) (Guangyu Pei et al., 1999; Iwata et al., 1999) is a proactive, hierarchical, multilevel cluster routing protocol. Nodes in HSR are divided into clusters, each cluster elects its cluster-head, and the cluster-heads then form the next level clusters, and so on, to construct a multilevel cluster topology. Each node then is assigned a hierarchical address, HID, according to its location in the hierarchical clusters. Using the hierarchical address, packets can be delivered from the source to destination. Figure 9.10 illustrates the multilevel clusters hierarchy.

In Figure 9.10, nodes are divided into four clusters, C0-1, C0-2, C0-3, and C-04, whose cluster-heads are node 1, 2, 3, and 4, respectively (Level 0). Node 1, 2, 3, and 4 then create the next level clusters, C1-1, C1-2, with cluster-head as node 1, 3, respectively (Level 1). Finally, node 1 and 3 form a cluster (Level 2). Nodes are assigned with a HID according its location in the hierarchy, such as ⟨1.1.5⟩ for node 5, ⟨3.3.6⟩ for node 6. When node 5 wants to send data to node 6, the data is sent to the cluster-head of top-level cluster of node 5, which is node 1. Node 1 then forwards it to node 3, which is the cluster-head of top-level cluster of node 6. Finally, node 3 can forward downward to node 6.

Since HID will change when cluster membership change, each node in HSR is assigned with another address, the logical address, whose type is ⟨subnet, host⟩. The subnet portion of the address is the logical cluster of the network. Each logical cluster has a home agent, which is responsible to map the logical address to HID of nodes in the logical cluster. Nodes in the logical cluster register their logical address and HID to its home agent. The logical cluster (subnet portion of the logical address) and the HID of the home agent will be registered at the top-level cluster, and broadcast to all nodes periodically. When a node wants to send packets to the destination using the logical address, it will first extract the subnet of the logical address, find the HID of the home agent that associates with the logical cluster, and send that packets to it. The home agent will receive the packets, find the HID of the destination, and forward the packets to the destination. Once the source and destination know each other's HIDs, they can communicate directly without the home agent.

In HSR, the overhead of periodic advertisement in table-driven routing protocols is reduced because the cluster membership maintenance is limited within a cluster, so it has better scalability than other proactive routing protocols, such as DSDV. But the route created in HSR is not an optimal path, and the complexity is higher because of the introduction of the home agent. Compared to reactive routing protocols, HSR has lower latency and overhead because the destination's address query doesn't flood the entire network.

- *Core Extraction-Based Distributed Ad-Hoc Routing* (CEDAR): CEDAR (Sinha et al., 1999a) uses a 2-level hierarchical substructure to compute routes in ad hoc networks. The computed routes do not necessarily follow the levels of the hierarchy, thus avoiding inefficiencies of most hierarchical routing protocols. The hierarchy is managed distributed using only local computation. The following is a brief description of the three key components of CEDAR:
- *Core Extraction*: A set of hosts is distributed and dynamically elected to form the *core* of the network by approximating a minimum dominating set of the ad hoc network using only local computation and local state. A dominating set is defined as a subset of nodes such that all other nodes of the ad hoc network are 1-hop away from at least one dominating set member. The dominators are referred to as core nodes in CEDAR. Each noncore node registers with at least one dominator in its 1-hop neighborhood. Each core node maintains the local topology of the nodes in its domain, and also performs route computation on behalf of these nodes. Periodic beaconing by the nodes is used to facilitate the election and maintenance of core nodes, and information accrual at the core nodes.
- *Link State Propagation*: Quality of service (QoS) routing in CEDAR is achieved by propagating the bandwidth availability information of stable links in the core graph. The basic idea is that the information about stable high bandwidth links can be made known to nodes far away in the network, while information about dynamic links or low bandwidth links should remain local.
- *Route Computation*: Route computation first establishes a core path from the dominator of the source to that of the destination. The core path provides the directionality of the route from the source to the destination. Using this directional information, CEDAR computes a route adjacent to the core path that can satisfy the QoS requirements.

9.2.3 Location-Based Routing Protocols

Protocols in this family take advantage of position information of nodes to direct the packet flow. There are two issues that need to be addressed for enabling the use of location information for routing. The first one is the availability of a location service, which keeps track of up-to-date locations of nodes and responds to location queries. The second one is how to design location-based routing protocols, which assume that the location information of the destination is known to the packet sender. This section presents some protocols from these two classes. DREAM, Quorum-based location service, GLS, and Homezone are location services, and LAR, GPSR, GRA, GEDIR, ZHLS, GRID, GeoTORA, GDSR, and GZRP are location-based routing protocols.

Distance Routing Effect Algorithm for Mobility (DREAM) (Basagni et al., 1998) provides location service for position-based routing. In this framework, each node maintains a position database that stores position information about other nodes in the network. It can therefore be classified as an *all-for-all* approach, which means all nodes work as the location service providers, and each node contains all other nodes' location information. An entry in the position database includes a node identifier, the direction and distance to the node, together with the time-stamp of entry creation. Each node regularly floods packets to update the position information maintained by the other nodes.

Since the accuracy of the position information in the database depends on its age, a node can control the accuracy of its position information available to other nodes by adjusting the frequency of sending position updates (temporal resolution), or by indicating how far a position update may travel before it is discarded (spatial resolution). The temporal resolution of sending updates is positively correlated with the mobility rate of a node. The spatial resolution is used to provide accurate position information in the direct neighborhood of a node and less accurate information for nodes farther away. Such disposition reflects the fact that "the greater the distance between two nodes, the slower they appear to be moving with respect to each other" (termed as the distance effect) (Basagni et al., 1998). The distance effect uses lower spatial resolution in areas farther away from the concerning node, and the costs of accurate position information for far away nodes can be reduced.

Quorum-Based Location Service (Haas and Liang, 1999) is derived from the theory of information replication in databases and distribution systems. In this framework, a subset of mobile nodes is chosen as the location databases. This subset is composed of several quorums, and each pair of these quorums has a nonempty intersection. A virtual backbone is constructed between the nodes in the subset, using a nonposition-based ad hoc routing mechanism.

A mobile node sends position update messages to the nearest backbone node. The backbone node then chooses a quorum of backbone nodes to store the position information. When another node intends to send packet to it, it sends a query message to the nearest backbone node first. The backbone node then relays the query to a quorum, which may be different from the former one. Since any two quorums have nonempty intersection, the sending node is guaranteed to get at least one response. If multiple responses are received, the receiver should choose the one representing the most up-to-date positions according to the time-stamp in the response messages.

From the description above, it is obvious that the quorum-based location service can be classified as *some-for-some*, which means some of the mobile nodes provide the location service, and each of them stores the position information for some mobile nodes. It can also be configured to work as all-for-all or as all-for-some, but the nature of some-for-some is essential.

Grid Location Service (GLS) (Li et al., 2000) divides the mobile ad hoc network into a hierarchy of grids with squares of increasing size. The largest square is called the order-n square, which covers the whole networks. Order-n square is partitioned into four order-$(\bar{n} - 1)$ squares, each of which can contain four order-$(n - 2)$ squares until the order-1 squares appear. In this way, a mobile node lies in exact one order-x $(x = 1, 2, \ldots, n)$ square, and any order-x has three neighbor squares under the same order-$(x + 1)$ square.

In GLS, mobile nodes are identified by their IDs. The IDs are distributed uniformly across the network, and nodes have only ID information about other nodes. The policy of selecting a node's location server is to choose its successor. In GLS, the "successor" is defined as the node with the least ID greater than the target ID in a circular manner. A node chooses three location servers for each level of the grid hierarchy, and each is its successor in that level of the grid. For example, in Figure 9.11, node 10 chooses nodes 25, 44, and 36 as its location server in order-2 square, and chooses 16, 77, and 98 as its location server in order-3 square. The same rule applies for 20, 23, and 80 in order-4 square. In the order-1 square containing the node, all nodes in that square are required to provide location service for it, and it knows about all other nodes in the square. In this way, the location servers for a node are relatively dense near the node but gets sparser as moving away from the node. Therefore, the GLS can be classified as *all-for-some*.

To perform a location query, the querying node sends a request to the target's successor in its square. If the receiving node does not know about the query target, it forwards the query to the target's successors in neighbor squares. This process is carried on until it reaches a location server of the destination. The server then sends back a reply using location-based forwarding to the querying node. For example, in Figure 9.11, node 99 intends to send a packet to node 10, it then inquires to node 35, which is a successor of node 10. Because node 35 is not a location server of node 10, it forwards the inquiry to node 7, 23, and 30, node 10's successors in node 35's neighbor squares. Since 23 is chosen by node 10 as its location server, node 23 will reply the inquiry.

Homezone (Giordano and Hamdi, 1999) is a framework in which each mobile node applies a well known hash function to the node identifier. The hash result is defined as the homezone position of the node.

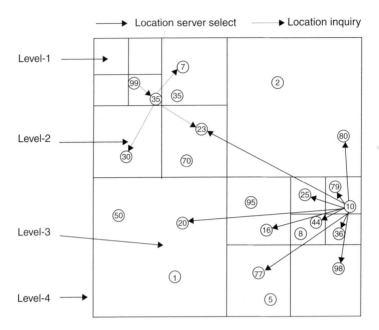

FIGURE 9.11　Example of GLS.

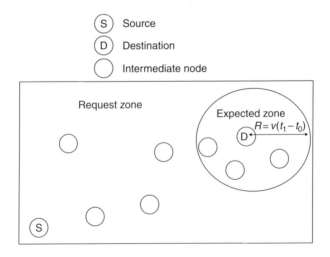

FIGURE 9.12　Expected zone and request zone.

Other nodes in the disk of radius R to the homezone position are required to provide location service for the target node. In this way, the position database can be found if the sender node and destination node agree on a common hash function without extra message exchanges. This approach can then be classified as an *all-for-some* one. If the homezones are sparsely distributed, the radius R should also be increased, which means trade-offs of the R within several location updates and queries.

Location-Aided Routing (LAR) protocol (Ko and Vaidya, 1998) assumes that the source S knows destination D's recent location (X_d, Y_d) and roaming speed v at time t_0. Then at current time t_1, S can define the *expected zone* in which node D is expected to be located in (Figure 9.12), which is a circle with the radius of $R = v(t_1 - t_0)$. The expected zone can also be shrunk if S knows more accurately about the mobility of D, or be expanded if the actual speed of D exceeds the anticipated speed or initial location of D at t_0 when it is unknown to S.

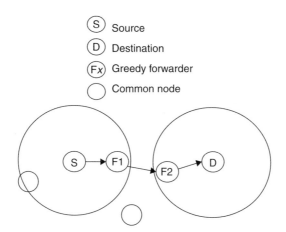

FIGURE 9.13 Greedy forwarding.

The LAR also uses flooding as the way to find routes, but in a restricted style. To reduce the unnecessary full network flooding, LAR defines a *request zone*. Only nodes inside the request zone will rebroadcast the RREQ packet. To increase the probability that the route request will reach destination D, the request zone should include the expected rectangular zone, as Figure 9.12 illustrates. When S issues the route request, it should piggyback necessary information about the request zone in the request packet. The node receiving the request only needs to compare its own coordinates to the request zone to decide whether it is inside the request zone.

After D receives the RREQ packet, it sends a RREP packet to S. Receiving the reply by S means the route is established. If the route cannot be discovered in a certain period, S can broadcast a new RREQ packet with coordinates of an expanded request zone. The expanded request zone should be larger than the previous one. It can be set as the whole network when necessary. Since the new request zone is larger, the probability of discovering a route is increased at the cost of more flooding traffic.

Greedy Perimeter Stateless Routing (GPSR) protocol (Karp and Kung, 2000) makes use of location information of all neighbors. Such information may be obtained by periodic beacon messages. The location of the destination should also be known in advance. The GPSR protocol does not need to discover a route prior to sending a packet. A node forwards a received packet directly based only on local information. In this routing protocol, two forwarding methods are used in GPSR: greedy forwarding and perimeter forwarding.

When a node intends to deliver a packet to another node, it picks one node from its neighbors, which is closest to the destination among itself and all of its neighbors, and then forwards the packet to it. Figure 9.13 shows an example of greedy forwarding. S is the source node, and D is the destination node. In this example, node R is the chosen neighbor. After receiving the packet, node R follows the same greedy forwarding procedure to find the next hop. This process is repeated until node D or a *local maximum node* is reached.

It is possible that the sending node cannot find a neighbor closer to the destination than itself. Such a node is called *local maximum node*. In the example in Figure 9.14, node S is a *local maximum node* because all its neighbors are farther from D than itself. In such a scenario, the greedy forwarding method cannot be applied. Instead, the perimeter forwarding method is used.

To use the perimeter forwarding, the *local maximum node* first needs to planarize the network topology graph. A graph is said to be a planar only if there are no two crossing edges. The graph may be transformed into a relative neighborhood graph (RNG) or a Gabriel graph (GG). Both are planar graphs. When the planarization is done, the *local maximum node* S forwards the packet according to a right-hand rule to guide the packet along the perimeter of a plane in a counterclockwise style, as Figure 9.14 illustrates.

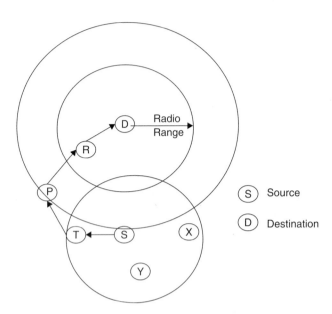

FIGURE 9.14 Perimeter forwarding.

When the packet reaches node R, where we are sure that we are closer to D (as opposed to the location of node T), the greedy forwarding is then applied again.

These two forwarding methods are used interchangeably until the packet reaches the destination node. GPSR only maintains the location information of all its neighbors, instead of entries for each destination, which makes it nearly a stateless routing protocol. Overall, the small per node routing state and small routing protocol message complexity are the advantages of GPSR.

Geographical Routing Algorithm (GRA) (Jain et al., 2001) integrates the greedy routing manner and the table-driven manner. Initially a GRA node only knows the position information of itself and its neighbors. But as time elapses, it will get more and more position information of other nodes. Before sending or forwarding a packet, a GRA node first checks its routing table. If the checking is successful, the packet is forwarded according to the entry found. If it is not the case, the packet is greedily forwarded, like in GPSR. It forwards the packet to the neighbor which is closest to the destination. If the packet reaches a local maximum, GRA uses a route discovery procedure with flooding to find a route to the destination, and then catches the new route in its routing table.

Geographic Distance Routing (GEDIR) (Lin and Stojmenovic, 1999) protocol also assumes that each node has the locations of its direct neighbors. Similar to GPSR, the GEDIR protocol also directly forwards packets to next hops without establishing routes in advance.

Compared to GPSR, there are two approaches of packet forwarding: based on distance and based on direction. In the distance-based approach, the packet is forwarded to the 1-hop neighbor that is nearest to the destination. However, in the direction-based approach, the packet is forwarded to the 1-hop neighbor whose direction is closest to the destination's direction. The direction here can be defined as the angle formed by the vector from the current node to the destination and to the next hop. The distance-based approach may route a packet to a *local maximum node*, while the direction approach may lead a packet into an endless loop. To resolve these problems, f-GEDIR (f means flooding) and c-GEDIR (c means the number of 1-hop downstream neighbors) were proposed. These enhancements are dedicated to help the packet leave the *local maximum node* or the loop. Furthermore, 2-hop GEDIR was proposed as another improvement, in which nodes need the location information of up to 2-hops. Before sending or forwarding a packet, a node selects a node which is closest to the destination among all up to 2-hop neighbors. If the picked node is a 1-hop neighbor, the packet is directly forwarded to it; otherwise, the packet is forwarded

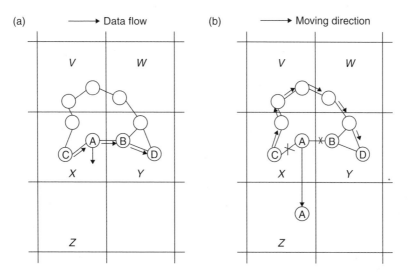

FIGURE 9.15 Example of zone-based routing. (a) Original connectivity between C and D. (b) Connectivity between C and D after A moves to zone Z.

to the node of a 1-hop neighbor of the picked node. In addition, flooding can be added to the protocol to find a route. GEDIR and its variants have all been proven to be loop-free.

Zone-Based Hierarchical Link State (ZHLS) is a routing protocol (Joa-Ng and Lu, 1999) that partitions the whole network area into squares in advance. Each mobile host knows this partition, so they know their own zone. There are two kinds of routing update in this protocol: intrazone and interzone. Local position change within a square triggers only local link state routing update in the square, while change of connectivity between squares triggers global routing update. For example, host A and host B connect square X and square Y, and is the only connection between these two squares (Figure 9.15[a]). In this scenario, A and B are called gateways of square X and square Y, respectively. Node C in zone X and node D in zone Y share the route C–A–B–D. When A moves from zone X to zone Y, it loses the connectivity with B, the connectivity between square X and square Y is also lost. This change will trigger the global routing information update. After that, C must resort to zone V and zone W to communicate with D (Figure 9.15[b]).

This table-driven approach requires each host to maintain an interzone routing table and an intrazone routing table. Before sending a packet, the host checks its intrazone routing table first. If a route is found, the packet is routed locally in the square. Otherwise, the host broadcasts a RREQ packet to other squares through the gateway of its square. The information needed here is only the square that the destination host belongs to. The packet is routed through interzone routing until it enters the destination square, then it will go through an intrazone routing until it reaches the destination host.

GRID (Liao et al., 2001) is, in fact, another zone-based routing protocol. It uses location information in all aspects of routing: route discovery, packet forwarding, and route maintenance. The network area is also partitioned into squares called *grids*. In each *grid* containing at least one mobile host, one host is elected as the leader. Packets routed between grids go through only leaders. In other words, only leaders forward packets.

During route discovery, a request zone is used to confine the scope of RREQ packets that are broadcast. In addition, since only leaders of grids are responsible for packet forwarding, the cost of broadcast flooding is highly reduced.

In packet forwarding, since the route is represented by a sequence of grid IDs, not host IDs, GRID provides great capacity of tolerance of host mobility. When the leader of some grid leaves its grid, a new leader is elected to replace the older one. In this way, the routing table only needs to record the next grid for a specific destination.

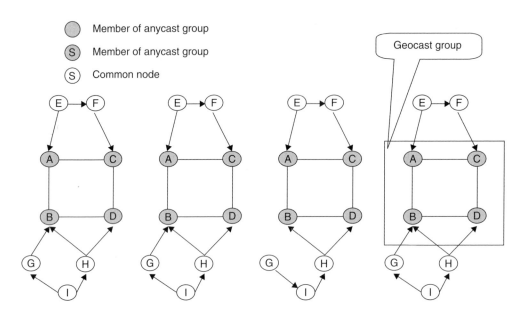

FIGURE 9.16 GeoTORA for anycast and geocast. (a) Example of GeoTORA for anycast. (b) Link between B and G breaks. (c) DAG rebuilding through link reversal. (d) Example of GeoTORA for geocast.

For route maintenance, the leader reelection provides the ability of maintaining the same route even if the old leader leaves its original grid. This process is called *handoff*; it is not possible for other routing protocols to do this job.

GeoTORA (Ko and Vaidya, 2000) is a variant of TORA introduced in (Park and Corson, 1997b). TORA is a unicast routing protocol, which uses destination-oriented directed acyclic graphs (DAGs) to acquire multiple paths to a destination, and uses "link-reversal" technique to maintain DAGs. GeoTORA first extends TORA to support anycast, and then it further extends anycast to support location-based routing.

Anycast is defined as transmission of packets to any node from a certain group. To support this function, GeoTORA requires each node to maintain a DAG for each anycast group, which has packets to send to. In this case, each member of the anycast group is called a sink. Since packets only need to reach any one of them, there is no logical direction links between any two sinks needed. Then packets are delivered following the DAG to reach any one sink. For example, suppose nodes A, B, C, and D form an anycast group in Figure 9.16(a), each other link generates a DAG for each node in the anycast group. It is worthwhile to mention that a node may not be able to reach all nodes in the anycast group, which can happen when a node in the anycast group lies in the critical path in Figure 9.16(a) for another node in the same anycast group. Nodes E and B is a good example for this. When the link between nodes B and G breaks in Figure 9.16(b), the route maintenance of TORA starts to work. Through link reversal between G and I, G reestablishes the DAG to reach the anycast group (Figure 9.16[c]).

Using the mechanism of anycast, GeoTORA further extends TORA to support geocast. To do this, all nodes within the destination region are considered as sinks, thus forming a special anycast group. Nodes A, B, C, and D in the rectangle in Figure 9.16(d) present a good example. The source node maintains a single DAG for a given geocast group. Source node first performs an anycast to the geocast group. When a group member receives a geocast packet, it floods it within the geocast region.

GPS-Based Route Discovery Optimizations (GDSR) (Boukerche and Sheetal, 2002; Boukerche and Sheetal, 2003) and GZRP (Boukerche and Rogers, 2001) are developed upon the reactive DSR (Johnson and Maltz, 1996) and hybrid ZRP (Haas and Pearlman, 1998, 2000; Pearlman and Haas, 1999; Zygmunt J. Haas, et al., 2002d), respectively. The GPS location information is used in these protocols to restrict the propagation direction of RREQ messages. In GDSR, nodes maintain the GPS coordinate information of their neighbor nodes and use the GPS screening angle to help the forwarding decision. To measure the GPS screening angle, the node uses two virtual lines: one from the previous-hop node to the current node, and another from this current node to next hop. These two lines form an angle, which is defined as the GPS screening angle. When a node receives a route query message from one of its neighbor nodes, it checks all other neighbor nodes to see whether the GPS screening angle is larger than a threshold value, say 90°. Then it will forward the route query message to each neighbor node with large enough GPS screening angle. In this way, the propagation of the query message is restricted to a relatively small scope, which means savings in route query messages. In GZRP, the authors use the same technique to reduce the number of route query messages in ZRP based on the GPS screening angle formed by previous-, current-, and next-hop nodes.

9.3   Broadcast Routing Protocols

Broadcast is a basic operation in ad hoc environment. Many routing protocols and applications are operated based on broadcast, such as AODV, DSR routing protocols, and data dissemination to all nodes in network. Therefore, a robust and efficient broadcasting algorithm is necessary in an ad hoc network environment.

The simplest and most trivial broadcasting algorithm is pure flooding. Every node that receives the broadcast message retransmits it to all its neighbors. The problem of pure flooding is that it produces many redundant messages, which may consume scarce radio and energy resources, and cause collision that is called *broadcast storm* problem (Ni et al., 1999). Therefore, the basic principle of designing an efficient and resource conservative broadcast algorithm is trying to reduce the redundant messages, which means to inhibit some nodes from rebroadcasting and the message can still be disseminated to all nodes in the network. In this section, we introduce the following broadcasting algorithms: SBA, DCB, and WMH, followed by discussion on some other schemes proposed for broadcast routing protocols.

Scalable Broadcast Algorithm (SBA) (Peng and Lu, 2000) is a simple, distributed broadcast algorithm that can reduce redundant messages in broadcast. The basic idea is: when a node receives a broadcast message, it will decide whether to rebroadcast the message or not according to whether all its neighbors have received the same message by previous broadcast. If all its neighbors have received the message before, then the node will not rebroadcast it, otherwise it will. Nodes only need to exchange their 1-hop neighbors' information with their neighbors to obtain the 2-hop neighbors' information, and can reduce unnecessary broadcast message to decrease energy consumption and save network bandwidth.

The SBA can be divided into two phases: local neighbors' maintenance and broadcasting. Local neighbors' maintenance is to let nodes know all their 2-hop neighbors. Nodes can exchange their 1-hop neighbors' information with their 1-hop neighbors by periodically advertising a "hello" message including their 1-hop neighbors. So every node in the network has its 2-hop neighbors list. When a node, u, receives a broadcast message, it will know where the message comes from, and also the source's neighbors. It can compare the source's neighbors, $N(s)$, with its neighbors, $N(u)$. Because all source's neighbors are supposed to receive the broadcast message, it is not necessary to broadcast for those nodes. Therefore, if u's neighbors are subset of source's neighbors, that is, $N(u) \subseteq N(s)$, then all u's neighbors should have received the message, and u does not need to rebroadcast it again.

If u's neighbors are all not covered by the source's broadcast, it will wait a random period before rebroadcasting it to reduce the opportunity of collision. In the waiting period, if u receives the same broadcast message from other nodes, it will compute which neighbors are covered. If all its neighbors have been covered at least by one broadcast before it rebroadcasts the message, it will cancel the rebroadcast. The random period T is calculated as a function of the maximum degree of u's neighbors and the degree of u:

$$T = f(\Delta \times [(1 + d_m(u))/(1 + d(u))])$$

Δ is a small constant, $d_m(u)$ is the maximum degree of u's neighbors, and $d(u)$ is the degree of u. $f(x)$ is a function that returns a random number between 0 and x. In this way, nodes with higher degree, which means having more neighbors, will have more opportunities to rebroadcast earlier than other nodes having fewer neighbors, and hence covers more nodes.

SBA is a distributed broadcast routing protocol that is adaptive to highly dynamic network. It requires only local topology knowledge to decide routing. Comparing to other broadcast algorithms such as probabilistic-based, counter-based, distance-based and location-based scheme [Ni et al., 1999], SBA is more reliable for disseminating messages to all nodes in the network. Though it is not optimal or approximated to optimal as minimum connected dominating set (MCDS) or other protocols using connected dominating set (CDS) as rebroadcast tree, it is very simple and efficient.

Double-Covered Broadcast (DCB) (Lou and Wu, 2004) is a broadcast algorithm designed to lower the broadcast overhead by reducing the number of forward nodes while providing a reliable transmission. Node that receives a broadcast packet selects some of its neighbors as forward nodes to rebroadcast the packet such that all its 2-hop neighbors are fully covered, and its 1-hop neighbors are covered by at least two forward nodes. Nodes overhear their forward nodes' rebroadcast packets as ACKs for confirmation of their receipt. If the node does not receive ACK from all its selected forward nodes in a period, it will rebroadcast the packet until all ACKs are received, or the maximum number of retry is reached.

The DCB constructs a CDS distributedly for broadcast. A dominating set is a subset of nodes such that each node in the network is either in the subset, or is a neighbor of a node in the subset. In DCB, nodes exchange their 1-hop neighbor node set with their 1-hop neighbors periodically, so every node knows its 2-hop neighbor node set. Let $N(v)$ denote the set of 1-hop neighbors of node v, and $N_2(v)$ be the set of 2-hop neighbors. If S is a node set, $N(S)$ represents the set containing all 1-hop neighbor nodes of nodes in S. When a source node u wants to broadcast a packet, it selects its forward node set $F(u)$ from the node set $X = N(u)$ to cover its 2-hop neighbors $U = N_2(u)$ using forward node set selection process (FNSSP) algorithm. The FNSSP algorithm works as follows:

1. Each node w in X calculates its degree, $|N(w) \cap U|$.
2. The node w_1 with the highest degree is selected as the forward node of w, and w_1 is removed from X and $N(w_1)$ is removed from U.
3. Repeat Steps 1 and 2 until U is empty.

If u is the node that receives the broadcast packet from node set $V(u)$, and is selected as the forward node to rebroadcast it, u selects its forward node set $F(u)$ using FNSSP algorithm, where $X = N(u) - V(u) - \bigcup_{\forall w \in V(u)} F(w)$ and $U = N_2(u) - N(V)(u)) - \bigcup_{\forall w \in V(u)} N(F(w))$. Here $F(u)$ is the 1-hop neighbor nodes of u, excluding the nodes from which the packet is received by u and the nodes which are forward nodes of nodes in $V(u)$, Figure 9.17 illustrates the node sets of X and U.

Once the forward node set is selected, node u broadcasts the packet piggybacked with the forward node set $F(u)$. Nodes that receive the packet can check if they are forward nodes or not by checking the forward node set in the packet. If node v is not a forward node, it does nothing. If node v is a forward node and receives the packet for the first time, it will compute its forward node set as described above. Otherwise it will just broadcast the packet as ACK without specifying the $F(v)$ in the packet.

When a node broadcasts a packet to all its neighbors, some neighbors may fail to receive the packet because of packets collisions, interference of radio signal, or node mobility. If the node that does not receive the packet is not a forward node, it still has another chance to receive the packet because DCB covers each node with at least two broadcast nodes. If the node is a forward node, the sender will retransmit the packet if it does not overhear the node's retransmission. But if the node moves away, the sender will never hear the retransmission from the moved node. In this case, the sender will remove that node from its neighbor node set and recalculate the forward nodes to cover the nodes that are supposed to be covered by the moved node.

There are other protocols, such as DP (Lim and Kim, 2001), MPR (Amir Qayyum *et al.*, 2002), TDP, and PDP (Lou and Wu, 2000), using similar concept of selecting dominating set of nodes, but with different scope of node set X and U or the way they select F, to forward packets and reduce the number of redundant

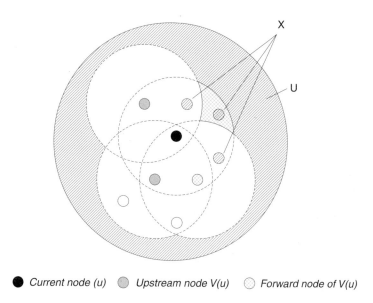

● *Current node (u)*　◉ *Upstream node V(u)*　○ *Forward node of V(u)*

FIGURE 9.17 Forward nodes and covered nodes.

transmission. But they may suffer from lack of robustness in broadcasting when the transmission error rate is high. DCB extends its capability to provide a more reliable broadcast mechanism by selecting forward nodes that make all nodes to be covered by at least two broadcast nodes, and retransmitting lost packets to forward nodes. But in highly dynamic network environments, the neighbor node sets information will become outdated quickly, which leads to unnecessary retransmission and missed packets, and increases the overhead of broadcast.

Wireless Multicast Tree with Hitch-Hiking (WMH) (Agarwal et al., 2004) is a broadcasting tree construction algorithm that takes advantage of the physical layer design, which uses the combining of partial signals. Using the combining of partial signals, a transmitting node can increase the signal strength to increase its coverage of the 2-hop neighbors, and reduce the required signal strength of its children to rebroadcast the packet. By calculating the increased power of the node and the reduced power of its children, WMH finds the transmission power of each node to reduce the total power consumption of broadcasting.

The concept of combining partial signals is used to increase the reliability of the communication. A transmission is either successful or failed according to the signal-to-noise ratio (SNR). If the ratio is too low, the receiver cannot decode the transmitting packet successfully, and the transmission will fail. But due to the combination of partial signals, multiple unsuccessful transmissions may be combined to form a successfully decoded message. Using this mechanism, WMH can increase a node's transmission power to increase the coverage of its 2-hop neighbors, thereby reducing the transmission power of its 1-hop neighbors to forward packets to their 1-hop neighbors. When the sum of the reduced power of its children is greater than the increased power of the node, the energy consumption is reduced.

The WMH first constructs a minimum spanning tree (MST) rooted at the source. Each node is assigned with a power level at which a node can fully cover its child nodes in MST. Starting from the source, WMH calculates a set of power levels such that each power level can fully cover all children nodes of one of its children nodes. As shown in Figure 9.18, node 5 calculates power levels 20, 20, and 10 for N1, N2, and N7, which is a child of node N3, N4, and N6, respectively. WMH then calculates all energy gains with each power level, and selects the power level that maximizes the energy gain. Energy gain is calculated as the sum of all reduced energy consumption of the children, minus the increased energy of the node. As in Figure 9.19, if N5 increases its power level to 10, the energy gain will be $(2 + 2 + 5) - 5 = 4$. If N5 increase its power level to 20, the energy gain will be $(5 + 5 + 5) - 15 = 0$. So N5 changes its power level to 10 because it has the maximum energy gain. When all children nodes are fully covered by the increased

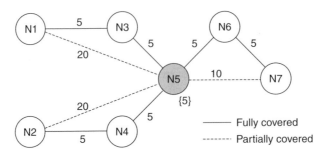

FIGURE 9.18   Power level calculation.

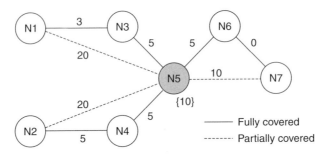

FIGURE 9.19   Energy gain by increasing the power level of the sender.

power node, such as node 7 in Figure 9.19, the child node, such as node 6, does not need to rebroadcast the packet. WMH repeats these steps to calculate the children's power level, and finally a broadcasting tree is constructed.

Using the WMH algorithm, the total energy consumption of broadcast can be reduced because redundant rebroadcast can be reduced and nodes may rebroadcast packets using lower transmission power. But the complexity of WMH is quite high because it has to construct a MST for each broadcasting node and has to find all the nodes' final power level, it is only suitable for highly static network environments.

In "The broadcast storm problem in a mobile ad hoc network" (Ni et al., 1999), the authors present some simple broadcast schemes that can reduce redundant rebroadcasting. Here we briefly summarize them as follows:

Probabilistic Scheme: Each node that receives a broadcast message will rebroadcast it with a probability P. Clearly, if P is 1, this scheme is equivalent to pure flooding.

Counter-Based Scheme: Each node that receives a broadcast message will wait for a period of time before rebroadcasting it. In this period, the node may receive the same broadcasting message from other nodes. It is clear that the more messages a node receives, the higher the probability that all its neighbors will be covered. Each node maintains a counter recording of how many times a message is received. If the counter exceeds a threshold, the node will not rebroadcast it.

Distance-Based Scheme: We can see from Figure 9.20 that if a node receives a broadcast message from a closer neighbor node, the extra coverage by this node's broadcast will be less than that a node receives from a farther neighbor node. Therefore, the node in distance-based scheme will determine whether or not to rebroadcast a message according to the distance of the source from which the message was received. Like in counter-based scheme, the node that receives a broadcast message will wait a for a period before rebroadcasting it. In this period, it is possible that it will receive the same message from other nodes. If the minimum distance of these nodes is smaller than a threshold, the node will not rebroadcast it.

Location-based Scheme: The location-based scheme is like the distance-based scheme, but instead of using the distance between the source and itself, nodes in location-based scheme uses nodes' location to calculate

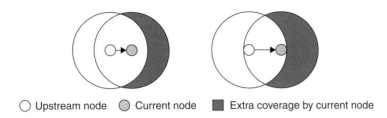

FIGURE 9.20   The extra coverage for different distance source.

the extra coverage by its rebroadcasting, which is more precise than only considering the distance between nodes. Nodes receiving multiple broadcasting messages can calculate the coverage by those broadcasts. If the extra coverage by its broadcasting is smaller than a threshold, the node will not rebroadcast it.

These protocols are simple and can be operated without the knowledge of network topology. The only problem is that they are not reliable. They cannot guarantee full coverage even in a collision-free and no-interference environment. Although with some positioning devices such as GPS, the information can be made more precise, it is still not easy to calculate the coverage.

9.4   Multicast Routing Protocols

There are many important applications in both mobile ad hoc networks and wired networks that need multicast, for example, online conference and multiplayer games. In wired networks, there are two main network layer multicast protocol families: per-source tree and core-tree. Many ad hoc multicast protocols also fall into these two families. Besides them, application multicast and stateless multicast protocols also play important roles in multicast protocol family. As for multicast routing protocols in mobile ad hoc networks, these protocols can be divided into five types like in wired networks: tree-based, mesh-based, backbone-based, stateless multicast, and application layer multicast. In this section, we will introduce some typical multicast routing protocols that fall into the following categories.

9.4.1   Forwarding Group Multicast Protocol

The forwarding group multicast protocol (FGMP) (Chiang et al., 1998) uses the forwarding group (FG) to confine the flooding to only a set of mobile nodes. Only nodes in FG are responsible for multicast packet rebroadcast. Each node in the FG needs to maintain a single flag and a timer. In this way, the storage overheads of intermediate nodes are greatly reduced, which leads to good scalability.

The core concept in FGMP is the FG. Nodes in FG use their forwarding flags to indicate whether to broadcast the received multicast packet or not. The timer is used to decide whether to relay the multicast packet. If the timer has expired without refreshment, the forwarding function is shut down.

To elect and maintain an efficient FG Chiang et al. (1998) proposed two approaches, via receiver advertising (RA) and via sender advertising (SA). In FGMP-RA, the multicast receiver periodically floods the whole networks with its member information, indicating its interested multicast group ID and its own ID. When a sender receives such a join request, it updates its multicast member table, recording ID information of all members with timers. Expired entries are deleted from the table. The sender then creates a forwarding table, encapsulating the multicast group ID, entries for each receiver and its next-hop to that receiver. The next-hop information is retrieved from the unicast routing table. The forwarding table is broadcast by the sender and received by all its neighbors. Only neighbors who can find themselves in the next-hop field of the forwarding table will process the message. It then becomes a member of the FG. Each neighbor in the next-hop list creates its own forwarding table through the entries where it itself is the next hop and its own unicast routing table, and then broadcast the new forwarding table. This process is carried on until the leaf node in the FG receives the forwarding table. The maintenance of FG is through the exchange of forwarding tables. It is important to mention that although nodes in

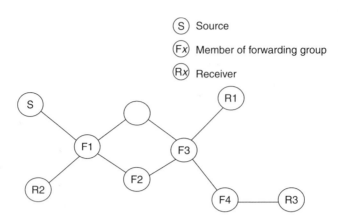

FIGURE 9.21 Typical topology of FGMP.

the FG create forwarding tables, they do not store these tables, instead, these tables are simply broadcast to their neighbors. In FGMP-SA, the multicast sender periodically floods the whole network with sender's information. Upon receiving such messages, group members broadcast locally "join table" messages, and they are propagated to the sender as in FGMP-RA.

Figure 9.21 illustrates a typical operating topology of FGMP. Circles marked by S is the sender, ones marked by Fx ($x = 1, \ldots, 4$) make up the FG, and others marked by Rx ($x = 1, 2, 3$) are receivers. Only F4 is the leaf node of the FG, since it does not need to forward multicast packet to other nodes in the FG.

9.4.2 On-Demand Multicast Routing Protocol

The on-demand multicast routing protocol (ODMRP) (Gerla and Chiang, 1999) is a mesh-based multicast routing protocol for mobile ad hoc networks. Because of the mesh-based nature, ODMRP can exploit redundant route when the shortest one is invalid. Therefore, ODMRP is robust to host mobility and failure of connectivity. In addition, since the multicast route and membership maintenance is totally on-demand, the protocol overhead is reduced.

Similar to other on-demand routing protocols, ODMRP comprises of a request phase and a reply phase. Upon sending a packet to a multicast group for the first time, the sender broadcasts a JOIN REQUEST message to the whole network. When a node receives a nonduplicate JOIN REQUEST message, it records the corresponding upstream node ID in a backward learning manner, and then rebroadcast the received JOIN REQUEST. When a multicast member receives the nonduplicate JOIN REQUEST message, it checks and updates its member table. If a valid entry exists, the receiver periodically broadcasts a corresponding JOIN TABLE message to its neighbor. When a neighbor receives the JOIN TABLE message, it checks it to see whether there is an entry in the message containing itself as a next node ID. If the searching succeeded, it can be sure that it is on the forward path to the multicast group. It then sets the FG_Flag and broadcasts its own JOIN TABLE message created from the entry matched before. This process is repeated until the source receives the JOIN TABLE message. As a result, a mesh connecting all the forwarding nodes and all the multicast group members is constructed. The source also periodically broadcast the JOIN REQUEST message to refresh the multicast routing mesh.

Figure 9.22 illustrates the mesh building process. In this figure, the node marked S denotes the source, the nodes marked R denote the multicast group members, and nodes marked F denote nonmember forwarding nodes; arrows with line tail denote JOIN REQUEST message, and arrows with dash tail denote JOIN REPLY message. From the figure, it can be seen that F1, R2, and R3 make up a mesh. At first, R2 receives multicast packets through R3. But when R2 moves out of the radio range of R3 but still can be reached by F1, R2 can reply to the JOIN REQUEST from F1 with new JOIN TABLE, then receives multicast packet from F1.

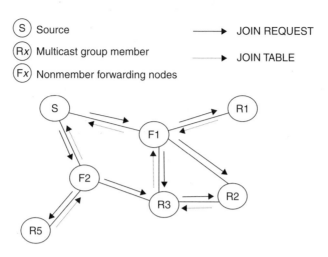

FIGURE 9.22   Multicast route setup process of ODMRP.

In ODMRP, no explicit control message is needed to leave the multicast group. To leave the group, the sender can simply stop sending JOIN REQUEST messages, while the receiver can achieve this by stopping its send JOIN TABLE messages.

From the description above, it is clear that the more complex the mesh, the more robust the protocol is to host mobility. In Reference (Gerla and Chiang, 1999), author prove that the packet delivery ratio, the number of control bytes transmitted per data byte delivered, and the number of data and control packets transmitted per data packet, are all improved with the increase of multicast group size, which exhibits good scalability for ODMRP.

9.4.3   Ad Hoc Multicast Routing Protocol

Ad hoc multicast routing protocol (AMRoute) (Xie et al., 2002) is an approach for robust IP multicast in mobile ad hoc networks. This multicast protocol is user-multicast tree based, and uses dynamic logical core to manage the tree, which tracks the group topology dynamics only, and is robust against the dynamic topology change in mobile ad hoc networks.

The user-multicast tree comprises of only senders and receivers, and both are required to join the group explicitly. Bidirectional tunnels are established between group members that are close to each other, and therefore, meshes are formed. Using a subset of all the meshes, AMRoute builds up a multicast tree periodically. Like core based tree (CBT), one group needs only one tree, which leads to improved scalability. AMRoute uses periodical signal message flooding in the tree to maintain the status, instead of flooding of data in distance vector multicast routing protocol (DVMRP) and protocol independent multicast-dense mode (PIM-DM). Unicast tunnels between group members are used as tree links to connect neighbors on the tree.

Each group has at least one core node. The logical core node is not a critical point for all data, and is not a preset one, which means the core node changes dynamically and the single point failure will be avoided. Only core nodes can initiate signaling messages to refresh the multicast tree. A node sets itself as a core when first joining a group, so that it can discover the group members, join the meshes and tree using tree creation message. If a core receives a tree creation message, it can infer that there is another core node in the tree. Then the core node resolution algorithm is run in a distributed manner by all group nodes to ensure that there is exactly one core in the tree. The basic criterion for the resolution is that a core node with the highest IP or ID wins. If a core node leaves the tree, other nodes in the tree will hear no tree creation messages for sometime. In this case, one node will designate itself as a new core, and initiate its tree creation message.

Simulation results show that AMRoute performs very well in reducing signaling traffic and join latency. In the data delivery rate, AMRoute performs very well even in dynamic networks.

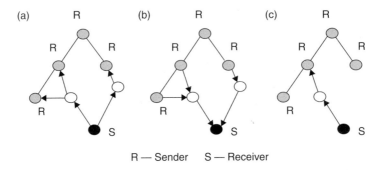

FIGURE 9.23   Multicast tree construction of MAODV: (a) RREQ propagation; (b) RREP propagation; (c) final multicast tree.

9.4.4 Multicast Operation of AODV Routing Protocol

The multicast operation of AODV routing protocol (MAODV) (Royer and Perkins, 1999) discovers multicast routes on demand. A node initiates a RREQ message joining a multicast group or when it has data to send to a multicast group but no established route to that group. Figure 9.23(a) illustrates the propagation of RREQ from node S. Only one member in the multicast group may respond to a join RREQ. If the RREQ is not a join request, any node with a fresh enough route (based on group sequence number) to the multicast group may respond to it. If a nonmember intermediate node receives a join RREQ, or if it receives a RREQ but it does not have a route to that group, it rebroadcasts the RREQ to its neighbors.

As the RREQ is flooded across the network, nodes set up pointers to establish the reverse route in their route tables. A node receiving an RREQ first updates its route table to reverse the route, and this entry may later be used to forward a response back to S. If received RREQ is a join RREQ, an additional entry is added to the multicast route table. This entry is not activated until the route becomes part of the multicast tree.

If a node receives a join RREQ for a multicast group, it may reply if it is a member of the multicast group's tree and its recorded sequence number for the multicast group is at least as great as that of the RREQ. The responding node updates its route and multicast route tables by placing the requesting node's next-hop information in the tables, and then unicasts a request response RREP back to S (Figure 9.23[b]). When nodes on the path to the source node S receive the RREP, they add both a route table entry and a multicast route table entry for the node from which they received the RREP, which forms the forward path.

After S broadcasts a RREQ for a multicast group, it usually receives more than one RREP. S keeps the received route reply with the greatest sequence number and shortest hop count to the nearest member of the multicast tree for a specified period of time while disregarding others. When this period is timeout, it enables the selected next hop in its multicast route table, and unicasts an activation message to this selected next hop. The next hop, on receiving this message, enables the entry for S in its multicast route table. If this node happens to be a member of the multicast tree, it does not forward the message anymore. However, if this node is not a member, it may have received one or more RREPs from its neighbors. In this case, it chooses and keeps the best next hop for its route to the multicast group, enables the corresponding entry in its multicast route table, and unicasts an activation message to the selected next hop, just as what S does before. This process continues until the final activation message reaches the node that sent the RREP. In this way, the redundant paths are cut to ensure that there is only one path to any tree node, and the multicast packets are forwarded along the selected paths. Figure 9.23(c) illustrated the final path selection for the case of Figure 9.23(a) and Figure 9.23(b).

In MAODV, the first member of the multicast group becomes the leader for that group, which manages the multicast group sequence number and broadcasts this number in group hello messages to the whole multicast group.

9.4.5 Core-Assisted Mesh Protocol

The core-assisted mesh protocol (CAMP) (Garcia-Luna-Aceves and Madruga, 1999) is a mesh-based multicast protocol. Instead of using a multicast tree, CAMP builds and maintains a mesh among the multicast group members and senders. The mesh is a subset of the network topology, which guarantees at least one route from each sender to each receiver. Besides the reachability, CAMP also ensures the path from the receiver to the sender is the shortest one. Such reverse shortest path also make packet flows conform to shortest paths from sources to receivers.

Because of the path redundancy, the impact of topology changes on packet delivery is greatly reduced. Further, if compared to core-based tree multicast protocols, the hop count between sources and destinations is also decreased.

In CAMP, one or more cores are used to limit the control traffic for joining group, and it need not to be a member of the mesh. In this way, CAMP eliminates the need for flooding, unless all cores are unreachable.

9.4.6 Ad Hoc Multicast Routing Protocol Utilizing Increasing id-NumberS

The ad hoc multicast routing protocol utilizing increasing id-numberS (AMRIS) (Wu and Tay, 1999) dynamically assigns each multicast group member a mem-id. The order of mem-ids is used to direct the multicast packet flow, while the gaps among the assigned mem-ids are used to repair connectivity. Each multicast group is initiated by a special node, namely Sid, which owns the smallest mem-id, and the mem-ids increases as nodes radiate away from Sid. Therefore, the mem-id is an indication of the height in the multicast tree. AMRIS is comprised of two main mechanisms: tree initialization and tree maintenance.

In the beginning of the tree initialization phase, the Sid is chosen from all senders. To construct a multicast tree, Sid broadcasts a NEW-SESSION message to its neighbors. The NEW-SESSION contains the mem-id of Sid, multicast session id, and routing metrics. All nodes receiving the NEW-SESSION message generate their own mem-id, which is greater than the received mem-id but not consecutive. The gaps are reserved for future connectivity repair. The receivers then overwrite the mem-id in the NEW-SESSION messages with their own, and broadcast their NEW-SESSION messages again. Chances are that a node receives more than one NEW-SESSION messages. To cope with such a situation, AMRIS requires a gap between receiving and broadcasting NEW-SESSION messages. If it has not rebroadcast its own NEW-SESSION message, it will keep the message with the best routing metrics and generate its mem-id based on that message.

If node X intends to join a multicast group, it first searches for neighbor nodes with smaller mem-ids than itself. These nodes are its potential parent. X then unicasts a JOIN_REQ message to Y, one of its potential parents. When Y receives the JOIN_REQ message, it first checks whether it is on the multicast tree. If it is the case, Y sends a JOIN_ACK message to X; otherwise, Y sends JOIN_REQ to find its own potential parent. This process continues until one receiving node becomes an actual parent. Then the JOIN_ACK message will be propagated through the reverse path to X. If finally the first attempt to contact a neighbor fails, X will do a 1-hop broadcast trying to locate a parent. If all 1-hop attempts fail, X will execute the branch reconstruction (BR) process in tree maintenance to join the tree.

The tree maintenance runs continuously in the background. When a link between two nodes breaks, the node with larger mem-id (child) is responsible for rejoining the tree. The rejoining is achieved through the BR process, which has two subroutines, BR1 and BR2. BR1 works just as a new joining process. BR2 is to cope with the situation when an isolated child cannot locate a neighbor as parent. In this case, the child broadcasts, instead of using unicast, a JOIN_REQ with hop limit R. When a node receives the broadcast JOIN_REQ, it checks whether it can satisfy the request. If so, it will send back JOIN_ACK. Since the isolated child may receive several such JOIN_ACK messages, the node replying with a JOIN_ACK message does not forward multicast traffic to that isolated child until it receives a JOIN_CONF message from it.

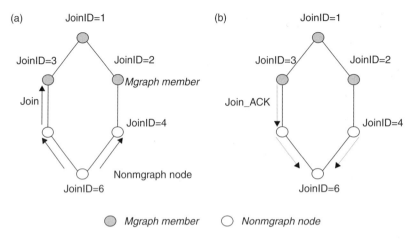

FIGURE 9.24   Joining process of MCEDAR.

9.4.7   Multicast Core-Extraction Distributed Ad Hoc Routing

The multicast core-extraction distributed ad hoc routing (MCEDAR) (Sinha et al., 1999c) extends the CEDAR routing protocol (Sivakumar et al., 1999) to support multicast. It merges the robustness of the mesh-based multicast routing protocols and approximate efficiency of tree-based ones.

CEDAR establishes a core of the network, and uses the core as the routing management infrastructure. The core propagates the link state to its nodes, and performs on-demand route computation using a fully distributed algorithm utilizing a purely local state where core nodes participate. Briefly, CEDAR consists of four key components: core extraction, core broadcast, link state propagation, and route computation. MCEDAR uses only the first two components of CEDAR and the core broadcast enables the tree-based multicast forwarding protocol on mesh infrastructure.

For each multicast group, MCEDAR extracts a subgraph of the core graph to function as the multicast routing infrastructure. The subgraph extracted is a mesh structure and is called mgraph, and the multicast forwarding is performed using core broadcast in mgraph. When a node intends to join a multicast group, it requests its core node to join the mgraph. Then the core node performs the join operation by core broadcasting a join request containing joinID, where the joinID represents the time when the core node joins the mgraph. When a nonmember core node receives the JOIN message, it forwards the request to its neighbors, in a manner conforming to core broadcast in the CEDAR routing algorithm. When a member core node receives the request, it replies it with JOIN_ACK only if its joinID is less that the incoming joinID, and then forwards the request. This joining process can be illustrated in Figure 9.24.

In packet forwarding, the core broadcast mechanism is also used when an mgraph member intends to forward a multicast packet to its nearby members, which generate a source-based tree automatically. MCEDAR made some further optimization to decouple the control infrastructure and the data forwarding infrastructure, which means packets do not need to follow the mgraph. This is done when JOIN_ACK is propagated back to the joiner, where intermediate nodes can compute optimized route from their domain to their children's domains upon receiving JOIN_ACK, add such information in the JOIN_ACK, and then notify the nodes they dominate to update their multicast routing table.

9.4.8   Adaptive Backbone-Based Multicast

The adaptive backbone-based multicast protocol (Jaikaeo and Shen, 2002) uses an adaptive dynamic backbone (ADB) to lessen the constraint of dominating property in traditional backbone-based multicast. In ADB, a node may choose a 1-hop away node as its core node. In this way, the number of core nodes in the network is reduced. The choice of how far away the core node lies depends on the current network status observed, such as link failure frequency.

To construct an ADB, each node in the network maintains the following data structures:

- *parent:* The ID of the upstream node toward the core (core node keeps its own ID), which is set and modified by the core selection process.
- *ntab:* A table containing all immediate neighbor's information, which is maintained by the neighbor discovery process.
- *nlff:* Normalized link failure frequency, which reflects the dynamic condition of the surrounding area.
- *degree:* The number of entries in the ntab table.
- *fwtab:* The core forwarding table maintaining a list of shortest path entries to the nearby cores, which is maintained by the core forwarding process.

Each node sets itself as a core node when it first joins the network. It then periodically broadcasts hello messages to discover its neighbors. Each hello message contains (*nodeId, coreId, hopsToCore, degree, cnlff*). The *nodeId* is the ID of sending node, the *coreId* refers to the core node that the sending node is associated with, *hopsToCore* tells the hop count to the associated core node, *degree* equals the number, and *cnlff* is the cumulative of every link from the associated core node to the sending node. When receiving hello message, nodes update their *ntab* and *nlff.*

After joining the network, a node should decide either to serve as a core node or choose a core node to join. ADB uses (*nlff*-1, *degree, id*) as the height metric to make such decisions, and *ntab* also contains the height metrics of neighbors. If its own height metric is higher than all its neighbors, it serves as a core node; otherwise, it chooses the neighbor with the highest metric to update its parent variable. From this point, further exchanges of hello message will help nodes find better parent and prevent hop count to core node or accumulated *nlff* from exceeding their thresholds. This process is called the core selection process. Each core node maintains a *fwtab* to direct the forwarding of multicast packets. This table is updated from the advertised information of nodes associated with other cores. In this way, each node does not need to have a neighbor node as a core node, which means in high mobility areas, the coverage of core nodes is small, while in low mobility areas, the coverage of core nodes is large. That is why such a backbone is named adaptive dynamic bone (ADB).

The multicast operation on ADB is a two-tier architecture, local multicast membership and forwarding management, and the counterpart in backbone. For local joining operation, cores manage the membership, and are local rendezvous points. Nodes intend to join the group send JOIN_REQUEST to their parents, and such process continues until requests reach the core node. Then all multicast packets are relayed by the core node to reach their destination receivers. For backbone multicast packet propagation, the core node encapsulates the packet into a CORE_FWD message, and forwards it according to the entry in *fwtab.* Receiving core node performs local multicast forwarding and backbone forwarding again, and discards duplicate packets received.

9.4.9 Differential Destination Multicast

The differential destination multicast (DDM) routing protocol (Ji and Corson, 2001) is small and sparse group oriented. It encodes the destination list in the packet header thus eliminating the multicast state maintenance in the whole network, and requires the underlying unicast routing protocols to work. When intermediate nodes receive a packet, they need to analyze the packet header to find the destination list, and forward the packet according to the list. Intermediate nodes also can work in a "soft state" mode, in which they keep the destination list of recent forwarded packets, and use them to direct future forwarding. When the destination list changes, upstream nodes will inform their downstream nodes, namely differential destination encoding.

The multicast group member list is managed by each sender. To join the multicast, receiver is required to have a JOIN/ACK exchange with the sender. Senders periodically refresh the multicast group membership by setting the refresh flag in some packets sent requiring receivers to respond with JOIN messages. If someone in the member list fails to respond with JOIN in a time-out period, the sender will erase it from the list, and update the downstream routers to update their caches if soft state mode is used.

Compared to other multicast routing protocol, DDM is suitable for small and sparsely distributed multicast group since it eliminates the high cost of tree or mesh maintenance, and the quality of the underlying unicast is critical for the operation of DDM.

9.4.10 Location-Guided Tree Construction Algorithms

The location-guided tree (LGT) construction algorithms (Chen and Nahrstedt, 2002) encapsulate the destination list in packet headers, and intermediate nodes analyze the header to decide where to forward the packet according to its unicast routing table. Like DDM, this stateless multicast routing protocol aims at small group multicast in mobile ad hoc networks. Two algorithms are designed to construct subtree when forwarding multicast packets, location-guided k-ary (LGK) tree and location-guided steiner (LGS) tree. Both of these utilize the geometric location information of the destination to approximate the network hop distance, and use such heuristics to construct packet distribution subtree.

The main assumptions of LGT are:

- Small multicast group.
- Each multicast group member is aware of other members and their geometry location.
- The network hop count to destination increases monotonically with the increase of geometry distance. This assumption is validated through simulations.

To construct a LGK tree, the sender selects the nearest k destinations as its children, then forwards to each child a copy of the destination. Each child then performs the same operation: group k nearest destination and send a copy to them. This process continues until all destinations receive the packet. Every time a packet runs through a destination, the node deletes its address from the destination list in the packet header, so that the tree is implicitly loop-free.

To construct a LGS tree, the sender initiates the tree as its only member. Then at each iteration, the nearest unconnected destination to the partially constructed tree is added to the tree until all destinations are added.

To inform other members the location information of each node, LGT uses two kinds of location and membership combined updates, in-band update and periodic update. The former includes update information in data packets, while the latter is used when there is no packet to send. To improve the performance of intermediate node, LGT also uses route caching at intermediate nodes, just as DDM.

Through simulation and analysis, authors conclude that LGS tree has lower bandwidth cost than LGK when the location information is up-to-date because it is greedier than LGK. But when the location information is out-dated, their bandwidth costs are similar, and LGK outperforms LGS since it has lower packet distribution delay and lower computation complexity.

9.4.11 Route-Driven Gossip

The route-driven gossip (RDG) (Luo et al., 2003) relies on partial member information, which is suitable for on-demand underlying unicast routing protocol such as DSR. The packet propagation is through a full gossip style, which plays the dynamics of mobile ad hoc networks with randomly packet distribution. Each member forwards packets to a random subset of the group, all of which it has routes to (gossiper-push), and the negative acknowledgment is piggybacked in multicast packets (gossiper-pull).

RDG has totally three sessions: *join, gossip,* and *leave*. The meanings of these are self-explanatory.

In the *join* session, the joining node floods the whole network with GROUPREQUEST messages trying to find routes to all members simultaneously. However, when one node receives this message, whether or not to send GROUPREPLY is controlled in a random fashion.

In the *gossip* session, each member of the group periodically sends gossip messages to a randomly selected subset of members it has routes to. Members receiving the gossip message update their own status information, and reply to gossiper-pull messages to make up the gap other nodes may have in their packet buffers. The *leave* session works almost the same way.

Based on RDG, topology-aware RDG (TA-RDG) is presented in the same paper. TA-RDG uses some information about the members to guide the gossip direction. Many metrics from underlying unicast routing protocols are useful here, such as the distance and route breakage frequency. Then one member may select some nearest members to have gossip with, which may increase the success probability of gossiping.

9.5 Summary

Routing protocols for unicast, multicast, and broadcast are key to ad hoc networks. Adaptations of routing protocols of the wired Internet have often failed for ad hoc networking environments. Various key routing protocols for enabling one-to-one (unicast), one-to-many (multicast) and one-to-all (broadcast) communication are outlined in this chapter.

A new paradigm of one-to-any communication, also known as *anycast*, has several useful applications for ad hoc networks. To make ad hoc networks more useful and practical, it must support various services possibly with the help of some "special" nodes that maintain relevant databases. These "special" nodes may have more energy, storage, and computing power. Such nodes may be used to store different types of information such as, location and ID of other nodes in the network, ID of nearest node connected to the Internet, various networks statistics, various parameters of other nodes in the network, and various parameters of network links. For such architectures, nodes can query or update the special nodes using *anycast*. Some of these ideas have been explored in (Choudhury and Vaidya, 2003).

Fluctuating channel conditions and frequent link disconnections due to node mobility pose significant challenges in supporting traffic requiring guarantees on specific end-to-end traffic parameters such as delay, loss-rate, and bandwidth. Some protocols such as CEDAR (Sinha et al., 1999b) and ticket-based routing protocols (Xiao et al., 2002) are specifically designed for providing soft guarantees on QoS parameters. QoS routing in ad hoc networks is a challenging research area, which is very significant for supporting real-time multimedia traffic.

Ad hoc networks are typically plagued with the problem of limited battery life. For extending the network lifetime of ad hoc networks, researchers have also explored the design of centralized as well as distributed protocols for minimum energy routing in ad hoc networks.

Further advances in the design of routing protocols for supporting energy conservation, low latency, and high throughput communication, are critically needed for making ad hoc networking a reality.

References

Agarwal, Manish, Cho, Joon Ho, Gao, Lixin, and Wu, Jie (2004). Energy efficiency broadcast in wireless ad hoc networks with hitch-hiking. In *Proceedings of the INFOCOM 2004*.

Amir Qayyum, Laurent Viennot, Anis Laouiti, (2002). Multipoint relaying for flooding broadcast messages in mobile wireless networks. In *Proceedings of the 35th HICSS*.

Basagni, Stefano, Chlamtac, Imrich, and Syrotiuk, Voilet R. (1998). A distance routing effect algorithm for mobility (dream). In *Proceedings of the MOBICOM 1998*, pp. 76–84.

Boukerche, Azzedine, Das, Sajal K., and Fabbri, Alessandro (2001). Message traffic control capabilities of the R-DSDV protocol in mobile ad hoc networks. In *Proceedings of the 4th ACM International Workshop on Modeling, Analysis and Simulation of Wireless and Mobile Systems*, pp. 105–112.

Boukerche, Azzedine and Rogers, Steve (2001). GPS query optimization in mobile and wireless networks. In *Proceedings of the 6th IEEE Symposium on Computers and Communications*, pp. 198–203.

Boukerche, Azzedine and Sheetal, Vaidya (2003). A performance evaluation of a dynamic source routing discovery optimization using GPS system. *Telecomunication Systems*, **22**: 337–354.

Boukerche, Azzedine and Sheetal, Vaidya (2002). A route discovery optimization scheme using GPS system. In *Proceedings of the 35th Annual Simulation Symposium*, pp. 20–26.

Boukerche, Azzedine and Zhang, Liqin (2003). A preemptive on-demand distance vector routing protocol for mobile and wireless ad hoc networks. In *Proceedings of the 36th Annual Simulation Symposium.*

Chen, Kai and Nahrstedt, Klara (2002). Effective location-guided tree construction algorithms for small group multicast in manet. In *Proceedings of the INFOCOM 2002*, Vol. 3, pp. 1180–1189.

Chiang, Ching-Chuan, Gerla, Mario, and Zhang, Lixia (1998). Forwarding group multicast protocol (FGMP) for multihop, mobile wireless networks. *ACM/Kluwer Journal of Cluster Computing: Special Issue on Mobile Computing*, **1**: 187–196.

Choudhury, Romit Roy and Vaidya, Nitin (2003). MAC-layer anycasting in wireless ad hoc networks. In *Proceedings of the 2nd Workshop on Hot Topics in Networks (HotNets II).*

Garcia-Luna-Aceves, J.J. and Madruga, Ewerton L. (1999). The core-assisted mesh protocol. *IEEE Journal on Selected Areas in Communications*, **17**: 1380–1394.

Gerla, Sung-Ju Lee Mario and Chiang, Ching-Chuan (1999). On-demand multicast routing protocol. In *Proceedings of the WCNC 1999*, Vol. 3, pp. 1298–1302.

Giordano, S. and Hamdi, M. (1999). Mobility management: the virtual home region. Technical Report.

Guangyu Pei, Mario Gerla, Xiaoyan Hong and Chiang, Ching-Chung (1999). A wireless hierarchical routing protocol with group mobility. In *Proceedings of the IEEE WCNC αϛ99.*

Haas, Z. and Pearlman, M. (1998). The performance of query control schemes for the zone routing protocol. In *Proceedings of the ACM SIGCOMM*, pp. 167–177.

Haas, Z. and Pearlman, M. (2000). The zone routing protocol (ZRP) for ad-hoc networks. Internet Draft, draft-ietf-manetzrp-03.txt.

Haas, Zygmunt J. and Liang, Ben (1999). Ad hoc mobility management with uniform quorum systems. *IEEE/ACM Transactions on Networking*, **7**: 228–240.

Iwata, Atsushi, Chiang, Ching-Chuan, Pei, Guangyu, Gerla, Mario, and Chen, Tsu-Wei (1999). Scalable routing strategies for ad hoc wireless networks. *IEEE Journal on Selected Area in Communications, Special Issue on Ad-Hoc Networks*, **17**(8): 1369–1379.

Jaikaeo, Chaiporn and Shen, Chien-Chung (2002). Adaptive backbone-based multicast for ad hoc networks. In *Proceedings of the ICC 2002*, Vol. 5, pp. 3149–3155.

Jain, Rahul, Puri, Anuj, and Sengupta, Raja (2001). Geographical routing using partial information for wireless ad hoc networks. *IEEE Personal Communications*, **8**(1): pp. 48–57.

Ji, Lusheng and Corson, M. Scott (2001). Differential destination multicast — a manet multicast routing protocol for small groups. In *Proceedings of the INFOCOM 2001*, Vol. 2, pp. 1192–1201.

Joa-Ng, Mario and Lu, I-Tai (1999). A peer-to-peer zone-based two-level link state routing for mobile ad hoc networks. *IEEE Journal on Selected Areas in Communications*, **17**: 1415–1425.

Johnson, D.B. and Maltz, D.A. (1996). *Dynamic Source Routing in Ad Hoc Wireless Networks.* Kluwer Academic Publishers, Dordrecht.

Karp, Brad and Kung, H.T. (2000). GPSR: greedy perimeter stateless routing for wireless networks. In *Proceedings of the MOBICOM 2000*, pp. 243–254.

Ko, Y. and Vaidya, N.H. (1998). Location-aided routing (LAR) mobile ad hoc networks. In *Proceedings of the ACM MOBICOM.*

Ko, Young-Bae and Vaidya, Nitin H. (2000). Geotora: a protocol for geocasting in mobile ad hoc networks. In *Proceedings of the ICNP 2000*, pp. 240–250.

Li, Jinyang, Jannotti, John, Couto, Douglas S.J. De, Karger, David R., and Morris, Robert (2000). A scalable location service for geographic ad-hoc routing. In *Proceedings of the MOBICOM 2000*, pp. 120–130.

Liao, Wen-Hwa, Tseng, Yu-Chee, and Sheu, Jang-Ping (2001). GRID: a fully location-aware routing protocol for mobile ad hoc networks. *Telecommunication Systems*, **18**: 61–84.

Lim, Hyojun and Kim, Chongkwon (2001). Flooding in wireless ad hoc networks. *Computer Communication Journal.* **24**(3–4): 353–363.

Lin, X. and Stojmenovic, I. (1999). GEDIR: Loop-free location based routing in wireless networks. In *Proceedings of the PDCS 1999*, pp. 1025–1028.

Lou, Wei and Wu, Jie (2000). On reducing broadcast redundancy in ad hoc wireless networks. In *Proceedings of the MOBIHOC.*

Lou, Wei and Wu, Jie (2004). Double-covered broadcast (DCB): a simple reliable broadcast algorithm in manets. In *Proceedings of the IEEE INFOCOM 2004*.

Luo, Jun, Eugster, Patrick Th., and Hubaux, Jean-Pierre (2003). Route driven gossip: probabilistic preliable multicast in ad hoc networks. In *Proceedings of the INFOCOM 2003*, Vol. 3, pp. 2229–2239.

Ni, S.Y., Tseng, Y.C., Chen, Y.S., and Sheu, J.P. (1999). The broadcast storm problem in a mobile ad hoc network. In *Proceedings of the 5th Annual ACM/IEEE International Conference on Mobile Computing and Networking (MOBICOM)*, pp. 151–162.

Park, V. and Corson, M.S. (1997a). A highly adaptive distributed routing algorithm for mobile wireless networks. In *Proceedings of the INFOCOM*, pp. 1405–1413.

Park, Vincent D. and Corson, M. Scott (1997b). A highly adaptive distributed routing algorithm for mobile wireless networks. In *Proceedings of the INFOCOM 1997*, Vol. 3, pp. 1405–1413.

Pearlman, M. and Haas, Z. (1999). Determining the optimal configuration of for the zone routing protocol. *IEEE Journal on Selected Areas of Communication (Special issue on ad-hoc networks)*, **17**: 1395–1414.

Peng, Wei and Lu, Xi-Cheng (2000). On the reduction of broadcast redundancy in mobile ad hoc networks. In *Proceedings of the 1st ACM International Symposium on Mobile Ad Hoc Networking and Computing*.

Perkins, C. and Bhagwat, P. (1994). Highly dynamic destination-sequenced distance vector routing (DSDV) for mobile computers. In *Proceedings of the ACM SIGCOMM*, pp. 234–244.

Perkins, C.E. (1999). Ad hoc on demand distance vector (AODV) routing. Internet Draft, draft-ietf-manet-aodv-04.txt.

Perkins, C.E. and Royer, E.M. (1999). Ad hoc on-demand distance vector routing. In *Proceedings of the 2nd IEEE Workshop on Mobile Computing Systems and Applications*, pp. 90–100.

Royer, Elizabeth M. and Perkins, Charles E. (1999). Multicast operation of the ad hoc on-demand distance vector routing protocol. In *Proceedings of the MOBICOM 1999*, pp. 207–218.

Sinha, P., Sivakumar, R., and Bharghavan, V. (1999a). CEDAR: a core-extraction distributed ad hoc routing algorithm. In *Proceedings of the IEEE INFOCOM*.

Sinha, P., Sivakumar, R., and Bharghavan, V. (1999b). CEDAR: a core-extraction distributed ad hoc routing algorithm. In *Proceedings of the IEEE INFOCOM*.

Sinha, Prasun, Sivakumar, Raghupathy, and Bharghavan, Vaduvur (1999c). Mcedar: multicast core-extraction distributed ad hoc routing. In *Proceedings of the WCNC 1999*, Vol. 3, pp. 1313–1317.

Sivakumar, Raghupathy, Sinha, Prasun, and Bharghavan, Vaduvur (1999). CEDAR: a core-extraction distributed ad hoc routing algorithm. *IEEE Journal on Selected Areas in Communications*, **17**: 1454–1465.

Toh, Chai-Keong (1996). A novel distributed routing protocol to support ad-hoc mobile computing. In *Proceedings of the IEEE 15th Annual International Phoenix Conference on Computers and Communications*.

Wu, C.W. and Tay, Y.C. (1999). AMRIS: A multicast protocol for ad hoc wireless networks. In *Proceedings of the MILCOM 1999*, Vol. 1, pp. 25–29.

Xiao, L., Wang, J., and Nahrstedt, K. (2002). The enhanced ticket based routing algorithm. In *Proceedings of the IEEE ICC*.

Xie, Jason, Talpade, Rajesh R., and Mcauley, Anthony (2002). AMRoute: ad hoc multicast routing protocol. *Mobile Networks and Applications*, **7**: 429–439.

Zygmunt J. Haas, Marc R. Pearlman, and Prince Samar (2002a). The bordercast resolution protocol (BRP) for ad hoc networks. Internet Draft, draft-ietf-manet-zone-brp-02.txt.

Zygmunt J. Haas, Marc R. Pearlman, and Prince Samar (2002b). The interzone routing protocol (IARP) for ad hoc networks. Internet Draft, draft-ietf-manet-zone-ierp-02.txt.

Zygmunt J. Haas, Marc R. Pearlman, and Prince Samar (2002c). The intrazone routing protocol (IARP) for ad hoc networks. Internet Draft, draft-ietf-manet-zone-iarp-02.txt.

Zygmunt J. Haas, Marc R. Pearlman, and Prince Samar (2002d). The zone routing protocol (ZRP) for ad hoc networks. Internet Draft, draft-ietf-manet-zone-zrp-04.txt.

10

Cluster-Based and Power-Aware Routing in Wireless Mobile Ad Hoc Networks

10.1 Introduction...**10**-217
10.2 Clusterhead Gateway Source Routing (CGSR) **10**-218
  GSR Infrastructure Creation Protocol • Gateways •
  Routing in CGSR • Disadvantages of CGSR
10.3 The VBS Infrastructure Creation Protocol.............**10**-220
  Routing in VBS • Disadvantages of VBS
10.4 The Power Aware Virtual Base Station (PA-VBS)
  Infrastructure Creation Protocol **10**-221
10.5 Warning Energy Aware Clusterhead (WEAC)
  Infrastructure Creation Protocol **10**-222
  Protocol Overview • Formal Description of the Protocol
10.6 The Virtual Base Station On-Demand (VBS-O)
  Routing Protocol.......................................**10**-227
10.7 Reservation and Polling **10**-229
  Exhaustive Polling Analysis • Partially Gated Polling Analysis
10.8 Network Model **10**-234
10.9 Power Analysis **10**-234
10.10 Summary **10**-236
References **10**-236

Tarek Sheltami
Hussein T. Mouftah

10.1 Introduction

Cluster-based protocols establish a dynamic wireless mobile infrastructure to mimic the operation of the fixed infrastructure in cellular networks. In cluster-based protocols [8–13], an ad hoc network is represented as a set of clusters. A clusterhead is selected based on an agreed upon rule [15–18]. On the contrary to on demand protocols, clustering protocols take advantage of their cluster structure, which limits the scope of route query flooding. Only clusterheads and gateways (nodes that lie within the transmission range of more than one clusterhead) are flooded with query packets, thus, reducing the bandwidth required by the route discovery mechanism. The aggregation of nodes into clusters under clusterhead control provides a convenient framework for the development of efficient protocols both at the medium access control (MAC) layer, for example, code separation among clusters channel access, bandwidth allocation, and at the network layer, for example, hierarchical routing [4]. Clustering provides an effective way to allocate wireless channels among different clusters. Across clusters, spatial reuse by using different spreading codes (i.e., code division multiple access (CDMA) [1,5]) is enhanced. Within a cluster a clusterhead controlled token protocol (i.e., polling) is used to allocate the channel among competing nodes. The token approach allows us to give priority to clusterheads in order to maximize channel utilization and minimize delay. A clusterhead should get more chances to transmit because it is in charge of broadcasting within the cluster and of forwarding messages between mobile hosts, which are not connected. However, in order to have CDMA as a MAC layer with more than one code, the nodes must be equipped with more than one transceiver, that is, a transceiver per code used, or they have to run an algorithm, which switches between codes, for example, if we have 16 codes (usually you can have 64), then this means that this algorithm will have to switch between the 16 codes. The same idea as frequency hopping [6], where nodes switch very fast between different frequencies in order to overcome interference. So, the probability that two nodes use the same frequency simultaneously while being neighbors becomes very low. Therefore, unless we have these transceivers or this algorithm, we cannot use CDMA with more than one code as a MAC layer. Even if we have an algorithm that can switch between codes, theoretically, we can switch between the codes, but how will the wireless nodes switch between the various channels with one transceiver not known.

Recently there has been interest in hierarchal architectures for ad hoc networks. A mobile infrastructure is developed in [7] to replace the wired backbone infrastructure in conventional cellular networks. Moreover, an adaptive hierarchical routing protocol is devised. The routing protocol utilizes the mobile infrastructure in routing packets from a source node to a destination node. The algorithm divides the wireless mobile ad hoc network into a set of clusters, each of which contains a number of mobile terminals (MTs), which are at most two hops away. Clusters have no clusterheads; instead they are only a logical arrangement of the nodes of the mobile network. Instead, a cluster center node is chosen. Cluster centers ensure that the distance between any two nodes in any cluster is at most two hops. The node with the highest connectivity [14] is chosen to be the center of the cluster. This, in fact, introduces a major drawback to the stability of the various clusters since in high-mobility networks; cluster re-formations frequently take place due to the continuous random movements of MTs. As the network changes slightly, the node's degree of connectivity is much more likely to change as well. Nodes that relay packets between clusters are called repeaters. Repeaters play a major role in performing the routing function. Maintaining and updating the cluster structure was not tackled. Also, the method by which cluster information is conveyed to the MTs of a cluster was not addressed. MTs are classified as follows:

Clusterhead: As it is named, the leader of the cluster.
Zone_MT: An MT supervised by a clusterhead.
Free_MT: An MT that is neither a clusterhead nor a zone_MT (i.e., it is not associated with a cluster).
Gateway or border mobile terminal (BMT): MT that lies between more than clusterhead or Free_MT, it can be clusterhead or zone_MT or free_MT.

10.2   Clusterhead Gateway Source Routing (CGSR)

A node or MT is elected to be a clusterhead. All nodes in the cluster communicate with each other through the clusterhead. Two algorithms can possibly be used to select the clusterhead:

- The smallest ID algorithm [16]
- The highest connectivity algorithm [17]

The disadvantage of the highest connectivity algorithm is mentioned above. Moreover, here are two major drawbacks in the smallest ID algorithm:

- The batteries of the nodes with small IDs drain much faster than those with large IDs.
- The network will be unstable, because each time a node with the smallest ID enters a cluster many actions are taken to elect it as a clusterhead.

The CGSR uses the least clusterhead change (LCC) clustering algorithm [18], which put more importance on a node becoming supervised by a clusterhead, rather than being a clusterhead. Clusterheads change due to either of the following conditions:

- When two clusterheads are within the transmission range of each other
- When a node becomes disconnected from any other cluster

10.2.1   GSR Infrastructure Creation Protocol

Initially, the smallest ID cluster algorithm (or highest connectivity cluster algorithm) is used to create clusters.

When a node in cluster 1 moves into cluster 2, the clusterhead of 1 and 2 will not be affected.

When a node (nonclusterhead) moves out of its cluster and does not enter another cluster, it becomes a clusterhead of itself.

When a clusterhead X, which has a smaller ID than clusterhead Y, is within the transmission range of clusterhead Y, clusterhead X will be a clusterhead, while clusterhead Y gives up its role as a clusterhead.

The nodes that clusterhead Y is in charge of will re-elect another clusterhead according to an agreed upon rule [16,17].

A common control code is used for initialization and configuration.

Each cluster has a different control code from its neighboring cluster.

10.2.2   Gateways

A node is defined as a gateway if it is within the range of more than one clusterhead. As in virtual base station (VBS), it is used to communicate between two or more clusterheads. However, it has to select the code used by each cluster. Chiang *et al.* [12] assume that a gateway can change its code after it returns the permission token or after receiving a message. They suggest that once a gateway is equipped with multiple radio interfaces, it can access multiple cluster channels by selecting corresponding codes for each radio. As a result, gateway conflict will be reduced. This solution is impractical. First of all, any node can be a gateway. Having said that, each node should be equipped with multiple radio interfaces so that it can access multiple cluster channels. This will increase the cost, weight, and level of complexity of the mobile unit. Second, if a node moves out of its cluster and enters another one, it will not be possible to pursue communications in that cluster.

10.2.3   Routing in CGSR

Each clusterhead is in charge of a set of nodes. There are two kinds of scheduling. Token scheduling, which is carried out by clusterheads, and code scheduling, which is carried out by gateways. In the CGSR routing protocol, each and every node in the ad hoc wireless node maintains two tables: a cluster member table, which maps the destination node address to its clusterhead address, and a routing table, which shows the next hop to reach the destination cluster. Both tables contain sequence numbers to eliminate stale routes and prevent looping.

10.2.4   Disadvantages of CGSR

There is a drawback due to the overhead associated with maintaining clusters. Specifically, each node needs to periodically broadcast its cluster member table and update its table based on the received updates. If two nodes, even while being in the same cluster, are using different codes, they will be unable to communicate with each other. This forms a so-called *pseudo link*. Chiang *et al.* do not state how the nodes merge with clusterheads. Do they send merge-request messages and then wait until they get merge-accept messages, in order to be one of the nodes the cluster in charge of? Or do they just send merge-request messages then consider themselves under the supervision of the clusterhead without getting an acknowledgment back from it? This decision affects the performance of the protocol.

10.3 The VBS Infrastructure Creation Protocol

In VBS infrastructure protocol [8–10] an MT is elected by one or more MTs to be their VBS based on an agreed rule, however, the smallest ID rule [16]. Every MT acknowledges its location by *hello* packets for a period of time, and it has a sequence number that reflects the changes that occur to that MT, and a my_VBS variable, which is used to store an ID number to the VBS in charge of that MT. An MTs my_VBS variable is set to the ID number of its VBS; however, if that MT is itself a VBS, then the my_VBS variable will be set to either 0, if it is serving other MTs, or to −1 if it is not serving any MT. A VBS collects complete information about all other VBSs and their lists of MTs and sends this information in its periodic hello messages. Zone_MTs accumulate information about the network from their neighbors between hello messages, and their network information is cleared after sending every hello message. MTs announce their ID number with their periodic hello message. An MT sends a merge-request message if it has a bigger ID number. The receiver of the merge-request message responds with an accept-merge message, increments its sequence number by one, and sets its my_VBS variable to 0. When the MT receives the accept-merge message, it increments its sequence number by one, and sets its my_VBS variable to the ID number of its VBS. If an MT hears from another MT, whose ID is smaller than its VBS, it sends a merge-request message to it. When it receives an accept-merge message, it increments its sequence number by one and updates its my_VBS field. The MT then sends a disjoin message; it removes its own sequence number by one. The timer may expire due to the motion of the VBS to a location at which the MT cannot hear its hello message, due to some interference at the time the hello message of the VBS was sent, or because the VBS is shut down by its user. In addition, every VBS maintains a timer from every MT, which it is in charge of from the last time it heard from it. If the timer expires, the VBS will remove the MT from its list of MTs, increment its sequence number by one, and check its new list of MTs. If its list of MTs becomes empty, it will set its my_VBS field to −1.

10.3.1 Routing in VBS

Each VBS is in charge of a set of MTs. Only certain nodes are eligible to acquire knowledge of the full network topology (VBS and BMT). Route requests are not flooded to the rest of the ad hoc network. If an MT wishes to send a packet to another MT in the network, first, it sends the packet to its VBS, which forwards the packet to the VBS in charge to the destination or the correct BMT. The sent packet contains the address of the destination. When the VBS receives the message, it searches the destination address in its table. If the destination is found, the VBS of the source MT will forward the packet to the VBS in charge of the destination. This is done by consulting the BMT field of that VBS. The message is then forwarded to the MT (or VBS) whose ID number is stored in the BMT field. The BMT, after receiving the message, forwards it to its own VBS. This process is repeated until the message reaches the destination. It is obvious that MTs are neither responsible for discovering new routes, nor maintaining existing ones. As a result this routing scheme eliminates the initial search latency, which degrades the performance of interactive and multimedia applications.

Based on simulation experiments done in Reference 10, the performance of CGSR is comparable with that of VBS in less stressful situations, that is, small number of mobile nodes. VBS; however, outperforms CGSR in more stressful situations with a widening performance gap. Also, based on simulation experiments done in Reference 8, VBS achieves packet delivery fractions between 90 and 100%. Ad hoc on-demand distance vector (AODV), on the other hand, achieves packet delivery fractions between 79 and 91%, while those achieved by distance state routing (DSR) are between 72 and 93.5%. Therefore, the difference is often significant, up to 30% higher. This is mainly because of the information stored in VBSs to aid in routing. By the time a packet is to be retransmitted, new information will be available at the VBS. Therefore, the probability of delivering packets to their destinations is high. Moreover, VBS outperforms both AODV and DSR in the case of 40 sources, in terms of the packet. Also, VBS always demonstrates a lower routing load than AODV and DSR. In fact, VBS routing load is very close to 0. In AODV and DSR, route discovery will have to be restarted if a route is no longer valid. This adds to the routing load of both protocols. VBSs do not store routes in their cache. They, instead, forward the packet to the current correct next hop to the destination. Therefore, VBSs will almost always have a next hop to forward the data packet to, which implies that retransmissions are relatively rare. This explains the very low routing load of VBS. VBS outperformed both AODV and DSR in Reference 2 in terms of routing load. Moreover, VBS outperformed CGSR in Reference 3, load balancing, throughput, and stability.

10.3.2 Disadvantages of VBS

Zone_MTs require the aid of their VBS(s) all the time, this results a very high MAC contention on the VBSs, so even neighboring nodes cannot communicate with each other, even though they can hear each other, which put unnecessary load on clusterheads. Moreover VBS suffers from the disadvantages of the lowest ID algorithm [16].

10.4 The Power Aware Virtual Base Station (PA-VBS) Infrastructure Creation Protocol

In this protocol [15], one of the MTs is elected to be in charge of some other MTs that are in its range of transmission. The elected MT is known as a VBS. Every MT acknowledges its location by *hello* packets for a period of time. Every MT has a *sequence number* that reflects the changes that occur to that MT and also used to save battery power whenever possible, and a my_VBS variable, which is used to store an ID number to the VBS in charge of that MT. An MTs my_VBS variable is set exactly the same way as that of VBS protocol [8–10]. An MT is elected by one or more MTs to be their VBS based on an agreed upon rule, the battery power level method in this protocol. MTs announce their ID number with their periodic hello message. An MT sends a merge-request message to another MT if it has a lower battery power level and the other MT power greater than or equal to THRESHOLD_1 (will be explained later). The receiver of the merge-request message responds with an accept-merge message, increments its sequence number by one, and sets its my_VBS variable to zero. When the MT receives the accept-merge message, it increments its sequence number by one, and sets its my_VBS variable to the ID number of its VBS. In Reference 15 the battery power level of each and every MT is characterized into one of the following three categories:

1. MT EL $\geq$ THRESHOLD_1: An MT is eligible to be a VBS and willing to accept other MTs to be under its supervision if they have a lower energy level (EL).
2. THRESHOLD_2 $\leq$ MT EL < THRESHOLD_1: An MT ignores any merge-request messages that are sent to it by other MTs. If the MT is serving as a VBS, it remains in its roll, but it adds no more MTs to its list of MTs. If the MT is not serving as a VBS or under the supervision of a VBS and the EL of its VBS is less than THRESHOLD_2, it sends a merge-request message to another MT that has greater EL, if the latter EL is greater than or equal to THRESHOLD_1.
3. MT EL < THRESHOLD_2: An MT will ignore any merge-request messages and send iAmNoLongerYourVBS message to all the nodes under its supervision, if it was serving other nodes.

The PA-VBS outperformed VBS in Reference 15, in terms of load balancing, and energy saving however, PA-VBS did not address the methods of selecting gateways. Moreover, clusterheads do not give their zone_MTs an implicit signal to look for another clusterhead; instead, MTs are free to look for another clusterhead when their clusterhead battery power goes below THRESHOLD_1, which may cause instability in the network.

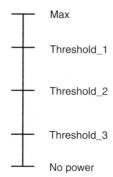

FIGURE 10.1   Three power levels WEAC.

10.5   Warning Energy Aware Clusterhead (WEAC) Infrastructure Creation Protocol

10.5.1   Protocol Overview

As in all cluster-based protocols, WEAC [19,20] the MTs, based on an agreed-upon policy, are elected to be in charge of all the MTs within their transmission ranges, or a subset of them. This can be achieved by electing one to be a clusterhead. Every MT acknowledges its location via *hello* packets, sometimes called beacon packets. The method of electing a clusterhead from a set of nominees is based on its EL (as it will be shown later). Another issue to be addressed is the handing of responsibilities of a clusterhead over from one clusterhead to another.

Every MT has a myCH variable. An MT myCH variable is set to the ID number of its clusterhead; however, if that MT itself is a clusterhead, then the myCH variable will be set to 0, otherwise it will be set to −1, indicating that it is a clusterhead of itself or a free node. A clusterhead collects complete information about all other clusterhead and their lists of MTs and broadcasts this information in its periodic hello messages. Zone_MTs, accumulate information about the network from their neighbors between hello messages, and they broadcast their neighbor_list to their neighbors in their hello packets. MTs announce their ID number with their periodic hello message. An MT sends a merge-request message to another MT if the latter has a higher EL and it should be more than or equal to THRESHOLD_1 (it will be explained later). The receiver of the merge-request message responds with accept-merge message and sets its myCH variable to zero. When an MT receives the accept-merge message it sets its myCH variable to the ID number of its clusterhead. The EL of each and every MT is characterized into one of the following four categories, as shown in Figure 10.1.

1. MT Energy Level ≥ THRESHOLD_1: An MT is eligible to be a clusterhead and willing to accept other MTs to be under its supervision if these MTs have a lower EL. If the MTs have the same EL, which is almost impossible, then the one with more number of neighbors wins. If myCH ≥ 0, no merge request will be sent by MTs, however, if myCH = −1, it will send a merge request to an MT with higher EL.

2. THESHOLD_2 ≤ MT Energy Level < THRESHOLD_1: An MT ignores any merge-request messages that are sent to it by other MTs. If the MT is serving as a clusterhead, it remains a clusterhead, but it adds no subscribers, however; as in the first point, if myCH ≥ 0 no merge request will be sent by MTs. If the myCH = −1, then it sends a merge-request message to an MT whose EL is greater than or equal to THRESHOLD_1.

3. THESHOLD_3 ≤ MT Energy Level ≤ THRESHOLD_2: If an MT is serving as a clusterhead, it sends a warning message to all MTs under its supervision, informing them to look for another clusterhead; nonetheless, they can remain with it till its EL drains to THRESHOLD_3. If the

myCH $= -1$, then it sends a merge-request message to an MT whose EL is greater than or equal to THRESHOLD_1.

4. MT Energy Level $\leq$ THRESHOLD_3: An MT ignores any merge-request messages and will send iAmNoLongerYourCH message to all the nodes under its supervision, if it was serving other nodes. If the myCH $= -1$, then it sends a merge-request message to an MT whose EL is greater than or equal to THRESHOLD_1.

10.5.2 Formal Description of the Protocol

This section contains the pseudo code for the algorithms executed by the MTs running the WEAC infrastructure-creation protocol. The hello() method is called by an MT periodically when it sends its hello message. MTs broadcast their current EL as part of their hello messages. Hello messages contain other useful pieces of information, such as the BMT flag, warningThreshold flag, and the iAmNoLongerYourCH flag. If the BMT flag is set to true, then it means that this node is a gateway, which is the only node eligible to acquire full knowledge of the network. The sendADisjoinMessage() method is executed by an MT when it finds a new clusterhead, based on the criterion described above, and when its current clusterhead sets its warningThreshold flag to true, that is, the current clusterhead EL is below THERSHOLD_2.

10.5.2.1 The WarningThreshold Flag

The warningThreshold flag is used by a clusterhead to notify the MTs, under its supervision, whether they can look for another clusterhead or not (Figure 10.2).

If it is set to false (i.e., its EL above THRESHOLD_2), its list of MTs know that they are not allowed to look for another clusterhead for at least one hello period. However, if the flag is set to true, its list of MTs can stay with their clusterhead, but they should look for another clusterhead or elect a new clusterhead (Figure 10.3).

10.5.2.2 The iAmNoLongerYourCH Flag

The iAmNoLongerYourCH flag is used by a clusterhead to convey to its list of MTs whether it can support them for another hello period or not (Figure 10.4).

If the flag is set to false (i.e., the warningThreshold is set to false), then the EL is above the warning threshold (THRESHOLD_2), its list of MTs learn that they will stay with their current clusterhead for at least one hello period, but if iAmNoLongerYourCH flag was set to true (i.e., its EL is below THRESHOLD_3), the MTs would have to look for another clusterhead or to elect a new clusterhead (Figure 10.5). The sendAMergeRequest() and sendMergeAccept() methods are called by an MT according to the mechanisms described in the previous section. Each node maintains a neighborList, which contains its neighbors, and broadcasts it in its hello message, in order to be used in routing decisions.

10.5.2.3 Merge and Disjoin Algorithms

After receiving a hello message, an MT sends a merge-request message to the issuer of the hello message, if at least one of following two conditions is satisfied:

```
1. if(myCH == 0)
2.     if((EL < THRESHOLD_2)&& (EL > Threshold_3))
3.         warningThreshold = true;
4.     elseif(EL > THRESHOLD_2) warningThreshold = false;
```

FIGURE 10.2   Using warningThreshold a VBS.

```
1.  if((myCH>0)&&(myCH == senderID))
2.      if(warningThreshold == true){
3.          if(thereIsAnMTWithEL>Threshold_1InMyNeighborList){
4.              if(itsEL > myEL){
5.                  sendAmergeRequestToThatMT(myID);
6.                  afterReceivingMergeAcceptMessage;
7.                  sendDisjoinMessageToMyOldCH;
8.                  deleteMyOldCHTimer;
9.                  startATimerForMyNewCH;
                }
            }
10.         else{
11.             electANewCH or stay with myCH till it goes<
                    Threshold_2;
12.             restartMyCHTimer();
13.             KeepSearchingForANewCH;
            }
        }
```

FIGURE 10.3   Using warningThreshold by an MT.

```
1.  if(myCH == 0)
2.      if(EL < Threshold_3)
3.          myListOfMTs.size = 0;
4.          iAmNoLongerYourCH = true;
5.          myCH = -1;
6.      else
7.          iAmNoLongerYourCH = false;
```

FIGURE 10.4   Using iAmNoLongerYourCH by a VBS.

```
1.  if((myCH > 0)&&(myCH == senderID)){
2.    if(iAmNoLongerYourCH == true){
3.          deleteMyCHTimer;
4.          myCH = -1;
      }
5.    else
6.          restartMyCHTimer();
    }
```

FIGURE 10.5   Using iAmNoLongerYourCH by an MT.

1. The MT is not a clusterhead, with the following constraints:
   (a) Its EL is less than the EL of the issuer of the hello message.
   (b) The EL of the sender is above THRESHOLD_1.
2. The MT is under the supervision of a clusterhead, with the following constraints:
   (a) The warning threshold of its clusterhead is true.

(b) Its EL is less than the EL of the hello message issuer.

(c) The EL of the sender is above THRESHOLD_1.

Figure 10.6 shows how the decision is made by an MT to determine whether to send a merge request to the issuer of the hello message, after receiving a hello message.

Upon receiving a merge-request message (Figure 10.7), an MT starts the merge-accept process, if and only if its EL is above THRESHOLD_1. If this condition is satisfied, the receiver of the merge request increments its list of MTs by 1, and sets its myCH variable to 0, indicating that it is a clusterhead, starts a timer for the sender (in order to activate the initiation of a timeout period), and sends a merge-accept message to the sender. If the receiver of the merger request was under the supervision of another cluster-head, upon receiving the merge request, it cancels the timer of its clusterhead and sends a disjoin message to it. Sometimes the EL of the receiver of the merge request, upon receiving it, is below THRESHOLD_1, or it might go out of the communication range of the sender before it receives the merge request. In this case the sender of the merge request assumes that its merge request was rejected, if it does not receive a feedback within a certain time.

As soon as an MT receives an accept message, it executes the steps shown in Figure 10.8 if and only if it is neither clusterhead nor an MT served by a clusterhead. It sets myCH variable to the ID of the sender of the merge-accept message, and starts a timer for its new clusterhead.

Figure 10.9 shows the series of actions taken by a clusterhead after receiving a disjoin message. First, it decrements its list of MTs by one, then cancels the timer of the sender of the disjoin message. If its list of MTs is empty, the clusterhead sets its myCH to -1. Figure 10.10 shows the mechanism of tables update for MTs under the supervision of clusterheads.

10.5.2.4 The Gateway Selection Algorithm

As mentioned earlier, each node broadcasts its neighbor list in its hello message. If the destination is the neighbor or the neighbor's neighbor, an MT sets the next hop to that neighbor; otherwise it sets its clusterhead as next hop, to connect to the backbone. It may happen that there are more than one neighbor that has an access to the destination; in this case, an MT runs the MT selection algorithm (Figure 10.11).

```
1. if (((receivedEL) >= THRESHOLD_1)&&(receivedEL > myEL))
      && ( (myCH == -1)
      || ((myCH > 0) && (myCHWarningThershold == true)))
2. sendMergeRequestMessageTo(issuerOfHello);
```

FIGURE 10.6 Merge request decision algorithm by an MT.

```
1. if (myEL > Threshold_1()){
2.   listOfMyMTs++;;
3.   startTimerForTheMergeRequestSender();
4.   if(myCH>0){
5.         cancelTimerOfMyOldCH();
6.         sendDisjoinMessageToMyOldCH();
      }
7.   if (myCH > 0 || myCH = -1)
8.         myCH = 0;
   }
```

FIGURE 10.7 Actions taken after receiving a merge-request message.

```
1. if (myCH == -1 ||(myCH > 0 && warningThresholdOfMyCH == true)){
2.     myCH = ID of the sender of the mergeAcceptMessage;
3.     storeLastPowerValueReportedByMyCH();
4.     startTimerForMyNewCH();
5.     if(myCH > 0){
6.         cancel the timer of my Old CH();
7.         sendADisjoinMessage(ID of myOldCH);
8.         }
    }
```

FIGURE 10.8   Actions taken after receiving a merge-accept message.

```
1. if (myCH == 0){
3.     listOfMyMTs--;
4.     cancelMTTimer(senderOFDisjoinMessage);
5.     if (listOfMyMTs == 0)
6.         myCH = -1;
        }
```

FIGURE 10.9   Actions taken after receiving a disjoin message.

```
1. if(myCH > 0)
2.    if(DEST my neighbor || DEST my neighbor's neighbor){
3.       set this neighbor as the next hop;
4.          if(more than one neighbor to the DEST)
5.          run MT selection algorithm(), (Figure 10.11)
6.    }
7.    else
8.       set myCH as the next hop
```

FIGURE 10.10   Table update by Zone_MTs.

```
1. if(there is more than one access to DEST){
2.    if(EL of BMT > THRESHOLD_2){
3.       getTheLeastNumberOfNeighbors();
4.       select the one with the least number of neighbors;
5.    }
6.    else
7.       choose the one with the highest EL;
   }
```

FIGURE 10.11   Gateway selection algorithm.

```
1. if(myCH ≤ 0)
2.    update routing table according to the received tables
3.    if(the DEST my neighbor || my neighbor's neighbor)
4.       set this neighbor as the next hop
5.          if(more than one neighbor to the DEST)
6.             run MT selection algorithm(); (Figure 10.11)
7.
8.
9.    else if(available route to the DEST)
10.      keep the current route till the next update;
11.   else
12.      set the DEST to null;
```

FIGURE 10.12   Table update by clusterheads and free MTs.

This algorithm selects MTs with the highest EL with the following two constraints:

1. If the EL of the MT $\geq$ THRESHOLD_2, select the MT with the least number of neighbors, in order to minimize the energy wasted by the network.
2. If the EL of the BMT is less than or equal to THRESHOLD_2, select the BMT with the highest EL.

10.5.2.5 Table Updates

If the MT is a clusterhead or a free_MT, the table update mechanism is slightly different (Figure 10.12). The first part (checking whether the destination is the next hop) is the same as the zone_MTs. If the destination is neither the neighbor nor the neighbor's neighbor, an MT updates its table according to the received routing tables. However, if there is more than a gateway to the destination, it runs the MT selection algorithm (Figure 10.11). If the destination is not accessible, it sets it to null.

WEAC protocol uses the concept of gateways or BMTs. As mentioned earlier, if the BMT flag is set to true, then that particular node is eligible to propagate routing tables. If the sender of the routing table has more than BMT within its range, it executes the algorithm (Figure 10.12).

10.6 The Virtual Base Station On-Demand (VBS-O) Routing Protocol

In this section we introduce the VBS-O routing protocol [11,13,19,20], which can be run on top of any mobile infrastructure protocol. As it is named, there are some on-demand features added to VBS-O routing protocol. Each VBS or clusterhead is in charge of a set of nodes in its neighbor. Only VBSs and free_MTs are eligible to acquire knowledge of the full network topology. Therefore, VBSs, BMSs, and free_MTs form the virtual backbone of the network. Since neighboring nodes, which are one hop away from each other, can hear each other, they are allowed to communicate with each other without the aid of their VBS(s), opposing VBS [8–10] and CGSR [10,12,13] routing protocols. Additionally, the new technique of broadcasting the neighbor list, used by WEAC protocol, speeds up even more packet delivery and improves the throughput. If an MT wishes to send a packet (see Figure 10.13), first it looks up its neighbor_list, if the destination is not found, it looks up the neighbor_lists of its neighbors (i.e., if the destination is the neighbor or the neighbor's neighbor, it sends the packet to that particular neighbor). If it has more than one access to the destination, it checks their battery power level, if it is more than THRESHOLD_2, it send the packet to the one with the least number of neighbors, otherwise it sends the packet to the one with the highest battery power. This reduces the MAC contention, balances the load, minimizes the energy wasted by the network, and as a result, enlarges the lifetime of the network. Moreover, this reduces the delay time of packets delivery between neighboring nodes in the network. However, if an MT wishes to send a

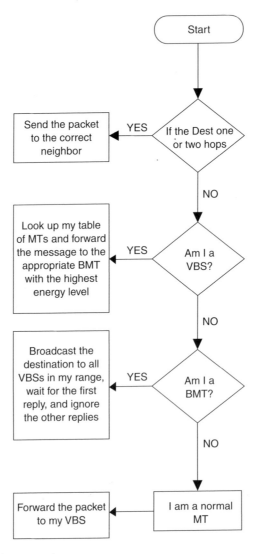

FIGURE 10.13   VBS-O routing protocol.

packet to another MT that is more than two hops, first, it sends the packet to its VBS. The VBS looks up its routing table and forwards the packet to the correct neighbor or BMT or the destination. At any time, if the destination is the neighbor or neighbor's neighbor, the packet will be forwarded to that neighbor. The sent packet contains the destination address. If the BMT is a zone_MT, it first performs the packet forwarding algorithm, if the destination is more than two hops, it broadcasts the destination to all VBSs and free_MTs in its transmission range. If one of them has an access to the destination, which is most probably the case, it replies to it. If the waiting time expires, it sends the packet to its VBS.

The VBS-O outperformed both of VBS and CGSR protocols in Reference [13] in terms of energy saving, average packet delivery time and throughput.

10.7 Reservation and Polling

In this section we derive formulas for the expected waiting time for each packet to be transmitted, provided that the link between the server [21–23], which is the VBS or clusterhead, and client, which is the MT that requires the assistant of the server, remains intact till the packet gets transmitted. We derived the formulas specifically for the VBS-O routing protocol.

Assumptions:

- The number of MTs $= m$, m is variable number.
- The transit time for the token from MT_i to MT_{i-1} (according to the clusterhead index) is random and depends on how busy the medium is, with mean $\overline{V_i}$ and second moment $\overline{V_i^2}$.
- The token can serve packets at an MT in different fashions (exhaustive and partially gated polling in this paper).

Let $\overline{X}$ = average transmission time, $\overline{X^2}$ = second moment of transmission time for all MTs. Arrival packets to any node is according to Poisson process with rate λ/m.

- X_i: Transmission time of ith arrival
- W_i: Delay in MAC queue of ith arrival
- R_i: Residual time of ith arrival
- N_i: Packets need to be transmitted before ith packet
- F_i: Packets need to be re-transmitted before ith packet
- Y_i: Total time token spends in full transit before ith packet gets transmitted
- V_i: Total time clusterhead in vacation before ith packet gets transmitted

10.7.1 Exhaustive Polling Analysis

For exhaustive polling, the total delay in queue:

$$W_i = R_i + \sum_{j=1}^{N_i} X_{ij} * B(j) + Y_i + F_i + V_i \tag{10.1}$$

where X_{ij} is the transmission time of jth packet that gets transmitted before ith arrival is transmitted,

$$B(i) = \begin{cases} 0, & \text{the carrier of the packet out of the transmission range of the clusterhead} \\ 1, & \text{the carrier of the packet within the transmission range of the clusterhead} \end{cases}$$

The expectation of the waiting time for the exhaustive polling is:

$$E[W_i] = E[R_i] + \overline{X} * E[N_i] + E[Y_i] + E[F_i] + E[V_i] \tag{10.2}$$

Now we will derive a formula of each of the components of the right-hand side:

Let

$$R = E[R_i] = \text{average residual time}$$

$$= \frac{1}{t} \lim_{t \to \infty} \int_0^t r(\tau)\, d(\tau), \, r(\tau) = \text{residual time at time } \tau \tag{10.3}$$

Let $V_{ij} = $ the transit time from MT_{i-1} to MT_i in the jth cycle, $t = $ the time that ends a cycle, $M(t) = $ the number of packets transmitted by time t, $L(t) = $ the number of cycles completed by time t.

Therefore,

$$\frac{1}{t}\int_0^t r(\tau)d\tau = \left[\underbrace{\frac{M(t)}{t}}_{=\lambda} * \underbrace{\frac{1}{M(t)} * \sum_{i=1}^{M(t)} \frac{X_i^2}{2}}_{=\frac{\overline{X_i^2}}{2}} * B(i) + \frac{L(t)}{t} * \underbrace{\frac{1}{L(t)} * \sum_{j=1}^{L(t)} \sum_{i=0}^{m-1} \frac{V_{ij}^2}{2}}_{=\sum_{i=0}^{m-1} \frac{\overline{V_i^2}}{2}} \right]$$

Let

$$C_j = \sum_{i=0}^{m-1} S_{ij} = \text{time in transit during } j\text{th cycle}$$

$$= \text{DIFS} + \text{SIFS} + \text{hello messages time} + \text{packet forwarding time}$$

$$+ \text{RTS/CTS periodes} + \cdots + \text{etc.}$$

$$\therefore \frac{1}{t} \sum_{i=1}^{L(t)} C_i \Rightarrow \underbrace{1 - \rho}_{\text{idle time}} (t \to \infty)$$

But

$$\frac{1}{L(t)} \sum_{i=1}^{L(t)} C_i = \overline{S}_0 + \overline{S}_1 + \overline{S}_2 + \cdots + \overline{S}_{m-1}, \text{ Law of Large Numbers (LLN)}$$

$$\therefore \frac{L(t)}{t} = \frac{1 - \rho}{\overline{S}_0 + \overline{S}_1 + \overline{S}_2 + \cdots + \overline{S}_{m-1}}, \text{ where } S_i \text{ is transit time of node } i$$

$$\therefore E[R_i] = \frac{\lambda * \overline{X}^2 * B(i)}{2} + \frac{(1 - \rho) * \sum_{i=0}^{m-1} \overline{S_i^2}}{2 * \sum_{i=1}^{m-1} \overline{S}_i} \tag{10.4}$$

Condition on which MT_ith arrival was in, and on which the token upon arrival.

Let

$$E_j = \{\text{the } i\text{th arrival of } MT_j\}$$

$$P(E_i) = \frac{1}{m} \text{(probability) and } \{E_j\}_{j=0}^{m-1} \text{ is partition}$$

$$A_f = \{\text{tokens is at } MT_f, \text{ when the } i\text{th packet arrival}\}$$

$$B_f = \{\text{token in a transit from } MT_{(f-1)} \bmod m \text{ to } MT_f\}, \text{ this is according to the index of the clusterhead.}$$

$$f = 0, 1, 2, \ldots, \text{number of MTs served by one VBS } (m-1).$$

$$\{A_f \cup B_f\}_{f=0}^{m-1} \text{ is also a partition.}$$

$$P(A_f) = \rho\frac{1}{m}$$

$$P(B_f) = (1-\rho)\frac{\overline{S}_f}{\overline{S}_0 + \cdots + \overline{S}_{m-1}}$$

$$E[Y_i] = \sum_{f=0}^{m-1}\sum_{j=0}^{m-1} E[Y_i \mid A_f \cup B_f, E_j] * P[(A_f \cup B_f) \cap E_j]$$

$$= \sum_{f=0}^{m-1}\sum_{j=0}^{m-1} E[Y_i \mid A_f \cup B_f, E_{(f+j)\bmod m}] * P(A_f \cup B_f) * P(E_{(f+j)\bmod m})$$

but at $j = 0$ there will be no full transit, so

$$E[Y_i] = \sum_{f=0}^{m-1}\sum_{j=1}^{m-1}(\overline{S}_{(f+1)\bmod m} + \cdots + \overline{S}_{(f+j)\bmod m}) * \left(\frac{\rho}{m} + \frac{(1-\rho)}{\sum_{i=0}^{m-1}\overline{S}_i}\right) * \frac{1}{m}$$

when $j = m - 1 : \overline{S}_{(f+1)\bmod m} + \cdots + \overline{S}_{(f+m-1)\bmod m}$

$$\therefore E[Y_i] = \sum_{f=0}^{m-1}\sum_{j=1}^{m-1}(m-j) * (\overline{V}_{(f+1)\bmod m}) * \left(\frac{\rho}{m} + \frac{(1-\rho)}{\sum_{i=0}^{m-1}\overline{S}_i}\right) * \frac{1}{m} \Rightarrow$$

$$= \sum_{j=1}^{m-1}(m-j)(\overline{S}_0 + \cdots + \overline{S}_{(m-1)}) * \frac{\rho}{m} * \frac{1}{m}$$

$$+ \underbrace{\frac{(1-\rho)}{\sum_{i=0}^{m-1}\overline{S}_i}}_{=m\overline{S}} * \frac{1}{m}\sum_{f=0}^{m-1}\sum_{j=1}^{m-1}(m-j) * (\overline{S}_{(f+1)\bmod m}) * \overline{S}_f$$

$$\because \overline{S} = \frac{1}{m}(\overline{S}_0 + \cdots + \overline{S}_{m-1}) \Rightarrow m\overline{S} = (\overline{S}_0 + \cdots + \overline{S}_{m-1})$$

$$E[Y_i] = \underbrace{\frac{\rho\overline{S}}{m}\sum_{j=1}^{m-1}(m-j)}_{=\frac{\rho\overline{S}}{m} * \frac{m(m-1)}{2}} + \frac{(1-\rho)}{m^2\overline{V}}\underbrace{\sum_{f=0}^{m-1}\sum_{j=1}^{m-1}(m-j) * (\overline{S}_{(f+1)\bmod m}) * \overline{S}_f}_{=X}$$

$$X = \sum_{f=0}^{m-1}\sum_{j=1}^{m-1}(m-j) * (\overline{S}_{(f+1)\bmod m}) * \overline{S}_f$$

Let us assume that $(f + j) \bmod m = f', f$ and $f' = 0, 1, 2, \ldots, m - 1$.
In both case of $f + j \leq m - 1$ and $f + j \geq m - 1, r = m - j, 1 \leq r \leq m - 1$

$$\because X = m \sum \sum_{f < f'} \overline{S}_f \overline{S}_{f'}$$

$$\therefore \sum \sum_{f < f'} \overline{S}_f \overline{S}_{f'} = \left[\left(\sum_{f=0}^{m-1} \overline{S}_f \right)^2 - \sum_{f=0}^{m-1} \overline{S}_f^2 \right] * \frac{1}{2}$$

$$\therefore E[Y_i] = \frac{\overline{S}(m - \rho)}{2} + \frac{(1 - \rho) \sum_{f=0}^{m-1} S_f^2}{2 \times m \times \overline{S}} \qquad (10.5)$$

which is the expected time the token spent in full transit before ith packet gets transmitted.

Regarding $E(N_i)$, which is the expected number of packets that need to be transmitted before ith packet, we have set some constraints:

We need to argue that we can apply Little's formula to get $E[N_i] = \lambda W * B(i)E[Y_i]$. Here is the argument:

1. The VBS sends the wireless token to its list of MT(s) at the correct time in order of their service start time.
2. From the point of view of VBS, everything is the same, but N_i, which is the number of packets in the system upon arrival, $E[N_i] =$ the average number of packets in node provided that both of the transmitting MT and the VBS stay in the transmission range of each other till the ith packet gets transmitted.

From the above argument we can conclude that:

$$E[N_i] = \lambda W * B(i)E[Y_i] = \frac{\overline{S}(m - \rho)}{2} - \frac{(1 - \rho) * \sum_{i=0}^{m-1} \overline{S}_i^2}{2 * m * \overline{S}} \qquad (10.6)$$

Regarding $E[F_i]$, it is simply the probability of re-transmission multiplied by the total number of packets need to be sent before the ith packet so:

$$E[F_i] = P(\text{re-transmission}) * E[N_i] \qquad (10.7)$$

To calculate the expected time the VBS on vacation is simply as the one in Equation (10.5).

$$E[V_i] = \frac{\overline{V^2}}{2 * \overline{V}} \qquad (10.8)$$

where m is the number of nodes and S is the transit time. Substituting Equations [(10.3)–(10.8)] in Equation (10.2) we get the average waiting time in the MAC queues of one packet provided that it stays in the queue till it gets transmitted is

$$W = \frac{\lambda * \overline{X^2} * B(i)}{2 * (1 - \rho)} + \frac{\overline{S} * (m - \rho)}{2 * (1 - \rho)} + \frac{\sigma_s^2}{2 * S} + \frac{\overline{V^2}}{2 * \overline{V} * (1 - \rho)} \qquad (10.9)$$

10.7.2 Partially Gated Polling Analysis

For the partially gated polling the average waiting time is

$$E[W_i] = \underbrace{E[R_i] + \overline{X} * E[N_i] + E[F_i] + E[V_i]}_{\text{same as exhaustive polling}} + \underbrace{E[Y_i]}_{\substack{\text{slightly} \\ \text{different}}}$$

From the previous analysis for exhaustive polling of $E[Y_i]$ with $j = 0$;

$$P[A_f \mid (A_f \cup B_f) \cap E_f] = \frac{P[A_f \cap \langle (A_f \cup B_f) \cap E_f \rangle]}{P[(A_f \cup B_f) \cap E_f]}$$

$$= \frac{P[A_f \cap (A_f \cup B_f)]}{P[A_f \cup B_f]}$$

$$= \frac{P[A_f]}{P[A_f \cup B_f]}$$

$$\sum_{f=0}^{m-1} E[Y_i \mid (A_f \cup B_f) \cap E_f] * P(A_f \cup B_f) * \frac{1}{m} E[Y_i \mid (A_f \cup B_f) \cap E_f] * P(A_f \cup B_f)$$

$$= E[Y_i \mid (A_f \cup B_f) \cap E_f \cap A_f] * P(A_f \mid (A_f \cup B_f) \cap E_f]$$

$$+ \underbrace{E[Y_i \mid (A_f \cup B_f) \cap E_f \cap B_f] * P(B_f \mid (A_f \cup B_f) \cap E_f]}_{= 0 \text{ for partially gated poling}}$$

$$E[Y_i \mid (A_f \cup B_f) \cap E_f] = E[Y_i \mid A_f \cap E_f]$$

$$= m\overline{V}$$

$$P[A_f \mid (A_f \cup B_f) \cap E_f] = \frac{P[A_f \cap \langle (A_f \cup B_f) \cap E_f \rangle]}{P[(A_f \cup B_f) \cap E_f]}$$

$$= \frac{P[A_f]}{P[A_f \cup B_f]}$$

Putting $j = 0$ in Equation (10.5) we get:

$$\sum_{f=0}^{m-1} E[Y_i \mid (A_f \cup B_f) \cap E_f] * P(A_f \cup B_f) * \frac{1}{m}$$

$$= \sum_{f=0}^{m-1} \frac{m\overline{V} * P[A_f]}{P[A_f \cup B_f]} * P[A_f \cup B_f] * \frac{1}{m} \qquad (10.10)$$

$$\therefore \sum_{f=0}^{m-1} E[Y_i \mid (A_f \cup B_f) \cap E_f] * P(A_f \cup B_f) * \frac{1}{m} = \overline{V} \sum_{f=0}^{m-1} P(A_f) = \rho\overline{V}$$

from Equations (10.9) and (10.10)

$$W = \frac{\lambda * \overline{X}^2 * B(i)}{2 * (1 - \rho)} + \frac{\overline{S}(m - \rho)}{2 * (1 - \rho)} + \frac{\sigma_s^2}{2 * S} + \frac{\overline{V}^2}{2 * \overline{V} * (1 - \rho)} + \frac{\rho * \overline{S}}{(1 - \rho)}$$

$$\left[W_{\text{partial}} = W_{\text{exhaustive}} + \frac{\rho * \overline{S}}{(1 - \rho)} \right]$$

10.8 Network Model

An ad hoc network is modeled as a graph $G = (N, L)$ [24–26], where N is a finite set of nodes and L is a set of undirected links. The routing protocol uses only bi-directional links.

Now, we are modeling VBS-O routing protocol.

- A node n_i has a set of neighbors, $NB_i = \{n_j \in N : (n_i, n_j) \in L\}$.
- The bandwidth partitioned into a set of time periods, $Tx = \{tx_1, tx_2, tx_3, \ldots, tx_m\}$, which represents the time when medium is occupied.
- The transmission schedule of a node n_i is defined as the set of periods TTx_i in which it transmits.
- The set of nodes Rx_i^k, which is n_i transmission targets set (receivers) in period Tx_k, $Tx_k \in TTx_i$, $Rx_i^k \in NB_i$.
- The set $RxTx_i = \{tx_k \in Tx : n_i \in Rx_j^k, n_j \in NB_i\}$, is the set of periods where node n_i is required to receive from its neighbors.
- Let $TN^k = \{n_i \in N, tx_k \in TTx_i\}$ be the set of nodes transmitting in period tx_k.

The period of TTx is the collection $\{TTx_i : n_i \in N\}$.

To ensure successful transmission:

If node n_i transmits in period tx_k, $(n_i \in TN^k)$ for every node $n_j \in Rx_i^k$, $NB_i \cap TN^k = \{n_i\}$ and $n_j \notin TN^k$.

In other words, when node n_i transmits to n_j in period tx_k, n_j itself does not transmit and n_i is the only transmitting neighbor of n_j in that time period.

10.9 Power Analysis

Given all the terminals initially have fully charged batteries, the maximum allowed energy dissipated per second so as to force the nodes to operate for h hours, provided that the network life time is at least h hours [24–26]:

1. A VBS V_i has a set of MTs, $MT_i = \{(n_i, V_i) \in N, n_i \in V_i, (n_i, V_i) \in L\}$, provided that n_i is within the transmission range of V_i
2. Given all the terminals initially have fully charged batteries, the maximum allowed energy dissipated per second so as to force the nodes to operate for h hours, provided that the network life time is at least h hours.

$$\text{Maximum flow threshold per second} = \frac{N * \text{battery\_capacity}}{h * 3600}$$

where N is the total number of nodes in the network and, h is the number of hours (network life time).

3. Each node in the network, which is served by VBS and requires the assistance of its VBS, should expect the following delays and constraints:
   (i) Sense the medium, if busy (DIFS + random back off).
   (ii) Interference of neighboring nodes.
   (iii) If VBS idle (sending packets, hello messages, . . ., etc).
   (iv) VBS above THRESHOLD_3.
   (v) VBS is still in the transmission range.
   (vi) For BER $= 10^{-6}$, differential quadrature phase shift keying (DQPSK) modulation: where E_b/N_o is the measure of signal-to-noise ratio for a digital communication system. It is measured at the input to the receiver and is used as the basic measure of how strong the signal is.

   We use energy per bit (E_b) to noise spectral density $(N_o)(E_b/N_o)$ and the carrier-to-noise ratio (C/N) to find out how much transmitter power we will need. We use DQPSK modulation scheme and transmit 2 Mbps with a carrier frequency of 2450 MHz. It will have a 30 dB fade margin and operate within a reasonable bit error rate (BER) at an outdoor distance of 100 m.

4. To the transmit power is to:
   - Determine E_b/N_o for the desired BER.
   - Convert E_b/N_o to C/N at the receiver using the bit rate.
   - Add the path loss and fading margins.

   We first decide what is the maximum BER that we can tolerate. For our example, we choose 10^{-6} figuring that we can retransmit the few packets that will have errors at this BER.

   From 3G Mobile Base Stations Standards, we find that for DQPSK modulation, a BER of 10^{-6} requires an E_b/N_o of 11.1 dB.

   Now we convert E_b/N_o to C/N using the equation:

   $$\frac{C}{N} = \frac{E_b}{N_o} * \frac{f_b}{Bw}$$

   where f_b is the bit rate and Bw is the receiver noise bandwidth.

   Since we now have the carrier-to-noise ratio, we can determine the necessary received carrier power after we calculate the receiver noise power.

   Noise power is computed using Boltzmann's equation:

   $$N = kTB$$

   where k is the Boltzmann's constant $= 1.380650 \times 10^{-23}$ J/K, T is the effective temperature in Kelvin, and B is the receiver bandwidth.

5. For BER $= 10^{-6}$, DQPSK modulation: $E_b/N_o = 11.1$ dB
6. Calculate E_b/N_o from:
   (a) $X = SNR * W/RATE$.
   (b) If $X <$ value of E_b/N_o found in (10.2), then reject it, . . ., else accept packet.
7. Receiver sensitivity: $Pw_{rx\_min} =$ receiver_noise_floor + SNR:
   (a) Receiver_noise_floor $= N +$ noise figure; $N = kTB$, noise figure $=15$ dB.
   (b) Path Loss $(PL) = 10 * \log_{10}(4\pi d/\lambda)$
   where d_{max} is the maximum possible distance between transceivers.
8. $Pw_{tx\_max} = Pw_{rx\_min} + PL +$ fade_margin.
9. $Pw_{rx} = Pw_{tx} \cdot \lambda_0^2/(4\pi d)^2$; Pw_{rx} is measured by receiver.
10. Compute d.
11. Find $Pw_{tx\_min\_prev}$ for d, Pw_{rx_min}:
    (a) Before an intermediate node (e.g., j) on the route forwards it to the next node toward the source (S), it enters $Pw_{tx\_min\_prev}$ calculated in (10.9).
    (b) Node i uses $Pw_{tx\_min\_prev}$ entered by j to calculate its $Pw_{tx\_min\_next}$.

12. $\displaystyle\sum_{e \in n_j} f(e) = f_1(Pw_{tx\_min\_next}) + \sum_{e \in n_j - \{j\}} f_2(Pw_{tx\_min\_next}, R_{e,j})$

13. $\therefore Pw_{tx\_min\_next}$ is used for energy consumption calculations by source node, and $Pw_{tx\_min\_prev}$ is used by the next node to the source to calculate its own $Pw_{tx\_min\_next}$.

10.10 Summary

This chapter presents an overview of ad hoc routing principles on cluster-based routing protocols. In cluster-based protocols, an ad hoc network is represented as a set of clusters. Clusterhead is elected based on an agreed upon rule. On the contrary to on demand protocols, cluster-based protocols take advantage of their cluster structure, which limits the scope of route query flooding. Only clusterheads and gateways are flooded with query packets, thus, reducing the bandwidth required by the route discovery mechanism. Delay models and clusterhead controlled token has been also presented in this chapter. An advantage of the exhaustive service policy is that, when a number of urgent frames are unexpectedly generated in a station with short transmission deadlines, the station can complete transmission of the frames before token passing. On the other hand, the policy may be a problem when an MT transmits frames is too long. Therefore, in general, the exhaustive token-controlled network is applied to a network with low traffic load. QoS issues can be implemented using clusterhead controlled token scheme. Power analysis is presented in the last part of the chapter. Adjusting the transmit power reduces interference. Moreover, it enlarges the lifetime of clusterheads, which means more stable clusters.

References

[1] I.F. Akyildiz, J. McNair, L. Carrasco, R. Puigjaner, and Y. Yesha, "Medium access control protocols for multimedia traffic in wireless networks," *IEEE Network*, 13: 39–47, 1999.

[2] S.R. Das, C.E. Perkins, and E.M. Royer, "Performance comparison of two on-demand routing protocols for ad hoc networks," in *Proceedings of the IEEE INFOCOM'2000*, Tel Aviv, Israel, pp. 3–12, March 2000.

[3] E.M. Royer and C.E. Perkins, "Multicast operation of the ad-hoc on-demand distance vector routing protocol," in *Proceedings of the ACM/IEEE MOBICOM'99*, Seattle, WA, pp. 207–218, August 1999.

[4] Mario Gerla and Jack Tzu-Chieh Tsai, "Multi-cluster, mobile, multi-media radio network," *ACM/Baltzer Journal of Wireless Networks*, 1: 255–265, 1995.

[5] K.S. Gilhousen and I.M. Jacobs *et al.* "On the capacity of a cellular CDMA system," *IEEE Transactions on Vehicular Technology*, 40: 303–312, 1991.

[6] Zoran Kosti and cacute, "Performance and implementation of dynamic frequency hopping in limited-bandwidth cellular systems," *IEEE Transactions on Wireless Communications*, 1: 28–36, 2002.

[7] M. Scott Corson and A. O'Neill, "An approach to fixed/mobile-converged routing," Institute for Systems Research Technical Report, TR 2000-5, Universtiy of Maryland, March 2000.

[8] A. Safwat and H. Hassanein, "Structured routing in wireless mobile ad-hoc networks," in *Proceedings of the IEEE International Symposium on Computers and Communications*, July 2001.

[9] H. Hassanein and A. Safwat, "Virtual base stations for wireless mobile ad-hoc communications: an infrastructure for the infrastructure-less," *The International Journal of Communications Systems*, 14(8): 763–782, 2001.

[10] T.R. Sheltami and H.T. Mouftah, "Comparative study of two clustering protocols," in *Proceedings of the IASTED International Conference Wireless and Optical Communications (WOC)*, Banff, Canada, pp. 251–256, June 2001.

[11] T.R. Sheltami and H.T. Mouftah, "Virtual base station on-demand," *ICWN'02*, Las Vegas, USA, pp. 421–425, June 24–27, 2002.

[12] C.-C. Chiang, H.-K. Wu, W. Liu, and M. Gerla, "Routing in clustered multihop, mobile wireless networks with fading channel," in *Proceedings of the IEEE Singapore International Conference on Networks (SICON)*, Singapore, pp. 197–211, April 1997.

[13] T.R. Sheltami and H.T. Mouftah, "Performance comparison of three clustering protocols in wireless mobile ad hoc networks," in *Proceedings of the Biennial Symposium on Communications*, Queen's University, pp. 131–135, June 2002.

[14] A. O'Neill, G. Tsirtsis, and S. Corson, "Edge mobility architecture," *Internet Draft*, October 1999.

[15] A. Safwat, H.S. Hassanein, and H. Mouftah, "Power-aware fair infrastructure formation for wireless mobile ad hoc communications," in *Proceedings of IEEE Globecom 2001*, 5: 2832–2836, 2001.

[16] A. Ephremides, J.E. Wieselthier, and Dennis J. Baker, "A design concept for reliable mobile radio networks with frequency hopping signaling," in *Proceedings of the IEEE*, 75: 56–73, 1987.

[17] M. Gerla and J.T.-C. Tsai, "Multicluster, mobile, multimedia radio network," *ACM-Baltzer Journal of Wireless Networks*, 1: 255–265, 1995.

[18] W. Liu, C.-C. Chiang, H-K. Wu, V. Jha, M. Gerla, and R. Bagrodia, "Parallel simulation environment for mobile wireless networks," in *Proceedings of the 1996 Winter Simulation Conference, WSC'96*, Coronado, CA, pp. 605–612, 1996.

[19] T.R. Sheltami and H.T. Mouftah, "A warning energy aware clusterhead," *IEEE Transaction on Wireless Communications*, submitted.

[20] T.R. Sheltami and H.T. Mouftah, "An efficient energy aware clusterhead formation infrastructure protocol for MANETs," *IEEE ISCC*, June 30–July 3, 2003, Kemer-Antalya, Turkey, pp. 203–208, 2003.

[21] T.R. Sheltami and H.T. Mouftah, "Clusterhead controlled token for virtual base station on-demand in MANETs," in *Proceedings of the Workshop on Mobile and Wireless Networks (MWN)*, Providence, Rhode Island, USA, pp. 716–721, 2003.

[22] T.R. Sheltami and H.T. Mouftah, "Two way of round robin MAC protocol for clusterheads in WEAC for MANETs," *IEEE CCECE 2003*, May 4–7, Montreal, Canada, pp. 371–374, 2003.

[23] T.R. Sheltami and H.T. Mouftah, "Average waiting time of clusterhead controlled token for virtual base station on-demand in MANETs," "Cluster Computing," accepted for publication.

[24] T.R. Sheltami and H.T. Mouftah, "Energy aware routing for virtual base stations on-demand routing protocol," *IEEE International Symposium on Wireless Networks (ISWSN'03)*, March 24–26, 2003, Dhahran, Saudi Arabia, 2003.

[25] T.R. Sheltami and H.T. Mouftah, "Power aware routing for the virtual base station on-demand protocol in MANETs," *The Arabian Journal for Science and Engineering*, 28(2C): 115–135, 2003.

[26] T. Sheltami and H.T. Mouftah, "Power issues on WEAC and VBS-O protocols for wireless mobile ad hoc networks, in *Proceedings of International Conference on Wireless Networks (ICWN)*, Monte Carlo Resort, Las Vegas, NV, June 21–24, 2004.

11

Broadcasting and Topology Control in Wireless Ad Hoc Networks

11.1 Introduction...**11**-239
11.2 Centralized Methods **11**-244
    Assumptions • Based on MST and Variations • Theoretical
    Analysis • Centralized Clustering
11.3 Localized Methods **11**-248
    Based on Distributed CDS • Localized Low Weight
    Structures • Combining Clustering and Low Weight •
    Flooding-Based Methods
11.4 Scheduling Active and Sleep Periods **11**-259
11.5 Conclusion and Future Research Directions...........**11**-260
References **11**-261

Xiang-Yang Li
Ivan Stojmenovic

11.1 Introduction

Network wide broadcasting in mobile ad hoc networks (MANET) provides important control and route establishment functionality for a number of unicast and multicast protocols. In this chapter, we present an overview of the recent progress of broadcast and multicast in wireless ad hoc networks. We discuss two energy models that could be used for broadcast: one is the nonadjustable power and the other is the adjustable power. If the power consumed at each node is not adjustable, minimizing the total power used by a reliable broadcast tree is equivalent to the minimum connected dominating set problem (MCDS), that is, minimize the number of nodes that relay the message, since all relaying nodes of a reliable broadcast form a connected dominating set (CDS). If the power consumed at each node is adjustable, we assume that the power consumed by a relay node u is $||uv||^{\beta}$, where real number $\beta \in [2, 5]$ depends on transmission environment and v is the farthest neighbor of u in the broadcast tree. For both models, we review several centralized methods that compute broadcast trees such that the broadcast based on them consumes the energy within a constant factor of the optimum if the original communication graph is unit disk graph. Since centralized methods are expensive to implement, we further review several localized methods that can approximate the minimum energy broadcast tree for nonadjustable power case. For adjustable power case, no localized methods can approximate the minimum energy broadcast tree within a constant factor, thus we review several currently best possible heuristics. Several local improvement methods and activity scheduling of nodes (active, idle, and sleep) are also discussed in this chapter.

Wireless Ad Hoc Networks: Due to its potential applications in various situations such as battlefield, emergency relief, environment monitoring, and so on, wireless ad hoc networks have recently emerged as a premier research topic. Wireless networks consist of a set of wireless nodes, which are spread over a geographical area. These nodes are able to perform processing as well as capable of communicating with each other by means of a wireless ad hoc network. With coordination among these wireless nodes, the network together will achieve a larger task both in urban environments and in inhospitable terrain. For example, the sheer numbers of wireless sensors and the expected dynamics in these environments present unique challenges in the design of wireless sensor networks. Many excellent researches have been conducted to study problems in this new field.

In this chapter, we consider a wireless ad hoc network consisting of a set V of n wireless nodes distributed in a two-dimensional plane. Each wireless node has an omni-directional antenna. This is attractive because a single transmission of a node can be received by many nodes within its vicinity which, we assume, is a disk centered at the node. We call the radius of this disk the *transmission range* of this wireless node. In other words, node v can receive the signal from node u if node v is within the transmission range of the sender u. Otherwise, two nodes communicate through multihop wireless links by using intermediate nodes to relay the message. Consequently, each node in the wireless network also acts as a router, forwarding data packets for other nodes. By a proper scaling, we assume that all nodes have the maximum transmission range equal to one unit. These wireless nodes define a *unit disk graph* UDG(V) in which there is an edge between two nodes if and only if their Euclidean distance is at most one.

In addition, we assume that each node has a low-power global position system (GPS) receiver, which provides the position information of the node itself. If GPS is not available, the distance between neighboring nodes can be estimated on the basis of incoming signal strengths. Relative coordinates of neighboring nodes can be obtained by exchanging such information between neighbors. With the position information, we can apply computational geometry techniques to solve some challenging questions in wireless networks.

Power-Attenuation Model: Energy conservation is a critical issue in wireless network for the node and network life, as the nodes are powered by batteries only. Each wireless node typically has a portable set with transmission and reception processing capabilities. To transmit a signal from a node to the other node, the power consumed by these two nodes consists of the following three parts. First, the source node needs to consume some power to prepare the signal. Second, in the most common power-attenuation model, the power needed to support a link uv is $||uv||^\beta$, where $||uv||$ is the Euclidean distance between u and v, β is a real constant between 2 and 5 dependent on the transmission environment. This power consumption is typically called *path loss*. Finally, when a node receives the signal, it needs to consume some power to receive, store, and then process that signal. For simplicity, this overhead cost can be integrated into one cost, which is almost the same for all nodes. Thus, we will use c to denote such constant overhead. In most results surveyed here, it is assumed that $c = 0$, that is, the path loss is the major part of power consumption to transmit signals. The power cost $p(e)$ of a link $e = uv$ is then defined as the power consumed for transmitting signal from u to node v, that is, $p(uv) = ||uv||^\beta$.

Broadcasting and Multicasting: Broadcasting is a communication paradigm that allows to send data packets from a source to multiple receivers. In one-to-all model, transmission by each node can reach *all* nodes that are within radius distance from it, while in the one-to-one model, each transmission is directed toward only one neighbor (using, for instance, directional antennas or separate frequencies for each node). The broadcasting in literature has been studied mainly for one-to-all model and we will use that model in this chapter. Broadcasting is also frequently referred to as *flooding*.

Broadcasting and multicasting in wireless ad hoc networks are critical mechanisms in various applications such as information diffusion, wireless networks, and also for maintaining consistent global network information. Broadcasting is often necessary in MANET routing protocols. For example, many unicast routing protocols such as dynamic source routing (DSR), ad hoc on demand distance vector (AODV), zone routing protocol (ZRP), and location aided routing (LAR) use broadcasting or a derivation of it to establish routes. Currently, these protocols all rely on a simplistic form of broadcasting called *flooding*, in which each node (or all nodes in a localized area) retransmits each received unique packet exactly one time. The main problems with flooding are that it typically causes unproductive and often harmful bandwidth congestion, as well as inefficient use of node resources. Broadcasting is also more efficient than sending multiple copies the same packet through unicast. It is highly important to use power-efficient broadcast algorithms for such networks since wireless devises are often powered by batteries only.

Recently, a number of research groups have proposed more efficient broadcasting techniques [1,2] with various goals such as minimizing the number of retransmissions, minimizing the total power used by all transmitting nodes, minimizing the overall delay of the broadcasting, and so on. Williams and Camp [2] classified the broadcast protocols into four categories: simple (blind) flooding, probability based, area based, and neighbor knowledge methods. Wu and Lou [3] classified broadcasting protocols based on neighbor knowledge information: global, quasi-global, quasi-local, and local. The global broadcast protocol, centralized or distributed, is based on global state information. In quasi-global broadcasting, a broadcast protocol is based on partial global state information. For example, the approximation algorithm in Reference 4 is based on building a global spanning tree (a form of partial global state information) that is constructed in a sequence of sequential propagations. In quasi-local broadcasting, a distributed broadcast protocol is based on mainly local state information and occasionally partial global state information. Cluster networks are such examples: while clusters can be constructed locally for most of the time, the chain reaction does occur occasionally. In local broadcasting, a distributed broadcast protocol is based on solely local state information. All protocols that select forward nodes locally (based on 1-hop or 2-hop neighbor set) belong to this category. It has been recognized that scalability in wireless networks cannot be achieved by relying on solutions where each node requires global knowledge about the network. To achieve scalability, the concept of localized algorithms was proposed, as distributed algorithms where simple local node behavior, based on local knowledge, achieves a desired global objective.

In this chapter, we categorize previously proposed broadcasting protocols into several families: centralized methods, distributed methods, and localized methods. Centralized methods calculate a tree used for broadcasting with various optimization objectives of the tree. In localized methods, each node has to maintain the state of its local neighbors (within some constant hops). After receiving a packets that needed to be relayed, the node decides whether to relay the packet only based on its local neighborhood information. Majority of the protocols are in this family. In distributed methods, a node may need some information more than a constant hop away to decide whether to relay the message. For example, broadcasting based on minimum spanning tree (MST) constructed in a distributed manner is a distributed method, but not localized method since we cannot construct MST in a localized manner.

Distributed or Localized Algorithms? Distributed algorithms and architectures have been commonly used terms for a long time in computer science. Unfortunately none of the already proposed approaches are applicable to wireless ad hoc networks. In order to address the needs of distributed computing in wireless ad hoc networks, one has to address how key goals, such as power minimization, low latency, security, and privacy, are affected by the algorithms used. Some common denominators are almost always present, such as high relative cost of communication to computation in wireless networks.

Due to the limited capability of processing power, storage, and energy supply, many conventional algorithms are too complicated to be implemented in wireless ad hoc networks. Thus, the wireless ad hoc networks require efficient distributed algorithms with low computation complexity and low communication complexity. More importantly, we expect the distributed algorithms for wireless ad hoc networks to be localized: each node running the algorithm only uses the information of nodes within a constant number of hops. However, localized algorithms are difficult to design or impossible sometimes. For example, we cannot construct the MST locally.

MAC Specification: Collision avoidance is inherently difficult in MANETs; one often cited difficulty is overcoming the hidden node problem, where a node cannot decide whether some of its neighbors are busy receiving transmissions from an uncommon neighbor. The 802.11 medium access control (MAC) follows a carrier sense multiple access/collision avoidance (CSMA/CA) scheme. For unicast, it utilizes a request to send (RTS)/clear to send (CTS)/data/acknowledgment (ACK) procedure to account for the hidden node problem. However, the RTS/CTS/Data/ACK procedure is too cumbersome to implement for broadcast packets as it would be difficult to coordinate and bandwidth expensive: a relay node has to perform RTS/CTS individually with all its neighbors that should receive the packets. Thus, the only requirement made for broadcasting nodes is that they assess a clear channel before broadcasting. Unfortunately, clear channel assessment does not prevent collisions from hidden nodes. Additionally, no resource is provided for collision when two neighbors assess a clear channel and transmit simultaneously. Ramifications of this environment are subtle but significant. Unless specific means are implemented at the network layer, a node has no way of knowing whether a packet was successfully reached by its neighbors. In congested networks, a significant amount of collisions occur leading to many dropped packets. The most effective broadcasting protocols try to limit the probability of collisions by limiting the number of rebroadcasts in the network. Thus, it is often imperative that the underlying structure for broadcasting is degree bounded and the links are at similar lengths. By using a power adjustment at each node, the collision of packets and contention for channel will be alleviated. Notice that, if the underlying structure for broadcasting is degree bounded, we can either use RTS/CTS scheme to avoid hidden node problem, or we can rebroadcast the dropped packets (such rebroadcast will be less since the number of intended receiving neighbors is bounded by a small constant).

Reliability: Reliability is the ability of a broadcast protocol to reach all the nodes in the network. It can be considered at the network or at the medium access layer. We will classify protocols according to their network layer performance. That is, assuming that MAC layer is ideal (every message sent by a node reaches all its neighbors), location update protocol provides accurate desired information to all nodes about their neighborhood, and the network is connected. Broadcast protocols can be *reliable* or *unreliable*. In a reliable protocol, every node in the network is reached, while in unreliable broadcast protocols, some nodes may not receive the message at all.

Message Contents: The broadcast schemes may require different neighborhood information, which is reflected in the contents of messages sent by nodes when they move, react to topological changes, change activity status, or simply send periodically update messages. For example, commonly seen *hello* message may contain (all or a subset of) the following information: its own ID, its position, one bit for dominating set status (informing neighbors whether or not the node itself is in dominating set), list of 1-hop neighbors, its degree. Other content is also possible, such as list of 1-hop neighbors with their positions, or list of 2-hop neighbors, or even global network information.

The broadcast message sent by the source, or retransmitted, may contain broadcast message only. In addition, it may contain various information needed for proper functioning of broadcast protocol, such as the same type of information already listed for *hello* messages, some constant bits of the system requirements (such as the maximum broadcast delay), or list of forwarding neighbors of current relaying node, informing them whether or not to retransmit the message.

Jitter and RAD: Suppose a source node originates a broadcast packet. Given that radio waves propagate at the speed of light, all neighbors will receive the transmission almost simultaneously. Assuming similar hardware and system loads, the neighbors will process the packet and rebroadcast at the same time. To overcome this problem, broadcast protocols jitter the scheduling of broadcast packets from the network layer to the MAC layer by some uniform random amount of time. This (small) offset allows one neighbor to obtain the channel first, while other neighbors detect that the channel is busy (clear channel assessment fails) and thus delay their transmissions to avoid collision. Since the node has to backup all received broadcast packets within *random assessment delay* (RAD), the RAD cannot be too large also. On the other hand, if RAD is small, this node may repeatedly broadcast the same packet, and thus cause the infinity loop of rebroadcast.

Many of the broadcasting protocols require a node to keep track of redundant packets received over a short time interval in order to determine whether to rebroadcast. That time interval, which were termed RAD [2], is randomly chosen from a uniform distribution between 0 and *Tmax* seconds, where Tmax is the highest possible delay interval. This delay in transmission accomplishes two things. First it allows nodes sufficient time to receive redundant packets and assess whether to rebroadcast. Second, the randomized scheduling prevents the collisions of transmission.

Performance Measurement: The performance of broadcast protocols can be measured by a variety of metrics. A commonly used metric is the number of message retransmissions with respect to the number of nodes. In case of broadcasting with adjusted transmission power (thus adjusted disk that the message can reach), the total power can be used as performance metrics. The next important metric is reachability, or the ratio of nodes connected to the source that received the broadcast message. Time delay or latency is sometimes used, which is the time needed for the last node to receive broadcast message initiated at the source. Note that retransmissions at MAC layer are normally deferred, to avoid message collisions. Some authors consider alternative more restricted indicator, whether or not the path from source to any node is always following a shortest path. This measure may be important if used as part of routing scheme, since route paths are created during the broadcast process.

Brief Literature Review: In the minimum energy broadcasting problem, each node can adjust its transmission power in order to minimize total energy consumption but still enable a message originated from a source node to reach all the other nodes in an ad hoc wireless network. The problem is known to be NP-complete. There exists a number of approximate solutions in literature where each node requires global network information (including distances between any two neighboring nodes in the network) in order to decide its own transmission radius. Three greedy heuristics were proposed in Reference 14 for the minimum-energy broadcast routing problem: MST, shortest-path tree (SPT), and broadcasting incremental power (BIP). It was shown that the total energy consumed by MST or BIP methods are no more than 12 times larger than the optimum [20]. Cartigny et al. [47] described a localized protocol where each node requires only the knowledge of its distance to all neighboring nodes and distances between its neighboring nodes (or, alternatively, geographic position of itself and its neighboring nodes). In addition to using only local information, the protocol is shown experimentally to be even competitive with the best-known globalized BIP solution [14], which is a variation of Dijkstra's shortest path algorithm. The solution [47] is based on the use of relative neighborhood graph (RNG) that preserves the network connectivity and is defined in a localized manner. The transmission range for each node is equal to the distance to its furthest RNG neighbor, excluding the neighbor from which the message came from. Localized energy efficient broadcast for wireless networks with directional antennas are described in Reference 62, and are also based on RNG. Messages are sent only along RNG edges, requiring about 50% more energy than BIP based [14] globalized solution. However, when the communication overhead for maintenance is added, localized solution becomes superior. Localized MST can replace RNG to improve energy efficiency, as proposed in References 46 and 63. Their simulations show the performance is comparable to that of BIP. Li et al. [6,7,9,10] recently proposed several methods with further improvements. They described several low weight planar structures (IMRG and $LMST_k$) that can be constructed by localized methods with total communication costs $O(n)$ and the simulations showed a significant improvement of energy consumption compared with [62,46]. Although the structures IMRG and $LMST_k (k \geq 2)$ have total edge length within a constant factor of the MST, the broadcasting based on these locally constructed structures could still consume energy arbitrarily large than the optimum, when we assume that the power needed to support a link uv is $||uv||^\beta$. It has been proved that the broadcasting based on MST consumes energy within a constant factor of the optimum, but MST cannot be constructed locally. They also showed that there is no structure that can be constructed locally and the broadcasting based on it consumes energy within a constant factor of the optimum.

Organization: The rest of the chapter is organized as follows. In Section 11.2, we discuss in detail several centralized broadcasting methods. Note that both minimizing the number of retransmissions and minimizing the total power used by transmitting nodes are NP-complete problems even for unit disk graphs. We thus discuss in detail several methods that can achieve constant approximation ratio in polynomial time. Since centralized methods are expensive to implement for wireless ad hoc networks, in Section 11.3, we then review several protocols that use only localized information. In Section 11.4, we discuss how to judiciously assign operation model to each node thus saving overall energy consumptions in the network. In Section 11.5, we conclude this chapter with discussion of some possible future works.

11.2　Centralized Methods

We assume that two energy models could be used for broadcast: one is nonadjustable power and one is adjustable power. If the power consumed at each node is not adjustable, minimizing the total power used by a reliable broadcast tree is equivalent to the MCDS problem, that is, minimize the number of nodes that relay the message, since all relaying nodes of a reliable broadcast form a CDS. If the power consumed at each node is adjustable, we assume that the power consumed by a relay node u is $||uv||^{\beta}$, where real number $\beta \in [2, 5]$ depends on transmission environment and v is the farthest neighbor of u in the broadcast tree. In the rest of the section, for these two energy models respectively, we reviewed several centralized methods that can build some broadcast tree whose energy consumption is within a constant factor of the optimum if the original communication graph is modeled by unit disk graph.

11.2.1　Assumptions

We first study the adjustable power model. Minimum-energy broadcast/multicast routing in a simple ad hoc networking environment has been addressed by the pioneering work in References 11–14. To assess the complexities *one at a time*, the nodes in the network are assumed to be randomly distributed in a two-dimensional plane and there is no mobility. Nevertheless, as argued in Reference 14, the impact of mobility can be incorporated into this static model because the transmitting power can be adjusted to accommodate the new locations of the nodes as necessary. In other words, the capability to adjust the transmission power provides considerable "elasticity" to the topological connectivity, and hence may reduce the need for hand-offs and tracking. In addition, as assumed in Reference 14, there are sufficient bandwidth and transceiver resources. Under these assumptions, centralized (as opposed to distributed) algorithms were presented by References 14–17 for minimum-energy broadcast/multicast routing. These centralized algorithms, in this simple networking environment, are expected to serve as the basis for further studies on distributed algorithms in a more practical network environment, with limited bandwidth and transceiver resources, as well as the node mobility.

11.2.2　Based on MST and Variations

Some centralized methods are based on optimization. The scheme proposed in Reference 18 is built upon an alternate search based paradigm in which the minimum-cost broadcast/multicast tree is constructed by a search process. Two procedures are devised to check the viability of a solution in the search space. Preliminary experimental results show that this method renders better solutions than BIP, though at a higher computational cost. Liang [15] showed that the minimum-energy broadcast tree problem is NP-complete, and proposed an approximate algorithm to provide a bounded performance guarantee for the problem in the general setting. Essentially they reduce the minimum-energy broadcast tree problem to an optimization problem on an auxiliary weighted graph and solve the optimization problem so as to give an approximate solution for the original problem. They also proposed another algorithm that yields better performance under a special case. Das et al. [16] proposed an evolutionary approach using genetic algorithms. The same authors also presented in Reference 19, three different integer programming models which can be used to find the solutions to the minimum-energy broadcast/multicast problem. The major drawback of optimization based schemes are, however, that they are centralized and require the availability of global topological information.

Some centralized methods are based on greedy heuristics. Three greedy heuristics were proposed in Reference 14 for the minimum-energy broadcast routing problem: MST, SPT, and BIP. The MST heuristic first applies the Prim's algorithm to obtain a MST, and then orient it as an arborescence rooted at the source node. The SPT heuristic applies the Dijkstra's algorithm to obtain a SPT rooted at the source node. The BIP heuristic is the node version of Dijkstra's algorithm for SPT. It maintains, throughout its execution, a single arborescence rooted at the source node. The arborescence starts from the source node, and new nodes are added to the arborescence one at a time on the minimum incremental cost basis until all nodes are included in the arborescence. The incremental cost of adding a new node to the arborescence is the minimum additional power increased by some node in the current arborescence to reach this new node. The implementation of BIP is based on the standard Dijkstra's algorithm, with one fundamental difference on the operation whenever a new node q is added. Whereas the Dijkstra's algorithm updates the node weights (representing the current knowing distances to the source node), BIP updates the cost of each link (representing the incremental power to reach the head node of the directed link). This update is performed by subtracting the cost of the added link pq from the cost of every link qr that starts from q to a node r not in the new arborescence. They have been evaluated through simulations in Reference 14, but little is known about their analytical performances in terms of the approximation ratio. Here, the approximation ratio of a heuristic is the maximum ratio of the energy needed to broadcast a message based on the arborescence generated by this heuristic to the least necessary energy by any arborescence for any set of points.

For a pure illustration purpose, another slight variation of BIP was discussed in detail in Reference 20. This greedy heuristic is similar to the Chvatal's algorithm [21] for the set cover problem and is a variation of BIP. Like BIP, an arborescence, which starts with the source node, is maintained throughout the execution of the algorithm. However, unlike BIP, many new nodes can be added one at a time. Similar to the Chvatal's algorithm [21], the new nodes added are chosen to have the minimal *average* incremental cost, which is defined as the ratio of the minimum additional power increased by some node in the current arborescence to reach these new nodes to the number of these new nodes. They called this heuristic as the broadcast average incremental power (BAIP). In contrast to the $1 + \log m$ approximation ratio of the Chvatal's algorithm [21], where m is the largest set size in the Set Cover Problem, they showed that the approximation ratio of BAIP is atleast $(4n/\ln n) - o(1)$, where n is the number of receiving nodes.

Wan et al. [20] showed that the approximation ratios of MST and BIP are between 6 and 12 and between $\frac{13}{3}$ and 12 respectively; on the other hand, the approximation ratios of SPT and BAIP are atleast $\frac{n}{2}$ and $(4n/\ln n) - o(1)$ respectively, where n is the number of nodes. We then discuss in detail of their proof techniques in the next section.

The iterative maximum-branch minimization (IMBM) algorithm was another effort [17] to construct power-efficient broadcast trees. It begins with a basic broadcast tree in which the source directly transmits to all other nodes. Then it attempts to approximate the minimum-energy broadcast tree by iteratively replacing the maximum branch with less-power, more-hop alternatives.

Both BIP and IMBM operate under the assumption that the transmission power of each node is unconstrained, that is, every node can reach every other node. Both algorithms are centralized in the sense that they require: (a) the source node needs to know the position/distance of every other node and (b) each node needs to know its downstream, on-tree neighbors so as to propagate broadcast messages. As a result, it may be difficult to extend both algorithms into distributed versions, as a significant amount of information is required to be exchanged among nodes.

11.2.3 Theoretical Analysis

Any broadcast routing is viewed as an arborescence (a directed tree) T, rooted at the source node of the broadcasting, that spans all nodes. Let $f_T(\mathbf{p})$ denote the transmission power of the node $\mathbf{p}$ required by T. For any leaf node $\mathbf{p}$ of $T, f_T(\mathbf{p}) = 0$. For any internal node $\mathbf{p}$ of T,

$$f_T(\mathbf{p}) = \max_{\mathbf{pq} \in T} ||\mathbf{pq}||^\beta,$$

in other words, the βth power of the longest distance between $\mathbf{p}$ and its children in T. The total energy required by T is $\sum_{\mathbf{p} \in P} f_T(\mathbf{p})$. Thus the minimum-energy broadcast routing problem is different from the conventional link-based MST problem. Indeed, while the MST can be solved in polynomial time by algorithms such as Prim's algorithm and Kruskal's algorithm, the minimum-energy broadcast routing problem cannot be solved in polynomial time unless $P = NP$ [11]. In its general graph version, the minimum-energy broadcast routing can be shown to be NP-hard [22], and even worse, it can not be approximated within a factor of $(1-\epsilon)\log\Delta$, unless $NP \subseteq DTIME[n^{O(\log\log n)}]$, where Δ is the maximal degree and ϵ is any arbitrary small positive constant. However, this hardness of its general graph version does not necessarily imply the same hardness of its geometric version. In fact, as shown later in the chapter, its geometric version can be approximated within a constant factor. Nevertheless, this suggests that the minimum-energy broadcast routing problem is considerably harder than the MST problem. Recently, Clementi et al. [11] proved that the minimum-energy broadcast routing problem is a NP-hard problem and obtained a parallel but weaker result to those of Reference 20.

Wan et al. [20] gave some lower bounds on the approximation ratios of MST and BIP by studying some special instances in Reference 20. Their deriving of the upper bounds relies extensively on the geometric structures of Euclidean MSTs. A key result in Reference 20 is an upper bound on the parameter $\sum_{e\in\text{mst}(P)}||e||^2$ for any finite point set P of radius one. Note that the supreme of the total edge lengths of mst(P), $\sum_{e\in\text{mst}(P)}||e||$, over all point sets P of radius one is infinity. However, the parameter $\sum_{e\in\text{mst}(P)}||e||^2$ is bounded from above by a constant for any point set P of radius one. They use c to denote the supreme of $\sum_{e\in\text{mst}(P)}||e||^2$ over all point sets P of radius one. The constant c is at most 12; see Reference 20. The proof of this theorem involves complicated geometric arguments; see Reference 20 for more detail. Note that for any point set P of radius one, the length of each edge in mst(P) is at most one. Therefore, for any point set P of radius one and any real number $\beta \geq 2$,

$$\sum_{e\in\text{mst}(P)}||e||^\beta \leq \sum_{e\in\text{mst}(P)}||e||^2 \leq c \leq 12.$$

The next theorem proved in Reference 20 explores a relation between the minimum energy required by a broadcasting and the energy required by the Euclidean MST of the corresponding point set.

Lemma 11.1 [20] *For any point set P in the plane, the total energy required by any broadcasting among P is at least $\frac{1}{c}\sum_{e\in\text{mst}(P)}||e||^\beta$.*

Proof. Let T be an arborescence for a broadcasting among P with the minimum energy consumption. For any non-leaf node $\mathbf{p}$ in T, let $T_\mathbf{p}$ be an Euclidean MST of the point set consisting $\mathbf{p}$ and all children of $\mathbf{p}$ in T. Suppose that the longest Euclidean distance between $\mathbf{p}$ and its children is r. Then the transmission power of node $\mathbf{p}$ is r^β, and all children of $\mathbf{p}$ lie in the disk centered at $\mathbf{p}$ with radius r. From the definition of c, we have

$$\sum_{e\in T_\mathbf{p}}\left(\frac{||e||}{r}\right)^\beta \leq c,$$

which implies that

$$r^\beta \geq \frac{1}{c}\sum_{e\in T_\mathbf{p}}||e||^\beta.$$

Let T^* denote the spanning tree obtained by superposing of all $T_\mathbf{p}$'s for non-leaf nodes of T. Then the total energy required by T is at least $\frac{1}{c}\sum_{e\in T^*}||e||^\beta$, which is further no less than $\frac{1}{c}\sum_{e\in\text{mst}(P)}||e||^\beta$. This completes the proof. ∎

Consider any point set P in a two-dimensional plane. Let T be an arborescence oriented from some mst(P). Then the total energy required by T is at most $\sum_{e\in T_\mathbf{p}}||e||^\beta$. From Lemma 11.1, this total energy is at most c times the optimum cost. Thus the approximation ratio of the link-based MST heuristic is at most c. Together with $c \leq 12$, this observation leads to the following theorem.

Theorem 11.2 [20] *The approximation ratio of the link-based MST heuristic is at most c, and therefore is at most 12.*

In addition, they derived an upper bound on the approximation ratio of the BIP heuristic. Once again, the Euclidean MST plays an important role.

Lemma 11.3 [20] *For any broadcasting among a point set P in a two-dimensional plane, the total energy required by the arborescence generated by the BIP algorithm is at most $\sum_{e \in \mathrm{mst}(P)} ||e||^{\beta}$.*

11.2.4 Centralized Clustering

We then study the nonadjustable power model case. The set of nodes that rebroadcast message in a reliable broadcasting scheme define a connected dominating set. A subset S of V is a *dominating set* if each node u in V is either in S or is adjacent to some node v in S. Nodes from S are called dominators, while nodes not is S are called dominatees. A subset C of V is a *connected dominating set* (CDS) if C is a dominating set and C induces a connected subgraph. Consequently, the nodes in C can communicate with each other without using nodes in $V-C$. A dominating set with minimum cardinality is called *minimum dominating set*, denoted by MDS. A connected dominating set with minimum cardinality is MCDS. A broadcasting based on connected dominating set only uses the nodes in CDS to relay the message. We first review several methods in the literature to build a CDS.

If every node cannot adjust its transmission power accordingly, then we need to find the MCDS to save the total power consumption of the broadcasting protocol. Unfortunately, the problem of finding CDS of minimal size is NP-complete even for unit disk graphs. Guha and Khuller [23] studied the approximation of the connected dominating set problem for general graphs. They gave two different approaches, both of them guarantee approximation ratio of $\Theta(H(\Delta))$. As their approaches are for general graphs, they do not utilize the geometry structure if applied to the wireless ad hoc networks. One approach is to grow a spanning tree that includes all nodes. The internal nodes of the spanning tree is selected as the final CDS. This approach has approximation ratio $2(H(\Delta) + 1)$. The other approach is first approximating the dominating set and then connecting the dominating set to a CDS. They [23] proved that this approach has approximation ratio $\ln \Delta + 3$.

One can also use the Steiner tree algorithm to connect the dominators. This straightforward method gives approximation ratio $c(H(\Delta) + 1)$, where c is the approximation ratio for the unweighted Steiner tree problem. Currently, the best ratio is $1 + \frac{\ln 3}{2} \simeq 1.55$, due to Robins and Zelikovsky [24].

By definition, any algorithm generating a maximal independent set is a clustering method. We first review the methods that approximates the maximum independent set (MIS), the MDS, and the MCDS. Hunt et al. [25] and Marathe et al. [26] studied the approximation of the maximum independent set and the minimum dominating set for unit disk graphs. They gave the first PTASs for MDS in UDG. The method is based on the following observations: a maximal independent set is always a dominating set; given a square Ω with a fixed area, the size of any maximal independent set is bounded by a constant C. Assume that there are n nodes in Ω. Then, we can enumerate all sets with size at most C in time $\Theta(n^C)$. Among these enumerated sets, the smallest dominating set is the MDS. Then, using the shifting strategy they derived a PTAS for the MDS problem.

Since we have PTAS for MDS and the graph VirtG connecting every pair of dominators within at most 3 hops is connected [27], we have an approximation algorithm (constructing a MST VirtG) for MCDS with approximation ratio $3 + \epsilon$. Notice that, Berman et al. [28] gave an $\frac{4}{3}$ approximation method to connect a dominating set and Robins and Zelikovsky [24] gave an $\frac{4}{3}$ approximation method to connect an independent set. Thus, we can easily have an $\frac{8}{3}$ approximation algorithm for MCDS, which was reported in Reference 29. Recently, Cheng et al. [30] designed a PTAS for MCDS in UDG. However, it is difficult to run their method efficiently in a distributed manner.

11.3 Localized Methods

11.3.1 Based on Distributed CDS

A natural structure for broadcasting is CDS. Many distributed clustering (or dominating set) algorithms have been proposed in the literature [31–34]. All algorithms assume that the nodes have distinctive identities (denoted by ID hereafter).

In the rest of section, we will interchange the terms cluster-head and dominator. The node that is not a cluster-head is also called *dominatee*. A node is called *white* node if its status is yet to be decided by the clustering algorithm. Initially, all nodes are *white*. The status of a node, after the clustering method finishes, could be *dominator* with color *black* or *dominatee* with color *gray*. The rest of this section is devoted for the distributed methods that approximates the MDS and the MCDS for UDG.

11.3.1.1 Clustering without Geometry Property

For general graphs, Jia et al. [35] described and analyzed some randomized distributed algorithms for the MDS problem that run in polylogarithmic time, independent of the diameter of the network, and that return a dominating set of size within a logarithmic factor from the optimum with high probability. Their best algorithm runs in $O(\log n \log \Delta)$ rounds with high probability, and every pair of neighbors exchange a constant number of messages in each round. The computed dominating set is within $O(\log \Delta)$ in expectation and within $O(\log n)$ with high probability. Their algorithm works for weighted dominating set also.

The method proposed by Das et al. [36,37] contains three stages: approximating the MDS, constructing a spanning forest of stars, and expanding the spanning forest to a spanning tree. Here the *stars* are formed by connecting each dominatee node to one of its dominators. The approximation method of MDS is essentially a distributed variation of the centralized Chvatal's greedy algorithm [21] for set cover. Notice that the dominating set problem is essentially the set cover problem, which is well studied. It is then not a surprise that the method by Das et al. [36,37] guarantees a $H(\Delta)$ for the MDS problem, where H is the harmonic function and Δ is the maximum node degree.

While the algorithm proposed by Das et al. [36,37] finds a dominating set and then grows it to a CDS, the algorithm proposed by Wu and Li [38] takes an opposite approach. They first find a CDS and then prune out certain redundant nodes from the CDS. The initial CDS $\mathbb{C}$ contains all nodes that have at least two nonadjacent neighbors. A node u is said to be *locally redundant* if it has either a neighbor in $\mathbb{C}$ with larger ID, which dominates all other neighbors of u, or two adjacent neighbors with larger ID, which together dominates all other neighbors of u. Their algorithm then keeps removing all locally redundant nodes from $\mathbb{C}$. They showed that this algorithm works well in practice when the nodes are distributed uniformly and randomly, although no any theoretical analysis is given by them both for the worst case and for the average approximation ratio. However, it was shown by Alzoubi et al. [31] that the approximation ratio of this algorithm could be as large as $\frac{n}{2}$.

Recently, Dai and Wu [39] have proposed a distributed dominant pruning algorithm. Each node has a priority, which can be simply its unique identifier or a combination of remaining battery, degree, or identifier. A node u is "fully covered" by a subset S of its neighboring nodes if and only if the following three conditions hold:

- The subset S is connected
- Any neighbor of u is neighbor of at least one node from S
- All nodes in S have higher priority than u

A node belongs to the dominating set if and only if there is no subset that fully covers it. The advantage of using CDS as defined in References 38 and 39 is that each node can decide whether or not it is in dominating set without any additional communication steps involved, other than those needed to maintain neighborhood information. The neighborhood information needed is either 2-hop neighbors knowledge, or 1-hop neighbor knowledge with their position.

Stojmenovic et al. [40] observed that distributed constructions of connecting dominating set can be obtained following the clustering scheme of Lin and Gerla [32]. Connecting dominating set consists of two types of nodes: clusterhead and border nodes (also called gateway or connectors elsewhere). The clusterhead nodes are decided as follows. At each step, all white nodes that have the lowest *rank* among all white neighbors are colored black, and the white neighbors are colored gray. The ranks of the white nodes are updated if necessary. The clustering method uses two messages which can be called *IamDominator* and *IamDominatee*. A white node claims itself to be a dominator if it has the smallest ID among all of its white neighbors, if there is any, and broadcasts *IamDominator* to its 1-hop neighbors. A white node receiving *IamDominator* message marks itself as dominatee and broadcasts *IamDominatee* to its 1-hop neighbors. The set of dominators generated by the above method is actually a maximal independent set. Here, we assume that each node knows the IDs of all its 1-hop neighbors, which can be achieved if each node broadcasts its ID to its neighbors initially. This approach of constructing MIS is well-known. The following rankings of a node are used in various methods: the ID only [33,32], the ordered pair of degree and ID [41], and an ordered pair of degree and location [40]. In Reference 42, Basagni et al., used a general *weight* as a ranking criterion for selecting the node as the clusterhead, where the weight is a combination of mentioned criteria and some new ones, such as mobility or remaining energy. After the clusterhead nodes are selected, border-nodes are selected to connect them. A node is a border-node if it is not a clusterhead and there are at least two clusterheads within its 2-hop neighborhood. It was shown by Reference 31 that the worst case approximation ratio of this method is also $\frac{n}{2}$, although it works well in practice.

11.3.1.2 Clustering with Geometry Property

Notice that none of the above algorithm utilizes the geometry property of the underlying UDG. Recently, several algorithms were proposed with a constant worst case approximation ratio by taking advantage of the geometry properties of the underlying graph. It is used to connect the clusterheads constructed as described above into a CDS with fewer additional nodes. During this second step of backbone formation, some *connectors* (also called *gateways*) are found among all the dominatees to connect the dominators. Then the connectors and the dominators form a CDS. Recently, Wan et al. [43] and Wu and Lou [3] proposed a communication efficient algorithm to find connectors based on the fact that there are only a constant number of dominators within k-hops of any node. The following observation is a basis of several algorithms for CDS. After clustering, one dominator node can be connected to many dominatees. However, it is well-known that a dominatee node can only be connected to at most *five* dominators in the unit disk graph model. Generally, it was shown in References 43, 27, and 3 that for each node v (dominator or dominatee), the number of dominators inside the disk centered at v with radius k-units is bounded by a constant $\ell_k < (2k+1)^2$.

Given a dominating set S, let VirtG be the graph connecting all pairs of dominators u and v if there is a path in UDG connecting them with at most 3 hops. Graph VirtG is connected. It is natural to form a connected dominating set by finding connectors to connect any pair of dominators u and v if they are connected in VirtG. This strategy is also adopted by Wan et al. [43] and Wu and Lou [3]. Notice that, in the approach by Stojmenovic et al. [40], they set any dominatee node as the connector if there are two dominators within its 2-hop neighborhood. This approach is very pessimistic and results in very large number of connectors in the worst case [31]. Instead, Wan et al. [4] suggested to find only one unique shortest path to connect any two dominators that are at most three hops away.

We briefly review their basic idea of forming a CDS in a distributed manner. Let $\prod_{UDG}(u, v)$ be the path connecting two nodes u and v in UDG with the smallest number of hops. Let us first consider how to connect two dominators within 3 hops. If the path $\prod_{UDG}(u, v)$ has two hops, then u finds the dominatee with the smallest ID to connect u and v. If the path $\prod_{UDG}(u, v)$ has three hops, then u finds the node, say w, with the smallest ID such that w and v are two hops apart. Then node w selects the node with the smallest ID to connect w and v. Wang and Li [27] and Alzoubi and coworkers [43] discussed in detail some approaches to optimize the communication cost and the memory cost.

The graph constructed by this algorithm is called a CDS graph (or *backbone* of the network). If we also add all edges that connect all dominatees to their dominators, the graph is called extended CDS, denoted by CDS'. Let *opt* be the size of the MCDS. It was shown [26] that the size of the computed maximal independent set has size at most $4 * opt + 1$. We already showed that the size of the connected dominating set found by the above algorithm is at most $\ell_3 k + k$, where k is the size of the maximal independent set found by the clustering algorithm. It implies that the found connected dominating set has size at most $4(\ell_3 + 1) * opt + \ell_3 + 1$. Consequently, the computed connected dominating set is at most $4(\ell_3 + 1)$ factor of the optimum (with an additional constant $\ell_3 + 1$). It was shown in References 4 and 27 that the CDS' graph is a sparse spanner in terms of both hops and length, meanwhile CDS has a bounded node degree.

11.3.1.3 Distributed Weighted CDS

In the previous section, we assumed that the power needed by every node is the same. Thus, to find the minimum power structure for broadcast is equivalent to find the MCDS. In this section, we will discuss the scenario when different nodes may have different power to send messages to their neighbors. Then, finding the minimum power structure for broadcast is equivalent to find the minimum cost weighted CDS. It is well-known that, for a general weighted network modeled by an arbitrary graph, we can find a weighted connected dominating set whose cost is no more than $O(\log n)$ times the optimum in a centralized manner. It is unknown how to achieve such in an efficient distributed manner and whether we can achieve better approximation ratio for networks with some special properties such as unit disk graphs, or the networks whose nodes' weights are smooth.

Wang and Li [44] recently presented the first distributed methods to address this problem and proved that their method has a better worst-case performances than all previous distributed methods. Before we review their methods here, we first give some necessary definitions that will be used in our presentation later.

Let $d_G(u)$ be the degree of node u in a graph G and d be the maximum degree of all wireless nodes. The average node degree is called *density* of the network. We assume that each wireless node u has a cost $c(u)$ of being in the backbone. Here the cost $c(u)$ could be the value computed based on a combination of its remaining battery power, its mobility, its node degree in the communication graph, and so on. We assume that smaller $c(u)$ means that the node is more suitable of being in the backbone. Let $\delta = \max_{i,j \in E} c(i)/c(j)$, where E is the set of communication links in the wireless network G. We call δ the cost *smoothness* of the wireless networks. When δ is bounded by some small constant, we say the node costs are *smooth*.

We call all nodes within a constant k hops of a node u in the communication graph G as the *k-local nodes* or *k-hop neighbors* of u, denoted by $N_k(u)$, which includes u itself. The *k-local graph* of a node u, denoted by $G_k(u)$, is the induced graph of G on $N_k(u)$, that is, $G_k(u)$ is defined on vertex set $N_k(u)$, and contains all edges in G with both end-points in $N_k(u)$. The independence number, denoted as $\alpha(G)$, of a graph G is the size of the maximum independent set of G. The *k-local independence number*, denoted by $\alpha^{[k]}(G)$, is defined as $\alpha^{[k]}(G) = \max_{u \in V} \alpha(G_k(u))$. It is well-known that for a UDG, $\alpha^{[1]}(G) \leq 5$ and $\alpha^{[2]}(G) \leq 18$. A subset C of V is a *minimum weighted connected dominating set* (MWCDS) if C is a CDS with minimum total cost among all CDSs.

The method in Reference 44 has the following two phases. The first phase (clustering phase) is to find a set of wireless nodes as the dominators[1]. The second phase is to find a set of nodes to connect these dominators to form the final backbone of the network. Notice that these two phases could interleave in the actual construction method. They separate them just for the sake of easy presentations.

We then first review their method [44] for constructing a CDS whose total cost is comparable with the optimum solution. Their method first constructs a MIS using node weights as the selection criterion. Then for each node v in MIS, we run local greedy set cover method on *local neighborhood* $N_2(v)$ to find some nodes $GRDY_v$ to cover all 1-hop neighbors of v. If $GRDY_v$ has a total cost smaller than v, then we

---

[1] We will interchange the terms cluster-head and dominator. The node that is not a cluster-head is also called *ordinary* node or *dominatee*. A node is called *white* node if its status is yet to be decided by the clustering algorithm. Initially, all nodes are white. The status of a node, after the clustering method finishes, could be *dominator* or *dominatee*.

use $GRDY_v$ to replace v as dominators, which will further reduce the cost of MIS. The method works as follows.

Algorithm 11.1(*Constructing Efficient Dominating Set*)

1. First assume that all nodes are originally marked WHITE.
2. A node u sends a message ItryDominator to all its 1-hop neighbors if it has the smallest cost among all its WHITE neighbors. Node u also marks itself PossibleDominator.
3. When a node v receives a message ItryDominator from its 1-hop neighbors, it marks itself Dominatee and sends a message IamDominatee to all its 1-hop neighbors.
4. When a node w receives a message IamDominatee from its neighbor v, node w removes node v from its list of WHITE neighbors.
5. Each node u marked with PossibleDominator collects the cost and ID of all of its two hop neighbors $N_2(u)$.
6. Using the greedy method for minimum weighted set cover (like the second method), Possible-Dominator node u selects a subset of its two hop-neighbors to cover *all* the 1-hop neighbors (including u) of node u. If the cost of the selected subset, denoted by $GRDY_u$, is smaller than the cost of node u, then node u sends a message YouAreDominator(w) to each node w in the selected subset. Otherwise, node u just marks itself Dominator.
7. When a node w received a message YouAreDominator(w), node w marks itself Dominator.

The second step of weighted connected dominating set formation is to find some *connectors* (also called *gateways*) among all the dominatees to connect the dominators. The connectors and the dominators form a CDS (or called backbone). Their method forms a CDS by finding connectors to connect any pair of dominators u and v if they are connected in VirtG. Their method uses the following data structures and messages:

1. $D_k(v)$ is the list of dominators that are k-hops away from a node v.
2. $P_k(v, u)$ is the least cost path from v to u using at most k-hops. Notice u and v may be less than k-hops away.
3. OneHopDominatorList($v, D_1(v)$): nodes $D_1(v)$ are the 1-hop dominators of node v.
4. TwoHopDominatorList($v, D_2(v)$): nodes $D_2(v)$ are the 2-hop dominators of node v.
5. TwoHopDominator($v, u, w, c(w)$): node u is a 2-hop dominator of node v and the path uwv has the least cost.

Algorithm 11.2 Connector Selection

1. Every dominatee node v broadcasts to its 1-hop neighbors the list of its one-hop dominators $D_1(v)$ using message OneHopDominatorList($v, D_1(v)$). When a node w receives OneHopDominator($v, D_1(v)$) from 1-hop neighbor v, it puts the dominator $u \in D_1(v)$ to $D_2(w)$ if $u \notin D_1(w)$. Update the path $P_3(w, u)$ as uvw if it has a smaller cost.
2. When a dominatee node w receives messages OneHopDominatorList from *all* its one-hop nodes, for each dominator node $u \in D_2(w)$, node w sends out message TwoHopDominator($w, u, x, c(x)$), where wxu is the least cost path $P_2(w, u)$.
3. When a dominator z receives a message TwoHopDominator($w, u, x, c(x)$) from its neighbor w, it puts u to $D_3(z)$ if $u \notin D_2(z)$, and updates the path $P_3(w, u)$ as $uwxz$ if $c(w) + c(x)$ has a less cost.
4. Each dominator u builds a virtual edge $\widetilde{uv}$ to connect each neighboring dominator v. The length of $\widetilde{uv}$ is the cost of path $P_3(u, v)$. Notice that here the cost of end-nodes u and v is not included. All virtual edges forms an *edge weighted* virtual graph VirtG in which all dominators are its vertices.
5. Run a distributed algorithm to build a MST on graph VirtG. Let VMST denote MST (VirtG).

6. For any virtual edge $e \in$ VMST, select each of the dominatees on the path corresponding to e as a connector.

The following theorems are proved to show that the constructed structure is indeed efficient for broadcast and also unicast. Please refer Reference 44 for the detail of the proofs.

Theorem 11.4 *For any communication graph G, our algorithm constructs a weighted connected dominating set whose total cost is no more than*

$$\min(\alpha^{[2]}(G)\log d, (\alpha^{[1]}(G) - 1)\delta + 1) + 2\alpha^{[1]}(G),$$

times of the optimum.

Specifically, when the network is modeled by a unit disk graph, we have the following corollary.

Corollary 11.5 *For homogeneous wireless networks, our algorithm constructs a weighted connected dominating set whose total cost is no more than* $\min(18\log d, 4\delta + 1) + 10$ *times of the optimum.*

For unicast, they proved that

Theorem 11.6 *For any communication graph, the unicast based on the constructed structure has a cost at most 3 times the cost on the original communication network.*

11.3.2 Localized Low Weight Structures

The centralized algorithms do not consider computational and message overheads incurred in collecting global information. Several of them also assume that the network topology does not change between two runs of information exchange. These assumptions may not hold in practice, since the network topology may change from time to time, and the computational and energy overheads incurred in collecting global information may not be negligible. This is especially true for large-scale wireless networks where the topology is changing dynamically due to the changes of position, energy availability, environmental interference, and failures, which implies that centralized algorithms that require global topological information may not be practical.

Flooding is also a good solution for the sake of scalability and simplicity. Several flooding techniques for wireless networks have been proposed each with respect to certain optimization criterion. However, none of them takes advantage of the feature that the transmission power of a node can be adjusted.

Some distributed heuristics are proposed, such as Reference 45. Most of them are based on distributed MST method. A possible drawback of these distributed method is that it may not perform well under frequent topological changes as it relies on information that is multiple hops away to construct the MST. Refer Reference 46 for more detail. The relative neighborhood graph, the Gabriel graph and the Yao graph all have $O(n)$ edges and contain the Euclidean MST. This implies that we can construct the minimum spanning tree using $O(n\log n)$ messages.

Localized minimum energy broadcast algorithms are based on the use of a locally defined geometric structure, such as RNG. RNG consists of all edges uv such that uv is not the longest edge in any triangle uvw. That is, uv belongs to RNG if there is no node w such that $uw < uv$ and $vw < uv$.

Cartigny et al. [47] proposed a localized algorithm, called RBOP [47] that is built upon the notion of RNG. In RBOP, the broadcast is initiated at the source and propagated, following the rules of neighbor elimination [40], on the topology represented by RNG. Simulation results show that the energy consumption could be as high as 100% as compared to BIP. However, the communication overhead due to mobility and changes in activity status in BIP are not considered, therefore RBOP is superior to BIP in dynamic ad hoc networks. Li and Hou [46], and Cartigny et al. [47] proposed another localized algorithm, which applies localized minimal spanning tree (LMST) instead of RNG as the broadcast topology. In LMST, proposed in Reference 48, each node calculates local MST of itself and its 1-hop neighbors. A node uv is in LMST if and only if u and v select each other in their respective trees. The simulations [46,47] show

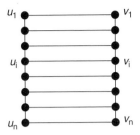

FIGURE 11.1 An instance of wireless nodes that every network structures described previously (except MST) have an arbitrarily large total weight.

that the performance of LMST based schemes is significantly better than the performance of RBOP, and with about 50% more energy consumption than BIP in static scenarios. Cartigny et al. [63] demonstrated that, when $c > 0$ in power-attenuation model where energy consumption for transmitting over an edge is $||uv||^{\beta} + c$, there exists an optimal "target" transmission radius, so that further energy savings can be obtained if transmission radii are selected near target radius.

However, as shown in Reference 6 (also by Figure 11.1), the total weights of RNG and LMST could still be as large as $O(n)$ times of the total weight of MST. Here, in Figure 11.1, $||u_i v_i|| = 1$ and $||u_i u_{i+1}|| = ||v_i v_{i+1}|| = \epsilon$ for a very small positive real number ϵ. Given a graph G, let $\omega_b(G) = \sum_{e \in G} ||e||^b$. Then $\omega_1(\text{RNG}) = \Theta(n) \cdot \omega_1(\text{MST})$ and $\omega_1(\text{LMST}) = \Theta(n) \cdot \omega_1(\text{MST})$.

In References 6 and 9, the authors described several low weight planar structures that can be constructed by localized methods with total communication costs $O(n)$. The energy consumption of broadcast based on those structures are within $O(n^{\beta-1})$ of the optimum, that is, $\omega_{\beta}(H) = O(n^{\beta-1}) \cdot \omega_{\beta}(\text{MST}), \omega_{\beta}(\text{LMST}_2) = O(n^{\beta-1}) \cdot \omega_{\beta}(\text{MST}), \omega_{\beta}(\text{IMRG}) = O(n^{\beta-1}) \cdot \omega_{\beta}(\text{MST})$ for any $\beta \geq 1$. This improves the previously known "lightest" structure RNG by $O(n)$ factor since in the worst case $\omega(\text{RNG}) = \Theta(n) \cdot \omega(\text{MST})$ and $\omega_{\beta}(\text{RNG}) = \Theta(n^{\beta}) \cdot \omega_{\beta}(\text{MST})$.

We will now review in detail these three structures.

11.3.2.1 Structure Based on RNG'

Although RNG is very sparse structure (the average number of neighbors per node is about 2.5), in some degenerate cases a particular node may have arbitrarily large degree. This motivated Stojmenovic [8] to define a modified structure where each node will have degree bounded by 6. The same structure was independently proposed by Li in Reference 6, with an additional motivation. Li proved that the modified RNG is the first localized method to construct a structure H with weight $O(\omega(\text{MST}))$ using total $O(n)$ local-broadcast messages. Note that, if each node already knows the positions and IDs of all its neighbors, then no messages are needed to decide which of its edges belong to (modified) RNG. Notice that, traditionally, the relative neighborhood graph will always select an edge uv even if there is some node on the boundary of $lune(u, v)$. Here $lune(u, v)$ is the intersection of two disks centered at nodes u and v with radius $||uv||$ respectively. Thus, RNG may have unbounded node degree, for example, considering $n - 1$ points equally distributed on the circle centered at the nth point v, the degree of v is $n - 1$. Notice that for the sake of lowing the weight of a structure, the structure should contain as less edges as possible without breaking the connectivity. Li [6] and Stojmenovic [8] then naturally extended the traditional definition of RNG as follows.

We need to make distinct edge lengths. This can be achieved by adding the secondary, and if necessary, the ternary keys for comparing two edges. Each node is assumed to have an unique ID. Then consider the record $(||uv||), \text{ID}(u), \text{ID}(v)$, where $\text{ID}(u) < \text{ID}(v)$ (otherwise u and v are exchanged for given edge). Two edges compare their lengths first to decide which one is longer. If same, they then compare their secondary key, which is their respective lower endpoint node's ID. If this is also same, then the ternary key resolves the comparison (otherwise we are comparing edge against itself). This simple method for making distinct edge length was proposed in Reference 48. The edge lengths, so defined, are then used in the regular definition of RNG. It is easy to show that two RNG edges uv and uw going out of the same node must have angle between them at least $\pi/6$, otherwise $vw < uv$ or $vw < uw$, and one of the two edges becomes the longest in the triangle and consequently could not be in RNG. Li denoted modified RNG structure by RNG′. Obviously, RNG′ is a subgraph of traditional RNG. It was proved in References 6 and 8 that RNG′ still contains a MST as a subgraph. However, RNG′ is still not a low weight structure.

Notice that it is well-known that the communication complexity of constructing a MST of a n-vertex graph G with m edges is $O(m + n \log n)$; while the communication complexity of constructing MST for UDG is $O(n \log n)$ even under the local broadcasting communication model in wireless networks. It was shown in Reference 6 that it is *impossible* to construct a low-weighted structure using only one-hop neighbor information.

The localized algorithm given in Reference 6 that constructs a low-weighted structure using only some two-hops information is as follows.

Algorithm 11.3 (*Construct Low Weight Structure H*)

---

1. All nodes together construct the graph RNG′ in a localized manner.
2. Each node u locally broadcasts its incident edges in RNG′ to its one-hop neighbors. Node u listens to the messages from its one-hop neighbors.
3. Assume node u received a message informing existence of edge $xy \in$ RNG′ from its neighbor x. For each edge $uv \in$ RNG′, if uv is the longest among uv, xy, ux, and vy, node u removes edge uv. Ties are broken by the label of the edges. Here assume that $uvyx$ is the convex hull of u, v, x, and y.
4. Let H be the final structure formed by all remaining edges in RNG′.

Obviously, if an edge uv is kept by node u, then it is also kept by node v. The following theorem was proved in Reference 6.

Theorem 11.7 [6] *The total edge weight of H is within a constant factor of that of the minimum spanning tree.*

This was proved by showing that the edges in H satisfy the *isolation property*. They [6] also showed that the final structure contains MST of UDG as a subgraph.

Clearly, the communication cost of Algorithm 11.3 is at most $7n$: initially each node spends one message to tell its 1-hop neighbors its position information, then each node uv tells its 1-hop neighbors all its incident edges $uv \in$ RNG′ (there are at most total $6n$ such messages since RNG′ has at most $3n$ edges). The computational cost of Algorithm 11.3 could be high since for each link $uv \in$ RNG′, node u has to test whether there is an edge $xy \in$ RNG′ and $x \in N_1(u)$ such that uv is the longest among uv, xy, ux, and vy. Then References 9 and 10 present some new algorithms that improve the computational complexity of each node while still maintains low communication costs.

11.3.2.2   Structure Based on LMST$_k$

The first new method in Reference 9 uses a structure called *local minimum spanning tree*, let us first review its definition. It is first proposed by Li et al. [48]. Each node u first collects its 1-hop neighbors $N_1(u)$. Node u then computes the MST($N_1(u)$) of the induced unit disk graph on its 1-hop neighbors $N_1(u)$. Node u keeps a directed edge uv if and only if uv is an edge in MST($N_1(u)$). They call the union of all directed edges of all nodes the *local minimum spanning tree*, denoted by LMST$_1$. If only symmetric edges are kept, then the graph is called LMST$_1^-$, that is, it has an edge uv iff both directed edge uv and directed edge vu exist. If ignoring the directions of the edges in LMST$_1$, they call the graph LMST$_1^+$, that is, it has an edge uv iff either directed edge uv or directed edge vu exist. They prove that the graph is connected, and has bounded degree 6. In Reference 9, Li et al. also showed that graph LMST$_1^-$ and LMST$_1^+$ are actually planar. Then they extend the definition to k-hop neighbors, the union of all edges of all MST($N_k(u)$) is the k *local minimum spanning tree*, denoted by LMST$_k$. For example, the two local minimum spanning tree can be constructed by the following algorithm.

Algorithm 11.4 (*Construct Low Weight Structure LMST₂ by 2-hop Neighbors*)

---

1. Each node u collects its 2-hop neighbors information $N_2(u)$ using a communication efficient protocol described in Reference 49.
2. Each node u computes the Euclidean minimum spanning tree $MST(N_2(u))$ of all nodes $N_2(u)$, including u itself.
3. For each edge $uv \in MST(N_2(u))$, node u tells node v about this directed edge.
4. Node u keeps an edge uv if $uv \in MST(N_2(u))$ or $vu \in MST(N_2(v))$. Let $LMST_2^+$ be the final structure formed by all edges kept. It keeps an edge if either node u or node v wants to keep it. Another option is to keep an edge only if both nodes want to keep it. Let $LMST_2^-$ be the structure formed by such edges.

In Reference 9, they prove that structures $LMST_2$ ($LMST_2^+$ and $LMST_2^-$) are connected, planar, low-weighted, and have bounded node degree at most 6. In addition, MST is a subgraph of $LMST_k$ and $LMST_k \subseteq RNG'$. Although the constructed structure $LMST_2$ has several nice properties such as being bounded degree, planar, and low-weighted, the communication cost of Algorithm 11.4 could be very large to save the computational cost of each node. The large communication costs are from collecting the two hop neighbors information $N_2(u)$ for each node u. Although the total communication of the protocol described in Reference 49 is $O(n)$, the hidden constant is large.

11.3.2.3 Combining RNG′ and LMST$_k$

We could improve the communication cost of collecting $N_2(u)$ by using a subset of two-hop information without sacrificing any properties. Define $N_2^{RNG'}(u) = \{w \mid vw \in RNG' \text{ and } v \in N_1(u)\} \cup N_1(u)$. We describe our modified algorithm as follows.

Algorithm 11.5 (*Construct Low Weight Structure IMRG by 2-hop Neighbors in RNG′*)

---

1. Each node u tells its position information to its 1-hop neighbors $N_1(u)$ using a local broadcast model. All nodes together construct the graph RNG' in a localized manner.
2. Each node u locally broadcasts its incident edges in RNG' to its 1-hop neighbors. Node u listens to the messages from its 1-hop neighbors.
3. Each node u computes the Euclidean minimum spanning tree $MST(N_2^{RNG'}(u))$ of all nodes $N_2^{RNG'}(u)$, including u itself.
4. For each edge $uv \in MST(N_2^{RNG'}(u))$, node u tells node v about this directed edge.
5. Node u keeps an edge uv if $uv \in MST(N_2^{RNG'}(u))$ or $vu \in MST(N_2^{RNG'}(v))$. Let $IMRG^+$ be the final structure formed by all edges kept. Similarly, the final structure is called $IMRG^-$ when edge $uv \in RNG'$ is kept iff $uv \in MST(N_2^{RNG'}(u))$ and $uv \in MST(N_2^{RNG'}(v))$. Here IMRG is the abbreviation of *Incident* MST *and* RNG *Graph*.

Notice that in the algorithm, node u constructs the local minimum spanning tree $MST(N_2^{RNG'}(u))$ based on the induced UDG of the point sets $N_2^{RNG'}(u)$. It is obvious that the communication cost of Algorithm 11.5 is at most $7n$.

It was shown that structures $IMRG^+$ and $IMRG^-$ are still connected, planar, bounded degree, and low-weighted. They are obviously planar, and with bounded degree since both structures are still subgraphs of the modified relative neighborhood graph RNG'. Clearly, the constructed structures are supergraphs of the previous structures, that is, $LSMT_2+ \subseteq IMRG^+$ and $LSMT_2^- \subseteq IMRG^-$, since Algorithm 11.5 uses less information than Algorithm 11.4 in constructing the local minimum spanning tree. It is proved in Reference 9 that Algorithm 11.5 constructs structures $IMRG^-$ or $IMRG^+$ using at most $7n$ messages. The structures IMRG$^-$ or IMRG$^+$ are connected, planar, bounded degree, and low-weighted. Both IMRG$^-$ and IMRG$^+$ have node degree at most 6.

Recall that until now there is no efficient localized algorithm that can achieve all following desirable features: bounded degree, planar, low weight, and spanner. It is still an open problem.

11.3.2.4 A Negative Result

In References 6, 9, and 10, Li et al. proposed several methods to construct structures in a localized manner such that the total edge lengths of these structures are within a constant factor of MST. They also showed that the energy consumption of broadcasting based on those structures are within $O(n^{\beta-1})$ of the optimum, that is, $\omega_\beta(H) = O(n^{\beta-1}) \cdot \omega_\beta(\text{MST}), \omega_\beta(\text{LMST}_2) = O(n^{\beta-1}) \cdot \omega_\beta(\text{MST}), \omega_\beta(\text{IMRG}) = O(n^{\beta-1}) \cdot \omega_\beta(\text{MST})$ for any $\beta \geq 1$.

They further showed that it is impossible to design a deterministic localized method that constructs a structure such that the broadcasting based on this structure consumes energy within a factor $O(n^{\beta-1})$ of the optimum. Assume that there is a deterministic localized algorithm to do so: it uses k-hop information of every node u to select the edges incident on u, and the energy consumption is no more than $O(n^{\beta-1})$ times of the optimum. They construct two set of nodes configurations such that the k-hop information collected in a special node u is same for both configurations. In addition, there is an edge uv in both UDGs such that if node u decides to keep edge uv (then edge uv is kept in both configurations), the energy consumption of one configuration is already more than $O(n^{\beta-1})$ times of the optimum; if node u decides to remove edge uv (then edge uv is removed in both configurations), then the structure constructed for another configuration is disconnected. See Reference 10 for more detail. This implies that the low-weighted structures are asymptotically optimum in terms the worst case energy consumption for broadcasting among *any* locally constructed topologies when assuming that the energy needed to support a link uv is proportional to $\|uv\|^\beta$.

11.3.3 Combining Clustering and Low Weight

Seddigh et al. [50] specify two more location based broadcasting algorithms that combine RNG and internal node concept (CDS) as follows. PI-broadcast algorithm applies the planar subgraph construction first, and then applies the internal nodes concept on the subgraph. The result is different from the internal nodes applied on the whole graph. IP-broadcast algorithm changes the order of concept application compared to the previous algorithm. Internal nodes are first identified in the whole graph, and then the obtained subgraph (containing only internal nodes) is further reduced to planar one by the RNG construction.

The solution in Reference 50 is for one-to-one communication model, where message sent from one node is received by only the targeted neighbor. In Reference 51, Li et al. combine the low-weighted structures and the connected dominating set for energy efficient broadcasting in traditional one-to-many (omnidirectional antenna) networks. Similarly, they proposed two approaches for combining them as in Reference 50. Notice that the constructed low-weighted structures are a subgraph of RNG, thus, they are still planar graphs. For simplicity, they also call these two combinations, *PI-broadcast* and *IP-broadcast*, respectively. They found that the energy consumption of the IP-broadcast schemes in dense networks is significantly less than that of the PI-broadcast schemes. The reason is that, in the IP-broadcast schemes, one retransmission by the internal nodes will be received by many noninternal nodes in dense networks, thus, energy consumption is reduced. Several localized improvement heuristics are also applied after the IP-broadcast or PI-broadcast schemes to further improve the energy consumption. Ingelrest and coworkers [63] in his master thesis also combined LMST with dominating sets, and target radius idea to derive new minimum energy broadcast protocols.

Following Figure 11.2 illustrates the different structures used for broadcasting. All figures are computed on a set of 500 nodes randomly distributed in a square region with side length 7500 m. Each node has transmission range 500 m. Here the CDS is generated using ID as criterion for selecting dominator/connector;

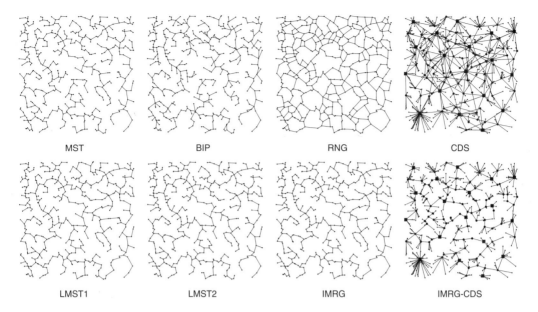

FIGURE 11.2 Different structures generated from a UDG used for broadcasting.

IMRG-CDS is the subgraph constructed by applying IMRG on the induced unit disk graph of all CDS nodes, that is, dominators and connectors. Our simulation results also confirms that CDS structure and IMRG-CDS consumes the least energy when the node power is nonadjustable, while the IMRG consumes the least energy when each node can adjust its power according to the longest incident link.

11.3.4 Flooding-Based Methods

11.3.4.1 Selecting Forwarding Neighbors

The simplest broadcasting mechanism is to let every node retransmit the message to all its 1-hop neighbors when receiving the first copy of the message, which is called *flooding* in the literature. Despite its simplicity, flooding is very inefficient and can result in high redundancy, contention, and collision. One approach to reducing the redundancy is to let a node only forward the message to a subset of 1-hop neighbors who together can cover the 2-hop neighbors. In other words, when a node retransmits a message to its neighbors, it explicitly asks a subset of its neighbors to relay the message.

In Reference 52, Lim and Kim proposed a broadcasting scheme that chooses some or all of its 1-hop neighbors as rebroadcasting node. When a node receives a broadcast packet, it uses a Greedy Set Cover algorithm to determine which subset of neighbors should rebroadcast the packet, given knowledge of which neighbors have already been covered by the sender's broadcast. The Greedy Set Cover algorithm recursively chooses 1-hop neighbors, which cover the most *uncovered* 2-hop neighbors and recalculates the cover set until all 2-hop neighbors are covered.

Călinescu et al. [53] gave two practical heuristics for this problem (they called selecting forwarding neighbors). The first algorithm runs in time $O(n \log n)$ and returns a subset with size at most 6 times of the minimum. The second algorithm has an improved approximation ratio 3, but with running time $O(n^2)$. Here n is the number of total 2-hop neighbors of a node. When all 2-hop neighbors are in the same quadrant with respect to the source node, they gave an exact solution in time $O(n^2)$ and a solution with approximation factor 2 in time $O(n \log n)$. Their algorithms partition the region surrounding the source node into four quadrants, solve each quadrants using an algorithm with approximation factor α, and then combine these solutions. They proved that the combined solution is at most 3α times of the optimum solution. They then gave two different algorithms for finding a disk cover when the 2-hop neighbors are restricted to one quadrant with approximation ratio $\alpha = 2$ and 1, respectively.

Their approach assumes that every node u can collect its 2-hop neighbors $N_2(u)$ efficiently. Notice that, the 1-hop neighbors of every node u can be collected efficiently by asking each node to broadcast its information to its 1-hop neighbors. Thus all nodes get their 1-hop neighbors information by using total $O(n)$ messages. However, until recently, it was not known how to collect the 2-hop neighbors information with $O(n)$ communications. The simplest broadcasting of 1-hop neighbors $N_1(u)$ to all neighbors u does let all nodes in $N_1(u)$ to collect their corresponding 2-hop neighbors. However, the total communication cost of this approach is $O(m)$, where m is the total number of links in UDG. Recently, Călinescu [49] proposed an efficient approach to collect $N_2(u)$ using the connected dominating set [43] as forwarding nodes. Assume that the node position is known. He proved that the approach takes total communications $O(n)$, which is optimum within a constant factor.

11.3.4.2 Gossip and Probabilistic Schemes

Probabilistic Scheme: The Probabilistic scheme from Reference 1 is similar to flooding, except that nodes only rebroadcast with a predetermined probability. In dense networks multiple nodes share similar transmission coverages. Thus, randomly having some nodes not rebroadcast saves node and network resources without harming delivery effectiveness. In sparse networks, there is much less shared coverage; thus, nodes will not receive all the broadcast packets with the Probabilistic scheme unless the probability parameter is high. When the probability is 100%, this scheme is identical to flooding. Cartigny and Simplot [65] applied probability, which is a function of the distance to the transmitting neighbor.

Counter-Based Scheme: Tseng et al. [1] show an inverse relationship between the number of times a packet is received at a node and the probability of that node being able to reach additional area on a rebroadcast. This result is the basis of their Counter-Based scheme. Upon reception of a previously unseen packet, the node initiates a counter with a value of one and sets a RAD (which is randomly chosen between 0 and Tmax seconds). During the RAD, the counter is incremented by one for each redundant packet received. If the counter is less than a threshold value when the RAD expires, the packet is rebroadcast. Otherwise, it is simply dropped. From Reference 1, threshold values above six relate to little additional coverage area being reached.

The over-riding compelling features of the counter-based scheme are its simplicity and its inherent adaptability to local topologies. That is, in a dense area of the network, some nodes will not rebroadcast; in sparse areas of the network, all nodes rebroadcast. The disadvantage of all counter and probabilistic schemes is that delivery is not guaranteed to all nodes even if ideal MAC is provided. In other words, they are not reliable.

11.3.4.3 Area-Based Decision

In either probabilistic schemes or the counter-based schemes a node decides whether to rebroadcast a received packets purely based on its own information. Tseng et al. [1] proposed several other criteria based on the additional coverage area to decide whether the node will rebroadcast the packet. These coverage-area based methods are similar to the methods of selecting forwarding neighbors, which tries to select a set of 1-hop neighbors sufficient to cover all its 2-hop neighbors. While area based methods only consider the coverage area of a transmission; they do not consider whether nodes exist within that area. Two coverage-area based methods are proposed in Reference 1: *Distance-Based Scheme* and *Location Based Scheme*.

In Distance-Based Scheme, a node compares the distance between itself and each neighbor node that has previously rebroadcast a given packet. Upon reception of a previously unseen packet, a RAD is initiated and redundant packets are cached. When the RAD expires, all source node locations are examined to see if any node is closer than a threshold distance value. If true, the node does not rebroadcast.

The location-based scheme uses a more precise estimation of expected additional coverage area in the decision to rebroadcast. In this method, each node must have the means to determine its own location, for example, a GPS. Whenever a node originates or rebroadcasts a packet it adds its own location to the header of the packet. When a node initially receives a packet, it notes the location of the sender and calculates the additional coverage area obtainable were it to rebroadcast. If the additional area is less than a threshold value, the node will not rebroadcast, and all future receptions of the same packet will be ignored. Otherwise, the node assigns a RAD before delivery. If the node receives a redundant packet during the RAD, it recalculates the additional coverage area and compares that value to the threshold. The area calculation and threshold comparison occur with all redundant broadcasts received until the packet reaches either its scheduled send time or is dropped.

We will review also some upcoming work related to dominating sets and broadcasting problem. In Reference 67, a beaconless broadcasting method is proposed. All nodes have the same transmission radius, and nodes are not aware of their neighborhood. That is, no beacons or hello messages are sent in order to discover neighbors prior to broadcasting process. The source transmits the message to all neighbors. Upon receiving the packet (together with geographic coordinates of the sender), each node calculates the portion of its perimeter, along circle of transmission radius, that is not covered by this and previous transmissions of the same packet. Node then sets or updates its timeout interval, which inversely depends on the size of the uncovered perimeter portion. If the perimeter becomes fully covered, the node cancels retransmissions. Otherwise, it retransmits at the end of timeout interval. The method is reliable, as opposed to other area based methods.

11.3.4.4 Neighbor Coverage-Based Decision

The method presented in the previous section were based on covering an area where nodes could be located. Instead of covering area, one could simply cover neighboring nodes, assuming their location, or existence of their link to a previous transmitting node, are known. The basic method was independently and almost simultaneously (August 2000) proposed in two articles [54,55]. The methods were called Neighbor Elimination by Stojmenovic and Seddigh [55], while a similar method, called Scalable Broadcast Algorithm, was proposed by Peng and Lu [54]. 2-hop neighbors information is used to determine whether a node will rebroadcast the packet. Suppose that a node u receives a broadcast data packet from its neighbor node v. Node u knows all the neighbors of node v, and thus all nodes that are common neighbors of them (already received the data from v). If node u has additional neighbors not reached by node v's broadcast, node u schedules the packet for delivery with a RAD. However, if node u receives a redundant broadcast packet from some other neighbors within RAD, node u will recalculate whether it needs to rebroadcast the packet. This process is continued until either the RAD expires and the packet is then sent, or the packet is dropped (when all its neighbors are already covered by the broadcasts of some of its neighbors).

Lipman et al. [66] described the following broadcasting protocol. Upon receiving a broadcast message(s) from a node h, each node i (that was determined by h as a forwarding node) determines which of its 1-hop neighbors also received the same message. For each of its remaining neighbors j (which did not receive a message yet, based on i's knowledge), node i determines whether j is closer to i than any 1-hop neighbors of i (that are also forwarding nodes of h) who received the message already. If so, i is responsible for message transmission to j, otherwise it is not. Node i then determines a transmission range equal to that of the farthest neighbor it is responsible for.

11.4 Scheduling Active and Sleep Periods

In ad hoc wireless networks, the limitation of power of each host poses a unique challenge for power-aware design. There has been an increasing focus on low cost and reduced node power consumption in ad hoc wireless networks. Even in standard networks such as IEEE 802.11, requirements are included to sacrifice performance in favor of reduced power consumption. In order to prolong the life span of each node and, hence, the network, power consumption should be minimized and balanced among nodes. Unfortunately, nodes in the dominating set in general consume more energy in handling various bypass traffic than nodes outside the set. Therefore, a static selection of dominating nodes will result in a shorter life span for certain nodes, which in turn result in a shorter life span of the whole network.

Wu et al. [56] study dynamic selection of dominating nodes, also called activity scheduling. Activity scheduling deals with the way to rotate the role of each node among a set of given operation modes.

For example, one set of operation modes is sending, receiving, idles, and sleeping. Different modes have different energy consumptions. Activity scheduling judiciously assigns a mode to each node to save overall energy consumptions in the networks and to prolong life span of each individual node. Note that saving overall energy consumptions does not necessarily prolong life span of a particular individual node. Specifically, they propose to save overall energy consumptions by allowing only dominating nodes (i.e., gateway nodes) to retransmit the broadcast packet. In addition, in order to maximize the lifetime of all nodes, an activity scheduling method is used that dynamically selects nodes to form a connected dominating set. Specifically, in the selection process of a gateway node, we give preference to a node with a higher energy level. The effectiveness of the proposed method in prolonging the life span of the network is confirmed through simulation. Source dependent forwarding sets appear to be more energy balanced. However, it was experimentally confirmed in Reference 57 that the difference in energy consumption between an idle node and a transmitting node is not major, while the major difference exists between idle and sleep states of nodes. Therefore the most energy efficient methods will select static dominating set for a given round, turning all remaining nodes to a sleep state. Depending on energy left, changes in activity status for the next round will be made. The change can therefore be triggered by changes of power status, in addition to node mobility. From this point of view, internal nodes based dominating sets provide static selection for a given round and more energy efficiency than forwarding set based method that requires all nodes to remain active in all the rounds.

In Reference 58, the key for deciding dominating set status is a combination of remained energy and node degree. Xu et al. [59] discuss the following sensor sleep node schedule. The tradeoff between network lifetime and density for this cell-based schedule was investigated in Reference 60. The given two dimensional space is partitioned into a set of squares (called cells), such as any node within a square can directly communicate with any nodes in an adjacent square. Therefore, one representative node from each cell is sufficient. To prolong the life span of each node, nodes in the cell are selected in a alternative fashion as a representative. The adjacent squares form a two dimensional grid and the broadcast process becomes trivial. Note that the selected nodes in Reference 59 make a dominating set, but the size of it is far from optimal, and also it depends on the selected size of squares. On the other hand, the dominating set concept used here has smaller size and is chosen without using any parameter (size of square, which has to be carefully selected and propagated with node relative positioning in solution [59]).

The Span algorithm [61] selects some nodes as coordinators. These nodes form dominating set. A node becomes coordinator if it discovers that two of its neighbors cannot communicate with each other directly or through one or two existing coordinators. Also, a node should withdraw if every pair of its neighbors can reach each other directly or via some other coordinators (they can also withdraw if each pair of neighbors is connected via possibly noncoordinating nodes, to give chance to other nodes to become coordinators). Since coordinators are not necessarily neighbors, 3-hop neighboring topology knowledge is required. However, the energy and bandwidth required for maintenance of 3-hop neighborhood information is not taken into account in experiments [61]. On the other hand, if the coordinators are restricted to be neighboring nodes, then the dominating set definition [61] becomes equivalent to one given by Wu and Li [38]. Next, protocol [61] heavily relies on proactive periodic beacons for synchronization, even if there is no pending traffic or node movement. The recent research on energy consumption [57] indicates that the use of such periodic beacons or hello messages is an energy expensive mechanism, because of significant start up cost for sending short messages. Finally, Blough and Santi [60] observed that the overhead required for coordination with SPAN tends to *explode* with node density, and thus counterbalances the potential savings achieved by the increased density.

11.5 Conclusion and Future Research Directions

In this chapter, we reviewed several methods for efficient broadcasting for wireless ad hoc networks. There are still many challenging questions left open for further research. So far, all the known theoretically good algorithms either assume that the power needed to support a link uv is proportional to $||uv||^\beta$ or is a fixed cost that is independent of the neighboring nodes that it will communicate with. In practice, the energy consumption of a node is neither solely dependent on the distance to its farthest neighbor, nor totally independent of its communication neighbor. For example, a more general power consumption model for a node u would be $c_1 + c_2 \cdot ||uv||^\beta$ for some constants c_1 and c_2 where v is its farthest communication neighbor in a broadcast structure. No theoretical result is known about the approximation of the optimum broadcast or multicast structure under this model.

Another important aspect of designing energy efficient protocols for broadcast and multicast that is often neglected in the literature is the physical constraints of the wireless communications. In most of the algorithms, it is assumed that the signal sent by a node will be received by all nodes at one shot it intends to send. However, in practice, it is not the case, which will make designing energy efficient broadcast and multicast protocols much harder. The first difficulty is that the sender needs to coordinate with the receivers so *all* receivers are ready to receive. It is often difficult, if not impossible, to do so. Then a natural question is how to set the threshold t such that if the number of receivers that are ready is more than t, node u sends the packets; otherwise node t will not send the packets. Clearly, the setting of the threshold will affect the total actual energy consumption of the broadcast and multicast protocols, in addition to affect the system performance and the stability of the networks. The second difficulty is that, even some receivers are ready to receive, the receivers cannot always decode the signal correctly due to the signal strength fluctuation. In other words, the link between two nodes is often probabilistic. We clearly should take this into account when we design energy efficient broadcast and multicast protocols.

References

[1] Y.-C. Tseng, S.-Y. Ni, Y.-S. Chen, and J.-P. Sheu, "The broadcast storm problem in a mobile ad hoc network," *Wireless Networks*, vol. 8, pp. 153–167, 2002, Short version in MOBICOM 99.

[2] B. Williams and T. Camp, "Comparison of broadcasting techniques for mobile ad hoc networks," in *Proceedings of the ACM International Symposium on Mobile Ad Hoc Networking and Computing (MOBIHOC)*, pp. 194–205, 2002.

[3] J. Wu and W. Lou, "Forward-node-set-based broadcast in clustered mobile ad hoc networks," *Wireless Communications and Mobile Computing*, vol. 3, pp. 155–173, 2003.

[4] Khaled Alzoubi, Peng-Jun Wan, and Ophir Frieder, "Message-optimal connected-dominating-set construction for routing in mobile ad hoc networks," in *Proceedings of the 3rd ACM International Symposium on Mobile Ad Hoc Networking and Computing (MOBIHOC'02)*, 2002.

[5] Khaled Alzoubi, Xiang-Yang Li, Yu Wang, Peng-Jun Wan, and Ophir Frieder, "Geometric spanners for wireless ad hoc networks," *IEEE Transactions on Parallel and Distributed Processing*, vol. 14, pp. 408–421, April 2003, Short version in IEEE ICDCS 2002.

[6] Xiang-Yang Li, "Approximate MST for UDG locally," in *Proceedings of the COCOON 2003*, Big Sky, MT, 2003.

[7] Xiang-Yang Li, Localized construction of low weighted structure and its applications in wireless ad hoc networks, ACM Wireless Network (WINET), 2004 (Accepted for Publication).

[8] Ivan Stojmenovic, "Degree limited RNG," in *Proceedings of the Workshop on Wireless Networks*, Cocoyoc, Mexico, 2003.

[9] Xiang-Yang Li, Yu Wang, Peng-Jun Wan, Wan-Zhen Song, and Ophir Frieder, "Localized low-weight graph and its applications in wireless ad hoc networks," in *Proceedings of the IEEE INFOCOM*, 2004.

[10] Xiang-Yang Li, Yu Wang, and Wen-Zhan Song, "Applications of k-local MST for topology control and broadcasting in wireless ad hoc networks," *IEEE Transaction on Parallel and Distributed Systems*, vol. 15, pp. 1057–1069, December 2004.

[11] A. Clementi, P. Crescenzi, P. Penna, G. Rossi, and P. Vocca, "On the complexity of computing minimum energy consumption broadcast subgraphs," in *Proceedings of the 18th Annual Symposium on Theoretical Aspects of Computer Science, LNCS 2010*, pp. 121–131, 2001.

[12] A. Clementi, P. Penna, and R. Silvestri, "On the power assignment problem in radio networks," *Electronic Colloquium on Computational Complexity*, 2001, To approach. Preliminary results in APPROX'99 and STACS'2000.

[13] L. M. Kirousis, E. Kranakis, D. Krizanc, and A. Pelc, "Power consumption in packet radio networks," *Theoretical Computer Science*, vol. 243, pp. 289–305, 2000.

[14] J. Wieselthier, G. Nguyen, and A. Ephremides, "On the construction of energy-efficient broadcast and multicast trees in wireless networks," in *Proceedings of the IEEE INFOCOM 2000*, pp. 586–594, 2000.

[15] Weifa Liang, "Constructing minimum-energy broadcast trees in wireless ad hoc networks," in *Proceedings of the ACM International Symposium on Mobile Ad Hoc Networking and Computing (MOBIHOC)*, pp. 112–122, 2002.

[16] Arindam K. Das, Robert J. Marks, Mohamed El-Sharkawi, Payman Arabshahi, and Andrew Gray, "Minimum power broadcast trees for wireless networks: an ant colony system approach," in *Proceedings of the IEEE International Symposium on Circuits and Systems*, 2002.

[17] Fulu Li and Ioanis Nikolaidis, "On minimum-energy broadcasting in all-wireless networks," in *Proceedings of the IEEE 26th Annual IEEE Conference on Local Computer Networks (LCN'01)*, 2001.

[18] R.J. Marks, A.K. Das, M. El-Sharkawi, P. Arabshahi, and A. Gray, "Minimum power broadcast trees for wireless networks: optimizing using the viability lemma," in *Proceedings of the IEEE International Symposium on Circuits and Systems*, pp. 245–248, 2002.

[19] Arindam K. Das, Robert J. Marks, Mohamed El-Sharkawi, Payman Arabshahi, and Andrew Gray, "Minimum power broadcast trees for wireless networks: integer programming formulations," in *Proceedings of the IEEE INFOCOM 2003*, 2003.

[20] Peng-Jun Wan, G. Calinescu, Xiang-Yang Li, and Ophir Frieder, "Minimum-energy broadcast routing in static ad hoc wireless networks," *ACM Wireless Networks*, 2002 (Preliminary version appeared in IEEE INFOCOM 2000).

[21] V. Chvátal, "A greedy heuristic for the set-covering problem," *Mathematics of Operations Research*, vol. 4, pp. 233–235, 1979.

[22] M.R. Garey and D.S. Johnson, *Computers and Intractability*, W.H. Freeman and Co., New York, 1979.

[23] Sudipto Guha and Samir Khuller, "Approximation algorithms for connected dominating sets," in *Proceedings of the European Symposium on Algorithms*, pp. 179–193, 1996.

[24] G. Robins and A. Zelikovsky, "Improved steiner tree approximation in graphs," in *Proceedings of the ACM/SIAM Symposium on Discrete Algorithms*, pp. 770–779, 2000.

[25] Harry B. Hunt III, Madhav V. Marathe, Venkatesh Radhakrishnan, S.S. Ravi, Daniel J. Rosenkrantz, and Richard E. Stearns, "NC-approximation schemes for NP- and PSPACE-hard problems for geometric graphs," *Journal of Algorithms*, vol. 26, pp. 238–274, 1999.

[26] Madhav V. Marathe, Heinz Breu, Harry B. Hunt III, S.S. Ravi, and Daniel J. Rosenkrantz, "Simple heuristics for unit disk graphs," *Networks*, vol. 25, pp. 59–68, 1995.

[27] Yu Wang and Xiang-Yang Li, "Geometric spanners for wireless ad hoc networks," in *Proceedings of the of 22nd IEEE International Conference on Distributed Computing Systems (ICDCS)*, 2002.

[28] P. Berman, M. Furer, and A. Zelikovsky, "Applications of matroid parity problem to approximating steiner trees," Technical Report 980021, Computer Science, UCLA, 1998.

[29] Khaled M. Alzoubi, *Virtual Backbone in Wireless Ad Hoc Networks*, Ph.D. thesis, Illinois Institute of Technology, 2002.

[30] X. Cheng, X. Huang, D. Li, W. Wu, and D.-Z. Du, "Polynomial-time approximation scheme for minimum connected set in ad hoc wireless network," *Networks*, vol. 42, pp. 202–208, 2003.

[31] Khaled M. Alzoubi, Peng-Jun Wan, and Ophir Frieder, "New distributed algorithm for connected dominating set in wireless ad hoc networks," in *Proceedings of the HICSS, Hawaii*, 2002.

[32] Chunhung Richard Lin and Mario Gerla, "Adaptive clustering for mobile wireless networks," *IEEE Journal of Selected Areas in Communications*, vol. 15, pp. 1265–1275, 1997.

[33] I. Chlamtac and A. Farago, "A new approach to design and analysis of peer to peer mobile networks," *Wireless Networks*, vol. 5, pp. 149–156, 1999.

[34] Alan D. Amis, Ravi Prakash, Dung Huynh, and Thai Vuong, "Max-min d-cluster formation in wireless ad hoc networks," in *Proceedings of the of the 19th Annual Joint Conference of the IEEE Computer and Communications Societies INFOCOM*, 2000, vol. 1, pp. 32–41, 2000.

[35] Lujun Jia, Rajmohan Rajaraman, and Torsten Suel, "An efficient distributed algorithm for constructing small dominating sets," in *Proceedings of the ACM PODC*, 2000.

[36] B. Das and V. Bharghavan, "Routing in ad-hoc networks using minimum connected dominating sets," in *Proceedings of the 1997 IEEE International Conference on Communications (ICC'97)*, 1997, vol. 1, pp. 376–380, 1997.

[37] R. Sivakumar, B. Das, and V. Bharghavan, "An improved spine-based infrastructure for routing in ad hoc networks," in *Proceedings of the IEEE Symposium on Computers and Communications*, Athens, Greece, June 1998.

[38] Jie Wu and Hailan Li, "A dominating-set-based routing scheme in ad hoc wireless networks," *The Special Issue on Wireless Networks in the Telecommunication Systems Journal*, vol. 3, pp. 63–84, 2001.

[39] F. Dai and J. Wu, "Distributed dominant pruning in ad hoc networks," in *Proceedings of the IEEE ICC*, 2003.

[40] Ivan Stojmenovic, Mahtab Seddigh, and Jovisa Zunic, "Dominating sets and neighbor elimination based broadcasting algorithms in wireless networks," *IEEE Transactions on Parallel and Distributed Systems*, vol. 13, pp. 14–25, 2002.

[41] F. Garcia, J. Solano, and I. Stojmenovic, "Connectivity based k-hop clustering in wireless networks," in *Telecommunication Systems*, vol. 22, pp. 205–220, 2003.

[42] S. Basagni, I. Chlamtac, and A. Farago, "A generalized clustering algorithm for peer-to-peer networks," in *Proceedings of the Workshop on Algorithmic Aspects of Communication*, 1997.

[43] Peng-Jun Wan, Khaled M. Alzoubi, and Ophir Frieder, "Distributed construction of connected dominating set in wireless ad hoc networks," in *Proceedings of the INFOCOM*, 2002.

[44] Yu Wang and Xiang-Yang Li, "Distributed low-cost weighted backbone formation for wireless ad hoc networks," (Submitted for Publication), 2004.

[45] Mario Cagalj, Jean-Pierre Hubaux, and Christian Enz, "Minimum-energy broadcast in all-wireless networks: Np-completeness and distribution issues," in *Proceedings of the ACM MOBICOM 02*, 2002.

[46] Ning Li and Jennifer C. Hou, "Blmst: a scalable, power efficient broadcast algorithm for wireless sensor networks," 2003, Submitted for publication, a version as UIUC Computer Science Department Technical Report.

[47] J. Cartigny, D. Simplot, and I. Stojmenovic, "Localized minimum-energy broadcasting in ad-hoc networks," in *Proceedings of the IEEE INFOCOM 2003*, 2003.

[48] Ning Li, Jennifer C. Hou, and Lui Sha, "Design and analysis of a mst-based topology control algorithm," in *Proceedings of the IEEE INFOCOM 2003*, 2003.

[49] Gruia Călinescu, "Computing 2-hop neighborhoods in ad hoc wireless networks," 2002, Manuscript.

[50] Mahtab Seddigh, J. Solano Gonzalez, and I. Stojmenovic, "Rng and internal node based broadcasting algorithms for wireless one-to-one networks," *ACM Mobile Computing and Communications Review*, vol. 5, pp. 37–44, 2002.

[51] Xiang-Yang Li, Kousha Moaveninejad, and Yu Wang, "Low weighted structures and internal node based broadcasting schemes for wireless ad hoc networks," Manuscript, 2003.

[52] H. Lim and C. Kim, "Multicast tree construction and flooding in wireless ad hoc networks," in *Proceedings of the ACM International Workshop on Modeling, Analysis and Simulation of Wireless and Mobile Systems (MSWIM)*, 2000.

[53] Gruia Călinescu, Ion Măndoiu, Peng-Jun Wan, and Alexander Zelikovsky, "Selecting forwarding neighbors in wireless ad hoc networks," in *Proceedings of the ACM DialM*, 2001.

[54] W. Peng and X. Lu, "On the reduction of broadcast redundancy in mobile ad hoc networks," in *Proceedings of the MOBIHOC*, 2000.

[55] I. Stojmenovic, and M. Seddigh, "Broadcasting algorithms in wireless networks," in *Proceedings of the International Conference on Advances in Infrastructure for Electronic Business, Science, and Education on the Internet SSGRR*, L'Aquila, Italy, July 31-Aug. 6, 2000.

[56] J. Wu, B. Wu, and I. Stojmenovic, "Power-aware broadcasting and activity scheduling in ad hoc wireless networks using connected dominating sets," *Wireless Communications and Mobile Computing*, vol. 3, 2003, pp. 425-438.

[57] L. M. Feeney and M. Nilson, "Investigating the energy consumption of a wireless network interface in an ad hoc networking environment," in *Proceedings of the IEEE INFOCOM 2001*, pp. 1548–1557, 2001.

[58] J. Shaikh, J. Solano, I. Stojmenovic, and J. Wu, "New metrics for dominating set based energy efficient activity scheduling in ad hoc networks," *IEEE Conference on Local Computer Networks/WLN*, Bonn, Germany, Oct. 20–24, 2003.

[59] Y. Xu, J. Heidemann, and D. Estrin, "Geography-informed energy conservation for ad hoc networks," in *Proceedings of the MOBICOM*, 2001.

[60] D. M. Blough and P. Santi, "Investigating upper bounds on network lifetime extension for cell-based energy conservation techniques in stationary ad hoc networks," in *Proceedings of the MOBICOM*, 2002.

[61] Benjie Chen, Kyle Jamieson, Hari Balakrishnan, and Robert Morris, "Span: an energy-efficient coordination algorithm for topology maintenance in ad hoc wireless networks," in *Proceedings of the Mobile Computing and Networking*, 2001, pp. 85–96.

[62] J. Cartigny, D. Simplot, and I. Stojmenovic, "Localized energy efficient broadcast for wireless networks with directional antennas," in *Proceedings of the IFIP Mediterranean Ad Hoc Networking Workshop (MED-HOC-NET 2002)*, Sardegna, Italy, 2002.

[63] J. Cartigny, F. Ingelrest, D. Simplot-Ryl, and I. Stojmenovic, "Localized LMST and RNG based minimum energy broadcast protocols in ad hoc networks," (Submitted for Publication).

[64] F. Ingelrest, D. Simplot-Ryl, and I. Stojmenovic, "Target transmission radius over LMST for energy-efficient broadcast protocol in ad hoc networks," (Submitted for Publication).

[65] J. Cartigny, and D. Simplot, "Border node retransmission based probabilistic broadcast protocols in ad hoc networks," *Telecommunication Systems*, 22, 2003, pp. 189–204.

[66] J. Lipman, P. Boustead, and J. Judge, "Efficient and scalable information dissemination in mobile ad hoc networks," in *Proceedings of the First International Conference on Ad-Hoc Networks and Wireless ADHOC-NOW*, 2002, pp. 119–134, 2002.

[67] J. Carle, D. Simplot, and I. Stojmenovic, "Area coverage based beaconless broadcasting in adhoc networks," (In Preparation).

12

Cross-Layer Optimization for Energy-Efficient Information Processing and Routing

12.1 Introduction..**12**-265
12.2 Data-Centric Paradigm.................................**12**-267
12.3 Collaborative Information Processing and Routing...**12**-267
12.4 Cross-Layer Optimization for Energy-Efficiency......**12**-269
  Motivation • Consideration for Collaborative Information Processing and Routing
12.5 Energy-Efficient Information Routing and Coding ...**12**-271
  Naive Data-Centric Routing Protocols • Analysis for Entropy-Aware Routing • Hybrid Routing Tree Algorithms • Network Flow-Based Routing Algorithm • From Information Coding Perspective
12.6 A Broad View of Cross-Layer Optimization for Energy-Efficiency **12**-276
  Hardware Layer • Physical Layer • MAC Layer • Routing Layer • Application Layer • Putting It All Together
12.7 Future Directions **12**-279
References **12**-280

Yang Yu
Viktor K. Prasanna

12.1 Introduction

With the advancements in micro-electro-mechanical systems (MEMS) and miniaturization techniques, networked sensor systems (NSSs) emerged as a new technology and research topic around six years ago [1–4]. The set of applications that were envisioned for NSSs include environment monitoring, habitat study, battlefield surveillance, and infrastructure monitoring. All of these applications require the sensing, processing, and gathering of information from the physical environments where the systems operate. The end users are interested in the content of the information, including the spatial and temporal specification of the information. Hence, the computation and communication are centered around the information itself, instead of the sensor nodes that hold the information. This feature is called *data-centric* computation and communication paradigm [5], as opposite to the address- or node-centric paradigm, which is exhibited in traditional Internet networks, mobile ad hoc networks, or parallel and distributed systems.

To deliver information to the end users in such a data-centric paradigm necessitates two fundamental operations: *information processing* and *information routing*. While information processing extracts useful information from the physical environment, information routing transports such information from sensor nodes across the network to the end users. These two operations are tightly related to each other. On one hand, information processing such as joint compression can be used to remove the redundancy among multiple data sources and hence reduce the data volume to be routed. On the other hand, routing provides the opportunities that data from multiple sources can meet at certain sensor nodes where joint compression can be performed. Hence, these two operations are often performed in a collaborative and mutually beneficial fashion.

This chapter deals with the critical problem of improving the energy-efficiency of the above operations. We take a cross-layer optimization perspective that has been widely used in NSSs because of two major reasons. First, information sharing across layers leads to a much larger optimization and tradeoff space, and hence better opportunities to achieve greater energy savings. For example, the application-level knowledge of joint entropy from multiple sources helps the design of more energy-efficient routing schemes [6–8]. Also, physical layer rate adaptation techniques can be used to achieve up to 90% energy savings based on the knowledge of routing schemes and application level delay constraint [9]. Moreover, one particular optimization technique can affect system performance at various layers. For example, media access control (MAC) layer sleep scheduling could affect signal interference at the physical layer [10], channel access at the MAC layer [11,12], network topology and routing selection at the routing layer [13–15], and sensing coverage at the application layer [16,17]. To avoid conflicts between optimization techniques applied at different layers, coordination among various optimization techniques is necessary. Both of the above two facts motivate a holistic optimization approach with simultaneous consideration of multiple layers and coordinated optimization across these layers.

While cross-layer optimization can be applied at various system layers to improve the energy-efficiency of collaborative information processing and routing, we are particularly interested in a set of techniques that are related to application level knowledge of joint entropy among source information. This is because the key reason of collaborative information processing and routing is to explore the benefits of joint data compression, which is inherently related to the joint entropy among source information. Specifically, we discuss these details from both routing and information coding perspectives.

From routing perspective, we first reveal the fundamental tradeoffs between Shortest Path Tree and Minimal Steiner Tree when joint data compression is considered. This tradeoffs motivates various studies on tree structures that approximate both Shortest Path Tree and Minimal Steiner Tree. Also, we discuss some results that address the problem by relating it to the maximal network flow problem. From information coding perspective, we show the theoretical results that joint coding and routing can provide a many-to-many communication solution that is supported by the network capacity. We further demonstrate the inherent tradeoffs between network traffic and scheduling delay for this technique with a linear array of sensor nodes.

Moreover, we also give a brief survey of techniques that have been proposed for the general goal of improving system energy-efficiency. Our survey is presented from several system layers, including the hardware layer, physical layer, MAC layer, routing layer, and application layer. These techniques together with the aforementioned techniques related to joint entropy knowledge provide a comprehensive view of enabling energy-efficient collaborative information processing and routing.

The rest of the chapter is organized as follows. In Section 12.2, we discuss several unique features of data-centric paradigm. The collaborative information processing and routing originated from this data-centric paradigm is described in Section 12.3. In Section 12.4, we list the cross-layer optimization techniques that have been widely used in NSSs. A detailed discussion of techniques that utilize application-level information entropy knowledge is presented in Section 12.6.4. This is followed by a brief survey of other existing optimization techniques in Section 12.6. Finally, we give a discussion on future directions and open issues in Section 12.7.

12.2 Data-Centric Paradigm

Data-centric computation and communication paradigm is a unique feature of NSSs. It differs from the address-centric paradigm in several ways. First, the typical goal of NSSs with data-centric paradigm is to gather specific information from multiples source nodes to a small number of sink nodes. Hence, the communication pattern normally resembles *a reversed-multicast tree*, instead of a more randomized peer-to-peer-based communication pattern in address-centric systems.

Second, most of data gathered from a physical environment are not of direct interest to the end users and they often show strong correlation. Therefore, it is necessary to process the data before they are transported to the end users. Such *in-network processing* includes signal and image processing to extract useful information out of the raw data, data compression to reduce communication load, and joint compression of data from multiple sources to eliminate redundancy among the data.

Third, the routing schemes that were originally developed for address-centric systems are based on the purpose of fulfilling each individual data routing request. Hence, they become unsuitable for data-centric systems, where the communication request of the whole system is application-specific, which may be as simple as to route data from all source nodes to a sink node. To investigate new routing schemes that are customized for NSS applications becomes important.

Fourth, since applications of an NSS are usually more specific and more well-defined, the sensor nodes within the system are more likely to work in a cooperative fashion, instead of a competing fashion. In this way, the aggregated resource of the system can be more efficiently utilized.

The difference between data- and address-centric paradigms can be effectively reflected by the user queries. For example, typical user queries or service requests in address-centric paradigm are "Load the web page at address...," "Transfer file A from host B to host C," etc. However, in data-centric paradigm systems, typical queries are "What is the average temperature at region A within time T?," "Are there any vehicles currently in region A?." For these queries, the users are interested purely in the information itself, but not the sensor nodes that generate or hold the information, neither the way how the information is transmitted to the end users.

The above unique features of data-centric paradigm have raised various hot research topics for NSSs, ranging from hardware components, networking techniques, to application algorithms. One of the most important topics is the way of processing and transporting information within the system so that the required functionality and performance metrics can be fulfilled. We discuss this topic in detail in the next section.

12.3 Collaborative Information Processing and Routing

In data-centric paradigm, the system operations shall be centered around the fundamental functionality: to deliver the required information to the end users. Apparently, this involves two operations:

1. *Information processing* includes sensing the environment and extracting useful information out of the raw data.
2. *Information routing* includes combining the information from difference sources across the network and routing the final set of information to the end users.

Note that in some sense these two operations are also applicable to address-centric paradigm. For example, users may want to extract information from a file on machine A and then transmit the result to

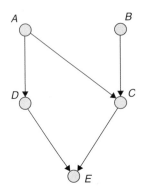

FIGURE 12.1   An example network scheme for collaborative information processing and routing.

machine B. However, in data-centric paradigm, these two operations are not sequentially and independently performed as they usually are in address-centric paradigm. On the contrary, these two operations are performed in a parallel fashion and are tightly related to each other.

On one hand, information processing can help reduce the data volume to route by processing raw data, or compressing the data, or removing redundancy among data through joint source compression. On the other hand, since source data are distributed across the network, many information processing techniques, especially joint source compression, rely on the data availability at sensor nodes that is largely determined by the routing scheme. Such a mutually beneficial relationship leads to tightly coupled design and implementation of information processing and routing, referred to as *collaborative information processing and routing.*

For example, consider the simple network scheme in Figure 12.1, where vertices A through D denote sensor nodes and edges denote valid communication links. In this example, there are two source nodes, A and B, and one sink node, E. The route from node B to E has one option, which is $B \rightarrow C \rightarrow E$, while the route from node A to E has two options: $A \rightarrow C \rightarrow E$ and $A \rightarrow D \rightarrow E$. If the two pieces of source information at nodes A and B are relatively independent, the routing selection for A is not crucial in terms of the communication cost. However, if the source information is highly correlated, we can take advantage of joint compression to reduce the data volume to be routed. Hence, the route $A \rightarrow C \rightarrow E$ will be preferred to enable joint compression at node C, given that the gain of reduced data volume to communicate over CE dominates the extra communication cost of choosing ACE instead of ADE, if any. This example clearly indicates the importance of coupling the selection of routing schemes with data compression techniques.

Since sensor readings from the physical world are usually highly correlated, the above joint compression technique is particularly useful to reduce data volume by removing redundancy among data from multiple sources. Such a joint compression is also referred to as *data aggregation* [5]. We will use the terms data aggregation and joint compression exchangeably hereafter.

Choosing an appropriate routing scheme with joint compression in general networks is much more challenging, since it is determined by many factors, including network topology, communication cost and reliability, computation cost, and the correlation among data. In many general settings, to choose the optimal routing scheme is NP-hard and hence heuristic solutions are preferred. For example, consider the problem with a simplified assumption that a perfect data aggregation can be achieved. This means the output of jointly compressing data from an arbitrary number of sources is exactly one unit data. To form a routing scheme from a given set of source nodes to a sink node under this assumption for minimizing the total communication cost in terms of hops is the Minimal Steiner Tree problem, which is known to be NP-hard.[1]

---

[1] Please note that the routing scheme does not necessarily need to be a tree in general cases. However, such a tree structure is the focus of many studies due to its simplicity.

In many real-life cases without perfect data aggregation, to choose an appropriate model for abstracting the data aggregation operation is difficult, since it heavily depends on the data correlation from multiple sources. Such a data correlation is determined by the nature of the observed phenomena as well as the physical situation of the operating field which affects the propagation of the phenomena. While several papers have proposed models to abstract the correlation among data from the physical world [7,18,19], a unified and widely accepted model is still lacking. Hence, several research efforts have been trying to study the inter-relationship between information processing and routing under certain simplifications. For example, the work by Cristescu et al. [8] assumes a fixed reduction in data volume for any source data which is jointly compressed by data from other sources. Some other models are discussed in Section 11.6.4.

Another important research direction in order to cope with the above challenge is to perform joint data compression from information coding perspective. For example, the Slepian–Wolf coding [20] can be used to code two correlated sources with a total data volume equal to the joint entropy of the two sources without explicit communication between the two sources. Indeed, the Slepian–Wolf coding is studied in Reference 8, which shows that the routing selection and joint data compression can be perfectly decoupled with such a coding scheme. While practical distributed source coding schemes for sensor networks are being developed [21], most existing works for data gathering are based on compression schemes with explicit communication [22].

Most of the existing works on are intended for a single sink node. When multiple sink nodes are considered, the concept of *network information flow* [23] can be used to design a joint routing and coding scheme that transports required information to all sources under a given network capacity. However, joint data compression is not considered in the original definition of network information flow. How to integrate these two concepts in the context of NSSs is an open challenge.

Most existing work on collaborative information processing and routing assumes a static system model in terms of communication reliability, network topology, and data correlation. The problem becomes even more challenging if we consider the possibly high dynamics in the above system parameters in real-life scenarios.

12.4 Cross-Layer Optimization for Energy-Efficiency

12.4.1 Motivation

A major concern for NSSs is energy-efficiency, which has been addressed by various techniques targeting hardware components [2,24], MAC scheduling policies [25,26], network organization [27,28], routing protocols [27,28], signal processing techniques [29,30], and application level algorithms [31,32]. It has also been realized that cross-layer optimization is of particular importance for saving energy in NSSs, since it enables a large space for tradeoffs and optimization of performance metrics across different layers [33–35].

Although cross-layer optimization has been widely studied in many contexts, there are unfortunately no formal definitions for cross-layer optimization. From a broad perspective, any hardware and software techniques applied at a specific system layer can be regarded as cross-layer optimization if they explicitly interact with the functionalities or optimization techniques at other system layers and in most cases, explicitly impact the system properties or performance at those layers.

The motivation for cross-layer optimization for energy-efficiency is multifold. First, system performance metrics in NSSs are often determined by multiple factors across several layers. For example, the performance of wireless communication is jointly determined by several factors across the physical, MAC, and routing layers, which is significantly different from wired communication. Also, an efficient routing scheme should take wireless communication conditions, network topology and connectivity, possibility of joint data compression, and application-level quality-of-service requirements into consideration. The optimization within each individual layer often leads to inefficient solution. Hence, a holistic approach that simultaneously considers different layers with cross-layer information sharing and coordinated optimization enables a much larger design and optimization space. Many research efforts on cross-layer optimization are based on this very basic hypothesis.

Second, optimization techniques applied at one particular layer often affects performance metrics at other layers. For example, sleep scheduling can affect signal interference at the physical layer, channel access at MAC layer, routing selection at the routing layer, and sensing coverage at the application layer. Isolating optimization techniques at each individual layer may cause conflicts in optimization goals or even counteracting solutions. It is therefore crucial to share information across the system stack and expose the effects of various optimization techniques to all layers so that coordinated optimization can be performed. This factor is however ignored by many research efforts, due to the inherent complexity to cope with multiple performance metrics at multiple layers.

Third, NSSs are usually application specific in terms of the required functionality. The generality of functionalities supported by strictly layered network and system structure becomes unnecessary compared with the layering overhead. Hence, a blurred boundary between layers or even the removal of unused layers helps build a lightweight and more efficient system.

Of course, the advantage of cross-layer optimization is not free. The large design and optimization space also leads to more challenging algorithm and system design and more complicated effects across various layers. However, given the application specific property of NSSs and fairly limited functionality of sensor nodes, these challenges are expected by most researchers to be tractable and worthwhile. Also, cross-layer optimization does not mean that layering is useless. Instead, we still need a layered structure so that a clean model is presented at each layer, which abstracts unnecessary details from lower layers. The key point in cross-layer optimization is the information sharing and coordinated optimization across the stack.

A common way to realize cross-layer optimization is to adjust the system performance by tuning low level hardware-based system knobs. One well-known example is shutdown or sleeping policy that tunes the awake time of the sensor nodes to adjust various upper-layer functionalities, including MAC layer scheduling [11,12], topology control [14,15], routing selection [13], and coverage for event detection [16,17]. Besides the motivation to reduce channel contention and to alleviate the scalability issue by keeping as a small number of awake sensor nodes as possible, the major principle in this case is to deliver "just enough" performance as required with a minimized resource usage, including energy. Other commonly used system knobs include voltage scaling that adjusts CPU computation time [36], power control that adjusts radio transmission radius [37], and rate adaptation that adjusts radio transmission speed [38].

12.4.2 Consideration for Collaborative Information Processing and Routing

It is quite natural to apply cross-layer optimization techniques into the context of collaborative information processing and routing. From one perspective, such techniques can be applied based on the following operating flow:

1. Data sensing at source nodes
2. Signal processing at source nodes
3. Joint information routing and compression across the network

We note that besides the last stage, the first two stages also require distributed and coordinated operations among sensor nodes in many cases. Although data sensing seems to be a localized operation that involves each sensor node as a basic function unit, the challenge lies in the fact that the aggregated sensing behavior of all sensor nodes usually needs to satisfy certain coverage requirement [16,17]. Since the sensing and computation capability of each sensor node is limited, signal processing usually requires the coordination a small group of sensor nodes in proximity to extract useful information from the raw data gathered by the sensor nodes, for example, the beamforming algorithm [30]. Hence, in most situations, the cross-layer optimization need to be performed in a distributed fashion.

From another perspective, cross-layer optimization techniques can be classified based on the layer where the techniques are applied. For our purpose, we divide the system into five layers: hardware layer, physical layer, MAC layer, routing layer, and application layer. Many cross-layer optimization techniques have

TABLE 12.1 Examples of Cross-Layer Optimization Techniques for
Energy-Efficient Collaborative Information Processing and Routing

System layer	Data sensing	Signal processing	Joint information routing and compression
Application		Energy-efficient signal processing, adaptive fidelity algorithms	Joint routing and coding, tunable compression
Routing			Energy-aware routing, entropy-aware routing
MAC			Radio sleep scheduling
Physical			Power control, rate adaptation, adaptive coding
Hardware	Low-power CPU, node sleep scheduling, voltage scaling		Low-power radio

been proposed at each of these layers with the general goal of improving the system energy-efficiency. In Table 12.1, we re-interpret them in the context of the above three operating stages of collaborative information processing and routing.

Since most of the above techniques are designed without specifically targeting energy-efficient collaborative information processing and routing, we are particularly interested in a subset of them that are explicitly related to application level knowledge of joint information entropy of source information. This is because the key motivation of collaborative information processing and routing is to explore the benefits by joint data compression, which is inherently related to the knowledge of joint information entropy of multiple source information.

Two classes of techniques have been studied to utilize this knowledge. The first class is routing techniques that take joint entropy into consideration, henceforth referred to as *entropy-aware routing*. Several papers including [5–8, 39–41] fall into this class. The second class is joint routing and coding from information theory perspective [8,18,42]. While both techniques are proposed for joint routing with information coding/compression, the first class of techniques assume the availability of certain data correlation models and (joint) data compression/coding schemes. On the other hand, the second class of techniques investigate concrete models for abstracting data correlation among sensor readings as well as coding schemes.

In the next section, we give a detailed description of algorithms and analysis that have been proposed for these two classes of techniques.

12.5 Energy-Efficient Information Routing and Coding

For our purpose, this section focuses on the inter-relationship between information processing and routing, rather than specific techniques for either routing or information processing and coding.

12.5.1 Naive Data-Centric Routing Protocols

The first batch of routing protocols that were adopted for NSSs are mostly based on protocols that were originally proposed for ad hoc networks, including extensions of AODV and DSR [15]. These routing protocols are still based on traditional address-centric peer-to-peer communication patterns instead of data-centric paradigm of NSS's. Hence, their applicability in the context of NSS's is quite limited.

Directed diffusion [27] is almost the first well-known protocol customized for information routing in data-centric paradigm. However, information processing is simply incorporated as a opportunistic by-product of routing in directed diffusion. While this might be sufficient for simple event monitoring applications where data volume is small, the lack of formal consideration of integrating information processing with routing makes the protocol suboptimal for applications with complex information processing and large data volume. Some other routing schemes, including geographic routing [43] and rumor routing [44] also fall into this category.

The LEACH protocol [45] adopts a two-tier clustering structure, where the information processing is performed at each cluster head and routing is simply divided into two stages: routing from sensor nodes to cluster heads and from cluster heads to the base station. This is, however, an empirical study that aims for energy-conservation by avoiding long distance communication but not really integrating information processing with routing.

12.5.2   Analysis for Entropy-Aware Routing

A formal analysis of the impact of data aggregation on routing in NSSs is presented by Krishnamachari et al. [5]. An intuitive theoretical bound is that in the case of perfect aggregation where every k pieces of information can be aggregated into a single piece of information, the amount of traffic to be routed can be reduced by a factor of at most k. This bound can be asymptotically achieved when all the source nodes are clustered in a small proximity that is far away from the sink node. Here, the value of k is usually referred to as the *aggregation factor* or *correlation factor* among data. However, to construct an optimal routing tree for minimizing the total amount of data transmission is NP-hard in general. The special case with perfect aggregation is exactly the Minimal Steiner Tree problem.

Along this line, the impact of k on two different routing schemes, the Shortest Path Tree and the Minimal Steiner Tree is investigated by Pattem et al. [7]. The performance metric is again the accumulated number of bits transmitted over each hop. Assuming that each source node generates one unit size of information, the study is motivated by two extreme cases. In the case when k is one, the correlation among source information is zero, which means there is no advantage for joint data compression. Hence, by selecting a shortest path from each source node to route its data to the sink node, the optimal solution is the Shortest Path Tree. In the case when k is infinity, the correlation among source information is so high that an arbitrary number of source information can be jointly compressed into one unit size of data. Therefore, exactly one unit size of data shall be routed on each edge of the tree, implying that the optimal solution is the Minimal Steiner Tree.

When k has a value in between the above two extreme cases, the optimal tree structure shall resemble a hybrid scheme of Shortest Path Tree and Minimal Steiner Tree. Since data correlation is usually high within a small area, a natural cluster-based scheme is to use a Minimal Steiner Tree structure within each cluster and a Shortest Path Tree to route compressed data from each cluster to the sink. While the optimal cluster size depends on the correlation factor, a surprising result is that in a grid-based scenario, a near-optimal cluster size can be analytically determined purely based on the network topology and is insensitive to the correlation factor.

12.5.3   Hybrid Routing Tree Algorithms

The idea of hybrid routing scheme in the previous section is conformed by several other results, under various assumptions to model the data correlation.

12.5.3.1   Shallow Light Tree

Cristescu et al. [8] assumes a simplified compression model, where the aggregation factor of a piece of information does not depend on the amount of side information, but only on its availability. Assume that each sensor node in the network generates one unit size of information. Whenever there is a side information transported to a source node, no matter the sources and size of this side information, the output of the source node after joint compression with the side information is a fixed value $\rho \in (0,1]$. Hence, the flow on the resulting routing tree can be separated into two classes — the flows on the path from all leaf nodes to the sink node equals one, while the flows on the path from all internal nodes to the sink node equals ρ.

To minimize the cost of a routing tree, we need to minimize the cost of routing information from all internal nodes to the sink, which prefers a Shortest Path Tree structure. On the other hand, we also need to minimize the cost of routing side information from a leaf node to the set of internal nodes that utilize this side information and the sink, which favors a Minimal Spanning Tree. Hence, the optimal solution lies between a Shortest Path Tree and a Minimal Spanning Tree. In fact, the Shallow Light Tree (STL) proposed by Bharat-Kumar and Jaffe [46] has the property that:

1. The total cost of the tree is bounded by a constant factor times the cost of a Minimal Spanning Tree.
2. The cost from any node in STL to the sink node is bounded by a constant factor times the shortest path between the node and the sink node.

This nice property can be used to prove that STL also provides a constant factor approximation of the considered problem.

12.5.3.2 Hierarchical Matching Tree

While STL provides an approximation for both Shortest Path Tree and Minimal Spanning Tree, the situation becomes different if the source nodes are a subset of the sensor nodes instead of all sensor nodes. In this case, we need to construct a tree structure that simultaneously approximates both Shortest Path Tree and Minimal Steiner Tree. Under a more general objective function that minimizes the sum of total cost of the tree and the accumulated length from each source node to the sink node where the cost and length can be defined based on two independent metrics, an approximation algorithms is proposed by Meyerson et al. [47] to achieve a $\log k$ performance bound, where k is the number of source nodes.

A simplified version of the algorithm in Reference 47 is used to solve the problem of transporting information from a set of source nodes to the sink node when the joint entropy of a set of source nodes is assumed to be a concave, but unknown function of the number of source nodes [6]. The network topology is assumed to be a complete graph with shortest path distance being the edge cost, if necessary. The algorithm constructs a hierarchical matching tree using an iterative method. In each iteration, a min-cost perfect matching is first performed, then a random node from each pair of nodes in the matching is removed from the network with half probability. The algorithm terminates when only one node is left in the network.

The above randomized algorithm provides a $\log k$ approximation to the optimal solution. It is noted that the assumption about concave data aggregation function essentially leads to an identical abstraction of the information routing problem with that of the single-source buy-at-bulk problem [48]. The key point is that the transmission cost spent on each edge is a concave function of the number of source nodes that use this edge to communicate to the sink.

12.5.3.3 Scale Free Aggregation Tree

Another randomized algorithm for information routing on a grid of sensor nodes is proposed by Enachescu et al. [39]. It is assumed that each sensor nodes has the readings of the physical world within l hops, referred to as the sensing area. Hence, the data redundancy among readings of two sensor nodes is modeled as the overlapped area between the sensing area of the two nodes, which essentially abstracts the effect of distance — a closer distance between two sensor nodes leads to more data redundancy. In this algorithm, the routing decision is simply based on the relative position of the node to the sink. Consider a sensor node horizontally x hops away from the sink, and vertically y hops away. After joint compression with the information received from all preceding nodes, the sensor nodes forwards its information toward the sink node horizontally with probability $x/(x + y)$ and vertically with probability $y/(x + y)$. The performance of such a randomized algorithm is proved to be a constant factor times the optimal cost, which is independent of l.

12.5.3.4 Routing with Tunable Compression

All the above works assume that the energy cost for (joint) data compression is negligible, which is reasonable for applications with small data volume. However, for computation-intensive applications such as the monitoring of complex systems and video surveillance, the energy for compressing large volume of data becomes comparable to communication energy. To avoid overpaying computation energy with maximum compression strategy, the concept of tunable compression is exploited by Yu et al. [41] for a balanced computation and communication energy cost.

Tunable compression basically explores the tradeoff between output size of data compression and the compression energy [49]. When tunable compression is considered, the inherent tradeoffs between Shortest Path Tree and Minimal Steiner Tree is intuitive. When computation cost is free and perfect data aggregation is possible, Minimal Steiner Tree is optimal. However, when computation cost is prohibitive or data correlation is zero, joint data compression becomes impossible and hence Shortest Path Tree is optimal. Again, a hybrid tree structure is desired to provide near-optimal performance in general cases. In fact, a poly-logarithmic approximation routing scheme can be developed based on probabilistic metric approximation [50,51].

12.5.4 Network Flow-Based Routing Algorithm

Moreover, some research efforts have investigated other routing substrates instead of a tree structure. Hong and Prasanna [40] proposed a distributed in-network processing algorithm that achieves maximal throughput under the assumption that the information processing is independent for all source nodes [40]. This assumption is particularly useful for decision based applications where every block of sensor readings is translated into a decision through proper signal processing algorithm and the size of the decision is assumed to be negligible compared with the size of sensor readings. The unique feature of this work is that all sensor nodes are eligible for processing the sensor readings from any other sensor nodes. Hence, the problem essentially requires a routing scheme to determine where the sensor readings shall be processed so that the total throughput of the system is maximized, subject to the limited processing and transmission capability and energy supply of sensor nodes.

Such a problem is solved after being converted to a maximal network flow problem. A completely decentralized algorithm is then proposed to find the optimal flow. A nice property of the algorithm is that it adapts to variations in the processing and transmission capability and energy supply of sensor nodes on-the-fly, without the need of re-synchronizing the system.

Some other results based on network flow model can be found in References 52 and 53.

12.5.5 From Information Coding Perspective

All the papers discussed in Sections 12.5.1 to 12.5.4 [5–8,27,39–41,43–45] assume that certain coding mechanisms are available to accomplish the data aggregation operation, hence the authors can focus on a relatively high level algorithm design or performance analysis. Nevertheless, there are also papers that try to understand the inter-relationship between information processing and routing from information coding perspective.

12.5.5.1 Joint Information Routing and Coding

Scaglione and Servetto [18] have considered tight coupling between information routing and coding for the problem of broadcast communication in an NSS subject to a given distortion constraint. The mission is to disseminate sensor readings at all sensor nodes throughout the network so that all sensor nodes have the same view of the entire sensing field. While such a problem is slightly different from the previously discussed scenarios with only a few sink nodes, it indicates the significance of collaborative information processing and routing from an information coding perspective.

It is first shown that without data aggregation, the whole network traffic for disseminating the information scales as $O(N \log N)$ (N being the number of sensor nodes), which is beyond the network capacity

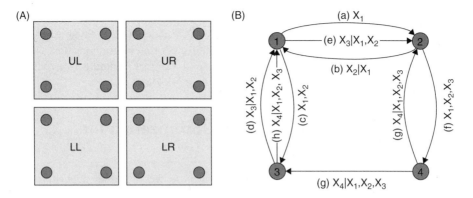

FIGURE 12.2 Illustration for joint information coding and routing. (A) Partition of the network into four blocks. (B) Packet exchanging within four nodes ($X_i \mid X_j$ is the entropy of X_i conditioning on X_j. The letter indicates the order of packet exchanging, that is, (a) is the first packet, (b) is the second packet), . . ., (h) is the eighth packet. Packets with the same letter are exchanged simultaneously.

with a scaling law of $O(\sqrt{N})$. This confirms the well-known result by Gupta and Kumar [54], which states that per-node throughput asymptotically vanishes as the network size grows to infinity. However, it is then argued that by exploiting the correlation among sensor readings during routing, it is possible to reduce the traffic to $O(\sqrt{N})$, and hence making it supportable by the network capacity.

The routing scheme is constructed using a divide-and-conquer algorithm. The network is divided into four partitions as shown in Figure 12.2(A). Assuming that all nodes in each partition have the complete view of the corresponding field, the information exchange among the four partitions can be illustrated with the case of four nodes (Figure 12.2[B]). Assuming that the readings at the four nodes are X_1, X_2, X_3, and X_4. After exchanging a total traffic of $3\mathcal{H}(X_1, X_2, X_3, X_4)$ bits in a delay of 8 time slots, every node has the complete information of the entire field. Moreover, it can be shown that under reasonable models of sensor readings, $O(\mathcal{H}(X_1, X_2, X_3, X_4)) \leq O(\sqrt{N})$, implying that there exists joint coding and routing schemes to perform the broadcast communication.

12.5.5.2 Scaling Laws of Transportation Traffic and Delay

The study in Reference 18 also reveals but without further exploration that there exist certain tradeoffs between the network traffic and time delay by exploiting various transmission scenarios other than the one shown in Figure 12.2(B). By considering the special case of a linear array of sensor nodes, a hierarchical joint coding and routing scheme can be used to achieve logarithmic scaling laws for both the transmission traffic and scheduling delay [42].

The scheme is constructed through two stages. In the first stage, the network is partitioned into groups with two sensor nodes. After information exchange within each group, adjacent groups are combined into a larger group within which, information exchanging between two representative nodes from the two original group is performed. After recursively performing group combination and information exchange for $\log N$ times, two nodes within the system have the sensor readings of the entire field while a hierarchical grouping structure is imposed on the network. Then in the second stage, the sensor readings are distributed in a top-down fashion of the hierarchical grouping structure, which is also performed in $\log N$ times.

12.5.5.3 Distributed Source Coding

When distributed source coding (e.g., Slepian–Wolf coding [20]) can be used for encoding the sensor readings, the problem of joint information routing with coding becomes trivial. This is because distributed source coding does not require explicit packet exchange among source nodes.

It is also argued by Cristescu et al. [8] that in this case, the problem of routing and coding can be decoupled. Moreover, the Shortest Path Tree turns out to be the optimal tree structure. Naturally, the optimal routing over this tree structure should have the property that the flow from the sensor node that is closest to the sink node must be the largest among the flow of all sensor nodes. Also, to ease the task of a global Slepian–Wolf coding, the authors proposed an approximation algorithm by grouping nodes into clusters and performing Slepian–Wolf coding within each cluster and Minimal Spanning Tree routing among clusters. In this case, the optimal clustering is conjectured to be NP-Hard.

12.6   A Broad View of Cross-Layer Optimization for Energy-Efficiency

We present this brief survey based on the system layers where the techniques are applied, including the hardware layer, physical layer, MAC layer, routing layer, and application layer. Note that although we have narrowed down our attention to only optimization techniques for energy-efficiency, this survey is by no means a complete list. Instead, we focus on the techniques that are listed in Table 12.1.

12.6.1   Hardware Layer

The most important hardware layer technique is the low-power circuit design for both CPU and radio modules, which has been addressed by a large body of literature [4,24]. Such low-power design gains energy-efficiency by: (1) developing dedicated low-power, low cost hardware modules for the expected low performance of sensor nodes, and (2) exploring tradeoffs between power consumption of the system and other performance metrics. The performance criteria for typical sensor nodes are around tens of MIPS for CPUs (e.g., 16 MIPS for the ATMEL ATmega128L processor [55] used in Berkeley Motes) and tens of Kbps for the radio modules, which are fairly low compared to the performance of usual PCs with commercial wireless LAN cards. This application-level requirement provides opportunities to design simple digital circuits, including digital signal processors (DSPs) and radio frequency (RF) circuits, with less power consumption.

We now discuss two important tradeoffs that have been explored at the hardware layer — the energy versus response time tradeoff and the energy versus delay tradeoff. The energy versus response time tradeoff is realized via shutting down the nodes in idle state [56,57], which is motivated by the well-known ACPI (Advanced Configuration and Power Interface) industry specification. Note that here we discuss the shutdown of the whole node; MAC layer radio sleep scheduling will be discussed in Section 12.6.3. The key points of the tradeoff lie in two aspects: (1) to select the appropriate shutdown mode based on the tradeoffs between shutdown duration and waking-up time/energy cost and (2) to determine the shutdown duration by exploring the tradeoffs between energy efficiency and possible event miss rate at application level. By converting the temporal event miss rate to the spatial percentage of coverage, the later tradeoff is also studied as the energy versus sensing coverage tradeoff [16,17].

Instead of reducing the operating duration of CPU by the shutdown technique, voltage scaling explores the energy versus delay tradeoff by running CPU at a lowered speed and longer operating duration with reduced supply voltage and operating frequency [36]. The key rationale is that the CPU power consumption is quadratically proportional to the supply voltage with delay being linearly inverse-proportional to the supply voltage, implying that the power-delay product increases with the supply voltage. Apparently, the energy versus delay tradeoff is meaningful for tasks with application-level real-time constraints, which are usually captured by setting a hard or soft deadline for each task. Various research efforts have proposed scheduling techniques for voltage scaling in uniprocessor systems [36,58–60] or multiprocessor systems [61–65].

A key question regarding the usefulness of the above CPU related techniques is the relative energy cost spent by CPUs compared to that of radio modules. One fact is that the energy cost to transmit one bit is typically around 500 to 1000 times greater than a single 32-bit computation [49,66]. Hence, for applications with simple functionality, striving for CPU energy-efficiency might not be worthwhile. However, we also envision the development of more advanced, computation-intensive NSSs applications within the near future. Moreover, it has been noted that for many high-end sensor nodes, the power consumption of CPU is around 30 to 50% of the total power consumption of the system [33], which motivates energy minimization for CPUs with complex applications.

12.6.2 Physical Layer

At the physical layer, energy-efficiency is often optimized using techniques such as power control, rate adaptation, and adaptive coding that have a direct impact on the signal strength and interference at receivers.[2] Such an impact eventually affects the network connectivity, topology, data transmission rate, and energy cost at the MAC layer, routing layer, and application layer. These complicated cross-layer effects make it difficult to isolate the tradeoffs involved with these techniques. Nevertheless, we can make a first-order classification that power control explores the tradeoff of energy versus connectivity and reliability, while rate adaptation and adaptive coding explore the tradeoff between energy and communication speed, or transmission delay.

The rationale behind these two tradeoffs can be explained using Shannon's law in wireless communication [67]. Consider an Additive White Gaussian Noise channel. This law states that the achievable communication rate is logarithmically proportional to the power at the receiver, which in turn is proportional to the transmission power at the receiver and decays with the transmission distance at a rate of d^α, where d is the radius and α is the pass loss exponent between 2 and 6. Hence, increment of either the communication radius or rate leads to increased transmission power. Following the principle of delivering just enough performance, we would like to decrease power while maintaining just enough communication radius and data rate.

Power control was originally proposed for single-hop multiuser systems like the cellular system to maintain a given level of signal-to-noise quality to compensate for fading effects, thermal noise, and more importantly, mutual interference in the shared radio spectrum [68,69]. In the context of NSSs where multihop communication prevails, power control has been mainly used for determining an appropriate communication radius, which can be either common for all sensor nodes [70] or not [37]. Many research efforts have proposed power control schemes to reduce the communication radius and hence the power consumption while achieving global network connectivity [37,70–74]. Also, the tradeoffs between energy and reliability through power control is studied [75].

Rate adaptation (sometimes also referred to as modulation scaling [76]) and adaptive coding were also originally proposed for cellular systems or local wireless networks with the goal of throughput optimization [77,78]. The use of these techniques for scheduling packet transmission over a given channel with delay constraint is studied in Reference 38, with an optimal off-line algorithm similar to the one in Reference 36. The problem is then extended to a star structure with multiple downstream links [79] or routing tree based structure [9].

Recently, many research efforts are trying to analyze and utilize the joint impact of these two techniques on energy-efficiency and throughput optimization, and many times together with MAC layer transmission scheduling and routing layer decisions [10,75,80–82]. As it has been realized that energy-efficiency depends on factors spanning multiple layers, this is becoming a promising and hot research direction.

12.6.3 MAC Layer

The way that MAC layer techniques affect the energy-efficiency is mainly through the adjustment of transmission scheduling and channel access. A common way to do that is via sleep scheduling [11,83–85] from long time scale perspective or time-division multiple access (TDMA) [12,86] from short time scale perspective. Similar to the shutdown technique of CPUs, sleep scheduling also explores the energy versus response time tradeoffs in wireless communication. During many studies, the response time is translated to network or application layer transmission delay or throughput.

---

[2]Power control is sometimes treated as a MAC layer technique since it affects the MAC layer topology.

The power aware multi-access protocol with signaling (PAMAS) protocol [83] is a simple incremental over the MACA protocol [87] by turning off the radios of nodes that cannot either transmit or receive given the current traffic in neighborhood. A more aggressive policy, S-MAC is proposed by Ye et al. [11], in which nodes determine their own sleep scheduling based on the sleep scheduling of neighboring nodes. To cope with the problem that the scheduling is predetermined in S-MAC, a more dynamic policy, T-MAC is proposed so that the scheduling of a node can be adapted on-the-fly based on transmission requests from neighbors [84]. While the above works are proposed for a general network topology, a scheduling policy dedicated to routing tree structure in NSSs is proposed in D-MAC [85]. The main advantage of D-MAC is that it facilitates a fully pipelined packet transmission over the routing tree by staggering the sleep scheduling of nodes.

Compared to the above adaptive sleep scheduling, TDMA provides a more strict, prespecified sleep scheduling. Tradeoffs between energy and delay as well as buffering size are studied in Reference 86. A novel performance metric of network diameter is studied by Lu et al. [12].

Another tradeoff explored by sleep scheduling is the energy versus topology tradeoffs, which in turn impacts concurrent transmission scheduling and channel access at the MAC layer and routing selection at the routing layer. Most of the research efforts along this line try to maintain a backbone style topology of the network such that the network remains connected with a minimum number of awake sensor nodes [14,15,88].

The Span protocol [14] uses a randomized method to elect so-called coordinator nodes to maintain a backbone connectivity with certain redundancy. The concept of virtual grid is proposed in GAF [15], the size of which is determined by the communication radius so that any nodes in two adjacent virtual grids can communicate with each other directly. The key point is then to ensure an active node in every virtual grid. In the STEM protocol [88], radios are proactivated by using either a paging channel with fixed duty cycle or a tone on a secondary channel, which provides extra means to explore the energy versus latency tradeoff.

12.6.4 Routing Layer

As previously mentioned, routing decision in NSSs is often jointly made with other factors, including connectivity determined by either power control at the physical layer or sleep scheduling at the MAC layer, and transmission scheduling at the MAC layer. Hence, many energy-efficient routing schemes are integrated as part of cross-layer optimization [10,75,80]. Besides the above fact, many routing schemes also explicitly consider details of performance metrics at lower layers. For example, the energy $\times$ delay metric is used to determine a data gathering substrate [89]; the expected number of transmissions (ETX) is used as a path metric for multihop transmission [90]; and the packet reception rate $\times$ distance metric is used to choose a forwarding node during routing [91]. In this context, various tradeoffs including energy versus transmission delay, number of hops, path length have been explored.

12.6.5 Application Layer

At the application layer, both energy-efficient signal processing algorithms [29,30] and adaptive-fidelity algorithms [92] have been studied for cross-layer optimization. The key idea is to trade the application-level information precision or accuracy for energy.

12.6.6 Putting It All Together

In Table 12.2, we illustrate the tradeoffs being explored by the aforementioned techniques. Note that it is often difficult, to clearly isolate different performance metrics involved in the tradeoffs. For example, transmission delay and reliability are closely related at both physical layer and routing layer, since transmission delay depends on both the time cost for each transmission and the expected number of re-transmissions, which is determined by reliability. Also, the radio sleep scheduling at MAC layer affects both transmission delay and throughput simultaneously. Hence, in many cases, it is necessary and helpful to understand the tradeoffs while taking multiple performance metrics into consideration.

TABLE 12.2 Examples of Cross-Layer Optimization Techniques and the Associated Tradeoffs

System layer	Techniques	Tradeoffs
Application	Energy-efficient signal processing	Energy versus information precision/accuracy
	Adaptive fidelity algorithms	Energy versus information precision/accuracy
	Joint routing and coding	Traffic versus delay
	Tunable compression	Energy versus output size
Routing	Energy-aware routing	Energy versus delay/reliability/path length
	Entropy-aware routing	Energy versus delay/routing complexity
MAC	Radio sleep scheduling	Energy versus delay/throughput/topology
Physical	Power control	Energy versus connectivity/topology/reliability
	Rate adaptation	Energy versus delay
	Adaptive coding	Energy versus delay/reliability
Hardware	Low-power circuit	Energy versus performance
	Node sleep scheduling	Energy versus response time/sensing coverage
	Voltage scaling	Energy versus delay

Moreover, the concrete interpretation of a single performance metric may vary across different levels. For example, delay at the application layer often refers to the end-to-end time duration for performing a specific task, such as, gathering information across the network. At the routing layer, delay usually refers to the time duration of transporting a packet over a path between two sensor nodes. At the physical layer, delay may refer to the time duration for packet transmission over a single link. However, link-wide delay at the physical layer and path-wide delay at the routing layer also affect system-wide delay at the application layer. Due to such an inherent relationship, we do not rigorously distinguish between these different interpretations.

Based on the table, we summarize two important issues in cross-layer optimization, which also refresh our motivation stated in Section 12.4.1.

First, it is worth noting that these optimization techniques are not independent. In fact, the behavior of certain techniques can change the optimization space and hence solution for other techniques. For example, the radio sleep scheduling at the MAC layer affects network topology, which in turn impacts the routing decision at routing layer. Also, the sleep scheduling affects channel access at MAC layer and hence signal interference at the physical layer, which is referenced while applying power control and rate adaptation techniques. Given such a complex inter-relationship, it is important to understand the impact of certain techniques across various stacks before applying it.

Second, one single performance metric observed by the users can be affected by techniques across different layers. For example, network topology is affected by both physical layer power control and MAC layer sleep scheduling. Also, application-level delay is co-determined by a series of techniques, including application layer joint routing and coding scheduling, routing layer decision, MAC layer sleep scheduling, physical layer packet transmission, and hardware layer CPU processing. If an application-level delay constraint is imposed by the user, the upfront question to explore the energy-delay at these different layers is how to break the application-level delay constraint into subconstraints for each individual layer. There is no easy answer for this question unless a cross-layer optimization technique can be developed to integrate the tradeoffs at different layers.

12.7 Future Directions

We envision the following list of future directions from perspectives spanning theoretical study, system design, and real implementation:

1. *Scaling laws for traffic, energy, and delay*: While certain efforts have been made to partially answer this question [18,19,42], this theoretical bound remains open. This bound is crucial to understand

the operating region or capacity of NSSs, including answers to questions such as "what is the highest frequency and what is the minimal energy cost for gathering information across the network?," which are fundamental to the design and implementation of any optimization techniques.

2. *Optimization for heterogeneous networks*: One of the trend in NSSs is to deploy heterogeneous sensor nodes in a mixed fashion. Sensor nodes with strong capability will serve as main platforms of signal processing and data routing, while sensor nodes with weak capability will serve as guarding nodes or sensing nodes. Intuitively, heterogeneous deployment is particularly suitable for cluster-based networks with each cluster centered around a strong sensor node. How to adapt existing optimization algorithms and techniques to handle such heterogeneous systems becomes an important and challenging research topic.

3. *Consideration for networks with mobile nodes*: Another trend in NSSs is the integration of mobile nodes into the traditionally static network [93]. The presence of mobile nodes will change the basic method for routing information across the network, implying a possible integration of existing results in mobile ad hoc networks together with coding techniques for correlated information source. Also, mobile nodes will affect the decomposition of system energy cost — a large portion of the system energy will be spent on moving the sensor nodes. This change will inevitably shift the research focus for energy-efficiency.

4. *Integration of and coordination among various techniques*: While most of the techniques discussed in this chapter have been developed with the goal of cross-layer optimization, it is hard to say that any of them has completely addressed the cross-layer effects of the proposed techniques. A comprehensive view is hence desired to assess the aggregated effects of these techniques, which is possible only after a systematic integration of various techniques.

5. *Formal and systematic design methodologies*: As we have stated before, the advantage of cross-layer optimization comes at the expense of increased complexity in design process. To cope with such a challenge as well as the increasingly application complexity, formal design methodologies become crucial. This requires the development of unified system models at various layers, which also provide systematic mechanisms for information sharing and coordinated optimization across layers. Intuitively, certain compromise between the design space and design complexity is expected.

6. *Validation in real-life systems*: The complicated inter-relationship among various optimization techniques can only be fully understood through experiments on real-life systems. However, a somewhat disappointing reality is that very few of the existing techniques have been validated through this step. We believe that the future of NSSs heavily depends on the advancements in real system implementation and commercially practical applications within these very few years.

References

[1] D. Estrin, R. Govindan, J. Heidemann, and S. Kumar, "Next century challenges: scalable coordination in sensor networks," in *Proceedings of the ACM/IEEE International Conference on Mobile Computing and Networking (MOBICOM)*, 1999, pp. 263–270.

[2] J. Hill, R. Szewczyk, A. Woo, S. Hollar, D. Culler, and K. Pister, "System architecture directions for networked sensors," in *Proceedings of the 9th International Conference on Architectural Support for Programming Languages and Operating Systems*, 2000.

[3] G.J. Pottie and W.J. Kaiser, "Wireless integrated network sensors," *Communications of the ACM*, vol. 43, pp. 551–558, May 2000.

[4] R. Min, M. Bhardwaj, S. Cho, A. Sinha, E. Shih, A. Wang, and A.P. Chandrakasan, "Low-power wireless sensor networks," in *Proceedings of the 14th International Conference on VLSI Design*, 2001, pp. 205–210.

[5] B. Krishnamachari, D. Estrin, and S. Wicker, "The impact of data aggregation in wireless sensor networks," in *Proceedings of the International Workshop on Distributed Event-Based Systems*, 2002.

[6] A. Goel and D. Estrin, "Simultaneous optimization for concave costs: single sink aggregation or single source buy-at-bulk," in *Proceedings of the ACM-SIAM Symposium on Discrete Algorithms*, 2003.

[7] S. Pattem, B. Krishnamachari, and R. Govindan, "The impact of spatial correlation on routing with compression in wireless sensor networks," in *Proceedings of the ACM/IEEE International Symposium on Information Processing in Sensor Networks*, 2004.

[8] R. Cristescu, B. Beferull-Lozano, and M. Vetterli, "On network correlated data gathering," in *Proceedings of the IEEE InfoCom*, 2004.

[9] Y. Yu, B. Krishnamachari, and V.K. Prasanna, "Energy-latency tradeoffs for data gathering in wireless sensor networks," in *Proceedings of the IEEE InfoCom*, Mar. 2004.

[10] R.L. Curz and A.V. Santhanam, "Optimal routing link scheduling and power control in multi-hop wireless networks," in *Proceedings of the IEEE InfoCom*, Apr. 2003.

[11] W. Ye, J. Heidemann, and D. Estrin, "An energy-efficient MAC protocol for wireless sensor networks," in *Proceedings of the IEEE InfoCom*, Mar. 2005.

[12] G. Lu, N. Sadagopan, B. Krishnamachari, and A. Goel, "Delay efficient sleep scheduling in wireless sensor networks," in *Proceedings of the IEEE InfoCom*, Mar. 2005.

[13] Y. Xu, J. Heidemann, and D. Estrin, "Adaptive energy-conserving routing for multihop ad hoc networks," University of Southern California/Information Sciences Institute, Technical Report 527, Oct. 2000.

[14] B. Chen, K. Jamieson, H. Balakrishnan, and R. Morris, "Span: an energy-efficient coordination algorithm for topology maintenance in ad hoc wireless networks," in *Proceedings of the ACM/IEEE International Conference on Mobile Computing and Networking (MOBICOM)*, 2001, pp. 85–96.

[15] Y. Xu, J. Heidemamij, and D. Estrin, "Geography-informed energy conservation for ad hoc routing," in *Proceedings of the ACM/IEEE International Conference on Mobile Computing and Networking (MO-BICOM)*, July 2001.

[16] D. Tian and N.D. Georganas, "A coverage-preserving node scheduling scheme for large wireless sensor networks," in *Proceedings of the ACM International Workshop on Wireless Sensor Networks and Applications (WSNA)*, Sep. 2002.

[17] Z. Abrams, A. Goel, and S. Plotkin, "Set k-cover algorithms for energy efficient monitoring in wireless sensor networks," in *Proceedings of the ACM/IEEE International Symposium on Information Processing in Sensor Networks*, Apr. 2004.

[18] A. Scaglione and S.D. Servetto, "On the interdependence of routing and data compression in multi-hop sensor networks," in *Proceedings of the ACM/IEEE International Conference on Mobile Computing and Networking (MOBICOM)*, Sep. 2002.

[19] D. Marco, E.J. Duarte-Melo, M. Liu, and D.L. Neuhoff, "On the many-to-one transport capacity of a dense wireless sensor network and the compressibility of its data," in *Proceedings of the ACM/IEEE International Symposium on Information Processing in Sensor Networks*, 2003, pp. 1–16.

[20] D. Slepian and J. Wolf, "Noiseless coding of correlated information sources," *IEEE Transactions on Information Theory*, vol. 19, no. T-19, pp. 471–480, 1973.

[21] A. Kashyap, L.A. Lastras-Montano, C. Xia and Z. Liu, Distributed source coding in dense sensor networks. Data compression conference pp. 13–22, 2005.

[22] C. Tang and C.S. Raghavendra, Bitplane coding for correlation exploitation in wireless sensor networks, ICC May 2005.

[23] R. Ahlswede, N. Cai, S.-Y.R. Li, and R.W. Yeung, "Network information flow," *IEEE Transactions on Information Theory*, vol. 46, pp. 1204–1216, 2000.

[24] J. Rabaey, J. Ammer, T. Karalar, S. Li, B. Otis, M. Sheets, and T. Tuan, "PicoRadios for wireless sensor networks: the next challenge in ultra-low power design," in *Proceedings of the International Solid-State Circuits Conference*, Feb. 2002.

[25] W. Heinzelman, A.P. Chandrakasan, and H. Balakrishnan, "An application specific protocol architecture for wireless microsensor networks," *IEEE Transactions on Wireless Communications*, vol. 1, pp. 660–670, 2002.

[26] M. Singh and V.K. Prasanna, "A hierarchical model for distributed collaborative computation in wireless sensor networks," in *Proceedings of the 5th Workshop on Advances in Parallel and Distributed Computational Models*, Apr. 2003.

[27] C. Intanagonwiwat, R. Govindan, and D. Estrin, "Directed diffusion: a scalable and robust communication paradigm for sensor networks," in *Proceedings of the ACM/IEEE International Conference on Mobile Computing and Networking (MOBICOM)*, 2000.

[28] D. Ganesan, R. Govindan, S. Shenker, and D. Estrin, "Highly-resilient, energy-efficient multi-path routing in wireless sensor networks," in *Proceedings of the ACM/IEEE International Conference on Mobile Computing and Networking (MOBICOM)*, 2001, pp. 251–254.

[29] S. Aldosari and J. Moura, "Fusion in sensor networks with communication constraints," in *Proceedings of the ACM/IEEE International Symposium on Information Processing in Sensor Networks*, Apr. 2004.

[30] G. Barriac, R. Mudumbai, and U. Madhow, "Distributed beamforming for information transfer in sensor networks," in *Proceedings of the ACM/IEEE International Symposium on Information Processing in Sensor Networks*, Apr. 2004.

[31] A. Sinha and A.P. Chandrakasan, "Operating system and algorithmic techniques for energy scalable wireless sensor networks," in *Proceedings of the 2nd International Conference on Mobile Data Management*, Jan. 2001.

[32] M. Singh and V.K. Prasanna, "Supporting topographic queries in a class of networked sensor systems," in *Proceedings of the Workshop on Sensor Networks and Systems for Pervasive Computing (PerSeNS)*, Mar. 2005.

[33] V. Raghunathan, C. Schurgers, S. Park, and M.B. Srivastava, "Energy-aware wireless microsensor networks," *IEEE Signal Processing Magazine*, vol. 19, pp. 40–50, Mar. 2002.

[34] S. Shakkottai, T.S. Rappaport, and P.C. Carlsson, "Cross-layer design for wireless networks," *IEEE Wireless Communication Magazine*, vol. 41, pp. 74–80, Oct. 2003.

[35] Y. Zhang and L. Cheng, "Cross-layer optimization for sensor networks," in *Proceedings of the New York Metro Area Networking Workshop*, Sep. 2003.

[36] F. Yao, A. Demers, and S. Shenker, "A scheduling model for reduced CPU energy," in *Proceedings of the Annual Symposium on Foundations of Computer Science (FOCS)*, 1995, pp. 374–382.

[37] V. Rodoplu and T.H Meng, "Minimum energy mobile wireless networks," *IEEE Journal of Selected Areas in Communication (JSAC)*, vol. 8, pp. 1333–1344, 1999.

[38] B. Prabhakar, E. Uysal-Biyikoglu, and A.E. Gamal, "Energy-efficient transmission over a wireless link via lazy packet scheduling," in *Proceedings of the IEEE InfoCom*, 2001.

[39] M. Enachescu, A. Goel, R. Govindan, and R. Motwani, "Scale free aggregation in sensor networks," in *Proceedings of the 1st International Workshop on Algorithmic Aspects of Wireless Sensor Networks (Algosensors)*, 2004.

[40] B. Hong and V.K. Prasanna, "Optimizing a class of in-network processing applications in networked sensor systems," in *Proceedings of the 1st IEEE International Conference on Mobile Ad Hoc and Sensor Systems (MASS)*, Oct. 2004.

[41] Y. Yu, B. Krishnamachari, and V.K. Prasanna, "Energy-efficient data gathering with tunable compression in wireless sensor networks," University of Southern California, Technical Report CENG-2004-15, 2004. http://halcyon.usc.edu/~yangyu/data/TR_CENG200415.pdf

[42] T. ElBatt, "On the scalability of hierarchical cooperation for dense sensor networks," in *Proceedings of the ACM/IEEE International Symposium on Information Processing in Sensor Networks*, Apr. 2003.

[43] B. Karp and H.T. Kung, "GPSR: Greedy perimeter stateless routing for wireless networks," in *Proceedings of the ACM/IEEE International Conference on Mobile Computing and Networking (MOBICOM)*, Aug. 2000.

[44] D. Bradinsky and D. Estrin, "Rumor routing algorithm for sensor networks," in *Proceedings of the ACM International Workshop on Wireless Sensor Networks and Applications (WSNA)*, Sep. 2002.

[45] W. Heinzelman, J. Kulik, and H. Balakrishnan, "Adaptive protocols for information dissemination in wireless sensor networks," in *Proceedings of the ACM/IEEE International Conference on Mobile Computing and Networking (MOBICOM)*, 1999.

[46] K. Bharat-Kumar and J. Jaffe, "Routing to multiple destination in computer networks," *IEEE Transactions on Computers*, pp. 343–351, 1983.

[47] A. Meyerson, K. Munagala, and S. Plotkin, "Cost-distance: two metric network design," in *Proceedings of the Annual Symposium on Foundations of Computer Science (FOCS)*, 2000.

[48] B. Awerbuch and Y. Azar, "Buy-at-bulk network design," in *Proceedings of the Annual Symposium on Foundations of Computer Science (FOCS)*, 1997.

[49] K. Barr and K. Asanović, "Energy aware lossless data compression," in *Proceedings of the 1st International Conference on Mobile Systems, Applications and Services*, May 2003.

[50] Y. Bartal, "Probabilistic approximations of metric spaces and its algorithmic applications," in *Proceedings of the Annual Symposium on Foundations of Computer Science (FOCS)*, 1997.

[51] J. Fakcheroenphol, S. Rao, and K. Talwar, "A tight bound on approximating arbitrary metrics by tree metrics," in *Proceedings of the Annual ACM Symposium on Theory of Computing (STOC)*, 2003.

[52] K. Kalpakis, K. Dasgupta, and P. Namjoshi, "Maximum lifetime data gathering and aggregation in wireless sensor networks," in *Proceedings of the IEEE International Conference on Networking (NETWORKS '02)*, pp. 685–696, Aug. 2002.

[53] Y.T. Hou, Y. Shi, and J. Pan, "A lifetime-aware flow routing algorithm for energy-constrained wireless sensor networks," in *Proceedings of the IEEE MILCOM*, 2003.

[54] P. Gupta and P.R. Kumar, "The capacity of wireless networks," *IEEE Transactions on Information Theory*, vol. 46, pp. 388–404, March 2000.

[55] "Atmel atmega1281 datasheet." http://www.atmel.com/dyn/resources/prod-documents/2467S.pdf

[56] M.B. Srivastava, A.P. Chandrakasan, and R.W. Brodersen, "Predictive system shutdown and other architectural techniques for energy efficient programmable computation," *IEEE Transactions on VLSI Systems*, vol. 4, pp. 42–55, Mar. 1996.

[57] A. Sinha and A.P. Chandrakasan, "Dynamic power management in wireless sensor networks," *IEEE Design and Test of Computers*, vol. 18, pp. 62–74, 2001.

[58] I. Hong, G. Qu, M. Potkonjak, and M.B. Srivastava, "Synthesis techniques for low-power hard real-time systems on variable voltage processors," in *Proceedings of the IEEE Real-Time Systems Symposium (RTSS)*, Dec. 1998.

[59] Y. Shin, K. Choi, and T. Sakurai, "Power optimization of real-time embedded systems on variable speed processors," in *Proceedings of the IEEE/ACM International Conference Computer-Aided Design*, 2000, pp. 365–368.

[60] H. Aydin, R. Melhem, D. Mossé, and P.M. Alvarez, "Determining optimal processor speeds for periodic real-time tasks with different power characteristics," in *Proceedings of the 13th Euromicro Conference on Real-Time Systems*, June 2001.

[61] F. Gruian and K. Kuchcinski, "LEneS: Task scheduling for low-energy systems using variable supply voltage processors," in *Proceedings of the Design Automation Conference (DAC)*, 2001, pp. 449–455.

[62] D. Zhu, R. Melhem, and B. Childers, "Scheduling with dynamic voltage/speed adjustment using slack reclamation in multi-processor real-time systems," in *Proceedings of the IEEE Real-Time Systems Symposium (RTSS)*, Dec. 2001.

[63] J. Luo and N.K. Jha, "Static and dynamic variable voltage scheduling algorithms for real-time heterogeneous distributed embedded systems," in *Proceedings of the VLSI Design*, Jan. 2002.

[64] Y. Zhang, X. Hu, and D.Z. Chen, "Task scheduling and voltage selection for energy minimization," in *Proceedings of the Design Automation Conference (DAC)*, 2002.

[65] Y. Yu and V.K. Prasanna, "Energy-balanced task allocation for collaborative processing in wireless sensor networks," in *Proceedings of the ACM Mobile Networks and Applications (MONET)*, vol. 10, pp. 115–131, 2005 (special issue on Algorithmic Solutions for Wireless, Mobile, Ad Hoc and Sensor Networks).

[66] M. Singh and V.K. Prasanna, "System level energy tradeoffs for collaborative computation in wireless networks," in *Proceedings of the IEEE IMPACCT Workshop*, May 2002.

[67] R.L. Cover and J.A. Thomas, *Elements of Information Theory*, John-Wiley and Sons Inc., 1991.

[68] C. Huang and R.D. Yates, "Rate of convergence for minimum power assignment algorithm in cellular radio systems," *Wireless Networks*, vol. 4, pp. 223–231, 1998.

[69] S. Hanly and D. Tse, "Power control and capacity of spread-spectrum wireless networks," *Automatica*, vol. 35, pp. 1987–2012, 1999.

[70] S. Narayanaswamy, V. Kawadia, R.S. Sreenivas, and P.R. Kumar, "Power control in ad-hoc networks: theory, architecture, algorithm and implementation of the COMPOW protocol," in *Proceedings of the European Wireless Conference — Next Generation Wireless Networks: Technologies, Protocols, Services and Applications*, Feb. 2002.

[71] R. Ramanathan and Rosales-Hail, "Topology control of multihop wireless networks using transmit power adjustment," in *Proceedings of the IEEE InfoCom*, 2000, pp. 404–413.

[72] E.M. Belding-Royer, P.M. Melliar-Smith, and L.E. Moser, "An analysis of the optimal node density for ad hoc mobile networks," in *Proceedings of the IEEE International Conference on Communications (ICC)*, 2001.

[73] R. Wattenhofer, L. Li, B. Bahl, and Y.M. Wang, "Distributed topology control for power efficient operation in multihop wireless ad hoc networks," in *Proceedings of the IEEE InfoCom*, 2001.

[74] M. Kubisch, H. Karl, A. Wolisz, L.C. Zhong, and J. Rabaey, "Distributed algorithms for transmission power control in wireless sensor networks," in *Proceedings of the IEEE Wireless Communications and Networking Conference (WCNC)*, Mar. 2003, pp. 16–20.

[75] B. Krishnamachari, Y. Mourtada, and S. Wicker, "The energy-robustness tradeoff for routing in wireless sensor networks," in *Proceedings of the IEEE International Conference on Communications (ICC)*, May 2003.

[76] C. Schurgers, O. Aberhorne, and M.B. Srivastava, "Modulation scaling for energy-aware communication systems," in *Proceedings of the International Symposium on Low Power Electronics and Design (ISLPED)*, 2001, pp. 96–99.

[77] T. Ue, S. Sampei, N. Morinaga, and K. Hamaguchi, "Symbol rate and modulation level-controlled adaptive modulation/TDMA/TDD system for high-bit rate wireless data transmission," *IEEE Transactions on Vehicular Technology*, vol. 47, pp. 1134–1147, Nov. 1998.

[78] G. Holland, N. Vaidya, and P. Bahl, "A rate-adaptive MAC protocol for multi-hop wireless networks," in *Proceedings of the ACM/IEEE International Conference on Mobile Computing and Networking (MOBICOM)*, 2001.

[79] A.E. Gamal, C. Nair, B. Prabhakar, E. Uysal-Biyikoglu, and S. Zahedi, "Energy-efficient scheduling of packet transmissions over wireless networks," in *Proceedings of the IEEE InfoCom*, 2002.

[80] R. Bhatia and M. Kodialam, "On power efficient communication over multi-hop wireless networks: joint routing, scheduling and power control," in *Proceedings of the IEEE InfoCom*, Mar. 2005.

[81] T. ElBatt and A. Ephremides, "Joint scheduling and power control for wireless ad hoc networks," *IEEE Transactions on Wireless Communications*, vol. 3, 2004.

[82] U.C. Kozat, I. Koutsopoulos, and L. Tassiulas, "A framework for cross-layer design of energy-efficient communication with QoS provisioning in multi-hop wireless networks," in *Proceedings of the IEEE InfoCom*, Mar. 2004.

[83] C.S. Raghavendra and S. Singh, "PAMAS — power aware multi-access protocol with signaling for ad hoc networks," *ACM SIGCOMM Computer Communication Review*, vol. 28, pp. 5–26, July 1998.

[84] T. van Dam and K. Langendoen, "An adaptive energy-efficient MAC protocol for wireless sensor networks," in *Proceedings of the ACM SenSys*, Nov. 2003.

[85] B.K.G. Lu and C. Raghavendra, "An adaptive energy-efficient and low-latency MAC for data gathering in sensor networks," in *Proceedings of the International Workshop on Algorithms for Wireless, Mobile, Ad Hoc and Sensor Networks*, Apr. 2004.

[86] K. Arisha, M. Youssef, and M. Younis, "Energy-aware TDMA-based MAC for sensor networks," in *Proceedings of the IEEE Workshop on Integrated Management of Power Aware Communications, Computing, and Networking (IMPACCT)*, May 2002.

[87] P. Karn, "MACA: a new channel access method for packet radio," in *Proceedings of the ARRL/CRRL Amateur Radio Computer Networking Conference*, Sep. 1990.

[88] C. Schurgers, V. Tsiatsis, S. Ganeriwal, and M.B. Srivastava, "Optimizing sensor networks in the energy-latency-density design space," *IEEE Transactions on Mobile Computing*, vol. 1, pp. 70–80, Jan. 2002.

[89] S. Lindsey, C.S. Raghavendra, and K. Sivalingam, "Data gathering algorithms in sensor networks using energy metrics," *IEEE Transactions on Parallel and Distributed Systems*, vol. 13, pp. 924–935, Sep. 2002.

[90] D.S.J. De Couto, D. Aguayo, J. Bicket, and R. Morris, "A high-throughput path metric for multi-hop wireless routing," in *Proceedings of the ACM/IEEE International Conference on Mobile Computing and Networking (MOBICOM)*, Sep. 2003.

[91] K. Seada, M. Zuniga, B. Krishnamachari, and A. Helmy, "Energy-efficient forwarding strategies for geographic routing in lossy wireless sensor networks," in *Proceedings of the ACM SenSys*, Nov. 2004.

[92] A. Sinha, A. Wang, and A. Chandrakasan, "Algorithmic transforms for efficient energy scalable computation," in *Proceedings of the IEEE International Symposium on Low Power Electronics and Design*, Aug. 2000.

[93] A. Howard, M.J. Mataricié, and G.S. Sukhatme, "Mobile sensor network deployment using potential fields: a distributed, scalable solution to the area coverage problem," in *Proceedings of the 6th International Symposium on Distributed Autonomous Robotics Systems*, June 2002.

13

Modeling Cellular Networks for Mobile Telephony Services

13.1 The Basic Model **13**-289
13.2 Modeling Mobility......................................**13**-291
    Arrival Process • Channel Holding Time • The Model •
    Performance Metrics
13.3 Channel Reservation....................................**13**-298
13.4 Model Extensions **13**-303
    Classes of Users • Services with Different Bandwidth • The
    Finite Population Case
13.5 A Model for Generally Distributed Dwell Times
    and Call Durations.....................................**13**-310
13.6 A Look at the Literature **13**-314
13.7 Conclusions and Outlook **13**-315
References **13**-316

Marco Ajmone Marsan
Michela Meo

In the last two decades, the advent of mobile telephony has changed the way we live and work, and has had an enormous impact on the telecommunications market, influencing the whole Information and Communications Technologies (ICT) sector.

The two most peculiar characteristics of mobile telephony lie in the cellular approach for the radio coverage of the area in which service is offered, and in the algorithms to support mobility. The cellular radio coverage allows the reuse of radio resources, thus increasing the size of the population that can be served; the support of mobility drastically expands the usage and the appeal of telephony services.

In a mobile telephony network, the area where service is offered is divided in small regions called *cells* (i.e., the origin of the term *cellular telephony*); each cell is served by a different *base station*, which provides the radio resources to carry the telephone traffic of the users in the cell. Depending on the considered cellular system, radio resources are shared between users according to different schemes, based on frequency, time, or code division, or, more frequently, a mix of these techniques. Regardless of the adopted technique, it is possible to define the elementary unit of radio resource, the *radio channel*, as the one that implements the *wireless local loop*, allowing the establishment of a circuit for a telephone call.

The fact that radio channels can be shared among users in a cell is quite a significant system improvement with respect to fixed telephony. Indeed, this drastically increases the local loop utilization, thus eliminating one of the main inefficiencies of traditional telephone networks.

Mobility implies that users can roam in the service area while accessing the telephony service. An active user, that is, a user that has established a telephony connection, and that roams from a cell to another, must execute a *handover* procedure, transferring the call from the radio channel in the old cell to a channel in the new cell, without interrupting the call.

Since a mobile telephony network is almost invariably far too complex to be analyzed and studied as a whole, its design and planning are typically decomposed into two tasks. The first task consists in finding a feasible allocation of the available radio channels to cells; this task is made difficult by the limited total number of radio channels and the many physical constraints. In the second task, cells are considered one by one. By taking into account the users' needs and behavior, the second task computes the number of channels to be activated in each cell, so as to obtain acceptable performance while providing the desired quality of service (QoS) to users.

A complete design and planning procedure may require several iterations between these two tasks. Thus, the models for the cell analysis (the second task mentioned above) should be as simple as possible, so that they can be efficiently solved and possibly introduced in complex tools for mobile telephony network design and planning. The focus of this chapter is on the development of such models.

For the development of models of cells, the following assumptions will be made:

- Radio channels are exclusively employed by one user at a time for the whole duration of the service.
- The dynamics, which drive the users behavior are:
  - The generation of new service requests.
  - The termination of a service in progress.
  - The users' movements in the service area. Movements translate into the generation of incoming handover requests (an active user enters the considered cell from a neighboring cell) and outgoing handover requests (a user with a call in progress moves from the considered cell to a neighboring cell).

Besides being as simple as possible, the cell models should also be reliable enough to accurately predict the behavior of the cell. In particular, models should provide answers to two classes of questions:

- *Performance evaluation.* Given the characteristics of the users' behavior and the number of channels in the cell, what is the performance of the cell? Among the considered performance metrics are: (i) the average number of active calls within a cell, which is an indirect metric of the revenues generated by the installed equipment; (ii) the probability that a new call cannot be established due to the lack of free radio channels, which must be kept small, specially if multiple operators offer mobile telephony services in the same area; (iii) the probability that a call in progress is dropped, which must be kept very low, in order to avoid user dissatisfaction.
- *Design.* Given the characteristics of the users' behavior and given some design objectives, how many radio channels should be activated in the cell in order to meet the design objectives? Typical design objectives are the guarantee that the new call blocking probability and the call dropping probability are below some specified target value.

In the next sections we discuss analytical models for performance evaluation and design of cells in a mobile telephony network. The first presented model is named the *basic model*. This model is extremely simple, especially for what concerns the description of the users' mobility, but it allows us to introduce the general modeling methodology and explain how such a model can be used for the cell performance evaluation and design. A more accurate description of the users' mobility is included in the model presented in Section 13.2. By allowing the investigation of the impact of mobility on the cell performance, this model points to the need for enhancements in the channel allocation policy. An enhanced allocation policy is then discussed in Section 13.3, that considers the case of channel reservation to incoming handovers. Further extensions of the cell model are described in Sections 13.4 and 13.5.

13.1 The Basic Model

The basic model for the performance analysis of a cell in a mobile telephony cellular network is based on the M/G/C/0 queue, which has already been used to study and dimension telephone systems for almost a century. Customers of the queue represent telephone calls: the arrival of a new customer corresponds to a request of a telephone call between two users, one of which resides in the cell under analysis. The request can either refer to a new call to be activated, or to an incoming handover request, that is, to a call that is already active in a neighboring cell and needs to be transfered to the cell under analysis. Servers represent the traffic channels available in the cell; usually, one traffic channel is requested to set up a telephone call. The customer's service time is the interval during which the telephone call uses the channel of the considered cell, and is usually referred to as the *channel holding time*. The lack of a waiting line in the queue is due to the typical policy adopted in telephony: a call is set up as long as resources are available; and it is rejected, or *blocked*, if no channel is free.

The following assumptions and notation will be adopted:

- Customers arrive at the queue according to a Poisson process with parameter λ.
- The customer service time S is distributed according to $F_S(t)$, with mean value $E[S]$.
- The queue load, or traffic intensity, is $\rho = \lambda E[S]$.
- The probability that the number of customers in the queue is equal to i is denoted by $\pi(i)$.
- The mean value of the number of customers in the queue is $E[A]$.
- The probability that a call is blocked is denoted by B.

The Poisson assumption for incoming traffic is the one that has been traditionally used for dimensioning telephone networks. While it is reasonable to assume that the collective behavior of the large population of users of the fixed telephony service can be accurately described by a Poisson process, the much smaller number of users in a cell may in some cases reduce the accuracy of this assumption. In addition, care is needed in the selection of the Poisson process parameter. Typically, the average peak hour traffic intensity was used for the design of fixed telephone networks; the rationale behind this choice is that a conservative design leading to some network over-dimensioning is preferable to the service deterioration under critical traffic conditions. When dimensioning cellular networks, however, the choice becomes more difficult. Indeed, one of the most critical aspects of cellular networks is the limited availability of radio channels, which makes the choice of over-dimensioning unfeasible. Moreover, the natural variability of telephone traffic is amplified by the user mobility: for example, the traffic intensity moves together with users in their commuting to work places in the morning, creating traffic peaks at rush hours at railway and public transportation stations; also, very high traffic peaks can be generated in correspondence of sport matches or rock concerts. Nonetheless, the Poisson assumption is the one that is normally adopted in most of the analytical models proposed in the literature for cellular networks; see Section 13.6 for a discussion of previous literature.

The choice of the distribution of the service times is another critical aspect of the model development, since it jointly accounts for the user mobility and for the duration of the service. As discussed in Section 13.6, many authors derived expressions or approximations for the service time distribution, based on different assumptions for the user behavior, and for the structure of the cellular network.

The solution of the M/G/C/0 queue provides the following expressions:

$$\pi(i) = \frac{\rho^i / i!}{\sum_{j=0}^{C} \rho^j / j!} \tag{13.1}$$

$$B = \pi(C) = \frac{\rho^C / C!}{\sum_{j=0}^{C} \rho^j / j!} \tag{13.2}$$

Equation (13.2) is the well-known Erlang-B formula and is often denoted by $B(\rho, C)$.

Procedure 1. Erlang-B

```
/* Input parameters:
     load: rho
     number of channels: C */
/* Output values:
     probabilities that i calls are active: vector pi
     blocking probability: B */

pi[0]=1; /* p is a vector with C+1 elements */
total=pi[0]; /* normalization constant to be computed */
for(i=1;i<=C;i++){
/* loop for the computation of the probabilities π(i)'s to be normalized */
    pi[i]=pi[i-1]*rho/i;
    total+=pi[i];
}
for(i=0;i<=C;i++) /* loop to normalize the vector p */
    pi[i]=pi[i]/total;
B=pi[C]; /* blocking probability */
```

FIGURE 13.1   Procedure for the computation of the Erlang-B formula.

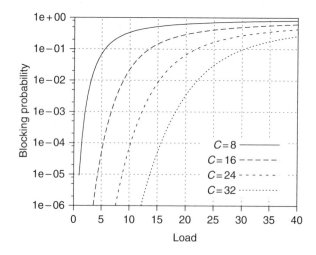

FIGURE 13.2   Basic model: blocking probability B versus the traffic load ρ, for different values of the number of available channels C.

Given a value of the load ρ and of the number of available channels C, the performance evaluation of the cell is directly obtained by the computation of (13.1) and (13.2). Equation (13.2), in particular, is a measure of the quality of service received by users. When B is large, a user's request undergoes a large blocking probability: the user cannot access the telephony service, and is thus dissatisfied. The procedure reported in Figure 13.1 provides guidelines for the efficient solution of the model.

Besides being efficient, the derivation of the term $\pi(i)$ from $\pi(i-1)$ in the algorithm of Figure 13.1 avoids the numerical problems, which may arise from the direct computation of factorials and powers.

The blocking probability is plotted in Figure 13.2 versus the cell traffic load, for different values of the number of available channels C. Clearly, the blocking probability decreases for increasing number of

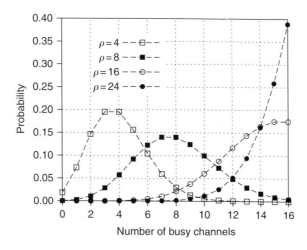

FIGURE 13.3 Basic model: distribution of the number of busy channels for different values of the traffic load, with $C = 16$.

channels, and increases with the traffic load. From the point of view of the system designer, this means that, in order to provide good QoS to users (i.e., low values of the blocking probability) a large investment is needed (i.e., the cell must be equipped with a large number of channels).

The distribution of the number of busy channels for $C = 16$ and different values of the traffic load is reported in Figure 13.3. Large values of the load make the mass of the probability distribution move toward large values, and the average number of busy channels $E[A]$ increases. This means that the radio resources are well exploited, at the cost of a worse service provided to users.

As already mentioned, the typical cell design problem consists in the computation of the minimum number of channels C which is necessary to guarantee that the blocking probability is smaller than a target value B^*, under given traffic conditions. The solution of this problem requires the inversion of the Erlang-B formula in (13.2), so that C can be expressed as a function of ρ and B. Unfortunately, no closed-form formula exists for this, and a numerical solution is needed. A procedure for the solution is shown in Figure 13.4.

In order to be efficient, the algorithm is based on the iterative computation of B for growing values of C. Starting from $B(\rho, 0) = 1$, $B(\rho, k)$ is computed from $B(\rho, k - 1)$. The procedure stops as soon as the blocking probability becomes smaller than the target B^*, and the corresponding value of k is the desired number of channels.

Figure 13.5 shows the results for different values of the constraint B^* versus increasing values of the load. The smaller the value of B^*, the larger the number of required channels: the increase of the number of required channels is the cost to be estimated in order to improve the quality of the provided service. For the same cases, the actual blocking probability perceived by users is plotted in Figure 13.6. As expected, the constraint on B is always satisfied; the oscillating behavior of the curves is due to the discretization on the number of radio channels, which must be an integer.

13.2 Modeling Mobility

Since mobility is a fundamental aspect of cellular systems, the inclusion of mobility in the models is a crucial issue. In the basic model that we presented in the previous section, mobility is not described in detail: no distinction is made between the generation of a new call and the arrival of an incoming handover, and the possibility that users roam out of the considered cell is implicitly accounted for only by the distribution of the channel holding time. In order to explicitly describe mobility, some enhancements to the basic model are necessary.

Procedure 2. Inverse of the Erlang-B

```
/* Input parameters:
     load: rho
     target blocking probability: Btarget */
/* Output value:
     number of channels: C */

C=0;
/* To be computed */
Num=Den=1; /* Numerator and denominator of the expression for B */
B=1; /* Initial value: blocking probability when C=0 */
while(B>Btarget){
   C+=1;
   Num=Num*rho/C; /* Numerator of the expression for B, for the new value of C */
   Den=Den+Num; /* Denominator of the expression for B, for the new value of C */
   B=Num/Den;
}
```

FIGURE 13.4   Procedure for the cell design through the basic model.

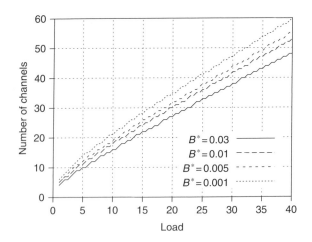

FIGURE 13.5   Basic model: number of channels needed to guarantee blocking probability smaller than the target B^* under a given traffic load.

13.2.1   Arrival Process

A distinction is introduced in the model between new calls and incoming handover requests, based on the following assumptions:

- The process of *new call* requests is Poisson, with parameter equal to δ.
- The process of *incoming handover* requests from other cells is Poisson with rate γ.

The aggregate process of connection requests is thus given by the superposition of two (independent) Poisson processes, which is a Poisson process with rate

$$\lambda = \delta + \gamma. \tag{13.3}$$

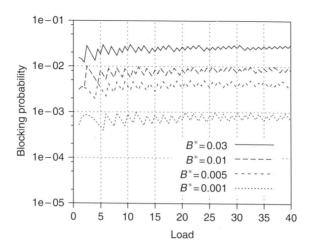

FIGURE 13.6 Basic model: actual blocking probability provided to users when the blocking probability is guaranteed to be smaller than the target B^* under a given traffic load.

When modeling a system, the value of the parameter δ, which is a measure of the fresh traffic, can be estimated in terms of the user's density and behavior, of the size of the cell, and so on; when available, δ can also be estimated by measurements on real systems. In the case of traditional telephony, for example, the new call request generation rate is usually estimated by measurements performed during the peak traffic hour. The value of γ, instead, depends on the user mobility, on the size and geometry of the cell and, above all, on the traffic intensity in neighboring cells. Indeed, given the user mobility, the flow of incoming handovers increases with the number of active connections in neighboring cells.

13.2.2 Channel Holding Time

For an accurate description of the channel holding time, it is necessary to distinguish between:

- *Call duration*
- *Dwell time*, which is the time spent by a user in a cell.

In the literature, it is often assumed that both the call duration and the dwell time are random variables with negative exponential distribution, and with parameters μ_d and μ_h, respectively.

The channel holding time is the time for which a call is active in the considered cell, that is, it is the interval between the activation of the call in the cell (because either a new call is set up or an incoming handover is performed successfully), and the channel release. Two events can be responsible for the channel release: (i) the call terminates, or (ii) the user moves to a neighboring cell and performs an outgoing handover.

Denoted by X the call duration, by D the dwell time, and by S the channel holding time. The distribution of S can be derived from that of X and D by considering the two following cases for the instant in which the channel holding time starts:

- *A new call is generated.* From the instant in which the call is set up, the channel holding time lasts until the channel release, which occurs as soon as either the call terminates or the user exits the cell. The channel holding time is given by

$$S_1 = \min(X, D) \tag{13.4}$$

Due to a well-known property of exponentially distributed random variables, the distribution of S_1 is negative exponential, with parameter $\mu = \mu_d + \mu_h$.

- *An incoming handover is successfully performed.* The channel holding time is the time between the transfer of the active call in the current cell to the channel release:

$$S_2 = \min(X^{(r)}, D) \tag{13.5}$$

where $X^{(r)}$ is the *residual* call duration, that is, the time left to complete the call when the user enters the considered cell. The memoryless property of exponential distributions makes $X^{(r)}$ distributed as X. Thus, $S_2 = \min(X, D)$, that is, S_2 is distributed as S_1.

By combining these two cases, we can conclude that the distribution of S is negative exponential with parameter $\mu = \mu_d + \mu_h$.

When modeling a system, the parameter μ_d is set by estimating the average call duration $1/\mu_d$; usually, a value of $1/\mu_d$ equal to 3 min is assumed. The parameter of the dwell time μ_h can instead be set based on the estimation of the average speed of the users and of the size of the cell.

Given the exponential assumptions for the call duration and the dwell time, some further considerations can be made about mobility. The probability that the channel holding time expires because of a handover, rather than a call completion, is given by

$$H = \frac{\mu_h}{\mu_d + \mu_h} \tag{13.6}$$

Moreover, in a network where all cells are equal, the distribution of the number of handovers per call is geometric with parameter H: the probability that a call requests exactly h handovers is given by

$$P\{\, h \text{ handovers } \} = H^h\,(1 - H) \tag{13.7}$$

The mean number of handovers per call is equal to $E[H] = 1/(1 - H)$.

13.2.3 The Model

Given the above assumptions, the model of the cell is now a M/M/C/0 queue, with load $\rho = \lambda/\mu$. Similar to the previous model, the solution is given by (13.1) and (13.2).

Since both γ and μ_h describe mobility by means of handovers, they should be carefully set so that they consistently model the same phenomenon. The behaviors of neighboring cells are indeed correlated by means of handovers: an incoming handover at a given cell is, at the same time, a channel release for a neighboring cell. An approach, which is commonly adopted in the literature [17], is to balance the incoming and the outgoing handover flow rates, so that, on average, the number of outgoing and incoming handovers per time unit is the same. The rational behind this assumption is that cells, which are close to each other tend to have the same average behavior, and, in particular, they have the same handover flow rates. This implies that, for a given cell, the average number of incoming handovers in the time unit, γ, is equal to the average number of outgoing handovers in the time unit. Let $\gamma^{(o)}$ denote the outgoing handovers flow rate. The handover flow balance relation is $\gamma = \gamma^{(o)}$.

In order to compute $\gamma^{(o)}$ from the M/M/C/0 model of the cell, consider the departure process from the queue. At steady state, the average number of departures in the time unit is equal to by $\lambda \cdot (1 - B)$. The fraction of departures due to a handover (rather than a call termination) is given by H, with H as in (13.6). By combining these two relations, the flow rate of outgoing handovers is given by

$$\gamma^{(o)} = \lambda\,(1 - B)H \tag{13.8}$$

$$= \frac{(\gamma + \delta)\,\mu_h\,(1 - B)}{\mu_d + \mu_h} \tag{13.9}$$

Procedure 3. Handover flow balance

```
/* Input parameters:
       new call arrival rate: delta
       parameter of the distribution of call duration: mu_d
       parameter of the distribution of dwell time: mu_h
       number of channels: C */
/* Output values:
       average incoming and outgoing handover flow: gamma
       blocking probability: B */

gamma=0; /* Initial guess for the incoming handover flow */
err=1; /* Initial error on the value of gamma */
eps=0.001; /* Desired accuracy on the evaluation of gamma */
while(err>eps){
    rho=(gamma+delta)/(mu_d+mu_h);
    B=ErlangB(rho,C);
    /* Apply now the flow balance equation */
    gammaOut=delta*mu_h*(1-B)/( (mu_d+mu_h)*(1-(1-B)*mu_h/(mu_d+mu_h)) );
    err=abs(gammaOut-gamma)/gammaOut;
    gamma=gammaOut;
}
```

FIGURE 13.7   Procedure for the solution of the model with handover flow balance.

By imposing that $\gamma = \gamma^{(o)}$ in (13.9), the flow balance relation becomes

$$\gamma = \frac{\delta \mu_h (1 - B)}{(\mu_d + \mu_h)[1 - (1 - B)\mu_h/(\mu_d + \mu_h)]} \tag{13.10}$$

Since B depends on γ and its expression is not invertible, a solution based on a fixed point approximation is needed for (13.10). A procedure that implements such solution is reported in Figure 13.7.

The above procedure consists in finding the crossing point of two functions. The first one is the Erlang-B, (13.2). Given C, the Erlang-B provides the blocking probability B as a function of the load ρ, which, in its turn, is function of γ. The second function is the flow balance Equation (13.10), which provides γ as a function of B. The existence and uniqueness of a crossing point between the two functions is guaranteed by their shape. The Erlang-B formula is a monotonically increasing function of γ, which starts from the value B_0 obtained for $\gamma = 0$, with $0 < B_0 < 1$, and tends to 1 for γ increasing from 0 to infinity. The outgoing handover flow given by (13.10) is monotonically decreasing with B, reaching 0 for $B = 1$. As an example, observe the curves in Figure 13.8 obtained for $C = 8$, $1/\mu_d = 180$ s, $\mu_h = 5\mu_d$ and $\lambda = \mu_d + \mu_h$. The procedure looks for the crossing point of the two curves, whose coordinates represent the desired value of γ and B.

13.2.4 Performance Metrics

The accurate description of the user mobility allows the computation of the *failure probability*, which is the probability that a call is not successfully terminated due to the failure of either a handover request or the initial set up at call generation, and of the *dropping probability*, which is the probability that a call, which is successfully set up at its generation is then forced to terminate before completion, due to a failed handover.

Assume that the behavior of neighboring cells is uncorrelated and that, on average, the blocking probability is the same. The failure probability U can be computed as the complement to 1 of the

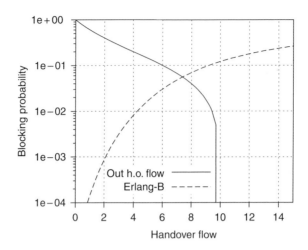

FIGURE 13.8   Convergence of the handover flow balance.

probability of successful completion, which can be expressed by considering the conditional probabilities that the call succeeds, given that it requires h handovers:

$$1 - U = P\{\text{access}\} \sum_{h=0}^{\infty} P\{\text{success} \mid h \text{ handovers}\} P\{h \text{ handovers}\} \tag{13.11}$$

where the term $P\{\text{access}\}$ accounts for the probability to successfully access the system at the call generation. It results,

$$U = 1 - \sum_{h=0}^{\infty} (1 - B)^{h+1} H^h (1 - H) \tag{13.12}$$

$$= 1 - \frac{(1 - B)(1 - H)}{1 - (1 - B)H} \tag{13.13}$$

Similarly, the dropping probability D can be derived as:

$$1 - D = \sum_{h=0}^{\infty} P\{\text{success} \mid h \text{ handovers}\} P\{h \text{ handovers}\} \tag{13.14}$$

from which,

$$D = 1 - \sum_{h=0}^{\infty} (1 - B)^h H^h (1 - H) \tag{13.15}$$

$$= 1 - \frac{(1 - H)}{1 - (1 - B)H} \tag{13.16}$$

From the solution of the model, it is now possible to study the impact of mobility on the performance of the system.

The dropping probability is plotted in Figure 13.9 versus the traffic load for different values of the average number of handovers per call. The nominal load on the x-axis is given by δ/μ_d, and represents

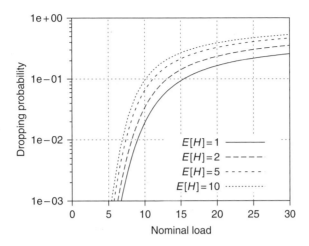

FIGURE 13.9 Model with handover flow balance: dropping probability versus nominal load; with $C = 16$ and different degrees of mobility.

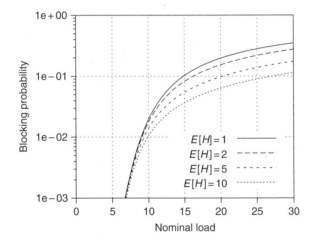

FIGURE 13.10 Model with handover flow balance: blocking probability versus nominal load; with $C = 16$ and different degrees of mobility.

the amount of work per cell required to satisfy the users' traffic. The dropping probability increases with mobility, since large numbers of handovers per call make handover failures more likely. Notice that the impact of mobility is remarkable. For example, consider the case of nominal load equal to 10. The system provides an acceptable value of the dropping probability (about 2%) under the assumption that $E[H] = 1$, while the dropping probability is as large as 10% if mobility is ten times higher. This implies that, depending on the environment in which the cellular network is operating (urban, suburban, highway), different choices for the cell dimensioning should be taken.

Under the same scenarios, Figure 13.10 shows the blocking probability. Observe that the blocking probability decreases as mobility increases, which, at a first sight may seem a counter-intuitive result. This behavior is a side effect of the fact that highly mobile terminals undergo a large probability of prematurely dropped calls: since a forced call termination reduces the time the user spends in the system, it also reduces the actual load in the cell, from which a smaller blocking probability results. This phenomenon is confirmed by Figure 13.11, which reports the actual load per cell, ρ: the reduction due to high mobility is clearly visible.

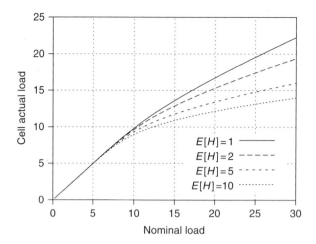

FIGURE 13.11   Model with handover flow balance: actual load versus nominal load; with $C = 16$ and different degrees of mobility.

Procedure 4. Dimensioning with handover flow balance

```
/* Input parameters:
      new call arrival rate: delta
      parameter of the distribution of call duration: mu_d
      parameter of the distribution of dwell time: mu_h
      target dropping probability: Dtarget */
/* Output value:
      number of channels: C */

C=0;
H=mu_h/(mu_d+mu_h);
D=1;
while(D>Dtarget){
   C+=1;
   B=ModelHandoverBalance(delta,mu_d,mu_h,C);
   D=1-(1-B)*(1-H)/(1-(1-B)*H);
}
```

FIGURE 13.12   Procedure for the cell design through the model with handover flow balance.

Similar to what was previously done, the model can be used to dimension the cell and to find the number of channels that are needed to satisfy a given QoS target. The QoS target can be expressed in terms of blocking probability and dropping probability, which is a measure closer to the actual quality perceived by the end user. As an example, in the procedure shown in Figure 13.12, the dimensioning aims at satisfying a given target dropping probability; the procedure calls Procedure 3 (named *ModelHandoverBalance* in the figure), which solves the model with handover flow balance.

13.3   Channel Reservation

Since radio resources are scarce and precious, they must be carefully exploited in order to provide the best possible QoS to users. In wired telephone systems, quality degradation is primarily due to blocking; that

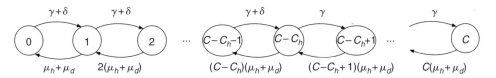

FIGURE 13.13   Markov chain model of the cell with C channels and C_h channels reserved to handovers.

is, the user cannot access the system due to the unavailability of resources. In cellular systems, besides blocking, dropping of a call in progress due to a handover failure may also occur. From the point of view of a user, this event is even worse than blocking, because it produces an interruption of a service, which is in progress. Thus, in an attempt to reduce handover failures, network operators usually adopt policies to favor handovers over new call requests; one of the simplest policies is *channel reservation*. Let C_h be the number of channels reserved to handovers out of the C channels available in the cell. While handovers are set up as long as a channel is free, a new call request is satisfied only if more than C_h channels are free; thus, new call and handover requests are blocked with different probability. By properly choosing C_h, the tradeoff between new call blocking probability and dropping probability can be decided.

Under the Markovian assumptions already made for the previous model, a continuous-time Markov chain can be developed to describe the behavior of the system. The state s of the Markov chain represents the number of busy channels, or active calls in the cell; s ranges between 0 and C. The state transition diagram of the Markov chain is reported in Figure 13.13. When $0 \leq s < C - C_h$, the arrival of both a new call or a handover request can be satisfied. In the Markov chain this is modeled by a transition from state s to state $s + 1$; the rate of the transition being given by the rate of the associated Poisson process, $\gamma + \delta$. When $C - C_h \leq s < C$, the number of available channels is smaller than C_h, new call requests are blocked and only handovers can enter the cell. Correspondingly, the Markov chain moves from s to $s + 1$ with rate γ. In state $s = C$ both new calls and handovers are blocked. From state s, with $s > 0$, the channel release is represented by the transition to state $s - 1$, which occurs with rate $s(\mu_h + \mu_d)$.

The steady-state probability $\pi(s)$ that s channels are busy is

$$
\pi(s) = \pi(0)
\begin{cases}
\dfrac{1}{s!}\left(\dfrac{\gamma + \delta}{\mu_h + \mu_d}\right)^s & \text{for } s \leq C - C_h \\[4mm]
\dfrac{1}{s!}\left(\dfrac{\gamma + \delta}{\mu_h + \mu_d}\right)^{C-C_h}\left(\dfrac{\gamma}{\mu_h + \mu_d}\right)^i & \text{for } s = C - C_h + i \quad \text{and} \quad 0 < i \leq C_h
\end{cases}
\tag{13.17}
$$

$$
\pi(0) = \left(\sum_{s=0}^{C-C_h}\frac{1}{s!}\left(\frac{\gamma + \delta}{\mu_h + \mu_d}\right)^s + \left(\frac{\gamma + \delta}{\mu_h + \mu_d}\right)^{C-C_h}\sum_{i=1}^{C_h}\frac{1}{(C - C_h + i)!}\left(\frac{\gamma}{\mu_h + \mu_d}\right)^i\right)^{-1}
\tag{13.18}
$$

The blocking probability of new calls is given by the probability that the system is in the state $s \geq C - C_h$,

$$
B_n = \sum_{s=C-C_h}^{C} \pi(s)
\tag{13.19}
$$

The blocking probability of a handover request is

$$
B_h = \pi(C)
\tag{13.20}
$$

which, as expected, is always larger than B_n.

As for the previous model, the incoming handover rate γ can be computed by imposing that at steady-state the incoming and outgoing handover flow rates are the same. The outgoing handover flow rate is given by

$$\gamma^{(o)} = H[\gamma(1 - B_h) + \delta(1 - B_n)] \tag{13.21}$$

The procedure for the derivation of the handover flow rate (skipped here for the sake of brevity) is simply obtained from Procedure 3 by properly changing the computation of the blocking probability.

The failure and dropping probabilities can now be derived as in (13.11) and (13.14), where $P\{$ access $\}$ is equal to $1 - B_n$ and $P\{$ success $\mid h$ handovers $\}$ is $(1 - B_h)^h$,

$$U = 1 - (1 - B_n) \sum_{h=0}^{\infty} (1 - B_h)^h H^h (1 - H) \tag{13.22}$$

$$= 1 - \frac{(1 - B_n)(1 - H)}{1 - (1 - B_h)H} \tag{13.23}$$

$$D = 1 - \frac{(1 - H)}{1 - (1 - B_h)H} \tag{13.24}$$

In Figure 13.14 and Figure 13.15, the new call blocking probability and dropping probability are plotted versus the nominal load for different values of the number of channels reserved to handovers when the cell has $C = 16$ channels and the average number of handovers per call is $E[H] = 5$. When $C_h = 0$, the new call blocking probability is significantly smaller than the dropping probability D, which is quite large, about 1% for nominal load as low as 7. By increasing the value of C_h, the dropping probability decreases, while the blocking probability increases. The choice of C_h is made so as to properly trade-off between these two performance metrics.

In order to do so, it can be convenient to provide a definition of the user satisfaction or dissatisfaction, which jointly accounts for both the new call blocking and dropping probabilities. Since, as already mentioned, from the point of view of the user, a forced termination of a call in progress is worse than a new call blocking, a simple metric of the user dissatisfaction can be given by: $B_n + \alpha D$, with $\alpha > 1$. In Figure 13.16 we plot the user dissatisfaction with $\alpha = 10$, versus increasing values of the number of channels reserved to handovers, C_h, when $C = 16$, and the nominal load is equal to 14. The optimal value of C_h is the one

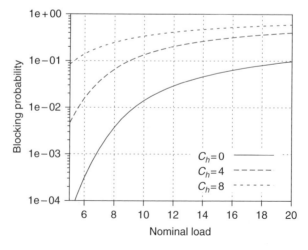

FIGURE 13.14 Model for channel reservation: new call blocking probability versus nominal load, for different values of C_h, with $C = 16$ and $E[H] = 5$.

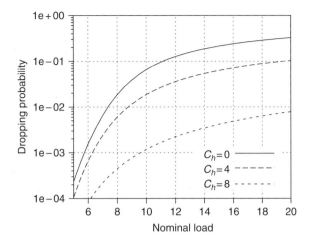

FIGURE 13.15 Model for channel reservation: dropping probability versus nominal load, for different values of C_h, with $C = 16$ and $E[H] = 5$.

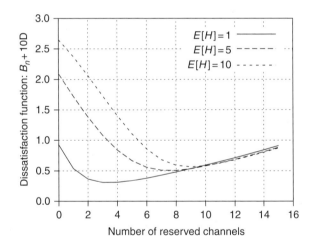

FIGURE 13.16 Model for channel reservation: user dissatisfaction versus the number of channels reserved to handovers, C_h, for different values of $E[H]$, with $C = 16$ and nominal load equal to 14.

which minimizes the user dissatisfaction. For example, $C_h = 3$ is the optimal choice when the mobility is low, that is, $E[H] = 1$; while a value as large as 9 should be chosen when $E[H] = 10$.

Similar results are plotted in Figure 13.17, for $E[H] = 5$ and different values of the load.

In the case of channel reservation, the cell dimensioning procedure has two outputs: the number of channels in the cell, C, and the number of channels reserved to handovers, C_h. QoS constraints can now be specified in terms of target values for some of the following metrics: the new call and the handover blocking probability, the dropping probability, the failure probability. Alternatively, QoS guarantees can be specified in terms of a maximum allowable value for the user dissatisfaction, where dissatisfaction can be defined, for example, as before, by $B_n + \alpha D$.

The procedures to find a solution to the dimensioning problem are based on the inversion of the model presented above; that is, on the solution of the model for different values of C and C_h. As an example, Figure 13.18 reports a procedure whose objective is to find C and C_h, which guarantee target values of the new call blocking probability and of the dropping probability.

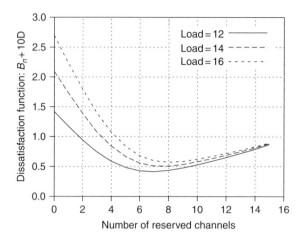

FIGURE 13.17 Model for channel reservation: user dissatisfaction versus the number of channels reserved to handovers, C_h, for different values of the nominal load, with $C = 16$ and $E[H] = 5$.

Procedure 5. Dimensioning with channel reservation

```
/* Input parameters:
    new call arrival rate: delta
    parameter of the distribution of call duration: mu_d
    parameter of the distribution of dwell time: mu_h
    target dropping probability: Dtarget
    target new call blocking probability: Btarget */
/* Output values:
    total number of channels: C
    number of channels reserved to handovers: Ch */

C=1;
Ch=0;
H=mu_h/(mu_d+mu_h);
B=1;
while(B>Btarget){
   C+=1;
   Ch=0;
   B=ModelHandoverBalance(gamma,mu_d,mu_h,C);
   D=1-(1-B)*(1-H)/(1-(1-B)*H);
   if(B<Btarget)
     while(D>Dtarget){
        Ch+=1;
        B=ModelHandoverBalance(delta,mu_d,mu_h,C);
        D=1-(1-B)*(1-H)/(1-(1-B)*H);
        }
   }
```

FIGURE 13.18 Procedure for the solution of the model of a cell with channels reserved to handovers.

The procedure works as follows. First, the minimum value of C that guarantees the target B_n is found, by setting C_h to 0. Then, given the current value of C, the smallest value of C_h is searched, such that the constraint on D is also met. Once C_h is found, the constraint on B is checked again. Indeed, while increasing C_h, the value of B_n too increases. If the constraint on B_n is not satisfied, the procedure is repeated with $C + 1$, and $C_h = 0$.

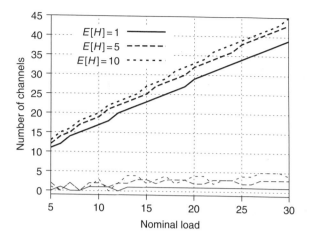

FIGURE 13.19 Model for channel reservation: dimensioning the cell to guarantee $D^* = 0.01$ and $B_n^* = 0.03$ versus nominal load, for different degrees of mobility.

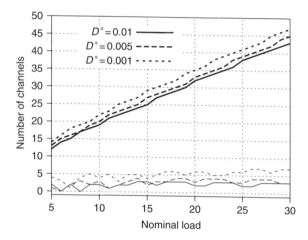

FIGURE 13.20 Model for channel reservation: dimensioning the cell to guarantee $B_n^* = 0.03$ and different values of D^* versus nominal load, with $E[H] = 5$.

Results are shown in Figure 13.19 for different degrees of mobility, under the constraint $D^* = 0.01$ and $B_n^* = 0.03$. For high levels of mobility, large values of C_h are typically requested. However, this behavior is not monotonic, as a side-effect of the discretization over the number of channels, which is an integer value.

Similar results are shown in Figure 13.20, by setting $E[H] = 5$, fixing the constraint on B_n^* to 0.03, and changing the constraint on D^*.

13.4 Model Extensions

The cell model can be further enhanced, so as to describe more complex scenarios. In this section we first describe a generalization that allows considering several different classes of users. Then, we exploit this generalization to also consider the case of services with different bandwidth requirements. Finally, we discuss the case of finite user population.

13.4.1 Classes of Users

The models presented so far describe the behavior of the whole user population by means of exponentially distributed random variables that model the general user mobility and call duration. When more accurate descriptions of the users behavior are needed, classes of users may be introduced with limited additional complexity. For example, the introduction of classes of users can be convenient for the following reasons.

1. *Multiple services.* Classes of users can be introduced in order to model the coexistence of different services, such as telephony and low-speed data transfer. Depending on the requested service, the typical duration of a connection changes, and classes of users are introduced in order to represent the different distributions of the call duration. Class i users request a service whose call duration is described by a random variable with negative exponential distribution and mean value $1/\mu_{d,i}$. In this case, user mobility does not depend on the requested service, thus, the user dwell time is modeled by a random variable with negative exponential distribution and mean value $1/\mu_h$ for all user classes, and the channel holding time for class i users is a random variable with negative exponential distribution and rate $\mu_i = \mu_{d,i} + \mu_h$. The arrival rate of requests typically depends on the considered service, and can be denoted by δ_i.

2. *Classes of mobility.* In some scenarios, classes of users may be introduced to model the fact that groups of users exhibit very different degrees of mobility. As an example, consider the case of cellular telephony provided in a urban scenario; pedestrian users are very slow, much slower than users on vehicles. The mobility of class i users is described by dwell times having negative exponential distribution with rate $\mu_{h,i}$, where $\mu_{h,i}$ is larger for faster users. In this case, call durations do not depend on mobility, and all classes of users have call durations with negative exponential distribution and rate μ_d. Thus, the channel holding time for class i users is a random variable with negative exponential distribution and rate $\mu_i = \mu_d + \mu_{h,i}$. The arrival rate of requests from class i users is denoted by δ_i.

Clearly, scenarios with classes of users exhibiting both different service times and different mobility can be also considered.

The model described in the previous section can be extended to deal with classes of users. Indeed, the system can be described by a M/M/C/0 queue with classes of customers or by a multidimensional continuous-time Markov chain.

Denote by N the number of classes of users, distinguished on the basis of call duration and user mobility and by δ_i, $\mu_{d,i}$, and $\mu_{h,i}$ the rate of new call arrivals, call duration, and dwell time for class i calls. Denoted also by γ_i the arrival rate of incoming handovers for active class i calls.

The state of the cell is defined by the vector $\bar{s} = (s_1, s_2, \ldots, s_N)$, where each component s_i represents the number of active class i calls. The model state space is

$$S = \left\{ \bar{s} \mid 0 \leq \sum_{i=1}^{N} s_i \leq C \right\} \tag{13.25}$$

The steady-state probabilities $\pi(\bar{s})$ for this system are computed as a product of the single-class solutions:

$$\pi(\bar{s}) = \pi(\bar{s}_0) \prod_{i=1}^{N} \frac{\rho_i^{s_i}}{s_i!} \quad \text{with } \pi(\bar{s}_0) = \pi(0,0,\ldots,0) = \left(\sum_{\bar{s} \in S} \prod_{i=1}^{N} \frac{\rho_i^{s_i}}{s_i!} \right)^{-1} \tag{13.26}$$

where ρ_i is the class i traffic intensity,

$$\rho_i = \frac{\delta_i + \gamma_i}{\mu_{d,i} + \mu_{h,i}} \tag{13.27}$$

The blocking probability for class i calls, B_i, is given by the sum of the probabilities of states in subset $\mathcal{S}_i$ in which all channels are busy with active calls,

$$B_i = \sum_{\bar{s} \in \mathcal{S}_i} \pi(\bar{s}) \qquad \text{with } \mathcal{S}_i = \left\{ \bar{s} \mid \sum_{j=1}^{N} s_j = C \right\} \tag{13.28}$$

The class i outgoing handover flow rate is,

$$\gamma_i^{(o)} = (\delta_i + \gamma_i)(1 - B_i)H_i \tag{13.29}$$

with

$$H_i = \frac{\mu_{h,i}}{\mu_{h,i} + \mu_{d,i}} \tag{13.30}$$

The value of $\gamma_i^{(o)}$ must be found by balancing incoming and outgoing handover flow rates. The handover flow rate balance is needed for each class of users.

The dropping and failure probabilities are computed from (13.13) and (13.16) by substituting B_i instead of B, and H_i instead of H.

The average number of active class i calls is given by

$$E[A_i] = \sum_{\bar{s} \in \mathcal{S}} s_i \pi(\bar{s}) \tag{13.31}$$

13.4.2 Services with Different Bandwidth

A further generalization of the previous model with classes of users is needed in order to describe the coexistence of services with different bandwidth requirements. Consider, for example, the case of offering the High-Speed Circuit-Switched (HSCS) service in a cellular GSM system; or, consider a third generation system in which circuit-switched services can be allocated with different bandwidth. While a telephone call requires the allocation of just one channel, when a high-speed service request is generated, more than one channel must be allocated to the request. Since services are circuit-switched, these channels are employed by the user for the whole call duration, and are released all together, only at the call termination or after an outgoing handover. From the point of view of the model, this means that customers may require a different number of channels to be served.

N classes of calls are distinguished, on the basis of three aspects: user mobility, call duration, and bandwidth requirements, which are expressed in terms of number of channels employed for the connection; let this number be C_i for class i calls. As before, the parameters which describe class i calls are denoted by δ_i, $\mu_{d,i}$, and $\mu_{h,i}$.

The state of the cell is defined by the vector $\bar{s} = (s_1, s_2, \ldots, s_N)$, where s_i represents the number of active class i calls. Since each call of class i requires C_i channels, the model state space is given by

$$\mathcal{S} = \left\{ \bar{s} \mid 0 \le \sum_{i=1}^{N} C_i \cdot s_i \le C \right\} \tag{13.32}$$

Compared to the previous model, the state space is now defined over a different set of states. In particular, what changes is the definition of the boundary conditions.

As an example, consider the case of a system in which two services are provided: telephony, which requires one channel per call, and data transfers, whose connections require two channels. The system is described by $N = 2$ classes of users, with bandwidth requirements specified by $C_1 = 1$ for telephony, and $C_2 = 2$ for data transfers. The Markov chain of the model is represented in Figure 13.21. In state $\bar{s} = (s_1, s_2)$, s_1 voice calls and s_2 data connections are active. Voice calls are generated with rate $\lambda_1 = \delta_1 + \gamma_1$ and are accepted if at least one channel is available, that is, if $s_1 + 2s_2 < C$. Data connections, instead, are generated with rate $\lambda_2 = \delta_2 + \gamma_2$, and are accepted if at least two channels are available, $s_1 + 2s_2 < C - 1$. Thus, in the states such that $C - 1 < s_1 + 2s_2 < C$ (such as $(C - 1, 0)$ in the figure) only voice calls are accepted. For simplicity, an even number is assumed for C in the figure, so that the maximum number of active data connections is equal to $C/2$ when no voice calls are active. From state (s_1, s_2), the termination rate of voice connections is equal to $s_1\mu_1$ with $\mu_1 = \mu_{d,1} + \mu_{h,1}$, and the termination rate of data connections is $s_2\mu_2$ with $\mu_2 = \mu_{d,2} + \mu_{h,2}$.

Coming back to the general case, the steady-state probabilities $\pi(\bar{s})$ are computed as the product of the single-class solutions as in (13.26) and (13.28); notice that the two cases differ in the state space, which is defined over a different set of possible states.

The blocking probability for class i calls, B_i, is computed as the sum of the probabilities of states in $\mathcal{S}_i$, in which less than C_i channels are free,

$$B_i = \sum_{\bar{s} \in \mathcal{S}_i} \pi(\bar{s}) \qquad \text{with } \mathcal{S}_i = \left\{ \bar{s} \mid \sum_{j=1}^{N} C_j \cdot s_j > C - C_i \right\} \tag{13.33}$$

In the above expression, the term $\sum_{j=1}^{N} C_j \cdot s_j$ provides the number of busy channels in state $\bar{s}$. Clearly, calls with larger bandwidth requirements (larger values of C_i) experience larger values of blocking probability.

The class i outgoing handover flow rate is given by (13.29) and it is used for balancing incoming and outgoing handover flow rates. Again, the dropping and failure probability are computed from (13.13) and (13.16) by substituting B_i instead of B, and H_i instead of H.

The average number of active class i calls is given by

$$E[A_i] = \sum_{\bar{s} \in \mathcal{S}} s_i \pi(\bar{s}) \tag{13.34}$$

while the average number of channels busy serving class i calls is given by $C_i E[A_i]$.

In order to assess the impact of large bandwidth requirements on the cell performance, consider the two-service case, whose Markov chain model is shown in Figure 13.21, with $C = 16$. The data connection request arrival rate is half the arrival rate of telephone call requests; the average call duration of the two services are the same, so that the nominal traffic intensity $\delta_i/(\mu_{d,i} + \mu_{h,i})$ for data is half the one for telephony; however, in terms of channel load, since data require twice the channels of voice, the two services have the same nominal load.

Figure 13.22 reports the dropping and blocking probabilities for voice and data services, versus the nominal channel load of each class. As expected, the data service experiences larger blocking and dropping probabilities. For the same scenario, Figure 13.23 shows the average number of active calls for voice and data transfers (thick lines) and the average number of busy channels for data connections (thin dashed line); clearly, in the case of voice, the number of busy channels and active calls are the same. Due to the larger blocking probability, the number of busy channels for data is much smaller than the number of busy channels for voice.

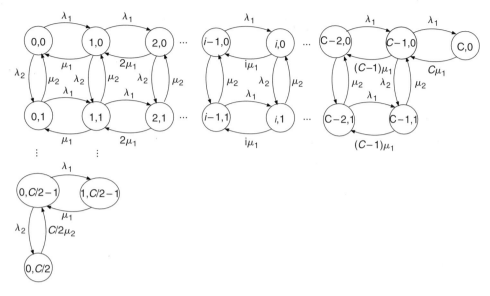

FIGURE 13.21 Markov chain model of the cell with C channels, two services with $C_1 = 1$ and $C_2 = 2$ channels per call.

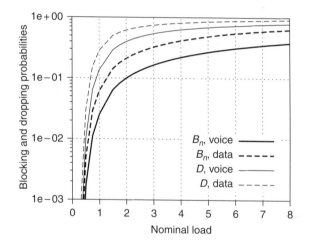

FIGURE 13.22 Model for voice and data: blocking and dropping probabilities versus nominal load, with $C = 16$ and $E[H] = 5$.

13.4.3 The Finite Population Case

The tendency to reduce the cell size in mobile telephony cellular systems is due to two main reasons. On the one hand, in urban environments with high population density, the need to increase the network capacity leads to reuse frequencies as much as possible by means of small cells. On the other hand, a typical scenario foreseen for third generation systems is the provisioning of high-bandwidth services in tiny cells (called picocells) to users with very low mobility, or no mobility at all.

When the cell size is small, the implicit infinite population assumption, on which the Poisson arrival process is based, becomes unrealistic. Indeed, when users in a cell are few, the fact that some users are active, significantly reduces the number of users that may issue new call requests. Thus, an accurate description of the small cell case requires the development of a finite population model.

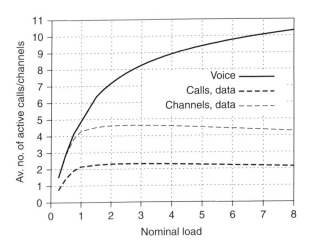

FIGURE 13.23　Model for voice and data: average number of busy channels and active calls versus nominal load, with $C = 16$ and $E[H] = 5$.

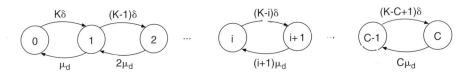

FIGURE 13.24　Markov chain model of the cell with C channels and finite population of K users.

Let K denote the population size, and let $1/\delta$ be the mean time between the instant in which a user terminates a call and the instant in which the same user issues the following request (we again adopt an exponential assumption). When i calls are active in the cell, $(K - i)$ are the users who may potentially generate a new call request; thus, $(K - i)\delta$ is the total call generation rate when i users are active in the cell. The average service time is equal to the average call duration $1/\mu_d$. The state-transition diagram of the Markovian model in this case is shown in Figure 13.24, where the state index corresponds to the number of active calls.

By denoting with ρ the ratio δ/μ_d, the steady-state probabilities for this model are given by

$$\pi(i) = \frac{K!}{i!(K - i)!}\rho^i \pi(0) = \binom{K}{i}\rho^i \pi(0) \tag{13.35}$$

with

$$\pi(0) = \left[\sum_{i=0}^{C}\binom{K}{i}\rho^i\right]^{-1} \tag{13.36}$$

Similarly to the basic model, the finite population model in Figure 13.24 does not explicitly account for handovers and mobility. While this is realistic for the picocell scenario mentioned earlier, in which users are not moving, mobility must be considered in the high population density scenario. A user entering the cell makes the population size increase, while a user roaming out of the cell makes it decrease. Although a detailed description of the system, calls for a model with variable population size, when K is significantly larger than C, K can be assumed to be constant with little loss of accuracy. Let γ be the arrival rate of

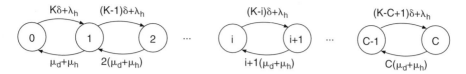

FIGURE 13.25 Markov chain model of the cell with C channels, finite population of K users and mobility.

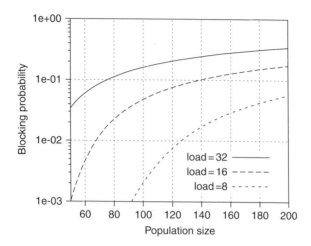

FIGURE 13.26 Finite population model: blocking probability versus population size, with $C = 16$ and $E[H] = 5$; the nominal load is given by $100 \cdot \delta / (\mu_d + \mu_h)$.

incoming handover requests. In state i, the total request arrival rate is given by $(K - i)\delta + \gamma$, and the service rate is $i(\mu_d + \mu_h)$. Assuming that both incoming and outgoing handover requests leave the population size unchanged, the Markovian model becomes as shown in Figure 13.25.

The steady-state probabilities in this case are given by

$$\pi(i) = \prod_{j=0}^{i-1} \frac{(N - j)\delta + \lambda}{(j + 1)(\mu_d + \mu_h)} \pi(0) \tag{13.37}$$

$$\pi(0) = \left[1 + \sum_{i=1}^{C} \prod_{j=0}^{i-1} \frac{(N - j)\delta + \lambda}{(j + 1)(\mu_d + \mu_h)} \right]^{-1} \tag{13.38}$$

The incoming handover rate γ can be derived by balancing incoming and outgoing handover rates by means of a fixed point procedure, similar to those previously described.

Results for the case of a small cell in a high population density environment (i.e., with mobility) are reported in Figure 13.26 and Figure 13.27 for the blocking and dropping probability, respectively. Curves are labeled with the nominal load, in the case of population size equal to 100. For example, the value of δ in the curve with load 16, is the one for which $100 \cdot \delta / \mu_d = 16$. Clearly, a large population size corresponds to a high load, and both blocking and dropping probability increase with load.

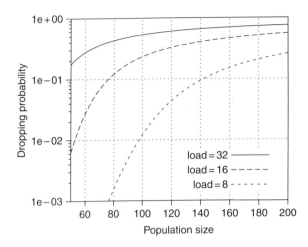

FIGURE 13.27 Finite population model: dropping probability versus population size, with $C = 16$ and $E[H] = 5$; the nominal load is given by $100 \cdot \delta/(\mu_d + \mu_h)$.

13.5 A Model for Generally Distributed Dwell Times and Call Durations

Consider now a more general case in which we depart from the exponential assumptions used so far. In this case, besides having N different classes of calls distinguished on the basis of: (i) bandwidth requirements, (ii) user mobility, (iii) new call arrival rate, and (iv) call duration, dwell times and call durations have general distribution.

Let $F_{D_i}(t)$ be the cumulative distribution function (CDF) of the random variable D_i, which describes the dwell time for class i calls; let $f_{D_i}(t)$ be the probability density function (pdf) of the same random variable. The duration of class i calls is specified by means of a random variable X_i, whose CDF and pdf are denoted by $F_{X_i}(t)$ and $f_{X_i}(t)$, respectively. Bandwidth requirements are expressed in terms of the number of channels required by calls; let this be a constant C_i for class i calls.

The arrival rate of new calls of class i in the cell is δ_i calls per time unit. Arrivals are still modeled by a Poisson process.

Let S_i be the random variable, which describes the channel holding time for class i users. The steady-state probability distribution of the number of active channels in the cell depends on the average service time $E[S_i]$ only, and is insensitive to higher moments of the distribution of S_i. Thus, an accurate estimate of $E[S_i]$ is needed in order to accurately model the cell. S_i is defined as $S_i = \min(D_i, X_i^{(r)})$, where $X_i^{(r)}$ is the residual call duration.

Since the exact characterization of $X_i^{(r)}$ is quite complex, the following approximate approach can be adopted. Let the random variable X_i be approximated by another random variable Y_i with a two-phase hyper-exponential distribution, which is much simpler to deal with, since its memory is completely captured by the phase variable. The pdf of Y_i, f_{Y_i}, is thus obtained as the composition of two exponential random variables $Y_{i,1}$ and $Y_{i,2}$,

$$f_{Y_i}(t) = \alpha_i f_{Y_{i,1}} + (1 - \alpha_i) f_{Y_{i,2}} = \alpha_i \mu_{i,1} e^{-\mu_{i,1} t} + (1 - \alpha_i) \mu_{i,2} e^{-\mu_{i,2} t} \qquad \text{with } t \geq 0 \qquad (13.39)$$

The parameters α_i, $\mu_{i,1}$, and $\mu_{i,2}$ are chosen so that the first two moments of f_{Y_i} and f_{X_i} coincide. This condition can be satisfied for all possible distributions of X_i whose coefficient of variation C_v is larger than one. For distributions with C_v smaller than 1, an Erlang distribution should be used instead. However, most of the works in the literature, which derive traffic models on the basis of real traffic measurements claim that the distribution of the call duration has typical values of C_V larger than 1. Therefore, this is the most interesting case.

Given (13.39), traffic class i can be represented with the combination of two (artificial) classes of traffic. The first one, say class $(i, 1)$, has exponentially distributed call duration with parameter $\mu_{i,1}$; the second class of traffic, $(i, 2)$, has exponentially distributed call duration with parameter $\mu_{i,2}$. By mapping each class of traffic i on a pair of classes $(i, 1)$ and $(i, 2)$, with exponential call durations, the model becomes much easier to solve. Indeed, as already mentioned, due to the memoryless property of negative exponential distributions, the residual call duration for class (i, n) traffic, with $n = 1, 2$, has the same distribution as the call duration: $Y_{i,n}^{(r)} \sim Y_{i,n}$. Then, approximating $X_i^{(r)}$ with $Y_i^{(r)}$ in (13.39),

$$S_{i,n} = \min(D_i, Y_{i,n}^{(r)}) \tag{13.40}$$

$$= \min(D_i, Y_{i,n}) \qquad \text{with } n = 1, 2 \tag{13.41}$$

Hence, the CDF of $S_{i,n}$, $F_{S_{i,n}}(t)$, is

$$F_{S_{i,n}}(t) = F_{Y_{i,n}}(t) + F_{D_i}(t) - F_{Y_{i,n}}(t) \cdot F_{D_i}(t) \tag{13.42}$$

from which $E[S_{i,n}]$ is easily computed. Notice that once a class i call is set up, C_i channels remain busy for a time described by S_i.

The call request arrival process at the cell is given by the superposition of the arrivals of new call requests and handover requests from neighboring cells.

The arrivals of new call requests follow a Poisson process with parameter δ_i for class i. Given the assumptions that were previously introduced, the class i calls Poisson arrival process can be mapped into the superposition of the two Poisson processes, which correspond to traffic classes $(i, 1)$ and $(i, 2)$; arrival rates are $\delta_{i,1} = \alpha_i \cdot \delta_i$ and $\delta_{i,2} = (1 - \alpha_i) \cdot \delta_i$. The overall arrival process of class i calls has rate $\delta_i = \delta_{i,1} + \delta_{i,2}$.

The arrival processes of handover requests from neighboring cells can be evaluated, as before, by balancing incoming and outgoing handover flow rates for each user class (i, n). For class (i, n), the request arrival process into each cell is the superposition of the new call generation process and the incoming handover process; the resulting total arrival rate is $\lambda_{i,n} = \delta_{i,n} + \gamma_{i,n}$.

The state of a cell in this case is defined by the vector $\bar{s} = \{s_{i,n}\}$ where each component $s_{i,n}$ represents the number of active class (i, n) calls, with $i = 1, 2, \ldots, N$ and $n = 1, 2$. Since each class i call requires C_i channels, the cell model state space is

$$S = \left\{ \bar{s} \mid 0 \le \sum_{n=1}^{2} \sum_{i=1}^{N} C_i \cdot s_{i,n} \le C \right\} \tag{13.43}$$

As in the previous multiple class models, the steady-state probabilities $\pi(\bar{s})$ can be computed as the product of the single-class solutions,

$$\pi(\bar{s}) = \pi(\bar{s}_0) \prod_{n=1}^{2} \prod_{i=1}^{N} \frac{\rho_{i,n}^{s_{i,n}}}{s_{i,n}!} \tag{13.44}$$

$$\pi(\bar{s}_0) = \pi(0, 0, \ldots, 0) = \left(\sum_{\bar{s} \in \mathcal{S}^{(m)}} \prod_{n=1}^{2} \prod_{i=1}^{N} \frac{\rho_{i,n}^{s_{i,n}}}{s_{i,n}!} \right)^{-1} \tag{13.45}$$

where $\rho_{i,n}$ is the class (i, n) traffic intensity: $\rho_{i,n} = (\lambda_{i,n} + \gamma_{i,n}) \cdot E[S_{i,n}]$.

The flow of calls that leave the cell (because of an outgoing handover or a call termination) has rate $E[A_{i,n}]/E[S_{i,n}]$, where $E[A_{i,n}]$ is the average number of active calls. Then, the outgoing handover flow rate is given by

$$\gamma_{i,n}^{(o)} = H_{i,n} \frac{E[A_{i,n}]}{E[S_{i,n}]} \tag{13.46}$$

where $H_{i,n}$ is the probability that a user with an active call in class (i, n) releases the radio resources due to a handover rather than a call termination,

$$H_{i,n} = P\{D_i \leq Y_{i,n}^{(r)}\} \tag{13.47}$$

$$= P\{D_i \leq Y_{i,n}\} \tag{13.48}$$

$$= \int_0^\infty F_{D_i}(t) f_{Y_{i,n}}(t)\, dt \tag{13.49}$$

$$= \int_0^\infty F_{D_i}(t) \mu_{i,n} e^{-\mu_{i,n} t}\, dt \tag{13.50}$$

As already explained, $\gamma_{i,n}^{(o)}$ is used in the iterative procedure for the solution of the model.

The blocking probability for class i calls, B_i, is given by the sum of the probabilities of states in subset $\mathcal{B}_i$ for which the number of free channels in the cell is not sufficient to satisfy a class i call:

$$B_i = \sum_{\bar{s} \in \mathcal{B}_i} \pi(\bar{s}) \quad \text{with } \mathcal{B}_i = \left\{ \bar{s} \mid \sum_{n=1}^{2} \sum_{j=1}^{N} C_j \cdot s_{j,n} > C - C_i \right\} \tag{13.51}$$

Observe that classes $(i, 1)$ and $(i, 2)$ have identical blocking probability B_i, since this probability depends on C_i only.

The average number of active class i calls in the cell is given by

$$E[A_i] = \sum_{\bar{s} \in \mathcal{S}} s_{i,n} \pi(\bar{s}) \tag{13.52}$$

Let the average number of busy channels in the cell be denoted by $E[N]$, and let $E[N_i]$ refer to the channels employed by class i users. Then

$$E[N_i] = C_i \sum_{\bar{s} \in \mathcal{S}} s_{i,n} \pi(\bar{s}) \tag{13.53}$$

$$E[N] = \sum_{i=1}^{C} E[N_i] \tag{13.54}$$

As an example, consider a scenario with $C = 16$ channels per cell and users requesting 1 channel per call. The call duration is distributed according to a Weibull distribution with mean value equal to 180 sec and squared coefficient of variation equal to 9.86. Based on their mobility, users can be divided in two classes ($N = 2$). All users have the same type of distribution of the dwell time, but users belonging to class 1 move faster than users in class 2, and thus have a smaller mean dwell time.

The average number of active calls for the two classes of users is plotted in Figure 13.28 versus the nominal load, for two different cases. In the first one, the distribution of the dwell time is hyper-exponential; the fast users mean dwell time is equal to 36 sec, while the slow users mean dwell time is equal to 180 sec; both distributions have squared coefficient of variation equal to 16. In the second case, the dwell time has

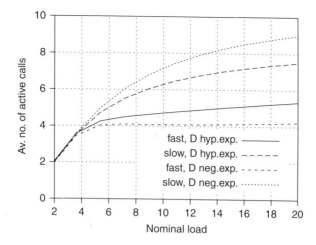

FIGURE 13.28 Model for general distributions: average number of active calls with Weibull distributed call duration, and dwell time distributed according either to a negative exponential or to a hyper-exponential distribution; two classes of users.

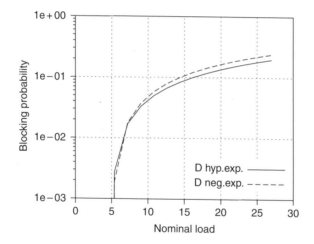

FIGURE 13.29 Model for general distributions: blocking probability with Weibull distributed call duration, and dwell time distributed according either to a negative exponential or to a hyper-exponential distribution; two classes of users.

negative exponential distribution, with the same mean values considered for the hyper-exponential case. Fast users (class 1) require a larger number of handovers per call, with respect to slow users, and thus are more frequently blocked. Indeed, the blocking probability for new call requests is the same for slow and fast users, but fast users are more frequently dropped. In other words, the reduced mobility of slow users makes them safer than fast users once their call has been established. Thus, the number of fast users' calls is smaller than that of slow users.

Under the same scenarios, Figure 13.29 plots the blocking probability versus the nominal load. Since the number of channels per call is the same for the two classes of users, that is, $C_1 = C_2 = 1$, all users have the same blocking probability. For the case of hyper-exponential dwell times, the blocking probability is smaller. This is due to the fact that distributions with large variance produce many short dwell times, and few quite long dwell times. Short dwell times correspond to calls, which request a large number of handovers, and thus undergo a large dropping probability, thus resulting in a lower system load.

13.6 A Look at the Literature

The central role that wireless telephony gained in the previous years within the world of telecommunications generated a vast amount of research and of literature in the field of performance modeling of cellular networks. Thus, in what follows, we will not provide a comprehensive overview of the literature in this field; rather, we will mention some modeling approaches, which are closely related to the simple models presented earlier.

As already mentioned, a very common modeling assumption is that the arrival process of new call requests is Poisson. However, in some cases this assumption is considered not to be realistic. Indeed, besides being difficult to properly select the arrival rate, as we discussed, the Poisson assumption for the service request arrival process may also be questionable in two common cases: small cell size and hierarchical cellular structure. In the first case, as we said, when the cell size is small, finite population models are more suitable to describe the user behavior. Indeed, when the cell size is small, the population of users, which may issue a service request is small too, so that the request arrival rate decreases with the number of active calls. This approach is followed by Reference 32, which models also the retrial phenomenon. The Poisson assumption is sometimes dropped also in the case of hierarchical cellular networks, in which the same area is covered by the superposition of two layers of cells of different size, which are usually called microcells and macrocells. Users try to access the radio resources of microcells, and if they do not succeed they overflow to macrocells. In this case, the overflow traffic requires some extra care in the model development, due to its high degree of burstiness. In Reference 12 the overflow traffic of users entering the macrocell from all underlying microcells is modeled as an Interrupted Poisson Process (IPP). Instead, in Reference 33 an IPP is used to model the overflow from each microcell; the aggregate overflow traffic from all microcells into the macrocell is described by the Markov Modulated Poisson Process (MMPP) which derives from the superposition of all IPPs. A similar approach is adopted also in Reference 14: the aggregate overflow traffic is modeled by a MMPP, which approximates the superposition of IPPs; the approximation is such that the two processes have the same values of the first three moments of the distribution of interarrival times. A different approach is proposed in Reference 18, where the overflow process is modeled by computing the average residual call service time. The network covering a highway is considered in Reference 30, where a deterministic fluid model as well as a stochastic model based on a nonhomogeneous Poisson arrival process are considered. Similarly, while the process of incoming handover requests is often assumed to be Poisson, in some papers different choices are made. In Reference 25 incoming handovers arrive according to a two-state MMPP derived from the analysis of a cluster of cells. In Reference 26 a technique is used to match the first two moments of the handover arrival process, in the case of networks with and without channel reservation.

Many papers are devoted to the derivation of accurate descriptions of the user behavior. Different models are derived based on different assumptions for the distribution of dwell times and call durations. Just to mention a few, in Reference 6 a general distribution of dwell times is assumed, while different kinds of distribution are considered for the call duration. Call arrivals are Poisson; the blocking probability is given, and the authors derive a number of performance metrics including the expected effective call duration. A similar approach is followed in Reference 7, by considering a general distribution of the call duration. In Reference 16 the case of general call duration and dwell time is considered when the policy of channel reservation to handovers is adopted. The results are obtained by introducing some approximations based on the Modified Offered Load techniques, and the optimal channel reservation rule is derived. In Reference 24 combinations of hyper-exponentials are employed for the dwell time, and a multidimensional Markov chain model is derived; in References 3 and 5 the hyper-Erlang distributions is used, instead. Phase-type distributions are employed in Reference 24, by expressing the distribution of both dwell times and call durations as sums of hyper-exponential distributions.

The distribution of dwell times can also be characterized from the cell shape and the users mobility in terms of speed and direction. In Reference 34 the distribution of dwell times and channel holding times, and the average number of handovers, are characterized based on general mobility parameters. It is shown that dwell times can be described by a Gamma distribution, and that channel holding times can be approximated by an exponential distribution. In Reference 2 dynamic channel allocation strategies are investigated, and a mobility model is derived based on the cell geometry. In Reference 9 it is shown that under some mobility model, the channel holding time follows an exponential distribution; however, this assumption may not always be true; for example, the assumption fails when only few direction changes occur in the user mobility. In Reference 23 geometric models are used to predict handover rates, and the relations between the handover rates and the user density are derived.

Handover flow rate balance was first proposed in Reference 11, and was then used in many other papers; see, for example [15,17,19,20,22,27,30]. In Reference 19 the handover flow rate balance is employed by assuming an Erlang distribution for the call duration and generally distributed dwell times. A Markovian approach (exponential assumptions are made for both call durations and dwell times) is used in Reference 27, where multidimensional birth–death processes are employed to model a hierarchical overlay scheme with channel reservation for handovers. In Reference 8 it is shown that an Erlang-B formula based on the equivalent traffic load can accurately predict blocking probability in a cellular network supporting user mobility.

Simple models were often used to derive schemes to favor handover requests over new call requests, so that by reducing the dropping probability a better service can be provided to users. Besides channel reservation to handovers, the possibility of queueing incoming handover requests when all channels are busy is also frequently considered. This technique takes advantage of the fact that adjacent cells overlap at the borders. Channel reservation and handover request queueing are investigated in Reference 11 under exponential assumptions. Schemes based on queueing handover requests are also investigated in Reference 29 under exponential assumptions by means of Markov chain models. A similar approach is employed in Reference 31 to study dynamic reordering of the queue of handover requests. Different channel allocation schemes to favor handovers over new call requests are also studied in Reference 21, by assuming generally distributed dwell times. The scenario of a hierarchical cellular network for highways is considered in Reference 4, where channel reservation schemes are also compared with schemes which exploit the hierarchical structure of the network. A Markovian modeling approach is also employed by Reference 10 for a scheme in which new calls, instead of handovers, are queued under heavy traffic conditions, so that handovers are favored by having access to some reserved channels, and, at the same time, new calls are not blocked but delayed, so that the carried traffic does not decrease. A Markov chain model is employed also in Reference 1 to study schemes based on the possibility to queue both new calls and handovers; the user impatience is modeled. The sub-rating technique to reduce dropping probability, which consists in dividing full-rate channels into lower-rate channels when handovers need extra resources, is investigated in Reference 20 by means of a Markovian model and handover flow balance.

A hierarchical cellular structure with different classes of user mobility is considered in Reference 13; the policy to choose whether to serve users with the microcell or macrocell channels is based on the cell channel occupancy and on the user mobility. A similar scenario is considered in Reference 22, where hierarchical cellular networks with classes of mobility are dimensioned so as to meet some QoS constraint on the blocking and dropping probability. In Reference 28 all the arrival processes are Poisson, while generally distributed dwell times and call durations are assumed in a hierarchical cellular network.

13.7 Conclusions and Outlook

The models presented in this chapter provide simple examples of approaches that can be exploited for the performance analysis and design of cells in mobile telephony cellular networks.

The basic model is extremely simple, yet, it allows the estimation of the cell behavior given the traffic load. By making the users' mobility description more accurate, the model can be easily improved, so that the impact of mobility on the cell performance can be evaluated, and dropping or blocking probabilities can be computed.

The simplicity of the models makes them flexible and improvable, thus becoming suitable for the description of different scenarios. Channel allocation policies can be easily introduced in the models; in particular, the case of channel reservation to handovers was presented as an example. Also, the models allow considering different classes of users; this is quite useful to expand the capabilities of the model. As examples we discussed scenarios with different classes of user mobility, with different types of services, and with generally distributed dwell times and call durations.

Thanks to their flexibility, the simple models that we presented can be extremely useful for cellular network design and planning. Indeed, they allow to get a first feeling of how a system performs, to compare channel allocation policies, to dimension the cell radio resources. Moreover, their simplicity makes the presented models easy to solve, and thus suitable for inclusion in complex tools for network design and planning.

References

[1] C.-J. Chang, T.-T. Su, Y.-Y. Chiang, "Analysis of a cutoff priority cellular radio system with finite queueing and reneging/dropping," *IEEE/ACM Transactions on Networking*, vol. 2, pp. 166–175, 1994.

[2] E. Del Re, R. Fantacci, G. Giambene, "Handover and dynamic channel allocation techniques in mobile cellular networks," *IEEE Transactions on Vehicular Technology*, vol. 44, pp. 229–237, 1995.

[3] Y. Fang, I. Chlamtac, "A new mobility model and its application in the channel holding time characterization in PCS networks," *IEEE Infocom '99*, New York, March 1999.

[4] S.A. El-Dolil, W.-C. Wong, R. Steele, "Teletraffic performance of highway microcells with overlay macrocell," *IEEE Journal on Selected Areas in Communications*, vol. 7, pp. 71–78, 1989.

[5] Y. Fang, I. Chlamtac, "Teletraffic analysis and mobility modeling of PCS networks", *IEEE Transactions on Communications*, vol. 47, pp. 1062–1072, 1999.

[6] Y. Fang, I. Chlamtac, Y.B. Lin, "Modeling PCS networks under general call holding time and call residence time distributions," *IEEE/ACM Transactions on Networking*, vol. 6, pp. 893–906, 1988.

[7] Y. Fang, I. Chlamtac, Y.B. Lin, "Call performance for a PCS network," *IEEE Journal on Selected Areas in Communications*, vol. 15, pp. 1568–1581, 1997.

[8] G.J. Foschini, B. Gopinath, Z. Miljanic, "Channel cost of mobility," *IEEE Transactions on Vehicular Technology*, vol. 42, pp. 414–424, 1993.

[9] R.A. Guerin, "Channel occupancy time distribution in a cellular radio system," *IEEE Transactions on Vehicular Technology*, vol. 35, pp. 89–99, 1987.

[10] R.A. Guerin, "Queuing-blocking system with two arrival streams and guard channels," *IEEE Transactions on Communications*, vol. 36, pp. 153–163, 1988.

[11] D. Hong, S. Rappaport, "Traffic model and performance analysis for cellular mobile radio telephone systems with prioritized and non-prioritized handoff procedures," *IEEE Transactions on Vehicular Technology*, vol. 35, pp. 77–92, 1986.

[12] L.-R. Hu, S.S. Rappaport, "Personal communication systems using multiple hierarchical cellular overlays," *IEEE Journal on Selected Areas in Communications*, vol. 13, pp. 1886–1896, 1995.

[13] B. Jabbari, W.F. Fuhrmann, "Teletraffic modeling and analysis of flexible hierarchical cellular networks with speed-sensitive handoff strategy," *IEEE Journal on Selected Areas in Communications*, vol. 15, pp. 1539–1548, 1997.

[14] X. Lagrange, P. Godlewski, "Teletraffic analysis of hierarchical cellular network," *IEEE 45th Vehicular Technology Conference*, vol. 2, Chicago, IL, pp. 882–886, July 1995.

[15] K.K. Leung, W.A. Massey, W. Whitt, "Traffic models for wireless communication networks", *IEEE Journal on Selected Areas in Communications*, vol. 12, pp. 1353–1364, 1994.

[16] W. Li, A.S. Alfa, "Channel reservation for handoff calls in a PCS network," *IEEE Transactions on Vehicular Technology*, vol. 49, pp. 95–104, 2000.

[17] Y. Lin, "Modeling techniques for large-scale PCS networks," *IEEE Communication Magazine*, vol. 35, pp. 102–107, 1997.

[18] Y.B. Lin, F.F. Chang, A. Noerpel, "Modeling hierarchical microcell/macrocell PCS architecture," *ICC'95*, Seattle, WA, June 1995.

[19] Y.-B. Lin, I. Chlamtac, "A model with generalized holding and cell residence times for evaluating handoff rates and channel occupancy times in PCS networks," *International Journal of Wireless Information Networks*, vol. 4, pp. 163–171, 1997.

[20] Y.B. Lin, A. Noerpel, D. Harasty, "The subrating channel assignment strategy for PCS handoffs," *IEEE Transactions on Vehicular Technology*, vol. 45, pp. 122–130, 1996.

[21] Y.B. Lin, S. Mohan, A. Noerpel, "Queueing priority channel assignment strategies for handoff and initial access for a PCS network," *IEEE Transactions on Vehicular Technology*, vol. 43, pp. 704–712, 1994.

[22] M. Meo, M. Ajmone Marsan, "Approximate analytical models for dual-band GSM networks design and planning," *Infocom 2000*, Tel Aviv, Israel, March 2000.

[23] S. Nanda, "Teletraffic models for urban and suburban microcells: cell sizes and handoff rates," *IEEE Transactions on Vehicular Technology*, vol. 42, pp. 673–682, 1993.

[24] P.V. Orlik, S.S. Rappaport, "A model for teletraffic performance and channel holding time characterization in wireless cellular communication with general session and dwell time distributions," *IEEE Journal on Selected Areas in Communications*, vol. 16, pp. 788–803, 1998.

[25] P.V. Orlik, S.S. Rappaport, "On the hand-off arrival process in cellular communications," *IEEE Wireless Communications and Networking Conference (WCNC)*, Chicago, IL, pp. 545–549, September 1999.

[26] M. Rajaratnam, F. Takawira, "Nonclassical traffic modeling and performance analysis of cellular mobile networks with and without channel reservation," *IEEE Transactions on Vehicular Technology*, vol. 49, pp. 817–834, 2000.

[27] S.S. Rappaport, L.-R. Hu, "Microcellular communication systems with hierarchical macrocell overlays: traffic performance models and analysis," *Proceedings of the IEEE*, vol. 82, pp. 1383–1397, 1994.

[28] K. Yeo, C.-H. Jun, "Modeling and analysis of hierarchical cellular networks with general distributions of call and cell residence times," *IEEE Transactions on Vehicular Technology*, vol. 51, pp. 1361–1374, 2002.

[29] C.H. Yoon, C.K. Un, "Performance of personal portable radio telephone systems with and without guard channels," *IEEE Journal on Selected Areas in Communications*, vol. 11, pp. 911–917, 1993.

[30] T.-S. Yum, K.L. Leung, "Blocking and handoff performance analysis of directed retry in cellular mobile systems," *IEEE Transactions on Vehicular Technology*, vol. 44, pp. 645–650, 1995.

[31] S. Tekinay, B. Jabbari, "A measurement-based prioritization scheme for handovers in mobile cellular networks," *IEEE Journal on Selected Areas in Communications*, vol. 10, pp. 1343–1350, 1992.

[32] P. Tran-Gia, M. Mandjes, "Modeling of customer retrial phenomenon in cellular mobile networks," *IEEE Journal on Selected Areas in Communications*, vol. 15, pp. 1406–1414, 1997.

[33] Y. Zhou, B. Jabbari, "Performance modeling and analysis of hierarchical wireless communications networks with overflow and take-back traffic," *IEEE 9th International Symposium on Personal, Indoor and Mobile Radio Communications*, vol. 3, pp. 1176–1180, 1998.

[34] M.M. Zonoozi, P. Dassanayake, "User mobility modeling and characterization of mobility patterns," *IEEE Journal on Selected Areas in Communications*, vol. 15, pp. 1239–1253, 1997.

14

Location Information Services in Mobile Ad Hoc Networks

14.1 Introduction...14-319
    Geographic Forwarding • Location Services
14.2 Proactive Location Database Systems 14-322
    Home Region Location Services • Quorum-Based Location
    Services
14.3 Proactive Location Dissemination Systems 14-330
    DREAM Location Service • Simple Location Service •
    Legend Exchange and Augmentation Protocol • Ants •
    Geographic Region Summary Service • Dead Reckoning: A
    Location Prediction Technique
14.4 Reactive Location Systems...........................14-336
    Reactive Location Service • Reactive Location Service (RLS′)
14.5 Conclusions and Future Research 14-337
Acknowledgments...14-337
References 14-337

Tracy Camp

14.1 Introduction

Compared to wired networks, routing a packet from a source to a destination in a mobile ad hoc network (MANET) is a much more difficult problem due to the interaction of three fundamental challenges. The first challenge is contention. Nodes in a MANET require wireless communication and the nature of wireless communication results in significant contention for the shared medium (the wireless channel). The second challenge is congestion. Compared to a wired network, a wireless network has decreased bandwidth, which results in a much higher congestion. Finally, and most importantly, there is a unique challenge in MANETs created by node mobility. Nodes in a MANET may move and cause frequent, unpredictable topological changes. This changing network topology is the key challenge that MANET routing protocols must overcome.

Several unicast routing protocols have been proposed for MANETs in an attempt to overcome the routing challenges. Example protocols include Dynamic Source Routing (DSR) [38,39], Ad hoc On-demand Distance Vector (AODV) [56,58], and the Zone Routing Protocol (ZRP) [26]; a review of these protocols is available in [60,66] and a performance comparison for a few of the protocols are in References 10, 17, 18, 37, and 57.

In an effort to improve the performance of unicast communication, some of the proposed MANET unicast routing protocols use location information in the routing protocol. A few of the proposed algorithms include the Location-Aided Routing (LAR) algorithm [44], the Distance Routing Effect Algorithm for Mobility (DREAM) [4], the Greedy Perimeter Stateless Routing (GPSR) algorithm [40], the Geographical Routing Algorithm (GRA) [35], the Terminode Remote Routing (TRR) protocol [5,6], the Scalable Location Update-based Routing Protocol (SLURP) [68], the Depth First Search (DFS) algorithm [64], Greedy-Face-Greedy (GFG) algorithm [19], and the GPS Zone Routing Protocol (GZRP) [9]. Reviews of position-based routing protocols for ad hoc networks are available in References 23, 24, 51, and 63.

Several benefits of position-based routing protocols exist:

1. *Reduced overhead*, since the establishment and maintenance of routes are (usually) not required in a protocol that uses location information for routing. In position-based routing protocols, each node maintains accurate neighborhood information and a rough idea of the destination's location.
2. *Scalability*, since routing protocols that do not use location information do not scale [35,46].
3. *Localized algorithm*, since a node determines which neighbor to forward a message only on its neighborhood and the destination's location.
4. *Higher performance*, as shown in the simulation results of several articles (e.g., References 13 and 44).

One of the main challenges of a position-based routing protocol is to learn an accurate position or location for a packet's destination. While some protocols (e.g., DREAM and LAR) include the exchange of location information as a part of its protocol, most position-based routing protocols assume a separate mechanism that provides location information. For example, knowledge about the location of a destination node is assumed available in all geographic forwarding routing protocols (e.g., References 20, 40, and 65). In fact, in the simulation results presented for these protocols, location information is typically provided to all mobile nodes without cost [20,40]. Thus, overhead in simulation results on protocols such as DREAM are much higher, because DREAM includes the task of maintaining location information on destination nodes.

In this chapter, we review all the proposed location information services that exist in the literature to date. We classify these location information services into three categories: proactive location database systems, proactive location dissemination systems, and reactive location systems.

As mentioned, in a position-based routing protocol, each node needs to determine its location in the ad hoc network. A node can determine its location if the Global Positioning System (GPS) is available. If GPS is unavailable, a relative coordinate system, such as the ones proposed in References 5, 11, 15, and 16 can be used. In a relative coordinate system, a node estimates its distance from a neighbor using signal strength. Absolute and relative location systems for indoor and outdoor environments are presented in Reference 30. Finally, if some nodes have the ability to determine their locations, De Couto and Morris [20] propose that nodes that are aware of their location serve as location proxies for location ignorant nodes.

The rest of this chapter is organized as follows: first, in Section 14.1.1, we detail one popular type of position-based routing algorithm (i.e., geographic forwarding). As mentioned, geographic forwarding algorithms require a separate mechanism (called a location service) to provide location information on nodes in the ad hoc network. In Section 14.1.2, we give an overview of the location services reviewed and discuss how we classify the reviewed location services in the rest of this chapter. The next three main sections (i.e., Sections 14.2 to 14.4) detail the location services reviewed. Finally, in Section 14.5, we state our conclusions and comment upon the future work that is needed on location services.

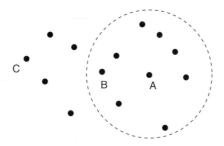

FIGURE 14.1   An example of the MFR scheme.

14.1.1 Geographic Forwarding

When location information is available, geographic forwarding can be used in place of establishing a route from a source node to a destination node. In geographic forwarding, which is a greedy scheme, each node forwards the packet to the neighbor closest to the packet's destination [21,31,65]; in other words, the packet is sent to a neighboring node that makes the most forward progress. (HELLO packets are used to maintain neighborhood information.) The goal of this strategy, which is called *Most Forward with fixed Radius* (MFR) in Reference 31, is to minimize the number of hops to the destination. Figure 14.1 shows an example of this greedy MFR scheme. If node *A* is forwarding a packet toward *C*'s location, node *A* chooses node *B* to forward the packet further.

We note that Hou and Li [31] evaluate three strategies in choosing a node to forward a packet, and two of these strategies use variable transmission power. The authors determined that a network can obtain higher throughput if a node adjusts its transmission power to reach the nearest node in a forward direction. In other words, fewer conflicts occur in this strategy than in the one that minimizes the number of hops to the destination.

If no neighbor node is closer to the packet's destination than the node currently holding the packet, then the packet has reached a "dead-end".[1] Several solutions have been proposed as a recovery strategy to a dead-end situation (e.g., References 7, 8, 19, 20, 21, 35, 40, 65, and 68). In Reference 21, a node recursively searches its neighbor's neighbors for a node that is closer to the destination than itself. If no node exists in a forward direction, Takagi and Kleinrock [65] propose that the packet is sent to the closest node in a backward direction. A node asks its neighbors for an alternative route to the destination in SLURP [68]. In the GPSR [40] protocol, a planar subgraph is created to route around the dead-end; this method was first proposed in References 7, 8, and 19. If a packet does not reach its destination via geographic forwarding, De Couto and Morris [20] propose the source node to randomly choose an intermediate node through which to forward packets; incorporating a stop at an intermediate node on the way to the destination may allow the packet to bypass a dead-end situation. Finally, a route discovery is initiated in the GRA protocol for a "dead-end" situation [35].

The authors of Reference 68 mention that precise coordinates on the destination are unknown when the first data packet is transmitted; instead, the source only knows the home region (see Section 14.2.1) where the destination is located. Thus, SLURP routes packets with MFR toward the center of the home region. Once a node inside the home region receives the packet, SLURP uses DSR [38,39] to determine a route to the destination within the home region.

14.1.2 Location Services

As mentioned, a location service is responsible for providing location information on nodes in the network. There are two general types of location services: proactive location services and reactive location

---

[1]If the network is dense, a "dead-end" situation should not occur [32,46].

services (RLSs). Proactive location services are those protocols that have nodes exchange location information periodically. Reactive location services query location information on an as and when needed basis.

Proactive location services can be further classified into two general types: location database systems and location dissemination systems [32]. In a location database system, specific nodes in the network serve as location databases for other specific nodes in the network. When a node moves to a new location, the node updates its location database servers with its new location; when location information for a node is needed, the node's location database servers are queried. In a location dissemination system, all nodes in the network periodically receive updates on a given node's location. Thus, when a given node requires location information on another node, the information is found in the node's location table. "A location dissemination system can be treated as a special case of a location database system, where every node in the network is a location database server for all the others" [32].

In comparing location database systems to location dissemination systems, we find that:

- A node maintains more state in a location dissemination system than in a location database system, since location information is maintained on all nodes in the network.
- An update in a location dissemination system is (usually) more costly than an update in a location database system, since an update is disseminated to all nodes in the network.
- A location dissemination system is more robust than a location database system, since location information is maintained by all nodes in the network.
- A location query in a location dissemination system is (usually) cheaper than a location query in a location database system, since the desired location information is available in a node's location table.

In Reference 51, location services are classified by:

1. How many nodes host the service (i.e., some specific nodes or all nodes).
2. How many positions of nodes a location server maintains (i.e., again, some specific nodes or all nodes).

As an example, in a traditional cellular network several dedicated location servers maintain location information about all nodes. Thus, a cellular network would be classified as a *some-for-all* approach. Typically, location database systems would be classified as an *all-for-some* approach and location dissemination systems would be classified as an *all-for-all* approach. In our discussion of location services, we refer to this classification as the *Mauve Classification*.

To maintain location information on other nodes in the network, we assume that each mobile node maintains a location table. This table contains an entry on every node in the network whose location information is known, including the node's own location information. A table entry contains (at least) the node identification, the coordinates of the node's location based on some reference system, and the time this location information was obtained. When a location request occurs, a node will first look into its location table for the information. If the information is not available in the table, a location database system will initiate a location query; if no result is returned, the node periodically transmits queries according to a time-out interval. On the other hand, a location dissemination system will (typically) flood a location request packet to all nodes in the MANET. A reply to this location request packet is transmitted by the node whose location was requested; nodes that receive a reply to a location request update their tables in a promiscuous manner.

We review proactive location database systems in Section 14.2. In Section 14.3 we present proactive location dissemination systems. Lastly, in Section 14.4, we present reactive location systems.

14.2 Proactive Location Database Systems

In this section, we present two types of proactive location database systems: home region location services and quorum-based location services. In each type, we review four protocols that have been proposed in the literature. As mentioned, in a location database system, specific nodes in the network serve as location databases for other specific nodes in the network. A node will update its location database servers when it moves to a new location; a node's location database servers are then queried when location information on the node is needed. According to the *Mauve Classification* [51], most location database systems would be classified as an *all-for-some* approach. That is, all nodes in the network serve as database servers, and each database server maintains location information for some specific nodes in the network.

14.2.1 Home Region Location Services

Several home region location services, which are similar to Mobile IP [55] and our cellular phone network, have been proposed [22,62,68]. In these protocols, each node in the network is associated with a home region. A home region (called virtual home region in References 5, 22, and 34 and home agent in Reference 62) is either defined by a rectangle or a location and a circle, with radius R, around that location. In Reference 62, the center of the circular home region for a node is defined by the node's initial location in the network; each node informs every other node about its initial location via a network-wide broadcast. In References 5, 22, and 34, the center of the circular home region is derived by applying a well-known hash function on the node's identifier. The ad hoc network in Reference 68 is logically arranged into equal-sized rectangular regions, and a static mapping, f, associates a node's identifier (ID) with a specific region. This many-to-one mapping is known by all nodes in the network [68].

If the number of nodes in the home region is too high, which leads to too much overhead, or too low, which leads to a lack of robustness, the radius of the home region is modified in References 5, 22, and 34. In other words, the radius R dynamically adapts with a goal of maintaining an approximate constant number of nodes in each home region.

All nodes within a given node's home region (i.e., all nodes currently within the defined circle or rectangle) maintain the location information for the node. Thus, when a node moves to a new location, it sends its current location to the nodes in its home region. In Reference 62, a location update is transmitted when the number of created or broken edge links reaches a fixed threshold. In Reference 5, three location update schemes are proposed: timer-based (i.e., periodic updates), distance-based (i.e., update when the node has moved more than a threshold), and predictive distance-based. In the predictive distance-based scheme, a node sends its location and velocity to the home region. The node transmits a location update when the difference between the predicted location and the node's actual location is greater than a threshold. Lastly, in Reference 68, a node transmits a location update when it moves from one home region to another.

Figure 14.2 gives an example of the rectangular home region approach. Suppose the home region of node D is $R4$. When D moves from $R3$ to $R6$, nodes A, B, and C will receive the location update packet.

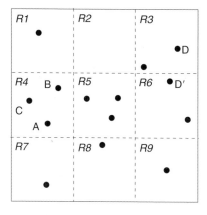

FIGURE 14.2   An example of the home region location service.

A location update packet travels to a home region via a geographical forwarding protocol. When a node inside the home region receives a location update packet, the node broadcasts the packet to all nodes within the home region. To decrease the overhead in this broadcast, Woo and Singh [68] suggests that a node include its neighbor list in each location update packet. When a node inside the home region receives a location update packet, it rebroadcasts the packet only if new neighbors will be reached. This technique is called Flooding with Self Pruning [49,67].

When a source wants to transmit a packet to a destination, the source queries the destination's home region for the destination's current location. Specifically, a location request packet is sent from the source to the center of the home region via a geographical forwarding protocol. Each node that receives this request inside the home region responds with the current location of the destination [5]. If a location cache is recent enough (i.e., 10 sec or less), an intermediate node responds to the location request in Reference 68. To avoid a reply storm, the location response is transmitted after a random waiting period based on the age of the cached entry.

The authors of Reference 68 give several details for their home region location service approach. For example, they propose a solution when no nodes are currently in a given node's home region (e.g., R2 in Figure 14.2). In this situation, all the regions surrounding the empty home region become home regions for nodes that map to the empty home region. Thus, if a home region is in the center of the network, the eight surrounding rectangular regions become proxies for the empty home region. For example, in Figure 14.2, regions R1, R3, R4, R5, and R6 all act as a home region for nodes whose identifiers map to R2. See Reference 68 for full details on the maintenance of home regions.

A fourth home region approach, the Hierarchical State Routing (HSR) protocol, has also been proposed [53,54]. The main difference between this scheme and the three ones previously discussed is that the HSR protocol assumes that the network is organized hierarchically. Thus a location update or request must travel along a hierarchical tree to the home region.

14.2.2 Quorum-Based Location Services

Several quorum-based location services have been developed and are based on replicating location information at multiple nodes that act as information repositories [2,27,28,41,46,48,61]. In other words, read and write quorums (i.e., subsets of the nodes) are defined in the network. When a node determines whether a location update is needed, it transmits the update to a write quorum. When a source node wants to transmit a packet to a destination node, it requests the location for the destination from a read quorum. The main challenge in a quorum-based system is to define the read or write quorums in such a way as to maximize the probability of a query success. That is, the goal is to define the quorums such that a read quorum for a node intersects the write quorum for any other node. Thus, up-to-date location information can be obtained for any given destination.

One simple solution to forming the read or write quorums is as follows: the write quorum for any node i is i and the read quorum for any node j is all nodes in the network. In this solution, the write quorum is of size 1 and the read quorum is of size n where n is the number of nodes in the network. Thus, while the cost of a location update is small, the cost of a location request is very large. In this section, we present quorum-based location services that attempt to minimize the combined cost of updates and queries.

14.2.2.1 Uniform Quorum System

The Uniform Quorum System (UQS) is presented in References 27, 28, and 48. During the initial setup of UQS, a subset of the network nodes are chosen that best serve as the network's virtual backbone (VB); for example, chosen nodes may be those that are uniformly distributed throughout the network. (See Reference 48 for implementation issues concerning the VB, for example, the maintenance of the VB in the presence of node disconnections.) Quorums are then defined as subsets of the VB nodes, such that any two quorums intersect.

To update its location, a node transmits its new location to the nearest VB node. The VB node then sends the location update to a randomly chosen quorum. In other words, there is no fixed association between the node and the quorum updated. The procedure for a location request is similar: a node transmits a location request to its nearest VB node, which then contacts a randomly chosen quorum for the location.

Haas and Liang propose a node updates its location in three different manners:

1. *VB change*: Whenever the node's nearest VB node changes, a location update occurs.
2. *Location request*: Whenever the node requests a location, a location update occurs in the queried quorum.
3. *Periodically*: Whenever a given time period has elapsed without an update, a location update occurs. This update will ensure time-stamps on location information are kept current.

Haas and Liang discuss several methods to generate quorums in Reference 28. Since the size of the quorum intersection is a configuration parameter, an implementor of UQS can decide the overhead to resiliency trade-off. That is, a larger quorum size means a larger cost for location updates and location requests; however, a larger quorum size also means a larger quorum intersection. Regarding *Mauve's Classification*, UQS can be configured to operate as an all-for-all, all-for-some, or some-for-some approach. Since the VB nodes are typically a subset of all network nodes and the nodes in a quorum are typically a subset of the VB nodes, UQS will most often be configured as a some-for-some approach.

14.2.2.2 A Column or Row Method

Another quorum-based location service is proposed in Reference 61. When a node determines a location update is needed,[2] the node transmits its new location to all nodes located in its *column*. A node's *column* is all the nodes to the north and south of the node's current location. In other words, the node transmits a location update packet that travels in both the north and south direction. While the thickness of the column can be configured, Stojmenovic [61] defines a column as one path for ease of understanding. All nodes on the path, and all neighbors of nodes on the path, receive the updated location information. To increase the column size to two, neighbor nodes would need to retransmit each location update.

When a source node initiates a location query for a destination, the location request is first sent to all neighbors within q hops of the source. If there is no reply to this request, the location request proceeds in the same manner as a location update. That is, a source node transmits a location request in its *row*, which traverses in an east–west direction. (Again, the thickness of the row can be configured.) The location request packet contains the time of the most recent location information known by the source. If a node that receives the location request has more recent location information, a location reply packet is sent. Once the easternmost or westernmost node is reached, the location request packet is transmitted via a geographic forwarding protocol toward the most recent location information on the destination. A second location request packet can be sent from the source node as well. This location request is transmitted toward the most recent location information on the destination using a geographic forwarding protocol.

One problem with this column or row method is illustrated in Figure 14.3. When node N transmits a location update, its northernmost node is currently A. If the location update terminates at A, then a search by source S, which follows the $X-Y-Z$ path, will not return the most recent location information on node N. To ensure an intersection between the north/south and east/west directions occurs, Stojmenovic [61] proposes that node A switches to FACE mode (see References 7 and 8) until another node, more northern, is found.

14.2.2.3 Grid Location Service

Another quorum-based location database system in an ad hoc network is called the Grid Location Service (GLS) [46]. The set of GLS location servers is determined by a predefined geographic grid and a predefined ordering of mobile node identifiers in the ad hoc network. There are three main activities in GLS, which we detail in the following discussion: location server selection, location query request, and location server update.

---

[2]Due to results in Reference 41, Stojmenovic [61] chose to transmit a location update when the number of created or broken edge links reaches a fixed threshold (see Section 14.2.2.5).

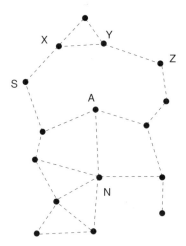

FIGURE 14.3 A problem with the column or row location service.

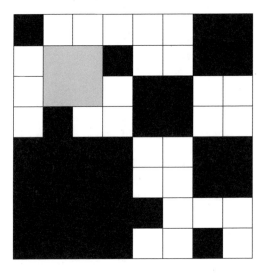

FIGURE 14.4 An example grid.

Initially, the area covered by the ad hoc network is arranged into a hierarchy of grids with squares of increasing size. The smallest square is called an order-1 square. Four order-1 squares make up an order-2 square, four order-2 squares make up an order-3 square, and so on. A few example squares of various orders are shown in Figure 14.4 with dark shading. (This figure is taken from Reference [46] ©2000 ACM, Inc. Reprinted by permission.) Specifically, five order-1 squares, three order-2 squares, one order-3 square, and one order-4 square are shown. An order-n square's lower left coordinates must be of the form $(a2^{n-1}, b2^{n-1})$ for integers a, b [46]. Thus, in Figure 14.4, the lower left coordinates of the lightly shaded square is $(1, 5)$; although this is an example of a 2×2 square, it is not an order-2 square since no integers a and b exist such that $(2a, 2b) = (1, 5)$.

Location Server Selection: A node chooses its location servers by selecting a set of nodes with IDs close to its own ID. Each of the chosen location servers have the least ID greater than the node's ID in that order square. Figure 14.5 provides an example of how node B selects its location servers; B's location servers are circled in the figure. (This figure is taken from Reference [46] ©2000 ACM, Inc. Reprinted by permission.) In this example, B (ID 17) determines which nodes will be its location servers by selecting nodes with IDs closest to its own. As mentioned, a node is defined as closest to B when it has the least ID greater than B.

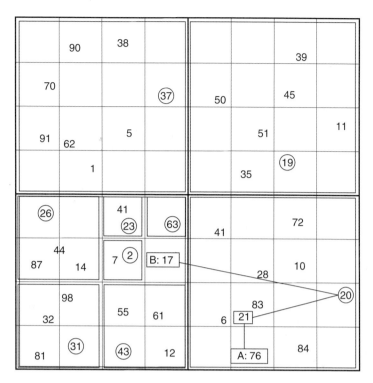

FIGURE 14.5   A GLS example.

In other words, the location server's ID is the smallest number that is greater than B's ID. For example, consider the grid to the left of B's grid. No ID exists that is greater than 17. Since the ID space is considered to be circular (i.e., wraps around), 2 is defined as closer to 17 than 7. B selects three location servers for each level of grid order square, which combine to make the next level of grid order square. For example, in Figure 14.5, B selects one location server from each order-1 square (e.g., 2, 23, and 63) that, along with its own order-1 square, will make an order-2 square. B then selects one location server from each order-2 square (e.g., 26, 31, and 43) that, along with its own order-2 square, will make an order-3 square. Each of the chosen location servers have the least ID greater than B in that order square.

Location Query Request: As a quorum-based location service, a node initiates a location query request when the node needs a location for a destination. Since each node knows all nodes within its order-1 square, the request is first sent to a potential location server for the destination desired in the requesting node's order-2 square. In other words, the location query request packet is forwarded, using geographic forwarding, to a node whose ID is the least greater than or equal to the ID of the destination within the order-2 square. That node then forwards the query using the same algorithm until it reaches a location server for the destination. This location server then forwards the query directly to the destination, which responds to the location query request with its most recent location.

Figure 14.5 illustrates the path of a query packet from A to B. A (ID 76) sends a location query to the node with ID 21, since A knows the location of node 21 and it is the least greater than B (ID 17) in that grid order square. In other words, no node with IDs between 17 and 21 exist in A's order-2 square. Node 21 then forwards the query packet using the same algorithm to the node in the next order square in the grid hierarchy, which is ID 20 in this example. Since node 20 is a location server for B in that grid order square, it knows B's location and is able to forward the query packet directly to B. Since the query packet contains A's location, B responds by sending its current location to A using geographic forwarding.

Location Server Update: Since GLS is a quorum-based location service, a location server update occurs when a node moves. Each node, acting as a location server for the nodes it serves, maintains two tables.

The location table holds the location of nodes that have selected this node as its location server; each entry contains a node's ID and geographic location. A node also maintains a location cache, which is used when a node originates a data packet; the cache holds information from the update packets that a node has forwarded. Because a node uses the routing table maintained by geographic forwarding for its order-1 square neighbors, it does not need to send GLS updates within its order-1 square.

When a node moves a given threshold, it must send an update packet to all of its location servers. To avoid excessive update traffic, the update frequency is calculated using a threshold distance and the location servers' square order [46]. The threshold distance is the distance the node has traveled since the last update. For example, suppose a node updates its location servers in order-2 squares when it moves a distance d; the node then updates its location servers in order-3 squares when it moves a distance $2d$. In other words, a node updates its location servers at a rate proportional to its speed, and the distant location servers are updated less frequently than the nearby location servers.

Before a node sends a data packet, it checks its location cache and location table to find the location of the destination. If it finds an entry for the destination, it forwards the data packet to that location. Otherwise, it initiates a location query using GLS, and stores the data packet in a buffer waiting for the query result. If no result is returned, the node periodically transmits queries according to a time-out interval. Once it gets the query result, it will use geographic forwarding to send the data packet.

Location Query Failures: There are two types of failure caused by node mobility [46]: a location server may have out-of-date information or a node may move out of its current grid. The solution for the first type of failure is to use the old location information. To overcome the second type of failure, which is more common, the moving node places a forwarding pointer in the departed grid. This forwarding pointer points to the grid the node has just entered. In other words, before a node moves out of a grid, it broadcasts its forwarding pointer to all nodes in the grid. Any node in this grid stores the forwarding pointer associated with the node that just left the grid; a node discards forwarding pointers when it departs a grid.

To share forwarding pointers with other nodes that have entered the order-1 square, a randomly chosen subset of forwarding pointers are transmitted with each HELLO packet [46]. (Table 14.1 shows the HELLO packet fields.) A node receiving a HELLO packet adds forwarding pointers to its own collection of forwarding pointers only if the broadcaster of the HELLO packet is in the same grid as the node's grid. A forwarding pointer allows a data packet to be forwarded to a grid that may contain the node. We note that a dense network is needed for the forwarding pointers to be effective and all simulation results presented in Reference 46 use dense networks. See References 25, 36, and 43 for further performance results on GLS.

14.2.2.4 Doubling Circles

The location service in Reference 2 is similar to GLS [46]. However, instead of arranging the network into squares of increasing size, [2] arranges the network into circles of increasing size centered on a node's location; the radius of each subsequent circle is 1 m larger than the radius of the previous circle.

Suppose a node was at location i when it transmitted its last location update. When the node moves outside a circle of radius r meters centered at location i, the node transmits a location update to all nodes within a circle of radius $(r + 1)$ meters centered at the node's new location. Figure 14.6 illustrates an example. When node A moves to A′, it transmits a location update packet to all nodes within the dashed

TABLE 14.1   HELLO Packet Fields

HELLO
Source ID   Source location   Source speed
Neighbor list: IDs and location
Forwarding pointers

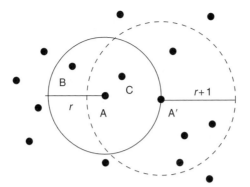

FIGURE 14.6   Example of location update in the doubling circles scheme.

circle of radius $(r+1)$. Multipoint Relaying [59,67], an efficient network-wide broadcast protocol, is used to limit the overhead associated with flooding a location update packet within a given circle.

Similar to other location services, the distance a location update packet propagates corresponds to the distance the node has moved since its previous update [2]. In other words, nodes close to a given destination will have more accurate location information on the destination than nodes farther away from the destination. Thus, when a packet is transmitted toward a destination, forwarding nodes redirect the packet with more up-to-date location information on the destination. For example, node B in Figure 14.6 forwards a packet for A to the center of the solid circle. When node C receives this packet, it redirects the packet to the center of the dashed circle.

14.2.2.5   Quorum-Based Details

As mentioned, the main challenge in a quorum-based system is to define the read/write quorums in such a way as to maximize the probability of a query success. This challenge is especially difficult in a MANET, since network partitions are possible. The authors of Reference 41 propose a quorum-based location service for information dissemination in MANETs with partitions. In Reference 41, the authors assume that the quorum sets are constructed *a priori*, that at least one node intersects any two quorum sets, and the quorum sets are known by every node in the network. Although the article does not propose a scheme to configure the quorums, it does attempt to answer three important questions:

1. When should a node transmit a location update?
2. Where should a node transmit a location update?
3. Which nodes should a node query for location information?

In Reference 41, four strategies are proposed to answer *When should a node transmit a location update?* Based on their performance evaluation, the "absolute connectivity-based policy" was found to be the best trigger update strategy. In this strategy, a node sends an update "when a certain *prespecified number* of links incident on it have been established or broken since the last update" [41]. The authors of Reference 41 argue that absolute distance-based updates and velocity-based updates are ineffective.

Three strategies are proposed to answer the remaining two questions: *Where should a node transmit a location update?* and *Which nodes should a node query for location information?* In all three strategies, a node uses reachability knowledge to determine which quorum set to choose for a location update or location request. That is, a node uses knowledge of unavailable nodes to select a quorum set in the hopes of minimizing a current or future location request failure. In their performance evaluation, the authors found the Eliminate_then_Select (ETS) Strategy was the best location update strategy and the Select_then_Eliminate (STE) Strategy was the best location request strategy in choosing a quorum set to update/query. In ETS, a node first eliminates all quorum sets that have at least one unreachable node. A quorum set is then randomly chosen from the remaining sets. The ETS strategy maximizes the number of nodes that receive an update. In STE, a node first randomly chooses a quorum set and then eliminates unreachable nodes from the set. The STE strategy maximizes the number of quorum sets to query.

14.3 Proactive Location Dissemination Systems

In this section, we present six proactive location dissemination systems. As mentioned, all nodes in the network periodically receive updates on a given node's location in a location dissemination system. Thus, location information should be available for every node in a given node's location table. According to the *Mauve Classification* [51], location dissemination systems would be classified as an *all-for-all* approach. That is, every node in the network maintains location information on every other node in the network.

14.3.1 DREAM Location Service

The DREAM Location Service (DLS) [12] is similar to a location service proposed by the authors of DREAM [4], a position-based routing protocol. Each Location Packet (LP), which updates location tables, contains the coordinates of the source node based on some reference system, the source node's speed, and the time the LP was transmitted. Each mobile node in the ad hoc network transmits an LP to nearby nodes at a given rate and to faraway nodes at another lower rate. The rate a mobile node transmits LPs adapts according to when the mobile node has moved a specified distance from its last update location. Since faraway nodes *appear* to move more slowly than nearby mobile nodes, it is not necessary for a mobile node to maintain up-to-date location information on faraway nodes. This phenomenon is termed the *distance effect* and is illustrated in Figure 14.7. By differentiating between nearby and faraway nodes, the overhead of LPs is limited.

DREAM controls the accuracy of the location information in the network via the frequency that it transmits LPs and how far the LP is allowed to travel [4]. However, details on implementing the proposed location service are lacking in Reference 4. The DLS solution for the transmission of LPs in Reference 12 is:

transmit nearby LP: $\dfrac{T_{\text{range}}}{\alpha} * \dfrac{1}{v} = \dfrac{T_{\text{range}}}{\alpha v}$

transmit far away LP: one for every X nearby LPs   or   at least every Y seconds,

where T_{range} is the transmission range of the mobile node, v is the average velocity of the mobile node, and α is a scaling factor. In DLS, a nearby node is a 1-hop neighbor; LPs to faraway nodes update all nodes in the network, including nearby nodes.

There are some similarities of DLS to the Internet standard Open Shortest Path First (OSPF) [52]. Specifically, both transmit information concerning the local environment to all nodes in the network on a periodic basis. There are, however, three significant differences between these two protocols. First, DLS differentiates between nearby nodes and faraway nodes; OSPF has no such differentiation. Second,

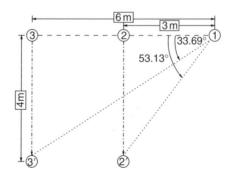

FIGURE 14.7   The distance effect: although nodes 2 and 3 have moved the same distance, from node 1's perspective, $\widehat{212'} > \widehat{313'}$.

a node using DLS will transmit its location information; a node using OSPF will transmit its neighbor information. Lastly, OSPF was not developed for an ad hoc network environment; thus, the rate of OSPF update packets is only based on time but not on the distance moved.

14.3.2 Simple Location Service

The Simple Location Service (SLS) is also a proactive dissemination location service, except this service only transmits location information to neighbors [12]. Specifically, each LP that updates location tables, contains the location of several nodes, the speed of each of these nodes, and the time the LP was transmitted. The rate a mobile node transmits LPs adapts according to location change:

$$\left(\frac{T_{\text{range}}}{\alpha}\right) * \left(\frac{1}{v}\right) = \frac{T_{\text{range}}}{\alpha v} \quad \text{or} \quad \text{at least every } Z \text{ seconds,}$$

where T_{range} is the transmission range of the node, v is the average velocity of the node, and α, which is a constant optimized through simulation, is a scaling factor. For a given T_{range}, the frequency a mobile node transmits LPs adapts according to the time used by the mobile node to move a specified distance from its last update location. The faster the node moves, the higher frequency the mobile node transmits LPs.

In SLS, each LP contains up to E entries from the node's location table [12]. These E entries are chosen from the table in a round robin fashion. That is, each LP transmission shares location information on several nodes in the MANET with the node's neighbors. As multiple LPs are transmitted, all the location information a node knows is shared with its neighbors. A node using SLS also periodically receives a LP from one of its neighbors. The node then updates its location table retaining the most recently received table entries. See Reference 12 for further details on SLS.

There are some similarities of SLS [12] to the Internet standard Routing Information Protocol (RIP) [29]. Specifically, both transmit tables to neighbors on a periodic basis. Differences between these two protocols include the following: SLS sends partial location tables compared to the total routing tables sent by RIP and, unlike RIP, a node using SLS will utilize its current location table in the calculation of its new location table. Lastly, RIP was not developed for an ad hoc network environment. Thus, SLS shares its tables with different neighbors as the network topology changes and, like OSPF, the rate of RIP update packets is only based on time but not on distance moved.

14.3.3 Legend Exchange and Augmentation Protocol

The Legend Exchange and Augmentation Protocol (LEAP) [36] is another proactive location dissemination system. LEAP consists of a legend (i.e., an explanatory list) that migrates (or leaps) from one node to another in a heterogeneous network. The legend's current state is the locations of nodes that have been previously collected. As the legend traverses the network, it collects the location information of each node, and distributes this information to all other nodes in the network.

In LEAP, each node in the ad hoc network periodically broadcasts a packet to its neighbors to announce its existence [36]. This broadcast packet or "hello" packet includes the current node's location information and the time in which the packet was sent.

There are two types of location tables in LEAP [36]. A local location table is stored in every node and includes location information on other nodes. The legend is a global location table and includes location information on all nodes as well as information to decide where to send the legend next. Initially, each node in the MANET has an empty local location table that can store n entries for the n nodes in the network. As mentioned in Section 14.2, each entry in a node's local location table has the following items for each node in the network:

- ID — node ID
- loc_info — location information for the node
- last_update_time — time-stamp for the location information

Each entry in the global location table (or legend) includes the previous three items and the following item:

- v_bit — boolean parameter to show whether the node has been visited by the legend

We note that while n local location tables exist, only one global location table (or legend) exists.

A local location table is updated in two ways [36]. First, every time a node receives a "hello" packet, the node updates corresponding entries for its neighbors in its local location table. Second, when the legend visits a node, both the global location table and the local location table are updated based on the time-stamps for the corresponding entries in the two tables. In other words, for each entry, the most recent entry of one table is stored in the other table. After this update procedure finishes, the legend migrates (or leaps) to another node. Thus, at this time, both the most recently visited node and the legend have identical location information for all visited nodes.

In the first implementation of LEAP [36], only one legend in the network exists. However, the authors state that they plan to consider the effect of multiple legends traversing a flat network. They also plan to develop a *hierarchical* legend-based service that provides current information (e.g., location information) on the state of nodes in a hierarchical network.

Initially, one node is selected to begin the single legend propagation [36]. As an initialization step, that node sets, for each node in the global location table, the visited bit to false and the current location of the node to undefined. The node then updates its current location and location of all known neighbors in the global location table (with a time-stamp), sets its visited bit in the global location table, and then sends the legend to the closest un-visited neighbor. If all neighbors of the current node have been visited, the node sends the legend to the closest un-visited node using the LAR protocol [44].

It is possible that the location information of some nodes are unknown by the currently visited nodes due to a partitioned network. Thus, LEAP cannot tell which node is the legend's closest un-visited node [36]. In this case, LEAP chooses the lowest ID node as the destination for the legend's next migration. If a partitioned network exists, the destination for the legend may not be reachable. When this problem occurs, the legend waits on the current node for a time-out period. When the time-out expires, the destination may now be reachable due to a change in the topology of the network. If the destination is still unreachable, LEAP repeats this wait procedure a second time. If the node is still unreachable, LEAP chooses a new destination.

After all nodes in the global location table are visited, the legend is paused at the last visited node for a time-out period [36]:

$$\left(\frac{T_{\text{range}}}{\alpha} \right) * \left(\frac{1}{v} \right) = \frac{T_{\text{range}}}{\alpha v},$$

where T_{range} is the transmission range of the nodes, v is the average velocity of the nodes, and α is a scaling factor. That is, the time the legend is paused corresponds to the movement of the nodes. If the nodes are moving quickly, then the legend propagates (almost) continuously; if the nodes are moving slowly, then the legend is often paused. In other words, there is no reason to propagate the legend if the locations of the nodes have barely changed. When the time-out expires, the visited bit of all nodes in the global location table is set to false. A new propagation of the legend then continues in the same manner.

In a real network environment, the legend may get lost during transmission [36]. A node sending the legend sets an acknowledgment timer to help in ensuring the reliable transmission of the legend. When node A forwards the legend packet to node B, node A sets a timer to receive a response from node B. If node B is a neighbor, the response is overhearing node B propagate the legend further. If node B is not a neighbor, then node B sends an acknowledgment packet to node A. Node B also sends an acknowledgment packet to node A if the legend is to be paused at node B. If node A receives a response from node B, then node A cancels the timer; otherwise, node A attempts to send the legend to node B a second time. If the timer expires a second time, node A chooses a new destination to send the legend packet.

A legend traversal example is illustrated in Figure 14.8. (This figure is taken from Reference 36.) We note that this figure is not a snapshot at a certain time. Instead, Figure 14.8 is a record of the position for

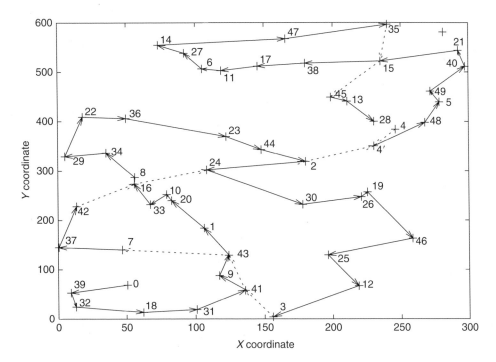

FIGURE 14.8 Example of the legend in LEAP migrating.

each node when it receives the legend. The legend begins its traversal at node 0 and ends its traversal at node 28. The dotted lines illustrate the use of LAR to transmit the legend to a nonneighbor node (e.g., node 3 sending the legend to node 7). As shown, LAR is needed to forward the legend three times in this example. The LAR case for node 4 is different from the other two LAR cases. After the legend visits node 42, which is at position (15,225), it determines node 4 as its next destination. However, it is impossible for node 42 to transfer any packet to node 4 as the network is partitioned. Thus, LEAP suspends the legend propagation for a time-out. When node 4 has moved from location 4 to 4′ (Figure 14.8), the network is again connected and node 42 can send the legend to node 4.

14.3.4 Ants

Mobile software agents, modeled on ant behavior, are used to collect and disseminate location information in GPSAL (the GPS/Ant-Like routing algorithm) [14]. An ant is an agent sent by a source node to a destination node in the network. The destination node may be the node farthest from the source, the node with the oldest location information in the source's routing table, or a random node. The ant packet contains a location table with the most recent location information known by the ant. The goal of each ant is to forage location information for a given destination node. Location information of intermediate nodes to the destination is also collected and shared by the ant.

As an ant traverses to its destination and back to its source, it shares more recent location information within the ant packet with the intermediate nodes it traverses. In addition, as the ant traverses to its destination and back to its source, it collects more recent location information within each intermediate node's location table. That is, similar to the legend traversing in LEAP [36], both the location table of an intermediate node and the ant packet are updated as the ant traverses the network.

In GPSAL, when a node determines it needs to send an ant to a destination node (in order to obtain more recent location information on the destination), the node uses its location table and a shortest-path algorithm to determine a route to the destination. The shortest-path route, comprised of a list of nodes and the corresponding time-stamps, is then included in the header of the ant packet similar to source routing in DSR [38,39]. The ant packet is then sent to the first node in the determined shortest-path route. This node shares and collects more recent location information (as previously discussed) and then updates the shortest-path route in the ant packet if needed. The ant packet is then sent to the next node in the shortest-path route. This process is repeated until the ant packet arrives at the destination node. The destination node then sends the ant packet back to the source node that created the ant packet in the same manner.

The main difference between LEAP [36] and the location service in GPSAL [14] is the method that is used to create/propagate the legend/ant. The legend in LEAP [36] is a global entity that traverses all nodes in the network; as the legend traverses, it collects and shares location information on all nodes in the network. An ant in GPSAL [14] traverses from a source node (that created the ant) to a destination node; only the location tables within the ant packet, intermediate nodes, and the source/destination nodes are updated.

14.3.5 Geographic Region Summary Service

The Geographic Region Summary Service (GRSS) is proposed in References 32 and 33. Like GLS, the area covered by the ad hoc network is arranged into a hierarchy of grids with squares of increasing size (see Figure 14.4). The location service is based upon two main functions. First, a node within the same order-0 square[3] transmits its location and neighborhood information to all other nodes in the same order-0 square. This task is accomplished via a link-state routing protocol. Thus, like GLS, full topology information is known within each order-0 square. Second, boundary nodes can transmit location information to nodes in adjacent squares; thus, boundary nodes in an order-i square share location information for all nodes within the order-i square to sibling order-i squares (four siblings of order-i make a square of order-$i + 1$) via "summary" LPs.

A summary packet is an aggregated list of location information [32,33]. Specifically, a summary packet contains a list of all nodes within the order-i square and the location of the order-i square's center. In other words, the hierarchical infrastructure created is used to generate and disseminate summaries that contain location information. The union of all received summary packets and the location information on a given node's order-0 square contains location information for all nodes in the network.

One advantage of GRSS is that a given node does not flood its location information in the entire network [32,33]. One disadvantage of GRSS is that, due to summary packets, location information may only be the center of the region a node resides in, instead of the exact location of the node. However, since the size of each region is approximately the radio range, all nodes in an order-0 square can be reached within two hops [32].

The authors of Reference 32 propose two types of summary packets: exact summary and imprecise summary. Exact summaries require more overhead than imprecise summaries, but then lead to fewer false positives. A false positive occurs when a node receives conflicting location information on another node, which causes a packet to be dropped. See Reference 32 for more details on GRSS.

14.3.6 Dead Reckoning: A Location Prediction Technique

A recently proposed location dissemination location system is the Dead Reckoning Method (DRM) [1,45]. In DRM, each node constructs a model of its movement, which is then disseminated with the node's current location. Other nodes are then able to predict the movement of every other node in the network. The basis for this technique is the prediction method discussed in Reference 47.

The authors of DRM mention that the movement model constructed by the nodes could be a deterministic or probabilistic first order or high order model [1]. In References 1 and 45, the following movement

---

[3]We note that GLS begins counting at one instead of zero.

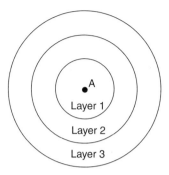

FIGURE 14.9   The layering method in DRM.

model is used. A node computes its velocity values (v_x and v_y) from two successive location samples, (x_1, y_1) and (x_2, y_2), taken at times t_1 and t_2:

$$v_x = \frac{x_2 - x_1}{t_2 - t_1} \quad \text{and} \quad v_y = \frac{y_2 - y_1}{t_2 - t_1}.$$

Once the velocity values are computed, the node floods the values, the node's current location, and the time the velocity values were computed. The packet transmitted is called a DRM update packet.

A node predicts another node's location via the following two equations [1,45]:

$$x_{\text{predict}} = x_{\text{location}} + (v_{x_{\text{model}}} * (t_{\text{current}} - t_{\text{model}})),$$

and

$$y_{\text{predict}} = y_{\text{location}} + (v_{y_{\text{model}}} * (t_{\text{current}} - t_{\text{model}})),$$

where ($x_{\text{location}}, y_{\text{location}}$), ($v_{x_{\text{model}}}, v_{y_{\text{model}}}$), and t_{model} are the values received in the received DRM update packet and t_{current} is the current time.

Each node periodically predicts its own location as well, which is compared against the node's actual value. That is, a deviation between a node's actual location and predicated location is calculated by:

$$d = \sqrt{(x_{\text{current}} - x_{\text{predicted}})^2 + (y_{\text{current}} - y_{\text{predicted}})^2}.$$

If its own predicted location deviates from its actual location by a given threshold, then the node transmits a new DRM update packet. Thus, if a node transmits an accurate movement model, very little location update cost exists in DRM [1,45].

In Reference 1, the authors mention an optimization to their dead reckoning technique that is based on the "distance effect" discussed in Section 14.3.1. Specifically, a node floods a DRM update packet if the calculated deviation is large; if the calculated distance is small, a node only transmits the DRM update packet to other nodes in close proximity of the node. In Reference 45, the authors detail a layering method, based on a given node's location (see Figure 14.9), to implement this optimization.

Once a node can predict all other node locations, Agarwal and Das [1] propose that each node computes the topology of the network. Dijkstra's shortest path algorithm can then be used to route packets from a source to a destination. In Reference 45, the authors use geographic forwarding to route packets.

14.4　Reactive Location Systems

The last type of location systems presented are reactive in nature. As mentioned in Section 14.1.2, RLSs query location information on an as and when needed basis. In this section, we present two reactive location systems proposed in the literature. According to the *Mauve Classification* [51], reactive location systems would be classified as an *all-for-some* approach. That is, every node in the network maintains location information on some other nodes in the network whose locations are needed.

14.4.1　Reactive Location Service

In the RLS protocol [12], when a mobile node requires a location for another node and the location information is either unknown or expired, the requesting node will first ask its neighbors for the requested location information (i.e., the time-to-live, or TTL, of the packet is set to zero). If the node's neighbors do not respond to the requested location information within a time-out period, then the node will flood a location request packet in the entire network.

When a node receives a location request packet and does not know the requested location information, the node propagates the flooded location request. If, however, a node receives a location request packet and the node's location table contains the requested location information, the node returns a location reply packet via the reverse source route obtained in the location request packet. In other words, each location request packet carries the full route (a sequenced list of nodes) that a location reply packet should be able to traverse in its header. Since IEEE 802.11 requires bidirectional links in the delivery of all nonbroadcast packets, RLS assumes bidirectional links. If bidirectional links are not available, this requirement can be removed via the manner proposed in the DSR protocol [10].

If feasible, each node using RLS should update its location table when a new LP is overheard or received. In other words, RLS suggests promiscuous mode operation is used by all nodes. (As noted in Reference 37, promiscuous mode operation is power consuming.) Lastly, entries in the location table are aged as the node associated with the entry moves; that is, an entry associated with a node that is moving quickly will age more quickly.

There are some similarities of RLS to LAR [44], DSR [10] and Location Trace Aided Routing (LOTAR) [69], three unicast routing protocols developed for a MANET. Specifically, all four protocols are reactive protocols that try to discover the required information on demand. In addition, all four protocols use the reverse source route to respond to a request for information. One significant difference between these four protocols is the following: RLS attempts to determine the location information, while LAR, DSR, and LOTAR attempt to determine full routes. Lastly, requesting desired information from neighbors first and allowing intermediate nodes to reply to a request are two features that both RLS and DSR have. Although not mentioned in LAR [44], the usefulness of these features for LAR are evaluated in Reference 13.

14.4.2　Reactive Location Service (RLS′)

Another RLS protocol, which was developed concurrently, is proposed in Reference 42 and then used for routing in a city environment in Reference 50. To avoid confusion, we refer to this second RLS protocol as RLS′. The differences between RLS and RLS′ are:

- While intermediate nodes are allowed to reply to location requests in RLS [12], a location request is forwarded until it reaches the node whose location is desired in RLS′ [42,50]. While cached replies are discussed in Reference 42, the authors chose not to implement them because they felt that the disadvantages (e.g., several location replies may be returned for one location request) outweighed the advantages.
- A location reply packet is returned via the reverse source route in RLS [12]; a location reply packet is returned via "the underlying routing protocol (e.g., greedy unicast routing, flooding, etc.)" in RLS′ [42].

- While several options for flooding a location request packet are discussed in RLS′ [42] (e.g., linear flooding and exponential flooding), the main flooding scheme evaluated is the same as the scheme used in RLS [12]. This scheme is called binary flooding in Reference 42.
- RLS′ includes radial flooding, which allows nodes farther away from a source to rebroadcast a location request packet before those closer to the source [42]. RLS does not include this feature.

14.5   Conclusions and Future Research

In this chapter, we review 16 different location information services that exist in the literature to date. An effective location service can be used to improve the performance and scalability of a routing protocol that requires location information (e.g., GPSR [40]). We classify the location information services into three categories: proactive location database systems, proactive location dissemination systems, and reactive location systems. Proactive location services are those protocols that have nodes exchange location information periodically; the information is either exchanged with a few select nodes or with all nodes. RLSs query location information on an as and when needed basis.

There are several areas of future investigation on location services that are needed. First, as mentioned in Reference 51, when a node is associated with a position, location privacy is difficult to achieve. Since none of the presented location services consider anonymity, future research in this area is essential.

In addition, more quantitative comparisons to discover the strengths and weaknesses of the various approaches are needed. Typically, a performance evaluation is given in a paper that proposes a location service. However, seldom are multiple location services quantitatively compared. The exceptions are:

- In Reference 51, the authors state the complexity and robustness of four location services (i.e., DLS, home regions, UQS, and GLS).
- A detailed comparison of DLS, SLS, and RLS exists in Reference 12.
- RLS′ with greedy location-based forwarding is compared to DSR and GLS with greedy location-based forwarding in Reference 42.
- A comparison of LEAP, GLS, RLS, and SLS is given in Reference 36.
- In Reference 45, a performance comparison exists for a geographical forwarding routing protocol that uses four different location services: DRM, GRSS, DLS, and SLS.

According to Reference 43, GLS is a promising distributed location service. However, due to its complexity, GLS requires a "deeper analysis and evaluation" [43]. See References 25, 36, and 43 for further performance results on GLS.

Acknowledgments

We thank several members (past and present) of Toilers (http://toilers.mines.edu) for the information presented herein, especially Jeff Boleng, Lucas Wilcox, Nalinrat Guba, and Xia Jiang. This work was supported in part by NSF Grants ANI-0073699 and ANI-0240588. Research group's URL is http://toilers.mines.edu.

References

[1] A. Agarwal and S.R. Das. Dead reckoning in mobile ad hoc networks. In *Proceedings of the IEEE Wireless Communications and Networking Conference (WCNC 2003)*, pp. 1838–1843, 2003.

[2] K.N. Amouris, S. Papavassiliou, and M. Li. A position-based multi-zone routing protocol for wide area mobile ad-hoc networks. In *Proceedings of the IEEE Vehicular Technology Conference (VTC)*, pp. 1365–1369, 1999.

[3] S. Basagni, I. Chlamtac, and V.R. Syrotiuk. Geographic messaging in wireless ad hoc networks. In *Proceedings of the IEEE Vehicular Technology Conference*, pp. 1957–1961, 1999.

[4] S. Basagni, I. Chlamtac, V.R. Syrotiuk, and B.A. Woodward. A distance routing effect algorithm for mobility (DREAM). In *Proceedings of the ACM/IEEE International Conference on Mobile Computing and Networking (MOBICOM)*, pp. 76–84, 1998.

[5] L. Blăzević, L. Buttyán, S. Čapkun, S. Giordano, J.-P. Hubaux, and J.-Y. Le Boudec. Self organization in mobile ad hoc networks: the approach of terminodes. *IEEE Communications Magazine*, 39: 166–174, 2001.

[6] L. Blăzević, S. Giordano, and J.-Y. Le Boudec. Self organized terminode routing. *Cluster Computing Journal*, 5: 205–218, 2002.

[7] P. Bose, P. Morin, I. Stojmenovic, and J. Urrutia. Routing with guaranteed delivery in ad hoc wireless networks. In *Proceedings of the ACM International Workshop on Discrete Algorithms and Methods for Mobile Computing and Communications (DIAL-M)*, pp. 48–55, 1999.

[8] P. Bose, P. Morin, I. Stojmenovic, and J. Urrutia. Routing with guaranteed delivery in ad hoc wireless networks. *ACM Wireless Networks*, 7: 609–616, 2001.

[9] A. Boukerche and S. Rogers. GPS query optimization in mobile and wireless ad hoc networking. In *Proceedings of the 6th IEEE Symposium on Computers and Communications*, pp. 198–203, 2001.

[10] J. Broch, D. Maltz, D. Johnson, Y. Hu, and J. Jetcheva. Multi-hop wireless ad hoc network routing protocols. In *Proceedings of the ACM/IEEE International Conference on Mobile Computing and Networking (MOBICOM)*, pp. 85–97, 1998.

[11] N. Bulusu, J. Heidemann, and D. Estrin. GPS-less low-cost outdoor localization for very small devices. *IEEE Personal Communications*, 7: 28–34, 2000.

[12] T. Camp, J. Boleng, and L. Wilcox. Location information services in mobile ad hoc networks. In *Proceedings of the IEEE International Conference on Communications (ICC)*, pp. 3318–3324, 2002.

[13] T. Camp, J. Boleng, B. Williams, L. Wilcox, and W. Navidi. Performance evaluation of two location based routing protocols. In *Proceedings of the Joint Conference of the IEEE Computer and Communications Societies (INFOCOM)*, pp. 1678–1687, 2002.

[14] D. Câmara and A. Loureiro. A novel routing algorithm for ad hoc networks. In *Proceedings of the 33rd Hawaii International Conference on System Sciences*, vol. 8, pp. 1–8, 2000.

[15] S. Čapkun, M. Hamdi, and J.P. Hubaux. GPS-free positioning in mobile ad-hoc networks. In *Proceedings of the 34th Hawaii International Conference on System Sciences*, vol. 9, pp. 1–10, 2001.

[16] S. Čapkun, M. Hamdi, and J.P. Hubaux. GPS-free positioning in mobile ad-hoc networks. *Cluster Computing Journal*, 5: 157–167, 2002.

[17] S. Das, R. Castañeda, and J. Yan. Simulation-based performance evaluation of routing protocols for mobile ad hoc networks. *ACM Mobile Networks and Applications*, 5: 179–189, 2000.

[18] S. Das, C. Perkins, and E. Royer. Performance comparison of two on-demand routing protocols for ad hoc networks. In *Proceedings of the Joint Conference of the IEEE Computer and Communications Societies (INFOCOM)*, pp. 3–12, 2000.

[19] S. Datta, I. Stojmenovic, and J. Wu. Internal node and shortcut based routing with guaranteed delivery in wireless networks. In *Proceedings of the IEEE International Conference on Distributed Computing Systems Workshop*, pp. 461–466, 2001.

[20] D. De Couto and R. Morris. Location proxies and intermediate node forwarding for practical geographic forwarding. Technical Report MIT-LCS-TR-824, MIT Laboratory for Computer Science, June 2001.

[21] G. Finn. Routing and addressing problems in large metropolitan-scale internetworks. Technical Report, USC/Information Sciences Institute, ISI/RR-87-180, March 1987.

[22] S. Giordano and M. Hamdi. Mobility management: the virtual home region. Technical Report SSC/1999/037, Ecole Polytechnique Federale de Lausanne, October 1999.

[23] S. Giordano and I. Stojmenovic. Position based ad hoc routes in ad hoc networks. In M. Ilyas, ed., *Handbook of Ad Hoc Wireless Networks*. CRC Press, Boca Raton, FL, 2003.

[24] S. Giordano, I. Stojmenovic, and L. Blăzević. Position based routing algorithms for ad hoc networks: a taxonomy. In X. Cheng, X. Huang, and D.Z. Du, eds., *Ad Hoc Wireless Networking*. Kluwer Academic Publishers, Dordrecht, 2003.

[25] N. Guba and T. Camp. Recent work on GLS: a location service for an ad hoc network. In *Proceedings of the Grace Hopper Celebration (GHC 2002)*, 2002.

[26] Z. Haas. A new routing protocol for reconfigurable wireless networks. In *Proceedings of the IEEE International Conference on Universal Personal Communications*, pp. 562–566, October 1997.

[27] Z.J. Haas and B. Liang. Ad hoc mobility management with randomized database groups. In *Proceedings of the IEEE International Conference on Communications (ICC)*, pp. 1756–1762, 1999.

[28] Z.J. Haas and B. Liang. Ad hoc mobility management with uniform quorum systems. *IEEE/ACM Transactions on Networking*, 7: 228–240, 1999.

[29] C. Hedrick. Routing Information Protocol. Request for Comments 1058, June 1988.

[30] J. Hightower and G. Borriello. Location systems for ubiquitous computing. *IEEE Computer*, 34: 57–66, 2001.

[31] T.-C. Hou and V. Li. Transmission range control in multihop packet radio networks. *IEEE Transactions on Communications*, 34: 38–44, 1986.

[32] P.-H. Hsiao. Geographical region summary service for geographical routing. *ACM SIGMOBILE Mobile Computing and Communications Review*, 5: 25–39, 2001.

[33] P.-H. Hsiao. Geographical region summary service for geographical routing. In *Proceedings of the ACM International Symposium on Mobile Ad Hoc Networking and Computing (MOBIHOC)*, pp. 263–266, 2001.

[34] J.-P. Hubaux, J.-Y. Le Boudec, S. Giordano, M. Hamdi, L. Blăzević, L. Buttyán, and M. Vojnović. Towards mobile ad-hoc WANs: terminodes. In *Proceedings of the IEEE Wireless Communications and Networking Conference (WCNC)*, pp. 1052–1059, 2000.

[35] R. Jain, A. Puri, and R. Sengupta. Geographical routing using partial information for wireless ad hoc networks. *IEEE Personal Communications*, pp. 48–57, February 2001.

[36] X. Jiang and T. Camp. An efficient location server for an ad hoc network. Technical Report, Department of Mathematics and Computer Sciences, Colorado School of Mines, MCS-03-06, May 2003.

[37] P. Johansson, T. Larsson, N. Hedman, B. Mielczarek, and M. Degermark. Routing protocols for mobile ad-hoc networks — a comparative performance analysis. In *Proceedings of the ACM/IEEE International Conference on Mobile Computing and Networking (MOBICOM)*, pp. 195–206, 1999.

[38] D. Johnson and D. Maltz. Dynamic source routing in ad hoc wireless networks. In T. Imelinsky and H. Korth, eds., *Mobile Computing*, pp. 153–181. Kluwer Academic Publishers, Dordrecht, 1996.

[39] D. Johnson, D. Maltz, Y. Hu, and J. Jetcheva. The dynamic source routing protocol for mobile ad hoc networks. Internet Draft: draft-ietf-manet-dsr-04.txt, November 2000.

[40] B. Karp and H.T. Kung. GPSR: Greedy perimeter stateless routing for wireless networks. In *Proceedings of the ACM/IEEE International Conference on Mobile Computing and Networking (MOBICOM)*, pp. 243–254, 2000.

[41] G. Karumanchi, S. Muralidharan, and R. Prakash. Information dissemination in partitionable mobile ad hoc networks. In *Proceedings of the IEEE Symposium on Reliable Distributed Systems*, pp. 4–13, 1999.

[42] M. Käsemann, H. Füßler, H. Hartenstein, and M. Mauve. A reactive location service for mobile ad hoc networks. Technical report, Department of Science, University of Mannheim, TR-02-014, November 2002.

[43] M. Käsemann, H. Hartenstein, H. Füßler, and M. Mauve. Analysis of a location service for position-based routing in mobile ad hoc networks. In *Proceedings of the 1st German Workshop on Mobile Ad-Hoc Networks (WMAN)*, pp. 121–133, 2002.

[44] Y. Ko and N.H. Vaidya. Location-aided routing (LAR) in mobile ad hoc networks. In *Proceedings of the ACM/IEEE International Conference on Mobile Computing and Networking (MOBICOM)*, pp. 66–75, 1998.

[45] V. Kumar and S.R. Das. Performance of dead reckoning-based location service for mobile ad hoc networks. *Wireless Communications and Mobile Computing Journal*, 4(2): 189–202, 2003.

[46] J. Li, J. Jannotti, D. De Couto, D. Karger, and R. Morris. A scalable location service for geographic ad hoc routing. In *Proceedings of the ACM/IEEE International Conference on Mobile Computing and Networking (MOBICOM)*, pp. 120–130, 2000.

[47] B. Liang and Z.J. Haas. Predictive distance-based mobility management for PCS networks. In *Proceedings of the Annual Joint Conference of the IEEE Computer and Communications Societies (INFOCOM)*, pp. 1377–1384, 1999.

[48] B. Liang and Z.J. Haas. Virtual backbone generation and maintenance in ad hoc network mobility management. In *Proceedings of the Annual Joint Conference of the IEEE Computer and Communications Societies (INFOCOM)*, pp. 1293–1302, 2000.

[49] H. Lim and C. Kim. Multicast tree construction and flooding in wireless ad hoc networks. In *Proceedings of the ACM International Workshop on Modeling, Analysis and Simulation of Wireless and Mobile Systems (MSWIM)*, pp. 61–68, 2000.

[50] C. Lochert, H. Hartenstein, J. Tian, H. Füßler, D. Hermann, and M. Mauve. A routing strategy for vehicular ad hoc networks in city environments. In *Proceedings of the IEEE Intelligent Vehicles Symposium*, pp. 156–161, 2003.

[51] M. Mauve, J. Widmer, and H. Hartenstein. A survey on position-based routing in mobile ad hoc networks. *IEEE Network*, 15: 30–39, 2001.

[52] J. Moy. OSPF version 2. Request for Comments 2178, July 1997.

[53] G. Pei and M. Gerla. Mobility management for hierarchical wireless networks. *Mobile Networks and Applications*, 6: 331–337, 2001.

[54] G. Pei, M. Gerla, X. Hong, and C. Chiang. A wireless hierarchical routing protocol with group mobility. In *Proceedings of the IEEE Wireless Communications and Networking Conference (WCNC)*, pp. 1538–1542, September 1999.

[55] C. Perkins. IP mobility support. Request for Comments 2002, October 1996.

[56] C. Perkins, E. Royer, and S. Das. Ad hoc on demand distance vector (AODV) routing. Internet Draft: draft-ietf-manet-aodv-07.txt, November 2000.

[57] C. Perkins, E. Royer, S. Das, and M.K. Marina. Performance comparison of two on-demand routing protocols for ad hoc networks. *IEEE Personal Communications*, 8: 16–28, 2001.

[58] C. Perkins and E.M. Royer. Ad-hoc on-demand distance vector routing. In *Proceedings of the 2nd IEEE Workshop on Mobile Computing Systems and Applications*, pp. 90–100, 1999.

[59] A. Qayyum, L. Viennot, and A. Laouiti. Multipoint relaying: an efficient technique for flooding in mobile wireless networks. Technical Report 3898, INRIA — Rapport de recherche, 2000.

[60] E. Royer and C.-K. Toh. A review of current routing protocols for ad-hoc mobile wireless networks. *IEEE Personal Communications Magazine*, pp. 46–55, April 1999.

[61] I. Stojmenovic. A scalable quorum based location update scheme for routing in ad hoc wireless networks. Technical Report TR-99-09, University of Ottawa, September 1999.

[62] I. Stojmenovic. Home agent based location update and destination search schemes in ad hoc wireless networks. In A. Zemliak and N.E. Mastorakis, eds, *Advances in Information Science and Soft Computing*, pp. 6–11. WSEAS Press, 2002. Also Technical Report, University of Ottawa, TR-99-10, September 1999.

[63] I. Stojmenovic. Position based routing in ad hoc networks. *IEEE Communications Magazine*, 40: 128–134, 2002.

[64] I. Stojmenovic, M. Russell, and B. Vukojevic. Depth first search and location based localized routing and QoS routing in wireless networks. In *Proceedings of the IEEE International Conference on Parallel Processing*, pp. 173–180, 2000.

[65] H. Takagi and L. Kleinrock. Optimal transmission ranges for randomly distributed packet radio terminals. *IEEE Transactions on Communications*, 32: 246–257, 1984.

[66] Y.C. Tseng, W.H. Liao, and S.L. Wu. Mobile ad hoc networks and routing protocols. In I. Stojmenovic, ed., *Handbook of Wireless Networks and Mobile Computing*. John Wiley & Sons, New York, pp. 371–392, 2002.

[67] B. Williams and T. Camp. Comparison of broadcasting techniques for mobile ad hoc networks. In *Proceedings of the ACM Symposium on Mobile Ad Hoc Networking and Computing (MOBIHOC)*, pp. 194–205, 2002.

[68] S.-C. Woo and S. Singh. Scalable routing protocol for ad hoc networks. *ACM Wireless Networks*, 7: 513–529, 2001.

[69] K. Wu and J. Harms. Location trace aided routing in mobile ad hoc networks. In *Proceedings of the International Conference on Computer Communications and Networks (ICCCN)*, pp. 354–359, 2000.

15

Geographic Services for Wireless Networks

15.1 Introduction..15-343
15.2 Geographic Routing 15-344
    Routing Mechanisms • Destination Location • Location
    Inaccuracy • The Effect of Link Losses on Geographic
    Routing
15.3 Geocasting..15-352
    Efficient Geocasting Protocols with Perfect Delivery
15.4 Geographic-Based Rendezvous 15-357
    Rendezvous Regions
15.5 Conclusions 15-360
References 15-361

Karim Seada
Ahmed Helmy

15.1 Introduction

In wireless ad hoc and sensor networks, building efficient and scalable protocols is a very challenging task due to the lack of infrastructure and the high dynamics. Geographic protocols, that take advantage of the location information of nodes, are very valuable in these environments. The state required to be maintained is minimum and their overhead is low, in addition to their fast response to dynamics. In this chapter, we present a state-of-the-art overview of geographic protocols providing basic functions such as geographic routing, geocasting, service location, and resource discovery. We introduce also some of our work on assessing and improving the robustness of geographic protocols to nonideal realistic conditions corresponding to the real-world environments.

Geographic protocols are very promising for multihop wireless networks. These protocols take advantage of the location information of nodes to provide higher efficiency and scalability. In wireless environments, the locations of nodes correspond to their network connectivity, which makes geographic protocols natural components in these environments and it is expected that they will become major elements for the development of these networks. For obtaining the location information, different kinds of localization systems exist such as GPS, infrastructure-based localization systems, and ad hoc localization systems.

Examples of multihop wireless networks are ad hoc networks and sensor networks. Ad hoc networks are infrastructureless dynamic networks that could be an extension or alternative to infrastructure wireless networks, especially in situations where it is difficult or time-critical to deploy an infrastructure such as in disaster recovery or military applications. Commercially it could also be used to build small fast networks for conferences and meetings, vehicle networks, rooftop networks, or to extend the services provided by the cellular infrastructure. Sensor networks are networks of small embedded low-power devices that can operate unattended to monitor and measure different phenomena in the environment. Sensor networks are suited for applications such as habitat monitoring, infrastructure protection, security, and tracking.

We consider basic geographic protocols at the network layer: geographic routing, geocasting and geographic rendezvous mechanisms. Geographic routing provides a way to deliver a packet to a destination location, based only on local information and without the need for any extra infrastructure, which makes geographic routing the main basic component for geographic protocols. With the existence of location information, geographic routing provides the most efficient and natural way to route packets comparable to other routing protocols. Geocasting is the delivery of packets to nodes within a certain geographic area. It is an extension to geographic routing where in this case the destination is a geographic region instead of a specific node or point. Geocasting is an important communication primitive in wireless ad hoc and sensor networks, since in many applications the target is to reach nodes in a certain region. In geographic-based rendezvous mechanisms, geographical locations are used as a rendezvous place for providers and seekers of information. Geographic-based rendezvous mechanisms can be used as an efficient means for service location and resource discovery in ad hoc networks. They can also provide efficient data dissemination and access in sensor networks.

In the rest of this chapter, we will go through the basic geographic mechanisms: routing, geocasting, and geographic rendezvous. In Section 15.2 we discuss geographic routing protocols and some important related problems: the determination of destination location, the effect of location inaccuracy, and the effect of lossy links. In Section 15.3 we present the different geocasting mechanisms. In Section 15.4 we explain several geographic rendezvous mechanisms used for service location, resource discovery, and data access. Finally, the conclusions are presented in Section 15.5.

15.2 Geographic Routing

Routing in ad hoc and sensor networks is a challenging task due to the high dynamics and limited resources. There has been a large amount of nongeographic ad hoc routing protocols proposed in the literature that are either proactive (maintain routes continuously) [48], reactive (create routes on-demand) [31,47,49] or a hybrid [21]. For a survey and comparison see References 52 and 10. Nongeographic routing protocols suffer from a huge amount of overhead for route setup and maintenance due to the frequent topology changes and they typically depend on flooding for route discovery or link state updates, which limit their scalability and efficiency.

On the other hand, geographic routing protocols require only local information and thus are very efficient in wireless networks. First, nodes need to know only the location information of their direct neighbors in order to forward packets and hence the state stored is minimum. Second, such protocols conserve energy and bandwidth since discovery floods and state propagation are not required beyond a single hop. Third, in mobile networks with frequent topology changes, geographic routing has fast response and can find new routes quickly by using only local topology information.

In the discussion of routing mechanisms in Section 15.2.1, we have the following assumptions:

- Each node knows its geographic location using some localization mechanism. Location-awareness is essential for many wireless network applications, so it is expected that wireless nodes will be equipped with localization techniques. Several techniques exist for location sensing based on proximity or triangulation using radio signals, acoustic signals, or infrared. These techniques differ in their localization granularity, range, deployment complexity, and cost. In general, many localization systems have been proposed in the literature: Global Positioning System (GPS), infrastructure-based localization systems [63,50], and ad hoc localization systems [11,53]. For an extensive survey of localization refer to Hightower and Borriello [27].
- Each node knows its direct neighbors locations. This information could be obtained by nodes periodically or on request broadcasting their locations to their neighbors.

- The source knows the destination location. In Section 15.2.2 we will discuss in detail how this information could be obtained.

15.2.1 Routing Mechanisms

In geographic routing, each node knows the location of its direct neighbors (neighbors within its radio range). The source inserts the destination location inside the packet. During packet forwarding, each node uses the location information of its neighbors and the location of the destination to forward the packet to the next-hop. Forwarding could be to a single node or to multiple nodes. Forwarding to multiple nodes is more robust and leads to multiple paths to the destination, but it could waste a lot of resources (energy and bandwidth) and thus forwarding to a single node is more efficient and it is the common approach among unicast protocols. A main component in geographic routing is greedy forwarding, in which the packet should make a progress at each step along the path. Each node forwards the packet to a neighbor closer to the destination than itself until ultimately the packet reaches the destination. If nodes have consistent location information, greedy forwarding is guaranteed to be loop-free.

Takagi and Kleinrock [60] is an early work that presented the most forward within R (MFR) routing model, where R is the transmission radius. In MFR, a node transmits to the neighbor that provides the maximum progress in the direction of the final destination, in order to minimize the number of hops between the source and the destination. The objective of that work was to obtain the optimum transmission radius in a content-based channel. In 1987, Finn [16] proposed Cartesian routing as a scalable routing solution to interconnect isolated LANs in the Internet. Each node forwards the packet to the neighbor closest to the destination among its neighbors that are closer to the destination. In References 28 and 44, Imielinski and Navas proposed integrating geographic coordinates into IP to enable the creation of location dependent services in the Internet. They presented a hierarchy of geographically aware routers that can route packets geographically and use IP tunnels to route through areas not supporting geographic routing. Geographically aware routers can determine which geographic areas they are servicing and based on that information and the packet destination area, each router, when it receives a packet, decides whether it services that destination area or it should forward the packet to its parent or to some of its children in the hierarchy.

In Reference 1, Akyildiz et al. used the reported geographic location of a mobile host to perform selective paging in cellular networks by paging a set of cells around that location. Among the earliest work to consider the geography for routing in ad hoc networks is LAR [35] by Ko and Vaidya, which uses the location information of nodes for route discovery and not for data delivery. LAR improves the performance of nongeographic ad hoc routing protocols by limiting discovery floods to a geographic area around the destination expected location. DREAM [5] is a routing protocol that uses the location information for data delivery in ad hoc networks. The packet is forwarded to all nodes in the direction of the destination. Based on the destination location and its velocity, the source determines an expected zone for the destination and forwards the packet to all nodes within an angle containing the expected zone. If the sender has no neighbors in the direction of the destination, a recovery procedure using partial flooding or flooding is invoked. Figure 15.1 shows an example for directional flooding, which could be used for route discovery in LAR or data delivery in DREAM. In Compass routing [38], a node forwards the packet to the neighbor whose edge has the closest slope to the line between that node and the destination; that is, the neighbor with the closest direction to the destination. Compass routing is not guaranteed to find a path if one exists.

Bose et al. [9] and GPSR [33] present the common form of greedy forwarding in ad hoc networks. Packets contain the position of the destination and nodes need only local information about their position and their immediate neighbors' positions to forward the packets. Each wireless node forwards the packet to the neighbor closest to the destination among its neighbors (within radio range) that are closer to the destination as shown in Figure 15.2.

Greedy forwarding is very efficient in dense uniform networks, where it is possible to make progress at each step. Greedy forwarding, however, fails in the presence of voids or dead-ends, when reaching a local maximum, a node that has no neighbor closer to the destination (Figure 15.3). In this case, it will fail

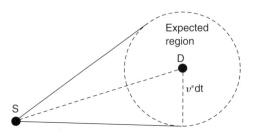

FIGURE 15.1   Source S sends the packet to all nodes in the direction of destination D expected region, where v is the velocity of D and dt is the time since last location update.

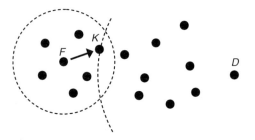

FIGURE 15.2   Greedy forwarding: node F forwards the packet to neighbor K, which is the neighbor closest to the destination D.

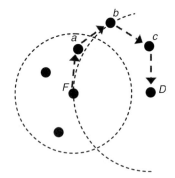

FIGURE 15.3   Greedy forwarding fails at node F when there are no neighbors closer to the destination D, although a path through a farther neighbor F-a-b-c-D exists.

to find a path to the destination, even though paths to the destination through farther nodes may exist. Previous protocols deals with dead-ends in different ways. In MFR [60], if no progress could be made in the forward direction, the dead-end node sends the packet to the least backward neighbor, which is the neighbor closest to the destination among its neighbors. This could cause looping and nodes need to detect when they get the same packet for a second time. Finn [16] proposed using limited flooding for a number of hops to overcome dead-ends. When a node is reached that has no neighbors closer to the destination, it sends a search packet for n hops away. Closer nodes to the destination reply back and the closest node to the destination among those nodes is chosen to forward the packet. The value of n is set based on the topology structure (estimated size of voids) and the desired degree of robustness. LAR and DREAM, which use directional flooding, did not provide specific mechanisms to deal with voids that stop the flood before reaching the expected zone. It is assumed that global flooding will be used as a recovery if directional flooding fails. In Reference 14 De Couto and Morris show a probabilistic approach that uses intermediate node forwarding to overcome dead-ends. When greedy forwarding fails, the source picks a random intermediate point and routes the packet though it to the destination. The random point is

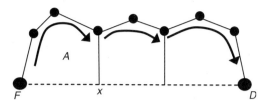

FIGURE 15.4 Face (perimeter) routing: the packets traverse planar faces between a node F and the destination D using the right-hand rule.

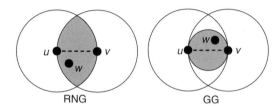

FIGURE 15.5 Local and distributed planarization algorithms. Node u removes the edge u–v from the planar graph, if a witness w exists.

picked randomly from an area between the source and the destination. The area is increased each time the routing fails.

The previous approaches for dead-end recovery do not guarantee that the packet reaches the destination if a path exists (unless global flooding is used, which causes large overhead). A local algorithm called Compass Routing II was presented in Reference 38, which guarantees that the packet reaches the destination. Compass Routing II, which becomes known as face routing or perimeter routing, works in planar[1] unit[2] graphs by traversing the faces intersecting the line between the source and the destination consecutively until reaching the destination as shown in Figure 15.4. Bose et al. [9] presented algorithms and proofs for extracting planar graphs from unit graphs and for face routing in the planar graphs to guarantee delivery. Due to the inefficient paths resulting from face routing, they proposed combining face routing with greedy forwarding to improve the path length. Face routing is used when greedy forwarding fails until a node closer to the destination is reached, then greedy forwarding could be resumed again. This way the algorithm will remain loop-free. In order for face routing to work correctly, a planar connectivity graph for the network needs to be constructed and so a planarization algorithm is required to create the planar graph. In Figure 15.5, RNG [62] and GG [17] are examples of algorithms that create a planar graph from the nonplanar physical topology by selecting a subset of the links and using only those links during face routing. A desirable feature in these algorithms is that they are local (a node needs to know only its own and neighbors' locations) and run in a distributed manner, so that each node can decide the links to include for planar routing using only local information independent of other nodes. The main idea of both algorithms is for a node to exclude an edge to a neighbor from the planar graph if there is another path through a different neighbor called *witness*. The witness should exist in a specific intersection area between the two nodes of the edge. In Reference 4, a variant of face routing is described that is more robust to irregular transmission ranges and can tolerate up to 40% of variation in the transmission range at the cost of a limited amount of extra overhead.

In summary, greedy forwarding alone *does not* guarantee the delivery of packets because of dead-ends (variously called local maxima or voids). Face routing on a planar graph theoretically does guarantee the delivery of packets. For improved performance, face routing is integrated with greedy forwarding and is used as a way to overcome dead-ends when greedy forwarding fails. Wireless network connectivity is in

---

[1] Planar graphs are graphs containing no cross links.
[2] In a unit graph a pair of nodes is connected if and only if the distance between them is below a certain threshold, which is the radio range in this case.

general nonplanar, this is why the planarization component is required to create a planar graph by using only a subset of the physical links during face routing. Face routing, similar to greedy forwarding, is also stateless and nodes need to keep only information about their direct neighbors in order to forward a packet, thus combined geographic protocols of greedy and face routing are stateless. Greedy forwarding coupled with face routing is the common efficient approach of the currently proposed geographic protocols.

GPSR [33] is a geographic routing protocol for wireless networks that combines greedy forwarding and face routing (perimeter routing). Packets contain the position of the destination and nodes need only local information about their position and their immediate neighbors' positions to forward the packets. Each node forwards the packet to the neighbor closest to the destination using greedy forwarding. When greedy forwarding fails, face routing is used to route around dead-ends until closer nodes to the destination are found. In Figure 15.4, node F is forwarding a packet using face routing to node D. Using the right-hand-rule the packet starts traversing face A, switching to other faces intersecting FD until reaching the face containing D. In Reference 33, packet-level simulations using 802.11 wireless MAC layer and comparisons with an ad hoc routing protocol, DSR, are provided. GOAFR [39] is another protocol proposed later that also combines greedy forwarding with face routing and is designed to be both asymptotically optimal and average-case efficient. GOAFR achieves worst-case optimality (analytically proved) of the path length by using limited elliptic regions for face routing and recursively increasing the ellipse size until finding a close-to-optimal path. This could improve the efficiency in low-density networks.

Other approaches for geographic routing have also been presented. In Gao et al. [18], a clustering algorithm is used to group nodes into clusters and then a planar graph called a restricted Delaunay graph (RDG) is built between the cluster-heads. RDG can be used as an underlying graph for geographic routing and it has the benefit that the path length between two nodes is a constant factor from the optimum length. Gao et al. [18] shows that routing on RDG graphs outperforms graphs built by RNG or GG, but maintenance for the clusters and graph is required. Terminode Routing [7] presents a different approach by dividing routing into two levels and using geographic routing for remote routing and a distance vector protocol for local routing. Geographic routing is used for routing to remote destinations to provide scalability in large mobile ad hoc networks, but as the packet arrives close to the destination (two hops away) local routing is used to avoid inconsistencies in the destination location. In Terminodes, a protocol called Anchored Geodesic Packet Forwarding is used for geographic routing, where the source defines a set of anchors (fixed geographic points) in the path to the destination. The goal of using anchored paths is to try to avoid obstacles and gaps by setting anchors accordingly. A packet is sent through the anchors to the destination; each node forwards the packet toward the next anchor in the list using a greedy approach, until the packet arrives to a node in proximity of this anchor, then the next anchor in the list is used and so on. A path discovery method is proposed to learn about anchors in the path. If no anchors are known, the destination location is used as the next anchor. Jain et al. [30] uses another approach, which is a mix of greedy forwarding and traditional ad hoc routing. Each node maintains a routing table containing its direct neighbors and their positions, which it uses for greedy forwarding. When a packet reaches a dead-end, a route discovery protocol is initiated to find a path to the destination. Each node along the path set an entry in its routing table for the next-hop to that destination. For route discovery no explicit algorithm is specified; flooding, depth first search, face routing, or distance vector routing could be used to learn the path.

In sensor networks communication is typically data-centric, which means communication between nodes is based on the content of data rather than the specific node identities. Messages are directed for named data instead of named nodes. Directed diffusion [29] is a data-centric communication approach presented for sensor networks. In directed diffusion data are named by attribute-value pairs and nodes interested in the data diffuse their interest to other nodes. Data can then be forwarded along the reverse path to the interested nodes. In Reference 25, different diffusion algorithms were discussed (e.g., push or pull) to design the protocol based on the application characteristics. Using geographic information to limit the diffusion by geographically scooping the messages was also presented. Geographical Energy Aware Routing (GEAR) [68] is an energy aware geographic protocol designed with the goal to increase the lifetime of sensor networks. GEAR uses energy aware metrics for neighbor selection in such a way that each node tries to balance the energy consumption among its neighbors using only local information by maintaining a cost function for each neighbor computed based on its location and an estimation for the energy consumed by that neighbor. Geographical Adaptive Fidelity (GAF) [65] uses the geographic information for energy conservation by building a geographical grid, such that only a single node needs to be turned on in each cell and other nodes are turned off. The cell size is set based on the radio range of nodes so that all nodes in the cell are equivalent from a routing perspective. Two-Tier Data Dissemination (TTDD) [67] provides a different way for data dissemination than directed diffusion. Instead of the sink propagating queries to all nodes and sources replying back; in TTDD each source builds a grid structure and sets its forwarding information at the nodes closest to the grid points, so that queries from the sink traverse only nodes in the local cell and some grid nodes towards the source. TTDD uses geographic greedy forwarding to construct and maintain the grid. This approach is beneficial when sinks are mobile, since location updates are propagates within the local cell and some grid nodes instead of the whole network.

The SPEED [24] is a geographic routing protocol designed for real-time communication in sensor networks. SPEED handles congestion and provides soft real-time communication by using feedback control and nondeterministic geographic forwarding. It also provides a different way to handle dead-ends similar to the way it handles congestion. Nondeterministic geographic forwarding is used to balance the load among multiple routes. A node computes a relay speed to each of its neighbors by dividing the advance in distance to the destination by the estimated delay to forward the packet to that neighbor. The node then forwards the packet to a neighbor closer to the destination that has a speed higher than a certain threshold with a probability based on that neighbor speed compared to other neighbors. If no neighbor has a speed higher than the desired speed, a neighborhood feedback loop determines whether to drop the packet or reroute it in order to reduce the congestion. Backpressure rerouting is used to avoid both congestion and dead-ends by sending a backpressure beacon to the upstream node. The upstream node will try to forward the packet on a different route or further backpressure will occur until a route is found. Dead-end recovery using this backpressure mechanism does not guarantee to find a path. SPEED considers also others functions such as geocast and anycast, which can be activated after the packet enters the destination area.

A related approach to geographic routing is trajectory-based forwarding [46], a method presented to route packets along curves in dense sensor networks. In this method, the trajectory is set by the source and intermediate nodes forward the packet to nodes close to the trajectory path.

15.2.2 Destination Location

In the previous section we have mainly focused on the routing problem and assumed that the packet destination location is known to the source. How the destination location is obtained is a separate problem that in many cases depends on the application. Most of the routing protocols discussed have not considered this problem explicitly. In many applications in ad hoc and sensor networks, the node ID itself is irrelevant and nodes are characterized by their location. In those applications packets do not need to be forwarded to specific nodes and a node close to the destination location or in a certain area around the destination can process the packet. For example, in sensor networks, queries may be sent to specific locations that the access point decides based on previous events and measurements. In geocasting, packets are sent toward regions and all nodes in the region can receive the packet.

In applications where the packet should be sent to a specific node, a mapping between the node ID and its current location is required. The source needs to obtain the destination current location before forwarding the packet, for example, by consulting a node location service. It is important for the location service to be efficient and at the same time consistent with node locations. A simple way to obtain node locations is by having nodes propagating their locations through the network and other nodes storing these locations. This approach causes large energy and bandwidth overhead, especially with node mobility, and the storage will be high since each node stores the locations of all other nodes, even if it may not need most of them. Another approach is to flood queries that search for the destination location and the destination can reply back with its current location. Approaches based on global flooding do not scale to large networks. DREAM [5] considered the problem of locating destinations and provided a solution based on location propagation. In order to limit the overhead, nodes propagate their locations based on two observations: the distance effect, where updates are propagated as a function of the distance between the node updating its location and the node receiving the update in such a way that closer nodes receive more updates and have more accurate information about a destination location. The second observation is that each node sets the frequency of location updates based on its mobility rate, so that low mobility nodes send fewer updates.

A different approach that avoids flooding is to use location servers that keep track of node locations. Nodes moving send only to these servers to update their locations and other nodes can query the servers to obtain the recent locations. In infrastructure-based networks (e.g. in Mobile IP and in cellular networks), centralized fixed well-known servers provide this service, but in ad hoc and sensor networks it is difficult to use centralized servers due to the lack of infrastructure and due to the topology changes and dynamics. In Terminodes [6], each node has a Virtual Home Region (VHR) that is known or can be computed by other nodes. Each node updates its location by sending the location update to its VHR. All nodes in the VHR will store that node location. Queries for a node location will be sent to the node VHR, where the nodes can reply back. In Reference 64 also, each node has a home region and all nodes in the home region store its location. GLS [41] presented a scalable distributed node location service for ad hoc networks. Each node updates a small set of location servers with its location. The node uses a predefined geographic hierarchy and a predefined ordering of node identifiers to determine its location servers. At each level of the hierarchy, the node chooses a location server as the node, from the corresponding region of that level, that has the closest ID to itself. Queries use the same hierarchy and identifier ordering to access a location server. In Reference 66, a geographic hierarchy is also used to map each node to location servers at different levels in the hierarchy such that the location is represented with different accuracy at each level. Instead of choosing a location server based on its node ID as in GLS, a mapping function is used to map the destination ID to one of the minimum partitions in the hierarchy and choose a node that covers this partition as the location server. The location of a node is stored by its location servers at different levels of accuracy, such that further location servers store approximate locations while closer servers have more accurate locations. This way a smaller number of location servers will need to be updated when the node moves, which reduces the overhead due to node mobility. Queries for the destination location will start with approximate regions and obtain more accurate information about the location as they get closer to the destination. A scheme that uses uniform quorums for mobility management is presented in Reference 22. A set of nodes in the network form a virtual backbone and each of these nodes stores a location database. The location databases are organized into quorums (sets of databases) in such a way that any two quorums have to intersect by having a shared number of databases. Location updates are sent to any quorum and stored by all its location databases. Due to the intersection between quorums, queries for a node location sent to any quorum should reach a database that maintains a location for that node.

General geographic rendezvous mechanisms could also be used for node location. In Reference 54, we have presented a geographic-based rendezvous architecture, *Rendezvous Regions*, which could be adjusted to provide a node location service. In Rendezvous Regions the network topology is divided into geographical regions. For a node location service, each region will be responsible for a set of nodes. Based on a hash-table-like mapping scheme, each node ID will be mapped to a region. Each node will store its location in the corresponding region and other nodes looking for its location could retrieve it from there. Inside each region, a few elected nodes are responsible for maintaining the information of the mapped nodes. The evaluations have shown that Rendezvous Regions is scalable, efficient and robust to node mobility, failures, and location inaccuracy. In Section 15.4, we will explain Rendezvous Regions in more detail.

15.2.3 Location Inaccuracy

Geographic routing protocols typically assumed the availability of accurate location information, which is necessary for their correct operation. However, in all localization systems an estimation error is incurred that depends on the system and the environment in which it is used. GPS is relatively accurate, but it requires visibility to its satellites and so is ineffective indoors or under coverage. In addition, the high cost, size, and power requirements make it impractical to deploy GPS on all nodes. Infrastructure-based localization systems [63,50] are mostly designed to work inside buildings and they either have a coarse-granularity of several meters or require a costly infrastructure. In ad hoc localization systems [11,53], nodes calculate their locations based on measurements to their neighbors or to other reference nodes in the environment. High localization errors can occur due to environmental factors affecting the location measurements such as obstacles. In addition, errors in a node location propagate to other nodes using it as a reference.

In Reference 23, simulation results were shown for the effect of localization errors on the performance of greedy forwarding. The conclusion was that routing performance is not significantly affected when the error is less than 40% of the radio range. Face routing is not considered in that work. In Reference 30, it is assumed that the system can deal with location errors, since a route discovery protocol is used when greedy forwarding fails. If a flooding route discovery approach is used, it will not be affected by the location errors at the cost of high discovery overhead, but if a route discovery approach based on location (e.g., face routing) is used, the route discovery itself could fail. Approaches that use the location for remote routing only and topology-based routing for local routing such as Terminodes [7] can tolerate inaccuracy in the destination location, but inaccuracy in intermediate nodes can still cause failures.

15.2.3.1 The Effect of Location Inaccuracy on Face Routing

As we mentioned, greedy forwarding coupled with face routing is an efficient approach that guarantees delivery and accordingly it is the most accepted approach among the currently proposed geographic protocols. In the absence of location errors it has been shown to work correctly and efficiently. In Reference 56, we provided a detailed analysis on the effect of location errors on complete geographic routing protocols consisting of greedy forwarding coupled with face routing. The methodology for this analysis is novel: using an elaborate, micro-level analysis of face routing protocols, we provided detailed scenarios in which the protocol correctness is violated when the location of a node is in error. We performed detailed analysis based on the protocol components to classify the errors and specify their conditions and bounds. We also performed extensive simulations to evaluate and quantify the effects of localization errors on a geographic routing protocol and a geographic-based rendezvous mechanism. Based on our analysis and error classification we introduced a simple and elegant protocol fix that eliminates the most likely protocol errors and we evaluated the efficacy of our fix. Our simulations show near-perfect performance for our modified geographic routing even in the presence of significant localization errors. This is the first work to point the different pathologies that can happen in planarization due to the violation of the unit graph assumption.

In other studies, we have also shown how location errors, caused by inconsistency of location dissemination [34] or node mobility [58] result in severe performance degradation and correctness problems in geographic routing protocols.

15.2.4 The Effect of Link Losses on Geographic Routing

In Reference 57, we provided energy-efficient forwarding strategies for geographic routing in lossy wireless sensor networks. Experimental studies have shown that wireless links in real sensor networks can be extremely unreliable, deviating to a large extent from the idealized perfect-reception-within-range models used in most common network simulation tools. The previous discussed protocols commonly employ a maximum-distance greedy forwarding technique that works well in ideal conditions. However, such a forwarding technique performs poorly in realistic conditions as it tends to forward packets on lossy links. Based on a realistic link loss model, we studied the distance-hop tradeoff via mathematical analysis and extensive simulations of a wide array of blacklisting/link-selection strategies; we also validated some strategies using real experiments on motes. Our analysis, simulations, and experiments, all show that the product of the packet reception rate (PRR) and the distance traversed toward destination is a very effective metric with and without ARQ. Nodes using this metric often take advantage of neighbors in the reception transitional region (high-variance links). Our results also show that reception-based strategies are in general better than distance-based, and we also provide blacklisting strategies that reduce the risk of routing disconnections.

In greedy forwarding, each node forwards a packet to the neighbor that is closest to the destination. The link quality of that neighbor may be very bad. The existence of such unreliable links exposes a key weakness in maximum-distance greedy forwarding that we refer to as the *weakest link* problem. At each step, the neighbors that are closest to the destination (also likely to be farthest from the forwarding node) may have poor links with the current node. These "weak links" would result in a high rate of packet drops, resulting in drastic reduction of delivery rate or increased energy wastage if retransmissions are employed. This observation brings to the fore the concept of neighbor classification based on link reliability. Some neighbors may be more favorable to choose than others, not only based on distance, but also based on loss characteristics. This suggests that a blacklisting/neighbor selection scheme may be needed to avoid "weak links." In Reference 57, we present and study in detail several blacklisting and neighbor selection schemes.

15.3   Geocasting

Geocasting is the delivery of packets to nodes within a certain geographic area. Perhaps the simplest way for geocasting is global flooding. In global flooding, the sender broadcasts the packet to its neighbors, and each neighbor that has not received the packet before broadcasts it to its neighbor, and so on, until the packet is received by all reachable nodes including the geocast region nodes. It is simple but has a very high overhead and is not scalable to large networks.

Imielinski and Navas [28,44] presented geocasting for the Internet by integrating geographic coordinates into IP and sending the packet to all nodes within a geographic area. They presented a hierarchy of geographically aware routers that can route packets geographically and use IP tunnels to route through areas not supporting geographic routing. Each router covers a certain geographic area called a service area. When a router receives a packet with a geocast region within its service area, it forwards the packet to its children nodes (routers or hosts) that cover or are within this geocast region. If the geocast region does not intersect with the router service area, the router forwards the packet to its parent. If the geocast region and the service area intersect, the router forwards to its children that cover the intersected part and also to its parent.

Ko and Vaidya [36] proposed geocasting algorithms to reduce the overhead, compared to global flooding, by restricting the forwarding zone for geocast packets. Nodes within the forwarding zone forward the geocast packet by broadcasting it to their neighbors and nodes outside the forwarding zone discard it. Each node has a localization mechanism to detect its location and to decide when it receives a packet, whether it is in the forwarding zone or not. The algorithms are the following:

- Fixed rectangular forwarding zone (FRFZ) (Figure 15.6): The forwarding zone is the smallest rectangle that includes the sender and the geocast region. Nodes inside the forwarding zone forward the packet to all neighbors and nodes outside the zone discard it.
- Adaptive rectangular forwarding zone (ARFZ) (Figure 15.7): Intermediate nodes adapt the forwarding zone to be the smallest rectangle including the intermediate node and the geocast region. The forwarding zones observed by different nodes can be different depending on the intermediate node from which a node receives the geocast packet.
- Progressively closer nodes (PCN) (Figure 15.8): When node B receives a packet from node A, it forwards the packet to its neighbors only if it is closer to the geocast region (center of region) than A or if it is inside the geocast region. Notice that this is different from geographic forwarding; in geographic forwarding a node forwards the packet to the neighbor closest to the region while here a node forwards the packet to all neighbors and *all* neighbors closer to the region forward it further.

Other variations of the FRFZ, ARFZ, and PCN mechanisms could also be used, for example, by increasing the area of the forwarding zone to include more nodes around the geocast region. These variations could improve the delivery rate at the expense of higher overhead, but they do not provide

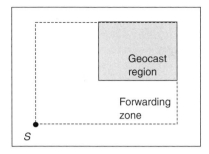

FIGURE 15.6   Fixed rectangular forwarding zone.

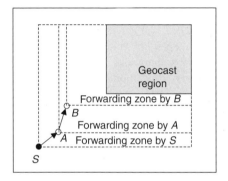

FIGURE 15.7   Adaptive rectangular forwarding zone.

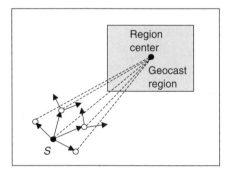

FIGURE 15.8   Progressively closer nodes: closer nodes to the region than the forwarding node forward the packet further and other nodes discard it.

guaranteed delivery. To reduce the overhead further, GeoTORA [37] uses a unicast routing protocol (TORA [47]) to deliver the packet to the region and then floods within the region. In Reference 59, the network is partitioned using the Voronoi diagram concept and each node forwards the packet to the neighbors whose Voronoi partitions (as seen locally by that node) intersect with the geocast region. The idea is to forward to a neighbor only if it is the closest neighbor to a point in the region.

Variations of global flooding and restricted flooding were presented that use some form of clustering or network divisions to divide the nodes [2,43], such that a single node only in each cluster or division needs to participate in the flooding. This approach can reduce the geocasting overhead by avoiding unnecessary flooding to all nodes at the cost of building and maintaining the clusters. Some approaches (e.g., mesh-based) [8,12] use flooding or restricted flooding only initially, to discover paths to nodes in the geocast region, then these paths are used to forward the packets.

Bose et al. [9] presented graph algorithms for extracting planar graphs and for face routing in the planar graphs to guarantee delivery for unicasting, broadcasting, and geocasting. For geocasting they provided an algorithm for enumerating all faces, edges, and vertices of a connected planar graph intersecting a region. The algorithm is a depth-first traversal of the face tree and works by defining a total order on the edges of the graph and traversing these edges. An entry edge, where a new face in the tree is entered, needs to be defined for each face based on a certain criteria. In order to determine the entry edges of faces using only local information and without a preprocessing phase, at each edge the other face containing the edge will need to be traversed to compare its edges with the current edge. This could lead to very high overhead.

15.3.1 Efficient Geocasting Protocols with Perfect Delivery

In Reference 55, we presented efficient and practical geocasting protocols that combine geographic routing mechanisms with region flooding to achieve high delivery rate and low overhead. The challenging problem in geocasting is distributing the packets to all the nodes within the geocast region with high probability but with low overhead. According to our study we noticed a clear tradeoff between the proportion of nodes in the geocast region that receive the packet and the overhead incurred by the geocast packet especially at low densities and irregular distributions. We presented two novel protocols for geocasting that achieve high delivery rate and low overhead by utilizing the local location information of nodes to combine geographic routing mechanisms with region flooding. We have shown that the first protocol, Geographic-Forwarding-Geocast (GFG), has close-to-minimum overhead in dense networks and that the second protocol, Geographic-Forwarding-Perimeter-Geocast (GFPG), provides guaranteed delivery without global flooding or global network information even at low densities and with the existence of region gaps. GFPG is based on the observation that by traversing all faces intersecting a region in a connected planar graph, every node of the graph inside the region is traversed. Our algorithm is efficient by using a combination of face routing and region flooding and initiating the face routing only at specific nodes. In the following section, we explain these protocols in more detail.

15.3.1.1 Geographic-Forwarding-Geocast

In geocast applications, nodes are expected to be aware of their geographic locations. GFG utilizes this geographic information to forward packets efficiently toward the geocast region. A geographic routing protocol consisting of greedy forwarding with perimeter (face) routing is used by nodes outside the region to guarantee the forwarding of the packet to the region. Nodes inside the region broadcast the packet to flood the region. An example is shown in Figure 15.9. In more detail, a node wishing to send a geocast creates a packet and puts the coordinates of the region in the packet header. Then it forwards the packet to the neighbor closest to the destination. The destination of geographic routing in this case is the region center. Each node successively forwards the packet to the neighbor closest to the destination using greedy

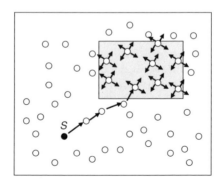

FIGURE 15.9 Sender S sends a geocast packet, geographic forwarding is used to deliver the packet to the region, then it is flooded in the region.

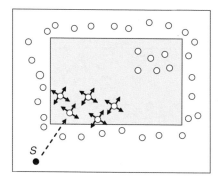

FIGURE 15.10 A gap (disconnection) in the geocast region. A packet flooded in the region cannot reach all nodes without going out of the region.

forwarding. When greedy forwarding fails, perimeter routing is used to route around dead-ends until closer nodes to the destination are found. Ultimately (in case there are nodes inside the region) the packet will enter the region. The first node to receive the geocast packet inside the region starts flooding the region by broadcasting to all neighbors. Each node inside the region that receives the packet for the first time broadcasts it to its neighbors and nodes outside the region discard the packet. For region flooding, smart flooding approaches [45] could also be used to reduce the overhead.

In dense networks without obstacles or gaps, GFG is sufficient to deliver the packet to all nodes in the region. In addition, since in dense networks geographic routes are close to optimal routes (shortest path), GFG has almost the minimum overhead a geocast algorithm can have, which mainly consists of the lowest number of hops to reach the region plus the number of nodes inside the region itself.

In order for GFG to provide perfect delivery (i.e., all nodes in the region receive the geocast packet), the nodes in the region need to be connected together such that each node can reach all other nodes without going out of the region. In dense networks normally this requirement is satisfied, but in sparse networks or due to obstacles, regions may have gaps such that a path between two nodes inside the region may have to go through other nodes outside the region as shown in Figure 15.10. In case of region gaps, GFG will fail to provide perfect delivery. GFPG overcomes this limitation.

15.3.1.2 Geographic-Forwarding-Perimeter-Geocast

We present an algorithm that guarantees the delivery of a geocast packet to all nodes inside the geocast region, given that the network as a whole is connected. The algorithm solves the region gap problem in sparse networks, but it causes unnecessary overhead in dense networks. Therefore, we present another practical version of the algorithm that provides perfect delivery at all densities and keeps the overhead low in dense networks. The practical version is not guaranteed as the original version, but the simulation results show that practically it still achieves perfect delivery.

This algorithm uses a mix of geocast and perimeter routing to guarantee the delivery of the geocast packet to all nodes in the region. To illustrate the idea, assume there is a gap between two clusters of nodes inside the region. The nodes around the gap are part of the same planar face. Thus, if a packet is sent in perimeter mode by a node on the gap border, it will go around the gap and traverse the nodes on the other side of the gap (see Figure 15.11). The idea is to use perimeter routing on the faces intersecting the region border in addition to flooding inside the region to reach all nodes. In geographic face routing protocols as GPSR, a planarization algorithm is used to create a planar graph for perimeter routing. Each node runs the planarization algorithm locally to choose the links (neighbors) used for perimeter forwarding. The region is composed of a set of planar faces with some faces totally in the region and other faces intersecting the borders of region. Traversing all faces guarantees reaching all nodes in the region.

We describe now the algorithm in more detail; please refer to Figure 15.11. Initially, similar to GFG, nodes outside of the geocast region use geographic forwarding to forward the packet toward the region.

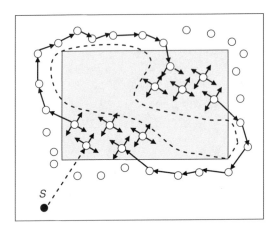

FIGURE 15.11 A mix of region flooding and face routing to reach all nodes in the region. Nodes around the gap are part of the same face. For clarity, here we are showing only the perimeter packet sent around the empty face, but notice that all region border nodes will send perimeter packets to their neighbors that are outside the region.

As the packet enters the region, nodes flood it inside the region. All nodes in the region broadcast the packet to their neighbors similar to GFG, in addition, all nodes on the border of the region send perimeter mode packets to their neighbors that are outside the region. A node is a region border node if it has neighbors outside the region. By sending perimeter packets to neighbors outside the region (notice that perimeter mode packets are sent only to neighbors in the planar graph not to all physical neighbors), the faces intersecting the region border are traversed. The node outside the region, receiving the perimeter mode packet, forwards the packet using the right-hand rule to its neighbor in the planar graph and that neighbor forwards it to its neighbor and so on. The packet goes around the face until it enters the region again. The first node inside the region to receive the perimeter packet floods it inside the region or ignores it if that packet was already received and flooded before. Notice that all the region border nodes send the perimeter mode packets to their neighbors outside of the region, the first time they receive the packet, whether they receive it through flooding, face routing, or the initial geographic forwarding. This way if the region consists of separated clusters of nodes, a geocast packet will start at one cluster, perimeter routes will connect these clusters together through nodes outside the region, and each cluster will be flooded as the geocast packet enters it for the first time. This guarantees that all nodes in the region receive the packet, since perimeter packets going out of the region will have to enter the region again from the opposite side of the face and accordingly all faces intersecting the region will be covered.

Due to the perimeter traversals of faces intersecting the region, GFPG will cause additional overhead that may not be required especially in dense networks, where as we mentioned GFG has optimal overhead by delivering the packet just to nodes inside the region. Ideally we would like perimeter routes to be used only when there are gaps inside the region such that we have perfect delivery also in sparse networks and minimum overhead in dense networks. In this section, we present an adaptation for the algorithm, in which perimeter packets are sent only when there is a suspicion that a gap exists. This new algorithm GFPG*, as the simulations show, practically has perfect delivery in all scenarios. In this algorithm each node inside the geocast region divides its radio range into four portions as shown in Figure 15.12(a) and determines the neighbors in each portion. This can be done easily, since each node knows its own location and its neighbors' locations. If a node has at least one neighbor in each portion, it will assume that there is no gap around it, since its neighbors are covering the space beyond its range and so it will not send a perimeter packet and will send only the region flood by broadcasting to its neighbors. If a node has no neighbors in a portion, then it sends a perimeter mode packet using the right-hand rule to the first neighbor counterclockwise from the empty portion as shown in Figure 15.12(b). Thus the face around the suspected void will be traversed and the nodes on the other side of the void will receive the packet. Notice that in this algorithm there is no specific role for region border nodes and that perimeter packets can be sent by any

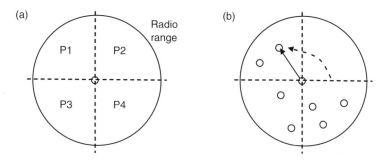

FIGURE 15.12 (a) A node divides its radio range into four portions. (b) If a node has no neighbors in a portion, it sends a perimeter packet using the right-hand rule to the first node counterclockwise from the empty portion.

node in the region, since the gap can exist and need to be detected anywhere. Therefore there are two types of packets in the region, flood packets and perimeter packets. Nodes have to forward perimeter packets even if that packet was flooded before. If a node receives a perimeter packet from the same neighbor for the second time, the packet is discarded, since this means that the corresponding face is already traversed. A node may receive the perimeter packet from different neighbors and thus forwards it on different faces. Our results show the improvement achieved by GFPG* in reducing the overhead at high densities. GFPG* does not guarantee delivery as GFPG, but our simulation results show that practically it has perfect delivery at all densities, in addition to close-to-minimum overhead at high densities. This is desirable for many types of high density applications. The evaluation results of our protocols are in Reference 55.

15.4 Geographic-Based Rendezvous

In geographic-based rendezvous mechanisms, a geographical location is used as a rendezvous place for providers and seekers of information. Geographic-based rendezvous mechanisms can be used as an efficient means for service location and resource discovery in ad hoc networks. They can also provide efficient data dissemination and access in sensor networks.

In wireless networks, the simplest form of data dissemination and resource discovery is global flooding. This scheme does not scale well. Other approaches that address scalability employ hierarchical schemes based on cluster-heads or landmarks [40]. These architectures, however, require complex coordination between nodes, and are susceptible to major re-configuration (e.g., adoption, re-election schemes) due to mobility or failure of the cluster-head or landmark, incurring significant overhead. GLS [41] provides a scalable location service by using a predefined geographic hierarchy and a predefined ordering of node identifiers to map nodes to their locations. GLS is presented for locating nodes and assumes that node identifiers are known. It is not clear that GLS could be extended efficiently to provide a general rendezvous-based mechanism. One way is to map keys to node identifiers and let the insertion and lookup use GLS to reach that node for storage and retrieval, respectively. A problem here is how nodes can guarantee that a node with that identifier exists and how to do reliable replication at multiple nodes. In addition, the path will be significantly longer, since the insertion or lookup has to find a location server first to get the node location and then it goes to the storage node. Another possibility is to use the key identifier itself to perform storage of the key-value pair in GLS servers similar to how node locations are stored. Since, in GLS the servers of a node are determined based on the node's location, the servers of a key will be determined based on the inserter location. This will create inconsistencies if multiple nodes can insert the same key.

Recently, some geographic-based rendezvous mechanisms have been proposed for data-centric storage in sensor networks. GHT [51] is a geographic hash table system that hashes keys into geographic *points*, and stores the key-value pair at the sensor node closest to the hash of its key. GHT requires nodes to know their exact geographic location and uses geographic routing to reach the destination. It uses

GPSR [33] for geographic routing, where it uses GPSR perimeter routing in a novel way to identify a packet home node (the node closest to the geographic destination). Packets enter perimeter mode at the home node (since no neighbor could be closer to the destination), and traverse the perimeter that encloses the destination (home perimeter) before returning back to home node. GHT uses a perimeter refresh protocol to replicate keys at nodes in the home perimeter. The perimeter refresh protocol refreshes keys periodically using also perimeter routing to deal with topology changes after failures or mobility. ARI [69] is another geographic-based rendezvous scheme for data-centric storage in sensor networks. In this scheme data are stored at nodes close to detecting nodes, and the location information of these storing nodes is pushed to some index nodes. The index nodes for a certain event type form a ring around a rendezvous location for that type. The idea of this scheme is that the nodes in the index ring capture storage and query messages passing through the ring. In order for the index nodes to do that, GAF [65] is used to divide the network into grids with a single node in each grid responsible for forwarding messages. Since GAF is based on the assumption that each node can only forward messages to the nodes in its neighboring grids, a message sent by a node outside of the ring-encircled region and destined to the index center, must pass some nodes on the ring. There are other variations of schemes providing data-centric storage in sensor networks. DIFS [20] is a system built on top of GHT to provide range searches for event properties in sensor networks. Another system, DIM [42] allows multidimensional range queries in sensor networks, which is useful in correlating multiple events. DIM uses a geographic embedding of an index data structure (multidimensional search tree) to provide a geographic locality-preserving hash that maps a multiattribute event to a geographic zone. The sensor field is logically divided into zones such that there is a single node in each zone. DIMENSIONS [19] provides multiresolution storage in sensor networks by using wavelet summarization and progressive aging of the summaries in order to efficiently utilize the network storage capacity.

In References 3 and 61, geographic curves are used for match-making between producers and consumers of content. The idea is for producers to send their advertisements along the four directions (north, south, east, and west) and for consumers to send queries also along the four directions. Nodes where advertisements and queries intersect will reply back to the consumers.

15.4.1 Rendezvous Regions

In Reference 54, we provided a scalable rendezvous-based architecture for wireless networks, called Rendezvous Regions (RR). The original RR idea borrowed from our earlier work on PIM-SM rendezvous mechanism [15] that uses consistent mapping to locate the rendezvous point (RP). However, a rendezvous *point* is insufficient in a highly dynamic environment as wireless networks. We first hinted at the RR idea in Reference 26, in the context of bootstrapping multicast routing in large-scale ad hoc networks, with no protocol details or evaluations. In Reference 54, we presented the detailed architecture for RR, with full description of the design and the mechanisms to deal with mobility, failures, and inaccuracies, and generalizing it to deal with resource discovery and data-centric architectures in general. A main goal in RR design is to target high mobility environments and this makes rendezvous regions more suitable than rendezvous points. RR is also based on our objective to design geographic systems that need only approximate location information. The use of regions affects many design details such as the server election, insertion, lookup, and replication. In RR, the network topology space is divided into rectangular geographical regions, where each region is responsible for a set of keys representing the data or resources of interest. A key, k_i, is mapped to a region, RR_j, by using a hash-table-like mapping function, $h(k_i) = RR_j$. The mapping is known to all nodes and is used during the insertion and lookup operations. A node wishing to insert or lookup a key obtains the region responsible for that key through the mapping, then uses geographic-aided routing to send a message to the region. Inside a region, a simple local election mechanism dynamically promotes nodes to be servers responsible for maintaining the mapped information. Replication between servers in the region reduces the effects of failures and mobility. By using regions instead of points, our scheme requires only approximate location information and accordingly is more robust to errors and imprecision in location measurement and estimation than schemes depending on exact location information. Regions also provide a dampening factor in reducing the effects of mobility, since no server changes are required as long as current servers move inside their region and hence the overhead due to mobility updates is quite manageable.

The network topology space is divided into geographical regions (RRs), where each region (e.g., RR_j) is responsible for a set of resources. The resource key space is divided among these regions, such that each resource key (K_i) is mapped to a region. The key-set to RR mapping ($KSet_i \leftrightarrow RR_j$) is known by all nodes.

The RR scheme can be built on top of any routing protocol that can route packets toward geographic regions. The only requirement of the routing protocol is to maintain approximate geographic information, such that given an insertion or lookup to a certain region, it should be able to obtain enough information to route the packet toward that region. In our design we use geocasts for insertions and anycasts for lookups. These design choices are simple to implement, robust to dynamics, and do not require tracking of nodes' locations. Following we describe the main components of our architecture:

Region detection: Using a localization mechanism, each node detects its location and accordingly its geographic region. When the node moves, it detects its new location and so it can keep track of its region. The node uses this information to forward packets toward regions, to detect packets forwarded to its region, and to potentially participate in server election in its region (if and when needed).

Server election: A simple local election mechanism is used inside the region to dynamically promote the servers. As the number of servers increases, the robustness to mobility and failures increases, but also the storage overhead increases. Servers are elected on-demand during insertions. When a data insertion operation is issued, the first node in the region that receives the insertion,[3] known as the *flooder*, geocasts the insertion inside the region. Each server receiving the insertion geocast sends an Ack back to the flooder. The flooder keeps track of the servers and if it does not get enough Acks (the minimum number of servers required), it geocasts again and includes a self-election probability, p, in the geocast message. Each node receiving the geocast elects itself with probability p and if it becomes a server, it replies to the flooder. If not enough Acks are received, the flooder increases p based on a back-off mechanism until the required number of servers reply or p reaches 1. When servers move out of the region or fail, new servers are elected in the same way. After the new servers are elected, they retrieve the stored keys from other servers.

Insertion: A node inserts a key, K, by first mapping the key to a rendezvous region, RR_i, where $K \in KSet_i \leftrightarrow RR_i$. The node generates a packet containing the region identifier, RR_i, in its header. Nodes routing the packet toward the region, check the region identifier to determine whether they are in or out of region. The first node inside RR_i to receive the packet, the *flooder*, geocasts the packet inside the region. Servers inside the region receive the geocast, store the key and data, then send Acks back to the flooder (Figure 15.13). The flooder collects the Acks and sends an Ack back to original sender. If no Ack is received by the sender, it timeouts and retransmits the insertion up to a fixed number of times.

Lookup: Lookups are similar to insertions except that nodes and previous flooders inside a region cache locations of the recent servers they hear from, and send the lookups directly to any of the servers (anycast). The server replies to the flooder and the flooder replies back to original sender (Figure 15.13). If the flooder receives no reply or if it has no cached servers, it geocasts the lookup inside the region.

Replication: Replication is inherent in this architecture, since several servers inside the region store the key and data. This adds extra robustness to failures and mobility. For additional robustness against severe dynamics such as group failures and partitions, multiple hash functions may be used to hash a key to multiple regions.

Mobility: Local movements of nodes and servers have negligible effect and overhead on our architecture as long as servers stay within their regions. The only condition we need to consider is when a server moves out of its region. The server checks its location periodically to detect when it gets out of its region, in order

---

[3]A node can identify that it is the first node in the region to receive the packet by a simple flag set in the packet header.

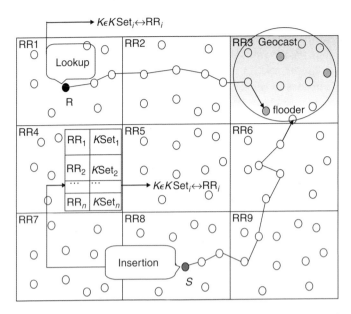

FIGURE 15.13 Rendezvous regions. Insertion: node S wishing to inset (or store) resource key K that belongs to $KSet_i$ gets the corresponding RR (in this case RR3) through the mapping ($KSet_i \rightarrow RR_i$). Node S then sends the resource information toward RR3, where it is geocast by the flooder and stored by the servers. Lookup: node R looking for a resource with key K that belongs to $KSet_i$ gets the corresponding RR (in this case RR3) through the mapping ($KSet_i \rightarrow RR_i$). R then sends the resource lookup toward RR3, where it is anycast to any server holding the information.

to send an insertion packet toward that region so that new servers are elected. The server then deletes its stored keys and is not a server anymore. It may or may not get elected again later in a new region.

Failures: Since each region contains several servers, and insertions and mobility may invoke new server elections, it is unlikely that independent reasonable failures will cause all servers to vanish. In order to avoid this case anyway, servers use a low-frequency periodic soft-state mechanism during silent (low traffic) periods, to detect failing servers and promote new servers. Each server runs a low-frequency timer, which is reset each time an insertion geocast is received. When the server times out, it geocasts a packet checking for other servers. Other servers reset their timers upon receiving this check and reply back demonstrating their existence. If not enough servers reply back, server election is triggered.

Bootstrap: One question remaining is how the mapping function is obtained. One option is to assume that it is pre-known or provided by out-of-band mechanisms. Another option is to use the same rendezvous mechanism, in order to provide a bootstrap overlay that publishes dynamic mappings. Using the mapping for a *well-known key,* a node sends request to a well-known region to obtain the mapping function of a set of services. These mappings however are not expected to change frequently. This introduces more flexibility for providing different mappings for different type of services and changing them when required.

15.5   Conclusions

We have presented an overview of geographic protocols for wireless ad hoc and sensor networks. It is obvious that utilizing the geographic information is vital for building scalable and efficient protocols in these environments. This study shows that there is a significant amount of work done in this area. Nevertheless, in order for geographic protocols to be implemented in the real-world, they need a higher degree of robustness to the realistic environmental conditions. In our work, we focus on this issue of assessing the robustness of geographic protocols to nonideal conditions corresponding to the real-world environments and designing new strategies and protocols that take these conditions into account. We pointed to some of our studies in this paper: the effect of inaccurate locations, lossy wireless links, robust geocasting, and Rendezvous Regions. In the present and future work, we are considering additional issues to improve the robustness of geographic protocols and move them closer to effective real-world deployment.

References

[1] I. Akyildiz, J. Ho, and Y. Lin. "Movement-based location update and selective paging for PCS networks". *IEEE/ACM Transactions on Networking*, 4(4), 629–638, August 1996.

[2] B. An and S. Papavassiliou. "Geomulticast: architectures and protocols for mobile ad hoc wireless networks". *Journal of Parallel and Distributed Computing*, 63, 182–195, 2003.

[3] I. Aydin, C.-C. Shen. "Facilitating match-making service in ad hoc and sensor networks using pseudo quorum." *IEEE ICCCN 2002*, 2002.

[4] L. Barrière, P. Fraigniaud, L. Narayanan, and J. Opatrny. "Robust position-based routing in wireless ad hoc networks with irregular transmission ranges." *Wireless Communications and Mobile Computing*, 2, 141–153, 2003.

[5] S. Basagni, I. Chlamtac, V. Syrotiuk, and B. Woodward. "A distance routing effect algorithm for mobility (DREAM)." *ACM MOBICOM 1998*, 1998.

[6] L. Blazevic, L. Buttyan, S. Capkun, S. Giordano, J. P. Hubaux, and J. Y. Le Boudec. "Self-organization in mobile ad-hoc networks: the approach of terminodes." *IEEE Communications Magazine*, 39(1), 118–124, June 2001.

[7] L. Blazevic, S. Giordano, J.-Y. Le Boudec. "Self-organized terminode routing." *Cluster Computing Journal*, 5(2), 205–218, 2002.

[8] J. Boleng, T. Camp, and V. Tolety. "Mesh-based geocast routing protocols in an ad hoc network." In *Proceedings of the IPDPS 2001*, 2001.

[9] P. Bose, P. Morin, I. Stojmenovic, and J. Urrutia. "Routing with guaranteed delivery in ad hoc wireless networks." In *Proceedings of the Workshop on Discrete Algorithms and Methods for Mobile Computing and Communications (DialM 1999)*, 1999.

[10] J. Broch, D. A. Maltz, D. B. Johnson, Y.-C. Hu, and J. Jetcheva. "A performance comparison of multi-hop wireless ad hoc network routing protocols." In *Proceedings of the ACM MOBICOM 1998*, 1998.

[11] N. Bulusu, J. Heidemann, D. Estrin, and T. Tran. "Self-configuring localization systems: design and experimental evaluation." *ACM Transactions on Embedded Computing Systems (TECS)*, 2003.

[12] T. Camp and Y. Liu. "An adaptive mesh-based protocol for geocast routing." *Journal on Parallel and Distributed Computing*: *Special Issue on Routing in Mobile and Wireless Ad Hoc Networks*, 63(2), 196–213, February 2003.

[13] Douglas S. J. De Couto, Daniel Aguayo, John Bicket, and Robert Morris. "A high-throughput path metric for multi-hop wireless routing." In *Proceedings of the ACM MOBICOM*, September 2003.

[14] D. De Couto and R. Morris. "Location proxies and intermediate node forwarding for practical geographic forwarding." MIT Technical Report, June 2001.

[15] D. Estrin, M. Handley, A. Helmy, P. Huang, and D. Thaler. "A dynamic bootstrap mechanism for rendezvous-based multicast routing." In *Proceedings of the IEEE INFOCOM 1999*, 1999.

[16] G. G. Finn. "Routing and addressing problems in large metropolitan-scale internetworks." Technical Report ISI/RR-87-180, Information Sciences Institute, March 1987.

[17] K. Gabriel and R. Sokal. "A new statistical approach to geographic variation analysis." *Systematic Zoology*, 18, 259–278, 1969.

[18] J. Gao, L. Guibas, J. Hershberger, L. Zhang, and A. Zhu. "Geometric spanners for routing in mobile networks." In *Proceedings of the ACM MOBIHOC 2001*, 2001.

[19] D. Ganesan, B. Greenstein, D. Perelyubskiy, D. Estrin, and J. Heidemann. "An evaluation of multi-resolution storage for sensor networks." In *Proceedings of the ACM SenSys 2003*, 2003.

[20] B. Greenstein, D. Estrin, R. Govindan, S. Ratnasamy, and S. Shenker. "DIFS: a distributed index for features in sensor networks." In *Proceedings of the IEEE SNPA 2003*, 2003.

[21] Z. Haas. "A new routing protocol for the reconfigurable wireless networks." In *Proceedings of the IEEE Conference on Universal Personal Communication*, pp. 562–566, 1997.

[22] Z. J. Haas and B. Liang. "Ad hoc mobility management with uniform quorum systems." *IEEE/ACM Transactions on Networking*, vol. 7, pp. 228–240, April 1999.

[23] T. He, C. Huang, B. Blum, J. Stankovic, and T. Abdelzaher. "Range-free localization schemes for large scale sensor networks." In *Proceedings of the ACM MOBICOM 2003*, 2003.

[24] T. He, J. Stankovic, C. Lu, and T. Abdelzaher. "SPEED: a stateless protocol for real-time communication in sensor networks." In *Proceedings of the International Conference on Distributed Computing Systems*, May 2003.

[25] J. Heidemann, F. Silva, and D. Estrin. "Matching data dissemination algorithms to application requirements." In *Proceedings of the ACM SenSys 2003*, 2003.

[26] A. Helmy. "Architectural framework for large-scale multicast in mobile ad hoc networks." In *Proceedings of the IEEE ICC*, April 2002.

[27] J. Hightower and G. Borriello. "Location systems for ubiquitous computing." *IEEE Computer*, August 2001.

[28] T. Imielinski and J. Navas. "GPS-based addressing and routing." In *Proceedings of the IETF RFC 2009*, November 1996.

[29] C. Intanagonwiwat, R. Govindan, and D. Estrin. "Directed diffusion: a scalable and robust communication paradigm for sensor networks." In *Proceedings of the ACM MOBICOM 2000*, 2000.

[30] R. Jain, A. Puri, and R. Sengupta. "Geographical routing using partial information for wireless ad hoc networks." *IEEE Personal Communications*, February 2001.

[31] D. B. Johnson and D. A. Maltz. "Dynamic source routing in ad-hoc wireless networks." *Mobile Computing*, 1996.

[32] B. Karp. "Challenges in geographic routing: sparse networks, obstacles, and traffic provisioning." Slides presented at the DIMACS workshop on Pervasive Networking, May 2001.

[33] B. Karp and H. T. Kung. "GPSR: greedy perimeter stateless routing for wireless networks." In *Proceedings of the ACM MOBICOM 2000*, 2000.

[34] Y. Kim, J. Lee, and A. Helmy. "Impact of location inconsistencies on geographic routing in wireless networks." In *Proceedings of the ACM MSWIM 2003*, 2003.

[35] Y. Ko and N. Vaidya. "Location-aided routing (LAR) in mobile ad hoc networks." In *Proceedings of the ACM MOBICOM 1998*, 1998.

[36] Y. Ko and N. Vaidya. "Flooding-based geocasting protocols for mobile ad hoc networks." *ACM/Baltzer Mobile Networks and Applications (MONET) Journal*, December 2002.

[37] Y. Ko and N. Vaidya. "Anycasting-based protocol for geocast service in mobile ad hoc networks." *Computer Networks Journal*, April 2003.

[38] E. Kranakis, H. Singh, and J. Urrutia. "Compass routing on geometric networks." In *Proceedings of 11th Canadian Conference on Computational Geometry*, August 1999.

[39] F. Kuhn, R. Wattenhofer, and A. Zollinger. "Worst-case optimal and average-case efficient geometric ad-hoc routing." In *Proceedings of the ACM MOBIHOC 2003*, 2003.

[40] S. Kumar, C. Alaettinoglu, and D. Estrin. "Scalable object-tracking through unattended techniques (SCOUT)." In *Proceedings of the IEEE ICNP 2000*, 2000.

[41] J. Li, J. Jannotti, D. Couto, D. Karger, and R. Morris. "A scalable location service for geographic ad hoc routing." In *Proceedings of the ACM MOBICOM 2000*, 2000.

[42] X. Li, Y. Kim, R. Govindan, and W. Hong. "Multi-dimensional range queries in sensor networks." In *Proceedings of the ACM SenSys*, November 2003.

[43] W. H. Liao, Y. C. Tseng, K. L. Lo, and J. P. Sheu. "GeoGRID: a geocasting protocol for mobile ad hoc networks based on GRID." *Journal of Internet Technology*, vol. 1, pp. 23–32, 2000.

[44] J. Navas and T. Imielinski. "Geographic addressing and routing." In *Proceedings of the ACM MOBICOM 1997*, 1997.

[45] S. Ni, Y. Tseng, Y. Chen, and J. Sheu. "The broadcast storm problem in a mobile ad hoc network." In *Proceedings of the ACM MOBICOM 1999*, 1999.

[46] D. Nicules and B. Nath. "Trajectory-based forwarding and its applications." In *Proceedings of the ACM MOBICOM 2003*, 2003.

[47] V. Park and M. Corson. "A highly adaptive distributed routing algorithm for mobile wireless networks." In *Proceedings of the IEEE INFOCOM 1997*, 1997.

[48] C. E. Perkins and P. Bhagwat. "Highly dynamic destination-sequenced distance vector routing for mobile computers." In *Proceedings of the ACM CCR*, October 1994.

[49] C Perkins and E. Royer. "Ad hoc on-demand distance vector routing." In *Proceedings of the 2nd IEEE Workshop on Mobile Computing Systems and Applications*, New Orleans, LA, February 1999, pp. 90–100.

[50] N. B. Priyantha, A. Chakraborty, and H. Balakrishnan. "The cricket location-support system." In *Proceedings of the ACM MOBICOM*, 2000.

[51] S. Ratnasamy, B. Karp, L. Yin, F. Yu, D. Estrin, R. Govindan, and S. Shenker. "GHT: a geographic hash table for data-centric storage." In *Proceedings of the ACM WSNA*, September 2002 and *ACM MONET* 2003.

[52] E. Royer and C. Toh. "A review of current routing protocols for ad-hoc mobile wireless networks." *IEEE Personal Communications Magazine*, April 1999, pp. 46–55.

[53] A. Savvides, C.-C. Han, and M. B. Srivastava. "Dynamic fine-grain localization in ad-hoc networks of sensors." In *Proceedings of the ACM MOBICOM 2001*, 2001.

[54] K. Seada and A. Helmy. "*Rendezvous Regions*: a scalable architecture for service location and data-centric storage in large-scale wireless networks." In *Proceedings of the IEEE/ACM IPDPS 4th International Workshop on Algorithms for Wireless, Mobile, Ad Hoc and Sensor Networks (WMAN)*, Santa Fe, New Mexico, April 2004.

[55] K. Seada and A. Helmy. "Efficient geocasting with perfect delivery in wireless networks." In *Proceedings of the IEEE Wireless Communications and Networking Conference (WCNC)*, Atlanta, Georgia, March 2004.

[56] K. Seada, A. Helmy, and R. Govindan. "On the effect of localization errors on geographic face routing in sensor networks." In *Proceedings of the IEEE/ACM 3rd International Symposium on Information Processing in Sensor Networks (IPSN)*, Berkeley, CA, April 2004.

[57] K. Seada, M. Zuniga, A. Helmy, and B. Krishnamachari. "Energy-efficient forwarding strategies for geographic routing in lossy wireless sensor networks." In *Proceedings of the ACM 2nd Conference on Embedded Networked Sensor Systems (SenSys)*, Baltimore, Maryland, November 2004.

[58] D. Son, A. Helmy, and B. Krishnamachari. "The effect of mobility-induced location errors on geographic routing in ad hoc networks: analysis and improvement using mobility prediction." In *Proceedings of the IEEE WCNC 2004*, 2004.

[59] I. Stojmenovic, A. P. Ruhil, and D. K. Lobiyal. "Voronoi diagram and convex hull-based geocasting and routing in wireless networks." In *Proceedings of the IEEE ISCC 2001*, 2001.

[60] H. Takagi and L. Kleinrock. "Optimal transmission ranges for randomly distributed packet radio terminals." *IEEE Transactions on Communications*, March 1984.

[61] J. Tchakarov and N. Vaidya. "Efficient content location in wireless ad hoc networks." In *Proceedings of the IEEE International Conference on Mobile Data Management (MDM)*, January 2004.

[62] G. Toussaint. "The relative neighborhood graph of a finite planar set." *Pattern Recognition*, 12, 261–268, 1980.

[63] A. Ward, A. Jones, and A. Hopper. "A new location technique for the active office." *IEEE Personal Communications*, October 1997.

[64] S. Woo and S. Singh. "Scalable routing protocol for ad hoc networks." *Wireless Networks*, January 2001.

[65] Y. Xu, J. Heidemann, and D. Estrin. "Geography-informed energy conservation for ad-hoc routing." In *Proceedings of the ACM MOBICOM 2001*, 2001.

[66] Y. Xue, B. Li, and K. Nahrstedt. "A scalable location management scheme in mobile ad-hoc networks." In *Proceedings of the IEEE Annual Conference on Local Computer Networks (LCN)*, November 2001.

[67] F. Ye, H. Luo, J. Cheng, S. Lu, and L. Zhang. "A two-tier data dissemination model for large-scale wireless sensor networks." In *Proceedings of the ACM MOBICOM 2002*, 2002.

[68] Y. Yu, R. Govindan, and D. Estrin. "Geographical and energy aware routing: a recursive data dissemination protocol for wireless sensor networks." *USC Technical Report*, 2001.

[69] W. Zhang, G. Cao, and T. La Porta. "Data dissemination with ring-based index for sensor networks." In *Proceedings of the IEEE ICNP 2003*, 2003.

16

Location Management Algorithms Based on Biologically Inspired Techniques

16.1 Introduction...**16**-365
16.2 Location Management Cost............................**16**-368
  Network Structure
16.3 Genetic Algorithm **16**-370
  Selection • Crossover • Mutation
16.4 Genetic Algorithm Framework **16**-374
16.5 Tabu Search **16**-375
  Tabu List • Search Intensification • Search Diversification
16.6 Tabu Search Framework...............................**16**-377
  Search Intensification • Search Diversification • Back to
  Normal Condition
16.7 Ant Colony Algorithm **16**-379
16.8 Ant Colony Algorithm Framework....................**16**-380
  Type-1 • Type-1 Rank Variant • Type-1 Tournament
  Variant • Type-2 • Type-2 Rank and Tournament Variant
16.9 Performance Comparison.............................**16**-385
  Stopping Conditions • Results
16.10 Summary **16**-389
References **16**-389

Riky Subrata
Albert Y. Zomaya

16.1 Introduction

One of the challenges facing mobile computing is the tracking of the current location of the user — the area of *location management*. In order to route incoming calls to appropriate mobile terminals, the network must from time to time keep track of the location of each mobile terminal.

Mobility tracking expends the limited resources of the wireless network. Besides the bandwidth used for registration and paging between the mobile terminal and base stations, power is also consumed from the portable devices, which usually have limited energy reserve. Furthermore, frequent signaling may result in degradation of quality of service (QoS), due to interferences. On the other hand, a miss on the location of a mobile terminal will necessitate a search operation on the network when a call comes in. Such an operation, again, requires the expenditure of limited wireless resources. The goal of mobility tracking, or location management, is to balance the registration and search operation, so as to minimize the cost of mobile terminal location tracking.

Two simple location management strategies are the *always-update* strategy, and the *never-update* strategy. In the always-update strategy, each mobile terminal performs a location update whenever it enters a new cell. As such, the resources used (overhead) for location update would be high. However, no search operation would be required for incoming calls. In the never-update strategy, no location update is ever performed. Instead, when a call comes in, a search operation is conducted to find the intended user. Clearly, the overhead for the search operation would be high, but no resources would be used for the location update. These two simple strategies represent the two extremes of location management strategies, whereby one cost is minimized and the other maximized. Most existing cellular systems use a combination of the above two strategies.

One of the common location management strategy used in existing systems today is the location area scheme [47,49,63]. In this scheme, the network is partitioned into *regions* or *location areas* (LA), with each region consisting of one or more cells (Figure 16.1). The never-update strategy can then be used within each region, with location update performed only when a user moves out to another region/location area. When a call arrives, only cells within the LA for which the user is in need to be searched. For example, in Figure 16.1, if a call arrives for user X, then search is confined to the 16 cells of that LA. It is recognized that optimal LA partitioning (one that gives the minimum location management cost) is in general an NP-complete problem [28].

Another location management scheme similar to the LA scheme is suggested in Reference 8. In this strategy, a subset of cells in the network is designated as the *reporting cells* (Figure 16.2). Each mobile terminal performs a location update only when it enters one of these reporting cells. When a call arrives, search is confined to the reporting cell the user last reported, and the neighboring bounded nonreporting cells. For example, in Figure 16.2, if a call arrives for user X, then search is confined to the reporting cell the user last reported in, and the non-reporting cells marked **P**. Obviously, certain reporting cells configuration leads to unbounded nonreporting cells, as shown in Figure 16.3. It was shown in Reference 8 that finding an optimal set of reporting cells, such that the location management cost is minimized, is in general an NP-complete problem. In Reference 31, a heuristic method to find near optimal solutions is described.

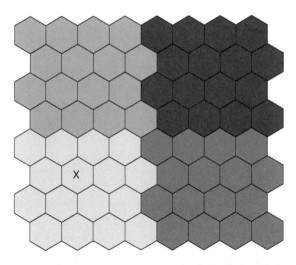

FIGURE 16.1   Regions representing location areas (LA) and individual cells (in this figure there are four LA, each consisting of 16 cells).

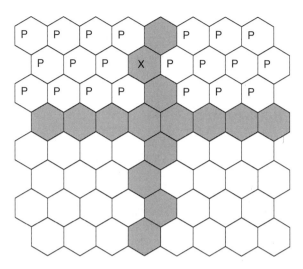

FIGURE 16.2 Network with reporting cells (shaded areas represent reporting cells).

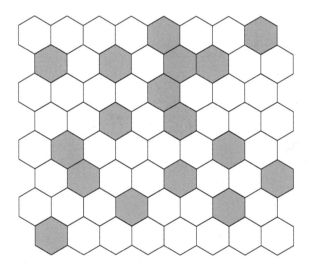

FIGURE 16.3 Network with reporting cells and unbounded nonreporting cells.

Biologically inspired techniques have been used to solve a wide range of complex problems in recent times. The power of these techniques stems from their capability in searching large search spaces, which arise in many combinatorial optimization problems, very efficiently. Several well-known biologically inspired techniques are presented and discussed for solving location management problems. Due to their popularity and robustness, genetic algorithm (GA), tabu search (TS), and ant colony algorithm (ACA) are discussed and presented to solve the reporting cells planning problem.

In a further discussion of the relatively new ACA, two types of ACA are presented — call it *type-1* and *type-2*, along with the introduction of a rank- and tournament-based selection variant for each type. The idea for the rank- and tournament-based selection variant comes from GA, in which rank and tournament selection has been used, and improves the algorithm's performances in several cases. Type-1 ACA is a more traditional ACA in the sense that each ant creates a new solution in each new cycle. In type-2 ACA, each ant *moves* from its current solution, to generate a new solution.

The next section is an overview of the location management problem and viewing it as an optimization problem with a description of the cost functions that can be used to lead to the best results. This is followed by the development of a GA, TS, and six different ACA to solve the location management problem. A performance comparison of the different algorithms for a number of test cases is then presented.

16.2   Location Management Cost

As noted earlier, location management involves two elementary operations of *location update* and *location inquiry*, as well as *network interrogation* operations. Clearly, a good location update strategy would reduce the *overhead* for location inquiry. At the same time, location updates should not be performed excessively, as it expends on the limited wireless resources.

To determine the *average cost* of a location management strategy, one can associate a cost component to each location update performed, as well as to each polling/paging of a cell. The most common cost component used is the wireless bandwidth used (wireless traffic load imposed on the network). That is, the wireless traffic from mobile terminals to base stations (and vice versa) during location updates and location inquiry. While there are also "fixed wire" network traffic (and database accesses and loads) between controllers within the network during location updates and location inquiry — "network interrogation," they are considered much cheaper (and much more "scalable") and is usually not considered.

The total cost of the above two cost components (location update and cell paging) over a period of time T, as determined by simulations (or analytically or by other means) can then be averaged to give the average cost of a location management strategy [55]. For example, the following simple equation can be used to calculate the total cost of a location management strategy:

$$\text{Total cost} = C \cdot N_{\text{LU}} + N_{\text{P}} \qquad (16.1)$$

where N_{LU} denotes the number of location updates performed during time T, N_{P} denotes the number of paging performed during time T, and C is a constant representing the cost ratio of location update and paging. It is recognized that the cost of location update is usually much higher than the cost of paging — up to several times higher [38], mainly due to the need to setup a "signaling" channel. In light of this fact, the value $C = 10$ is used [28,62].

16.2.1   Network Structure

Most, if not all, today's wireless network consists of cells. Each cell contains a base station, which is wired to a fixed wire network. These cells are usually represented as hexagonal cells, resulting in six possible neighbors for each cell (see Figure 16.4). For further information on wireless networks and communications, refer to 9, 12, 50, 51, and 54. Note, however, the use of GA, TS, and ACA are not restricted to hexagonal or uniformly distributed cellular network.

In the reporting cells location management scheme, some cells in the network are designated as reporting cells. Mobile terminals do location update (update their positions) upon entering one of these reporting cells.

When calls arrive for a user, the user has to be located. Some cells in the network, however, may not need to be searched at all, if there is no path from the last location of the user to that cell, without entering a *new* reporting cell (a reporting cell that is not the last reporting cell the user reported in). That is, the reporting cells form a "solid line" barrier, which means a user will have to enter one of these reporting cells to get to the other side. For example, in Figure 16.4, a user moving from cell 4 to cell 6 would have to enter a reporting cell. As such, for location management cost evaluation purposes, the cells that are in bounded areas are first identified, and the maximum area to be searched for each cell (if a call arrives for that cell) is calculated (described later).

Suppose we call the collection of all the cells that are reachable from a reporting cell i without entering another reporting cell as the *vicinity* of reporting cell i; by definition, the reporting cell i is also included.

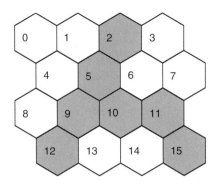

FIGURE 16.4   Network with reporting cells (shaded areas represent reporting cells).

Then the *vicinity value* (number of cells in the vicinity) of a reporting cell i is the maximum number of cells to be searched, when a call arrives for a user whose last location is known to be cell i. As an example, in Figure 16.4, the *vicinity* of reporting cell 9 includes the cells 0, 1, 4, 8, 13, 14, and cell 9 itself. The *vicinity value* is then 7, as there are seven cells in the vicinity.

Each nonreporting cell can also be assigned a vicinity value. However, it is clear that a nonreporting cell may belong to the vicinity of several reporting cells, which may have different vicinity values. For example, in Figure 16.4, cell 4 belongs to the vicinity of reporting cells 2, 5, 9, and 12, with vicinity values 8, 8, 7, and 7 respectively. For location management cost evaluation purposes, the maximum vicinity value will be used. As such, in this case, the vicinity value of eight is assigned to cell 4.

Each cell i is further associated with a *movement* weight, and *call arrival* weight, denoted w_{mi} and w_{ci}, respectively. The movement weight represents the frequency or total number of movement into a cell, while the call arrival weight represents the frequency or total number of call arrivals within a cell. Clearly, if a cell i is a reporting cell, then the number of location updates occurring in that cell would be dependent on the movement weight of that cell. Further, because call arrivals result in a search/paging operation, the total number of paging performed would be directly related to the call arrival weight of the cells in the network. As such, the total number of location updates and paging (performed during a certain time period T) can be represented by the following formulas:

$$N_{\text{LU}} = \sum_{i \in S} w_{mi}$$

$$N_{\text{P}} = \sum_{j=0}^{N-1} w_{cj} \cdot v(j) \tag{16.2}$$

where N_{LU} denotes the number of location updates performed during time T, N_{P} denotes the number of paging performed during time T, w_{mi} denotes the movement weight associated with cell i, w_{cj} denotes the call arrival weight associated with cell j, $v(j)$ denotes the vicinity value of cell j, N denotes the total number of cells in the network, and S denotes the set of reporting cells in the network.

Using Equations (16.1) and (16.2) gives the following formula to calculate the location management cost of a particular reporting cell's configuration:

$$\text{Total cost} = C \cdot \sum_{i \in S} w_{mi} + \sum_{j=0}^{N-1} w_{cj} \cdot v(j) \tag{16.3}$$

where C is a constant representing the cost ratio of location update and paging, as described earlier. The value $C = 10$ is used.

In the earlier equation, the vicinity value $v(j)$ calculation will undoubtedly take most of the computing time. The total cost shown above can also be divided by the total number of call arrivals, giving the *cost per call arrival*. Such "normalization" does not affect the relative location management costs of different reporting cells configuration, and is what has been done in the experiments.

16.3 Genetic Algorithm

A genetic algorithm (GA) [27,44,46,61] is a biologically inspired optimization and search technique developed by Holland [36]. Its behavior mimics the evolution of simple, single celled organisms. It is particularly useful in situations where the solution space to be searched is huge, making sequential search computationally expensive and time consuming.

Genetic algorithm is a type of guided random search technique, able to find "efficient" solutions in a variety of cases. Efficient here is defined as a solution that, though they may not be the absolute optimal, is within the constraints of the problem, and is found in a reasonable amount of time and resources. To put another way, the effectiveness or quality of a GA (for a particular problem) can be judged by its performance against other known techniques — in terms of solutions found, and time and resources used to find the solutions. That is, much like everything else in life, we judge by relative performance or benchmark — have we done better than our competitors? It should also be pointed out that in many cases (much like everything else in life) the absolute optimal is not known. In this respect, GA has shown itself to be extremely effective in problems ranging from optimizations to machine learning [19,39,40,44,46,56–58,64,65].

Some common terminology used in GA is described below. The relationships between the different "units" used in the GA system are shown in Figure 16.5.

- *Gene*: A basic entity in GA, a single gene represents a particular feature or characteristic of *individuals* in the GA system.
- *Chromosome*: An *individual* in the GA system, consisting of a *string* (or collection) of genes. Each chromosome represents a *solution* (usually an encoding of the solution), or a parameter set, for the particular problem.
- *Population*: Collection of chromosomes.

The gene in GA takes on a value from an alphabet of size $r \in Z^+$. Each chromosome/string then consists of a series of n genes:

$$\tilde{s} = \{b_1, b_2, \ldots, b_n\} \tag{16.4}$$

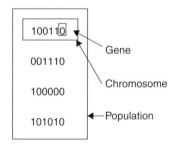

FIGURE 16.5   Genetic algorithm system. Population is a collection of chromosomes, and chromosome is a collection of genes.

resulting in a solution space of size r^n. The most common, and generally most effective [27] string representation is the binary alphabet $\{0, 1\}$. In this case, each gene takes on a possible value of 0 or 1:

$$b_i \in \{0, 1\}, \qquad i = 1, 2, \ldots, n \tag{16.5}$$

whereby the solution space is now of size 2^n. In most cases, the use of binary gene (resulting in a binary string) results in longer chromosome length (to represent the parameter set, or solutions, to the problem). At the same time, this results in more genes available to be "exploited" by the GA, resulting in better performances in many cases [27,46].

In GA, parameter sets (solutions) for a given problem is encoded into the chromosome. While the encoding is arbitrary, it should be chosen so that the mapping of the solution (or parameter set) to the chromosome is unique. Once the solutions, or parameter sets of the problem is appropriately encoded to the chromosome, a population of chromosomes is generated, and "evolved." A basic GA is described in the algorithm below.

Algorithm 16.1 (BASIC GA):

Input: Parameters for the GA.
Output: Population of chromosomes, P. Each of these chromosomes represents a solution to the problem.
Begin
       Initialize the population, P.
       Evaluate P.
       While stopping conditions not true **do**
              Apply *Selection* in P to create P_{mating}.
              Crossover P_{mating}.
              Mutate P_{mating}.
              Replace P with P_{mating}.
              Evaluate P.
End.

At the initial stage, a population of chromosomes P of a specified size $m \in Z^+$ is created and initialized. At initialization, the value of each gene in each chromosome is usually generated randomly. This initial population is usually denoted as P^0, and subsequent generations as P^1, P^2, and so on, until P^{final}. P^{final} denotes the final population of chromosomes, when the stopping conditions are met and the algorithm terminates.

After initialization, the population goes through an "evolution loop." Ideally, as each generation evolves, the "fitness" of the chromosomes in the population of the current generation should be better than the chromosomes of the previous generation. This "evolution" continues until the *stopping conditions* are met. The stopping conditions is generally one or a combination of the following conditions:

- Specified number of iterations/generations has been reached.
- Specified number of iterations/generations has passed since the best solution was found.
- Maximum computation time allowed reached.
- Chromosomes similarity — the population consists of many "similar" (or the same) chromosomes. That is, the GA has *converged* to a solution.
- Solution(s) within an acceptable error margin achieved.

At each generation, the fitness of all the chromosomes in the population is evaluated. If there are m chromosomes in the population, this results in m number of fitness calculation at each generation. As such, the stopping condition(s) of a GA should be set so that the total number of fitness calculation is much less than the size of the solution space. Otherwise, the algorithm is inefficient, as an exhaustive, brute-force search — whereby all possible solutions are examined, and is guaranteed to give the best possible solution (the global optimal), and without the added computation time for the GA.

The evolution of the GA population from one generation to the next is usually achieved through the use of three "operators" that are fundamental in GA: *selection, crossover,* and *mutation.*

16.3.1 Selection

Selection is the process of selecting chromosomes (in the population) from the current generation, for "processing" to the next generation. For this purpose, highly fit chromosomes/individuals are usually given a higher chance of being selected more often, thereby producing more "offspring" for the next generation. One common technique to achieve this objective is the fitness-proportionate selection.

In this selection method, the number of times an individual is selected for reproduction is proportional to its fitness value, as compared to the average fitness of the population. One simple method is to set the *expected* "count" of each individual to the following:

$$E(i) = \frac{f_i}{\bar{f}} \tag{16.6}$$

where $E(i)$ denotes the expected count of an individual i, f_i denotes the fitness value associated with individual i, and $\bar{f}$ denotes the average fitness of the population.

One can implement the above fitness-proportionate selection method using a *biased* roulette wheel, whereby highly fit individuals is given a bigger slice of the wheel. The wheel can then be "spun" to select individuals for "reproduction." Mathematically, the selection probability for each individual i can be expressed as:

$$p_i = \frac{f_i}{\sum_{j=1}^{n} f_j} \tag{16.7}$$

where n denotes the population size (total number of individual in the population), and f_i denotes the fitness value associated with individual i.

16.3.1.1 Elitism

Elitism is usually used in conjunction with the other selection methods. In its simplest form, the best σ individuals are "cloned" and guaranteed to be present, in its current form, in the next generation. Such method is used to ensure that the best individuals are not lost, either because they are not selected in the selection procedure, or because their genes have been changed by crossover and mutation. The use of elitism improves the performance of the genetic algorithm in many cases [10,32,33,41,42,46,48,53].

16.3.2 Crossover

Once individuals/chromosomes have been selected (for reproduction), crossover is applied to the chosen individuals. The crossover operator usually operates on two individuals/parents, to produce two children. The crossover operator is one of the distinguishing features of a GA, and its use ensures that characteristics of each parent are inherited in the children.

Common crossover methods that are readily used for binary gene include *one-point* crossover and *two-point* crossover, and *uniform* crossover. Which of these crossover methods is the most appropriate to use depends on the particular problems, in particular the chromosome encoding (to represent the problem set), and fitness function used. Note though the two-point crossover and uniform crossover is commonly used in recent GA applications [46].

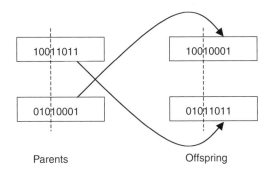

FIGURE 16.6   One-point crossover. The crossover point is at the "third" position.

16.3.2.1 One-Point Crossover

In the one-point crossover, a crossover point r_1 is randomly generated. This crossover point ranges in $[1, n-1]$, where n is the length of the chromosome. For example, in Figure 16.6, the crossover point is at the third position, dividing the chromosome into two segments. The first segment consists of three genes, and the second consists of five genes. The first offspring then gets the first segment of the first parent, and the second segment of the second parent. Similarly, the second offspring gets the first segment of the first parent, and the second segment of the first parent (see Figure 16.6).

16.3.2.2 k-Point Crossover

The one-point crossover can be extended to the two-point and k-point crossover. In the k-point crossover, k crossover points $r_1, r_2, \ldots, r_k$ are generated, each with value in $[1, n-1]$, and satisfying the condition $r_1 < r_2 < \cdots < r_{k-1} < r_k$. Each offspring then gets the segments of each offspring alternately. For example, a four-point crossover of two parents 00000000 and 11111111, with crossover points of 1, 2, 3, 4 will produce two offspring 01010000 and 10101111.

16.3.2.3 Uniform Crossover

Unlike the k-point crossover, in which "exchanges" of genes between the parents (to produce offspring) take place only at designated segments, in uniform crossover gene exchanges can occur at any position on the chromosome. For example, if we have two parents 000000 and 111111, then to produce two offspring we go through each of the six genes of one of the parent. For each gene, there is a probability p_s (e.g, $p_s = 0.7$) that this gene will be exchanged with the other parent. As such, if an exchange happens at position 2 and 5, then the two offspring produced will be 010010 and 101101.

16.3.3 Mutation

While the crossover operator works on a pair (or more) of chromosomes/individuals to produce two (or more) offspring, the mutation operator works on each individual offspring. The mutation operator helps prevent early convergence of the GA by changing characteristics of chromosomes in the population. Such changes in the chromosomes also results in the GAs ability to "jump" to far away solutions, hopefully to unexplored areas of the solution space. Clearly though, the *mutation rate* of the chromosomes should not be set too high, as too much mutations has the effect of "changing" the guided random search of the GA to a purely random search.

The simplest type of mutation commonly used is a *bit-flip*. In the case of binary gene, each gene in the chromosome has a probability p_m (e.g., $p_m = .001$) of being "flipped" from a "0" to a "1," or from a "1" to a "0." In the case of nonbinary, integer gene, whereby each gene can take on some integer values, flipping involves changing the gene's value to a value *other* than the current value.

16.4 Genetic Algorithm Framework

The GA used for discovering efficient solutions to the reporting cell problem is shown below.

Algorithm 16.2 (GA FOR REPORTING CELL PLANNER):

Input: Parameters for the GA.
Output: Population of solutions, P. Each of these solutions can be used as a reporting cell configuration.
Begin
      Initialize the population, P.
      Evaluate P.
      While stopping conditions not true **do**
          Select *Elite* in P consisting of k ($1 \leq k \leq$ *population size*) best individuals.
          Apply *Selection* from individuals in P to create P_{mating}, consisting of (*population size* — k) individuals.
          Crossover P_{mating}.
          Mutate P_{mating}.
          Replace the worst (*population size* — k) individuals in P with (the whole individuals of) P_{mating}.
          Evaluate P.
          If escape condition true **then** *Escape*.
End.

The parameters for the GA are shown in Table 16.1 First, a population of solution is created. In our case, a population of 100 chromosomes is used. Each chromosome consists of binary string representing a reporting cell configuration. For an 8×8 network, the chromosome will have length of 64. Each binary gene, which represents a cell in the network, can have a value of either "0," representing a nonreporting cell, and a "1," representing a reporting cell.

At initialization, the configuration for each chromosome is randomly generated. This initial population is then evaluated. After the initial evaluation, the GA goes into the "evolution loop." The GA continues to evolve until the stopping conditions are met. Such conditions include maximum computation time, rule similarity, number of iterations/generations since the best solution was found, and number of iterations/generations to be run, to name a few.

At each generation, the chromosomes/rules in the population is sorted in descending order, according to their fitness. In this experiment, the best five rules are kept, and survive to the next generation intact. To this end, *tournament selection* is applied to the 100 chromosomes in the current population, with probability $p_s = 0.8$, to create a mating pool of 95. In this tournament selection, two individuals are chosen at random from the current population P. The two individuals then "compete" in a virtual tournament, whereby the fitter individual has a p_s chance of winning. The winner is then given the right to "mate," proceeding in this case to the mating pool. This process is repeated until a specified number of individuals, in this

TABLE 16.1   Genetic Algorithm
Parameters

Population m	100
Best cloned σ	5
Tournament selection p_s	0.8
Two-point crossover p_c	0.8
Gene mutation p_m	0.001
Max stale period T_{stale}	20
Escape mutation p_{mm}	0.2

case 95, have been selected to the mating pool. It should be noted that the same individual can be selected more than once, leading to the individual "mating" more than once.

To the chromosomes in the mating pool, *two-point crossover* with probability $p_c = 0.8$, and *gene mutation* operator of probability $p_m = .001$ is applied. These 95 chromosomes are then copied to the current population P, which, together with the best five chromosomes, forms the next generation of chromosomes.

Finally, the "escape" condition is checked. Here, the escape condition is the number of generation since the last improvement in solution. If this number exceeds the maximum stale period $T_{stale} = 20$, an escape operator is applied. In the escape operator, the population undergoes a massive gene mutation of $p_{mm} = .2$.

16.5   Tabu Search

Tabu search (TS) [21,22,26] was first proposed in its current form by Glover [20]. It has been successfully applied to a wide range of theoretical and practical problems, including graph coloring, vehicle routing, job shop scheduling, course scheduling, and maximum independent set problem [1–5,7,34,35,56,57].

One main ingredient of TS is the use of *adaptive memory* to guide problem solving. One may argue that *memory* is a necessary component for "intelligence," and intelligent problem solving. TS uses a set of strategies and learned information to "mimic" human insights for problem solving, creating essentially an "artificial intelligence" unto itself — though problem specific it may be.

Tabu search also has links to evolutionary algorithms through its "intimate relation" to scatter search and path re-linking [23,25]. Numerous computational studies over the years have shown that tabu search can compete, and in many cases surpass the best-known techniques [24,35]. In some cases, the improvements can be quite dramatic.

In its most basic sense, a tabu search can be thought of as a local search procedure, whereby it "moves" from one solution to a "neighboring" solution. In choosing the next solution to move to, however, TS uses memory and extra knowledge endowed about the problem. A basic TS algorithm is shown below.

Algorithm 16.3 (BASIC TABU SEARCH):

Input: Parameters for the TS.
Output: A feasible solution to the problem.
Begin
       Generate an initial solution s.
      While stopping conditions not true **do**
           Select next solution neighboring s.
           Update memory.
End.

For a given problem to be solved using TS, three things need to be defined: a set V of feasible solutions (for the problem), a *neighborhood structure* $N(s)$ for a given solution $s \in V$, and a *tabu list* (TL).

The $N(s)$ defines all possible neighboring solutions to move to from a solution s. The choice of the neighborhood structure is one of the most important factors influencing the performance of a TS algorithm. Clearly, the neighborhood structure should be chosen so that there is a "path" to the optimal solution, from any solution s. But also, if the choice of the $N(s)$ results in the "landscape" of the solution space to consist of many large mountains, and deep troughs, then it would be very difficult for the algorithm to find an optimal solution, particularly because TS "moves" locally from solution to solution.

As shown in Algorithm 16.3, an initial solution s is chosen from the set of feasible solution V. This initial solution is usually chosen randomly. Once an initial solution is chosen, the algorithm goes into a loop that terminates when one or more of the stopping conditions are met — much like a genetic algorithm.

In the next step of Algorithm 16.3, the next solution s^{i+1} is selected from the neighbors of the current solution s^i, $s^{i+1} \in N(s^i)$. In a basic TS, all possible solutions $s \in N(s^i)$ is considered, and the best solution is chosen as the next solution s^{i+1}. Note, however, the use of such selection rule may result in the algorithm going in circles. At the very worst, the algorithm will go back and forth between two solutions. To avoid this, memory is incorporated into TS.

16.5.1 Tabu List

The simplest way such a memory can be used is to remember the last k solutions that have been visited, and to avoid these solutions. That is, the algorithm keeps a *list* of tabu solutions. The length of this TL may be varied, and a longer list will prevent cycles of greater length — a list of length k will prevent cycles of length $\leq k$. However, it may be impractical (e.g., time consuming to compare the solutions) to incorporate such memory. As such, *simpler* type of memories is usually incorporated, one of which is remembering the last k *moves* made. A move for a solution s^i can be viewed as "changes" applied to the solution s^i, to get to a new solution $s^{i+1} \in N(s^i)$. Generally, moves are defined so that they are *reversible*: for a move m from s^i to s^{i+1}, there is a *reverse* move m^{-1} from s^{i+1} to s^i.

Clearly though, such a list is not a perfect replacement for a solution list. While a move list is much easier to use than a solution list, in many cases information is lost in that one cannot know for certain if a solution has in fact been visited in the last k moves. That is, the use of a move list does not guarantee that no cycle of length $\leq k$ will occur. In other cases, its use results in a restrictive search pattern — an unvisited solution may be ignored because a move to that solution is in the TL. Still, in other cases, the use of a move list results in both a loss of information, and a more restrictive search pattern.

To partially overcome the restriction imposed by using a move list (for the TL), a TS usually incorporates *aspiration conditions*, whereby a tabu move (a move that is in the TL) is selected if it satisfies one or more of the aspiration conditions. One simple aspiration condition that is commonly used in TS is the following. If a tabu move leads to a solution that is better than the best solution found so far, then it should be selected.

Finally, one other important feature of TS is that of *search intensification* and *search diversification*. For this purpose, a more elaborate memory structure that takes into account the *recency, frequency,* and *quality* of solutions and moves made so far is used.

16.5.2 Search Intensification

In search intensification, exploration is concentrated on the vicinity, or "neighbors" of solutions that have historically been found to be good. That is, search is concentrated around the *elite* solutions. Such search may be initiated when there is evidence that a region (on the solution space) may contain "good" solutions.

During search intensification, a *modified* fitness function is used to "encourage" moves or solution's features that have historically been found to be good. For example, a "reward" value may be added to the fitness function, so that certain moves or solution's features become more attractive. Further, a modified neighborhood structure may also be used, in order to effectively combine good solution's features.

Search intensification is usually carried out over a few numbers of iterations. If the process is not able to find a better solution than the best one found so far, a search diversification is usually carried out.

16.5.3 Search Diversification

The search diversification process, unlike search intensification, attempts to spread out the search process to unvisited regions, and encourages moves or solution's features that are *significantly* different from those that have been found. For this purpose, a modified fitness function is used, whereby *penalties* are applied to certain moves or solution's features. For example, penalties may be given to frequently made moves, or common solution's features.

16.6 Tabu Search Framework

In this TS implementation, two TLs are used, along with two types of search intensification and two types of search diversification, one of which is a global restart operator.

First, as TS can be considered as a local search procedure, a $N(s)$ needs to be defined for a given solution S (where S represents a suitable reporting cell configuration, as described earlier). Here, the N_S of a solution S is defined to consist of a "0 to 1" move and a "1 to 0" move for each cell. If a cell is at state "0," then the only allowed move for that cell is "0 to 1." Similarly, if a cell is at state "1," the allowed move is "1 to 0." The TS implementation for the reporting cell problem is graphically depicted in Figure 16.7.

During the exploration process, the best move is selected, and is then put into a tabu list, to prevent *excessive* "cycling" of solutions. Moves table are used instead of solutions table due to the impracticality of using solutions table.

Here, two TLs are used, one for the "0 to 1" move, and another for the "1 to 0" move, for each cell. Moves are kept in the tabu list for a period of between TL_{min} and TL_{max}. The exact period for each move is chosen randomly and uniformly from $[TL_{min}, TL_{max}]$. Further to this is an *aspiration condition*, whereby a move that's tabu will be accepted if it improves the best solution found so far.

To complement the exploration process, two types of search *intensification*, and search *diversification* are implemented. The inclusion of each type *improves* the performance of tabu search for the reporting cell problem.

16.6.1 Search Intensification

Two types of search intensification are implemented, which will be referred here as *type-1* and *type-2*. In search intensification, a modified fitness function is used to decide the next move to be made.

In the first type, moves with a history of good "score" are rewarded, so that such moves are likely to be chosen during the type-1 intensification period. To do this, a "score" is kept for each move made: the best move for the current iteration that gives a better solution than the solution from the previous iteration is considered good; the difference in value is then recorded. The "reward" for a move is then the average of the calculated sum. Note here that the "reward" value is always "positive," as only moves that results in a better solution triggers a recording. For moves that always give worse solutions, the reward value would be zero (no recording).

At the start of this search intensification period, the TL is reset, and the period that moves stay in the tabu list is shortened to $[TL_{I,min}, TL_{I,max}]$. The type-1 search intensification is triggered when the best solution found is repeated. This intensification procedure would normally last for T_{I1} iterations (unless the *Back to Normal condition* is satisfied).

In the second type of search intensification, moves with a history of bad "score" are penalized, so that these moves are less likely to be chosen during this period. To achieve this, a scoring scheme similar to the one described earlier is used. Here, however, a move that results in a worse solution triggers a recording, whereby the difference in value is recorded. The penalty for a move is then the average of the calculated sum. As shown in Figure 16.7 this intensification procedure is executed when the solution is not improved by the type-1 intensification procedure. This procedure lasts for T_{I2} iterations, unless disturbed by an external action (*Back to Normal condition*).

16.6.2 Search Diversification

Here, two types of search diversification procedure are implemented, *mild*, and *hot* diversification.

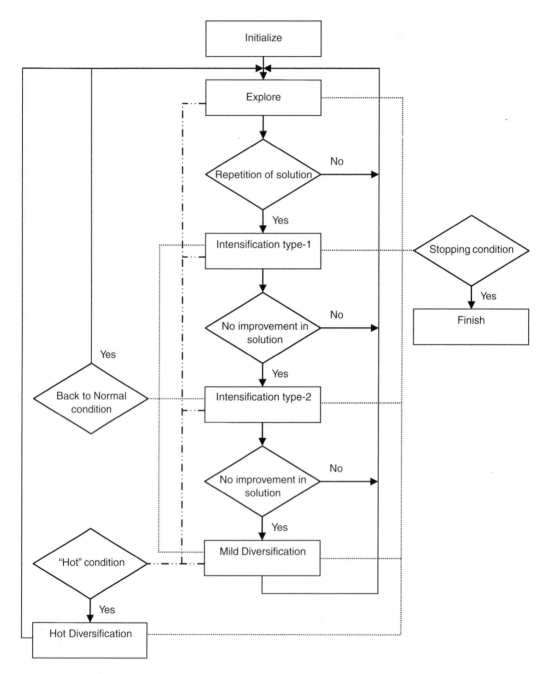

FIGURE 16.7   Tabu search procedure.

In the mild diversification procedure, a modified fitness function is again used. Here, frequently made moves are penalized, so that such moves are less likely to be chosen during this diversification procedure. To do this, the ratio of the move to the total moves performed is calculated. The penalty (for a move) is then calculated as the fitness value × ratio. This procedure is invoked when type-2 search intensification is not able to improve the solution (Figure 16.7). The procedure lasts for a maximum period of T_{Dm}.

Finally, the hot diversification procedure is global in nature, and is executed when the solution found is not improved within a period T_{stale}. The procedure resets the TL, and a new random solution is generated.

TABLE 16.2 Tabu Search Parameters

Tabu period TL_{min}	2
Tabu period TL_{max}	5
Tabu period $TL_{I,min}$	1
Tabu period $TL_{I,max}$	1
Intensification period T_{I1}	10
Intensification period T_{I2}	10
Diversification period T_{Dm}	10
Max stale period T_{stale}	20

Further to this, the move counts used in the mild diversification procedure are also reset. In the event that the conditions for type-1 search intensification and hot diversification are both satisfied, the type-1 search intensification takes precedence. Further, at the beginning of each search intensification and mild diversification, the countdown for T_{stale} is reset.

16.6.3 Back to Normal Condition

Each of the type-1, type-2 search intensification, and mild diversification procedure would normally last for T_{I1}, T_{I2}, and T_{Dm} respectively. However, the procedure will exit and go back to normal exploration if a better solution is found. The parameters for the TS are shown in Table 16.2.

16.7 Ant Colony Algorithm

Ant colony algorithm (ACA) [16–18] was first introduced in Reference 13 as a distributed optimization tool. It was first used to solve the traveling salesman problem (TSP), but has been successfully used to solve other combinatorial problems including vehicle routing, network routing, graph coloring, quadratic assignment, and the shortest common super-sequence problem [11,14,30,43,45,56,57].

The ACA was inspired by the behavior of real ant colonies. Each ant possesses only limited skill and communication capability. However, through these simple, local interactions between individual ants, a global behavior emerges, whereby the colony takes on a complex autonomous existence of its own. Of particular interest is the way real ants find the shortest path between a food source and its nest.

The foraging behavior of ants [37,52,59] has captured the imaginations of many researchers. Different species and colonies of ants use different methods of searching for foods, although the majority of the ant species forage "collectively": once a food source is found, the individual ant would "inform" the other ants in the colony of the presence of food — *recruitment*. Again, the primary means of recruitments differs in different species of ants. Some species of ants communicate *directly*, whereby other ants would follow the ant back to the food source (group recruitment). Other species of ants communicate *indirectly*, through the use of a chemical called *pheromone* (mass recruitment). In some ant species, the ants lay these chemical pheromones while *exploring* from the nest (exploratory trail), as well as while going back to the nest from a food source (recruitment trail). It is these species of ants that inspired and form the basis of the ACA.

Each ant, in isolation, explores the surrounding (searching for food) in an almost random manner. Once it found a food source, however, it will usually walk in a straight line from the food source back to the nest, depositing the chemical pheromone along the way. In this way, a pheromone *trail* is laid, from the food source to the nest. Ants use different methods to find its way back to the nest, including visual cues, memory, and smell. When another ant stumbles upon this pheromone scent, it is more likely to follow this pheromone trail to the food source. Upon returning to the nest (from the food source), more pheromone would be deposited on the trail by the new ant. As the pheromone level builds up on this trail, even more ants would choose this trail, resulting in even higher pheromone build up. Through this positive feedback

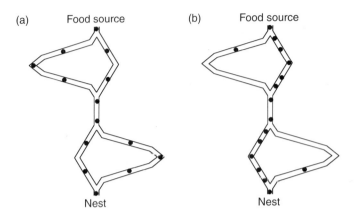

FIGURE 16.8 Two-bridge experiment. Each bridge consists of two paths, one shorter and one longer. (a) In the absence of pheromone, ants take on the branches with equal probability. (b) Higher pheromone levels on the shorter branches result in most ants taking the shorter branches [29].

(autocatalytic) process the ant colony as a whole is able to "collectively" forage the landscape for food. This autocatalytic "behavior" also assists the ants in finding the shortest path from a food source to the nest, as illustrated further.

An experiment involving the Argentine ant, *Iridomyrmex humilis*, was done in References 15 and 29. Previously, the Argentine ant has been shown to lay the same chemical pheromone both when leaving the nest to explore and when returning to the nest from a food source, and tend to follow pheromonal cues more than visual ones [6,60]. Using this knowledge, a set of experiments is conducted to investigate the self-organizing foraging behavior of the Argentine ant.

In a set of experiment [29], two bridges are set up on the path between the ant's nest and a food source (Figure 16.8). Each bridge consists of two unequal paths, of which one is shorter than the other. Initially, there is no pheromone on the paths, and to reduce the possibility of the ants using "environmental factors" (e.g., visual cues) to choose a path, the experiment is performed in the dark.

Initially, the ants will choose the different paths with equal probability (Figure 16.8[a]). The ants that take on the shortest path (the two shorter branches of the bridge), however, will arrive at the food source quicker than the ants that take on the other longer paths. On their return trip (returning from the food source to the nest), higher pheromone levels will be present on the shorter branches, making them more attractive. As these ants walk on the shortest path, even more pheromones will be deposited on the two shorter branches, resulting in even higher pheromone levels on the shorter branches. As a result of this positive feedback process, very soon essentially all of the ants will choose the shortest path (Figure 16.8[b]).

16.8 Ant Colony Algorithm Framework

Here, two types of ACA are implemented, *type-1* and *type-2*, along with two extra variants for each type, to solve the reporting cell planning problem. The ACA utilizes elitism whereby the elite ant deposits extra pheromones. TL is also used for the type-2 ACA.

16.8.1 Type-1

The ACA used for discovering efficient solutions to the reporting cell problem is shown in the following page.

Algorithm 16.4 (ACA for reporting cell planner):

Input: Parameters for the ACA.
Output: Solutions representing suitable reporting cell configurations.
Begin
        Initialize population and pheromone.
        While stopping conditions not true **do**
            For each ant:
                Generate solution.
                Lay pheromone.
                Update pheromone.
            If escape condition true **then** *Escape*.
End.

In type-1, each ant generates a new solution each cycle. It does this by crawling through the binary tree that represents all the possible permutation of the reporting cell problem (Figure 16.9). For a 4 × 4 network, the tree would have a depth of 17 nodes (including the starting node).

At each node of the tree, the ant has two choices (0 or 1). Which path the ant actually takes is based on pheromone levels, and heuristic values.

16.8.1.1 Pheromone Trail

Clearly, the binary tree is huge even for a small network, and laying trail on each individual path leads to an enormous pheromone trail, which is very impractical to use. Furthermore, the build-up of pheromone, or rather, the nonexistence of it, would be a problem.

So instead, moves are used for the trail, whereby pheromone trail is defined for a "to 0" move, and a "to 1" move, for each cell. Each move, for a cell, indicates whether the ant changes the cell to a 0, or a 1. In terms of the binary tree, this means each "to 0" and "to 1" move from each node at a given depth shares the same move trail.

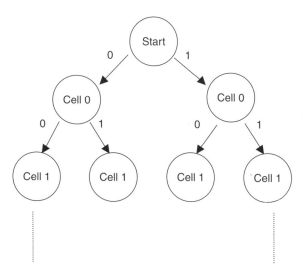

FIGURE 16.9 Binary tree for the reporting cell problem.

16.8.1.2 Local Heuristic

For a particular cell c, a "to 1" move uses the following heuristic:

$$\eta_{c,1} = \frac{\sum_{i=0}^{N-1} w_{ci}}{\sum_{i=0}^{N-1} w_{ci} - w_{cc}} \tag{16.8}$$

where w_{ci} denotes the call arrival weight associated with cell i, w_{cc} denotes the call arrival weight associated with cell c, and N denotes the total number of cells in the network.

Similarly, a "to 0" move associated with a cell c uses the following heuristic:

$$\eta_{c,0} = \frac{\sum_{i=0}^{N-1} w_{mi}}{\sum_{i=0}^{N-1} w_{mi} - w_{mc}} \tag{16.9}$$

where w_{mi} denotes the movement weight associated with cell i, and w_{mc} denotes the movement weight associated with cell c.

Clearly, the heuristic value given above is only a rough guide. However, the heuristic value only guides the ant during the initial phases of the run (during the early cycles), where the pheromone levels are negligible.

16.8.1.3 Transition Function

Each ant decides which path/move to take based on the pheromone levels and heuristic value of each move. It chooses with higher probability moves with higher pheromone levels and heuristic value. The probability that an ant k at node c chooses a move i is given by the following equation:

$$p_{c,i}^k = \frac{[\tau_{c,i}]^\alpha \cdot [\eta_{c,i}]^\beta}{\sum_{j \in N_c} [\tau_{c,j}]^\alpha \cdot [\eta_{c,j}]^\beta} \tag{16.10}$$

where N_c denotes a list of all possible moves from cell c, $\tau_{c,i}$ denotes the pheromone levels associated with move i for cell c — move(c, i), $\eta_{c,i}$ denotes the heuristic value associated with move(c, i), and α and β are positive constants specifying the relative weights of pheromone levels and heuristic values.

16.8.1.4 Pheromone Update

Once all the ants have generated a solution, the pheromone levels of each move i for cell c — move(c, i) are updated using the following equation:

$$\tau_{c,i} = \rho \cdot \tau_{c,i} + \Delta\tau_{c,i} + \Delta\tau_{c,i}^* \tag{16.11}$$

In the above equation, $\rho \in (0, 1]$ represents the pheromone evaporation rate, and is used to control the rate of convergence of the ant colony algorithm. $\Delta\tau_{c,i}$ represents the amount of pheromone to be added on move(c, i), and is given by:

$$\Delta\tau_{c,i} = \sum_{k=1}^{m} \Delta\tau_{c,i}^k \tag{16.12}$$

where m denotes the total number of ants in the population. The amount of pheromone added by ant k on move(c, i) is given by:

$$\Delta\tau_{c,i}^k = \begin{cases} \dfrac{Q}{L_k}, & \text{if move}(c, i) \in T_k \\ 0, & \text{otherwise} \end{cases} \tag{16.13}$$

where Q is a positive constant, T_k denotes a list of moves that need to be made to get the solution S_k, and L_k denotes the fitness value associated with solution S_k.

Finally, $\Delta\tau^*_{c,i}$ represents *elitism* that is implemented in this type-1. In this elitist strategy, the best solution found so far deposits extra pheromones. Such strategy has been discussed widely — both in GA and recently in ant algorithms, and improves the performance in many cases. The extra pheromone is implemented using the following equation:

$$\Delta\tau^*_{c,i} = \begin{cases} \sigma\dfrac{Q}{L^*}, & \text{if move}(c,i) \in T_* \\ 0, & \text{otherwise} \end{cases} \tag{16.14}$$

where σ is a positive constant, T^* denotes a list of moves that need to be made to get the best solution found so far – S^*, and L^* denotes the fitness value associated with solution S^*.

The above equation is equivalent to having extra σ elitist ants (that uses the best path) depositing pheromones. Such strategy improves the performance of the ACA , as applied to the reporting cell problem.

16.8.1.5 Escape Operator

The escape operator is invoked when solution is not improved within a specified maximum stale period T_{stale}. When invoked, the operator resets the pheromone levels associated with all moves for all cells, to its initial level τ_{initial}.

The parameter values used in this experiment are indicated in Table 16.3.

Note that for $Q \gg \tau_{\text{initial}}$, the Q term basically cancels out in Equation (16.10). That is, $Q \cong 1$.

16.8.2 Type-1 Rank Variant

In this variant, as well as using the elitist strategy outlined above, some *ranking* is introduced. The ranking is implemented as follows:

Once all the ants have generated a solution, they are ranked based on their fitness. The first n of these ants deposits extra pheromones, based on their ranks. The modified pheromone update formula is as shown:

$$\tau_{c,i} = \rho \cdot \tau_{c,i} + \Delta\tau_{c,i} + \Delta\tau^*_{c,i} + \Delta\tau^+_{c,i} \tag{16.15}$$

where $\Delta\tau^+_{c,i}$ represents the extra pheromones added by the extra n ants, and is given by the following formula:

$$\Delta\tau^+_{c,j} = \sum_{r=1}^{n} \Delta\tau^r_{c,i} \tag{16.16}$$

TABLE 16.3 Ant Colony Algorithm type-1 Parameters

Population m	50
Pheromone control α	0.9
Heuristic control β	5
Pheromone evaporation ρ	0.5
Initial pheromone $\tau_{initial}$	0.01
Pheromone amount Q	1
Elitist ants σ	6
Max stale period T_{stale}	25

where n denotes the number of ranked ants, and r denotes the ant's ranking, with $r = 1$ assigned to the fittest ant. The amount of pheromone added by ant of rank r is given by:

$$\Delta \tau_{c,i}^r = \begin{cases} R(r) \cdot \dfrac{Q}{L_r}, & \text{if move}(c, i) \in T_r \\ 0, & \text{otherwise} \end{cases} \qquad (16.17)$$

In the above, $R(r)$ represents a modifier function with maximum value at $R(1)$, and minimum value at $R(n)$. Ideally, $R(1)$ should be less than σ. In this experiment, the value $n = 8$ (the first eight ants are considered) is used, and $R(r) = (9 - r)/2$.

As will be shown later, the addition of $\Delta \tau_{c,i}^+$ improves the ant colony algorithm's performance for the reporting cell problem. Further, exclusions of the $\Delta \tau_{c,i}$ or $\Delta \tau_{c,i}^*$ term leads to a decrease in performance.

16.8.3 Type-1 Tournament Variant

This variant is similar to the rank variant described above, but *tournament* selection, instead of *ranking*, is used. After sorting, the first n ants compete in a tournament for the right to lay extra pheromones. In this way, which ants deposit more pheromones are randomly decided (instead of fixed like the rank variant above), though biased toward the fitter ants. That is, $R(r)$ in the ranking variant is now a "probabilistic" function, biased toward the fitter ants.

The tournament selection is similar to genetic algorithm's tournament selection, and $p_s = 0.8$ is used. Here, the first eight ants compete in a tournament that lasts twenty rounds. Specifically, in this tournament selection, two individuals are chosen at random from the first eight ants. The two individuals then "compete" in a virtual tournament, whereby the fitter individual has a p_s chance of winning. The winner then lays extra pheromones. This process is repeated for 20 rounds. It should be noted that the same individual can be selected more than once, leading to multiple pheromone deposit by the individual.

16.8.4 Type-2

Here, the ants walk on the solution space. The neighborhood N_S of a solution S consists of a "0 to 1" move and a "1 to 0" move for each cell. If a cell is at state "0," then the only allowed move for that cell is "0 to 1." Similarly, if a cell is at state "1," the allowed move is "1 to 0."

In the initial cycle, a random solution is created. The ant then chooses a path to move to (leading to another solution, as each ant walks on the solution space). The pheromone levels and heuristic values again govern which path/moves the ant chooses.

This implementation is similar to the TS implementation described earlier. However, while TS chooses the best path, here a probabilistic transition function is used, like simulated annealing. However, unlike simulated annealing, here the transition function is "guided" by pheromone levels and heuristic values.

To further enhance the performance, a TL is also included for each ant — to minimize the possibility of the ant going in circle. The TL implemented here is similar to the TL in TS with recent moves made by the ant kept in the TL. Each move is kept in the TL for a period of between $\text{TL}_{\min}$ and $\text{TL}_{\max}$.

The probability that an ant k chooses a move i for a cell c is given by the following modified equation:

$$p_{c,i}^k = \frac{[\tau_{c,i}]^\alpha \cdot [\eta_{c,i}]^\beta}{\sum_{\text{move}(d,j) \in N_{S_k}^k} [\tau_{d,j}]^\alpha \cdot [\eta_{d,j}]^\beta} \qquad (16.18)$$

where $N_{S_k}^k \subseteq N_{S_k}$ represents all possible moves from solution S_k, excluding moves that are in the ant's tabu list (TL_k). Certainly, only moves that are valid — move$(c, i) \in N_{S_k}^k$ are considered by each ant k.

TABLE 16.4   Ant Colony Algorithm
Type-2 Parameters

Population m	50
Pheromone control α	1
Heuristic control β	5
Pheromone evaporation ρ	0.5
Initial pheromone $\tau_{initial}$	0.01
Pheromone amount Q	1
Elitist ants σ	6
Tabu period TL_{min}	2
Tabu period TL_{max}	5
Max stale period T_{stale}	25
Mutation p_m	0.05

Pheromone laying and heuristic values are the same as type-1. Here, however, move "to 1" and "to 0" is replaced by move "0 to 1" and "1 to 0," respectively. Note also that here the global heuristic of fitness value can be used as the heuristic value (instead of the heuristic value of type-1). However, the number of computations to be performed would go up dramatically.

Further, while in type-1 a new solution is generated at each cycle, here the ants *move* from its current solution to create a new solution, which may take several cycles to generate a solution "dissimilar" from its current solution. As such, pheromones are added by an ant k, only when the new solution *improves* upon the previous solution.

The parameter values used are indicated in Table 16.4.

As before, the escape operator is invoked when solution is not improved within a specified maximum stale period T_{stale}. Here, the operator resets the pheromone levels associated with all moves for all cells, to its initial level $\tau_{initial}$. Further, each ant's solution S_k undergoes a bit mutation with probability $p_m = .05$. For an 8×8 network, this translates to probability $= .038$ that a solution S_k remains intact (does not undergo mutation). Finally, each ant's TL (TL_k) is also reset.

16.8.5 Type-2 Rank and Tournament Variant

Type-2 rank and tournament variant is conceptually similar to its type-1s equivalent. The same rank and tournament parameter values are also used.

16.9   Performance Comparison

Three cellular networks of 16, 36, and 64 cells are used to analyze the different algorithms presented earlier. The three cellular networks are of size 4×4, 6×6, and 8×8, and are in the form as shown in Figure 16.10. For location management cost evaluation purposes, the data set for each of the three networks are shown in Tables 16.5–16.7.

16.9.1 Stopping Conditions

The number of iteration allowed for each algorithm is shown in Tables 16.8–16.10.

In the solution's fitness calculation (Equation [16.3]), the vicinity value calculation, $v(j)$, involves a relatively large amount of computation. As such, for the simulation, it is assumed that calculation of the solution's fitness will take most of the computation time. The fitness calculation is also the one computation that is common among the different algorithms. As such, the number of fitness calculation allowed is used as the stopping condition for the different algorithms. For the 8×8 network (Table 16.10), the number of solution calculation allowed is 16,000, which is a fraction of the total possible solution of 2^{64}.

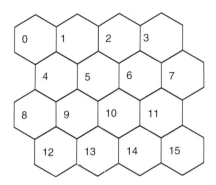

FIGURE 16.10   A network of size 4 × 4.

TABLE 16.5   Data Set for a 4 × 4 Network

Cell	w_{ci}	w_{mi}	Cell	w_{ci}	w_{mi}
0	517	518	8	251	445
1	573	774	9	224	2149
2	155	153	10	841	1658
3	307	1696	11	600	952
4	642	1617	12	25	307
5	951	472	13	540	385
6	526	650	14	695	1346
7	509	269	15	225	572

TABLE 16.6   Data Set for a 6 × 6 Network

Cell	w_{ci}	w_{mi}	Cell	w_{ci}	w_{mi}	Cell	w_{ci}	w_{mi}
0	714	1039	12	238	507	24	328	16
1	120	1476	13	964	603	25	255	332
2	414	262	14	789	1479	26	393	1203
3	639	442	15	457	756	27	370	1342
4	419	1052	16	708	695	28	721	814
5	332	1902	17	825	356	29	769	747
6	494	444	18	462	1945	30	17	146
7	810	1103	19	682	1368	31	265	904
8	546	1829	20	241	1850	32	958	359
9	221	296	21	700	1131	33	191	1729
10	856	793	22	23	236	34	551	190
11	652	317	23	827	1622	35	467	1907

In terms of computation time, this translates to a *maximum* allowed computation time of two seconds (on a P3-733).

For GA, the solution's fitness is calculated at each generation for each chromosome (although a few chromosomes do not undergo a change and their fitness is not recalculated). As the GA has a population of 100, for the 8 × 8 network this translates to a maximum generation of 160. Similarly, for the ACA, as the colony has a population of 50, and fitness of each ant's solution is calculated at each cycle, this translates to a maximum cycle of 320. Finally, TS chooses the best solution at each iteration. For the 8 × 8 network, this translates to 64 fitness calculation at each iteration (moves that are in the tabu list are still calculated, due to the *aspiration* condition — tabu moves are accepted if it improves the best solution found so far).

TABLE 16.7   Data Set for an 8 × 8 Network

Cell	w_{ci}	w_{mi}	Cell	w_{ci}	w_{mi}	Cell	w_{ci}	w_{mi}
0	968	533	22	104	101	43	820	436
1	745	907	23	881	539	44	362	672
2	827	515	24	694	655	45	356	822
3	705	1965	25	793	131	46	637	1912
4	902	1336	26	955	1227	47	626	1402
5	498	1318	27	126	450	48	345	524
6	807	1292	28	268	470	49	135	1400
7	62	1789	29	96	1081	50	175	393
8	331	541	30	285	1714	51	596	1272
9	212	1071	31	368	308	52	677	1197
10	787	1759	32	952	121	53	283	462
11	664	1416	33	367	1410	54	139	548
12	938	1413	34	132	1011	55	307	500
13	719	1224	35	439	1298	56	272	113
14	794	484	36	134	1634	57	931	47
15	543	1892	37	153	1750	58	38	1676
16	184	626	38	612	1948	59	896	1017
17	787	104	39	216	662	60	164	1307
18	319	1408	40	878	700	61	78	499
19	25	1256	41	957	765	62	303	1451
20	934	1637	42	363	756	63	578	1606
21	414	1950						

TABLE 16.8   Calculation Allowed for 4 × 4 Network

4 × 4 network
Fitness calculation: 4000
Max computation time: <1 sec

Method	Iteration
Genetic algorithm	40
Tabu search	250
Ant colony algorithm type-1	80
Ant colony algorithm type-2	80

TABLE 16.9   Calculation Allowed for 6 × 6 Network

6 × 6 network
Fitness calculation: 8000
Max computation time: 1 sec

Method	Iteration
Genetic algorithm	80
Tabu search	222
Ant colony algorithm type-1	160
Ant colony algorithm type-2	160

As such, it is allowed a maximum of 250 iterations. Note that the actual number of fitness calculation for each algorithm may be slightly higher, due to the "escape" operator of each algorithm. However, such "escape" only happens a few times during a run — 2 or 3 (or not at all), and is only a minute fraction of the total fitness calculation performed.

TABLE 16.10 Calculation Allowed for
8 × 8 Network

8 × 8 network Fitness calculation: 16,000 Max computation time: 2 sec	
Method	Iteration
Genetic algorithm	160
Tabu search	250
Ant colony algorithm type-1	320
Ant colony algorithm type-2	320

TABLE 16.11 Results for the 4 × 4 Network

4 × 4 network
Optimum value: 12.252

Method	Average value	Deviation %	Min value	Deviation %	Max value	Deviation %
GA	12.253	0.006	12.252	0.000	12.373	0.986
TS	12.252	0.000	12.252	0.000	12.252	0.000
AC_1	12.252	0.000	12.252	0.000	12.252	0.000
$AC_{1,rank}$	12.252	0.000	12.252	0.000	12.252	0.000
$AC_{1,tournament}$	12.252	0.000	12.252	0.000	12.252	0.000
AC_2	12.252	0.000	12.252	0.000	12.252	0.000
$AC_{2,rank}$	12.252	0.000	12.252	0.000	12.252	0.000
$AC_{2,tournament}$	12.252	0.000	12.252	0.000	12.252	0.000

TABLE 16.12 Results for the 6 × 6 Network

6 × 6 network
Optimum value: 11.471

Method	Average value	Deviation %	Min value	Deviation %	Max value	Deviation %
GA	11.511	0.343	11.471	0.000	12.030	4.867
TS	11.471	0.000	11.471	0.000	11.471	0.000
AC_1	11.472	0.007	11.471	0.000	11.573	0.883
$AC_{1,rank}$	11.471	0.000	11.471	0.000	11.471	0.000
$AC_{1,tournament}$	11.471	0.000	11.471	0.000	11.471	0.000
AC_2	11.491	0.172	11.471	0.000	12.129	5.737
$AC_{2,rank}$	11.477	0.051	11.471	0.000	11.740	2.341
$AC_{2,tournament}$	11.474	0.026	11.471	0.000	11.740	2.341

16.9.2 Results

Each algorithm is run 200 times for each test network. Results of the different algorithms are shown in Tables 16.11–16.13. In the tables, *Average value* is the average of the solution's value obtained from the 200 runs. *Deviation* is the percentage deviation from the optimal value.

Two values are of interest from the tables — the *average value*, and the *maximum value*. The average value represents the expected fitness of solutions obtained using the particular algorithm, while the maximum value represents the "worst-case behavior" of the algorithm. Each algorithm was run for 200 times, to get an accurate representation of the actual average value, and more importantly, the worst-case behavior — to see if the algorithm can unexpectedly give a poor performance. After all, one does not expect to run the algorithm many times just to get a "good" solution — otherwise the many run needed to get a good solution can be considered a "one run."

TABLE 16.13   Results for the 8 × 8 Network

8 × 8 network Optimum value: 11.471 Method	Average value	Deviation %	Min value	Deviation %	Max value	Deviation %
GA	14.005	1.619	13.782	0.000	14.671	6.454
TS	13.791	0.071	13.782	0.000	13.999	1.580
AC_1	14.107	2.361	13.801	0.141	14.407	4.539
$AC_{1,rank}$	13.860	0.569	13.782	0.000	14.007	1.637
$AC_{1,tournament}$	13.868	0.625	13.782	0.000	14.051	1.952
AC_2	14.393	4.434	13.901	0.866	15.373	11.545
$AC_{2,rank}$	14.032	1.821	13.782	0.000	14.462	4.939
$AC_{2,tournament}$	14.045	1.915	13.782	0.000	14.417	4.610

As can be seen from the results above, tabu search performs better than the other algorithms, for the three test networks, followed closely by the ant colony algorithms. Further, it can be seen that the use of ranking and tournament selection in ant colony algorithms improves the algorithms' performance.

16.10   Summary

This chapter discussed the use of GA, TS, and ACA to solve the reporting cells planning problem. As a further discussion of the relatively new ACA, two types of ACA were presented, along with the introduction of a rank-based and tournament-based selection variant for each type. It was shown that the three methods can be effectively used for reporting cells planning. While TS search shows the best performance, followed closely by ACA, results show the three methods are able to find optimal, or near optimal solutions to the reporting cells planning problem. Furthermore, for the reporting cells planning problem, the addition of ranking and tournament selection in the ant colony algorithm results in significant improvement in the algorithm's performance.

References

[1]   H.S. Abdinnour and S.W. Hadley, "Tabu search based heuristics for multi-floor facility layout," *International Journal of Production Research*, vol. 38, pp. 365–83, 2000.

[2]   M.A. Abido, "A novel approach to conventional power system stabilizer design using tabu search," *International Journal of Electrical Power & Energy Systems*, vol. 21, pp. 443–54, 1999.

[3]   P.J. Agrell, M. Sun, and A. Stam, "A tabu search multi-criteria decision model for facility location planning," in *Proceedings of the 28th Annual Meeting Decision Sciences Institute*, vol. 2, pp. 908–10, 1997.

[4]   M.A.S. Al, "Commercial applications of tabu search heuristics," in *Proceedings of the IEEE International Conference on Systems, Man, and Cybernetics*, vol. 3, pp. 2391–95, 1998.

[5]   V.R. Alvarez, E. Crespo, and J.M. Tamarit, "Assigning students to course sections using tabu search," *Annals of Operations Research*, vol. 96, pp. 1–16, 2000.

[6]   S. Aron, J.M. Pasteels, and J.L. Deneubourg, "Trail-laying behaviour during exploratory recruitment in the argentine ant, Iridomyrmex humilis (Mayr)," *Biology of Behaviour*, vol. 14, pp. 207–17, 1989.

[7]   G. Barbarosoglu and D. Ozgur, "A tabu search algorithm for the vehicle routing problem," *Computers & Operations Research*, vol. 26, pp. 255–70, 1999.

[8]   A. Bar-Noy and I. Kessler, "Tracking mobile users in wireless communications networks," *IEEE Transactions on Information Theory*, vol. 39, pp. 1877–86, 1993.

[9]   U. Black, *Emerging Communications Technologies*, 2nd ed., New Jersey: Prentice Hall, 1997.

[10]   R.R. Brooks, S.S. Iyengar, and J. Chen, "Automatic correlation and calibration of noisy sensor readings using elite genetic algorithms," *Artificial Intelligence*, vol. 84, pp. 339–54, 1996.

[11] B. Bullnheimer, R.F. Hartl, and C. Strauss, "An improved ant system algorithm for the vehicle routing problem," *Annals of Operations Research*, vol. 89, pp. 319–28, 1999.

[12] M.F. Catedra and J.P. Arriaga, *Cell Planning for Wireless Communications*. Boston, MA, Artech House, 1999.

[13] A. Colorni, M. Dorigo, and V. Maniezzo, "Distributed optimization by ant colonies," in *Proceedings of the First European Conference on Artificial Life*, pp. 134–42, 1991

[14] D. Costa and A. Hertz, "Ants can colour graphs," *Journal of the Operational Research Society*, vol. 48, pp. 295–305, 1997.

[15] J.L. Deneubourg, S. Aron, S. Goss, and J.M. Pasteels, "The self-organizing exploratory pattern of the Argentine ant," *Journal of Insect Behavior*, vol. 3, pp. 159–68, 1990.

[16] M. Dorigo and C.G. Di, "Ant colony optimization: a new meta-heuristic," in *Proceedings of the 1999 Congress on Evolutionary Computation*, vol. 2, pp. 1470–77, 1999.

[17] M. Dorigo, C.G. Di, and L.M. Gambardella, "Ant algorithms for discrete optimization," *Artificial Life*, vol. 5, pp. 137–72, 1999.

[18] M. Dorigo, V. Maniezzo, and A. Colorni, "The ant system: optimization by a colony of cooperating agents," *IEEE Transactions on Systems, Man and Cybernetics*, vol. 26, pp. 29–41, 1996.

[19] M.A.C. Gill and A.Y. Zomaya, "A cell decomposition-based collision avoidance algorithm for robot manipulators," *Cybernetics and Systems*, vol. 29, pp. 113–35, 1998.

[20] F. Glover, "Future paths for integer programming and links to artificial intelligence," *Computers & Operations Research*, vol. 13, pp. 533–49, 1986.

[21] F. Glover, "Tabu search. I," *ORSA Journal on Computing*, vol. 1, pp. 190–206, 1989.

[22] F. Glover, "Tabu search. II," *ORSA Journal on Computing*, vol. 2, pp. 4–32, 1990.

[23] F. Glover, J.P. Kelly, and M. Laguna, "Genetic algorithms and tabu search: hybrids for optimization," *Computers & Operations Research*, vol. 22, pp. 111–34, 1995.

[24] F. Glover and M. Laguma, "Tabu search," in *Handbook of Combinatorial Optimization*, vol. 3, D. Du and P.M. Pardalos (Eds.), Dordrecht: Kluwer Academic Publishers, pp. 621–757, 1999.

[25] F. Glover, M. Laguma, and R. Marti, "Fundamentals of scatter search and path relinking," *Control & Cybernetics*, vol. 29, pp. 653–84, 2000.

[26] F. Glover, E. Taillard, and W.D. de, "A user's guide to tabu search," *Annals of Operations Research*, vol. 41, pp. 3–28, 1993.

[27] D.E. Goldberg, *Genetic Algorithms in Search, Optimization and Machine Learning*, Reading, MA, Addison-Wesley publishing company inc, 1989.

[28] P.R.L. Gondim, "Genetic algorithms and the location area partitioning problem in cellular networks," in *Proceedings of the IEEE 46th Vehicular Technology Conference*, vol. 3, pp. 1835–38, 1996.

[29] S. Goss, S. Aron, J.L. Deneubourg, and J.M. Pasteels, "Self-organized shortcuts in the Argentine ant," *Naturwissenschaften*, vol. 76, pp. 579–81, 1989.

[30] L. Guoying and L. Zemin, "Multicast routing based on ant-algorithm with delay and delay variation constraints," in *Proceedings of the IEEE Asia Pacific Conference on Circuits and Systems Electronic Communication Systems*, pp. 243–46, 2000.

[31] A. Hac and X. Zhou, "Locating strategies for personal communication networks, a novel tracking strategy," *IEEE Journal on Selected Areas in Communications*, vol. 15, pp. 1425–36, 1997.

[32] K. Hatta, S. Wakabayashi, and T. Koide, "Adapting genetic operators and GA parameters based on elite degree of an individual in a genetic algorithm," *Transactions of the Institute of Electronics, Information and Communication Engineers*, vol. 9, pp. 1135–43, 1999.

[33] K. Hatta, S. Wakabayashi, and T. Koide, "Adaptation of genetic operators and parameters of a genetic algorithm based on the elite degree of an individual," *Systems and Computers in Japan*, vol. 32, pp. 29–37, 2001.

[34] A. Hertz and W.D. de, "Using tabu search techniques for graph coloring," *Computing*, vol. 39, pp. 345–51, 1987.

[35] A. Hertz, E. Taillard, and D. Werra, "Tabu Search," in *Local Search in Combinatorial Optimization*, E. Aarts and J.K. Lenstra (Eds.), Chichester: John Wiley & Sons Ltd., pp. 121–36, 1997.

[36] J.H. Holland, *Adaptation in Natural and Artificial Systems*, Ann Arbor: University of Michigan Press, 1975.

[37] B. Holldobler and E.O. Wilson, *The Ants*, Berlin: Springer Verlag, 1990.

[38] T. Imielinski and B.R. Badrinath, "Querying locations in wireless environments," in *Proceedings of the Wireless Communications Future Directions*, pp. 85–108, 1992.

[39] N. Kapoor, M. Russell, I. Stojmenovic, and A.Y. Zomaya, "A genetic algorithm for finding the page number of interconnection networks," *Journal of Parallel and Distributed Computing*, vol. 62, pp. 267–83, 2002.

[40] C.L. Karr and L.M. Freeman, *Industrial Applications of Genetic Algorithms*, Boca Raton, FL, CRC Press, 1999.

[41] H. Kawanishi and M. Hagiwara, "Improved genetic algorithms using inverse-elitism," *Transactions of the Institute of Electrical Engineers of Japan*, pp. 707–13, 1998.

[42] M. Laumanns, E. Zitzler, and L. Thiele, "On the effects of archiving, elitism, and density based selection in evolutionary multi-objective optimization," in *Proceedings of the 1st International Conference Evolutionary Multi Criterion Optimization (EMO 2001)*, pp. 181–96, 2001.

[43] V. Maniezzo and A. Colorni, "The ant system applied to the quadratic assignment problem," *IEEE Transactions on Knowledge and Data Engineering*, vol. 11, pp. 769–78, 1999.

[44] Z. Michalewicz, *Genetic Algorithms + Data Structures = Evolution Programs*, Berlin: Springer-Verlag, 1994.

[45] R. Michel and M. Middendorf, "An island model-based ant system with lookahead for the shortest supersequence problem," in *Proceedings of the Parallel Problem Solving from Nature PPSN V 5th International Conference*, pp. 692–701, 1998.

[46] M. Mitchell, *An Introduction to Genetic Algorithms*, Cambridge, MA: MIT Press, 1996.

[47] S. Okasaka, S. Onoe, S. Yasuda, and A. Maebara, "A new location updating method for digital cellular systems," in *Proceedings of the 41st IEEE Vehicular Technology Conference*, pp. 345–50, 1991.

[48] G. Parks, J. Li, M. Balazs, and I. Miller, "An empirical investigation of elitism in multiobjective genetic algorithms," *Foundations of Computing and Decision Sciences*, vol. 26, pp. 51–74, 2001.

[49] D. Plassmann, "Location management strategies for mobile cellular networks of 3rd generation," in *Proceedings of the IEEE 44th Vehicular Technology Conference*, vol. 1, pp. 649–53, 1994.

[50] T.S. Rappaport, *Cellular Radio and Personal Communications: Selected Readings*, Piscataway, NJ: IEEE, 1995.

[51] T.S. Rappaport, *Wireless Communications: Principle and Practice*, Inglewood Cliffs, NJ: Prentice Hall, 1996.

[52] M. Resnick, *Turtles, Termites, and Traffic Jams: Explorations in Massively Parallel Microworlds*. Cambridge, MA: MIT Press, 1994.

[53] J. Seijas, C. Morato, and G.J.L. Sanz, "Genetic algorithms: two different elitism operators for stochastic and deterministic applications," in *Proceedings of the 4th International Conference Parallel Processing and Applied Mathematics (PPAM 2001)*, pp. 617–25, 2002.

[54] G.L. Stuber, *Principles of Mobile Communication*. Boston: Kluwer Academic, 1996.

[55] R. Subrata and A.Y. Zomaya, "Location management in mobile computing," in *Proceedings of the ACS/IEEE International Conference on Computer Systems and Applications*, pp. 287–89, 2001.

[56] R. Subrata and A.Y. Zomaya, "Artificial life techniques for reporting cell planning in mobile computing," in *Proceedings of the Workshop on Biologically Inspired Solutions to Parallel Processing Problems (BioSP3)* (published in the CD-ROM proceedings of the IEEE 16th international parallel and distributed processing symposium), pp. 203–10, 2002.

[57] R. Subrata and A.Y. Zomaya, "A Comparison of three artificial life techniques for reporting cell planning in mobile computing," *IEEE Transactions on Parallel and Distributed Systems*, vol. 14, pp. 142–53, 2003.

[58] R. Subrata and A.Y. Zomaya, "Evolving cellular automata for location management in mobile computing networks," *IEEE Transactions on Parallel and Distributed Systems*, vol. 14, pp. 13–26, 2003.

[59] J.H. Sudd, *An Introduction to the Behavior of Ants*. London: Edward Arnold, 1967.

[60] S.E. Van Vorhis Key and T.C. Baker, "Observations on the trail deposition and recruitment behaviors of the Argentine ant, Iridomyrmex humilis (Hymenoptera: Formicidae)," *Annals of the Entomological Society of America*, vol. 79, pp. 283–88, 1986.

[61] M.D. Vose, *The Simple Genetic Algorithm: Foundations and Theory*. Cambridge, MA: MIT Press, 1999.

[62] H. Xie, S. Tabbane, and D.J. Goodman, "Dynamic location area management and performance analysis," in *Proceedings of the 43rd IEEE Vehicular Technology Conference Personal Communication Freedom Through Wireless Technology*, pp. 536–9, 1993.

[63] K.L. Yeung and T.S.P. Yum, "A comparative study on location tracking strategies in cellular mobile radio systems," in *Proceedings of the IEEE Global Telecommunications Conference*, pp. 22–8, 1995.

[64] A.Y. Zomaya, C. Ward, and B. Macey, "Genetic scheduling for parallel processor systems: comparative studies and performance issues," *IEEE Transactions on Parallel and Distributed Systems*, vol. 10, pp. 795–812, 1999.

[65] A.Y. Zomaya and H.T. Yee, "Observations on using genetic algorithms for dynamic load-balancing," *IEEE Transactions on Parallel and Distributed Systems*, vol. 12, pp. 899–911, 2001.

17

Location Privacy

17.1 Introduction...17-393
17.2 Location-Aware Communication Devices 17-394
17.3 Attacks on Location Privacy...........................17-395
First-Hand Communication • Second-Hand
Communication • Observation • Inference
17.4 Solutions...17-396
Policies • A Privacy Awareness System (pawS) for Ubiquitous
Computing Environments • Framework for Security and
Privacy in Automotive Telematics • Concepts for Personal
Location Privacy Policies • Protecting First-Hand
Communication • Approaches Using MIX-Nets • ANODR:
Anonymous on Demand Routing • MIXes in Mobile
Communication Systems • Mix Zones • The Cricket
Location-Support System • PlaceLab • Temporal and
Spatial Cloaking • The Blocker Tag
17.5 Summary 17-409
References 17-409

Andreas Görlach
Andreas Heinemann
Wesley W. Terpstra
Max Mühlhäuser

17.1 Introduction

The proliferation of portable electronic devices into our day-to-day lives introduced many unresolved privacy concerns. The principal concern in this paper is that these devices are being increasingly equipped with communication capabilities and location awareness. While these features present a wide array of new quality-of-life enhancing applications, they also present new threats. We must be careful that the potential quality-of-life lost through the surrender of private information does not overwhelm the benefits.

An important question is how much privacy protection is necessary. Perfect privacy is clearly impossible as long as communication takes place. Therefore, research aims at minimizing the information disclosed. The required level of this protection is not a matter of technology; different people have different privacy needs. Nevertheless, technology should not *force* society to accept less privacy.

The major privacy concern with mobile devices equipped with communication ability is that they can reveal the location of their bearers. This concern is in itself not new; people can recognize each other. What is new is the increased scope of the problem due to automated information gathering and analysis. Poorly designed mobile devices enable anyone to obtain another's location.

If we allow automation to create an effective public record of people's locations, discrimination against minorities will be impossible to control. AIDS patients could be identified by the offices of doctors they visit, Alcoholics Anonymous members by their group meetings, and religious groups by their churches.

This chapter will present an overview of the state-of-the-art in location privacy. In Section 17.2, mobile devices, which possess both location awareness and communication ability will be examined. Section 17.3 lists attacks by which an invader can obtain your private location information. Existing countermeasures and safeguards are detailed in Section 17.4. These include high level schemes such as policies that operate like contracts, and lower-level solutions that reduce information disclosure. Among the latter are anonymous routing algorithms, schemes for hiding within a group, methods to determine location by the device itself without revealing them to third parties, and frequency modulation techniques to hinder triangulation.

17.2 Location-Aware Communication Devices

Many technologies can determine the location of an individual. This section provides an overview of what technologies are presently deployed and which are coming in the near future.

One of the earliest systems designed for location tracking is the global positioning system (GPS) [15]. This system uses satellites to help devices determine their location. The GPS works best outdoors where it has line-of-sight to the satellites and few obstructions. For commercial products, resolution to within 4 m is achievable. The GPS is widely deployed and integrated, especially in map applications. Although GPS devices do not transmit, they are being increasingly integrated into PDAs and other devices which do.

For indoor use, the Active Badges [32] from AT&T Laboratories Cambridge were developed. These are small devices worn by individuals that actively transmit an identifier via infrared. This information is received by sensors deployed in the environment. This system provides essentially room-level resolution and has problems following individuals due to the infrequency of updates. The environment consolidates this information and can provide the current location of an individual.

A later refinement, the Bat [33], increased the detected resolution. With the increased resolution, the Bat can be used to touch virtual hot spots. Their work reports accuracy as good as 4 cm. These refined devices used ultrasonic pings similar to bat sonar. However, once again the environment measures the Bat's location as opposed to real bats who learn about their environment.

The Cricket Location-Support System [24] takes a similar approach. It uses radio and ultrasonic waves to determine distance and thus location. Like the Cambridge Bat, resolution to within inches is possible. As opposed to the similar Cambridge work, beacons are placed in the environment as opposed to on individuals. The Cricket devices carried by individuals listen to their environment in order to determine their location. In this way, the device knows its location, while the environment does not.

An approach to location sensing, which does not require new infrastructure is taken by Carnegie Mellon University [30]. Here, the existing wireless LAN is observed by devices to recognize their location. By passively observing the signal strengths of various base stations, a device can determine its location. Though there are no requirements for new infrastructure, there is a training overhead. During training a virtual map of signals is created, which is used by the devices to determine their location.

Cell phones can be abused to provide location information. Although not originally intended for this purpose, the E-911 [25] requirements in the United States forced cell phone providers to determine customer location when they dialed an emergency phone number. Although this practice was clearly beneficial, the technology has since spread. The underlying problem is the omnipresent possibility of performing triangulation (with varying accuracy, though).

In the near future radio frequency identification (RFID) [14] will be found in many consumer goods. Intended as a replacement for barcodes, these tiny devices are placed in products to respond to a wireless query. Unlike barcodes, RFIDs are distinct for every item, even those from the same product line. This allows companies to determine their inventory by simply walking through the shelves and automatically recording the observed products.

17.3 Attacks on Location Privacy

In a successful privacy attack, some party obtains unauthorized information. Individuals intend that some informations about themselves should be available to others, and that the rest remain private. The means by which the individual's preferences were circumvented is the attack vector.

The main privacy concern with regards to ubiquitous computing is that many new *automated* attack vectors become possible. Loosely categorized, automated digital devices obtain information either through communication, observation, or inference. In this section the attack vectors available in each of these channels will be explored.

17.3.1 First-Hand Communication

An attacker obtains private information through first-hand communication when an individual unwittingly provides it directly to the attacker. In a world with ubiquitous computing, the threat of disclosure via accident or trickery is significant. All digital devices of a given type, by virtue of being homogeneous, make the same mistakes — and do not learn from them. The designers of the Windows file sharing protocol never intended it to be used to obtain people's names. Nevertheless, Windows laptops will happily reveal their owner's name to anyone who asks it. Due to a bug in bluetooth phones, attackers may often trick the phone into revealing its address book and phone number [23]. By asking a device with known location for owner information, both of these attacks pinpoint the owner's location, among other things. Naturally, these attacks can be built into an automated device.

Many ubiquitous devices are also chatty. The Bats and Active Badges broadcast their location information for all to hear. WLAN cards periodically emit traffic which includes their unique multiple access control, MAC ID. Devices providing their exact location information to location-based services also seems overly permissive. At the bare minimum, these problems must be addressed.

A unique characteristic of digital devices is their potential for brain-washing. Manufacturers may choose to place secret spyware in their products[1] as a means to recoup financial losses. Furthermore, a vulnerability may allow an attacker to completely assume control of the device, and thus obtain a live location feed. For devices where the location information is known to the infrastructure, the threat of a system vulnerability is magnified.

17.3.2 Second-Hand Communication

Also known as gossip, attacks via second-hand communication relay information from one party to another unauthorized party. The primary difference between these attacks and first-hand attacks is that the individual no longer controls the information. Fortunately, in the human scenario, talking about individuals behind their backs requires some effort. Unfortunately, aggregation and spreading of this information in a digital system is significantly easier.

This behavior has already been observed in the Internet where Doubleclick regularly sells personal habit and preference information. It seems naïve to assume that the much finer grained information available from ubiquitous devices will not similarly be sold. Services are already available for individuals to locate their friends via the cell phone networks [2].

17.3.3 Observation

Attackers may also obtain information by configuring devices to observe their environment. The most obvious problem is the deployment of many nearly-invisible cameras in the environment. However, there are other risks that are more feasible to launch with the current technology.

A widespread and very common attack that can be launched against mobile communications-equipped devices is triangulation. By measuring timing delays in a signal, the attacker can determine the location

---

[1]For example, the Kazaa Media Desktop.

of the device. This is similar to how the Bat operates, only using electromagnetic waves instead of sound waves. Triangulation is almost always possible within certain constraints and heavily varying accuracy. What we will not take into account within this chapter are approach coming from electrical engineering since they often require a deep understanding of signal processing issues.

17.3.4 Inference

One of the fears about automated privacy invasion is the compilation of a profile. After gathering large amounts of information via communication and observation, an automated system combines these facts and draws inferences. Given enough data, the idea is to build a complete picture of the victim's life.

From a more location-centric point of view, location information could be processed to obtain useful information for discrimination. If a person regularly visits the location of a group meeting, she is probably a member of that group. In the consumer arena, the fact that an individual shops at a particular store at regular intervals may be a useful information for price discrimination [6].

Tracking an individual's location through time may also enable an attacker to link information to the individual. For example, if an individual's car regularly sends out *totally anonymous* weather requests, it might still be possible for a weather network to track the car by correlating the observed request locations. Later, when the individual buys gas at an affiliate's gas station, the network can link the individual's name and bank account to the tracked car. Now, the network can deduce information such as where the person shops, lives, and works; who the person regularly visits; etc.

17.4 Solutions

In the literature there exist several approaches to protect the location of a user. Most of them try to prevent disclosure of unnecessary information. Here one explicitly or implicitly controls what information is given to whom, and when. For the purpose of this chapter, this information is primarily the identity and the location of an individual. However, other properties of an individual such as interests, behavior, or communication patterns could lead to the identity and location by inference or statistical analysis.

In some cases giving out information cannot be avoided. This can be a threat to personal privacy if an adversary is able to access different sources and link the retrieved data. Unwanted personal profiles may be the result. To prevent this, people request that their information be treated confidentially. For the automated world of databases and data mining, researchers developed policy schemes. These may enable adequate privacy protection, although they similarly rely on laws or goodwill of third parties.

17.4.1 Policies

In general, all policy-based approaches must trust a system. If a system betrays a user, his privacy might be lost. Here, the suitable counter-measure is a nontechnical one. With the help of legislation the privacy policy can be enforced.

All policy-based systems have the drawback that a service could simply ignore the individual's privacy preferences and say, "To use this service you have to give up your privacy or go away." This certainly puts the user in a dilemma and he will probably accept these terms as he wants to use the service.

17.4.2 A Privacy Awareness System (pawS) for Ubiquitous Computing Environments

In Reference 22 Langheinrich proposes the *pawS* system. *pawS* provides users with a privacy *enabling* technology. This approach is based on the platform for privacy preferences project (P3P) [11], a framework that enables the encoding of privacy policies into machine-readable XML. Using a trusted device, the user negotiates his privacy preferences with the UbiCom environment.

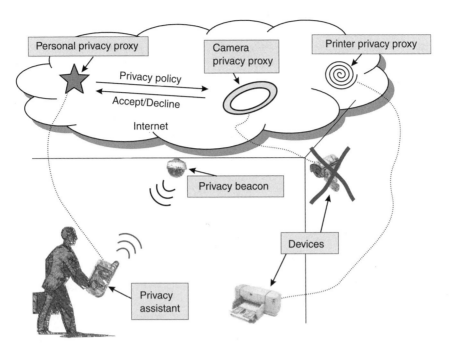

FIGURE 17.1   Privacy management system.

This is depicted in Figure 17.1. A user equipped with a *privacy assistant (PA)* is able to detect a *privacy beacon* upon entering a UbiCom environment. The privacy beacon announces the available services (e.g., a printer, a video camera) with a reference (e.g., a URL) to their data collection capabilities and policies. The PA in turn contacts the user's personal privacy proxy located somewhere on the Internet, which contacts the corresponding service proxies at their advertised addresses and asks for their *privacy policies*. These are compared to the user's privacy preferences and a result may be to decline usage of a tracking service, which will result in disabling the video camera.

Regarding adequate security, anonymity, and pseudonymity as useful tools to support a privacy preserving infrastructure, Langheinrich's design includes four more principles. See Reference 21 for an elaborated discussion. We sum-up his ideas here.

Notice: Since future UbiCom environments will be ideally suited for unnoticed operation, monitoring, and tracking, the author proposes some kind of announcement mechanism that allows users to notice the data collection capabilities in their environment. This announcement mechanism should be standardized and agreed on. The privacy beacon from Figure 17.1 serves as an example.

Choice and consent: To preserve privacy, a future UbiCom system should offer a user the choice to allow or deny any kind of data collection. A system should respect this choice and operate only on user consent. Since a user will not always explicitly express his will, his choice and the derived consent should be expressed in machine-readable policies (cf. personal privacy proxy in Figure 17.1).

Proximity and locality: Locality information for collected data should be used by the system to enforce access restriction. This restriction may be based on the location of a user who wants to use the data, for example, a user is able to record an audio stream at a meeting while attending, but not from the opposite side of the world (proximity). Also the audio stream should not be disseminated through the network (locality).

Access and recourse: A system should give users easy access to collected personal information (e.g., using a standardized interface). Also, they should be informed about any usage of their data. Thus, abuse would be noticed.

Following these principles, Langheinrich's goal is to allow people and systems who *want* to respect a user's privacy to behave in such a way. This should help to build a lasting relationship based on mutual trust and respect. His goal is not to try to provide perfect protection for personal information that is hardly achievable anyway.

17.4.3 Framework for Security and Privacy in Automotive Telematics

A framework for security and privacy in automotive telematics, that is, embedded computing and tele-communication technology for vehicles, is described by Duri et al. [12]. The primary goal of their framework is to enable building telematics computing platforms that can be trusted by users and service providers.

Their work is motivated by a *pay-per-use*-insurance scenario, that is, a user will only be charged when he/she uses the car. Insurance rates will be based upon miles driven, whether the driving is done in an urban or suburban area and it is taken into account how familiar the driver is with the visited area. In addition, a user can choose from several privacy policies. Polices range from *don't-give-away-any-personal-information* to *personal-information-might-be-sold-to-third-parties* kind of polices. The insurance company could then offer a greater discount when more personal information is disclosed by the user.

To enable this kind of pay-per-use program, detailed information about the usage, driving speed, driving area, etc. has to be collected, aggregated according to the chosen policy and delivered to the insurance company. Figure 17.2 shows the proposed system components. The central and crucial component is called the data protection manager (DPM) to handle sensitive data. The key advantage in this approach is that every data access passes through the DPM. This simplifies verifying that data accesses comply with a user's privacy policies.

Their proposed framework employs three key concepts to build a trusted system. First, it uses a defense-in-depth approach to build a secure platform from the ground up. Second, the framework enables data aggregation close to the source and third, the framework uses user defined privacy policies for obtaining user consent before data collection and usage. We will explain these concepts in more detail:

Defense in depth: By this the authors understand a bottom-up approach to build a system that is trust-worthy for both the data subject (the user) and a service provider. This approach comprises several layers of hardware and software. Each layer provides its own security functions and rely on the directly underlying layer. At the bottom, tamper-resistant hardware, for example, smart cards for key storage and secure

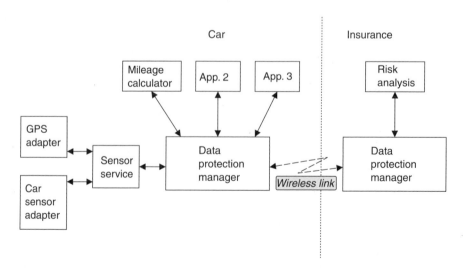

FIGURE 17.2   Overview system components.

cryptographic coprocessors, are deployed. On top of this, the platform should allow for secure configuration, update, and booting of the system (see Reference 4). The operating system itself must also provide security features, such as access control, secure communication protocols, encrypted file systems, and so on (for more details see References 1 and 3).

Data aggregation close to source: Looking at Figure 17.2, we see that a *Mileage calculator* is part of the car and data to and from this component is controlled by the DPM. With this design, data aggregation is as close to the source as possible, that is, raw data collected by the *GPS adapter* and *Car sensor adapter* is delivered to the Mileage calculator via the DPM. The Mileage calculator carries out some suitable aggregation and returns the results to the DPM. It is the responsibility of the DPM to pass only aggregated on to the insurance company. This way, no raw data leaves the car and misuse is prevented.

User privacy policies: To enforce user privacy policies the DPM includes a privacy enabled resource manager (PERM) (not shown in Figure 17.2). The PERM handles any request for private data. Such a request includes application credentials, privacy policy concerning data. The PERM verifies application credentials and then compares application privacy policies with a user's privacy policy to determine whether to grant or deny access. For details on this see Reference 8.

The framework and key concepts proposed by Duri et al. help system designers and developers to implement automotive systems that control the sharing of private data according to policies agreed to by the owner of the data.

17.4.4 Concepts for Personal Location Privacy Policies

Snekkenes [31] presents concepts, which may be useful when constructing tools to enable individuals to formulate a personal location privacy policy. Snekkenes's idea is that the individual should be able to adjust the accuracy of his location, identity, time, and speed, and therefore have the power to enforce the need-to-know principle. The accuracy is dependent on the intended use of the data, and the use in turn is encoded within privacy policies.

The question, Snekkenes aims to answer is *Who should have access to what location information under which circumstances?* This concept is applicable to location tracking, that is, position data is calculated by an external entity and therefore not under direct control of a user. As an example, a GSM/UMTS network operator is able to locate the cell of a mobile phone. The proposed system components are depicted in Figure 17.3. Their roles are defined as follows (in brackets, we explain the interaction according to the sequence numbers):

Initiator: Entity that would like and/or accept that a service is produced, for example, a user. (In Step 1, this entity expresses to the requestor to ask for a service production.)

Requestor: Entity that makes a request to the service provider for a service to be produced. (In Step 2, the requestor asks a service provider to produce a service.)

Service provider: Entity that combines location data with other data to produce some service. (In Step 3, this entity asks a location provider to provide the location information of an initiator.)

Location provider: Entity that provides location data. Any release of data should be subject to a policy. (First, in Step 4, the location provider looks up in the policy custodian directory where to find the policy for the initiator. Second, in Step 5, it provides the necessary data to the policy custodian to enforce a policy.)

Policy custodian directory: Directory to look up where to find a policy. This works similar to the *Domain Name Service* of the Internet.

Policy custodian: Entity that stores and possibly also enforces a policy. There may be many policy custodian available for different groups of users.

Policy: Statement of what can be released to whom and when. This entity is stored at the policy custodian.

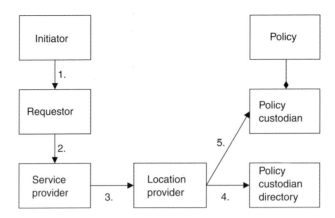

FIGURE 17.3   System components and sequence of interaction.

```
Owner: John Smith
LocatedObject: John's cellular phone
DefaultObservationResponse: (B,B,B,B)
  Request:
  (Requestor = Consumer) AND
  (Initiator = Consumer) AND
  IsFriendOrRelative(Consumer) AND
  ((London) LESS_ACCURATE (RawObservation(John's cellular phone,
                                       Location(CurrentTime)
                                       )

                          )
  )
ObservationResponse:

  (CurrentTime, London, John Smith, B)
```

FIGURE 17.4   Policy example.

Consumer: Entity to which the service is presented for consumption. (Not explicitly labelled in figure, might be same as requestor).

Having introduced the system components, we will now present how to blur tracking information. The basic datum in tracking is the *observation* of an object. This observation for an object, *O*, is defined by its *time, location, ID*, and *speed*:

$$\text{Obs}_O = (\text{ObservationTime}, \text{ObjectLocation}, \text{ObjectID}, \text{ObjectSpeed})$$

For all aspects a user policy expresses the observation accuracy to be released. In Figure 17.4 the policy expresses *My friends and relatives should know whether or not I'm in London* for a user John Smith, but not his exact location. The `DefaultObservationResponse: (B,B,B,B)` defines the most inaccurate response possible.

The work of Sneekkenes also includes fragments of a language for formulating personal location privacy policies as shown in the example above. Performance and scalability issues in a system integrating his concepts are ongoing research.

17.4.5   Protecting First-Hand Communication

Most approaches address the problem of information disclosure. Many different ideas have been proposed to prevent unnecessary information from becoming known to a third party.

17.4.6   Approaches Using MIX-Nets

In the following section we will present a number of approaches using concepts such as Onion Routing [26] and MIX-Nets [9]. These concepts need a short introduction, which we give with the next few paragraphs.

From now on a network working the described way will be called a *MIX network*, its nodes are MIXes, and the data structure is a cryptographic *onion*. The term onion is a good image to reflect the way in which the original plaintext is processed.

MIX networks usually rely on the availability of public key cryptography The sending node needs to know the public key of the receiver as well as those of all intermediary nodes. The outgoing data is now encrypted with the public key of its supposed destination. Additionally one encryption iteration is done for every node on the way to this destination.

This scheme has the following properties. First, exactly one node per step is able to decrypt the data with its private key. Second, since the routing information is within the encrypted part, any outgoing packet of a node cannot be correlated to a previously received packet by an external observer.[2] Third, this means — except some very advanced attacks — it is not possible to follow one specific packet through a MIX network without controlling *all* nodes the packet passes through.

A cryptographic onion looks like this:

$$K_n(R_n, K_{n-1}(\ldots K_2(R_2, K_1(R_1, K_a(R_0, T), A), M_1), \ldots), M_{n-1}) \qquad (17.1)$$

where A is the destination with its public key K_a, T the to be delivered message. M_x denotes MIX x, K_x its public key, and random nonces R_x.

A nonce is a piece of data supposed to be used only once as the name already suggests. Its purpose is preventing attacks such as replay. It is a good idea to choose nonces at random.

17.4.7   ANODR: Anonymous on Demand Routing

Introduction: With the scenario of a battlefield in mind, Kong and Hong described in Reference 20 their scheme ANODR. This is a routing protocol addressing the problems of route anonymity and location privacy.

The intention is that packets in the network cannot be traced by any observing adversary. Aside from the observation of actively sending devices an external observer cannot figure out who is communicating with whom. This is due to the fact that no correlation can be determined between any two observed network packets.

Prior to one node's ability to send a message to another, a route must be established through route discovery. This route discovery is achieved by broadcasting and forwarding packets. The sender of a message is anonymous because it is impossible to judge whether a node is actually sending a message it generated or is simply forwarding a packet as part of a route.

Route discovery: In order to send data in ANODR a route discovery phase must precede the actual sending phase. The task of the route discovery is to create route pseudonyms. The individual nodes can be identified by the MAC addresses of their network interfaces. For sender anonymity the sending node has the possibility to use a broadcast address as source address. Since we speak about wireless networks the actual transmission is broadcast anyway.

---

[2]This statement refers to the contents of a sent packet. Of course, it is possible to correlate two packets if there was only one packet coming in and only one leaving one node. These issues have also been taken into account by the MIX network community.

Trapdoor one-way functions: The route pseudonyms can be seen as local trapdoors. Trapdoors in turn refer to the concept of cryptographic trapdoor one-way functions.

A function

$$f : \quad X \longrightarrow Y$$

is a *one-way function* it is "easy" to compute an image for every element $x \in X$, but computational infeasible to find preimages given any element $y \in Y$. A function f is a *trapdoor one-way function* if f is a *one-way function* and it becomes feasible to find preimages for $y \in Y$ given some *trapdoor information*. These functions are only used in the route discovery phase. In ANODR the trapdoors base on shared secrets among all nodes of the network. Details about the used trapdoor functions are not detailed within this chapter but can be found in Reference 20.

Initiation of the route discovery: The route discovery is initiated by a RREQ packet which looks like as follows:

$$\langle RREQ, seqnum, tr_{dest}, onion \rangle$$

with *seqnum* being a globally unique sequence number, tr_{dest} a cryptographic trapdoor as defined above and only to be opened by its destination, and *onion*, a cryptographic onion as defined in Equation (17.1). After reaching its destination the RREQ packet is being answered by a RREP packet. With the received onion, a locally unique route pseudonym N, the proof of the global trapdoor opening pr_{dest}, the RREP packet is assembled:

$$\langle RREP, N, pr_{dest}, onion \rangle$$

For best anonymity guarantee and performance, the *onion* is in ANODR a *trapdoor boomerang onion*[3] and looks like this:

$$TBO_X = K_X(N_X, K_{X-1}(N_{X-1}, K_{X-2}(\ldots)))$$

In this route discovery scheme each forwarding node X adds a random nonce N_X to the boomerang onion. The onion is then encrypted using a random symmetric key K_X. The route request packet RREQ is broadcasted locally to the next hop form, which the RREP packet will come back after the destinated bounced the onion. Only X will then be able to decrypt its formerly encrypted onion layer and broadcast it back.

Data transfer: The actual data transfer operates on the precomputed route pseudonyms. A source wraps its data with the according route pseudonym. Each receiving node looks up the pseudonym of the packet in its forwarding table and either discards the packet (no match) or replaces the route pseudonym by its pseudonym found in the table (match). Again, the packet is broadcasted locally and the procedures is repeated until the packet reaches its destination.

Route maintenance: Table entries of the route pseudonyms are affected by both a timeout T_{win} and nodes being defunct. For the defunct nodes, table entries are being recycled. An error packet $\langle RERR, N' \rangle$ where N' is the route pseudonym associated with N^4 of the not responding node is locally broadcasted.

---

[3]Kong and Hong examined other schemes as well, but the trapdoor boomerang onion showed the best results in terms of network overhead and data packet latency.

[4]Might that be a possibility to de-anonymize by selectively jamming nodes and analyzing the answers of the surrounding nodes?

17.4.8 MIXes in Mobile Communication Systems

It is easy for cellular networks like GSM to track their mobile subscribers. Location information is required in order to route calls appropriately. Avoiding this by simply broadcasting is not an option because of the limited bandwidth in current cellular networks.

In Reference 13 this issue has been investigated and a two-part solution is proposed. One part handles the storage and processing of the location information of the mobile subscriber within the central databases of the mobile services provider. The other takes into account routing issues and proposes solutions to prevent observers from doing traffic analysis.

How current cellular networks work: Networks for mobile telephones are organized as cells. There is a Base Transceiver Station for each cell broadcasting the calls into the cell. The network centrally stores the mapping of a mobile subscriber number (MSISDN) to the cell in which the mobile station (MS)[5] is situated. An explicit handover of two base transceiver stations (when the MS moves from one cell to another) updates the location information in the central database of the network.

Protection of location data: Federrath et al. changed the routing scheme in a way that the network does not know anymore the location information of device corresponding to a certain mobile subscriber number. This implies that the routing information must be created by the device itself. The home location register (HLR),[6] still has a database with mappings of MSISDNs and the location information of the mobile device. But now the location information is stored in a covered form. In addition to this the network assigns a temporary mobile subscriber identity (TMSI)[7] to the mobile device.

The requirements on the *mobile device* are:

- Capabilities to compute several cryptographic primitives
- Storage of cryptographic keys
- Storage of the TMSI

In order to receive phone calls the mobile device must first register its location within the network. The network is usually the infrastructure of one (or more) telephone services provider(s). All nodes within this network can act as MIX nodes. To reach its destination a data packet or a phone call is being routed through a certain subset of theses nodes.

Here is a short summary of the procedure necessary to register with the HLR:

Encrypt the TMSI for the first MIX,

$$c_{M_1}(k_{M_1}, \text{TMSI})$$

where c_{M_1} is the public key of the first MIX, k_{M_1} a symmetric key, which has been generated by the MS and will be used by M_1 in a later step to encrypt data for the MS.

As mentioned in the introduction on MIX networks earlier, the route through MIX networks must be predetermined. So this step needs to be recursively repeated for all intermediary MIXes. The nodes within the network of the mobile services provider[8] realize these MIXes. Routing information is included, too.

Then — out of the location information (LAI) — the *untraceable return address* {LAI}[9] is calculated.

$$\{\text{LAI}\} := A_{M_n}, c_{M_n}(k_{M_n}, A_{M_{n-1}}, c_{M_{n-1}}(\ldots))))$$

where A_{M_x} denotes the address of a MIX x.

---

[5]This is in most cases simply the cell phone.

[6]The database where the information about location of a certain cell phone is stored. Each phone is represented by a telephone number. Thus the HLR contains the mapping of the number and the location.

[7]The GSM networks use *International Mobile Subscriber Identities* to identify mobile device. Note, that this is independent of the — with the SIM card assigned — telephone number.

[8]Or at least a subset of the nodes.

[9]The curly brackets denote encrypted data.

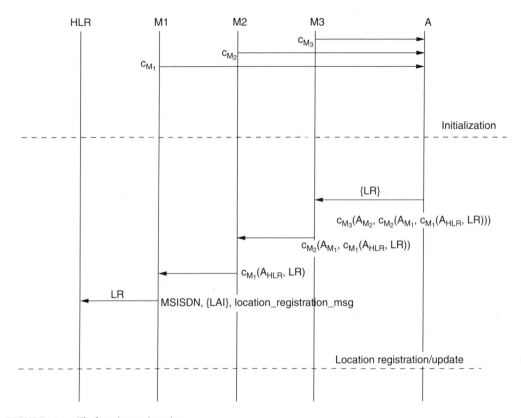

FIGURE 17.5   The location registration.

The creation of a {LAI} requires a number of cryptographic computations and consumes bandwidth. Especially the transmission and computation of LAI is an issue to be considered since the underlying cryptographic scheme requires continuous change of LAI. In order to reduce the load caused by this scheme Federrath et al. introduce improvements in terms of efficiency.

We just computed the covered form of the LAI. Now the location record (LR) for the central database needs to be created:

$$LR := MSISDN, \{LAI\}, location\_registration\_msg$$

This LR, again, is protected[10] by encryption in the same way as the LAI.

$$LR := A_{M_n}, c_{M_n}(A_{M_{n-1}}, c_{M_{n-1}}(\cdots(A_{M_1}, c_{M_1}(A_{HLR}, LR))))$$

It should be noted that an indeterministic cryptosystem[11] is required in the approach.

Figure 17.5 shows the location registration process.

The call setup: Figure 17.6 visualizes the call setup. In this case the caller submits the number (MSISDN) of the party to be called as usual. The incoming call is routed to the central database (HLR) containing the necessary routing information. This database contains the mapping of MSISDN and LAI. But now the HLR only has the covered form of the LAI (denoted by LAI) as defined earlier. It only sees encrypted data and the address of the first MIX M_n (or M_3 in the example of the Figure 17.6). There is no further

---

[10]To prevent attacks such as traffic analysis by observers.

[11]An indeterministic cryptosystem encrypts equal plaintexts to different cryptotexts.

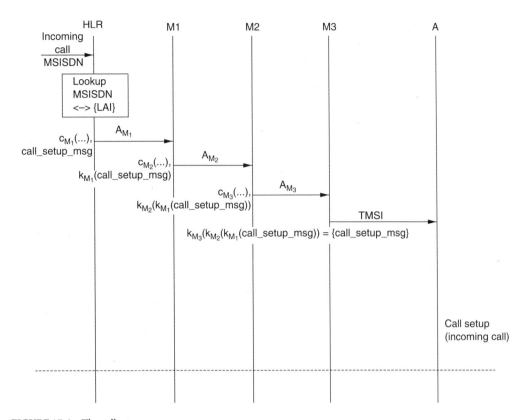

FIGURE 17.6   The call setup.

information available to the HLR. The encrypted LAI is now being sent to the first MIX together with the call setup message with information of the incoming call.

The first MIX decrypts the received part of the LAI. After decryption this MIX knows the next MIX to which the data needs to be forwarded. In addition to this it has the symmetric key previously sent within the LAI from the MS to the HLR. This key is now used to encrypt the call setup message (*call_setup_msg*). The remaining part of the LAI is encrypted with the public key of the next MIX and thus unreadable for this MIX. This procedure is repeated until finally the last MIX broadcasts the encrypted call setup messages within the cell in which the MS is.

After reception of the encrypted message the cell phone can easily decrypt it using the symmetric keys k_{M_i}, $i \in \{1, \ldots, n\}$. These are derived directly from the TMSI.

17.4.9 Mix Zones

The concept of mixes has been used by Beresford and Stajano to develop *mix zones* (see Reference 7 for a description of the approach, threat model, practical experiments, and evaluation). A mix zone is the geographical equivalent of a mix node in traditional mix networks. A proxy is introduced to pseudonimize users within the mix zones and handle application requests for location data. While the applications, which process the location data are seen as attackers, the infrastructure, that is, the middleware containing the proxy and the sensors collecting location information of users, is a trusted entity. In order to evaluate their solution they also defined two metrics to measure location privacy.

The mixing process: Entities within the system are:

- Users, who — from time to time — use location-based services
- A sensing infrastructure usually knowing about the current position of the users

- A pseudonimizing proxy whose role is to do the mixing process
- Several applications to which the spatiotemporal information about a user might be unrevealed

A mix zone is defined "for a group of users as a connected spatial region of maximum size in which none of these users has registered any application callback." By registering such an application callback a user leaves the mix zone and enters an application zone.

A first step for the protection of location privacy is to replace real names by pseudonyms. This prevents simple user profiling by the accessed services. Since it is possible to uncover pseudonyms and get real names by simply following them and identifying offices or homes, short term pseudonyms have been implemented. This addresses the problem of observation. In addition to these measures it will be much harder for services to collude against user by combining their databases (second-hand communication).

Anonymity measure: Location privacy is measured based on *anonymity sets* and *entropy*. The anonymity set is defined as "the group of people visiting the mix zone during the same time period." Parameters influencing the anonymity level are multifarious. The relation of the spatial size of the mix zone to the average movement speed of users within matters. A strong correlation is true for ingress position and egress position of the mix zone varying with the zone's geography. This issue is covered by the entropy, a concept originally developed by Shannon [29] and adopted to mixes by Serjantov and Danezis in Reference 28. If preceding and subsequent mix zones are known for users it is possible to calculate — out of all possible pairs of preceding and subsequent zones — probabilities. These conditional probabilities display the (un-)certainty with which an observer can make assumptions about users.

The anonymity level concerning entropy can be expressed by a number h, which Beresford and Stajano compute as follows:

$$h = -\sum_i P(m_i|M) \cdot \log P(m_i|M)$$

where,

$$P(m_i|M) := \frac{P(m_i \wedge M)}{P(M)} = \frac{P(m_i)}{\sum_i P(m_i)}$$

and M is the set of possible mappings m_i of ingress and egress events.

17.4.10 The Cricket Location-Support System

Cricket [24] provides a location support system. The authors distinguish their solution from location tracking systems such as the BAT [33] and Active Badge [32]. The primary difference is that the environment provides the necessary equipment for devices to determine their own location. The environment does not determine the location of mobile devices themselves.

Cricket is primarily designed to operate indoors. The GPS infrastructure already provides a location support system for outdoor situations. The indoor scenario is considerably more difficult. Walls and ceilings prevent long-range signals from penetrating, requiring that infrastructure be scattered throughout the building. This in turn leads to administration and management problems for the system. Maintaining privacy must contend with the infrastructural costs.

For this chapter, the most interesting property about Cricket is that the mobile devices determine their own position. Even more interestingly, the *listeners*, which are attached to mobile devices are reported to cost only US$10. The cost-effective nature of cricket's privacy makes it a very appealing solution.

Cricket *listeners* listen to *beacons*, which are placed throughout the building. These beacons do not track the listeners, rather they broadcast location information periodically, which the listeners receive. This places the burden of computation on the listeners, which must calculate their location inspite of the difficulties involved with indoor transmission.

Cricket uses a hybrid approach to keep the location finding problem simple. Beacons transmit both a RF signal (containing data) and an ultrasonic pulse. Since RF signal travels at the speed of light, it reaches the mobile device after an effectively fixed delay depending on the hardware and not the distance. The ultrasonic pulse, however, travels more slowly through the air. By measuring the delay of the pulse, the listener can determine its distance from the beacon.

Combining the distance information from several beacons allows listeners to determine their location. The distance calculation is performed by an inexpensive 10 MHz PIC micro-controller contained in the listener. This micro-controller has only 68 bytes of RAM and 1024 words of ROM. The string contained in the RF signal and the calculated distance is passed from the listener on to the mobile device via a serial port. The mobile device then combines information from multiple beacons to determine its location.

To prevent confusion from multiple beacons, the Cricket system continues the RF transmission for long enough that the ultrasonic pulse occurs simultaneously with the RF signal. This allows the weak listener processor to correlate the two and identify collisions. The authors assume that the penetration of RF is better than ultrasonic; therefore, the RF signal will always be present whenever the ultrasonic pulse is, guaranteeing correctness. These measures allow Cricket to be deployed with such low-grade hardware.

In all, Cricket provides a very convincing and elegant solution. It allows mobile devices to obtain their location without the supporting infrastructure performing tracking. It is cost effective. Finally, deployment of Cricket appears to be flexible and require little maintenance.

17.4.11 PlaceLab

Similar to Cricket, PlaceLab [27] discovers location information on the mobile device itself. Unlike Cricket, PlaceLab aims to provide an extremely low-cost solution by reusing existing hardware and infrastructure. The general idea of PlaceLab is to observe wireless access points and use this information to determine the mobile device's location.

The Cricket system requires special beacons deployed within the environment and a special listener device on the mobile devices. Each of these components costs US$10, but another problem is that these devices are not available and not widely deployed. The GPS is quite well deployed, however, the cost of GPS devices is quite high. PlaceLab, like many other systems [5,10,17], uses standard wireless LAN hardware in mobile devices to recognize wireless access points; thus, it is effectively zero-cost to add to mobile clients and to deploy. Unfortunately, the resolution of PlaceLab is on the order of city blocks.

In a paper devoted to privacy issues [18], PlaceLab presents the Place Bar. This bar shows the current location of the user in the web browser's menu bar. Different granularity can be selected to enable different services. The intention is to make the trade-off of privacy versus utility visible to the user and also simple.

One of the interesting aspects addressed is that of wireless access point (WAP) privacy. Access points broadcast periodically a basic service set identifier (BSSID), which is in the format of a MAC address. This is the token used by PlaceLab to determine location. To map MAC addresses to locations, PlaceLab requires a database of WAPs. This presents a potential privacy concern for the WAP operators.

As a step toward solving this problem, the authors suggest to store data in a special way:

$$\text{Database}(F(P)) = E_P(\text{Position})$$

Here P is the access point's MAC address, F is a one-way hash function that cannot be inverted, and E_P a symmetric encryption function using P for the encryption key. When a mobile agent sees an access point it can hash the MAC to obtain a key into the database. If the key has associated data, the position can be decrypted using the MAC address. Without knowing the MAC address, the wireless agent learns nothing.

One of the problems with this proposal is that there are too few MAC addresses. Therefore, a brute force attack trying all possible MAC addresses is feasible. To solve this problem, the authors propose to concatenate two MAC addresses in order to increase the key length. The two MAC addresses would be obtained from adjacent WAPs.

17.4.12 Temporal and Spatial Cloaking

In Reference 16, Gruteser and Grunwald propose a mechanism called *cloaking* that conceals a user within a group of k people. This work is primarily concerned with telematic applications, where vehicles use location-based services. The main contribution is an algorithm for perturbing disclosed location information.

Only one problem is addressed in this work: how best to disclose location information. How to send messages anonymously is left open, relying upon mix networks as discussed in previous sections. Determination of location information is assumed to be local like Cricket. Information is only disclosed through the use of services. Attacks using triangulation of radio signals from the vehicles are left untreated. The authors consider such attacks to be infeasible since the E-911 service had difficulty attaining a 125 m resolution. Finally, the work assumes the existence of a trusted anonymizing service.

A user is defined to be *k-anonymous* if, and only if, they are indistinguishable from at least $k - 1$ other users. The owner of the vehicle presumably sets his privacy requirement. Gruteser and Grunwald assume that a reasonable value for k is between 5 and 10.

The perturbations allowed by the authors include spatial discretization, and delayed transmission. Rather than disclosing the x and y spatial coordinates at the exact time t, intervals $x \in (x_1, x_2)$, $y \in (y_1, y_2)$, and $t \in (t_1, t_2)$ are disclosed. By widening the intervals, more users could have been the source of the disclosed information.

The authors believe there are three primary axes for location disclosure. The frequency of disclosure, the accuracy of time information, and the accuracy of location information. Different location-based services will have different requirements for each of these axes.

Three scenarios are considered. Driving conditions (weather) are disclosed by vehicles to highway operators. Here, neither the location information nor the time need be very accurate; however, the highway operators require frequent updates. City officials may gather statistics about hazardous locations from vehicles. The location of the hazardous location must be very accurate, but since statistics are being calculated, the information is not time sensitive nor requiring frequent updates. A mapping service requires only a rough location and very infrequent requests; however, the information needs to be provided in a timely manner.

We now describe the quad-tree algorithms employed by Gruteser and Grunwald to perturb location information. The first algorithm provides anonymity by cloaking the physical location. The algorithm starts with the area (in space) covered by the anonymizing service. This area is repeatedly divided into four quadrants. As long as the quadrant containing the user has at least k other vehicles, recurse. Otherwise, return the remaining area and the exact time. The second algorithm provides anonymity by cloaking the time of disclosure. The algorithm recurses through physical quadrants until a discrete area meeting the required resolution is obtained. Then the algorithm delays until there have been k vehicles passing through the area. Finally, disclose the physical area and a time interval. The interval ends with the current time, and starts at the time of the request minus a random cloaking factor.

17.4.13 The Blocker Tag

Introduction: A special case among pervasive devices are RFIDs. People carrying objects that contain RFIDs might not even be aware of the existence of these devices because of their size[12] and passive nature. A second property of RFIDs is their inability to do any intensive computation like encryption.

Juels, Rivest, and Szydlo examined several possible solutions in Reference 19 ranging from destruction of tags to regulation (i.e., policies). Since the authors find disadvantages in all of the examined solutions, they present their own approach, which is the development of a special tag: the blocker tag. This tag blocks attempts of readers to identify RFIDs. In order to not block desired RFIDs and to temporarily enable reading of RFIDs, the blocking process can be done selectively.

---

[12]The smallest RFIDs are currently only of 0.4 mm × 0.4 mm size.

The tree-walking singulation protocol: RFID reader devices depend on a special procedure to handle multiple tags. As it is difficult to read many RFID tags in parallel, the readers read the tags sequentially. Aside from space division, frequency division, and code division multiple access techniques there are time division multiple access procedures to which the tree-walking singulation protocol belongs.

For the tree-walking singulation protocol the reader executes a recursive search. It repeatedly queries sub-trees; only the contained tags respond. The reader device can easily determine if some tags answered or none. By repeatedly dividing tree, the device can eliminate sub-trees where there are no tags. When there is only one tag left within a sub-tree the reader can request the whole ID of this tag.

Operation: The blocker tag only works with the tree-walking singulation protocol. Reader-tag combinations based on other protocols such as space division multiple access (SDMA) are not affected. Physical properties of the tag response are also not affected; specially designed readers could distinguish the different answers from a tag and the blocker.

The concept of the blocker tag is easy: it simply simulates all possible 2^k serial numbers of RFIDs. By applying this trick the singulation process necessarily fails.

17.5 Summary

The solutions we have seen can be categorized into policies and minimizing information disclosure. These approaches aim to address threats in the areas of first- and second-hand communication, observation, and inference.

Policies seem to work well wherever consent underlies the transaction. For example, when information is to be provided to a service, an agreement can be reached regarding the further distribution of the information. If no agreement can be reached, then the individual will be unwilling to use the service, but the service will likewise not obtain the information or any associated remuneration. Similarly, the individual can negotiate terms about how his information may be used; this can mitigate attacks based on inference, though not prevent them.

There is no consent in observation. This means that policies can not be applied to these attacks since the individual is in no position to negotiate. Here, legal safeguards and countermeasures are required. Unfortunately, there is currently insufficient discourse between technical and legal experts.

Accuracy reduction techniques and onion routing apply primarily to first-hand communication problems. These schemes aim at reducing the amount of confidential information disclosed first-hand. We discussed a variety of techniques, which obscure the location information, the timestamp of the transaction, and the identity of the individual.

References

[1] Bastille Linux. http://www.bastille-linux.org/
[2] Mobiloco — Location based services for mobile communities. http://www.mobiloco.de/
[3] Security-enhanced linux. http://www.nsa.gov/selinux/
[4] W.A. Arbaugh, D.J. Farbner, and J. Smith. A secure and reliable bootstrap architecture. In *Proceedings of the Privacy Enhancing Technologies Workshop 2002.* IEEE Computer Society, 1997.
[5] P. Bahl, A. Balachandran, A. Miu, W. Russell, G.M. Voelker, and Y.-M. Wang. PAWNs: satisfying the need for secure ubiquitous connectivity and location services. *IEEE Wireless Communications Magazine, Special Issue on Future Wireless Applications,* pp. 40–48, February 2002.
[6] J. Bailey. Internet price discrimination: self-regulation, public policy, and global electronic commerce, Telecommunications policy research conference (TPRC), online at: http://www.tprc.org/agenda98.htm, 1998.
[7] A.R. Beresford and F. Stajano. Location privacy in pervasive computing. *PERVASIVE Computing, IEEE CS and IEEE Communications Society,* 3: pp. 46–55, 2003.

[8] K. Bohrer, X. Liu, D. Kesdogan, E. Schonberg, and M. Singh. Personal information management and distribution. In *Proceedings of the 4th International Conference on Electronic Commerce Research (ICECR-4)*, 2001.

[9] D. Chaum. Untraceable electronic mail, return addresses, and digital pseudonyms. *Communications of the ACM*, 24: pp. 84–88, 1981.

[10] K. Cheverst, N. Davies, K. Mitchell, and A. Friday. Experiences of developing and deploying a context-aware tourist guide: the guide project. In *Proceedings of the 6th Annual International Conference on Mobile Computing and Networking*, pp. 20–31, ACM Press, 2000.

[11] L. Cranor, M. Langheinrich, M. Marchiori, M. Presler-Marshall, and J. Reagle. The platform for privacy preferences 1.0 (P3P1.0) specification. http://www.w3.org/TR/P3P/

[12] S. Duri, M. Gruteser, X. Liu, P. Moskowitz, R. Perez, M. Singh, and J.-M. Tang. Framework for security and privacy in automotive telematics. In *International Conference on Mobile Computing and Networking*, pp. 25–32, ACM Press, 2002.

[13] H. Federrath, A. Jerichow, and A. Pfitzmann. MIXes in mobile communication systems: location management with privacy. In *Proceedings of the Information Hiding*, pp. 121–135, 1996.

[14] K. Finkenzeller. *RFID — Handbook, 2nd Edn.* Wiley & Sons LTD, 2003.

[15] I.A. Getting. The global positioning system. *IEEE Spectrum*, 30: pp. 36–47, December 1993.

[16] M. Gruteser and D. Grunwald. Anonymous usage of location-based services through spatial and temporal cloaking. In *MobiSys*, pp. 31–42. USENIX, 2003.

[17] J. Hightower and G. Borriello. Location systems for ubiquitous computing. *Computer*, 34: pp. 57–66, 2001.

[18] J. Hong, G. Borriello, J. Landay, D. McDonald, B. Schilit, and D. Tygar. Privacy and security in the location-enhanced world wide web. In *Proceedings of the Ubicomp 2003*, October 2003.

[19] A. Juels, R.L. Rivest, and M. Szydlo. The blocker tag: selective blocking of RFID tags for consumer privacy. V. Atluri, ed. In *8th ACM Conference on Computer and Communications Security*, pp. 103–111, ACM Press, New York, 2003.

[20] J. Kong and X. Hong. ANODR: Anonymous on demand routing with untraceable routes for mobile ad-hoc networks. In *Proceedings of the 4th ACM International Symposium on Mobile Ad Hoc Networking and Computing*, pp. 291–302, ACM Press, 2003.

[21] M. Langheinrich. Privacy by design — principles of privacy-aware ubiquitous systems. In G.D. Abowd, B. Brumitt, and S.A. Shafer eds, *Ubicomp*, vol. 2201 of *Lecture Notes in Computer Science*, pp. 273–291, Springer, Berlin, 2001.

[22] M. Langheinrich. A privacy awareness system for ubiquitous computing environments. In G. Borriello and L.E. Holmquist, eds, *Ubicomp*, vol. 2498 of *Lecture Notes in Computer Science*, pp. 237–245, Springer, Berlin, 2002.

[23] A. Laurie. Serious flaws in bluetooth security lead to disclosure of personal data. http://www.thebunker.net/security/bluetooth.htm, 2003.

[24] N.B. Priyantha, A. Chakraborty, and H. Balakrishnan. The cricket location-support system. In *Mobile Computing and Networking*, pp. 32–43, 2000.

[25] J.H. Reed, K.J. Krizman, B.D. Woerner, and T.S. Rappaport. An overview of the challenges and progress in meeting the E-911 requirement for location service. *IEEE Communications Magazine*, 5: pp. 30–37, April 1998.

[26] M.G. Reed, P.F. Syverson, and D.M. Goldschlag. Anonymous connections and onion routing, vol. 4, pp. 44–54. *IEEE Computer society press*, 1997.

[27] B. Schilit, A. LaMarca, G. Borriello, W. Griswold, D. McDonald, E. Lazowska, A. Balachandran, J. Hong, and V. Iverson. Challenge: ubiquitous location-aware computing and the place lab initiative. In *Proceedings of the 1st ACM International Workshop on Wireless Mobile Applications and Services on WLAN (WMASH 2003)*. ACM Press, September 2003.

[28] A. Serjantov and G. Danezis. Towards an information theoretic metric for anonymity. P. Syverson and R. Dingledine, eds, In *Proceedings of the Privacy Enhancing Technologies*, LNCS, San Francisco, CA, 2002. http://petworkshop.org/2002/program.html.

[29] C.E. Shannon. A mathematical theory of communication. *Bell System Technical Journal*, 27: pp. 623–656, pp. 379–423, July and October 1948.

[30] A. Smailagic, D.P. Siewiorek, J. Anhalt, D. Kogan, and Y. Wang. Location sensing and privacy in a context aware computing environment. In *Proceedings of the Pervasive Computing*, 2001.

[31] E. Snekkenes. Concepts for personal location privacy policies. In *Proceedings of the 3rd ACM Conference on Electronic Commerce*, pp. 48–57, ACM Press, New York, 2001.

[32] R. Want, A. Hopper, V. Falcão, and J. Gibbons. The active badge location system. *ACM Transactions on Information Systems*, 10: pp. 91–102, 1992.

[33] A. Ward, A. Jones, and A. Hopper. A new location technique for the active office. *IEEE Personal Communication*, 4: pp. 42–47, 1997.

III

Resource Allocation and Management in Wireless Networks

Radio Resource Management Algorithms in Wireless Cellular Networks

18.1 Introduction...18-415
    Limitations of Cellular Radio Systems • Radio Resource
    Management
18.2 Handoff and Mobility Management....................18-418
    Overview • Handoff Types • Desirable Features of Handoff
    • Handoff Algorithms
18.3 Effective Call Admission Control Algorithms for
    Resource Management 18-423
    Overview • Classification of Call Admission Control • Call
    Admission Control Algorithms
18.4 Management of Power in Cellular Systems:
    Mechanisms and Algorithms..........................18-430
    Overview • Power Control Mechanisms • Power Control
    Algorithms
18.5 Conclusion and Future Work 18-433
References 18-434

Nidal Nasser
Hossam Hassanein

18.1 Introduction

The general goal of emerging mobile and personal communication services is to enable communication with a person, at any time, at any place, and in any form. Different techniques are employed in different systems to achieve this goal. A central technique, and thus of great practical importance, in providing mobile user access to the fixed communication networks is based on wireless cellular networks. A *wireless cellular network* (WCN) provides a radio access for users within a specific service area and allows the dynamic relocation of mobile terminals. As shown in Figure 18.1, the cellular architecture comprises two levels — a stationary level and a mobile level. The stationary level consists of fixed base stations (BSs) that are interconnected through a fixed network that is referred to as the backbone network. The mobile

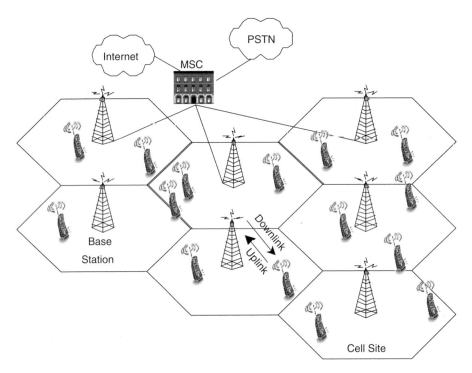

FIGURE 18.1　The cellular architecture.

level consists of mobile terminals (MTs) that communicate with the BSs via wireless links. They can both transmit and receive information. We will refer to a MT as a mobile (or user). The geographic area within which mobiles can communicate with a particular BS, and which also represents the BSs radio coverage, is referred to as a *cell*. Neighboring cells overlap with each other, thus ensuring continuity of communication. The BSs are connected through fixed links to a *mobile switching center* (MSC), which is a local switching exchange with additional features to handle mobility management requirements of a cellular system. Moreover, MSCs interconnect with the public switched telephone network (PSTN) and the Internet. Mobiles communicate among themselves, as well as with fixed networks, through the BSs and the backbone network. Transmission from the BS to the MT is called *downlink (forward link)* and the transmission from the MT to the BS is called *uplink (reverse link)*. The network access points, that is, BSs, of the mobiles may change as the users travel from one location to another.

Radio resources in WCNs such as radio (frequency) spectrum, transmit powers and BSs are generally limited due to the physical and regulatory restrictions, as well as the interference-limited nature of WCNs. Thus, to provide communication services with high capacity and good quality of service (QoS) requires powerful methods for sharing the radio spectrum in the most efficient way. Spectrum sharing methods are called *multiple access techniques*. A multiple access technique is a definition of how the radio spectrum is divided into channels and how channels are allocated to users of the system. The objective of multiple access techniques is to provide communication services with sufficient bandwidth when the radio spectrum is shared with many simultaneous users. A channel can be thought of a portion of the radio spectrum that is temporarily allocated for a specific purpose, such as someone's phone call. The most common multiple access techniques are *frequency division multiple access* (FDMA), *time division multiple access* (TDMA), and *code division multiple access* (CDMA). The division of FDMA, TDMA, and CDMA channels into time-frequency plane is illustrated in Figure 18.2.

Consider a bandwidth B from a radio spectrum that allocated to serve mobile users simultaneously. In FDMA, B is divided into narrowband channels by the frequency domain. Each mobile is allocated one channel for the entire duration of the call as shown in Figure 18.2. FDMA was used in the first generation (1G) of the cellular systems such as advanced mobile phone service (AMPS) systems. TDMA enhances

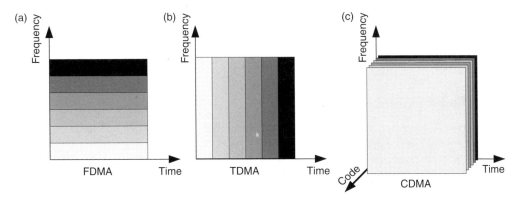

FIGURE 18.2 Multiple access techniques: (a) FDMA (b) TDMA (c) CDMA.

FDMA by further dividing the bandwidth into channels by the time domain as well. A channel in the frequency domain is divided among multiple users. Each user is allocated a slot in the channel for some time slots. In each slot only one user is allowed to either transmit or receive. TDMA is used as the access technology for global system for mobile communications (GSM), which is representative of the second generation (2G) of cellular systems. Unlike FDMA and TDMA, CDMA transmission does not work by allocating channels for each user. Instead, CDMA utilize the entire bandwidth (B) for transmission of each user. Multiple access is achieved by assigning each user a distinguished spreading code (chip code). This chip code is used to transform a user's narrowband signal to a much wider spectrum prior to transmission in a manner known as a spread spectrum transmission. The receiver correlates the received composite signal with the same chip code to recover the original information-bearing signal. Commercial cellular systems utilizing CDMA technique are the narrowband IS-95 and the third generation (3G) UMTS/IMT-2000 wideband CDMA (WCDMA).

18.1.1 Limitations of Cellular Radio Systems

WCNs are relatively complex systems and many issues must be addressed in their design, regarding both the stationary level and the mobile level. Unlike wired networks, which assume low error rates and stationary users, there are several factors in WCNs, which make it very difficult to provide QoS guarantees:

- Resources in WCNs are scarcer than in wired networks. Wireless links, in general, provide much lower bandwidths than the wired counterpart. This disparity is expected to hold in future even though rapid progress is being made for high-speed wireless transmission, due mainly to the physical limitation of the wireless media.
- Wireless channels are inherently unreliable and prone to location-dependent, time-varying, and bursty errors due to noise, multipath fading, shadowing, and interferences. Even with channel coding, diversity-combining, and power control techniques, their unreliability is much higher than that of wired links.
- Users tend to move around during a communication session causing handoffs between adjacent cells. The current trend in cellular networks to reduce cell size (i.e., picocells or microcells [21]) in order to accommodate more mobile users in a given area will make it even more difficult to deal with the mobility-related problems.
- The MTs equipped with communications ability can reveal the location of their bears, which prohibits privacy. Location privacy is an important factor to support QoS such as security guarantees.

Due to these distinct characteristics of WCNs, it is necessary to develop mechanisms tailored to support QoS for mobile users that run wireless services. In this chapter, we are concerned with the first three factors to provide QoS guarantees for mobile users. Last factor is discussed in more details in the previous chapter.

To overcome the aforementioned limitations of wireless cellular systems and maintain service continuity with QoS guarantees to the wireless services users, it is important to utilize system resources efficiently and provide service differentiation according to mobile user's traffic profile. Radio resource management (RRM) module in the WCN system is responsible for efficient utilization of air interface resources and to guarantee certain QoS level for different users according to their traffic profiles. In the next section, we discuss the RRM module in more detail.

18.1.2 Radio Resource Management

Radio Resource Management (RRM) is a set of algorithms that control the usage of radio resources. It is located in the MT and the BS. RRM functionality is aimed to maximize the overall *system capacity* in the cellular network. A common definition for *capacity* is the maximum traffic load that the system can accommodate under some predefined service quality requirements. In order to study effective resource management algorithms, it is necessary to understand and define the conditions that limit the cellular capacity. These conditions are related to the services characteristics (voice, video, or data), the propagation channel variations, the power control operation, and the user mobility patterns. All factors affecting the system capacity are included in the system load functions, which are derived for both uplink and downlink. The basic RRM algorithms can be classified as follows: Handoff and mobility management algorithms, call admission control (CAC) algorithms, and power control algorithms.

If a new call is admitted to access the network, then the CAC algorithm will make a decision to accept or reject it according to the amount of available resources versus users QoS requirements, and the effect on QoS of exiting calls that may occur as a result of the new call. If the call is accepted the following has to be decided: transmission (bit) rate, BS, and channel assignment and transmission power. Most of these resources have to be dynamically controlled during the transmission. For example, the BS assignment has to be changed as the MT moves further away from the BS. The handoff algorithm takes care of the re-assignment of BSs. When moving closer to the BS, the same received signal strength (RSS) can be upheld for a lower transmitted power. Thus, efficient power control algorithm is needed to reduce the transmission power and to keep the interference levels at a minimum in the system to provide the required QoS and to increase the system capacity. Since available bandwidth (radio resource spectrum) in cellular communication is limited, it is important to utilize it efficiently. For this reason, frequencies are reused in different cells in the system. This can be done as long as different users are sufficiently spaced apart, to ensure that interference caused by transmission by other users will be negligible.

Traditionally all these algorithms have been studied separately. Recently, however, there has been a move toward the investigation of combined radio resource algorithms [1–3,18]. In this chapter, we review some of these algorithms.

This chapter is organized as follows. Section 18.2 reviews different algorithms for handoff and mobility management in WCNs. Some of the existing call decision-based admission control algorithms are described and compared in Section 18.3. Management of power for WCNs are introduced, explained, and studied in Section 18.4. Finally, Section 18.5 presents the chapter conclusion and future work.

18.2 Handoff and Mobility Management

18.2.1 Overview

In wireless cellular systems users are allowed to move freely in the terrain while maintaining continuous contact with the system. When a mobile user travels from one cell to another during the call duration, the call should be transferred to the new cell's BS. Otherwise, the call will be dropped because the link with the current BS becomes too weak as the mobile recedes. Indeed, this ability for transferring is a design issue in wireless cellular system and is called *handoff*. Before the introduction of this feature, a call was simply dropped when the distance between the mobile and the serving BS became too large. Efficient handoff algorithms enhance the capacity and guarantee QoS in cellular systems [4].

Figure 18.3 shows a simple handoff scenario in which an MT travels from cell A to cell B. Initially, the MT is connected to BS A (①). The overlap between the two cells is the handoff region in which the mobile may

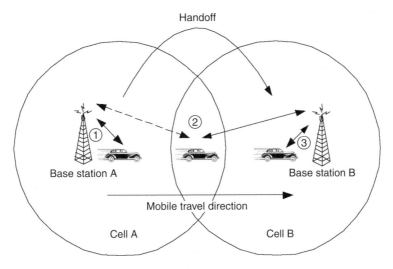

FIGURE 18.3 Handoff scenario in cellular systems.

be connected to either BS A or BS B (②). At a certain time while the MT is traversing the overlap region, the mobile is handed off from BS A to BS B. When the MT is close to BS B, it remains connected to BS B (③).

The overall handoff algorithm can be thought of as having two distinct phases [5]: the initiation phase (in which the decision about handoff is made) and the execution phase (in which either a new channel is assigned to the MT or the call is forced to terminate).

Handoff may be caused by factors related to radio link, network management, or service options [6,7]:

- *Radio Link Related Causes:* Radio link related causes reflect the quality perceived by users. Some of the major variables affecting the service quality are RSS, and signal to interference ratio (SIR). Insufficient RSS and low SIR reduce the service quality. A situation where handoff is requested due to insufficient RSS is when the MT approaches the cell boundary and, hence, the RSS drops below a predetermined threshold. Another situation is when the MT is inside a "signal strength hole" within a cell, that is, the signal is too weak to be detected. SIR results in handoffs when it drops a result of increasing CCI. Bit error rate (BER) can be used to estimate SIR.
- *Network Management Related Causes:* The network may handoff a call to avoid congestion in a cell.
- *Service Options Related Causes:* When an MT asks for a service that is not provided at the current BS, the network may initiate a handoff so that the desired service can be offered from the neighboring BSs [6].

Network management and service related handoffs are infrequent and easy to tackle. Radio link related handoffs are the ones that are most commonly encountered and most difficult to handle.

18.2.2 Handoff Types

A handoff made within the currently serving cell (e.g., by changing the frequency) is called an *intracell* handoff. A handoff made from one cell to another is referred to as an *intercell* handoff. Handoff may be hard or soft. *Hard handoff* (HHO) is "break before make" meaning that the connection to the old BS is broken before a connection to the candidate BS is made. HHO occurs when handoff is made between disjointed radio systems, different frequency assignments, or different air interface characteristics or technologies [8]. *Soft handoff* (SHO) is "make before break" meaning that the connection to the old BS is not broken until a connection to the new BS is made. That is, more than one BS would simultaneously connected to the MT. For example, in Figure 18.3, both the BSs will be connected to the MT in the handoff region.

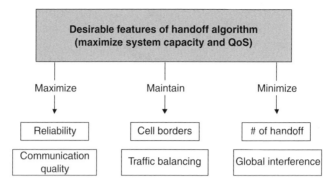

FIGURE 18.4   Desirable features of handoff algorithm.

18.2.3   Desirable Features of Handoff

An efficient handoff algorithm can achieve many desirable features by trading different operating characteristics. Some of the major desirable features of handoff algorithm [4,6,9–14]. are described further (refer to Figure 18.4):

- A handoff algorithm should be reliable. This means that the call should have a good quality after handoff. SIR and RSS help determine the potential service quality of the neighbor BS.
- A handoff algorithm should be fast so that the user does not experience service degradation or interruption. Service degradation may be due to a continuous reduction in signal strength or an increase in CCI. Service interruption may be due to a "break before make" approach of HHO being exercised in the network. Note that the delay in the execution of a handoff algorithm adds to the network delay at the MSC. Fast handoff also reduces CCI since it prevents the MT from going too far into the new cell.
- A handoff algorithm should maintain the planned cellular borders to avoid congestion, high interference, and use of assigned channels inside the new cell. Each BS can carry only its planned traffic load. Moreover, there is a possibility of increased interference if the MT moves well into another cell site while still being connected to a distant BS. This is because the co-channel distance is reduced and the distant BS tends to use a high transmit power to serve the MT.
- A handoff algorithm should balance traffic in adjacent cells, eliminating the need for borrowing channels from neighboring cells that have free channels; simplifying cell planning and operation, and reducing the probability of new call blocking.
- The number of handoffs should be minimized. Excessive handoffs lead to heavy handoff processing loads and poor communication quality. This results from two elements. First, the more handoff attempts, the greater the chances that a call will be denied access to a channel, resulting in a higher handoff call dropping probability; Second, a high number of handoff attempts results in more delay in the MSC processing of handoff requests, which will cause signal strength to decrease over a longer time period to a level of unacceptable quality. In addition, the call may be dropped if sufficient SIR is not achieved. Handoff algorithm requires network resources to connect the call to a new BS. Thus, minimizing the number of handoffs reduces the switching load. Unnecessary handoffs should be prevented, especially when the current BS might be able to provide the desired service quality without interfering with other MTs and BSs.
- The global interference level should be minimized by the handoff algorithm. Transmission of minimum power and maintenance of planned cellular borders can help achieve this goal.

18.2.4   Handoff Algorithms

The rest of this section looks at handoff algorithms. These can be categorized as conventional handoff algorithms and handoff prioritization algorithms as shown in Figure 18.5.

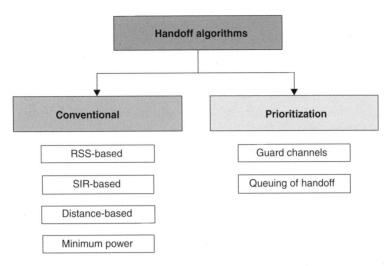

FIGURE 18.5   Classification of handoff algorithms.

18.2.4.1   Conventional Handoff Algorithms

Several variables have been proposed and used as handoff criteria in the design and implementation of handoff algorithms. The handoff criteria discussed here include RSS, SIR, distance, and transmit power.

RSS-Based Algorithms: The RSS criterion is simple, direct, and widely used. Many systems are interference limited, meaning that signal strength adequately indicates the signal quality, and this is the motivation behind having a signal strength-based decision scheme.

According to the RSS criterion in Reference 14, the handoff algorithm works as follows. The BS that receives the strongest signal from the MT is connected to the MT. The advantage of this algorithm is its ability to connect the MT with the strongest BS. However, the disadvantage is the excessive handoffs due to shadow fading variations associated with the signal strength. *Shadow fading* is a phenomenon that occurs when a mobile moves behind an obstruction and experiences a significant reduction in signal power. Another disadvantage is that the MTs connection is retained to the current BS even if it passes the planned cell boundary as long as the signal strength is above a predetermined threshold. A variation of this basic RSS algorithm incorporates delta, a minor increment. For such an algorithm, a handoff is made if the RSS from neighboring BS exceeds the RSS from the current BS by a predetermined delta. Thus, the number of unnecessary handoffs is reduced but dropouts can be increased since it can also prevent necessary handoffs by introducing a delay in handoff [13]. A necessary balance between the number of handoffs and delay in handoff needs to be achieved by appropriate delta and signal strength averaging. The averaging should consider the MT speed and shadow fading [5]. The shadow fading is characterized by a Gaussian distribution with zero mean and a certain standard deviation (which depends on the environment). A scheme for estimating the shadow fading standard deviation based on squared deviations of the RSS at the MT is proposed in References 26 and 27.

SIR-Based Algorithms: An advantage of using SIR as a criterion in the handoff algorithm is that SIR is a parameter common to voice quality, system capacity, and dropped call rate. BER is often used to estimate SIR. When the actual SIR is lower than the desired SIR, radio link becomes poor, and the rate of dropped calls increases. Unfortunately, SIR may oscillate due to propagation conditions and may cause a ping-pong effect (in which the MT repeatedly switches between the adjacent BSs). Another disadvantage is that even though BER is a good indicator of link quality, bad link quality may be experienced near the serving BS, and handoff may not be desirable in such situations [15]. In an interference-limited environment, deterioration in BER does not necessarily imply the need for an intercell handoff. In some cases, an intracell handoff may be sufficient [16].

The SIR is a measure of communication quality. The algorithm in Reference 7 makes a handoff when the current BSs SIR drops below a threshold and neighboring BS can provide sufficient SIR. The lower SIR may be due to high interference or low carrier power. In either case, handoff is desirable when SIR is low. Although, SIR-based handoff algorithms help to prevent handoffs near cell boundaries, they demand high transmit power [17].

Distance-Based Algorithms: Distance criterion can help preserve planned cell boundaries. The distance can be estimated based on signal strength measurements [19], delay between the signals received from different BSs [20], etc. Distance measurement was shown to improve the handoff performance [11].

Based on this criterion the algorithm in Reference 20 connects the MT to the nearest BS. The relative distance measurement is obtained by comparing propagation delay times. This criterion allows handoff at the planned cell boundaries, giving better spectrum efficiency compared to the signal strength criterion [14]. However, it is difficult to plan cell boundaries in a microcellular (with a cell radius is <1 km) system due to complex propagation characteristics. Thus, the advantage of distance criterion over signal strength criterion begins to disappear for smaller cells due to inaccuracies in distance measurements. A RSS-based algorithm gives less interference probability compared to a relative distance-based algorithm. In particular, when a line of sight (LOS) path exists between the current and distant BS and the MT, the current BS gives stronger signal strength compared to the nearer non-line of sight (NLOS) BS. In such cases, the RSS criterion can avoid interference; relative distance criterion experiences more interference.

Minimum Power Algorithms: Transmit power can be used as a handoff criterion to reduce the power requirement, reduce interference, and increase battery life.

A minimum power handoff algorithm that minimizes the uplink transmit power by searching for a suitable combination of a BS and a channel is suggested in Reference 17. First, the channel that gives minimum interference at each BS is found. Then, the BS that has a minimum power channel is determined. This algorithm was shown to reduce call dropping and to minimize the uplink transmit power.

Reference 11 uses a power budget criterion to ensure that the MT is always assigned to the cell with the lowest path loss, even if the thresholds for signal strength or signal quality have not been reached. This criterion results in the lowest transmit power and a reduced probability of CCI.

18.2.4.2 Handoff Prioritization Algorithms

In cellular systems, calls may be classified as new calls and handoff calls. A new call occurs when a user requests a new connection, while a handoff call occurs when an active user moves from one cell to another. An example of new and handoff calls is illustrated in Figure 18.6. Call dropping occurs when a call in progress is forcefully terminated due to lack of available resources in the new cell. On the other hand, call blocking takes place when a new call may not be served. Call dropping is less desirable than call blocking. Hence, handoff calls are given higher priority over new calls.

One of the ways to reduce the handoff dropping rate is to prioritize handoff. Handoff algorithms that try to minimize the number of handoffs give poor performance in heavy traffic situations [22]. In such situations, a significant handoff performance improvement can be obtained by prioritizing handoff. Two basic algorithms of handoff prioritization, guard channels, and queuing, are explained next.

Guard Channels Algorithms: A Guard Channels (GC) algorithm improves the probability of successful handoffs by reserving a fixed or dynamically adjustable number of channels exclusively for handoffs. For example, priority can be given to handoffs by reserving N channels for handoffs among C ($N < C$) channels in the cell [23]. The remaining $(C - N)$ channels are shared by both new calls and handoff calls. A new call is blocked if the number of channels available is greater than $(C - N)$. Handoff fails if no channel is available in the candidate cell. However, this algorithm has the risk of underutilization of spectrum. Variations of the GC algorithm such as fractional guard channel (FGC) algorithm can help reduce this problem. The FGC is based on probabilistic analysis. The reader can refer to the next chapter for the detailed mathematical analysis for both the GC and the FGC.

Queuing of Handoff Algorithms: Queuing is a way of delaying handoff [7]. The principle idea of the queuing of handoff algorithm is that the MSC queues the handoff requests instead of denying access if

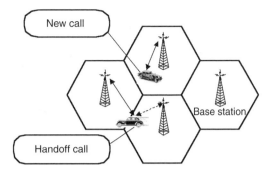

FIGURE 18.6 New call and handoff call.

the candidate BS is busy. The probability of a successful handoff can be improved by queuing handoff requests at the cost of increased call blocking probability since new calls are not assigned a channel until all the handoff requests in the queue are served. Detailed studies for queuing both new and handoff calls are discussed in Chapter 19, which show better performance results in terms of new call blocking probability and handoff call dropping probability. Queuing is possible due to the overlap region between the adjacent cells in which an MT can communicate with more than one BS.

If handoff requests occur uniformly, queuing is not needed; queuing is effective only when handoff requests arrive in groups and when traffic is high. Queuing is very beneficial in macrocells (with cell radius exceeding 35 km in rural areas) since the MT can wait for handoff before signal quality drops to an unacceptable level. However, the effectiveness of queuing decreases for microcells due to stricter time requirements. The combination of queuing and channel reservation can be employed to obtain better performance [24].

Handoff Priority Algorithms: Reference 22 investigates the performance of different handoff priority algorithms using a simulation model that incorporates transmission and traffic characteristics. The priority scheme of GSM has been evaluated. Simulation results show that the queuing and channel reservation schemes improve the dropout performance significantly, and that the priority algorithms provide up to 16% further improvement.

Reference 25 presents a handoff prioritization algorithm to improve the service quality by minimizing handoff failures and spectrum utilization degradation. The algorithm operates as follows. If all the channels are occupied, new calls are blocked while handoff requests are queued. The handoff queue is dynamically reordered based on the measurements. The performance of the proposed handoff priority algorithm has been evaluated through simulations and compared with nonprioritized call handling algorithm and the first in first out (FIFO) queuing algorithm. The algorithm was shown to provide a lower probability of dropped handoff calls and a small reduction in delay compared to the FIFO algorithm under all traffic conditions [25]. However, the algorithm improves the handoff dropping probability at the cost of an increase in new call blocking probability. The priorities are defined by the RSS at the MT from the current BS. The degradation rate in service due to queuing depends on the velocity of the MT, and the proposed algorithm considers this degradation rate.

18.3 Effective Call Admission Control Algorithms for Resource Management

18.3.1 Overview

The CAC is an algorithm that manages radio resources in order to adapt to traffic variations. CAC is always performed when a mobile initiates communication in a new cell either through a new call or handoff. Furthermore, admission control is performed when a new service is added during an active call. CAC makes a decision to accept or reject a new call according to the amount of available resources versus user QoS requirements, and the effect on the QoS of existing calls that may occur as a result of the new call.

A connection is accepted if resources are available and the requested QoS can be met, and if other existing connections and their agreed upon QoS will not be adversely affected. Moreover, the admission control algorithm ensures that the interference created after adding a new call does not exceed a pre-specified threshold.

The purpose of an admission control algorithm, as also mentioned in Chapter 19, is to regulate admission of new users into the system, while controlling the signal quality of the already serviced users without leading to call dropping. The admission control algorithm will then balance between high capacity and interference. Another goal of admission control is to optimize the network revenue. This can, for example, be done by maximizing the instantaneous reward achievable when a new service request arrives. The reward associated with each QoS level is assumed to increase with the amount of resources required for the service. A few simple models for setting rewards based on the number of required channels are discussed in the literature [28–32,37,38]:

- *Threshold Policy:* The threshold policy attempts to improve throughput by rejecting some of the arriving service requests from one class (even though there are free resources), in order to improve the chances for a request from a preferred or disadvantaged class.
- *Multi-level Policy:* The multi-level policy is based on the assumption that if users cannot acquire the necessary resources in order to obtain their highest QoS level, they are willing to accept an admission at a lower level (requiring less resources), rather than totally being denied service.
- *Redistribution Policy:* The redistribution policy is an extension of the multi-level policy. Instead of just trying to lower the QoS level of the arriving user, it also considers lowering the service of some of the already admitted users. An optimization that maximizes the future reward would be best, but in practice such policy is difficult to implement due to the uncertainty in determining the remaining session length of users.

The CAC needs to be performed separately for the uplink and downlink. This is especially important if the traffic is highly asymmetric. Typical criteria for admission control are call blocking (system capacity measurement) and call dropping (system quality measurement). Blocking occurs when a new user is denied access to the system. Call dropping means that a call of an existing user is terminated. Call dropping is considered to be more costly than blocking.

18.3.2 Classification of Call Admission Control

To describe admission control algorithms it is common to characterize them by means of being either:

- Measurement-based
- Interactive or noninteractive
- Distributed or nondistributed
- Predictive or nonpredictive

Measurement-based CAC algorithms use measurement mechanisms to determine the actual current traffic load to make the admission decision. In cellular networks the most common measurements are the SIR, cell traffic load and BS power. Noninteractive algorithms make an instantaneous decision on whether or not to accept a request to the system based on previously measured interference values or received power values; that is, when the measured interference (or currently received power) on some channel (cell) is too high, admission is denied. Interactive algorithms let the new user interact with the system and monitor/predict how the system will be affected if the user is accepted. These involve gradually increasing the power of a new user until they are admitted. A distributed algorithm takes into consideration status information for other BSs than the one that the request is made to. The nonpredictive feature means that no predictions are made on future traffic conditions. The predictive feature tries to make a prediction on future traffic conditions.

TABLE 18.1 CAC Algorithms with Different Characteristics

Scheme	Reference	Measurement-based	Interactive	Distributed	Predictive
Absolute guarantee	[33]			✓	✓
Looking around	[35]	✓		✓	✓
Admission control for downlink packet transmission in WCDMA	[37]	✓			
Efficient interactive call admission control in power-controlled mobile systems	[38]	✓	✓		
The received power call admission control	[39]	✓			
The predictive call admission control	[40]	✓		✓	✓
The modified predictive call admission control	[41]	✓		✓	✓

18.3.3 Call Admission Control Algorithms

Usually CAC algorithm is defined as a combination of the above characteristics. In this section, a number of CAC algorithms are described. A classification of the algorithms is shown in Table 18.1.

In addition to the above classification, authors in the next chapter classify the CAC schemes into two major categories, static approaches and dynamic approaches. In this chapter, however, we will discuss only the schemes that are listed in Table 18.1.

The Absolute Guarantee (AG) Algorithm: The AG algorithm [33] is based on per-connection reservation that guarantees no handoff drop for any existing connection. This is possible by checking the bandwidth in all cells, which the mobile requesting a new connection will traverse, then reserving the required bandwidth in each of those cells. Therefore, this admission algorithm involves per-connection bandwidth reservation in each cell. In this scheme each mobile should inform the wired network (MSC) or the corresponding BS of the mobility specification that is composed of the cells the mobile will traverse during the lifetime of the requested connection. It is assumed the availability of the mobility specification as in Reference 34.

For the mobility specification, M_{sp}, of a newly-requested connection, which consists of a set of cells, and its required bandwidth, b_{new}, admission control and per-connection bandwidth reservation are carried out as follows [33]:

1. For each cell i in the mobility specification, M_{sp}, by adding b_{new} to the existing used bandwidth by other users, check that the total used bandwidth remains below a designed threshold value.
2. If the above test is positive, for each cell i in the mobility specification, M_{sp}, the connection is admitted.

It has been shown using simulation that there are no handoff drops. As a conclusion, even though this algorithm shows perfect results, it is unfeasible to implement since it is impossible to know exactly a certain user's destination in advance.

The Looking Around Algorithm: The call admission decision in Reference 35 is based on the effective traffic loads, which are calculated for both the target cell and the neighboring cells. Its goal to reduce dropped handoff calls in CDMA cellular systems.

It is assumed that a target cell in a WCN can be partitioned into two zones [36]: the core zone (CZ) and the soft handoff zone (SHZ). With respect to the BS in the target cell, the area immediately adjacent to the SHZ is called the neighborhood zone (NZ), see Figure 18.7.

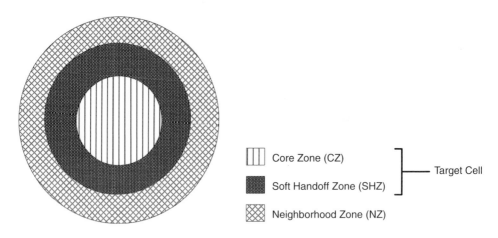

FIGURE 18.7   A concentric geometry for a target cell and its neighborhood area.

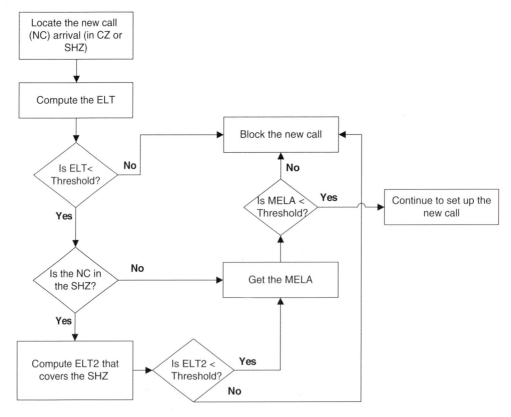

FIGURE 18.8   The CAC algorithm for the new call arrivals in the target cell.

Figure 18.8 shows the CAC algorithm for new calls. When a new call (NC) arrives, the BS at the target cell first checks whether the NC is in the CZ or the SHZ. One way to obtain this is by measuring the signal strength. The effective load for the target cell (ELT) is then calculated through the following equation:

$$\text{ELT} = k_c + w_s \cdot k_s + w_n \cdot k_n \tag{18.1}$$

where k_c (k_s) is the number of calls in CZ (SHZ) before a possible admission of a new call; k_n is the number of calls in the NZ; w_s and w_n are weights. If the target cell is already saturated, the NC is blocked.

Otherwise, if the NC is in the SHZ, the effective load for the other cell (ELT2) that covers the SHZ is also calculated. Again, the threshold is compared with ELT2, if no saturation point is reached for the other cell, the algorithm continues with the final check for the neighborhood cells. The call is accepted if the maximum effective load for the adjacent cells (MELA) is below the threshold.

It has been shown in Reference 35 that the Looking Around algorithm can significantly reduce the probability for dropped calls. However, this comes with the drawback of increasing the blocking probability.

The Admission Control Algorithm for Downlink Packet Transmission in WCDMA: Reference 37 suggests an admission control algorithm for downlink packet transmission in WCDMA. The transmitted packets are classified into different types of services. It assumes two types of traffic classes: real-time, requiring guaranteed QoS, and nonreal-time, where best-effort packet delivery is sufficient. Both new and handoff calls are handled by this algorithm with higher priority to handoff requests. The admission decision is based on mean power and mean traffic load in the target BS.

Upon arrival, calls are put in their respective queues according to the type of service required. Calls in the same queue are processed in a FIFO mechanism. Among new arrivals as well as handoffs, a higher class user has higher priority. The admission algorithm is also nonpreemptive which implies that a higher class user awaiting admission cannot preempt a lower priority user already admitted in the system. A user of traffic class k ($k = 1, 2$) is admitted to cell j ($j = 0, 1, \ldots, 6$) provided the following criteria are met:

$$\text{Power} \quad \overline{P_j(t)} < P_T \tag{18.2}$$

$$\text{Traffic} \quad \frac{\sum_{i=1}^{k} \overline{N_{ij}}}{\mu_j} < D_k \tag{18.3}$$

where $\overline{P_j(t)}$ is the mean output power from BS j averaged over a certain time interval (W) and P_T is a predefined power level or maximum power budget; similarly, $\overline{N_{ij}}$ is the mean number of transport blocks of traffic class i in BS j averaged over W, μ_j is the service rate of BS j in terms of transport blocks that can be transmitted per time frame; D_k is the maximum tolerable delay of traffic class k, $k = 1$ or 2. This scheme improves the system performance at the expense of a slight increase in NC blocking.

The Efficient Interactive Call Admission Control Algorithm in Power-Controlled Mobile Systems: The admission control algorithm proposed in Reference 38 lets new mobiles interact with the system. It admits a new mobile at the moment it arrives in a controlled manner where the SIR of existing calls are adequately protected and make the final accept/reject after a number of power-updating iterations. When a mobile seeks admission in power-controlled cellular systems, interactive CAC guides the evolution of the power transmitted by the new mobile in order to protect the transmission quality of ongoing calls from dropping below a desired level. In addition, it eventually decides whether the new mobile should be accepted or rejected without errors. However, existing interactive call admission algorithms incur high network signaling costs and suffer from slow speed in the accept/reject decision, which often renders the implementation of these algorithms impractical.

The Received Power Call Admission Control (RPCAC) Algorithm: When employing the RPCAC algorithm in Reference 39 NC at a BS are blocked if the total amount of received power for that BS exceeds a certain threshold-value, Z_k^t. It is a noninteractive, nondistributed uplink scheme, which is designed for CDMA system, and when used in such a system the performance is very good. The algorithm can be simply stated as:

$$\text{Admit at base station } k \text{ if and only if } Z_k(P) \le Z_k^t$$

where $Z_k(P)$ is the total amount of received power at BS k. The scheme is based on the fact that the total amount of received power can be expressed in terms of cell load. *Cell load* is defined as the ratio between the number of users assigned to a cell and the maximum capacity of the cell. As the number of users requesting a radio link from a BS increases, the received power by the BS will also increase.

Assuming that each user i has an identical SIR requirement, $\gamma = \gamma_i$, and perfect uplink PC (Perfect uplink PC implies that the received power at the BS is the same for all users). The SIR, γ, of a user assigned to BS k with received power P_k is

$$\gamma = P_k/((N_k - 1)P_k + I_k + n_k) \tag{18.4}$$

where N_k is the number of users assigned, I_k is the outer-cell interference (i.e., the interference from adjacent cells), and n_k is the receiver noise. Since CDMA systems are only interference limited, maximum capacity of a cell is achieved when there is no outer-cell interference and no receiver noise.

Therefore, with $I_k = n_k = 0$, the maximum capacity for BS k is

$$N_{max} = 1 + (1/\gamma) \tag{18.5}$$

Let L_k denote the efficient number of users caused by outer-cell interference, such as

$$L_k = I_k/P_k \tag{18.6}$$

From Equations (18.4) and (18.6) the received power P_k can then be written as

$$P_k = n_k/((1/\gamma) + 1 - N_k - L_k) \tag{18.7}$$

In CDMA systems, the number of users that each BS can support depends on the load of the adjacent cells. Therefore, the cell load, X_k, of BS k is

$$X_k = (N_k + L_k)/N_{max} = (N_k + L_k)/(1/\gamma) + 1 \tag{18.8}$$

Then, X_k can be written as the ratio of the sum of all active user's received powers to the total received power,

$$X_k = (N_k P_k + I_k)/(N_k P_k + I_k + n_k) = (Z_k - n_k)/Z_k \tag{18.9}$$

where

$$Z_k = N_k P_k + I_k + n_k \tag{18.10}$$

denotes the total received power at the BS k. By using Equation (18.5) one can determine how loaded the cell should be allowed to be before blocking NCs. If the total amount of received power is fairly stable, this algorithm will not admit users that will degrade system performance, but if the received power is severely fluctuating, we can suspect that this algorithm will not perform very well.

The Predictive Call Admission Control (PCAC) Algorithm: The PCAC algorithm [40] is an uplink control algorithm designed for *multi-service* CDMA systems.

The basic idea behind PCAC algorithm is to predict future traffic conditions when the admission for an NC is considered. The *resource criterion* is defined by what is known as the *equivalent bandwidth*. The algorithm was initially designed to take into account post admission traffic variations due to handoff and call departures during the call set up time T. Here, traffic variations due to the characteristics of the different services as well as the handoff and departure are taken into account. A user's *equivalent bandwidth, W,* is defined as its SIR, γ, times its relative bit rate, R, [40] to be

$$W = \gamma R \tag{18.11}$$

The algorithm operates as follows:

1. A request is made at a certain cell, for a certain bit rate, that is, equivalent bandwidth.
2. The BS has an estimate of used equivalent bandwidth and tries to compute how this value will change when admitting a new user. In this calculation the algorithm tries to predict future values of the equivalent bandwidth, and takes this into account when making a decision.
3. If this newly predicted value is less than a certain threshold, the user is admitted. Otherwise the user will be blocked.

The request to enter cell C_0 of a new call requiring bandwidth W_d can be accepted if the following conditions are satisfied for the uplink:

$$\text{First condition} \qquad W_d + \sum_{j=1}^{\lambda_h T} W_{hj}^0 + \sum_{j=1}^{k_{c0}} \check{W}_{aj}^0 p_{aj}^0 \leq \rho W_T \qquad (18.12)$$

where λ_h represents the average call handoff rate, W_{hj}^i is the equivalent bandwidth of user j handed off to cell C_i, $\check{W}_{aj}^i$ is the predicted value of the jth active user's equivalent bandwidth, p_{aj}^i is the probability of user j remaining in the cell C_i during time T and K_{C0} is the number of active users in cell C_i at the time of call admission. The second term represents the equivalent bandwidth consumed by users arriving in the home cell due to handoff during the interval T. The third term is the equivalent bandwidth taken by users already present in the home cell. The variable ρ is the CDMA system frequency reuse efficiency, which is defined as the ratio of interference from mobiles within a given cell to the total interference from all cells, $\rho = 1/(1+f)$. Thus, ρW_T denotes users in the home cell. Call duration is exponentially distributed with mean holding time $1/\mu_j$ and hence $p_{aj} = e^{-\mu j T}$.

$$\text{Second condition} \qquad \phi_i W_d \leq \text{REB}_i \qquad (18.13)$$

where REB_i is the cell's remaining equivalent bandwidth for cell C_i and $\phi_i W_d$ denotes the equivalent interference to cell C_i contributed by the new user. Depending on the distance between cells, propagation environment, and path loss exponent, the factor, ϕ_i, varies significantly. This condition makes the PCAC a distributed algorithm.

The Modified Predictive Call Admission Control (MPCAC) Algorithm: The MPCAC in Reference 41 combines the predictive feature of PCAC algorithm [40] and the absolute, momentary feature of RPCAC algorithm [39]. Instead of only looking at where the traffic situation is heading (as in PCAC), MPCAC will also look at the current condition. This is achieved by first using the PCAC to check if the traffic conditions in the near future, for both the up- and down-link, will be higher than accepted. The RPCAC is then executed for the uplink. If there are sufficient resources left at the time, the user will be admitted. Otherwise, the user is rejected.

In this manner, the algorithm compares the demand with resources at both the present and future times. The algorithm is valid for the uplink as well as for the downlink. This is because one might suspect that the assumption that the uplink will be the limiting link will not be valid in a more complex traffic situation than the old circuit-switched speech traffic. For instance, when browsing the web and when sending/receiving e-mails most of the data (statistically) will be sent on the downlink.

The request to enter cell C_0 of a new call requiring bandwidth $W_{d,\text{up}}$ for the uplink and $W_{d,\text{down}}$ for the downlink, can be accepted if the following conditions are satisfied:

$$W_{d,\text{up}} + \sum_{j=1}^{k_{c0}} \check{W}_{aj,\text{up}}^0 p_{aj}^0 \leq \rho W_{T,\text{up}} \qquad (18.14)$$

$$W_{d,\text{down}} + \sum_{j=1}^{k_{c0}} \check{W}_{aj,\text{down}}^0 p_{aj}^0 \leq \rho W_{T,\text{down}} \qquad (18.15)$$

where the predicted amount of equivalent bandwidth, for both the up- and down-link, summed with the demanded equivalent bandwidth must be lower than a predefined total amount of equivalent bandwidth offered from the cell,

$$Z_k(P) \leq Z_k^t \qquad (18.16)$$

which, as for RPCAC, means that the amount of total received power at the BS must be lower than a certain, predefined level.

18.4   Management of Power in Cellular Systems: Mechanisms and Algorithms

18.4.1   Overview

In wireless networks, power control (PC) is one of the most important system requirements, and it is analyzed for cellular networks based on FDMA and TDMA [42–45], and for CDMA cellular networks [46–62]. Downlink PC applies to BSs and uplink PC to mobiles. In most modern systems, both BS and mobiles have the capability of real-time (dynamic) adjustment of their transmit powers.

In FDMA and TDMA, PC is applied to reduce co-channel interference within the cellular system that arises from frequency reuse. In CDMA systems, PC allows users to share resources of the system fairly among themselves. In the case where PC is not used, all mobiles transmit signals toward the BS with the same power without taking into account the fading and the distance from the BS. *Fading* is defined as signal propagation effects that may disrupt the signal and cause errors. If a mobile's signal is received at the BS with a very low level of received power (i.e., the mobile is far from the cell site or using an unusually high attenuation channel), a high level of interference is experienced by this mobile and its performance (bit error rate) will be degraded (high BER). On the other hand, if the received power level is too high (i.e., the mobile station is close to the BS or using an unusually low attenuation channel), the performance of this mobile is acceptable, but it increases interference with other mobiles that are using the same channel, and may result in unacceptable performance to other mobiles. This is known as the *"near far problem."* Therefore the main purposes of power control are:

1. To maintain all user's signal energies received at the BS nearly equal in the spread spectrum shared [59]. This combats the near far problem.
2. To make the received power level less dependent on the fading and shadowing effects of the transmission channel [60].

18.4.2   Power Control Mechanisms

In order to properly manage the PC in CDMA, the system uses one of the two well-known PC mechanisms [61]:

- Open-Loop Power Control (OLPC)
- Closed-Loop Power Control (CLPC)

18.4.2.1   Open-Loop Power Control

The OLPC adjusts the transmitted power according to its estimation of the channel — it does not attempt to obtain feedback information. The major benefit of OLPC is to provide a very rapid response over a period of just a few microseconds (i.e., it does not wait for feedback information). This is desirable for cases of sudden change in channel conditions, such as a mobile behind a building. It adjusts the mobile transmit level and thus prevents the mobile transmitter power from exceeding some threshold with respect to the downlink received power level. Prior to any transmission, the mobile measures the total power received by the BS. The measured power provides an indication of the propogation loss to each user. Then, the mobile

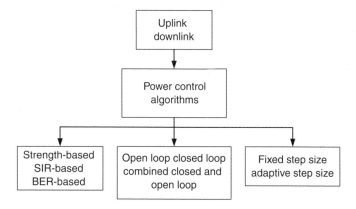

FIGURE 18.9 Classification of power control techniques.

adjusts its transmission power so that it is inversely proportional to the total signal power it receives. The OLPC may not very accurate though, since it does not have employ any feedback mechanisms for better effectiveness.

18.4.2.2 Closed-Loop Power Control

This is a more effective long-term PC method. The BS provides continuous feedback to each mobile so that the mobile varies its power accordingly. Each BS measures the received signal power P_m (or SIR_m in case of SIR-based CLPC) from each mobile. The measured P_m is compared to the desired power level P_d for that mobile and a PC adjustment command is sent accordingly. If the measured value is above or below the desired nominal threshold, then a one bit command is sent to adjust power by ΔP dB. The transmitter adjusts its power up or down, relative to the open loop estimate. The CLPC is a form of "fine tuning" on the open loop power estimate. It should be fast enough to keep up with the fast fading. Hence, it is the crucial component of any effective scheme to combat fading.

18.4.3 Power Control Algorithms

Several algorithms to control the power in wireless cellular systems were proposed, especially for CDMA systems. Most of the existing algorithms utilize either SIR or transmit power as a reference point in the PC decision-making process. PC algorithms can be classified in many different ways, as shown in Figure 18.9 [40]. One of the possible classifications is the following:

- Power Control for Uplink (Reverse link)
- Power Control for Downlink (Forward link)

18.4.3.1 Power Control Algorithms for the Uplink

The PC for the uplink usually combines algorithm consisting of closed-loop and open-loop PCs. The following is some of the proposed algorithms in the literature.

Power Control Algorithms for 3G WCDMA System: PC of the 3G WCDMA system is studied in Reference 5. Two algorithms are proposed based on received quality level (i.e., SIR) and PC update rate, which has a fixed value.

The WCDMA air interface is organized in frames of 10 msec duration. A frame contains 15 time slots and each slot includes one PC command (up or down), which gives a PC update rate of 1500 b/sec. The transmitted power has a fixed value during a given time slot. PC in WCDMA is usually CLPC, which is a combination of outer and inner closed-loop control. The inner (also called fast) CLPC adjusts the transmitted power in order to keep the received SIR equal to a given target. This SIR target is fixed according to the received block error rate (BLER) or BER. The setting of the SIR target is done by the outer loop PC in order to match the required BLER [51]. The BLER target is a function of the service that is carried. Ensuring that the lowest possible SIR target is used results in greater network capacity.

The inner CLPC measures the received quality, defined as the received SIR, and sends commands to the transmitter for the transmitted power update. In order to estimate the received SIR, the receiver estimates the received power of the connection. The obtained SIR estimate, SIR_{est}, is then used by the receiver to generate PC commands according to the 3GPP specifications [51].

In Algorithm 1 of Reference 51, the transmitted power is updated at each time slot. It is increased or decreased by a fixed value:

- If $SIR_{est} > SIR_{target}$ then request a transmit power decrease
- If $SIR_{est} < SIR_{target}$ then request a transmit power increase

Simulation has shown that Algorithm 1 cannot fully react to fast changes in the signal level (deep fades) at a reasonable update rate. Thus, Algorithm 2 of Reference 51 is proposed as a variation of Algorithm 1 to overcome this tardiness. In Algorithm 2, the transmitted powers may be updated each five time slots, which simulates smaller power update steps. It was shown that Algorithm 2 performs better than Algorithm 1 in reacting to fast changes in the signal level.

The PC step size is a parameter of the fast (inner) CLPC. The power update step size may be chosen according to the average mobile speed and other operating environment parameters.

The Adaptive-Step Power Control (ASPC) Algorithm: An evolution of the PC algorithm of the 3G WCDMA system is proposed in Reference 53. The algorithm is based on a modification of the transmitted power update step size. Instead of the presently-suggested fixed value [54], the step size is modified dynamically in order to guarantee more adapted power variations.

Frequent variations occur in the wireless network due to communication setup and termination, mobiles movements, and propagation channel changes. Radio channel modification may be due to multipath fading and to obstacles movements. Thus, fast PC is needed. In WCDMA, one PC command is transmitted per time slot. The value of the fixed step [51] may be too small or too large for the transmitted power adjustment needed. In Reference 53, the fixed power update step is replaced by an adaptive step in the inner closed-loop one bit PC of 3G WCDMA system. The algorithm is based on the following principle. If the transmitter detects several simultaneous up commands (increase power), the step is increased. This is also done for several simultaneous down commands (decrease power). The update step is decreased if an alternative succession of up and down appears, showing that the update step is probably too large. Using this basic principle, the algorithm shows that the quicker convergence of the proposed ASPC (with regard to the present version of PC in WCDMA) may give a capacity increase.

Strength and SIR-Combined Power Control Algorithm: One of the ways to alleviate the possible effects of positive feedback in PC-based on SIR measurements is to use strength-based and SIR-based PC algorithms in a combined manner [52]. The *Positive feedback* problem arises in the situation when one mobile under instructions from the BS has to raise its transmit power in order to deliver a desirable SIR to the BS. However, this increase in the MTs power also results in an increase in interference to other mobiles such that these other mobiles are consequently forced to increase their transmit power accordingly. Both the received power and the received SIR are measured at BSs. If both quantities are lower or higher than the threshold, it is clear that the associated mobile should be instructed to power up in the first case (lower quantities) or to power down in the second case (higher quantities). In the case that SIR is in satisfactory condition and that the power is below the threshold, the mobile does not need to take any action because SIR reflects better system performance such as QoS and capacity. In the case that the received power is satisfactory and the SIR is below the threshold, the mobile needs to raise its transmitting power but in a more careful manner because of a potential positive feedback situation since the power of the signal already exceeds its threshold. One way to solve this problem is to take a number of the most recent states when the power is satisfactory and SIR is not. The higher that number, the less amount of power the mobile should be asked to increase. It has been shown in Reference 52 that this combined PC technique yields better performances than PC-based solely on SIR measurements.

18.4.3.2 Power Control Algorithms for the Downlink

The purpose of the downlink PC is to reduce power for users that are either stationary, relatively close to the BS, less impacted by multipath fading and shadowing effects, or experiencing minimal interference from other cells. Therefore, additional power can be given to users that are either in a more difficult environment or far away from the BS and experiencing high error rates.

A simple PC algorithm that does not take into account fading is considered in Reference 64, where the transmitted power from the BS to the MT is proportional to the distance between the BS and the MT. It has been shown a poor performance is observed in the absence of PC. In Reference 65, a simple analytical model is developed for the neighboring cell interference experienced by users as a function of distance from their home BS, not taking fading into consideration. Using this model and the previously mentioned distance PC, the optimum radial distance dependent PC is designed to provide a uniform service to all users and approximately increases the capacity compared to the no PC case. This is a very optimistic result since fading and shadowing were not considered. An optimal downlink PC relationship in the presence of fading is presented in Reference 56, where the expression for the average BER of a mobile located at some distance from the BS is given analytically. This expression for the desired BER is the optimality criterion that is equivalent to maximization in capacity. It has been shown that the performances with optimal PC law are improved and the number of admitted users is increased by a factor of two. However, this optimal PC law is given implicitly as the solution of an equation rather than in closed form and is hence impractical.

18.5 Conclusion and Future Work

The rapid growth in demand for mobile communication has led to intense research and development efforts toward a new generation of cellular systems. The new systems must be able to provide QoS guarantees, support a wide range of services, and improve the system capacity. In order to support high user densities in cellular cells, while maintaining high quality in the wireless links, RRM is essential.

Handoff, CAC and PC schemes are important RRM algorithms, which significantly increase the capacity and guarantee high QoS in cellular systems. In this chapter, we have studied several RRM algorithms.

Handoff is a process of transferring a mobile user's call or flow from one BS to another. It is thus an integral component of cellular communications. Efficient handoff algorithms enhance system capacity and service quality in a cost-effective manner. In this study, handoff algorithms are categorized as conventional handoff algorithms and handoff prioritization algorithms. It is shown that both types of algorithms reduce the call dropping rate.

A new call requires a new radio link. Admission control algorithm makes a decision to accept or reject a new call according to the amount of available resources with respect to the user's QoS requirements, and the negative impact on QoS of existing calls that may occur as a result of the new call. A connection is accepted only if sufficient resources are available, the requested QoS requirement can be met, and if other existing connections and their agreed upon QoS will not be adversely affected. In the literature, different algorithms, such as dividing the cell into zones and traffic prioritization, are used to guarantee a certain degree of link quality for existing users as well as keeping the call dropping and blocking probabilities low. Our study classifies the CAC algorithms as follows:

- Measurement-based
- Interactive or noninteractive
- Distributed or nondistributed
- Predictive or nonpredictive

The PC is essential to RRM in cellular systems. The purpose of different PC algorithms is to find the trade-off between change of transmit power level and interference. In a way, PC algorithms try to keep the interference levels in the system at a minimum to provide the required QoS and to increase the system capacity.

We have presented a review of the proposed up- and down-link PC algorithms. First, PC algorithms were classified according to different criteria. Then, different PC algorithms for the CDMA up- and down-links were presented and their properties were studied. Also, PC algorithms in 3G wireless communication systems were presented.

Over the past two decades, various RRM algorithms have been introduced. Nevertheless, future RRM algorithms have to be intelligent enough to dynamically adapt to the changing traffic and interference levels of the cellular network.

References

[1] Sudhir Dixit, Yile Gud, and Zoe Antonion, "Resource management and quality of service in third-generation wireless networks," *IEEE Communication Magazine*, vol. 39, 2001, pp. 125–133.

[2] C.W. Sung and W.S. Wong, "Power control and rate management for wireless multimedia CDMA systems," *IEEE Transactions on Communications*, vol. 49, 2001, pp. 1215–1226.

[3] R. Verdone, and A. Zanella, "Performance of received power and traffic driven handover algorithms in urban cellular networks," *IEEE Wireless Communications Magazine*, vol. 9, 2002, pp. 60–71.

[4] G.P. Pollini, "Trends in handover design," in *IEEE Communications Magazine*, vol. 34, 1996, pp. 82–90.

[5] G. Corazza, D. Giancristofaro, and F. Santucci, "Characterization of handover initialization in cellular mobile radio networks," in *Proceedings of the 44th IEEE VTC*, vol. 3, 1994, pp. 1869–1872.

[6] M. Anagnostou and G.C. Manos, "Handover related performance of mobile communication networks," in *Proceedings of the 44th IEEE VTC*, vol. 1, 1994, pp. 111–114.

[7] W.C.Y. Lee, *Mobile Cellular Telecommunications*. McGraw Hill, 2nd Ed., New York, 1995.

[8] V.K. Garg and J.E. Wilkes, *Wireless and Personal Communications Systems*. Prentice-Hall Inc., 1996.

[9] W.C.Y. Lee, *Mobile Communications Design Fundamentals*. John Wiley & Sons Inc., 2nd Ed., New York, 1993.

[10] E.A. Frech and C.L. Mesquida, "Cellular models and handoff criteria," in *Proceedings of the 39th IEEE VTC*, vol. 1, 1989, pp. 128–135.

[11] W. Mende, "Evaluation of a proposed handover algorithm for the GSM cellular system," in *Proceedings of the 40th IEEE VTC*, 1990, pp. 264–269.

[12] D. Munoz-Rodriguez and K. Cattermole, "Multicriteria for handoff in cellular mobile radio," *IEE Proceedings*, vol. 134, 1987, pp. 85–88.

[13] G.H. Senarath and D. Everitt, "Comparison of alternative handoff strategies for micro-cellular mobile communication systems," in *Proceedings of the 44th IEEE VTC*, vol. 3, 1994, pp. 1465–1469.

[14] T. Kanai and Y. Furuya, "A handoff control process for microcellular systems," in *Proceedings of the 38th IEEE VTC*, 1988, pp. 170–175.

[15] G. Liodakis and P. Stavroulakis, "A novel approach in handover initiation for microcellular systems," in *Proceedings of the 44th IEEE VTC*, vol. 3, 1994, pp. 1820–1823.

[16] S. Chia, "The control of handover initiation in microcells," in *Proceedings of the 41st IEEE VTC*, 1991, pp. 531–536.

[17] C.-N. Chuah and R.D. Yates, "Evaluation of a minimum power handoff algorithm," in *Proceedings of the IEEE International Symposium in Personal Indoor-Outdoor Mobile Radio Communication*, Toronto, Canada, Sep. 1995, pp. 814–818.

[18] C.-N. Chuah, R.D. Yates, and D. Goodman, "Integrated dynamic radio resource management," in *Proceedings of the 45th IEEE VTC*, vol. 2, 1995, pp. 584–588.

[19] M. Hata and T. Nagatsu, "Mobile location using signal strength measurements in cellular systems," *IEEE Transactions on Vehicular Technology*, vol. 29, 1980, pp. 245–252.

[20] G.D. Ott, "Vehicle location in cellular mobile radio systems," *IEEE Transactions on Vehicular Technology*, vol. 26, 1977, pp. 36–43.

[21] B. Jabbari, G. Colombo, A. Nakajima, and J. Kulkarni, "Network issues for wireless communications," *IEEE Communications Magazine*, vol. 33, Jan. 1995, pp. 88–99.

[22] G.H. Senarath and D. Everitt, "Performance of handover priority and queuing systems under different handover request strategies for microcellular mobile communication systems," in *Proceedings of the 45th IEEE VTC*, vol. 2, 1995, pp. 897–901.

[23] P. Gassvik, M. Cornefjord, and V. Svensson, "Different methods of giving priority to handoff traffic in a mobile telephone system with directed retry," in *Proceedings of the 41st IEEE VTC*, May 1991, pp. 549–553.

[24] D. Giancristofaro, M. Ruggieri, and F. Santucci, "Queuing of handover requests in microcellular network architectures," in *Proceedings of the 44th IEEE VTC*, vol. 3, 1994, pp. 1846–1849.

[25] Tekinay and B. Jabbari, "A measurement-based prioritization scheme for handovers in mobile cellular networks," *IEEE Journal on Selected Areas in Networks*, vol. 10, Oct. 1992, pp. 1343–1350.

[26] B. Eklundh, "Channel utilization and blocking probability in a cellular mobile telephone system with directed retry," *IEEE Transactions on Communications*, vol. COM-34, Apr. 1986, pp. 329–337.

[27] A. Sampath and J.M. Holtzman, "Adaptive handoffs through estimation of fading parameters," in *Proceedings of the IEEE International Conference on Communications*, vol. 2, May 1994, pp. 1131–1135.

[28] E. Altman, G. Koole, and T. Jimenez, "On optimal call admission control in a resource-sharing system," *IEEE Transactions on Communications*, vol. 49, 2001, pp. 1659–1668.

[29] S. Spitler and D.C. Lee, "Optimality of soft-threshold policy for call admission control with packet loss constraint," in *Proceedings of the IEEE International Conference on Communications*, vol. 8, Helsinki, Finland, June 2001, pp. 2354–2358.

[30] T. Kwon, S. Kim, Y. Choi, and M. Naghshineh, "Threshold-type call admission control in wireless/mobile multimedia networks using a prioritized adaptive framework," *IEE Electronics Letters*, vol. 36, Apr. 2000, pp. 852–854.

[31] S. Kim, T. Kwon, and Y. Choi, "Call admission control for prioritized adaptive multimedia services in wireless/mobile networks," in *Proceedings of the 51st IEEE VTC*, vol. 2, Tokyo, May 2000, pp. 1536–1540.

[32] Y. Guo and B. Aazhong, "Call admission control in multi-class traffic CDMA cellular system using multiuser antenna array receiver," in *Proceedings of the 51st IEEE VTC*, vol. 1, Tokyo, May 2000, pp. 365–369.

[33] Sunghyun Choi and Kang G. Shin, "A comparative study of bandwidth reservation and admission control schemes in QoS-sensitive cellular networks," *ACM Wireless Networks*, vol. 6, 2000, pp. 289–305.

[34] A.K. Talukdar, B.R. Badrinath, and A. Acharya, "On accommodating mobile hosts in an integrated services packet network," in *Proceedings of the IEEE INFOCOM*, vol. 3, April 1997, pp. 1046–1053.

[35] Y. Ma, J.J. Han, and K.S. Trivedi, "Call admission control for reducing dropped calls in code division multiple access (CDMA) cellular systems," in *Proceedings of the IEEE INFOCOM*, New York, March 2000, pp. 1481–1490.

[36] J.C.Jr. Liberti and T.S. Rappaport, "Analytical results for capacity improvements in CDMA," *IEEE Transactions on Vehicular Technology*, vol. 43, Aug. 1994, pp. 680–690.

[37] M. Kazmi, P. Godlewski, and C. Cordier, "Admission control strategy and scheduling algorithms for downlink packet transmission in WCDMA," in *Proceedings of the 52nd IEEE VTC*, vol. 2, Sep. 2000, Boston, USA, pp. 674–680.

[38] D. Kim, "Efficient interactive call admission control in power-controlled mobile systems," *IEEE Transactions on Vehicular Technology*, vol. 49, May 2000, pp. 1017–1028.

[39] C.Y. Huang and R.D. Yates, "Call admission in power controlled CDMA systems," in *Proceedings of the 46th IEEE VTC*, vol. 3, May 1996, pp. 1665–1669.

[40] S. Sun and W.A. Krzyman, "Call admission policies and capacity analysis of a multi-service CDMA personal communication system with continuous and discontinuous transmission," in *Proceedings of the 48th IEEE Vehicular Technology Conference*, vol. 1, May 1998, pp. 218–223.

[41] Bjrn Hjelm, "Admission control in future multi-service wideband direct-sequence CDMA (WCDMA) systems," in *Proceedings of the 52nd IEEE VTC*, vol. 3, Sept. 2000, pp. 1086–1093.

[42] J. Blom and F. Gunnarsson, "Power control in cellular radio systems," Licentiate Thesis, Department of Electrical Engineering, Linköpings Universitet, Sweden, 1998.

[43] J. Zender, "Transmitter power control for co-channel interference management in cellular radio systems," in *Proceedings of the 4th WINLAB Workshop*, New Brunswick, NF, USA, Oct. 1993.

[44] S.A. Grandhi and J. Zander, "Constrained power control in cellular radio systems," *Wireless Personal Communications*, vol. 1, 1995, pp. 824–828.

[45] M. Andersin, Z. Rosberg, and J. Zander, "Distributed discrete power control in cellular PCS," in *Proceedings of the Workshop. Multiaccess, Mobility and Teletraffic for PCS, MMT'96*, Paris, France, May 1996.

[46] K. S. Gilhousen et al., "On the capacity of a cellular CDMA systems," *IEEE Transactions on Vehicular Technology*, vol. 40, May 1991, pp. 303–312.

[47] H. Ji and C-Y Huang, "Non-cooperative uplink power control in cellular radio systems," *Wireless Networks*, vol. 4, Apr. 1998, pp. 233–240.

[48] A. Chockalingam and L.B. Milstein, "Open loop power control performance in DS-CDMA networks with frequency selective fading and non-stationary base stations," *Wireless Networks*, vol. 4, Apr. 1998, pp. 249–261.

[49] A. Chockalingam et al., "Performance of CLPC in DS-CDMA cellular systems," *IEEE Transactions on Vehicular Technology*, vol. 47, 1998, pp. 774–789.

[50] F. Vatalaro et al., "CDMA cellular systems performance with imperfect power control and shadowing," in *Proceedings of the 46th IEEE VTC*, vol. 2, Atlanta, USA, May 1996, pp. 874–878.

[51] 3G TS 25.214 v4.1.0 (2001–06), "Physical layer procedures (FDD) (Release 4)" 3rd Generation Partnership Project; Technical Specification Group Radio Access Networks, June 2001.

[52] Y.-J. Yang and J.-F. Chang, "A strength and SIR combined adaptive power control for CDMA mobile radio channels," in *Proceedings of the IEEE 4th International Symposium on Spread Spectrum Techniques and Applications*, vol. 3, 1996, pp. 1167–1171.

[53] L. Nuaymi, X. Lagrange, and P. Godlewski, "A power control algorithm for 3G WCDMA system," in *Proceedings of the European Wireless 2002*, Florence, Italy, Feb. 2002.

[54] S.V. Hanly and D. Tse, "Power control and capacity of spread-spectrum wireless networks," *Automatica*, vol. 35, Dec. 1999, pp. 1987–2012.

[55] D.M. Novakovic and M.L. Dukic, "An analysis of integrated power control and advanced receivers algorithms for reverse DS-CDMA link," *IEEE 6th International Symposium on Spread Spectrum Techniques and Applications*, vol. 2, New Jersey, USA, Sep. 2000, pp. 771–775.

[56] M. Zorzi and L.B. Milstein, "Power control on the forward link in cellular CDMA," in *Proceedings of the International Zurich Seminar'94*, Mar. 1994, pp. 391–399.

[57] M. Zorzi, "Optimization of a class of power control laws for the forward-link in a cellular CDMA system," in *Proceedings of the IEEE Globecom'94*, vol. 3, San Francisco, USA, Nov. 1994, pp. 1712–1716.

[58] M. Zorzi, "Simplified forward-link power control law in cellular CDMA," *IEEE Transactions on Vehicular Technology*, vol. 43, Nov. 1994, pp. 1088–1093.

[59] A.J. Viterbi, A.M. Viterbi, and Ephraim Zehavi, "Performance of power controlled wideband-terrestrial digital communication," *IEEE Transactions on Vehicular Technology*, vol. 41, May 1991, pp. 559–569.

[60] Po-Rong Chang, "Adaptive fuzzy power control for CDMA mobile radio systems," *IEEE Transactions on Vehicular Technology*, vol. 45, May 1996, pp. 225–236.

[61] H. Kaaranen, A. Ahtiainen, L. Laitinen, S. Naghian, and V. Niemi, *UMTS Networks: Architecture, Mobility and Services*, Wiley, 2001

[62] K.S. Gilhousen, I.M. Jacobs, R. Padovani, A.J. Viterbi, and C.E. Wheatly, " On the capacity of cellular CDMA system," *IEEE Transactions on Vehicular Technology*, vol. 40, May 1991, pp. 303–312.

[63] Dejan M. Novakovic and Miroslav L. Dukic, University of Belgrade, "Evolution of the Power Control Techniques for DS-CDMA Toward 3G Wireless Communication Systems," *IEEE Communications Surveys and Tutorials*, 4th quarter 2000.

[64] W.C.Y. Lee, "Overview of cellular CDMA," *IEEE Transactions on Vehicular Technology*, vol. 40, May 1991, pp. 291–302.

[65] R.R. Gejji, "Forward-link-power control in CDMA cellular systems," *IEEE Transactions on Vehicular Technology*, vol. 41, Nov. 1992, pp. 532–536.

19

Resource Allocation and Call Admission Control in Mobile Wireless Networks

19.1 Introduction..**19**-439
19.2 Major Challenges **19**-441
Increased Handoff Frequency • Multiple Service Types •
Traffic Asymmetry
19.3 Static Approaches **19**-442
Guard Channel • Fractional Guard Channel • Queueing
Schemes • Admission Control for Multiple Traffic Classes •
Remarks on Static CAC Approaches
19.4 Dynamic Approaches **19**-447
Distributed Call Admission Control • Stable Dynamic Call
Admission Control • Admission Control Using Shadow
Cluster • Dynamic Multiple-Threshold Bandwidth
Reservation • Call Admission for Asymmetric Traffic •
Local Predictive Resource Reservation • Remarks on
Dynamic CAC Approaches
19.5 Conclusions **19**-455
Acknowledgments..**19**-455
References **19**-455

Xiang Chen
Wei Liu
Yuguang Fang

19.1 Introduction

Recent years have witnessed an ever-increasing popularity of wireless networks, as a result of wireless technological advances and rapid development of handheld wireless terminals [27, 28, 6, 24, 32, 33]. Since the mid-1980s, cellular mobile communications have been experiencing explosive growth and by the end of 2000, the number of wireless subscribers has exceeded 700 million. It is expected that the number will top two billion by 2007 [16].

As the number of users is increasingly growing, the wireless network should serve as many users as possible, given the limited resources (e.g., bandwidth). On the other hand, as various service types, such as voice, video, and data, are being offered, quality of service (QoS) is highly demanded. Obviously, these

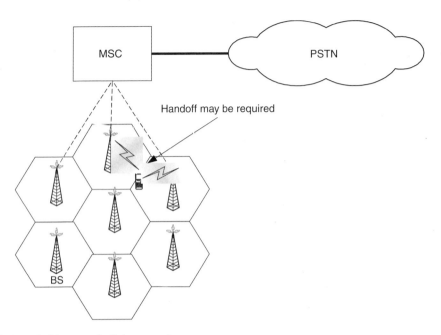

FIGURE 19.1   Architecture of cellular networks.

two requirements are competing against each other. To strike a good balance between these two competing requirements, resource allocation plays a key role. Essentially, resource allocation is responsible for efficient utilization of network resources while providing QoS guarantees to various applications. However, this goal is not easy to achieve, since resource allocation is confronted with more difficulty such as error-prone wireless channel, limited bandwidth, and mobility in mobile wireless networks than in wired networks.

As shown in Figure 19.1, a wireless cellular network is typically organized into geographical areas called cells so that the same radio channels may be reused in other cells some distance away [32]. A base station (BS) serves mobile users in the cell it covers. To communicate with other users, a mobile user must issue to the BS a new connection request, in which a set of desirable QoS requirements are explicitly or implicitly specified. The BS then will make the admission decision based on the premise that the QoS of currently existing connections will not be sacrificed. And at the same time, the new connection, if admitted, should be guaranteed the set of QoS requirements in its duration, since granting an admission is equivalent to such a guarantee. As a result, resource allocation is enforced by the BS through constructing a connection admission strategy, which is aimed to maximize network utilization and satisfy the QoS requirements of each connected service.

When designing a connection admission control (CAC) scheme, we should take several issues into consideration. First, we need to give handoff connection requests higher priority than new connection requests. As we know, a handoff request occurs when a user engaged in a call connection moves from one cell to another. To keep the QoS contract agreed during the connection setup stage, the network should provide uninterrupted service to the previously established connection. However, if the new cell does not have enough resources, the ongoing connection will be forced to terminate before normal completion. Since mobile users are more sensitive to the termination of an ongoing connection than the blocking of a new call connection, handoff call connections are usually given higher priority over new call connections. Second, since the various services offered by the network have inherently different traffic characteristics, their QoS requirements may differ in terms of bandwidth, delay, and connection dropping probabilities. It is the network's responsibility to assign different priorities to these services in accordance with their QoS demands and traffic characteristics. Finally, when there are multiple types of services coexisting in the network, it is critical that the network can provide fairness among those services in addition to satisfying their specific QoS requirements. Thus, the network needs to fairly allocate network resources among different users such that differentiated QoS requirements can be satisfied for each type of service independent of the others. To address these concerns, we focus on handoff prioritized call admission control schemes in this chapter. Accordingly, we are interested in two connection-level QoS metrics, namely, the new call blocking probability (CBP) and the handoff call dropping probability (CDP). In addition, the system utilization is also considered.

The quest for efficiently allocating network bandwidth among different traffic classes while giving handoff calls higher priority than new call requests over wireless networks has been courting extensive efforts. Consequently, a magnitude of CAC schemes have been proposed. Among various approaches, there are two major categories: static approaches and dynamic approaches. In light of this classification, this chapter is organized as follows. We introduce some major challenges in the next section. In Section 19.3, we present static CAC approaches, whereas dynamic approaches are given in Section 19.4. Finally, concluding remarks are given in Section 19.5.

19.2   Major Challenges

As the wireless networks evolve from 2G FDMA/TDMA technology to 3G CDMA technology [13] and the carried traffic changes from voice service to multimedia service, such as voice, video, and data, we need to deal with several challenges when designing appropriate CAC schemes for optimal resource allocation. In the following, we pinpoint some major challenges.

19.2.1   Increased Handoff Frequency

As the number of subscribers is increasingly growing and the bandwidth required by some applications such as data or video is much larger than that for traditional voice service, the number of channels assigned to a cell becomes insufficient. Thus, the overall system capacity needs to be increased. Given that the frequency spectrum for the system is largely unchanged, the cell size has to be reduced so that more efficient frequency reuse can be achieved, which results in the evolution from macro-cell to micro-cell or pico-cell. While helping increase system capacity, this reduction in cell size increases the frequency of handoffs as users engaged in a call are more likely to move across the cell boundary during the call's lifetime. Since dropping handoff calls is more objectionable than blocking new call requests, it is more challenging to guarantee a desirable handoff dropping probability with the presence of increased handoff frequency.

It is worth noting that unlike hard handoff in FDMA/TDMA cellular networks, which forces a mobile user to break the connection to the current base station before building a new connection to a new base station, soft handoff in CDMA networks allows a mobile user to simultaneously connect to several base stations in neighboring cells, since all the cells use the same frequency. When a user moves from one cell to another, he can simply drop the connection whose communication quality degrades to a certain level while maintaining the connection with good quality.

19.2.2   Multiple Service Types

A variety of applications, such as voice, video, and data, are expected to be supported with QoS guarantee in the next generation wireless networks. Due to their different traffic characteristics, they may have different QoS requirements in terms of delay, delay jitter, bit error rate (BER).

In Universal Mobile Telecommunications Service (UMTS) [1], four traffic classes are supported:

- *Conversational*: This class of traffic has stringent requirements on delay and delay jitter, although they are not very sensitive to BER. Typical applications include voice and video telephony.
- *Streaming*: Real-time streaming video belongs to this class. Usually, it is less sensitive to delay or delay jitter than the conversational class.

- *Interactive*: Applications belonging to this class include web browsing and database retrieval. Typically, the response time they require should be within a certain range and the BER should be very low since the payload content should be preserved.
- *Background*: It may include data services like email or file transfer. For these services, the destination can tolerate delays ranging from seconds to minutes. However, data to be transferred has to be received error-free.

19.2.3 Traffic Asymmetry

As the traffic types evolve from voice-centric to multimedia, traffic asymmetry between downlink and uplink has become a distinguished feature of next generation wireless networks [18, 20, 33]. Among all the four classes mentioned earlier, whereas conversational traffic is symmetric or nearly symmetric, streaming video or web browsing are very asymmetric, usually requiring larger bandwidth for downlink than for uplink. To efficiently handle traffic asymmetry, not only the bandwidth in a system should be asymmetrically allocated accordingly, but also the call admission schemes should explicitly take this into account, especially when current schemes fail to do so.

19.3 Static Approaches

In this section, we present some representative static CAC schemes. Before we proceed, we introduce the channel assignment strategies used in current cellular networks. Typically each cell in a network is assigned an amount of resources (wireless channels or bandwidth). Depending on whether the amount of the assigned resources is fixed, a channel assignment scheme can be classified as either fixed channel allocation (FCA) or dynamic channel allocation (DCA) [19]. In FCA, the number of channels is fixed so a new or handoff call will be blocked if there is no free channel available in the cell in which the call is originated, whereas in DCA, the number of channels assigned to one cell is not fixed and varies depending on co-channel interference limitations. More discussions about both FCA and DCA, including comparisons, can be found in Reference 19 and the references therein. For simplicity reason, all the schemes introduced below assume FCA is adopted.

19.3.1 Guard Channel

In this scheme [14], a number of channels called *guard channels* are reserved for handoff calls. Let C denote the total number of channels in a cell. Among C channels, C_h channels are guard channels, meaning that only handoff calls can access them. The rest $C - C_h$ channels are shared between new calls and handoff calls. Thus, the admission control is the following. When there are more than C_h idle channels, both new call requests and handoff call requests can be accepted. If the number of free channels is less than or equal to C_h, only handoff call requests can be accepted and all new calls are blocked. For simplicity of analysis, it is assumed that the arrival process for both new calls and handoff calls are Poissonian with the arrival rate λ and λ_h, respectively. The channel holding time for new calls and handoff calls follow the same exponential distribution with mean $1/\mu$. Therefore, we can model the cell occupancy as a continuous time Markov chain with C state, where the system state is the number of busy channels. The state-transition diagram is shown in Figure 19.2.

According to Figure 19.2, the state equations are:

$$P_j = \begin{cases} \dfrac{\lambda + \lambda_h}{j\mu} P_{j-1}, & \text{for } j = 1, 2, \ldots, C - C_h \\[3mm] \dfrac{\lambda_h}{j\mu} P_{j-1}, & \text{for } j = C - C_h + 1, \ldots, C \end{cases} \tag{19.1}$$

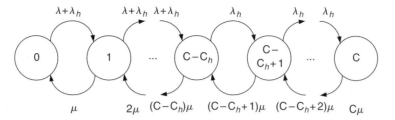

FIGURE 19.2   State transition diagram for GC [14].

Meanwhile, the normalization equation is

$$\sum_{j=0}^{\infty} P_j = 1 \tag{19.2}$$

By solving these equations, we obtain the probability distribution as follows:

$$P_0 = \left[\sum_{k=0}^{C-C_h} \frac{(\lambda + \lambda_h)^k}{k!\mu^k} + \sum_{k=C-C_h+1}^{C} \frac{(\lambda + \lambda_h)^{C-C_h}\lambda_h^{k-(C-C_h)}}{k!\mu^k} \right]^{-1} \tag{19.3}$$

$$P_j = \begin{cases} \dfrac{(\lambda + \lambda_h)^j}{j!\mu^j} P_0, & \text{for } j = 1, 2, \ldots, C - C_h \\[3mm] \dfrac{(\lambda + \lambda_h)^{C-C_h}\lambda_h^{j-(C-C_h)}}{j!\mu^j} P_0, & \text{for } j = C - C_h + 1, \ldots, C \end{cases} \tag{19.4}$$

As we know, a new call request will be rejected if the number of busy channels is equal to or greater than $C - C_h$. Thus, the new call blocking probability, denoted by P_{nb}, is the sum of the probabilities of all the states that are equal to or larger than $C - C_h$.

$$P_{nb} = \sum_{C-C_h}^{C} P_j \tag{19.5}$$

Since handoff requests cannot be accepted into the network if and only if all the channels are busy, we obtain the handoff call dropping probability P_{hd}:

$$P_{hd} = P_C \tag{19.6}$$

19.3.2 Fractional Guard Channel

Based on the guard channel described earlier, Ramjee et al. [30] proposed a fractional guard channel policy (FGC). The basic idea is the following. Unlike the guard channel scheme, in which a number of channels are exclusively reserved for use of handoff calls, FGC reserves a nonintegral number of guard channels for handoff calls by accepting new calls with some probability that depends on the current network status, specifically, the current channel occupancy. With similar assumptions made for guard channel, the state transition diagram is obtained and shown in Figure 19.3, where $\beta_i, 1 \le i \le C$, is defined as the new call acceptance probability when there are $i - 1$ busy channels in the cell.

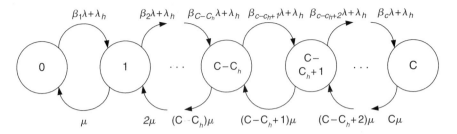

FIGURE 19.3   State transition diagram for FGC [30].

Let $\alpha = \lambda_h/(\lambda + \lambda_h)$ and $\rho = (\lambda + \lambda_h)/\mu$. Also let $\gamma_i = \alpha + (1 - \alpha)\beta_i, 1 \le i \le C$. Then, the stationary probability distribution can be calculated as

$$P_0 = \frac{1}{\sum\limits_{j=0}^{C} (\rho^j \prod\limits_{i=1}^{j} (\gamma_i/j!))} \tag{19.7}$$

$$P_j = \frac{\rho^j \prod\limits_{i=1}^{j} \gamma_i}{j!} P_0, \qquad \text{for } 1 \le j \le C \tag{19.8}$$

Given the state probabilities and the new call acceptance probability β_i, the new call blocking probability is

$$P_{nb} = \sum_{j=0}^{C} (1 - \beta_{j+1})P_j, \qquad \text{where } \beta_{C+1} = 0 \tag{19.9}$$

and the handoff dropping probability is

$$P_{hd} = P_C \tag{19.10}$$

Furthermore, FGC is shown to be optimal for two optimization problems, that is, the problem of minimizing the new call blocking probability with a hard constraint on the handoff call dropping probability and the problem of minimizing the number of channels with hard constraints on both the new call blocking probability and the handoff call dropping probability. The authors also proved that the guard channel is optimal for the problem of minimizing a linear objective function of the new call blocking probability and the handoff call dropping probability

19.3.3 Queueing Schemes

In the original guard channel scheme, the handoff call dropping probability is improved. However, this is achieved at the price of increased new call blocking probability. As a result, the overall carried traffic of the entire network is reduced. To reduce the damage to the carried traffic, the blocked new calls can be queued. On the other hand, handoff calls can also be queued. In References 14, 34, 3, and 10, some queueing schemes are studied.

19.3.3.1 Queueing Handoff Calls

The scheme in Reference 14 is the same as the original guard channel scheme, except that a handoff request is queued if it finds out that there is no idle channel. However, no new calls are queued. The calls in the queue are served in a first-come-first-served (FCFS) manner. The time for a handoff call to stay in queue equals the time for which the handoff call is in the handoff area. The handoff area is defined as the area in which the received signal power level from a base station drops below the handoff threshold, but is still

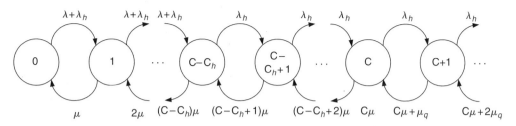

FIGURE 19.4 State transition diagram for GC with queueing [14].

higher than the minimum receiver threshold. Normally, it varies with the cell size, mobile user's traveling speed and direction, and the characteristics of geography. To make the analysis tractable, we assume the dwell time of a handoff call in the queue is exponentially distributed with mean $1/\mu_q$. The state transition diagram is shown in Figure 19.4.

Similarly, by solving the state equations and the normalization equation, the probability distribution P_j is the following.

$$
P_0 = \left[\sum_{k=0}^{C-C_h} \frac{(\lambda + \lambda_h)^k}{k! \mu^k} + \sum_{k=C-C_h+1}^{C} \frac{(\lambda + \lambda_h)^{C-C_h} \lambda_h^{k-(C-C_h)}}{k! \mu^k} \right.
$$
$$
\left. + \sum_{k=C+1}^{\infty} \frac{(\lambda + \lambda_h)^{C-C_h} \lambda_h^{k-(C-C_h)}}{C! \mu_q^C \prod_{i=1}^{k-C} (C_u + i\mu_q)} \right]^{-1}
\tag{19.11}
$$

$$
P_j = \begin{cases}
\dfrac{(\lambda + \lambda_h)^j}{j! \mu^j} P_0, & \text{for } 1 \le j \le C - C_h \\[3mm]
\dfrac{(\lambda + \lambda_h)^{C-C_h} \lambda_h^{j-(C-C_h)}}{j! \mu^j} P_0, & \text{for } C - C_h + 1 \le j \le C \\[3mm]
\dfrac{(\lambda + \lambda_h)^{C-C_h} \lambda_h^{j-(C-C_h)}}{C! \mu_q^C \prod_{i=1}^{j-C} (C_u + i\mu_q)} P_0, & \text{for } j \ge C + 1
\end{cases}
\tag{19.12}
$$

Thus, the new call blocking probability is the sum of the probabilities of states that are equal to or larger than $C - C_h$.

$$
P_{nb} = \sum_{C-C_h}^{C} P_j
\tag{19.13}
$$

For handoff calls, things are somewhat complicated. The probability that a handoff call will be dropped is the probability that the handoff call will leave the handoff area before it comes to the top of the queue and gets a channel. So the handoff dropping probability is expressed as follows:

$$
P_{hd} = \sum_{k=0}^{\infty} P_{C+k} P_{hd|k}
\tag{19.14}
$$

where $P_{hd|k}$ represents the conditional probability that a handoff call will be dropped given that it enters the queue in position $k + 1$. Through mathematic analysis, we obtain that

$$
P_{hd|k} = \frac{(k+1)\mu_q}{C\mu + (k+1)\mu_q}
\tag{19.15}
$$

Rather than using the FCFS queueing discipline for the queued handoff calls, Tekinay and Jabbari [34] adopted a non-preemptive dynamic priority discipline for the handoff queue in their measurement based prioritization scheme (MBPS). At first, the latest handoff request, if no free channel is available, is put at the end of the queue. The power level of all the handoff calls in the queue is monitored continuously, and the priority of the handoff calls are dynamically updated depending on their measured power levels. A queued handoff call will be assigned a higher priority if its measured power level is decreasing from the handoff threshold to the minimum receiver threshold. Whenever there is a free channel, the handoff call with the highest priority gets served. As a result, this scheme can further reduce the dropped handoff calls.

19.3.3.2 Queueing New Calls

Guerin [10] studied a cellular system in which guard channel policy is used. As before, there are two kinds of incoming traffic, that is, new calls and handoff calls. The call admission criteria are the same as original guard channel scheme except that the new calls will be queued if they cannot be accepted into the system when arriving. In other words, the handoff calls can access all the available channels with no restriction, whereas new calls are not rejected even when there are less than C_h available channels. Through complicated mathematical analysis, the author showed that the total carried traffic can be substantially increased and the dropping probability of handoff calls are only slightly increased by queueing of new calls.

19.3.3.3 Queueing Both New and Handoff Calls

In Reference 3, Chang et al. investigated the performance of a system that adopts the guard channel scheme and allows both new and handoff calls to be queued in buffers with finite size. Further, a queued new call may leave the system due to caller impatience and a queued handoff call may be dropped if the user move out of the handoff area before a successful completion. It was concluded that for both new and handoff calls, we can find appropriate queue sizes in such a system; moreover, there exists an optimal guard channel threshold for minimizing the overall blocking probability.

19.3.4 Admission Control for Multiple Traffic Classes

Epstein and Schwartz [7] studied admission control of multimedia traffic in wireless networks where FCA is used. Two kinds of traffic are considered, namely narrowband (NB) and wideband (WB) traffic. To guarantee the QoS requirements in terms of the new call blocking probability and the handoff dropping probability, three kinds of admission control policies, that is, complete sharing (CS), complete partitioning (CP), and hybrid policies that lie between CS and CP, are evaluated. The study revealed several conclusions.

- Although CS can achieve high channel utilization, it cannot provide high priority to handoff traffic. In addition, it favors NB users over WB users because WB traffic needs more bandwidth to be admitted.
- Since CP divides the total bandwidth in a cell into subgroups, which correspond to different traffic classes, different traffic classes can be admitted with different priority levels, and as a result, their specific QoS requirements can be satisfied. However, this is achieved at the expense of reduced channel utilization.
- With appropriate reservation algorithms, hybrid policies can adapt to desired QoS, and hence achieve the minimized cost function that is defined based on the blocking or dropping probability and can be considered as the overall performance.

Note that besides [7], there are several works focusing on the performance analysis of call admission schemes for integrated voice and data [36, 22, 12] or multiple-class cellular networks [4].

19.3.5 Remarks on Static CAC Approaches

As can be seen earlier, the key idea behind the static schemes is that handoff calls are granted higher priority by exclusively reserving some channels for them, although queueing handoff calls that cannot be admitted immediately also helps achieve this prioritization.

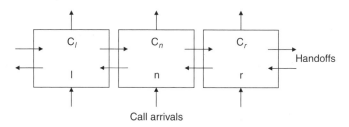

FIGURE 19.5 One-dimensional cellular system [26].

These schemes are static in the sense that the admission control parameters, such as the size of reserved channels or bandwidth, or the length of the queue, are determined according to *a priori* knowledge of the traffic pattern. And once determined, they cannot be changed during the course of system operation.

In conclusion, these schemes have advantages as well as drawbacks. On the positive side, they are simple and mathematically tractable. By deriving the important performance metrics like the new call blocking probability and the handoff dropping probability as shown earlier, they provide us with a benchmark for further improvements. Furthermore, they are effective. As long as the control parameters are set properly, the schemes can fulfill the QoS requirements. This is verified through performance evaluation in the original papers. In addition, Lin et al. [25] observed that giving higher priority to handoff calls would dramatically improve the forced termination probability of the system without seriously degrading the new call blocking probability. However, because of their static nature, they are unable to cope with dynamically changing network environments, where traffic patterns change dramatically and users exhibit high mobility. As a matter of fact, this is the case for a micro-cell system with densely populated mobile users. As a result, the static schemes become inefficient in system utilization and cannot provide QoS guarantee, thereby giving way to dynamic and adaptive approaches.

19.4 Dynamic Approaches

Dynamic call admission control approaches are the focus of this section. By *dynamic*, we mean these schemes are capable of reacting to network dynamics on a real-time basis.

19.4.1 Distributed Call Admission Control

In all the previous schemes, call admission control is made at each base station where only local state information is used. However, because of user mobility, the call occupancy and the traffic load between adjacent cells are correlated. Thus, it is necessary for each cell to exercise admission control in light of the state information in neighboring cells. Motivated by this observation, Naghshineh and Schwartz developed a distributed call admission control scheme (DCA) for wireless networks in Reference 26. At the beginning of each estimation period T, each cell exchanges state information with its neighbors. In this following, a one-dimensional cellular system shown in Figure 19.5 is used for demonstration.

We assume each cell has N channels. We denote C_n as the cell where a call admission request is made, and C_r and C_l as C_n's adjacent cells. We further denote by p_s the probability that a mobile user remains in the same cell, and p_m the probability that a mobile user moves out of its current cell during T.

If P_i is the probability that a cell has i ongoing calls, the overload probability of a cell is expressed as $P_0 = \sum_{N+1}^{\infty} P_i$. Since a handoff call will be dropped only if there is no free channel in the target cell, the handoff dropping probability can be approximately modeled as P_0. Thus, the admission control should make sure that P_0 is always less than or equal to P_{QoS}, a predefined QoS parameter. Accordingly, the admission decision for a newly admitted call at time t_0 must satisfy two requirements:

1. At time $t_0 + T$, the overload probability of cell C_n is smaller than P_{QoS}, given both the possible incoming handoff calls from C_r or C_l to C_n and the possible outgoing handoff calls from C_n to its neighboring cells.

2. At time $t_0 + T$, the overload probability of cell C_r or C_l is also smaller than P_{QoS}, given possible handoff calls and newly admitted calls.

It is assumed that at time t_0 the number of calls at cell C_n, C_r, C_l is n, r, l, respectively. Then the probability that there are i_n calls in the same cell C_n at time $t_0 + T$ is $B(n, i_n, p_s)$, where $B(i, k, p)$ is defined as Binomial distribution:

$$B(i, k, p) = \begin{pmatrix} i \\ k \end{pmatrix} p^k (1 - p)^{i-k} \tag{19.16}$$

In a similar way, the probability that i_l calls will be handed off from cell C_l to C_n is $B(n, i_l, p_m/2)$; the probability that i_r calls will be handed off from cell C_r to C_n is $B(n, i_r, p_m/2)$. Therefore, the probability distribution of the number of calls in C_n at time $t_0 + T$, denoted by $P_{n,t_0+T}(k)$, is the convolution sum of the three Binomial distributions. If we use a Gaussian distribution to approximate a Binomial distribution $B(i, k, p)$, the mean is ip and the variance is $ip(1 - p)$. Then, $P_{n,t_0+T}(k)$ can be written as

$$P_{n,\,t_0+T}(k) \approx G(np_s + (l + r)p_m/2, \sqrt{(np_s(1 - p_s) + (l + r)p_m(1 - p_m/2)/2)}) \tag{19.17}$$

Next, the overload probability can be easily obtained as

$$P_O = \sum_{N+1}^{l+n+r} P_{n,\,t_0+T} \approx Q \left(\frac{N - (np_s + (l + r)p_m/2)}{\sqrt{(np_s(1 - p_s) + (l + r)p_m(1 - p_m/2)/2)}} \right) \tag{19.18}$$

where $Q(.)$ is the tail of a Gaussian distribution with mean 0 and variance 1. Letting $P_O = P_{QoS}$, we can solve for n get n_1, the admission threshold that satisfies the first admission requirement. Using the similar method, we can solve the two equations for the second admission requirement and accordingly get two additional admission threshold, n_2 and n_3. Therefore, the final admission threshold is calculated by $n = \min(n_1, n_2, n_3)$.

Simulation showed that, compared with the guard channel scheme, DCA can achieve a lower handoff dropping probability without increasing the new call blocking probability. However, it has some limitations. First, the Gaussian approximation is not accurate, since each cell has a finite capacity. Second, it neglects the probability that in one estimation period, a call could hand off more than once. Finally, only single class traffic is considered, which discourages its use in networks supporting multimedia. It is interesting to note that this scheme was later extended to support multiple traffic classes in Reference 8.

19.4.2 Stable Dynamic Call Admission Control

To overcome the deficiencies of DCA, Wu et al. proposed a novel stable dynamic call admission control mechanism (SDCA) [35]. Similar to DCA, SDCA conducts admission control based on periodic control. Over each control interval, the scheme calculates the average handoff dropping probability. By letting this probability equal to the QoS parameter, namely the maximum allowable handoff dropping probability, and solving the equation, this scheme derives an *acceptance ratio a*, rather than deriving a hard admission threshold as in Reference 26. By accepting each new call in the coming control interval with probability a, this scheme spreads all new calls over the control interval, thereby avoiding suddenly overloading the network at the beginning of each interval during congestion, as is so often the case when a hard admission threshold is used, and leading to more effective and stable control.

Mainly, SDCA has several desirable features:

- The time-dependence of handoff dropping probability is considered as the average dropping probability is computed over a control interval. Thus, this increases the precision as compared to a single-value approximation.

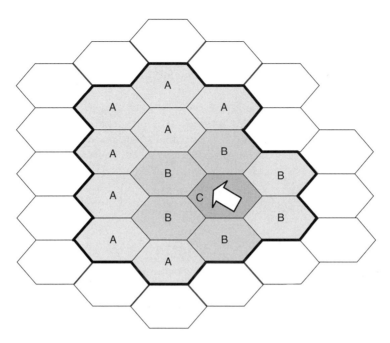

FIGURE 19.6 A shadow cluster formed for an active mobile terminal in cell C. A denotes the non-neighboring cells and B the neighboring cells [21].

- Since the estimation of the dropping probability is based on the solution to the evolution equation of the occupancy distribution, it takes into account the finite capacity of each cell. Obviously, this is better than the Gaussian approximation.
- The scheme also accounts for the possibility of multiple handoffs over a single control interval, as the estimation of the dropping probability considers the call transition probabilities between nearest, second nearest, and third nearest neighboring cells.
- To further reduce the signaling overhead incurred due to periodical state information exchanges among neighboring cells, a local estimation algorithm was also proposed. In this algorithm, all the control parameters are calculated locally using the exponential smoothing technique. More details can be found in Reference 23, 35.

Through extensive simulations, this scheme was shown to be able to achieve high channel utilization while keeping the handoff dropping probability below a predetermined bound.

19.4.3 Admission Control Using Shadow Cluster

In Reference 21, the *shadow cluster* concept is used to estimate future resource requirements and to perform admission control for wireless multimedia networks. As we know, to continuously provide services for mobile terminals that are engaged in an ongoing call, the system should have knowledge about the movement of mobile terminals and their resource requirements. Based on this observation, the concept of a shadow cluster, which is at the core of this scheme, is formed. Essentially, every mobile terminal with an active connection is exerting an influence on the cells (and their corresponding base stations) in the vicinity of it current location and along its travel direction. When the mobile terminal moves, this influence region will move as well, just as its shadow does. In this sense, a mobile terminal's shadow cluster is said to be formed by all the base stations currently being influenced by it, as shown in Figure 19.6, where the size of the shadow cluster and the level of influence (or the "darkness" of the shade) are determined by bandwidth requirements, current call holding time and priority, and the mobile terminal's trajectory and velocity.

To implement a shadow cluster, every base station, for every active mobile terminal currently under its control, must inform those base stations that belong to the mobile terminal's shadow cluster about the mobile terminal's future location probability, which is termed *active mobile probability*.

Let x be an active mobile terminal under the control of base station i. We also assume the time is quantized into time slots with equal length as $t = t_1, t_2, \ldots, t_m$. Then the active mobile probability of x, denoted by $P_{x,i,j}(t)$, is represented as $P_{x,i,j}(t) = [P_{x,i,j}(t_1), P_{x,i,j}(t_2), \ldots, P_{x,i,j}(t_m)]$, where x is currently in cell i and will be active in cell j at time $t_1, t_2, \ldots, t_m$. As time passes, cell i will recompute this probability for x and send it to all base stations within x's shadow cluster, denoted by $K(x)$. Note that the probability $P_{x,i,j}(t)$ can also be interpreted as the minimum percentage of the total amount of resources currently being used by mobile terminal x that base station i recommends to base station j to make available at times $t = t_1, t_2, \ldots, t_m$ if x moves to cell j.

With the active mobile probabilities, base station i can determine the number of bandwidth units (BU's) to be used at times $t = t_1, t_2, \ldots, t_m$ by

$$C_{u_i}(t) = C_{u_i}(t_0) - C_{u_i}^*(t) + C_{u_i}^\#(t) \tag{19.19}$$

where $C_{u_i}(t_0)$ is the initial number of used BU's, $C_{u_i}^*(t)$ is the estimate of the number of BU's that will be freed by the active terminals that either end their calls or move to other cells by time t, and $C_{u_i}^\#(t)$ is the estimate of the number of BU's that will be used due to handoffs from neighboring cells whose shadow cluster contains cell i.

Let $c(x)$ denote the number of BU's x is using and X_i denote the set of all active mobile terminal in cell i, then $C_{u_i}^*(t)$ can be obtained as

$$C_{u_i}^*(t) = \sum_{x \in X_i} [1 - P_{x,i,i}(t)] c(x) \tag{19.20}$$

and $C_{u_i}^\#(t)$ can be obtained as

$$C_{u_i}^\#(t) = \sum_{x \notin X_i, i \in K(x)} P_{x,j,i}(t) c(x) \tag{19.21}$$

Accordingly, the estimate of free BU's at base station i at time $t_1, t_2, \ldots, t_m$ is

$$C_{f_i}(t) = \begin{cases} C_i - C_{u_i}(t), & \text{if } C_i - C_{u_i}(t) \geq 0 \\ 0, & \text{otherwise} \end{cases} \tag{19.22}$$

where C_i is the total BU's in cell i.

Based on the calculated resource demands, the call admission algorithm is composed of the following steps in sequence.

1. Assume a call request is issued at time t_0. Base station i will check if $c(x) \leq C_{f_i}(t_0)$. If not, the request is simply rejected. If multiple call requests are issued at the same time and the free BU's can accommodate at least one of such requests, continue.

2. For each mobile terminal x that makes a call request, base station i generates a shadow cluster $K(x)$ and informs its neighbors about the *preliminary active mobile probabilities*.

3. Based on the preliminary active mobile probabilities from its own requests and from requests in neighboring cells, base station i computes *availability estimates* $\Delta_i(t)$ for future time $t = t_1, t_2, \ldots, t_m$ as the following:

$$\Delta_i(t) = \frac{C_{f_i}(t)}{\sum_{x \in X_i} P_{x,i,i}(t) c(x) + \sum_{x \notin X_i} P_{x,j,i}(t) c(x)} \tag{19.23}$$

Next, the availability estimates are distributed among base stations that sent data about their own current connection requests earlier.

4. Upon receiving the availability estimates from its neighbors, base station i computes survivability estimates $\Lambda_i(x, \hat{t}_x)$ for each mobile terminal x making requests in cell i as:

$$\Lambda_i(x, \hat{t}_x) = \frac{1}{\hat{t}_x} \sum_{t=t_1}^{\hat{t}_x} \left[\Delta_i(t) P_{x,i,i}(t) + \sum_{j \in K(x)} \Delta_j(t) P_{x,i,j}(t) \right] \tag{19.24}$$

where $\hat{t}_x$ is the known average call duration time for mobile terminal x.

5. An *acceptance value* is returned by a function $\Omega[\Lambda(x, \hat{t}_x), D_{dp}(x)]$, where $D_{dp}(x)$ is the maximum dropping probability allowed for the call originated from mobile terminal x. A positive acceptance value indicates that the request can be accepted.

6. Finally, all the call requests that have a positive acceptance value are accepted in a descending order. Note this may terminate when either the base station accepts all requests with a positive acceptance value or there are no more free BU's.

Several comments are in order. First, since the formation of the shadow cluster concept involves network traffic load, resource requirements of each active terminal, user mobility patter, and so forth, it demonstrates how such network dynamics can be utilized together to provide a better QoS. As a matter of fact, it opens an avenue for research along this direction [1, 15, 29]. Second, it requires frequent information exchanges among neighboring and non-neighboring cells, thereby incurring significant communication and processing overhead. Third, it can be easily seen that its performance depends largely on the accurate and detailed knowledge of user mobility pattern, which may not be available at the present time and hard to collect.

19.4.4 Dynamic Multiple-Threshold Bandwidth Reservation

We proposed a dynamic multiple-threshold bandwidth reservation (DMTBR) scheme for mobile multimedia networks [5]. In the system, two kinds of traffic are considered: real-time traffic and nonreal-time traffic. The objective of the proposed scheme is twofold. The scheme first provides QoS provisioning by keeping the CDP below the predefined bound even under network congestion situation, that is:

$$\begin{cases} P_{d,\text{rt}} \leq \text{QoS}_\text{rt} \\ P_{d,\text{nrt}} \leq \text{QoS}_\text{nrt} \end{cases} \tag{19.25}$$

where $P_{d,\text{rt}}$ ($P_{d,\text{nrt}}$) is the CDP for real-time (nonreal-time) traffic, and QoS_rt (QoS_nrt) is the predefined QoS requirement, that is, the maximum allowable handoff dropping probability, for real-time (nonreal-time) traffic. Second, in a fair manner, it maintains the relative priorities among real-time traffic and nonreal-time traffic in terms of the CBP according to their traffic profiles and instantaneous traffic situations:

$$\frac{P_{b,\text{rt}}}{P_{b,\text{nrt}}} = \frac{OL_\text{rt}}{OL_\text{nrt}} \times \frac{W_\text{rt}}{W_\text{nrt}} \tag{19.26}$$

where OL_rt (OL_nrt) is the offered traffic load, which can be calculated as the product of traffic arrival rate, the connection holding time, and the normal bandwidth requirement; and W_rt (W_nrt) is the traffic priority weight for real-time (nonreal-time) traffic, which can be set during the negotiation between the user and the network operator. Note this is a generalized concept of relative priority [8].

Three bandwidth reservation thresholds, G_1, G_2, and G_3 ($G_1 < G_2 < G_3$), are used. As shown in Figure 19.7, G_1 is the threshold for real-time handoff connections only, G_2 is the threshold for both real-time and nonreal-time handoff connections; G_3's role, however, is not fixed in that it could be the threshold for new real-time traffic against new nonreal-time traffic, or the threshold for new nonreal-time traffic

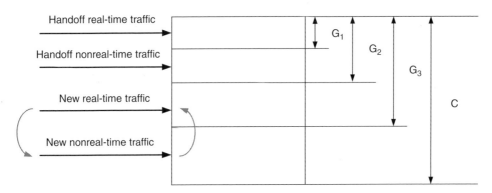

FIGURE 19.7   The bandwidth reservation thresholds in DMTBR.

against new real-time traffic, depending on the instantaneous relative priority status for the two traffic classes. The thresholds are calculated in a two-step manner: we first estimate the initial value of each threshold using a method similar to Reference 31, and then adapt it according to the traffic conditions and QoS status.

To satisfy the predefined QoS requirements in case of network overload due to traffic business, we also proposed a cooperative mechanism, in which each cell will inform all of its neighbors to throttle the admissions of new connections if it discovers that the QoS requirements for the handoff dropping probability is in danger of violation. Upon receiving such notifications, the neighboring cells admit new connection requests with a certain probability generated online, which is called the *probability of throttling new connections* (similar to that in References 30 and 9).

In summary, this scheme has the following features:

- It gives differential priorities to both new and handoff connections with different types of services by keeping multiple bandwidth reservation thresholds, based on which connection admission decisions are made.
- It maintains the relative priorities and fairness among traffic classes by taking user QoS profiles and real traffic conditions into account.
- It generalizes the concept of relative priorities and fairness among traffic classes. By doing so, we can further reduce the CBPs of some real-time services while not seriously deteriorating the QoS of nonreal-time services.

19.4.5 Call Admission for Asymmetric Traffic

As traffic becomes asymmetric between uplink and downlink, call admission should be tailored for this trend. Jeon and Jeong proposed a call admission scheme to handle traffic asymmetry in Reference 7.

To differentiate handoff calls and new calls, and to differentiate calls of different classes, the call admission scheme relies on the use of guard bandwidth, similar to guard channel. In this scheme, it is assumed that there are L classes of calls in the system. A class i ($0 \leq i \leq L - 1$) call has higher priority than a class j ($0 \leq j \leq L - 1$) call if $i < j$. A handoff call, regardless of its class, has higher priority over new calls of any class. Each cell has a total uplink and downlink bandwidth of W_u and W_d. For a class i call, the required uplink and downlink bandwidth are B_i^u and B_i^d, respectively.

For a class i handoff call during time period $(t_m, t_{m+1}]$, the uplink guard bandwidth $G_{h,i}^u(t_m)$ is

$$G_{h,i}^u(t_m) = \left(\sum_{k=0}^{i-1} \hat{\lambda}_k(t_m) T_k B_k^u \right) \Delta_i \tag{19.27}$$

where $\hat{\lambda}_k$ is the estimated handoff call arrival rate for class k, T_k is the mean connection holding time of class k calls in the cells. Since $\hat{\lambda}_k(t_m) T_k B_k^u$ represents the mean uplink bandwidth required by class k handoff calls, $\sum_{k=0}^{i-1} \hat{\lambda}_k(t_m) T_k B_k^u$ is the total uplink bandwidth required by handoff calls with higher priority than i. Thus, the guard bandwidth $G_{h,i}^u(t_m)$ is set to be proportional to the total required bandwidth with coefficient Δ_i ($0 < \Delta_i < 1$). Similarly, the downlink guard bandwidth $G_{h,i}^d(t_m)$ for a class i handoff call can be estimated as:

$$G_{h,i}^d(t_m) = \left(\sum_{k=0}^{i-1} \hat{\lambda}_k(t_m) T_k B_k^d \right) \Delta_i \tag{19.28}$$

The uplink guard bandwidth, denoted by $G_{n,i}^u(t_m)$, for a new class i call during time period $(t_m, t_{m+1}]$ can be set as follows:

$$G_{n,i}^u(t_m) = \left(\sum_{k=0}^{L-1} \hat{\lambda}_k(t_m) T_k B_k^u + \sum_{j=0}^{i-1} \hat{\Lambda}_j(t_m) T_j B_j^u \right) \Delta_i \tag{19.29}$$

where $\hat{\Lambda}_j$ is the estimated new call arrival rate for class j.

It can be seen that in Equation (19.29), since class i new calls has lower priority than all handoff calls and class j ($j < i$) new calls, the guard bandwidth includes the required bandwidth of class j new calls ($j < i$), which is represented by the $\sum_{j=0}^{i-1} \hat{\Lambda}_j(t_m) T_j B_j^u$, and the required bandwidth for handoff calls of all classes, as represented by $\sum_{k=0}^{L-1} \hat{\lambda}_k(t_m) T_k B_k^u$. Similarly, the downlink guard bandwidth $G_{n,i}^d(t_m)$ for a class i new call can be estimated as:

$$G_{n,i}^d(t_m) = \left(\sum_{k=0}^{L-1} \hat{\lambda}_k(t_m) T_k B_k^d + \sum_{j=0}^{i-1} \hat{\Lambda}_j(t_m) T_j B_j^d \right) \Delta_i \tag{19.30}$$

Note in the above, both the estimated handoff call arrival rate, $\hat{\lambda}_k$, and new call arrival rate, $\hat{\Lambda}_k$, can be calculated using the exponential moving average method [11].

Assuming that at the moment of a call arrival in period $(t_m, t_{m+1}]$, there are n_i number of class i ongoing calls in a cell, and the guard bandwidth for both handoff calls and new calls of class i are obtained, a class i handoff call will be accepted if and only if

$$\begin{cases} W_u - \displaystyle\sum_{k=0}^{L-1} n_k B_k^u \geq B_i^u + G_{h,i}^u(t_m) \\ W_d - \displaystyle\sum_{k=0}^{L-1} n_k B_k^d \geq B_i^d + G_{h,i}^d(t_m) \end{cases} \tag{19.31}$$

and a class i new call will be accepted if and only if

$$\begin{cases} W_u - \displaystyle\sum_{k=0}^{L-1} n_k B_k^u \geq B_i^u + G_{n,i}^u(t_m) \\ W_d - \displaystyle\sum_{k=0}^{L-1} n_k B_k^d \geq B_i^d + G_{n,i}^d(t_m) \end{cases} \tag{19.32}$$

Simulation results showed that, when this scheme is used for systems where arriving traffic is asymmetric between uplink and downlink, asymmetric bandwidth allocation outperforms symmetric bandwidth allocation in terms of the handoff dropping probability and the new call blocking probability, as well as system bandwidth utilization.

19.4.6 Local Predictive Resource Reservation

As can be seen, most of the dynamic call admission schemes [5, 8, 21, 31] entail periodical and timely real-time information exchanges among neighboring cells, which may be difficult and increase system complexity and costs due to the reduced cell size and increased user mobility. Thus, Zhang et al. [37] presented two new methods that only use local information to predict the resource demands of and determine resource reservation levels for future handoff calls. Another desirable feature of these methods is that they can deal with some general traffic cases that other methods cannot, which is attributable to the fact that the prediction is directly based on modeling the instantaneous resource demands, in comparison with previous prediction schemes that first model the factors that affect the resource demands and then derive the resource demands from the model of the factors.

The first method is developed on the basis of Wiener process, which is a Markov process whereby only the present value is needed for predicting the future. Specifically, if we assume the amount of resources required by handoff calls in a cell at time t is a stochastic process denoted by $R(t)$ and assume Δt is the prediction time interval, $R(t)$ can be modeled as:

$$\Delta R = R(t) - R(t - \Delta t) = \mu \Delta t + \alpha \delta \sqrt{\Delta t} \tag{19.33}$$

where α is a random variable following a normal distribution with mean 0 and standard deviation 1, μ and δ are constant parameters that are also referred to as the *expected drift rate* and the *standard deviation rate* of ΔR. It can be seen that ΔR is a normally distributed random variable with mean $\mu \Delta t$ and standard deviation $\delta \sqrt{\Delta t}$. For any given time interval τ, μ, and δ can be estimated as follows [2]:

$$\mu = \frac{\sum_{i=0}^{k-1} (r(t - i\tau) - r(t - i\tau - \tau))}{k\tau} = \frac{r(t) - r(t - k\tau)}{k\tau} \tag{19.34}$$

$$\delta = \frac{1}{\sqrt{\tau}} \sqrt{\frac{\sum_{i=0}^{k-1} (r(t - i\tau) - r(t - i\tau - \tau) - \mu\tau)^2}{k}} \tag{19.35}$$

where $r(t)$ is the sample value of $R(t)$ and k is the number of intervals used for estimation. Note that τ does not need to be the same as Δt.

Next, given the maximum allowable handoff dropping probability P_{hd}, we can set the reservation level L such that $\text{Prob}(\Delta R \le L) = 1 - P_{hd}$. Therefore, we statistically guarantee that the handoff dropping probability in the next time interval Δt is at most P_{hd}.

The second method is based on time series, or more precisely, based on the *auto regressive moving average* (ARMA) model [4]. An ARMA(p, q) process $\{X_t\}$ is defined as:

$$X_t - \alpha_1 X_{t-1} - \cdots - \alpha_p X_{t-p} = Z_t + \beta_1 Z_{t-1} + \cdots + \beta_q Z_{t-q} \tag{19.36}$$

where $\{Z_t\}$ are uncorrelated random variables with zero mean and variance σ^2. It is argued that in practice, ARMA($p, 0$) or namely AR(p), ($p = 1, 2,$ or 3) is sufficient to model ΔR_t, which is defined as $R(t) - R(t - 1)$. In fact, this method is generalization of the first method, since it takes into account the correlation between future handoff resource requirements and present and past handoff resource requirements. If no correlation is considered, it reduces to the first method.

It is verified through simulations that these two methods can accurately predict the future resource demands of handoff traffic and yield a comparable performance compared to other handoff prioritized schemes based either on local prediction method or on collaboration method that requires information exchanges among cells.

19.4.7 Remarks on Dynamic CAC Approaches

Before concluding this section, we make several important remarks about the dynamic CAC schemes. First, unlike their static counterparts described in previous section, dynamic schemes conduct adaptive admission control in a collaborative manner. Typically, base stations in different cells collect cell-specific status information and communicate it with others. Therefore, each base station has a more comprehensive view about changes in traffic pattern and user mobility, and can predict future resource demands of handoff requests more accurately. Consequently, it can take proper actions in admission control, thereby achieving better QoS provisioning and higher system utilization than it does in the static schemes. Second, information exchange enables distributed control, which helps handle traffic irregularity. However, these benefits are achieved at the expense of increased control overhead and system complexity. In the regard, dynamic schemes based on local prediction, such as [37], are highly needed. We believe this research direction is worth further exploration. Another downside is the dynamic schemes are not amenable to analysis. It is thus hard to precisely describe their behavior and quantify the performance gain in realistic scenarios.

19.5 Conclusions

Give the limited bandwidth and increasing number of users in wireless communication networks, resource allocation plays an increasingly important role in efficiently utilizing network resource while providing QoS. To better understand how and why current resource allocation and call admission control schemes work, in this chapter we survey some representative research works in this area. We first point out the major challenges facing the design of resource allocation and call admission control, such as high frequency of handoff, diverse QoS requirements, and unbalanced traffic. Then, we present several representative CAC approaches proposed for current and next generation mobile wireless networks, which largely can be classified into two categories, that is, static CAC schemes and dynamic CAC schemes. Compared with static schemes, dynamic schemes can adapt to changing traffic load thereby improving resource utilization and QoS provisioning. However, the latter is more complicated and incurs more overhead than the former in that a base station typically needs to collect and process more information about user mobility and traffic patterns in both the cell where it resides and the neighboring cells. How to take advantage of dynamic call admission control with low overhead is believed to be the focus of future research.

Acknowledgments

This work was supported in part by the U.S. Office of Naval Research under grant N000140210464 (Young Investigator Award) and under grant N000140210554, and by the U.S. National Science Foundation under grant ANI-0093241 (CA-REER Award) and under grant ANI-0220287.

References

[1] 3G Partnership Project. http:// www.3gpp.org/.

[2] 3G Partnership Project 2, http://www.3gpp2.org/.

[3] A. Aljadhai and T.F. Znati, Predictive mobility support for QoS provisioning in mobile wireless networks, *IEEE J. Select. Areas Commun.*, vol. 19, 2001.

[4] P.J. Brockwell and R.A. Davis, *Time Series: Theory and Methods*, 2nd ed. New York: Springer Verlag, 1991.

[5] C.-J. Chang, T.-T. Su, and Y.-Y. Chiang, Analysis of a cutoff priority cellular radio system with finite queueing and reneging/dropping, *IEEE/ACM Transactions on Networking*, vol. 2, pp. 166–175, 1994.

[6] C.C. Chao and W. Chen, Connection admission control for mobile multiple-class personal communications networks, *IEEE J. Select. Areas Commun.*, vol. 15, 1997.

[7] X. Chen, B. Li, and Y. Fang, A dynamic multiple-threshold bandwidth reservation (DMTBR) scheme for QoS provisioning in multimedia wireless networks, *IEEE Trans. Wireless Commun.*, to appear

[8] D.C. Cox, Wireless personal communications: what is it?, *IEEE Personal Communication*, 1995.

[9] B. Epstein and M. Schwartz, Reservation strateies for multi-media traffic in a wireless environment. In *Proceedings of the IEEE VTC*, 1995.

[10] B.M. Epstein and M. Schwartz, Predictive QoS-based admission control for multiclass traffic in cellular wireless networks, *IEEE J. Select. Areas Commun.*, vol. 18, pp. 523–534, 2000.

[11] Y. Fang and Y. Zhang, Call admission control schemes and performance analysis in wireless mobile networks, *IEEE Trans. Vehicular Technol.*, vol. 51, pp. 371–382, 2002.

[12] R. Guerin, Queueing-blocking system with two arrival streams and guard channels, *IEEE Trans. Commun.*, vol. 36, pp. 153–163, 1988.

[13] G.C. Goodwin and R.L. Payne, *Dynamic System Identification: Experiment Design and Data Analysis.* New York: Academic, 1977.

[14] Y.R. Haung, Y.B. Lin, and J.M. Ho, Performance analysis for voice/data integration on a finite-buffer mobile system, *IEEE Trans. Vehicular Technol.*, vol. 43, 2000.

[15] H. Holma and A. Toskala, *WCDMA for UMTS: Radio Access for Third Generation Mobile Communications.* 2nd ed, New York: Wiley, 2002.

[16] D. Hong and S.S. Rappaport, Traffic model and performance analysis for cellular mobile radio systems with prioritized and non-prioritized handoff procedures, *IEEE Trans. Vehicular Technol.*, vol. VT-35, pp. 77–92, 1986. See also: *CEAS Technical Report* No. 773, June 1, 1999, College of Engineering and Applied Science, State University of New York, Stony Brook, NY 11794, USA.

[17] J. Hou and Y. Fang, Mobility-based call admission control schemes for wireless mobile networks, *Wireless Commun. Mobile Comput.*, vol. 1, pp. 269–282, 2001.

[18] In-Stat/MDR, http://www.instat.com, 2003.

[19] W.S. Jeon and D.G. Jeong, Call admission control for mobile multimedia communications with traffic asymmetry between uplink and downlink, *IEEE Trans. Vehicular Technol.*, vol. 50, 2001.

[20] D.G. Jeong and W.S. Jeon, CDMA/TDD system for wireless multimedia services with traffic unbalance between uplink and downlink, *IEEE J. Select. Areas Commun.*, vol. 17, 1999.

[21] I. Katzela and M. Naghshineh, Channel assignment scheme for cellular mobile telecommunication systems: a comprehensive survey, *IEEE Personal Communication*, 1996.

[22] D. Kim and D.G. Jeong, Capacity unbalance between uplink and down-link in spectrally over-laid narrowband and wideband CDMA mobile systems, *IEEE Trans. Vehicular Technol.*, vol. 49, pp. 1086–1093, 2000.

[23] D. Levine, I. Akyildiz, and M. Naghshineh, A resource estimation and call admission algorithm for wireless multimedia networks using shadow cluster concept, *IEEE/ACM Trans. Networking*, vol. 5, pp. 1–12, 1997.

[24] B. Li, L. Li, B. Li, and X. Cao, On handoff performance for an integrated voice/data cellular system, *ACM Wireless Networks*, vol. 9, pp. 393–402, 2003.

[25] B. Li, L. Yin, K.Y. Wong, and S. Wu, An effective and adaptive bandwidth allocation scheme for mobile wireless networks using an on-line local estimation technique, *ACM Wireless Networks*, vol. 7, pp. 107–116, 2001.

[26] V.O.K. Li and X. Qiu, Personal communication systems (PCS), *Proc. of the IEEE*, vol. 83, 1995.

[27] Y.B. Lin, S. Mohan, and A. Noerpel, Queueing priority channel assignment strategies for PCS hand-off and initial access, *IEEE Trans. Vehicular Technol.*, vol. 43, 1994.

[28] M. Naghshineh and M. Schwartz, Distributed call admission control in mobile/wireless networks, *IEEE J. Select. Areas Commun.*, vol. 16, pp. 711–717, 1996.

[29] C. Oliver, J.B. Kim, and T. Suda, An adaptive bandwidth reservation scheme for high-speed multimedia wireless networks, *IEEE J. Select. Areas Commun.*, vol. 16, pp. 858–874, 1998.

[30] R. Ramjee, D. Towsley, and R. Nagarajan, On optimal call admission control in cellular networks, *ACM Wireless Networks*, vol. 3, pp. 29–41, 1997.

[31] P. Ramanathan, K.M. Sivalingam, P. Agrawal, and S. Kishore, Dynamic resource allocation schemes during handoff for mobile multimedia wireless networks, *IEEE J. Select. Areas Commun.*, vol. 17, pp. 1270–1283, 1999.

[32] T.S. Rappaport, *Wireless Communications*: *Principles and Practice*, 2nd ed, New York: Prentice-Hall, Inc. 2002.

[33] T.S. Rappaport, A. Annamalai, R.M. Buehrer, and W.H. Tranter, Wireless communications: past events and a future perspective, *IEEE Communications Magazine*, May 2002.

[34] S. Tekinay and B. Jabbari, A measurement based prioritization scheme for handovers in cellular and microcellular networks, *IEEE J. Select. Areas Commun.*, vol. 10, 1992.

[35] L. Yin, B. Li, Z. Zhang, and Y.-B. Lin, Performance analysis of a dual-threshold reservation (DTR) scheme for voice/data integrated mobile wireless networks, *Proc. IEEE WCNC*, 2002.

[36] T. Zhang, E. Berg, J. Chennikara, P. Agrawal, J.C. Chen, and T. Kodama, Local predictive resource reservation for handoff in multimedia wireless IP networks, *IEEE J. Select. Areas Commun.*, vol. 19, 2001.

20

Distributed Channel Allocation Protocols for Wireless and Mobile Networks

20.1 Introduction...**20**-460
20.2 A Mobile Cellular Communication Network..........**20**-460
    An Architecture for a Mobile Cellular Network • Channel Allocation
20.3 Classification of Channel Allocation Algorithms......**20**-462
    Fixed Channel Allocation Protocols • Dynamic Channel Allocation Protocols • Hybrid Channel Allocation Protocols
20.4 Biologically Inspired Channel Allocation Protocols...**20**-463
    Simulated Annealing-Based Dynamic Channel Allocation Protocols • Neural Network-Based Dynamic Channel Allocation Protocol • Genetic-Based Dynamic Channel Allocation Protocol
20.5 Algorithms Based on Borrowing for Dynamic Channel Allocation **20**-466
    Basic Borrowing Algorithm with Reassignment • Dynamic Channel Allocation Based on Borrowing from the Richest, First Available • DCA Algorithm Based on Dynamic Load Balancing Strategy
20.6 Algorithms Based on the Mutual Exclusion Paradigm for Dynamic Channel Allocation **20**-468
    Channel Allocation versus Mutual Exclusion • The Prakash–Shivaratri–Singhal Algorithm • The Choy and Singh Algorithm • The Cao and Singhal Algorithm • The Boukerche–Hong–Jacob Algorithm
20.7 Algorithms Based on Graph Coloring for Dynamic Channel Allocation **20**-470
    Deterministic Distributed Approach • Randomized Distributed Approach
20.8 Channel Allocation for QoS Enhancement **20**-471
    Cao's QoS-Enhanced Channel Allocation Scheme

20.9 Channel Allocation for Fault-Tolerant Enhancement .**20**-472
      Prakash, Shivaratri, and Singhal's Fault-Tolerant
      Enhanced Channel Allocation Scheme • Cao and Singhal's
      Fault-Tolerant Enhanced Channel Allocation Scheme

      Acknowledgment..**20**-473
      References **20**-473

Azzedine Boukerche

Tingxue Huang

Tom Jacob

20.1 Introduction

Technological advances coupled with the proliferation of mobile devices among wireless and mobile users require efficient resource management and reuse of the scarce radio spectrum allocated to wireless and mobile communication systems. Resource management in mobile radio communication networks consists of several components: waveform (channel) selection, transmitter power management, antenna pattern choice, access port (base station) selection, and the channel allocation criteria [40]. In earlier chapters, several important aspects of resource management for wireless and mobile communication systems have been discussed. In this chapter, we focus upon the channel allocation problem for wireless and mobile communication systems. We present an overview of different channel allocation algorithms and discuss the general idea behind each major channel allocation scheme. Then, we discuss the major channel allocation protocols in each category.

The remainder of this chapter is organized as follows: In Section 20.2, we review some of the related concepts of mobile radio cellular networks and discuss the main issues in channel allocation. In Section 20.3, we discuss our classification of well-known channel assignment algorithms and summarize their main characteristics. In Section 20.4, we present major biologically inspired dynamic channel allocation algorithms that have been developed for wireless and mobile networks, and in Section 20.5, we discuss well-known borrowing-based dynamic channel allocation algorithms. Section 20.6 examines major mutual exclusion-based dynamic channel allocation protocols, and in Section 20.7, we present two dynamic channel allocation schemes based on graph coloring, followed by a discussion of how these protocols could be extended to tolerate the failure of mobile as well as static nodes, and further enhance the quality of services by reusing channels. Last, but not the least, the conclusion follows in Section 20.8.

20.2 A Mobile Cellular Communication Network

20.2.1 An Architecture for a Mobile Cellular Network

A typical architecture for a mobile cellular network is shown in Figure 20.1. Generally speaking, the coverage area of such a network is partitioned into many adjacent regions called *cells*. In real life, the size and shape of the cells are not regular because of variable signal strength and the presence of natural obstacles. In theory, all cells can be simplified and represented as equal-sized hexagons. In each cell, there is one and only one *base station* (BS) that serves a number of mobile users, referred to as *mobile hosts* (MHs). Neighboring BSs are grouped into sets, each controlled by a *base station controller* (BSC) that executes some centralized functions. Several BSs and BSCs are served by a *mobile switching center* (MSC), which manages a large coverage area. While most research system models for mobile cellular networks assume wired links between BSs and their BSC, wired links between BS/BSC and MSC, and wireless links between MHs and BSs, only a few research system models assume that all of the links are wireless. That is to say, there are few models of a pure wireless network.

Some operations are enforced on a call made to or from a cell. If all channels along the path between the source MH and the destination MH are available, the call is *setup*. If no channel on some part of the communication path is available, the call is *blocked*. If the MH moves from one cell to a neighboring cell

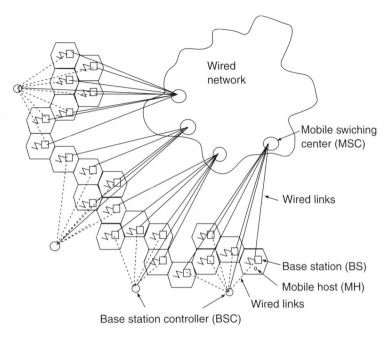

FIGURE 20.1   An architecture for a mobile radio network.

while the call is in progress, the MH must be connected to the destination BS through a *handoff*. During the handoff, the channel that is occupied by this MH in the source cell is released, and the neighboring cell has to reassign a channel for the call to continue. However, if the neighboring cell lacks an available channel during the *handoff* phase, the call will simply be *dropped*.

20.2.2 Channel Allocation

Communication channels are the most important resources in a mobile cellular network. A usable radio spectrum (bandwidth) is assigned to a given mobile cellular network. The spectrum is divided into a set of noninterfering communication channels using some basic division techniques such as frequency division (FD), time division (TD), or code division (CD). These transmission protocols are known as frequency, time, or code division multiple access (FDMA, TDMA, or CDMA, respectively) [10]. Nevertheless, regardless of the technique one chooses to use, scarce bandwidth places several constraints on the design of any wireless or mobile communication system. Indeed, in the same cell, a channel cannot be allocated simultaneously to two calls. This is referred to as the *co-channel* constraint (CCC). If the channels are not all orthogonal to each other, then two adjacent channels cannot be used simultaneously in the same cell. This is known as an *adjacent-channel* constraint (ACC). Since the coverage areas of two neighboring cells may overlap, the co-channel and adjacent channel restraints must be enforced throughout the overlapping cells. This is called the *co-cluster* constraint. Anytime these constraints are violated, channel interference happens, and the mobile communication system suffers from poor performance as a consequence.

In order to utilize communication channels efficiently, the same channel can be used in two cells that are sufficiently distant from each other. This minimum reuse distance is called the *co-channel reuse distance* (R_u), and the *co-channel reuse ratio* [38] is defined as

$$\frac{R_u}{R_c} = \sqrt{3N_c}$$

where R_u is the distance between the two closest co-channel cells, R_c the cell radius, and N_c the number of cells per cluster.

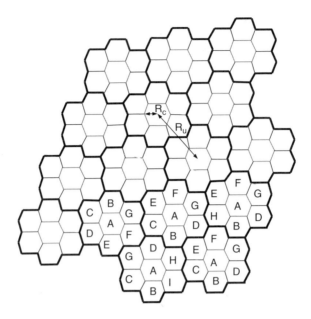

FIGURE 20.2   Co-channel reuse.

To reuse communication channels, several channel allocation protocols have been proposed [1,3–5,7,11–16,30–32,36,39]. Their basic idea is to group cells into clusters according to the co-channel reuse distance. In Figure 20.2, we illustrate a typical cluster of seven cells, a block of channels assigned to each cluster, with the cells in each cluster using different channels. A set of co-channel cells must lie in different clusters with a minimum separation equal to the reuse distance R_u.

20.3   Classification of Channel Allocation Algorithms

The channel allocation algorithms listed earlier can be divided into three different categories according to the method used to choose a channel for assignment: *Fixed Channel Allocation* (FCA), *Dynamic Channel Allocation* (DCA), and *Hybrid Channel Allocation* (HCA). We discuss each category and describe representative channel allocation protocols for that category.

20.3.1   Fixed Channel Allocation Protocols

Fixed Channel Allocation protocols have been extensively studied in the past [1,4,7,11,16,32]. In FCA, communication channels are statically allocated to individual cells with approximately an equal number of channels to each cell. However, because of the nonuniform distribution of calls, in which some cells have light traffic load (cold cells) whereas others have high traffic intensity (hot cells), this approach has proven to exhibit low performance and a high blocking rate. Assigning channels based on the communication requests in each cell might improve the performance of the mobile cellular network, but requires a predictable and relatively stable call distribution in each cell. For unpredictable call distributions, the FCA strategy and its variants have proven to be inefficient. Consequently, a variety of dynamic channel allocation algorithms have emerged.

20.3.2   Dynamic Channel Allocation Protocols

Dynamic Channel Allocation strategies were introduced to solve the problem of nonuniform distribution of communication requests across the different cells of the cellular network. Unlike the FCA paradigm,

DCA schemes assume that all of the channels can be used by any cell and a base station (BS). In other words, communication channels and cells have no fixed relationships.

The DCA schemes have been studied extensively in the recent years. Accordingly, several schemes have been proposed. One can classify them into call-by-call DCA schemes [31] and adaptive DCA schemes [3,31]. While the first scheme uses only current channel usage information, the second uses both present and previous channel usage.

The DCA protocols can also be classified based upon their implementations, that is, centralized versus distributed. In the centralized DCA schemes [12–15,30], all channels are kept in a central pool and a centralized controller is responsible for allocating these channels. When a communication request is received, a channel from the central pool is assigned to the request for temporary use. The key question of centralized DCA schemes is how to select one of the candidate channels for assignment. Several cost functions have been investigated in order to get a better channel selection. First available and locally optimized dynamic assignments are two of the many approaches.

On the other hand, with the distributed DCA schemes, the allocation algorithm is performed in each BS, using local information from the cells controlled by the BS. Several distributed DCA protocols have been proposed, differing in the way they use local information, that is, cell-based local information [21,22,30,31] and signal strength information [1,19,35]. When compared to the centralized DCA schemes, known for their bottleneck problem, distributed DCA schemes have the advantage of dealing with BS failure more efficiently.

20.3.3 Hybrid Channel Allocation Protocols

Hybrid Channel Allocation schemes were proposed as a result of simulations [14,24] and analysis [23], that indicated that the performance of FCA and DCA schemes change depending on the traffic load in the mobile cellular network. Under a heavy uniform traffic load, FCA schemes have shown better performance than DCA, whereas under low and moderate traffic intensity, DCA strategies have better performance than FCA, particularly with nonuniform traffic. HCA schemes were introduced in order to take advantages of both FCA and DCA techniques.

In HCA schemes [15,24,36,39], the channels of a mobile cellular system are partitioned into *fixed* and *dynamic* sets. The channels included in the *fixed* set are assigned to each cell using the FCA scheme, whereas the *dynamic* set of channels is shared by all BSs. The channel allocation procedure for the dynamic set may follow any of the DCA strategies.

20.4 Biologically Inspired Channel Allocation Protocols

For obvious reasons, there has been a growing and continued interest in developing efficient solutions and heuristics for the dynamic channel allocation problem. Many such solutions are based on biological (such as cellular automata and neural networks), evolutionary (such as genetic algorithms), and physical or natural phenomena (such as simulated annealing). Some of these solution techniques can be classified under a general paradigm, called *biocomputing*, It also turns out that almost all of these approaches are inherently distributed in nature. Consequently, many distributed algorithms have been proposed for real-world application problems, including DCA for mobile networks.

In this section, we present three well-known biologically inspired dynamic channel allocation protocols based on simulated annealing, neural networks, and genetic algorithms, respectively. In this section, we briefly review the algorithms.

20.4.1 Simulated Annealing-Based Dynamic Channel Allocation Protocols

Simulated annealing (SA), introduced by Kirkpatrick et al. [26], is a powerful method for optimizing functions defined over complex systems. It is based on ideas from statistical mechanics and motivated by an analogy to the behavior of physical systems in the presence of a heat bath.

While greedy algorithms, and other simple iterative improvement techniques accept a new configuration of lower cost and reject the costly states, SA escapes from local minima by sometimes accepting higher cost arrangements with a probability determined by the simulated "temperature." When compared to other approaches to solving combinatorial optimization problems, SA leads to efficient heuristic algorithms and can easily handle multiple, potentially conflicting goals in the same problem.

Simulated annealing has been successfully applied to resolve the channel allocation problem [17] in a way similar to solving combinatorial optimization problems. The DCA SA-based algorithm is first expressed as an optimization problem with the objective of avoiding interference to efficiently serve the expected mobile hosts' traffic by using configuration space S, cost function (C), and neighborhood structure (N) information. Then, an appropriate cooling schedule is selected and executed via the traditional annealing process to allocate m channels in a mobile radio network of n cells according to the following steps:

Step 1: Choose an initial configuration space S_{start} and an initial temperature t_0 by hand. Since the m radio channels are to be assigned to n radio cells, the initial configuration space is defined as a binary matrix (S_{ij}) of dimension $m \times n$ with the following elements:

$$S_{ij} = \left\{ {}^0_1 \text{ if channel } i, \text{where } i = 1, \dots, m, \text{is}{{\text{not used}}\atop{\text{used}}} \right\} \qquad \text{in radio cell } j \text{ where } j = 1, \dots, n.$$

Step 2: Construct a neighborhood structure $N(s)$ of the current configuration space S by performing the following two transitions. First, one channel i is used or dropped by one cell j. Second, one unused channel i_1 replaces one used channel i_2 in one cell j.

Step 3: Propose a new configuration space S' from the neighborhood structure $N(s)$.

Step 4: Compute the cost function

$$C(S) = C_{interference} + C_{traffic}$$

where the $C_{interference}$ represents a penalty associated with channel interference and $C_{traffic}$ represents the penalty associated with traffic violations.

Step 5: Compute the acceptance probability

$$\min \{1, \exp(-(C(S') - C(S))/t_k\}.$$

If the acceptance probability is between 0.7 and 0.9, then assign S the new value S'.

Step 6: If the equilibrium has not been reached **go to Step 2**.

Step 7: If a substantial improvement in cost cannot be expected, then the *final temperature* has been reached. **Stop.**

Step 8: Update the temperature level t_k.

$$t_k = t_{k-1} \cdot \exp(-(\lambda t)/\sigma). \qquad \textbf{Go to Step 2.}$$

20.4.2 Neural Network-Based Dynamic Channel Allocation Protocol

To avoid channel interference and meet channel demand, neural networks have been used in recent years. In this chapter, we present a neural network solution proposed by Kunz [27].

The DCA algorithm based on neural networks uses one neuron u_{ij} for each channel i at each BS j. To allocate the m channels, the neural network-based DCA protocol consists of the following basic steps [27]:

Step 1: Set an arbitrary initial value for the neuron matrix ($\{u_{ij}\}$).

Step 2: Compute the interference matrix ($\{\text{inter}f(j, j')\}$), where $\text{inter}f(j, j') = 1$, if the BSs j, j' interfere; otherwise $\text{inter} f(j, j') = 0$.

Step 3: Set weights $T_{ij,i'j'}$ and external input I_{ij} according to interference and channel demand.

$$T_{ij,i'j'} = -A\delta_{ii'}\mathrm{interf}(j,j') - B\delta_{jj'}(\delta_{ii'-1} + \delta_{ii'+1}) - C\delta_{jj'} + D\delta_{ii'}\delta_{jj'}$$

$$I_{ij} = C\mathrm{traf}(j)$$

where A, B, C, and D are nonnegative parameter values, $\delta_{ii} = 1$ and $\delta_{ij} = 0$.

Step 4: Execute the iterative operation of the corresponding energy function E until it reaches a constant value. $V_{ij} := (1 + \tanh(\lambda u_{ij}))/2$, where λ is a parameter.

$$E = 0.5A \sum_{j=1}^{n} \sum_{\substack{j'=1 \\ \mathrm{interf}(j,j')=1}}^{m} \sum_{i=1}^{n} V_{ij}V_{ij'} + B\sum_{j=1}^{n}\sum_{i=1}^{m-1} V_{ij}V_{i+1j} + 0.5C\sum_{j=1}^{n}\left(\sum_{i=1}^{m} V_{ij} - \mathrm{traf}(j)\right)^2$$

$$- D\sum_{j=1}^{n}\sum_{i=1}^{m} V_{ij}^2$$

Step 5: Check whether the final output V_{ij} of each neuron u_{ij} is at its maximum or minimum. If not, **go to Step 3**. If yes, **stop**.

Finally, if the output V_{ij} of one neuron takes the maximum value, the channel i is permitted for use at BS j. On the contrary, if the output V_{ij} takes the minimum value, the channel i is prohibited for use by BS j.

20.4.3 Genetic-Based Dynamic Channel Allocation Protocol

Genetic algorithms simulate genetic phenomena. The DCA algorithm based on the genetic paradigm borrows important genetic processes such as reproduction, crossover, and mutation from biology. The offspring inherits good characteristics from its parents, making its fitness to the new environment be greater than that of either parent.

Kim et al. [25] has proposed an efficient DCA scheme based upon the genetic paradigm while considering the following three constraints: the CCC, the ACC, and the co-site constraint (CSC). The proposed scheme assigns m_i channels to the ith cell by taking the following steps.

Step 1: Initialize the *population* by setting its size. For each *string* S_r, the channel assignment is first performed in the cell with the largest number of calls, next in the one with the next largest number of calls, and so on, until all cells have been assigned frequencies. Any two frequencies assigned to a cell must have the interval γ, which is the minimal frequency interval satisfying the three compatibility constraints.

Step 2: Calculate the energy function E_{S_r} and the fitness function F_{S_r} of each *string* S_r.

$$E_{S_r} = E_{CSCS_r} + E_{ACS_r} = (\text{energy for CSC}) + (\text{energy for ACC and CCC})$$

$$F_{S_r} = \frac{1/E_{S_r}}{\sum_{n=1}^{P}(1/E_{S_r})}$$

Step 3: Generate the roulette wheel slot of each *string* S_r, whose size is proportional to the ratio of its fitness to the total amount of fitness in the population.

Step 4: Select P pairs of *strings* into a mating pool to reproduce the next generation of *population*. The *string* is selected if a random number between 0 and 1 generated for each selection is within the range of its roulette wheel slot.

Step 5: The *strings* in the mating pool are mated under a crossover operation at random. There are three crossover techniques: uniform crossover, one point crossover, and two point crossover.

Step 6: Find a new search space through mutation, avoiding getting trapped in a local minimum. The five mutation techniques can be considered.

Step 7: If the maximum number of iterations is reached or if $E_{S_r} = 0$, **stop**. Otherwise, **go to Step 2**.

The genetic algorithm cannot guarantee convergence, so the maximum number of iterations must be set. However, it is more attractive than simulated annealing and neural networks in that it can find an acceptably good globally optimal solution very quickly.

20.5 Algorithms Based on Borrowing for Dynamic Channel Allocation

In this section, we investigate three widely known algorithms based on the borrowing paradigm for DCA. These include the basic borrowing algorithm with reassignment, the borrow first available scheme, and a dynamic load balancing strategy.

20.5.1 Basic Borrowing Algorithm with Reassignment

The basic borrowing algorithm was proposed by Engel and Peritsky [18]. It is one of the earliest developed algorithms based on the borrowing mechanism. The channels assigned to a cell i are classified as *standard* and *nonstandard* channels, where the standard channels are only used by the cell i and the nonstandard channels are borrowable. Borrowed channels must be returned after the call drops.

The borrowing-based channel allocation algorithm maintains for each cell a series of lists about the usage of channels, such as available standard channels, available nonstandard channels, and used standard channels. When a call is initiated, a nonstandard channel, which has the minimum probability of blocking neighboring cells will be chosen if the standard channels are exhausted. When a call that occupies a standard channel is dropped, an appropriate used nonstandard channel that can reduce the call blocking probability will be freed by shifting its call to the standard channel. When a call is initiated, a reorganization of the channels is performed if all standard and nonstandard channels are unavailable.

20.5.2 Dynamic Channel Allocation Based on Borrowing from the Richest, First Available

Anderson [2] has proposed three borrowing algorithms for dynamic channel assignment. In these schemes, among the earliest DCA techniques, a set of channels are assigned to a cell as its *nominal* channels. Likewise, a set of cells is assigned to a channel as its *nominal* cells.

When setting up a call, the cell first looks to see if a nominal channel is available. If not, it will compute the number of channels available for borrowing from each adjacent cell. If no neighboring channel is available, this call is blocked. Otherwise, a channel is borrowed from the cell that has the most available for borrowing. When a borrowing cell finishes with a borrowed channel, that channel must be returned to its owner.

The second algorithm is much more complex. We define the *worst case* to be the nominal cell that has the fewest nominal channels available after the proposed borrowing. The idea is that the candidate borrowed channel has to be chosen in such a way that it maximizes the available nominal channels in the worst-case, nominal, interferable cell. If a cell has no nominal channel available for a call, it will check the adjacent cells one by one. For each nominal channel of a adjacent cell, it first checks whether the channel is available. Then, all the nominal cells of the available channel are examined for interferability. Within those interferable cells, the minimum number of nominal channels $min(i)$ is recorded (note: i denotes the potential borrowed channel of the adjacent cell). Finally, the potential borrowed channel with the maximum $min(i)$ is borrowed.

The third algorithm is relative simple. It simply chooses the first available channel instead of trying to optimize when borrowing. All channels are first divided into several channel sets. These channel sets are nominally allocated to cells at the regular reuse interval. Channel sets are searched in a prescribed sequence for setting up a call, with the search continuing until an available channel is found or the search is terminated. If no available channel is found, the call is blocked.

20.5.3 DCA Algorithm Based on Dynamic Load Balancing Strategy

The first algorithm presented by Das et al. [16] is based on a dynamic load balancing strategy. Its idea is to reassign channels according to dynamic load distribution. All cells are classified into two categories: hot cells and cold cells. A parameter *degree of coldness* d_c is defined as the ratio of the number of available channels to the number of initial channels assigned to the cell. A mobile radio network has a fixed threshold parameter h calculated once by a Markov model. If $d_c \leq h$, the particular cell is hot, otherwise it is cold. As the time passes, the degree of coldness and state of a cell will change. A hot cell can borrow channels from a cold cell or from another hot cell.

The mobile users in a cell are divided into three types: *new, departing,* or *other* according to holding time and distance from the BS. A user is *new* if it has been in the current cell for a short period. If a user is close to the boundary of a cell so that its received signal strength (RSS) is weakened, it is a *departing* user. A user who is neither "new" nor "departing" is classified as *other*. The state of the mobile users in a cell is an important factor in selective borrowing. To borrow channels, each hot cell has to compute the value *NumDpart*, which stores the number of users that are departing from this cell.

The channel borrowing algorithm defines a selection criteria function for borrower and lender pairs, given by

$$F(B, L) = f(d_c(L), D(B, L), H(B, L))$$

where B and L stand for a borrower and a lender, respectively. The parameter $d_c(L)$ represents the coldness of the lender. The value $D(B, L)$ is the cell distance between the borrower B and lender L. The value $H(B, L)$ denotes the number of hot co-channel cells of the cold lender cell L. These lender cells are also nonco-channel cells of the borrower cell B. The cold cell whose parameters maximize the value of $F(B, L)$ is selected as a lender.

The channel borrowing algorithm can run on a concentric architecture like MSC or on a distributed architecture. Regardless of the architecture, the algorithm runs once for each borrower cell by taking the following steps:

Step 1: Request the array *NumDepart* from the hot cells.

Step 2: Examine the neighboring cells of the borrower cell B. Choose the cold ones and the hot ones with nonzero *NumDepart* as probable lender cells.

Step 3: Order the probable lender cells L in decreasing order using the function $F(B, L)$.

Step 4: For each cell i in the listed order, borrow channels until either the function $F(B, L)$ is not maximum or the number of borrowed channels is equal to *NumDepart*. Lock each lent channel to avoid interference.

Step 5: If the required number of channels has not been borrowed and the list of ordered cells is not exhausted, **go to Step 4**. Otherwise, if the required channels have been borrowed, **terminate**; else continue **next step**.

Step 6: Calculate $F(B, L)$ for all cold cells L in the frequency reuse pattern of B, excluding those already considered above.

Step 7: Borrow a channel from the cell L with the maximum $F(B, L)$, lock this channel and recompute the function $F(B, L)$.

Step 8: **Repeat Step 7** until the required number of channels has been borrowed.

20.6 Algorithms Based on the Mutual Exclusion Paradigm for Dynamic Channel Allocation

In this section, we investigate four widely known algorithms for DCA. These are Prakash et al. [32], Choy and Singh [11], Cao and Singhal [7] and finally the Distributed Dynamic Resource/Channel Allocation Algorithm (DDRA) [4]. All are based on the mutual exclusion paradigm.

20.6.1 Channel Allocation versus Mutual Exclusion

Mutual exclusion has long been a research topic in operating systems and distributed computing. It is used to provide an efficient access control mechanism for processes to access shared resources without any conflict or deadlock. For example, operating systems use mutual exclusion to allow exclusive access of processes to a database. Mutual exclusion has also been used in distributed computing to control access to a shared data section.

The channel allocation problem can be treated as a mutual exclusion problem, where two adjacent cells must use the same communication channels exclusively. Bandwidth and communication channels are limited, so that when many communication sessions happen simultaneously, it is very possible that two neighboring cells choose a same channel from the limited channel set. Solving this conflict is similar to the mutual exclusion problem.

There are two basic approaches to using mutual exclusion to solve the co-channel interference problem. The first approach adopts the use of a *time-stamp*. When a BS requests a channel from its neighboring BSs, it sends a time-stamped REQUEST. If two requests need the same channel, the one with the earlier time-stamp acquires the channel. This method is applied by the Prakash–Shivaratri–Singhal and Cao–Singhal algorithms.

The second method adopts the use of another variable *competition*. When two sessions request the same channel, their *competition* values are compared. The greater one wins. The loser increments its *competition* by one in order to improve its competition ability in the next round. This solution is used by DDRA.

The third method sets a priority for each BS. If two BSs request the same channel, the one with the greater priority acquires it. This method is adopted by the Choy–Singh algorithm.

20.6.2 The Prakash–Shivaratri–Singhal Algorithm

The first DCA algorithm [32] we present is the distributed DCA algorithm of Prakash, Shivaratri, and Singhal. The authors use a system model that consists of a set of MBSs and MHs connected by a completely wireless network. The model also assumes that the MBS–MH short-hop communication channels and MBS–MBS backbone channels change with time. The models assumed by the other algorithms assume there are changes only in the MBS–MH short-hop communication channels. The communication channels in the system are divided into two disjoint sets: Spectrum$_b$ for backbone communication and Spectrum$_s$ for short-hop communication. The MBSs in the network use independent but similar algorithms for backbone and short-hop channel allocation.

A mobile base station MBS$_i$ in the network maintains three kinds of local information: the set of short-hop channels allocated to its cell, Allocate$_i$, the set of used channels, Busy$_i$, and the set of candidate channels for possible transfer, Transfer$_i$. Initially, the three sets are empty. MBS$_i$ chooses a channel from Spectrum$_s$ or Spectrum$_b$. As channels are allocated and exchanged, MBS$_i$ updates the three local sets, and exchanges information about these sets with neighboring MBSs.

MBS$_i$ allocates a channel according to the following steps:

Step 1: MBS$_i$ calculates the expression Allocate$_i$ − Busy^i − Transfer^i to check whether there is a unused local channel. If there is, it can assign that channel to the communication session.

Step 2: If there is no available local channel, *MBS$_i$* sends time-stamped REQUEST messages to the neighboring MBSs. After receiving REPLY messages from each neighbor, MBS$_i$ takes the union of its own Allocate, Busy, and Transfer set and its neighbors' Allocate, Busy, and Transfer sets and computes the interference set, Interfere$_i$. Then, MBS$_i$ calculates the free set Free$_i$ = Spectrum$_s$ − Interfere^i. If Free$_i$ is empty, the channel request is dropped. Otherwise, MBS$_i$ performs Step 3.

Step 3: *MBS$_i$* selects channel k from Free$_i$ to be transferred and sends TRANSFER(k) messages to all neighbors. If all neighboring cells reply with AGREED messages, MBS$_i$ assigns channel k to the communication session. Otherwise, if any of the neighbors replies with a REFUSE message, MBS$_i$ tries the next channel in Free$_i$. If all channels fail, the communication request is dropped.

This distributed DCA algorithm has two characteristics. First, the mutual exclusion mechanism is based on the communication requests' time-stamp. The earlier the request's time-stamp, the higher the priority. The second characteristic is that the newly acquired channel is kept with the same MBS, even after completion of the communication session, that is, it is stored in the Allocate set until it is transferred.

Based on the simulation results presented in Reference 32, the algorithm shows good performance under smooth change of uniform and nonuniform loads. However, with a heavy nonuniform load with sudden changes, the dropping rate and blocking rate of the algorithm are both high and the response time is longer.

20.6.3 The Choy and Singh Algorithm

Our second distributed DCA algorithm [11] was proposed by Choy and Singh. The algorithm uses the dining philosophers algorithm as its kernel, with a mutual exclusion mechanism that is based on each MBS's color value, representing its priority. The greater the color value, the higher the priority. Initially, each MBS is assigned a color, and all neighboring MBSs have different colors. When two neighboring MBSs request the same channel simultaneously, the one with the higher color value takes the channel.

In addition, the algorithm divides the set of n_c channels into n_g groups. Moreover, n_g instances of a dining philosophers algorithm are run independently. When an MBS receives a communication request, it executes the following steps to allocate a channel to the request:

Step 1: The MBS tries to satisfy a user request with a local unused channel. If it fails, it takes Step 2.
Step 2: The MBS will transfer channels through executing the dining philosophers algorithm. The set of channels in the group is examined for any possible transfers. Therefore, other pending requests can also be served simultaneously.

The authors studied the performance of the algorithm through a set of experiments based on uniform and nonuniform cases. These experiments compared two approaches of borrowing channels: an optimistic and a pessimistic approach. The results show the optimistic approach performs well under light loads but not under heavy loads, whereas the pessimistic approach showed better performance under heavy loads.

20.6.4 The Cao and Singhal Algorithm

The third algorithm we have selected is the adaptive distributed DCA algorithm proposed by Cao and Singhal [7]. Its basic idea and implementation procedures are similar to the Prakash algorithm, but there are some important differences. Here, we concentrate on this algorithm's unique characteristics.

Initially, a fixed number of primary channels are assigned to each cell. For cell C_i, its primary channel set is designated by P_i. Compared with the Prakash–Shivaratri–Singhal algorithm, which assumes each cell's original allocated channel set is empty, the Cao–Singhal algorithm performs better under light load traffic because each cell can satisfy its communication requests using local channels without consulting with its neighboring cells.

This algorithm depends on the principle that a cell borrows a channel from its "richest" interference neighbor. This principle prevents the case where a neighbor uses up its primary channels soon after lending channels. Therefore, some extra borrowing operations are avoided, a short response time is obtained, and message complexity is kept low.

This algorithm has a unique mechanism called intrahandoff. In contrast to interhandoff, where a MH is assigned a new channel when it moves from one cell to the other, intrahandoff enables an MH to release its current channel and be assigned a new channel within the same cell, based on interference information. Intrahandoff can achieve better channel reuse through minimizing co-channel interference. However, it increases the message overhead.

20.6.5 The Boukerche–Hong–Jacob Algorithm

The Distributed Dynamic Resource/Channel Allocation (DDRA) was proposed by Boukerche et al. [4]. Compared to the previous dynamic channel assignment algorithms, which make use of co-channel interference, the DDRA algorithm adopts the co-group interference approach. The set of all channels is divided into several groups, with the number of channel groups equal to the number of BSs in a cluster. Any two adjacent BSs always hold different groups, an idea obtained from the three-coloring theorem.

The DDRA algorithm uses a variable *competition* to solve the mutual Exclusion problem. The greater the value of *competition*, the higher the priority. If two *competition* numbers are the same, the winner is selected randomly. When a communication request is initiated in cell C_i, BS_i allocates a channel to this request by implementing the following steps:

Step 1: BS_i picks a group g_j that it has not yet visited and sends a REQUEST message to all neighboring BSs.

Step 2: BS_i receives REPLY messages from all neighbors. If any of the neighbors sends back a REJECT message, BS_i cannot use the group g_j. In that case, BS_i increments the value of *competition* by one in order to enhance its competition ability in the next round. The algorithm then goes back to Step 1. If no group can be used, then BS_i drops this communication request.

Step 3: If all neighbors reply with a AGREE message, BS_i searches for a free channel in g_j. If it finds one, BS_i changes the value of *competition* to zero and assigns the channel to the request. At the same time, a BLOCK message for this channel is sent all neighboring BSs. Otherwise, the algorithm goes back to Step 1.

The authors compare the performance of their algorithm with the optimistic and pessimistic algorithms through a set of simulations. Because DDRA makes use of co-group interference and does not borrow any free channel from a neighboring BS, the simulation results show that low message complexity and low channel acquisition time can be obtained.

20.7 Algorithms Based on Graph Coloring for Dynamic Channel Allocation

Some authors have modeled the channel allocation problem as a generalized list coloring problem. To avoid channel interference, the same channel cannot be used in two neighboring cells. For the list chromatic number of a graph coloring, two neighboring nodes have different colors, making it feasible to allocate channels using graph theory. There are two approaches presented by Garg et al. [20]: a deterministic distributed approach and a randomized distributed approach.

20.7.1 Deterministic Distributed Approach

This approach also makes use of the idea of mutual exclusion, but because it employs graph coloring, we classify it as a graph coloring algorithm. First, to serve as a priority scheme, an initial vertex coloring of the graph is assumed, where D colors are used. The smaller the color, the higher the priority. Some notation is introduced before we review how it works. $Busy_i$ and $Free_i$ denote the set of busy colors and free colors in node v_i, respectively (notice: a node of a graph means a cell or a BS and a color stands for a frequency). $Occupied_{ij}$ of node v_i is the size of $Busy_j$ for node v_j (v_is neighbors). l_{ij} represents the last synchronization message received from the neighbor v_j.

Step 1: Send a message $sync_1$ to each neighbor v_j when all $l_{ij} \neq sync_1$. This is the first doorway.

Step 2: Send a message $sync_2$ to each neighbor v_j when all $l_{ij} \neq sync_2$. This is the second doorway.

Step 3: Wait until node v_i has the highest priority.

Step 4: Choose the requested frequencies S from $Free_i$. Then send a confirmation message $con f(S)$ to each neighbor. Wait for all acknowledge messages from neighbors.

Step 5: Implement $Busy_i = Busy_i \bigcup S$ and $Free_i = Free_i - S$ when all neighbors acknowledge. Finally, send a message $sync_3$ to each neighbor.

20.7.2 Randomized Distributed Approach

This scheme assumes that the list coloring is much larger than the total number of colors needed by a node. When the node v_i requests several colors, it first randomly picks some colors from $Free_i$. It then checks with its neighbors. Two doorways are avoided by this approach. Colors are acquired as follows:

Step 1: Randomly pick colors S from $Free_i$. Calculate $Busy_i = Busy_i \bigcup S$ and $Free_i = Free_i - S$.

Step 2: Send a message $pick(S)$ to each neighboring cell and wait for all replies.

Step 3: For each color $f \in S$ do if some neighbor has replied "No" for f then calculate $Busy_i = Busy_i - f, Free_i = Free_i \bigcup f$ and $S = S - f$.

Step 4: Finally, send a message $con f(S)$ to each neighbor v_j.

20.8 Channel Allocation for QoS Enhancement

Each channel allocation scheme discussed earlier is designed to utilize bandwidth efficiently but does not take into account any Quality of Service (QoS) guarantees. In multimedia applications, different types of media have different QoS requirements. For example, such media as video and audio require a minimum transmission rate and maximum delay constraints (real-timely) but do not demand error-free transmission. However, such media as data, for example, transmitted software or an "email" message, requires error-free transmission but do not have real-time constraints. Because of this, some QoS-enhanced channel allocation schemes have been proposed.

20.8.1 Cao's QoS-Enhanced Channel Allocation Scheme

The inter-cell handoff is a frequent operation in a mobile cellular network. If a handoff fails, the ongoing call is dropped, so the handoff operation heavily affects the connection dropping rate (CDR), an important QoS metric. In general, to keep the connection dropping rate below its target value, each cell reserves a fixed number of channels or makes reservations based solely on prediction. These strategies may not work well because they waste a large amount of wireless bandwidth, while failing to reduce the CDR below the target value. Cao's scheme enhances the QoS by integrating distributed channel allocation with adaptive handoff management [6].

The DCA scheme was introduced in the previous section. Here, we review how to implement the QoS enhancement based on the distributed channel allocation scheme. In cell C_i, A_i and R_i stand for the number of available and reserved channels, respectively.

Step 1: When a new connection in cell C_i requires m channels, the connection is admitted if $A_i \geq m + R_i$. If $A_i < m + R_i$, attempt to borrow $R_i + m - A_i$ channels through the DCA. If borrowing fails, this connection is blocked.

Step 2: When a handoff requires m channels, the handoff succeeds if $A_i \geq m$; otherwise, it fails. When $A_i < R_i$, attempt to borrow $R_i - A_i$ channels through the DCA.

Step 3: In all other cases, cell C_i calculates its DCR, six neighboring cells' DCR and the average DCR. When the DCR of C_i is greater than the target value, increase the R_i and notify neighboring cells to stop borrowing channels from cell C_i. Concurrently, block β percent of the connections, which may go to cell C_j, whose DCR is high. Otherwise, when the DCR of C_i is less than a predefined percentage, then decrease the value of R_i and inform the neighboring cells to resume borrowing channels from cell C_i. Simultaneously, stop blocking the connections that were going to cell C_j.

20.9 Channel Allocation for Fault-Tolerant Enhancement

All of the DCA schemes we have discussed so far require the exchange of channel usage information between neighboring BSs. Some implementations cannot continue until a response message from the neighboring BS is obtained. Thus, if a BS fails or a link between neighboring BS fails, some BSs will fall into infinite waiting. Hence, it is necessary to enhance fault-tolerance for the DCA. We shall list two channel allocation schemes for enhanced fault-tolerance in the following section.

20.9.1 Prakash, Shivaratri, and Singhal's Fault-Tolerant Enhanced Channel Allocation Scheme

The DCA scheme of Prakash, Shivaratri, and Singhal was reviewed in the previous section. The channel allocation algorithm for fault-tolerant enhancement proposed by Prakash et al. [33] is based on that DCA scheme. Here, we review how this algorithm enhances fault-tolerance based on failures.

Case 1: Mobile Host fails. If, during an ongoing communication session, the MH fails, the BS (BS_i) removes the occupied channel from the set Busy$_i$ and adds it to the set Free$_i$, considering it simply to be the completion of a communication session.

Case 2: The link between neighboring BSs fails. The underlying network protocol can handle this failure. If the direct wired link between two neighboring BSs fail, an usable alternate wired replacement path will be found if one exists. Even if all the wired links fail, the 8 kb/s wireless control channel is still usable.

Case 3: BS_i fails. In this case, resolution is relatively complicated. Each BS sets a timer for each information exchange. If there is a time-out after sending a REPLY message, BS_j assumes it has received a REPLY message from BS_i such that Allocate$_i$, Busy$_i$ and Transfer$_i$ are empty sets. If there is a time-out after sending a TRANSFER(k) message, BS_j assumes that it receives an AGREED(k) message from BS_i. After knowing BS_i has failed, BS_j removes BS_i from its list of neighbors and does not send any more messages to BS_i. When the fault has been repaired, BS_i sends a notification message to all neighboring cells, which add BS_i to their list of neighbors.

20.9.2 Cao and Singhal's Fault-Tolerant Enhanced Channel Allocation Scheme

On the basis of Cao and Singhal's DCA scheme, they presented a channel allocation scheme for fault-tolerance enhancement [8]. This fault-tolerance enhanced scheme deals with the same three cases of failures as Prakash, Shivaratri, and Singhal: MH failure, BS failure, and link failure. Most resolutions are similar except for the failure of a BS.

For the failure of a BS, the difference lies on the different system model. The Prakash, Shivaratri, and Singhal system model assumes each cell has six neighboring cells and that a channel can be reused as long as two cells are not neighbors. However, Cao and Singhal's system model assumes the co-channel reuse distance is $3\sqrt{3}R$ (R is the radius of a cell), meaning that each cell has up to 30 neighboring cells. This scheme introduces the following property: if two cells C_i and C_j of a cluster request the same channel r, they have at least one common cell, which is an interference-nominated cell of channel r. According to this property, a borrower does not have to receive response messages from all interference neighboring cells to make a decision. The failure of a neighboring BS may not affect other BS's borrowing. Therefore, the channel allocation scheme is fault-tolerance enhanced.

Acknowledgment

This work was partially supported by NSERC and Canada Research Chair (CRC) Program, Canada Foundation for Innovation Funds, and OIT/Distinguished Researcher Award.

References

[1] Y. Akaiwa and H. Andoh. Channel segregation — a self organized dynamic allocation method: application to TDMA/FDMA microcellular system. *Journal on Selected Areas in Communication*, 11: 949–954, 1993.

[2] L.G. Anderson. A simulation study of some dynamic channel assignment algorithms in a high capacity mobile telecommunications system. *IEEE Trans. Commun.*, COM-21: 1294–1301, 1973.

[3] R. Beck and H. Panzer. Strategies for handover and dynamic channel allocation in micro-cellular mobile radio telephone systems. *IEEE VTC*, 1: 178–185, 1989.

[4] A. Boukerche, S. Hong, and T. Jacob. An efficient dynamic channel allocation algorithm for mobile telecommunication systems. In *Proceedings of the ACM/Kluwer Mobile Networks and Applications (MONET)*, 2000. Vol. 7, pp. 115–126, 2002.

[5] A. Boukerche, K. El-Khatib, and T. Huang. A performance comparison of dynamic channel and resource allocation protocols for mobile cellular Networks. In *Proceedings of the 2nd International ACM Workshop on Mobility Management and Wireless Access*, pp.

[6] G. Cao. Integrating distributed channel allocation and adaptive handoff management for QoS-sensitive cellular networks. *Wireless Networks*, 9: 131–142, 2003.

[7] G. Cao and M. Singhal. An adaptive distributed channel allocation strategy for mobile cellular networks. *J. Parallel Dist. Comput.*, 60: 451–473, 2000.

[8] G. Cao and M. Singhal. Distributed fault-tolerant channel allocation for cellular networks. *IEEE J. Selected Areas Commun.*, 18: 1326–13337, 2000.

[9] K.M. Chandy and J. Misra. The drinking philosophers problem. *ACM Trans. Programming Languages Syst.*, 6: 632–646, 1984.

[10] I. Chlamtac and J.Y. Lin. *Wireless Networks*. John Wiley & Son, New York, 2000.

[11] M. Choy and A. Singh. Efficient distributed algorithm for dynamic channel assignment. In *Proceedings of 7th IEEE International Symposium on Personal, Indoor and Mobile Radio Communication*, 1996.

[12] D.C. Cox and D.O. Reudink. Dynamic channel assignment in high capacity mobile communications systems. *Bell Syst. Tech. J.*, 50: 1833–1857, 1971.

[13] D. Cox and D.O. Reudink. A comparison of some channel assignment strategies in large mobile communication systems. *IEEE Trans. Commun.*, 20: 190–195, 1972.

[14] D. Cox and D.O. Reudink. Dynamic channel assignment in two dimension large-scale radio systems. *Bell Syst. Tech. J.*, 51: 1611–1628, 1972.

[15] D. Cox and D.O. Reudink. Increasing channel occupancy in large scale mobile radio systems: dynamic channel reassignment. *IEEE Trans. Vehicular Technol.*, VT-22: 218–222, 1973.

[16] S.K. Das, S.K. Sen, and R. Jayaram. A dynamic load balancing strategy for channel assignment using selective borrowing in cellular mobile environment. *ACM/Baltzer J. Wireless Networks*, 3: 333–347, 1997.

[17] M. Duque-Anton, D. Kunz, and B. Ruber. Channel assignment for cellular radio using simulated annealing. *IEEE Trans. Vehicular Technol.*, 42: 14–21, 1993.

[18] J.S. Engel and M.M. Peritsky. Statistically-optimum dynamic server assignment in systems with interfering servers. *IEEE Trans. Commun.*, COM-21: 1287–1293, 1973.

[19] Y. Furuya and Y. Akaiwa. Channel segregation: A distributed channel allocation scheme for mobile communication systems. *IEICE Trans.*, 74: 1531–1537, 1991.

[20] N. Garg, M. Papatriantafilou, and P. Tsigas. Distributed list coloring: how to dynamically allocate frequencies to mobile base stations. In *Proceedings of the 8th IEEE Symposium on Parallel and Dist. Processing (SPDP'96)*, October 1996.

[21] C.L. I and P.H. Chao. Local packing-distributed dynamic channel allocation at cellular base station. In *Proceedings of the IEEE GLOBECOM '93*, 1993.

[22] C.L. I and P.H. Chao. Distributed dynamic channel allocation algorithms with adjacent channel constraints. *PIMRC*, B2.3: 169–175, 1994.

[23] S. Jordan and A. Khan. *Optimal Dynamic Allocation in Cellular Systems.* 1993.

[24] T. J. Kahwa and N. Georganas. A hybrid channel assignment scheme in large scale cellular-structured mobile communication systems. *IEEE Trans. Commun.*, COM26: 432–438, 1978.

[25] J.-S. Kim, S.H. Park, P.W. Dowd, and N.M. Nasrabadi. A dynamic information-structure mutual exclusion algorithm for distributed systems. *Wireless Personal Commun.*, (3): 273–286, 1996.

[26] S. Kirkpatrick, C.D. Gelatt, and M.P. Vecchi. Optimization by simulated annealing. *Science*, 220: 671–680, 1983.

[27] D. Kunz. Channel assignment for cellular radio using neural networks. *IEEE Trans. Parallel Vehicular Technol.*, 40: 188–193, 1991.

[28] L. Lamport. Time, clocks and the ordering of events in a distributed system. *Commun. ACM*, 26: 1978.

[29] M. Maekawa. An algorithm for mutual exclusion in decentralized systems. *ACM Trans. Comput. Syst.*, 145–159, 1985.

[30] K. Okada and F. Kubota. On dynamic channel assignment in cellular mobile radio systems. *Proc. IEEE Int. Symp. Circ. Syst.*, 2: 938–941, 1991.

[31] K. Okada and F. Kubota. On dynamic channel assignment strategies in cellular mobile radio systems. *IEICE Trans. Fundamentals*, 75: 1634–1641, 1992.

[32] R. Prakash, N. Shivaratri, and M. Singhal. Distributed dynamic fault-tolerant channel allocation for mobile computing. *IEEE Trans. Vehicular Technol*, 48: 1874–1888, 1999.

[33] R. Prakash, N.G. Shivaratri, and M. Singhal. Distributed dynamic fault-tolerant channel allocation for mobile computing. *IEEE Trans. Vehicular Technol.*, 48: 1874–1888, 1999.

[34] G. Ricart and A.K. Agrawala. An optimal algorithm for mutual exclusion in computer networks. *Commun. ACM*, 24: 9–17, 1981.

[35] M. Serizawa and D. Goodman. Instability and deadlock of distributed dynamic channel allocation. In *Proceedings of the 43rd IEEE VIC*, pp. 528–531, 1993.

[36] J. Sin and N. Georganas. A simulation study of a hybrid channel assignment scheme for cellular land-mobile radio systems with erlang-c service. *IEEE Trans. Commun.*, COM-9: 143–147, 1981.

[37] M. Singhal. A dynamic information-structure mutual exclusion algorithm for distributed systems. *IEEE Trans. Parallel Dist. Syst.*, 3: 121–125, 1992.

[38] S.H. Wang. *Channel Allocation for Broadband Fixed Wireless Access Networks.* Ph.D. dissertation, Cambridge University, May 2003.

[39] W. Yue. Analytical methods to calculate the performance of a cellular mobile radio communication system with hybrid channel assignment. *IEEE Trans. Vehicular Technol.*, VT-40: 453–459, 1991.

[40] J. Zander. Trends in resource management future wireless networks. *Future Telecommunications Forum, FTF 99, Beijing, China*, December 1999.

IV

Mobility and Location Management in Wireless Networks

21

Adaptive Mobility Management in Wireless Cellular Networks

21.1 Introduction **21**-477
    Chapter Organization
21.2 Taxonomy of Location Management Schemes **21**-480
    Paging Schemes • Static Update Schemes • Dynamic
    Update Schemes • What's Missing?
21.3 User Mobility **21**-483
    Network Topology • Movement History • Mobility Model
21.4 Location Uncertainty and Entropy **21**-488
    Basic Terminologies • Comparison of Models
21.5 The LeZi-Update Algorithm **21**-490
    Update Procedure • Incremental Parse Tree • Update Cost
    and Message Complexity • Profile-Based Paging
21.6 Tradeoff between Update and Paging **21**-496
21.7 Conclusion **21**-497
Acknowledgments **21**-498
References **21**-498

Sajal K. Das

21.1 Introduction

The major driving forces behind the rapidly growing field of mobile computing include the tremendous development in wireless networking access technologies and powerful portable devices over the recent years, in addition to node mobility that accounts for the flexibility of communicating and computing on the move. However, most of the challenging issues and problems in this field arise due to the fact that the underlying environment is inherently uncertain and extremely resource poor [14]. As an example, wireless communication channels are bandwidth-limited, error-prone and have time-varying stochastic nature, implying their availability is unpredictable. The uncertainty due to node (user) mobility has some fundamental impacts, as it induces uncertainty in the network topology and hence routing, for example. Moreover, traffic load and resource demands in wireless mobile networks are also uncertain.

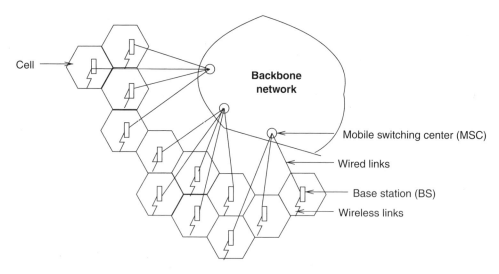

FIGURE 21.1   Overview of cellular architecture.

This chapter investigates issues related to mobility with a goal to devise techniques to keep the associated uncertainties under control. We assume the wireless network has some infrastructure where base stations (BSs) in cellular networks or access points in wireless LANs provide communication links to the mobile terminals. We will mainly focus on cellular networks, in which given geographic area is divided into a set of *cells*, each serviced by a BS. Several BSs are often wired to a *BS controller* (BSC), and a number of BSCs are further connected to a *mobile switching center* (MSC) forming a tree-like cluster. The communications between the MSCs, BSCs, and BSs are via wired networks, while those between the BSs and the mobile users are via air interfaces. The MSCs act as the gateway to the wireline backbone networks such as PSTN, ISDN, and the Internet. Figure 21.1 illustrates the hierarchical architecture of cellular networks.

Seamless and ubiquitous connectivity is the main driving force behind mobile computing and the growing demand for wireless Internet. However, tracking down a mobile user is a nontrivial task. This is because the freedom of user mobility creates an *uncertainty* about the exact location of the registered mobile terminal (in short, *mobile*). The network (or system) has to work against this uncertainty for successful delivery of calls. Furthermore, this has to be done effectively and efficiently (i.e., with as little information exchange as possible) for every user. Transfer of any more message than necessary results in redundant signaling, and hence wastage of scarce wireless bandwidth and energy (battery power) of the mobile.

Depending on whether a mobile terminal is actively communicating or in standby mode, we can distinguish (i) in-session mobility management, and (ii) out-of-session mobility management. Case (i), also known as *hand-off* management, deals with mechanisms by which calls/sessions are kept alive while the terminal moves from cell to cell, thereby changing the point of attachment. It is relatively easier to handle. On the other hand, case (ii), commonly known as *location management*, deals with mechanisms of keeping track of mobiles in standby mode. This is more intricate to tackle optimally, as the network has to route incoming sessions within a time bound to avoid connection failure. The rest of this chapter will deal with the location management problem.

The mobility/location management problem in wireless cellular networks involves two key components: paging and update. At one end of the spectrum, one can propose a solution to this problem with the help of what is called *paging* from the system's perspective. On a call arrival, the network initiates a search for the target mobile, by simultaneously polling all the cells where it can possibly be found. The MSC broadcasts a *page* message over a designated *forward control channel* via the BSs in a set of cells. All the mobiles listen to the page message and only the target sends a response message back over a *reverse channel*. In the worst case, the system may have to page all the cells in the entire service area. Clearly, the paging-only location tracking solution is very inefficient due to excessive signaling traffic of paging messages if an exhaustive search is to be performed for each and every call. The limited number of paging channels are bound to overload as the call volume grows.

At the other end of the spectrum, one can design a solution, which requires a mobile to report every time it moves from cell to cell. This reporting, called *location update* or *registration* mechanism, starts with an update message sent by the mobile over a *reverse channel*, which is followed by some traffic that takes care of related database maintenance operations at the network side. Again it is easy to see that this naive approach leads to enormous signaling overhead due to update messages, and hence impractical.

In reality, cellular networks employ location management schemes, which lie in between the above two extreme solutions. In other words, a mobile is made to report its whereabouts at a coarser granularity than cell-level, by means of some threshold-based mechanisms. This effectively limits the search space for paging at a later point of time, and thus puts an upper bound on the amount of *location uncertainty*. Hence, practical location management involves a combination of update and paging with a goal to reduce their combined cost. A critical analysis of the literature reveals that almost all of the existing solutions are based on heuristics that minimize one of the component's costs more than the other, but can never guarantee optimal cost for mobility tracking. Given that the paging and update costs are interdependent rather than independent, the true complexity and characterization of the location management problem was not known until the fundamental work by Bhattacharya and Das [6,7], who tackled this important problem from an *information-theoretic* standpoint. This outlook provides the insight to design an adaptive algorithmic framework for mobility tracking, which is *optimal* in terms of both update and paging costs. The objective of their update scheme is to *learn* user mobility pattern with optimal message exchange. Learning endows the paging mechanism with a predictive power, which reduces the average paging cost.

This chapter takes a more fundamental approach of characterizing the complexity of the mobility tracking problem under an information-theoretic framework. The use of information theory is motivated by the observation that the signaling costs due to update and paging can be traced to the randomness, or *uncertainty*, associated with an individual user's mobility pattern. More precisely, Shannon's entropy measure is identified as a basis for quantifying this uncertainty [11,39] and thus comparing user mobility models. We then develop an adaptive model-independent framework for estimating and managing the underlying uncertainty and relate it to the entropy or information content of the user's movement process.

Entropy captures the uncertainty of a probabilistic source, which is also the information content of the message (e.g., in number of bits) it generates. The performance of the mobility tracking algorithm can be measured too in the number of bits exchanged between the mobile and the system database during updating or paging. Since no algorithm can track the mobile by exchanging any less information on an average than the uncertainty created by its mobility, the proposed algorithm is optimal when these two amounts are the same.

By building and maintaining a compressed dictionary of user's *path* updates (as opposed to the widely used "position" updates), our on-line algorithmic framework is simple, optimally adapts and learns user's mobility patterns. It also allows for flexible tradeoff between location update and paging costs.

21.1.1 Chapter Organization

Section 21.2 provides a taxonomical classification of the existing location management schemes, bringing out important results on paging and update mechanisms. This section also analyzes the motivation behind the development of an information-theoretic framework for mobility management. Section 21.3 describes general models for the cellular network and individual user mobility. Section 21.4 shows how to compare mobility models in terms of uncertainty, and discusses the richness and limitation of the models. Section 21.5 gives an overview of an adaptive, predictive scheme called "LeZi-update," designed around the well-known universal lossless compression techniques such as the Lempel-Ziv family. Subsequently, Section 21.6 develops an optimal tradeoff between update and paging. Section 21.7 concludes the chapter with some discussions on ongoing research.

21.2 Taxonomy of Location Management Schemes

This section reviews various paging mechanisms as well as static and dynamic update strategies.

21.2.1 Paging Schemes

As mentioned earlier, the naive approach to location tracking is to page the mobile simultaneously either in the entire network, or in a part as dictated by the last update message. As pointed out in Reference 31, this may be overkill. In reality, paging must be completed within an allowable *delay constraint*, and there exists the possibility of sequentializing the procedure. An alternative considered there is to split the service area into *location areas* (LAs), where all the cells within an LA are paged simultaneously. When a call arrives, the LAs are paged sequentially for the mobile following an ordering termed the *paging strategy*. A very intuitive result derived in Reference 31 states that, under steady-state location probability distribution of a mobile user, the optimal paging strategy (in terms of the mean cost of paging) with no delay constraint should page the LAs in the order of decreasing probability values. Clearly, a uniform distribution is the worst adversary, because no additional improvement is obtainable by changing paging strategy. An algorithm to find the optimal paging sequence under given delay constraints is also presented in Reference 31. However, the worst case still ends up with systemwide paging and incurring high cost.

To counteract the worst case complexity issue in pure paging-based strategies, most location management schemes rely on time to time *location updates* by the mobile. The idea is to enforce an upper bound on the uncertainty by restricting the paging domain within the vicinity of the last known location. The cost of location update, however, adds a new component to the location management cost. The update cost over a time period is proportional to the number of times the mobile updates, whereas the paging cost depends on the number of calls placed and the number of cells paged while routing each of those calls. The tradeoff between the two components is evident from the fact that, the more frequently the mobile updates (incurring higher update cost) the less is the number of paging attempts required to track it down.

From an operational point of view, it seems that paging is more fundamental than updating in a cellular system. Yet, it is somewhat interesting to observe that the majority of the research on location management has actually focussed on update schemes, assuming some obvious version of paging algorithm. The most compelling reason for this is the lack of a single unquestionable probability model for user mobility that reflects all kinds of movements seen in practice. As pointed out in Reference 46, not enough studies have been conducted in realistic movement profiles of mobiles that take care of speed and directional variations. Usually, pedestrian and vehicular movement are treated separately. This makes the mobility models inappropriate for PCS networks [9] and third-generation systems [22].

Most paging algorithms proceed with a high reliance factor on the latest update information. The mobile's last known position and its surroundings are considered to be the most probable current position — the probability decreasing in an omni-directional way with increasing distance. This is the underlying assumption for the popular *cluster paging* [26] and *selective paging* [1,2,17]. The directional bias in user movement has been taken into account in References 4 and 8 by associating a number of states for each cell under a Markov model. Only while designing paging algorithms, user profiles are considered in Reference 28 which assumed that probabilistic information about a profile is readily available either with the user or at the billing database.

21.2.2 Static Update Schemes

The following two approaches to location update, based on either partitioning of cells into LAs or selection of a few designated reporting cells, have been characterized as static or global techniques [4]. These are *static* schemes because the cells at which a mobile updates are fixed. They are also *global* in the sense that all or a batch of mobiles always originate their update messages from the same set of cells.

21.2.2.1 LA Partitioning

The update scheme most widely adopted by current cellular systems (such as GSM [25]), follow the idea presented in Reference 31. The service area under an MSC is partitioned into LA, that are formed by nonoverlapped grouping of neighboring cells. A mobile must update whenever it crosses an LA boundary. Its location uncertainty is reduced by limiting the search space to the set of cells under the current LA. All the cells under the LA are paged upon a call arrival, resulting in an assured success within a single step. The BSs must broadcast the LA-id (along with the cell-id) to aid the mobiles in following the update protocol. Consequently, a global LA assignment is induced for all subscribers. An obvious drawback of this scheme is that the update traffic originates only in the boundary cells of the LAs, thereby reducing the bandwidth availability for other calls.

It has been shown in Reference 44 how to partition cells into optimal LAs. Assuming that the relative cost of paging versus update and how the paging cost varies with LA size are known, their solution is dictated by the call arrival rate and mobility of the user. Two variants of the method are proposed. The pure *static* variant uses the average call arrival rate and mobility index for all users. Thus, a mobile behaving as a potential outlier would transmit frequent uninformative update messages by crossing LA boundaries. In contrast, a somewhat *dynamic* variant of this primarily static scheme allows different LA assignments for different mobiles, based on individual call arrival and mobility patterns. This alleviates the localization problem of the update traffic to some extent, although at the cost of the computational complexity involved in choosing, maintaining, and uploading a wide range of LA maps to individual mobiles. This was further extended in Reference 20 to optimize on the signaling costs that reflects the mobile's direction of movement and regional cell characteristics.

21.2.2.2 Reporting Center Selection

An alternative approach due to Reference 3 does not impose any partition on the cellular map, but designates some cells as *reporting cells* where the mobiles must update upon entering. On arrival of a call, the mobile is paged in the vicinity of the reporting cell it has last updated at. Choosing an optimal set of reporting cells for a general cellular network has been shown to be NP-complete. Optimal or near-optimal solutions for special types of cellular topologies (e.g., tree, ring, and grid), and an approximation technique for solutions to general graphs have been presented. But the scheme is built upon a number of simplifying assumptions such as square shape for cells and LAs, and a fluid flow model of user mobility.

The basic drawbacks of a static or global scheme, even if reduced by this approach, still lingers. For example, a user can generate uninformative update messages by hopping in and out of reporting cells. As suggested by Sen *et al.* [38], considering per-user mobility is the first step toward dealing with these problems. Following the spirit of reporting center selection, they impose a selective update scheme tuned to individual users over an LA-based cellular network. Such a technique lies in between the static update schemes and the dynamic ones, described next.

21.2.3 Dynamic Update Schemes

Under these schemes, the mobile updates based only on the user's activity, but not necessarily at prede-termined cells. These are *local* in the sense that mobiles can make the decision whether to update or not without gathering any global or design specific information about the cell planning. Three major schemes fall under the dynamic category, viz., (i) distance-based, (ii) movement-based, and (iii) time-based, which are named by the kind of threshold used to initiate an update.

21.2.3.1 Distance-Based

Under the *distance-based* scheme [1,4,17,23], the mobile is required to track the Euclidean distance from the location of the previous update and initiates a new update if the distance crosses a specified threshold, d. Although the distance would ideally be specified in terms of a unit such as mile or kilometer, it can also be specified in terms of the number of cells between the two positions. So the location uncertainty is reduced by limiting the search space to the set of cells falling within a circular region of radius d around the last known cell. One can think of these cells effectively forming an LA centered around the last known position of that specific user. However, it is difficult to measure the Euclidean distance between two locations, even if the traversed distance is made available from the vehicle. Irrespective of its characterization among local ones, implementing a distance-based scheme calls for some global knowledge about the cell map. In order for the mobile to identify the cells within the distance threshold, the system has to load these cell-ids as a response to its update message. In Reference 23 a scheme is described to find an optimal threshold $\hat{d}$ by minimizing the expectation of the sum of update and paging costs until the next call. Considering the evolution of the system in between call arrivals under a memoryless movement model, an iterative algorithm based on dynamic programming is used to compute the optimal threshold distance. A similar approach due to Reference 17 computes the optimal threshold distance, $\hat{d}$, under a two-dimensional random walk user mobility model over hexagonal cell geometry. The random walk is mapped onto a discrete-time Markov chain, where the state is defined as the distance between the mobile's current location and its last known cell. The residing area of the user contains $\hat{d}$ layers of cells around its last known cell, which is partitioned into paging areas according to the given delay constraint.

21.2.3.2 Movement-Based

The *movement-based* scheme [2,4] is essentially a way of over-estimating the Euclidean distance by traversed distance, when the distance is considered only in terms of the number of cells. The mobile needs to count the number of cell boundary crossings and update when the count reaches a certain threshold, M. It is truly a local scheme since there is no longer any need for any knowledge about cell neighborhood. The penalty paid is an increased number of updates that it triggers counting local movements between different cells, even though the distance threshold is not being crossed.

21.2.3.3 Time-Based

Under the *time-based* [4,32] scheme, the mobile sends periodic updates to the system. The period or time threshold, T, can easily be programmed into the mobile using a hardware or software timer. While this makes it a truly local and attractive solution for implementation, the cost due to redundant updates made by stationary mobiles has to be accommodated. A mobile is paged by searching all possible cells reachable by the user within the elapsed time from the last known cell. Thus the search space evolves as a function of the elapsed time, user mobility, and possibly the last known cell.

In Reference 4, the time-based, movement-based, and distance-based update schemes were compared in terms of the paging cost with varying update rates. Two types of user movement models, viz., independent and Markovian, are used on a ring cellular topology. It has been observed that the distance-based scheme performs the best consistently. The result is intuitive because the distance threshold directly puts an upper bound on the location uncertainty, whereas in both time and movement-based schemes the uncertainty imposed by user mobility interacts with it.

21.2.4 What's Missing?

A critical analysis of the above discussions lead to the following observations:

1. The location management problem has been formulated by the vast majority of researchers as some static optimization problem to minimize either paging or update cost, or a combination thereof. The inputs to the optimization problem are often some probability values that need to be computed from statistics gathered from system snapshots. In the static version of the optimization problem, an algorithm works off-line to provide a solution that remains in effect until the next run. This interval is very critical to the performance of the technique, but is not provided by the static optimization formulation. Educated guesses are used in most cases. Formulating the problem as a dynamic optimization provides a better insight into its stationary behavior and applicability of static versions. Better yet, it may lead to the design of a *on-line algorithm* to solve the problem.

2. Unlike the resource management problems in cellular networks, the location tracking problem is user oriented by definition. Naturally, it would be wise to make use of personal mobility and

calling profiles of individual subscribers for optimization purposes. Both the update and paging techniques should be user specific to add the "personal" touch to the *personal communication service*. *Knowledge representation* and *learning* of the user profile are thus two key factors.

3. The effectiveness of sequencing the paging process critically depends on early success of the deployed paging strategy. A large number of failed paging attempts would not only result in more call drops, but also cause overloading on paging channels. The essence of designing a good paging strategy is to enhance the predictability of a mobile's behavior making use of the user profile.

4. Since the sole purpose of the update mechanism is to aid the paging process, there is no reason to treat them as two independent components of location management cost. As opposed to deciding on a paging strategy first and then optimizing on the update strategy, one can come up with a collaborative pair of paging and update policies. In other words, the update mechanism needs to keep the system better informed about a user's mobility, sending the maximum possible information in a compact form and avoiding redundancy as far as possible.

Before proceeding further, let us re-interpret three fundamental questions originally posed in References 31 and 32:

1. *Complexity:* Given a user mobility model, what is the least amount of effort necessary on an average, to track the mobile? What metric do we use to quantify the effort and what unit do we use?

2. *Optimality:* Given that the user mobility model is known to both the user and the system, how to pick an optimal paging strategy, an optimal update scheme, or an optimal combination of paging and update policies?

3. *Learning:* How can the time-varying dynamic location probabilities be efficiently estimated based on measurements taken from user motion? How to pick the right mobility model that does not impose fundamental restrictions in terms of richness?

21.3   User Mobility

The user mobility model plays too important a role in designing location management schemes. The underlying model can potentially influence the analysis and simulation results so much that observations and conclusions would merely reflect the nature of the model itself. Thus, there are reasons to be skeptical and cautious about extrapolating theoretical claims to practice. For example, Markov models are often constructed with states dictated by the current cell in a regular geometric cellular architecture. Transition probabilities for a typical user to move into a specific neighboring cell are arbitrarily assumed. This is far from being practical.

21.3.1   Network Topology

Structured graph models for a cellular network have been quite popular among researchers engaged in solving the location management problem. Circular, hexagonal, or square areas are often used to model the cells, while various regular graph topologies such as rings, trees, one- and two-dimensional grids are used to model their interconnection. These models do not accurately represent a real cellular network, where the cell shapes may vary depending on the antenna radiation pattern of the BSs, and a cell can have an arbitrary (but bounded) number of neighbors. Although structured graphs help in the early stage of frequency planning, they over-simplify the mobility tracking problem. Since enforcing strict assumptions on the topology has a potential to weaken the model, we refrain from making any assumption about either the geometry or the interconnections between cells.

With a GSM-like deployment in mind, let us consider an LA-based system. The network can be represented by a bounded-degree, connected graph $G = (\vartheta, \varepsilon)$, where the node-set ϑ represents the LAs

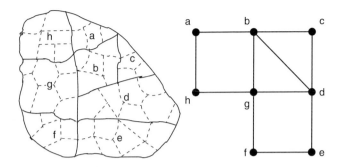

FIGURE 21.2   Model of a realistic cellular system.

TABLE 21.1   User Movement between 9:00 a.m. and 9:00 p.m.

	a.m.			p.m.							
Time	11:04	11:32	11:57	3:18	4:12	4:52	5:13	6:11	6:33	6:54	$\cdots$
Crossing	$a \rightarrow b$	$b \rightarrow a$	$a \rightarrow b$	$b \rightarrow a$	$a \rightarrow b$	$b \rightarrow c$	$c \rightarrow d$	$d \rightarrow c$	$c \rightarrow b$	$b \rightarrow a$	$\cdots$

and the edge-set ε represents the neighborhood (roads, highways, etc.) between pairs of LAs. Figure 21.2 shows a service area with eight LAs, viz., a, b, c, d, e, f, g, h, and the corresponding graph representation.

The general graph representation is of course not restricted to only LA-based cell planning. Some existing systems use paging strategies that work with a finer granularity, and choose the individual cells to be paged next. In such cases, our model would give rise to a graph with a larger node-set $\vartheta = \{a_1, a_2, \ldots, b_1, b_2, \ldots, c_1, c_2, \ldots\}$, where $a_1, a_2, \ldots$ are cells in the LA a and so on. We introduce a more general term *zone* to refer to a node in our network model. A zone can be an LA or a cell depending on the system. The important thing is that the current zone-id should always be made available to a mobile within by frequent and periodic broadcast messages from the BSs.

21.3.2 Movement History

The real power of an adaptive algorithm comes from its ability to learn. As and when the events unfold, a learning system observes the history to use it as a source of information. Let us now look at the movement history of a typical user within the service area shown in Figure 21.2. For simplicity, we only record the movement at the coarse granularity of LAs. This means that the user must be in one of the eight zones, viz., a, b, c, d, e, f, g, h, at any point in time. Suppose the service was turned on at 9:00 a.m., when the mobile initially registered. Table 21.1 shows the time (limited to the precision of a minute) at which the LA boundaries were crossed. Current time is 9:00 p.m. An update message reports the current zone to the system. Consequently, all that the system obtains from an update scheme is a sequence of zone-id's. The information carried by this sequence about the user's mobility profile depends heavily on the underlying update scheme and how it is interpreted by the system.

Let us compare and contrast the sequences generated by the time-based and movement-based schemes. Table 21.2 shows the zone sequences generated. For the time-based scheme, two values of the threshold T have been considered, viz., 1 and $\frac{1}{2}$ h. Both do a good job in capturing the fact that the user resides in zones a and b for a relatively longer time as compared to zones c and d. On the other hand, smaller values of T must be chosen to trace in detail the routes taken by the user. The choice of $\frac{1}{2}$ h performed better than that of 1 h in that respect, yet missed to detect the zone c in the round trip $a \rightarrow b \rightarrow c \rightarrow d \rightarrow c \rightarrow b \rightarrow a$. In contrast, the movement-based scheme is more apt to capture routes. Two choices of the threshold M, viz., 1 and 2, have been used to illustrate its effectiveness. Clearly, $M = 1$ captures movements in the finest detail. Yet, both are lacking in their capability to convey the relative durations of residence.

Based on these observations we are now motivated to use a combination of time-based and movement-based schemes to make the movement history more informative. The last row of Table 21.2 shows the zone

TABLE 21.2   Zone Sequence Reported by Various Update Schemes

Time-based ($T = 1$ h)	*aaabbbbacdaaa . . .*
Time-based ($T = \frac{1}{2}$ h)	*aaaaabbbbbbbbbaabcddcaaaaa . . .*
Movement-based ($M = 1$)	*abababcdcba . . .*
Movement-based ($M = 2$)	*aaacca . . .*
Time- and movement-based ($T = 1$ h, $M = 1$)	*aaababbbbbbaabccddcbaaaa . . .*

sequence obtained if a dual mode update is deployed. This scheme generates an hourly update starting at 9:00 a.m., as well as whenever a zone boundary crossing is detected. In the rest of the paper we assume such an update scheme and use this sequence *aaababbbbbbaabccddcbaaaa . . .* for illustrative purposes. However, without any loss of generality, we view the *movement history* as just a sequence of zones reported by successive updates.

Definition 21.1 *The* movement history *of a user is a string "$v_1 v_2 v_3 \cdots$" of symbols from the alphabet ϑ, where ϑ is the set of zones under the service area and v_i denotes the zone-id reported by the ith update. Consequently, v_is are not necessarily distinct.*

21.3.3 Mobility Model

A clear understanding of the distinctions between the *movement history* and the *mobility model* is necessary before an appropriate definition of the latter can be given. The movement history of a user is deterministic, but a matter of the past. The mobility model, on the other hand, is probabilistic and extends to the future. The tacit assumption is that a user's movement is merely a reflection of the patterns of his/her life, and those can be learned over time in either off-line or on-line mode. Specifically, the patterns in the movement history correspond to the user's favorite routes and habitual duration of stay. The essence of learning lies in extracting those patterns from history and storing them in the form of knowledge. Learning aids in decision making when re-appearance of those patterns are detected. In other words, learning works because "history repeats itself."

Characterization of mobility model as a probabilistic sequence suggests that it can be defined as a stochastic process, while the repetitive nature of identifiable patterns adds stationarity as an essential property. Prior works referenced in this chapter also assume this property either explicitly or implicitly.

Definition 21.2 *The* mobility model *of a user is a stationary stochastic process $V = \{V_i\}$, such that V_i assumes the value $v_i \in \vartheta$ in the event that the ith update reports the user in zone v_i. The joint distribution of any subsequence of V_is is invariant with respect to shifts in the time axis, that is,*

$$\Pr[V_1 = v_1, V_2 = v_2, \ldots, V_n = v_n] = \Pr[V_{1+l} = v_1, V_{2+l} = v_2, \ldots, V_{n+l} = v_n] \quad (21.1)$$

for every shift l and for all $v_i \in \vartheta$. The movement history *is a trajectory or sample path of V.*

An important question is why such a general mobility model is not as popular as the restrictive models so abundant in the literature. The most likely reason is that the general model allows nothing to be assumed to start the analysis. The model has to be learned. Given that the user mobility model defined just as a stationary process, learning is possible if and only if one can construct a universal predictor or estimator for this process. Here we illustrate an approach for building models from history. Rigorous treatments on these topics can be found in References 15, 30, 31, and 43.

Let us first get a better feel of how commonly used models interpret the movement history and end up imposing restrictions during the learning phase.

21.3.3.1 Ignorant Model

The ignorant model disbelieves and disregards the information available from movement history. Due to the lack of knowledge, it assigns equal residence probabilities to all the eight zones in Figure 21.2. In other words, $\pi_a = \pi_b = \pi_c = \pi_d = \pi_e = \pi_f = \pi_g = \pi_h = \frac{1}{8} = 0.125$. The assumption of uniform probability distribution suffers from the consequence that no single paging strategy can be adjudged better than another in terms of average paging cost [31].

21.3.3.2 The IID Model

The IID model assumes that V_is are independent and identically distributed. Using the relative frequencies of the symbols as estimates of residence probabilities, we obtain the residence probabilities as $\pi_a = \frac{10}{23} \approx 0.435$, $\pi_b = \frac{8}{23} \approx 0.348$, $\pi_c = \frac{3}{23} \approx 0.13$, $\pi_d = \frac{2}{23} \approx 0.087$, and $\pi_e = \pi_f = \pi_g = \pi_h = 0$.

21.3.3.3 Markov Model

The simplest possible Markov model assumes that the process S is a time-invariant Markov chain, defined by

$$\Pr[V_k = v_k \mid V_1 = v_1, \dots, V_{k-1} = v_{k-1}]$$
$$= \Pr[V_k = v_k \mid V_{k-1} = v_{k-1}] \tag{21.2}$$
$$= \Pr[V_i = v_i \mid V_{i-1} = v_{i-1}] \tag{21.3}$$

for any arbitrary choice of k and i.

Zones e, f, g, and h, never being visited, acquire zero probability mass. The effective state space is thus reduced to the set $\{a, b, c, d\}$. The one-step transition probabilities

$$P_{i,j} = \Pr[V_k = v_j \mid V_{k-1} = v_i], \qquad \text{where } v_i, v_j \in \{a, b, c, d\}$$

are estimated by the relative counts. From Figure 21.3, the probability transition matrix $\mathbf{P} = ((P_{i,j}))$ is given by

$$\mathbf{P} = \begin{bmatrix} \frac{2}{3} & \frac{1}{3} & 0 & 0 \\ \frac{3}{8} & \frac{1}{2} & \frac{1}{8} & 0 \\ 0 & \frac{1}{3} & \frac{1}{3} & \frac{1}{3} \\ 0 & 0 & \frac{1}{2} & \frac{1}{2} \end{bmatrix}.$$

The Markov chain is finite, aperiodic, and irreducible, and thus *ergodic*. Let $\mathbf{\Pi} = [\pi_a \ \pi_b \ \pi_c \ \pi_d]^T$ be the steady-state probability vector. Solving for $\mathbf{\Pi} = \mathbf{\Pi} \times \mathbf{P}$ with $\pi_a + \pi_b + \pi_c + \pi_d = 1$, we obtain $\pi_a = \frac{9}{22} \approx 0.409$, $\pi_b = \frac{4}{11} \approx 0.364$, $\pi_c = \frac{3}{22} \approx 0.136$, $\pi_d = \frac{1}{11} \approx 0.091$. Also, $\pi_e = \pi_f = \pi_g = \pi_h = 0$.

21.3.3.4 Finite-Context Model

The ignorant model is incapable of learning and cannot lead to any adaptive technique. The IID model takes the first step toward learning from movement history. For example, $\langle a, b, c, d, e, f, g, h \rangle$ is an optimal unconditional paging strategy that can be derived based on the IID model. However, if we know that the mobile has made the last update in zone d, neither a nor b is a more likely candidate for paging in

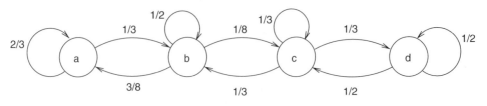

FIGURE 21.3 Markov model (order-1) for movement profile.

comparison to c or d. Unfortunately, the IID model does not carry any information about the symbols' order of appearance and falls short in such situations. The Markov model carries a little more information about the ordering, at least to the extent of one symbol context. To be more precise, let us adopt the terminology *order-1 Markov model* to refer to this particular model. The same nomenclature designates the IID model as the *order-0 Markov model*. To maintain this convention, the concept of order has to be extrapolated even to the negative domain without much real meaning. We call the ignorant model an *order-(-1) Markov model*.

The construction of higher order Markov models is illustrated by enumerating all order-2 contexts in the sequence "*aaababbbbbaabccddcbaaaa*" and the symbols that appear in those contexts with their respective frequencies. Table 21.3 enumerates the contexts with symbol frequencies for orders 0, 1, and 2. An entry of "$v \mid w(f)$" implies that the symbol $v \in \vartheta$ appears with frequency f in the context $w \in \vartheta^*$, where ϑ^* is the regular grammar notation for a sequence of zero or more symbols from set ϑ. Null contexts have not been shown explicitly. In other words, f is the number of matches of "$w.v$" in the history, where the *dot* represents concatenation. A dictionary of such contexts can be maintained in a compact form by a *trie* or *digital search tree*, as shown in Figure 21.4. Every node represents a context, and stores its last symbol along with the relative frequency of its appearance at the context of the parent node. Naturally, a node can have at most $|\vartheta|$ children. The root, at level 0, represents a null context. Level l stores the necessary statistics for an order-$(l-1)$ Markov model.

It is now intuitively clear how higher order Markov models carry more detailed information about the ordering of symbols in a sequence. Applied to the update sequences, they capture a lot of information about the user's favorite routes. For example, according to the order-1 model, the probability of the route "*abcbcd*" being taken is $\frac{9}{22} \times \frac{1}{3} \times \frac{1}{8} \times \frac{1}{3} \times \frac{1}{8} \times \frac{1}{3} = \frac{1}{4224} \approx 2.37 \times 10^{-4}$. However small this may seem, such a zigzag route is highly unlikely to be taken by any sensible person, and should have been assigned a zero probability. Under the order-2 model, this route turns out to be an impossible event, as expected.

TABLE 21.3 Contexts of Orders 0, 1, and 2 with Frequencies

Order-0	Order-1		Order-2		
$a(10)$	$a\mid a(6)$	$b\mid c(1)$	$a\mid aa(3)$	$a\mid ba(2)$	$a\mid cb(1)$
$b(8)$	$b\mid a(3)$	$c\mid c(1)$	$b\mid aa(2)$	$b\mid ba(1)$	$d\mid cc(1)$
$c(3)$	$a\mid b(3)$	$d\mid c(1)$	$a\mid ab(1)$	$a\mid bb(1)$	$d\mid cd(1)$
$d(2)$	$b\mid b(4)$	$c\mid d(1)$	$b\mid ab(1)$	$b\mid bb(3)$	$b\mid dc(1)$
	$c\mid b(1)$	$d\mid d(1)$	$c\mid ab(1)$	$c\mid bc(1)$	$c\mid dd(1)$

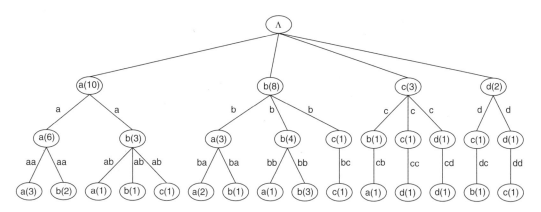

FIGURE 21.4 A trie for all contexts up to order-2 in the sequence "*aaababbbbbaabccddcbaaaa* . . . ."

The possibility of Markov models based on higher order finite contexts adds hope as well as confusion to the pursuit of a general mobility model. Does an increase in the order of context always make the model richer? There should be a limit to the richness, because the sought after general model has to be the richest. At what order k, do we stop? Should we use an order-k model only, or consider all models of orders 0 through k inclusive? And, what is the figure of merit for richness anyway?

21.4 Location Uncertainty and Entropy

It seems impossible to answer the fundamental questions posed in the previous section without a quantitative comparison of the candidate models. Of course, we need a fair basis for such comparison, and uncertainty is a potential choice. The rule of thumb is that the lower the uncertainty under a model, the richer the model is. Consequently, we need to formalize the notion of uncertainty we have been talking about since the beginning.

21.4.1 Basic Terminologies

We have defined user mobility earlier as a stochastic process $\mathcal{V} = \{V_i\}$, where the V_is form a sequence of random variables. The traditional information-theoretic definitions of *entropy* and *conditional entropy* of random variables, as well as *entropy rate* of a stochastic process are given below [11,39].

Definition 21.3 *The entropy $H_b(X)$ of a discrete random variable X, with probability mass function $p(x), x \in \mathcal{X}$, is defined by $H_b(X) = -\sum_{x \in \mathcal{X}} p(x) \log_b p(x)$. The limiting value "$\lim_{p \to 0} p \log_b p = 0$" is used in the expression when $p(x) = 0$. The base of the logarithm depends on the unit used. As we usually measure information in terms of bits, using $b = 2$ we write*

$$H(X) = -\sum_{x \in \mathcal{X}} p(x) \log p(x) \tag{21.4}$$

Since $p(x) \in [0, 1]$, we see that $H(X) \geq 0$.

Definition 21.4 *The joint entropy $H(X, Y)$ of a pair of discrete random variables X and Y with a joint distribution $p(x, y)$, where $x \in \mathcal{X}$ and $y \in \mathcal{Y}$, is defined by*

$$H(X, Y) = -\sum_{x \in \mathcal{X}} \sum_{y \in \mathcal{Y}} p(x, y) \log p(x, y) \tag{21.5}$$

Also the conditional entropy $H(Y \mid X)$ is defined as

$$H(Y \mid X) = \sum_{x \in \mathcal{X}} p(x) \, H(Y \mid X = x)$$

$$= -\sum_{x \in \mathcal{X}} p(x) \sum_{y \in \mathcal{Y}} p(y \mid x) \, \log p(y|x) \tag{21.6}$$

$$= -\sum_{x \in \mathcal{X}} \sum_{y \in \mathcal{Y}} p(x, y) \log p(y \mid x) \tag{21.7}$$

Definition 21.5 *The per symbol entropy rate $H(\mathcal{V})$ for a stochastic process $\mathcal{V} = \{V_i\}$, is defined by*

$$H(\mathcal{V}) = \lim_{n \to \infty} \frac{1}{n} H(V_1, V_2, \dots, V_n) \tag{21.8}$$

if the limit exists. The conditional entropy rate $H'(\mathcal{V})$ for the same process is defined by

$$H'(\mathcal{V}) = \lim_{n \to \infty} H(V_n | V_1, V_2, \ldots, V_{n-1}) \tag{21.9}$$

if the limit exists.

21.4.2 Comparison of Models

First we try to compare different models based on the two entropy rates $H(\mathcal{V})$ and $H'(\mathcal{V})$. Let us consider the three special cases of context models of orders -1, 0, and 1. Let us use the notation $Pr[V_1 = v_1, V_2 = v_2, \ldots, V_n = v_n] = p(v_1, v_2, \ldots, v_n)$ for all $v_i \in \vartheta$.

Order-(-1) model: V_is are independent and uniformly distributed, that is, with distribution $p(v) = \pi_v = \frac{1}{8}, \forall v \in \vartheta$. Due to independence, $p(v_n | v_1, \ldots, v_{n-1}) = p(v_n)$ and therefore $p(v_1, v_2, \ldots, v_n) = \{p(v_1)\}^n$. Thus, $H(\mathcal{V}) = H'(\mathcal{V}) = \log 8 = 3$ bits.

Order-0 model: V_is are independent and identically distributed with distribution $p(v) = \pi_v$, $\forall v \in \vartheta$. Due to independence, $p(v_n | v_1, \ldots, v_{n-1}) = p(v_n)$ and therefore $H(\mathcal{V}) = H'(\mathcal{V}) = \sum_{v \in \vartheta} -p(v) \log p(v) = \sum_{v \in \vartheta} -\pi_v \log \pi_v = \frac{10}{23} \log \frac{23}{10} + \frac{8}{23} \log \frac{23}{8} + \frac{3}{23} \log \frac{23}{3} + \frac{2}{23} \log \frac{23}{2} \approx 1.742$ bits, from Equation (21.6).

Order-1 model: V_is form Markov chain, which means that $p(v_n | v_1, \ldots, v_{n-1}) = p(v_n | v_{n-1}) = P_{v_i, v_j}$. Substituting $p(v) = \pi_v$ in Equation (21.6), we get,

$$H'(\mathcal{V}) = -\sum_i \pi_i \left(\sum_j P_{i,j} \log P_{i,j} \right)$$

$$= \frac{9}{22} \left(\frac{2}{3} \log \frac{3}{2} + \frac{1}{3} \log 3 \right) + \frac{4}{11} \left(\frac{3}{8} \log \frac{8}{3} + \frac{1}{2} \log 2 + \frac{1}{8} \log 8 \right)$$

$$+ \frac{3}{22} \left(3 \times \frac{1}{3} \log 3 \right) + \frac{1}{11} \left(2 \times \frac{1}{2} \log 2 \right) \approx 1.194 \text{ bits.}$$

The π_v values come from the solution of $\mathbf{\Pi} \times \mathbf{P} = \mathbf{\Pi}$.

A few observations are in order. First, at most three bits are sufficient to span the message space of eight alternatives. All three bits of uncertainty exist in order-(-1) model. Thus the model itself is not at all informative, and is the one that maximizes entropy rate. Order-0 and order-1 models show gradual improvement in richness with decreasing entropy rate. Second, the two kinds of entropy rate $H(\mathcal{V})$ and $H'(\mathcal{V})$ are equal due to the independence in order-(-1) and order-0 models. The question is whether this is true for models of order-1 or higher. An important result that follows from the earlier definitions clarifies this issue [11].

Result 21.1 *For any set $\{V_1, V_2, \ldots, V_k\}$ of k discrete random variables with a joint distribution*

$$p(v_1, v_2, \ldots, v_k) = Pr[V_1 = v_1, V_2 = v_2, \ldots, V_k = v_k], \qquad \forall i \, (v_i \in \vartheta)$$

the joint entropy $H(V_1, V_2, \ldots, V_k)$ is given by

$$H(V_1, V_2, \ldots, V_k) = \sum_{i=1}^{k} H(V_i | V_1, V_2, \ldots, V_{i-1}) \tag{21.10}$$

Suppose the random variables V_is are taken from $\mathcal{V}$. Equation (21.10) reveals an interesting relationship when $k - 1$ is the highest order context under consideration. Substituting $k = 3$, for example, we get

$$H(V_1, V_2, V_3) = H(V_1) + H(V_2 | V_1) + H(V_3 | V_1, V_2)$$

The additive terms on the right-hand side consists of the information carried by levels 1, 2, and 3 (root at level zero) of the trie in Figure 21.4. Higher order context models are thus more information-rich as compared to the lower order ones. Another way to look at it is that the lower order models mislead the algorithm designer by projecting an under-estimate of uncertainty. To see that the trie holds all the necessary parameters to compute $H(V_i|V_1, V_2, \ldots, V_{i-1})$, we expand using Equation (21.6) to find

$$H(V_i|V_1, V_2, \ldots, V_{i-1}) = \sum_{\vartheta^{i-1}} p(v_1, \ldots, v_{i-1})$$

$$\times \left\{ \sum_{\vartheta} p(v_i|v_1, \ldots, v_{i-1}) \log p(v_i|v_1, \ldots, v_{i-1}) \right\}. \tag{21.11}$$

The probabilities $p(v_1, \ldots, v_{i-1})$ and $p(v_i|v_1, \ldots, v_{i-1})$ are estimated from the relative frequencies preserved in the trie. Since the conditional entropy computation for order-i requires the joint distribution for all orders up to $(i-1)$, we need to maintain models of all orders up to a suitably large value. Note that order-1 model is an exception due to the simplicity in computing the vector $\mathbf{\Pi}$ from matrix $\mathbf{P}$.

As a part of illustration, let us compute the conditional entropies for contexts of orders 0, 1, and 2 from Figure 21.4. We have, $H(V_1) = 1.742$ as before. However, $H(V_2|V_1) = \frac{10}{23}(\frac{2}{3}\log\frac{3}{2} + \frac{1}{3}\log 3) + \frac{8}{23}(\frac{3}{8}\log\frac{8}{3} + \frac{1}{2}\log 2 + \frac{1}{8}\log 8) + \frac{3}{23}(3 \times \frac{1}{3}\log 3) + \frac{2}{23}(2 \times \frac{1}{2}\log 2) \approx 1.182$. The deviation from the value 1.194 obtained before is due to the fact that we consulted the steady-state distribution from order-0 instead. Similarly, $H(V_3|V_1 V_2) = \frac{6}{23}(\frac{1}{3}\log 3 + \frac{2}{3}\log\frac{3}{2}) + \frac{3}{23}(3 \times \frac{1}{3}\log 3) + \frac{3}{23}(\frac{2}{3}\log\frac{3}{2} + \frac{1}{3}\log 3) + \frac{4}{23}(\frac{1}{4}\log 4 + \frac{3}{4}\log 43) \approx 0.707$. Finally, $H(V_1, V_2, V_3) \approx 3.631$ and by taking the running average, we arrive at an estimate of $H(V) = (1.742 + 1.1818 + 0.707)/3 \approx 1.21$. As the order increases, this running average estimate should converge to the true value of the per-symbol entropy rate, if it exists.

A natural question arises regarding the highest order k, which dictates the height of the trie. The answer comes from the following result [11].

Result 21.2 *For a stationary stochastic process* $V = \{V_i\}$, *the conditional entropy* $H(V_n|V_1, \ldots, V_{n-1})$ *is a decreasing function in n and has a limit* $H'(V)$.

Clearly, the marginal improvement in model richness starts to die out soon. Intuitively, the largest meaningful order has something to do with largest chain of dependency observed in the movement history. This may come from the longest route taken by the user or the longest stay at a zone (reported by the time-based update scheme only). This becomes evident when we look at the left-hand side of Equation (21.10). This is the joint entropy of a sequence of the k random variables. According to Equation (21.8), the per-symbol entropy rate would then represent the running average of conditional entropy rates, giving rise to the following result [11]:

Result 21.3 *For a stationary stochastic process* V, *both the limits in Equations (21.8) and (21.9) exist and are equal, that is,*

$$H(V) = H'(V) \tag{21.12}$$

For a universal model, all we now need is a way to automatically arrive at the appropriate order dictated by the input sequence. Fortunately, this is exactly what a class of compression algorithms achieves.

21.5 The LeZi-Update Algorithm

In Section 21.2 we reasoned why the time-based and movement-based update schemes are more localized as compared to the distance-based scheme. Let us recall that this was due to the system's involvement in re-computing and uploading the mobile's new neighborhood after each update. The mobile has to maintain a list of neighboring cells in a cache until the next update. Proponents of the movement-based schemes have also touted a caching scheme such that the new cell-id is cached on every boundary crossing and entry into a cell already in cache is not counted toward reaching the threshold M. This is supposed to avoid some unnecessary updates when cells are re-visited.

Our proposed update scheme essentially uses an enhanced form of caching. Discussions on richness of models in Section 21.3 reveals the need for gathering statistics based on contexts seen in the movement history. While the zone-ids are treated as character symbols, these contexts must be treated as phrases. A dictionary of such phrases must replace the cache. Whereas just doubling or tripling the threshold only eliminates some primary updates, our algorithm tries to hold them back and send them together in a merged way. Being motivated by the dictionary-based LZ78 compression algorithm, originally proposed by Ziv and Lempel [45], it assumes the name "LeZi-update" (pronounced "lazy update").

It is worth mentioning that an analogy exists in the field of data compression [5], which says that a message cannot be compressed beyond the entropy of its source without losing any part of it. Optimality is achieved if the length of the compressed message approaches the entropy, that is, the redundancy in the encoding approaches zero. Motivated by this duality, we have designed an update scheme around a compressor–decompressor duo. A family of efficient on-line algorithms for text compression is known to show asymptotically optimal performance [45]. Some versions of these algorithms are hardware-implementable and are in commercial use (e.g., V.42bis compression standard for modems) [43], which indicates practical viability. It has already been understood that a good data compressor must also be a good predictor, because one must predict future data well to compress them. Such techniques based on compression have earlier been used as predictors in caching and prefetching of database and hypertext documents [12,41], as well as for instruction prefetching and branch prediction in processor architecture [16]. Here we have molded them into efficient update and paging techniques for a universal user mobility model in a cellular network environment.

21.5.1 Update Procedure

Before describing the algorithm, let us clarify a very important point: LeZi-update is not meant to replace the threshold-based dynamic update schemes. Rather, it is supposed to reduce the update cost by working as an add-on module to the underlying update scheme. The responsibility of generating the movement history "$v_1 v_2 v_3 \cdots$" still lies with the primary update scheme as before. Let us identify this process of data acquisition as *sampling*. However, a real update message is not generated for each *sampled symbol*. The LeZi-update algorithm captures the sampled message and tries to process it in chunks, thereby delaying the actual update for some sampled symbols. When it finally triggers an actual update, it reports in an encoded form the whole sequence of sampled symbols withheld since the last reporting. In effect, the movement history "$v_1 v_2 v_3 \cdots$" reaches the system as a sequence "$C(w_1) C(w_2) C(w_3) \cdots$," where the w_is are nonoverlapping segments of the string "$v_1 v_2 v_3 \cdots$" and $C(w)$ is the encoding for segment w. The *prime requirement for LeZi-update* (following LZ78) is that *the w_is must be distinct*.

The system's knowledge about the mobile's location always lags by at most the gap between two updates. The uncertainty increases with this gap, yet larger gaps reduce the number of updates. A natural action would be to delay the update if the current string segment being parsed has been seen earlier, that is, the path traversed since the last update is a familiar one. Although the gap goes on increasing, it is expected that the information lag will not affect paging much if the system can somehow make use of the profile generated so far. This prefix-matching technique of parsing is the basis of the LZ78 compression algorithm, which encodes variable length string segments using fixed-length dictionary indices, while the dictionary gets continuously updated as new phrases are seen. In Figure 21.5 and Figure 21.6, we outline this greedy parsing technique from the classical LZ78 algorithm, as used in our context.

The mobile acts as the *encoder*, while the system takes the role of the *decoder*. It must however not be overlooked that the LeZi-update really makes a paradigm shift from the existing *zone-based* to a new *path-based* update messaging. Re-designing the message format would thus be necessary.

```
initialize dictionary := null
initialize phrase w := null
loop
  wait for next symbol v
  if(w.v in dictionary)
    w := w.v
  else
    encode <index(w),v>
    add w.v to dictionary
    w := null
  endif
forever
```

FIGURE 21.5   Encoder at the mobile.

```
initialize dictionary := null
loop
  wait for next codeword <i,s>
  decode phrase := dictionary [i].s
  add phrase to dictionary
  increment frequency for every
          prefix of phrase
forever
```

FIGURE 21.6   Decoder at the system.

21.5.2 Incremental Parse Tree

The LZ78 algorithm emerged out of a need for finding some *universal variable-to-fixed* length coding scheme, where the coding process is interlaced with the learning process for the source characteristics. The key to the learning is a de-correlating process, which works by efficiently creating and looking up an explicit dictionary. The algorithm in Reference 45 parses the input string "$v_1, v_2, \ldots, v_n$" where $v \in \vartheta$, into $c(n)$ distinct substrings $w_1, w_2, \ldots, w_{c(n)}$ such that for all $j \geq 1$, the prefix of substring w_j (i.e., all but the last character of w_j) is equal to some w_i, for $1 \leq i < j$. Because of this *prefix property*, substrings parsed so far can be efficiently maintained in a trie [21].

Figure 21.7 shows the trie formed while parsing the movement history "*aaababbbbbaabccddcbaaaa . . .*" as "*a, aa, b, ab, bb, bba, abc, c, d, dc, ba, aaa, ...*" Commas separate the parsed phrases and indicate the points of updates. In addition to representing the dictionary, the trie can store statistics for contexts explored, resulting in a *symbolwise* model for LZ78. A New dictionary entry can only be created by concatenating a single symbol v to a phrase w already in it. Thus $c(n)$ also captures the storage requirement for maintaining the dictionary at both the mobile and system side.

It is easy to see that as the process of incremental parsing progresses, larger and larger phrases accumulate in the dictionary. Consequently estimates of conditional probabilities for larger contexts start building up. Intuitively, it would gather the predictability or richness of higher and higher order Markov models. Since there is a limit to the model richness for stationary processes, the Lempel–Ziv symbolwise model should eventually converge to the universal model. A result from Reference 15 states that:

Result 21.4 *The symbolwise model created by the incremental parsing asymptotically outperforms a Markov model of any finite order and attains the finite-state predictability. At any point, the effective number of states in the incremental parsing model is $O(c(n))$ and the equivalent Markov order is $O(\log c(n))$. Moreover, for stationary ergodic sources, it attains the predictability of the universal model.*

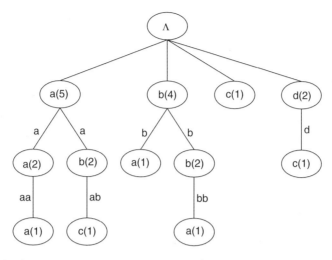

FIGURE 21.7   Trie for the classical LZ symbolwise model.

```
initialize dictionary := null
loop
   wait for next codeword <i,s>
   decode phrase := dictionary [i].s
   add phrase to dictionary
   increment frequency for every prefix of
            every suffix of phrase
forever
```

FIGURE 21.8   Enhanced decoder at the system.

For our running example, the largest order we see is 2. However, not all the order-2 contexts get detected. The first reason behind this is that the algorithm remains unaware about the contexts that cross over phrase boundaries. Unfortunately, the algorithm has to work on one phrase at a time. The second reason is that the decoding algorithm, as is, logs the statistics only for the prefixes of the decoded phrases. None of them influence the asymptotic behavior, but the rate of convergence gets affected to a considerable extent. A simple modification on the decoder (system side) as in Figure 21.8 improves the performance if the dictionary is augmented by the suffixes of the decoded phrases. The enhanced trie for the symbolwise model is shown in Figure 21.9. To appreciate the effect, let us compute the conditional entropies for both the classical and the enhanced symbolwise models. For the classical one, $H(V_1) = \frac{5}{12} \log \frac{12}{5} + \frac{1}{3} \log 3 + \frac{1}{12} \log 12 + \frac{1}{6} \log 6 \approx 1.784$, and $H(V_2|V_1) = \frac{5}{12}(2 \times \frac{1}{2} \log 2) + \frac{1}{3}(\frac{1}{3} \log 3 + \frac{2}{3} \log \frac{3}{2}) \approx 0.723$ bits. An estimate for $H(\mathcal{V})$ is $(1.784 + 0.723)/2 = 1.254$ bits. Conditional entropy for all order-2 contexts turns out to be zero. This is also true for the enhanced one. However, we still have, $H(V_1) = \frac{10}{23} \log \frac{23}{10} + \frac{8}{23} \log \frac{23}{8} + \frac{3}{23} \log \frac{23}{3} + \frac{2}{23} \log \frac{23}{2} \approx 1.742$ bits. Then, $H(V_2|V_1) = \frac{10}{23}(\frac{3}{5} \log \frac{5}{3} + \frac{2}{5} \log \frac{5}{2}) + \frac{8}{23}(2 \times \frac{2}{5} \log \frac{5}{2} + \frac{1}{5} \log 5) \approx 0.952$ bits. The estimate for $H(\mathcal{V})$ is $(1.742 + 0.952)/2 = 1.347$ bits.

21.5.3 Update Cost and Message Complexity

Had the update cost been measured in terms of the volume of update messages in bits, the cost minimization problem can be easily identified with compressing the movement history. However, only the number of updates contributes to the update cost. In an off-line scenario where both the dictionary and the input string are given, it is possible to find out a parsing that minimizes the number of parsed phrases using

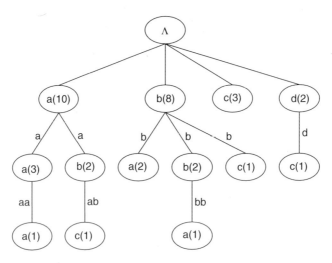

FIGURE 21.9   Trie for the enhanced LZ symbolwise model.

either a breadth-first search or dynamic programming. But, when the history is generated on-line and the dictionary is built adaptively, we do not have that luxury. The literature has an asymptotic optimality result for Lempel–Ziv parsing in terms of the total volume of update messages, while only an upper bound on the growth rate can be derived for the number of updates [11].

Result 21.5 *For a stationary ergodic stochastic process $\mathcal{V} = \{V_i\}$, if $l(V_1, V_2, \ldots, V_n)$ is the Lempel–Ziv codeword length associated with $V_1, V_2, \ldots, V_n$, then*

$$\limsup_{n\to\infty} \frac{1}{n}[l(V_1, V_2, \ldots, V_n)] = H(\mathcal{V}) \tag{21.13}$$

with probability 1.

Result 21.6 *The number $c(n)$ of phrases in a distinct parsing of a string "$v_1, v_2, \ldots, v_n$" satisfies*

$$c(n) = O\left(\frac{n}{\log n - \log\log n}\right), \tag{21.14}$$

where the base of the logarithms is $|\vartheta|$.

The O-notation in the above equation has the usual meaning that $f(n) = O(g(n))$ if and only if there exist positive constants c and n_0 such that $0 \leq f(n) \leq c\,g(n)$ for all $n \geq n_0$.

Result 21.5 essentially implies that LZ78 approaches optimality asymptotically for stationary sources. This coveted property is inherited by LeZi-update, which replaces n location-updates of the primary update algorithm by $c(n)$ path-updates. Assuming that the update messages handle the path-updates with no extra cost, the improvement in update cost from Equation (21.14) is $\Omega(\log n - \log\log n)$, where $f(n) = \Omega(g(n))$ if and only if there exist positive constants c and n_0 such that $f(n) \geq c\,g(n) \geq 0$ for all $n \geq n_0$. According to the algorithms in Figures 21.5 and Figure 21.6, the LeZi-update adds only one entry to the dictionary per update. So, the size of the dictionary after n samplings is also given by $c(n)$. The index needed for encoding would have $O(\log c(n)) = O(\log n - \log\log n)$ bits.

21.5.4 Profile-Based Paging

The location database of every user holds a trie, which is the symbolwise context model corresponding to the enhanced Lempel–Ziv incremental parse tree. Each node except for the root preserves the relevant

TABLE 21.4   Phrases and their Frequencies at Contexts "*aa*," "*a*," and "Λ"

aa (order-2)	*a* (order-1)	Λ (order-0)		
$a\|aa(1)$	$a\|a(2)$	$a(5)$	$ba(2)$	$d(1)$
$\Lambda\|aa(2)$	$aa\|a(1)$	$aa(2)$	$bb(1)$	$dc(1)$
	$b\|a(1)$	$ab(1)$	$bba(1)$	$\Lambda(1)$
	$bc\|a(1)$	$abc(1)$	$bc(1)$	
	$\Lambda\|a(5)$	$b(3)$	$c(3)$	

statistics that can be used to compute total probabilities of contexts as well as conditional probabilities of the symbols based on a given context. A path from the root to any node w in the trie represents a context, and the sub-trie rooted at w reveals the conditional probability model given that context. The paths from the root to the leaves in the Lempel–Ziv trie represent the largest contexts, which are not contained in any other contexts.

Here we describe the underlying principles behind the probability assignments, which are motivated by the principles used in the *prediction by partial match* (PPM) family of text compression schemes [5,10]. However, while the PPM techniques for text compression are concerned with the probability of the next symbol, we are more interested in the probability of occurrence of the symbols (zones) on the path segment to be reported by the next update. These segments are the sequences of zones generated when traversing from the root to the leaves of the sub-trie representing the current context. The estimated conditional probabilities for all the zones at the current context constitutes the conditional probability distribution, based on which zones are ranked for paging. Instead of relying completely on the conditional probability estimates given the context of a specific order, a PPM-style *blending* of these distributions is desirable [5,10]. A family of paging algorithms may emerge based on the choice of these blending strategies.

We use the enhanced trie in Figure 21.9 to illustrate a blending strategy known as *exclusion*, which is often used by the PPM schemes. Assume that the call has arrived, and no LeZi-style path update message has been received since receiving "*aaa*" at 9 p.m. The contexts that can be used are suffixes of "*aaa*," with the exception of itself, viz., "*aa*" (order-2), "*a*" (order-1), and "Λ" (order-0). First, we need to find all possible paths that can be predicted at these contexts. A list of all such paths are shown in Table 21.4, with their respective frequencies. The unconditional probabilities of occurrence of these phrases are then computed by blending. Without going into mathematical details, let us show how to compute of a couple of such probability estimates. The phrase "*a*," for example, appears in the contexts of all the orders 0, 1, and 2. We start from the highest order, that is, the context "*aa*." The phrase "*a*" occurs only once out of three possible occurrences of this context, the other two producing null prediction. Thus we can predict the phrase "*a*" with probability $\frac{1}{3}$ at context "*aa*" and fall back to the lower order with probability $\frac{2}{3}$. Now "*a*" occurs twice at the context "*a*," out of a total ten occurrences of the context. Thus, "*a*" can be predicted with probability $\frac{1}{5}$ at the order-1 context. Due to five occurrences out of the ten producing null predictions, we need to escape to lower order with probability $\frac{1}{2}$. Finally, "*a*" shows up five times out of twenty-three possible phrases including the null phrase, leading to a probability value $\frac{5}{23}$. The blended probability of occurrence for phrase "*a*" is thus $\frac{1}{3} + \frac{2}{3}\{\frac{1}{5} + \frac{1}{2}(\frac{5}{23})\} = 0.5319$. Since the phrase is made of only one symbol "*a*," the whole probability mass gets assigned to it.

To see a little variation, let us also compute the blended probability of the phrase "*bba*." This does not occur in either contexts of order 1 or 2. The probability of escaping the contexts of these two orders by null prediction is $\frac{2}{3} \times \frac{1}{2}$. However, "*bba*" occurs only once among 23 possible phrases seen at a null context. Thus the net probability of occurrence of "*bba*" is $\frac{2}{3}\{\frac{1}{2}(\frac{1}{23})\} = 0.0145$. Because there is only one "*a*" as opposed to two "*b*"s in "*bba*," the individual probabilities of symbols "*a*" and "*b*" within phrase "*bba*" are computed as $\frac{1}{3} \times 0.0145 = 0.0048$ and $\frac{2}{3} \times 0.0145 = 0.0097$, respectively. The probability of individual symbols are thus computed based on relative weights of symbols on these phrases as shown in Table 21.5. These are precisely the location probabilities of the mobile terminal in different cells. Paging order is determined by arranging the cells in decreasing order of these location probabilities.

TABLE 21.5   Probabilistic Prediction of Individual Symbols on Path until Next Update

Phrase	Pr[Phrase]	a	b	c	d
a	$\frac{1}{3} + \frac{2}{3}\{\frac{1}{5} + \frac{1}{2}(\frac{5}{23})\} = 0.5391$	0.5391	0.0000	0.0000	0.0000
aa	$\frac{2}{3}\{\frac{1}{10} + \frac{1}{2}(\frac{2}{23})\} = 0.0957$	0.0957	0.0000	0.0000	0.0000
ab	$\frac{2}{3}\{\frac{1}{2}(\frac{1}{23})\} = 0.0145$	0.0073	0.0073	0.0000	0.0000
abc	$\frac{2}{3}\{\frac{1}{2}(\frac{1}{23})\} = 0.0145$	0.0048	0.0048	0.0048	0.0000
b	$\frac{2}{3}\{\frac{1}{10} + \frac{1}{2}(\frac{3}{23})\} = 0.1104$	0.0000	0.1104	0.0000	0.0000
ba	$\frac{2}{3}\{\frac{1}{2}(\frac{2}{23})\} = 0.0290$	0.0145	0.0145	0.0000	0.0000
bb	$\frac{2}{3}\{\frac{1}{2}(\frac{1}{23})\} = 0.0145$	0.0000	0.0145	0.0000	0.0000
bba	$\frac{2}{3}\{\frac{1}{2}(\frac{1}{23})\} = 0.0145$	0.0048	0.0097	0.0000	0.0000
bc	$\frac{2}{3}\{\frac{1}{10} + \frac{1}{2}(\frac{1}{23})\} = 0.0812$	0.0000	0.0406	0.0406	0.0000
c	$\frac{2}{3}\{\frac{1}{2}(\frac{3}{23})\} = 0.0435$	0.0000	0.0000	0.0435	0.0000
d	$\frac{2}{3}\{\frac{1}{2}(\frac{1}{23})\} = 0.0145$	0.0000	0.0000	0.0000	0.0145
dc	$\frac{2}{3}\{\frac{1}{2}(\frac{1}{23})\} = 0.0145$	0.0000	0.0000	0.0073	0.0073
Sum		0.6662	0.2018	0.0962	0.0218

Probabilistic prediction is not necessarily the only advantage of building the Lempel–Ziv tries. Search can also be done on the entire trie or a context sub-trie using breadth-first, depth-first, or a combined strategy. Geographical information about the neighborhood of LAs comes as a by-product of constructing these tries, especially when the movement-based update scheme with $M = 1$ is used for sampling. A breadth-first search pages the mobile in the neighboring zones where it is known to have gone before. This is really a pruned version of paging tier by tier [26]. On the other hand, a depth first search can be spawned biased by the weights of the paths emanating from the node and finishing in a leaf, which gives preference to movement in specific directions. An ordering of zones is always prepared for paging, irrespective of the criterion used for this ordering. This list, of course, is guaranteed not to include a zone never visited before, based on the protocol that the mobile will update if the user ends up moving to a new location.

21.6   Tradeoff between Update and Paging

While the LeZi-Update strategy achieves near-optimal location management, it does not allow mobile users to arbitrarily tradeoff between update and paging costs. For example, if two devices, one a WatchPad and the other a laptop, exhibit the same movement pattern, they will have identical update costs, even though the WatchPad is clearly more resource-constrained than the laptop and would possibly like to reduce its update cost even further. Moreover, LeZi-Update does not explicitly exploit the registration based location management infrastructure currently deployed in cellular PCS networks which generally cluster a group of cells into *registration areas* (RAs), such that a mobile's location uncertainty is confined into its last reported RA. We now outline a framework from [35] that permits devices to customize the tradeoff between their location update and paging costs, and yet exploits the existing concept of cellular clusters or RAs.

As the update cost in LeZi-Update is designed to approach the entropy of the mobile node's (MN's) movement pattern, it is clear that no further reduction in update cost is feasible without compromising on the accuracy of the reported information. This leads us to investigate the use of *lossy quantization* algorithms with the LeZi-Update framework, such that the network does not necessarily receive the mobile's true movement history, but a *reasonably close* approximation to it. From a theoretical perspective, we are looking for algorithms that result in nonzero distortion D, and consequently are lower-bounded

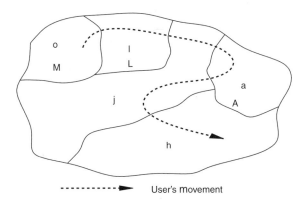

FIGURE 21.10 Organization of cells into RAs.

by a rate-distortion value $R(D)$ that is even lower than the entropy $H(X)$. Intuitively, we aim at reducing the update cost even further by allowing some loss of information.

We restrict ourselves to two simple variants of LeZi-Update in this chapter, details of which are available in Reference 35. The modified protocols are motivated by a fundamental result [14] that a *cascaded combination of scalar quantization and entropy coding usually results in a communication overhead that is fairly close to the lower bound of rate-distortion theory*. In this enhanced framework, the network still receives the update from the MN as a sequence of symbols. Now, however, *each symbol represents an RA*, rather than an individual cell. The MN essentially has quantized its location coordinates, revealing its cell-level location information at the coarser granularity of the RA. Figure 21.6 shows the organization of the cells into three different RAs, say A, J, and M are consisting of cells $(a, h), (j, l)$, and (o), respectively.

In the first enhanced approach, called LeZi–RA, the MN's movement pattern is first quantized into a sequence of RAs on which the $LZ78$-based incremental parsing technique is applied. Then the quantized, lossy RA sequence becomes: "$A, J, J, J, M, M, J, A, A, A, A, J, J, J, M, M, J, A, A, J, J, J, M, M, J, A, A, J, M, J.$" The incremental parsing results in "$A, J, JJ, M, MJ, AA, AAJ, JJM, MJA, AJ, JJMM, JAA, AJM, J \ldots.$" In the second approach, called RA–LeZi, the run-length of the quantized RA sequence is ignored and a new quantized sequence is generated only when the mobile changes its current RA. Thus, the lossy quantization and its subsequent parsing of the original sequence S_1 now results in the sequence "$A, J, M, J, A, J, M, J, A, J, M, J, A, J, M$" and set of phrases "$A, J, M, JA, JM, JAJ, MJ, AJ, J \ldots.$" The $LZ78$-based LeZi-Update scheme is then applied on these modified RA-level sequences, rather than the original cell-level movement sequence (Figure 21.10).

At the network's end, the paging process is, however, identical for both these schemes. The residence probability of every symbol (RA) in the trie is computed as before, and is *equally distributed* among its constituent cells. The paging strategy then polls the RAs according to their *normalized residence probabilities* (residence probability divided by RA size). With each RA, the cells may either be paged simultaneously or sequentially in random order. The increased uncertainty in the paging process arises from the fact that the quantization into RA-level information loses the details about the MNs precise movement pattern *within* an RA. Naturally, we can expect the paging costs with these two quantized protocols to be higher than LeZi-Update. However, as the rate-distortion theory shows, the paging costs would be the minimum possible, given the reduced location update costs.

21.7 Conclusion

We have identified the *uncertainty* due to user mobility as the complexity of the location management problem in a cellular environment. Therefore, *entropy* is the natural measure for comparing mobility tracking algorithms. As entropy is oftentimes quantified in terms of the number of bits, we arrive at a very intuitive definition of location uncertainty as the least amount of message that needs to be exchanged so that the exact location is known. This provides a common ground for comparing various techniques that are proposed in the literature.

The main difference between our approach and the traditional line of research on location management in cellular PCS networks lies in the perspective. We formulate mobility tracking as a *dynamic optimization* problem as opposed to the existing static formulation specified in terms of some known input parameters. The scope of our study extends to the estimation of those parameters from data provided by the underlying *sampling mechanism*. The update schemes proposed in the literature so far are essentially a characterization of these sampling mechanisms. While sampling is not the primary focus of this chapter, we have pointed out that one can obtain highly informative samples by combining the time-based and movement-based schemes. The important issue is that the sampled data may have enough redundancy, and need not be sent "as is" in the update messages. Characterizing the user movement data as a stochastic process, we have identified *stationarity* to be the sufficient criterion for *learning* the mobility profile. Using entropy as the basis for comparing models, we have illustrated the existence and properties of a *universal model* when the profile is stationary.

We have adopted the *LZ78 compression* algorithm as the basis of our update scheme. The choice was guided by two factors, viz., the existence of an explicit symbolwise context model created by incremental parsing, and the tendency of this model to asymptotically converge to a universal one. As seen in lossless data-compression [5], *piecewise stationarity* of the sampled sequence should help LZ78 to achieve a high level of performance. The use of pure LZ78 in this paper is motivated by its simplicity in establishing the theoretical basis. Practical implementation would call for a fancier LZ78 variant which must work efficiently with limited memory and should have the capability of forgetting beyond the recent past. We have also outlined how a paging strategy is to be derived out of such a model.

This optimal location tracking scheme is recently extended to multisystem, heterogenous cellular networks. The details of this strategy can be found in Reference 24 where we have introduced the concept of *cost entropy* and *session-state information* to capture the different signalling costs and mobile's active/idle state. Centralized, distributed, and heuristic-based location update and paging strategies, with different level of coordination among the sub-networks are formulated. A new near-optimal greedy paging strategy for heterogeneous network is subsequently proposed in Reference 34. Finally, the proposed scheme is likely to be of immediate interest to wireless service providers. By maintaining global dictionaries along with individual user profile decoded from updates, it will be possible to predict group behavior. In wireless data networks, this can lead to more efficient bandwidth management and quality of service (QoS) provisioning based on reservation [36,37].

Acknowledgments

This work is supported by NSF grants IIS-0326505 and IIS-0121297. The authors are also grateful to Diane Cook, Abhishek Roy and Archan Misra for collaborations on extensions to and applications of the LeZi-Update scheme.

References

[1] I. F. Akyildiz and J. S. M. Ho, "Dynamic mobile user location update for wireless PCS networks," *Wireless Networks*, 1: 187–196, 1995.

[2] I. F. Akyildiz and J. S. M. Ho, "Movement-based location update and selective paging for PCS networks," *IEEE/ACM Transactions on Networking*, 4: 629–638, 1995.

[3] A. Bar-Noy and I. Kessler, "Tracking mobile users in wireless communication networks," *IEEE Transactions on Information Theory*, 39: 1877–1886, 1993.

[4] A. Bar-Noy, I. Kessler, and M. Sidi, "Mobile users: To update or not to update?," *Wireless Networks*, 1: 175–185, 1995.

[5] T. C. Bell, J. G. Cleary, and I. H. Witten, *Text Compression*, Prentice Hall, New York, 1990.

[6] A. Bhattacharya and S. K. Das, "LeZi-update: An information-theoretic approach to track mobile users in PCS networks," in *Proceedings of the Annual International Conference on Mobile Computing and Networking*, 1–12 August 1999.

[7] A. Bhattacharya and S. K. Das, "LeZi-Update: An Information-theoretic approach for personal mobility tracking in PCS networks," *ACM-Kluwer Wireless Networks* 8: 121–137, 2002.

[8] Y. Birk and Y. Nachman, "Using direction and elapsed-time information to reduce the wireless cost of locating mobile units in cellular networks," *Wireless Networks*, 1: 403–412, 1995.

[9] T. X. Brown and S. Mohan, "Mobility management for personal communication systems," *IEEE Transactions on Vehicular Technology*, 46: 269–278, 1997.

[10] J. G. Cleary and I. H. Witten, "Data compression using adaptive coding and partial string matching," *IEEE Transactions on Communications*, 32: 396–402, 1984.

[11] T. M. Cover and J. A. Thomas, *Elements of Information Theory*, John Wiley, New York, 1991.

[12] K. M. Curewitz, P. Krishnan, and J. S. Vitter, "Practical prefetching via data compression," in *Proceedings of the ACM International Conference on Management of Data (SIGMOD)*, pp. 257–266, May 1993.

[13] S. K. Das and C. Rose, "Coping with uncertainty in mobile wireless networks," in *Proceedings of the Personal, Indoor and Mobile Radio Communications (PIMRC)*, Barcelona, Spain, September 2004 (Invited Paper).

[14] N. Farvardin and J. Modestino, "Optimum quantizer performance for a class of non-gaussian memoryless sources," *IEEE Transactions on Information Theory*, May 30: 485–498, 1984.

[15] M. Feder, N. Merhav, and M. Gutman, "Universal prediction of individual sequences," *IEEE Transactions on Information Theory*, 38: 1258–1270, 1992.

[16] E. Federovsky, M. Feder, and S. Weiss, "Branch prediction based on universal data compression algorithms," in *Proceedings of the Annual International Symposium on Computer Architecture*, pp. 62–72, June–July 1998.

[17] J. S. M. Ho and I. F. Akyildiz, "Mobile user location update and paging under delay constraints," *Wireless Networks*, 1: 413–425, 1995.

[18] S. Kahan, "A model for data in motion," in *Proceedings of the Annual ACM Symposium on Theory of Computing*, pp. 267–277, 1991.

[19] J. N. Kapur and H. K. Kesavan, *Entropy Optimization Principles with Applications*, Academic Press, New York, 1992.

[20] S. J. Kim and C. Y. Lee, "Modeling and analysis of the dynamic location registration and paging in microcellular systems," *IEEE Transactions on Vehicular Technology*, 45: 82–90, 1996.

[21] G. G. Langdon, "A note on Ziv-Lempel model for compressing individual sequences," *IEEE Transactions on Information Theory*, 29: 284–287, 1983.

[22] G. L. Lyberopoulos, J. G. Markoulidakis, D. V. Polymeros, D. F. Tsirkas, and E. D. Sykas, "Intelligent paging strategies for third generation mobile telecommunication systems," *IEEE Transactions on Vehicular Technology*, 44: 543–553, 1995.

[23] U. Madhow, M. L. Honig, and K. Steiglitz, "Optimization of wireless resources for personal communications mobility tracking," *IEEE/ACM Transactions on Networking*, 3: 698–707, 1995.

[24] A. Misra, A. Roy, and S. K. Das, "An information theoretic framework for optimal location management in multi-system 4G wireless networks", in *Proceedings of the IEEE INFOCOM*, March 2004.

[25] M. Mouly and M.-B. Pautet, *The GSM System for Mobile Communications*, 1992.

[26] D. Munoz-Rodrguez, "Cluster paging for traveling subscribers," in *Proceedings of the IEEE Vehicular Technology Conference*, pp. 748–753, May 1990.

[27] D. Plassmann, "Location management strategies for mobile cellular networks of 3rd generation," in *Proceedings of the IEEE Vehicular Technology Conference*, pp. 649–653, June 1994.

[28] G. P. Pollini and C.-L. I, "A profile-based location strategy and its performance," *IEEE Journal on Selected Areas in Communications*, 15: pp. 1415–1424, 1997.

[29] J. Rissanen and G. G. Langdon, "Universal modeling and coding," *IEEE Transactions on Information Theory*, 27: 12–23, 1981.

[30] J. Rissanen, "A Universal data compression system," *IEEE Transactions on Information Theory*, 29: 656–664, 1983.

[31] C. Rose and R. Yates, "Minimizing the average cost of paging under delay constraints," *Wireless Networks*, 1: 211–219, 1995.

[32] C. Rose, "Minimizing the average cost of paging and registration: A timer-based method," *Wireless Networks*, 2: 109–116, 1996.

[33] A. Roy, S. K. Das, and A. Misra, "Exploiting information theory for adaptive mobility and resource management in future cellular networks," *IEEE Wireless Communications* (Special Issue on Mobility and Resource Management), 11: 59–65, 2004.

[34] A. Roy, A. Misra, and S. K. Das, "The optimal paging problem for multi-system heterogenous wireless networks," in *Proceedings of the WiOpt*, Cambridge, UK, pp. 94–103, March 2004.

[35] A. Roy, A. Misra, and S. K. Das, "A rate-distortion framework for information theoretic mobility management", in *Proceedings of IEEE ICC*, June 2004.

[36] A. Roy, S. K. Bhaumik, A. Bhattacharya, K. Basu, D. J. Cook, and S. K. Das, "Location aware resource management in smart homes," in *Proceedings of First IEEE International Conference on Pervasive Computing and Communications* (PerCom'03), Ft Worth, TX, pp. 481–488, March 2003.

[37] A. Roy, S. K. Bhaumik, K. Basu, and S. K. Das, "Resource reservation for location-oriented multimedia in smart homes," in *Proceedings of the 3rd International Workshop on Mobile Multimedia Communications* (MoMuc), Munich, Germany, October 2003.

[38] S. Sen, A. Bhattacharya, and S. K. Das, "A selective location update strategy for PCS users," *Wireless Networks*, 5: 313–326, 1999.

[39] C. E. Shannon, "The mathematical theory of communication," *Bell System Technical Journal*, 27: 379–423, 623–659, 1948, reprinted in *The Mathematical Theory of Communications by C. E. Shannon and W. Weaver*, University of Illinois Press, pp. 29–125, 1949.

[40] J. F. Traub and A. G. Werschulz, *Complexity and Information*, Cambridge University Press, Cambridge, UK, 1998.

[41] J. S. Vitter and P. Krishnan, "Optimal prefetching via data compression," *Journal of the ACM*, 43: 771–793, 1996.

[42] M. J. Weinberger, J. J. Rissanen, and M. Feder, "A universal finite memory source," *IEEE Transactions on Information Theory*, 41: 643–652, 1995.

[43] T. A. Welch, "A technique for high-performance data compression," *IEEE Computer*, 17: 8–19, 1984.

[44] H. Xie, S. Tabbane, and D. Goodman, "Dynamic location area management and performance analysis," in *Proceedings of the IEEE Vehicular Technology Conference*, pp. 533–539, May 1993.

[45] J. Ziv and A. Lempel, "Compression of individual sequences via variable-rate coding," *IEEE Transactions on Information Theory*, 24: 530–536, 1978.

[46] M. M. Zonoozi and P. Dassanayake, "User mobility modeling and characterization of mobility patterns," *IEEE Journal on Selected Areas in Communications*, 15: 1239–1252, 1997.

22

Mobility Control and Its Applications in Mobile Ad Hoc Networks

22.1 Introduction..**22**-501
22.2 Background and Motivation **22**-502
22.3 Existing Work **22**-503
    CDS Formation • Topology Control • Mobility Control
22.4 Link Availability.....................................**22**-506
22.5 View Consistency **22**-509
    View Consistency Issue • Consistent Local View •
    Conservative Local View • Smart Neighbor Discovery
22.6 Simulation Results **22**-514
22.7 Conclusion **22**-516
Acknowledgment..**22**-516
References **22**-516

Jie Wu
Fei Dai

22.1 Introduction

Most existing localized protocols in mobile ad hoc networks (MANETs) use local information for decentralized decision among nodes to achieve certain global objectives. These objectives include determining a small connected dominating set (CDS) for virtual backbone and topology control by adjusting transmission ranges of nodes. Because of asynchronous sampling of local information at each node, delays at different stages of protocol handshake, and movement of mobile nodes, local view captured at different nodes may be *outdated* and *inconsistent*, and may not reflect the actual network situation. The former may incur "broken" links that fail to keep the given global constraint such as global domination and connectivity; the latter may cause "bad" decisions, which in turn will ultimately cause the failure of the constraint. This chapter reviews several *mobility control* techniques that handle outdated and inconsistent local views. These techniques are illustrated using several well-known protocols in data communication and topology control.

The remainder of this chapter is organized as follows. Section 22.2 introduces the topology control problem and our motivation. Section 22.3 reviews related work in CDS formation, topology control, and mobility control. Section 22.4 discusses the broken link issue and introduces a solution using large actual transmission ranges to create a buffer zone. Section 22.5 illustrates bad decisions caused by inconsistent local views, and introduces three view consistency mechanisms. Simulation results showing the effectiveness of the proposed topology control methods are presented in Section 22.6. Section 22.7 concludes this chapter and discusses some open issues.

22.2　Background and Motivation

In MANETs, all nodes cooperate to achieve a global task, such as data gathering, communication, and area monitoring. A MANET is usually modeled as a *unit disk graph* where two nodes are connected if and only if their geographical distance is smaller than a given transmission range (as shown in Figure 22.1(a)). To design protocols that are simple and quick to converge, many protocols in MANETs rely on localized algorithms. A deterministic distributed algorithm is *localized* if its execution on each node relies on local information (e.g., 2-hop topology information or 1-hop location information) only, without any sequential propagation of partial computation results; otherwise, it is *nonlocalized*. Collectively, nodes running the localized algorithm achieve some desirable global objectives. Two popular applications of the localized algorithm are (1) determining a CDS for efficient routing [10,26,35,49] and (2) selecting an appropriate transmission range of each node for topology control [22,29,42,43].

A CDS is a subset such that each node in the system is either in the set or the neighbor of a node in the set. In addition, the induced graph of the CDS is connected. Nodes in the CDS are called *dominators* and those outside *dominatees*. As dominators form a backbone of the MANET, any dominatee can switch to the sleep mode for energy saving without causing network partition. Other applications of CDS include area coverage and efficient broadcasting. Finding a minimum CDS is known as NP-complete. Most practical approaches in MANETs use localized algorithms to find a small CDS. In a typical localized CDS protocol, each node v uses local information to determine its status, dominator, or dominatee, based on its 2-hop topology information, including v's neighbors and neighbors' neighbors. Figure 22.1(b) shows a CDS constructed via a localized algorithm [10]. In this diagram the connections between dominatees are not shown.

In MANETs, in order to reduce energy consumption and signal interference, it is important to select an appropriate transmission power for each node, while still satisfying certain global constraints such as connectivity. This task, called topology control, is usually implemented via localized algorithms. In such an algorithm, each node first selects only a few *logical neighbors* among its *physical neighbors* (i.e., nodes within its normal transmission range), and then sets its actual transmission range to the distance of the farthest logical neighbor. Localized topology control algorithms usually rely on 1-hop location information of the current node v, including ids and (x, y) coordinations of v's physical neighbors. Some algorithms require less information where the distance or angle of arrival (AoA) information is used instead of location information. Figure 22.1(c) shows the result of a localized topology control algorithm [22], where both the average number of neighbors and transmission range are reduced significantly, while the network is still connected.

Compared with their nonlocalized counterparts, localized algorithms are lightweight, fast to converge, and resilient to node movement. However, without a mobility control mechanism, global domination and connectivity may still be compromised by node movement. In most existing localized algorithms, each node in a MANET emits a periodic "Hello" message to advertise its presence and its position (if needed) at a fixed interval Δ. "Hello" intervals at different nodes are asynchronous to reduce message collision. Each node extracts neighborhood information from received "Hello" messages to construct a local view of its vicinity (e.g., 2-hop topology or 1-hop location information). In a MANET with mobile nodes, the limited sample frequency, asynchronous "Hello" intervals, and delays at different stages of protocol handshake will cause a mismatch between the *virtual network* constructed from the collection of local views sampled at different nodes and the *actual network*. This mismatch will cause the link availability issue, where a neighbor in a virtual network is no longer a neighbor in the actual network, because the

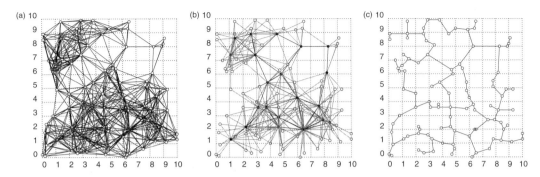

FIGURE 22.1 Virtual networks constructed via localized algorithms. The original MANET has 100 nodes and a transmission range of 2.5. (a) The original network. (b) Connected dominating set. (c) Topology control.

virtual network is constructed from outdated information. Therefore, special mechanisms are needed to address the following issue:

- *Delay and mobility management*: How protocols deal with imprecise neighborhood information caused by node mobility and various delays introduced at different stages of protocol handshake.

One solution in Reference 47, as will be discussed later in detail, uses two transmission ranges to address the link availability issue. First, a transmission range r is determined based on the selected protocol. This transmission range is either the same as the "Hello" message range r' as in the CDS protocol or shorter than r' as in the topology control protocol. The actual transmission uses a long transmission range set to $r + l$. The difference, l, between these two ranges is based on the update frequency and the speed of node movement.

The mismatch between the virtual- and actual-network will cause a more serious problem: inconsistent local views. Inconsistent local views may cause "bad" decisions that fail to keep the global constraint such as global domination and connectivity. Again, special mechanisms are needed to address the following issue:

- *Node synchronization and consistent local view*: How each node knows when to sample its local view. How each node collects and uses the local information in a consistent way.

We will examine three different approaches that address the consistency issue: enforcing consistent views, making conservative decisions, and smart neighbor discovery. The first approach was initially proposed in Reference 48 to construct consistent 1-hop location information for mobility control. This approach can also be extended to support the construction of 2-hop information. The second approach was originally proposed in Reference 47 for CDS formation. The same principle can be used in topology control. The third approach was first proposed in Reference 11 and extended in Reference 47 for mobility control in CDS-based efficient broadcasting.

The main objective of this chapter is to expose to the reader the challenging issue related to mobility control. Through discussion on the effect of mobile nodes on several important protocols, we present some problems, provide possible solutions, and discuss several open issues.

22.3 Existing Work

Before introducing our mobility control schemes, we shall first have a brief look at existing CDS formation and topology control protocols. Here we focus on localized algorithms based on accurate neighborhood information, which have no or limited mobility control. This section also examines some preliminary work on the mobility control issue.

22.3.1　CDS Formation

Both *centralized* and *decentralized* algorithms have been proposed to form a CDS in MANETs. Centralized algorithms [12,41] usually yield a smaller CDS than decentralized algorithms, but their application is limited due to the high maintenance cost. Decentralized algorithms can be further divided into cluster-based algorithms and localized algorithms. Cluster-based algorithms [2,5,14,23] have a constant approximation ratio in unit disk graphs and relatively slow convergence. In localized algorithms [1,9,26,28,32,33,46,49], the status of each node depends on its 2- or 3-hop information only. A localized algorithm takes constant steps to converge, produces a small CDS on average, but has no constant approximation ratio.

Wu and Li [49] devised the marking process and two pruning rules (called Rule 1 and Rule 2) that select a few nodes to form a CDS in an undirected graph. Wu [46] extended this scheme for CDS formation in directed graphs. This scheme is further extended by Dai and Wu [10] via a more efficient pruning rule. Rieck et al. [28] extended Wu and Li's marking process to form a *d*-hop CDS, and suggested to use Rule 1 only to preserve shortest paths among any pair of source and destination nodes. Since Rule 1 alone is not so efficient in reducing the number of dominators, Rieck's scheme usually yields a larger CDS.

Chen et al. [9] presented an approach similar to the marking process (called Span) to select a few *coordinators* for energy efficient routing. Ideally coordinators form a CDS such that other nodes can switch to the energy saving mode without compromising the routing capability of the network. A node *v* becomes a coordinator if it has two neighbors that are not directly connected, indirectly connected via one intermediate coordinator, or indirectly connected via two intermediate coordinators. Before a node changes its status from noncoordinator to coordinator, it waits for a backoff delay which is computed from its energy level and 2-hop neighborhood topology. The backoff delay can be viewed as a priority value, such that nodes with shorter backoff delay have a higher chance of becoming coordinators. Span cannot ensure a CDS when two coordinators simultaneously change back to noncoordinators.

Qayyum et al. [26] put forward an efficient broadcast scheme called multipoint relaying (MPR). In MPR, each node designates a few 1-hop neighbors (MPRs) to cover its 2-hop neighbor set. All MPRs form a CDS, which was used as a backbone in the optimized link state routing (OLSR) protocol. Adjih et al. [1] suggested a method to reduce the backbone size. Based on the new rule, not all MPRs are dominators; a MPR becomes a dominator only when it happens to be designated by a node with the lowest id in its 1-hop neighbors. Meanwhile, to maintain the connectivity, any node with the lowest id among its neighbors becomes a dominator automatically.

Sinha et al. [32] proposed the *core*-based scheme. Each node designates a node with the highest priority among its neighbors (including itself) as a its dominator (called core). The set of cores form a dominating set, but not necessarily a CDS. To get a CDS, each core designates several gateway nodes to connect neighboring cores within 3 hops [33]. This scheme was extended by Amis et al. [3] to form a *d*-hop dominating set.

22.3.2　Topology Control

Many topology control schemes have been proposed to use a small actual transmission range to conserve energy and bandwidth consumption while maintaining network connectivity. Centralized algorithms [25,27,44] construct optimized solutions based on global information and, therefore, are not suitable in MANETs. Probabilistic algorithms [7,24,27] adjust transmission range to maintain an optimal number of neighbors, which balances energy consumption, contention level, and connectivity. However, they do not provide hard guarantees on network connectivity. Most localized topology control algorithms use nonuniform actual transmission ranges computed from 1-hop location information. Here the 1-hop neighborhood refers to physical neighbors under the normal transmission range.

The relative neighborhood graph (RNG) [38] is a geometrical graph used to remove edges (i.e., reduce the number of neighbors) while maintaining network connectivity. An edge (u, v) is removed if there exists a third node w such that $d(u, v) > d(u, w)$ and $d(u, v) > d(v, w)$, where $d(u, v)$ is the Euclidean distance between u and v. In localized topology control protocols [8,31], each node determines its logical neighbor set based on positions of 1-hop neighbors. Two nodes u and v are logical neighbors if and only if edge (u, v) exists in RNG. The Gabriel graph [16] is a special case of RNG, where the third node w is restricted to the disk with diameter uv.

Rodoplu and Meng [29] gave another method of reducing the number of edges while maintaining network connectivity and, in addition, preserving all minimum-energy paths. A minimum-energy path between two nodes u and v is defined as the shortest path between u and v, using transmission power as edge cost. An edge (u, v) can be removed if there exists another node w, such that the 2-hop path (u, w, v) consumes less energy than direct transmission. Li and Halpern [20] extended this scheme by using k-hop ($k \geq 2$) paths to remove more edges and, at the same time, reduce the computation overhead.

In both protocols [20] and [29], instead of selecting logical neighbors from the normal 1-hop neighbor set, each node collects location information of nodes within a small *search region* to conserve control message overhead. The radius of the search region is iteratively enlarged until logical neighbors in the search region can cover the entire normal 1-hop neighborhood; that is, each position outside of the search region can be reached via a k-hop path through a selected logical neighbor, and the k-hop path is more energy-efficient than direct transmission. If the search region is the entire 1-hop neighborhood, the Li and Halpern's algorithm is equivalent to constructing a local shortest path tree (SPT) where only neighbors of the root in SPT become logical neighbors.

In cone-based topology control (CBTC) [21,43], the logical neighbor set $\{w_1, w_2, \ldots, w_k\}$ of node v is selected to satisfy the following condition: if a disk centered at v is divided into k cones by lines vw_i ($1 \leq i \leq k$), the angle of the maximal cone is no more than α. It was proved in Reference 21 that, when $\alpha \leq 5\pi/6$, CBTC preserves connectivity, and, when $\alpha \leq 2\pi/3$, the corresponding symmetric subgraph (a subgraph after removing all unidirectional edges) is connected. Several optimizations are also proposed in Reference 21 to further reduce the number of logical neighbors and transmission range. Bahramgiri et al. [4] extended CBTC to provide k-connectivity with $\alpha \leq 2\pi/3k$. Similar to the minimum-energy algorithms, CBTC uses dynamic search regions to reduce control overhead. Furthermore, CBTC requires only direction information instead of accurate location information.

A similar but different scheme is based on Yao graph [42], where a disk centered at node v is evenly divided into k cones, and a logical neighbor is selected from each cone. It was proved that Yao graph is connected with $k \geq 6$. Yao graph with $k = 6$ can be viewed as a special case of CBTC with $\alpha = 2\pi/3$, but not vice versa.

Li et al. [22] proposed to build a local minimal spanning tree (MST) at each node based on 1-hop location information and select neighbors in MST as logical neighbors. This scheme guarantees connectivity, is easy to implement, and has a constant upper bound (6) on the number of logical neighbors of each node.

22.3.3 Mobility Control

We use the name mobility control to denote mechanisms in localized algorithms that compensate inaccurate neighborhood information and maintain the global constraints. Previous work on topology control is very limited. This section addresses those preliminary attempts to handle the mobility issue.

Locally formed CDS are frequently used in efficient broadcast protocols, where dominators become forward nodes to rely the broadcast packet. In their simulation study, Williams and Camp [45] showed that localized broadcast protocols suffer low delivery ratio in mobile and congested networks. In a later simulation study [11], Dai and Wu concluded that the major cause of the low delivery ratio is contention and mobility. In order to reduce contention, fewer forward nodes should be used to reduce redundancy. However, low redundancy makes a broadcast protocol more vulnerable to packet losses caused by mobility. Therefore, an adaptive approach is desirable that has low redundancy under low mobility, and increases redundancy only when the mobility level rises. Although several probabilistic broadcast protocols [18,39] have been proposed by trading between efficiency with delivery ratio, it is difficult to establish a direct connection between forwarding probability and node mobility.

It was shown in Reference 24 that connectivity in probabilistic topology control algorithms is barely affected by mobility. Blough et al. [7] showed that connectivity is preserved with high probability (95%) if every node keeps nine logical neighbors. In our approach, the logical neighbor set and transmission range are first computed from the neighborhood information of each individual node, and then adjusted to balance the mobility. Compared with the uniform optimal node degree in probabilistic algorithms, our approach requires fewer logical neighbors on average. Although the node degree in Reference 7 can be further reduced, it is not clear whether the resultant topology is still resilient to mobility after optimization.

Very little work has been done in maintaining an accurate neighbor set in ad hoc networks. One exception is [19], where a *stable zone* and a *caution zone* of each node has been defined based on a node's position, speed, and direction information obtained from GPS. Specifically, stable zone is the area in which a mobile node can maintain a relatively stable link with its neighbor nodes since they are located close to each other. Caution zone is the area in which a node can maintain an unstable link with its neighbor nodes since they are relatively far from each other. The drawback of this approach is that it is GPS-based, which comes with a cost, as will be discussed in the next section. In addition, there is no rigorous analysis on the impact of mobility on the selection of these two zones.

Several papers [6] addressed the issue of how long two nodes will remain in close enough proximity for a link between them to remain active. Several routing protocols, associativity-based routing (ABR) [37] and signal stability-based adaptive routing (SSA) [13], have been proposed that select *stable links* to construct a route. In Reference 36, GPS information is used to estimate the expiration time of the link between two adjacent hosts. Recently, several studies have been done on the effect of mobility on routing path [30]. However, no broadcast protocol uses the notion of stable link to evaluate the stability of neighbor set in order to better decide the forwarding status of each node.

In general, the capacity of ad hoc networks is constrained by the mutual interference of concurrent transmissions between nodes. The mobility of nodes adds another dimension of complexity in the mutual interference. Epidemic routing [40] is a scheme to provide eventual delivery in occasionally partitioned networks. It exploits rather than overcomes mobility to achieve a certain degree of connectivity. Basically, when two nodes are moving close to each other within their mutual transmission range, these two nodes will exchange data (based on a probabilistic model) that does not appear in the other node. This scheme extends the Infostation model [15,17,34], where one or several neighbors are selected as intermediate nodes to store the data received from the source and to relay the data when they eventually roam into the transmission range of the destination. The difference is that Infostation corresponds to a 2-hop routing process (from source to destination via one intermediate node) whereas epidemic routing is a general multiple-hop routing process.

22.4　Link Availability

The broken link problem caused by outdated local views is relatively easy to understand and deal with. In this section, we discuss the link availability issue and our solution to this problem. The view consistency issue is more complex and will be discussed in the next section.

In MANETs, because of asynchronous "Hello" messages and various protocol handshake delays, neighborhood information and position used in decision making may be outdated. For example, a previously sampled neighbor can move out of transmission range during the actual transmission. In order to apply existing protocols without having to redesign them, the notion of *buffer zone* is used in Reference 47, where two circles with radii r and $r + l$ are used (see Figure 22.2), r corresponds to the transmission range determined by a selected protocol, whereas $r + l$ corresponds to the actual transmission range used. $l = d \times 2t$ is defined as a buffered range depending on the moving speed t of mobile nodes and the maximum delay d. To simplify the discussion, both "Hello" intervals and moving patterns/speeds are homogeneous, and hence, l is uniform for each node.

The above requirement of buffered range guarantees link availability in the worst case situation. However, probabilistic study in Reference 47 reveals that the worst case rarely happens. In MANETs

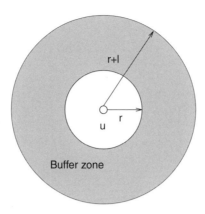

FIGURE 22.2 The notion of buffer zone with different transmission ranges.

with very high moving speed (t), it is either impossible or too expensive to use such a large l. Both probabilistic analysis and simulation results in Reference 47 show that link availability is preserved in most cases with a buffered range much smaller than $d \times 2t$. There is a wide range of potential trade-offs between efficiency and connectivity.

Specifically, suppose r' is the normal "Hello" message range. A typical CDS protocol works as follows:

1. Select $r = r'$ for neighborhood information exchange.
2. Apply the selected localized CDS protocol to determine the status of each node.
3. Use $r + l$ for each dominator in the actual transmission.

Step 2 of the above process varies from protocol to protocol. Here we use Wu and Li's marking process and Rules 1 and 2 [49] to illustrate:

At each node u:

- Marking Process: u is marked true (i.e., becomes a dominator) if there are two unconnected neighbors.
- Rule 1: u is unmarked (i.e., becomes a dominatee) if its neighbor set is covered by another node with a higher id.
- Rule 2: u is unmarked if its neighbor set is covered jointly by two connected nodes with higher id's.

In Rules 1 and 2, we say u's neighbor set, $N(u)$, is *covered* by one or two *covering* nodes, if every node w in $N(u)$ is either a covering node or a neighbor of a covering node. Using the above marking process with Rules 1 and 2, marked nodes form a CDS. Figure 22.3 shows a sample ad hoc network with 9 nodes. Node r is unmarked by the marking process because its neighbors u and z are directly connected. Node w is unmarked by Rule 1 because its neighbor set is covered by node x. Here node id x is higher than w according to the alphabetical order. Node u is unmarked by Rule 2 because its neighbor set is covered by two connected nodes x and z. Clearly, the marked nodes v, x, and z form a CDS of the sample network.

Originally, Rules 1 and 2 use only marked nodes as covering nodes, and involve overhead in communicating dominating set status. Stojmenovic et al. [35] showed that unmarked nodes can also be covering nodes, and there is no need to exchange dominating set status. Dai and Wu's [10] proposed a generalized rule (called Rule k) to construct a smaller CDS. Based on the generalized rule, u is unmarked if its neighbor set is covered by several connected nodes with higher id's. The number of the covering nodes is allowed to be more than two. The generalized rule can unmark more nodes, and is easier to check than Rules 1 and 2.

When the network is static or local views are consistent (say, all nodes see only solid line) in Figure 22.4, both nodes u and v are marked after the marking process. u will be unmarked using Rule 1.

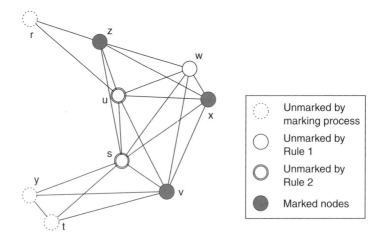

FIGURE 22.3   Wu and Li's CDS algorithm. Black nodes are marked (i.e., in the CDS). Nodes unmarked by the marking process have dotted lines; those unmarked by Rule 1 have single solid lines; those by Rule 2 have double solid lines.

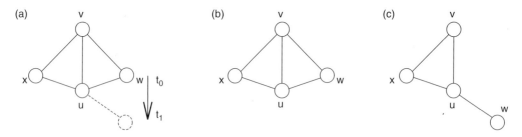

FIGURE 22.4   Both nodes u and v are unmarked due to inconsistent local views sampled at nodes u and v. (a) w's position at t_0 and t_1. (b) u's decision as an unmarked node at t_1. (c) v's decision as unmarked at t_1.

A typical topology control protocol works as follows:

1. Select r' to collect neighborhood information.
2. Apply a selected localized topology control protocol to select $r(u), r(u) \leq r'$, for node u to cover its farthest logical neighbor.
3. Use $r(u) + l$ for the actual transmission.

We use Li, Hou, and Sha's topology control algorithm based on local MST [22] to illustrate Step 2 of the above process.

At each node u:

1. Build a local MST using Prim's algorithm based on 1-hop location information. The resultant MST covers all 1-hop neighbors of u.
2. Select neighbors in MST as logical neighbors of u.
3. Set the transmission range of u to its distance to the farthest logical neighbor.

When the network is static or local views are consistent in Figure 22.5(a), all nodes build an identical MST that includes two links (u, v) and (w, v). Node u has one logical neighbor v and sets its range to 4. Node w has one logical neighbor v and sets its range to 5. Node v has two logical neighbors u and w and sets its range to 5 to reach the farthest node w.

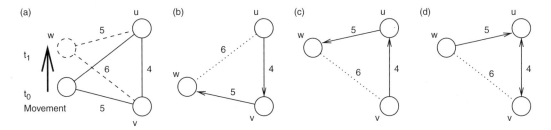

FIGURE 22.5 Node w becomes unreachable from nodes u and v due to inconsistent local views sampled at nodes u and v. (a) w's position at t_0 and t_1. (b) LMST(u) built at time t_0. (c) LMST(v) built at time t_1. (d) Network topology at time t_1.

When the network contains mobile nodes, such as node w in Figure 22.5, the transmission range of each node is increased to maintain the link availability. For example, if it is known that the maximum relative movement between two nodes during one "Hello" interval is $l = 2$, then the actual transmission range of nodes u, v, and w are adjusted to 8, 7, and 8, respectively. Therefore, link (v, w) is still available even if node w moves upward and the distance between v and w becomes 6. It is also observed in Reference 47 that the buffered range $l = 2$ is conservative and not always necessary. The probability is high that all links can be maintained with a smaller l.

22.5   View Consistency

In this section, we first explain what are inconsistent local views and consequences of using inconsistent views in localized algorithms. Then we discuss three possible solutions to overcome the inconsistent view problem. In the first solution, a certain synchronization mechanism is used to ensure consistent views during a given time period. In the second solution, each node makes conservative solutions based on recent history information. The third solution relies on an aggressive fault detection mechanism, which precautions neighbors about unreliable links. Due to the early warning, most "bad" decisions of neighbors can be avoided.

22.5.1   View Consistency Issue

We use two localized algorithms as examples to demonstrate how inconsistent local views cause "bad" decisions in MANETs. Both algorithms, Wu and Li's marking process [49] for CDS construction, and Li, Hou, and Sha's topology control algorithm based on local MST [22], have been introduced in the last section.

In the CDS construction example (as shown in Figure 22.4), we assume that node w moves southward. Link (v, w) exists at time t_0 and is broken at time t_1. We also assume t_0 and t_1 belong to two intervals. Since link (v, w) is two hops away from node u, when node u decides its status, it uses the outdated information (lagging by one interval) that link (v, w) still exists. The local view u is shown in Figure 22.4(b). Based on Rule 1, node u is unmarked because its neighbor set is covered by node v. However, when node v decides its status, it has the fresh information that link (w, v) is broken since it is adjacent to the link (as shown in Figure 22.4(c)). Based on the marking process, the only two neighbors of v, x, and u, are connected, so node v is also marked false. As a consequence, none of the nodes in the network are marked.

In the topology control example (as shown in Figure 22.5), assume node u's view reflects the topology at t_0 (as shown in Figure 22.5(b)) whereas node v's view corresponds to the topology at t_1 (as shown in Figure 22.5(c)). This happens when the recent "Hello" message from w is sent at t, where $t_0 < t < t_1$. In this case, u has only one logical neighbor v, and v has only one logical neighbor u. Based on the protocol, a link is selected only if both end nodes select each other. As a result, only one link (u, v) exists after the topology control (as shown in Figure 22.5(d)). A network partition occurs.

In the earlier examples, individual nodes make "bad" decisions based on inconsistent local views. Two views are inconsistent if their common parts do not match. In the CDS example, link (v, w) exists in node u's view but not in node v's view. In the topology control example, w is closer to v in u's view but closer to u in v's view.

22.5.2 Consistent Local View

One solution to the above problem is to enforce consistent views such that the same information "version" is used by all nodes in making their decisions. First we consider the problem of collecting consistent 1-hop information via loosely synchronous "Hello" messages. Originally, each node receives "Hello" messages from its 1-hop neighbors, and updates its local view upon the arrival of every "Hello" message. If all nodes have synchronized clocks, this scheme actually works. In the topology control example, if both nodes u and v make their decisions at t_0, they will agree that v is closer to w; at t_1, they will agree that u is closer. Here we omit the propagation delay and assume that a "Hello" message is received by all neighbors at the same time. However, it is impossible to have totally synchronized clocks in a MANET without centralized control. If u makes its decision slightly earlier than v, and w's "Hello" message arrives after u's decision and before v's decision, then the two nodes have inconsistent views. This inconsistency cannot be avoided no matter how small the asynchrony is.

The traditional solution for this problem is to build local views only once at the beginning of each "Hello" interval. As shown in Figure 22.6(a), each "Hello" interval is divided into three time periods $\Delta = h + s + d_1$. Because of asynchronous clocks, different nodes may start their "Hello" intervals at different times. That is, some nodes have "faster" clocks than other nodes. However, we assume the difference between two clocks is bounded by s. In the construction of consistent views, each node sends its "Hello" message during period h, waits for a period s, and conducts normal activities (e.g., sending data packets) in period d_1. As the h period of the "slowest" node ends before the s period of the "fastest" node, every node receives all "Hello" messages before the end of its s period. Local views built in the end of s are consistent. It is safe to route data packets in period d_1 based on these local views.

This scheme can be extended to build 2-hop information. As shown in Figure 22.6(b), each "Hello" interval is divided into five periods $\Delta = h_1 + s_1 + h_2 + s_2 + d_2$. Normally $h_1 = h_2$ and $s_1 = s_2$. Here we assume the clock difference is bounded by both s_1 and s_2. Each node first advertises its 0-hop information (i.e., its id and location) in period h_1, builds 1-hop information at the end of period s_1, and then advertises the newly constructed 1-hop information in period h_2. At the end of period s_2, every node constructs its consistent local view, which is ready for use in period d_2. The drawback of this scheme is that two "Hello" messages are sent during each interval Δ, and the effective communication period d_2 is further reduced.

Then we consider another situation with asynchronous "Hello" messages. The traditional solution relies on the assumption that the maximal difference among local clocks, s, is predictable and $s \leq \Delta$. In a totally asynchronous system, $s = \Delta$ and the above simple approach cannot be applied. Note that even if $s < \Delta$ at a particular network, delays accumulate unless some clock synchronization protocol is

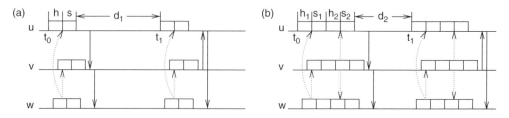

FIGURE 22.6 Build consistent local views at the beginning of each "Hello" interval. Dotted lines represent "Hello" messages. Solid lines represent data packets. (a) Consistent 1-hop information. (b) Consistent 2-hop information.

applied. Although various solutions exist to adjust clock values, frequent clock synchronization is costly. When maintaining (partially) synchronous "Hello" interval becomes too expensive or impossible, we propose using timestamped asynchronous "Hello" messages to enforce application specific consistent local views.

The basic idea is to maintain a sequence number i_v at each node v, and attach i_v to each "Hello" message from v. The sequence number serves as a timestamp. Consistent local views are obtained from "Hello" messages with the same timestamps. This can be done by carrying a timestamp in each data packet. The timestamp is chosen by the originator of the data packet, and all nodes relaying this packet must determine their logical neighbors based on information of the same version (i.e., with the same timestamp). In this scheme, each node keeps several local views, each local view corresponding to a recently used timestamp. Similarly, several logic topologies coexist in the same network. Each logic topology corresponds to a timestamp and is connected. The logic time (i.e., the timestamp of the latest local view) of the originator of the data packet is used as a selector. It indicates in which logical topology this data packet is traveling.

In Figure 22.5, suppose the first "Hello" message from node w has timestamp 0, and the second one has timestamp 1. When the above method is applied, two parallel logic topologies exist. The logical topology corresponding to timestamp 0 includes two bidirectional links (u, v) and (v, w). The logic topology corresponding to timestamp 1 includes (u, v) and (u, w). When a data packet p is sent from u to w, the source node u selects a recent timestamp and forwards p on the corresponding logical topology. If p has timestamp 0, it is first forwarded to v. Based on v's local view with timestamp 0, w is a logical neighbor of v, and p is forwarded along the logical link (v, w). If p has timestamp 1, it is sent to w directly via logical link (u, w). In both cases, p arrives safely at its destination.

The above approach can tolerate a larger "time skew" among different local views and, therefore, involves less synchronization overhead. Let Δ' be the maximal time difference between any two "Hello" messages with the same timestamp. This scheme still works when $\Delta' > \Delta$. However, the buffer zone should be adjusted accordingly to compensate for the "older" local views. If the local clock speed is different in each node, Δ' may accumulate during a long time period, and the buffer zone may be very large. A certain degree of synchronization is still required to maintain a relatively small Δ'. For example, randomly selected nodes can initiate periodical broadcast processes that help each node adjust its timestamp. This synchronization process need not be frequent when each node has a relatively accurate clock. If all local clocks are accurate, only one broadcasting is sufficient.

22.5.3 Conservative Local View

In the previous section, we discussed two methods of enforcing consistent local views. Both methods require a certain degree of internode synchronization, which introduces extra overhead. When maintaining consistent local views becomes too expensive or impossible, another approach called *conservative local view* [48] can be applied, which makes conservative decisions based on inconsistent views. No synchronization is necessary. A conservative decision is one that maintains the global property with the penalty of lower efficiency. That means selecting more logical neighbors in a topology control algorithm and marks more nodes as dominators in a CDS formation process. We use Wu and Li's marking process as an example to illustrate the conservative approach.

In Wu and Li's marking process, a node v may be incorrectly unmarked if (1) v no longer views a node w as its neighbor, and (2) another node u still views w as v's neighbor and unmarks itself based on this view. As the broken link (v, w) is first detected by v and then propagated to u via periodical "Hello" messages, local views of nodes u and v are inconsistent for a short period. During that period, u and v may be unmarked simultaneously, and the CDS is temporarily compromised. In order to prevent conditions (1) and (2) from happening together, each node must use a conservative local view, instead of its most recent local view, to make conservative decisions. In this case, the conservative local view $\text{View}_c(v)$ of node v is constructed from k most recent local view $\text{View}_1(v), \text{View}_2(v), \ldots, \text{View}_k(v)$ based on the following rule: a link (u, w) exists in $\text{View}_c(v)$ if and only if (1) (u, w) exists in the most recent local views $\text{View}_1(v)$, or (2) $u = v$ and (u, v) exists in at least one recent local view $\text{View}_i(v)$ for $1 \leq i \leq k$. That is, a broken link is preserved longer in the conservative views of its two end nodes than in those of all other nodes.

As shown in Figure 22.7, after node v detects a broken link (v, w), it will keep a virtual link corresponding to the broken link in its local view for a short time period. Based on this conservative view (as shown in Figure 22.7(c)), v is still a dominator. Note that the virtual link (v, w) is still available during this time period, if v uses a large actual transmission range to create a buffer zone, as discussed in Section 22.2. The virtual link stays in v's view until all other nodes have removed this link from their views. When 2-hop information is used, link (v, w) exists in local views of v's 1-hop neighbors and w's 1-hop neighbors, which will remove link (v, w) from their local views after receiving a "Hello" message from v or w. Node v will send its next "Hello" message within a "Hello" interval (Δ). Node w may detect the broken link and send its "Hello" message later than v, but the difference is bounded by Δ. Therefore, it is safe to remove the virtual link (v, w) for v's local view after 2Δ.

This approach can also be applied to other localized CDS and topology control algorithms. However, the conservative decisions are different from algorithm to algorithm, and the construction of conservative views depends on the specific localized algorithm. For example, in Li, Hou, and Sha's local MST algorithm, a conservative view of node v can be defined as follows: given k most recent local views $\text{View}_1(v), \text{View}_2(v), \ldots, \text{View}_k(v)$, which contain distance values $d_i(u, w)(1 \leq i \leq k)$ between any two nodes u and w within v's transmission range (including v), their distance in the conservative view is (1) $\max_i d_i(u, w)$, if $u \neq v$ and $w \neq v$, and (2) $\min_i d_i(u, w)$ otherwise. That is, the virtual distance between v and a neighbor w in its conservative view may be smaller than the actual distance, and the virtual distance between two neighbors may be larger than the actual distance. As shown in Figure 22.8, when conservative local views are used, both nodes u and v select w as a logical neighbor, and the network after topology control is connected.

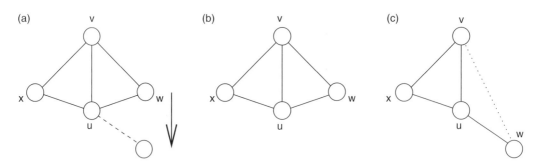

FIGURE 22.7   Based on the conservative view, node v is still marked for a little while after it detected the broken link (v, w). The dotted line represents a virtual link in v's view. (a) w's position at t_0 and t_1. (b) u's decision as an unmarked node at t_0. (c) v's decision as marked at t_1.

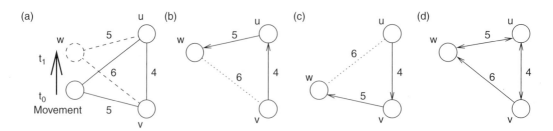

FIGURE 22.8   Network connectivity is preserved using conservative views. (a) w's position at t_0 and t_1. (b) u's conservative view. (c) v's conservative view. (d) Network topology.

22.5.4 Smart Neighbor Discovery

A simple mobility control mechanism was proposed in Reference 11 to make conservative local views by maintaining two, instead of one, neighbor sets at each node. Most localized CDS formation and CDS-based broadcast algorithms use a neighbor discovery mechanism that monitors periodical "Hello" messages from neighbors. When a node receives a "Hello" message, the sender's id is added to its current neighbor set. This id stays in the neighbor set as long as "Hello" messages from the same sender are received regularly. Otherwise, a link failure is detected and the corresponding node id is removed from the current neighbor set. In order to reduce the number of false alerts caused by collisions of "Hello" messages, the link failure detection mechanism is usually designed to tolerate two consecutive losses of "Hello" messages. A node v will consider another node w as a neighbor as long as it has received at least one "Hello" message from w during the last three "Hello" intervals. Therefore, the 1-hop neighbor information is usually outdated, and the protocol adapts poorly to node movement. On the other hand, an aggressive failure detector is not safer. For example, if node v in Figure 22.9(c) removes node w from its neighbor set by mistake, it will be unmarked based on the false information, leaving w uncovered.

In the dual neighbor set solution [11], two neighbor sets, one private and another public, are maintained using different link failure detection standards. After node v misses the first "Hello" message from node w, w is removed from node v's public neighbor set. The public neighbor set is promptly advertised to other nodes so that a graceful handover can be arranged in advance. Meanwhile, w remains in v's private neighbor set until losses of three consecutive "Hello" messages. The status of v is determined based on its private neighbor set, which remains a dominator during the handover period. For example, when link (v, w) fails in Figure 22.9(a) due to the movement of w, it is first removed from the public neighbor set of v, which is advertised to node u. Therefore, node u is marked shortly after the node movement, as shown in Figure 22.9(b). On the other hand, as node w remains in v's private neighbor set, its local view is as in Figure 22.9(c), based on which v is still marked for a smooth handover. Simulation results in Reference 11 show that this enhancement is very effective in improving reliability, where the size of CDS increases automatically to compensate for increased mobility level.

The above "smart" neighbor discovery scheme is extended in Reference 47, where signal strength is used for early detection of unstable links. Here the "Hello" message is transmitted with the actual transmission range $r + l$. When a node v receives a "Hello" message, its distance to the sender w is estimated based on the incoming signal strength. The id w is always put into v's private neighbor set, but will be removed from v's public neighbor set if the distance between v and w is larger than r. This link failure detection scheme is more aggressive and eliminates the possibility of simultaneous withdrawal in Figure 22.4. In this case, node u is warned about the failure of link (v, w) and becomes a dominator before the failure actually happens. In the same time, node v remains a dominator until the link fails. Therefore, node w is always covered by either node u or node v or both. This scheme is also useful in sparse networks, where using a large "Hello" transmission range is necessary to avoid network partition.

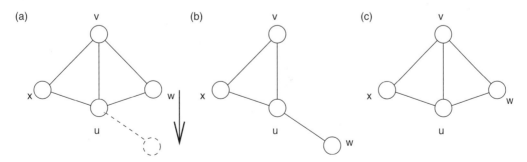

FIGURE 22.9 Because node w is removed from node v's public neighbor set while it is still in v's private neighbor set, both nodes u and v are marked shortly after t_1. (a) w's position at t_0 and t_1. (b) u's position as marked at t_1. (c) v's position as marked at t_1.

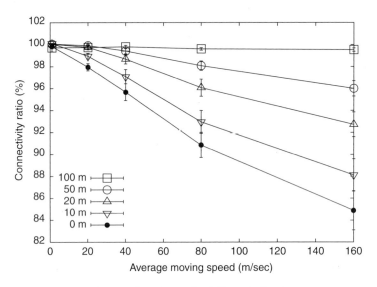

FIGURE 22.10   Connectivity ratio of a CDS algorithm under different buffered ranges.

22.6 Simulation Results

The mobility control mechanisms for localized CDS and topology control algorithms can be evaluated using a popular network simulator $ns-2$. Simulation results show that, under moderate mobility, most existing CDS and topology control algorithms can no longer maintain a connected dominating set or connected topology in a MANET. On the other hand, global domination and connectivity can be preserved with high probability, when the mobility control mechanisms are used to enhance link availability and view consistency.

Figure 22.10 and Figure 22.11 show sample results from simulations of a CDS algorithm and a topology control algorithm. For results of more algorithms, the readers can refer to [48,47]. In the simulation, 100 nodes are randomly placed in a 900×900 m^2 area. The normal transmission range r' is 250 m. The "Hello" interval is 1 sec. The mobility pattern follows the random waypoint model, where each node selects a random destination and moves to this destination with a random speed. After a node arrives at a destination, it pauses for a little while, and then selects a new destination and continues moving. This simulation uses a pause time of zero and an average moving speed varying from 1 to 160 m/sec. An ideal medium access control (MAC) layer without collision and contention is used to isolate the effect of mobility from other factors. The network connectivity is measured in terms of the connectivity ratio, which is defined as the ratio of pairs of connected nodes to the total number of pairs. In the simulation of a CDS algorithm, links between dominatees are removed before computing the connectivity ratio. That is, dominatees must be connected via dominators.

Figure 22.10 shows the connectivity ratio of Dai and Wu's general rule [10], which is an extension of Wu and Li's marking process and Rules 1 and 2 [49]. Different buffer zones are used to solve the link availability problem. In the original algorithm (0 m), the connectivity ratio drops rapidly as the average moving speed increases. When a small (20 m) buffer zone is used to tolerate broken links, the delivery ratio improves significantly under low (1 m/sec) to moderate (40 m/sec) mobility. With a 100 m buffer zone, the algorithm has almost 100% connectivity ratio under very high (160 m/sec) mobility. The effect of inconsistent views is not obvious in this case, because the buffer zone also increases redundancy in the CDS. A CDS with higher redundancy can tolerate more "bad" decisions caused by inconsistent views.

Figure 22.11(a) shows the connectivity ratio of Li, Hou, and Sha's topology control algorithm based on local MST [22]. Various buffer zones are used to tolerate the inaccurate node location information caused by mobility. When there is no buffer zone (0 m), the connectivity ratio is very low (10%) under an average

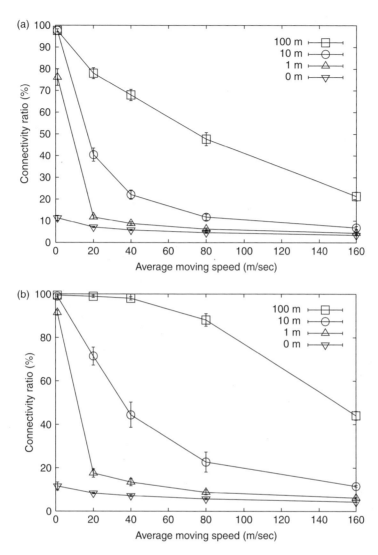

FIGURE 22.11 Connectivity ratio of a topology control algorithm under different buffered ranges. (a) With inconsistent local views. (b) With consistent local views.

moving speed of 1 m/sec. The connectivity ratio increases to 75% after a 1 m buffer zone is used, and 98% for a 10 m buffer zone. Obviously, the topology connectivity can be enhanced via a buffer zone with an appropriate width. On the other hand, using a buffer zone alone does not solve the entire problem. Hundred percent connectivity ratio is not achieved even with very low mobility level. Moderate and high mobility causes low connectivity ratio.

Figure 22.11(b) shows the effect of combining the buffer zone technique with consistent views. After individual nodes construct consistent local views, the connectivity ratio is improved significantly. When using a 20 m buffer zone in MANETs with a 10 m/sec average moving speed, the connectivity ratio is 40% without consistent views, and 70% with consistent views. When using a 100 m buffer zone under a 40 m/sec average moving speed, the connectivity ratio reaches 98% with consistent views, while the original connectivity ratio without consistent views is only 70%.

Overall, simulation results confirm that the global connectivity can be compromised by both link availability and view consistency issues. Both issues can be overcome with mobility control mechanisms, and the global property can be preserved with high probability and relatively small overhead.

22.7 Conclusion

We have addressed issues related to mobility control in mobile ad hoc networks. To illustrate the importance of the negative impact of mobile nodes on various protocols, we focus on two localized protocols, one for CDS construction and the other for topology control. It has been shown that most existing protocols on CDS construction and topology control will generate incorrect results in the presence of mobile nodes. We discuss two major problems in mobility control: link availability and view consistency, and provide several solutions. Mobility control in MANETs is still in its infancy. Many open issues exist:

- How does mobility affect protocols at other layers?
- Can approaches for view consistency in distributed systems be applied in mobile ad hoc networks?
- How should various kinds of cost and efficiency trade-off be done?

More efforts are needed to address these issues before various protocols can be applied in MANETs with mobile nodes.

Acknowledgment

This work was supported in part by NSF grants CCR 0329741, ANI 0073736, and EIA 0130806.

References

[1] C. Adjih, P. Jacquet, and L. Viennot. Computing connected dominated sets with multipoint relays. Technical Report 4597, INRIA-Rapport de recherche, Oct. 2002.

[2] K.M. Alzoubi, P.-J. Wan, and O. Frieder. Distributed heuristics for connected dominating sets in wireless ad hoc networks. *Journal of Communications and Networks*, 4: 22–29, 2002.

[3] A.D. Amis, R. Prakash, T.H.P. Vuong, and D.T. Huynh. Max–min d-cluster formation in wireless ad hoc networks. In *Proceedings of the IEEE INFOCOM 2000*, pp. 32–41, 2000.

[4] M. Bahramgiri, M. Hajiaghayi, and V.S. Mirrokni. Fault-tolerant and 3-dimensional distributed topology control algorithm in wireless multi-hop networks. In *Proceedings of the ICCCN*, pp. 392–397, 2002.

[5] S. Basagni. Distributed clustering for ad hoc networks. In *Proceedings of the 1999 International Symposium on Parallel Architectures, Algorithms, and Networks (I-SPAN'99)*, pp. 310–315, June 1999.

[6] S. Basagni, I. Chlamtac, A. Farago, V.R. Syroutiuk, and R. Talebi. Route selection in mobile multimedia ad hoc networks. *Proceeding of the IEEE MOMUC*, Nov. 1999.

[7] D. Blough, M. Leoncini, G. Resta, and P. Santi. The K-Neigh protocol for symmetric topology control in ad hoc networks. In *Proceedings of the MobiHoc*, pp. 141–152, June 2003.

[8] J. Cartigny, D. Simplot, and I. Stojmenovic. Localized minimum-energy broadcasting in ad hoc networks. In *Proceedings of the IEEE INFOCOM*, pp. 2210–2217, 2003.

[9] B. Chen, K. Jamieson, H. Balakrishnan, and R. Morris. Span: an energy-efficient coordination algorithm for topology maintenance in ad hoc wireless networks. *ACM Wireless Networks Journal*, 8: 481–494, 2002.

[10] F. Dai and J. Wu. Distributed dominant pruning in ad hoc wireless networks. In *Proceeding of the ICC*, p. 353, May 2003.

[11] F. Dai and J. Wu. Performance comparison of broadcast protocols for ad hoc networks based on self-pruning. In *Proceedings of the IEEE WCNC*, Mar. 2004.

[12] B. Das, R. Sivakumar, and V. Bharagavan. Routing in ad hoc networks using a spine. In *Proceedings of the 6th IEEE International Conference on Computers Communications and Networks (IC3N)*, pp. 1–20, Sep. 1997.

[13] R. Dube, C. Rais, K. Wang, and S. Tripathi. Signal stability based adaptive routing (ssa) for ad hoc mobile networks. *IEEE Personal Communications* 4: 36–45, 1997.

[14] A. Ephremides, J.E. Wjeselthier, and D.J. Baker. A design concept for reliable mobile radio networks with frequently hopping signaling. *Proceedings of the IEEE*, 71: 56–73, 1987.

[15] R.H. Frenkiel, B.R. Badrinath, J. Borras, and R.D. Yates. The infostations challenge: balancing cost and ubiquity in delivering wireless data. *IEEE Personal Communication*, 7: 66–71, 2000.

[16] K.R. Gabriel and R.R. Sokal. A new statistical approach to geographic variation analysis. *Systematic Zoology*, 18: 259–278, 1969.

[17] M. Grossglauser and D.N.C. Tse. Mobility increases the capacity of ad-hoc wireless networks. In *Proceedings of the INFOCOM*, pp. 1360–1369, 2001.

[18] Z.J. Haas, J.Y. Halpern, and L. Li. Gossip-based ad hoc routing. In *Proceedings of the IEEE INFOCOM*, vol. 3, pp. 1707–1716, June 2002.

[19] W.-I. Kim, D.H. Kwon, and Y.-J. Suh. A reliable route selection algorithm using global positioning systems in mobile ad-hoc networks. In *Proceedings the ICC*, June 2001.

[20] L. Li and J.Y. Halpern. Minimum energy mobile wireless networks revisited. In *Proceeding of the ICC*, pp. 278–283, June 2001.

[21] L. Li, J.Y. Halpern, V. Bahl, Y.M. Wang, and R. Wattenhofer. Analysis of a cone-based distributed topology control algorithm for wireless multi-hop networks. In *Proceedings of the PODC*, pp. 264–273, Aug. 2001.

[22] N. Li, J.C. Hou, and L. Sha. Design and analysis of and MST-based topology control algorithm. In *Proceedings of the INFOCOM*, vol. 3, pp. 1702–1712, Mar./Apr. 2003.

[23] C.R. Lin and M. Gerla. Adapting clustering for mobile wireless networks. *IEEE Journal on Selected Areas in Communications*, 15: 1265–1275, 1996.

[24] J. Liu and B. Li. MobileGrid: capacity-aware topology control in mobile ad hoc networks. In *Proceedings of the ICCCN*, pp. 570–574, Oct. 2002.

[25] E.L. Lloyd, R. Liu, M.V. Marathe, R. Ramanathan, and S.S. Ravi. Algorithmic aspects of topology control problems for ad hoc networks. In *Proceedings of the MobiHoc*, pp. 123–134, June 2002.

[26] A. Qayyum, L. Viennot, and A. Laouiti. Multipoint relaying for flooding broadcast message in mobile wireless networks. In *Proceedings of the HICSS-35*, Jan. 2002.

[27] R. Ramanathan and R. Rosales-Hain. Topology control of multihop wireless networks using transmit power adjustment. In *Proceedings of the INFOCOM*, pp. 404–413, Mar. 2000.

[28] M.Q. Rieck, S. Pai, and S. Dhar. Distributed routing algorithm for wireless ad hoc networks using d-hop connected dominating sets. In *Proceedings of the 6th International Conference on High Performance Computing in Asia Pacific Region*, Dec. 2002.

[29] V. Rodoplu and T.H. Meng. Minimum energy mobile wireless networks. *IEEE Journal of Selected Areas in Communications*, 17: 1333–1344, 1999.

[30] N. Sadagopan, F. Bai, B. Krishnamachari, and A. Helmy. Paths: analysis of path duration statistics and their impact on reactive manet routing protocols. In *Proceedings of the MobiHoc*, June 2003.

[31] M. Seddigh, J. Solano, and I. Stojmenović. RNG and internal node based broadcasting in one-to-one wireless networks. *ACM Mobile Computing and Communications Review*, 5: 37–44, Apr. 2001.

[32] P. Shinha, R. Sivakumar, and V. Bharghavan. CEDAR: a core-extraction distributed ad hoc routing algorithm. *IEEE Journal on Selected Areas in Communications, Special Issue on Ad Hoc Networks*, 17: 1454–1465, Aug. 1999.

[33] P. Sinha, R. Sivakumar, and V. Bharghavan. Enhancing ad hoc routing with dynamic virtual infrastructures. In *Proceedings of the IEEE INFOCOM 2001*, pp. 1763–1772, 2001.

[34] T. Small and Z.J. Haas. The shared wireless infostation model: a new ad hoc networking paradigm (or where there is a whale, there is a way). In *Proceedings of the MOBIHOC*, pp. 233–244, June 2003.

[35] I. Stojmenovic, M. Seddigh, and J. Zunic. Dominating sets and neighbor elimination based broadcasting algorithms in wireless networks. *IEEE Transactions on Parallel and Distributed Systems*, 13: 14–25, 2002.

[36] W. Su, S.-J. See, and M. Gerla. Mobility prediction and routing in ad hoc wireless networks. *International Journal of Network Management*, 11: 3–30, 2001.

[37] C. Toh. Associativity-based routing for ad hoc mobile networks. *IEEE Personal Communications*, 4: 36–45, 1997.

[38] G. Toussaint. The relative neighborhood graph of finite planar set. *Pattern Recognition*, 12: 261–268, 1980.

[39] Y.-C. Tseng, S.-Y. Ni, Y.-S. Chen, and J.-P. Sheu. The broadcast storm problem in a mobile ad hoc network. *Wireless Networks*, 8: 153–167, 2002.

[40] A. Vahdat and D. Becker. Epidemic routing for partially connected ad hoc networks. Technical Report CS-200006, Duke University, Apr. 2000.

[41] P.J. Wan, K. Alzoubi, and O. Frieder. Distributed construction of connected dominating set in wireless ad hoc networks. In *Proceedings of the IEEE INFOCOM*, vol. 3, pp. 1597–1604, June 2002.

[42] Y. Wang, X. Li, P. Wan, and O. Frider. Distributed spanners with bounded degree for wireless ad hoc networks. *International Journal of Foundations of Computer Science*, 14: 183–200, 2003.

[43] R. Wattenhofer, L. Li, V. Bahl, and Y.M. Wang. Distributed topology control for power efficient operation in multihop wireless ad hoc networks. In *Proceedings of the INFOCOM*, pp. 1388–1397, Apr. 2001.

[44] J.E. Wieselthier, G.D. Nguyen, and A. Epheremides. On the construction of energy-efficient broadcast and multicast trees in wireless networks. In *Proceedings of the INFOCOM*, pp. 585–594, Mar. 2000.

[45] B. Williams and T. Camp. Comparison of broadcasting techniques for mobile ad hoc networks. In *Proceedings of the MobiHoc*, pp. 194–205, June 2002.

[46] J. Wu. Extended dominating-set-based routing in ad hoc wireless networks with unidirectional links. *IEEE Transactions on Parallel and Distributed Systems*, 13: 866–881, 2002.

[47] J. Wu and F. Dai. Mobility management and its applications in efficient broadcasting in mobile ad hoc networks. In *Proceedings of the INFOCOM*, Mar. 2004.

[48] J. Wu and F. Dai. Mobility-sensitive topology control in mobile ad hoc networks. In *Proceedings of the IPDPS*, pp. 28a, Apr. 2004.

[49] J. Wu and H. Li. On calculating connected dominating set for efficient routing in ad hoc wireless networks. In *Proceedings of the DiaLM*, pp. 7–14, 1999.

23

Self-Organization Algorithms for Wireless Networks

23.1 Why Self-Organization? **23**-519
23.2 Chapter Outline...**23**-520
23.3 Communication Algorithms **23**-521
23.4 Density Control Algorithms **23**-524
23.5 Clustering-Based Algorithms **23**-527
23.6 Localization Algorithms **23**-531
Acknowledgment...**23**-531
References **23**-531

Carlos M.S. Figueiredo
Eduardo F. Nakamura
Antonio A.F. Loureiro

23.1 Why Self-Organization?

A wireless sensor network (WSN) consists of sensor nodes connected among themselves by a wireless medium to perform distributed sensing tasks. These networks are expected to be used in different applications such as environmental and health monitoring, surveillance, and security. An important aspect of WSNs comes from having many sensors generating sensing data for the same set of events. The integration of sensing, signal processing, and data communication functions allows a WSN to create a powerful platform for processing data collected from the environment.

A WSN must be able to operate under very dynamic conditions and, in most cases, in an unattended mode leading to an autonomous network. This scenario becomes more complex when we consider that there will be probably sensor networks that comprise tens to hundreds of thousands sensor nodes. The design of algorithms for WSNs must consider the hardware limitations of sensor nodes, the physical environment where the nodes will operate, and the application requirements. A communication protocol must be designed to provide a robust and energy-efficient communication mechanism.

The algorithms and protocols for this kind of network must be able to enable network operation during its initialization and during both normal and exception situations. A vision to reach this autonomy is through the concept of self-organization, which is defined in Reference 1 as "the spontaneous creation of a globally coherent pattern out of local interactions."

Local interactions will be probably based on local rules to achieve a global goal. Note that the local rules assigned to each sensor may be different depending on its hardware characteristics, node location, traffic pattern, security, and other attributes associated with the application. The ultimate goal of these local rules is to design a self-organizing WSN.

Self-organization is a concept well-known in the literature and has been employed in different areas such as physics, chemistry, biology, and mathematics [2–4]. To illustrate the application of this concept, we cite some examples [5]:

- In physics, where the term seems to be first applied, one of its uses refers to structural phase transformations such as spontaneous magnetization and crystallization. Another use refers to complex systems made up of small, simple units connected to each other.
- In chemistry, self-organization includes self-assembly, which denotes a reunion of molecules without the help of an external entity in oscillating chemical reactions and autocatalytic networks.
- In biological systems, there are many phenomena that have been described as "self-organizing" such as the formation of lipid bilayer membranes, the self-maintaining nature of systems from the cell to the whole organism (homeostasis), the process of how a living organism develops and grows (morphogenesis), the creation of structures by social animals (e.g., social insects as bees, ants, and termites), and the origin of life itself from self-organizing chemical systems.
- In mathematics and computer science, we find the term related to cellular automata, evolutionary computation, artificial life, swarms of robots, and neural networks.

Particularly in computer science, a more specific view is needed. For example, in Reference 6 a self-organizing system is defined as a "system which creates its own organization." An organization is a "structure with function, where structure means that the components of a system are arranged in a particular order and function means that this structure fulfills a purpose." This definition can be applied to a WSN and, in this case, we could define a self-organizing wireless network as a network that has the capacity to maintain all the network functions (e.g., communication, and collaboration), adapts to perturbations (topological or environmental changes), saves resources (e.g., energy, bandwidth etc.), along its lifetime without the need of external interventions.

In contrast to other areas, such as in physics, chemistry, and biology, in which the laws of the nature control the local interactions, in WSNs we need to devise simple local interaction rules codified in distributed algorithms to achieve the desired goal. In general, this is not an easy task, especially if efficiency is required due to network constraints such as energy, bandwidth, and processing capacity. Despite this fact, some general features of these algorithms can be identified as follows:

- The entire system must have a common goal to be obtained by the cooperation of individual elements, which leads to a well-defined behavior.
- There is no centralized entity that controls the system organization. This must be done according to fully distributed algorithms.
- The organization occurs from local interactions. Thus, elements must act together to reach a common goal.
- The system must be dynamic. The state of a node must change based on its inputs (network perceptions) and the local interactions. The system must adapt itself to reach a stable global organization.
- The nodes must preserve their limited resources and the self-organization cannot sacrifice the goal of the application running on the sensor network. Thus, the dynamics of the system from local interactions must be objective and precise with simple established rules.

23.2 Chapter Outline

This chapter presents some algorithms that can guide the development of new applications and algorithms for wireless networks. We concentrate on algorithms for ad hoc and sensor networks, due to their natural characteristics of autonomic behavior (see References 7 and 8). The algorithms are divided in classes that correspond to the main activities of these networks: communication, clustering, density control, and localization algorithms.

23.3 Communication Algorithms

Communication is a basic functionality in any data communication network and was the first class of algorithms to be developed considering the self-organizing approach, especially in ad hoc and sensor networks. The problem is to create a communication infrastructure — basically the establishment of links and routes between nodes — and maintain this infrastructure in the presence of topological changes.

In most existing ad hoc networks, the channel access is done by a contention method, like the 802.11 MAC protocol. This method does not require a special organization for channel access. When a node wants to transmit a frame it simply verifies if the channel is in use and if so does not transmit the frame thereby controlling possible collisions. Another approach consists in organizing the channel access in collision-free links, mainly based on time divisor multiple access (TDMA) schemes. An advantage of this approach is the possibility to save energy in more restricted networks such as sensor networks. The task of channel assignment to links in a large network is a very difficult problem to be performed by a centralized entity, especially in a dynamic scenario. Thus, algorithms that possess the self-organizing feature have been studied in the sensor network domain. The protocol called self-organizing medium access control for sensor networks (SMACS) [9] is an algorithm that allows the sensor nodes to discover their neighbors and to establish a schedule for transmissions and receptions for communicating among them. The pseudo-code of the SMACS algorithm is depicted in Algorithm 23.1.

This algorithm shows the negotiation to establish a communication between nodes based on the TDMA schedule. A node wakes up at random times and listens to the channel for invitation messages. After some time, if a message is not received, the node decides to send an invitation (INV_i) by broadcasting it to the network and expects some response (represented by $Invitation()$ from lines 4–8). The negotiation between nodes are controlled by the variables $Expecting_i$ and $Expected_i$ that are reinitiated if the receiving process reaches a timeout ($Receive.timeout()$). A node that receives an invitation message from a node will send a response (R_INV_i) to it in a random time and stays expecting a link establishment from it (lines 11–15). The node that invites others can receive a response from more than one node. In this case, it will choose the first one, ignoring the others, and sends a confirmation message with its schedule information ($CONF_i(Frame_i)$). This is shown in lines 16–19. In lines 20–26, the invited node will compare this information with its own schedule scheme ($slot\_pair_i = compare(F, Frame_i)$) to determine a free pair of slots to assign to a link formed between the nodes ($assign(slot\_pair_i, Frame_i)$). This information is sent back to the node that initiated the invitation (through $R\_CONF_i(slot_pair_i)$ message), which is added to its schedule scheme (lines 27–31).

Some details were suppressed from the description, like the use of multiple channels to increase the possibilities of having schedule schemes and to avoid collisions. However, it illustrates the activity of the organization of the link layer. Another algorithm can be found in Reference 10.

Currently, there are several proposals for routing protocols focusing on both ad hoc [11–14] and sensor networks [15–18]. Naturally they consider some self-organizing aspect due to the ad hoc nature of these networks that do not use a fixed infrastructure. Thus, routes must be discovered and maintained based on local cooperation. An interesting example to be considered is the ad hoc on-demand distance vector algorithm (AODV) [13] that creates routes only on demand, that is, when the source nodes need them. Basically, the AODV protocol has two phases: the first one is the route discovery, based on a flooding process started by the node that wants to send a message; and the second phase is the route maintenance, realized by all nodes of a discovered route to update it. This basic behavior is presented in Algorithm 23.2 and it is also found in reactive algorithms like the Push Diffusion [19].

The path discovery, represented by "$Discovery(n_j \in N)$" in line 3, is executed by a node that wants to send a message to another node (n_j) and does not have a valid route to it. This is done by a broadcast of a route request message ($RREQ_i(j)$), uniquely identified, to all neighbors of the source node (line 5).

Algorithm 23.1 (*SMACS — Stationary MAC and StartUp Procedure*)

```
 1:  Frameᵢ  ←  {};
 2:  Expectingᵢ  ←  false;
 3:  Expectedᵢ  ←  nil;

 4:  Invitation()
 5:  begin
 6:     Expectingᵢ  ←  true;
 7:     Send INVᵢ to all nⱼ ∈ Neigᵢ;
 8:  end

 9:  Receive(msgᵢ such that originᵢ(msgᵢ) = (nᵢ, nⱼ))
10:  begin
11:     if msgᵢ = INVⱼ and Expectingᵢ = false then
12:        Expectingᵢ  ←  true;
13:        Expectedᵢ  ←  j;
14:        Send R_INVᵢ to nⱼ in a random time;
15:     end if
16:     if msgᵢ = R_INVⱼ and Expectingᵢ = true then
17:        Expectedᵢ  ←  j;
18:        Send CONFᵢ(Frameᵢ) to nⱼ;
19:     end if
20:     if msgᵢ = CONFⱼ(F) and Expectedᵢ = j then
21:        slot_pairᵢ  ←  compare(F, Frameᵢ);
22:        assign(slot_pairᵢ, Frameᵢ);
23:        Send R_CONFᵢ(slotₚairᵢ) to nⱼ;
24:        Expectingᵢ  ←  false;
25:        Expectedᵢ  ←  nil;
26:     end if
27:     if msgᵢ = R_CONFⱼ(S) and Expectedᵢ = j then
28:        assign(S, Frameᵢ);
29:        Expectingᵢ  ←  false;
30:        Expectedᵢ  ←  nil;
31:     end if
32:  end

33:  Receive.timeout()
34:  begin
35:     Expectingᵢ  ←  false;
36:     Expectedᵢ  ←  nil;
37:  end
```

This message is forwarded by every intermediate node (broadcast in line 18) until the required destination (verification of line 12) or an intermediate node with a valid route to it (verification of line 15) is reached. During the process of forwarding request messages, intermediate nodes record a list of packets received and their immediate destinations (*ForwardedListᵢ*), avoiding the forwarding of request copies and establishing a reverse path toward the source (lines 10 and 11). Once a request is received by the destination or an intermediate node with a valid route, they will respond by sending a route reply message ($RREP_k(d)$) back to the node from which it first received the request (lines 13 and 16). As this is done toward the reverse path, intermediate nodes record the response packets and their immediate destinations in a list (*RequisitionListᵢ*) indicating the active route to the communication between the source and the destination

Algorithm 23.2 *AODV — Ad Hoc On-Demand Distance Vector Algorithm*

```
 1:  ForwardedList_i  ←  {};
 2:  RequisitionList_i  ←  {};

 3:  Discovery(n_j ∈ N)
 4:  begin
 5:     Send RREQ_i(j) to all n_k ∈ Neig_i;
 6:  end

 7:  Receive(msg_i(d) such that origin_i(msg_i) = (n_i, n_j))
 8:    begin
 9:      if msg_i(d) = RREQ_k(d) then
10:        if not(msg_i(d) ∈ ForwardedList_i) then
11:           Put msg_i(d) in FowardedList_i;
12:           if i = d then
13:              Send RREP_i(k) to n_j;
14:           else
15:              if ∃ RREP_d(k) ∈ RequisitionList_i such that
                     (NOW − timestamp_i(RREP_d(k))) < VALID_INTERVAL AL then
16:                 Send RREP_d(k) to n_j;
17:              else
18:                 Send msg_i(d) to all n_j ∈ Neigh_i;
19:              end if
20:           end if
21:        end if
22:      end if
23:      if msg_i(d) = RREP_k(d) then
24:         Put msg_i(d) in RequisitionList_i;
25:         if i ≠ d then
26:            Send msg_i(d) to n_j such that
                  n_j = origin_i(RREQ_d(k)) and RREQ_d(k) ∈ ForwardedList_i;
27:         end if
28:      end if
29:    end
```

(lines 23–28). Associated with each entry in the node list is a timestamp to verify the validity of routes (comparison between the elapsed time "$NOW — timestamp_i(RREP_d(k))$" and a predefined time interval "$VALID\_INTERVAL$" in line 15).

Another simple and efficient solution adopted in WSNs is a routing tree that has been evaluated in studies such as SAR [20], TD-DES [21], and One-phase pull diffusion [22]. The tree structure is created and maintained by a special node (the sink in WSNs), which can be elected or defined by an external entity, in a proactive fashion. This solution is presented in Algorithm 23.3.

The special node starts the process by broadcasting a control message to the network ($TREE\_BUILD_i$) in periodical intervals (represented by $TreeBuildTimer.timeout()$ in lines 3–6). When a node initially receives the building message, it identifies the sender as its parent ($Parent_i$) and broadcasts the building message to all its neighbors (lines 14–20). Duplicated messages received from other neighbors, maintained in a list ($ForwardedList_i$), are discarded. Whenever a node has data to be transmitted (generated or forwarded), it will send it directly to its parent, following the path of the formed tree (lines 9–13).

Another approach to self-organize the routing process of a wireless network is to automatically change the algorithm used in the data communication according to the network perceptions.

Algorithm 23.3 Routing Tree Algorithm

```
 1:  Parent_i  ←  nil;
 2:  ForwardedList_i  ←  {};

 3:  TreeBuildTimer.timeout()
 4:  begin
 5:      Send TREE_BUILD_i to all n_j ∈ Neig_i;
 6:  end

 7:  Receive(msg_i such that origin_i(msg_i) = (n_i, n_j))
 8:    begin
 9:      if msg_i = DATA_k for some n_k ∈ N then
10:        if Parent_i ≠ nil then
11:          Send msg_i to Parent_i;
12:        end if
13:      end if
14:      if msg_i = TREE_BUILD_k for some n_k ∈ N then
15:        if not (msg_i ∈ ForwardedList_i) then
16:          Parent_i  ←  n_j;
17:          Send msg_i to all n_l ∈ Neig_i;
18:          Put msg_i in ForwardedList_i;
19:        end if
20:      end if
21:    end
```

Multi [23] is an example of such an algorithm that encompasses the behavior of a reactive algorithm, similar to the behavior of AODV and a proactive algorithm like the Routing Tree. The idea is to monitor the network traffic and to compare it with a threshold. When the traffic is low it is better to use a reactive strategy, allowing to save resources, and, on the other hand, when the traffic is high it is better to use a proactive strategy, building the entire routing infrastructure in a predictive way. The pseudo-code for this protocol is shown in Algorithm 23.4.

The algorithm shows that the special node periodically verifies (*VerificationTimer.timeout()* in lines 5–13) the traffic pattern (*traffic_i*) and compares it with a given threshold (*TRAFFIC_THRESHOLD*). Based on its value, the algorithm decides to use a routing protocol based either on AODV or on Tree routing. The mode of operation of a node is determined by the type of control message received (*Mode_i* = *proactive*, *reactive*), shown in lines 29–36. Whenever a node has data (*DATA_k*) to be transmitted (generated or forwarded), it will send it preferably by the tree, otherwise the path established by reactive responses (*RREP_k*) are taken (lines 16–28). In Reference 23, it is shown that resource savings can be achieved with this autonomous adaptive scheme.

23.4   Density Control Algorithms

Wireless sensor networks are expected to operate for periods of time varying from weeks to years in an autonomous way. The success of this vision depends fundamentally on the amount of energy available at each sensor node in the network. In many applications sensor nodes may not be easily accessible because of the locations where they are deployed or owing to the large scale of such networks. In both cases, network maintenance for energy replenishment becomes impractical. Furthermore, in case a sensor battery needs to be frequently replaced the main advantages of a WSN are lost, that is, its operational cost, freedom from wiring constraints, and possibly more important, many sensing applications may become impractical.

Algorithm 23.4 *Multi Algorithm*

```
 1:  traffici ← 0;
 2:  ForwardedListi ← {};
 3:  Parenti ← nil;
 4:  Modei ← reactive;

 5:  VerificationTimer.timeout()
 6:  begin
 7:      if traffici ≥ TRAFFIC_THRESHOLD then
 8:          {ni will send tree build messages.}
 9:      else
10:          {ni will react only for existing sources nodes.}
11:      end if
12:      traffici ← 0;
13:  end

14:  Receive(msgi such that origini(msgi) = (ni, nj))
15:  begin
16:      if msgi = DATAk from some nk ∈ N then
17:          if not (msgi ∈ ForwardedListi) then
18:              if Modei = proactive then
19:                  Send msgi to Parenti;
20:              else
21:              if Modei = reactive then
22:                  Send msgi to nk such that origini(RREPk) = (nk → ni)y;
23:              end if
24:          end if
25:          Put msgi in ForwardedListi;
26:          traffici ← traffici + 1;
27:      end if
28:  end if
29:  if msgi = RREPk then
30:      {Proceed like AODV Algorithm}
31:      Modei ← reactive;
32:  end if
33:  if msgi = TREE_BUILDk then
34:      {Proceed like Routing Tree Algorithm}
35:      Modei ← proactive;
36:  end if
37:  end
```

The design of algorithms and protocols for these networks has to consider the energy consumption of a sensor node. In particular, it is very important to pay attention to the communication aspects of a given algorithm or protocol for a WSN. The node transceiver consumes energy not only during transmission and reception, but also in channel listening (idle state). In general, the more expensive states in terms of energy consumption are transmission followed by reception, but the energy spent in the idle state cannot be neglected in the network lifetime. A possible strategy to save energy is to turn off the sensor radio. When this happens, the node is no longer capable of communicating with other nodes and, thus, there is one active node less changing the network topology and density. A distributed algorithm that implements this approach is geographic adaptive fidelity (GAF) [24], which self-configures redundant nodes into small

Algorithm 23.5 (*GAF — Geographic Adaptive Fidelity*)

```
 1:  State_i  ←  discovery;
 2:  ActiveNeigh_i  ←  false;
 3:  Radio_i  ←  off;

 4:  DiscoveryTimer.timeout()
 5:  begin
 6:     State_i  ←  discovery;
 7:     Radio_i  ←  on;
 8:     ActiveNeigh_i  ←  false;
 9:     Send DISC_i(GridID_i, Energy_i) to all n_j ∈ Neig_i;
10:     Wait Td seconds;
11:     if ActiveNeigh_i = true then
12:         Sleep();
13:     else
14:         Active();
15:     end if
16:  end

17:  Sleep()
18:  begin
19:     State_i  ←  sleeping;
20:     Radio_i  ←  off;
21:     DiscoveryTimer.schedule(Ts);
22:  end

23:  Active()
24:  begin
25:     State_i  ←  active;
26:     Radio_i  ←  on;
27:     DiscoveryTimer.schedule(Ta);
28:  end

29:  Receive(msg_i(g, e) such that origin_i(msg_i) = (n_i, n_j) and msg_i = DISC_j)
30:  begin
31:     if g = GridID_i and e > Energy_i then
32:         ActiveNeigh_i  ←  true;
33:         if State_i = active then
34:             DiscoveryTimer.reset();
35:             Sleep();
36:         end if
37:     end if
38:  end
```

groups based on their location and controls their duty cycles to save energy. Its pseudo-code is presented in Algorithm 23.5.

The GAF protocol addresses the problem of controlling the network density by dividing the network into small "virtual" grids. A virtual grid is defined as a set of regions in which all nodes inside a grid can communicate among themselves. The size of a virtual grid can be easily derived from the radio range of a node. This definition allows all nodes of a grid to be equivalent for routing purposes. Therefore, if one node is active the others can sleep. Using the GAF protocol, a node can be in one of three possible states:

sleeping, discovery, or active (represented by $State_i$). Initially a node starts in the discovery state with its radio turned on ($Radio_i = on$) and exchanges discovery messages ($DISC_i$) with its neighbors to find out about the other nodes in the same grid (lines 4–16). This process takes Td seconds. A node detects if there is an active node (controlled by $ActiveNeigh_i$) in its grid (same $GridID_i$ that can be determined locally with its own location coordinates and known grid size) if it receives a discovery message from a node with a higher residual energy ($Energy_i$). This is shown in lines 31–37. If there is an active node in its grid (verification in line 11), the node moves to the sleeping state, turns its radio off, and reinitiates the discovery process in Ts seconds ($Sleep()$ in lines 17–22). Otherwise, if there is no active node in the grid, the node goes to the active state, turns its radio on, and reinitiates the discovery process after Ta seconds ($Active()$ in lines 23–28). A node in the active state can go to the sleeping state when it determines that another grid node is in the active state. In this case, the timers are reset ($DiscoveryTimer.reset()$ in line 34) and the radio is turned off.

This algorithm depends on localization systems to determine the node coordinates and grids. Sometimes location information is not available and for these cases there are other approaches such as cluster-based energy conservation algorithm (CEC) [24]. Other proposals can be found in References 25–27.

23.5 Clustering-Based Algorithms

The clustering of a network is the process of dividing it into hierarchical levels. A cluster is composed of common nodes that is coordinated by a node with a special role and called cluster-head. Cluster-heads can also be organized into other hierarchies. In ad hoc and sensor networks, this solution is frequently used because it can save energy and bandwidth. Some of the proposals presented in the literature that use clustering algorithms can be found in References 16 and 28–31.

The organization of wireless networks in clusters is an interesting solution to group related nodes for some collaborative task. This strategy may have several advantages: it is probably easier to devise a self-organization algorithm considering a small set of nodes; it is possible to minimize the number of message transmissions since communication occurs from a common node to its cluster-head and from the cluster-head to the sink node; the network complexity can be treated at the grid level and considering the set of grids; and finally, given all these points, it is possible to have a more scalable solution.

A clustering approach for data dissemination in WSNs is the Low-energy adaptive clustering hierarchy (LEACH) protocol [16]. LEACH is a self-organizing algorithm that creates local clusters with a node acting as cluster-head. The algorithm promotes a random rotation of cluster-heads to distribute the energy consumption in the network. Basically, each node elects itself as a cluster-head at any time with a given probability. A cluster-head broadcasts its operational state to the other nodes in the network. Each node determines as to which cluster it wants to belong by choosing the cluster-head with the better metric (e.g., minimum communication energy). The pseudo-code of LEACH is shown in Algorithm 23.6.

The operation of LEACH is divided in rounds. Each round has a set-up phase, where the clusters are organized, and a steady-state phase, where the network operates normally. The set-up phase is subdivided in the Advertisement and Confirmation phases. In the Advertisement phase, represented by "Advertisement.timeout()" in lines 6–23, a node decides whether it wants to be a cluster-head or not in that round (controlled by $Round_i$). This decision is based on its desired percentage from the network (represented by P) and the number of times the node has been a cluster-head so far. Thus, a node chooses a random number between 0 and 1 ($rand_i = Uniform(0, 1)$) and compares this with a threshold (T_i) established according to the condition that the node has not been a cluster-head in the last $1/P$ rounds (controlled by the difference between $Round_i$ and $Last\_Round\_CH_i$). This is shown in lines 9–13. If the random number is less than the threshold (verification in line 15), the node will be a cluster-head ($Cluster_i = i$) and this process guarantees that it will be at some point in $1/P$ rounds. The self-elected cluster-heads send an advertisement (ADV_i message) to all the network nodes ($n_j \in N$) by broadcast (the algorithm assumes that all nodes are reachable). In lines 26–29, the noncluster-head nodes receive an advertisement of all cluster-head nodes. At this moment, each noncluster-head node

Algorithm 23.6 *LEACH — Low-Energy Adaptive Clustering Hierarchy*

```
 1:  Cluster_i  ←  nil;
 2:  Round_i  ←  0;
 3:  Better_Signal_i  ←  0;
 4:  set_Member_i  ←  {};
 5:  Last_Round_CH_i  ←  −1/P;

 6:  Advertisement.timeout()
 7:  begin
 8:      Round_i  ←  Round_i + 1;
 9:      if Round_i − Last_Round_CH_i > 1/P then
10:          T_i  ←  ────────P────────;
                   1−P*(Round_i mod(1/P))
11:      else
12:          T_i  ←  0;
13:      end if
14:      rand_i  ←  Uniform(0, 1);
15:      if rand_i < T_i then
16:          Cluster_i  ←  i;
17:          Last_Round_CH_i  ←  Round_i;
18:          set_Member_i  ←  {};
19:          Send ADV_i to all n_j ∈ N;
20:      else
21:          Better_Signal_i  ←  0;
22:      end if
23:  end

24:  Receive(msg_i such that origin_i(msg_i) = (n_i, n_j))
25:  begin
26:      if msg_i = ADV_j and signal_i(msg_i) > Better_Signal_i then
27:          Better_Signal_i  ←  signal_i(msg_i);
28:          Cluster_i  ←  j;
29:      end if
30:      if msg_i = MEMBER_j(i) and Cluster_i = i then
31:          set_Member_i  ←  set_Member_i ∪ n_j;
32:      end if
33:  end

34:  Confirmation.timeout()
35:  begin
36:      if Cluster_i ≠ i then
37:          Send MEMBER_i(Cluster_i) to all n_j ∈ N;
38:      end if
39:  end
```

decides as to where the cluster-head will belong to based on its received signal strength (comparison between $signal_i(msg_i)$ and $Better\_Signal_i$). After this phase, in the Confirmation phase shown in lines 34–39, each node must inform its cluster-head that it will be a member of that cluster-head. This is done by sending a broadcast message ($MEMBER_i(Cluster_i)$) to the cluster-head. The cluster-head, in turn, includes the originator of the member message in a list of nodes of its cluster (set_Member_i), as shown in lines 30–32.

Algorithm 23.7 (*Expanding Ring Approach*)

```
 1: Cluster_i ← nil;
 2: Depth_i ← 1;
 3: set_Member_i ← {};
 4: Parent_i ← nil;

 5: Cluster_SetUp.timeout()
 6:   begin
 7:     Parent_i ← i;
 8:     Cluster_i ← i;
 9:     if size_i(set_Member_i) < MAX_BOUND then
10:         Send ADV_i(Depth_i) to all n_j ∈ Neig_i;
11:     end if
12:     Depth_i ← Depth_i + 1;
13:   end

14: Receive(msg_i(d) such that origin_i(msg_i) = (n_i, n_j))
15:   begin
16:     if msg_i(d) = ADV_c(d) then
17:         if d = 1 then
18:             if Cluster_i = nil then
19:                 Cluster_i ← c;
20:                 Parent_i ← n_j;
21:                 Send MEMBER_i(c) to n_j;
22:             end if
23:         else
24:             if Cluster_i = c then
25:                 Send ADV_c(d − −) to all n_k ∈ Neig_i;
26:             end if
27:         end if
28:     end if
29:     if msg_i(c) = MEMBER_j(c) then
30:         if i ≠ c then
31:             Send MEMBER_j(c) to Parent_i;
32:         else
33:             set_Member_i ← set_Member_i ∪ n_j;
34:         end if
35:     end if
36:   end
```

Note that this is a simple solution that does not impose a limit on the cluster size. Thus its structure may not be efficient due to resource and communication restrictions. Additionally, the solution requires direct access to all nodes of the network, which may be difficult to achieve in practice. An expanding ring search solution [31] can be used as a complementary approach to build a cluster as shown in Algorithm 23.7.

The Expanding Ring algorithm proceeds in rounds and in every round the cluster-head acquires nodes distant one hop more than the previous one, starting with depth 1. Each round is represented by an execution of "*Cluster_SetUp.timeout*()" (lines 5–13) that is executed while the limit of the cluster is not reached ($size_i(set\_Member_i) < MAX\_BOUND$). An advertisement message with a depth scope ($ADV_i(Depth_i)$) is sent to the neighbors of the cluster-head. This message is propagated in a multi-hop way and in every hop the depth is decreased (lines 16–28). A node that does not belong to a cluster and

Algorithm 23.8 *SPA — Self-Positioning Algorithm*

```
 1:  Dᵢ[j][j]  ←  nil for all nⱼ ∈ N;
 2:  Cᵢ[j]  ←  (nil, nil) for all nⱼ ∈ N;

 3:  Beacon.timeout()
 4:  begin
 5:    Send BEACONᵢ to all nⱼ ∈ Neigᵢ;
 6:  end

 7:  Information.timeout()
 8:  begin
 9:    Send ADVᵢ(Dᵢ[i][ ]) to all nⱼ ∈ Neigᵢ;
10:  end

11:  Receive(msgᵢ(d) such that originᵢ(msgᵢ) = (nᵢ, nⱼ))
12:  begin
13:    if msgᵢ = BEACONⱼ then
14:        Dᵢ[i][j]  ←  TOAᵢ(BEACONⱼ);
15:    end if
16:    if msgᵢ = ADVⱼ(Dⱼ[j][ ]) then
17:        Dᵢ[j][n]  ←  Dⱼ[j][n] for all n ≥ 0;
18:    end if
19:  end

20:  LCS()
21:  begin
22:    Cᵢ[j]  ←  (nil, nil) for all nⱼ ∈ N;
23:    Cᵢ[i]  ←  (0,0);
24:    Choose p, q ∈ Neighᵢ such that
```
$$D_i[p][q] > 0 \text{ and } \gamma_i = \arccos\frac{D_i[i][p]\hat{}2 + D_i[i][q]\hat{}2 - D_i[p][q]\hat{}2}{2 \times\ D_i[i][p] \times\ D_i[i][q]} \neq 0$$
```
25:    Cᵢ[i]  ←  (0,0);
26:    Cᵢ[p]  ←  (Dᵢ[i][p], 0);
27:    Cᵢ[q]  ←  (Dᵢ[i][q] × cos γᵢ, Dᵢ[i][q] × cos γᵢ)
28:    foreach nⱼ ∈ Neighᵢ do
29:    begin
30:        if Cᵢ[j] = (nil, nil) and
              Dᵢ[i][j] > 0 and
              ∃nₓ, nᵧ such as (Cᵢ[x], Cᵢ[y] ≠ (nil, nil)
              and Dᵢ[x][j], Dᵢ[y][j] > 0) then
31:            Cᵢ[j]  ←  triangulisationᵢ(i, x, y);
32:        end if
33:    end
34:  end
```

---

receives an advertisement with depth of 1 will be a member of the cluster defined by the node that sent the advertisement ($Cluster_i = c$) and will set the neighbor from which it received the message as its father ($Parent_i = n_j$) thereby, forming a tree. Finally, a confirmation message ($MEMBER_i(c)$ in line 21) is sent through the tree to the cluster-head, which includes the originator of the member message in a list of nodes of its cluster (set_Member_i) as shown in lines 29–35.

Other proposals for clustering in ad hoc and sensor networks can be found in the literature, including more elaborated solutions for multilayer organization and role-assignment (e.g., Reference 30).

23.6 Localization Algorithms

Many applications and activities in ad hoc and sensor networks depend on the information of node localization. Some examples are routing, like in greedy perimeter stateless routing (GPSR) [32], where packets are forwarded to neighbors in the physical direction of the destination, in density control algorithms, like the GAF algorithm [24], and applications that need to relate events with their localizations. Some proposals assume that information about node positioning is given by the GPS system. However, there are cases when the GPS signal is not available (e.g., indoor environment) or simply it is not possible to have a GPS receiver due to cost or size restrictions. An alternative solution is to use a distributed algorithm that allows nodes to find their relative positions inside the network area using range measurements among them to build a network coordinate system. An algorithm with this functionality is called self-positioning algorithm (SPA) [33]. The process of creating a local coordinate system (LCS) is presented in Algorithm 23.8

The algorithm shows the establishment of a LCS for every node i, where its coordinates will be the reference ($C_i[i] = (0,0)$). The algorithm operates detecting its neighbors through the use of beacon messages ($BEACON_i$ in lines 3–6) and calculating the distance between them by a method like time of arrival ($TOA_i()$ in lines 13–15). Every node sends this information to its neighbors ($ADV_i(D_i[i][\ ]$) in lines 7–10). Thus, the node knows its two-hop neighbors and some of the distances between its one- and two-hop neighbors (information stored in $D_i[j][j]$ in lines 16–18). The process of establishing an LCS (represented by $LCS()$ in lines 20–34) starts by choosing two nodes $p, q \in Neigh_i$ such that the distance between them is known and larger than zero ($D_i[p][q] > 0$) and such that the nodes i, p, and q do not lie on the same line ($gamma_i \neq 0$). This is summarized in line 24. The coordinates of the nodes are defined by triangle rules such that node p lies on positive x-axis and node q has a positive y component (lines 25–27). The position of the other nodes $n_j \in Neigh_i$ can be computed by triangularization if the positions of at least two other nodes (n_x, n_y) and the distance of them to n_j are known ($C_i[x], C_i[y] \neq (nil, nil)$ and $D_i[x][j], D_i[y][j] > 0$). This is shown in lines 28–33. The original work of SPA also describes how to adjust a node in LCS to a whole network coordinate system by electing a node that will be the center of it.

Acknowledgment

This work has been partially supported by CNPq — Brazilian Research Council under process 55.2111/02-3.

References

[1] Francis Heylighen. The science of self-organization and adaptivity. *The Encyclopedia of Life Support Systems*. EOLSS Publishers, Oxford, UK, 2002.

[2] W. Ross Ashby. Principles of the self-organizing dynamic system. *Journal of General Psychology*, 37: 125–128, 1947.

[3] Stuart Kauffman. *Origins of Order: Self-Organization and Selection in Evolution*. Oxford University Press, Oxford 1993.

[4] Scott Camazine, Jean-Louis Deneubourg, Nigel R. Franks, James Sneyd, Guy Theraulaz, and Eric Bonabeau, eds. *Self-Organization in Biological Systems*. Princeton University Press, Princeton, NJ, 2001.

[5] Self-organization. Wikipedia: The free encyclopedia. Access: October 2004. [Online] Available: http://en.wikipedia.org/wiki/Self-organization.

[6] Francis Heylighen and Carlos Gershenson. The meaning of self-organization in computing. *IEEE Intelligent Systems, Trends and Controversies. Self-Organization and Information Systems*, May/June 2003.

[7] Ljubica Blazevic, Levente Buttyán, Srdan Capkun, Silvia Giordano, Jean-Pierre Hubaux, and Jean-Yves Le Boudec. Self-organization in mobile ad-hoc networks: the approach of terminodes. *IEEE Communications Magazine*, June 2001.

[8] Loren P. Clare, Gregory J. Pottie, and Jonathan R. Agre. Self-organizing distributed sensor networks. In *Proceedings of the SPIE Conference on Unattended Ground Sensor Technologies and Applications*, pp. 229–237, April 1999.

[9] K. Sohrabi, J. Gao, V. Ailawadhi, and G. Pottie. Protocols for self-organization of a wireless sensor network. *IEEE Personal Communications*, 7: 16–27, 2000.

[10] Wei Ye, John Heidemann, and Deborah Estrin. An energy-efficient mac protocol for wireless sensor networks. In *Proceedings of the IEEE Infocom*, pp. 1567–1576, New York, USA, June 2002. USC/Information Sciences Institute, IEEE.

[11] David B. Johnson and David A. Maltz. Dynamic source routing in ad hoc wireless networks. In Imielinski and Korth, eds, *Mobile Computing*, vol. 353. Kluwer Academic Publishers, Dordrecht, 1996.

[12] Charles Perkins and Pravin Bhagwat. Highly dynamic destination-sequenced distance-vector routing (DSDV) for mobile computers. In *Proceedings of the ACM SIGCOMM'94 Conference on Communications Architectures, Protocols and Applications*, pp. 234–244, 1994.

[13] C. Perkins. Ad-hoc on-demand distance vector routing, 1997.

[14] Vincent D. Park and M. Scott Corson. A highly adaptive distributed routing algorithm for mobile wireless networks. In *Proceedings of the INFOCOM (3)*, pp. 1405–1413, 1997.

[15] Chalermek Intanagonwiwat, Ramesh Govindan, and Deborah Estrin. Directed diffusion: a scalable and robust communication paradigm for sensor networks. In *Proceedings of the 6th ACM International Conference on Mobile Computing and Networking (MobiCom'00)*, pp. 56–67, Boston, MA, USA, August 2000.

[16] Wendi Heinzelman, Anantha Chandrakasan, and Hari Balakrishnan. Energy-efficient communication protocols for wireless microsensor networks. In *Proceedings of the Hawaiian International Conference on Systems Science (HICSS)*, Maui, Hawaii, USA, January 2000.

[17] Deepak Ganesan, Ramesh Govindan, Scott Shenker, and Deborah Estrin. Highly-resilient, energy-efficient multipath routing in wireless sensor networks. *ACM SIGMOBILE Mobile Computing and Communications Review*, 5: 11–25, 2001.

[18] Joanna Kulik, Wendi Heinzelman, and Hari Balakrishnan. Negotiation-based protocols for disseminating information in wireless sensor networks. *Wireless Networks*, 8: 169–185, 2002.

[19] John Heidemann, Fabio Silva, and Deborah Estrin. Matching data dissemination algorithms to application requirements. In *Proceedings of the 1st International Conference on Embedded Networked Sensor Systems (SenSys'03)*, pp. 218–229, ACM Press. Los Angeles, CA, USA, November 2003.

[20] K. Sohrabi, J. Gao, V. Ailawadhi, and G. Pottie. Protocols for self-organization of a wireless sensor network. *IEEE Personal Communications*, 7: 16–27, 2000.

[21] Ugur Cetintemel, Andrew Flinders, and Ye Sun. Power-efficient data dissemination in wireless sensor networks. In *Proceedings of the 3rd ACM International Workshop on Data Engineering for Wireless and Mobile Access*, pp. 1–8. ACM Press, New York, 2003.

[22] John Heidemann, Fabio Silva, and Deborah Estrin. Matching data dissemination algorithms to application requirements. In *Proceedings of the 1st ACM Conference on Embedded Networked Sensor Systems (SenSys 2003)*, November 2003.

[23] Carlos M.S. Figueiredo, Eduardo F. Nakamura, and Antonio A.F. Loureiro. Multi: a hybrid adaptive dissemination protocol for wireless sensor networks. In *Proceedings of the 1st International Workshop on Algorithmic Aspects of Wireless Sensor Networks (Algosensors 2004)*, Volume 3121 of *Lecture Notes in Computer Science*, pp. 171–186, Springer. Turku, Finland, July 2004.

[24] Y. Xu, S. Bien, Y. Mori, J. Heidemann, and D. Estrin. Topology control protocols to conserve energy in wireless ad hoc networks. In *Submitted for review to IEEE Transactions on Mobile Computing*. CENS Technical Report 0006, January 2003.

[25] Curt Schurgers, Vlasios Tsiatsis, Saurabh Ganeriwal, and Mani Srivastava. Optimizing sensor networks in the energy-latency-density design space. *IEEE Transactions on Mobile Computing*, 1: 70–80, 2002.

[26] Alberto Cerpa and Deborah Estrin. Ascent: adaptive self-configuring sensor networks topologies. In *Proceedings of the 21st International Annual Joint Conference of the IEEE Computer and Communications Societies (INFOCOM 2002)*, vol. 3, pp. 1278–1287, New York, NY, USA, June 2002.

[27] Benjie Chen, Kyle Jamieson, Hari Balakrishnan, and Robert Morris. Span: an energy-efficient coordination algorithm for topology maintenance in ad hoc wireless networks. *ACM Wireless Networks*, 8: 2002.

[28] Elaine Catterall, Kristof Van Laerhoven, and Martin Strohbach. Self-organization in ad hoc sensor networks: an empirical study. In *Proceedings of the 8th International Conference on Artificial Life*, pp. 260–263. MIT Press, Cambridge, MA, 2003.

[29] Rajesh Krishnan and David Starobinski. Message-efficient self-organization of wireless sensor networks. In *Proceedings of the IEEE WCNC 2003*, pp. 1603–1608, March 2003.

[30] Manish Kochhal, Loren Schwiebert, and Sandeep Gupta. Role-based hierarchical self organization for wireless ad hoc sensor networks. In *Proceedings of the 2nd ACM International Conference on Wireless Sensor Networks and Applications*, pp. 98–107. ACM Press, New York, 2003.

[31] C.V. Ramamoorthy, A. Bhide, and J. Srivastava. Reliable clustering techniques for large, mobile packet radio networks. In *Proceedings of the 6th Annual Joint Conference of the IEEE Computer and Communications Societies (INFOCOM '87)*, San Francisco, CA, USA, March/April 1987.

[32] Brad Karp and H.T. Kung. Gpsr: greedy perimeter stateless routing for wireless networks. In *Proceedings of the 6th Annual International Conference on Mobile Computing and Networking*, pp. 243–254. ACM Press, New York, 2000.

[33] S. Capkun, M. Hamdi, and J. Hubaux. Gps-free positioning in mobile ad-hoc networks. In *Proceedings of the 34th Annual Hawaii International Conference on System Sciences (HICSS-34)-Volume 9*, page 9008. IEEE Computer Society Press, Washington, 2001.

24

Power Management in Mobile and Pervasive Computing Systems

24.1 Introduction **24**-535
24.2 Research Trends for Mobile and Pervasive
Computing **24**-536
24.3 Battery Technology **24**-538
24.4 Power Management Approaches for Mobile Devices .**24**-540
Power Breakdown of Typical Mobile Devices • General
Scheme for HW Components Power Management • Reactive
Policies • Workload Modifications • Discussion
24.5 Power Management in Infrastructure-Based
Networks **24**-548
MAC-Layer Policies • Upper-Layer Policies
24.6 Power Management and Pervasive Computing **24**-559
Power Management in Single-Hop Ad Hoc Networks •
Power Management in Multiple-Hop Ad Hoc Networks
24.7 Conclusions and Open Issues **24**-568
Acknowledgments **24**-569
References **24**-569

G. Anastasi
M. Conti
A. Passarella

24.1 Introduction

The proliferation of mobile computing and communication devices is producing a revolutionary change in our information society. Laptops, smart-phones, and PDAs, equipped with wireless technologies, support users in accomplishing their tasks, accessing information, or communicating with other users anytime, anywhere. Projections show that in few years the number of mobile connections and the number of shipments of mobile terminals will grow yet by another 20 to 50% [WWRF]. With this trend, we can expect that the total number of mobile Internet users will soon exceed the number of fixed-line Internet users.

Currently, most of the connections among mobile devices and the Internet occur over fixed infrastructure-based networks, which are extended with a wireless last hop. For example, cellular networks (GSM, GPRS, or UMTS) provide a wireless link between mobile phones and a base station, while laptops connect to the Internet via Wi-Fi Access Points (i.e., inside Wi-Fi hotspots). This networking paradigm is throughout referred to as *infrastructure-based* or *single-hop hotspot scenario* (see Figure 24.1[a]). In the

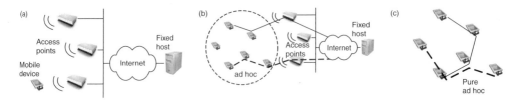

FIGURE 24.1 Mobile and pervasive networking scenarios. (a) Single-hop hotspot scenario, (b) ad hoc improved hotspot scenario, and (c) pure multi-hop ad hoc scenario.

near future the networking environments for mobile devices are expected to include *multi-hop ad hoc networks* (Figure 24.1[b] and Figure 24.1[c]). In the ad hoc networking paradigm, a group of mobile devices self-organize to create a network, without the need of any predeployed infrastructure. Applications running on those devices are able to communicate as in the standard Internet paradigm. The most straightforward application of the ad hoc paradigm is using ad hoc technologies to extend Wi-Fi hotspots. For example, in Figure 24.1[b], mobile devices inside a hotspot build an ad hoc network. Devices that are not in the Access Points' transmission range can nevertheless access the Internet by relying on the ad hoc network. Finally, the ad hoc technology is very likely to fuel the development of new network applications, explicitly designed for pure infrastructureless environments (Figure 24.1[c]).

The networking paradigms depicted in Figure 24.1 will be an important step toward the pervasive computing scenario envisaged by Mark Weiser [W91]. In such scenario the environment around the users will be saturated with devices, which will wirelessly interact among them, and with the users, to automatically adapt the environment to the users' needs (e.g., smart houses, smart offices, . . .) [IG00,ECPS02]. Many technical problems have to be fixed for this scenario becoming a reality. Among them, one of the most critical is *power management at mobile devices*[1]. To allow users' mobility, devices must be battery-supplied. It is common experience that current mobile devices (laptops, PDAs, etc.) can operate just for few hours before the battery gets exhausted. Even worse, the difference between power requirements of electronic components and battery capacities is expected to increase in the near future [S03]. In a nutshell, power management for mobile devices is a must for the development of mobile and pervasive computing scenarios, and each (hardware or software) component of a mobile device should be designed to be energy efficient. The networking subsystem is one of the critical components from the power management standpoint, as it accounts for a significant fraction of the total power consumption (around 10% for laptops, and up to 50% for small hand-held devices, such as PDAs).

This chapter provides an up-to-date state of the art of power management techniques for mobile and pervasive computing environments, and highlights some open issues. Section 24.2 provides an overview of the main power management approaches that have been proposed for such environments and introduces the research trends that will be analyzed in detail in the rest of the chapter.

24.2   Research Trends for Mobile and Pervasive Computing

The limited resources of mobile devices (e.g., limited CPU and memory) constitute the driving force of mobile and pervasive computing research [FZ94]. The energy available in a portable device is probably the most critical resource. In principle, either increasing the battery capacity, or reducing the power consumption, could alleviate mobile-device energy-related problems. However, as shown in Section 24.3, projections on progresses in battery technologies show that only small improvements in the battery capacity are expected in next future [S03,DWBP01]. If the battery capacity cannot be improved significantly, it is vital to manage power utilization efficiently, by identifying ways to use less energy preferably with no impact on the applications.

In Section 24.4, we provide a general framework for power management in a mobile device. These strategies can operate at different layers of the mobile-system architecture, including hardware, operating

---

[1]In this chapter the sentences *power consumption* and *energy consumption* are used interchangeably.

system, network protocols, and applications. At the operating system level techniques for hard-disk management [HLS96], CPU scheduling [LS97] and screen blanking [LS97] have been proposed. Techniques implemented at the application level include disconnected operations [S93], remote task execution [FPS02], use of agents [PS98], and the exploitation of the application semantic[2] [J00]. Techniques for adapting the application behavior to the changing level of energy also fall in this category [FS99].

Among the components of a mobile device that contribute to power consumption (CPU, video, hard-disk, network interface, etc.), the impact of the network interface becomes more and more relevant as the mobile-device size decreases. In a laptop the percentage of power drained by the wireless interface is about 10% of the overall system consumption [KK00,ACL03]. This percentage grows up to 50% when we consider small-size mobile devices like a PDA [KK00,ANF03]. This difference can be justified if we consider that small-size portable computers frequently have no hard disk and limited computational resources. On the other hand, the wireless interface provides almost the same functionalities in a laptop or a desktop PC. It is thus extremely important to design power-efficient network protocols and applications.

In principle, there are several approaches to reduce the power consumed by the network interface. In Section 24.5 we survey the most relevant power management policies for reducing the network-interface power consumption in the infrastructure-based scenario (Figure 24.1[a]). These techniques are presented in great detail, since the vast majority of today mobile computing environments fall in this category.

Power management schemes may be implemented at different layers of the network architecture [JSAC01]. Some researchers have focused on the impact of the transmission layer errors on the power consumption [RB96,ZR96]. When the bit error rate of the wireless channel is too high a transmitted message will be almost certainly corrupted, and this will cause power wastage. In this case, it is wise to defer the transmission. Other works suggest limiting the number of transmissions, as transmissions needs more power than receptions. The above strategies apply well in a cellular environment where the power consumption is asymmetric (the power consumed in the receiving mode is significantly less than in the transmitting mode). However, they are not suitable for a wireless local area network (WLAN) environment (i.e., Wi-Fi hotspot) where the network interface consumes approximately the same power in transmit, receive, and idle status [SK97,F04]. In this case the only effective approach to power saving consists in switching off the network interface (or, alternately, putting it in a sleep mode) when it is not strictly necessary. In addition, data-transfer phases should be as short as possible in order to reduce the time interval during which the wireless interface must be active (the ideal case is when data transfers are done at the maximum throughput allowed by the wireless link). The effectiveness of this approach was pointed out in several research works [SK97,KK98,ACL03].

Single-hop wireless Internet access represents the first step in the direction of pervasive computing [W91]. However, in a pervasive computing environment, the infrastructure-based wireless communication model is often not adequate as it takes time to set up the infrastructure network, while the costs associated with installing infrastructure can be quite high. In addition, there are situations where the network infrastructure is not available, cannot be installed, or cannot be installed in time. The answer to these problems is represented by the *ad hoc networking paradigm*, in which mobile devices build a multi-hop network, without relying on any predeployed infrastructure. Providing the needed connectivity and network services in these situations requires new networking solutions [C04]. In Section 24.6 we discuss novel power management issues generated by wireless networks developed for pervasive computing environments.

To summarize, in this chapter we survey the solutions proposed to minimize power consumption in mobile devices at different levels of the system architecture. Special emphasis is devoted to techniques aimed at reducing the power consumption in networking activities. We focus primarily on techniques for infrastructure-based wireless networks. However, we also consider power management solutions conceived for infrastructureless network technologies (ad hoc and sensor networks) that will have an important role for providing mobile and wireless communications in pervasive computing environments.

---

[2]For example, the application compresses the data before exchanging them, or finds some tradeoffs between performance and power consumption.

24.3 Battery Technology

The spectrum of mobile computing devices is quite large. They range from laptops to sensor nodes that have a size of few tens of mm^3 [DWBP01]. Hereafter, we divide mobile devices in three classes: (i) high-end devices (such as laptops), (ii) medium-size devices (such as PDAs and smart phones), and (iii) small-size devices (such as sensors).

Power requirements highly depend on the device type, and on the task the device is designed for. For example, laptops require rechargeable, high-capacity batteries, but dimension and weight are not primary concerns. On the other hand, sensor nodes are typically not reachable after the sensor network is deployed (e.g., they may monitor radioactive or polluted zones, they may be put below buildings to monitor seismic waves, etc.). Therefore, sensor nodes are designed to work unattended until the battery is completely exhausted. Thus, batteries for sensor nodes are typically not rechargeable, and must be very small [PSS01]. To summarize, due to the high variety of devices, the spectrum of battery technologies used in mobile devices is wide.

Today, the most widely used technology is lithium-ion cells [PSS01,DWBP01]. These batteries are available in many different form factors. Moreover, they can be restored at nearly full capacity for many recharge cycles. At the state of the art, lithium batteries are used as an off-the-shelf component, and can be found in devices of almost any class. For example, they are used in:

- *laptops*: Capacity around 4000 mAh, volume around 200 cm^3 [LBS]
- *PDAs*: Capacity around 900 mAh, volume around 50 cm^3 [IPAQ03]
- *sensors*: Capacity around 500 mAh, volume around 10 cm^3 [CMD,PSS01]

The tradeoff between capacity and size guarantees a reasonable lifetime to devices such as laptops, PDAs, or smart phones. However, for wearable computers and sensor networks, form factors of commercial batteries might not be sufficient. Current research efforts are devoted to miniaturize sensor nodes and wearable devices as much as possible. For example, the SmartDust project of the University of California at Berkeley has developed prototype sensor nodes as small as few tens of mm^3 [DWBP01]. In this scenario, borderline technologies are currently investigated, in order to provide sufficient energy in so small form factors. Not only technologies are investigated aimed at improving the energy density, but also solutions are proposed where devices scavenge energy from the environment around them, without being equipped with traditional batteries. Some examples are highlighted in the following:

- *Vanadium and molybdenum oxide batteries* [HDD98]: This technology allows to produce thin films with capacities comparable to standard lithium cells.
- *Ultra-capacitors* [MT]: These components store less capacity that lithium cells, but can provide high power, due to lower source impedance.
- *Solar cells* [DWBP01]: This technology exploits the well-known photovoltaic effect to produce energy. Therefore, it can be utilized only when some source of lighting is available. Moreover, the surface needed and the rate at which energy is produced may be severe limiting factors [S03].
- *Vibrations* [MMA+01]: Ambient vibrations can be converted in electrical energy by means of micro-electromechanical systems (MEMS). Experimental prototypes have shown to produce between 5 and 8 μW with a chip of about 6 mm^2, resulting in performance similar to solar cells (around 1 μW [DWBP01]).
- More exotic systems have been proposed based, for example, on exploiting heat from micro-rocket engines [TMCP01].

Despite this spectrum of activities, researchers in the mobile networking area agree that battery capacities are today one of the main limiting factors to the growth of this computing paradigm [ECPS02,PSS01,CR00]. More important, in the near future the difference between battery capacities and energetic requirements is likely to become even deeper. The work in [S03] collects the main performance

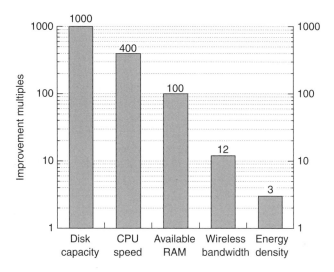

FIGURE 24.2   Increase of components' capabilities from 1990 to 2000 [S03].

characteristics of several components of a mobile device from 1990 to 2000, pointing out the improvements achieved in this period by each component, see Figure 24.2. For easy presentation, improvements are expressed as multiples of the performance in the year 1990 (please note the log scale on the y-axis).

It is worth noting that battery improvements are far slower than electronic-component improvements. In addition, the increase of components' performance usually produces an increase in energy requirements. For example, the power, P, drained by a processor is proportional to the frequency of its clock (i.e., $P \propto C \cdot V^2 \cdot f$, where C is the characteristic capacitance of the transistor technology, V is the supply voltage, and f is the clock frequency). To summarize, Figure 24.2 indicates that a *wise power management is a key enabling factor for the deployment of mobile networks*.

To design effective power management policies, some batteries' properties must be taken into account. The most important are: the *Rate Capacity Effect* and the *Relaxation Effect* [PCD+01,CR00]. The energy that a battery can provide depends not only on its capacity, but also on the profile of the current that it supplies (throughout referred to as I_d). If I_d is constant and greater than a certain threshold (referred to as the *rated current* of the battery), the energetic efficiency can be dramatically low. Specifically, it is well known that the voltage of a battery drops down while it is supplying current. Eventually, a *cut-off voltage* is reached, after which the battery is no longer able to provide energy. It has been shown that, if I_d is constant over time, the more I_d grows beyond the rated current, the less energy is provided before reaching the cut-off voltage [PCD+01,PSS01,DN97] (i.e., the lower is the energetic efficiency). This phenomenon is known as the *rate capacity effect* [PCD+01]. A battery is composed by an anode, a cathode, and an electrolyte. According to the diffusion phenomenon, positive ions flow from the anode to the cathode through the electrolyte, thus balancing the electrons flow on the (outer) circuit supplied by the battery. Specifically, during discharge, positive ions are produced by oxidation at the anode. Ions travel through the electrolyte and arrive at the cathode surface. The cathode contains reactions sites, where positive ions undergo reduction, becoming inactive. At low discharge currents, inactive sites are distributed almost uniformly in the cathode. But, at high discharge currents, inactive sites become concentrated at the outer surface of the cathode, making internal reduction sites inaccessible. In this case, the battery is declared discharged but many reaction sites at the cathode have not yet been used. Furthermore, unavailability of positive ions at the cathode surface contributes to reduce the battery efficiency. In detail, during discharge, ions that undergo reduction at the cathode are replaced by ions traveling through the electrolyte. At low discharge currents, the concentration of ions at the anode–electrolyte interface, and at the cathode–electrolyte interface, is almost uniform. On the other hand, at high discharge currents, the diffusion process in the electrolyte becomes unable to transport enough positive ions at the cathode. As a result, the concentration of positive ions in the electrolyte becomes higher at the anode than at the cathode, and hence the battery voltage decreases. To face this problem, many authors propose to drain current in *pulsed mode*. That way, much higher energetic efficiency can be achieved [PCS02,PCD+01,PSS01]. Specifically, a (constant) current I_d is drained for a time interval t_{load}, then the battery is left idle for a time interval t_{idle}, and so on. During t_{idle} uniform concentration of positive ions at the anode and cathode is restored, and hence the battery recovers the voltage lost during the previous t_{load}. This phenomenon is known as the *relaxation effect*. By exploiting this effect, if a pulsed scheme is used, the energy provided before reaching the cut-off voltage is higher than in the case of continuous drain. Obviously, the performance improvements depend on the battery capacity, and on the ratio between t_{load} and t_{idle} [PCS02].

To summarize, for the deployment of mobile computing environments, smart power managers are required as (i) improvements in battery capacity are significantly lower than the increase of power requirements of electronic components, and (ii) the capacity of batteries can be fully exploited only by properly interleaving idle and active phases.

24.4   Power Management Approaches for Mobile Devices

At a very high level of abstraction, a computer can be seen as a set of hardware and software components servicing user requests. Hardware components need energy to operate. Instead, software components do not consume energy by themselves, but they impact on the power consumption as they control the activity of hardware components.

Great efforts have been devoted to design and manage hardware components in a power-efficient way. The most widespread approach consists in focusing on a particular component (e.g., the hard disk, the network interface, etc.), and deploying a *dedicated power manager* guaranteeing that the component provides the required services, while optimizing the power spent to provide them. For example, laptop screens are usually blanked after a pre-specified inactivity interval, and are resumed when new activity is detected (e.g., a mouse movement or a key press). That is, a power manager is "attached" to the screen, monitoring its activity, and putting it in a low-power operation mode when it is not used for a while. In principle, power managers can be deployed either in hardware or in software. However, due to their inherent flexibility, software managers are usually preferred. It must be finally pointed out that — in general — care should be taken in the additional power consumption of the manager itself.

Hereafter, we survey the power management approaches presented in the literature. To the purpose of this presentation, in Section 24.4.1 we introduce the breakdowns of the power consumption of hardware components in typical mobile devices.

24.4.1   Power Breakdown of Typical Mobile Devices

In the past years, several studies have highlighted the power drained by various components of laptops [LS98,US96]. Specifically, [US96] shows that the CPU, the wireless network interface, the hard disk and the display require approximately 21, 18, 18, and 36% of the total power, respectively. Similar results are also presented in [LS98]. More recently, several works confirm the large impact on the overall system consumption due to the wireless interface [ACL03,KK98], CPU [CBB+02,PS01] and hard disks [BBD00].

The power breakdown of devices such as PDAs is shown in [SBGD01]. Usually, such systems do not have hard disks, and use flash memories to store persistent data. The wireless network interface consumes about 50% of the total power, while the display and the CPU consume about 30 and 10%, respectively. The great impact of the network interface was previously highlighted also in [KK98].

Finally, the power breakdown of typical sensor nodes is highlighted in [DWBP01,RSPS02,ACF+04]. This type of computing devices drastically differ from both laptops and PDAs, and hence also the power breakdown is different. Furthermore, the power consumption may greatly depend on the specific task implemented by the sensor network [DWBP01]. Typically, the wireless network interface and the sensing subsystem are the most power-hungry components of a sensor node. On the contrary, the CPU accounts for a minor percentage of the power consumption.

To summarize, several hardware components of a computing device need power managers. In the past years, many researchers have focused mainly on hard disks and CPUs, as they account for a great portion of the total power consumption in laptops. More recently, increasing interest has been devoted to the networking subsystem, as it is the main source of power consumption for medium and small-size mobile devices (e.g., PDA). In the following of this section we survey power-management approaches that apply to different hardware subsystems. Then, in Sections 24.5 and 24.6, power-management approaches tailored to the networking subsystem will be discussed in depth.

24.4.2 General Scheme for HW Components Power Management

Benini et al. [BBD99] present a very useful scheme to understand how power management works, see Figure 24.3. To better understand Figure 24.3, let us define the concept of power-manageable component (PMC). A PMC is a hardware component that provides a well-defined kind of service. For example, hard disks, network interfaces, CPUs can be seen as power-manageable components. A fundamental requirement of a PMC is the availability of different *operating modes*. Each mode is characterized by a specific power consumption and performance level. For example, a hard disk shows high power consumption when disks are spinning, and data can be accessed very quickly. Instead, when disk heads are parked, the power consumption is much lower, but the access time becomes very high, since disks must be spun up again before accessing data.

In Figure 24.3, a PMC is represented as a *service provider* (SP). Each item in the *queue* (Q) is a request for an SP service. A *service requester* (SR) — representing, in an aggregate way, the SP users — puts items in the Queue. The *power manager* (PM) observes the status of the SR, Q, and SP (i.e., the status of the system), and issues commands to the SP, dynamically switching it among its operating modes. The algorithm used by the PM to decide the SP operating modes is referred to as *policy*. The goal of a policy is to make the SP handle items in the Q by: (i) spending the minimum possible energy, and (ii) still providing an acceptable quality of service (QoS) to users. It must be pointed out that transitions between different SP modes usually have a cost. Specifically, each transition requires a time interval during which the SP consumes energy but is not able to service any request. Furthermore, the more the power consumption of the low-power mode decreases, the more the time interval to switch to a high-performance mode increases. For each low-power mode M, a *break-even time* (T_{BE}^M) can be defined. When the SP is put in low-power mode M, it should not be switched back to the high-power mode before T_{BE}^M sec. Otherwise, the energetic advantage of using the low-power mode is overwhelmed by the transition cost [BBD00]. Hence, the SP can be modeled as a server with vacations. During vacation periods, the SP operates in low-power modes and it is not able to service requests in the Queue. The vacation lengths depend on the particular mode the SP is operating in, and on the policy implemented by the PM. Specifically, for each

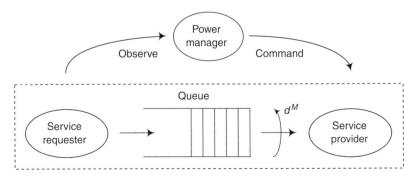

FIGURE 24.3   High-level system model [BBD99].

power mode M, T_{BE}^M represents the minimum vacation length. The PM may increase the vacation length, based on the observed system status. Therefore, if d^M denotes the vacation length related to the operating mode M, we can express d^M as follows:

$$d^M = T_{BE}^M + f_d(\text{policy}) \tag{24.1}$$

where $f_d(\,)$ denotes the delay introduced by the selected power management policy.

It is worth noting that a typical drawback of any power management policy is a negative impact on the QoS perceived by SP users. With reference to Equation (24.1), energy saving increases with the increase of d^M. However, d^M also impacts on the delay introduced by power management to service times, thus decreasing the QoS. For example, when a hard disk is spun down, the user must wait for a long time at the first access. Therefore, the PM design must trade-off between these contrasting requirements.

In the original model [BBD99], the system represented in Figure 24.3 is completely isolated, that is, the behavior of SR, SP, and PM neither affects nor is affected by any other computer component. Though this model has been exploited to analyze several power-saving strategies [BBD00,BD02], it is worth enriching it, to achieve a more complete characterization of power management for mobile devices. Specifically:

- The PM could observe not only the status of the SR, Q, and SP, but also the status of other computer components (i.e., interactions between different computer components could be taken into consideration).
- The PM could manage not only the SP, but also the SR, and Q. That is, not only the SP *behavior* can be controlled based on the requests performed by the SR, but the *workload* generated by the SR could be shaped in order to reduce the power consumption of the SP.

As a final remark, it must be noted that PMs considered in this chapter are sometimes referred to as *external* PMs. These PMs are used with SPs that allow an external controller to put them in different operating modes. External PMs are usually designed separately from SPs, and take into account the status of the entire system (at least, SR, Q, and SP) to save energy. On the other hand, *internal* PMs switch SPs between different operating modes based only on the internal state of the SP itself. They are usually implemented in hardware, and are designed together with the SP (i.e., they are part of the SP itself). Note that operating modes used by internal and external PMs differ. For example, hard disks by IBM define an idle mode to be used by external PMs; this idle mode is actually a wrapper for three possible idle modes, which can be set only by an internal PM, implemented in the hard-disk circuitry [BBD00]. Using internal PMs is one of the power-saving methodologies that are used in the chip design of SPs. For the sake of space, hereafter we omit considering this kind of techniques (the interested reader can refer to [BD02]).

24.4.2.1 Overview of Power Management Strategies

The core of any power management system is defining the behavior of PMs. Research efforts usually follow two complementary approaches. With reference to Figure 24.3, the first approach focuses on dynamically driving the SP in the various operating modes, based on the knowledge of the system status (i.e., the SP, SR, and Q status). This approach is usually referred to as *dynamic power management* (DPM). Solutions of this type range from simple timeout policies, to policies based on dynamically estimating the profile of requests issued by the SR. These solutions do not modify the profile of SR requests (in order to optimize the power consumption of the SP), but they dynamically react to the requests profile by selecting the appropriate mode of the SP. From this standpoint, this approach can be seen as a *reactive approach*. Reactive policies are surveyed in Section 24.3.

A second approach focuses on modifying the workload issued to the SP, that is, PM modifies the profile of SR requests. Policies based on this approach are aware of the impact that the workload shape has on the SP power consumption, and hence modify the workload to optimize the performance. This approach can be seen as a *proactive approach*. For example, some compilers modify the execution order of instructions to reduce transitions between 0s and 1s on the buses' lines. Obviously, it must be assured that the SP provides the same results after processing either the original or the modified workload.

As the reactive and proactive approaches are orthogonal, some solutions exploit a combination of them. Hereafter, we refer to these solutions as *mixed policies*. Mixed policies implement algorithms that dynamically select the SP operating mode, based on the workload shape. If possible, they also modify the workload in such a way that low-power operating modes could be exploited more efficiently than with the original workload. Section 24.4.4 is devoted proactive and mixed policies.

24.4.3　Reactive Policies

Reactive policies dynamically drive the SP behavior by deciding on-line the best operating mode for the SP, given the current status of the mobile device. Before analyzing the different types of reactive policies, it is worth spending few words on how different operating modes are implemented.

24.4.3.1　Dynamic Voltage and Frequency Scaling

Dynamic voltage scaling (DVS) and dynamic frequency scaling (DFS) are techniques used to implement low-power operating modes, and hence they are the enabling technologies for dynamic power managers. Moreover, some PMs explicitly scale voltage and frequency as a power-saving technique.

The motivation behind DFS is that the power consumption of an electronic component is proportional to $V^2 \cdot f$, where V is the supply voltage and f is the clock frequency. Many researchers highlight that DFS alone may not be sufficient to save energy. Specifically, the time needed to perform a given task is proportional to $1/f$, making the energy spent independent of the operating frequency ($E \propto V^2$). However, Poellabauer and Schwan [PS02] show that in some cases just arranging the operating frequency allows saving energy. Specifically, Poellabauer and Schwan [PS02] focus on dedicated CPUs periodically executing real-time event handlers. If a handler execution finishes before its deadline, the CPU remains idle till the new execution. Hence the energy spent can be expressed as $E = E_\mathrm{h} + E_\mathrm{idle}$, where E_h is the energy spent to execute the handler and E_idle is the energy spent when the CPU is idle. In this case, if the frequency is reduced in such a way that the handler execution finishes immediately before the next execution, the energy spent becomes equal to E_h.

The best performance, in terms of power saving, is achieved when DVS and DFS are used together. When the CMOS technology is used, clock frequency and supply voltage are strictly tied. Specifically, the supply voltage reduces by decreasing the transistors operating frequency [BB95]. Since the energy depends quadratically on V, the energy saving achieved in this way can be quite large [LBD02,PS01].

The most aggressive forms of DFS and DVS are achieved by completely freezing the clock (i.e., $f = 0$), or powering the circuitry down (i.e., $V = 0$). Usually, not the entire SP is frozen or powered down, but only some subsystems. The different operating modes of SPs are typically implemented in this way. For example, 802.11 WLAN cards use a sleep mode, where only synchronization-related circuitries are active, while the rest of the card is powered down. It must be pointed out that clock freezing achieves lower energy saving than powering down, because transistors still drain a static current, known as leakage current. On the other hand, transition times related to clock freezing are typically negligible, while powering up subsystems may take as long as few seconds, depending on the particular SP [BBD99].

24.4.3.2　Taxonomy and Examples of Reactive Policies

The research on Reactive Policies is a very hot topic. A possible taxonomy of proposed policies is presented in Figure 24.4 [BBD00]. We can distinguish between predictive and stochastic policies. Predictive policies estimate the (short-term) behavior of the SR, and exploit this information to decide whether it is convenient for the SP to go into low-power states. As shown in Figure 24.4, predictive policies can be further subdivided in several classes. On the other hand, stochastic policies are not based on estimates of SR future behavior. Instead, SP and SR are modeled by means of finite state machines (FSM), and the power-management policy defines the transitions between the states of the SP machine. In the following of this section we describe the main features of each power-management class shown in Figure 24.4.

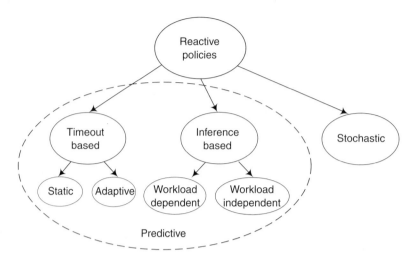

FIGURE 24.4 Taxonomy of reactive policies.

24.4.3.3 Predictive Policies

The fundamental assumption behind these policies is that some correlation exists between the past history of SR requests and the near-future requests. This correlation is exploited to predict the near-future shape of the workload, and hence to decide the operating mode of the SP. Without loss of generality, to understand how policies of each class operate, we can consider a SP having only two operating modes, a high-power high-performance operating mode, and a low-power low-performance operating mode.[3]

Timeout-Based Policies: This is the simplest type of reactive policies. Timeouts are typically used to decide when the SP must switch to the low-power operating mode. Specifically, if T_{TO} is the timeout value, the SP is forced in the low-power mode T_{TO} seconds after the last served request. An example of such techniques is the policy used by several operating systems to blank the screen.

This approach is effective only if the following equation holds

$$p(T > T_{TO} + T_{BE} \mid T \geq T_{TO}) \approx 1 \qquad (24.2)$$

where T denotes the length of an idle phase, that is, the time during which the Queue remains empty. As said before, after the transition to a low-power operating mode is decided, the SP should not switch back to the high-power mode before T_{BE} sec, in order to save energy. Therefore, the probability of having an idle phase longer than $T_{TO} + T_{BE}$, provided that it is longer than T_{TO}, must be close to one; otherwise, this approach may lead to a greater power consumption than letting the SP always in the high-power mode. The main concern of timeout policies is the choice of the T_{TO} value. Typically, by increasing the timeout value T_{TO}, the probability in the left-hand-side of Equation (24.2) increases. On the other hand, the SP remains in the high-power mode for T_{TO} sec after the last request, and this significantly increases power consumption.

Several works use *fixed* timeout values, and propose algorithms for setting the appropriate T_{TO} value. These solutions can be seen as *static timeout-based* policies, as they do not change the timeout value over time. Static-timeout policies have been the first class of reactive policies studied in the literature, and have been mainly applied to hard disks [DKM94,GBS+95, KMMO94]. More recently, Kravets and Krishnan [KK00] proposed to use timeout-based policies to manage the wireless network interface (see Section 24.5 for details). Today, the vast majority of policies implemented in commercial operating systems belong to

---

[3]This is the most common case in the real world, and the discussion can be easily extended to include more general scenarios.

this class. However, several works have shown that better performance can be achieved if the T_{TO} value is dynamically adjusted, based on the past history of idle phases. These policies are referred to as *adaptive timeout-based* policies. For example, Krishnan et al. [KLV95] maintain a list of possible timeouts, together with a list of "scores" showing how well each timeout would have performed in the past history. When an idle phase starts, the timeout with the highest score is chosen. Similarly, Helmbold et al. [HLS96] use a list of candidate timeouts and a list of corresponding scores. When an idle phase occurs, the timeout value is evaluated as the weighted sum of the candidate timeouts, using the scores as weights. Finally, Douglis et al. [DKB95] record the number of transitions to low-power operating modes that have occurred in the recent past. When this number becomes too high, it means that the T_{TO} value is too short with respect to the inter-arrival times between requests (i.e., new requests arrive very soon after the SP has been put into low-power operating mode). In this case, the T_{TO} value is increased. On the other hand, the T_{TO} value is decreased when the number of transitions into low-power operating modes becomes too low (i.e., the SP is almost always in the high-power operating mode).

Inference-Based Policies: Two main drawbacks of timeout policies have been highlighted in the literature [BBD00]: (i) they waste power waiting for the timeout to expire; and (ii) they always pay a performance penalty when a new request arrives and the SP is operating in low-power mode. To overcome these limitations, several works try to estimate the *length* of the next idle phase, and manage the SP accordingly. Once an estimate is available, the SP is forced in the low-power mode if the estimate is greater than T_{BE}. Then, at the expiration of the estimate, it is resumed to manage the next request. The work in [SCB96] proposes two methods for predicting the length of an idle phase, both based on past idle and active phases' lengths. The first one uses a nonlinear regression, based on the lengths of the last k idle and active phases. The form of the regression equation, as well as k, is chosen by means of offline data analysis. The second one is based only on the length of the last active phase, say T_{active}. If T_{active} is below a certain threshold, the next idle phase is assumed to be longer than T_{BE}. This policy is tailored to a particular type of workload, that is, workloads where short active phases are followed by long idle phases. Furthermore, the value of the threshold is chosen using an offline data analysis. These solutions require an *a priori* knowledge of the statistical characteristics of the workload, that is, they are *workload dependent*. As knowledge about the workload could not be available at design time, Hwang and Wu [HW97] propose a *workload-independent* policy. In this work, the length of the next idle phase is estimated by means of a smoothed average algorithm, similar to the one used by the Transmission Control Protocol (TCP) to estimate the Round Trip Time. Specifically, if $T_{est}^{(n)}$ is the estimate at time n, and T_{idle} is the actual length of the last idle phase, T_{est} is updated as $T_{est}^{(n+1)} \leftarrow \alpha \cdot T_{idle} + (1 - \alpha) \cdot T_{est}^{(n)}$, where $\alpha \in [0, 1]$ is the smoothing factor. Chung et al. [CBBD99] propose a workload-independent technique, tailored to hard-disk management and based on adaptive learning trees. Workload dependent and independent policies have been proposed and extensively evaluated for managing wireless network interfaces [ACGP03b,ACGP03c] (see Section 24.5).

24.4.3.4 Stochastic Policies

Stochastic policies model the SP and the SR as FSMs. The FSM representing the SR models the SR behavior. Hence, the meaning of its states and transitions between states depends on the specific SR. On the other hand, the SP states usually represent the different operating modes of the device, and transitions between states are driven by the PM policy. Specifically, if P is the set of the SP states, R is the set of the SR states, and C is the set of the possible commands used by the PM to drive the SP, then the policy is represented by a function f_p: $(P \times R) \rightarrow C$. These models provide an analytical framework to define optimal power-management policies. Furthermore, tradeoffs between power saving and performance penalties are included in the framework by representing performance penalties as constraints of the optimization problem. The optimization problem can be solved offline, and provides the *optimal* power-management policy, that is, the policy that achieves the greatest power saving under the problem constraints [BBD00]. It must be pointed out that the policy optimality relies on the accuracy of the SR and SP models. Hence, when the policy is used to manage real workloads and SP, its performance can be worse than expected [SBGD01].

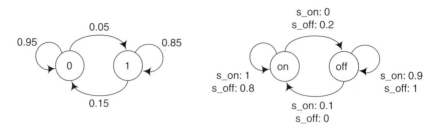

FIGURE 24.5   Example of Markov models of a service requester (left) and a service provider (right) [BBD00].

A first class of stochastic policies uses discrete-time Markov chains to model both the SR and the SP. In this type of model the time is slotted, and the state of the system is sampled at the beginning of each slot. For example, Figure 24.5 reproduces a scheme of such a model [BBD00]. The SR is modeled as a two-state Markov chain: state 0 means "no request is issued by SR," while state 1 means "a new request is issued by SR." The example in Figure 24.5 represents a bursty workload: if a new request is issued during the slot n, another request will be issued during the slot $n + 1$ with probability 0.85. Also the SP is represented by a two-state Markov chain: the SP has a low-power operating mode (off state), and a high-power operating mode (on state). Transitions between states are driven by commands issued by the PM (s_on means "switch on" and s_off means "switch off"). For example, let us focus on the on state, and let us assume that, at the beginning of slot n, the SP receives the s_off command. Then, at the beginning of slot $n + 1$, the SP will still be in the on state with probability 0.8, while it will be in the off state with probability 0.2. These probabilities model the transition times between the SP operating modes. Finally, the Power-Manager policy is an unknown function that, at the beginning of each slot, gives a command to the SP, based on the current SP and SR state. The optimization problem obtained from the system of Figure 24.5 is solved under different performance-penalty constraints by Benini et al. [BBD00]. Then, the performance-penalty versus power-consumption curve is plotted, showing that the stochastic approach provides a very useful tool for PM designers.

Chung et al. [CBB+02] highlight that the analytical approach used by Benini et al. [BBD00] relies on the assumption that the stochastic processes representing both the SR and the SP are known *a priori*, and are stationary (i.e., transition probabilities do not change over time). Unfortunately, these assumptions may be completely wrong, especially with respect to the SR. Therefore, [CBB+02] extends the previous model to the case of unknown stationary and nonstationary environments. In the former case, the parameters of the unknown *stationary* Markov model are estimated on-line, and then refined by monitoring the behavior of the real system. The system steady-state assumption guarantees that estimated parameters tend to the real ones. As the optimal policy depends on these parameters, it should be theoretically recomputed at each time slot. Since this approach is practically unfeasible, Chung et al. [CBB+02] compute offline optimal policies for different values of the system parameters, and then use an online interpolation technique to approximate the best policy. In the case of nonstationary environments, the technique used to estimate system parameters must be changed. To this end, [CBB+02] uses a window-based system, which provides estimates at run-time, based on the evolution of the system in the recent past. The interpolation technique is still used to compute the best policy.

All the above policies are time-slotted: at each time slot the system is observed and the PM sends a command to the SP. This can lead to power wastage, as commands might be computed too often [SBGD01]. Several works [SBGD01,QP00,QP99] overcome this drawback by using time-continuous, event-driven models, where commands are computed only when new events occur in the system. Moreover, Simunic et al. [SBGD01] highlight that using Markov chains (both in time-slotted and time-continuous models) means modeling transition times between SP and SR states by means of memoryless distributions (i.e., geometric or exponential). However, these assumptions are rarely met in real systems, resulting in poor performances in terms of power saving. To cope with this problem, in [SBGD01] two analytical approaches are proposed. Both are event-driven, and can be used with any distribution of transition times. The first one is based on the renewal theory and models the system by a finite state machine. PM commands are computed only when the system is in a particular state of this FSM. Under these hypotheses, the optimal policy can still be derived as the solution of an optimization problem. The second approach is based on time-indexed semi-Markov decision processes (TISMDP), and still models the system by means of a FSM. It is more complex than the previous one, but can deal with more general systems, where PM commands are computed in different states of the FSM. Even in this case, the optimal policy is the solution of an optimization problem. Finally, it should be noted that both approaches require an *a priori* knowledge of the distribution of transition times.

24.4.4 Workload Modifications

Instead of just driving the SP in different operating modes, the PM could also manage the SR to reduce the SP power consumption. These policies can be roughly divided between *proactive policies* and *mixed policies*. Proactive policies (see, e.g., Panigrahi et al. [PCD+01] and Kandemir et al. [KVIY01]) do not exploit possible low-power operating modes of the SP. Instead, they use some knowledge about the impact of different workload shapes on power consumption. Based on this information, PMs drive SR and Q. On the other hand, mixed policies [ACGP03b,PS01] manage SR and Q in order to better exploit the use of the low-power operating modes of the controlled SP.

24.4.4.1 Proactive Policies

Proactive policies include power-aware optimizations of software components. These policies can be classified depending on when code optimization is performed: before or after compilation. *Software synthesis* includes all policies in which the source code is transformed before compilation. Some researchers focus on exploiting computational kernels, that is, code blocks being executed frequently [BGS94,W96]. Kernels are then specialized having in mind power consumption. An example of this technique is procedure inlining, in which function calls are substituted by the body of the functions to eliminate the overhead related to invocations. A similar approach is based on value profiling and partial evaluation. The main idea is looking for function calls with frequent parameter values, and generating specialized versions of these functions for recurring parameter values [CFE99].

Compilers can also introduce power-aware optimizations. Some authors propose algorithms for instructions selection [TMW94] or registers allocation [MOI+97]. Others works reduce the power drained by bus lines (when switching between successive instructions) by instruction reordering [STD94]. Finally, techniques that optimize the code to reduce the energy spent by memories have also been investigated [KVIY01].

Lorch and Smith [LS98] highlight that well-known techniques such as caching and prefetching can be seen as proactive policies. Li et al. [LKHA94] show that energy saving can be achieved by properly sizing the cache of hard disks. Furthermore, the benefit of prefetching on energy saving can be highlighted by considering file systems such as Coda [S93].

Proactive policies have also been proposed to optimize the battery access patterns [PCD+01,CR00]. In these papers the SP is the battery, and the main idea is exploiting the relaxation effect (see Section 24.3) to extend the battery lifetime. Panigrahi et al. [PCD+01] focus on the networking subsystem of a general operating system. It is shown that very different battery lifetimes are achieved when the activities (and, hence, the battery accesses) required to process incoming and outgoing packets are scheduled in different ways. Chiasserini and Rao [CR00] highlight that lithium polymer cells can be shaped to provide a mobile device with an *array* of independent cells of a reasonable size. This opportunity is exploited by dynamically scheduling the cell that supplies the computer. The scheduler lets each cell relaxing for a sufficient amount of time.

24.4.4.2 Mixed Policies

These policies are often applied to CPU scheduling. The aim is to arrange dynamically tasks in the ready queue in order to maintain the clock frequency and the supply voltage as low as possible (see

Section 24.4.3.1). This methodology has been investigated for the first time by Weiser et al. [WWDS94], and then several researchers have used it. The works in [PS01,PS02,KL00,G01,MACM00,SC00] use this approach for real-time environments. For example, [PS02] focuses on periodic tasks. When an instance of the task must be executed, a set of estimates is generated indicating the execution times at each available frequency. The clock is then set at the minimum frequency meeting the task deadline. These estimates are based on offline measurements and previous executions of the same task. Finally, execution is frozen periodically, the estimates are updated, and the clock frequency is modified, if necessary. On the other hand, Flautner, et al. [FRM01], and Lorch and Smith [LS01] consider non real-time environments, with the aim of maintaining good interactive performance while reducing the CPU power consumption. For example, [LS01] assigns soft-deadlines to non real-time tasks, and schedules the CPU using these deadlines. Soft-deadlines are assigned based on estimations of tasks execution times derived from the history of previously executed (similar) tasks.

Mixed policies have also been proposed for managing wireless network interfaces. The works in [ACGP02, ACGP03b] focus on Web applications, and propose a power saving network architecture (see Section 24.5 for details). Among others, the goal is to concentrate data transfers in big bursts, in order to fully exploit the wireless link bandwidth. When a burst is completely transferred, the network interface is shut down. In [ACGP03b] it is shown that modifying the workload leads to about 10% additional energy with respect to a policy that lets the workload unmodified. Finally, policies have been proposed that drastically modify the networking applications in order to change the workload issued to SP. Examples can be found in [J00,PEBI01].

24.4.5 Discussion

To summarize, let us draw some qualitative comparison among the presented approaches. Timeout-based policies are very simple, and can be easily implemented. However, they usually exhibit poor performance, even if timeouts are dynamically adjusted. Candidate scenarios for such solutions are systems with long inactivity periods, where approximate timeouts are sufficient to detect idle phases.

Inference-based policies are very flexible, since they can work entirely on-line, without any assumption about the statistical characteristics of the SR [ACGP03c]. Moreover, when awareness of the workload shape is available, they can be customized. For example, the work in [ACGP03b] manages a WLAN card in the presence of Web traffic, and allows a power saving of about 90% with practically negligible performance penalties. In a similar scenario, stochastic policies achieve a power saving of approximately 70%, with higher performance penalties [SBGD01]. A drawback of predictive policies is that, in many cases, they are just heuristics, and hence it is hard to fully understand their behavior.

Finally, the main advantage of the stochastic approach is that it allows developing a general analytical framework. Thus, optimal policies can be derived, and a clear understanding of the PM behavior can be achieved. On the other hand, the stochastic approach usually needs quite strong assumptions about the statistical behavior of the real environment. When these conditions are not met, stochastic policies may perform poorly [SBGD01].

Proactive policies are the only way to save power when the SP is not a PMC, that is, when it cannot operate in different modes. On the other hand, in presence of PMC, mixed policies join the advantages of both reactive and proactive approaches. For example, in [ACGP03b] a mixed policy tailored to the wireless interface is compared with a reactive policy. Both policies exploit low-power modes of the wireless interface during inactivity periods. Moreover, the mixed policy also modifies the workload of the interface by using caching and prefetching techniques. Experimental results show that the mixed policy consumes about half of the power spent by the reactive one.

24.5   Power Management in Infrastructure-Based Networks

This section introduces some power management policies for wireless interfaces in infrastructure-based wireless networks (Figure 24.1[a]). According to the classification introduced in Section 24.4.2.1, all

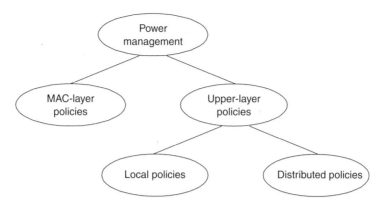

FIGURE 24.6 A classification of power-management policies used in infrastructure, wireless networks, based on the layer of the network architecture where they operate.

power management policies presented below are mainly *reactive* policies. However, some of them include mechanisms (e.g., caching, prefetching, etc.) that modify the original traffic shape in order to reduce power consumption at the mobile device. Therefore, according to the same classification, these techniques can be regarded as *mixed* policies.

A further classification can be based on the layer of the network architecture where these policies operate. From this standpoint, power management policies used in infrastructure-based networks can be divided in *MAC-layer* and *upper-layer* (e.g., transport or application) policies. In addition, as shown in Figure 24.6, upper-layer policies may be either *local* or *distributed*. Local policies are implemented locally at the mobile device, while distributed policies are implemented partially at the mobile device, and partially at the access point (AP). To the best of our knowledge, medium access control (MAC)-layer strategies are always distributed policies. MAC- and upper-layer policies are not necessarily alternative policies. On the other hand, they are often used jointly.

24.5.1 MAC-Layer Policies

24.5.1.1 IEEE 802.11 Power Saving Mode (PSM)

Currently, IEEE 802.11 [802.11] (also known as Wi-Fi) is the most mature wireless technology, and Wi-Fi products are largely available on the market. Readers can refer to [C04] for a survey on general aspects of this technology. In this Section we focus on power management issues. IEEE 802.11 wireless cards can operate in two different modes: *active* and *sleep* (or *doze*) mode. In the active mode, the wireless interface is able to exchange data and can be in one of the following states: *receive, transmit,* or *idle*. The power consumption in the active mode is very little influenced by the specific operating state and, hence, it is typically assumed as almost constant (e.g., 750 mW for Enterasys Networks RoamAbout interfaces [KB02]). In the sleep mode, the wireless interface is not able to exchange data but its power consumption is, at least, one order of magnitude lower than in the active mode (e.g., 50 mW for Enterasys Networks RoamAbout interfaces [KB02]).

The IEEE 802.11 standard includes a *power saving mode* (PSM) — operating at the MAC layer — whose objective is to let the wireless interface in the active mode only for the time necessary to exchange data, and to switch it into the sleep mode as soon as there are no more data to exchange. In infrastructure-based 802.11 WLANs (e.g., in Wi-Fi hotspots) power management is achieved by exploiting the central role of the AP. According to the IEEE 802.11 terminology, mobile devices registered with the same AP are said to form a basic service set (BSS). Each mobile device belonging to the BSS has to inform the AP whether or not it is using PSM. The AP buffers data destined for mobile devices that are using PSM. Furthermore, every *Beacon Interval* (usually 100 msec), the AP broadcasts a special frame, called *Beacon*, to mobile devices

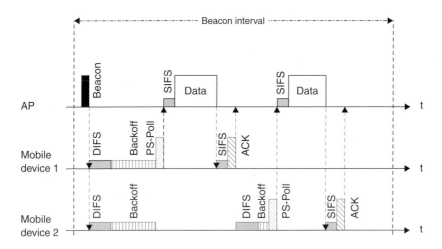

FIGURE 24.7   Wi-Fi PSM operations.

in the BSS. Mobile devices operating in PSM are synchronized with the AP,[4] and wake up periodically to listen to the Beacon. This special frame includes a *traffic indication map* (TIM) that specifies which mobile devices have frames buffered at the AP. Mobile devices indicated in the TIM remains active for the rest of the Beacon period to receive buffered frames. All the other ones go to sleep until the beginning of the next Beacon Period.

Frames buffered at the AP can be downloaded as shown in Figure 24.7. The mobile device sends a special frame, *PS-Poll*, to the AP using the standard distributed control function (DCF) procedure. Upon receiving a PS-Poll, the AP sends the first DATA frame to the mobile device and waits for the related ACK frame. If there are still data to be sent, the AP sets the *more data* bit in the DATA frame. To download the next DATA frame, the mobile device must send another PS-Poll. When the mobile device realizes that there are no more data to fetch, it puts the wireless interface in the sleep mode. As soon as the mobile device has a data frame to send it switches back to the active mode and performs the standard DCF procedure.

PSM achieves the goal of saving energy especially when the network activity is low. In [KB02, ACGP04a] it is shown that using PSM in a standard TCP/IP architecture allows saving up to 90% of the energy spent by the wireless interface. However, this result refers to a scenario where there is a single mobile device in the BSS. Since PSM relies upon the standard DCF procedure, mobile devices served by the same AP have to contend with each other to access the wireless medium and download buffered messages. Therefore, the PSM ability to save energy depends on the number of mobile devices in the BSS (e.g., in a Wi-Fi hotspot). In [ACGP04a] it is shown that with 10 mobile devices in the BSS the power saving achieved by PSM reduces to 60 to 70%.

An additional major limitation of PSM is that it implements a static power-management policy, that is, the Beacon Period remains constant, irrespectively of the network activity. This static behavior may result in significant performance degradations in some specific scenarios [KB02,ANF03], and, in some cases, may even increase the overall power consumption [ANF03]. Specifically, the effects described below have been observed. In all cases it is assumed that the AP is serving a single mobile device. Things are expected to be even worse when several mobile devices operate under the same AP.

Increase in the Round Trip Time (RTT) of a TCP connection: When using PSM the wireless interface is put in the sleep mode whenever the mobile device is not sending or receiving data. If the mobile device has data to send (e.g., a TCP SYN, or an ACK segment, or a TCP segment containing a Web request, etc.) it immediately wakes up the wireless interface. However, the wireless interface is switched back to the sleep mode as soon as data have been transmitted. When the response data arrive from the server, they

---

[4]The Beacon frame contains parameters for clock synchronization.

must be buffered at the AP until the next Beacon occurs. This delay increases the RTT of the connection. Specifically, the first RTT is increased by a value that depends on the time instant at which the TCP SYN segment was sent by the mobile device.[5] Subsequent initial RTTs are *rounded up* to the nearest 100 msec (even if the actual RTT is 5 msec the observed RTT is in the order of 100 msec) [KB02]. This can be explained as follows. Since TCP segments addressed to the mobile device are transmitted at the beginning of the next Beacon Period, the mobile device always responds with TCP ACKs immediately after the Beacon, and the TCP connection becomes synchronized with the PSM Beacon Period.

Increase in the execution time of interactive request/response applications: Applications involving many successive request-response transactions, like file operations on a network file system (NFS), may be negatively affected by the RTT increase described earlier. NFS is based on remote procedure calls (RPCs) and issues requests in sequence, that is, the next RPC request is issued after receiving the response for the previous one. According to the above remarks, RPCs experience a delay of approximately 100 msec.[6] If we consider the simple `ls` command to list files in a directory (NFS makes two RPCs for each file), the delay experienced in PSM may be up to 16 to 32 times greater than that experienced without power management [ANF03]. Similar considerations apply to other file and directory operations. Therefore, applications and programs containing a large number of NFS operations may experience a significant increase in their execution times [ANF03].

The above results show that the PSM sleep interval is too large especially when the connection RTT is small. In this case PSM may degrade significantly the performance of interactive request/response applications [KB02]. In addition, it may cause two paradoxical phenomena that are discussed below.

Increase in the energy consumption of interactive request/response applications: PSM may even increase the power consumption of interactive request/response applications. It may be worthwhile to recall here that the energy consumed by the wireless interface is only a fraction of the total energy consumed by the system. PSM actually reduces the energy consumed by the wireless interface but, as shown earlier, it may increase considerably the execution times. Since the total energy consumed by the system is the integral of power over time, it may happen that the energy consumed by the system in the additional execution time is greater than the energy saved by PSM [ANF03]. As mentioned earlier, in a palmtop computer the power consumption of the wireless interface is approximately 50% of the total consumption, while it is about 10% in a laptop computer [KK00,ACL03,ANF03]. In the latter case, reducing the energy consumed by the wireless interface while increasing, at the same time, the execution time of the application may result in an increase of the overall system consumption. Therefore, power management should take into consideration the overall power characteristics of the entire system.

Performance inversion: The PSM may also cause a performance inversion so that TCP applications achieve higher throughput with a slower wireless link. Since the TCP connection becomes synchronized with the PSM Beacon, at the beginning of a Beacon period the amount of data buffered at the AP is equal to the TCP window size (assuming that there is sufficient bandwidth between the server and the AP). If the AP finishes transmitting the buffered data before new data arrive from the server, the mobile device enters the sleep mode until the next Beacon, and the throughput of the TCP connection is limited to one window per Beacon period. When the TCP window size increases beyond a certain threshold, new data arrives at the AP when the AP itself is still transmitting and, hence, the wireless interface remains continuously active. The threshold value is given by the product of the wireless link bandwidth and the actual RTT between the server and the mobile device. Therefore, a lower wireless bandwidth saturates sooner and prevents the wireless interface from entering the sleep mode. This performance-inversion effect shows that when moving, for example, from IEEE 802.11b to IEEE 802.11g products, the performance of TCP applications might degrade [KB02].

---

[5]For instance, if the actual RTT is 120 msec, then the observed RTT is 150 msec if the TCP SYN was sent 50 msec after the Beacon arrival, while it is 210 msec if the TCP SYN was sent 90 msec after the Beacon arrival.

[6]It is assumed that the delay between the NFS server and the AP is small with respect to the Beacon period.

From the above remarks it clearly emerges that the 100 msec sleep interval of PSM can be *too large* to provide good performance, especially to interactive applications that are sensitive to the RTT of the connection. In principle, the problem could be overcome by considering a shorter sleep interval. On the other hand, many network applications exhibit a bursty behavior characterized by data-transfer phases interleaved by idle periods without any network activity. For example, a Web user typically reads information for several seconds after downloading them. Although there is no network activity during idle periods, PSM wakes up the wireless interface every 100 msec resulting in a large amount of energy wastage. From this standpoint, the sleep interval of PSM is *too short* to minimize power consumption during long idle periods.

24.5.1.2 Adaptive PSM

The problem with IEEE 802.11 PSM is its static nature that does not allow it to adapt to the network conditions and the application requirements. Ideally, power management should maintain the wireless interface active when there are packets to transfer, and disable it during idle periods. Cisco Aironnet 350 cards partially implement this policy. They appear to disable PSM (thus remaining continuously active) when more than one packet destined for the mobile device is waiting at the AP. PSM is re-enabled after approximately 800 msec without receiving a packet [ANF03]. Throughout, this protocol will be referred to as adaptive PSM. Other adaptive power-management policies are presented below.

24.5.1.3 Bounded Slowdown (BSD)

Starting from the evidence that the 100 msec PSM sleep interval is too large especially for interactive request/response applications, and too short for minimizing power consumption during long inactive periods, Krashinsky and Balakrishnan [KB02] propose the BSD protocol that dynamically adapts its behavior to the network conditions. The target of this protocol is to minimize the power consumption while guaranteeing that the RTT does not increase by more than a given percentage. To achieve this goal the wireless interface remains active for a short period after sending a request, and adaptively listens to a subset of Beacons when the network is idle. Staying in the active state reduces communication delays but increases power consumption. On the other hand, listening to fewer Beacons saves energy but may increase communication delays.

Formally, let R be the RTT of the TCP connection between the mobile device and the remote server when the wireless interface is continuously active (i.e., without power management). The goal of the BSD protocol is to minimize power consumption while limiting the observed RTT to $(1 + p) \cdot R$, where $p > 0$ is a specified parameter. The protocol works as follows. After sending any packet (e.g., a request, or a TCP ACK) the mobile device maintains the wireless interface in the active mode for a time interval T_{awake}. Then, it enters a loop during which, at each step, performs the following actions: (i) sleeps for a period T_{sleep} and (ii) wakes up and polls the AP for receiving newly arrived packets (if any). The loop is exited as soon as the mobile device has a new packet to send. When this occurs, the algorithm is restarted (on the other hand, packets received from the AP have no effect on the algorithm). At each iteration, the sleep interval T_{sleep} is obtained as the time elapsed since the (last) packet was sent by the mobile device, multiplied by p. Since T_{sleep} is the maximum additional delay introduced by the BSD protocol, the observed RTT is, at most, equal to $R + T_{\text{sleep}} = (1 + p) \cdot R$.

To implement the BSD protocol in a realistic environment, for example, in IEEE 802.11 cards, T_{awake} and T_{sleep} need to be synchronized with the Beacon period $T_{\text{bp}}(T_{\text{bp}} = 100$ msec in IEEE 802.11 PSM). Specifically, T_{sleep} must always be *rounded down* to a multiple of T_{bp}. The first time the wireless interface is put in the sleep mode T_{sleep} is equal to T_{bp} and, hence, the RTT might increase by, at most, T_{bp}. Therefore, T_{awake} must be equal to T_{bp}/p. Figure 24.8 shows the evolution of PSM and BSD when $p = 0.5$. In this example $T_{\text{awake}} = T_{\text{bp}}/p = 100/0.5 = 200$ msec [KB02].

A simulation analysis of BSD, based on Web traffic traces is performed in [KB02]. A single mobile device communicating with the AP is considered. It is shown that, compared to PSM, BSD reduces average page retrieval times by 5 to 64%, and (at the same time) the power consumption by 1 to 14%. However, the

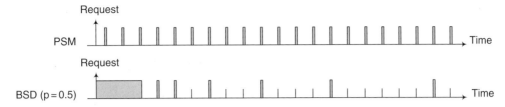

FIGURE 24.8   Evolution of PSM and BSD with $p = 0.5$ [KB02].

benefits of using BSD instead of PSM are more evident when the RTT of the TCP connection is smaller than the Beacon period (i.e., 100 msec).

The BSD protocol has some important drawbacks. Since the wireless interface may be disabled even for very long times, this policy is nice for a mobile device operating as a client in request/response applications (e.g., Web, e-mail, file transfer, etc.). It is not appropriate when the mobile device acts as a server, or in a peer-to-peer scenario. In these cases, long disconnections may prevent the mobile device from receiving packets (e.g., a service request) from other devices. Furthermore, the proposed protocol has no knowledge about the application behavior. This limits the potentialities of the protocol. Finally, to be used in Wi-Fi cards the BSD protocol requires the mobile device to be able to skip a dynamically varying number of Beacons. This may not be easy to implement in current devices.

24.5.2   Upper-Layer Policies

Both adaptive PSM and BSD are modified versions of the IEEE 802.11 PSM trying to adapt to the network activity. Therefore, these proposals cannot be used in Wi-Fi hotspots where mobile devices are equipped with fully-compliant IEEE 802.11 wireless cards. In addition, they operate at the MAC layer and, thus, have no knowledge about the application behavior. Upper-layer solutions discussed hereafter overcome these problems as they can be used, in principle, with any wireless interface. In addition, they can exploit application-specific information.

In the next section we will introduce several upper-layer policies: some of them are local policies, while some others are distributed policies.

24.5.2.1   Application-Driven Power Management

The simplest form of upper-layer power management is an application-specific policy that disables the wireless interface (i.e., put it in the doze mode) whenever there are no more data to be transferred. If we look at the typical behavior of a (nonreal-time) network application such as Web browsing — shown in Figure 24.9 — we can observe that *active times*, during which data transfer occur, alternate with *inactive times* characterized by no network activity. Furthermore, data transfers inside the same active time are separated by short idle periods called *active off times*. During both inactive times and active off times the wireless interface is idle. However, active off times are generated by the interaction between the mobile device and the remote server, while inactive periods are related to the user activity (they are also called user think times). Inactive times are usually much longer than active off times, and are typically greater than 1 sec [BC98]. Therefore, simply disabling the wireless interface during inactive phases can save a large amount of energy.

Based on this evidence, two simple application-layer policies are investigated in [SK97]. They are local policies and are specifically tailored to e-mail and Web-browsing applications (currently, the most popular applications for handheld devices), respectively. For e-mail applications, the wireless interface is usually in the doze mode. It wakes up periodically to check for possible messages, and remains active only for the amount of time necessary to send and receive e-mail messages to and from the mail server. For Web-browsing applications, the wireless interface is disabled after a certain period without any network activity and re-enabled when a new request is generated by the application.

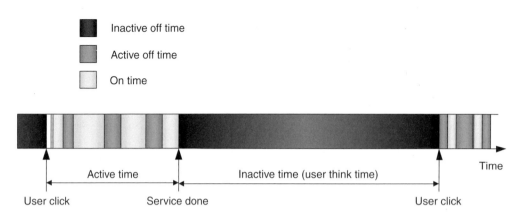

FIGURE 24.9   Typical behavior of a network application.

Simulation results show that the power consumption can be reduced to the minimum power necessary to send and receive e-mail messages, while for web browsing the power consumption can be reduced by approximately 75% with a negligible impact on the QoS perceived by the user [SK97,ACGP04b]. However, the proposed solutions require some application modifications. Furthermore, a software module that coordinates the activity of different network applications is required: the wireless interface can be disabled only when all the applications are not transferring data, and must be enabled as soon as any application has data to transmit. Again, suspending the wireless interface for long time periods may be suitable only when the mobile device acts as a client, while it is not appropriate when the mobile device has to respond to external requests (e.g., when the mobile device acts as a server or in a peer-to-peer environment).

24.5.2.2   Self-Tuning Power Management (STPM)

Anand et al. [ANF03] propose a Self-Tuning Power Management (STPM) protocol that extends the idea of application-driven power management to any network applications. The proposed solution not only takes into account the application behavior, but also adapts to (i) the network activity; (ii) the characteristics of the wireless interface; and (iii) the system on which the protocol is running. The goal is to have a flexible power management that optimizes the power consumption, based on the energy characteristics of both the wireless interface and the entire system, while guaranteeing good performance to delay-sensitive applications. To achieve this goal, STPM switches the wireless interface between the different operation modes that are supported (active, doze, PSM), depending on several conditions. Specifically, the protocol is built upon the following design principles:

- *Power management should know the application behavior.* This is important to take appropriate decisions since different applications have different behaviors.
- *Power management should be predictive.* If it is possible to know in advance that a large number of successive data transfers will occur back-to-back it might be convenient remaining continuously active to improve performance. The wireless interface must be disabled (or put doze mode) only at the end of the overall data transfer.
- *Power management should differentiate network data transfers.* In particular, it should discriminate between *foreground* data transfers that are delay-sensitive, and *background* data transfers that are not.
- *Power management should reach a performance/energy tradeoff.* This tradeoff should depend on the context where the user's activity is carried on. If the battery of the mobile device is fully charged, a better performance could be provided at the cost of greater power consumption. If the battery is nearly exhausted performance should be sacrificed to save power.

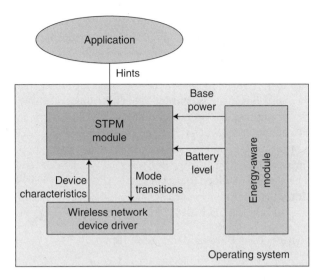

FIGURE 24.10 STPM software architecture [ANF03].

- *Power management should adapt to the operating conditions.* The goal of power management should be to minimize the power consumed by the entire system, not only the power consumed by the wireless interface.

Figure 24.10 shows a possible implementation of the STPM protocol as a kernel module. Applications provide the STPM module with hints about their behavior. Specifically, before each data transfer, the application specifies whether the forthcoming data transfer is a foreground or background activity and, optionally, the expected amount of data to be transferred. At the end, the application informs the STPM that the data transfer has been completed. To implement a predictive power management, STMP monitors the number of data transfers that are closely correlated in time so that they can be assumed to form a cluster of transfers, or *run*. STPM uses an empirical distribution of run lengths to calculate the number of expected data transfers in the current run, given the number of data transfers already observed. Based on this information, it decides whether it is convenient to remain continuously active or switching to PSM (or to doze mode).

Applications also specify the maximum additional delay (due to power management) they are willing to tolerate for incoming packets. Information about the power consumption of the mobile device (base power), as well as hints for deciding the right performance/energy tradeoff, can be provided either by the user or the operating system itself (as shown in Figure 24.10).

Finally, to take appropriate decisions, STMP needs to know the energy characteristics of the wireless interface. Wireless cards differ in the types and number of power saving modes they support, energy costs to transit from one mode to another, and transitions times. In the implementation proposed in [ANF03], STPM uses a code component that includes the energy characteristics of the specific wireless card used by the mobile device. This code component is loaded by STPM in the same way that the operating system loads the device driver for a given component.

Based on the above information, STPM decides when it is convenient to switch the wireless interface from one operational mode to another (e.g., from PSM to active, or vice versa). Assuming for simplicity that the wireless card only supports two operational modes, that is, PSM and active, STPM calculates the total costs (in terms of energy and times) of performing the data transfer in each mode. It compares the results obtained and decides to switch to the active mode if it predicts that performing the data transfer staying continuously active will both save energy and improve performance. If STPM estimates that a transition to the active mode will either reduce power consumption or improve performance but will not reach both objectives, the decision is based on information (provided by either the user or the operating system) about the relative tradeoff between performance and energy conservation. Similarly, STPM decides to switch back to PSM when no data transfer is in progress, and it predicts that the expected benefits of reduced power consumption during the remaining idle period will exceed the performance and energy costs of starting the next data transfer in the active mode (see [ANF03] for details).

In [ANF03] it is shown that, compared to PSM, STPM reduces the power consumption of a palmtop computer operating on a distributed file system by 21% and, simultaneously, reduces the filesystem interactive delay by 80%. In addition, the proposed solution is flexible as it is able to adapt to different systems (e.g., laptops and palmtops) and wireless interfaces with different energy characteristics. However, in the presence of Web-browsing applications (i.e., the most popular application, along with e-mail, for handheld computers) it is not clear which is the performance of STPM. A major SMTP drawback is that application clients need to be modified to provide the STPM module with specific hints about the application behavior. Although these modifications usually consists in adding few lines of code, this aspect prevents using legacy network applications with STPM.

24.5.2.3　Communication-Based Power Management

According to the classification provided at the beginning of Section 24.5, both application-driven power management and STPM are *local* policies because the power management algorithm is confined inside the mobile device. On the other hand, MAC-layer policies described in Section 24.5.1 implement a distributed power management that relies on the AP to buffer incoming packets during mobile-device disconnection periods. When using a local approach, the mobile device remains disconnected from the rest of the system when the wireless interface is disabled. This may prevent the mobile device from receiving external packets. Hence, the wireless interface can be disabled only when the mobile device does not expect to receive any packet from the network (e.g., during Think Times in Web browsing applications). However, even if all data transfers are controlled by the mobile device itself, by looking at Figure 24.9, the wireless interface can be disabled only during Inactive Times (or User Think Times), while it must be continuously active (or in PSM) during Active Times (when the mobile device can receive packets from the server). This may result in power wastage especially when the server is busy and idle times between consecutive data transfers (i.e., Active Off Times in Figure 24.9) are not short. Policies presented hereafter overcome this problem by taking a distributed approach similar to IEEE 802.11 PSM. In addition, they are implemented at the transport or application layer and can, thus, exploit hints from the applications.

Kravets and Krishnan [KK98,KK00] propose a communication-based power management protocol that operates at the transport layer but allows applications to control the power management parameters. The protocol is based on the Indirect-TCP model [BB97], that is, the TCP connection between the mobile device and the remote server is split into two TCP connections: one from the mobile device to the AP, and another from the AP to the remote server. Furthermore, the protocol is designed as a master–slave protocol where the role of master is played by the mobile device. The mobile device decides when to suspend/resume the wireless interface, and tells the AP when data transmissions can occur. Data arriving at the AP, while the mobile device is disconnected, are buffered and delivered later, when the mobile device reconnects.

Ideally, the wireless interface should be disabled as soon as the application enters a phase of inactivity, that is, during user thinks times or between active off times (see Figure 24.9). However, unless the application sends an explicit notification, the power management protocol is not able to know when the inactivity phase exactly starts. Therefore an *inactivity timer* is used that is reset after each transmission or reception. If the inactivity timer expires, the mobile device informs the AP that it is going to sleep and disables the wireless interface. The wireless interface will be re-enabled either when the application generates new data to send, or upon expiration of a *wakeup timer*. Then, the mobile device sends a WAKE_UP message to the AP, along with any new data, and waits for receiving buffered packets (if any).

A key role in the above protocol is played by two timing parameters, the *timeout period* (for the inactivity timer) and the *sleeping duration* (for the wake_up timer). As highlighted in Section 24.4, if the timeout period is too short, the wireless interface may go to sleep too early, causing an increase in the application response time. If it is too large, the wireless interface remains active for long periods of time, thus wasting energy unnecessarily. Similarly, if the sleeping duration is too small, the wireless interface is almost always active, thus consuming more than necessary. If it is too long, the additional delay introduced by the power management protocol may degrade the QoS perceived by the user. Clearly, the appropriate sleeping duration depends on the application. Therefore, the power management protocol should set the sleeping duration by exploiting hints provided by the application. Furthermore, if more than one application is running on the mobile device, the power management protocol must consider hints from all applications and set the timing parameters according to the more stringent requirements. Finally, the power management protocol should be adaptive by varying the timeout period and sleep duration according to the application activity. Compared to the solution based on fixed timeouts, adaptive power management improves power saving and reduces additional delays introduced by power management [KK00].

24.5.2.4 Application-Dependent Adaptive Power Management

Adaptation is the key idea also in [ACGP02,ACGP03a,ACGP03b]. These authors take an approach similar to the one in [KK00]. Specifically, they propose a policy based on the Indirect-TCP model [BB97] that (i) operates just below the application; (ii) exploits the characteristics of the application itself; and (iii) buffers at the AP packets addressed to the mobile device during disconnection periods (distributed policy). The novelty of this approach is its ability to adapt to the application behavior without requiring any modification of the application itself. Specifically, this approach can be implemented either in an *application-dependent* way or in an *application-independent* way.

The *application-dependent* approach is proposed in [ACGP02,ACGP03b] by developing a power management algorithm which is application specific (e.g., tailored to Web-browsing, e-mail or some other application). The AP is aware of some (off-line) statistics about the traffic profile generated by the application (e.g., in Web-browsing applications, expected number of embedded files in a web page, expected size of html and embedded files, etc.). In addition, the AP also continuously monitors the network activity in order to obtain run-time estimates of the throughput available along the connection with the remote server.

To better explain this policy in the following we will make reference to Web-browsing applications. Upon receiving a Web-page request from the mobile device, the AP derives an estimate of the main-file transfer time (i.e., the time needed to download the main html file from the remote Web server), and decides whether (or not) it is convenient for the mobile device to disconnect in the meanwhile. This decision takes into account: (i) energy and delay costs associated with suspending and resuming the wireless interface at the mobile device and (ii) maximum additional delay the application is willing to tolerate. If disconnection is convenient, the AP suggests this decision to the mobile device along with the time interval it should sleep. The mobile device suspends the wireless interface for the notified sleeping time[7]. While the mobile device is disconnected, the AP stores the main html file and, automatically, prefetches embedded files, if any. When the mobile device reconnects, the AP delivers all the available data (main file and embedded files, if available). If more data have to be downloaded, the AP also derives an estimate for the residual transfer time. If disconnection is convenient, the AP suggests the mobile device to disconnect.

From the above description it clearly emerges that the AP dynamically generates updated estimates for the residual transfer time, and based on these estimates, suggests the mobile device either to disconnect or to remain active. Furthermore, the AP acts as an agent for the mobile device, that is, it performs some actions on behalf of the mobile device to improve power saving and reduce the delay. Specifically, it includes mechanisms for file caching and prefetching. In practice, the AP can be seen as a Web proxy with power management capabilities.

Results obtained by considering Web-browsing applications in a real environment (i.e., by using a real Web server accessed by the client through a real Internet connection) have shown that the application-dependent power management protocol is able to achieve power saving not only during inactivity periods

---

[7]The mobile device may reconnect before the sleeping interval has expired if a new request is generated by the client-side application.

(i.e., during User Think Times) but also during the data transfer phase, that is, during the download of Web pages. Furthermore, by continuously monitoring the network connection between the AP and the Web server, it is able to dynamically adapt to network and server conditions. In [ACGP03b] it is shown that the application-dependent protocol reduces significantly the overall time during which the mobile device should be connected with respect to the application-driven protocol (this protocol switches off the wireless interface only during User Think Times). In addition, the performance of the application-dependent protocol is not far from an optimal (unfeasible) algorithm that switches on the wireless interface exactly when data becomes available at the AP, and disable it immediately after the data transfer.

24.5.2.5 Application-Independent Adaptive Power Management

The major drawback of the application-dependent power management protocol is that it must be tailored to a specific application. Therefore, a specific power management module is required for each network application. Furthermore, when there are more applications running on the mobile device, power management modules related to the various applications need to coordinate themselves in such way that the wireless interface is actually disabled only when *all* power management modules have decided to suspend it, and is re-enabled as soon *any* of them decides to resume it.

To overcome these difficulties an *application-independent* power management protocol is proposed in [ACGP03a]. This protocol autonomously becomes aware of the application's requirements and communication pattern, and exploits this knowledge to achieve power saving. Specifically, the protocol continuously monitors the application trying to learn its behavior and adapt to it. The protocol can be used with any network application, since no hints are required from the application. Furthermore, legacy applications can be used unmodified.

To take appropriate decision about when to suspend and resume the wireless interface, the application-independent power management protocol relies on an algorithm that provides estimates of both packet inter-arrival times (i.e., time intervals between consecutive packets) and inactivity periods' duration. The variable-share update algorithm [HW95], customized to the specific problem, is used to this purpose.

Basically, the Variable-Share Update algorithm works as follows (see Figure 24.11). Let I be the range of possible values for a variable y to be estimated. To predict the y value, the variable-share update algorithm relies upon a set of *experts*. Each expert x_i provides a (fixed) value within the range I, that is, a value that y can potentially assume. The number of experts to be used, as well as their distribution among the range I, are input data for the algorithm. Each expert x_i is associated with a weight w_i, a real number that measures the reliability of the expert (i.e., how accurately the expert has estimated y in the past). At a given time instant, an estimate of y is achieved as the weighted sum of all experts, using the current w_i value as the weight (i.e., reliability) for the expert x_i. Once the actual value of the variable y becomes known, it is compared with the estimate provided by the experts to evaluate their reliability and update the weight associated with each of them. The update policy requires two additional parameters, α and η, and a loss function L. This function provides a measure of the deviation of each expert from the actual

Parameters: $\eta > 0, 0 \leq \alpha \leq 1, n$ (number of experts)

Variables: x_i (experts), w_i (weights), y (actual variable's value), $\hat{y}$ (estimated variable's value)

Initialization: $w_i = 1/n \quad \forall \, i = 1, \ldots, n$

Prediction: $\hat{y} = \sum_{i=1}^{n} w_i x_i \Big/ \sum_{i=1}^{n} w_i$

Loss Update: $w_i' \stackrel{\Delta}{=} w_i e^{-\eta L(y, x_i)}$

Variable share:
$$\begin{cases} pool = \sum_i [1 - (1-\alpha)^{L(y,x_i)}] w_i' \\ w_i = (1-\alpha)^{L(y,x_i)} w_i' + \frac{1}{n-1} \\ \{pool - [1 - (1-\alpha)^{L(y,x_i)}] w_i'\} \end{cases}$$

FIGURE 24.11 Variable-share update algorithm.

value of the variable, and its values must lie in [0,1]. According to [HLS96], $\alpha = 0.08$ and $\eta = 4$ are used in the application-independent protocol. Finally, $L(x_i, y) = |x_i - y|/\max_i |x_i - y|$ is used as error function (see [HW95] for details).

The variable-share update algorithm has to provide estimates for both packet inter-arrival times and inactivity periods. These variables span different range of values since inter-arrival times are typically smaller than inactivity periods. Two different sets of experts, with nonoverlapping intervals for expert values, are thus used. The first interval ranges in $[0, P_{max}]$, where P_{max} is the maximum expected inter-arrival time. Idle periods greater than P_{max} are assumed to belong to inactivity periods. The second interval ranges in $(P_{max}, T_{max}]$, where T_{max} is the largest estimated inactivity period. This means that the mobile device will poll the AP for any incoming packets, at most, every T_{max}. Therefore, T_{max} is also the maximum additional delay introduced by the power management protocol to the first packet of a data transfer.

The authors of [ACGP04b] evaluated the application-independent protocol in a realistic prototype system (real Internet environment) considering different applications running in the mobile device. In particular, three scenarios are considered. In the first and second scenario there is a single active network application: Web and e-mail, respectively. In the third scenario, Web and e-mail are simultaneously active. The experimental results show that the application-independent protocol is able to save a considerable amount of energy irrespectively of the specific application running in the mobile device. In the mixed-traffic (Web + e-mail) scenario the protocol reduces the overall time during which the wireless interface should be on of about 72% with respect to the case without power management. In addition, this reduction is obtained without a significant impact on the QoS perceived by the user. In the mixed-traffic scenario the average additional delay introduced by the power management protocol is, on average, 0.41 sec for downloading a Web page and 1.9 sec for checking, downloading, and uploading e-mail messages. In the other single-application scenarios analyzed the performance of the application-independent protocol are even better.

24.6   Power Management and Pervasive Computing

The 1990s have started the mobile computing age. Mobile users can use their cellular phone to check e-mail and browse the Internet. Travellers with portable computers can access Internet services from airports, railway stations, and other public locations. In the same period, the hardware and software progresses provided the basic elements (PDAs, wearable computers, wireless networks, devices for sensing and remote control, etc.) for implementing a new computing paradigm, *pervasive* or *ubiquitous computing*, that will further extend the users possibility to access information or communicating with other users at anytime, anywhere, and from any device [W91].

The nature of ubiquitous devices[8] makes wireless communication the easiest solution for their interconnection. WLANs constitute the basis for these environments. However, the heterogeneity of devices to be interconnected (ranging from small sensors and actuators to multimedia PDAs) results in a large spectrum of requirements that cannot all be covered by WLANs. From one extreme, we have very simple devices for which small size, low cost, and low power consumption are the main constraints, while a high data rate is not necessary. On the other side, we have devices that need to exchange multimedia information for which a high data rate is a must. Also coverage requirements may vary significantly (from few meters to an entire building). This generates a strong push for new classes of networks, referred to as *body area network* (BAN) and *personal area network* (PAN), to cover all the pervasive-computing communication requirements [C03].

A BAN constitutes the solution for connecting devices distributed on the human body (e.g., head-mounted displays, microphones, earphones, processors, and mass storage) that form a wearable computer

---

[8]The devices are often located in inaccessible places or are spread on the human body.

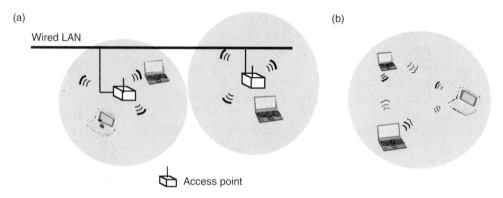

FIGURE 24.12   The WLAN configurations. (a) Infrastructure-based and (b) ad hoc.

[CH03]. A BAN is a network with a transmission range of a human body, that is, 1 to 2 m. Short-range, low-power wireless communications provide an efficient solution to BAN implementation.

While a BAN is devoted to the interconnection of one-person wearable devices, a PAN enables the interconnection of the BANs of persons close to each other, and the interconnection of a BAN with the environment around it [C03]. A PAN communicating range is typically up to 10 m and, of course, wireless communications constitute the basis for PANs.

To cope with these new communication requirements, IEEE 802 Committees are currently developing standard specifications that constitute the basis for wireless communications in pervasive computing environments. Specifically, in addition to the IEEE 802.11 standards for WLANs [802.11], the IEEE 802.15 working groups [802.15] are developing the specifications for short-range wireless communications among computers, peripherals, and (wearable) consumer electronic devices [C04]. These standards provide suitable solutions for both wireless PANs and BANs. A common feature of these network technologies is the ability to self-configure: a network exists as soon as two or more devices communicate on the same physical channel. No pre-existing infrastructure is required.

In mobile computing environments, most of the connections among wireless devices are achieved via fixed infrastructure WLAN networks (see Figure 24.1[a] and Figure 24.12[a]). The drawbacks of infrastructure-based networks are the costs and time associated with purchasing and installing the infrastructure. These costs and delays may not be acceptable for dynamic environments where people and vehicles need to be temporarily interconnected in areas without a pre-existing communication infrastructure (e.g., inter-vehicular and disaster networks), or where the infrastructure cost is not justified (e.g., in-building networks, specific residential communities networks, etc.). In these cases, infrastructureless or ad hoc networks can provide a more efficient solution. An ad hoc network is a peer-to-peer network formed by a set of stations within the range of each other that dynamically configure themselves to set up a temporary network (see Figure 24.12[b]). Ad hoc networks are created, for example, when a group of people come together, and use wireless communications for some computer-based collaborative activities; this is also referred to as *spontaneous networking* [FAW01]. In the ad hoc configuration, no fixed controller (AP) is required, but a controller is dynamically elected among all the stations participating to the communication.

It is worth remembering that while 802.11 networks are generally implemented as infrastructure-based networks, the 802.11 standards also enable the construction of peer-to-peer WLANs. In this case, the IEEE 802.11 stations communicate directly without requiring the intervention of a centralized AP. Ad hoc networks (hereafter referred to as single-hop ad hoc networks) interconnect devices that are within the same transmission range. This limitation can be overcome by exploiting the multi-hop ad hoc (MANET) technology. In a MANET, the users' mobile devices are the network, and they must cooperatively provide the functionality usually provided by the network infrastructure (e.g., routers, switches, servers). In a MANET, no infrastructure is required to enable information exchange among users' mobile devices.

Nearby terminals can communicate directly by exploiting, for example, WLAN technologies. Devices that are not directly connected communicate by forwarding their traffic via a sequence of intermediate devices.

Ad hoc networks inherit the traditional problems of wireless infrastructure-based networks discussed in the previous section. To these problems, the (multi-hop) ad hoc nature and the lack of fixed infrastructure add a number of characteristics, complexities, and design constraints that are specific of ad hoc networking, such as [CMC99]: multi-hop routing, dynamically changing network topologies, etc. Among these, power management is a key issue. The lack of fixed network elements, directly connected with a power supply, increases the processing and communication load of each mobile device, and this heavily impacts on the limited power of mobile devices. This becomes even a bigger issue in MANETs where a mobile device must spend its energy to act as a router forwarding the other mobile devices' packets [CCL03].

Hereafter, we will first discuss the power-management problem in single-hop ad hoc networks, and then we will address power-management issues in multi-hop ad hoc networks.

24.6.1 Power Management in Single-Hop Ad Hoc Networks

Currently, the widespread use of IEEE 802.11 cards makes this technology the most interesting off-the-shelf enabler for ad hoc networks [ZD04]. For this reason, hereafter, we will concentrate on power management policies applied to 802.11 technology. A more general discussion on power management for ad hoc networks can be found in [WC02,F04].

Power-management policies for IEEE 802.11 networks operate at the MAC layer with the aim to maximize the mobile device battery lifetime without affecting the behavior of the high-level protocols. Two types of policies, operating with different time scales, have been designed/investigated for 802.11 networks. The first class operates on small time scales (say milliseconds) and tries to optimize the behavior of the CSMA/CA protocol used by 802.11 by avoiding power wastage during packet transmissions. The latter class extends to infrastructureless 802.11 networks the policy developed for infrastructure-based networks (see Section 24.5.1.1): mobile devices operating in PSM spend most of the time in the sleep mode (also referred to as *doze* mode), and periodically wake up to receive/transmit data.

24.6.1.1 Power Management During Transmission

The power-management policies belonging to this class are designed to avoid transmitting when the channel is congested, and hence there is a high collision probability [BCD01,BCG02]. Power saving is obtained by minimizing the energy required to successfully transmit a packet. This is achieved in two steps:

1. The development of an analytical model to characterize the system-parameter values (mainly the backoff parameters) that correspond to the minimum power consumption.
2. The enhancement of 802.11 backoff mechanism to tune (at run time) the backoff window size to minimize the energy required for a successful packet transmission given the current status of the network, that is, number of active mobile devices, length of the messages transmitted on the channel, etc.

As far as point (1), in [BCG02] was developed an analytical model for the energy drained by the wireless interface of a tagged mobile device to successfully transmit a packet. Specifically, the model provides a characterization of the power consumption of the tagged device by studying the system behavior in the interval (referred to as *virtual transmission time*) between two consecutive successful transmissions of the tagged device itself, which corresponds to a system renewal interval. As shown in Figure 24.13, the virtual transmission time can be partitioned into: idle periods, collisions, and successful transmissions. Collisions and successful transmissions can be further partitioned into two subsets depending on the tagged device involvement. This distinction is useful because the tagged-device power consumption is different in tagged collisions (i.e., collisions where the tagged device is involved) and nontagged collisions.

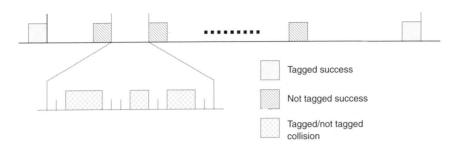

FIGURE 24.13   Channel structure during a virtual transmission time.

It can be shown that the system regenerates with respect to the starting point of a virtual transmission time. Hence, the system power efficiency, from the tagged-device wireless-interface standpoint, is the ratio between the average energy it uses to successfully transmit a message divided by the average energy it consumes in a virtual transmission time:

$$\rho_{energy} = \frac{PTX \cdot \ell}{E[Energy_{virtual\_transmission\_time}]},$$

where l is the average time required to the tagged device to transmit a message, PTX denotes the power consumption (per time unit) of the network interface when it is in the transmit state (PTX $\cdot$ l is the average energy required to the tagged device to successfully transmit the payload of a message), and $E[Energy_{virtual\_transmission\_time}]$ is the overall energy consumed by the tagged device during a virtual transmission time, and includes all the protocol overheads (idle slots, collisions, etc.).

The optimal energy consumption for the tagged device corresponds to the maximum value of ρ_{energy}, say ρ_{energy}^{MAX}. As $E[Energy_{virtual\_transmission\_time}]$ depends on the current status of the network, ρ_{energy}^{MAX} can be expressed as $\rho_{energy}^{MAX} = f(network\ status)$. The function $f()$, derived in [BCG02], shows that, to achieve the minimum energy consumption, the parameters of the backoff algorithm must be dynamically updated. In [BCD01] it is presented and evaluated a simple and effective extension of the 802.11 backoff algorithm to dynamically tune its parameters thus allowing the optimal energy consumption. This is achieved by exploiting a distributed mechanism, named *asymptotically optimal backoff* (AOB), which dynamically adapts the backoff window size to the current load [BCG04]. AOB guarantees that an IEEE 802.11 WLAN asymptotically (i.e., for a large number of active mobile devices) achieves its optimal behavior. The AOB mechanism adapts the backoff window to the network contention level by using two simple load estimates that can be obtained with no additional costs or overheads [BCG04].

In [BCG02] it has been shown that, for 802.11b technology, the tuning of the network interface for achieving the minimal energy consumption almost coincides with the optimal channel utilization. This behavior is associated with the energy consumption model of WLANs interface in which transmit, receive, and idle states are almost equivalent from a power consumption standpoint. This implies that, for 802.11b networks, energy savings can be achieved by optimizing at the same time also the other parameters that characterize the QoS perceived by the users, such as bandwidth, and MAC delay.

24.6.1.2 Power Saving Mode (PSM) in IEEE 802.11 Ad Hoc Networks

In the IEEE 802.11 standards' terminology, ad hoc networks are named *independent basic service set* (IBSS). An IBSS enables two or more IEEE 802.11 mobile devices to communicate directly without requiring the intervention of a centralized AP. Synchronization (to a common clock) of mobile devices belonging to an IBSS is sufficient to receive or transmit data correctly [ACG04]. In an IBSS, no centralized controller exists and Beacon frames, which contain timing information, are managed through a distributed process. The mobile device that initializes the IBSS defines the Beacon interval. At the beginning of the Beacon interval, all mobile devices wake up and randomly contend to transmit the synchronization Beacon. Specifically, all mobile devices in the IBSS schedule the transmission of a Beacon frame at a random time just after

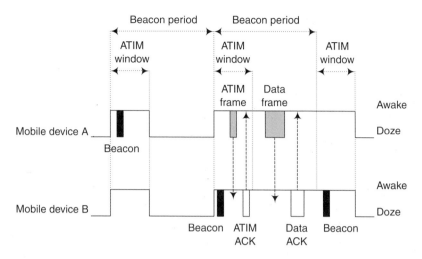

FIGURE 24.14 A data exchange between stations operating in PS mode in an 802.11 ad hoc network.

the target time identified by the Beacon interval. After the first successful transmission of the Beacon, all the other transmissions are cancelled. Mobile devices synchronize themselves with the first Beacon they receive. A more detailed description of IBSS setting up and management can be found in [ACG04].

For power management, a fixed-length ad hoc traffic indication map (ATIM) window is associated to the IBSS. All mobile devices must stay awake for an ATIM window after each Beacon. During the ATIM window, each mobile device sends an *ad hoc traffic indication map* (ATIM) to every other mobile device for which it has pending traffic. Each mobile device that receives an ATIM message responds with an ATIM acknowledgment.[9] At the end of the ATIM window, mobile devices that have neither sent nor received ATIM announcements go back to sleep. All other mobile devices remain awake throughout the remainder of the Beacon Interval, in order to send and receive the announced traffic.

Figure 24.14 shows the operations in an IBSS with two mobile devices, A and B. Two consecutive Beacon Intervals are considered in which the Beacon is issued by mobile devices A and B, respectively. In addition, in the second interval mobile device A has a message to be delivered to mobile device B. In this case, an ATIM frame and an ATIM ACK are exchanged during the ATIM window interval. At the end of this interval, mobile devices A and B remain awake to complete the data exchange.

The efficiency of IBSS power management mechanisms highly depends on the values selected for the Beacon intervals, the offered load and the length of the ATIM window. Results presented in the literature point out performance problems. Specifically, simulation results presented in [WESW98] indicate that power saving is obtained only at quite moderate loads; as offered load increases the saving declines substantially. The offered load also influences the choice of the parameter (e.g., the ATIM interval), hence the authors recommend adopting an adaptive ATIM window.

24.6.2 Power Management in Multiple-Hop Ad Hoc Networks

Multi-hop ad hoc networking introduces a new metric for measuring the power saving: the network lifetime [E02]. In an infrastructure-based wireless network, power management policies are aimed at minimizing the mobile-device power consumption. This metric is not appropriate for ad hoc networks where mobile devices must also cooperate to network operations to guarantee the network connectivity. In a MANET, a selfish mobile device that remains most of the time in the doze mode, without contributing to routing and forwarding, will maximize its battery lifetime but compromise the lifetime of the network.

---

[9]These transmissions occur following the basic 802.11 access mechanisms (e.g., DCF). To avoid contention with data traffic, only Beacons, frames' announcements, and acknowledgments are transmitted during the ATIM window.

On the other hand, policies aimed at maximizing the network lifetime must have a network-wide view of the problem. The idea is that, when a region is dense in terms of mobile devices, only a small number of them need to be turned on in order to forward the traffic. To achieve this, a set of mobile devices is identified which must guarantee network connectivity (to participate in packet forwarding), while the remaining mobile devices can spend most of the time in the doze mode to minimize power consumption. Mobile devices participating in packet forwarding may naturally exhaust their energy sooner, thus compromising the network connectivity. Therefore, periodically, the set of active mobile devices is recomputed by selecting alternative paths in a way that maximizes the overall network lifetime. Identifying the network's dominating sets is a typical goal of a global strategy. A dominating set is a subset of mobile devices such that each mobile device is in the set, or it has a neighbor in that set. Dominating sets, if connected, constitute the routing and forwarding backbone in the multi-hop ad hoc network. As the computation of the minimal dominating set is computationally unfeasible, in the literature several distributed algorithms exist to approximate suitable dominating sets, see for example [DB97,WL99,WDGS02,CJBM02,XHE01,XHE00].

The SPAN [CJBM02] is a distributed algorithm to construct dominating sets using local decisions to sleep, or to join the routing backbone. Mobile devices participating in the backbone are named *coordinators*. Coordinators are always in an active mode, while noncoordinator mobile devices are normally in the doze mode, and wake up to exchange traffic with the coordinators. Periodically, the coordinators' set is recomputed. The effectiveness of SPAN depends on the power consumption in the idle and sleep state: SPAN performance improves with the increase of the idle-to-sleep power-consumption ratio [CJBM02]. SPAN integrates with the 802.11 PSM, thus guaranteeing that noncoordinator mobile devices can receive packets that are buffered by the coordinators while they are sleeping.

The physical position of mobile devices (obtained, for example, via GPS) is used in the GAF algorithm to construct the routing and forwarding backbone. A grid structure is superposed on the network, and each mobile device is associated with a square in the grid using its physical position. Inside the square only one mobile device is in the nonsleeping mode [XHE01].

AFECA [XHE00] is an asynchronous distributed algorithm for constructing a routing backbone. Mobile devices alternate between active and sleep modes. In principle, a mobile device remains in the sleep mode for a time proportional to the number of its neighbors, thus guaranteeing, in average, a constant number of active mobile devices.

Controlling the power of the transmitting mobile device is the other main direction for power management in ad hoc networks. In wireless systems, the existence (or lack) of a link between two mobile devices mainly depends (given the acceptable bit error rate) on the transmission power and the transmission rate. By increasing the transmission power the number of feasible links is increased, but, at the same time, this increases the power consumption and the interference [E02]. Recently, several studies focused on controlling network topology by assigning per-device transmit powers that guarantee network connectivity and minimize the transmit power [NKSK02,WLBW01,RRH00,F04,R04]. The algorithmic aspects of topology control problems are discussed in [LLM+02].

Transmission power is highly correlated with power consumption. It determines both the amount of energy drained from the battery for each transmission, and the number of feasible links. These two effects have an opposite impact on the power consumption. By increasing the transmission power we increase the per-packet transmission cost (negative effect), but we decrease the number of hops to reach the destination (positive effect) because more and longer links become available. Finding the balance is not simple. On one hand, we have to consider the fact that signal strength at a distance r from the sender has nonlinear decay, specifically $S(r) = S \cdot r^{-\alpha}$ ($\alpha \in [2, 4]$), where S is the amplitude of the transmitted signal [E02]. This implies that, from the transmission standpoint, to cover the sender-to-receiver distance a multi-hop path generally requires less power. On the other hand, on a multi-hop path the delay (due to the multiple hops), as well as the processing energy (to receive and locally process a packet) increase.

The trade-off between minimum transmission power, and number of hops, further complicates the design of routing algorithms. A large part of recent works on power efficiency in ad hoc networks is concentrated on routing [RM99,SL00,RRH00], where the transmitting power level is an additional variable in the routing protocol design [F01].

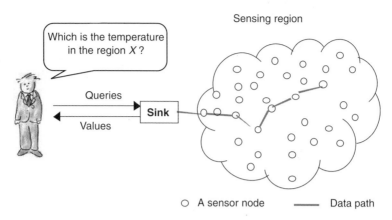

FIGURE 24.15   A wireless sensor network.

Power-efficient routing has been addressed from two different perspectives: (i) energy is an expensive, but not a limited resource (battery can be recharged/replaced), or (ii) the energy is finite. The former case applies to (mobile) ad hoc networks in general, while the latter is a suitable model for sensor networks. In case (i), power consumption must be minimized; typically, this translates in the following target: *minimize the total power consumed per packet to forward it from source to destination.* Minimizing per-packet energy does not maximize network lifetime, as residual energy of the mobile devices is not taken into consideration. On the other hand, in case (ii), the energy is a hard constraint [E02], and the maximum lifetime is the target.

Minimum-energy routings minimize the energy consumed to forward a packet from the source to the destination [LH01,RM99,GCNB03]. Similarly to proactive routing algorithms, [LH01,RM99] try to find minimum energy routes for all mobile devices; on the other hand, PARO [GCNB03] behaves as a reactive algorithm by minimizing the energy consumption of ongoing flows. In PARO, mobile devices intermediate to the source–destination pair elect themselves to forward packets, thus reducing the aggregate transmission power consumed by network devices.

On-line maximum-lifetime routing is a complex problem [LAR01]. In [CT00], for a static network with known and constant flows, the maximum lifetime routing is modeled as a linear programming problem. The solution of this model provides the upper bound on the network lifetime that is used to analyze the effectiveness of the algorithms. For a single power level, an optimal algorithm is presented; while, for the general case, the authors present an algorithm that selects routes and adjusts the corresponding power levels achieving a close to the optimal lifetime.

A balance between minimum-energy and maximum lifetime is the target of the CMMBCR policy [T01]. CMMBCR applies a conditional strategy that uses the minimum energy route, if the mobile devices residual energy is greater than a given threshold. Otherwise, a route that maximizes the minimum residual energy is selected.

24.6.2.1   Power Management in Sensor Networks

Sensor networks constitute basic elements of emerging pervasive environments. Sensor nodes are small devices with computing, communication, and sensing capabilities that can be used for various purposes. Typical sensing tasks could be temperature, light, sound, etc. Different kinds of "infrastructures" can be deployed to deliver the collected information from the sensing field to the place this information is elaborated. In recent years, advances in simple, low power, and efficient wireless communication equipments made wireless sensor networks the most interesting way to deliver the sensed data to the device(s) in charge to collect and elaborate them. However, due to the technologic constraints, single-hop wireless communications are often not efficient from the power consumption standpoint, and multi-hop communications are typically used to deliver the collected information. Specifically, a wireless sensor network is typically organized as shown in Figure 24.15. The sensor nodes are densely (and randomly) deployed inside the area in which a phenomenon is being monitored. Each sensor node delivers the collected data to one (or more) neighbor node, one hop away. By following a multi-hop communication paradigm, data are routed to the sink and through this to the user. Therefore, multi-hop ad hoc techniques constitute the basis also for wireless sensor networks. However, the special constraints imposed by the unique characteristics of sensing devices, and the application requirements, make solutions designed for multi-hop wireless networks generally not suitable for sensor networks [ASSC02]. First of all, power management is a "pervasive" issue in the overall design of a sensor network. Sensor networks utilize on-board batteries with limited energy that cannot be replenished in most application scenarios [S03]. This strongly affects the lifetime of a sensor network, and makes power awareness a critical factor in the design of sensor network protocols. Furthermore, sensor networks produce a shift in the networking paradigm from a node-centric to a data-centric view. This introduces a new dimension to power management. The aim of a sensor network is to collect information about events occurring in the sensor field. As sensor nodes are densely deployed several sensor nodes collect the same information and hence the importance of (the information collected by) each node is reduced with respect to MANET nodes. Indeed, while in MANETs the correct delivery of all data is important, the delivery of the information collected by each sensor node may tolerate a lower reliability. Moreover, while in node-centric networks the information is not manipulated inside the network (at the most packets can be discarded if resources are not available for forwarding them), in sensor networks intermediate devices — along the path from the sender (a sensor device) to the receiver (the sink) — can exploit the correlation in the observed data to reduce (via compression and aggregation) the amount of information flowing into the network, thus increasing the power efficiency. This is based on the observation that, in sensor devices, from the power consumption standpoint, processing is a much cheaper operation than transmitting [DWBP01,S03].

In a MANET only protocols operating at layer 1 to 3 of the OSI reference model participate to the operations inside the network (higher layers are active only in the sender and receiver devices); on the other hand, in a sensor network also higher layers[10] contribute to power-efficient data delivery, making power management a "cross-layer" issue [CMTG04]. Tiny DB [MFHH03] represents the extreme case, where almost all the networking functions are implemented at the application layer.

The interested readers can find an excellent and comprehensive coverage of sensor networks in a recent survey [ASSC02]. Hereafter, we will briefly discuss issues related to power management. Specifically, we will survey power management policies at different layers of the OSI reference model.

For Physical and Data Link layers the power efficiency questions are similar to those addressed in wireless networks: how to transmit in a power efficient way bits and frames, respectively, to devices one-hop away. These problems include identifying suitable modulation schemes, medium access control policies [VL03,YHE02], efficient FEC schemes, etc., see [ASSC02,KW03]. Of course the solutions of these problems are strongly affected by the sensor-device resources' constraints. The proposed solutions are generally independent from the applications however, recently, some authors [VA03] proposed to apply data-centric policies also at the MAC layer. The basic idea is to exploit spatial correlation among neighboring nodes to reduce the number of transmissions at the MAC layer.

Power efficiency at network layer has several aspects in common with MANETs. Specifically, it deals with identifying paths for deliver the information (from the source to the receiver) that maximize the network lifetime. This involves the solutions of problems such as: the control of the network topology, and the selection of the paths to be used for data delivery given a network topology. Similarly to MANETs, topology control algorithms either exploit power control techniques to construct energy efficient sub-graphs of the sensor network (see Section 24.6.2), or re-organize the network in energy efficient clusters (see below).

Topology management techniques, like GAF and SPAN (see Section 24.6.2), can be applied to sensors networks for minimizing the number of devices that must be active to guarantee a connected topology.

---

[10]Mainly the presentation and the application layers where compression and aggregation of data are applied to reduce redundancy in delivered data.

Furthermore, in a sensor network, additional savings can be obtained by observing that devices spend most of their time waiting for events without forwarding traffic, and hence, the network interface of active devices can remain in a sleep state for most of the time. This feature is exploited by *Sparse Topology and Energy Management* (STEM) [STGM02] to implement an energy efficient topology management scheme for sensor networks. STEM, when necessary, efficiently wakes up devices from a deep sleep state with a minimum latency. In addition, this scheme can be integrated with topology management techniques, like GAF and SPAN thus achieving compounded energy savings.

Clustering was introduced in 1980s to provide distributed control in mobile radio networks [BEF84]. Inside the cluster one device is in charge of coordinating the cluster activities (*clusterhead*). Beyond the clusterhead, inside the cluster, we have: *ordinary nodes* that have direct access only to this one clusterhead, and *gateways*, that is, nodes that can hear two or more clusterheads [BEF84]. As all nodes in the cluster can hear the clusterhead, all inter-cluster communications occur in (at most) two hops, while intra-cluster communications occurs through the gateway nodes. Ordinary nodes send the packets to their clusterhead that either distributes the packets inside the cluster, or (if the destination is outside the cluster) forwards them to a gateway node to be delivered to the other clusters. Only gateways and clusterheads participate in the propagation of routing control/update messages. In dense networks this significantly reduces the routing overhead, thus solving scalability problems for routing algorithms in large ad hoc networks. Several clustering algorithms have been proposed for wireless sensor networks, for example, see [BC03] and the references herein. On of most popular is the *Low Energy Adaptive Clustering Hierarchy* (LEACH). LEACH includes distributed cluster formation, local processing to reduce global communications, and randomized rotation of cluster-heads to evenly distribute the energy load among the sensor nodes in the network [HCB00].

The routing and forwarding of information inside a sensors network are data-centric functions through which data is disseminated to and from the sensors node. For example, the query "tell me the temperature in the region X" needs to be disseminated to sensor nodes of a region X (see Figure 24.15). At the same time, data coming from the region X have to be delivered to the user(s) issuing the query. Simple techniques such as flooding and gossiping can be used to disseminate the data inside the sensor network [ASSC02]. However, these techniques waste energy resources sending redundant information throughout the network. Several application-aware algorithms have been devised to efficiently disseminate information in a wireless sensor network. These algorithms are based on the publishing/subscribe paradigm. Nodes publish the available data that are then delivered only to nodes requesting them. One of the most popular approaches is *Directed Diffusion* [IGE03], which delivers data only to nodes that have expressed interest to them. Similarly, SPIN sends data to sensor nodes only if they are interested [HKB01]. In addition, SPIN has access to the current energy level of the node and adapts the protocol behavior on the energy remaining in the node.

Dissemination algorithms achieve additional energy savings through in-network data processing based on data aggregation (or data fusion). Data aggregation is useful when sensed data come from sensor nodes that have overlapping sensing regions, or belong to the query region X. By combining data values (and descriptions) at intermediate nodes, the amount of information to be delivered is minimized thus contributing to energy savings. For example, if a user queries "tell me the highest temperature in the region X" not all temperature samples need to be delivered, but an intermediate node needs to forward only the data related to the highest sample among those it received.

The sensor networks' data-centric nature combined with strong resources' limitation make TCP not suitable for the sensor network domain. Indeed, inside a sensor network, different reliability levels and different congestion control approaches may be required depending on the nature of the data to be delivered. The transport layer functionalities must be therefore designed in a power-aware fashion to achieve the requested service level, and minimize the energy consumption, at the same time. This implies the use of different policies for the forward path (from sensor nodes toward the sink) and the reverse path (from the sink toward sensor nodes).

In the forward path an event-reliability principle needs to be applied, that is, not all data need to be correctly delivered but, by exploiting spatial and temporal correlations, the transport protocol must guarantee the correct delivery of a number of samples sufficient for correctly observing (at the user side) the monitored event. Typically, sensor networks operate under light loads, but suddenly become active in response to a detected event and this may generate congestion conditions. In [WEC03] an event-driven congestion control policy is designed to manage the congestion in the nodes-to-sink path by controlling the number of messages that notify a single event. Indeed, the transport protocol should guarantee that, when an event is detected, the user correctly receives enough information. In [SAA03] the concept of event-driven transport protocol introduced in [WEC03] is enhanced to guarantee reliable event detection with minimum energy expenditure.

The reverse path typically requires a very high reliability as data delivered toward the sensors contain critical information (e.g., queries and commands or programming instructions) delivered by the sink to control the activities of the sensor nodes. In this case more robust and power-greedy policies must be applied (see e.g., [WCK02]).

Sensor nodes in the sensing region X are typically set up to achieve in a cooperative way a predefined objective (e.g., monitoring the temperature in region X). This is achieved by distributing tasks to be performed on the sensor nodes. Therefore a sensor network is similar to a distributed system on which, at the same time, multiple applications are running. Each application is composed by several tasks that run on different (sensor) nodes. Starting from this view of a sensor network, in [BS03] the authors propose middleware-layer algorithms to manage, in a power-efficient way, a set of applications that may differ for the energy requirements and users' rewards. Specifically, the authors propose an admission control policy that, when an application starts, decides (given its energy costs and users' rewards) to accept and reject it to maximize the users' rewards. A policing mechanism is adopted, at runtime, to control that applications conform to the resource usage they declared at the admission stage.

24.7 Conclusions and Open Issues

In this chapter we have presented the main research trends in the field of power management for mobile and pervasive computing systems. Since mobile devices are supplied by batteries, we have first presented techniques to improve the battery capacity. We have highlighted that just relying on such improvements, without designing power management strategies, will greatly limit the use of mobile devices. Therefore, the remaining part of the chapter concentrates on power management policies. We have presented a general framework for mobile-device power management. Within this framework, several approaches have been presented which can be applied to any component of a mobile device. Since networking activities account for a significant portion of the total power consumption, we have then focused on power management policies designed for the networking subsystem.

Most of the current mobile computing environments fall into the infrastructure-based or single-hop hotspot scenario (e.g., 2.5/3G cellular networks and Wi-Fi hotspots) in which a fixed network is extended by means of a wireless last-hop to provide connectivity to mobile devices. We have surveyed the main power management policies proposed for such scenario. Specifically, we have focused on power management solutions for nonreal-time applications. To the best of our knowledge, less attention has been devoted to power management in the presence of real-time applications [CV02,CH03,MCD+03,SR03], and this is still an important open issue. Finally, we have presented power management techniques for mobile devices in single and multi-hop ad hoc networks. This networking scenario is fundamental, since it is the basis of pervasive computing environments.

In our opinion, the design of cross-layer power-management policies is a very promising direction, which should be further explored. Cross-layering can reduce the traffic on ad hoc networks by sharing information that is of interest to different layers [CMTG03]. Moreover, information collected at a particular layer (e.g., a route failure) can be exploited by several layers to tune the protocol behavior (e.g., the transport protocol can avoid sending data until a new route is found). Such cross-layer optimizations are not explicitly designed for power management, but reduce the power consumption as a side effect.

On the other hand, cross-layer power-management policies could be designed, that exploit information residing at different layers in the protocol stack to manage communications on the wireless links. We have highlighted that the most appropriate way to save power is putting the wireless interface of mobile devices in low-power operating modes whenever possible. In ad hoc networks, the decisions on when to switch the wireless interface in the different operating modes is typically based on MAC [VL03,YHE02] or routing-level information [CJBM02,HCB00]. In a cross-layer approach also information about the application behavior could be used to schedule the time intervals during which nodes are available for communications. Moreover, such an approach could be exploited also in infrastructure-based environments. So far, little attention has been devoted to such policies [VA03,ANF03].

We also believe that experimental validations of power management solutions are strongly required. Usually, performance evaluations rely just on simulations. However, simulation models tend to drastically approximate the complex physical behavior of wireless links. Therefore, experimental measurements typically show several features of network protocols that are completely unknown from simulations [ABCG04,LNT02].

Acknowledgments

This work was partially funded by the Information Society Technologies program of the European Commission, Future and Emerging Technologies under the IST-2001-38113 *MobileMAN* project, and partially by the Italian Ministry for Education and Scientific Research in the framework of the projects *Internet: Efficiency, Integration and Security* and *FIRB-VICOM*.

References

[802.11]   Web site of the IEEE 802.11 WLAN: http://grouper.ieee.org/grups/802/11/main.html

[802.15]   Web site of the IEEE 802.15 WPAN: http://www.ieee802.org/15/pub/main.html

[ABCG04]   G. Anastasi, E. Borgia, M. Conti, and E. Gregori, "Wi-Fi in ad hoc mode: a measurement study," in *Proceedings of the IEEE Conference on Pervasive Computing and Communications (PerCom 2004)*, March 2004.

[ACF+04]   G. Anastasi, M. Conti, A. Falchi, E. Gregori, and A. Passarella, "Performance measurements of mote sensor networks," in *Proceedings of the ACM/IEEE Symposium on Modeling, Analysis and Simulation of Wireless and Mobile System (MSWIM 2004)*, Venice, Italy, October 4–6, 2004.

[ACG04]   G. Anastasi, M. Conti, and E. Gregori, "IEEE 802.11 ad hoc networks: protocols, performance and open issues," Chapter 3, *Ad hoc Networking*, S. Basagni, M. Conti, S. Giordano, and I. Stojmenovic, (Eds.), IEEE Press and John Wiley and Sons, Inc., New York, 2004.

[ACGP02]   G. Anastasi, M. Conti, E. Gregori, and A. Passarella, "A power saving architecture for web access from mobile devices," in *Proceedings of the IFIP Networking Conference (Networking'02)*, Pisa (I), Lecture Notes in Computer Science, 2345, May 2002, pp. 240–251.

[ACGP03a]   G. Anastasi, M. Conti, E. Gregori, and A. Passarella, "Balancing energy saving and QoS in the mobile internet: an application-independent approach," in *Proceedings of the Hawaii International Conference on System Science (HICSS 2003)*, Minitrack on Quality of Service Issues in Mobile and Wireless Networks, Hawaii, January 6–9, 2003.

[ACGP03b]   G. Anastasi, M. Conti, E. Gregori, and A. Passarella, "Performance comparison of power saving strategies for mobile web access," *Performance Evaluation*, Vol. 53, 2003, pp. 273–294.

[ACGP03c]   G. Anastasi, M. Conti, E. Gregori, and A. Passarella, "A performance study of power saving policies for Wi-Fi hotspots," *Computer Networks*, Vol. 45, June 2004, pp. 295–318.

[ACGP04a]   G. Anastasi, M. Conti, E. Gregori, and A. Passarella, "Saving energy in Wi-Fi hotspots through 802.11 PSM: an analytical model," in *Proceedings of the Workshop on Modeling and Optimization in Mobile, Ad Hoc and Wireless Networks (WiOpt 2004)*, University of Cambridge (UK), March 24–26, 2004.

[ACGP04b]   G. Anastasi, M. Conti, E. Gregori, and A. Passarella, "Experimental analysis of an application-independent energy management scheme for Wi-Fi hotspots," in *Proceedings of the IEEE Symposium on Computers and Communications (ISCC 2004)*, Alexandria, Egypt, June 29–July 1, 2004.

[ACL03]   G. Anastasi, M. Conti, and W. Lapenna, "A power saving network architecture for accessing the internet from mobile devices: design, implementation and measurements," *The Computer Journal*, Vol. 46, 2003, pp. 3–15.

[ANF03]   M. Anand, E. Nightingale, and J. Flinn, "Self-tuning wireless network power management," in *Proceedings of the ACM Mobicom 2003*, San Diego CA, September 14–19, 2003.

[ASSC02]   I.F. Akyildiz, W. Su, Y. Sankarasubramaniam, and E. Cayirci, "Wireless sensor networks: a survey," *Computer Networks*, Vol. 38, 2002, pp. 393–422.

[BB95]   T.D. Burd and R.W. Brodersen, "Energy efficient CMOS microprocessor design," in *Proceedings of the 28th IEEE Annual Hawaii International Conference on System Sciences (HICSS'95)*, January 1995.

[BB97]   A. Bakre, and B.R. Badrinath, "Implementation and performance evaluation of indirect TCP," *IEEE Transactions on Computers*, Vol. 46, March 1997, pp. 260–278.

[BBD00]   L. Benini, A. Bogliolo, and G. De Micheli, "A survey of design techniques for system-level dynamic power management," *IEEE Transactions on VLSI Systems*, Vol. 8, June 2000, pp. 299–316.

[BBD99]   L. Benini, A. Bogliolo, and G. De Micheli, "Dynamic power management of electronic systems," in *System-Level Synthesis*, A. Jerraya and J. Mermet (Eds.), Klumer, 1999, pp. 263–292.

[BC03]   S. Bandyopadhyay and E.J. Coyle, "An energy efficient hierarchical clustering algorithm for wireless sensor networks," in *Proceedings of IEEE INFOCOM 2003*, 2003.

[BC98]   P. Barford and M. Crovella, "Generating representative web workloads for network and server performance evaluation," in *Proceedings of ACM SIGMETRICS'98*, Madison, WI, June 1998, pp. 151–160.

[BCD01]   L. Bononi, M. Conti, and L. Donatiello, "A distributed mechanism for power saving in IEEE 802.11 wireless LANs," *ACM/Kluwer Mobile Networks and Applications Journal*, Vol. 6, 2001, pp. 211–222.

[BCG02]   R. Bruno, M. Conti, and E. Gregori, "Optimization of efficiency and energy consumption in p-persistent CSMA-based wireless LANs," *IEEE Transactions on Mobile Computing*, Vol. 1, 2002, pp. 10–31.

[BCG04]   L. Bononi, M. Conti, and E. Gregori, "Runtime optimization of IEEE 802.11 wireless LANs performance," IEEE *Transactions on Parallel and Distributed Systems*, Vol. 15(1), January 2004, pp. 66–80.

[BD02]   L. Benini and G. De Micheli, "Energy-efficient system-level design," in *Power-Aware Design Methdologies*, J. Rabaey and M. Pedram (Eds.), Kluwer, Dordrecht, 2002, pp. 473–516.

[BEF84]   D.J. Baker, A. Ephremides, and J.A. Flynn, "The design and simulation of a mobile radio network with distributed control," *IEEE Journal on Selected Areas in Communications*, Vol. 2, 1984, pp. 226–237.

[BGS94]   D. Bacon, S. Graham, and O. Sharp, "Compiler transformation for high-performance computing", *ACM Computing Survey*, Vol. 26, 1994, pp. 345–420.

[BS03]   A. Boulis and M.B. Srivastava, "Node-level energy management for sensor networks in the presence of multiple applications," *Wireless Networks*, Vol. 10(6), November 2004, pp. 737–746.

[C03]   M. Conti, "Body, personal, and local wireless ad hoc networks," *Handbook of Ad Hoc Networks*, M. Ilyas, (Ed.), CRC Press, New York, Chapter 1, 2003.

[CH03]   S. Chandra, "Wireless network interface energy consumption implications for popular streaming formats," *Multimedia Systems*, Vol. 9, 2003, pp. 185–201.

[C04]   M. Conti, "Wireless communications and pervasive technologies," *Smart Environments: Technologies, Protocols and Applications*, Diane Cook and Sajal K. Das (Eds.), John Wiley & Sons, New York, Chapter 4, 2004.

[CBB+02]   E.-Y. Chung, L. Benini, A. Bogliolo, Y.-H. Lu and G. De Micheli, "Dynamic power management for nonstationary service requests," *IEEE Transactions on Computers*, Vol. 51, 2002, pp. 1345–1361.

[CBBD99]   E.-Y. Chung, L. Benini, A. Bogliolo, and G. De Micheli, "Dynamic power management using adaptive learning tree," in *Proceedings of the IEEE International Conference on Computer Aided Design*, November 1999, pp. 274–279.

[CCL03]   I. Chlamtac, M. Conti, and J. Liu, "Mobile ad hoc networking: imperatives and challenges," *Ad Hoc Network Journal*, Vol. 1, January–February–March 2003, pp. 13–64.

[CFE99]   B. Calder, P. Feller, and A. Eustace, "Value profiling and optimization," *Journal of Instruction-Level Parallelism*, Vol. 1, 1999.

[CJBM02]   Benjie Chen, Kyle Jamieson, Hari Balakrishnan, and Robert Morris, "Span: an energy-efficient coordination algorithm for topology maintenance in ad hoc wireless networks," *ACM/Kluwer Wireless Networks*, Vol. 8, 2002, pp. 481–494.

[CMC99]   M.S. Corson, J.P. Maker, and J.H. Cernicione, "Internet-based mobile ad hoc networking," *IEEE Internet Computing*, 1999, pp. 63–70.

[CMD]   Crossbow MICA2DOT Datasheet, available at http://www.xbow.com/Products/ Product_pdf_files/Wireless_pdf/6020-0043-03_A_MICA2DOT.pdf

[CMTG04]   M. Conti, G. Maselli, G. Turi, and S. Giordano, "Cross-layering in mobile ad hoc network design," *IEEE Computer*, Vol. 37, 2004, pp. 48–51.

[CR00]   C.F. Chiasserini and R.R. Rao, "Energy efficient battery management," in *Proceedings of the IEEE INFOCOM 2000*, 2000.

[CT00]   Jae-Hwan Chang and Leandros Tassiulas, "Energy conserving routing in wireless ad-hoc networks," in *Proceedings of the IEEE INFOCOM*, March 2000, pp. 22–31.

[CV02]   S. Chandra and A. Vahadat, "Application-specific network management for energy-aware streaming of popular multimedia formats," in *Proceedings of the Annual Technical Conference, USENIX*, June 2002.

[DB97]   B. Das and V. Bharghavan, "Routing in ad-hoc networks using minimum connected dominating sets," in *Proceedings of the IEEE International Conference on Communications (ICC'97)*, June 1997.

[DKB95]   F. Douglis, P. Krishnan, and B. Bershad, "Adaptive disk spin-down policies for mobile devices," in *Proceedings of the 2nd Usenix Symposium on Mobile and Location-Independent Computing*, April 1995, pp. 121–137.

[DKM94]   F. Douglis, P. Krishnan, and B. Marsh, "Thwarting the power-hungry disk," in *Proceedings of the Usenix Technical Conference*, 1994, pp. 292–306.

[DN97]   M. Doyle and J.S. Newman, "Analysis of capacity-rate data for lithium batteries using simplified models of the discharge process," *Journal of Applied Electrochemistry*, Vol. 27, 1997, pp. 846–856.

[DWBP01]   L. Doherty, B.A. Warneke, B.E. Boser, and K.S.J. Pister, "Energy and performance considerations for smart dust," *International Journal of Parallel Distributed Systems and Networks*, Vol. 4, 2001, pp. 121–133.

[E02]   A. Ephremides, "Energy concerns in wireless networks," *IEEE Wireless Communications*, Vol. 9, 2002, pp. 48–59.

[ECPS02]   D. Estrin, D. Culler, K. Pister, and G. Sukhatme, "Connecting the phys-
            ical world with pervasive networks," *IEEE Pervasive Computing*, Vol. 1, 2002,
            pp. 59–69.

[F01]   L.M. Feeney, "An energy-consumption model for performance analysis of routing pro-
            tocols for mobile ad hoc networks," *ACM/Kluwer Mobile Networks and Applications
            (MONET)*, Vol. 6, 2001, pp. 239–249.

[F04]   L. Feeney, "Energy efficient communication in ad hoc wireless networks," in *Ad Hoc
            Networking*, S. Basagni, M. Conti, S. Giordano, and I. Stojmenovic (Eds.), IEEE Press and
            John Wiley and Sons, Inc., New York, 2004.

[FAW01]   L. Feeney, B. Ahlgren, and A. Westerlund, "Spontaneous networking: an application-
            oriented approach to ad hoc networking," *IEEE Communications Magazine*, Vol. 39(6),
            June 2001, 176–181.

[FPS02]   J. Flinn, S.Y. Park, and M. Satyanarayanan, "Balancing performance, energy, and quality in
            pervasive computing," in *Proceedings of the IEEE International Conference on Distributed
            Computing Systems (ICDCS'02)*, Wien, Austria, July 2002.

[FRM01]   K. Flautner, S. Reinhardt, and T. Mudge, "Automatic performance-setting for dynamic
            voltage scaling," in *Proceedings of the ACM International Conference on Mobile Computing
            and Networking (MOBICOM 2001)*, Rome, Italy, July 2001.

[FS99]   J. Flinn and M. Satyanarayanan, "Energy-aware adaptation for mobile applications," in
            Proceedings of the ACM Symposium on Operating System Principles, December 1999.

[FZ94]   G.H. Forman and J. Zahorjan, "The challenges of mobile computing," *IEEE Computer*,
            Vol. 27, 1994, pp. 38–47.

[G01]   F. Gruian, "Hard real-time scheduling for low energy using stochastic data and DVS
            processors," in *Proceedings of the International Symposium on Low Power Electronics and
            Design (ISPLED 2001)*, Huntington Beach, CA, August 2001.

[GBS+95]   R. Golding, P. Bosh, C. Staelin, T. Sullivan, and J. Wilkes, "Idleness is not sloth," in
            Proceedings of the Usenix Technical Conference, January 1995, pp. 201–212.

[GCNB03]   J. Gomez, A. Campbell, M. Naghshineh, and C. Bisdikian, "PARO: supporting dynamic
            power controlled routing in wireless ad hoc networks," *ACM/Kluwer Wireless Networks*,
            Vol. 9(5), September 2003, pp. 443–460.

[HCB00]   W. Heinzelman, A. Chandrakasan, and H. Balakrishnan, "Energy-efficient communication
            protocol for wireless microsensor networks," in *Proceedings of the 33rd Hawaii International
            Conference on System Sciences (HICSS'00)*, January 2000.

[HDD98]   J.H. Harreld, W. Dong, and B. Dunn, "Ambient pressure synthesis or aerogel-like vanadium
            oxide and molybdenum oxide," *Materials Research Bulletin*, Vol. 33, 1998, pp. 561–567.

[HKB01]   W.R. Heinzelman, J. Kulik, and H. Balakrishnan, "Adaptive protocols for information
            dissemination in wireless sensor networks," in *Proceedings of the 5th Annual International
            Conference on Mobile Computing and Networking*, Seattle, WA, August 2001, pp. 174–185.

[HLS96]   D.P. Helmbold, D.E. Long, and B. Sherrod "A dynamic disk spin-down technique for
            mobile computing," in *Proceedings of the 2nd Annual ACM International Conference on
            Mobile Computing and Networking*, NY, November 1996, pp. 130–142.

[HW95]   M. Herbster and M.K. Warmuth, "Tracking the best expert," in *Proceedings of the 12th
            International Conference on Machine Learning*, Tahoe City, CA, Morgan Kaufmann, 1995,
            pp. 286–294.

[HW97]   C.-H. Hwang and A. Wu, "A predictive system shutdown method for energy saving of
            event-driven computation," in *Proceedings of the International Conference on Computer
            Aided Design*, November 1997, pp. 28–32.

[IG00]   T. Imielinski and S. Goel, "DataSpace: querying and monitoring deeply networked
            collections in physical space," *IEEE Personal Communications*, October 2000, pp. 4–9.

[IGE03]   C. Intanagonwiwat, R. Govindan, D. Estrin, J. Heidemann, and F. Silva, Directed diffusion
            for wireless sensor networking, *IEEE Transaction on Networking*, Vol. 11, 2003, pp. 2–16.

[IPAQ03]   Hp iPAQ Pocket PC h2200 Series Datasheet, available at http://www.hp.com/hpinfo/
             newsroom/press_kits/2003/ipaq/ds_h2200.pdf.

[J00]   A. Joshi, "On proxy agents, mobility, and web access," *ACM/Baltzer Mobile Networks and
             Applications*, Vol. 5, 2000, pp. 233–241.

[JSAC01]   J.C. Jones, K. Sivalingam, P. Agarwal, and J.C. Chen, "A survey of energy efficient network
             protocols for wireless and mobile networks," *ACM/Kluwer Wireless Networks*, Vol. 7, 2001,
             pp. 343–358.

[KB02]   R. Krashinsky and H. Balakrishnan, "Minimizing energy for wireless web access with
             bounded slowdown," in *Proceedings of the ACM International Conference on Mobile
             Computing and Networking (MOBICOM 2002)*, 2002.

[KK00]   R. Kravets and P. Krishnan, "Application-driven power management for mobile commu-
             nication," *ACM/Baltzer Wireless Networks*, Vol. 6, 2000, pp. 263–277.

[KK98]   R. Kravets and P. Krishnan, "Power management techniques for mobile communication,"
             in *Proceedings of the 4th Annual ACM/IEEE International Conference on Mobile Computing
             and Networking (MOBICOM'98)*, 1998.

[KL00]   C.M. Krishna and Y.H. Lee, "Voltage-clock-scaling techniques for low power in hard
             real-time systems," in *Proceedings of the IEEE Real-Time Technology and Applications
             Symposium*, Washington, D.C., May 2000, pp. 156–165.

[KLV95]   P. Krishnan, P. Long, and J. Vitter, "Adaptive disk spindown via optimal rent-to-buy in
             probabilistic environments," in *Proceedings of the International Conference on Machine
             Learning*, July 1995, pp. 322–330.

[KMMO94]   A. Karlin, M.S. Manasse, L.A. McGeoch, and S. Owicki, "Competitive randomized
             algorithms for non-uniform problems," *Algorithmica*, Vol. 11, 1994, pp. 542–571.

[KVIY01]   M. Kandemir, M. Vijaykrishnan, M. Irwin, and W. Ye, "Influence of compiler optimizations
             on system power," *IEEE Transactions on VLSI Systems*, Vol. 9, 2001, pp. 801–804.

[KW03]   Holger Karl and Andreas Willig, "A short survey of wireless sensor networks," TKN
             Technical Report TKN-03-018.

[LAR01]   Qun Li, Javed Aslam, and Daniela Rus, "Online power-aware routing in wireless ad-hoc
             networks," in *Proceedings of the ACM International Conference on Mobile Computing and
             Networking (MOBICOM 2001)*, Rome, Italy, 2001, July 16–21.

[LBD02]   Y.-H. Lu, L. Benini, and G. De Micheli, "Dynamic frequency scaling with buffer insertion
             for mixed workloads," *IEEE Transactions on CAD-ICAS*, Vol. 21, 2002, pp. 1284–1305.

[LBS]   Laptop Battery Store site, at http://shop.store.yahoo.com/batteryjuice/

[LH01]   L. Li and J.Y. Halpern, "Minimum-energy mobile wireless networks revisited," in
             Proceedings of the IEEE International Conference on Communications (ICC01), 2001,
             pp. 278–283.

[LKHA94]   K. Li, R. Kumpf, P. Horton, and T. Anderson, "A quantitative analysis of disk drive power
             management in portable computers," in *Proccedings of the 1994 Winter Usenix Conference*,
             January 1994, pp. 279–291.

[LLM+02]   E.L. Lloyd, R. Liu, M.V. Marathe, R. Ramanathan, and S.S. Ravi, "Algorithmic aspects of
             topology control problems for ad hoc networks," in *Proceedings of the ACM International
             Symposium on Mobile Ad hoc Networking & Computing*, Lausanne, CH, 2002.

[LNT02]   Henrik Lundgren, Erik Nordström, and Christian Tschudin, "Coping with communication
             gray zones in IEEE 802.11b based ad hoc networks," in *Proceedings of the 5th International
             Workshop on Wireless Mobile Multimedia (WoWMoM 2002)*, September 2002.

[LS01]   J.R. Lorch and A.J. Smith, "Improving dynamic voltage scaling with pace," in *Proceedings
             of the ACM Sigmetrics 2001 Conference*, Cambridge, MA, June 2001, pp. 50–61.

[LS97]   J.R. Lorch and A.J. Smith, "Scheduling techniques for reducing processor energy use in
             macOS," *ACM/Baltzer Wireless Networks*, Vol. 3(5), 1997, pp. 311–324.

[LS98]   J.R. Lorch and A.J. Smith, "Software strategies for portable computer energy management,"
             IEEE Personal Communications, Vol. 5, 1998, pp. 60–73.

[MACM00]　D. Mosse, H. Aydin, B. Childers, and R. Melhem, "Compiler-assisted dynamic power-aware scheduling for real-time applications," in *Proceedings of the Workshop on Compiler and Operating Systems for Low Power (COLP 2000)*, Philadelphia, PA, October 2000.

[MCD+03]　S. Mohapatra, R. Cornea, N. Dutt, A. Nicolau, and N. Venkatasubramanian, "Integrated power management for video streaming to mobile handheld devices," in *Proceedings of the ACM International Conference on Multimedia 2003*, Berkeley, CA, November 2–8, 2003.

[MFHH03]　S.R. Madden, M.J. Franklin, J.M. Hellerstein, and W. Hong, "The design of an acquisitional query processor for sensor networks," in *Proceedings of the SIGMOD*, June 2003.

[MMA+01]　S. Meninger, J.O. Mur-Miranda, R. Amirtharajah, A.P. Chandrakasan, and J.H. Lang, "Vibration-to-electric energy conversion," *IEEE Transactions on VLSI Systems*, Vol. 9, 2001, pp. 64–76.

[MOI+97]　H. Mehta, R. Owens, M. Irwin, R. Chen, and D. Ghosh, "Techniques for low energy software," in *Proceedings of the International Symposium on Low Power Electronics and Design*, August 1997, pp. 72–75.

[MT]　Maxwell Technologies, at http://www.maxwell.com/

[NKSK02]　S. Narayanaswamy, V. Kawadia, R.S. Sreenivas, and P.R. Kumar, "Power control in ad-hoc networks: theory, architecture, algorithm and implementation of the COMPOW protocol," in *Proceedings of the European Wireless 2002*, February 2002, pp. 156–162.

[PCD+01]　D. Panigrahi, C.F. Chiasserini, S. Dey, R.R. Rao, A. Raghunathan, and K. Lahiri, "Battery life estimation of mobile embedded systems," in *Proceedings of the International Conference on VLSI Design*, 2001.

[PCS02]　B.J. Prahbu, A. Chockalingam, and V. Sharma, "Performance analysis of battery power management schemes in wireless mobile devices," in *Proceedings of the IEEE Wireless Communication and Networking Conference (WCNC'02)*, 2002.

[PEBI01]　S.H. Phatak, V. Esakki, R. Badrinath, and L. Iftode, "Web&: an architecture for non-interactive web," in *Proceedings of the IEEE Workshop on Internet Applications, (WIAPP'01)*, July 2001.

[PS01]　P. Pillai and K.G. Shin, "Real-time dynamic voltage scaling for low-power embedded operating systems," in *Proceedings of the 18th Symposium on Operating Systems Principles (SOSP'01)*, 2001.

[PS02]　C. Poellabauer and K. Schwan, "Power-aware video decoding using real-time event handlers," in *Proceedings of the 5th ACM International Workshop on Wireless Mobile Multimedia (WoWMoM'02)*, September 2002, pp. 72–79.

[PS98]　E. Pitoura and G. Samaras, "*Data Management for Mobile Computing*," Kluwer Academic Publishers, Dordrecht, 1998.

[PSS01]　S. Park, A. Savvides, and M.B. Srivastava, "Battery capacity measurement and analysis using lithium coin cell battery," in *Proceedings of the ACM International Symposium on Low Power Electronic Design (ISLPED '01)*, August 2001.

[QP00]　Q. Qiu and M. Pedram, "Dynamic power management of complex systems using generalized stochastic petri nets," in *Proceedings of the Design Automation Conference*, June 2000, pp. 352–356.

[QP99]　Q. Qiu and M. Pedram, "Dynamic power management based on continuous-time Markov decision processes," in *Proceedigns of the Design Automation Conference*, June 1999, pp. 555–561.

[R04]　R. Ramanathan, "Antenna beamforming and power control for ad hoc networks," in *Ad Hoc Networking*, S. Basagni, M. Conti, S. Giordano, and I. Stojmenovic (Eds.), IEEE Press and John Wiley and Sons, Inc., New York, 2004.

[RB98]　M. Rulnick and N. Bambos, "Mobile power management for wireless communication networks," *Wireless Networks*, Vol. 3(1), March 1997, pp. 3–14.

[RM99]   Volkan Rodoplu and Teresa H.-Y. Meng "Minimum energy mobile wireless networks," *IEEE Journal on Selected Areas in Communications*, Vol. 17, 1999, pp. 1333–1344.

[RRH00]   R. Ramanathan and R. Rosales-Hain, "Topology control of multi-hop wireless networks using transmit power adjustment," in *Proceedings of the IEEE INFOCOM 2000*, Tel Aviv, Israel, March 2000.

[RSPS02] V. Raghunathan, C. Schurgers, S. Park, and M.B. Srivastava, "Energy-aware wireless microsensor networks," *IEEE Signal Processing Magazine*, Vol. 19(2), March 2002, pp. 40–50.

[S03]   T.E. Starner, "Powerful change part 1: batteries and possible alternatives for the mobile market," *IEEE Pervasive Computing*, Vol. 2(4), 2003, pp. 86–88.

[S93]   M. Satyanarayanan et al., "Experience with disconnected operation in a mobile computing environment," in *Proceedings of the Usenix Mobile and Location-Independent Computing Symposium*, August 1993, pp. 11–28.

[SAA03]   Y. Sankarasubramaniam, O.B. Akan, and I.F. Akyildiz, "ESRT: event-to-sink reliable transport for wireless sensor networks," in *Proceedings of the ACM (MobiHoc'03)*, Annapolis, MD, June 2003.

[SBGD01] T. Simunic, L. Benini, P. Glynn, and G. De Micheli, "Event-driven power management," *IEEE Transactions on CAD-ICAS*, Vol. 20, 2001, pp. 840–857.

[SC00]   V. Swaminathan and K. Chakrabarty, "Real-time task scheduling for energy-aware embedded systems," in *Proceedings of the IEEE Real-Time Systems Symposium*, Orlando, FL, November 2000.

[SCB96]   M. Srivastava, A. Chandrakasan, and R. Brodersen, "Predictive system shutdown and other architectural techniques for energy efficient programmable computation," *IEEE Transactions on VLSI Systems*, Vol. 4, 1996, pp. 42–55.

[SK97]   Stemm, M. and Katz, R.H., "Measuring and reducing energy consumption of network interfaces in handheld devices," *IEICE Transactions on Fundamentals of Electronics, Communications and Computer Science*, Vol. 80, 1997, pp. 1125–1131. (Special Issue on Mobile Computing).

[SL00]   I. Stojmenovic and X. Lin, "Power-aware localized routing in wireless networks," in *Proceedings of the IEEE Symposium Parallel and Distant Processing System*, May 2000.

[SR03]   P. Shenoy and P. Radkov, "Proxy-assisted power-friendly streaming to mobile devices," in *Proceedings of the SPIE — Volume 5019, Multimedia Computing and Networking 2003*, January 2003, pp. 177–191.

[STD94]   C. Su, C. Tsui, and A. Despain, "Saving power in the control path of embedded processors," *IEEE Design and Test of Computers*, Vol. 11, 1994, pp. 24–30.

[STGM02] C. Schurgers, V. Tsiatsis, S. Ganeriwal, and M. Srivastava, "Optimizing sensor networks in the energy-latency-density design space," *IEEE Transactions on Mobile Computing*, Vol. 1, 2002, pp. 70–80.

[T01]   C.-K. Toh, "Maximum battery life routing to support ubiquitous mobile computing in wireless ad hoc networks," *IEEE Communications Magazine*, Vol. 39, 2001, pp. 138–147.

[TMCP01] D. Teasdale, V. Milanovic, P. Chang, and K.S.J. Pister, "Microrockets for smart dust," *Smart Materials and Structures*, Vol. 10, 2001, pp. 1145–1155.

[TMW94]   V. Tiwari, S. Malik, and A. Wolfe, "Power analysis of embedded software: a first step towards software power minimization," *IEEE Transactions on VLSI Systems*, Vol. 2, 1994, pp. 437–445.

[US96]   S. Udani and J. Smith, "The power broker: intelligent power management for mobile computing," Technical Report MS-CIS-96-12, Department of Computer Information Science, University of Pennsylvania, May 1996.

[VA03]   M.C. Vuran and I.F. Akyildiz, "Correlation-based collaborative medium access in wireless sensor networks," November 2003. http://www.ece.gatech.edu/research/labs/bwn/publications.html.

[VL03]   T. van Dam and K. Langendoen, "*An Adaptive Energy-Efficient MAC Protocol for Wireless Sensor Networks,*" ACM SenSys, November 2003.

[W91]   M. Weiser, "*The Computer for the Twenty-First Century,*" Scientific American, September 1991.

[W96]   M. Wolfe, "*High Performance Compilers for Parallel Computing.*" Reading, MA, Addison-Wesley, 1996.

[WC02]   IEEE Wireless Communications, Special issue on "Energy-Aware Ad Hoc Wireless Networks," Vol. 9, August 2002.

[WCK02]   C.Y. Wan, A.T. Campbell, and L. Krishnamurthy, "PSFQ: a reliable transport protocol for wireless sensor networks," in *Proceedings of the WSNA'02,* Atlanta, GA, September 2002.

[WDGS02]   Jie Wu, Fei Dai, Ming Gao, and Ivan Stojmenovic, "On calculating poweraware connected dominating sets for efficient routing in ad hoc wireless networks," *IEEE/KICS Journal of Communications and Networks,* Vol. 4, 2002, pp. 59–70.

[WEC03]   C.-Y. Wan, S.B. Eisenman, and A.T. Campbell, "CODA: congestion detection and avoidance in sensor networks," in *Proceedings of the ACM SENSYS 2003,* November 2003.

[WESW98]   Hagen Woesner, Jean-Pierre Ebert, Morten Schlager, and Adam Wolisz, "Power saving mechanisms in emerging standards for wireless LANs: the M level perspecitve," *IEEE Personal Communications,* Vol. 5, 1998, pp. 40–48.

[WL99]   J. Wu and H. Li, "On calculating connected dominating set for efficient routing in ad hoc wireless networks," in *Proceedings of the 3rd International Workshop on Discrete Algorithms and Methods for Mobile Computing and Communications,* Seattle, WA, August 1999.

[WLBW01]   Roger Wattenhofer, Li Li, Paramvir Bahl, and Yi-Min Wang, "Distributed topology control for wireless multihop ad-hoc networks," in *Proceedings of the IEEE Infocom,* April 2001, pp. 1388–1397.

[WWDS94]   M. Weiser, B. Welch, A. Demers, and S. Shenker, "Scheduling for reduced CPU energy," in *Proceedings of the 1st Symposium on Operating Systems Design and Implementation (OSDI),* November 1994, pp. 13–23.

[WWRF]   Wireless World Research Forum (WWRF): http://www.ist-wsi.org.

[XHE00]   Ya Xu, John Heidemann, and Deborah Estrin, "Adaptive energy conserving routing for multihop ad hoc networks," Technical Report 527, USC/Information Sciences Institute, October 2000.

[XHE01]   Ya Xu, John Heidemann, and Deborah Estrin, "Geography-informed energy conservation for ad hoc routing," in *Proceedings of the 7th Annual International Conference on Mobile Computing and Networking,* July 2001, pp. 70–84.

[YHE02]   W. Ye, J. Heidemann, and D. Estrin, "An energy-efficient MAC protocol for wireless sensor networks," in *Proceedings of the 21st International Annual Joint Conference of the IEEE Computer and Communications Societies (INFOCOM 2002),* New York, USA, June 2002.

[ZD04]   G. Zaruba and S. Das, "Off-the-shelf enablers of ad hoc networks," in *Ad Hoc Networking,* S. Basagni, M. Conti, S. Giordano, and I. Stojmenovic (Eds.), IEEE Press and John Wiley and Sons, Inc., New York, 2004.

[ZR96]   M. Zorzie and R.R. Rao, "Energy constrained error control for wireless channels," in *Proceedings of the IEEE GLOBECOM '96,* 1996, pp. 1411–1416.

Collision Avoidance, Contention Control, and Power Saving in IEEE 802.11 Wireless LANs

25.1 Introduction 25-577
25.2 The Role of Wireless Protocols Design 25-578
      Adaptive Protocols and Cross Layering • WLANs and
      MANETs
25.3 The IEEE 802.11 Standard 25-581
      Distributed Foundation Wireless Medium Access Control
25.4 Background and Wireless Assumptions for MAC
      Protocols 25-583
      Wireless Signals • Collision Domains, Capture Effect • Half
      Duplex Channels • Collision Detection • Full-Duplex Links
25.5 Evolutionary Perspective of Distributed
      Contention-Based MAC 25-586
      Distributed Contention-Based MAC Protocols • Collision
      Detection in Wireless Systems • Collision Avoidance •
      Collision Resolution Protocols • Contention Control in IEEE
      802.11 DCF • Contention and Collision Avoidance in
      Multi-Hop Scenarios
25.6 Power Saving Protocols 25-600
      Power Saving Solutions at the MAC Layer • The MAC
      Contention under the Power Saving Viewpoint •
      Sleep-Based Solutions • Power Control Solutions • IEEE
      802.11 Power Saving
25.7 Conclusion 25-606
References 25-606

Luciano Bononi
Lorenzo Donatiello

25.1 Introduction

Ethernet is one of the predominant network technologies for supporting local area network (LAN) communication [71]. In recent years the proliferation of portable and laptop computers required LAN technology to support wireless connectivity [9,26,28,45,68]. Mobile and wireless solutions for communication have been designed to allow mobile users to access information anywhere and at anytime [45]. The well-known Internet services (as an example, WWW, e-mail) are considered the most promising killer applications pushing for wireless Internet technologies, services, and infrastructures deployment on behalf of network service providers and private customers. The integration of the Internet communication with the innovative and challenging last-mile wireless connectivity is required to support full services and protocols' integration among the two worlds. A variety of networking solutions, services, protocols, standards, technologies, and new applications have been proposed to meet the wireless Internet goals. One of the most successful technologies in recent years is based on the IEEE 802.11 Standard. The IEEE 802.11 Standard defines the Physical, the medium access control (MAC), and link layer control (LLC) sub-layers of the wireless technology that is widely deployed in today's wireless LANs. Specifically, at the MAC layer, the IEEE 802.11 Standard defines two coordination functions: the centralized point coordination function (PCF) and the distributed coordination function (DCF). Due to the principles adopted for the design, the IEEE 802.11 DCF MAC protocol could be referenced to as "Ethernet on the air" solution for a wireless LAN.

The aim of this chapter is to illustrate the evolution in the design and the main issues originating and characterizing the protocols as the basis of the distributed, random-access implementation of the IEEE 802.11 DCF MAC protocol. In general, it is interesting to determine the motivations for a consistent redesign work and the leading role of protocols in the resolution of many peculiar issues for the wireless scenario. Basically, this chapter will introduce and analyze the problems considered in the protocol design that are determined by the wireless scenarios and, more specifically, the evolution of protocols that inherited most of the design concepts from their wired counterpart. The chapter describes the new protocol-adaptation solutions proposed to control new critical system factors and limitations such as time- and space-contention under distributed control, channel bandwidth limitation, host mobility and variable hosts density, battery energy limitation, and power control.

The chapter is structured in the following way: in Section 25.2, we introduce general concepts about the role of protocols and new protocol-design concepts for the wireless LANs and Mobile Ad Hoc Network (MANETs); in Section 25.3, we provide a short illustration of the IEEE 802.11 Standard, as the basis for successive discussions; in Section 25.4, we analyze the new assumptions to characterize the wireless scenario, at the MAC layer; in Section 25.5, we illustrate a summary of the evolution of distributed, contention-based (random-access) MAC protocols up to the skeleton of the recent IEEE 802.11 DCF design by focusing on the distributed time- and space-contention control and collision avoidance solutions; in Section 25.6, we illustrate the MAC protocol design issues devoted to the power saving; and in Section 25.7, we draw the chapter conclusions.

25.2   The Role of Wireless Protocols Design

Wireless medium problems and resource restrictions in wireless systems made the "anywhere and at anytime connectivity" difficult to obtain. Some of the problems to be solved include environment obstacles and interference, user mobility and dynamic network topologies, variable user-density, low channel bandwidth, frequent link failures, energy limitations, and overheads reduction. Under the protocols' design viewpoint these problems have been dealt with at many layers in the OSI protocol stack. At the physical layer, suitable technologies for transmission, reception, and coding are required. At upper layers, protocols' design plays an important role: protocols define the way resources are used, shared, and also protocols define the way the information is coded, fragmented, delivered, reordered, assembled, checked, and validated. Protocols also determine which services the system can support and the Quality of Service (QoS) guaranteed to the users' applications.

It results in the opening up of a vast area of research in the investigation of the role of protocols and distributed algorithms for management of networks. New limiting constraints given by the wireless scenarios have resulted in the need for consistent research work to realize optimal tuning of the protocols and algorithms derived from those adopted in wired networks. One of the challenging tasks for researchers in recent years has been (and still is) the need to overcome the wireless weaknesses by maintaining the inertial definition of management protocols, and architectures, for the intercommunication of wireless systems with their wired counterparts. The need to maintain system and service architectures, and the protocols' definitions derived from the wired networks counterpart, has brought adaptive solutions on the wireless side. The concept of *adaptation* applied to existing protocols has become the alternative to a complete redesign of the wireless protocols. By designing *adaptive protocols* the system integration should be as much transparent as possible to the final users, devices, and service providers, both on the wired and on the wireless side.

25.2.1 Adaptive Protocols and Cross Layering

It is widely recognized that the dynamic nature of the wireless link demands fast and low-cost adaptation on the part of protocols [8,14,15,28]. As an example, due to wireless channels characteristics, mobility of users and frequent fluctuations of radio parameters may stress the adaptive behavior of protocols. Therefore, the study of tuning knobs of adaptive protocols is an important issue already in the protocol design. It is also required to understand problems such as excessive overheads, stability, and fairness problems that one might encounter with adaptive protocols.

All network layers will require the ability to adapt to variable channel conditions, perhaps implemented through some form of channel state estimation and tracking. Solutions have been proposed for an appropriate suite of adaptive, event-driven protocols that pass state-information across all the layers in an effort to cope with this variability. Little is known about this new approach in the protocol design, and considerable research is required here, although a large payoff potential is expected [13,15]. As an example, dealing with new assumptions of wireless scenarios and the effects of such new assumptions on the *medium access control* (MAC) protocol design, this chapter illustrates the evolutionary design perspective of the class of distributed, contention-based (random-access) MAC protocols. A distributed, contention-based MAC protocol is the basic access scheme in today's IEEE 802.11 MAC definition.

Recently, the need for adaptive behavior of protocols, based on the information exchange between the OSI protocol layers, has evolved to the idea of a collapse of the OSI layering structure for the wireless world (i.e., the *cross-layering* principle). Motivations and criticisms emerged that made the cross-layering principle in the design of protocols double-edged: the need for adaptive behavior of protocols based on the exchange of information among the protocol layers is quite clear and consolidated. On the other hand, a warning on the risk of unstructured and "spaghetti-design" principles for wireless scenarios, and the correlated risks for cyclic design solutions and unstable protocols, were recently discussed in the literature [49].

25.2.2 WLANs and MANETs

Wireless LANs (WLANs) are going to integrate and replace wired LANs at a fast rate because the technology solutions are becoming less expensive and acceptable under the performance viewpoint. The WLAN infrastructure is based on static access points (APs) serving and managing local (mobile) hosts. If hosts leave the WLAN area they should register to a new AP, if any.

On the other hand, new classes of wireless networks, such as the MANETs or infrastructure-less networks have no counterpart in today's networks. MANETs are composed of a set of mobile hosts (MHs), possibly connected to each other in a best-effort way, through one (single-hop) or more (multi-hop) communication links. The transmission range of an MH is limited, and the topology of the network is dynamic, such that multi-hop communication is necessary for hosts to communicate with each other. Basic assumptions in current wired networks, including the notions of a quasi-permanent fixed topology and stable links, might not apply to such new networks. The dynamic nature and topology of the MANETs challenges current MAC and routing techniques, and requires a more autonomous style of *peer-to peer* (P2P) network management than today's centralized, stationary systems.

The problem of how such a network self-organizes and responds to host mobility, channel interference, and contention requires solutions. Moreover, the rapidly variable number of hosts in such a network highlights the need for a level of scalability not commonly present in most of the existing approaches to network management. For example, a burst of users with mobile devices that moves, at a given time instant, in the same meeting room may generate a sharp increase in the traffic of the corresponding WLAN. Although the number of hosts that can be connected to a WLAN may be large, wireless links will continue to have lower capacity than wired links, hence congestion is more problematic. The protocol scalability influences the QoS perceived by the users and the resources' utilization (mainly battery energy and channel bandwidth). Since energy and bandwidth are limited resources in wireless networks, there should be a focus on protocols able to reduce their use.

By focusing on the MAC layer, the objective is to make the most effective use of the limited energy and channel spectrum available, while supporting the distributed services that adequately meet the QoS requirements of the users' applications accessing the communication network. Accordingly, researchers will have to gain a better understanding about how to design the MAC, data link, network, and transport protocols, and their interactions, for these networks [4]. Furthermore, the multiple access technique in use on any wireless subnetwork should allow flexible and efficient internetworking with both WLANs and MANETs. In this way, MANETs could be adopted to extend the WLAN coverage areas (e.g., WLAN hot-spots) [26]. At the upper layers, protocols should allow heterogeneous systems to communicate, and permit integration between the wireless and the wired backbone network.

The success of WLANs is connected to the development of networking products that can provide wireless network access at a competitive price. A major factor in achieving this goal is the availability of appropriate networking standards. WLANs experienced an explosive growth and user demand in recent years. The IEEE 802.11 Standard (Wi-Fi) technology has become a *de facto* standard for the MAC layer in such networks. This fact led the research in the field of WLANs and MANETs to be mainly focused on IEEE 802.11-based MAC sloutions.

25.2.2.1 The MAC level perspective

In a wireless network, given the broadcast nature of the wireless transmission, mobile hosts can access a shared channel with their wireless transmissions which may be detected by all neighbor hosts within a given range. In WLANs and MANETs, the MAC protocol is the main management component that determines the efficiency in sharing the limited communication bandwidth of the wireless channel. At the same time, it manages the congestion situations that may occur inside the network. The MAC protocol definition and tuning is essential in providing an efficient resource allocation and power saving, among competing hosts. The MAC layer definitions play an important role also for QoS, real-time services, unicast and multicast delivery, load distribution, fairness, and priority-based channel access.

Centralized MAC protocols are based on the support of a centralized coordinator, as an example, a base station or AP coordinating the channel accesses. The centralized access scheme can support QoS, priority schemes, asymmetric channel scheduling among coordinated hosts, but it may suffer system dynamics like mobility, load changes and could result in complex management and resource waste. The need for a central coordinator is a strong assumption that can be acceptable in infrastructure-based and static WLAN systems. On the other hand, it is not a reliable choice in infrastructure-less and mobile ad hoc networks.

Distributed MAC protocols realize less critical implementations, defined to work under peer-to-peer management conditions thereby resulting in easy implementation without a need for coordinating hosts. For this reason, common choices for WLANs and MANETs are based and realized by distributed MAC protocols. Unfortunately, distributed MAC protocols, have been demonstrated to suffer low scalability, low efficiency, and QoS problems, under high communication load. Among the distributed MAC protocols, two classes of protocols can be identified: reservation-based schemes and contention-based (random-access) schemes. Reservation-based schemes are realized with the assumption that hosts should agree on the order and duration of their respective channel accesses before to try any access. In centralized schemes, this policy can be easily implemented by the central coordinator, which collects requests from the wireless hosts and generates a shedule of the channel accesses. Channel access is governed by the central coordinator by means of a polling-authorization mechanism. In distributed schemes, this policy is much more difficult to realize due to the absence of the central coordinator. Two common approaches can be adopted to realize the reservation-based access in a distributed way: the explicit reservation (i.e., the static list approach) and the implicit reservation (i.e., token-based approach). The explicit reservation approach is based on the creation of a static ordered list of hosts and the duration of their respective accesses. It can be adopted when the number of hosts and the traffic requirements are stable. Such an approach is really unpractical in wireless and mobile systems because it cannot adapt to the system dynamics, and it may result in a waste of resources. In the token-based approach, a message called *token* circulates and is captured in mutually exclusive way by hosts organized in a cyclic sequence. The host receiving the token owns the right to transmit for a given time after which it must pass the token to the next host in the list. This access scheme has been considered for wireless systems, but the risk to lose the token, and the need for distributed failure-tolerance and management issues, made the implementation of this choice quite complex and unpractical for wireless networks.

The distributed, random-access (or contention-based) MAC protocols have been considered as a good compromise to balance the ease of system management, resources' utilization, and performances, in many wireless systems. The idea behind such MAC protocols is to define distributed protocols as event-based algorithms randomly spreading the accesses of hosts in an effort to obtain system stability, acceptable resources' utilization, and performances, as the aggregate behavior of hosts. The events governing the distributed, contention- based MAC protocols are represented by the limited feedback information perceived by the network interface of every host. In next sections we provide a historical perspective of proposals based on different assumptions about the feedback information that could be exploited by hosts. The efficient implementation of distributed MAC managements in MANETs would require every MH to obtain the maximum information regarding the neighbor hosts, if any. This information could be adopted by clustering and routing protocols, and in the MAC contention control as well. As we see in the next sections, information gathering may be a complex activity in WLANs and MANETs, and it is subject to many biasing effects. The increase in the confidence level of the information obtained at the MAC layer is subject to inverse trade-offs with resources' utilization, power control, and power saving principles.

A complete taxonomy of the possible MAC protocols and management techniques proposed in recent years for the wireless scenario is out of the scope of this chapter. In this chapter, we provide the reader with a historical perspective and a state-of-the-art illustration of examples and solutions (with some milestones) that have been proposed in the field of the distributed MAC layer protocols for wireless and mobile ad hoc networks.

25.3   The IEEE 802.11 Standard

In this section, we present the essential information of the IEEE 802.11 Standard for WLANs, which is required for the analysis of some problems in the congestion reaction and power saving mechanisms proposed for the Standard protocols. The IEEE Standard 802.11-1997 and successive releases specify a single MAC sublayer and three physical layer specifications: frequency hopping spread spectrum (FHSS), direct sequence spread spectrum (DSSS), and infrared (IR) [32]. The physical layer is out of the scope of this chapter. Two projects are currently ongoing to develop higher-speed PHY extensions to 802.11 operating in the 2.4-GHz band (project 802.11b, handled by TGb) and in the 5-GHz band (project 802.11a handled by TGa), see Reference 33. Many other task groups have been created to study and design solutions for many issues. As an example, the IEEE 802.11 TGe is studying MAC layer solutions oriented to MAC enhancements for QoS and access differentiation mechanisms [33].

The IEEE 802.11 WLAN is a single-hop, infrastructure network. In addition, it is emerging as one of the most promising technologies for realizing multi-hop MANETs. The current definition of the IEEE 802.11 MAC is not ideal under this viewpoint: the following sections illustrate that further analysis and enhancements are required to capture more system characteristics under the optimized protocols' design at the MAC and LLC layers. Specifically, support for multi-hop communication, flow synchronization, power control, and power saving would require additional work at the MAC layer [88]. Anyway, the

IEEE 802.11 Standard can be considered the prototype standard definition, and the basis for prototype implementation of MANETs. It has become a reference both for practical implementations and for the research in this field. In the following section, we first provide an overview of distributed contention control management in IEEE 802.11 networks (specifically the DCF).

25.3.1 Distributed Foundation Wireless Medium Access Control

In the IEEE 802.11 systems considered in this chapter, the *distributed foundation wireless medium access control* (DFWMAC) *Protocol* includes the definition of two access schemes, coexisting in a time-interleaving super-frame structure [32]. The DCF is the basic access scheme and it is a distributed, random-access MAC protocol for asynchronous, contention-based, distributed accesses to the channel. On top of the DCF, an optional PCF is defined as the scheme to support infrastructure-based systems based on a central coordinator (i.e., AP) for centralized, contention-free accesses.

Stations can operate in both configurations, based on the different coordination function:

- *Distributed Coordination Function (ad hoc network)*: The MHs exchange data like in a P2P communication, and there is no need for fixed infrastructures to be installed. The DCF is completely distributed, and the channel access is contention-based. Stations in such a configuration realize an independent basic service set (IBSS). Two or more IBSS in communication via an intermediate wireless station realize the multi-hop communication between different IBSS.
- *Point Coordination Function (infrastructure network)*: The MHs communicate to APs that are part of a static distribution system (DS). An AP serves the stations in a basic service set (BSS) by implementing a centralized control of the system. The access method is similar to a reservation-based polling system and exploits the coordinator AP to determine the transmission scheduling of many MHs.

The basic access method in the IEEE 802.11 MAC protocol is the DCF that is a *carrier sense multiple access with collision avoidance* (CSMA/CA) MAC protocol. In few words, this means that an MH listens to the channel before it transmits. Since this protocol is the main focus of this work, it is separately considered in detail in the following sections.

Before taking control of the channel and transmitting, each MH in the IBSS (including the AP, if any) is associated with an amount of idle time (i.e., the Inter Frame Space (IFS)) that the MH must wait while performing a carrier sensing of the channel state. If an MH senses transmission of another station on the channel during its own IFS, it defers its own transmission. The result is that the MH with the shortest IFS is allowed to take control of the channel. The basic assumption of this access scheme is that the state of the channel is consistent for all the stations sharing the same channel. As we illustrate in next sections, this is not always guaranteed, due to hidden terminals in the space domain. At least three possible levels of priority are considered, respectively associated to three IFS intervals: in increasing order, short interframe space (SIFS), PCF interframe space (PIFS), and DCF interframe space (DIFS).

The SIFS is the shortest interframe space, and it is implicitly assigned to an MH which is expected to transmit immediately by the context of the communication process: for example, to send an acknowledgment for a received frame, or to send a frame after a polling signal is received by the AP.

The PCF is supported on the top of the DCF by exploiting the PIFS. The PIFS is the intermediate IFS (SIFS < PIFS < DIFS), and it is assigned to the Access Point (AP) only. Whenever the AP wants to take control of the channel, it waits up until the first idle PIFS time appears on the channel, and immediately transmits to take control of the channel. The AP gains control of the channel and maintains it until it leaves a DIFS of idle time to elapse on the channel (see Figure 25.1). The assumption under the PCF management is that only one AP is coordinating the channel, and that there is no contention present among the APs.

The DIFS is the longest interframe space, and it is the enabler IFS to start the DCF phase. After every transmission, all MHs under the DCF control must wait for a DIFS idle time on the channel before they start the contention for the channel (see Figure 25.1). Contention-based accesses are performed

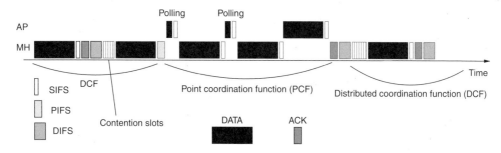

FIGURE 25.1   IEEE 802.11 DFWMAC superframe structure.

by peer-MHs by adopting collision avoidance and contention control schemes. This short description of interframe spaces is sufficient for the considerations that are presented in this chapter, but it is not exhaustive. Interested readers can address Reference 32 for further details.

In the next sections we concentrate our analysis on the DCF collision avoidance and contention control, which may be affected in significant way by the congestion problem due to distributed random-access characteristics. The PCF may be affected by the contention problem as well, in an indirect way. The transmission requests from the MHs to the AP are performed in DCF frames and are subject to contention-based accesses. In other words, the DCF access scheme is considered the basic access scheme in IEEE 802.11 networks, hence its optimization is a relevant research activity for both DCF and PCF access schemes.

25.4   Background and Wireless Assumptions for MAC Protocols

In this section, we summarize some of the assumptions and characteristics that have been considered for the design, tuning, and implementation of distributed, random-access MAC protocols in wireless scenarios. Most of the following assumptions can be considered as the new issues in the MAC protocol design, which made most the successful solutions for the wired scenarios to become unpractical for the wireless scenarios.

25.4.1   Wireless Signals

Wireless signals can be used to code the information in many ways for transmission. Coding techniques are out of the scope of this chapter, and more information on this topic can be found in Reference 63. From the physical viewpoint, wireless signals are electromagnetic waves that propagate in the open space all around from their sources (like the light from a lamp). This phenomenon denotes the physical "broadcast nature" of the wireless transmission, that is, signals cannot be restricted on a wire, but they cover all the area around the transmitter. It is quite clear how this assumption is to be considered in the MAC design, which is devoted to manage the channel capture. The way the wireless signals propagate can be described in many ways by adopting *propagation models*. Propagation models may define the combined effects of the medium characteristics, environment obstacles, and transmission power of the signal source (the wireless transmitter). The transmission power of wireless signals (Ptx) is subject to natural decay when the signal is propagated on the medium. The more the distance d from the transmitter, the lower the residual power (Prx) for the signal being detected by a receiver (in the order of $Prx = Ptx/d^k, k \geq 2$) (see Figure 25.2). In general, network interfaces can be managed to transmit signals with a variable transmission power Ptx. In the receiving phase, network interfaces can be summarily described by means of the *receiving threshold Rth* and a *carrier-sense thresold Cth* [63] (see Figure 25.2). For every transmitter–receiver pair, if the receiving power perceived for the ongoing transmission is greater than Rth, then a data communication is possible between the sender and the receiver. If the receiving power is greater than Cth and lower Rth it would be sufficient for detection and carrier-sensing. Signals lower than Cth would be considered simple interference. To allow a communication (i.e., a communication link establishment) between a sender

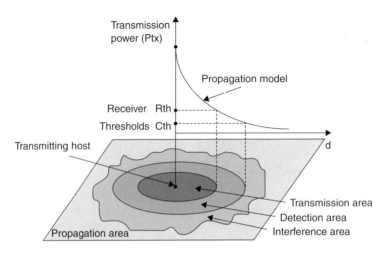

FIGURE 25.2 Transmission power, propagation, and coverage areas.

and a receiver it would be required to increase the transmission power of the sender, or to reduce their relative distance d, depending on the receiver's sensitivity thresholds. A common assumption required for most of the MAC and LLC protocols is to obtain bidirectional links: the communication must be possible on both directions, from sender to receiver and vice versa. This assumption must be considered also when heterogeneous devices, with different sensitivity thresholds, and different transmission power levels coexist in the same scenario.

A *carrier sensing* (CS) mechanism can be briefly described as the physical capability of the network interface to detect transmission and to send a signal to the MAC layer indicating the event: "some signal is being detected on this channel." Reception and carrier sensing events can be detected by the network devices and can be made available to MAC protocols to support local management of the transmission.

Given any transmitter–receiver pair, the *transmission (coverage) area* of a wireless transmitter (see Figure 25.2) is the area where the wireless signal propagates and can be correctly detected and decoded (i.e., transmission is possible with few or no errors due to interference). It should be noted that this area depends (i) on the transmission power of the transmitter, (ii) on the propagation characteristics of the medium, (iii) on the reception threshold (sensitivity) of the receiving network interface (Rth), and (iv) on the amount of interference (noise) caused by other transmission and by environment factors [63]. Only the receiver (i.e., not the sender) knows if the transmission has been received or detected, hence the transmission area cannot be completely determined by the transmitter's properties only.

The *detection area* (see Figure 25.2) of a wireless transmission device is the area outside the *transmission area* where the signal propagates, and where it can be detected by a carrier sensing mechanism (see below), without being necessarily decoded (i.e., $Cth \leq Prx \leq Rth$). In the detection area, a mobile receiver can sense the wireless medium as busy without being able to decode the received signals. The existence of this area in the system, for each transmitter, should be considered for the evaluation of detailed carrier sensing and MAC level effects, such as exposed terminals, hidden terminals, capture effects (described in the following section) and so on.

The *interference area* of a wireless transmission device is defined as the area where the residual wireless signal propagates without being detected or decoded by any receiver. Interference adds noise to any ongoing transmission whose intended receivers are within the interference range. The cumulative effect of multiple interferences might add errors to the bits of information transmitted to receivers.

25.4.2 Collision Domains, Capture Effect

For a tagged receiver X, a *collision domain* could be defined as the set of the transmitters whose coverage areas may contain X on a single shared communication channel. Two or more concurrent transmissions

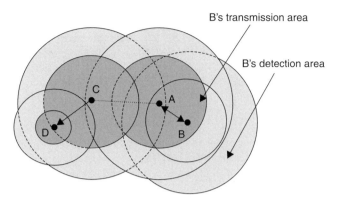

FIGURE 25.3 Example of a collision domain.

in the same collision domain would cause an interference problem for the tagged receiver X. Given the collision domain example shown in Figure 25.3, D can only receive C, C and A can detect each other but cannot receive their respective transmissions, and A and B can receive (and detect) each other. Interference areas are not shown in Figure 25.3: it could be assumed that every host is in the interference range of each other. A *collision* happens on a given receiver when two or more portions of concurrent transmissions overlap each other in the time and space domains by causing decoding errors for the information received. If the received signal power of one of the colliding signals is much more greater than others, it may happen that the receiver is able to capture such a transmission anyway (e.g., A could capture B's data while C transmits). In this case, a *capture effect* of signals can be exploited at the MAC layer. Otherwise, collisions of concurrent signals transmitted on the same collision domain may cause a destructive interference for detected signals on the receiver. The goals of an MAC protocol policy is to avoid such collisions and to adapt the density of transmissions, that is, the contention level. The contention level can be thought as the risk to cause a collision if a transmission is performed. In the next sections, we illustrate other problems and issues for the MAC design.

25.4.3 Half Duplex Channels

Once the link existence is established at the physical transmission level, the MAC protocol should manage the link properties inherited from the physical layer.

A single wireless communication device, that is, a wireless network interface (NI), can transmit and receive data in separate time windows. There is no practical way for a single NI to transmit and receive at the same time on the same wireless channel. Hence, a single wireless communication channel between any two hosts, A and B can be used only in *half-duplex mode* for communications, that is, to send data from A to B or vice versa, but not both ways at the same time (see Figure 25.3). In other words, an NI cannot "listen" to receive communications while it is transmitting on the same channel. This behavior could be technically possible at the hardware level (e.g., by using two NIs), but the result would be nullified because the transmitted signal would be, in most cases, much more powerful in the proximity of the device than every other received signal. This is due to the physical propagation laws for wireless signals.

25.4.4 Collision Detection

A collision detection (CD) technique is based on the concept that a transmitter can listen to the outcome of a channel while it is transmitting a signal on the same channel. If the transmitter hears a collision overlapping its own ongoing transmission, then it can immediately interrupt the transmission. The half-duplex characteristic of wireless channels is one of the most critical and limiting assumptions to be considered in the design of MAC protocols for the wireless scenario. As a consequence, CD techniques based on the "listen while transmitting" concept and adopted in wired LANs MAC protocols (e.g., IEEE 802.3 and Ethernet) cannot be implemented on a single wireless channel. The only way to obtain a similar function in wireless scenarios would be by adopting a couple of network interfaces and a couple of channels: while A sends data to B on the DATA channel, if a collision is detected, B may send a jamming signal to A on the CD channel by causing an early stop in the DATA transmission.

25.4.5 Full-Duplex Links

A bidirectional full-duplex link can be obtained by adopting duplexing techniques such as time division duplex (TDD) or frequency division duplex (FDD). TDD creates a logical abstraction of a full-duplex link by splitting the transmission and reception phases over consecutive, nonoverlapped time intervals on a single half-duplex (physical) channel. The FDD technique consists of adopting two physical channels: one for transmission and one for reception. In most WLANs and MANETs, logical (bidirectional) links are commonly defined as time division duplex channels. All data transmissions and receptions have to be in the same frequency band, since there are no "bridge" hosts (or base stations) to translate the transmissions from one physical channel to another. This usually requires strict time synchronization in the system and MAC protocols to be defined accordingly [17].

In the following section, we describe the principles for distributed MAC protocols proposed for WLAN and MANET scenarios. It is worth noting that three important and leading factors determine the MAC protocol definition: time, space, and power (energy). In few words, a distributed MAC protocol should locally manage the time schedule of transmissions, depending on the variable traffic load requirements in such a way so as to avoid collisions on the receivers and to exploit to the maximum degree the spatial-reuse of the limited channel resource. On the other hand, any reduction of energy consumed for transmission and reception is another critical point for battery-based devices at the MAC layer. In background to the factors cited above the effect of the mobility of users resulting in highly dynamic collisions domains and contention levels should be considered.

25.5 Evolutionary Perspective of Distributed Contention-Based MAC

This section describes the evolution of proposals in the field of distributed, contention-based (multiple-access) MAC protocols for the wireless scenario. The list of proposals is not exhaustive due to space reasons. It is an incremental illustration of the milestones and the evolution of the protocols leading to the IEEE 802.11 DCF MAC design. Specifically, given the focus of the chapter, this perspective is oriented to the contention control and power saving issues in the distributed contention-based MAC protocol design.

25.5.1 Distributed Contention-Based MAC Protocols

The following MAC protocols have been incrementally defined to reduce the vulnerability of distributed, contention-based transmissions over the same wireless collision domain. The *vulnerability* of a tagged frame f being transmitted to a receiver Y could be defined as the maximum size of the time window (containing the tagged frame f) during which another frame may be transmitted on the collision domain of Y, by originating a collision. Given that the structure of collision domains is dynamically determined by properties of current transmissions and topology of mobile hosts, the vulnerability problem involves both the time domain and the space domain (who transmits when? And who transmits where, in the collision domain of the candidate receiver Y?).

25.5.1.1 The Aloha MAC protocol

The pioneering MAC protocol for distributed, multiple-access wireless transmission of data frames (called packets) is the *Aloha* protocol [1]. In this protocol the CS concept is still not considered, that is, every MH is not assumed to "listen" to the channel before transmitting. The MAC protocol policy is simple: data in the

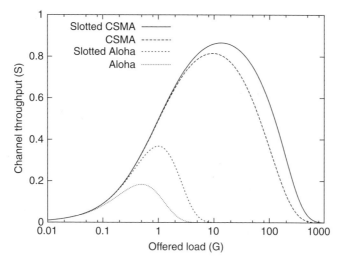

FIGURE 25.4 Analytical investigation of contention-based MAC throughput.

buffer queues is transmitted immediately, whenever it is ready. If the transmission originates a collision, CD is not possible, and the transmission attempt is completed up to the end of the data frame over half-duplex channels. After the transmission, an *Acknowledgment* (Ack) is sent back (e.g., on a separate channel) to ascertain a successful transmission. The transmitter waits for a maximum amount of time (Ack timeout) for the Ack. The bidirectional "Data + Ack" structure of the frame transmission is conceived to realize the abstraction of a reliable logical link layer. In case of an unsuccessful transmission (i.e., missing acknowledgment after the Ack timeout), a new transmission attempt is required till a corresponding Ack is received. In order to avoid synchronization of retransmission attempts which create new collisions, every retransmission is scheduled after a pseudo-random waiting time.

The vulnerability period for each frame (by assuming constant frame size) in the Aloha access scheme is two times the average frame size (expressed in time units) [51]. By assuming (i) independent Poisson-distributed arrivals of frames' transmissions (with constant size) and (ii) collisions as the only source of transmission errors, the expected average channel utilization is upper bounded by only 18% of the channel capacity (i.e., the maximum channel throughput) [51]. In other words, by increasing the transmission load offered by independent MHs (load G is the frame size multiplied by the Poisson inter-arrival rate), the probability of a collision increases, and the MAC policy would not be able to provide an average channel utilization greater than 18% (see Figure 25.4). This theory result well describes the scalability limits of this MAC policy under the time vulnerability and distributed contention control viewpoint.

25.5.1.2 The Slotted-Aloha MAC protocol

The *Slotted Aloha* protocol is introduced to limit the vulnerability of each frame, with respect to Aloha. The new assumption is that time must be divided in *frame slots* (with fixed frame size), each one able to contain a frame transmission. This protocol is similar to Aloha, but a quantized synchronization of MHs is assumed such that every transmission starts only at the beginning of a *frame slot*. In this way, the vulnerability period is limited by the single frame slot where the transmission is performed. Analytical models demonstrate that the average channel utilization is upper bounded by 36% of the channel capacity, that is, two times the Aloha value (see Figure 25.4). This theory result shows that the scalability of this MAC policy is better than Aloha, but is still far from the optimality.

25.5.1.3 The pure CSMA MAC protocol

The carrier sense multiple access (CSMA) concept [51] has been introduced to further reduce the frame vulnerability. In the CSMA MAC, every host "listens" to the channel before transmitting: if the channel is busy then the host defers the transmission to a later time (this is called nonpersistent carrier sensing), otherwise the host transmits immediately. The advantage of this policy is that ongoing transmissions in the range of the candidate transmitter can be detected immediately before the transmission, hence the candidate transmitter could avoid causing collisions. Unfortunately, if the propagation delays are significant with respect to the frame size, the performance of CSMA would be negatively influenced. The propagation delay temporarily hides the new transmissions to other potential transmitters, whose respective transmissions may cause collision on the intended receivers depending on the structure of collision domains. For this reason, the vulnerability of the frames is demonstrated to be two times the maximum propagation delay $(2 * \tau)$ of wireless signals, between any two distant transmitters. As an example, the transmission of a transmitter X starting at time t_x is exposed to the risk of collisions with possible transmissions started at time $t_y \geq (t_x - \tau)$ by any transmitter Y (whose transmission still has not been detected by X). Similarly, the transmission from X is exposed to the risk of collision with any possible transmitter Z whose transmission will start at time $t_z \leq (t_x + \tau)$ (i.e., before z can detect the ongoing transmission from X). The propagation delay τ is usually considered orders of magnitude lower than the size of the typical fames [51]. This is more probable for common WLAN and MANET scenarios. The upper bound of the average CSMA throughput is defined as high as 80% of the channel capacity (see Figure 25.4).

25.5.1.4   The Slotted-CSMA MAC protocol

By applying the slot-based concept to CSMA, the *Slotted-CSMA* protocol is proposed as a further enhancement of CSMA [51]. A *minislot* is defined as the upper bound of the propagation delay between different transmitters in the system (τ). In the Slotted-CSMA protocol, every host with a frame to transmit "listens" to the channel at the beginning of the current *minislot*. If the channel is busy, then it defers the transmission to later time, or else it transmits immediately at the beginning of the next minislot. The advantage of this policy is that the beginning of candidate transmissions are synchronized up to a minislot-quantized time. New transmissions originated by transmitter X are guaranteed to be detected, at worst, at the beginning of the next minislot by all the transmitters in the space proximity of X. In such a way, the vulnerability of the frame is demonstrated to be reduced to a single minislot time (τ). If the propagation delays are significant with respect to the frame size, the performance of Slotted-CSMA would degrade. The upper bound for the Slotted-CSMA throughput gives better values than the CSMA channel capacity, under the same scenarios (see Figure 25.4)

25.5.2   Collision Detection in Wireless Systems

All the MAC policies described so far are assumed to receive, directly from the intended receiver, an Ack-based feedback indication of the successful transmission. The Ack is usually provided as a short reply frame, and it can be exploited to realize the link layer concept of "reliable link" between a transmitter and a receiver.

In some systems, and in early wireless MAC proposals, Acks were sent on separate control channels. In recent solutions, the Ack transmission is piggybacked by the data receiver immediately following the end of the data reception, over the single, shared, half-duplex communication channel. In this way, at the MAC/LLC layer, the receiver could immediately exploit the contention won by the transmitter for sending the Ack frame (i.e., a new contention is not required since the shared channel has been already successfully captured by the sender).

Different policies and definitions of the MAC and LLC layers can be defined by assuming to receive the explicit motivation for unsuccessful transmissions (e.g., if a frame was received with errors, if it was subject to collision, etc.). Anyway, the LLC layer is out of the scope of this chapter. The knowledge of the reasons for the unsuccessful transmissions has been considered in the literature, and analysis has shown that the more information feedback is provided on the cause of a contention failure (i.e., collision, number of colliding MHs, bit error due to interference, etc.), the more performances and adaptive behavior can be obtained by the MAC protocol. Unfortunately, in most scenarios, the only feedback information obtained after a transmission attempt is the existence of Ack frames within a timeout period.

As mentioned before, in wireless systems, CSMA and slotting techniques can be exploited to reduce the vulnerability of frames being transmitted. On the other hand, when the frame transmission starts, there is no easy way to detect early if a collision is occurring at the receiver MH. Under these assumptions, CSMA/CD schemes such as Ethernet and IEEE 802.3 cannot be exploited in wireless MAC protocols. In early wired LANs, researchers suggested solutions based on CSMA techniques, where the implementation of collision detection mechanisms in wireless scenarios has been investigated, for example, see References 57,55,65,69. Given the characteristics of wireless systems, the only practical way to obtain the equivalent of collision detection is the adoption of separate signalling channels and multiple NIs. This would require twice the channel bandwidth, power consumption, and NIs than other mechanisms. We see in the following sections how researchers defined new MAC policies for the wireless world that could be considered almost equivalent to the collision detection schemes, under the channel reservation and channel utilization viewpoint.

25.5.3 Collision Avoidance

The Aloha and CSMA MAC protocols were thought for the reduction of the vulnerability time of the distributed, contention-based frame transmissions, without recourse to CD mechanisms. The Collision Avoidance (CA) techniques have been designed in order to create the same conditions provided by CD, still using a single shared channel and a single NI. This translates in the following concept: under CA schemes, if a contention-based transmission would result in a collision to the intended receiver, then the amount of energy and channel occupancy wasted by the colliding transmissions should be as much limited as possible, like in the CD approach.

Basically, CA can be thought as a preliminary spatial reservation of the collision domain between a sender and a receiver, in order to preserve the whole successive data transmission. The preliminary spatial reservation can be performed by resolving the channel contention among multiple transmitters in the neighborhood of both the sender and the receiver. The cost of this reservation should be comparable with the cost of collisions under the CD schemes. Before of illustrating the proposals for collision avoidance, we are going to define the most representative problems to be considered at the MAC layer by introducing a little more detail on the space-contention viewpoint.

25.5.3.1 Hidden and exposed terminals

The dynamic topology of wireless ad hoc networks, the adoption of shared channels for transmissions, and a carrier-sensing-based policy for the MAC protocol implementation may bring some MHs to experience the *hidden terminal* and the *exposed terminal* problems. These problems happen for an MH receiving the concurrent transmissions of at least two other neighbor MHs, respectively hidden to each other. In such scenarios, any time-overlapping of concurrent transmissions on the receiver may result in a *collision* that has a destructive effect. It is worth noting that collisions happen only on the receivers, and may depend on many local factors such as the threshold levels (sensitivity) for reception, the interference, and the residual energy perceived on the receiver's network interface.

As an example, let us suppose A is within the *detection area* of C and vice versa, B is within the *transmission area* of A and vice versa, and C is outside the *detection area* of B (see Figure 25.5). In this scenario C senses the transmission of A, that is, it senses the channel as busy, but it cannot decode the transmissions. This condition illustrates the *exposed terminal* condition for C with respect to A [17]. Exposed terminals are terminals (like C) whose transmission is not allowed (e.g., by a CSMA MAC policy over C's collision domain) due to exposure to irrelevant transmissions, (e.g., from A to B). The exposed terminals' problem is the cause of a limitation of the possible channel reuse.

Another problem is given by *hidden terminals*: due to shadowing effects and limited transmission ranges, a given terminal (like B) could start a transmission toward another terminal A (B and A are within each other's transmission area, see Figure 25.5), while A is receiving signals from a terminal C hidden

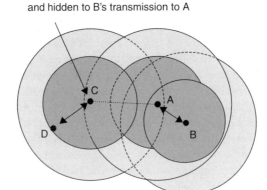

FIGURE 25.5   Example of hidden and exposed terminal.

to B. This means that A cannot complete any reception due to the destructive collision of signals from the respectively hidden terminals B and C. It may happen that A could detect and isolate one of the colliding transmissions, (e.g., from B to A); in this case, we obtain a *capture effect* of transmission from B to A, despite C's interference and collision. A discussion of details for hidden and exposed terminals and capture effect can be found in References 17 and 61.

All CSMA-based solutions are implemented by transmitters [51], while collisions happen on the receivers. This asymmetry makes the hidden terminal problem a practical counter-example showing that CSMA is not sufficient for transmitters to avoid collisions in the system. In scenarios with hidden terminals, the throughput of CSMA may fall to low channel capacity. The hidden terminal problem was discussed quite early, in 1975, by Tobagi and Kleinrock [75], who proposed the busy tone multiple access (BTMA) protocol. In BTMA, a separate signalling channel is used to send a busy tone whenever an MH detects a transmission on the data channel. In other words, information about the (local) occupancy of the channel at each receiver is forwarded by a busy tone signal to neighbor MHs. Every MH should perform a CSMA technique on the signalling channel before sending a frame. The major drawback of the BTMA is that it would require a separate channel and a couple of network interfaces to perform carrier sensing on the busy-tone channel while transmitting on the data channel.

25.5.3.2 The Request-to-Send and Clear-to-Send Scheme

A more convenient and simple way to limit the hidden terminal problem, among wired terminals connected to a centralized server, was suggested in the split-channel reservation multiple access protocol (SRMA) in 1976 [76]. The solution is based on the handshake of short messages Request-to-send/Clear-to-send (RTS/CTS) between senders (i.e., terminals) and receivers (i.e., the server) over three separate channels (RTS, CTS, and DATA channels, respectively). The RTS/CTS scheme was originally designed to manage the efficient transmission scheduling between senders and receivers without originating interference among hidden terminals on the server [76].

Some time later, the RTS/CTS scheme was interpreted and adopted in quite a different way with respect to SRMA. Basically, RTS/CTS signalling was performed on a single transmission channel, as originally proposed for wired LANs, in the CSMA/CA scheme by Colvin in 1983 [24]. The idea is to send a short RTS frame (by applying the CSMA MAC channel access scheme) to the intended receiver before a data transmission. If the receiver correctly receives the RTS, then it immediately responds with a CTS back to the transmitter. In this way the double successful transmission of both RTS and CTS should reserve the channel (i.e., the collision domain) around both the sender and the receiver against hidden transmitters.

The principle of this solution, modified in opportune way, has been successively adopted, in the years following its proposal, in many protocols and standards. Collision avoidance based on RTS/CTS is an optional function included in many current IEEE 802.11 DCF implementations.

25.5.3.3 Multiple Access Collision Avoidance

RTS/CTS was first introduced in wireless systems by the multiple access collision avoidance (MACA) protocol [48]. MACA is a random-access MAC protocol proposing to limit the hidden terminal problem by taking a step back with respect to the carrier-sensing approach. In MACA, it was considered that the relevant contention is at the receiver side, not at the sender side. This fact suggested that the carrier sensing approach on the transmitter is not fully appropriate for the purpose of implementing collision avoidance. Carrier sensing at the sender would provide information about potential collisions on the sender, but not at the receiver side. Since the carrier sensing mechanism implemented on the transmitter side cannot ensure against hidden terminals, and leads to exposed terminals, the radical proposal in MACA is to ignore carrier sensing as the main indicator for driving channel accesses. The idea is to adopt a slot-based approach and to bet on the preliminary contention of two short frames, RTS and CTS, before sending the (long) DATA frame. RTS and CTS are adopted in the preliminary attempt to reserve the coverage area (i.e., the collision domain) between the sender and the receiver. In MACA, both RTS and CTS are 30 bytes long, and they contain the information about the expected duration of the channel occupancy for the following data transmission. The RTS and CTS duration defines the "slot time" for quantizing transmissions.

The main critical assumption in this definition is the perfect symmetry of links, that is, both the sender and the receiver must detect and receive their respective transmissions. The transmitting MH sends the RTS to the receiver as a broadcast message. Upon reception to the RTS, if the receiver is not deferring due to a previous reservation, the receiver immediately responds with the CTS. The CTS also contains information about the expected duration of the channel occupancy. Any MH different from the sender receiving the CTS (meaning that the MH is in the critical range for the receiver) would be informed about the transmission duration, and it would enter the deferring phase for that interval. If the sender receives the CTS, it also receives the indication that the receiver is within the transmission range, and that the channel should have been reserved successfully. This indicates that the data transmission could start with a good probability to be successful.

All MHs receiving the RTS, different from the intended sender and receiver, should listen and wait to hear the CTS reply. If they are not receiving the CTS reply, they could assume they are not in the critical range of the receiver (i.e., the receiver is not exposed) and hence they could start their own transmissions by increasing in this way the spatial reuse of the channel. It is worth noting that acknowledgments after the data transmission are not expected on the sender side in MACA.

MACA is considered a first solution for terminals "exposed" with respect to the sender's RTS. In this scheme, the advantage in the channel capture is based on a lower risk to waste channel bandwidth, if a collision occurs, than the risk given by collisions of two or more long data frames. If CTS is not received within a timeout period, the sender assumes a collision occurred and reschedules a new RTS transmission. To select the "slot time" for the new RTS transmission the sender should adopt a contention control scheme (like the Backoff protocol, see Section 25.5.3.4).

Other solutions for collision avoidance are based on reversing of the RTS/CTS handshake. In MACA by invitation (MACA-BI) [74] and in many receiver initiated multiple access (RIMA) versions [79], the RTS/CTS scheme is initiated by candidate receivers sending ready-to-receive (RTR) frames to the neighbors. In this way the collision avoidance scheme reduces the overhead required in some scenarios.

25.5.3.4 The backoff protocol

The backoff protocol is a *contention control* protocol that is frequently associated with contention-based collision avoidance and slotted access schemes. We anticipate here a sketched presentation, even if collision resolution and contention control protocols, and more specifically the Binary Exponential Backoff (BEB) protocol, are described in Sections 25.5.4, 25.5.5, and 25.5.5.1. Whenever a collision is detected by the sender (e.g., by a missing Ack or a missing CTS) this event can be considered as an indication of a high contention level on the channel. A time-spreading and randomization of retransmission attempts is required to reduce the contention, that is, to reduce the collision risk due to the choice of the same slot. The backoff scheme realizes the adaptive contention reduction based on the experience of collisions after frame transmissions. For every new frame transmission, the backoff mechanism is reinitialized. The first transmission attempt will be performed in one of the next (future) slots selected with pseudo-random uniform distribution in the interval [0, $CW\_Size\_min - 1$], where CW_Size_min is the minimum value assigned to an integer CW_Size parameter. The CW_Size and the related slot interval [0, $CW\_Size - 1$] are increased after each collision, up to a maximum value CW_Size_MAX. This incrementally reduces the probability that colliding hosts select another busy slot for transmission. The CW_Size interval is reset to the minimum CW_Size_min after a successful transmission. In the backoff protocol defined in MACA, the CW_Size is doubled after every collision (this is called a Binary Exponential Backoff, BEB), $CW\_Size\_min = 2$ and $CW\_Size\_max = 64$.

25.5.3.5 MACA for Wireless

MACA for Wireless (MACAW) [6] is a modified version of MACA where additional wireless scenario's assumptions play an important role. Note that MAC protocols should deliver high network utilization together with fair access to every MH (i.e., no "one MH takes all" solution). In Reference 6 the unfairness problem of the BEB is described: local experience of collisions on one sender could make it reaching high CW_Size, while other senders could keep winning the contention within CW_Size_min slots. This may translate into unfair throughput levels for peer hosts at the MAC layer. The suggested solution to this problem is to insert in the frames' header the CW_Size value: every MH receiving a frame would locally copy the CW_Size value by obtaining a more fair access distribution density. The multiplicative increase and linear decrease (MILD) algorithm is applied to the CW_Size parameter to avoid wild fluctuations. Also, the concept of "stream-oriented" fairness, instead of station-oriented fairness, is taken into account in Reference 6. As an example, imagine two contending stations: the first one with a single frame queue, and the second one with many frame queues associated to many different running applications. The "contention pressure" on the channel given by the MAC protocols of the two stations is the same, even if the "flows pressure" on the MAC protocols is not fair. The new idea is to adopt one backoff queue per stream with local scheduling and resolution of virtual collisions of frames within the local host. In this way the density of accesses on the channel is not the same for all the hosts, but it is a function of the virtual contention among flows inside each host. Recently, this idea has been reconsidered in the IEEE 802.11e design, leading to a distributed implementation of differentiated accesses to the channel, for flows with different priority levels. New special frames are defined in MACAW to propose new solutions for the synchronization problems and for making the receiver able to contend for bandwidth even in the presence of high congestion [6]. Some optimizations of the BEB and CW_Size-copying algorithm have been proposed (i) based on the observation that the "copying" algorithm works well only in uniform contention scenarios and (ii) based on the assumption to know the motivation for RTS and CTS problems, if any. MACAW introduces the assumption that channel contention in wireless scenarios is location dependent, and some kind of collective cooperation should be adopted in order to allocate the media access fairly. The MAC protocol should propagate contention information, instead of assuming every MH is able to discover such information on a local basis. Finally the MAC protocol should propagate synchronization information about contention periods in order to allow every device to contend in an effective way, for example, by exploiting contention initiated on the receiver side (RRTS).

In MACAW, by following the suggestion given by Tobagi and Kleinrock [77], Appletalk [69], and the early IEEE 802.11 working groups, immediate acknowledgement is introduced after the RTS–CTS–DATA exchange of information. In this way, if RTS–CTS–DATA–ACK fails, immediate retransmission can be performed if the frame was not correctly received at the link layer, for some reason. This is assumed by the sender if the ACK is not received, even if the CTS is received. The retransmission management anticipated at the MAC/LLC layer gives great advantages with respect to a transport-layer retransmission management in wireless networks. This is because the frequent frame errors due to characteristics of the wireless scenario (mainly the high risk of bit error and interference and collisions) can be detected early and managed with respect to end-to-end congestion management reactions implemented at the transport layer.

The immediate ACK required from the receiver to complete the transmission sequence makes the sender to act as a receiver during the RTS–CTS–DATA–ACK transmission scheme. The solution proposed in MACA for reducing exposed terminals presents, now, a drawback for MACAW because concurrent transmitters could interfere with the reception of ACKs. This limits the spatial reuse of the channel that was obtained by the RTS/CTS policy in MACA (where a sender may transmit anyway, if it was receiving another sender's RTS but not the corresponding CTS). It is worth nothing that both MACA and MACAW are not based on the carrier sensing activity at the transmitter before the transmission of the RTS. Also, at least a double propagation-delay time of idle-channel space should be required between the time the channels becomes idle after a transmission and the time of a new RTS transmission, in order to allow for the full reception of ACKs [6].

25.5.3.6 Floor Acquisition Multiple Access

Floor acquisition multiple access (FAMA) [31,30] is a refinement of MACA and MACAW. FAMA completes the wireless adaptation of MACA and MACAW protocols to the wireless scenario by introducing (i) carrier sensing (CSMA) on both senders and receivers, before and after every transmission, in order to acquire additional information on the channel capture, (ii) nonpersistence in the CSMA access scheme (if the channel is found to be busy, a random wait is performed before a new carrier sensing), (iii) lower bound of the size for RTSs and CTSs based on worst case assumptions on the propagation delays and processing time, and (iv) RTS size shorter than CTS (called CTS dominance) to avoid hidden collisions among RTS and CTS. It is worth noting that, after MACAW, the frame transmission is considered complete when the RTS–CTS–DATA–ACK cycle is completed. The need for ACK reception on the sender to complete the handshake implies that both receiver *and* sender must be protected against hidden terminals (as mentioned before, the main drawback of hidden terminals is the collision that may happen on a terminal acting as a receiver).

Since CD is not practical in (single channel) wireless scenarios, in FAMA the carrier sensing approach is extended to both the sender (for RTS) and the receiver (for CTS). The sender and the receiver aim to reserve the "floor" around them, in order to protect the DATA reception on the receiver, and the ACK reception on the sender, against their respective hidden terminals. This conservative approach may give a reduction of long collisions and link layer transmission delays, hence a better utilization of scarce resources such as channel bandwidth and battery energy. The illustration of sufficient conditions to lead RTS/CTS exchange a "floor" acquisition strategy is provided (with and without carrier sensing) in Reference 30.

25.5.3.7 Analysis of collision avoidance schemes

To summarize, the RTS/CTS mechanism has many interesting features and a couple of drawbacks.

First, let us describe the interesting features of the RTS/CTS mechanism: its adoption guarantees that, in most cases, the transmission will be worthwhile because a successful RTS/CTS handshake ensures that (i) the sender successfully captured the channel in its local range of connectivity, (ii) the receiver is active, (iii) the receiver reserved the channel in its local range of connectivity (not necessarily the same area of the sender) and it is ready and able to receive data from the sender, and (iv) the RTS/CTS exchange would allow the sender and receiver to tune their transmission power in an adaptive way (hence, by saving energy and reducing interference). Recent studies have shown that the hidden terminal problem seems to have a lower practical impact than expected (at least by the amount of work on this topic) [85]. On the other hand, as a conservative approach, the RTS/CTS solution is the basis for many other research proposals. Specifically, RTS/CTS exchange could be considered as a milestone solution for MAC in wireless multi-hop scenarios such as in MANETs. Ongoing activities are based on the adoption of directional antennas to implement directional collision avoidance schemes. The idea is to adopt directional antenna beams to reserve the channel over small area sectors between the sender and the receiver. In this way, less energy can be used and more spatial reuse of channel can be obtained. Directional MAC protocols and directional collision avoidance schemes are ongoing research activities.

In ideal conditions (i.e., when the contention is low), the preliminary additional transmissions of RTS and CTS frames, before a DATA frame, would require additional bandwidth and energy than the amount strictly required. One possible solution to this drawback, adopted in IEEE 802.11 networks, is to set a RTS/CTS_threshold defining the lower bound of frame sizes that require the adoption of RTS/CTS exchange. If at least one transmitter needs to send a long frame, whose size exceeds the RTS/CTS threshold, then an RTS message would be adopted to avoid a long-collision risk. The RTS/CTS overhead is not worthwhile if any possible collision would not be exceeding the predefined threshold. With this scheme, the RTS/CTS goal is twofold: (i) a channel reservation is performed to limit hidden terminals and (ii) long collisions can be avoided.

The second drawback is given by a set of worst case scenarios where the adoption of RTS/CTS would not guarantee the successful transmission, due to collisions among RTS and CTSs, and due to the characteristics of interference and propagation of wireless signals [85]. For a description of such worst case scenarios, see, for example, References 6,22,31,70 and 85. In Reference 85, the analytical and simulation-based evaluation of the RTS/CTS solution for ad hoc networks has been performed by assuming IEEE 802.11 DCF MAC protocol. The relevant contribution in this work is given by the analytical investigation of transmission ranges, detection ranges, and interference ranges (in open space) by assuming a two-way ground reflection propagation model [63]. Another important issue being considered is the difference existing between the reception (or detection) thresholds and the interference threshold in current wireless network interfaces [85]. The analysis showed that RTS/CTS handshake would leave a consistent area around the receivers where potential transmitters may not receive any RTS or CTS. Such potential transmitters would not activate their "deferring phases" (i.e., Virtual Carrier Sensing). The interference generated by such transmitters would be sufficient to generate collisions on the receivers despite the fact that they successfully exchanged RTS/CTS [85]. The proposed solution to enchance the RTS/CTS scheme was the Conservative CTS-Reply (CCR) scheme: a quite simple modification of the standard RTS/CTS solution. A conservative RTS_threshold power level is defined as that level that should be reached by the RTS signal on the receiver side in order to allow the receiver to send the CTS back [85]. With this assumption, data exchange is activated only if the transmitter is received with high power, and if capture effect is probable despite any possible interference.

25.5.4 Collision Resolution Protocols

The IEEE 802.11 DCF taken as a reference in this chapter is based on a slotted CSMA/CA MAC protocol. The slot size is kept as low as possible, and it is defined as a function of (i) the maximum propagation delay in the collision domain and (ii) the time required to switch the NI from carrier sensing to transmission phases. While the CA scheme tries to reduce the risk of a collision caused by hidden terminals, the contention control and collision resolution schemes (which may be considered as secondary components of the collision avoidance) are defined in order to reduce the risk of a new collision after a previous one.

In the following we will consider only collisions caused by the selection of the same transmission slot on behalf of more than one transmitter in the range of the receiver. We assume that collision avoidance (RTS/CTS) is performed in background against hidden terminals.

The collisions become more probable if the number of users waiting for transmission on a given collision domain is high, that is, if the channel contention is high. The collision resolution protocols can be defined, similarly to the contention control protocols, to reduce the probability of collision as low as possible, in an adaptive way, with respect to the variable load in the collision domain.

Tree-based collision resolution mechanisms have been suggested in References 16 and 78. A good survey of such protocols can be found in References 46 and 35, and in the related bibliography. The set of k contending MHs is assumed to belong to the initial set $S0$: every MH randomly selects one slot for transmission among the next R slots (with fixed integer value R), called a "round" or a "contention frame." Whenever the feedback information indicating a collision is perceived, all the colliding MHs randomly split into two or more subsets. Every subset will try a new retransmission in a separate subset of R slots (i.e., separate rounds), hence by reducing the contention. Further splitting is performed after every new collision by originating a tree-like structure of subsets thereby giving the name to this mechanism. The main problem with tree-based schemes is to adapt parameters like the number of slots R in each round, and the number of split subsets, in order to maximize the channel utilization, and to minimize the energy consumption and the collision-resolution delay. Different assumptions about the amount of information perceived by the collisions were used to define many tree-based schemes. As an example, by assuming to detect the number of colliding MHs by the residual energy detected, the round length R could be tuned in an adaptive way [50,37]. In neighborhood-aware contention resolution protocol (NCR) [5], some contention resolution protocols were based on the assumption that ever MH knows the IDs of neighbors within two-hops. Such assumptions are quite strong, and in general, collision resolution schemes have not been considered as a practical choice in IEEE 802.11 WLANs and MANETs, based on the CSMA/CA with contention control protocol.

Under light load conditions, collision avoidance and collision resolution protocols achieve the same average throughput of FAMA protocols [35]. A good description and comparison of collision avoidance and collision resolution schemes such as ICRMA, RAMA, TRAMA, DQRUMA, DQRAP, and CARMA can be found in Reference 35.

25.5.5 Contention Control in IEEE 802.11 DCF

From previous sections the reader would have gathered sufficient background knowledge to begin the analysis of the IEEE 802.11 DCF contention control and collision reaction.

The DCF access method is based on CSMA/CA MAC protocol. This protocol requires that every station performs a carrier sensing activity to determine the state of the channel (idle or busy) before transmitting. This allows each station to start a transmission without interfering with any other ongoing transmission. If the medium is idle for an interval that exceeds the DIFS, the station executes its CA access scheme. If the medium is busy, the transmission is deferred until the ongoing transmission terminates, then after the DIFS, a CA mechanism is adopted.

The IEEE 802.11 CA mechanism is based on the (optional) RTS/CTS exchange. Positive acknowledgments are employed to ascertain a successful transmission. Immediately following the reception of the data frame, the receiver initiates the transmission of an acknowledgment frame (Ack), after a SIFS time interval (see Figure 25.1).

If a collision occurs, this event is interpreted in the backoff protocols as an indication of a high level of contention for the channel access. To react to the high contention level that caused the collision, a variable time-spreading of the scheduling of next accesses is performed. Hence, in a contention-based MAC protocol, a channel waste is caused both by collisions and by the idle periods introduced by the time-spreading of accesses (i.e., by idle slots). The reduction of the idle periods generally produces an increase in the number of collisions. Hence, to maximize the channel and energy utilization the MAC protocol should balance the tradeoff between the two conflicting costs [12,11,15,34]. Since the optimal tradeoff changes dynamically, depending on the network load, and on the number of mobile users, the MAC protocol should be made adaptive to the contention level on the channel [20,34,42].

The distributed reaction to collisions in the IEEE 802.11 CSMA/CA MAC is based on a *contention control* scheme realized by the BEB protocol [32,40,42]. The contention control makes the probability of collision as low as possible, in an adaptive way, with respect to the variable load in the collision domain. Backoff protocols have been already sketched with the description of the backoff protocol introduced by MACA in Section 25.5.3.4.

25.5.5.1 The BEB Protocol

In this section we analyze the characteristics of the BEB protocol adopted as a contention control mechanism in the IEEE 802.11 DCF. By assuming that a collision occurred because of the selection of the same slot by at least two contending MHs, a backoff protocol is adopted to control the contention level by exploiting the frame history regarding successes or collisions [42]. Specifically, given the system assumptions, each user is not assumed to have any knowledge about other users' successes or collisions, or even about the number of users in the system.

The objectives of the IEEE 802.11 backoff scheme are (i) a distribution (the most uniform as possible) of transmissions over a variable-sized time window and (ii) a small access delay under light load conditions. A station selects a random interval, named *backoff interval*, that is used to initialize a *backoff counter*. When the channel is idle, the time is measured in constant-length (Slot_Time) units indicated as "slots." The backoff counter is decreased as long as the channel is sensed idle for a Slot_Time, it is stopped when a transmission is detected on the channel, and it is restarted when a DIFS is sensed idle on the channel. A station transmits as soon as its own backoff counter is zero. The backoff counter is an integer number of slots whose value is randomly and uniformly selected in the backoff interval $(0, CW\_Size - 1)$, where CW_Size is a local parameter that defines the current *Contention Window* size in each single station. Specifically, the backoff value is defined by the following expression [32]:

$$Backoff\_Counter(CW\_Size) = int(Rand() * CW\_Size)$$

where $Rand()$ is a function that returns pseudo-random numbers uniformly distributed in $[0..1]$. The name BEB is originated by the CW_Size parameter definition, which depends on the *number of unsuccessful transmissions* (NUT) already performed for a given frame. In Reference 32 it is defined that the first transmission attempt for a given frame is performed by adopting CW_Size equal to the minimum value CW_Size_min (by assuming low contention). After each unsuccessful (re)transmission of the same frame, the station doubles CW_Size until it reaches the maximal value fixed by the standard, that is, CW_Size_MAX, as follows:

$$CW\_Size(NUT) = min(CW\_Size\_MAX, CW\_Size\_min * 2^{NUT-1}).$$

where NUT (default value: 1) is the counter of the transmission attempts performed for a given frame. When the transmission of a given frame is successful, then the mechanism is restarted by assigning $NUT = 1$, even if a new frame is ready for transmission. In this way there is no "memory effect" of the contention level perceived for a given frame in successive transmissions. The $CW\_Size\_min = [16, 32]$ and $CW\_Size\_MAX = 1024$ in IEEE 802.11 DCF [32]. If the fixed maximum number of transmission attempts is reached, for a given frame, a "link failure" is indicated to the upper layers.

Analytical investigation of stability and characteristics of various backoff schemes have been presented in References 34,36,40, and 42.

25.5.5.2　Analysis of IEEE 802.11 Contention Control

The increase of the CW_Size parameter value after a collision is the reaction that the 802.11 standard DCF provides to make the access mechanism adaptive to channel conditions. By analyzing the behavior of the 802.11 DCF mechanism, under various contention levels (i.e., the number of active stations with continuous transmission requirements), some problems could be identified. Figure 25.6 shows simulation data regarding the channel utilization of a standard 802.11 system under DCF control, with respect to the variable contention level, that is, the number of active stations. The parameters adopted in the simulation, presented in Table 25.1, refer to the FHSS implementation [32]. The RTS/CTS mechanism is off, and a single static collision domain is assumed to capture the contention effect.

Figure 25.6 shows simulation results indicating that the channel utilization is negatively influenced by the increase of the contention level. These results can be explained because in the IEEE 802.11 backoff algorithm, a station selects the initial size of the contention window by assuming a low level of contention in the system. This choice avoids long access delays when the load is light. Unfortunately, this choice causes efficiency problems in burst-arrival scenarios, and in congested systems, because it concentrates the accesses in a small time window by causing a high collision probability. In high-contention conditions each station reacts to the contention on the basis of the collisions so far experienced while transmitting a given frame. Every station performs its attempts blindly, with respect to the contention level, with a late collision reaction performed (i.e., by increasing *CW_Size*). The number of collisions so far experienced is reflected in the size of the *CW_Size*, whose value can be considered as a local estimate of the contention level. Adaptation of the *CW_Size* is obtained by paying the cost of collisions, and the channel is not available for the whole time required to transmit the longest one among colliding frames. The carrier sensing mechanism protects the vulnerability of frames, but it does not give any preliminary indication about the contention level. Furthermore, after a successful transmission, the *CW_Size* is collapsed to the minimum value without maintaining any knowledge of the current contention level estimate. To summarize, the IEEE 802.11 backoff mechanism has some drawbacks. The increase of the *CW_Size* is obtained by paying the cost of many collisions, and each collision does not provide a significant indication of the actual contention level. Due to stochastic variability in the slot selection, collisions could occur even with few stations, so the contention indication obtained could be overestimated. After a successful transmission no state information indicating the actual contention level is maintained.

Several authors have investigated the enhancement of the IEEE 802.11 DCF MAC protocol to increase its performance when it is used in WLANs (i.e., a typical single collision domain) and MANETs (i.e., multi-hop collision domains) [2,7,60,88]. Unfortunately, in a real scenario, a station does not have an exact knowledge of the network and load configurations but, at most, can estimate it. In References 21 and 27, the tuning of the Standard's parameters has been studied via a performance analysis. In References 8

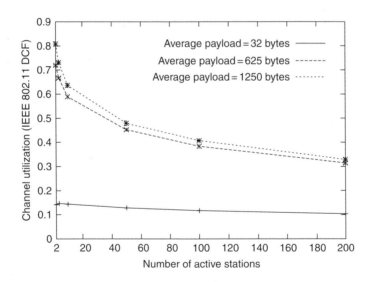

FIGURE 25.6 Channel utilization of the IEEE 802.11 DCF under variable contention level.

TABLE 25.1 IEEE 802.11 DCF Physical Parameters (FHSS Implementation)

Parameter	Value
Contention level	2 to 200 stations
Nominal channel bitrate	2 Mb/sec
CW_Size_min	16
CW_Size_MAX	1024
Header size	136 μSec(34 Bytes)
Payload size	variable(Geometric distribution)
Acknowledgment size	200 μ sec (50 Bytes)
Slot Time	50 μ sec
SIFS	28 μ sec
DIFS	128 μ sec
Propagation time	< 1 μ sec

and 83, solutions have been proposed for achieving a more uniform distribution of the accesses in the BEB scheme. The most promising direction for improving backoff protocols is to obtain the network status through channel observation [38,41,10]. A great amount of work has been done to study the information that can be obtained by observing the system's parameters [36,64,81]. For the IEEE 802.11 MAC protocol, some authors have proposed an adaptive control of the network congestion by investigating the number of users in the system [8,14,15]. This investigation would be time-expensive, hence difficult to obtain and subject to significant errors, especially in high contention situations[14].

In Reference 10 a simple mechanism named distributed contention control (DCC) is proposed to exploit the information obtained by the carrier sensing mechanism as a preliminary contention level estimation to be adopted in the contention control mechanism. The *slot utilization* observed during the carrier sensing phases (i.e., the ratio of nonempty slots observed during the backoff) has been demonstrated to be a better indicator of the contention level than the spotted collision events. In Reference 10, the slot utilization estimate is proposed to be adopted in a probabilistic mechanism (DCC) by extending the backoff protocol. The DCC mechanism defers scheduled transmissions in an adaptive way, on the basis of the local contention level estimate and local priority parameters (with no need for priority-negotiations). Implementation details of DCC, stability analysis, and performance results can be found in Reference 10.

The Asymptotically Optimal Backoff (AOB) mechanism proposed in Reference 12 tunes the backoff parameters to the network contention level by using two simple and low-cost estimates (as they are obtained by the information provided by the carrier sensing mechanism): the *slot utilization*, and the *average size of transmitted frames*. AOB is based on the results derived by exploiting the analytical model of the IEEE 802.11 protocol presented in Reference 15, and the enhancement of the DCC mechanism presented in Reference 10. In Reference 12 it was shown that, for any average length of the transmitted frames, there exists a value for the *slot utilization* that maximizes the protocol capacity, indicated as *optimal slot utilization*. In addition, in Reference 12 the analytical model presented in Reference 15 has been extended to show that the optimal value for the slot utilization is almost independent on the contention level and the network configuration (i.e., the number active stations). This fact is really important because it would relax the need to estimate the number of users in the system by simply estimating the slot utilization. Moreover, a simple definition of a tuning function that is adopted in the AOB mechanism to control the channel contention in congested scenarios is defined in Reference 12. In AOB, by exploiting a rough and low cost estimate of the average size of transmitted frames, the channel utilization converges to the optimal value when the network is congested, and no overheads are introduced in a low contention scenario. To achieve this goal, AOB schedules the frames' transmission according to the IEEE 802.11 backoff algorithm but adds an additional level of control before a transmission is enabled. Specifically, a transmission already enabled by the standard backoff algorithm is postponed by AOB in a probabilistic way. The probability to postpone a transmission depends on the network congestion level. The postponed transmission is rescheduled as in the case of collision, that is, the transmission is delayed of a further backoff interval, as if a virtual collision occurred. This simple feedback mechanism could be implemented to extend the Standard IEEE 802.11 contention control without any additional hardware and converges to the near-to-optimal channel utilization. Additional interesting features of the AOB mechanism are definition of priority-based contention control without the need for negotiations, good stability, and good tolerance to estimation errors. More details about these points can be found in References 10 and 12.

25.5.6 Contention and Collision Avoidance in Multi-Hop Scenarios

Another MAC problem is the "self-contention" arising in IEEE 802.11 MANETs between multi-hop flows of frames that share a common area of transmission (i.e., the same collision domain). This problem has been marginally addressed at the MAC layer in the literature [89,88,2,60], and some proposals are documented at the higher layers, for example, inter-stream contention in transport [58] and routing layers [25], intra-stream contention at the link layer [29] and transport layer [25,73]. The problem is due to the unawareness of the generalized MAC protocols (e.g., like in the IEEE 802.11 DCF) with respect

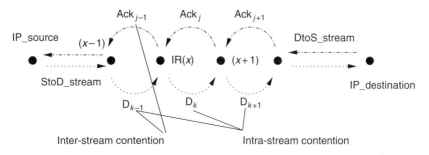

FIGURE 25.7   Self-contention of MAC frames.

to the transport layer session and multi-hop flows a MAC frame belongs to. As a result, MAC frames related to IP packets belonging to the same transport flows (both IPsender-to-IPreceiver and vice versa) may contend for the local channel resource without any synchronization by increasing the risk of mutual collisions and end-to-end delays. This problem may result in low goodput at the transport layer when multi-hop communication is implemented at the wireless MAC layer (like in MANETs).

According to Reference 89, we define a *TCP stream* as a sequence of IP packets routed from the transport layer *IP_source* to *IP_destination*. A TCP connection typically consists of a couple of streams: the data packets from the source to destination (StoD_stream) and the Ack packets from the destination to the source (DtoS_stream). Every MAC frame (e.g., D_k) encapsulating a (portion of) IP packet that belongs to a TCP stream is forwarded in the chain of intermediate receivers by using the RTS–CTS–DATA–ACK double handshake (the handshake is not shown for any D_k and Ack_j, in Figure 25.7, for readability). MAC frames are subject to two types of self-contention at the MAC layer: *intra-stream* and *inter-stream* self-contention. Intra-stream self-contention is determined by MAC frames belonging to the same TCP stream: if an intermediate receiver $IR(x)$ (i.e., the xth MH in the multi-hop chain at the MAC layer) receives a MAC frame D_{k-1} by the $IR(x-1)$, it would need to forward that frame to the next intermediate receiver $IR(x+1)$ by contending for a new channel access. This new channel access for D_k by $IR(x)$ would contend for the local channel with any frame D_{k-1} and D_{k+1} belonging to the same TCP stream (intra-stream) to be sent by neighbor IRs (see Figure 25.7). In most cases, the transport layer implements reliable end-to-end connections (e.g., as it happens with TCP, R-TCP, R-UDP). This implies that a DtoS_stream of acknowledgments would be usually transmitted on a reverse routing path of the StoD_stream of data frames (see Figure 25.7). Inter-stream self-contention at the host x is thus determined by the Ack frames coming from $IR(x+1)$ to $IR(x)$ (Ack_j in the DtoS_stream), contending with the Data frames going from $IR(x-1)$ to $IR(x)$ (D_{k-1} in the StoD_stream). The lack of any synchronization mechanism at the MAC layer for the (many) opposite streams is the cause for contention problems in multi-hop communication, resulting in the increasing end-to-end delays and collision rates. Any synchronization scheme would be required to adopt dynamic scheduling policies, given the highly variable set of parameters in such scenarios (host mobility, variable transmission power, hosts topology and routing, variable loads). On the other hand, self-contention is a MAC layer problem, and a distributed access scheme such as IEEE 802.11 DCF would be devoted to solve this kind of problem by leaving untouched the upper layers, if possible [89]. In Reference 89, two solutions have been sketched: *quick-exchange* and *fast-forward*. The quick-exchange solution is designed to alleviate the inter-stream self-contention: the idea is to exploit the channel capture obtained by a StoD_stream frame D_k from $IR(x)$ to $IR(x+1)$, to piggyback also possible DtoS_stream Ack_j frames from $IR(x+1)$ to $IR(x)$. In this way a new channel capture is not required and once the channel is captured by the sender or the receiver, the channel is not released since both streams' transmissions have been performed. The fast-forward solution works in the direction of favoring the multi-hop transmission of intra-stream frames: the idea is to create a hybrid MAC-layer acknowledgment frame for MAC data frames (not to be confused with transport layer Acks shown in the figure). Hybrid-Acks transmitted by $IR(x)$ to $IR(x-1)$ would work as implicit RTS toward the $IR(x+1)$ for the current MAC data frame. The hybrid-Ack sent by $IR(x)$ would be a broadcast frame (like RTS) with additional information to identify the intended receiver of its "acknowledgment" interpretation IR$(x-1)$, and the intended receiver of its "RTS" interpretation IR$(x+1)$. Hosts receiving the hybrid-Ack would consider it as an RTS request coming from IR(x), and they would set their virtual carrier sensing accordingly. Investigation of such mechanisms and proposals are still ongoing activity [89,88,2,60].

25.6   Power Saving Protocols

Wireless networks are typically viewed as bandwidth-constrained with respect to wired networks. However, for the portion of a wireless network consisting of battery-powered MHs, a finite energy capacity may be the most significant bottleneck, and energy utilization should be considered as a primary network-control parameter [4,45,52,56,66,84]. Projections on the expected progress in battery technology show that only a 20% improvement in the battery capacity is likely to occur over the next 10 years [68]. If the battery capacity cannot be improved, it is vital that energy utilization is managed efficiently by identifying ways to use less power with no impact on the applications, on the management efficiency, and on the resources' utilization.

Base stations may typically be assumed to be power-rich, whereas the MHs they serve may be power-constrained. Thus, this asymmetry should be accounted for in the network protocol design at all levels by offloading complexity from the MHs to the base stations as much as possible. Again, the problem may be more difficult in MANETs as the entire network may be energy-constrained: protocol complexity must be uniformly distributed throughout the network and be kept as low as possible. Solutions based on the exploitation of dynamic clustering protocols based on MH resources could be adopted to differentiate the energy drain required for network management in MANETs.

The low-power concept impacts protocol design at all levels of network control, including the MAC layer [4,56,84]. Due to the characteristics of wireless transmissions and wireless devices, the radio and network interface activities are among some of the most power consuming operations to perform [56,72]. To save energy, the first idea is to minimize the "on" time of network interfaces, that is, by switching the NI in *sleep mode* [72]. On the other hand, in WLAN and MANET scenarios, portable devices often need to transmit and receive data frames required both by applications and by distributed and cooperative management protocols. Techniques based on *synchronized sleep periods* are relatively easy to define in systems where the system area is "centrally" controlled by a given base station, as in infrastructure WLANs. However, in systems relying on *asynchronous, distributed control algorithms* at all network layers (including the MAC), such as in MANETs, the participation in the control algorithms prohibits usage of static sleep-schedules, and hence more sophisticated methods are required.

Several studies have been carried out in order to define mechanisms, system architectures, and software strategies useful for power saving (PS) and energy conservation in wireless LANs [4,56]. Transmitter power control strategies for minimizing power consumption, mitigating interference, and increasing the channel capacity have been proposed, and the design aspects of power-sensitive wireless network architectures have been investigated [4,66,90]. The impact of network technologies and interfaces on power consumption has been investigated in depth in References 72 and 18. The power saving features of the emerging standards for wireless LANs have been analyzed in References 84 and 18.

Multidimensional tradeoffs between energy usage and various performance criteria exist. One may choose to drain more energy to decrease latency, to increase throughput, to achieve a higher level of QoS support, or to save energy by mitigating interference, or some combination thereof [11–13,15]. From an energy-usage perspective, it may be better to be a less efficient in the usage of the wireless spectrum, for example, by adopting separate signalling channels.

The adaptive behavior of the MAC protocol can best be accomplished by integrating the information provided by lower and higher levels in the protocol stack (e.g., user profile information, battery level indication, channel tracking information). Again, the MANETs and the multi-hop scenarios are considered the most challenging scenarios under these viewpoints.

In infrastructure networks and in reservation-based access schemes the power saving topic is still considered a hot topic by researchers, even if many assumptions are less critical. For this reason we will mainly illustrate the distributed approaches for power saving, and we present some of the power saving strategies at the MAC level. Specifically, we focus on the distributed, contention-based access for WLANs and MANETs, and on the CSMA/CA access mechanism adopted in the IEEE 802.11 DCF Standard [32].

25.6.1　Power Saving Solutions at the MAC Layer

Three categories of solutions for power saving and energy conservation have been considered at the MAC layer in MANETs and WLANs: transmission power control, low-power sleep management, and power-aware contention control.

Transmission power control: The idea behind power control is to use the minimum transmission power required for correct reception to the destination. Given the positive acknowledgment required to complete a frame transmission in CSMA/CA schemes, the transmission power control must be considered on both sides: sender-to-receiver and vice versa. Transmission power control strongly impacts factors such as bit error rates, achievable transmission bitrate, network topology, and host connectivity (i.e., hosts' density related to the contention level). Solutions have been proposed to deal with power control and its influence at the link layer to determine network topology properties [43,62,82,44]. Also, network throughput can be influenced by power control because of the differences in the frequency reuse and in the spatial reuse of channels [53]. When transmitters use less power to reach the destination host, the collision domains are limited, and multiple transmissions could be performed concurrently over separate collision domains [44, 53]. Limited collision domains would allow the same channel (i.e., frequency band) to be used among multiple, separate collision domains. This is an important result in multi-hop wireless networks [53]. On the other hand, a high transmission power may contribute to maintain high signal-to-noise ratio (SNR) resulting in high bitrate coding techniques exploited on a wide-range area. The negative effect is that high transmission power would also contribute to the increase in the in-band interference among signals resulting in low signal-to-interference-and-noise ratio (SINR), low bit-rates coding techniques, and high Bit Error Rate (BER).

Low-Power sleep mode: Many wireless devices support a low-power *sleep* or *doze* mode, as the opposite of the *active* mode. The sleep mode is considered in the IEEE 802.11 standard as a way to reduce the energy drain of NIs. Many investigations of the wireless NI consumption have shown that a significant amount of energy consumed in a wireless host is due to the wireless NI activity. Many levels of power consumption can be identified depending on the NI's state [18]. The active mode for NIs includes the transmission, reception, and carrier sensing phases. When the NI is in the transmission phases, the amount of energy consumed is significant (in the order of mW). In carrier sensing and in reception phases, the amount of energy consumed is lower than in the transmission phases, but it is still significant. In current devices, the transmission phases can be considered at least twice more power-consuming than the reception (and carrier sensing) phases [72,18]. In doze or sleep phases the NI's energy consumption is limited to the minimum (both carrier sensing and radio components are switched off), and the energy drain is orders of magnitude lower than in active states [18]. These observations indicate that, in order to reduce the energy consumption by the NI, it would be useful to reduce the whole time the NI is in active state, that is, in carrier sensing, reception, or transmission phases. When communication is not expected from or to a given host, it could switch the NI into sleep mode to save energy. Unfortunately, most NIs require a significant time (many microseconds) and a significant burst of energy to switch back from sleep to active state. This is the reason why it would not be always convenient to switch the NI in idle state as soon as the channel is idle for a short time. Sleep time management has been considered in many research proposals. The main challenges are given by the need for continuous carrier sensing to realize the MAC protocol functions, and by the need to receive asynchronous frames, which could be sent while the receiver's NI is sleeping. Keeping the NI in the doze state also limits the neighbors' discovery and neighbors' information maintenance at the basis of many network-management protocols. In infrastructure networks, the NIs wake up periodically to check for buffered packets from the AP or to receive beacon frames [80,84,67,70,23,56]. Centralized buffering schemes give the advantage that many transmissions and receptions can be clustered as consecutive ones by increasing the average duration of sleep phases and by reducing the rate of state switches [18]. Many MAC protocols for infrastructure networks have been compared under the power saving viewpoint in Reference 18. The sleep-synchronization scheme may be quite complicated in multi-hop networks as shown in the following paragraphs. Recently, solutions have been proposed to switch off the network interface of wireless devices by exploiting dynamic, cluster-based infrastructures among peer mobile hosts. Other solutions exploit information derived from the application layer (e.g., user think times in interactive applications [3]).

Power aware Collision Avoidance and Contention Control: A previous discussion about these topics has illustrated the need to adapt access delays and the risk of collisions. Advantages obtained by the optimal tuning of the contention control and collision avoidance under the channel utilization viewpoint could be reflected in the reduction of energy wasted on collisions and carrier sensing activities, see for example, References [11 and 18].

The *Power Aware Routing:* This topic is out of the scope of this chapter being located at the network layer. The main solutions considered at this level are based on the filtering of forwarding hosts based on the remaining energy and transmission power reduction [39]. This approach is cited in this context since it may be considered as a power saving policy in multi-hop forwarding techniques at the MAC layer and in cross-layer hybrid solutions for routing at the MAC layer (like with label-switching). Many other solutions, for example, SPAN [19], GAF [86], AFECA [87], that guarantee a substantial degree of network connectivity (at the network layer) are based on the dynamic election of coordinator hosts, and on local and global information like energy, GPS position, mobility, and connectivity (i.e., hosts density). Such choices have an effect on the MAC and physical layers since only coordinator hosts never sleep while trying to adjust their transmission power in order to maintain a fully connected network. In this way the contention for channel access can be under control because a reduced number of hosts try to forward frames in the high density areas. The problem of the "broadcast storm" in the flooding-based solutions for routing is quite similar to the "self-contention" problem of multi-hop frame-flows in wireless broadcast channels that was considered in previous sections.

25.6.2   The MAC Contention under the Power Saving Viewpoint

For contention-based MACs like the CSMA/CA protocols, the amount of power consumed by transmissions is negatively affected by the congestion level of the network. By increasing the congestion level, a considerable power is wasted due to collisions. To reduce the collision probability, the stations perform a variable time-spreading of accesses (e.g., by exploiting backoff protocols), which results in additional power consumption due to carrier sensing performed over additional idle periods. Hence, CSMA/CA and contention control protocols suffer a power waste caused by both transmissions resulting in a collision and the amount of carrier sensing (active channel-state detection time) introduced by the time-spreading of the accesses. It is worth noting that collisions may cause a power waste involving more than one transmitter. Some kind of transmission policy optimization could be performed by evaluating the risk (i.e., the costs/benefits ratio) of transmission attempts being performed, given the current congestion conditions and the power-consumption parameters of the system. As an example, the power saving criterion adopted in Reference 11 is based on balancing the power consumed by the network interface in the transmission (including collisions) and the reception (or idle) phases (e.g., physical carrier sensing). Since these costs change dynamically, depending on the network load, a power-saving contention control protocol must be adaptive to the congestion variations in the system. Accurate tuning of the adaptive contention-based access was designed by considering different (parameter-based) levels of energy required by the network interface's transmission, reception, and idle (doze) states in Reference 11. The model and tuning information were adopted to implement the power-save distributed contention control (PS-DCC) mechanism in Reference 11. PS-DCC can be adopted on top of IEEE 802.11 DCF contention control by driving the contention control to converge to the near-to-optimal network-interface power consumption. In addition, the PS-DCC power-saving strategy balances the need for high battery lifetime with the need to maximize the channel utilization and QoS perceived by the network users in IEEE 802.11 DCF systems [11,56].

25.6.3 Sleep-Based Solutions

Given the open broadcast nature of the wireless medium, any ongoing transmission is potentially over-heard by all the neighbor hosts within the communication range. Thus all active neighbor hosts consume power by receiving frames even though the frame transmission is not addressed to them. The latter point has been faced in some cases; for example, the IEEE 802.11 networks try to reduce the amount of physical carrier sensing activity (called clear channel assessment, CCA) by exploiting a virtual carrier sensing based on network allocation vectors (NAVs) [32]. NAVs are local timers counting the time to the expected end of the ongoing transmission. If any ongoing transmission is not addressed to the receiving host, its NAV can be initialized to the duration of the ongoing transmission. If the transmission duration is long enough to consider worthwhile the transition to the sleep state, then the NI is switched off only to be reactivated when the NAV expires. The information to set the NAV timers can be obtained in frame headers and in preliminary RTS and CTS messages adopted for collision avoidance.

Many power saving protocols have been proposed for MANETs and WLANs to allow mobile hosts to switch to sleep mode, depending on the role of hosts and energy availability (e.g., power line or battery based). In infrastructure based networks, under the IEEE 802.11 PCF, the sleep mode can be exploited by demanding the base station to broadcast transmission scheduling lists. The problem here can be considered quite easy to solve because the base station can act as a central coordinator for hosts. The base station buffers the frames sent to sleeping hosts, and it periodically sends beacon frames at fixed intervals containing the information about the timeline of scheduled pending transmissions. Administrated (slave) hosts sleep most of the time, and wake up just in time to receive and send their information to the base station. This management approach based on the master–slave role of host has also been introduced in Bluetooth piconets and in cluster-based architectures for MANETs by exploiting the hosts asymmetry and by demanding administration roles to the best candidates.

Many more problems arise in the distributed sleep coordination schemes required for MANETs and multi-hop wireless networks. Usually, proposed solutions for power saving assume fully connected networks (a strong assumption for multi-hop systems) and overall synchronization of clocks. An example is given by the IEEE 802.11 timing synchronization function (TSF) in the PCF scheme and its DCF version that is presented in the next section. Another critical issue, related to the wireless scenario, is the mobility of hosts resulting in variable network topology, variable contention level, and variable channel-traffic loads. The heterogeneous characteristics of hosts in MANETs are other problems that should be considered in the design of power saving mechanisms at the MAC layer. To sum up, unpredictable mobility, multi-hop communication, difficult clock-synchronization mechanisms, and heterogeneous power supplies (power line vs. battery based) are some of the most critical design assumptions to be considered in power saving schemes for MANETs [70]. The absence of clock-synchronization mechanisms is the main problem in distributed scenarios because it would be hard to predict if and when the receiver host would be ready to receive. A sleeping host can be considered a missing host. Nonetheless, neighbor discovery in highly dynamic scenarios, with hosts mobility and sleeping policies, is even more critical. The need for asynchronous sleep-based protocols and solutions has been discussed in Reference 80.

Other solutions have been proposed for MANETs in distributed and multi-hop scenarios. In power aware multi-access protocol with signalling (PAMAS) [70] a separate signalling channel is adopted to discover and to manage the state of neighbor hosts. PAMAS is based on the MACA definition, where the collision avoidance based on RTS/CTS messages (on the signalling channel) is considered as a power saving solution. PAMAS was designed by assuming fully connected scenarios and busy tones were thought in order to stop neighbor hosts from getting involved in ongoing transmissions, hence by switching-off their network interfaces to save energy. Every host is required to solve locally the problem of the NI activation in order to be able to receive frames when required. The proposal is to adopt a sequence of channel probes on a separate control channel to properly determine the reactivation time.

In Reference 23, different sleep patterns are defined to differentiate between hosts' sleeping-periods based on residual energy and QoS needs. A technological solution called remote activated switch (RAS) is required to wake up the sleeping hosts by sending them a wake-up signal. In this scheme, the sleep management is passive (i.e., it is controlled by senders) instead of active, (i.e., managed by NAVs).

25.6.4   Power Control Solutions

Dealing with power control, many similar solutions appear in the literature such as SmartNode [59], power controlled multiple access (PCMA) MAC protocol [53], and many others cited in Reference 47. The common idea adopted in such schemes was called the *basic power control scheme*. The idea is to exploit dynamic power adjustment between the senders and the receivers by exploiting the RTS/CTS handshake as a common reference. The RTS and CTS frames are sent with the maximum nominal transmission power, and the adjustment is performed for the data transmission relative to the residual power detected by the counterpart. This approach becomes quite critical with heterogeneous devices with different nominal power levels. In Reference 47, a modification of the basic power adjustment schemes was proposed based on periodic pulses of the transmission power during the data transmission. This scheme was thought as a way to contrast the throughput degradation due to the risk of hidden terminals during the data transmission that cannot be avoided by the RTS/CTS handshake.

In Reference 54, the COMPOW protocol was proposed as a distributed policy to find the minimum COMmon POWer for transmissions leading to a sustainable degree of host connectivity and bidirectional links.

25.6.5   IEEE 802.11 Power Saving

The IEEE 802.11 Standard supports two power modes for MHs: *active* MHs can transmit and receive at any time, and *power-saving* MHs may be sleeping and wake up from time to time to check for incoming packets.

All the power saving schemes explained below are based on the PCF access scheme defined for infrastructure systems based on APs, and the DCF access scheme, referred to as the basic access scheme for ad hoc networks.

In infrastructure networks, indicated as a BSS of hosts, the existence of the AP station managing the point coordinated channel access (PCF) is assumed. The AP is in charge of monitoring the state of each MH, and a MH should always refer to the AP for registration requests, transmission requests, and state changes. The AP is in charge of the synchronization of sleep and active periods between the stations. The synchronization is achieved by means of the timing synchronization function (TSF), that is, every MH would get synchronized by the clock indicated by the AP in special frames called *beacon frames*. Periodically, the AP sends beacon frames to start a *beacon interval*. Beacon frames should be monitored by MHs, that is, they should wake up in time to receive beacons. Every beacon contains a traffic indication map (TIM) indicating the list of MH's identifiers with buffered traffic on the AP: such MHs should stay active in order to receive the buffered traffic in the current beacon interval. For buffered broadcast frames, the AP sends a delivery TIM (DTIM) message (indicating that every MH should stay active) and immediately starts with the broadcast frames' transmission.

In ad hoc networks, supported by the IBSS structure of hosts, the existence of the AP as a centralized coordinator cannot be assumed. This requires the power saving management to be implemented as a distributed policy. The MH initiating the IBSS assumes the role of the synchronization coordinator, and the synchronization approach is still based on beacon frames. How the IBSS is started and initialized is out of the scope of this chapter, see Reference 32 for details. Every station in the IBSS assumes to receive a beacon message within a nominal amount of time, that is, the *Beacon Period* proposed by the IBSS initiator. Local TSF timers are used to obtain a weak synchronization of distributed beacon intervals. At

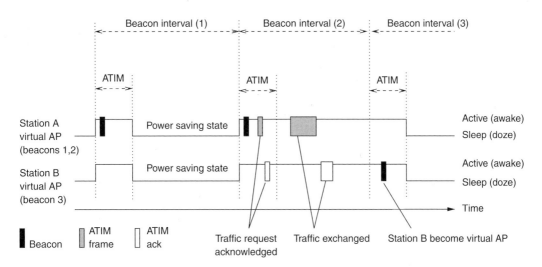

FIGURE 25.8   IEEE 802.11 TSF and power saving.

the beginning of a beacon interval, every MH listens for beacon frames while decrementing a randomly initialized backoff counter. If the backoff counter expires before it hears any beacon frame, the MH itself sends a beacon frame indicating its local TSF timer value. In this way, if any problem occurred, the IBSS initiator can be replaced on the fly (see station B in Figure 25.8). Every MH receiving a beacon frame will compare the TSF indicated in the beacon with its local TSF timer. If the beacon-TSF value is later than the local TSF, the MH initializes its TSF timer to the beacon-TSF value. In this way, a weak synchronization scheme similar to the scheme adopted in infrastructure systems can be maintained, and local time is guaranteed to monotonically advance on every MH.

During the DCF, an MH can send a *PS-poll* (by contending for the channel) to the AP when it is ready to receive buffered frames. If the PS-poll is correctly received, the AP transmits the buffered frames. MHs can sleep most of the time, and periodically wake-up during short ad hoc TIM (ATIM) time-windows, that are located at the beginning of each beacon interval (see Figure 25.8). The assumption here is that all MHs in the ad hoc network would have synchronized ATIM windows, where they can exchange *ATIM frames* by notifying each other about buffered frames (see Figure 25.8). ATIM frames in ad hoc scenarios have the purpose to inform neighbor hosts about the pending traffic. ATIM and data frames are sent, subject to standard contention rules (i.e., the DCF CSMA/CA and BEB rules), within the ATIM window and after the ATIM window respectively. An MH receiving a unicast ATIM frame will immediately acknowledge the ATIM frame and will wait for the transmission of buffered packets after the end of the ATIM window. If the ATIM sender does not receive the Ack, it will try again later in the next ATIM window. Broadcast ATIM frames need no acknowledgment, and they can be sent under DCF contention rules at the end of the ATIM window. During the ATIM window, only RTS/CTS, Ack, Beacon, and ATIM frames can be sent.

Such a distributed power saving mode is designed for single-hop networks. In multi-hop scenarios, the global ATIM window synchronization becomes a problem because of the increasing propagation delays, clock drifts among multiple hosts, and temporary network partitions. This is even worse when the network scales to many hosts [80]. The discovery of neighbor hosts under power saving mode is not trivial because the host mobility would change the neighbors set of every host, and because during the sleeping times hosts are unable to receive and to transmit beacon messages. On the other hand, beacon messages contending for transmission over small time windows would have high collision probability.

In Reference 84, the simulation analysis of the IEEE 802.11 power saving mechanism at the MAC layer has been performed. In Reference 80, the proposal was to insert more beacons in every ATIM window, suggesting that beacons should be adopted not only for clock synchronization, but also for discovering neighbors and for self-advertising. Another proposal was to design the ATIM windows such that overlapping awake intervals are guaranteed even with maximum clock drift and worst scenario assumptions [80]. A definition and analysis of three power-saving-oriented beaconing algorithms for IEEE 802.11 MAC have been proposed in Reference 80: dominating-awake-interval, periodically-full-awake-interval, and quorum-based protocols. The relationships among beaconing process, neighbors discovery delay, and power saving characteristics have been investigated. Some of the proposed solutions are more appropriate for highly mobile and low mobility scenarios, respectively. In general, solutions should be adaptive dealing with system mobility (both predictable and un-predictable), multi-hop communication, variable traffic loads, and weak clock synchronization.

25.7 Conclusion

In WLANSs and MANETs the MAC protocol is the first protocol-candidate to support the management issues of mobile hosts in a highly dynamic scenario. The MAC protocols also influence the utilization of scarce resources such as the channel bandwidth and the limited battery energy.

In this chapter, we illustrated the motivations and the assumptions that were the basis of the new design, adaptation, and tuning of existing and new distributed MAC protocols for the wireless networks scenario. Some assumptions and the main problems and limiting constraints of the wireless communication channels have been sketched as a background information. The evolutionary perspective of distributed, contention-based (random-access) MAC protocols has been presented to illustrate, in incremental way, the problems considered and solutions proposed thereby leading to current IEEE 802.11 standard definition. The illustration of the time and space contention control in IEEE 802.11 DCF with a discussion of related problems and solutions has been shown. Specifically, contention problems and solutions have been illustrated for single-hop WLANs and for multi-hop MANETs. Finally, a perspective of power saving solutions to be considered at the MAC layer has been presented. Many prototype solutions have been described and referenced. Anyway, the research in this field can be considered still in preliminary phase. The design and tuning of stable, fair, low-overhead and adaptive distributed MAC protocols supporting multi-hop communication, contention control, and power saving for WLANs and MANETs still require additional research work. The aim of this chapter is to give a self-contained roadmap for orientation in this problem by illustrating the state of the art and some milestone solutions that can be considered as the background for additional research in this field.

References

[1] N. Abramson, "The ALOHA System — Another Alternative for Computer Communications," in *Proceedings of the Fall Joint Computer Conference AFIPS*, 1970.

[2] A. Acharya, A. Misra, and S. Bansal, "A Label-switching Packet Forwarding Architecture for Multi-hop Wireless LANs," in *Proceedings of the ACM WoWMoM 2002*, 09/2002, Atlanta, Georgia, USA.

[3] G. Anastasi, M. Conti, E. Gregori, and A. Passarella, "Power Saving in Wi-Fi Hotspots: an Analytical Study," in *Proceedings of the PWC 2003*, Venice, September 2003.

[4] N. Bambos, "Toward Power-Sensitive Network Architectures in Wireless Communications: Concepts, Issues and Design Aspects," IEEE Personal Communications, June 1998, pp. 50–59.

[5] L. Bao and J.J. Garcia-Luna Aceves, "A New Approach to Channel Access Scheduling for Ad Hoc Networks," in *Proceedings of the Sigmobile: ACM SIG on Mobility of Systems, Users, Data and Computing*, 2001, pp. 210–221.

[6] V. Bharghavan, A. Demers, S. Shenker, and L. Zhang, "MACAW: A Media Access Protocol for Wireless LAN's," in *Proceedings of the ACM SIGCOMM'94*, London, 1994, pp. 212–225.

[7] V. Bharghavan, "Performance Evaluation of Algorithms for Wireless Medium Access," in *Proceedings of the IEEE Performance and Dependability Symposium '98*, Raleigh, NC. August 1998.

[8] G. Bianchi, L. Fratta, and M. Olivieri, "Performance Evaluation and Enhancement of the CSMA/CA MAC protocol for 802.11 Wireless LANs," in *Proceedings of the PIMRC 1996*, October 1996, Taipei, Taiwan, 1996, pp. 392–396.

[9] K. Bieseker, "The Promise of Broadband Wireless," IT Pro November/December 2000, pp. 31–39.

[10] L. Bononi, M. Conti, and L. Donatiello, "Design and Performance Evaluation of a Distributed Contention Control (DCC) Mechanism for IEEE 802.11 Wireless Local Area Networks," Journal of Parallel and Distributed Computing (JPDC), Vol. 60, 2000, pp. 407–430.

[11] L. Bononi, M. Conti, and L. Donatiello, "A Distributed Mechanism for Power Savings in IEEE 802.11 Wireless LANs," ACM Mobile Networks and Applications (ACM MONET) Journal, Vol. 6, 2001, pp. 211–222

[12] L. Bononi, M. Conti, and E. Gregori, "Run-Time Optimization of IEEE 802.11 Wireless LANs performance," IEEE Transaction on Parallel and Distributed Systems (IEEE TPDS), Vol. 15, 2004, pp. 66–80.

[13] R. Bruno, M. Conti, and E. Gregori, "WLAN technologies for mobile ad-hoc networks," in Proceedings of the HICSS-34, Maui, Hawaii, January 2001, pp. 3–6.

[14] F. Cali', M. Conti, and E. Gregori, "Dynamic IEEE 802.11: design, modeling and performance evaluation," IEEE Journal on Selected Areas in Communications, 18, 2000, pp. 1774–1786.

[15] F. Cali', M. Conti, and E. Gregori, "Dynamic Tuning of the IEEE 802.11 Protocol to Achieve a Theoretical Throughput Limit," IEEE/ACM Transactions on Networking, Vol. 8, 2000, pp. 785–799.

[16] J.I. Capetanakis, "Tree Algorithms for Packet Broadcast Channels," IEEE Transactions on Information Theory, Vol. 25, 1979, pp. 505–515.

[17] A. Chandra, V. Gumalla, and J.O. Limb, "Wireless Medium Access Control Protocols," IEEE Communications Surveys, Second Quarter, 2000.

[18] J.-C. Chen, K.M. Sivalingam, P. Agrawal, and S. Kishore, "A Comparison of MAC protocols for Wireless LANs Nased on Battery Power Consumption," in *Proceedings of the INFOCOM'98*, San Francisco, CA, March 1998, pp. 150–157.

[19] B. Chen, K. Jamieson, H. Balakrishnan, and R. Morris, "Span: an energy-efficient coordination algorithm for topology maintenance in ad hoc wireless networks," ACM Wireless Networks Journal, 8, 2002, pp. 85–96.

[20] K.C. Chen, "Medium Access Control of Wireless LANs for Mobile Computing," IEEE Networks, 9-10/1994, 1994.

[21] H.S. Chhaya, "Performance Evaluation of the IEEE 802.11 MAC protocol for Wireless LANs," Master Thesis, Graduate School of Illinois Institute of Technology, May 1996.

[22] H.S. Chhaya and S. Gupta, "Performance Modeling of asynchronous data transfer methods in the IEEE 802.11 MAC protocol," ACM/Balzer Wireless Networks, Vol. 3,1997, pp. 217–234.

[23] C.F. Chiasserini and R.R. Rao, "A Distributed Power Management Policy for Wireless Ad Hoc Networks," in *Proceedings of the IEEE Wireless Communication and Networking Conference*, 2000, pp. 1209–1213.

[24] A. Colvin, "CSMA with Collision Avoidance," Computer Communications, Vol. 6, 1983, pp. 227–235.

[25] C. Cordeiro, S.R. Das, and D.P. Agrawal, "COPAS: Dynamic Contention Balancing to Enhance the Performance of TCP over Multi-hop Wireless Networks," in Proceedings of the 10th ICCCN 2003, October 2002, pp. 382–387.

[26] M.S. Corson, J.P. Maker, and J.H. Cerincione, "Internet-based Mobile Ad Hoc Networking," Internet Computing, July–August 1999, pp. 63–70.

[27] B.P. Crow, "Performance Evaluation of the IEEE 802.11 Wireless Local Area Network Protocol," Master thesis, University of Arizona, 1996.

[28] G.H. Forman and J. Zahorjan, "The challenges of mobile computing," IEEE Computer, April 1994, pp. 38–47.

[29] Z. Fu, P. Zerfos, H. Luo, S. Lu, L. Zhang, and M. Gerla, "The Impact of Multihop Wireless Channel on TCP Throughput and Loss," in Proceedings of the Infocom 2003, 2003.

[30] C.L. Fullmer and J.J. Garcia-Luna-Aceves, "Floor Acquisition Multiple Access (FAMA) for Packet Radio Networks," in *Proceedings of the ACM Sigcomm'95*, Cambridge, MA, 1995.

[31] C.L. Fullmer, Collision Avoidance Techniques for Packet radio Networks. Ph.D. Thesis, UC Santa Cruz, 1998.

[32] IEEE 802.11 Std., IEEE Standard for Wireless LAN Medium Access Control (MAC) and Physical Layer (PHY) Specifications, June 1999.

[33] http://grouper.ieee.org/groups/802/11/main.html

[34] R.G. Gallagher, "A perspective on multiaccess channels," IEEE Transactions on Information Theory, Vol. IT-31, 1985, pp. 124–142.

[35] R. Garces, "Collision Avoidance And Resolution Multiple Access", Ph.D. Thesis, University of California at Santa Cruz, 1999.

[36] L. Georgiadis and P. Papantoni-Kazakos, "Limited Feedback Sensing Algorithms for the Packet Broadcast channel," IEEE Transactions on Information Theory, Vol.IT-31, 1985, pp. 280–294.

[37] L. Georgiadis and P. Papantoni-Kazakos, "A Collision Resolution Protocol for Random Access Channels with Energy Detection," IEEE Transactions on Communications, Vol. 30, 1982, pp. 2413–2420.

[38] M. Gerla, L. Kleinrock, "Closed loop stability control for S-Aloha satellite communications," *Proceedings of the Fifth Data Communications Symposium*, September 1977, pp. 2.10–2.19.

[39] J. Gomez, A. T. Campbell, M. Naghshineh, and C. Bisdikian, "Power-aware Routing in Wireless Networks," in *Proceedings of the MoMuC 1999*, San Diego, 1999, pp. 114–123.

[40] J. Goodman, A.G. Greenberg, N. Madras, and P. March, "Stability of Binary Exponential Backoff," Journal of the ACM, Vol. 35, 1988, pp. 579–602.

[41] B. Hajek and T. Van Loon, "Decentralized Dynamic Control of a Multiaccess Broadcast Channel," IEEE Transaction on Automotion Control, Vol. 27, 1982, pp. 559–569.

[42] J. Hastad, T. Leighton, and B. Rogoff, "Analysis of Backoff Protocols for Multiple Access Channels," SIAM Journal of Computing, Vol. 25, 1996, pp. 740–774.

[43] L. Hu, "Topology Control for Multihop Packet Radio Networks," IEEE Transactions on Communication, Vol. 41, pp. 1474–1481, 1993.

[44] C.F. Huang, Y.C. Tseng, S.L. Wu, and J.P. Sheu, "Increasing the Throughput of Multihop Packet Radio Networks with Power Adjustment," in *Proceedings of the ICCCN*, 2001.

[45] T. Imielinsky and B.R. Badrinath, "Mobile Wireless Computing: Solutions and Challenges in Data Management," Communication of ACM, Vol. 37, 1994, pp. 18–28.

[46] A.J.E.M. Janssen and M.J.M. de Jong, "Analysis of Contention-Tree Algorithms," in *Proceedings of the IEEE Informatiion Theory Workshop*, Killarney, Irelapd, 1998.

[47] E.-S. Jung and N.H. Vaydia, "A Power Control MAC Protocol for Ad Hoc Nerworks," in *Proceedings of the MOBICOM 2002*, September 2002, Atlanta, GA.

[48] P. Karn, "MACA — A new Channel Access Method for Packet Radio," in *Proceedings of the 9th Computer Nerworking Conference*, September 1990.

[49] V. Kawadia and P.R. Kumar, "A Cautionary Perspective on Cross Layer Desingn," Submitted for publication, 2003 ($http://black1.csl.uiuc.edu/prkumar/$).

[50] S. Khanna, S. Sarkar, and I. Shin, "An Energy Measurement based Collision Resolution Protocol," in *Proceedings of the ITC 18*, Berlin, Germany, September 2003.

[51] L. Kleinrock and F.A. Tobagi "Packet Switching in Radio Channels: Part I - Carrier Sense Multiple-Access modes and their throughput-delay characteristics," IEEE Transactions on Communications, Vol. Com-23, 1975, pp. 1400–1416.

[52] R. Kravets and P. Krishnan, "Power Management Techniques for Mobile Communication," in *Proceedings of the Fourth Annual ACM/IEEE International Conference on Mobile Computing and Networking (MOBICOM'98)*.

[53] J.P. Monks, J.-P. Ebert, A. Wolisz, and W.-M.W. Hwu, "A Study of the Energy Saving and Capacity Improvment Potential of Power Control in Multi-Hop Wireless Networks," in *Proceedings of the 26th LCN and WLN*, Tampa, FL, November, 2001.

[54] S. Narayanaswamy, V. Kawadia, R.S. Sreenivas, and P.R. Kumar, "Power Control in Ad Hoc Networks: Theory, Architecture, Algorithm and Implementations of the COMPOW Protocol," in *Proceedings of the European Wireless Conference*, February 2002.

[55] W.F. Lo and H.T. Mouftah, "Carrier Sense Multiple Access with Collision Detection for Radio Channels," in *Proceedings of the IEEE 13th International Communication and Energy Conference*, 1984, pp. 244–247.

[56] J.R. Lorch, A.J. Smith, "Software Strategies for Portable Computer Energy Management," IEEE Personal Communications, June 1998, pp. 60–73.

[57] R.M. Metcalfe, D.R. Boggs, "Ethernet: Distributed Packet Switching for Local Computer Networks," Communication of ACM, Vol. 19, 1976, pp. 395–404.

[58] C.Parsa, J.J. Garcia-Luna Aceves, "Improving TCP Performance over Wireless Networks at the Link Layer," ACM Mobile Networks and Applications, Vol. 5, 2000, pp. 57–71.

[59] E. Poon and B. Li, "SmartNode: Achieving 802.11 MAC Interoperability in Power-efficient Ad Hoc Networks with Dynamic Range Adjustments," in *Proceedings of the IEEE International Conference on Distributed Computing Systems (ICDCS 2003)*, May 2003, pp. 650–657.

[60] D. Raguin, M. Kubisch, H. Karl, and A. Wolisz, "Queue-driven Cut-through Medium Access in Wireless Ad Hoc Networks," in *Proceedings of the IEEE Wireless Communications and Networking Conference (WCNC)*, Atlanta, Georgia, USA, 03/2004.

[61] B. Ramamurthi, D.J. Goodman, and A. Saleh, "Perfect capture for local radio communications," IEEE Journal on Selected Areas in Communication, Vol. SAC-5, June 1987.

[62] R. Ramanathan and R. Rosales-Hain, "Topology Control of Multihop Wireless Networks using Transmit power adjustment," in *Proceedings of the IEEE INFOCOM 2000*, pp. 404–413.

[63] T.S. Rappaport, *Wireless Communications: Principles and Practice*, 2nd ed, Prentice Hall, Upper Sadle River, NJ, 2002.

[64] R.L. Rivest, "Network Control by Bayesian Broadcast," IEEE Transactions on Information Theory, Vol. IT-33, 1987, pp. 323–328.

[65] R. Rom, "Collision Detection in Radio Channels," Local Area and Mutiple Access Networks, 1986, pp. 235–249.

[66] M. Rulnick and N. Bambos, "Mobile Power Management for Wireless Communication Networks," Wireless Networks, Vol. 3, 1996, pp. 3–14.

[67] A.K. Salkintzis and C. Chamzas, "Performance Analysis of a downlink MAC protocol with Power-Saving Support," IEEE Transactions on Vehicular Technology, Vol. 49, May 2000, pp. 1029–1040.

[68] S. Sheng, A. Chandrakasan, and R.W. Brodersen, "A Portable Multimedia Terminal," IEEE Communications Magazine, December 1992.

[69] G. Sidhu, R. Andrews, and A. Oppenheimer, *Inside Appletalk*, Addison-Wesley, Reading, MA, 1989.

[70] S. Singh and C.S. Raghavendra, "PAMAS — Power Aware Multi-Access protocol with Signalling for Ad Hoc Networks," ACM Communications Review, July 1998, Vol. 28, pp. 5–26.

[71] W. Stallings, *Local & Metropolitan Area Networks*, 5th ed, Prentice Hall, New York, 1996, pp. 356–383.

[72] M. Stemm and R.H. Katz, "Measuring and Reducing Energy Consumption of Network Interfaces in Hand-Held Devices," in *Proceedings of the 3rd International workshop on Mobile Multimedia Communications (MoMuC-3)*, Princeton, NJ, September 1996.

[73] K. Sunderasan, V. Anantharaman, H.-Y. Hsieh, R. Sivakumar, "ATP: A reliable transport protocol for ad hoc networks," in *Proceedings of the ACM MOBIHOC 2003*, June 2003.

[74] F. Talucci and M. Gerla, "MACA-BI (MACA by Invitation): A Wireless MAC Protocol for High Speed Ad Hoc Networking," in Proceedings of the IEEE ICUPC'97, Vol. 2, San Diego, CA, October 1997, pp. 913–917.

[75] F.A. Tobagi and L. Kleinrock, "Packet Switching in Radio Channels: Part II — The Hidden Terminal Problem in Carrier Sensing Multiple Access and Busy Tone Solution," IEEE Transactions on Communications, Vol. Com-23, 1975, pp. 1417–1433.

[76] F.A. Tobagi and L. Kleinrock, "Packet Switching in Radio Channels: Part III — Polling and (Dynamic) Split Channel Reservation Multiple Access," IEEE Transactions on Computer, Vol. 24, 1976, pp. 832–845.

[77] F.A. Tobagi and L. Kleinrock, "The Effect of Acknowledgment Traffic on the Capacity of Packet Switched Radio Channels," IEEE Transactions on Communications, Vol. Com-26, 1978, pp. 815–826.

[78] B.S. Tsybakov, V.A. Mikhailov, "Random Multiple Access of Packets: Part and Try Algorithm," Problems in Information Transmission, Vol. 16, 1980, pp. 305–317.

[79] A. Tzamaloukas, "Sender- and Receiver-Initiated Multiple Access Protocols for Ad Hoc Networks," Ph.D. Thesis, University of California at Santa Cruz, 2000.

[80] Yu-Chee Tseng, Chin-Shun Hsu, and Ten-Yueng Hsieh, "Power-Saving Protocols for IEEE 802.11-Based Multi-hop Ad Hoc Networks," in Proceedings of INFOCOM'02, New York, June 2002, pp. 200–209.

[81] J.N. Tsitsiklis, "Analysis on a multiaccess Control Scheme," IEEE Transactions on Automation Control, Vol. AC-32, 1987, pp. 1017–1020.

[82] R. Wattenhofer, L. Li, P. Bahl, and Y.M. Wang, "Distributed Topology Control for Power Efficient Operation in Multihop Wireless Ad Hoc Networks," in *Proceedings of the IEEE INFOCOM 2001*, pp. 1388–1397.

[83] J. Weinmiller, H. Woesner, J.P. Ebert, and A. Wolisz, "Analyzing and tuning the Distributed Cordination Function in the IEEE 802.11 DFWMAC Draft Standard," in *Proceedings of the International Workshop on Modeling, MASCOT 96*, San Jose, CA, 1996.

[84] H. Woesner, J.P. Ebert, M. Schlager, and A. Wolisz, "Power-Saving Mechanisms in Emerging Standards for Wireless LANs: The MAC Level Perspective," IEEE Personal Communications, June 1998, pp. 40–48.

[85] Kaixin Xu, M. Gerla, Sang Bae, "How effective is the IEEE 802.11 RTS/CTS Handshake in Ad Hoc Networks?" in *Proceedings of the GLOBECOM 2002*, Vol. 1, 2002, pp. 72–76.

[86] Ya Xu, John Heidemann, and Deborah Estrin, "Geography-Informed Energy Conservation for Ad Hoc Routing," in *Proceedings of the Seventh Annual ACM/IEEE International Conference on Mobile Computing and Networking*, Rome, Italy, July 2001.

[87] Ya Xu, John Heidemann, and Deborah Estrin "Geography-Informed Energy Conservation for Ad Hoc Routing," in *Proceedings ACM MOBICOM 2001*, Rome, Italy, pp. 70–84, July 2001.

[88] S. Xu and T. Saadawi, "Revealing the problems with 802.11 medium access control protocol in multi-hop wireless ad hoc networks," Computer Networks, Vol. 38, 2002, pp. 531–548.

[89] Z. Ye, D. Berger, P. Sinha, S. Krishnamurthy, M. Faloutsos, and S.K. Tripathi, "Alleviating MAC layer Self-Contention in Ad Hoc Networks," in Proceedings of the ACM Mobicom 2003 Poster session.

[90] M. Zorzi and R. R. Rao, "Energy Management in Wireless Communications," in *Proceedings of the 6th WINLAB Workshop on Third Generation Wireless Information Networks*, March 1997.

V

QoS in Wireless Networks

26

QoS Enhancements of the Distributed IEEE 802.11 Medium Access Control Protocol

26.1 Introduction **26**-613

26.2 IEEE 802.11 Distributed Coordination Function **26**-615
Basic Access Mode • RTS–CTS Access Mode

26.3 QoS Leverages of Distributed MAC Protocols.........**26**-617

26.4 QoS Enhancement Proposals **26**-619
Predetermined Static Differentiation • Distributed Priority
Access • Distributed Fair Scheduling • Distributed
Reservation-Based Access

26.5 IEEE 802.11e EDCF.....................................**26**-627

26.6 Summary **26**-628

References **26**-629

Kenan Xu
Hossam Hassanein
Glen Takahara

26.1 Introduction

In recent years, the proliferation of portable computing and communication devices, such as laptops, PDAs, and cellular phones has pushed a revolution toward ubiquitous computing. As an alternative technique to last-hop wireless access, wireless local area networks (WLANs) have gained growing popularity worldwide. Being both cost-efficient and simple, WLANs provide high-bandwidth packet transmission service over the free-of-charge industrial, scientific, and medical (ISM) frequency band.

First approved in 1997, the IEEE 802.11 [Wg97] is a widely accepted and commercialized WLAN standard. It specifies the WLAN operation on the physical (PHY) and medium access control (MAC) layers. According to the IEEE 802.11 standard, a WLAN may be constructed in either of the two architectures, namely, infrastructure-based or ad hoc. An infrastructure-based network consists of one central controller or coordinator and a number of distributed mobile terminals. The controller, known as the access point in the standard, has the responsibility of controlling the operation of the network, such as authentication, association/disassociation, coordinating the media access among the wireless nodes, synchronization, and power management. In contrast, an ad hoc WLAN has no central coordinator/controller. All the wireless, possibly mobile, nodes are self-configured to form a network without the aid of any established infrastructure. The mobile nodes handle the necessary control and fulfill networking tasks in a distributed, peer-to-peer fashion. Due to ease of deployment and flexibility, ad hoc based WLANs have attracted great interest in both industry and academia.

Being the essential component of the IEEE 802.11 standard, a distributed contention-based MAC protocol, known as the distributed coordination function (DCF), is defined to regulate the media access procedure in an ad hoc WLAN. Using the DCF scheme, all wireless stations contend to access the wireless medium following the slotted carrier sense multiple access with collision avoidance (CSMA/CA) mechanism. Under CSMA/CA every station determines its packet transmission time instance by executing a two-stage control procedure. The station will first sense the channel idle for a fixed period of time. In order to avoid collisions, the station will defer the transmission with an extra random backoff time after the fixed time period. Retransmission of collided packets is arranged by a binary exponential backoff (BEB) algorithm.

DCF-based ad hoc WLANs have the advantage of easy deployment and they are robust against node failure and node mobility. Moreover, they are flexible enough to be extended to multi-hop communication by adding routing intelligence to mobile nodes so that packets can be delivered to destinations more than one-hop away.

However, the DCF function does not include any explicit quality of service (QoS) support. All packets, regardless of their belongings to various traffic flows requiring different QoSs or the network status, are treated and served in exactly the same manner. Thus, the packet delay and session throughput cannot be predicted. The lack of predictability and controllability largely constrain the ad hoc WLANs to best effort applications only. A possible solution to alleviate this problem is devising QoS extensions to the basic best effort MAC, which we call QoS MAC protocols. QoS MAC allows mobile terminals to access the physical channel in a differentiated and controllable manner, in order that the specific QoS requirements of individual flows are satisfied.

The design of QoS MAC for ad hoc based WLANs is a challenging task for two main reasons. First, wireless channels are time-varying and error prone. Using radio as the transmission carrier, the propagation of electromagnetic waves is strongly influenced by the surrounding environment and is limited by three major factors, namely, reflection, diffraction, and scattering. The variation of the environment due to nodal mobility and change of surroundings, combined with sources of interference, leads to a complex and time-varying communication environment. Second, the absence of the central control entity makes ad hoc WLAN QoS design even harder since most existing QoS mechanisms and algorithms, such as fair queuing and scheduling, require the support of a central controller.

In recent years, numerous research efforts have been reported in the literature to enhance DCF to enable QoS provisioning. Constrained by the nature of distribution of ad hoc WLANs, most of the QoS MAC research efforts take the design objective as *soft* QoS, which means that the QoS objective is satisfied in a statistical manner. For example, higher priority packets will have better chance to access the media earlier on average over the long term, but the priority for a specific packet at a specific instant is not guaranteed. In other words, soft QoS means QoS differentiation rather than QoS guarantees.

This chapter provides a taxonomy and comprehensive review of the research efforts dedicated to QoS enhancements of the IEEE 802.11 DCF. To better understand the various protocols and algorithms, we categorize the proposals based on their design objectives and specifications and illustrate each of the categories with examples.

The remainder of this chapter is organized as follows. In Section 26.2, we will briefly review the operation of the IEEE 802.11 DCF. Section 26.3 will discuss the QoS leverages that exist in DCF-like protocols and how these leverages affect the packet access priority. In Section 26.4, we will present the QoS algorithms and protocols in four categories, namely, predetermined static differentiation, dynamic priority access, distributed fair scheduling, and distributed reservation access. Section 26.5 outlines the IEEE 802.11e

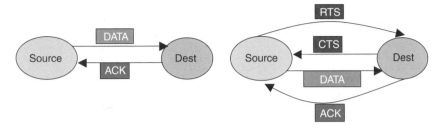

FIGURE 26.1   Two-way and four-way handshake.

EDCF, which is a QoS extension scheme in the process of standardization. Section 26.6 concludes the chapter and provides a discussion of future directions of QoS provisioning research in WLANs.

26.2   IEEE 802.11 Distributed Coordination Function

The IEEE 802.11 standard defines a contention-based MAC component, known as the DCF. The DCF defines three interframe space (IFS) intervals: the short IFS (SIFS), the point coordination function IFS (PIFS), and the DCF IFS (DIFS). Each of them defines a fixed time period.

The SIFS is the shortest interframe spacing and is used for the transmission of essential control frames in DCF. The PIFS is used by a node operating as an access coordinator in PCF mode to allocate priority access to the medium. It is one slot time longer than the SIFS. Nodes operating in DCF mode transmit data packets using the DIFS. It defines the fixed length of time that nodes have to wait before a random countdown procedure in order to access the channel eventually. The IEEE 802.11 DCF works as a listen-before-talk scheme, and operates in either basic access mode or RTS–CTS mode.

26.2.1   Basic Access Mode

As depicted in Figure 26.1, the basic access mode is a two-way handshaking mechanism between the transmitter and the receiver, which uses the data packet and an ACK packet.

In the basic access mode, a station starts its transmission immediately after the medium is sensed to be idle for an interval of length DIFS. If the medium becomes busy during the sensing period, the station defers until an idle period of DIFS length is detected and then further counts down a random backoff period before transmission. The backoff period lasts for the product of one time slot and an integral number. This integral number is referred to as the backoff value and is uniformly chosen in the range $(0, CW - 1)$, where CW stands for a runtime parameter called the Contention Window. The backoff value decrements every slot time when the channel is sensed idle until it reaches zero. The node then starts to transmit the queued packet when the backoff value counts down to zero. Upon proper reception of one data packet, the receiver responds with an ACK message after a SIFS time period. The DATA–ACK exchange is referred to as two-way handshaking. If the channel becomes busy before the backoff reaches zero, for example, due to the transmission of other nodes, the backoff procedure stops and the backoff value is frozen. The node will wait for another idle period of DIFS length, and resume the backoff procedure with the remaining backoff value.

If two or more nodes transmit at the same time, the transmissions are destined to collide with each other. When collisions take place, the nodes involved will timeout waiting for ACK packets and execute a recovery procedure. The recovery procedure first invokes a BEB algorithm to compute a new CW value. According to the BEB algorithm, CW is initially set to CW_{min}, a parameter that may be agreed upon among all operating nodes. CW is doubled every time a collision happens (implied by ACK timeout in IEEE 802.11) until it reaches the parameter CW_{max}. After CW reaches CW_{max}, CW will remain

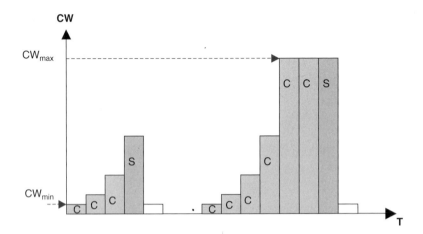

FIGURE 26.2   The BEB algorithm: first packet is transmitted successfully after retrying three times; the second packet is transmitted successfully after retrying six times.

unchanged for further retransmissions. CW is reset to CW_{min} when finally the retransmission succeeds. Figure 26.2 illustrates the BEB algorithm. After generating the new CW using the BEB algorithm, the new backoff value is drawn from $(0, CW - 1)$ using the same algorithm. The same backoff procedure applies afterwards.

The DCF basic access mechanism exchanges data using the two-way handshaking; hence the data is transmitted right after the node finishes the backoff procedure. However, since two or more nodes may happen to finish the random backoff procedure at the same time, a collision may occur and it would last for more than the length of the whole packet (packet length plus timeout) due to the lack of detection ability in the wireless nodes. So a collision will cause a large amount of resource waste. Moreover, the basic access mechanism makes no effort to deal with the "hidden terminal" problem [Bha94], which results in a large fraction of collision cases. The RTS–CTS mechanism addresses both of these problems by "acquiring the floor" before transmitting.

26.2.2 RTS–CTS Access Mode

The RTS–CTS access mode executes the same carrier sensing and random backoff procedure as in the basic mode. It also uses the same BEB algorithm to regulate the adjustment of the contention window. It enhances the basic access mode in two respects.

First, the RTS–CTS access mode enhances the basic mode two-way handshaking with RTS–CTS–DATA–ACK four-way handshaking (see Figure 26.3). When a node finishes the preceding carrier sensing and random countdown, instead of transmitting the data packet immediately, the node sends out a request-to-send (RTS) control packet to the intended receiver. Upon correct reception of the RTS message in an idle state, the receiver responds to the sender with a clear-to-send (CTS) control packet. After receiving the CTS message, the sender then starts transmitting the real data packet, and waits for the ACK packet as the confirmation of a successful data exchange.

Four-way handshaking decreases the chance of collision significantly since the RTS packet has much smaller length than the data packet. Even though collisions may sometimes occur, the resource wastage is minimized since collisions last only for the length of the RTS packet. Moreover, the RTS–CTS exchange can solve the hidden terminal problem. By setting the network allocation vector (NAV), which is a field of RTS/CTS control message, as the length of the ongoing packet transmission or larger, the nonintended-destination nodes, which overheard the RTS or CTS, are prevented from transmitting until the end of the current transmission. Thus the nodes within the radio transmission range of the receiver would not start transmissions that would cause collisions at the receiver of an ongoing transmission. The same benefit

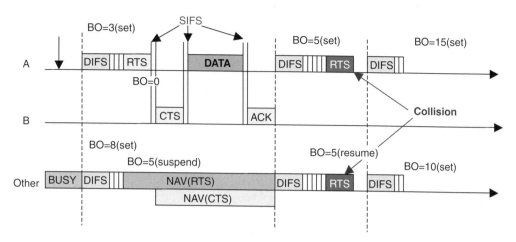

FIGURE 26.3 An example of virtual carrier sense and operation of DCF in RTS/CTS mode. The first transmission is successful from node A to B. Other nodes are prohibited to transmit by virtual sensing.

applies to the transmitter since the ACK message would not be damaged by the transmission from its neighbors.

The DCF provides an efficient random access mechanism for WLANs. However, QoS is not considered in the original design. All packets, of any user or any application are treated in exactly the same manner, follow the same transmission procedure and use the same control parameters. DCF, therefore, does not support service differentiation nor does it provide QoS guarantees.

26.3   QoS Leverages of Distributed MAC Protocols

As the core mechanism of DCF function, the CSMA protocol has been proven to be an efficient distributed MAC protocol, especially under light to medium traffic load [Kle75]. However, DCF does not support any level of QoS guarantee, such as access delay bounds, or QoS differentiation, such as bandwidth sharing with weighted fairness. No matter what is the type of the head of line (*HOL*) packet (data or voice), a backlogged mobile node competes for a transmission opportunity by executing the same carrier sensing and backoff procedure using identical parameters. Thus, in a certain time instance, it is unpredicted which station or type of station will win the next access. Thus, if the transmission of a voice packet collides with another concurrent transmission, the contention window of the voice station will be increased. Consequently, the delay sensitive voice station will wait for a longer period before gaining access to the channel.

In order to support QoS differentiation in WLANs, it is intuitive to differentiate the parameters of DCF algorithm according to QoS requirements and fluctuation of the traffic load. Recall that mobile stations running the DCF have to sense the channel idle for a fixed time period (interframe spacing) followed by some randomly generated time period (backoff interval) before accessing the channel. The basic idea of QoS differentiation is setting divergent protocol parameters per priority class such that the waiting time of higher priority packets is shorter than that of lower priority packets. Since the distinct parameters will lead to the differentiated service quality, we refer to them as QoS leverages.

The first leverage, known as DIFS in DCF (AIFS in IEEE 802.11e EDCF, and PIFS in black burst (BB)), is the fixed length of time that the node has to sense the media as idle before backing off. With the different designs, the differentiated DIFSs will typically make QoS effects in two manners, absolute priority or relative priority.

For example, Aad and Castelluccia [Aad01a,b] proposed one distinct DIFS length for each traffic priority of total two, so that the DIFS of a low priority flow will not be less than the sum of the DIFS length and the

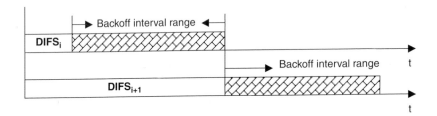

FIGURE 26.4 Differentiated DIFSs for absolute priorities.

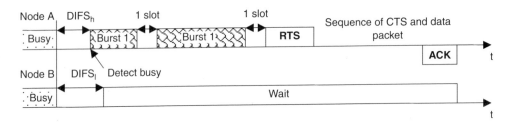

FIGURE 26.5 Differentiated DIFSs and jamming signal for absolute priorities.

contention window size of a high priority flow. Consequently, a low priority flow will have no chance to access the channel if a high priority flow is backlogged (see Figure 26.4). A major difficulty with this design is the proper selection of distinct DIFS values. A large difference between DIFSs will result in capacity waste due to the channel idle state when there is very little high-priority traffic; while a small difference between DIFSs (inferring that CW is small) will lead to excessive network contention if there are a large number of competing flows of same priority. Therefore, the optimal value is largely affected by the volume of traffic load of each level. The proposal in [Ban02b] considered a different design approach for absolute priority by introducing a jamming signal between the carrier sensing period and the random backoff period. It uses two distinct DIFSs for two levels of traffic flows. The shorter DIFS, denoted by $DIFS_h$, is used for high-priority flows, while the longer DIFS, denoted by $DIFS_l$, which is one time slot longer than $DIFS_h$, is used for low-priority flows. When a high-priority node tries to access the channel, it has to first sense the channel idle for a time period of $DIFS_h$ length. Afterwards, the station will jam the channel with a burst of energy for a random length of time. The backlogged low-priority flows, which have to sense the channel idle for a time period of length $DIFS_l$, will sense the jamming signal and backoff. Thus a high-priority flow will always be able to access the channel before the low-priority flows (see Figure 26.5).

The IEEE 802.11e enhanced DCF (EDCF) [Wg99] is an example of using the differentiated DIFSs (called AIFSs in EDCF) for relative priority differentiation. The length difference between two AIFSs is an integer number of time slots, with the shortest AIFS equal to DIFS in DCF. Since the long AIFS is usually less than the sum of the short AIFS plus the contention window (see Figure 26.6), the packet using the short AIFS has a greater chance to gain an earlier transmission opportunity than a packet using the long AIFS. However, such privilege is not guaranteed for any specific access instance.

Besides, there are three QoS leverages affecting the backoff procedure, that is, the contention window boundary values, CW_{min} and CW_{max}, and the contention window enlargement factor, which is fixed as two for DCF. As described in Section 26.2, CW_{min} is the initial contention window value. The enlargement factor regulates the rate of contention window enlargement when a collision happens – the contention window will be increased by the time as enlargement factor for each consecutive collision until it reaches CW_{max}. Further QoS differentiation can be achieved using both the DIFS and backoff parameter leverages than by using the DIFS leverage alone.

If two flows with the same DIFS compete with each other for a transmission opportunity, the high-priority flow should have smaller CW_{min}, smaller CW_{max}, and smaller enlargement factor. During the

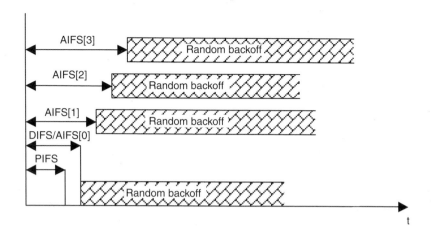

FIGURE 26.6 The IEEE 802.11 EDCF uses differentiated AIFS.

competition for a given transmission opportunity, the high priority flow will more likely choose a smaller backoff counter and finish the backoff earlier. Upon a collision, the contention windows of both flows will be increased approximately proportional to the enlargement factor. Thus, the higher priority flow will have a smaller contention window for the next competition. Note that the differentiation on backoff parameters can only provide statistical QoS differentiation, but no QoS bound. In contrast, the proposal in [Vai00], instead of regulating the backoff value indirectly through the three backoff parameter, calculates and assigns the backoff value directly according to the "finish tag," calculated by using a distributed self-clocked fair queuing (SCFQ) transmission algorithm.

The persistence factor is another optional QoS leverage. Ozugur et al. [Ozu02,Ozu99,Ozu98] tried to address fairness mechanisms in multi-hop WLAN networks using a combination of p-persistent CSMA and DCF. In the proposal, which can also be used to provide service differentiation in single-hop WLANs, a station sends the RTS packet with probability p after the backoff period, or backs off again with probability $1 - p$ using the same contention window size. We refer to the probability p as persistence factor. It is obvious that a station using a large value p will retain higher access priority. Ozugur et al. discussed two methods, the connection-based media access method and the time-based balanced media access method, to calculate the persistence factor value for every link so that fairness can be maintained between flows.

Lastly, some proposals, such as [Sob99,Ban02b], attach an extra priority-based contention resolution stage on the end of carrier sensing. In specific, these proposals use jamming signals of differentiated length as the QoS leverage. It is helpful to think of the length of the jamming signal as the "negative" of the backoff interval since longer jamming signals correspond to higher access priority. The operation of a MAC protocol using jamming signals is typically as follows. After sensing the channel idle for a period of length DIFS, the mobile station will send a burst of jamming signals onto the channel. Afterwards, the mobile station will listen to the channel. If the channel is sensed busy, then the station will quit the competition and try for the next round. Otherwise, the station will continue with its access protocol until it gets access to the channel or quits. For QoS purposes, the length of the jamming signal should reflect the priority of the HOL packet in the queue. The well-known BB MAC protocol [Sob99] is one example of using differentiated jamming signals.

26.4 QoS Enhancement Proposals

The Quality of Service has been one of the major concerns of distributed MAC protocol design for WLANs since the early 1990s. In [Bha94], the unfairness property of a distributed multiple access protocol, multiple

TABLE 26.1   Taxonomy of QoS-Based DCF Extensions

Categories	Examples
Predetermined static differentiation	[Den99,Den00,Kan01b,Ver01,Bar01]
Distributed priority access	BB [Sob99], DPS [Kan01a]
Distributed fair packet scheduling	DFS [Vai00], DWFQ [Ban02a], FS-FCR [Kwo03]
Distributed reservation access	DBASE [She01b], GAMA [Mui97]

access collision avoidance (MACA), was discussed in detail. Since then, researchers have made a lot effort to extend and enhance CSMA/CA-based MAC protocols so that they can meet various QoS objectives in a distributed manner.

In this section, we present and discuss these QoS enhancement algorithms and protocols. According to the different QoS objectives (as the group name implies), we categorize them into four groups as shown in Table 26.1. In the rest of this section, they are explained in detail with examples.

For the sake of clarity and simplicity, we assume each station carries only one flow of a certain priority/type. Thus, the words "station" and "flow" will be used interchangeably. Little modification in algorithm description would be made when multiple flows are present in one station. Interested readers can refer to the corresponding references.

26.4.1 Predetermined Static Differentiation

Proposals in this category are relatively simple and fundamental compared to those of other categories. The major contribution of these researches is mostly on the illustration and evaluation of the effect of the QoS leverages discussed in Section 26.3, either individually or with multiple leverages combined together. The research does not define any quantitative QoS objectives nor employ any traffic isolation mechanism. Thus, the traffic load fluctuation of one flow will cause changes on the QoS that other flows experience. In most cases, the QoS performance is illustrated by measuring the output metrics under a heterogeneous traffic load.

In [Den99], Deng et al. proposed a protocol enhancement with two priority classes. Therein, the high priority source stations choose the backoff value from the interval $[0, 2^{i+1} - 1]$ while low priority source stations choose the backoff value from the interval $[2^{i+1}, 2^{i+2} - 1]$, where i is the number of consecutive times a station attempts to send a frame. Thus, the initial backoff values are in disjoint intervals. It is worthy to note that more priority levels can be obtained by modifying the backoff algorithms in a similar way [Den00].

Likewise, Kang and Mutka [Kan01b] proposed three formulas to calculate the backoff interval length for three traffic priorities. A high priority node chooses the backoff interval from the range $[0, A - 1]$, where A is a fixed integer. A medium priority node uses the original BEB algorithm to maintain the contention window value. However, it chooses the backoff value from $[A, A + \lceil CW/B \rceil]$ instead of $[0, CW - 1]$, where CW is the current contention window and B is another fixed integer. A low priority node also operates the original BEB algorithm, and it chooses the backoff value from $[A, A + CW]$. The research in [Kan01b] evaluated the QoS performance when $A = 8$, and $B = 3$. All proposals above use different backoff values to achieve differentiated service. Other research efforts taking a similar approach include [Ver01,Bar01].

The predetermined differentiated parameter approach cannot lead to quantitatively controllable QoS differentiation among flows as the traffic load fluctuation of one flow will affect the QoS experienced by other flows, since there is no traffic isolation mechanism involved. Moreover, static CW sizes cannot balance the tradeoff between collisions and the excess waiting time. To solve these difficulties, a practical QoS scheme should take into account the following three issues (1) a quantitative definition of the QoS

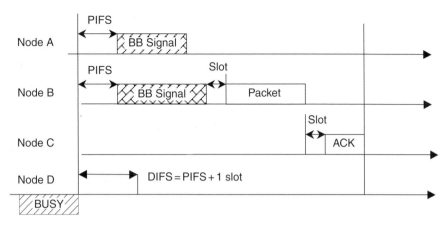

FIGURE 26.7   Illustration of BB algorithm: differentiated IFSs and jamming signals.

objectives; (2) as the traffic load fluctuates, the QoS leverages should be dynamically calculated and adjusted so as to maintain the QoS objectives; (3) some traffic isolation mechanism, such as multiple output queues, can help prevent a change of traffic load of one class from causing a violation of the QoS objectives of other classes. In the following three sections, we take a close look at some examples that attempt to incorporate the aspects above.

26.4.2 Distributed Priority Access

An important QoS objective of DCF-like protocols is the support of distributed prioritized access. For example, a real-time voice packet should have higher access priority to the channel compared with a best-effort data packet. As another example in the earliest deadline first (EDF) paradigm, the packets are scheduled to be served in an increasing order of packet delivery deadline. The challenge is to satisfy such requirements in a distributed contention-based infrastructureless environment, viz. ad hoc WLANs. In the following paragraphs, we discuss how BB [Sob99] provides two-level priority access and how distributed priority scheduling (DPS) [Kan01a] implements EDF scheduling, both in a distributed manner.

The BB protocol was proposed to provide two-level priority channel access to voice and data flows in an ad hoc network. The BB scheme employs two QoS leverages: the IFS and the attached jamming signal after carrier sensing. Consider the example in Figure 26.7. Stations A and B are two backlogged voice stations, and station D is a backlogged data station. Stations A and B sense the idle channel for a fixed period of time of length PIFS. After this, both A and B will send a jamming signal pulse, whose length is proportional to the queuing time of the HOL packet in the respective station. In our case, the jamming signal from station A is longer than that from station B. Meanwhile, the data station accesses the channel using the standard DCF procedure. Since voice stations have a short IFS, in Figure 26.7, station D will sense the jamming signal during the sensing period DIFS. After finishing sending their jamming signals, the voice stations A and B once again sense the channel. Station A senses the channel as busy and therefore backs off for the next transmission opportunity, while station B senses the channel as idle for the sensing period of one time slot and thus begins transmission.

The BB mechanism provides QoS differentiation in two respects. First, voice packets always have absolute priority over the data packets, since a data station will be blocked by the BB signal sent by voice stations after the PIFS sensing period. Second, the differentiated lengths of the jamming signals provide a further level of priority differentiation among voice packets, with voice packets that have been waiting longer having higher priority.

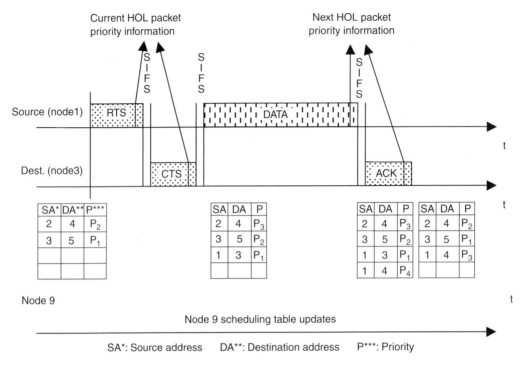

FIGURE 26.8   Illustration of scheduling tables in DPS.

In [Kan01a], Kanodia et al. proposed a distributed an priority scheme called DPS to realize centralized EDF in a distributed manner. EDF schedules packets in increasing order of priority indices, where, typically, the priority index of a packet is calculated as the sum of the packet's arrival time and its delay bound.

In order to decentralize the EDF scheme, two issues have to be addressed explicitly, that is, the packet priority information sharing and a mapping function from scheduling order to the backoff parameters. In an infrastructureless WLAN, the priority index of a packet must become known to the dispersed wireless nodes. Therefore, an information collection and sharing mechanism has to be employed in order that nodes can determine, in a collaborative manner, the next packet that would be transmitted in a (hypothetical) centralized and ideal dynamic priority scheduler. DPS solves the priority information sharing issue by exploiting the broadcast nature of a wireless channel. Specifically, a node broadcasts the priority information by piggybacking the priority index of the HOL packet on the RTS/CTS/DATA/ACK dialogue, while other nodes will store the overheard priority information locally in a scheduling table. Figure 26.8 shows an example of DPS priority information broadcasting. Assume node 1 gains the channel access opportunity for the next packet transmission to node 3. First, the priority index of the current HOL packet at node 1 will be piggybacked on the RTS message. If the RTS is received correctly, node 3 will respond with a CTS packet on which the same priority message will be attached. The neighboring node 9 overhears the RTS/CTS dialogue and inserts the HOL packet priority index into its scheduling table. Afterwards, node 1 starts to transmit the data packet to node 3, and node 3 finishes the dialogue by sending back an ACK packet. Again, the priority information of the next HOL packet in node 1 will be broadcast. This time, the HOL packet would be the packet following the current data packet since it is expected that the current packet will be transmitted properly. The priority information on the ACK has another usage of canceling the entry for the currently transmitted packet, such as the transition from the third table to the fourth table in Figure 26.8.

As well, the DPS scheme comprises an algorithm of mapping priority information to the appropriate backoff interval, which reflects the packet priority. In the DPS scheme a node calculates a packet's backoff interval as follows:

$$f(l, r) = \begin{cases} \text{Unif}[0, 2^l \text{CW}_{min} - 1], & r = 1, \quad l < m \\ \alpha \text{CW}_{min} + \text{Unif}[0, \gamma \text{CW}_{min} - 1], & r > 1, \quad l = 0 \\ \text{Unif}[0, 2^l \gamma \text{CW}_{min} - 1], & r > 1, \quad l \geq 1 \end{cases} \quad (26.1)$$

where r is the rank of the node's packet in its own scheduling table, m is the maximal number of attempts before the packet is discarded, l $(0 < l < m)$ represents the number of retransmission attempts, and α and γ are two static parameters.

It should be noted that the DPS scheme implicitly assumes all network nodes are synchronized so that the priority indices will be consistent. However, this is not a feasible assumption in a multi-hop network. A solution to this problem is to calculate the priority index as the delay bound minus the queuing delay. The new priority index bears the same logic in indicating the priority of the packet; that is, the smaller the value is, the higher the priority. If the value becomes negative, the packet should be discarded.

26.4.3 Distributed Fair Scheduling

Another approach to QoS provisioning in WLANs is distributed fair scheduling (DFS). The objective of fair scheduling can be described by an idealized fluid flow model as defined in Equation (26.2). Therein, r_i denotes the bandwidth used by flow i and ϕ_i denotes the weight of the flow; that is, the service share of each flow should be proportional to the flow weight during any period of time. It has been proven that fair scheduling provides end-to-end delay bounds for any well-behaved traffic flows [Par93].

$$\frac{r_i}{\phi_i} = \frac{r_j}{\phi_j} \quad \forall i, \quad \forall j \quad (26.2)$$

There is a critical difference between fair scheduling in a wired network and that in a single-hop ad hoc WLAN. For the former, the fairness model applies to multiple flows of a single node, while for the latter the fairness paradigm applies to the flows of multiple nodes since these flows share the same bandwidth. In other words, WLAN fairness scheduling tries to allocate bandwidth among flows belonging to dispersed nodes in the network, complying with the fairness objective defined in Equation (26.2). The distribution of flows makes the design of a fair scheduling algorithm in WLANs a challenging task.

We first remark that none of the existing packet scheduling algorithms can be directly deployed in ad hoc WLAN to achieve fair scheduling due to the lack of a centralized coordinator for the distributed flows. As the priority information sharing is essential to the DPS schemes, the sharing of the service share/progress information of individual flows is a critical aspect of the distributed implementation of fair scheduling. This is because the goal of *a distributed implementation of fair scheduling paradigm* is to make the same service order decisions as a hypothetical centralized scheduler would. The service progress information varies with the schemes in terms of the specific contents, such as the finish tag in [Vai00,Kwo03], the service label in [Ban02a], and the successful/collided transmissions and idle period in [Qia02]. Also, the service progress information sharing mechanisms differ in terms of the methods of information acquisition and dissemination, such as constantly monitoring the network activities [Qia02], or piggybacking the information on data or control packets [Vai00,Kwo03,Ban02a]. Next, we discuss two distributed fair scheduling proposals.

Distributed fair scheduling was proposed by Vaidya in [Vai00]. Most of the centralized scheduling algorithms, such as weighted fair queuing (WFQ), need to know whether other flows are backlogged or not in real-time. The real-time flow backlog status update will generate a large amount of traffic load to the network. In order to eliminate such overhead, DFS chooses to emulate the self-clocked fair queuing (SCFQ) scheduling model. SCFQ only requires a minimal amount of status information sharing, which is the finish tag of the last transmitted packet. The reader interested in the details of the SCFQ algorithm should consult the reference [Ban02a].

The DFS scheme is a SCFQ implementation based on the IEEE 802.11 MAC. The key idea has three aspects (1) calculate the packet finish tag as the centralized SCFQ scheduler does; (2) piggyback the finish tag on the control packet and data packet, while other nodes retrieve the finish tag from the overheard control/data packet, and update their virtual time; (3) translate the finish tag of the packet into the backoff interval value, so that the packet with the smaller finish tag will have a smaller interval value and therefore be served earlier. The actual channel access should follow the original DCF function defined in the IEEE 802.11 standard. Below is a brief summary of how these ideas are implemented in DFS.

Each node i maintains a local virtual clock $v_i(t)$, where t is the actual time and $v_i(0) = 0$. Let P_i^k represent the kth packet arriving at node i.

- Each transmitted packet is tagged with its finish tag.
- When node i either overhears or transmits a packet with finish tag Z at time t, it sets its virtual clock v_i equal to $\max(v_i(t), Z)$.
- The start and finish tags for a packet are only calculated when the packet becomes the HOL packet of the node. Assume packet P_i^k reaches the front of node i. The packet is stamped with start tag S_i^k, calculated as $S_i^k = v_i(f_i^k)$, where f_i^k denotes the real time when packet P_i^k reaches the front of the ith flow. The finish tag, F_i^k, is calculated as in Equation (26.3). Note that L_i^k is the length of packet P_i^k and the scaling factor in Equation (26.3) is not an essential component of the SCFQ algorithm itself. A well-chosen value will help the translation procedure as shown below.

$$F_i^k = S_i^k + \text{Scaling\_Factor}^* \frac{L_i^k}{\phi_i}$$

$$= v(f_i^k) + \text{Scaling\_Factor}^* \frac{L_i^k}{\phi_i} \qquad (26.3)$$

- The objective of the next step is to choose a backoff interval such that a packet with a smaller finish tag will ideally be assigned a smaller backoff interval. This step is performed at time f_i^k, that is, when the packet becomes the first packet in the queue. Specifically, node i picks a backoff value B_i for packet P_i^k as a function of F_i^k and the current virtual time $v_i(f_i^k)$, as $B_i = \lfloor F_i^k - v(f_i^k) \rfloor$ slots. Now, replacing F_i^k with Equation (26.3), the above expression reduces to $B_i = \lfloor \text{Scaling\_Factor}^* L_i^k / \phi_i \rfloor$. Finally, to reduce the probability of collisions, B_i above is randomized as $B_i = rand(0.9, 1.1)^* B_i$.

   The DFS can achieve long-term weighted fairness among the nodes in the network, while short-term fairness will be violated occasionally due to the randomization at the last step and the possible collisions.

Distributed weighted fair queuing (DWFQ) is a heuristic distributed fairness scheduling scheme proposed by Banchs in [Ban02a]. DWFQ is similar to DFS in several respects. First, both schemes are based on the DCF contention-based algorithm. Second, based on the overheard and local service share information, the nodes compute the random backoff interval parameter locally in order to achieve the service ratios among flows defined in Equation (26.2).

On the other hand, DWFQ has a few characteristics deviated from DFS. Instead of emulating the existing centralized scheduling scheme, SCFQ, in DFS, DWFQ takes the feedback-control-based approach. In DWFQ, a node compares its service progress with others, and adjusts its backoff parameters accordingly. As an inherent property of the feedback control paradigm, the adjustment is always based on past information, which leads to a gross implementation of fair scheduling. Moreover, instead of calculating a backoff interval, DWFQ controls the backoff time indirectly through the contention window. In other words, a DWFQ node calculates its contention window value based on comparison of its service share and other nodes' shares, and then the backoff interval is generated from the contention window value. The service share of node i is represented by the normalized service rate, calculated as $L_i = \gamma_i / \phi_i$, where γ_i is the estimated (moving average) bandwidth experienced by flow i and ϕ_i is its weight. The operation of the DWFQ algorithm is as follows:

- Node i updates its estimated throughput γ_i whenever a new packet is transmitted using a smoothing algorithm as

$$\gamma_i^{\text{new}} = (1 - e^{-t_i/K})\frac{l_i}{t_i} + e^{-t_i/K}\gamma_i^{\text{old}} \qquad (26.4)$$

  where l_i is the packet length and t_i is time interval since the last transmission, and K is a constant.
- With the above definition of the normalized service rate, L_i, the resource distribution expressed in Equation (26.4) can be achieved by imposing the condition that L_i should have the same value for all flows, that is, $L_i = L \ \forall i$.
- Every packet in DWFQ will have its flow's service rate label L_i, piggybacked with its transmission. Other stations in the WLAN domain can retrieve the corresponding label from the packet. CW adjustment is based on comparing the local rate, L_{own}, and the observed rate of every other station, L_{rcv}. It is calculated recursively for every L value received. Let p denote the CW scaling factor. Then for each observed packet:

$$\text{If}\,(L_{\text{own}} > L_{\text{rcv}})\ \text{then}\ p = (1 + \delta_1)p$$

$$\text{Else If (queue\_empty)}\ \textit{then}\ p = (1 + \delta_1)p$$

$$\text{Else}\ p = (1 - \delta_1)p$$

$$p = \min\{p, 1\}$$

$$\text{CW} = p \cdot \text{CW}_{802.11}$$

where $\delta_1 = 0.01 \times |(L_{\text{own}} - L_{\text{rcv}}/L_{\text{own}} + L_{\text{rcv}})|$ and $\text{CW}_{802.11}$ refers to the CW size maintained by the station from the BEB algorithm.

Although the specific formulas and parameters are somehow arbitrary, the outline of the algorithm above reflects its design objectives. First, if the local service rate is higher than the overheard service rate, then the CW scaling factor is increased, which will slow down the local transmission rate. Second, if the local service rate is lower because the queue is empty, the scaling factor is also increased. Third, if the local service rate is lower and the queue is not empty, the scaling factor is decreased.

26.4.4 Distributed Reservation-Based Access

The algorithms discussed earlier only provide soft (statistical) QoS due to the stochastic nature of the contention-based access operations. It is never guaranteed that a packet of high priority or with a smaller finish tag (requiring earlier service) will gain the next transmission opportunity against a packet of low priority or with a larger finish tag. The algorithm also does not guarantee timely medium access within the delay bound. The ideal access order is preserved in a statistical manner.

Reservation-based algorithms [Mui97,She01b] try to provide delay-bounded transmission service to real-time applications in a distributed manner. When a new real-time flow arrives, the flow will first reserve its access opportunities by sending a "join the transmission group" request. Requests are granted with transmission slot(s) or transmission sequence number if there is enough resources are still available. Afterwards, the transmission opportunity is secured for the duration of the flow. Since the group transmission repetition interval is bounded by a predetermined value, the access delay and bandwidth will be bounded for any group-member flow. The basic operation of reservation-based algorithms is outlined below:

- Time axis is segmented into recurring cycles. Each cycle contains a contention period and a group-transmission period, during which one or more stations transmit real-time packets without collisions.
- There is a limit on the length of the group-transmission period, which translates to a bounded time interval between occurrences of a transmission period.
- A new flow competes for the right to be added to the transmission group by executing a CSMA/CA-based access procedure and appending the join request on an RTS message.
- The destination will decide whether enough resources are available for the new flow. If this is the case, a CTS will be sent back by the destination, appended with the grant information.
- In a single-hop network, both RTS and CTS messages will be heard by all the stations. Since the resource request and grant details are included in the RTS and CTS message, every station can track all the flows and the resource availability. As an option, the first station of each transmission group can broadcast its knowledge about the traffic flows and their resource reservation at the beginning of every group transmission period to ensure that every station has the same view of the network status.
- Once a station is a member of the transmission group, it is able to transmit a collision-free real-time packet during each cycle. As long as a station has data to send, it maintains its position, which is represented by a sequence ID (SID) in the group. In some cases, a station also maintains the total number of active stations, referred to as AN.
- During the group transmission period, every station should listen to the channel in order to track the group transmission progress. Each station will know when to transmit its next packet by comparing its own SID with the current progress. The station also knows whether a flow has finished its session and quit the transmission group by detecting idles for the timeout period. In the latter case, the SID number of every station after the finished flow should be adjusted.

The basic operation provides delay-bounded service to synchronous real-time traffic flows with homogenous traffic characteristics, such as packet arrival rate. However, it behaves awkwardly if the traffic volume fluctuates with time, such as with variable bit rate (VBR) flows.

The DBASE [She01b] proposes a set of bandwidth allocation/sharing mechanisms to solve this problem. For allocation purposes, every station declares its bandwidth requirement for the next cycle in every transmitted MAC packet. If the required bandwidth is smaller than or equal to the average bandwidth (a preset threshold value corresponding to traffic type), the bandwidth is granted. On the other hand, if the required bandwidth is greater than the average bandwidth, only average bandwidth will be granted. Note that some stations may have bandwidth requirement lower than average bandwidth for the next cycle due to traffic fluctuation. Furthermore, some flows may finish transmission and quit the transmission group, which is detected by a series of idle slots exceeding a timeout. Spare resources will be tracked by all stations, and is shared by stations which have greater-than-average bandwidth requirements in a fair manner.

Although the reservation-based distributed QoS MAC protocols have the ability to guarantee delay-bounded service to synchronized packet flows, the performance of such protocols will deteriorate in an error-prone wireless communication environment. For example, in the DBASE protocol, a station determines its next transmission opportunity according to its SID, which is contained in the reservation frame (RF) generated and broadcast by the contention free period generator (CFPG) at the beginning of each cycle. Any failure of reception of the RF by a mobile station will be misunderstood as the absence of CFPG. In consequence, the corresponding node will regard itself as the CFPG. The conflicted *RFs* might rise in the network, which may lead to the violated transmission sequences.

26.5 IEEE 802.11e EDCF

The IEEE 802.11 working group has long recognized the need of a QoS MAC for future QoS provisioning in WLANs. A standard draft, known as IEEE 802.11e [Wg03], has been under serious investigation and consideration.

The enhanced distribution coordination function (EDCF) is a QoS extension to the DCF function. Under EDCF, each node operates up to four prioritized queues in order to provide differentiated QoS to the different flows of traffic. These independent queues, as illustrated in Figure 26.9, also called virtual stations in the IEEE 802.11e standard, execute the same random access procedure as a DCF node, but with differentiated, access category (AC) specific parameters. Using AC-specific parameters, EDCF can provide the differentiated QoS to various data flows.

In EDCF, there are three AC-specific parameters. The first is the arbitration interframe space (AIFS). The AIFS is the counterpart of the DIFS in DCF, and its length is a function of the access category, having the form as $AIFS = DIFS + k * slot\_time$, where $k \in [0, 7]$. Also, EDCF uses AC-specific CW_{min} and CW_{max} to regulate the backoff and retransmission procedure. The contention window is originally set to CW_{min}. Upon the occurrence of packet collision, CW is calculated as

$$CW[AC] = (CW[AC] + 1)^* 2 - 1 \tag{26.5}$$

Enlargement of the CW is limited by CW_{max}. So if the new CW calculated using Equation (26.5) is greater than CW_{max}, the new CW would be CW_{max}.

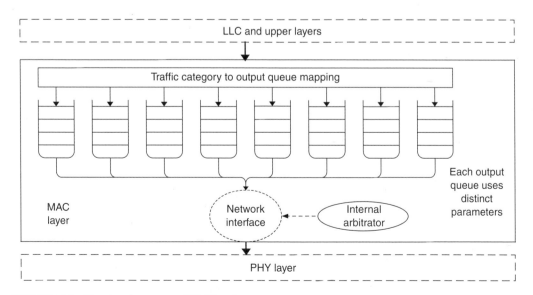

FIGURE 26.9 The virtual structure of EDCF nodes.

During the operation of WLANs, the value of AIFS, CW_{min} and CW_{max} are announced by the QoS access point (QAP) via periodically broadcasted beacon frames. The QAP can also adaptively tune these EDCF parameters based on the network traffic conditions.

In EDCF, packets are delivered through multiple virtual stations within one node, and each virtual station instance is parameterized with AC-specific parameters. A virtual station independently contends for a transmission opportunity as follows. It first senses the channel being idle for an AIFS. After for the idle period as AIFS, the virtual station sets a backoff counter to a random number drawn from the interval [0, CW]. If the channel is determined busy before the counter reaches zero, the backoff has to wait for the channel being idle for an AIFS again, before continuing to count down the backoff value. Upon a transmission collision, CW is enlarged as stated above and a new backoff value is generated randomly from [0, CW] again. If the counters of two or more parallel virtual stations in a single station reach zero at the same time, a scheduler inside the station avoids the virtual collision. The scheduler grants the transmission opportunity to the virtual station with highest priority, out of the virtual stations that virtually collided within the station. All other virtual stations will invoke another random backoff procedure. Yet collisions may occur between transmissions by different physical nodes.

The EDCF provides prioritized channel access by running the separate virtual stations in one physical station. Each virtual station follows the same random access procedure but with differentiated parameters. With well-designed mapping mechanisms from application data flows to MAC layer traffic categories, integration of data traffic of different QoS characteristics is feasible.

The EDCF is a compound QoS differentiation platform for WLANs, which is not bounded to any specific QoS paradigm, as stated in Section 26.4. In other words, EDCF is a building block to a complete QoS (*soft or absolute*) provisioning framework. Some research [Gri02,Man02,Rom02,Rom03,Xu03] has been conducted to investigate the performance of QoS support under EDCF. The simulation results in [Gri02,Man02,Rom02,Rom03] have illustrated the QoS differentiation effect of EDCF under various traffic conditions. The analytical evaluation in [Xu03] shows that the performance of certain flows are affected by a number of factors, including priority class of the flow, priority classes of other flows, parameters of each priority class, and the traffic load of other flows.

26.6 Summary

This chapter discusses QoS issues in ad hoc WLANs, with emphasis on QoS-based MAC algorithms and protocols. Existing ad hoc WLANs do not provide QoS packet delivery at any level. Typically, an ad hoc WLAN uses the IEEE 802.11 DCF to control the media access operations of the distributed wireless nodes. DCF is a discrete-time version of CSMA/CA protocol and serves efficiently in ad hoc WLANs without the help of a centralized controller. Moreover, DCF treats all packets in exactly the same manner, without distinguishing the packet service requirement, by enforcing the identically-parameterized carrier sensing and backoff procedure. Therefore, in any time scale, the network bandwidth is shared by all flows in a statistically fair manner. However, the lack of controllability and QoS differentiation renders DCF inefficient for QoS provisioning.

As a critical part to QoS provisioning in *WLANs*, QoS MAC design has to address two major and unique characteristics: the intrinsically error-prone and time-varying wireless communication environment, and the absence of the central controller. Specifically, the former factor leads to the unreliable delivery of data/control packets and the latter factor makes existing centralized QoS algorithms, such as weighted fairness scheduling, not applicable in a distributed ad hoc WLANs.

Despite of these challenges, a number of research studies have been conducted to enhance IEEE 802.11 DCF with QoS capability. This chapter reviews such studies beginning with highlighting a number of QoS leverages, that is, differentiable CSMA/CA parameters, discussing the usage and impact of each of these leverages. Based on their QoS objectives, we classify QoS-enhancement proposals into four categories, namely, predetermined static differentiation, distributed priority access, distributed weighted fair scheduling, and distributed reservation access. Each of the categories is discussed in detail with examples.

As has been discussed in the chapter, the design of MAC protocol for ad hoc WLANs that provide QoS guarantees is very challenging. Indeed, none of the existing schemes can achieve such a goal. Although distributed reservation access proposals declaim the capability of providing delay-bound service to real-time traffic, the implicit assumptions of reliable wireless links and perfect information exchange is impractical in real life. Therefore, most of research efforts have taken the approach soft QoS, which means that the QoS objective is satisfied in a statistical manner. That is, higher priority packets will have better chance to access the media earlier on average over the long term, but the priority for a specific packet at a specific instant is not guaranteed.

It is, therefore, our view that future research in QoS-based MAC design in ad hoc WLANs should concentrate on architectures, algorithms, and mechanisms that enable QoS guarantees as opposed to relative QoS differentiation. This will facilitate the deployment of wireless multimedia services over WLANs and help provide seamless integration of wireless cellular and wireless local networks.

References

[Aad01a] I. Aad and C. Castelluccia, "Differentiation mechanisms for IEEE 802.11," in *Proceedings of the IEEE INFOCOM 2001*, Anchorage, AL, USA, April 2001.

[Aad01b] I. Aad and C. Castelluccia, "Introducing service differentiation into IEEE 802.11," in *Proceedings of the IEEE ISCC 2000*, Antibes, France, July 2000.

[Ban02a] A. Banchs and X. Perez, "Distributed weighted fair queuing in 802.11 wireless LAN," in *Proceedings of the IEEE ICC 2002*, New York, USA, May 2002.

[Ban02b] A. Banchs, M. Radimirsch and X. Perez, "Assured and expedited forwarding extensions for IEEE 802.11 Wireless LAN," in *Proceedings of the IEEE/ACM IWQoS 2002*, Miami Beach, FL, USA, May 2002.

[Bar01] M. Barry, A. T. Campbell and A. Veres, "Distributed control algorithms for service differentiation in wireless packet networks," in *Proceedings of the IEEE INFOCOM 2001*, Anchorage, AL, USA, April 2001.

[Bha94] V. Bharghavan, "MACAW: a media access protocol for wireless LANs," in Proceedings of the *ACM SIGCOMM 1994*, London, UK, September 1994.

[Den99] D.J. Deng and R.S. Chang, "A Priority scheme for IEEE 802.11 DCF access method," *IEICE Transactions on Communications* Vol. E82-B, 1999, pp. 96–102.

[Den00] D.J. Deng and R.S. Chang, "A Nonpreemptive Priority-Based Access Control Scheme for Broadband Ad Hoc Wireless Local Networks", IEEE Journal on selected areas in communication, Vol. 18, No. 9, September 2000, 1731–1739.

[Gri02] A. Grilo and M. Nunes, "Performance evaluation of IEEE802.11e," in Proceedings of the *IEEE PIMRC 2002*, Lisbon, Portugal, September 2002.

[Kan01a] V. Kanodia, C. Li, A. Sabharwal, B. Sadeghi and E. Knightly, "Distributed multi-hop scheduling and medium access with delay and throughput constraints," in *Proceedings of the ACM MOBICOM 2001*, Rome, Italy, July 2001.

[Kan01b] S. Kang and M.W. Mutka, "Provisioning service differentiation in ad hoc networks by modification of the backoff algorithm," in *Proceedings of the IEEE ICCCN 2001*, Scottsdale, Arizona, USA, October 2001.

[Kle75] L. Kleinrock and F.A. Tobagi, "Packet switching in radio channels: part I — carrier sense multiple access modes and their throughput-delay characteristics," *IEEE Transactions on Communications*, Vol. 23, 1975, pp. 1400–1416.

[Kwo03]   Y. Kwon, Y. Fang and H. Latchman, "A novel MAC protocol with fast collision resolution for wireless LANs," in *Proceedings of the IEEE INFOCOM 2003*, San Francisco, CA, USA, March 2003.

[Man02]   S. Mangold, S. Chio, P. May, O. Klein, G. Hiertz and L. Stibor, "IEEE 802.11e wireless LAN for quality of service," in *Proceedings of the European Wireless*, Florence, Italy, February 2002.

[Moh03]   P. Mohapatra, J. Li and C. Gui "QoS in mobile ad hoc networks," *IEEE Transactions on Wireless Communications*, Vol. 10, 2003, pp. 44–52.

[Mui97]   A. Muir and J.J. Garcia-Luna-Aceves, "Supporting real-time multimedia traffic in a wireless LAN," in *Proceedings of the ACM/SPIE Multimedia Computing and Networking Conference (MMNC) 1997*, San Jose, California, February 1997.

[Ozu98]   T. Ozugur, M. Naghshineh, P. Kermani, C.M. Olsen, B. Rezvani and J.A. Copeland, "Balanced media access methods for wireless networks," in *Proceedings of the ACM MOBICOM 1998*, Dallas, TX, USA, October 1998.

[Ozu99]   T. Ozugur, M. Naghshineh, P. Kermani and J.A. Copeland, "Fair media access for wireless LANs," in *Proceedings of the IEEE GLOBECOM 1999*, Rio de Janeiro, Brazil, December 1999.

[Ozu02]   T. Ozugur, "Optimal MAC-layer fairness in 802.11 networks," in *Proceedings of the IEEE ICC, 2002*, New York, May 2002.

[Par93]   A.K. Parekh and R.G. Gallager, "A generalized processor sharing approach to flow control in integrated services networks: the single-node case," *IEEE/ACM Transactions on Networking*, Vol. 1, pp. 344–357, 1993.

[Qia02]   D. Qiao and K.G. Shin, "Achieving efficient channel utilization and weighted fairness for data communications in IEEE 802.11 WLAN under the DCF," in *Proceedings of the IEEE IWQoS 2002*, Miami Beach, Florida, USA, May 2002.

[Rom02]   L. Romdhani, Q. Ni and T. Turletti, "AEDCF: enhanced service differentiation for IEEE 802.11 wireless ad hoc networks," *INRIA* Research Report No. 4544, 2002.

[Rom03]   L. Romdhani, Q. Ni and T. Turletti, "Adaptive EDCF: enhanced service differentiation for IEEE 802.11 wireless ad-hoc networks," *IEEE WCNC* Differentiation for IEEE 802.11 Wireless Ad-Hoc Networks, in *Proceedings of the IEEE WCNC 2003*, New Orleans, LO, USA, March 2003.

[She01b]   S.T. Sheu and T.F. Sheu, "A bandwidth allocation/sharing/extension protocol for multimedia over IEEE 802.11 ad hoc wireless LANs," *IEEE Journal in Selected Areas in Communication*, Vol. 19, 2001, pp. 2065–2080.

[Sob99]   J.L. Sobrinho and A.S. Krishnakumar, "Quality-of-service in ad hoc carrier sense multiple access wireless networks," *IEEE Journal on Selected Areas in Communication*, Vol. 17, 1999, pp. 1353–1368.

[Vai00]   N.H. Vaidya, P. Bahl and S. Gupta, "Distributed fair scheduling in a wireless LAN," in *Proceedings of the ACM MOBICOM 2000*, Boston, MA, USA, August 2000.

[Ver01]   A. Veres, A.T. Campbell, M. Barry and L.H. Sun, "Supporting service differentiation in wireless packet networks using distributed control," *IEEE Journal on Selected Areas in Communication*, Vol. 19, 2001, pp. 2081–2093.

[Wg97]   IEEE 802.11 WG, "IEEE Std 802.11-1997. Information Technology — Telecommunications and Information Exchange Between Systems — Local and Metropolitan Area Networks — Specific Requirements — Part 11: Wireless LAN Medium Access Control (MAC) and Physical Layer (PHY) Specifications," 1997.

[Wg99]   IEEE 802.11 WG, "ISO/IEC 8802-11:1999(E) IEEE Std 802.11, 1999 Edition. International Standard [for] Information Technology-Telecommunications and Information Exchange Between Systems — Local and Metropolitan Area Networks — Specific Requirements — Part 11: Wireless LAN Medium Access Control (MAC) and Physical Layer (PHY) Specifications," 1999.

[Wg03]   IEEE 802.11 WG, "Draft Supplement to STANDARD FOR Telecommunications and Inform-
          ation Exchange Between Systems — LAN/MAN Specific Requirements — Part 11: Wireless
          LAN Medium Access Control (MAC) and Physical Layer (PHY) specifications: Medium Access
          Control (MAC) Quality of Service(QoS) Enhancement," IEEE P802.11E/D6.0, November
          2003.

[Xu03]   K. Xu, Q. Wang and H. Hassanein, "Performance analysis of differentiated QoS supported by
          IEEE802.11e enhanced distributed coordination function (EDCF) in WLANs," in *Proceedings
          of the IEEE GLOBECOM 2003*, San Francisco, CA, USA, December 2003.

27

QoS Support in Mobile Ad Hoc Networks

27.1 Introduction..27-633
27.2 The QoS Models in IP Networks 27-634
Integrated Services • Differentiated Services • IntServ Over DiffServ • Multi-Protocol Label Switching • Related Issues
27.3 The QoS Models in Ad Hoc Networks 27-637
27.4 The QoS Signaling 27-637
Resource ReSerVation Protocol • The INSIGNIA
27.5 The QoS Routing Protocols 27-640
Core-Extracted Distributed Ad Hoc Routing • Ticket-Based Probing • The QoS over AODV • Bandwidth Protocol • Trigger-Based Distributed Routing
27.6 Related Work 27-643
27.7 Summary 27-643
References 27-644

Mohammed Al-Kahtani
Hussein T. Mouftah

27.1 Introduction

Since the Internet was introduced to the general public at the beginning of the 1990s, its traffic has been rapidly increasing with an explosively fast growth of its users. As we already know, the current Internet was implemented to offer what is called "best effort" service, in which all data is treated equally without any reservations or priority levels. With the growth of the Internet and its emergence with the *multimedia applications*, it has become necessary to change this approach. Since *real time applications* are delay sensitive and need more bandwidth, they should be treated in a different fashion. Increasing the network resources is not the solution; the solution is how to use it in a more efficient manner. Unfortunately, most of the present Internet services are still unable to provide the required communication quality. Therefore, the next item on the Internet's agenda is to add the features and characteristics that were missing from its original design such as the provision of the quality of service (QoS) [1,2].

In the last 10 years, wireless technology, as a new trend in the communication industry, has reached most locations on the face of the earth. This convergence between the wireless networks and the Internet makes new challenges for the Internet evolution as well as the wireless networks. As it is already known, the wireless network is not able to achieve performance similar to that of the wired network because of the bandwidth, quality, and power limitations, which face that technology. QoS parameters for typical applications are bandwidth, packet delay, packet loss rate, and jitter. Bandwidth is the most obvious difference between the wired and the wireless domains, the wireless being much slower than the wired one. The wireless network typically experiences longer delays than the wired. The wireless link is easily affected by fading and interference so it is considered to be poor and unreliable with its quality changing with time. Wireless communication is also affected by the fact that the mobile units run by batteries. Therefore, delivering hard QoS guarantees in the wireless networks is very difficult [3,4].

Quality of service [1–4], with respect to computer networks, is a set of service requirements provided to certain traffic by the network to meet the satisfaction of the user of that traffic. QoS requires a capability in the network to be able to treat some selected traffic with better or different service than what other traffic receives. The QoS mechanism does not create an additional bandwidth for the selected network, instead, it uses the available bandwidth in a way that the network can provide the maximum requested QoS with the maximum bandwidth available for that traffic during the session. The network traffic is selected based on classification of packets so that some packet flows can be treated better than the others.

In the Internet networks, the only service model traditionally provided was the best-effort service. This was acceptable when the connectivity and data delivery were the main targets. The continuous growth of Internet services, including the convergence of real time application like voice over IP (VoIP) and multimedia, QoS is required to guarantee their operation. Simply increasing the available bandwidth will not be enough to meet every application requirements, since the Internet traffic increases in proportion to bandwidth as fast as it is added. Assuming that the network's bandwidth capacity is sufficient. However, it is a well-known reality that the Internet delays are a common occurrence. Applications like multimedia or ones involving live interaction cannot tolerate large delays. The nature of these new applications place additional demands on the network. To meet these demands, the network must be enhanced with new technologies that can offer the capabilities to provide or control its quality of services.

Recently, because of the rising popularity of real-time applications and potential commercial needed of MANETs, QoS support has become an unavoidable task in this kind of network. Its dynamic nature not only makes the routing fundamentally different from the fixed networks, but also its QoS support. The major challenge facing the QoS support in MANETs is the quality of the networks is varying with time. Therefore, providing the traditional QoS models, integrated services (IntServ) and differentiated Services (DiffServ), is insufficient. They require accurate link state and topology information, which is so difficult to maintain in this time varying and low capacity resource environments. However, QoS of MANET should benefit from the concepts and features of the existing models in order to come up with a model that can satisfy such networks. In this section, we review the current research on QoS support in MANETs. This includes; QoS models, resource reservations signaling, QoS routing, and some other related works.

27.2 The QoS Models in IP Networks

The IETF has defined different IP QoS mechanisms or architectures [1]. The most prominent among them are the IntServ [5] and the DiffServ [6]. These services differ in their QoS strictness, which describes how tightly the service can be bound by specific bandwidth, delay, jitter, and loss characteristics. This section briefly examines these mechanisms with focus on the wireless and mobile applications.

27.2.1 Integrated Services

The philosophy behind IntServ [1,5] is that routers must be able to reserve resources for individual flows to provide QoS guarantees to end-users. The IntServ QoS control framework supports two additional classes of services besides "best effort":

- *Guaranteed service* provides quantitative and hard guarantees such as; no packet loss from data buffers to ensure loss-less transmission and upper bound end-to-end delay and jitter. This service is useful for hard time real-time applications that are intolerant of any datagram arriving after their playback time.

- *Controlled load service* promises performance as good as in an unloaded datagram network but it provides no quantitative assurance. This service is intended to support a broad class of applications that are highly sensitive to overloaded conditions.

Both services must ensure that adequate bandwidth and packet processing resources are available to satisfy the level of service requested. This must be accomplished through active admission control. In the IntServ, the most needed component is the signaling protocol, which is used to set up and tear down reservations. Resource ReSerVation Protocol (RSVP) [7] is used by the IntServ as a signaling protocol. To achieve its goals, IntServ requires per-flow state to be maintained throughout the network. Unfortunately, the best-known problem with the per-flow approach is its poor scalability because it incurs huge storage and processing overhead at routers in the presence of thousands of flows. Also it is well-known that reservations need to be regularly refreshed, which consumes valuable resources, especially in bandwidth scarce environments such as the wireless. Other problems are that the call admission procedures and the routing scheduling schemes are complex and rely heavily on the per-flow state.

27.2.2 Differentiated Services

DiffServ [1,6], on the other hand, aggregate multiple flows with similar traffic characteristics and performance requirements into a few classes. This approach pushes complex decision making to the edges routers or Ingress routers, which results less processing load on core routers and thus, faster operation. According to this mechanism, QoS information is carried in-band within the packet in the type of service (TOS) field in IPv4 header or differentiated service (DS) field in IPv6. Classification, rate shaping, and policing are done at the edge routers and packets are mapped onto service levels through static service level agreements. Per-hop queuing and scheduling behaviors, or simply per-hop behaviors (PHBs), are defined through which number of edge-to-edge services might be built. Two service models have been proposed [1,6]:

- *Assured service* is intended for customers that need reliable services from the service providers. The customers themselves are responsible for deciding how their applications share the amount of bandwidth allocated.
- *Premium service* provides low-delay and low-jitter service, and is suitable for Internet telephony, video-conferencing, and creating virtual lease lines for virtual private networks (VPNs).

DiffServ appears to be a good solution to part of the QoS problem as it removes the per-flow state and scheduling that leads to scalability problems in the IntServ architecture. Therefore, it is simpler than the IntServ and does not require end-to-end signaling. However, it provides only static QoS configuration, typically through service level agreements, as there is no signaling of the negotiation of QoS. Also, it cannot guarantee the end-to-end QoS.

27.2.3 IntServ Over DiffServ

It would be very interesting to study this framework for IntServ operation in the future over DiffServ networks [8]. This solution tries to conjugate benefits of both the IntServ and the DiffServ approaches to QoS provisioning. In fact, it achieves scalability due to the DiffServ aggregation in the core network, while keeping the advantages of end-to-end signaling by employing the IntServ reservation in the edges of the network. The reference architecture includes DiffServ region in the middle of two IntServ regions. The advantage of this solution is that it allows hop-by-hop call admission and per-flow reservation at the edge of the network where low densities make this the most practical way to achieve good-quality guarantees. In the core of the network, the scalable solution of DiffServ scheduling can be used where hard guarantees in QoS can be made on the basis of more probabilistic judgments. This approach has been proposed by the integrated services over specific link layer (ISSLL) working group at the IETF.

In this architecture, at the edge routers, all policy related processing such as classification, metering, marking, etc., would take place. Resource provisioning at the edge cloud would be done with RSVP and

IntServ. At the core, DiffServ might be used to keep the number of traffic aggregates at a manageable level. The basic philosophy of the architecture is to remove as many computation intensive functions as possible from the backbone routers and push these functions toward the edge routers. That way, the core routers would be free to do high-speed forwarding of the packets and they would remain simple to manage.

The mapping at the edge router (ER) between the RSVP-based reservations and the DiffServ priority levels is the basic requirement and assumption in this approach. The ER is the router placed at the boundary between the IntServ access network and the DiffServ core. Differently from core routers, it is RSVP aware and stores per-flow states. Nevertheless, it is capable of managing packets both on a micro-flow basis and on an aggregate basis. The choice between the two possibilities depends on the role of the ER. When it acts as ingress ER, that is, for packets from the originating IntServ network to the DiffServ core it forwards packets in an aggregate fashion on the outgoing interface, while it still handles micro-flows on the incoming interface. Instead, when it behaves as egress ER, that is, for packets from the DiffServ core to the destination IntServ network, it is able to distinguish micro-flows on the outgoing interface. Note, however, that the distinction between ingress and egress edge router depends only on the direction of the data stream. This means that the same ER may be ingress ER for a flow and egress ER for another one in the opposite direction.

27.2.4 Multi-Protocol Label Switching

In the conventional IP routing, the packets are forwarded independently in each router. Its forwarding algorithm is quite slow because of the processing time of the IP header. Also its route selection is based on the shortest path algorithm. This restricts the capabilities to manage the resources of the network.

Multi-protocol label switching (MPLS) [9] as a new technology tries to overcome the problems of traditional IP routing. MPLS was originally presented as a way of improving the forwarding speed of routers, but it has capabilities that enable an operator to provide service differentiation. MPLS is a combination of switching and routing. It combines some of the features of ATM with IP and the basic idea is that routers perform flow classification, tunneling, and QoS marking at the edge of the domain, while switches do high-speed forwarding in the core. MPLS introduces the concept of connection-oriented into the Internet. The MPLS nodes will route based on topology not traffic and they will set-up flows on an end-to-end basis rather than hop-by-hop, which is done by IP protocol. The key idea is that control plan is separated from forwarding plane and forwarding is based on label swapping, which replaces the incoming label with a new outgoing label on each packet. MPLS is inherently more scalable than VC-based technologies and because label swapping is so general, forwarding can take place over circuits, LANs and nonbroadcast multiple access networks like ATM. Indeed, MPLS is becoming widely used as a network management or traffic engineering mechanism.

In an MPLS domain, a label is assigned to each packet at the ingress node of the domain. Then the packet is forwarded according to the label until it is received at the egress node of the domain. That means, the full IP header analysis is done only once at the ingress node, which simplifies and accelerates the routing process. Each label can be associated with a forwarding equivalence class (FEC), which means a group of packets that can be treated in the same manner. The binding between a label and an FEC is distributed by label distribution protocol (LPD) [10]. IETF defines three LPD: LDP, CR-LDP [11], and RSVP-TE [12], which will be described later.

The MPLS supports two types of route selection: hop-by-hop and explicit routing. The strength of MPLS is in provisioning of traffic engineering [13,14]. The use of explicit routes as an example, gives the ability to manage the network resources and support new services.

27.2.5 Related Issues

The QoS mechanisms typically manage the router behavior at the network layer in order to achieve the requested quality. This might work well if the link behaved well or if the packets might not experience any significant impact on the delay, jitter, or loss. However, this is not true with the wireless link layer. Low signal-to-noise ratio (SNR), burst errors, link fading, and interference are the main sources that lead to data loss and increase the error rates in the wireless links. To overcome this problem, radio links employ both forward and backward error correction schemes. For example, forward error correction (FEC) and automatic repeat request (ARQ). Also interleaving the bits of the packets can be used to overcome the burst errors and fading problems. However, all these techniques may cause delay problems for real-time applications [4].

The design principles that state the interaction between the link layer and the higher layers should be respected. Higher layers should not communicate directly with the link layer and the protocols should not be designed to the requirements or capabilities of the link layer. Application and transport layer protocols should not have "wireless aware" releases. Nevertheless, the error correction techniques should be used wherever possible and the errors should be corrected as much as possible within the delay budget. Furthermore, network providers should assume that a significant proportion of any delay budget should be reserved for use within a wireless link [4].

Regarding the impact of the mobility on the QoS of an existing connection, it is important to discuss two key issues [4]. First, that the requested QoS is made available to a connection during its lifetime with a high probability, in the absence of network failure, when a network accepts the connection to a fixed endpoint. This cannot be said for the mobile environment when the connection changes frequently. The second issue is, that during the route recovery there is a period of time which the end-to-end connection data path is incomplete. The extent to which this disruption affects the application performance depends on the nature of the application and the period of the disruption.

27.3   The QoS Models in Ad Hoc Networks

A flexible QoS model for MANET (FQMM) [15] is the first QoS model designed especially for MANET. It takes the characteristics of MANETs into account and combines the advantages of both the IntServ and DiffServ. FQMM defines three types of nodes: an ingress node that sends data, an egress node that is a destination, and an interior node that forwards data to other nodes. Obviously, each node may have multiple roles. The provisioning in FQMM to allocate and determine the resources at the mobile nodes, is a hybrid IntServ (per-flow) and DiffServ (per-class) scheme. Therefore, the traffic of the highest priority is given per-flow treatment, while other traffic is given per-class differentiation. The traffic conditioner, to mark, discard, and shape the packets is based on the traffic profile, and is placed on the ingress node. FQMM assumes that the smaller proportion of the traffic belongs to the highest priority class. Therefore, the scalability problem of IntServ is not an issue in FQMM. However, the DiffServ problem still exists because making a dynamically negotiated traffic profile is very difficult.

27.4   The QoS Signaling

The QoS signaling is used to set up, tear down, and renegotiate flows, and reserve and release resources in the networks [4]. It is useful for coordinating traffic handling techniques like shaping, policing, and marking. Signaling is important to configure uniform successful end-to-end QoS service handling across the network. True end-to-end QoS requires that all the network elements like switches, routers, and firewalls appropriately support QoS. The coordination of these components is done by QoS signaling. Therefore, QoS signaling information must be carried, interpreted, and processed by all the networks elements. Using the scarce resources in an efficient manner while maintaining strong service guarantees, assumes that some reservation signaling will be required for real-time applications.

The signaling can be fall into two types [4]: in-band signaling, when it is carried with the data, and out-band signaling, where it is separated from the data in explicit control packets. The in-band signaling ensures that the information is always carried to each router that the data may visit, which is useful when routes change frequently, as in mobile networks. However, the important point here is the fact that it requires an overhead, which in some application may reach 10%. Out-band signaling, as used in the telephone networks, is more easily transported to devices not directly involved in data transmission, as in admission control servers.

The signaling may be soft in which case, the reservation needs to be refreshed frequently. This makes it resilient to node failures. Conversely, a hard-state protocol can minimize the amount of signaling. However, care needs to be taken to remove the reservation that is no longer required.

27.4.1 Resource ReSerVation Protocol

The Resource ReSerVation Protocol (RSVP) [1,4,7] is a mechanism for signaling QoS across the network. It is a key element of both IntServ and ISSLL approaches, which were described earlier. RSVP is an out-band signaling system that operates in a soft-state mode. It is also possible to operate it in a near-hard-state mode across any section of a network. Two types of messages, PATH and RESV, are used in RSVP to set up resource reservation states on the nodes along the path between the sender and the receiver. The PATH messages are generated by the sender and propagated through the network to the receiver, gathering information about the network. Each node that processes the message records the flow identifier and the address of the previous RSVP-enable IP router. Once this message reaches the intended receiver, the receiver responds with a RESV message, which actually chooses the required QoS and establishes the reservation. Therefore, RSVP is known as a receiver-initiated resource reservation. This message is propagated back along the path that stored in the selected PATH message. If the sufficient resources are available, a soft reservation state is established in the router. Otherwise, an error message is generated and sent back to the receiver.

In RSVP, maintaining a reservation when the MN moves between regions is a challenge. A scheme is needed to define how smooth this transition should be since it effects the QoS reservation. The disruption of service significantly impacts the QoS. In the presence of micro-mobility, the situation will get worse. Efforts are underway to enhance the RSVP protocol in the mobile networks for transporting real-time applications. According to the original RSVP signaling protocol, it cannot be applied directly in the MANETs because it is not aware of mobility. When the MN moves and then the topology changes, the prior reservations are no longer available. The impact of mobility in the last-hop mobile IP networks has been investigated and resolved by RSVP Tunnel [16], mobile RSVP (MRSVP) [17], and hierarchical MRSVP (HMRSVP) [18]. However, these techniques are not suitable for the MANETs due to the different concepts.

The RSVP Tunnel is proposed in Reference 16 to resolve the RSVP messages invisibility problem. The idea is to establish nested RSVP sessions between the entry and the exit points of the tunnel. RSVP messages, without encapsulation, may be sent to establish a QoS quarantined communication path between the tunnel entry and exit points. Thus, all routers on the path of tunnel can reserve the desired resources for the receiver if sufficient resources are available. Using this nested RSVP session, RSVP Tunnel can actually resolve the invisibility problem of the RSVP signaling in the presence of IP-in-IP encapsulation. Although RSVP Tunnel has resolved the first problem stated earlier, the second problem regarding the lack of resource reservations during mobile movements still exists. To obtain good service guarantees, the MN needs to make advanced reservations at all locations that it may visit during the reservation session. This is called the MRSVP.

Talukdar et al. [17] proposed MRSVP to achieve the desired QoS, for the real time application in the IntServ networks, that requested by the MN regardless its mobility. The MRSVP proposal suggests making reservations in advance at all locations where the MN is expected to visit during the session. The MN will make an active reservation to its current location and passive reservations to each location in its mobility specification (MSPEC). The MSPEC indicates the set locations where the MN may visit in the near future. The difference between the active and the passive reservation is that in active reservation data flows through the path where the resources are reserved while in passive reservations, the resources are reserved but no data from the session flows through the path. Whenever the MN moves from one location to another, the active reservation from the sender to its previous location will be tuned to a passive reservation and the passive reservation to its new location will be tuned into an active reservation.

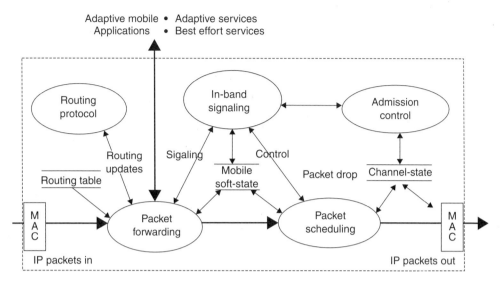

FIGURE 27.1   Flow management model in INSIGNIA protocol.

In this way, the needed resources in the new region can be retrieved quickly because it is already reserved. Although the seamless handover for the QoS guarantees can be retained using MRSVP, it wastes too much bandwidth in making advance resource reservations. This excessive resources waste may degrade system performance significantly.

The hierarchical MRSVP [18] is proposed to integrate RSVP with the mobile IP registration protocol and make advance resources only when the handover delay tends to be long. HMRSVP adopts the hierarchical mobile IP concept and makes advance reservation only when the MN visits an overlapped area of the boundary cells of two regions. Therefore, the setup time for the resources reservation path for an intra-region handover becomes short.

27.4.2 The INSIGNIA

The first signaling protocol designed especially to deliver real-time application in the MANETs is INSGNIA [19,20]. It is a lightweight protocol in terms of bandwidth consuming because it is in-band signaling, which supports fast flow reservation and restoration. The signaling control information is carried in the IP option of every packet, which is called INSIGNIA option. Since it is per-flow based protocol, like RSVP, each flow state is managed over an end-to-end session in response to topology and end-to-end QoS condition and in soft-state method. The wireless flow management model at the mobile node includes a packet forwarding module, a routing module, an INSIGNIA module, an admission control module, a packet scheduling module, and a medium access controller module as shown in Figure 27.1. In this figure the position and the function of INSIGNIA in this model is illustrated as well. Coordinating with the admission control module, INSIGNIA allocates bandwidth to the flow if the resource requirement can be satisfied. Otherwise, if the required resources are not available, the flow will be degraded to best-effort service. For fast responding to the topology changes and end-to-end QoS conditions, INSIGNIA uses QoS reports to inform the sources node of the status of the real-time flows. The destination node actively monitors the received flows and calculates QoS statistical results such as loss rate, delay, and throughput, etc. The reports are sent to the source node periodically. By using this kind of feedback information, the source node can take corresponding actions to adapt the flows to observed network conditions.

Coordinating with other network components, INSIGNIA is an effective signaling protocol and can efficiently deliver real-time applications in MANETs. However, since the flow state information should be kept in the mobile nodes, the scalability problem may hinder its deployment in future similar to RSVP.

Also, in some cases, it is hard to implement some functionallity with in-band signaling. For example, when the established data flow is unidirectional from the source to the destination, sending the feedback control messages back to the source with in-band signaling with the session of the flow is impossible.

27.5 The QoS Routing Protocols

Many authors have addressed, as mentioned earlier, the routing issues in MANETS from best-effort service point of view. Although, these protocols attempt to minimize the control and database maintenance overhead, they do not meet real-time QoS requirements. We believe that the routing protocols have to be adopted to tackle the delay and bandwidth constrains of real-time applications in MANET. At the same time the buffer and signaling overhead should be managed in a proper way to optimize the resource utilization. In general, the notion of QoS has been proposed to capture the qualitatively or quantitatively defined performance contract between the service provider and the user applications. The QoS requirement of a connection is given as a set of constraints. The basic function of QoS routing is to find a network path that satisfies the given constraints [21,22]. In addition, most QoS routing algorithms consider the optimization of resource utilization. The problem of QoS routing is difficult because multiple constraints often make the routing problem intractable and the dynamics of network state makes it difficult to gather up-to-date information in a large network. Moreover, the QoS routing protocols should work together with resources management to establish paths through the network that meet end-to-end QoS requirements. Therefore, the QoS routing is an essential part of the QoS architecture. Before any connection established or any resources reserved, a feasible path between a source–destination pair should be discovered. QoS routing is a routing mechanism under which paths for flows are determined on the basis of some knowledge of resources available in the network as well as the QoS requirements of the flows itself. QoS routing is difficult in MANETs because of the following reasons [23,24]:

- The QoS routing overhead is too high, in these already limited bandwidth environments, the MN should have a mechanism to store and update the link state information. Therefore, balancing the benefit of QoS routing against the bandwidth consumption is needed.
- Maintaining the precise link state information is very difficult because of the dynamic nature of the MANETs.
- Establishing a feasible path does not mean the required QoS should be ensured. The reserved resources may not be guaranteed because of the mobility caused route disconnection and power reduction of the MN. Therefore, QoS routing should rapidly find a feasible new route to recover the service.

Although, QoS routing in MANETs is relatively scarce due to the above reasons, some promising works have been done and show a good performance. Below we introduce CEDAR, ticket-based probing, QoS over AODV, bandwidth routing, and trigger-based distributed routing as examples to illustrate how to deal with the above difficulties.

27.5.1 Core-Extracted Distributed Ad Hoc Routing

The core-extracted distributed ad hoc routing (CEDAR) [25] protocol has been proposed as a QoS routing protocol for networks consisting of tens to hundreds of nodes. CEDAR dynamically establishes a *core* of the network, and then incrementally propagates link state of stable high bandwidth links to the nodes of the core. Route computation is on-demand and is performed by core hosts using local state only. CEDAR does not compute optimal routes because of the minimalist approach to state management, but the trade-off of robustness and adaptation for optimality is believed to be well justified in ad-hoc networks. Following is a brief description of the three key components of CEDAR:

- *Core Extraction*: CEDAR does core extraction in order to extract a subset of nodes in the network that would be the only one that performs state management and route computation. The core extraction is done dynamically by approximating a minimum dominating set of the ad hoc network using only local computation and local state. Each host in the core then establishes a virtual link (via a tunnel) to nearby core hosts (within the third neighborhood in the ad hoc network). Each core host maintains the local topology of the hosts in its domain and also performs route computation on behalf of these hosts. The core computation and core management upon change in the network topology are purely local computations to enable the core to adapt efficiently to the dynamics of the network.

- *Link State Propagation*: While it is possible to execute ad hoc routing algorithms using only local topology information at the core nodes, QoS routing in CEDAR is achieved by propagating the bandwidth availability information of stable links in the core. The basic idea is that the information about stable high-bandwidth links can be made known to nodes far away in the network, while information about dynamic links or low bandwidth links should remain local. The propagation of link-state is achieved through slow-moving "increase" waves (which denote increase of bandwidth) and fast-moving "decrease" waves (which denote decrease of bandwidth). The key questions to answer in link state propagation are: When should an increase/decrease wave be initiated? How far should a wave propagate? How fast should a wave propagate?

- *Route Computation*: Route computation first establishes a core path from the domain of the source to the domain of the destination. This initial phase involves probing on the core and the resultant core path is cached for future use. The core path provides the directionality of the route from the source to the destination. Using this directional information, CEDAR iteratively tries to find a partial route from the source to the domain of the farthest possible node in the core path (which then becomes the source for the next iteration) which can satisfy the requested bandwidth, using only local information. Effectively, the computed route is a concatenation of the shortest–widest (the least hop path among the maximum bandwidth paths) paths found locally using the core path as the guideline or a failure as reported. Upon a link failure, CEDAR initially attempts to recompute an admissible route at the point of failure. As long term solution, source-initiated recomputation is used.

27.5.2 Ticket-Based Probing

The authors in [26] developed a generic distributed QoS routing framework based on selective probing. Using this framework, they proposed a distributed multi-path QoS routing scheme foe MANET, called *ticket-based probing*, which is designed to work with imprecise state information. It allows the dynamic trade-off between routing performance and overhead. For ad-hoc networks, they devised a source-initiated routing algorithm and a destination-initiated routing algorithm for the bandwidth-constrained routing problem. These algorithms do not require global state or topology information and use only the local state maintained at every node. Imprecise state information can be tolerated and multiple paths are searched simultaneously to find the most feasible and the best path.

The basic idea of this scheme is that the tickets are utilized to limit the number of the candidate paths during the route discovery. A ticket is the permission to search a single path. When a source needs a path to reach the destination, it first issues a routing message, called a probe, to the destination. A probe is required to carry at least one ticket, but it may consist of more. Therefore, the maximum number of the searched paths is bounded by the tickets issued by the source. At the intermediate node, a probe with more than one ticket is allowed to split into multiple tickets. Then, based on its available state information, it decides whether the received probe should split and to which neighbors the probe(s) should be forwarded. Once the destination node receives the probe message, a possible path is found.

In ticket-based probing scheme, three-level path redundancy is utilized in case of route failures. The path redundancy scheme establishes multiple routes for the same connection. For the highest level of redundancy, resources are reserved along multiple paths and every packet is routed along each path. In the second level, resources are reserved along multiple paths; however, only one is used as the primary path while the rest serve as backup. In the third level, resources are only reserved on the primary path even multiple path are selected.

27.5.3 The QoS over AODV

The ad hoc on-demand distance vector (AODV) [27] routing protocol has been introduced for best-effort routing in MANET. AODV builds routes using a route request/route reply query cycle. When the source needs to find a path to reach a new destination, it broadcasts a RREQ packet across the network. Nodes receiving this packet update their information for the source node and set up backward pointers to the source node in the route tables. In addition to the source node's IP address, current sequence number, and broadcast ID, the RREQ also contains the most recent sequence number for the destination of which the source node is aware. A node receiving the RREQ may send a route reply (RREP) if it is either the destination or has a route to the destination with corresponding sequence number greater than or equal to that contained in the RREQ. If this is the case, it sends a RREP back to the source. Otherwise, it rebroadcasts the RREQ. Nodes keep track of the RREQs source IP address and broadcast ID. If they receive a RREQ, which they have already processed, they discard the RREQ and do not forward it.

As the RREP propagates back to the source, nodes set up forward pointers to the destination. Once the source node receives the RREP, it may begin to forward data packets to the destination. If the source later receives a RREP containing a greater sequence number or contains the same sequence number with a smaller hop-count, it may update its routing information for that destination and begin using the better route.

To provide QoS support, a minimal set of QoS extensions has been specified for the RREQ and RREP messages [26]. Specifically, a mobile node may specify one of the two services: a maximum delay and minimum bandwidth. Before broadcasting RREQ or unicasting RREP, node must be capable to meet the QoS constraints. When the node detects that the required QoS is no longer available, it must send an ICMP QoS_LOST message pack to the source. This protocol, which is called E-AODV, addresses the bandwidth and delay guarantee requirements. However, its reactive nature does not help minimize the service disruptions due to nodal mobility.

27.5.4 Bandwidth Protocol

In Reference 28, the authors proposed an algorithm for bandwidth calculation and reservation based on the destination sequenced distance vector (DSDV) routing algorithm for multimedia support in multihop wireless networks. The proposed bandwidth routing scheme depends on the use of a CDMA over TMDA medium access scheme. The transmission scale is organized as frames, each containing a fixed number of time slots, and a global clock or time synchronized mechanism is used. The path bandwidth between a source and destination is defined as the number of free time slots between them. Bandwidth calculation needs information about the available bandwidth on each link along the path as well as resolving the scheduling of free slots. Since this problem is an NP-problem, a heuristic approach is needed, which is detailed in Reference 29. The purpose of the proposed QoS routing protocol is to derive a route to satisfy bandwidth requirement for QoS constraint. At a source node, the bandwidth information can be used to accept a new call or delay it. In addition, the authors introduced the "standby" routing method to support fast rerouting when a path is broken. When the primary path fails, the secondary route becomes the primary and another secondary is discovered.

27.5.5 Trigger-Based Distributed Routing

In Reference 30, the authors presented an on-demand and yet proactive routing algorithm, called trigger-based distributed routing (TDR), to deal with the link failures due to nodal mobility in MANETs. They try to provide real-time QoS support while keeping the network overhead low. Thus, the flow state for each session is maintained in distribution fashion at active nodes. In case of imminent link failure in the active route, alternate route searching overhead is kept limited by localizing the reroute queries to within certain neighbors of the node along of the source–destination active route. For cost efficiency, rerouting is attempted from the location of an imminent link failure, which is donated as intermediate node initiated rerouting (INIR). If INIR fails to keep the flow state disruption to a minimum, rerouting is attempted from the source node, which is termed as source initiated rerouting (SIRR). The TDR scheme keeps the size of the nodal database small by maintaining only the local neighbored information. In addition, it makes use of selective forwarding of routing requests based on GPS information. However, GPS information requires additional communication among the nodes.

From the network point of view, the TDR scheme is an on-demand algorithm, as the rerouting routine is triggered as an active node based on the level and trend of variation of its receiving power from the downstream active node. Hence the name of "trigger-based" routing. On the other hand, from the user application point of view, it is a proactive algorithm as, ideally, the traffic experiences no break in the logical route during the session, thus making it suitable for dealing with real-time traffic. TDR also is "distributed" in the sense that any active node participating in a session can make its own routing decision.

27.6 Related Work

Recently, providing a desirable QoS in MANETs became an important design objective. It has been studied and investigated by different researchers and several proposals have been published to address how the QoS can be supported in MANETs. Nevertheless, QoS support in MANETs still remains an open problem. In Reference 31, the authors introduce a new definition of QoS for MANET. This definition, instead of being able to guarantee hard QoS, is based on the adaptation of the application to the network quality while the routing process is going on. They suggested a cross-layer QoS model based on this definition to support adaptively and optimization across multiple layers of the protocol.

Since an efficient resource discovery method is needed to support QoS in MANETs, the authors in Reference 32 proposed a framework that can provide a unified solution to the discovery of resources and QoS-aware selection of resources providers. The key entities of this framework are a set of self-organized discovery agents. These agents manage the directory information of resources using hash indexing. They also dynamically partition the network into domains and collect intra- and inter-domain QoS information to select appropriate providers. They also monitor QoS information continuously and predict path QoS on behalf of other nodes to reduce overall cost and improve accuracy.

Moreover, since the resource reservation is one of the important QoS issues, the concept of neighborhood reservation (NR) is proposed in Reference 33. When a QoS flow reserves bandwidth in a node, the corresponding bandwidth reservation will be done in all its neighborhood nodes simultaneously. This may reduce the impact of channel contention at the target node by its surrounding nodes. However, the authors assumed that all nodes are motionless, which it is not an acceptable assumption in the MANET.

Sometimes packets are replicated along predetermined multiple paths to the destination to seek communication reliability. Alternatively, data traffic is distributed along disjoint or meshed multiple paths — called selective forwarding. In Reference 34, the authors studied the resource usage in these schemes; packet replication and selective forwarding. They have shown that packet by packet throughput with packet replication is higher than that in selective forwarding. This is because sending a packet along multiple error-prone paths, rather than along a chosen route, surely increases the chance of successful arrival of at least a copy of the packet. However, for successfully routing of a message, packet replication has substantially high network resource requirements such as channel bandwidth and battery power. Therefore, they observed that meshed multiple path routing with selective forwarding has the overall superior performance.

27.7 Summary

Due to user mobility, battery power limitations and wireless link characteristics in the MANET, end-to end connections are frequently interrupted and they need to be rerouted in the network while preserving their QoS requirements. Therefore, the existence of a route between the source–destination pair is not guaranteed and then the MANET becomes a challenging task. However, providing the QoS in MANETs is necessary because the convergence between the wireless networks and the multimedia applications. This chapter has presented the current researches related to providing QoS support in MANETs. We have presented the traditional QoS models in computer networks with focus on the wireless and mobile applications. We have then described the first QoS model designed especially for MANET. Also, we have discussed the proposed protocols and algorithms for signaling and QoS routing components.

The QoS protocols for MANETs should be scalable and capable of efficiently utilizing scarce network resources. Therefore, because of the bandwidth and power limitations in MANET, minimizing network and QoS overheads should be a key objective for most network researchers and designers. In addition, the topology changes could also affect the available bandwidth and resources' reservations. Therefore, minimizing the influence of the host mobility should be considered to support and provide the QoS in MANETs.

References

[1] G. Armitage, *Quality of Service in IP Networks*, 1st ed. Pearson Higher Education, New York, 2000.

[2] Z. Wang, *Internet QoS: Architectures and Mechanisms for Quality of Service*, 1st ed. Morgan Kaufmann, San Fransico, CA, 2001.

[3] R. Lloyd-Evans, *QoS in Integrated 3G Networks*, 1st ed. Artech House, Norwood, MA, 2002.

[4] D. Wisely, P. Eardly, and Louise Burnees, *IP for 3G: Networking Technologies for Mobile Communications*, West Sussex, PO, England, John Wiley & Sons Ltd, New York, 2002.

[5] R. Braden, D. Clark, and S. Shenker, "Integrated Services in the Internet Architecture: An Overview," RFC 1633, IETF, June 1994.

[6] S. Blake, D. Black, M. Carlson, E. Davies, Z. Wang, and W. Weiss, "An Architecture for Differentiated Services," RFC 2475, IETF, Dec. 1999.

[7] R. Braden, L. Zhang, S. Berson, S. Herzog, and S. Jamin, "Resource ReSerVation Protocol (RSVP)," RFC 2205, IETF, Sep. 1997.

[8] Y. Bernet, P. Ford, R. Yavatkar, F. Baker, L. Zhang, M. Speer, R. Braden, B. Davie, J. Wroclawski, and E. Felstaine, "A Framework for Integrated Services Operation over DiffServ Networks," RFC 2998, Nov. 2000.

[9] E. Rosen, A. Viswanathan, and R. Callon, "Multi-Protocol Label Switching Architecture," RFC 3031, IETF, Jan. 2001.

[10] L. Andersson, P. Doolan, N. Feldman, A. Fredette, and B. Thomas, "LDP Specification," RFC 3036, IETF, Jan. 2001.

[11] B. Jamoussi, Ed., Constraint-Based LSP Setup Using LDP, RFC 3212, Jan. 2002.

[12] D. Awduche, L. Berger, D. Gan, T. Li, V. Srinivasan, and G. Swallow, "RSVP-TE: Extensions to RSVP for LSP Tunnels," RFC 3209, IETF, Dec. 2001.

[13] Awduche et al., "Overview and Principles of Internet Traffic Engineering," RFC3272, May 2002.

[14] Awduche et al., "Requirements for Traffic Engineering over MPLS," RFC2702, Sep. 1999.

[15] H. Xiao, W. K. G. Seah, A. Lo, and K. C. Chua, "A flexible quality of service model for mobile ad-hoc networks," in *Proceedings of the IEEE VTC2000-Spring*, Vol. 1, pp. 445–449, May 2000, Tokyo, Japan.

[16] A. Terzis, M. Srivastava, and L. Zhang, "A simple QoS signaling protocol for mobile hosts in integrated services internet," IEEE Proceedings, Vol. 3, 1999.

[17] A. Talukdar, B. Badrinath, and A. Acharya, "MRSVP: a resource reservation protocol for an integrated services packets networks with mobile hosts," in *Proceedings of the ACTS Mobile Summit'98*, June 1998.

[18] C. Tseng, G. Lee, and R. Liu, "HMRSVP: a hierarchical mobile RSVP protocol," *in Proceedings of the 21st International Conference on Distributed Computing Systems Workshops (ICDCSW '01)*, April 16–19, 2001.

[19] B. Lee, G.S. Ahn, X. Zhang, and A.T. Campbell, "INSIGNIA: an IP-based quality of service framework for mobile ad hoc networks", *Journal of Parallel and Distributed Computing, Special issue on Wireless and Mobile Computing and Communications,* Vol. 60, pp. 374–406, 2000.

[20] G-S Ahn, A.T. Campbell, S-B Lee, and X. Zhang, "INSIGNIA," IETF Internet Draft, <draft-ietf-manet-insignia-01.txt>, Oct. 1999.

[21] E. Crawely, R. Nair, B. Rajagopalan, and H. Sandrick, "A Framework for QoS Based Routing in the Internet," RFC 2386, Aug. 1998.

[22] M. Mirhakkak, N. Schultz, and D. Thompson, "Dynamic QoS for Mobile Ad Hoc Networks," in *Proceedings of the 1st ACM International Symposium on Mobile Ad Hoc Networking & Computing,* Boston, Massachusetts, pp. 137–138, 2000.

[23] D.D. Perkins and H.D. Hughes, "A survey on quality-of-service support for mobile ad hoc networks," *Wireless Communications and Mobile Computing,* 2002.

[24] D. Chalmers and M. Sloman, "A survey of QoS in mobile computing environments," *IEEE Communications Surveys,* 1999.

[25] P. Sinha, R. Sivakumar, and V. Bharghavan, "CEDAR: a core extraction distributed ad hoc routing algorithm," in *Proceedings of the IEEE Infocom'99,* New York, March 1999.

[26] S. Chen, and K. Nahrstedt, "Distributed quality of service routing in ad hoc networks," *IEEE Journal in Selected Areas in Communication,* Vol. 17, pp. 1488–1505, 1999.

[27] Perkins, C. et al., "Ad Hoc On-Demand Distance-Vector Routing (AODV)," IETF draft, 14 July 2000. [referred 25.9.2000] <http://www.ietf.org/internet-drafts/draftietf-manet-aodv-06.txt.

[28] C. Perkins, E. Royer, and S.R. Das, "Quality of Service for Ad Hoc On-Demand Distance Vector (AODV) Routing," Internet Draft, July 2000.

[29] C. R. Lin, J-S Liu, "QoS routing in ad hoc wireless networks," *IEEE Journal in Selected Areas in Communication,* Vol. 17, pp. 1426–1438, 1999.

[30] S. De, S. Das, H. Wu, and C. Qiao, "Trigger-based distributed QoS routing in mobile ad hoc network," *Mobile Computing and Communications,* Review, Vol. 6, July 2002.

[31] N. Nikaein and C. Bonnet, "A glance at quality of service models in mobile ad hoc networks", in *Proceedings of the DNAC 2002 16th Conference of New Architectures for Communications,* France, Paris, 2002.

[32] J. Liu, Q Zhang, B. Li, W. Zhu, and J. Zhang, "A unified framework for resource discovery and QoS-aware provider selection in ad hoc networks," *Mobile Computing and Communications Review,* Vol. 6, Jan. 2002.

[33] G. Wang, Y. Shu, Y. Fan, and O.W.W. Yang, "An efficient resource reservation mechanism in ad hoc networks," in *Proceedings of the CCECE 2003,* Montreal, May, 2003.

[34] S. De and C. Qiao, "Does packet replication along mulitpath really help?," in *Proceedings of the IEEE International Conference on Communications 2003,* Alaska, USA, pp. 1069–1073, May 2003.

28

QoS Provisioning in Multi-Class Multimedia Wireless Networks

28.1 Introduction and Background 28-647
Admission Control • Multimedia Connections
28.2 The QoS Provisioning — Standard Strategies 28-650
Cellular Networks • The LEO Satellite Networks
28.3 Novel QoS Provisioning Schemes in
Cellular Networks.......................................28-654
Our Simulation Model • El-Kadi's Rate-Based Bandwidth
Borrowing Scheme • A Max–Min Fair Bandwidth Borrowing
Scheme
28.4 The QoS Provisioning in LEO Satellite Networks 28-667
Mobility and Traffic Model • The PRAS Bandwidth
Reservation Strategy • The PRAS Call Admission Strategy •
Simulation Results
28.5 Directions for Further Research 28-671
References 28-674

Mona El-Kadi Rizvi
Stephan Olariu

28.1 Introduction and Background

We are witnessing an unprecedented demand for wireless networks to support both data and real-time multimedia traffic. While best-effort service suffices for datagram traffic, the usability of real-time multimedia applications is vastly improved if the underlying network can provide adequate quality of service (QoS) guarantees.

Indeed, QoS provisioning for wireless networks is becoming more important by the day. Cellular phones are now ubiquitous all over the world. Because the physical infrastructure costs of wireless networks are so much less expensive than those of wired networks, in many developing countries, the wireless coverage, in area and population, is greater than the wired coverage. With more users, there is a growing demand for more services. People want to access services like email, instant messaging, web browsing, and even video streaming and video conferencing, as well as the more common services like telephony and paging, from anywhere or while on the move.

Many cell phone handsets and network providers in the United States can provide users access to telephony, paging, instant messaging, and trivial web browsing on the same device. Some new phones add the ability to do more complex web browsing by combining a small PDA with a phone handset. And users that have more powerful computers equipped with wireless network access want to use the same, possibly high-bandwidth, applications they are accustomed to using with wired network access, such as video conferencing and other real-time networked applications.

QoS provisioning in single-class traffic wireless networks is a more complicated task than it is for wired networks due to physical constraints, bandwidth constraints, and user mobility. Clearly, providing QoS guarantees to users of wireless networks that handle different types or classes of service is an even more complex task, and it is currently a busy research area [8].

Some well-known QoS measures, such as delay, bandwidth, jitter, and error rate, are important in both wired and wireless networks. The chance that a new connection request will be denied, the *call blocking probability* (CBP), is an important QoS measure in wireless networks due to limited bandwidth. And the chance that an in-progress connection will be forcibly terminated by the network, the *call dropping probability* (CDP), is important due to user mobility. In [32], Kalyanasundaram proposes a hierarchical classification of QoS measures:

- *Packet level:* QoS at this level includes packet dropping probability, maximum packet delay, and maximum jitter.
- *Call level:* At this level several authors consider only CBP and CDP. We consider that supplied bandwidth also belongs at this level.
- *Class level:* QoS at this level describes the requirements for and relationships between the call-level QoS for different classes of traffic.

Whether traffic is classified according to its call-level and packet-level QoS needs, or according to the price paid for service, or by some other means, class-level QoS will describe the prioritization between the different classes of traffic.

The main goal of this chapter is to provide an extensive survey of call-level and class-level QoS provisioning in cellular and low earth orbiting (LEO) satellite networks. We consider the end-user needs in relation to the service providers' needs, that is, minimizing the CDP and CBP while keeping the bandwidth utilization as high as possible. And, we consider the end-user needs with respect to each other, within and between traffic classes, that is, making our algorithms *fair*. The following two chapters in this book, namely Chapter 26 titled: QoS Enhancements of the Distributed IEEE 80211 Medium Access Control Protocol and Chapter 27 titled QoS Support in Mobile Ad Hoc Networks are addressing two other important aspects of QoS provisioning in wireless networks. Specifically, Chapter 26 discusses enhancements to the IEEE standard 80211 MAC protocol to enable it to provide some elements of QoS at the MAC layer. The original version of IEEE 80211 was notorious for lacking in the areas of fairness and QoS provisioning. Later enhancements have addressed this deficiency to various degrees. Chapter 27 addressed a companion problem — namely that of providing QoS support in mobile ad hoc networks (MANETs). Unlike cellular and satellite networks that constitute the main topic of our chapter, MANET is a rapidly deployable network that does not rely on a preexisting infrastructure. As such, MANET finds applications to search and rescue operation, law enforcement efforts and on-the-fly collaborative environments, for example, multimedia classrooms where the participating students, equipped with laptops, self-organize into a short-lives MANET for the purpose of exchanging class notes or other teaching related material.

28.1.1 Admission Control

Admission control and bandwidth allocation schemes can offer wireline networks the ability to provide their users with such guarantees. Due to host mobility, scarcity of bandwidth, and an assortment of channel impairments, the QoS provisioning problem is far more challenging in wireless networks than in their wireline counterparts. For example, a mobile host may be admitted into the network in a cell where its needs can easily be met, but the mobile host may eventually move to a cell that has little or no resources to offer. Since the user's itinerary and the availability of resources in various cells is usually not known in advance, global QoS guarantees are very hard to provide [1,3,10,18,19,27,28,31–33,37,44,48,49,52].

Admission control refers to the task of deciding if a connection should be admitted into, and supported by, the network. Admission control is necessary for real-time, continuous media connections since the amount of resources requested by these connections may not match the level of resources available at the time of connection setup [45,46,49]. Admitting a connection into the network is tantamount to a contract between the network and the connection: on one hand the network guarantees that a certain level of resources will be maintained for the duration of the connection. On the other hand, the connection is expected not to request additional resources over and above those negotiated at connection setup. The agreed-upon amount of resources that the network guarantees to a connection is commonly referred to as QoS. Traditional QoS parameters include bandwidth, end-to-end delay, and jitter. However, there are some QoS parameters that are specific to wireless networks.

It is typical in most admission schemes to deny service to a new connection whose requests for resources cannot be met by the network. In such a case, the connection[1] is said to be *blocked*. In cellular networks, an important QoS parameter is the CBP, denoting the likelihood that a new connection request will be denied admission into the network. A similar situation arises when an established connection in one cell attempts to migrate into a neighboring cell (i.e., a handoff is attempted). If the new cell cannot support the level of resources required by the connection, the handoff is denied and the connection is dropped. The CDP expresses the likelihood that an existing connection will be forcibly terminated during a handoff between cells due to a lack of resources in the target cell. The CBP and CDP together offer a good indication of a network's quality of service in the face of mobility. An additional important consideration is the degree to which the network makes effective use of bandwidth — unquestionably its most scarce resource. This parameter, referred to as *bandwidth utilization*, expresses the ratio between the amount of bandwidth used by various applications admitted into the network and either the total bandwidth requested or the total bandwidth available, whichever is smaller. Keeping the CBP and CDP low, while at the same time maximizing bandwidth utilization is one of the most challenging tasks facing protocol designers [38,39,44].

28.1.2 Multimedia Connections

The traditional admission control process outlined earlier is, in many cases, too conservative and pessimistic. Indeed, multimedia applications are known to be able to tolerate and adapt to transient fluctuations in QoS [11,47,51]. This adaptation is typically achieved by the use of an adjustable-rate codec or by employing hierarchical encoding of voice and video streams [11]. The codec, along with appropriate buffering before play-out, can allow applications to gracefully adapt to temporary bandwidth fluctuations with little or no perceived degradation in overall quality. The graceful adaptation of applications to transient fluctuations in QoS is fundamental in wireless networks, where QoS provisioning is a very challenging task. As we shall demonstrate in this chapter, the additional flexibility afforded by this ability to adapt can be exploited by protocol designers to significantly improve the overall performance of wireless systems.

As we briefly mentioned, once a connection is admitted into the network, resources must be *allocated*, at the negotiated level, for the duration of the connection. It is important to realize that in a cellular network where the user may move through the network traversing a sequence of cells, this commitment cannot be only local to the cell in which the connection originated. If the connection is to be maintained after the user crosses the boundary between neighboring cells (i.e., after a handoff), the network must guarantee an appropriate level of resources in each new cell that the user traverses. Without detailed knowledge about the intended destination of each connection, honoring this commitment is a very difficult task [12,15,16,26,30,34,35,55].

While the strategies of enqueing handoff requests and that of rearranging channels, especially in dynamic channel allocation environments, are worthwhile and can reduce the CDP, in this chapter we only

---

[1]We will follow common practice and refer to connections as "calls."

survey QoS provisioning schemes that *reserve* resources in cells on behalf of mobile hosts in anticipation of their arrival. Being simple and natural the resource reservation problem has recently received well-deserved attention. We refer the reader to References 44 and 49 for surveys of recent literature.

There are, essentially, two approaches to resource reservation: with fixed reservation a certain percentage of the available resources in a cell are permanently reserved for handoff connections; by contrast, with statistical reservation resources are reserved using a heuristic approach. These heuristic approaches range from allocating the maximum of the resource requirements of all connections in neighboring cells, to reserving only a fraction of this amount [35,44].

The remainder of the chapter is organized as follows. Section 28.2 reviews well classic QoS provisioning strategies in cellular and LEO satellite networks. Section 28.3 discusses the details of two recent borrowing-based QoS provisioning schemes in cellular networks. Specifically, we take a close look at a rate-based borrowing scheme and at a max–min fair borrowing scheme. Next, Section 28.4 presents a recent predictive handoff management and call admission control scheme for QoS provisioning in LEO satellite networks. Finally, Section 28.5 offers concluding remarks and offers direction for further research.

28.2   The QoS Provisioning — Standard Strategies

The main goal of this section is to review standard QoS provisioning strategies in both cellular and LEO satellite networks.

28.2.1   Cellular Networks

Figure 28.1 illustrates the cellular system model we adopt in this chapter. The geographic area of interest is assumed to be tiled by a collection of regular hexagons referred to as cells [6,7,9,12,13,17,38]. The wireless communication in a cell is supported by a base station (BS). In turn, the BSs are connected to each other by wireline links. Several BSs are connected to a mobile switching center (MSC) that acts as a gateway from the cellular network to existing wireline networks, the Internet and the PSTN. In addition, with the active participation of the mobile hosts and the MSC, the BSs are instrumental in initiating and finalizing handoffs [12,13,38]. The mobile hosts in a cell communicate directly with the corresponding BS that has the responsibility of handling all demands for service originating in the cell. In particular, the BS is in charge of negotiating QoS parameters, of performing admission control, and of reserving resources for ongoing connections. This could mean denying access to new connections in order to provide an acceptable level of service to active connections. Every new connection would like to be accepted regardless of the current demand. The BS must be able to manage the load efficiently to permit continual cellular network traffic. Thus, an efficient load balancing strategy should be an important part of the communication technology used in a network. When a MH moves out of the range of one BS and into the area served by a neighboring BS, it is called a *handoff* or *handover*.

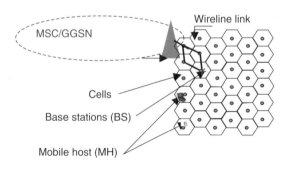

FIGURE 28.1   Architecture of a cellular network.

Because of cochannel interference, the entire frequency spectrum assigned to a network cannot be used in each and every cell. There has been much research [43] into how to allocate channels among the cells in order to maximize the available bandwidth for user requests. The simplest is the *fixed* channel allocation, in which all the channels of the spectrum are distributed statically to the BSs in a pattern that reduces or prevents cochannel interference. Importantly, each cell has a fixed number of channels to accommodate new and handoff calls [1]. In *dynamic* channel assignment schemes, channels are assigned to cells as they are needed [2]. *Flexible* channel assignment schemes are hybrids of fixed and dynamic assignment, where a set of channels is distributed statically and the rest of the channels are kept in a pool to be distributed as the need arises [2,12,17]. Dynamic channel assignment will clearly outperform fixed assignment in times when a few cells are highly loaded, because highly loaded fixed assignment cells will exhaust all their available bandwidth with no method to obtain more. However fixed assignment will perform better under uniformly high load conditions, unless the dynamic assignment scheme can produce the optimal configuration of fixed assignment as it progresses.

Many dynamic channel assignment schemes are complex and require a central controller. But in References 6 and 50, a distributed dynamic allocation scheme is proposed. A BS makes its channel assignment decisions itself, after conferring with neighboring BSs that are within cochannel interference distance of it. In Reference 17, a most interesting borrowing scheme is proposed. A fixed channel assignment is made initially, then heavily loaded cells may borrow free channels from their neighbors when needed.

To provide greater frequency utilization, cells may be divided into concentric regions. In the region closest to the BS, channels can be used at low power without interfering with the same channels being used in the central regions of neighboring cells. Although these types of schemes increase the available bandwidth, they are complex to manage and introduce an imbalance between resources available for calls depending on the region in which they originate. In Reference 47, an algorithm is proposed to solve the problem that fewer channels are available in the region farthest from the BS, and create a uniform CBP across an entire cell. Dividing a cell into regions also introduces the possibility of handoff being necessary when an MH moves from the inner to the outer region of a cell.

Handoff is a problem for QoS provisioning because the new region or cell that an MH is entering may not have the resources available to grant the same level of QoS that the hosts had originally negotiated or may not be able to continue the host's call at all. And as cells become smaller (as in micro- and pico-cellular networks), handoffs will occur much more frequently. Handoff detection is in itself a research issue. As an MH moves away from one BS and toward a neighboring BS, the signal from one will get progressively weaker and from the other progressively stronger, until in some overlap area, they become equal for a time. But other things, such as terrain, can also affect signal strength.

Reducing the CDP, or the handoff failure rate, is more important than reducing the CBP, because we assume that it is more disturbing to users to have an ongoing call cut off, than to have a call attempt denied. Dynamic and flexible channel allocation schemes aid in lowering the CDP because channels can be allocated or borrowed when handoffs arrive and find no available bandwidth. A common technique for reducing dropped calls is to reserve some portion of the bandwidth in a cell for use only by handoffs [15,44,57]. The simplest method of bandwidth or channel reservation is static, where a fixed set of channels, or fixed percentage of a cell's bandwidth, is set aside for handoffs. Selecting the amount of bandwidth to reserve is a trade-off between the CBP and CDP — the larger the reservation, the less bandwidth available for new calls. Dynamic or adaptive reservation uses information about the network or the MHs to vary the amount of reserved bandwidth with the current conditions in order to reach the most optimal balance between CBP and CDP.

In Reference 57, the traffic statistics of neighboring cells are polled to determine how much bandwidth should be reserved for handoffs. Knowledge or prediction of the mobility characteristics (i.e., speed and direction) of the MHs can improve adaptive bandwidth reservation schemes. In Reference 15, the authors use traffic history to predict MH behavior, assuming that MHs entering a cell from the same direction as other MHs have done in the past will likely go in the same direction as their predecessors and at a similar speed, because of roads. The *shadow cluster concept*, introduced in Reference 35, refers to the group of cells surrounding a MHs current cell, which has the highest probability of being visited by the MH in the near future. This *region of influence* for each MH is calculated based on the probable call duration and reports sent by the MH to the BS containing position, speed, and heading information. Aljadhai and Znati [3] propose a scheme based on the shadow cluster concept, but which uses historical information about an MHs movement to compute probabilities for its future movements, rather than relying on the MHs to report. This method is of course less precise, but likely more feasible. Aljadhai and Znati [3] also extend the algorithms in Reference 35 to apply to multiple classes of traffic.

Call admission control schemes help a BS decide whether or not to grant service to a connection request. They are like bandwidth reservation, in that by using information about network conditions, they may deny a connection request even though the call may be serviceable in the short term, in order to leave resources available for demands expected later. Agarwal et al. [2] propose an admission control method that assumes single-class traffic requiring one channel per MH. The area at the perimeter of a cell is designated as the *handoff zone*, and the area just inside that is the *pre-handoff zone*. In this scheme, a new call is assigned a channel only if no MHs are in the cell's handoff zone and if the number of MHs in the pre-handoff zone is less than the number of available channels. In Reference 38, neighboring cells periodically exchange information about their state. When a new call request arrives, a cell uses its knowledge about its neighborhood to decide whether or not to admit the new call, with the goal of limiting the CDP to a given value. This scheme has been criticized for its complexity [16]. A hybrid of the *weighted sum scheme* and the *probability index scheme* [48] has been shown to compare favorably to a static bandwidth reservation scheme and to the scheme in Reference 38. The weighed sum scheme makes admission decisions based on the weighed sum of the number of ongoing calls in the neighboring cells, with the weight based on the distance between the cell and its neighbor. The probability index scheme estimates the dropping probability for the new call and compares it to a threshold value in order to make an admission decision. Recent work by Sadeghi and Knightly [52] proposes an admission control scheme called the *Virtual Bottleneck Cell*, which forms cells into clusters and avoids resource control at the user level in order to be highly scalable.

Most of the schemes we have discussed so far consider just single-class traffic, using one channel or bandwidth unit per call. A few researchers have proposed generic algorithms and models for multi-class traffic of n types [10,26,37]. Others have modeled systems where each call may request different QoS, but traffic is mainly classified based on whether it is real-time or nonreal-time [18,32,39,44]. Das [18] makes just this distinction in a framework he proposes for QoS at the packet and at the call level. The same classification is made by Naghshineh [39], where an end-to-end QoS provisioning framework is proposed. These researchers suggest that some real-time applications have the ability to adapt themselves to fluctuations in the QoS. For example, video can be sent in *layers* of quality, so that if supplied bandwidth is reduced, layers can be dropped and the video quality lowered, without suffering a loss in the continuity of the video stream. Nonreal-time applications, such as email or ftp, cannot tolerate loss, but are able to tolerate significant changes in supplied bandwidth.

Later in this chapter we will describe the algorithms proposed in Reference 44 in detail. At this point, we note briefly that it is assumed that traffic falls into two classes, *Class I* designating real-time and *Class II* designating nonreal-time, and that successful handoffs are more important to Class I users than Class II. The schemes in Reference 44 specify that for a Class I connection to be accepted in a cell, bandwidth must be available to be reserved for it in the neighboring cells. In one scheme, a fixed amount is reserved in a cell depending on the number of connections requesting a reservation. In another version, the amount reserved is equal to the maximum of all the current reservation requests for the cell. When a new connection is attempted in a cell, the connection class, the desired bandwidth, and the minimum acceptable bandwidth are specified. If the desired bandwidth cannot be granted, the connection is blocked. Class I calls are blocked if reservations cannot be made for them. For handoff connection requests, only Class I connections have the cell's reserved bandwidth at their disposal. A Class I handoff must be granted at least its minimum bandwidth, and reservations must be made for it in neighboring cells, or it is dropped. A Class II handoff will succeed if there is at least one unit of bandwidth available for it.

28.2.2 The LEO Satellite Networks

Due to various economic constraints, terrestrial wireless networks provide communication services with a limited geographic coverage. Recently, in response to increasing demands for truly global coverage needed by personal communication services (PCS), a number of LEO satellites have been deployed [13,14,23,30]. Being deployed at low altitudes ranging from 500 km to about 2000 km above the surface of the Earth, LEO satellites feature low propagation delays and low power requirements, allowing hand-held devices to interface directly with a LEO system. It is anticipated that LEO satellite systems will feature low cost-per-minute utilization charges, making them ideally suited to support PCS [20–22,29,31,40–42].

The LEO satellites rotate around the Earth at constant speed. Thus, in order to provide continuous coverage to MHs in a geographical area, a larger number of LEO satellites are required. The coverage area of a satellite — a circular area of the surface of the Earth — is referred to as its *footprint*. By using an antenna with a honeycomb arrangement of spotbeams, the satellite footprint can be partitioned into slightly overlapping circular individual cells, sometimes referred to as *spotbeams*. A spotbeam controller can be considered similar to a ground-based cellular system's BS. But unlike a cellular BS, a LEO satellite is moving at a very high velocity. Because LEO satellites are deployed at low-altitude, Kepler's third law implies that these satellites must traverse their orbits at very high speed. Thus, in LEO systems a connection established in one of the cells of the satellite's footprint is likely to experience a large number of handoffs during its lifetime as it crosses over into several spotbeams (Figure 28.2). Consequently, LEO satellite systems require sophisticated handoff management and call admission control protocols.

A handoff of a connection from spotbeam to spotbeam is an *intra-satellite handoff*, and one from satellite to satellite is an *inter-satellite handoff* [4,13,53,56,58]. Handoff management is a key part of QoS provisioning in LEO satellite networks. While providing significant advantages over their terrestrial counterparts, LEO satellite networks present protocol designers with an array of daunting challenges, including handoff, mobility, and location management [21,22,40,45,46].

The LEO satellite networks are expected to support real-time interactive multimedia traffic and must be able, therefore, to provide their users with QoS guarantees. One of the important QoS parameters is the CDP, which quantifies the likelihood that an on-going connection will be force-terminated due to an unsuccessful handoff attempt.

In order to analyze the problem of intra-satellite handoffs, it is common practice to model a satellite network like a cellular network in which the MHs are all moving in straight lines in the same direction from spotbeam to spotbeam at a constant speed that is equal to the satellite's orbital velocity [21,22]. Some researchers are applying the techniques designed for cellular systems directly, albeit with a loss in efficiency, because the specific characteristics of the LEO networks are not considered. Del Re et al. [21] proposes a dynamic channel allocation scheme and applies the *handoff queuing* technique.

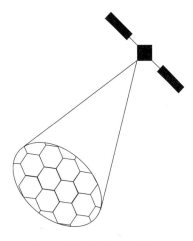

FIGURE 28.2   The footprint of a LEO satellite subdivided into spotbeams.

The need for mobility prediction does not exist in the satellite model because the motion of the MHs is defined by the motion of the satellites. This simplifies some of the cellular schemes, or provides them better performance through exploiting the knowledge of the speed and direction of the MHs. For example, in Reference 32, the *channel-sharing* scheme is applied to satellite networks. In this scheme, each adjacent pair of cells, $\{n, n + 1\}$, forms a *meta-cell* that has a set of channels assigned to it, thus allowing some users to go from cell n to $n + 1$ carrying the same channels with them. A fixed channel allocation scheme partitions the channels according to the correct reuse distances, then a call admission scheme determines which channels to assign a connection request from a MH in order to improve its chances of carrying them through its next spotbeam handoff. Applied to a land-based cellular network, the scheme from Reference 32 is much more complex.

Many researchers have developed resource reservation techniques that use the knowledge of the MH mobility to improve their efficiency [13,14,24,54].

28.3 Novel QoS Provisioning Schemes in Cellular Networks

As discussed earlier, much of the work in QoS for cellular networks has considered only single-class traffic, ignoring the more complex multi-class traffic case. But in light of the current and expected future demands for wireless networks, it is essential to create QoS provisioning schemes that address networks carrying a mixture of traffic classes. Multimedia applications are known to be able to tolerate and adapt to transient fluctuations in QoS [11,51]. This adaptation is typically achieved by the use of an adjustable-rate codec or by employing hierarchical encoding of voice and video streams [11]. In this section we review two schemes first proposed in References 25 and 36 that take advantage of the adaptability of some the classes of traffic in order to improve the QoS provided by the network in terms of CBP and CDP while maintaining good resource utilization.

It is well-known [17,18] that it is more far acceptable to deny a new connection request than to terminate an in-progress connection. The numerous strategies proposed to lower the CDP include channel rearrangement, handoff queuing, and channel reservation. Channel rearrangement and handoff queuing, techniques are effective; however the schemes proposed in References 25 and 36 fall into the category of reservation schemes. There are, essentially, two approaches to resource reservation:

- *fixed reservation:* Where a certain percentage of the available resources in a cell are permanently reserved for handoff connections.
- *statistical reservation:* Where resources are reserved using a heuristic approach. These approaches range from allocating the maximum of the resource requirements of all connections in neighboring cells, to reserving only a fraction of this amount [35,44].

Each of our schemes combines a resource reservation scheme with a companion fair borrowing scheme.

In order to set the stage for the description of the algorithms in References 25 and 36, we now review the bandwidth allocation and reservation schemes proposed in Reference 44 in more detail. We chose these schemes as a benchmark since they are arguably better than other comparable bandwidth allocation and reservation schemes found in the literature [44].

When an MH requests a new connection in a given cell, it provides the following parameters:

- The desired class of traffic (either I or II)
- The desired amount of bandwidth for the connection
- The minimum acceptable amount of bandwidth, that is the smallest amount of bandwidth that the source requires in order to maintain acceptable quality, for example, the smallest encoding rate of its codec

One of the significant features of the call admission control and bandwidth reservation schemes in Reference 44 is that in order to admit the connection, bandwidth must be allocated in the originating cell and, at the same time, bandwidth must be reserved for the connection in all the neighboring cells.

Specifically, for a new connection to be admitted in a cell, the cell must be able to allocate the connection with its desired bandwidth. For Class I connections, the call will be blocked unless the desired bandwidth can be allocated to it in the original cell, and some bandwidth can be reserved for it in each of its six neighboring cells.

During a handoff, an established Class I connection is dropped if its minimum bandwidth requirement cannot be met in the new cell or if appropriate reservations cannot be made on its behalf in the new set of neighboring cells. However, Class II traffic has no minimum bandwidth requirement in the case of a handoff, and a call will be continued if there is any free bandwidth available in the new cell.

The schemes presented in Reference 44 use statistical reservation techniques based on the number of connections in neighboring cells, the size of the connections in neighboring cells, the predicted movement of MHs, and combinations of these factors. It is worth noting that the reservation schemes in Reference 44 keep the dropping probability for Class I connections very low, since the MH should find bandwidth reserved for it, regardless of the cell to which it moves. But bandwidth may be wasted in the neighboring cells (the host can only move to one neighbor), and the blocking probability in those cells may increase because unused bandwidth is being kept in reserve. In general, the schemes described in Reference 44 favor minimizing the CDP at the expense of the CBP and give Class I traffic precedence over Class II traffic.

It is clear that keeping a small pool of bandwidth always reserved for handoffs, as in Reference 44, yields low CDP. However, in the schemes of References 25 and 36, the size of the reserved pool is not determined by requests from neighboring cells, but is *fixed* at a certain percentage of the total amount of bandwidth available in the cell. We found that this produced results similar to the best results reported in Reference 44, without the overhead of communication between neighboring BSs to request and release reservations. It was also reported in Reference 48 that simple fixed reservation can outperform more complex statistical techniques. To further reduce the CDP in our scheme, we treat the reserved pool very carefully. We do not allow bandwidth from the reserved pool to be allocated to incoming handoffs unless the bandwidth is needed to meet the minimum bandwidth requirements of the connection. Like Oliviera et al. [44], our scheme gives precedence to Class I connections; Class II traffic does not make use of the reserved bandwidth.

28.3.1 Our Simulation Model

In order to evaluate the performance of the QoS provisioning schemes discussed in this section we have designed and implemented a sophisticated simulator for cellular networks. Written in Java, it simulates a set of cells networked together with direct communication from a cell to any of its neighbors, or from a cell to the network controller. Traffic is introduced to each cell in the form of MH objects which have some predefined characteristics pertaining to movement (i.e., speed and direction) and QoS needs (i.e., service class, required bandwidth, connection duration, etc.). The MHs communicate with the cells, negotiating QoS for new connections, and attempting handoffs when needed. The cells collect statistics about the service they are providing, and the MHs collect statistics about the service they are receiving.

To fairly compare and contrast the schemes in this section with other schemes from the literature, we have used the traffic types and characteristics given in Reference 44 as input to our simulations, and modeled traffic behavior just as described there, with the exception of the handoffs. In Reference 44, a handoff occurs during a connection with some given probability, and that probability decreases exponentially with each successive handoff during the connection. We have chosen a different approach that seems more realistic. We give each MH a speed characteristic specifying the amount of time that will be spent in each cell during a call. Thus, longer calls are likely to experience more handoffs than shorter ones.

As in Reference 44, the traffic offered to the cellular system is assumed to belong to two classes:

1. *Class I traffic* — real-time multimedia traffic, such as interactive voice and video applications
2. *Class II traffic* — nonreal-time data traffic, such as email or ftp

TABLE 28.1   Traffic Characteristics for our Simulation Model

Class	AVG. BPS (kbps)	MIN. BPS (kbps)	MAX. BPS (kbps)	AVG. Call (Sec)	MIN. Call (Sec)	MAX. Call (Sec)
Class I	30	30	30	180	60	600
Class I	256	256	256	300	60	1800
Class I	3000	1000	6000	600	300	18000
Class II	10	5	20	30	10	120
Class II	256	64	512	180	30	36000
Class II	5000	1000	10000	120	30	1200

Table 28.1 shows the exact characteristics of the traffic used in our simulations. Each of the six types occurs with equal probability. This set represents an estimation of what the traffic mix in a future multimedia wireless network will look like Reference 44.

28.3.2   El-Kadi's Rate-Based Bandwidth Borrowing Scheme

El-Kadi et al. [25] proposed a rate-based scheme, henceforth referred to as EBS, that attempts to allocate the desired bandwidth to every multimedia connection originating in a cell or being handed off to that cell. The novelty of EBS is that in case of insufficient bandwidth, in order not to deny service to a requesting connection (new or handoff), bandwidth will be borrowed, on a temporary basis, from existing connections. EBS guarantees that no connection will give up more than its *fair share* of bandwidth, in the sense that the amount of bandwidth borrowed from a connection is proportional to its tolerance to bandwidth loss.

The EBS has the following interesting features:

1. It guarantees that the bandwidth allocated to a real-time connection never drops below the minimum bandwidth requirement specified by the connection at call setup time. This is very critical to ensuring that the corresponding application can still function at an acceptable level.
2. It guarantees that if bandwidth is borrowed from a connection, it is borrowed in small increments, allowing time for application-level adaptation.
3. It is *fair* in the sense that if bandwidth is borrowed from one connection, it is also borrowed from the existing connections. Specifically, if borrowing is necessary in order to accommodate a requesting connection (new or handoff), every existing connection will give up bandwidth in proportion to its tolerance to bandwidth loss. This motivated us to refer to our scheme as *rate-based* fair.
4. Finally, the borrowed bandwidth is returned to the connections as soon as possible. Thus, the degradation in the QoS is transient and limited to a minimum.

28.3.2.1   Cell and Connection Parameters

Each cell maintains a pool of bandwidth reserved for Class I handoffs which, initially, represents r percent of the total bandwidth. At setup time, each connection specifies to the cell in which it originates a *maximum bandwidth M* (termed the *desired bandwidth*) and a *minimum bandwidth m* as illustrated in Figure 28.3. The difference between these two values is the *bandwidth_loss tolerance* (BLT) of the connection. Thus,

$$\text{BLT} = M - m,$$

we note that for constant bit rate (CBR) connections $M = m$, indicating no BLT and, thus, BLT $= 0$.

Each cell maintains a local parameter, f, $(0 \leq f \leq 1)$, which represents the fraction of the BLT that a connection may have to give up, in the worst case. This fraction is the *actual borrowable bandwidth* (ABB) of the connection. Thus,

$$\text{ABB} = f \times \text{BLT} = f(M - m).$$

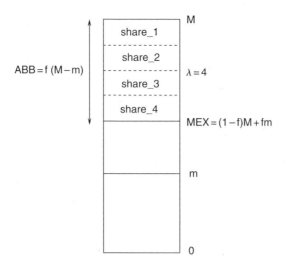

FIGURE 28.3 Illustrating the main connection parameters.

By accepting a new call, the cell agrees that the supplied bandwidth will not fall below a certain level that we call the *minimum expected* (MEX) bandwidth that the connection is guaranteed to receive during its stay in its starting cell. By definition, MEX $= M -$ ABB. It is worth noting that MEX $\geq m$. Simple computation shows that MEX is a weighted average of M and m in the sense that

$$\text{MEX} = (1 - f) \cdot M + f \cdot m.$$

To prevent borrowing from producing noticeable changes in a connection's QoS, we introduce another cell parameter, λ. The ABB is divided into λ *shares*, each share being equal to $(M - m)/\lambda$. This provides the basis for a method of borrowing bandwidth gradually from a set of connections whose allocated resources may be quite different. A cell is said to be operating at level L ($0 \leq L \leq \lambda$), when all its ongoing connections have had L (or more) shares borrowed from them.

It is important to note, however, that it is possible for a connection to be missing more than L shares after a handoff, due to the sacrifices made to prevent call dropping. However, our scheme attempts to restore bandwidth to handoff connections as soon as it becomes available.

28.3.2.2 Fairness of EBS

We now introduce a further connection parameter that we call *adaptivity* (AD), which underlies the EBS. Specifically, for a given connection, AD is the ratio between the connection's bandwidth loss tolerance and the maximum bandwidth that the connection can use.

$$\text{AD} = \frac{\text{bandwidth\_loss\_tolerance}}{\text{desired\_bandwidth}} = \frac{M - m}{M}. \tag{28.1}$$

It is worth noting that the higher the AD the more adaptive the connection, and the lower the probability of a forced termination in case of a handoff. Notice, again, that for CBR connections the adaptivity is 0.

Consider an arbitrary cell operating at level L. Recall that this implies that every connection in the cell has given up L of its shares. Consider an arbitrary connection C with desired and minimum bandwidth M and m, respectively. Since the cell operates at level L connection C must have lost L of its shares operating at an effective bandwidth of

$$M - L \times \frac{\text{ABB}}{\lambda}.$$

The *loss ratio* (LR) of connection C is the ratio between the amount of bandwidth borrowed from C and the maximum bandwidth M specified by C at setup time. In other words

$$\text{LR} = \frac{L \times (\text{ABB}/\lambda)}{M}. \tag{28.2}$$

Direct manipulations of Equation (28.2) reveal that

$$\text{LR} = \frac{Lf}{\lambda} \times \frac{M - m}{M} = \frac{Lf}{\lambda} \times \text{AD}, \tag{28.3}$$

since for a given cell, and a given point in time, Lf/λ is a constant, Equation (28.3) shows that the connection will give up an amount of bandwidth proportional to its adaptivity.

Let C' be a arbitrary connection in the same cell as C, and let $\text{LR}(C)$ and $\text{LR}(C')$ be the corresponding loss ratios. Then Equation (28.3) allows us to write

$$\frac{\text{LR}(C)}{\text{LR}(C')} = \frac{(Lf/\lambda) \times \text{AD}(C)}{(Lf/\lambda) \times \text{AD}(C')} = \frac{\text{AD}(C)}{\text{AD}(C')}.$$

Thus, the ratio of the loss ratios of two connections is invariant to L and is only a function of the adaptivity of the connections. This is the sense in which we consider EBS to be fair.

28.3.2.3 New Call Admission Protocol

When a new call requests admission into the network in a cell operating at level L, the cell first attempts to provide the connection with an amount of bandwidth equal to its desired bandwidth minus L shares of its ABB, that is

$$M - L \cdot \frac{\text{ABB}}{\lambda} = \left(1 - \frac{Lf}{\lambda}\right) \cdot M + \frac{Lf}{\lambda} \cdot m. \tag{28.4}$$

If the amount of bandwidth specified in Equation (28.4) exceeds the amount of bandwidth available, the cell tests to see if the call could be admitted if the cell progressed to level $L + 1$. If transition to level $L + 1$ will provide enough bandwidth to admit the call, the bandwidth is borrowed, the level is incremented, and the call is admitted; otherwise, the call is blocked. When the cell is operating at level $L = \lambda$, no more borrowing is allowed. It is important to note that EBS never borrows from CBR connections or from connections that have already lost more than L shares.

Every time bandwidth becomes available in a cell due to a connection releasing its bandwidth allocation, the cell will attempt to make a transition to the next lower level. As a result, the available bandwidth is returned to the connections that have lost bandwidth due to borrowing. All fluctuations in a connection's allocated bandwidth are gradual as only one share can be borrowed or returned at a time.

28.3.2.4 Handoff Management

The handoff admission policies differentiate between Classes I and II connections. The reserved bandwidth is used only for Class I connections, which are admitted only if their minimum bandwidth needs can be met. When a Class I connection requests admission into a cell as a handoff, the cell checks to see if the minimum bandwidth requirement can be met with the sum of the available free and reserved bandwidth in the cell. If such is the case, the call is admitted into the cell and given bandwidth from the free bandwidth up to its desired level minus L shares. The connection is given bandwidth from the reserved bandwidth pool only if it is needed to reach its minimum requirement. If the minimum cannot be met using the free and reserved bandwidth, the cell tests to see if scaling to level $L + 1$ would free up enough bandwidth to admit the call. If so, the cell scales the other calls in the cell and provides the handoff call with bandwidth according to the guidelines described above.

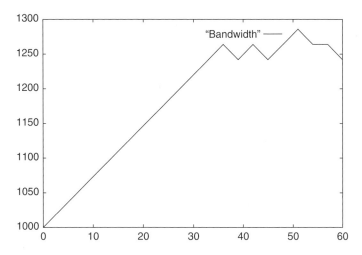

FIGURE 28.4   Illustrating the recovery of a Class I handoff.

On the other hand, Class II traffic will only be dropped if there is no free bandwidth left in the cell at all. The reserved pool is not available to these connections, because, as in Reference 44, we assume that Class II traffic is able and willing to incur a possibly substantial fluctuation in service rather than be disconnected. Calls that have suffered a lowering of bandwidth due to a handoff will eventually be brought back to a reasonable level as their new cell has free bandwidth to give them. This is in sharp contrast to the schemes presented in Reference 44, which have no facility to improve connections that have been degraded due to a handoff.

28.3.2.5   How Well Does a Handoff Do?

Recall that a handoff Class I connection may be cut down to its minimum in order to avoid dropping the call. In addition, EBS specifically disallows borrowing from connections that are below the cell level L. When bandwidth becomes available, EBS attempts to bring all Class I connections to the cell level L. In particular, this means that handoff connections are expected to *recover* from a bandwidth loss incurred at handoff time.

Figure 28.4 is illustrating this recovery process by plotting the bandwidth allocated to a Class I handoff connection over time. At time 0 the connection is admitted into the cell at its minimum acceptable level. In roughly 35 time units (sec, in our simulation) the bandwidth has been replenished to the cell level. That the connection has reached the cell level L is evident from the fact that the connection is borrowed[2] from.

28.3.2.6   Simulation Results

In order to evaluate the performance of EBS, we implemented and simulated two other schemes for comparison. First, we implemented a request-based statistical reservation scheme from Reference 44, termed the uniform and bandwidth-based model. According to this scheme, when reservations are made on behalf of a connection in neighboring cells, an equal amount of bandwidth is reserved in each neighboring cell, with no consideration of the most likely cell to which the host might travel. A cell does not reserve the sum of all the bandwidth it is asked to reserve, but just the largest of all the current requests.

We also simulated a simple scheme that reserves 5% of the total bandwidth in each cell for handoffs. New calls are admitted into the network if their desired bandwidth can be met, otherwise they are blocked. Class I handoffs are admitted if at least their minimum bandwidth requirements can be met. They are

---

[2]In Figure 28.4 this shows as a small decrease in bandwidth.

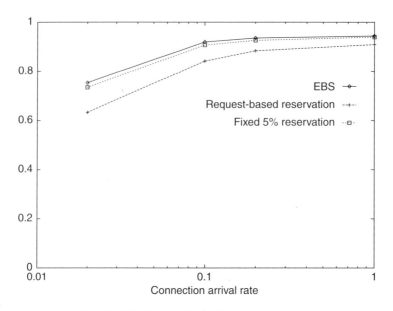

FIGURE 28.5   A comparison of bandwidth utilization by the three schemes.

only given enough bandwidth from the reserved pool to meet their minimum if there is too little free bandwidth available. Class II handoffs are admitted if there is any free bandwidth in the cell.

To simulate EBS, we used a fixed reservation pool representing 5% of the total bandwidth. We set f to 0.5, thus permitting borrowing up to half of the BLT. And we set λ to 10, so that each call had 10 shares to give.

For the results shown in the coming figures, the speed of the MHs was set to a host spending from 1 to 15 min in a cell, with an average of 5 min per cell. Each cell had 30 Mbps of bandwidth. The network was a hexagonal grid of size 6 × 6 consisting of 36 cells. Traffic was provided to each cell at the level being measured.

Figure 28.5 compares the values of bandwidth utilization for the request-based reservation scheme from Reference 44, for a fixed reservation scheme with $r = 5\%$, and for EBS with $r = 5\%, \lambda = 10$, and $f = 0.5$, so that at most half of a call's BLT can be borrowed. For the fixed reservation scheme and the rate-based borrowing scheme, at the maximum connection rate, the bandwidth utilization comes close to equaling the bandwidth outside of the reserved pool. The results for the request-based reservation scheme are worse than for the other two, because we did not implement a cap on the size of the reserved pool.

Figure 28.6 and Figure 28.7 show, respectively, the CDP for Class I traffic alone and for Classes I and II traffic combined. EBS outperforms the other two schemes in both cases. In fact, the dropping probability for Class I connections is very close to zero. The motivation, of course, for favoring Class I connections by giving them exclusive use of the handoff reserves, is that real-time connections would suffer an actual loss by being dropped. We assume that a Class II application, although inconvenienced by being dropped, would be able to resume its transmission at a later time, without any significant loss. Despite this, Class II traffic fares significantly better under EBS than under the others; it is especially important that EBS returns bandwidth to connections that have suffered cuts during a handoff.

Next, Figure 28.8 and Figure 28.9 illustrate, respectively, the CBP for Class I traffic alone and for Classes I and II traffic combined. They demonstrate how borrowing allows a significant improvement in the CBP while also improving the dropping probability. As with CDP, the combined traffic also fares worse than Class I traffic alone in terms of CBP. However this is not due to any bias in the algorithms, but rather to the characteristics of the traffic being simulated. The Class II traffic requires more bandwidth on average.

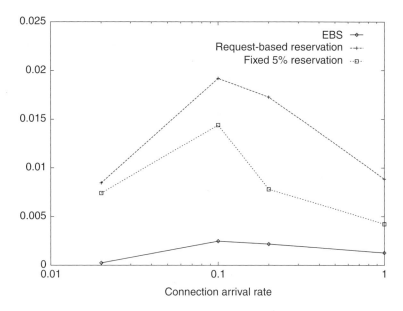

FIGURE 28.6 Call dropping probabilities for Class I traffic.

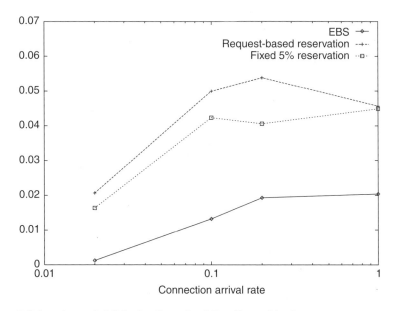

FIGURE 28.7 Call dropping probabilities for Classes I and II traffic combined.

28.3.3 A Max–Min Fair Bandwidth Borrowing Scheme

The second scheme we review is also a reservation and borrowing scheme, but it is based on max–min fairness, rather than the proportional, rate-based fairness described earlier. This scheme, proposed by Malla et al. [36], will be referred to henceforth as MBS. Max–min fairness is reviewed in the following section.

28.3.3.1 Max–Min Fairness

When the amount of bandwidth requested by the connections in a cell exceeds the total bandwidth available in the cell, it is unavoidable that some of the connections will receive less than their desired

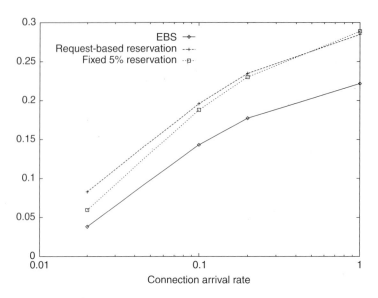

FIGURE 28.8   Call blocking probabilities for Class I traffic.

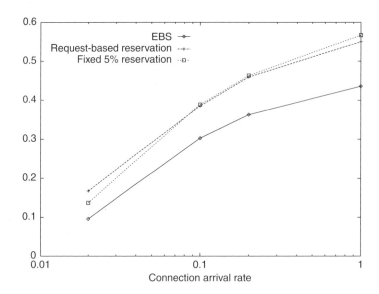

FIGURE 28.9   Call blocking probabilities for Classes I and II traffic combined.

amount of bandwidth. It is, however, important that the bandwidth allocation be fair in some sense. One of the best-known fair allocation schemes is the classic max–min fairness. An allocation is max–min fair if there is no way to give more bandwidth to a connection without decreasing the allocation of a connection of lesser or equal bandwidth [5].

For an illustration, consider a cell that has a total of 20 units of bandwidth. At some point there are four active connections A, B, C, and D using, respectively, 5, 4, 3, and 2 units of bandwidth. Referring to Figure 28.10, assume that a new connection E, requesting 7 units, has been accepted into the cell. Clearly, the total amount of bandwidth requested by these connections is 21 units, exceeding the capacity of the cell. How should the bandwidth be partitioned between the connections in a fair manner?

The key idea in max–min fairness is that if n connections need to partition b units of bandwidth, then each is guaranteed its equal share of b/n units. Connections that require at most their equal share

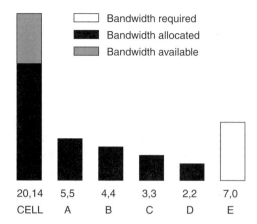

FIGURE 28.10   Insufficient bandwidth to satisfy new connection.

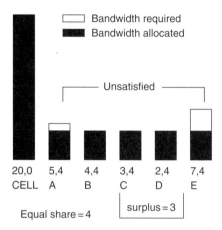

FIGURE 28.11   Allocation of the equal share to all connections.

are granted their desired bandwidth and are referred to as *satisfied*. For definiteness, assume that of the n connections, m are satisfied and the total bandwidth requested by the m satisfied connections is S units. The residual bandwidth, $R = mb/n - S$, is partitioned among the remaining connections, each receiving $R/(n - m)$ units.

Notice that, in the current example, the addition of the new connection E brings the total number of connections to 5, and consequently, each is guaranteed 4 units of bandwidth. As illustrated in Figure 28.11, connections B, C, and D are satisfied, while connections A and E are not. Since connections C and D are requesting less than their equal share, there are 3 units of residual bandwidth. As shown in Figure 28.12, the residual bandwidth is now partitioned between the unsatisfied connections A and E, each receiving 1.5 units of additional bandwidth. With this new allocation, connection A becomes satisfied, leaving E as the only unsatisfied connection.

Finally, the residual 0.5 units of bandwidth are now given to connection E. The final max–min fair bandwidth allocation is shown in Figure 28.13. Notice that connections A, B, C, and D are satisfied and that the only unsatisfied connection is the relatively bandwidth-intensive connection E.

As illustrated by the example above, max–min fairness attempts to maximize the bandwidth allocation to the connections requesting the least amount of bandwidth. The max–min algorithm considers that each connection is entitled to an equal share of the limited bandwidth. Some connections request less bandwidth than others. In the end, connections with a low bandwidth demand receive their desired amount of

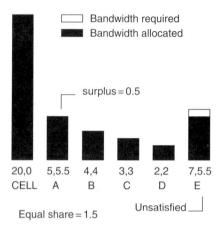

FIGURE 28.12   Redistribution of the residual bandwidth.

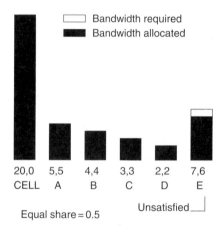

FIGURE 28.13   Final redistribution of the residual bandwidth.

bandwidth, thus becoming satisfied, while the remainder of the bandwidth is equally apportioned among the unsatisfied connections.

28.3.3.2 The MBS Scheme

Similar to the EBS scheme discussed in Section 28.3.2, MBS keeps a pool of bandwidth reserved for handoffs. This reserved pool is used for Class I handoffs only, with the assumption that real-time connections have more stringent QoS requirements than Class II connections. It has been shown already that this reservation technique significantly reduces the CDP without adversely affecting the CBP. In order to also improve the CBP, our scheme allows for the temporary borrowing of bandwidth from existing connections in order to accommodate new and handoff connections.

The parameters that must be specified in a connection request are the connection class and the desired, minimum and *expected* levels of bandwidth. The expected amount falls between the desired and the minimum and represents a comfortable working level for the application. In fact, in MBS, the expected bandwidth is guaranteed to a connection while it remains in its initial cell. The difference between the desired and expected amounts of a connection is the ABB, and the cell may borrow some of this bandwidth from an existing connection in order to accommodate other incoming connections.

A new connection, whether Class I or II, is accepted into a cell only if its expected bandwidth is less than or equal to the total bandwidth of the cell (not including the reserved handoff pool) divided by the number of connections in the cell. This requirement is actually a key feature of MBS, because it can cause a new connection to be blocked even if there is enough bandwidth for it in the cell, in order not to give it an unfairly large amount of bandwidth. If all the connections in a cell are functioning at their desired levels, and a new connection can also be accommodated at its desired level, it is simply admitted. If a connection cannot be given its expected amount using all the borrowable and free bandwidth in the cell, it is rejected. However, if it can be accommodated at its expected level, then the sum of the borrowable amounts of each connection plus any free bandwidth in the cell is divided equally among all the connections. Specifically, the equal share in a cell is determined by the formula below:

$$\text{Connections} = \text{active Connections} + 1$$

$$\text{EqualShare} = \frac{\text{total Bandwidth}}{\text{connections}}$$

If a new mobile user requests a connection, the cell will accept or deny the connection based on the following algorithm:

```
if (expected bandwidth of new user <= equalShare)
    if (expected bandwidth of new user
            <= borrowable bandwidth of all users)
        accept connection
    else
        block call
else
    block call
```

Handoff management differs for Classes I and II connections. In the case of Class II connections, we assume that they want to be continued, even if the bandwidth allocated is very small, since they do not have stringent QoS requirements. Therefore, Class II handoffs are not dropped as long as there is some free bandwidth in the new cell; clearly, their stated minimum may not be honored. The reserved bandwidth pool is for the exclusive use of Class I handoffs, and not available to Class II connections. However, a Class I handoff is only given bandwidth from the reserved pool if its minimum cannot be met using the borrowable and free bandwidth in the cell.

When a connection terminates, the freed bandwidth is used first to replenish the reserved pool, with the leftover allocated to connections that are functioning below their expected level. Finally, any residual bandwidth is distributed in a max–min fair manner among the connections that are functioning below their desired level.

28.3.3.3 Simulation Results

The simulation environment used for MBS is exactly the same one used for EBS. In fact, the two algorithms were compared against each other directly in a simulator subjecting both to exactly the same traffic.

Figure 28.14 and Figure 28.15 show, respectively, the CDP for Class I traffic alone, and for Classes I and II traffic, combined. The results show that the max–min scheme outperforms the other scheme in both cases. By blocking slightly more bandwidth-intensive connections, MBS makes more bandwidth available to both new and handoff connections. Somewhat surprisingly, the more strict call admission regimen of MBS does not adversely impact the overall CBP, as illustrated in Figure 28.16 and Figure 28.17. We note here that the significant difference in performance between MBS and EBS can be partially attributed to the traffic mix assumed in the simulation model. This mix features connections whose bandwidth requirements differ dramatically. However, this is anticipated to be a realistic scenario in future wireless multimedia networks [44].

It is important to note that all the desirable characteristics of MBS do not adversely impact the network bandwidth utilization. This is shown in Figure 28.18.

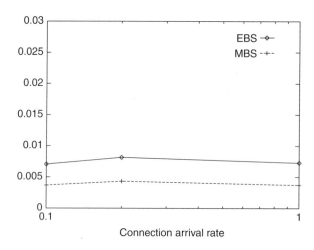

FIGURE 28.14   Call dropping probabilities for Class I traffic.

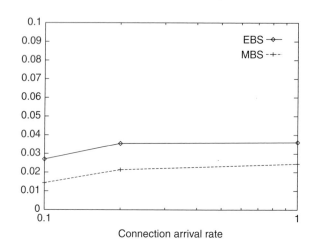

FIGURE 28.15   Call dropping probabilities for Classes I and II traffic combined.

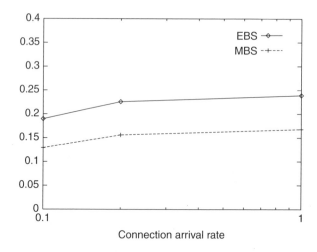

FIGURE 28.16   Call blocking probabilities for Class I traffic.

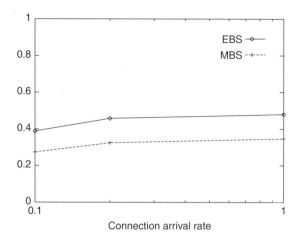

FIGURE 28.17   Call blocking probabilities for Classes I and II traffic combined.

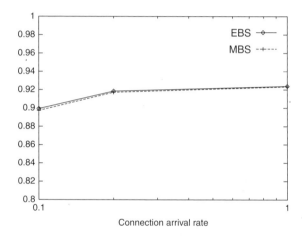

FIGURE 28.18   A comparison of bandwidth utilization by EBS and MBS.

The bandwidth utilization of MBS is just slightly worse than that of EBS. We note here that the better performance of the MBS in terms of CDP and CBP more than compensates for the slight degradation in bandwidth utilization. As noted before, this comes at the price of denying admission into the system to those connections that are requesting an inordinate amount of resources. Since bandwidth-intensive connections have a high probability of being dropped in handoff situations, denying them admission is likely to improve both CBP and CDP.

One of the shortcomings of MBS is that occasionally the users can be subjected to significant bandwidth fluctuations. While high-quality codecs can dampen the effects of these changes, it would be desirable to smooth out, to the largest extent possible, these bandwidth fluctuations. In this respect, EBS is superior to MBS as it borrows and returns bandwidth in a graded manner.

28.4   The QoS Provisioning in LEO Satellite Networks

Due to the high velocity of LEO satellites, in LEO systems, a connection established in one of the cells of a satellite's footprint is likely to experience a large number of handoffs during its lifetime as it passes through several spotbeams. Consequently, LEO satellite systems require sophisticated handoff management and call admission control protocols. In this section we present the details of the predictive handoff management and admission control strategy for multimedia LEO satellite networks, that we will refer to as PRAS, proposed by El-Kadi et al. [24]. A key ingredient of PRAS is an adaptive resource reservation protocol. Simulation results have confirmed that PRAS offers very low call dropping probability, while at the same time keeping resource utilization high.

Predictive handoff management and admission control strategy involves some processing overhead. However, this overhead does not affect the MHs hosts as all the processing is handled by the satellite. Consequently, the scheme scales to a large number of users.

28.4.1 Mobility and Traffic Model

The LEO system model we have adopted for simulating PRAS is based on the well-known Iridium satellite system, where each satellite rotates around the Earth in a polar orbit [22]. We follow common practice and assume that the speed of individual MHs is negligible with respect to the orbital velocity of the satellite [13,22]. Consequently, MH trajectories are straight lines. The footprint of a satellite is partitioned into spotbeams, each approximated by a regular hexagon.

In order to model MH behavior we make the following assumptions:

- MHs move at constant speed, essentially equal to the orbital speed of the satellite, and they cross each cell along its maximum diameter.
- The time t_s it takes a MH to cross a cell is $t_s = $ (cell_width/speed).
- A MH remains in the cell where the connection was initiated for t_f time, where t_f is uniformly distributed between 0 and t_s; thus, t_f is the time until the first handoff request, assuming that the call does not end in the original cell.
- After the first handoff, a fixed time t_s is assumed between subsequent handoff requests until call termination.

Referring to Figure 28.19, when a new connection is requested in cell N, it is associated with a *trajectory*, consisting of a list $N, N+1, N+2, \ldots, N+i, \ldots$ of cells that the connection may visit during its lifetime. For a generic call C, we let H be the random variable denoting the holding time of C. We assume that H is exponentially distributed with mean $(1/\mu)$.

Assume that C was accepted in cell N. After t_f time units, C is about to cross into cell $N+1$. Let p_f be the probability of this first handoff request. Clearly,

$$p_f = \Pr[H > t_f] = \frac{1}{t_s} \int_0^{t_s} e^{-\mu t} dt = \frac{1 - e^{-\mu t_s}}{\mu t_s}. \tag{28.5}$$

Due to the memoryless property of the exponential distribution, the probability of the $(k+1)$th, $(k \geq 1)$, handoff request is

$$\Pr[H > t_f + k \cdot t_s | H > t_f + (k-1) \cdot t_s]$$

$$= \frac{\Pr[H > t_f + k \cdot t_s]}{\Pr[H > t_f + (k-1) \cdot t_s]} = \frac{e^{-\mu(t_f + k \cdot t_s)}}{e^{-\mu(t_f + (k-1) \cdot t_s)}} = e^{-\mu t_s}$$

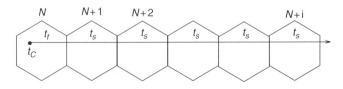

FIGURE 28.19   Illustrating the trajectory of a connection.

which, as expected, is independent of k. Consequently, we will let

$$p_s = e^{-\mu t_s} \tag{28.6}$$

denote the probability of a *subsequent* handoff request. It is important to note that t_f, t_s, p_f, and p_s are mobility parameters that can be easily evaluated by the satellite using its on-board processing capabilities.

When a MH requests a new connection C in a given cell, it provides the following parameters:

- The desired class of traffic for C (either I or II)
- M_C the desired amount of bandwidth for the connection
- If the request is for a Class I connection the following parameters are also specified:
  (a) m_C the minimum acceptable amount of bandwidth, that is the smallest amount of bandwidth that the source requires in order to maintain acceptable quality, for example, the smallest encoding rate of its codec.
  (b) θ_C the largest acceptable call dropping probability that the connection can tolerate.
  (c) $1/\mu_C$ the mean holding time of C.

28.4.2 The PRAS Bandwidth Reservation Strategy

The PRAS handoff admission policies distinguish between Classes I and II. Class I handoffs are admitted only if their minimum bandwidth requirements can be met. However, Class II handoff requests will be accepted as long as there is some bandwidth left in the cell. Thus, bandwidth reservation only pertains to Class I handoffs. The basic idea of PRAS is to reserve for each accepted Class I connection a certain amount of bandwidth in each cell along its trajectory.

Let p_h denote the handoff failure probability of a Class I connection, that is, the probability that a handoff request is denied for lack of resources. Let S_i denote the event that a Class I connection C admitted in cell N goes successfully through i handoffs and will, therefore, show up in cell $N + i$. It is easy to confirm that the probability of S_i is

$$\Pr[S_i] = p_f \cdot (1 - p_h) \cdot [p_s \cdot (1 - p_h)]^{i-1}. \tag{28.7}$$

Equation (28.7) suggests the following natural reservation strategy: in preparation for the arrival of connection C, an amount of bandwidth equal to

$$B_{N+i} = m_C \cdot \Pr[S_i], \tag{28.8}$$

will be reserved in cell $N + i$, $(i \geq 1)$, during the time interval $I_{N+i} = [t_C + t_f + (i-1)t_s, t_C + t_f + it_s]$, where t_C is the time C was admitted into the system.

It is worth noting that PRAS is lightweight: since the mobility parameters t_f and t_s are readily available, and since the trajectory of connection C is a straight line, the task of computing for every $i > 1$ the amount of bandwidth B_{N+i} to reserve as well as the time interval I_{N+i} during which B_{N+i} must be available is straightforward.

The bandwidth reservation strategy discussed earlier is meant to ensure that the parameter p_h is kept as low as possible. We emphasize here that p_h and CDP are *not* the same: while p_h quantifies the likelihood that an arbitrary Class I handoff request is denied, CDP has a long-term flavor denoting the probability that a Class I connection will be dropped at some point during its lifetime. It is straightforward to show that

$$\text{CDP} = p_f \cdot p_h + p_f \cdot (1 - p_h) \cdot p_s \cdot p_h + \cdots$$

$$= p_f \cdot p_h \sum_{i=0}^{\infty} [p_s(1 - p_h)]^i = \frac{p_f \cdot p_h}{1 - p_s(1 - p_h)} \tag{28.9}$$

where the first term in Equation (28.9) is the probability that the call will be dropped on the first handoff attempt, the second term denotes the probability that the call will be dropped on the second attempt, and so on.

We now point out a strategy for ensuring that for an accepted Class I connection C the negotiated CDP is maintained below the specified threshold θ_C. Thus, the goal is to ensure that CDP $< \theta_C$. By Equation (28.9) this amounts to insisting that

$$\frac{p_f \cdot p_h}{1 - p_s(1 - p_h)} < \theta_C.$$

Solving for p_h we get,

$$p_h < \frac{\theta_C \cdot (1 - p_s)}{p_f - \theta_C \cdot p_s}. \qquad (28.10)$$

All the quantities in the right-hand side of Equation (28.10) are either specified by the connection or can be determined by the satellite from the mean holding time of C by using Equations (28.5) and (28.6). Thus, in order to enforce the CDP commitment, the satellite keeps track of the handoff failure probability p_h for Class I connections. If p_h is close to the value of the right-hand side of Equation (28.10) new calls are temporarily blocked.

28.4.3 The PRAS Call Admission Strategy

The PRAS call admission strategy involves two criteria. The first call admission criterion, which is local in scope, applies to both Classes I and II connections, attempting to ensure that the originating cell has sufficient resources to provide the connection with its desired amount of bandwidth.

The second admission control criterion, which is global in scope, applies to Class I connections only, attempting to minimize the chances that, once accepted, the connection will be dropped later due to a lack of bandwidth in some cell into which it may handoff.

Consider a request for a new Class I connection C in cell N at time t_C and let t_f be the estimated residence time of C in N. Referring to Figure 28.20, the key observation that inspired our second criterion is that when C is about to handoff into cell $N + i, (i \geq 1)$, the connections resident in $N + i$ are likely to be those in region A of cell N and those in region B of cell $N + 1$. More precisely, these regions are defined as follows:

- A connection is in region A if at time t_C its residual residence time in cell N is less than t_f
- A connection is in region B if at time t_C its residual residence time in cell $N + 1$ is larger than t_f

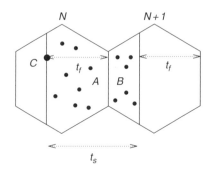

FIGURE 28.20   Illustrating the sliding window concept for call admission.

As illustrated in Figure 28.20 these two regions define a *window* of size t_s anchored at t_C. Let R_{N+i} be the projected bandwidth that has been reserved in cell $N + i$ for the connections in this window. The task of estimating R_{N+i} can be handled by the satellite in, essentially, two ways. First, if enough on-board computational power is available, such a sliding window is maintained for each cell in the footprint. On the other hand, in the presence of reduced on-board capabilities R_{N+i} can be estimated as follows. Let D_N and D_{N+1} be the total projected bandwidth requirements for Class I connections currently in cells N and $N + 1$, respectively. It is natural to approximate R_{N+i} by the following *weighted average* of D_N and D_{N+1}:

$$R_{N+i} = \frac{t_f}{t_s} \cdot D_N + (1 - \frac{t_f}{t_s}) \cdot D_{N+1}. \tag{28.11}$$

To justify Equation (28.11) let S denote the area of a cell. Clearly, the areas covered by regions A and B are, respectively, $(t_f/t_s) \cdot S$ and $(1 - (t_f/t_s)) \cdot S$. Now, assuming that D_N and D_{N+1} are uniformly distributed in cells N and $N + 1$, respectively, it follows that the projected bandwidth reserved for Class I connections in A and B is, respectively, $(t_f/t_s) \cdot D_N$ and $(1 - (t_f/t_s)) \cdot D_{N+1}$ and their sum is exactly R_{N+i}.

Observe that if the residual bandwidth available in cell $N + i$ once R_{N+i} has been committed is less than the projected bandwidth needs of connection C, it is very likely that C will be dropped. In the face of this bleak outlook connection C is not admitted into the system. Thus, the second admission criterion acts as an additional safeguard against a Class I connection being accepted only to be dropped at some later point.

28.4.4 Simulation Results

Our goal was to assess the performance of PRAS in terms of keeping the CDP low and bandwidth utilization high. For this purpose, we have compared PRAS to a fixed-rate scheme that sets aside 5% of the bandwidth for exclusive use of Class I handoffs.

In our simulation model we adopted some of the parameters of the Iridium system [22] where the radius of a cell is 212.5 km and the orbital speed of the satellites is 26,000 km/h. This implies that $t_s \approx 65$ sec. Residence times in the originating cells are uniformly distributed between 0 and 65 sec.

We have simulated a one-dimensional array of 36 cells each with a static bandwidth allocation of 30 Mbps. Thus, in the fixed reservation scenario, an amount of bandwidth equal to 1.5 Mbps (i.e., 5% of 30 Mbps) is set aside for Class I handoffs.

The results of the simulation are shown in Figures 28.21 to 28.23.

First, Figure 28.21 is plotting the call dropping probability for Class I connections against call arrival rate. The graphs in the figure confirm that PRAS offers a very low CDP, outperforming the fixed reservation policy.

Next, Figure 28.22 illustrates the combined call dropping probability of Classes I and II traffic against call arrival rate.

It is well-known that the goals of keeping the call dropping probability low and that of keeping the bandwidth utilization high are conflicting. It is easy to ensure a low CDP at the expense of bandwidth utilization and similarly, it is easy to ensure a high bandwidth utilization at the expense of CDP [13, 35]. The challenge, of course, is to come up with a handoff management protocol that strikes a sensible balance between the two. As Figure 28.23 shows, PRAS features a high bandwidth utilization in addition to keeping the call dropping probability low.

28.5 Directions for Further Research

Our simulation results have shown that the borrowing schemes presented in Sections 28.3.2 and 28.3.3 are very promising new tools for QoS provisioning in cellular networks. Some other researchers have briefly mentioned the possibility, but no one has explored it in detail or introduced a specific algorithm. Despite the slightly better performance of MBS over EBS we believe that EBS is superior due to its property of

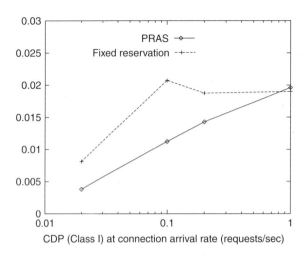

FIGURE 28.21   Call dropping probability for Class I traffic.

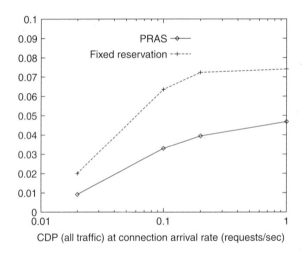

FIGURE 28.22   Combined call dropping probability for Classes I and II traffic.

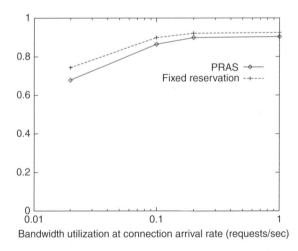

FIGURE 28.23   Illustrating bandwidth utilization.

changing a connection's bandwidth in small increments. But even with this feature, bandwidth borrowing subjects connections to possibly frequent fluctuations in the amount of bandwidth they are provided. In the schemes we have used for comparison and most others proposed in the literature, the only fluctuation in the bandwidth provided would occur due to a handoff.

Using EBS in a simulation run at a rate of one connection per second, we found that the bandwidth supplied to a connection fluctuated an average of once every 10 sec. It would be desirable to refine the borrowing algorithm to *smooth* out this fluctuation by adjusting the requirements for borrowing, and, in particular, for returning bandwidth, without significantly affecting the very low CDP and CBP achieved by the scheme.

- *Research Topic 1: Smooth* the fluctuations in supplied bandwidth caused by the rate-based borrowing algorithm of EBS.

Bandwidth borrowing also decreases the probability that calls will always be provided their desired amount of bandwidth. In simulation we noticed that the calls that lost bandwidth during a handoff were always steadily replenished; however, they were often operating at a level below their desired. In the case of adaptive multimedia applications it makes sense to introduce a new QoS parameter that specifies the fraction of time that a connection spends operating below its desired or expected bandwidth level. IT would be interesting to incorporate such a QoS parameter into cellular networks.

- *Research Topic 2:* Measure and control the time that connections spend operating at levels below expected.

Our extensive simulations have shown that the values chosen for r (the fixed reservation pool size), f (the fraction of the BLT value made available for borrowing), and λ (the number of shares) all have a marked impact on the performance results. Future research will involve finding optimal values for these parameters, understanding how they relate to each other and to the QoS parameters, and determining whether they can be adjusted dynamically to further increase network performance.

- *Research Topic 3:* Analyze the effects of r, f, and λ, and evaluate the merits of making them adapt to network conditions.

In LEO satellite networks, the exact knowledge of the speed and direction of the MHs makes predictive reservation more successful than simple fixed reservation. For the cases when a handoff call still does not find enough resources, it makes sense that the rate-based borrowing scheme would work well. We intend to add the borrowing scheme to the predictive reservation scheme for LEO satellite systems.

- *Research Topic 4:* Incorporate rate-based borrowing in reservation scheme for LEO networks.

As discussed earlier, EBS, MBS, and PRAS have dealt only with admission control and resource reservation in networks where a fixed channel allocation scheme is assumed (i.e., each cell has the same amount of bandwidth). It is likely that other methods of reducing the CDP and CBP, such as handoff queuing dynamic channel allocation would be complementary to these schemes. It would be interesting to add a dynamic resource allocation and (inter-cell) borrowing algorithm to the intra-cell borrowing one. We believe that this would improve performance on all fronts — lowering the CDP and CBP, as well as providing even more *smoothing* if free resources are borrowed from neighboring cells first, and borrowing from local connections happens only as a last resort.

- *Research Topic 5:* Enhance intra-cell resource borrowing with the addition of dynamic channel allocation and inter-cell borrowing.

The concept of bundling a set of streams into a single multimedia session has been well-researched for wireline networks and just recently explored in wireless networks. We believe that the idea of session-level QoS negotiation would fit very well with our provisioning schemes. A session would identify its component streams and their attributes, allowing for decisions at handoff time such as dropping one substream temporarily in order to continue the whole session. Borrowing could be allowed from one substream but not from another, in order to preserve the best possible overall QoS for the session.

- *Research Topic 6:* Provide QoS guarantees to multimedia sessions.

All algorithms reviewed in this chapter divide traffic into just Class I and Class II, real-time, and nonreal-time. Class I is given priority over Class II. It would be of interest to consider a general set of classes, that are distinguished from one another not by traffic type but by a user- or network-defined priority. The most obvious example of this is priorities based on price. In which case, for example, a customer with a telnet connection who insisted on a very low CDP and on maintaining consistent bandwidth across handoffs could pay for that privilege, while a customer with a video connection might be willing to risk a high dropping probability in order to save money.

- *Research Topic 7:* Extend existing schemes to apply to n classes of traffic distinguished by user- or network-specified priority rather than traffic type.

References

[1] A. Acampora and M. Naghshineh, Control and quality-of-service provisioning in high-speed microcellular networks, *IEEE Personal Communications*, 1, 1994, 36–43.

[2] P. Agarwal, D.K. Anvekar, and B. Narendran, Channel management policies for handovers in cellular networks, *Bell Labs Technical Journal*, 1, 1996, 97–110.

[3] A.R. Aljadhai and T.F. Znati, Predictive mobility support for QoS provisioning in mobile wireless environments, *IEEE Journal on Selected Areas in Communications*, 19, 2001, 1915–1930.

[4] P. Berge, A. Ferreira, J. Galtier, and J.-N. Petit, A probabilistic study of inter-satellite links load in polar orbit satellite constellations, *Telecommunication Systems*, 18, 2001, 123–135.

[5] D. Bertsekas and R. Gallager, *Data Networks*, Prentice-Hall, New York, 1992.

[6] A. Boukerche, H. Song, and T. Jacob, An efficient dynamic channel allocation algorithm for mobile telecommunication systems, *ACM Mobile Networks and Applications*, 7, 2002, 115–126.

[7] A. Boukerche, K. El-Khatib, and T. Huang, A performance comparison of dynamic channel and resource allocation protocols for mobile cellular networks, in *Proceedings of the 2nd International ACM Workshop on Mobility and Wireless Access Protocols (MobiWac)*, Philadelphia, PA, September 2004,

[8] A. Boukerche and H. Owens, Media synchronization QoS packet scheduling scheme in wireless and mobile networks, *ACM/Kluwer Mobile Networks and Applications*, 10, 2005, 233–249.

[9] G. Cao and M. Singhal, Distributed fault tolerant channel allocation for cellular networks, *IEEE Journal of Selected Areas in Communications*, 18, 2000, 1326–1337.

[10] C. Chao and W. Chen, Connection admission control for mobile multiple-class personal communications networks, *IEEE Journal on Selected Areas in Communications*, 15, 1997, 1618–1626.

[11] S. Chen and K. Nahrstedt, Distributed quality-of-service routing in ad-hoc networks, *IEEE Journal on Selected Areas in Communications*, 17, 1999, 1488–1505.

[12] M.-H. Chiu and M.A. Bassiouni, Predictive schemes for handoff prioritization in cellular networks based on mobile positioning, *IEEE Journal on Selected Areas in Communications*, 15, 2000, 510–522.

[13] S. Cho, Adaptive dynamic channel allocation scheme for spotbeam handover in LEO satellite networks, in *Proceedings of the 50th IEEE Vehicular Technology Conference (VTC'2000)*, Boston, MA, September 2000, pp. 1925–1929.

[14] S. Cho, I.F. Akyildiz, M.D. Bender, and H. Uzunalioglu, A new spotbeam handover management technique for LEO satellite networks, in *Proceedings of the IEEE GlobeCom'2000*, San Francisco, CA, November 2000.

[15] S. Choi and K.G. Shin, Adaptive bandwidth reservation and admission control in QoS-sensitive cellular networks, *IEEE Transactions on Parallel and Distributed Systems*, 13, 2002, 882–897.

[16] S. Choi and K.G. Shin, A comparative study of bandwidth reservation and admission control schemes in QoS-sensitive networks, *ACM/Balzer Wireless Networks*, 6, 2000, 289–305.

[17] S.K. Das, S.K. Sen, and R. Jayaram, A dynamic load-balancing strategy for channel assignment using selective borrowing in a cellular mobile environment, *ACM/Balzer Wireless Networks*, 3, 1997, 333–347.

[18] S.K. Das, M. Chatterjee, and N.K. Kakani, QoS provisioning in wireless multimedia networks, in *Proceedings of the Wireless Communications and Networking Conference (WCNC '99)*, New Orleans, LO, September 1999.

[19] S.K. Das, R. Jayaram, N.K. Kakani, and S.K. Sen, A call admission and control scheme for quality-of-service (QoS) provisioning in next generation wireless networks, *ACM/Baltzer Journal on Wireless Networks*, 6, 2000, 17–30.

[20] E. Del Re, R. Fantacci and G. Giambene, Performance evaluation of different resource management strategies in mobile cellular networks, *Telecommunication Systems*, 12, 1999, 315–340.

[21] E. Del Re, R. Fantacci, and G. Giambene, Efficient dynamic channel allocation techniques with handover queuing for mobile satellite networks, *IEEE Journal on Selected Areas in Communications*, 13, 1995, 397–405.

[22] E. Del Re, R. Fantacci, and G. Giambene, Characterization of user mobility in low earth orbit mobile satellite systems, *ACM/Balzer Wireless Networks*, 6, 2000, 165–179.

[23] B. El-Jabu and R. Steele, Cellular communications using aerial platforms, *IEEE Transactions on Vehicular Technology*, 50, 2001, 686–700.

[24] M. El-Kadi, S. Olariu, and P. Todorova, Predictive resource allocation in multimedia satellite networks, in *Proceedings of the IEEE GLOBECOM 2001*, San Antonio, TX, November 2001.

[25] M. El-Kadi, S. Olariu, and H. Abdel-Wahab, A rate-based borrowing scheme for QoS provisioning in multimedia wireless networks, *IEEE Transactions on Parallel and Distributed Systems*, 13, 2002, 156–166.

[26] B.M. Epstein and M. Schwartz, Predictive QoS-based admission control for multi-class traffic in cellular wireless networks, *IEEE Journal on Selected Areas in Communications*, 18, 2000, 523–534.

[27] E.C. Foudriat, K. Maly, and S. Olariu, An architecture for robust QoS provisioning for mobile tactical networks, in *Proceedings of the IEEE MILCOM'2000*, Los Angeles, CA, October 2000.

[28] E.C. Foudriat, K. Maly, and S. Olariu, H3M — a rapidly-deployable architecture with QoS provisioning for wireless networks, in *Proceedings of the 6th IFIP Conference on Intelligence in Networks (SmartNet'2000)*, Vienna, Austria, September 2000, 282–304.

[29] J. Galtier, Geographical reservation for guaranteed handover and routing in low earth orbit constellations, *Telecommunications Systems*, 18, 2001, 101–121.

[30] A. Iera, A. Molinaro, S. Marano, and M. Petrone, QoS for multimedia applications in satellite systems, *IEEE Multimedia*, 6, 1999, 46–53.

[31] S. Kalyanasundaram, E.K.P. Chong, and N.B. Shroff, An efficient scheme to reduce handoff dropping in LEO satellite systems, *ACM/Balzer Wireless Networks*, 7, 2001, 75–85.

[32] S. Kalyanasundaram, *Call-Level and Class-Level Quality-of-Service in Multiservice Networks*, Ph.D. Thesis, Purdue University, 2000.

[33] S. Kalyanasundaram, E.K.P. Chong, and N.B. Shroff, Admission control schemes to provide class-level QoS in multiservice networks, *Computer Networks*, 35, 307–326.

[34] A. Kuzmanovic and E. Knightly, Measuring service in multi-class networks, in *Proceedings of the IEEE INFOCOM 2001*, Anchorage, AL, April 2001.

[35] D. Levine, I. Akyildiz, and M. Naghshineh, A resource estimation and call admission algorithm for wireless multimedia networks using the shadow cluster concept, *IEEE/ACM Transactions on Networking*, 5, 1997, 1–12.

[36] A. Malla, M. El-Kadi, S. Olariu, and P. Todorova, A fair resource allocation protocol for multimedia wireless networks, *IEEE Transactions on Parallel and Distributed Systems*, 14, 2003, 63–71.

[37] J. Misic, S.T. Chanson, and F.A. Lai, Admission control for wireless multimedia networks with hard call level quality of service bounds, *Computer Networks* 31, 1999, 125–140.

[38] M. Naghshineh and M. Schwartz, Distributed call admission control in mobile/wireless networks, *IEEE Journal of Selected Areas in Communications*, 14, 1996, 711–717.

[39] M. Naghshineh and M. Willebeek-LeMair, End-to-end QoS provisioning in multimedia wireless/ mobile networks using an adaptive framework, *IEEE Communications Magazine*, November 1997, 72–81.

[40] H.N. Nguyen, J. Schuringa, and H. van As, Handover schemes for QoS guarantees in LEO-satellite networks, in *Proceedings of the IEEE International Conference on Networks (ICON'2000)*, Singapore, September 2000, 393–398.

[41] H.N. Nguyen, S. Olariu, and P. Todorova, A two-cell lookahead call admission and handoff management scheme for multimedia LEO satellite networks, in *Proceedings of the 35th Hawaii International Conference on System Sciences (HICSS-35)*, Big Island, Hawaii, HI, January 2002.

[42] N.H. Nguyen, S. Olariu, and P. Todorova, A novel mobility model and resource allocation strategy for multimedia LEO satellite networks, in *Proceedings of the IEEE WCNC*, Orlando, FL, March 2002.

[43] S. Oh and D. Tcha, Prioritized channel assignment in a cellular radio network, *IEEE Transactions on Communications*, 40, 1992, 1259–1269.

[44] C. Oliviera, J. Kim, and T. Suda, An adaptive bandwidth reservation scheme for high speed multimedia wireless networks, *IEEE Journal of Selected Areas in Communications*, 16, 1998, 858–874.

[45] S. Olariu and P. Todorova, Resource management in LEO satellite networks, *IEEE Potential*, 22, 2003, 6–12.

[46] S. Olariu, S.A. Rizvi, R. Shirhatti, and P. Todorova, Q-Win — A new admission and handoff management scheme for multimedia LEO satellite networks, *Telecommunication Systems*, 22, 2003, 151–168.

[47] S. Papavassiliou, L. Tassiulas, and P. Tandon, Meeting QoS requirements in a cellular network with reuse partitioning, *IEEE Journal on Selected Areas in Communications*, 12, 1994, 1389–1400.

[48] J.M. Peha and A. Sutivong, Admission control algorithms for cellular systems, *ACM/Balzer Wireless Networks*, 7, 2001, 117–125.

[49] H.G. Perros and K.M. Elsayyed, Call admission control schemes: a review, *IEEE Communications Magazine*, 34, 1996, 82–91.

[50] R. Prakash, N.G. Shivaratri, and M. Singhal, Distributed dynamic fault-tolerant channel allocation for mobile computing, *IEEE Transactions on Vehicular Technology*, 48, 1999, 1874–1888.

[51] S.V. Raghavan and S.K. Tripathy, *Networked Multimedia Systems*, Prentice-Hall, New York, 1998.

[52] B. Sadeghi and E.W. Knightly, Architecture and algorithms for scalable mobile QoS, *ACM Wireless Networks Journal (Special issue on wireless multimedia)*, 9, 2003, 7–20.

[53] P.T.S. Tam, J.C.S. Lui, H.W. Chan, C.C.N. Sze, and C.N. Sze, An optimized routing scheme and a channel reservation strategy for a low earth orbit satellite system, in *Proceedings of the IEEE Vehicular Technology Conference (VTC'1999)*, Amsterdam, The Netherlands, September 1999, 2870–2874.

[54] P. Todorova, S. Olariu, and H.N. Nguyen, A selective look-ahead bandwidth allocation scheme for reliable handoff in multimedia LEO satellite networks, in *Proceedings of the 2nd European Conference on Universal Multiservice*, Colmar, France, April 2002.

[55] L. Trajković and A. Neidhardt, Effect of traffic knowledge on the efficiency of admission-control policies, *ACM Computer Communication Review*, 1999, 5–34.

[56] H. Uzunalioglu, A connection admission control algorithm for LEO satellite networks, in *Proceedings of the IEEE International Conference on Communications (ICC'99)*, Vancouver, Canada, June 1999, pp. 1074–1078.

[57] O.T.W. Yu and V.C.M. Leung, Adaptive resource allocation for prioritized call admission over an ATM-based wireless PCN, *IEEE Journal on Selected Areas in Communications*, 15, 1997, 1208–1225.

[58] A.H. Zaim, George N. Rouskas, and Harry G. Perros, Computing call blocking probabilities in LEO satellite networks: the single orbit case, in *Proceedings of the 17th International Teletraffic Congress (ITC 17)*, December 2001, Salvador da Bahia, Brazil, pp. 505–516.

VI

TCP Studies in Wireless Networks

29

TCP over Wireless Networks

29.1 Introduction..29-679
29.2 Reno TCP Congestion Control 29-680
29.3 Characteristics of Wireless Media 29-681
29.4 Wireless TCP Mechanisms 29-681
    End-to-End Mechanisms • Split Connection Mechanisms •
    Local Recovery Mechanisms
29.5 Comparison of Wireless TCP Mechanisms 29-688
    Interaction Between TCP and the Link Layer • Interaction
    Between TCP and the Network Layer • Retransmission
    Approaches and ACKs • Flow Control • Loss Detection and
    Dependence on Timers • Effect of Path Asymmetry • Effect
    of Wireless Link Latency on RTT • State Maintained •
    Robustness and Overhead • Preserving End-to-End
    Semantics • Suitability to Different Network Environments
    • Adaptation to Long Fading Periods • Energy Efficiency
29.6 The Internet Engineering Task Force (IETF) Efforts ..29-692
    TCP over Satellites
29.7 Summary and Open Issues..............................29-694
References 29-696

Sonia Fahmy
Ossama Younis
Venkatesh Prabhakar
Srinivas R. Avasarala

29.1 Introduction

Because of the low loss nature of wired links, TCP assumes that packet losses occur mostly due to congestion. TCP reacts to congestion by decreasing its congestion window [1], thus reducing network utilization. In wireless networks, however, losses may occur due to the high bit-error rate of the transmission medium or due to fading or mobility. TCP still reacts to losses according to its congestion control scheme, thus unnecessarily reducing the network utilization. This chapter discusses proposals to address this problem.

A typical wireless network system model is depicted in Figure 29.1. Although we depict the wireless hop as the last hop, and depict the mobile unit to be the TCP receiver most of the time, this chapter is not limited to considering that particular scenario. This model can represent both wireless local area networks (WLANs) such as 802.11 networks, and wireless wide area networks (WWANs) such as CDPD and GPRS networks. In ad hoc networks, on the other hand, there is no required infrastructure (such as base stations (BSs)), and nodes can organize themselves to establish communication routes. Satellite networks allow various configurations with potentially multiple wireless hops. Our focus in this chapter

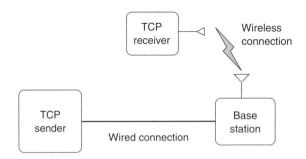

FIGURE 29.1   Typical model of a wireless TCP connection.

is on configurations where the wireless link is the last hop, though we also classify approaches where this is not necessarily the case.

Handling wireless losses may or may not be transparent to the sender. In other words, the sender implementation may be modified to be aware of losses due to wireless links, or local recovery can be used to handle such losses. The latter method shields the sender from knowing about the wireless link. A special case of this method terminates each TCP connection at the wireless interface (the base station), which in turn uses some other reliable connection to connect to the destination. This solution is referred to as a split-connection.

In this chapter, we compare and contrast a number of wireless TCP proposals. We find that some TCP sender adaptations work well for certain environments only, while others are generally useful. Local recovery works best when the wireless link delay is short, and split connection schemes allow application/transport level optimizations, but exhibit high overhead and lack flexibility and robustness.

The remainder of the chapter is organized as follows. Section 29.2 briefly reviews basic TCP congestion control. Section 29.3 discusses the characteristics of wireless media that affect performance of TCP. Section 29.4 classifies the main mechanisms proposed to solve the wireless transport problem. Section 29.5 presents a comparative study of the various mechanisms. Section 29.6 summarizes standardization efforts at the IETF. Finally, Section 29.7 concludes the chapter.

29.2   Reno TCP Congestion Control

A TCP connection starts off in the **slow start** phase [2]. The slow start algorithm uses a variable called congestion window ($cwnd$). The sender can only send the minimum of $cwnd$ and the receiver advertised window, which we call $rwnd$ (for receiver flow control). Slow start tries to rapidly reach equilibrium by opening up the window very quickly. The sender initially sets $cwnd$ to 1 (RFC 2581 [1] suggests an initial window size value of 2 and RFC 2414 [3] suggests $\min[4 \times \text{MSS}, \max(2 \times \text{MSS}, 4380 \text{ bytes})]$), and sending one segment. (MSS is the maximum segment size.) For each ACK that the sender receives, the $cwnd$ is increased by one segment. Increasing by one for every ACK results in an exponential increase of $cwnd$ over round trips, as shown in Figure 29.2.

TCP uses another variable ssthresh, the slow start threshold. Conceptually, ssthresh indicates the "right" window size depending on current network load. The slow start phase continues as long as $cwnd$ is less than ssthresh. As soon as $cwnd$ crosses ssthresh, TCP goes into **congestion avoidance**. In congestion avoidance, for each ACK received, $cwnd$ is increased by $1/cwnd$ segments. This is *approximately* equivalent to increasing $cwnd$ by one segment in one round trip (an additive increase), if every segment is acknowledged, as shown in Figure 29.2.

The TCP sender assumes congestion in the network when it times out waiting for an ACK. ssthresh is set to $\max(2, \min[cwnd/2, rwnd])$ segments, $cwnd$ is set to one, and the system goes to slow start [1]. If a TCP receiver receives an out of order segment, it immediately sends back a duplicate ACK (DUPACK) to the sender. The **fast retransmit** algorithm uses these DUPACKs to make retransmission decisions. If the sender receives n DUPACKs ($n = 3$ was chosen to prevent spurious retransmissions due to out-of-order delivery), it assumes loss and retransmits the lost segment without waiting for the retransmit timer to

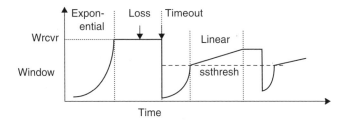

FIGURE 29.2   TCP congestion control.

go off. It also updates ssthresh. **Fast recovery** keeps track of the number of DUPACKs received and tries to estimate the amount of outstanding data in the network. It inflates *cwnd* (by one segment) for each DUPACK received, thus maintaining the flow of traffic. The sender comes out of fast recovery when the segment whose loss resulted in the duplicate ACKs is acknowledged. TCP then deflates the window by returning it to ssthresh, and enters the congestion avoidance phase. Variations on TCP congestion control include NewReno, SACK, FACK, and Vegas, presented in References 4 to 8.

29.3   Characteristics of Wireless Media

Certain characteristics of wireless media affect TCP performance. These characteristics include:

- *Channel losses*: Signals carried by wireless media are subject to significant interference from other signals, and subsequently, losses due to modification of bits while frames are being transmitted. These losses are difficult to recover from at the link layer despite the presence of error correction techniques. Retransmission can be performed at the link layer or at the transport layer (TCP). TCP performance is affected by frequent link layer losses, since TCP assumes all losses occur due to congestion, and invokes the congestion control algorithms upon detecting any loss.
- *Low bandwidth*: Bandwidth of wireless links may be low, which can sometimes result in buffer overflows at the base station and high observed round trip times (RTTs) [9].
- *Signal fading*: Interference from physical factors like weather, obstacles, unavailability of channels, overlapping areas of different cells, and mobility can result in signal fading and blackouts. Such blackouts can exist for prolonged periods of time.
- *Movement across cells*: Mobility of a wireless host involves addressing connection handoff. In addition to the link layer state that has to be handed off, the base station may maintain transport-layer connection state, which may need to be handed off. Signal fading may also occur when a host moves across cells, as discussed earlier.
- *Channel asymmetry*: In many cases, the sending entity (e.g., base station) gets more transmission time/bandwidth than the receiving entity (e.g., mobile device). This can lead to TCP acknowledgments being queued for transmission at the link layer of the receiving entity. Therefore, longer round trip times are observed by the TCP sender and traffic becomes more bursty.
- *Link latency*: Wireless links may exhibit high latencies. When such latencies are a significant fraction of the total round trip times observed by TCP, the retransmission timeouts of TCP are set to high values, which subsequently affects TCP performance. Such conditions occur in wireless WANs and satellite networks. In addition, high variance in the measured round trip times impacts the TCP RTT estimation algorithm [9].

29.4   Wireless TCP Mechanisms

Several mechanisms have been proposed to enhance the performance of TCP over wireless links. The proposed approaches can be classified as end-to-end, local recovery, or split connection. We briefly

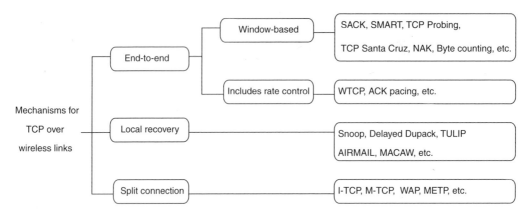

FIGURE 29.3   Categorization of mechanisms for TCP over wireless links.

discuss the various mechanisms below and compare them in the later sections. Figure 29.3 illustrates the three categories and sample schemes in each category.

29.4.1   End-to-End Mechanisms

End-to-end mechanisms solve the wireless loss problems at the transport layers of the sender and receiver. We examine the main proposals in this section. Other alternatives being standardized are discussed in Section 29.6.

29.4.1.1   Selective Acknowledgments

TCP selective acknowledgments (SACK) [10,6,7] uses the options field of TCP to precisely inform the sender which segments were not received. This enables the sender to retransmit only those segments that are lost, thereby not unduly wasting network bandwidth. TCP SACK was proposed as a solution to recover from multiple losses in the same window without degrading the throughput of the connection. The same idea can be applied to networks with wireless hops to aid the sender in recovering from noncongestion related losses. TCP still performs congestion control as in References 6 or 7 upon detecting packet loss even if it is on the wireless link as there is no way to deduce where the loss occurred.

29.4.1.2   SMART Retransmissions

A simple method to aid retransmissions (SMART) [11] decouples flow and error control. Each ACK packet from the receiver carries both the standard cumulative ACK and the sequence number of the packet that initiated the ACK. This informs the sender of the packet that is lost, so that the sender can selectively retransmit. SMART uses a different heuristic to detect lost *retransmissions*. SMART also avoids the dependence on timers for the most part except for the worst case when all packets and all ACKs are lost. The scheme uses two different buffers and windows: an error-control window at the receiver for buffering the out-of-sequence packets, and a flow-control window at the sender for buffering unacknowledged packets. Thus, error control and flow control are decoupled.

29.4.1.3   TCP Probing

With TCP probing [12], when a data segment is delayed or lost, the sender, instead of retransmitting and reducing the congestion window size, enters a probe cycle. A probe cycle entails exchanging probe segments between the sender and receiver to monitor the network. The probes are TCP segments with header options and no payload. This helps alleviate congestion because the probe segments are small compared to the retransmitted segments. The cycle terminates when the sender can make two successive RTT measurements with the aid of receiver probes. In case of persistent errors, TCP decreases its congestion window and threshold. But for transient random errors, the sender resumes transmission at the same window size it used before entering the probe cycle.

29.4.1.4 TCP Santa Cruz

TCP Santa Cruz [13], like TCP probing, makes use of the options field in the TCP header. The congestion control algorithm is based on relative delays that packets experience with respect to one another in the forward direction — an idea that was first introduced in TCP Vegas [18]. Therefore, ACK losses and reverse path congestion do not affect the throughput of a connection. The scheme also improves the error recovery mechanism by making better RTT estimates than other TCP implementations. Retransmissions are included in making RTT estimates to take into consideration the RTT during congestion periods. TCP Santa Cruz uses selective acknowledgments (SACK).

29.4.1.5 Negative Acknowledgments

Negative ACKs (NAKs) can be included in the options field of the TCP header to explicitly indicate which packet has been received in error so that retransmission of that packet can be initiated quickly. This is under the assumption that a corrupted packet can still reach the destination and that the source address of the packet is still known. The sender, on receiving a NAK, can retransmit the packet without modifying the congestion window size. RTT measurement from the retransmitted packets is ignored to avoid inflating the RTT estimate.

29.4.1.6 Wireless TCP (WTCP99)

Wireless TCP (WTCP99) [9] uses rate-based transmission control — not a self-clocking window-based scheme like TCP. (We call this WTCP99 to distinguish it from a number of earlier proposals, which also used the name WTCP.) The transmission rate is determined by the receiver using a ratio of the average inter-packet delay at the receiver to the average inter-packet delay at the sender. The sender transmits its current inter-packet delay with each data packet. The receiver updates the transmission rates at regular intervals based on the information in the packets, and conveys this information to the sender in the ACKs. WTCP99 computes an appropriate initial transmission rate for a connection based on a packet-pair approach. This is useful for short-lived connections in wireless wide area networks (WWANs) where round trips are long. WTCP99 achieves reliability by using selective ACKs [10]. No retransmission timeouts are triggered as it is difficult to maintain good round trip time estimates. Instead, the sender goes into "blackout mode" when ACKs do not arrive at sender-specified intervals. In this mode, the sender uses probes to elicit ACKs from the receiver, similar to TCP probing [12].

29.4.1.7 ACK Pacing

ACK pacing [14] is a rate based approach to ACK generation at the receiver. ACK pacing results in rate-controlled sender packets, and hence avoids bursty traffic that can result in packet losses, delays, and lower throughput. Pacing, however, does not help distinguish between congestion losses and wireless losses.

29.4.1.8 TCP Westwood

TCP Westwood (TCPW) [15] estimates the effective bandwidth at the TCP sender, without any modifications to the TCP receiver. Bandwidth estimation is based on the returning ACKs. A duplicate ACK (DUPACK) updates the bandwidth estimate (BWE) since it notifies the sender that a data packet was received at the destination, even though the packet was out-of-order. When n DUPACKs are received, the slow start threshold is updated as follows:

$$\text{ssthresh} = \frac{\text{BWE} \times \text{RTTmin}}{\text{SS}} \qquad (29.1)$$

where RTTmin is the minimum measured RTT, and SS is the TCP segment size. The congestion window is set to ssthresh if it currently exceeds the new ssthresh value. Thus, the connection speed remains close to network capacity if BWE is high. This approach is called "faster recovery." In the case of timer expiration, ssthresh is set to minimum (Equation [29.1]; 2), and the congestion window is set to 1 as in TCP Reno. The slow start and congestion avoidance phases are also similar to TCP Reno, and therefore TCPW is "TCP-friendly." Fairness among users cannot be guaranteed if bottlenecks exist on the backward links, since DUPACKs of certain connections are dropped. TCPW is unsuitable for networks with load that is highly dynamic, since measurements may not reflect the current network state.

29.4.1.9   pTCP

pTCP [16] was specifically designed for mobile hosts with multiple heterogeneous interfaces. The basic assumption in this work is that a mobile host wishes to use its interfaces simultaneously for a single application connection. pTCP stripes data across these interfaces in order to achieve bandwidth aggreg-ation at the receiver. Both application-layer and link-layer data striping approaches are demonstrated to be inferior to transport layer approaches due to reasons such as (1) fluctuating bandwidth and data rates, (2) increased application complexity, and (3) presence of multiple congestion control schemes that may be used at the different interfaces. The basic philosophy of pTCP mandates decoupling reliability and congestion control, striping data based on the congestion window of the individual pipes, reassigning windows dynamically, and supporting different congestion control mechanisms. Mobility and handoffs may reduce the effectiveness of this approach because some pipes may be stalled during handoffs. pTCP may exhaust host resources due to maintaining significant state. It also has the disadvantage of not being TCP-friendly.

29.4.1.10   TCP-Peach

TCP-Peach [17] is primarily targeted at satellite networks and avoids problems such as the long duration of the slow start phase caused by the long satellite propagation delays, the low rate imposed on the sender due to the long RTT, and the high error rate of the wireless environment. TCP-Peach uses a *Sudden Start* mechanism instead of slow start. In *Sudden Start*, a number of dummy segments are sent during each RTT (in addition to data segments) and the congestion window grows with returning ACKs for these dummy segments as well as data. These dummy segments have lower priority than data and therefore intermediate routers can drop them in times of congestion. TCP-Peach also proposes a *Rapid Recovery* algorithm for handling packet losses that are not due to congestion. This algorithm is executed after the execution of the *Fast Recovery* algorithm when n DUPACKs are received. In the *Rapid Recovery* phase, n dummy segments are sent to probe the availability of network resources. According to the number of received ACKs for the dummy segments, the sender can determine whether the loss was due to congestion or channel errors.

A problem with TCP-Peach is that it requires awareness from the sender, receiver, and intermediate routers (for priority drop of dummy segments). This limits its deployability. Another subtle point is that its approach for increasing the congestion window using dummy segments is aggressive, and thus not TCP friendly.

29.4.1.11   Explicit Bad State Notification (EBSN)

Losses on the wireless link of a connection may cause timeouts at the sender, as well as unnecessary retransmissions over the entire wired/wireless network. The EBSN scheme [18] proposes sending EBSN messages from the base station to the sender whenever the base station is unsuccessful in transmitting a packet over the wireless network. This scheme is *not* end-to-end, but we discuss it here because it requires sender support like most end-to-end schemes.

EBSN receipt at the sender re-starts the TCP timer and prevents the sender from decreasing its window when there is no congestion. Although this scheme requires modifications to the TCP implementation at the sender, the required changes are minimal, no state maintenance is required, and the clock granularity (timeout interval) has little impact on performance.

29.4.1.12   Explicit Loss Notification (ELN) Strategies

Like EBSN, ELN is not purely end-to-end. ELN has been proposed to recover from errors that occur when the wireless link is the first hop [19]. A base station is used to monitor TCP packets in either direction. When the receiver sends DUPACKs, the base station checks if it has received the packet that triggered the DUPACKs. If not, then the base station sets an ELN bit in the header of the ACK to inform the sender that the packet has been lost on the wireless link. The sender can then decouple retransmission from congestion control, that is, it does reduce the congestion window. In contrast to Snoop [20] (discussed further), the base station need not cache any TCP segments in this case, because it does not perform retransmission.

29.4.1.13   Receiver-Based Congestion Control

The rationale behind receiver-based congestion control is two-fold. First, since the mobile host has better knowledge of the channel conditions than the sender, it is more practical in terms of scalability and deployment to let the receiver make congestion control, power management, and loss recovery decisions. This relieves the sender from having to be aware of all possible congestion control implementations optimized for each wireless technology. Second, mobile hosts are becoming multi-homed, and a host may belong to multiple domains. With multiple heterogeneous interfaces, it is preferable that the receiver control the type, timing, and amount of data to receive. This is important for: (1) seamless handoffs (no packet losses), (2) server migration (minimum state transfer among senders), and (3) bandwidth aggregation (mobile host coordination of its heterogeneous interfaces to gain bandwidth). In addition, receiver-based protocols are less sensitive to path asymmetry [21]. Receiver-based transport protocols, however, may require extra processing and energy at the mobile host.

Examples of receiver-based transport protocols include [21–24]. Spring et al. [23] proposed receiver-based bandwidth sharing among connections running on low bandwidth access links. To achieve this, TCP connections are prioritized according to the application. The receiver is more capable of setting these priorities than the sender, and it is also more aware of the bandwidth on the wireless access link. The receiver controls the bandwidth share of each connection by manipulating the receive socket buffer size, thus overriding the advertised window of each open socket. This indirectly manipulates the sliding window of the sender. The main disadvantage of this approach is that it requires knowledge of link capacity. In addition, long-lived data transfers may be starved in this approach because short-lived transfers and interactive applications have higher priority.

The BandWidth Sharing System (BWSS) [22] handles the problem of last hop access links that are bottlenecked due to limited capacity. As in Reference 23, bandwidth sharing among multiple TCP flows is based upon user preferences. The receiver estimates and manipulates the RTT of each flow and controls the advertised window. This approach provides weighted bandwidth shares to users, which makes it suitable for interactive applications.

The Reception Control Protocol (RCP)[21] delegates the responsibility of flow control, congestion control, and reliability to the receiver. The receiver sends the first request for data (REQ) to the sender with the initial sequence numbers. The sender replies by sending the first data segment, and transmission proceeds similarly afterwards. RCP gives flexibility for server migration and seamless handoffs, TCP friendliness, intelligent loss recovery, scalable congestion control, and efficient power management.

TCP-Real [24] hands the receiver the responsibility of error detection and classification, and decision making in case of packet loss. The congestion avoidance technique is slightly different from TCP. The congestion window is modified based on an estimate of current channel contention. This reduces unnecessary transmission gaps and optimizes TCP-Real for handling time-constrained applications. TCP-Real goodput is unaffected by path asymmetry, and its rate is TCP-friendly.

29.4.2   Split Connection Mechanisms

In this class of mechanisms, the TCP connection is split at the base station. TCP is still used from the sender to the base station, whereas either TCP or some other reliable connection-oriented transport protocol is used between the base station and the receiver. The TCP sender is only affected by the congestion in the wired network and hence the sender is shielded from the wireless losses. This strategy has the advantage that the transport protocol between the base station and mobile node can make use of its knowledge of the wireless link characteristics, or even of the application requirements. The problem with these proposals, however, is their efficiency, robustness, and handoff requirements.

29.4.2.1 Indirect TCP (I-TCP)

Indirect TCP (I-TCP), proposed in 1994–1995, was one of the earliest wireless TCP proposals. With I-TCP, a transport layer connection between a mobile host and a fixed host is established as two separate connections: one over the wireless link and the other over the wired link with a "mobile support router" at the junction [25]. Packets from the sender are buffered at the mobile support router until transmitted across the wireless connection. A handoff mechanism is proposed to handle the situation when the wireless host moves across different cells. A consequence of using I-TCP is that the TCP ACKs are not end-to-end thereby violating the end-to-end semantics of TCP.

29.4.2.2 Mobile TCP (M-TCP)

Mobile TCP (M-TCP) also uses a split connection based approach, but tries to preserve end-to-end semantics [26]. M-TCP adopts a three level hierarchy. At the lowest level, mobile hosts communicate with mobile support stations in each cell, which are in turn controlled by a "supervisor host." The supervisor host is connected to the wired network and serves as the point where the connection is split. A TCP client exists at the supervisor host. The TCP client receives each segment from the TCP sender and passes it to an M-TCP client to send it to the wireless device. Thus, between the sender and the supervisor host, standard TCP is used, while M-TCP is used between the supervisor host and the wireless device. M-TCP is designed to recover quickly from wireless losses due to disconnections and to eliminate serial timeouts. TCP on the supervisor host does not ACK packets it receives until the wireless device has acknowledged them. This preserves end-to-end semantics, preserves the sender timeout estimate based on the whole round trip time, and handles mobility of the host with minimal state transformation.

29.4.2.3 Mobile End Transport Protocol (METP)

METP proposes the elimination of the TCP and IP layers from the wireless hosts and replacing them with an METP designed specifically to run directly over the link layer [27]. This approach shifts the IP datagram reception and reassembly for all the wireless hosts to the base station. The base station also removes the transport headers from the IP datagram. The base station acts as a proxy for TCP connections. It also buffers all the datagrams to be sent to the wireless host. These datagrams are sent using METP and a reduced header, thus providing minimal information about source and destination addresses and ports, and connection parameters. In addition, the base station uses retransmissions at the link layer to provide reliable delivery at the receiver. It does not require any change in the application programs running on the wireless host as the socket API is maintained as in a normal TCP/IP stack. However, when handoff occurs, all the state information needs to be transferred to the new base station.

29.4.2.4 Wireless Application Protocol (WAP)

In 1997, the WAP forum was founded by Ericsson, Nokia, Motorola, and Phone.Com (formerly Unwired Planet). The WAP Forum consolidated into the open mobile alliance (OMA) in 2002 [28].

Like METP, WAP includes a full protocol stack at the receiver, as well as gateway support. The WAP protocol layers are shown in Figure 29.4. The top WAP layer, the Wireless Application Environment (WAE), establishes an environment to allow users to build applications that can be used over a wide variety of wireless systems. WAE is composed of user agents such as browsers, text editors, date book, or phone book. WAE also includes scripting, high-level programming languages, and image formats. WAE uses languages such as WMLScript (similar to JavaScript) and WML (similar to HTML).

The wireless session protocol (WSP) handles communication between the client and proxy or server. WSP opens a session of communication between client and server, exchanges encoded data, exchanges requests and replies, and supports several asynchronous transmission modes of data.

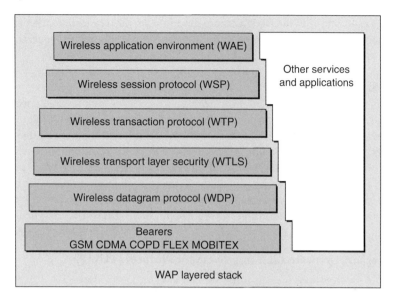

FIGURE 29.4   Wireless application protocol layers.

The wireless transaction protocol (WTP) handles transactions, re-transmission of data, and separation and concatenation of data. The protocol has a separate interface that manages and referees the WTP layer and the settings of the handheld device. This management application is known as the WTP Management Entity. For WTP to work, the following must hold: (1) The handheld device is within the coverage area of base agent; (2) The handheld device is turned on and is reliable; (3) Resources are adequate, for example, CPU and memory; and (4) WTP settings are correct.

The wireless transport layer security (WTLS) is the layer that handles security of data and validity of data between two communicating entities. To transport data, WTLS needs the source address and port number to identify the message creator, and from where the message is being sent, the destination address and port number to which data is being sent, and of course, the data itself. WTLS has a connection interface, which provides a connection protocol between client and server.

The wireless data protocol (WDP) acts as the communication layer between the upper level protocols (WTLS, WTP, and WSP), and the bearer services. The function of WDP is to provide a stable environment so that any of the underlying bearers can operate using WAP. WDP can be adapted to different bearers with different services. However, the services offered by WDP remain constant, thus providing a continuous interface to the upper layers of the WAP stack.

29.4.3   Local Recovery Mechanisms

The rationale behind local recovery is that wireless losses are local and hence recovery from them should be local. The mechanisms discussed below entail detecting wireless losses at the base station or at the receiver and using local retransmissions and sometimes forward error correction (FEC) for recovery. This class of mechanisms can be considered as a compromise between the above two categories (end-to-end and split connections) [29].

29.4.3.1   The Snoop Protocol

The Snoop protocol caches TCP packets at the base station [20]. State is maintained for all the TCP connections traversing the base station to the wireless device. The link layer is TCP-aware, thereby making it possible to monitor TCP segments and infer packet losses on the wireless link from duplicate ACKs (DUPACKs). Packet losses are also inferred through local timeouts. Lost TCP segments are retransmitted on the wireless link. This is possible because all TCP segments are cached until ACKs for them arrive from the wireless device. Additionally, the DUPACKs are suppressed at the base station in order to prevent the sender from unnecessarily invoking congestion control algorithms. This method appears to work well in wireless local area networks (WLANs)[29].

29.4.3.2 Delayed Duplicate ACKs

Delayed duplicate ACKs (DDA) recovers from wireless losses using link level retransmissions, while *delaying* the DUPACKs at the receiver to prevent the sender from retransmitting [30]. If the retransmitted packets are received after a time interval d, the DUPACKs are discarded; otherwise they are allowed to return to the sender. DDA, unlike Snoop, is *TCP-unaware* in that it relies on link level (not TCP-level like Snoop) segments and ACKs. This has the advantage that it can work even when the IP payload is encrypted, such as with IPSEC. However, due to the delaying of all receiver DUPACKs by a time d, the sender may not invoke congestion control algorithms like fast retransmit and recovery (see Section 29.2) and may time-out with both congestion and wireless losses. Thus, DDA works best when congestion loss is minimal.

29.4.3.3 Transport Unaware Link Improvement Protocol (TULIP)

Like DDA, TULIP recovers from wireless losses via link level retransmissions [31]. The protocol is TCP-unaware (though it is aware of reliability requirements). DUPACK generation at the receiver is avoided by only delivering in-order frames to the receiver TCP/IP. Lost packets are detected through a bit vector that is returned as a part of the link layer ACK, which helps immediate loss recovery. TULIP is designed for efficient operation over half-duplex radio channels. It introduces a media access control (MAC) acceleration feature to accelerate the return of link layer ACKs in order to improve throughput.

29.4.3.4 TCP interaction with MAC layer (MACAW)

MACAW [32] investigates the interaction between TCP and MAC layer backoff timers, which may cause unfairness and capture conditions. The authors show these effects are very pronounced in CSMA and FAMA MAC protocols. They propose adding link layer ACKs and a less aggressive backoff policy to improve performance of TCP over these MAC protocols. The new MAC layer protocol is called MACAW.

29.4.3.5 AIRMAIL

AIRMAIL [33] is a popular link layer protocol designed for indoor and outdoor wireless networks. It provides a reliable link layer by using local retransmissions and FEC at the physical layer. The mobile terminal combines several ACKs into a single event-driven ACK, and the base station sends periodic status messages. This, however, entails that no error correction can be done until the ACKs arrive, which can cause TCP to time out if the error rate is high.

29.4.3.6 TCP–DCR

TCP with delayed congestion response (TCP–DCR) [34] delays making congestion control decisions for some time period τ after a packet loss is encountered. During this period τ, local link layer recovery mechanisms are employed. After τ elapses, persistent loss is assumed to be due to congestion, and consequently TCP congestion control is employed. Setting the τ parameter for TCP–DCR is difficult since it depends on network dynamics. Waiting for some time before making congestion control decisions may worsen the congestion state of the network. Therefore, TCP–DCR is more suitable for environments where errors are mostly due to channel errors.

29.5 Comparison of Wireless TCP Mechanisms

In this section, we compare and contrast wireless TCP mechanisms based on interaction among protocol stack layers, division of responsibilities among entities, complexity, and performance under various conditions.

29.5.1   Interaction Between TCP and the Link Layer

End-to-end mechanisms typically do not need any support from the link layer. On the other hand, approaches based on recovering from a wireless loss locally need substantial support at the link layer. The Snoop protocol [35] monitors and caches TCP segments, and performs local retransmissions of the TCP segments, which necessitates the link layer to be TCP aware. TULIP [31] relies on link layer retransmission and MAC acceleration to improve the throughput of the wireless link. Both TULIP and DDA do not require base station TCP-level support. AIRMAIL [33] relies on event-driven acknowledgments and reliable retransmissions at the link layer. Mobile End Transport Protocol (METP) [27], which uses a split connection, also relies on a reliable retransmission and efficient demultiplexing of the reduced header at the link layer. MACAW [32] uses link layer ACKs and a less aggressive backoff policy to improve the performance of TCP over the link layer.

29.5.2   Interaction Between TCP and the Network Layer

Typically, mechanisms to improve TCP in wireless networks do not need support from the IP layer. With METP [27], however, the IP and TCP layers are transferred from the wireless device to the base station. METP is directly implemented over the link layer and uses link layer reliability mechanisms and reduced headers to improve the throughput in the wireless link. Hence, this requires modifying the network layer at the base station and the wireless device. WAP [28] also modifies the protocol stack on the mobile device, as well as the gateway.

29.5.3   Retransmission Approaches and ACKs

End-to-end mechanisms perform retransmissions at the TCP sender. TCP SACKs [10] are used by the TCP sender to selectively retransmit lost segments. SMART retransmissions [11] infer the packets lost from the additional (receiver-transmitted) sequence number of the packet that caused the cumulative ACK. This eliminates the need for a bit vector in the ACK to specify lost packets. TCP probing [12] enters a probe cycle upon detecting a lost packet instead of retransmitting. This helps in alleviating congestion while continuing to obtain round trip measurements. TCP Santa Cruz [13] uses an ACK window from the receiver, similar to SACK. In addition, it uses retransmitted packets to obtain the more accurate round trip time measurements. TCP Westwood [15] uses duplicate ACKs to estimate the effective bandwidth of a TCP connection at congestion time.

    Local recovery mechanisms perform retransmissions only on the wireless link. The Snoop protocol [30] performs retransmissions of the whole TCP segments on the wireless link upon receipt of DUPACKs at the base station. TULIP [31], delayed DUPACKs [30], and AIRMAIL [33] depend on *link layer* retransmissions as the base station used need not be TCP-aware. TULIP uses a bit vector in link layer ACKs, similar to SACK at the transport layer.

    Split connection mechanisms like I-TCP [25] and M-TCP [26] use a separate connection between the base station and the wireless device. So the retransmissions are done by the transport protocol between the wireless device and the base station as appropriate. Mobile End Transport Protocol [27] uses link layer retransmissions to recover from losses.

29.5.4   Flow Control

Among the approaches we have examined, only WTCP99 [9] and ACK pacing (to some extent) [14] use rate based flow control. In WTCP99, the transmission rate is controlled by the receiver using inter packet delay. The motivation behind this idea is that the round trip time calculation for a wireless WAN is highly susceptible to fluctuations. All the other schemes use window based flow control. With ACK pacing, the sender transmission of packets is controlled by the rate at which ACKs arrive. This avoids bursty traffic and reduced throughput.

29.5.5 Loss Detection and Dependence on Timers

SMART [11] tries to avoid timeouts. The sender discovers lost packets by the additional cumulative ACK information. SMART uses RTT estimates to detect loss of retransmissions. Timers are used only to deal with the worst case where all packets and all ACKs are lost. In schemes like EBSN [18], NAK and ELN [19], where the wireless channel state is explicitly conveyed to the sender, the timeout for the transmitted packets can be reset on the arrival of loss notification. TCP Santa Cruz [13] measures RTTs for retransmissions to provide better RTT estimates. This eliminates the need for timer-backoff strategies where the timeout is doubled after every timeout and retransmission.

In the Snoop approach [20], the base station detects wireless losses from DUPACKs on the reverse path. In situations when wireless channel losses are very high and loss detection is hampered, it uses a timer, set to a fraction of the wireless RTT to perform local retransmissions. Using coarse TCP timers and fine link layer timers avoids retransmission contention between the two layers. However, unnecessary link layer transmissions may result in out-of-order packets at the receiver which generates DUPACKs, resulting in retransmissions.

29.5.6 Effect of Path Asymmetry

Cumulative ACK schemes used in most TCP implementations offer robustness from ACK losses occurring on the reverse path from the receiver to the sender. In cases where ACKs follow slower or more congested links than data packets, RTT estimates at the sender may be inflated and do not reflect the forward path state.

This can create a problem for TCP Westwood [15] since it relies on returning DUPACKs for bandwidth estimation, and consequently slow start threshold and congestion window size updates. This can also be problematic to TCP-Peach [17], since it may affect the order and RTT of dummy segments, thus resulting in inaccurate estimation of network resources.

TCP Santa Cruz [13] avoids this problem by making use of only forward delay measurements for congestion control. WTCP99 [9] makes use of the ratio of inter packet delay at the receiver and the sender for congestion control and, therefore, avoids the RTT inflation problem.

29.5.7 Effect of Wireless Link Latency on RTT

Local recovery schemes such as References 20 and 34 use either TCP or link layer retransmissions. Such schemes operate under the assumption that the wireless link latency is small compared to latency of the entire connection. In cases where the wireless link latency is considerably high, there will be contention between TCP retransmissions from the sender and local retransmissions, resulting in reduced throughput.

29.5.8 State Maintained

End-to-end mechanisms do not need any state at any intermediate node. Some local recovery mechanisms need to maintain state while others do not. The Snoop protocol [35] places the complexity at the base station, so TCP segments need to be cached until acknowledged by the receiver. In Explicit Loss Notification strategies for recovering from wireless losses when the wireless link is the first hop [19], the TCP segments need not be maintained by the base station as no retransmission is performed. However, information like sequence numbers of the packets that pass through the base station need to be maintained to detect the losses. Other protocols like TULIP [31], Delayed DUPACKs [30], and AIRMAIL [33] rely on link level retransmissions and state, as the protocols are TCP-unaware. Split connection mechanisms require the TCP packets be stored at the base station to be transmitted on a separate connection across the wireless link.

29.5.9 Robustness and Overhead

Copying is required in split connection-based approaches. This is because the connection is terminated at the base station and hence the packets destined for the wireless host need to be buffered at the base station until they are transmitted on the wireless link. Overhead is incurred by the kernel when copying the packets across buffers for different connections [29]. The difference between this and TCP state in Snoop, for example, is that in the latter, the state is soft. A failure of the base station in Snoop only temporarily affects the throughput of the connection. However, if the base station fails in split connection-based approaches, the receiving TCP entity might not receive the packets that have been transmitted by the sending entity, and subsequently ACKed by the base station. M-TCP avoids this through delaying ACKs to the sender until packets are received at the mobile node. However, split connection approaches generally suffer from robustness, efficiency, and flexibility problems.

29.5.10 Preserving End-to-End Semantics

End-to-end mechanisms clearly preserve end-to-end semantics. Local recovery mechanisms also preserve end-to-end semantics except that DUPACKs are suppressed as in the Snoop protocol [35] or delayed as in DDA [30]. Split connection approaches typically terminate the connection at the base station [36,27], thereby violating the end-to-end semantics of TCP in the strict sense. The M-TCP approach [26], however, attempts to maintain end-to-end semantics while splitting the connection at the base station by acknowledging packets back on the wired network only when the ACKs arrive on the wireless link.

Another important point is that TCP-aware local recovery, such as Snoop [20], reads TCP headers and thus needs to be adapted when the IP payload is encrypted such as with IPSEC. DDA and TULIP, as well as approaches like MACAW and AIRMAIL, are not TCP aware, and can work with encrypted IP payloads.

29.5.11 Suitability to Different Network Environments

Some protocols work very well for certain types of network environments. TULIP [31] is tailored for the half-duplex radio links and provides a MAC acceleration feature to improve the throughput in wireless links. The Snoop protocol [35] and delayed duplicate ACKs work best in local area network environments where wireless round trip times are small and fading losses do not occur frequently. In fact, Snoop [35] has little benefit if the link layer already provides reliable in-order delivery, or if the wired link delay is small. WTCP99 [9] is tuned to perform well in wireless WANs with low bandwidths, large round trip times, asymmetric channels, and occasional blackouts and signal fading.

TCP Westwood [15] is more suitable for lightly loaded networks since the congestion window update mechanism may be too aggressive under heavy traffic and high error rates. In addition, congestion on the reverse paths may affect the bandwidth estimation of some connections, thus resulting in unfair bandwidth sharing among them. TCP-Peach [17] is also more suitable for lightly loaded networks because of its active probing approach using dummy segments. This requires extra processing from the hosts and imposes extra load on the network that may not be acceptable during high data loads.

29.5.12 Adaptation to Long Fading Periods

WTCP99 [9] adapts to long fading periods effectively. This is because the motivation for its design stems from wireless WANs, which exhibit fading. WTCP99 uses rate based control, inter-packet delay as a congestion metric, and blackout detection to adapt to long fading periods. Snoop [35] and similar protocols adapt to fading by making use of local timers in absence of DUPACKs. TCP probing [12] also adapts to long fading periods because it enters a probe cycle upon detection of a packet loss. The cycle terminates only after the sender makes two successive RTT measurements. pTCP [16] also handles blackouts by performing redundant striping of the first MSS of data in a congestion window that has suffered a timeout. These stripes are then assigned to different pipes.

29.5.13 Energy Efficiency

Energy Efficiency is extremely important for mobile devices, especially when re-charging the battery is impossible or time-consuming. In most of the techniques we have presented in this chapter, energy efficiency was not considered or quantified. Recent work has compared the energy efficiency of TCP Tahoe,

Reno, and SACK [37,38]. Although SACK typically achieves higher throughput, it is less energy-efficient than other TCP variants in environments where packet losses are bursty and correlated (assuming low idle-power consumption). Thus, at high error rates, it is preferable to halve the window size than to reduce it. The same intuition and argument about SACK energy inefficiency applies to aggressive rate-based techniques, such as WTCP99 [9] or window-based techniques, such as TCP Westwood [15].

In receiver-based protocols, such as RCP [21] and TCP-Real [24], more energy may be consumed at the mobile host due to its increased responsibilities. A careful study is needed to determine the tradeoffs among accuracy of power control decisions and energy consumption at the receiver.

Split connection mechanisms involve a third party, the base station, which typically has abundant energy and thus its energy consumption in serving the split connections is negligible. A recent study presented in Reference 39 quantifies the relationship between the transmission range of wireless devices and two performance parameters of TCP, namely, throughput and energy consumption. The study shows that a larger number of hops (i.e., using smaller transmission ranges) result in decreased energy-efficiency. Reducing the number of hops, however, also reduces the TCP goodput, especially if the link error is high. Thus, a network designer must consider this tradeoff when selecting the mechanism to use for the transport layer.

29.6   The Internet Engineering Task Force (IETF) Efforts

The performance implications of link characteristics (PILC) working group at the IETF has recommended certain changes to help TCP adapt to (1) asymmetry, (2) high error rate, and (3) low speed links.

Asymmetry in network bandwidth can result in variability in the ACK feedback returning to the sender. Several techniques can mitigate this effect, including using *header compression* [40]; reducing ACK frequency by taking advantage of cumulative ACKs; using TCP congestion control for ACKs on the reverse path; giving scheduling priority to ACKs over reverse channel data in routers; and applying backpressure with scheduling.

The TCP sender must also handle infrequent ACKs. This can be done by bounding the number of back-to-back segment transmissions. Taking into account cumulative ACKs and not number of ACKs at the sender can also improve performance. This scheme is called *byte (versus ACK) counting* [41] because the sender increases its congestion window based on the number of bytes covered (ACKed) by each ACK.

In addition, reconstructing the ACK stream at the sender; router ACK filtering (removing redundant ACKs from the router queue); or ACK compaction/expansion (conveying information about discarded ACKs from the compacter to the expander) can be used.

For *high error rate* links, experiments show that approaches such as explicit congestion notification (ECN) [42], fast retransmit and recovery and SACK are especially beneficial. Explicit loss notification, delayed duplicate acknowledgments (Section 29.4.3.2), persistent TCP connections, and byte counting are a few of the open research issues in this area. TCP-aware performance enhancing proxies (PEPs), such as split connection mechanisms and Snoop can also be used.

For *low speed* links, in addition to *compressing the TCP header and payload* [40,43], several changes to the congestion avoidance algorithm are recommended. First, hosts that are directly connected to low-speed links should advertise small receiver window sizes to prevent unproductive probing for nonexistent bandwidth.

Second, maximum transmission units (MTUs) should be carefully selected to not monopolize network interfaces for human-perceptible amounts of time (e.g., 100–200 msec) and to allow delayed acknowledgments. Large MTUs which monopolize the network interfaces for long periods are likely to cause the receiver to generate an ACK for every segment rather than delayed ACKs. Using a smaller MTU size will decrease the queuing delay of a TCP flow compared to using larger MTU size with the same number of packets in the queue.

Third, the receiver advertised window size, *rwnd*, should be carefully selected. Dynamic allocation of TCP buffers (or *buffer auto-tuning*) [44] based on the current effective window can be used.

Finally, binary encoding of web pages, such as with WAP (Section 29.4.2.4) can be used to make web transmissions more compact. Many of the suggested solutions mentioned in this section are still in the research phase.

RFC 2757 [45] provides a good summary of various TCP over wireless WAN proposals.

29.6.1 TCP over Satellites

In addition to bandwidth asymmetry, restricted available bandwidth, intermittent connectivity, and high error rate due to noise, Geostationary Earth Orbit (GEO) satellite links are characterized by very high latency. This is because such satellites are usually placed at an altitude of around 36,000 km, resulting in a one-way link delay of around 279 msec, or a round trip delay of approximately 558 msec. This results in a long feedback loop and a large delay bandwidth product. For Low Earth Orbit (LEO) and Medium Earth Orbit (MEO) satellites, the delays are shorter, but inter-satellite links are more common and round trip delays are variable.

Allman et al. [46] recommend in RFC 2488 several techniques to mitigate the effect of these problems. These techniques include using path MTU discovery; FEC; TCP slow start, congestion avoidance, fast retransmit, fast recovery, and selective acknowledgments (SACK). The TCP window scaling option must also be used to increase the receiver window (*rwnd*) to place a larger upper bound on the TCP window size. The algorithms companion to window scaling, including protection against wrapped sequence space (PAWS) and round trip time measurements (RTTM) are also recommended.

A number of additional mitigations are still being researched, and are summarized in RFC 2760 [47]. These include using larger initial window sizes (to eliminate the timeout observed with delayed ACKs at startup and reduce transmission time for short flows); transaction TCP (eliminates the TCP 3-way handshake with every connection); using multiple TCP connections for a transmission; pacing TCP segment transmissions; persistent TCP connections; byte counting at the sender; ACK filtering; ACK congestion control; explicit loss notification; using delayed ACKs only after slow start; setting the initial

TABLE 29.1   Mitigations for Special Link Characteristics

Technique	Satellites	Asymmetry	High error rate	Low Speed
Path MTU discovery	Recommended			Recommended small
Forward error correction	Recommended			
Transaction TCP	Recommended			
Larger initial window	Recommended			
Delayed ACK after slow start	Recommended			
Estimating ssthresh	Recommended			
Advertised receiver window	Large (window scaling)			Small/auto-tuned
Byte counting	Recommended	Recommended	Recommended	
Explicit loss notification	Recommended		Recommended	
Explicit congestion notification	Recommended		Recommended	
Multiple connections	Recommended			
Pacing segments	Recommended			
Header and payload compression	Recommended	Recommended		Recommended
Persistent connections	Recommended		Recommended	
ACK congestion control	Recommended	Recommended		
ACK filtering	Recommended	Recommended		
ACK compaction		Recommended		
ACK scheduling priority and backpressure		Recommended		

TABLE 29.2   Important TCP Congestion Control-Related RFCs

RFC	Describes
2001	Slow start, congestion avoidance, fast retransmit, and fast recovery
2018	Selective acknowledgments (SACK)
2309	Random early detection (RED)
2414–6	Increasing initial window size
2481	Explicit congestion notification
2488	TCP over satellite enhancements
2525	Known TCP implementation problems
2581	Slow start, congestion avoidance, fast retransmit, fast recovery, idle periods, ACK generation
2582	NewReno
2757	Long thin networks, for example, wireless WANs
2760	Ongoing TCP satellite research
2861	Congestion window validation (decay during idle periods)
2883	SACK extensions (use of SACK for acknowledging duplicate packets)
2884	Performance of explicit congestion notification
2914	Congestion control principles
2923	MTU discovery
2988	Retransmission timer (RTO) computation
3042	Limited transmit option (recovery with a small window or multiple losses)
3048	Reliable multicast transport
3124	The Congestion Manager (CM)
3390	Increasing TCP initial window
3448	TCP Friendly Rate Control (TFRC)
3465	Congestion control with byte counting
3517	Conservative selective ACK-based recovery
3522	Detecting if TCP has entered loss recovery unnecessarily
3649	High-speed TCP for large congestion windows
3708	Duplicate selective ACKs, Stream Control Transmission Protocol (SCTP), and Transmission Sequence Numbers (TSN)
3714	Congestion control for voice traffic in the Internet
3742	Modified slow start for TCP with large congestion windows
3782	NewReno modifications to TCP fast recovery

slow start threshold (ssthresh) to the delay bandwidth product as in Reference 5; header compression; SACK, FACK, random early detection (RED) at routers, ECN, and TCP-friendly control.

Table 29.1 summarizes the current research issues with link characteristics. A blank entry means the technique was not mentioned in the relevant RFC/draft. Note that slow start, congestion avoidance, fast retransmit, and fast recovery are required and SACK is recommended for all cases. Table 29.2 summarizes the RFCs describing TCP congestion control.

29.7   Summary and Open Issues

Table 29.3 gives a summary of the mechanisms discussed in this paper and their requirements and implications. Clearly, reliable transport for wireless networks has been an active topic of research since 1994. Several new protocols and TCP modifications have been proposed to improve performance over

TABLE 29.3 Summary of Mechanisms for TCP over Wireless Links

Scheme	Type	Sender support	Receiver support	BS support	State at BS	Retransmit at	LL support	Flow control	BS Crash impact	Preserves E2E
SACK	E2E	Yes	Yes	No	No	Sender	No	WB	No	Yes
SMART	E2E	Yes	Yes	No	No	Sender	No	WB	No	Yes
TCP-P	E2E	Yes	Yes	No	No	Sender	No	WB	No	Yes
TCP-SC	E2E	Yes	Yes	No	No	Sender	No	WB	No	Yes
NAK	E2E	Yes	Yes	No	No	Sender	No	WB	No	Yes
Byte-C	E2E	Yes	No	No	No	Sender	No	WB	No	Yes
WTCP99	E2E	Yes	Yes	No	No	Sender	No	RB	No	Yes
ACK-P	E2E	No	Yes	No	No	Sender	No	WB/RB	No	Yes
TCPW	E2E	Yes	No	No	No	Sender	No	WB	No	Yes
pTCP	E2E	Yes	Yes	No	No	Sender	No	WB	No	Yes
TCP-Peach	E2E+	Yes	Yes	Minimal	No	Sender	No	WB	No	Yes
EBSN	E2E+	Yes	No	LL	No	Sender	No	WB	No	Yes
ELN	E2E+	Yes	No	TL	Minimal	Sender	Yes	WB	No	Yes
R-based	E2E	No except RCP, TCP-Real	Yes	No	No	Sender	No	WB	No	Yes
I-TCP	SC	No	Yes	TL/AL	Yes	BS	No	WB	Yes	No
M-TCP	SC	No	Yes	TL/AL	Yes	BS	No	WB	Yes	Yes
METP	SC	No	Yes	TL/AL	Yes	BS	Yes	WB	Yes	No
WAP	SC	No	Yes	TL/AL	Yes	BS	No	WB	Yes	No
Snoop	LR	No	No	TL/LL	TL/LL	BS	Yes	WB	Small	Yes
DDA	LR	No	LL	LL	LL	LL-BS	Yes	WB	Minimal	Yes
TULIP	LR	No	LL	LL	LL	LL-BS	Yes	WB	Minimal	Yes
AIRMAIL	LR	No	LL	LL	LL	LL-BS	Yes	WB	Minimal	Yes
MACAW	LR	No	LL	LL	LL	LL-BS	Yes	WB	Minimal	Yes
TCP-DCR	LR	Yes	LL	LL	LL	LL-BS	Yes	WB	Minimal	Yes

TCP-P = TCP probing, TCP-SC = TCP Santa Cruz, byte-C = Byte counting, ACK-P = ACK pacing, TCPW = TCP Westwood, R-based = receiver-based, TCP-DCR = TCP delayed congestion response, WB = Window based, RB = Rate based, AL = application level, TL = transport level, and LL = link level, BS = Base station.

wireless links. We have presented a taxonomy of these approaches and compared them in terms of complexity, ease of deployment, and performance under different conditions.

Local recovery mechanisms work well when the wireless link latency is small compared to the RTT of the connection. For local recovery mechanisms that use link layer retransmissions, it is important that the granularity of the link layer timers be very fine compared to the TCP timers. Otherwise, contention among the two retransmissions can reduce throughput. Another important observation is that out-of-order delivery of link layer retransmissions at the receiver [29] may trigger duplicate ACKs and TCP retransmissions by the sender. TULIP [31] avoids this problem by preventing out-of-order delivery. In addition, hiding wireless losses from the sender by suppressing DUPACKs may cause inaccurate RTT measurements at the sender. This is especially important in cases where the wireless link latency is comparable to the RTT.

Split connection schemes shield the sender from wireless losses, and allow application and transport level optimizations at the base station, which is aware of the wireless link characteristics. However, split connection schemes do not preserve end-to-end semantics, and hence they are not robust to base station crashes. Moreover, the schemes are inefficient and involve significant copying overhead and handoff management.

Modifications proposed to TCP for lossy, asymmetric, low speed or high delay links are being standardized by IETF. It is important to study the interactions among these mechanisms to ensure no adverse effects occur. For example, some straightforward mechanisms, such as byte counting, result in bursty traffic, and should be used with caution, even though they are very useful for satellite environments.

A number of the TCP sender adaptations work well for certain environments only, while others, such as selective ACKs, are generally useful. Mechanisms designed for specific environments must be carefully examined if they are to be universally deployed. For example, WTCP99 [9] is well-suited for wireless wide area networks. TULIP [31] is tailored for half-duplex radio links, while MACAW [32] is designed for CSMA and FAMA links. TCP-Peach [17] works best in satellite networks or environments with very long RTT values. pTCP [16] is tailored to multi-homed hosts. Receiver-based transport protocols can serve interactive applications better than sender-based protocols.

References

[1] M. Allman, V. Paxson, and W. Stevens, "TCP congestion control," RFC 2581, April 1999, also see http://tcpsat.lerc.nasa.gov/tcpsat/papers.html.

[2] V. Jacobson, "Congestion avoidance and control," in *Proceedings of the ACM SIGCOMM*, August 1988, pp. 314–329, ftp://ftp.ee.lbl.gov/papers/congavoid.ps.Z.

[3] M. Allman, S. Floyd, and C. Partridge, "Increasing TCP's initial window," in *Proceedings of the RFC 2414*, September 1998.

[4] S. Floyd and T. Henderson, "The NewReno modification to TCP's fast recovery algorithm," *RFC 2582*, April 1999, also see http://www.aciri.org/floyd/tcp_small.html.

[5] J. Hoe, "Improving the start-up behavior of a congestion control scheme for TCP," in *Proceedings of the ACM SIGCOMM*, August 1996, pp. 270–280, http://www.acm.org/sigcomm/ccr/archive/1996/conf/hoe.ps.

[6] K. Fall and S. Floyd, "Simulation-based comparisons of Tahoe, Reno, and SACK TCP," *ACM Computer Communication Review*, July 1996, pp. 5–21, ftp://ftp.ee.lbl.gov/papers/sacks.ps.Z.

[7] M. Mathis and J. Mahdavi, "Forward acknowledgment: refining TCP congestion control," in *Proceedings of the ACM SIGCOMM*, August 1996, also see http://www.psc.edu/networking/papers/papers.html.

[8] L. Brakmo, S. O'Malley, and L. Peterson, "TCP vegas: new techniques for congestion detection and avoidance," in *Proceedings of the ACM SIGCOMM*, August 1994, pp. 24–35, http://netweb.usc.edu/yaxu/Vegas/Reference/vegas93.ps.

[9] P. Sinha, N. Venkitaraman, R. Sivakumar, and V. Bharghavan, "WTCP: a reliable transport protocol for wireless wide-area networks," in *Proceedings of the ACM MobiCom '99*, Seattle, Washington, August 1999.

[10] M. Mathis, J. Madhavi, S. Floyd, and A. Romanow, "TCP selective acknowledgement options," *RFC 2018*, October 1996.

[11] S. Keshav and S.P. Morgan, "SMART retransmission: performance with random losses and overload," in *Proceedings of the IEEE INFOCOM '97*, Kobe, Japan, April 1997.

[12] V. Tsaoussidis and H. Badr, "TCP-probing: towards an error control schema with energy and throughput performance gains," Technical Report TR4-20-2000, Department of Computer Science, SUNY Stony Brook, April 1999, also appears in *Proceedings of ICNP 2000*.

[13] C. Parsa and J.J. Garcia-Luna-Aceves, "Improving TCP congestion control over internets with heterogeneous transmission media," in *Proceedings of the 7th Annual International Conference on Network Protocols*, Toronto, Canada, November 1999.

[14] A. Aggarwal, S. Savage, and T. Anderson, "Understanding the performance of TCP pacing," in *Proceedings of the 2000 IEEE Infocom Conference*, Tel-Aviv, Israel, March 2000.

[15] S. Mascolo, C. Casetti, M. Gerla, M.Y. Sanadidi, and R. Wang, "TCP Westwood: bandwidth estimation for enhanced transport over wireless links," in *Proceedings of the ACM MobiCom*, Italy, July 2001.

[16] H.-Y. Hsieh and R. Sivakumar, "A transport layer approach for achieving aggregate bandwidths on multi-homed mobile hosts," in *Proceedings of the ACM MobiCom*, September 2002.

[17] I.F. Akylyidiz, G. Morabito, and S. Palazzo, "TCP-Peach: a new congestion control scheme for satellite IP networks," *IEEE/ACM Transactions on Networking*, 9, 307–321, 2001.

[18] B.S. Bakshi, P. Krishna, N.H. Vaidya, and D.K. Pradhan, "Improving performance of TCP over wireless networks," in *Proceedings of the 17th International Conference on Distributed Computing Systems '97*, Baltimore, MD, May 1997.

[19] H. Balakrishnan and R.H. Katz, "Explicit Loss Notification and wireless web performance," in *Proceedings of the IEEE Globecom Internet Mini-Conference*, Sydney, Australia, November 1998.

[20] H. Balakrishnan, S. Seshan, E. Amir, and R.H. Katz, "Improving TCP/IP performance over wireless networks," in *Proceedings of the 1st ACM Conference on Mobile Computing and Networking*, Berkeley, CA, November 1995, http://daedalus.cs.berkeley.edu/publications/mcn.ps.

[21] H.-Y. Hsieh, K.-H. Kim, Y. Zhu, and R. Sivakumar, "A receiver-centric transport protocol for mobile hosts with heterogeneous wireless interfaces," in *Proceedings of the ACM MobiCom*, San Diego, CA, September 2003.

[22] P. Mehra, C. De Vleeschouwer, and A. Zakhor, "Receiver-driven bandwidth sharing for TCP," in *Proceedings of the IEEE INFOCOM*, San Francisco, CA, April 2003.

[23] N. Spring, M. Chesire, M. Berryman, V. Sahasranaman, T. Anderson, and B. Bershad, "Receiver-based management of low bandwidth access links," in *Proceedings of the IEEE INFOCOM*, March 2000.

[24] V. Tsaoussidis and C. Zhang, "TCP-Real: receiver-oriented congestion control," *Computer Networks*, 40, 477–497, 2002.

[25] Ajay Bakre and B.R. Badrinath, "Handoff and system support for indirect TCP/IP," in *Second USENIX Symposium on Mobile and Location-Independent Computing Proceedings*, Ann Arbor, MI, April 1995.

[26] K. Brown and S. Singh, "M-TCP: TCP for mobile cellular networks," *ACM Computer Commnication Review*, 27, 19–43, 1997.

[27] K.Y. Wang and S.K. Tripathi, "Mobile-end transport protocol: an alternative to TCP/IP over wireless links," in *INFOCOM*, San Francisco, CA, March/April 1988, p. 1046.

[28] "The Open Mobile Alliance," http://www.openmobilealliance.org/tech/affiliates/wap/wapindex. html, 2004.

[29] H. Balakrishnan, V.N. Padmanabhan, S. Seshan, and R.H. Katz, "A comparison of mechanisms for improving TCP performance over wireless links," *IEEE/ACM Transactions on Networking*, 5, 756–769, 1997, http://www.ccrc.wustl.edu/~ton/dec97.html.

[30] Miten N. Mehta and Nitin Il. Vaidya, "Delayed duplicate acknowledgements: a proposal to improve performance of TCP on wireless links," Interim Report, Texas A&M University, College Station, TX, December 1997.

[31] C. Parsa and J.J. Garcia-Luna-Aceves, "TULIP: A link-level protocol for improving TCP over wireless links," in *Proceedings of the IEEE Wireless Communications and Networking Conference 1999*, New Orleans, Louisiana, September 1999, also appears in the MONET journal in 2000.

[32] Mario Gerla, Ken Tang, and Rajiv Bagrodia, "TCP performance in wireless multihop networks," in *Proceedings of the IEEE WMCSA*, New Orleans, LA, February 1999.

[33] E. Ayanoglu, S. Paul, T.F. LaPorta, K.K. Sabnani, and R.D. Gitlin, "AIRMAIL: a link-layer protocol for wireless networks," *ACM Wireless Networks*, 1, 47–60, 1995.

[34] S. Bhandarkar, N. Sadry, A.L.N. Reddy, and N. Vaidya, "TCP-DCR: a novel protocol for tolerating wireless channel errors," in *IEEE Transactions on Mobile Computing*, to appear.

[35] H. Balakrishnan, S. Seshan, and R.H. Katz, "Improving reliable transport and handoff performance in cellular wireless networks," *ACM Wireless Networks*, 1, 469–481, 1995.

[36] A. Bakre and B.R. Badrinath, "I-TCP: indirect tcp for mobile hosts," in *Proceedings of the 15th International Conference on Distributed Computing Systems*, Vancouver, BC, May 1995.

[37] M. Zorzi and R. Rao, "Energy efficiency of TCP in a local wireless environment," *Mobile Networks and Applications*, 6, 265–278, 2001.

[38] H. Singh and S. Singh, "Energy consumption of TCP Reno, Newreno, and SACK in multihop wireless networks," in *ACM SIGMETRICS*, Marina Del Rey, CA, June 2002.

[39] S. Bansal, R. Gupta, R. Shorey, and A. Misra, "Energy-efficiency and throughput for TCP traffic in multi-hop wireless networks," in *Proceedings of the IEEE INFOCOM*, New York, April 2004.

[40] V. Jacobson, "Compressing TCP/IP headers for low-speed serial links," RFC 1144, February 1990.

[41] M. Allman, "TCP byte counting refinements," *ACM Computer Communication Review*, 29, 14–22, 1999.

[42] K. Ramakrishnan and S. Floyd, "A proposal to add explicit congestion notification (ECN) to IP," *RFC 2481*, January 1999.

[43] A. Shacham, R. Monsour, R. Pereira, and M. Thomas, "IP-payload compression protocol," *RFC 2393*, December 1998.

[44] Jeffrey Semke, Jamshid Mahdavi, and Matthew Mathis, "Automatic TCP buffer tuning," *ACM Computer Communication Review*, 28, 315–323, 1998.

[45] G. Montenegro, S. Dawkins, M. Kojo, V. Magret, and N. Vaidya, "Long thin networks," *RFC 2757*, January 2000.

[46] M. Allman, D. Glover, and L. Sanchez, "Enhancing TCP over satellite channels using standard mechanisms," *RFC 2488*, January 1999.

[47] M. Allman, S. Dawkins, D. Glover, J. Griner, D. Tran, T. Henderson, J. Heidemann, S. Ostermann, K. Scott, J. Semke, and J. Touch, "Ongoing TCP research related to satellites," *RFC 2760*, February 2000.

30

TCP Developments in Mobile Ad Hoc Networks

30.1 Introduction..**30**-699
30.2 MANET Characteristics **30**-700
    Wireless Channel Characteristics • Mobile Node
    Characteristics • Routing Issues
30.3 Investigation of TCP Performance in MANETs **30**-705
30.4 Adapting TCP for MANETs **30**-707
    End-to-End Modifications • Point-to-Point Modifications
30.5 Conclusions **30**-718
References **30**-720

Stylianos Papanastasiou
Mohamed Ould-Khaoua
Lewis M. MacKenzie

30.1 Introduction

Mobile ad hoc networks (MANETs) have recently become a popular area of scientific research coinciding with the advent of affordable and powerful mobile devices utilizing common protocols such as IEEE 802.11 [23] and Bluetooth [20]. These devices or "nodes" can act both as hosts and routers, communicating through wireless links to form "on-the-fly" networks with potentially high dynamic and unpredictable topologies. Existing infrastructure or centralized administration is not assumed and consequently MANETs are particularly applicable in settings where these factors are in short supply, such as military or disaster relief operations [26].

The Transmission control protocol (TCP) has become the de facto reliable connection oriented transport protocol over IEEE 802.11 ad hoc networks due to its wide use on the Internet. Across its several variants, TCP implements a sliding window, additive increase multiplicative decrease (AIMD) congestion control mechanism; this operates by monitoring for signs of congestion and when necessary throttles the send rate using the automatic repeat request (ARQ) paradigm and cumulative acknowledgments (ACKs) [24]. TCP is normally used over IP, an unreliable network protocol, and implements end-to-end communication between source and destination hosts. Although TCP is independent of the physical medium used to accommodate communications, certain assumptions are evident in its design and these generally reflect the characteristics of the wired networks dominant at the time of its creation. Unfortunately, these assumptions do not hold over multihop wireless networks and several studies have shown that TCP exhibits suboptimal performance because of the characteristics of the new environment [18,45,49].

Essentially, any type of loss detected by TCP, either through the expiration of its retransmission timer or through the reception of three duplicate ACKs, is interpreted as a sign of congestion. At such an occurrence, TCP reduces its sending rate to avoid congestion collapse in the network. The interpretation of packet loss as a congestion indication is valid in wired LANs since very few packet losses can be traced to other causes. In wireless communications, in general, and MANETs in particular, however, a packet drop may be attributed to other causes such as the high bit error rate of the wireless medium, route failures due to mobility, network partitioning or buffer overflow, and spatial congestion along the communications path. Traditional TCP agents cannot distinguish between the different causes, and reduce the sending rate; however, this is not always the appropriate action and performance can degrade as a result. Furthermore, the IEEE 802.11 protocol contains a distributed medium access mechanism, which is inherently unfair and even more so toward TCP traffic because of a particular interplay with TCP's exponential backoff of the retransmission time-out (RTO) mechanism when a packet loss occurs [50].

The above mentioned problems are present in other forms of wireless communications; however, while relevant solutions have been proposed, these normally use preexisting infrastructure available in those environments. Such adaptations have been proposed for infrastructure wireless LANs [6], mobile cellular [36], and satellite networks [2]. However, the challenges faced by MANETs although largely similar in nature (e.g., the high bit error rate of the wireless medium), cannot be dealt with by using similar strategies, since any proposed solution must be distributed and assumes no fixed infrastructure.

In this survey, we classify the proposed adaptations to TCP in MANETs according to their point of deployment. Hence, we broadly categorize them as *end-to-end* or *point-to-point* depending on whether they need be deployed only at the end points or at intermediate nodes as well. Further, each approach is explicitly defined as *single layer* or *cross-layer*. In single layer approaches, modifications are confined in one layer in the network stack, while cross-layer approaches use signalling information between layers to improve TCP performance. The behavior of TCP in MANETs cannot be fully discussed without referring to general characteristics of the routing protocols that operate in dynamic topologies. However, a detailed description of ad hoc routing protocols falls outside the scope of this report and interested readers are advised to consult [37] for further information on this topic. Henceforth, some basic understanding of routing protocols in MANETs is assumed.

The rest of this chapter is organized as follows. Section 30.2 provides an overview of the unique characteristics of MANETs. The summary includes concise descriptions of wireless channel characteristics, mobile node characteristics, and the routing challenges in a MANET setting. A survey of existing studies on TCP performance in MANETs is provided in Section 30.3. Section 30.4 contains a review of proposed adaptations to improve TCP performance in MANETs. Finally, Section 30.5 concludes the chapter and outlines future research prospects.

30.2 MANET Characteristics

MANETs provide protocol designers with unique challenges as they differ substantially from traditional wired networks. A primary cause of divergence is to be found in the unique properties of the wireless medium itself; because of the way the transmitted signal radiates, wireless links are more susceptible to packet losses due to errors and improper transmission timing, which may lead to spatial contention. Another difference is that the mobile nodes participating in MANETs are typically limited in power reserves and CPU capacity and this militates against complex implementations and solutions. Finally, MANETs often exhibit highly dynamic topologies, which consequently lead to frequent and sudden connectivity changes due to potentially continuous mobility of nodes. This may rapidly render older routes obsolete while allowing new ones to form in their place. Routing protocols have been tuned to deal with these issues, which nonetheless have a major impact on the transport layer. A brief overview is presented below.

30.2.1 Wireless Channel Characteristics

The wireless channel in MANETs shares many of the properties of wired-infrastructure LANs but also possesses certain unique features, which are derived from the distributed nature of the medium access mechanism.

1. *Signal attenuation*: As the transmitted signal spreads out from the aerial in all directions it quickly attenuates as distance increases. Hence, the intensity of the electromagnetic energy at the receiver decreases and, at some point, the signal-to-noise (SNR) ratio may become low enough for the receiver to decode the transmission successfully.

   In wireless communications and for an omnidirectional transceiver, three ranges can be identified. These are, from the sender's perspective:

   Transmission Range (R_{tx}): The range within which a transmitted packet can be successfully received by the intended receiver, that is, within this range the SNR is high enough for a packet to be decoded by the receiver.

   Carrier Sensing Range (R_{cs}): The range within which the transmitter triggers carrier sense detection. When carrier sense detection is triggered the medium is considered busy and the sensing node defers transmission.

   Interference Range (R_i): The range within which an intended receiver may be subject to interference from an unrelated transmission, thereby suffering a loss. This range largely depends on the distance between the sender and the interfering node.

Those ranges are related to one another; $R_{tx} < R_i \leq R_{cs}$, as the energy required for a signal to be decoded is much more than what is needed to cause interference. The interference and transmission ranges depend on the signal propagation model and the sensitivity of the receiver, assuming that power constraints apply and all transmitters transmit at the maximum allowed power level. Interested readers may refer to [29,47] for more details.

2. *Multipath fading*: Multipath fading occurs because of different versions of the same signal arriving at different times at the receiver. These versions effectively follow different paths, with different propagation delays, from the transmitter due to multiple reflections off intervening obstacles. The superposition of these randomly phased components can make the multipath phenomenon a real problem especially if there are many reflective surfaces in the environment and the receiver is situated in a fringe area of reception.

3. *High bit error rates*: Due to the nature of signal propagation, the wireless medium potentially exhibits a high bit error rate (BER). The link layer frame format in IEEE 802.11 networks is similar to 802.3 (Ethernet) frames [44] and uses the same 48-bit medium access control (MAC) address fields. The format also includes the IEEE 802.3 32-bit CRC polynomial-based error detection mechanism. However, the protection offered by this scheme only extends to data actually traveling on point-to-point links. It is still possible, though somewhat unlikely, for corrupt data to be accepted by the receiver, but this is offset by the adoption of error discovery, implemented at higher layers. Furthermore, MANETs are often used in lossy environments, served by low bandwidth links as opposed to wired networks where the BER is very low [43]. Hence, packet losses due to errors are more frequent in MANETs than in wired LANs. Traditional transport and routing protocols used in wired networks exhibit suboptimal behavior under such conditions.

4. *Hidden and exposed terminals*: Consider the scenario illustrated in Figure 30.1(a). Node A is transmitting to node B. Node C cannot hear the transmission and since its carrier sense function detects an idle medium, it will not defer transmission to D and a collision will be produced at node B. Node A is hidden with respect to node C (and vice versa). This problem is offset in 802.11 by using a short packet exchanges of request-to-send (RTS), clear-to-send (CTS) frames. This is a two-way handshake whereas the source terminal transmits the RTS to the destination, which then replies with a CTS frame. If there is no reply, transmission is then deferred as presumably the medium around the destination area is busy. If a CTS reply is received then DATA transmission follows. Since the duration of the transmission is included in

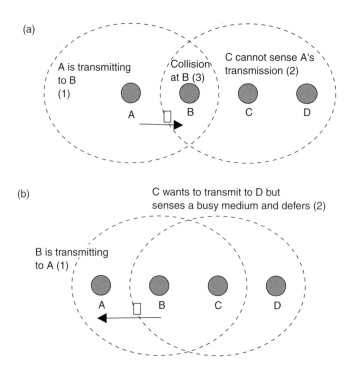

FIGURE 30.1 Illustration of the hidden and exposed terminal effects: (a) the hidden terminal effect; (b) the exposed terminal effect.

the RTS/CTS exchange, neighboring nodes defer their transmissions for the time the medium is occupied. Point-to-point transmission is reliable since the DATA frame is followed by an ACK transmission from the destination if the frame is successfully received.

The exposed terminal effect occurs when a station that wants to transmit senses a busy medium and defers transmission even though its transmission would not have interfered with the other sender's transmission. An instance of the exposed terminal effect is demonstrated in Figure 30.1(b). In this instance, node B is transmitting to node A. Node C senses node B's signal and defers transmission. However, it need not have done so as C's transmission does not reach node A and would not have interfered with B's transmission at the location of the intended destination (node A). Node C is the exposed terminal in this case. Note that both the hidden and exposed terminal effects are related to the transmission range. As the transmission range increases the hidden terminal effect becomes less prominent because the sensing range increases. Nonetheless, the exposed terminal effect then becomes more prominent as a greater area is "reserved" for each transmission.

In the above examples, the transmission (R_{tx}), interference (R_i), and carrier sense (R_{cs}) ranges are all assumed to be equal. However, several research efforts have concentrated on the effects of interference on the hidden and exposed terminal effects [10,11,48–50], when $R_i > R_{tx}$. As such, Xu et al. in Reference 47 have shown that when the distance d between the source and destination nodes is $0.56 * R_{tx} \leq d \leq R_{tx}$, where R_{tx} is the transmission range of the sender, the effectiveness of the RTS/CTS exchange declines rapidly.

5. *Spatial contention and reuse*: Unlike wired networks where the links are fixed and do not interact with each other (there is typically little interference between physical cables) wireless networks operate differently. Assuming omnidirectional antennas, when a node transmits, it "reserves" the area around it for its transmission, that is, no other transmission should take place because it will result in a collision and waste of bandwidth. The spatial reuse refers to the number of concurrent transmissions that may occur so without interfering with each other. It is the responsibility of the MAC protocol to ensure that transmissions are coordinated in such a way so as to maximize spatial reuse.

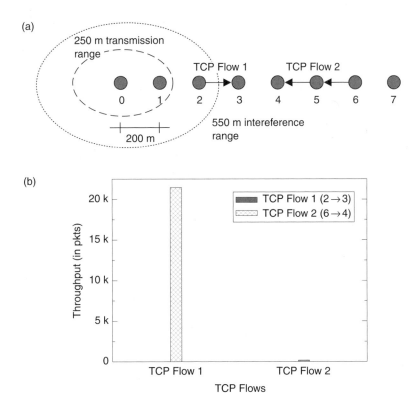

FIGURE 30.2 Illustration of the capture effect: (a) string topology and connectivity pattern demonstrating the capture effect; (b) TCP Flow 1 starves TCP flow 2, even though flow 2 starts first.

For instance, in Figure 30.2 communication between nodes $0 \rightarrow 1$ and $4 \rightarrow 5$ may happen simultaneously. Then communication of the other nodes could happen concurrently in turn, as long as each pair is 4 hops apart from the other. Since at most two pairs can transmit at the same time without affecting each other, the spatial reuse of this string topology is 2. However, for any topology its spatial reuse is an optimal level of concurrency; it is not always achievable and often this many nodes transmitting simultaneously will result in packet drops due to interference. This is referred to as spatial contention and can become the main cause of packet drops when a path is long enough as noted in [18]. This is in contrast with wired networks were packet drops are mainly caused by buffer overflows at the routers.

6. *Capture effect (Interplay of TCP with 802.11)*: In wired networks, TCP has a well-documented bias against long flows [15]. In 802.11 multihop networks, the bias is much stronger and is manifested in the form of the *channel capture effect*. In short, if two TCP connections are located within the vicinity of each other and therefore interfere with one another, this effect favors the session that originated earlier or the one that flows over fewer hops. The favored session often starves the other almost completely and no data transport may be possible for the mistreated session until the other one has completed all of its data transmission.

The bias is rooted in the exponential back off of the 802.11 distributed coordination function (DCF) mechanism, which is inherently unfair and is further augmented by TCP's own exponential back off mechanism. We only present an example below but interested readers may refer to References 33, 48, and 50 for further details.

Consider a string topology as in Figure 30.2(a) with two TCP flows, specifically Flow 1 $(2 \rightarrow 3)$ and Flow 2 $(6 \rightarrow 4)$. Flow 2, which spans over 2 hops starts 2 sec before Flow 1 while both deliver File Transfer Protocol (FTP) traffic to their respective destinations. Figure 30.2(b) shows the throughput in packets for each connection at the end of a 200 sec transfer, over dynamic source routing (DSR) [27]. Clearly, Flow 1 (2 → 3) occupies the majority of the available bandwidth while Flow 2 (6 → 4) is treated with severe unfairness. The problem of unfairness and channel capture is an important research challenge and we outline existing proposed solutions in Section 30.4.

30.2.2 Mobile Node Characteristics

The mobile nodes that participate in a MANET operate under limitations, which need to be taken into account by the various protocols if they are to be applicable in such an environment. The MANET devices may be as diverse as laptops, PDAs, or even desktop systems. To be mobile, devices will usually have limited power reserves and may possibly operate with limited processing capacity. These restrictions and the way they affect protocol operations are discussed briefly in this section.

In general, power is a scarce and valuable commodity in MANETs and its consumption is therefore just as important a measure as throughput, latency, and other metrics when evaluating MANET protocols at any layer. Several methods have been proposed to conserve power at all possible levels, including the physical layer, the operating system, and the applications. An overview of approaches to power conservation is included in Reference 28.

From a TCP perspective, power savings are best achieved by minimizing redundant retransmissions whenever possible [34]. The savings in this case are twofold. The source conserves power by transmitting fewer packets but also every forwarding node in the path benefits since fewer unnecessary retransmissions occur.

Another factor to take into account is the possibly restricted CPU capacity at each node. Routing algorithms, in particular, are designed to be simple as they operate in environments where processing capacity is limited [41]. It follows that any adjustments proposed to the TCP protocol should be of as low complexity as possible, so that any CPU time costs do not outweigh gains in throughput or latency. In addition, heavy CPU usage places extra demands on the power supply of the device, which makes processor-intensive modifications even more costly. Surprisingly, although power considerations are a key focus of routing and clustering techniques in MANETs, proposed TCP modifications are not always examined with respect to their overall power and CPU demands.

30.2.3 Routing Issues

MANETs are characterized by significant node mobility that induces highly dynamic topologies and potentially may result in partitioned networks. In this section, network partitioning and the effects of routing failures are discussed.

1. *Network Partitioning:* Network partitioning occurs when, due to mobility, nodes which were able to communicate directly or through the cooperation of other nodes at time T_1 are unable to do so at a later time, T_2, since there is no longer a usable path between them. It is further possible that at a later time, say T_3, the nodes have placed themselves in such a position that the network is again connected and every node can reach every other one, either directly or indirectly. The concept is illustrated in Figure 30.3.

TCP is not engineered to deal with network partitioning as it does not normally occur in wired networks. The exponential back off of TCP's RTO mechanism, in effect, facilitates exponentially delayed probing of a valid path as follows. In Figure 30.3, assume that the TCP source is communicating as normal with the destination. At time T_2 there is a network partitioning and so ACKs do not reach the source and packets are not being forwarded to the destination. A new packet is sent every time the RTO timer expires. If this packet reaches the destination, TCP continues with normal transmission and the route is utilized. However, since the RTO is doubled after every time out, those "probing" packet transmissions take longer each time. Hence, if the disconnection persists for consecutive RTOs, there might be long periods of inactivity during which the network is connected again, but TCP is in the back off state. Several approaches have been proposed to solve this problem and an overview is presented in Section 30.4.

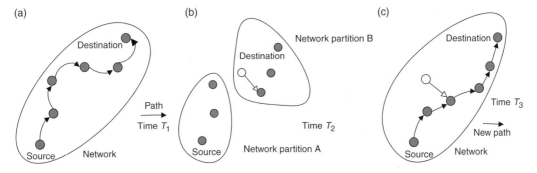

FIGURE 30.3 Illustration of network partitioning: (a) An ad hoc network; (b) network partitioning; (c) network is connected again.

2. *Routing failures:* In MANETs, unlike wired networks, routing failures occur frequently due to the mobile nature of the participating nodes. Route failures may also occur when repeated point-to-point transmissions fail, either due to high BER, which causes packet corruption, or because of the effects of spatial contention. When a routing breakage occurs, the routing protocol attempts to discover an alternate path, but the duration of the restoration period largely depends on the nodes' mobility, the mechanism of the routing protocol itself, and the network traffic characteristics.

The effects of route failures on TCP operation resemble those of network partitioning as discussed previously. If the route takes some time to restore, TCP enters its back off state and sends "probes" for a restored route at increasingly longer time intervals. Hence, the route might be restored for quite sometime but TCP remains idle until it launches the retransmitted packet after the RTO expiration. Another problem is the round trip time (RTT) calculation. After the route has been reestablished, the RTT measurements required for the RTO calculation reflect the new route. However, since the RTO is calculated using a weighted average of new and old RTT measurements, for some time after the route restoration the RTO value contains a mixed estimate of the old and new route characteristics. These problems have been noted in the literature and various methods have been proposed to overcome them. An overview is presented in Section 30.4.

30.3 Investigation of TCP Performance in MANETs

The vast majority of TCP evaluation studies over multihop wireless ad hoc networks have been carried out with simulations and experimental testbeds. The unique characteristics of the wireless medium and its interplay with existing TCP mechanisms, as described in Section 30.2, makes the development of analytic models of TCP behavior extremely difficult and it would be fair to say that an aspiration that still remains unmet.

The first evaluation of TCP performance under different routing algorithms over MANETs was conducted by Ahuja et al. in Reference 1 for several routing protocols, namely, the ad hoc on-demand distance vector (AODV) [41] routing protocol, the dynamic source routing (DSR) [27] protocol, the dynamic destination-sequenced distance-vector routing (DSDV) [42] protocol, and the signal stability-based adaptive (SSA) [13] routing protocol were considered. The work showed the detrimental effect of mobility on TCP Tahoe as well as the relative merits of SSA routing. Notably, SSA incorporates considerations of the signal strength along the path during the route discovery phase so that the most stable route is chosen rather than just the shortest one.

An investigation of TCP Reno over AODV, DSR, and the adaptive distance vector (ADV) [32] routing protocols is included in Reference 14. The ADV routing protocol is shown to maximize TCP throughput, while AODV is found to be the second best performer under various mobility conditions. Further, the degrading effect of stale cache entry usage is noted for DSR. The authors also propose the use of a heuristic called the fixed-RTO, which dramatically improves TCP throughput in AODV and DSR as it aids TCP in utilizing restored routes quickly, without resorting to feedback from the routing protocol. An overview of this technique is included in Section 30.4.1.2.

The effects of interference on TCP, as noted in Section 30.2.1, have been widely studied in literature. In Reference 49, Xu and Saadawi examine the throughput of TCP Tahoe, Reno, Newreno, SACK, and Vegas over multihop chain topologies in an open space environment under the 802.11 protocol. These authors demonstrate that the Tahoe TCP and its variants (Reno, Newreno, and SACK) exhibit throughput instability in such topologies as interference causes packet drops, which are interpreted as congestion losses. TCP Vegas does not suffer from this problem and further investigation has revealed that the other variants may regain throughput stability by limiting the maximum congestion window (cwnd) to four segments. Further, the viability of using the delayed-ACK modification is discussed in the same work and is shown to lead to a 15 to 32% throughput improvement.

The work of Xu and Saadawi is augmented in Reference 18 through further studies by Zhenghua et al. In particular, the authors in Reference 18 note that for the string, cross, and grid topologies there is an optimal cwnd size, which maximizes throughput by improving spatial reuse. Since TCP continuously increases its cwnd size until packet loss is detected, it typically grows beyond this optimal size and, moreover, normally operates at an average window size larger than the optimal one, thereby causing spatial contention. The authors compute analytically the optimal window size for each of the above mentioned topologies. Finally, they propose two link layer modifications called "link random early detection (RED)" and "adaptive pacing" to aid optimal spatial reuse. An overview of these is provided in Section 30.4.2.5.

The issue of TCP throughput instability and more effective spatial reuse is formally addressed in References 10 and 11 by Chen et al. They show that by relaxing MAC layer assumptions, that is, not 802.11 specific, the bandwidth delay product in multihop MANET paths cannot exceed the round-trip hop count (RTHC) of the path * packet size. In the case of the 802.11 protocol, this bound is shown to reach no more than one-fifth of the RTHC. According to the new definition of the bandwidth delay product the authors then propose an adaptive mechanism, which sets the maximum cwnd according to the route hop count, noting an 8 to 16% throughput improvement. In Reference 11 the performance merit of TCP-pacing, which evenly spaces a windows worth of packets over the current estimated round-trip time, is also evaluated but no worthwhile performance improvement is noted. In Reference 39, Papanastasiou and Ould-Khaoua note fairness problems caused by the adaptive maximum cwnd setting interacting with background traffic and propose a new technique called slow congestion avoidance (SCA), which decreases the standard rate of increase of TCP during the congestion avoidance (CA) phase and provides similar performance improvements as in Reference 10 but delivers much better throughput when there is background traffic in the network or moderate node mobility. An overview of these techniques is included in Sections 30.4.1.7 and 30.4.1.8.

Finally, the issue of fairness amongst TCP flows is addressed in References 48 and 50. It is demonstrated that the interplay between the exponential backoff mechanism used by TCP in the presence of congestion and the 802.11 ad hoc protocol causes severe unfairness and can starve flows that are subject to interference from other TCP flows. The unfairness problem persists even in the case when both wireless and wired hops are utilized in the TCP path as shown in Reference 46.

In Reference 48, Xu et al. address this problem of fairness by expanding the notion of a RED queue [16] to account for the "neighborhood" of a node, that is, those other nodes that may interfere with its transmissions. To contain overhead, only 1-hop information is exchanged and link layer information is utilized to form a queue at each node, which reflects the network congestion state. Packets may be dropped from this queue to slow down the sending rate of favored flows and allow others to utilize the link. The authors further note that there is a trade-off between total throughput and fairness, which is worth considering given that the starvation of TCP flows may be extreme in some cases. An overview of this technique is included in Section 30.4.2.6.

Concluding, Anastasi et al. [5] have conducted measurements on an actual testbed and have verified simulation results by showing that interference becomes a serious problem in ad hoc networks when TCP traffic is considered. Moreover, the authors note that there was sufficiently high variability in channel conditions at different times during their experiments, which made comparison of results difficult. Finally, it is observed that certain aspects of real wireless transmissions are not effectively captured in simulation such as the different sending rates of the preamble, the RTS/CTS, and the DATA frames in 802.11 networks, as well as the variability of the transmission and physical sensing ranges even in the same session.

30.4 Adapting TCP for MANETs

We categorize the proposed improvements to TCP in MANETs into two categories. The *End-to-End* modifications require alterations only at the sender and the receiver, while the *Point-to-Point* modifications need apply to intermediate nodes as well. For each proposed solution it is noted whether it is cross-layer (modifications to more than one layer) or single layer (self-contained in one layer). In theory a single layer, end-to-end solution is preferable since this maintains layering semantics intact. However, in practice, cross-layer changes can sometimes yield significant benefits. The trade-off between formal adherence to the layering concept and practical performance is illustrated in many of the techniques surveyed below.

30.4.1 End-to-End Modifications

The end-to-end modifications can be categorized according to the problem they address. The explicit link failure notification (ELFN), Fixed-RTO, and TCP-DOOR modifications aim to reduce the effect of route breakages on TCP, as described in Section 30.2.3. On the other hand, TCP HACK and other mechanisms that deal with the high BER of the shared medium are designed to distinguish between error and congestion-induced losses so that the proper action may be taken by the TCP agent. Finally, the Dynamic delayed ACK, the Adaptive CWL Setting, and the SCA techniques try to help the TCP agent avoid spatial contention as discussed in Section 30.2.1.

30.4.1.1 ELFN Approaches

The idea of the ELFN [22] is based on providing feedback to the TCP agent about route failure, so that consequent packet losses are not interpreted as signs of congestion. Such a notification can be realized through the use of a "host unreachable" internet control message protocol (ICMP) message or by piggybacking the required information on a source-bound route failure message generated by the routing protocol, such as a route error (RERR) message in AODV or DSR. Those feedback providing messages would contain the relevant sender and receiver addresses and ports and the current TCP sequence number.

Whenever the TCP-sender receives an ELFN message (or a RERR message with the required information) it enters a "stand-by" mode where its retransmission timer is frozen. A probing packet is sent during this stand-by period and, if an acknowledgment is received, TCP restores its retransmission timers and resumes operations as normal. An illustration of ELFN's principal operation is shown in Figure 30.4. The authors in Reference 22 also mention that it is possible to explicitly notify the source of the route reestablishment but they do not evaluate such a mechanism in their study. The feedback mechanism utilizes cross-layer feedback but maintains compatibility with unaltered nodes (if a node cannot interpret the ICMP message correctly, normal routing functions are not affected).

In the subsequent evaluation of TCP-F, the authors find that a value of 2 secs for the probing interval works best and propose as a suggestion for future study, that the interval value be expressed as a function of the RTT. Furthermore, they raise the question as to whether the cwnd or the RTO should be restored when a route is reestablished. Through simulation experiments, they find that using the "frozen" RTO and cwnd values after the route has been repaired improves on throughput, rather than setting the cwnd to 1 and the RTO to 3 sec (as recommended in Reference 40). However, since this technique can be used on flows in MANETs that may have an end-point on the Internet, more study is needed before allowing the initial RTO and cwnd values to deviate from their standard settings. Future research prospects in this direction include evaluating ELFN techniques with modern proactive and reactive protocols as well as estimating the feasibility of embedding the required feedback information in RERR packets generated by the routing agent.

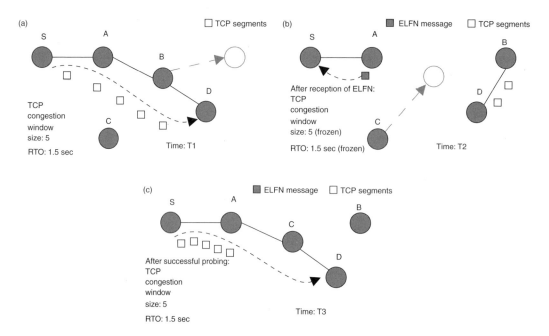

FIGURE 30.4   Illustration of the ELFN operation: (a) normal TCP operation; (b) link failure notification of TCP agent; (c) after successful probing TCP restores frozen state variables.

30.4.1.2   Fixed RTO

The fixed-RTO technique [14] is a heuristic to distinguish between route failure and congestion-induced losses without relying on feedback from other layers or mobile nodes.

The only modification made to TCP concerns the exponential back off of the RTO timer. When consecutive time-outs occur, if the missing ACK is not received before the second RTO expires, a route failure is assumed. As such, the missing segment is retransmitted following the normal TCP rules but the RTO is not doubled, instead remains the same until the retransmitted packet is acknowledged. This, in turn, denotes that an alternate route has been discovered.

The modification allows for restored routes to be utilized quickly by the TCP agent through its persistence. The authors in Reference 14 note dramatic improvements in TCP throughput for the DSR, AODV, and the ADV routing protocols (up to 75%). The technique provides a "quick fix" that can be used to mitigate to a certain extent the effect of route failures on TCP [10]. However, the fixed-RTO method breaks the operation principle of the RTO mechanism, which ensures protection against congestion collapse and, as such, more study is required before its use can be justified in a real MANET environment. This is especially the case, since a MANET can be a gateway to the Internet, where it is important to maintain the congestion control principles of TCP. In Reference 14 this limitation is implicitly recognized and the authors recommend the use of the fixed-RTO mechanism only in wireless networks. However, on the plus side, this modification is clearly a single layer, end-to-end self-contained solution.

30.4.1.3   TCP-DOOR

The TCP with detection of out-of-order and response (TCP-DOOR) [45] deals with the problem of route failure induced losses but instead of using special feedback packets for signaling like ELFN or TCP-BuS (presented in Section 30.4.2), it uses out-of-order (OO) delivery events as a heuristic to detect link failure.

The main idea is that route failures contribute to packet reordering, which can then provide clues to TCP about a link breakage event, so that it does not back off unnecessarily because it has misidentified congestion. An example of this is shown in Figure 30.5 with gratuitous replies and DSR. The proposed modifications are strictly end-to-end and do not assume any feedback from other layers. Specifically, since

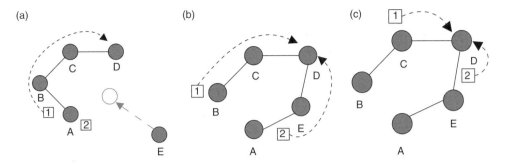

FIGURE 30.5   Out-of-order delivery due to route alteration: (a) Packet 1 sent through route A–B–C–D; (b) packet 2 sent through route A–E–D; (c) out-of-order delivery: Packet 2 reaches D before packet 1.

reordering can occur in both directions, that is, for ACKs and data segments, there are two OO detection mechanisms, a sender and a receiver one. These are explained briefly below:

- *Sender OO detection*: Detecting the OO nonduplicate ACKs is straightforward as the sequence numbers are monotonically increasing. The sender, however, cannot detect OO delivery of duplicate ACKs using the standard TCP segment format, so a 1-byte optional extension is added to express a monotonically increasing sequence for these as well. This is named the ACK duplication sequence number (ADSN) and enables the sender to detect OO duplicate ACKs.
- *Receiver OO detection*: The receiver cannot easily identify OO delivery of data packets because TCP retransmissions may cause a packet with a lower sequence number (but transmitted later) to arrive after one with a higher sequence number (thus falsely implying a reordering if the sequence numbers alone are taken into account). To overcome this difficulty, the authors in Reference 45 propose the adoption of a 2-byte option called the TCP packet sequence number (TPSN), which increases monotonically even for retransmissions and can aid in the detection of reordering. Another alternative would be the use of the TCP time-stamp [25] option, which is the typical solution to the "retransmission ambiguity" problem.

Two courses of action are proposed once reordering has been detected. The first is to disable temporarily congestion control, that is, keep the RTO and the cwnd constant for some time after the OO event has occurred. Second, instant recovery during the CA phase is recommended. The rationale is that if some time, T, before the OO packets started arriving, a congestion-indication event (RTO expiration or three dupACKs) occurred, then the latter was most likely triggered by path failure and could be rapidly reversed.

The two detection methods are examined in Reference 45 in combination with the two courses of action in five different combinations or configurations using DSR Routing. The evaluation revealed that instant recovery was more effective than temporarily disabling the congestion control, but the two methods bring about optimal throughput when used in unison (up to approximately 50%). There was no significant difference in the results when using sender and receiver OO detection, which implies that there is correlation between data and ACK reordering as a result of route failure, and this, in turn, suggests that either method would work as well as the other. However, it is not clear how the technique would be useful in the context of multipath routing or how much packet reordering occurs as a result of link breakages.

30.4.1.4 Error-Congestion Loss Distinction

From an end-to-end perspective, there have been several attempts to distinguish error from congestion losses in wireless communications. TCP-Veno [17], like TCP-Vegas [8], uses an estimator of the expected and actual throughput but still maintains the RENO TCP congestion avoidance (CA) mechanism with only a slight modification during the CA phase. If the Vegas estimator denotes imminent congestion in the CA phase then TCP-Veno increases the cwnd by 1 full-sized segment every other RTT, but does not decrease the cwnd, unlike Vegas. If the Vegas estimator does not denote congestion and a packet loss occurs, then Veno assumes it is due to errors and only decreases the cwnd by 1/5 instead of Reno's 1/2. For heavy loss environments, the authors in Reference 17 report improvements in throughput up to 80% compared to TCP Reno. Live measurements over an 802.11 infrastructure network have shown improvements of up to 35% but the technique has not been tested in a MANET environment.

In general, approaches that aim to distinguish congestion-induced losses from error-induced ones such as TCP-Aware [7] and TCP-Veno [17] can work reasonably well in infrastructure wireless networks but it is questionable whether they can be adapted to work in multihop 802.11 ad hoc networks. In such environments, the bandwidth-delay product is a different concept to that in infrastructure networks, due to the added factor of spatial contention [11].

Overall, distinguishing congestion losses from other type of losses in MANETs, from an end-to-end point of view, is a complex problem that may be unsolvable without explicit notification. As such, solutions used in infrastructure wireless networks are not directly applicable to MANETs.

30.4.1.5 TCP HACK

The header checksum option (HACK) [43] for TCP is an end-to-end modification proposed for infrastructure wireless networks that deals with errors and effectively solves the ambiguity between congestion and packet error drops. The reason this is treated separately from the error detection mechanisms mentioned in the previous section is that it is directly applicable to multihop wireless networks since it relies on explicit notification, instead of making assumptions about the less well-understood bandwidth-delay product in MANETs.

The technique is implemented as a TCP option. An extra header checksum is added, which complements the normal TCP packet checksum and also caters for the IP header (without the options), as shown in Figure 30.6(a). If the received packet at the destination has an invalid overall checksum, it is not dropped but, instead, the header checksum is checked. If the header is intact then the source of the packet can be identified and informed that one of its packets contains errors. It should therefore not initiate CA due to the loss. A special ACK, as shown in Figure 30.6(b), is produced and sent to the source informing it of the packet to be resent, whereupon the sender complies and transmits the packet without halving the cwnd.

The scheme is orthogonal to other TCP modifications such as SACK, can be incrementally deployed and is shown to be particularly effective in the case of bursty errors. A similar scheme and further analysis are also introduced in Reference 21, which provides further evidence for the validity of the idea. However, there is no evaluation of such a proposal in a multihop scenario and its applicability under such conditions remains an open research challenge.

30.4.1.6 Dynamic Delayed ACK

According to the TCP standard [3], the receiver can delay producing the cumulative ACK upon successful reception of a TCP segment, so long as the segment is not OO. The idea is that if the next segment contains a consecutive sequence number, a single cumulative ACK can be produced that acknowledges both segments. This delayed ACK should be generated at least for every two full-sized segments and within a 500 msec period in order to avoid ACK-clocking related problems.

In this approach [4], the possibility of producing delayed ACKs for more than two received segments is explored in string topologies. When the delay coefficient is set to higher values, fewer ACKs are produced on the return path, less spatial contention is observed, and performance increases become evident. The authors propose adjusting the delay coefficient parameter d with the sequence number of the TCP packets. The modifications are confined to the TCP sink (destination) and are completely end-to-end implementable. In particular, three thresholds are defined, $l1$, $l2$, and $l3$ such that $d = 1$ for packets with

(a)	1 byte	1 byte	2 bytes
	Kind = 14	Length = 4	1's complement checksum of TCP header and pseudo-IP header

(b)	1 byte	1 byte	16 bytes
	Kind = 15	Length = 6	32-bit sequence number of corrupted segment to resend

FIGURE 30.6 TCP HACK options format: (a) TCP header checksum option; (b) TCP header checksum ACK option.

sequence N numbers smaller than $l1$, $d = 2$ for packets with $l1 \leq N < 12$, $d = 3$ for $l2 \leq N < 13$, and $d = 4$ for $l3 \leq N$. As more and more segments appear in order the delay coefficient increases. Subsequent evaluation of the technique reveals dramatic improvements in delay, packet losses, and throughput when compared with classic Newreno in string topologies. However, ACK clocking issues and evaluation in realistic topologies require further investigation.

30.4.1.7 Adaptive CWL Setting

The adaptive congestion window limit (CWL) strategy [10] sets a limit on the maximum cwnd size so as to allow better spatial reuse of the medium. Consider the string topology in Figure 30.7 with the denoted transmission and interference ranges (which actually reflect Lucent WavcLan II transceivers in an open space environment [30]).

In this particular topology, and without considering any extra modifications to the standard 802.11 MAC mechanism, two packets can be transmitted at the same time, that is, two segments may be in the pipe simultaneously without interfering with one another. Two outstanding segments in the pipe are allowed by the TCP agent when its cwnd is two segments. In Reference 10, the authors approach the issue of the bandwidth-delay product and demonstrate that, under most circumstances, regardless of the MAC protocol used, it cannot exceed the RTHC*size of data packet. This is true so long as the channel does not hold multiple "back-to-back" packets per transmission, that is, after each transmission the sender has to recompete for channel access. For the 802.11 MAC, this limit becomes tighter, being set to one-fifth of the RTHC.

The authors utilize string topologies and DSR routing to derive the optimal cwnd for different hop counts. A feedback mechanism can then be used to inform the TCP agent of the route length, acquiring this information from the routing protocol; TCP then sets its maximum cwnd, according to the hop count. For instance, as the cwnd sets the amount of segments that TCP will allow outstanding in the pipe, a setting of 2 hops would be ideal in the case presented in Figure 30.7. Performance improvements of up to 16% are reported in throughput. This solution is an interlayer one, as it requires feedback from the routing protocol to function; however the changes concern only the sender.

In the experiments conducted in Reference 10, neither background traffic nor competing TCP connections are considered. As such, the solution is directed toward single TCP connections in a MANET, and this limits its applicability. The discovery of optimal cwnd settings, according to the number of active TCP connections, remains a topic of future research.

30.4.1.8 Slow Congestion Avoidance

In Reference 39, the SCA modification to TCP is proposed to help alleviate the spatial contention issue that arises from the unique properties of the wireless channels in MANETs. Unlike the adaptive CWL approach [10], the maximum cwnd limit is not affected, but instead the rate of increase during the CA phase of TCP is altered.

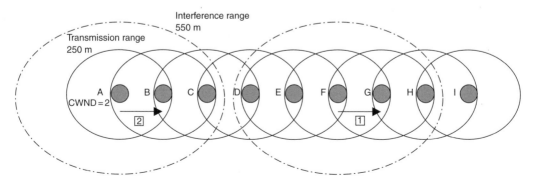

FIGURE 30.7 A chain topology with spatial reuse of 2 hops.

The method works as follows. The CA phase of TCP generates a linear increase in the sending rate by enlarging the cwnd by one full-sized segment every RTT. This increase is triggered by the TCP ACK clock, whereby each segment acknowledge triggers the dispatching of one or more segments. The SCA TCP agent maintains an SCA parameter P, which controls a delay in the increase of the cwnd during the CA phase. The cwnd still increases linearly but the increment occurs for every P RTTs. So a parameter value of two means that an increase in the cwnd occurs for every other RTT. This simple adjustment has a profound effect on TCP throughput for string topologies (up to 14 hops) delivering improved throughput ranging between 32 and 71% over "vanilla" Reno. In short, the idea is that TCP spends more time in an optimal cwnd region (optimal meaning in terms of the spatial reuse factor) even though it does eventually overshoot this window and spatial contention sets in. However, even when this happens, there are fewer packets in flight since cwnd increases less aggressively; thus the effects are less severe than "vanilla" TCP.

The SCA technique is evaluated in References 38 and 39 in a fashion similar to adaptive CWL. Specifically, the effectiveness of a range of parameters is considered for different hop counts under DSR [27] and AODV [41] routing. The improvement in throughput generally stabilizes or flattens at around an SCA parameter of 10, implying that this is an all-round "good" value usable for any hop count. If the parameter is set to 10, there is no need for feedback from the routing layer as the hop-count information is then redundant. Performance evaluation shows gains of up to 24% in dynamic MANET scenarios.

The penalty occurred when the technique is activated unnecessarily (i.e., when there are not enough hops in the path for TCP to be affected by interference from its own flow) is found to be small. However, in principle, the approach could be modified to use feedback from the routing layer and then to be activated only when necessary. In summary, SCA is an end-to-end modification, which may or may not require cross-layer feedback depending on whether a slight throughput penalty (<3%) is important for the sender or not.

Compared to adaptive CWL, the SCA method is not affected as severely by background traffic and performs much better in the presence of other interfering flows when the fixed RTO [14] technique is not used. However the issues of responsiveness and fairness were not evaluated in the study of Reference 39 and may yet prove significant since the bandwidth-delay product of MANETs is different to wired networks and not well understood [11]. More research is needed on this topic.

30.4.2 Point-to-Point Modifications

The point-to-point modifications of TCP involve nodes other than the two endpoints of the communication. Such modifications are harder to deploy as typically every node must implement any proposed changes, which can be difficult to guarantee in a MANET context. Partial implementation of a modification may not always be possible as compatibility issues with existing mechanisms may arise. According to the problem area targeted, the point-to-point modifications may be categorized in a similar fashion to the end-to-end ones; TCP-F, TCP-BuS, and proactive route selection deal with route breakages and their effect on TCP. ATCP is concerned with the high loss rate characteristic of the wireless channel. LRED, Neighborhood RED, COPAS, Split TCP, and the nonwork conserving scheduling algorithm, deal with the "capturing effect" and poor spatial reuse of TCP in 802.11 networks as described in Section 30.2.1.

30.4.2.1 TCP-Feedback

TCP-feedback (TCP-F) [9] is an additive modification to the TCP standard, which aims to mitigate the negative effects of frequent route breakages in MANETs. By producing feedback to the source from the point of route failure, TCP can distinguish between packet losses caused by changes in topology as nodes move about and those caused by congestion. The technique requires feedback from the routing algorithm concerning route failure and reestablishment and can only be deployed network-wide. Thus it is not a purely end-to end modification (unlike ELFN as mentioned in Section 30.4.1).

TCP-F works as follows. Whenever the routing algorithm detects disruption in a route, a route failure notification (RFN) packet is sent from the point of failure to the source, notifying the nodes in the path of the invalid route. If any of the intermediate nodes in the route are aware of an alternate route then that route is used and the RFN is discarded. However, if the source receives the RFN then it enters a "snooze state" where it stops transmission, freezes the TCP state variables and starts a time-out route failure timer. When a path is reestablished, the source receives a route reestablishment notification (RRN), unfreezes the TCP state variables and resumes transmission. The RRN can be produced by an intermediate node if it learns of a new route to the destination (like the local repair mechanism in AODV). Should the RRN be lost, the route failure timer ensures that the TCP agent does not remain in a snooze state indefinitely but falls back to the usual TCP behavior.

The evaluation of the proposed modifications in Reference 9 is carried out on a rough approximation of a multihop network where frequent topology changes and a long path (10 hops) is assumed. The positive results of the changes are two fold: first, TCP does not remain in a back off state for long periods of time, waiting for the RTO timer to expire before realizing that the routing protocol has reestablished a route; second, slow-start can be avoided since the state variables assume their prefrozen values as soon as the route is reestablished. However, there could be an adverse effect if the newly discovered route is congested, and TCP has incorrect information about the condition of the channel, since this might add further to the congestion. The authors do not view this as a problem, and rely on TCP's subsequent reaction to congestion to provide a fair division of the bandwidth as well as to provide protection against congestion collapse.

30.4.2.2 TCP-BuS

TCP with buffering capability and sequence information (TCP-BuS) [31] is another feedback-based approach to deal with the effects of route failures in MANETs, like TCP-F [9]. The proposed TCP modification is tightly coupled with the associativity-based routing (ABR) protocol whose principal operation, as a reactive routing protocol, closely resembles AODV.

Whenever a route alteration occurs, notification to the TCP agent is provided via "explicit route disconnection notification" (ERDN) and "explicit route successful notification" (ERSN) messages. Whenever there is a route failure, the source receives the ERDN from the point of failure, freezes transmission, stores the TCP state, including window sizes and timer values, and waits for an ERSN message. When the route is reestablished the source receives an ERSN message and resumes transmission. The key factor in this technique is that packets are buffered along the path. The TCP agent is aware, through the ERSN and ERDN messages, which packets are so buffered. Therefore, after the route has been reestablished, the TCP agent assumes that these packets are already on the path and will be delivered to the destination, and so does not retransmit them. Moreover, since TCP is aware of which packets are in the pipe and will be forwarded to the destination, it can update the time-out timers to take into account the extra amount of time necessary to reestablish the route.

A crucial requirement of the algorithm is the reliable transmission of the ERSN message and the authors propose two ways of achieving this goal. First, probing can be used as in the case of ATCP, which then triggers the transmission of an ERSN from the node at the point of route reestablishment. Second, passive acknowledgments can be used to ensure reliable point-to-point transport of the message. Specifically, after having transmitted the ERSN, the node then listens to the channel to see if the message is propagated further. If this does not happen for some period of time, then the message is retransmitted. This technique requires omnidirectional antennas at the transceivers. Overall, TCP-BuS proposes point-to-point changes, which depend on interlayer feedback.

The subsequent performance evaluation is performed on a custom-written discrete event simulator, which nonetheless portrays a multihop wireless network. No specific mobility model is taken into consideration but instead the authors opt to simulate the effect of route failures through some (unspecified) method and study their algorithm against TCP-F and "vanilla" TCP (Reno, presumably). They report an increase in throughput against TCP-F and "vanilla" TCP in the range of 10% and 30%, respectively. Essentially, TCP-BuS is an improvement over the TCP-F method, which relies heavily on routing feedback but its evaluation is incomplete as it has not been studied with respect to modern routing protocols or widely accepted mobility models. Finally, the requirement of buffering of packets at intermediate nodes during route breakages may strain the limited resources of some mobile devices.

30.4.2.3 Proactive Route Selection and Maintenance

Preemptive routing [19] is a cross-layer approach, which ensures that routing failures are dealt with in a proactive fashion for on-demand ad hoc routing algorithms. Specifically, when a path is likely to be broken a warning is issued to the source node indicating the likelihood of the breakage. Then the source may seek an alternate, more stable path, potentially avoiding the disconnection altogether. This can mitigate the negative effects of route failures on TCP, as discussed in Section 30.2.3.

Traditionally, routing protocols react to route breakages after they have occurred. In Reference 19, the goal is to detect, as reliably as possible, imminent route failure and discover an alternate path. In order to detect a potential path breakage before it happens, the authors in Reference 19 use signal strength indication of received packets at each node. Assuming that the signal strength measurements are a good indication of the distance of the receiver from the transmitter, if the measured amount declines below a certain threshold then it could be assumed that after a short time because of the destination moving away from the source, the route will be broken; hence a warning packet should be sent to the source to inform it to look for a different route.

In the actual implementation, when the received signal strength falls below the minimum threshold, the receiver starts pinging the sender to ascertain that the signal is in fact weakening and that the fluctuation is not just temporary. If the measurements of these minimum packet exchanges are, on average, below the minimum signal strength threshold a route warning is sent to the source, which then initiates the protocol's route-discovery procedure by flooding RREQs. Nodes only reply to those RREQs that are above the minimum signal strength threshold, thereby ensuring that the new path is stable. After the route discovery phase is over, the new route may be used instead of the old one.

The proposed changes are evaluated with both AODV and DSR routing. It is observed that TCP latency improves by up to 40% and throughput by up to 10% in single flows. Evaluation with multiple TCP flows is more difficult as the results are skewed by the TCP "capturing effect." As potential areas for future study, the authors mention studying preemptive routing for table-driven protocols, conducting experiments using a wider range of scenarios, and employing other measures of the quality of the path such as link age and path congestion.

30.4.2.4 Ad Hoc TCP

The ad hoc TCP (ATCP) [35] proposal entails the creation of a thin layer between TCP and IP, which utilizes feedback from the network layer. In particular, the BER and transient partitioning problems, as mentioned in Section 30.2.3, as well as reordering caused by multipath routing are dealt with this approach.

The ATCP relies on ICMP "destination unreachable" messages as well as "explicit congestion notification" (ECN) for its operation. The ECN notification consists of a bit set in the TCP packet that notifies the TCP agent of congestion along the path and is used by queueing management systems such as RED [16]. When ATCP is in its normal state it monitors the number of duplicate ACKs received for a segment. If a third duplicate ACK is received, ATCP intercepts the ACK and puts TCP into "persist" mode, which means that its state is frozen, while ATCP enters the "loss" state. ATCP reacts in the same fashion if an RTO event is about to occur. In the loss state, ATCP transmits unacknowledged segments from the TCP send buffer until an ACK arrives for a previously unacknowledged segment. At that time ATCP returns to its normal state. The loss state is designed to deal with losses caused by packet errors.

If an ACK has the ECN flag on, ATCP moves into its "congested" state until the TCP agent has sent a new segment, at which time ATCP returns to its normal state. During the congested state ATCP relies solely on TCP's CA, that is, it does nothing. The congested state is thus designed to deal with network congestion and, for this reason, the classic TCP behavior is left unaltered.

Finally, when an ICMP "destination unreachable" message is received, the TCP sender is placed in persist mode and sends probing packets to detect when the route might be reestablished. ATCP enters the "disconnected" state and returns to the "normal" state only when a duplicate ACK (or data segment) is received, indicating such reestablishment. ATCP takes care to reset TCP's cwnd to one segment when it enters the disconnected state, so that TCP can reestimate the available bandwidth of the new route. The

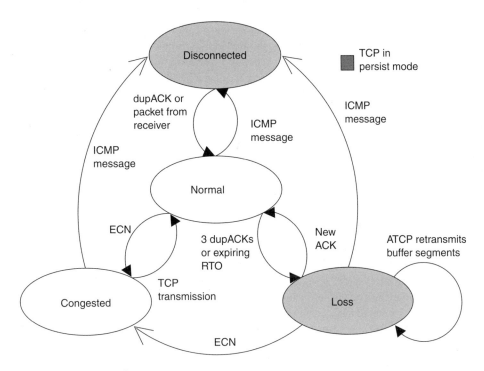

FIGURE 30.8   State diagram for ATCP.

proposed modifications are applied to all nodes, each of which is required to produce ICMP messages when route failures occur, utilizing cross-layer information.

A full state diagram of ATCP is shown in Figure 30.8. The authors in Reference 35, created an actual implementation of ATCP and studied it across a range of specialized scenarios reflecting high BER, reordering, and network congestion. The results show an increase in throughput by a factor of 2 to 3. However, the behavior of ATCP was not examined in a large MANET (only five nodes were employed in the actual testing) and no real routing algorithm was used. Furthermore, ATCP needs to maintain additional timers at the sender and this overhead is not taken into account in the study. Finally, its reliance on ECN producing routers or nodes is a requirement, which might be difficult to realize in the inherently, heterogeneous MANET environments, where some nodes may not be capable or aware of ECN.

30.4.2.5 LRED and Adaptive Pacing

The distributed link RED (LRED) [18] algorithm is a link layer modification to the 802.11 standard, which aims to reduce spatial contention and allow better utilization of the shared media. This helps TCP traffic as it is sensitive to packet drops due to spatial contention because it interprets such a loss as a sign of congestion. The method is inspired by the well-known RED [16] active queue management technique used on the Internet.

With these modifications, each node monitors the number of transmission retries at the link-layer and packets are discarded with a probability based on the average number of retries. If the average number of retries is smaller than a threshold then it is assumed that the node is rarely hidden and so no packets in the buffer are dropped. When it gets higher, a dropping probability is introduced that is constrained by a minimum and a maximum value. There is no differentiation between flows since all traffic transmitted by the node adds to spatial contention and is of interest. It is possible to use this technique with ECN marking of TCP packets so as to inform a sender that it should reduce its cwnd without dropping any packets. The technique is contained in the link layer but if the ECN capability is used it requires cross-layer modifications (as TCP packets are marked according to link layer feedback).

In the same work [18], the Adaptive Pacing modification is introduced, which works with the 802.11 MAC protocol to help coordinate transmissions so that spatial contention is dealt with. The only change in the protocol is allowing the node, when necessary, to back off for an interval time, which is an additional packet transmission time above the normal deferral period (a random back off plus one packet transmission time). To identify as to when such an additional time period is needed, the technique is coupled with LRED monitoring. When the average number of retries exceeds the minimum threshold as set in LRED, adaptive pacing is activated. The proposed modifications are confined to the MAC layer.

Both techniques are evaluated in string, cross, and grid topologies but not in mobile scenarios. The observed TCP throughput increase by using both techniques is up to 20% and there is a decidedly positive effect on fairness. Further evaluation in realistic topologies would be useful, as the simulation experiments in Reference 18 were set up in a way as to make the effects of spatial contention prominent and this might not be the case in realistic mobile situations. Finally, since the modifications involve changes in the MAC layer, issues of incremental deployment and interoperability might arise in real implementations when nodes do not operate with identical MAC protocols (e.g., LRED-enhanced and "plain" 802.11).

30.4.2.6 Neighborhood RED

Neighborhood RED (NRED) [48] is another cross-layer approach, which aims to increase TCP fairness in MANETs. Although feedback from the link layer is utilized, no modifications are assumed at the MAC layer, which makes the method easily deployable.

The idea is based on RED queueing [16] as used in wired networks. However, instead of monitoring its local queue, each node maintains the concept of a distributed neighborhood queue. Through experiments it is shown that such a queue concept is necessary because contention in MANETs occurs in a region rather than a single node or router. The neighborhood of a node is defined to be the transmitting node of all other nodes which may interfere with that node's transmissions. For the purposes of the study only neighbors up to 2 hops distant are considered even though the authors accept that nodes farther away may interfere with the source's transmission because of the great discrepancy that occurs sometimes between interference and transmission ranges.

Obtaining information for neighbors up to 2 hops away that might affect transmission requires a lot of signaling overhead as information must be exchanged with each node via its immediate neighbors. Therefore, a simplified expression of the distributed neighborhood queue is adopted. To monitor other nodes' queue sizes passive measurements are used. Congested nodes issue a "neighborhood congestion notification" packet, which aids neighbors to estimate their drop probability. A RED-like algorithm is used to estimate which packets should be discarded to aid fairness between flows in the neighborhood. The proposed solution is a cross-layer one and requires changes at every node, even though it can be deployed in a mixed environment.

This technique is validated over static topologies and evaluated in static and mobile scenarios. It is shown that although throughput might suffer when using NRED, fairness between flows improves radically and there is no starvation of connections. However, a thorough evaluation of this technique is still needed to ascertain its effectiveness under various mobility conditions. Finally, NRED's usefulness in a gradually deployment scenario is a prospect for future study.

30.4.2.7 Nonwork Conserving Scheduling Algorithm

The nonwork-conserving scheduling algorithm proposed in Reference 51 is a replacement for the first-in first-out (FIFO) work-conserving scheduling scheme use in multihop ad hoc networks. The technique aims to improve fairness for TCP traffic and avoid the "capturing" effect, where a TCP flow may starve competing flows even though the aggregate throughput could be considered sufficient.

This method introduces a timer at the MAC level, which must expire before a packet is released for transmission. Effectively, there is a delay d before a packet is sent for transmission. This delay is calculated as the sum of three parts:

$$delay = D_{tr-delay} + D_{delay} + D_{random}$$

$D_{tr-delay}$ is an estimate of the time it would take to transmit the packet if no contention were present and is calculated as the packet length divided by the bandwidth of the channel. D_{random} is a random value uniformly distributed between 0 and D_{delay} and which is used to avoid synchronization effects. Finally, D_{delay} is a delay determined by the recent queue output rate, that is, by counting the number of bytes the queue outputs in every fixed interval. In short, as the queue output in the fixed interval increases, D_{delay} increases. Routing packets are treated as high priority as in typical work-conserving scheduling schemes and are transmitted without waiting for the delay value.

Various simulation scenarios that highlight the unfairness of the standard work-conserving scheduling scheme in MANETs demonstrate the effectiveness of the new modifications. In particular, the approach clearly increases fairness in the scenarios considered in Reference 51 but also leads to a slight decrease in aggregate throughput. The scheme is completely contained in the MAC layer and does not rely on cross-layer feedback. However, it requires changes to the firmware of the transceiver as the delay is so fine grained that it must be hardware-assisted. As a prospect for future study, it would be useful to determine the technique's effectiveness in realistic mobile scenarios. Moreover, the effects of incremental deployment would have to be studied: solutions that require changes in the firmware of the transceiver might not be applied to all hosts in a MANET and hence a mixed firmware/MAC access environment is likely.

30.4.2.8 Contention-Based Path Selection

The contention-based path selection (COPAS) [12] is a routing algorithm modification proposed to alleviate the TCP "capture" effect and help lift spatial contention from paths. To do this it takes into account how "busy" the medium is around each node and uses different paths for forward and reverse traffic. The technique can be applied to any of the reactive routing protocols that use the standard route request (RREQ), route reply (RREP) cycles but in Reference 12 it is evaluated over the DSR 27 protocol.

In short, during the RREQ cycle of the path discovery phase, the destination waits for some time to collect as many paths as possible. It then chooses two, using path disjointness and minimal route congestion as selection criteria, and responds with two RREP messages. Congestion is measured along the route by having each node monitor its MAC layer behavior, in particular a weighted average of its back offs due to activity in the medium. A higher average implies a busier medium.

For instance, in Figure 30.9 there are three equally disjoint routes namely {Source → D → E → F → Destination}, {Source → A → B → C → Destination}, and {Source → G → H → I → J → Destination}. After the criterion of congestion has been evaluated, the latter two are chosen as the DATA and ACK paths, respectively.

Using separate paths for ACKs and DATA helps alleviate spatial contention as the two do not compete over the same path for usage of the medium. Since congestion is a dynamic property, COPAS maintains snapshots of the path contention by piggybacking information on packets in the forward and reverse paths. If the route experiences contention that exceeds a certain threshold, a new, less contented route is sought out. COPAS requires alterations at the routing layer and needs cross-layer feedback to maintain the contention information at the nodes.

In Reference 12 the proposed modifications are evaluated over static topologies using DSR. The throughput improvement of TCP Reno reaches 90% but there is a slight delay penalty for end-to-end delivery as COPAS tends to "go around" the shortest route. Future research into COPAS will need to evaluate it under realistic conditions of mobility. It will also have to examine whether the method can yield cumulative performance improvement in TCP when coupled with orthogonal techniques such as ELFN or Fixed-RTO (both are described in Section 30.4.1).

30.4.2.9 Split TCP

Split TCP [33] is a modification which aims to alleviate various TCP problems that arise with connections along long paths. In particular, since it is more likely for long paths to experience link failures (due to mobility), short connections are advantaged over long ones. Furthermore, flows along long path suffer from the effects of hidden terminal interference and exhibit unfairness when mixed with shorter flows. The split TCP scheme separates a long connection into several shorter localized segments through the use

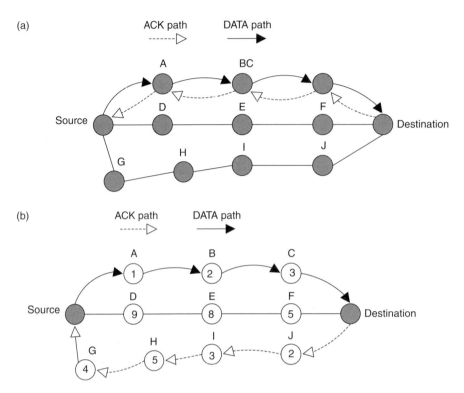

FIGURE 30.9 Example of COPAS routing path selection: (a) DSR routing uses the same path for ACKs and DATA; (b) COPAS uses disjoint paths and takes into account network contention.

of proxies as interfacing agents at each. The placement of the proxies is fixed by the interproxy distance parameter and determines which nodes will be acting as proxies.

Each proxy intercepts TCP packets, buffers them, and acknowledges their receipt to the source (or previous proxy) by using local acknowledgments (LACKs). Whenever a LACK is received, the transmitting proxy purges the packet from its buffer as the reliable delivery of the packet to the next segment is confirmed. Note that the cumulative ACK operation of TCP is still used to maintain end-to-end reliability. Proxy nodes are also responsible for the flow and congestion control between the short segments and obey the usual TCP rules. In particular, the source node maintains a congestion and an end-to-end window. The cwnd changes according to the LACKs received by the next segment, while the end-to-end cwnd is governed by the rate of arrival of the end-to-end ACKs from the destination. The cwnd is always a subwindow of the end-to-end window and the authors suggest limiting end-to-end ACKs to approximately one for every 100 packets as the possibility of nondelivery from proxies is expected to be small. An example of split TCP is shown in Figure 30.10. Split TCP proposes changes that do not require cross-layer information but have to be implemented at every node in the network.

Simulation experiments show that an interproxy distance between 3 and 5 hops has a decidedly positive effect on throughput (5 to 30%) and fairness. However, large buffer space may be required for the proxies to work effectively and additional computation effort is needed at each chosen interfacing node. This could have detrimental effect on the battery reserves of these devices.

30.5 Conclusions

This chapter has presented the latest developments on addressing TCP issues over MANETs. Since TCP was originally designed for wired LANs, several studies have shown that it exhibits suboptimal performance

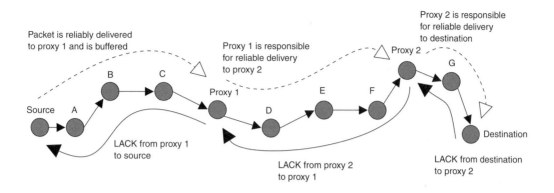

FIGURE 30.10 Split TCP over an 11 hop string topology with an interproxy distance of 3.

in MANET environments. The main reason for this is the conventional TCP assumption that all perceived packet losses are due to congestion. TCP's reaction to such events is a rapid reduction of the sending rate. However, in MANET environments, packet losses may be caused by other factors but TCP, being unable to distinguish among them, invariably reacts by simply slowing down transmission, which is not necessarily the right course of action. For instance, in the case of packet drops due to errors, the correct action would be to retransmit quickly the corrupt segment and continue as normal. Further, the main symptom watched for by TCP to identify packet loss, namely, the reception of three duplicate ACKs, is not always a valid heuristic in MANETs where frequent link failures and multipath routing make duplicate ACKs as part of the networks normal rather than pathological behavior.

In order to address these issues, several modifications have been proposed in the literature. These can be categorized into solutions deployable in an end-to-end fashion, affecting only the TCP sender and receiver, and those deployed point-to-point, which involve alterations at intermediate nodes as well as, possibly, the end points. Modifications may also be categorized as single layer or cross-layer depending on whether they are confined to a single or cross-layer of the network stack. Additionally, point-to-point solutions may either utilize explicit notification to the end points (as in TCP-ELFN) or gather implicit information without using any such signalling (e.g., neighborhood RED). In principle, an end-to-end transport layer solution is preferable as modifications are confined to end points, but it proves quite challenging to achieve this in practice since available information is limited at the transport layer. Hence, cross-layered, point-to-point modifications although more difficult to deploy, tend to yield greater improvements than end-to-end layered ones.

In conclusion, although TCP behavior over MANETs has received much attention from the research community in recent times, there is still a significant scope for further work, since the ultimate goal is convergence of wired and wireless networks. It is crucial in heterogeneous environments, such as MANETs, to examine the viability of incremental deployment for each of the proposed solutions. In particular, the interaction of modified with traditional TCP agents needs be taken into account to ensure that they can interoperate properly and that the unmodified agents are not disadvantaged. Moreover, since several of the proposed mechanisms are orthogonal to one another, it is interesting to examine whether the combination of different solutions results in aggregated improvements in performance. Finally, it has been suggested that TCP may not be the best protocol to use in MANETs since its link-monitoring mechanisms (through round-trip time measurements) while quite well suited for wired environments are not so suitable for multihop wireless conditions. As such, a viable option might be the introduction of new connection oriented, transport protocols, which provide more feedback about wireless conditions and include provisions for identification of the cause of packet losses.

The need for real testbed measurements is evident in much of the existing literature. Most TCP proposals in MANETs are evaluated through simulations or limited testbeds with few participating nodes. The advantages of real-life implementation and evaluation are twofold: they increase confidence in the simulators used, possibly revealing any flaws, which might exist in the tools; and they help validate previous work. Unfortunately, wide-scale examination of existing solutions in realistic MANET environments is still lacking. Finally, as the various factors affecting TCP behavior in MANETs become better understood it is hoped that mathematical models will be developed to reflect TCP behavior under various mobility and wireless channel conditions.

References

[1] A. Ahuja, S. Agarwal, J.P. Singh, and R. Shorey. Performance of TCP over different routing protocols in mobile ad-hoc networks. In *Proceedings of the IEEE Vehicular Technology Conference (VTC 2000)*, pp. 2315–2319, May 2000.

[2] Ian F. Akyildiz, O.B. Akan, Chao Chen, Jian Fang, and Weilian Su. InterPlaNetary Internet: state-of-the-art and research challenges. *Computer Networks*, 43: 75–112, 2003.

[3] M. Allman, V. Paxson, and W. Stevens. *TCP Congestion Control*. Internet Draft, http://www.ietf.org/rfc/rfc2581.txt, April 1999. Request For Comments.

[4] Eitan Altman and Tania Jimnez. Novel delayed ACK techniques for improving TCP performance in multihop wireless networks. In *Proceedings of the Personal Wireless Communications*, vol. 2775, pp. 237–250. Springer-Verlag, Heidelberg, September 2003.

[5] Giuseppe Anastasi, Eleonora Borgia, Marco Conti, and Enrico Grego. IEEE 802.11 ad hoc networks: performance measurements. In *Proceedings of the 23rd International Conference on Distributed Computing Systems Workshops (ICDCSW'03)*, pp. 758–763, May 2003.

[6] Hari Balakrishnan, Venkata N. Padmanabhan, Srinivasan Seshan, and Randy H. Katz. A comparison of mechanisms for improving TCP performance over wireless links. *IEEE/ACM Transactions on Networking*, 5 : 756–769, 1997.

[7] S. Biaz and N. Vaidya. Discriminating congestion losses from wireless losses using interarrival times at the receiver. In *Proceedings of the IEEE Symposium on Application-Specific Systems and Software Engineering and Technology (ASSET'99)*, pp. 10–17, March 1999.

[8] Lawrence S. Brakmo, Sean W. O'Malley, and Larry L. Peterson. TCP vegas: new techniques for congestion detection and avoidance. In *Proceedings of the Conference on Communications Architectures, Protocols and Applications*, pp. 24–35. ACM Press, New York, 1994.

[9] Kartik Chandran, Sudarshan Raghunathan, S. Venkatesan, and Ravi Prakash. A feedback based scheme for improving TCP performance in ad-hoc wireless networks. In *Proceedings of the 18th Annual International Conference on Distributed Computing Systems*, pp. 472–479, May 1998.

[10] Kai Chen, Yuan Xue, and Klara Nahrstedt. On setting TCP's congestion window limit in mobile ad hoc networks. In *Proceedings of the 38th Annual IEEE International Conference on Communications (ICC 2003)*, pp. 1080–1084, May 2003.

[11] Kai Chen, Yuan Xue, Samarth H. Shah, and Klara Nahrstedt. Understanding bandwidth-delay product in mobile ad hoc networks. *Computer Communications*, 27: 923–934, 2004.

[12] C. Cordeiro, S.R. Das, and D.P. Agrawal. COPAS: dynamic contention-balancing to enhance the performance of TCP over multi-hop wireless networks. In *Proceedings of the 10th International Conference on Computer Communication and Networks (IC3N)*, pp. 382–387, October 2002.

[13] R. Dube, C.D. Rais, Kuang-Yeh Wang, and S.K. Tripathi. Signal stability-based adaptive routing (SSA) for ad hoc mobile networks. In *Proceedings of the IEEE Personal Communications*, vol. 4, pp. 36–45, February 1997.

[14] Thomas D. Dyer and Rajendra V. Boppana. A comparison of TCP performance over three routing protocols for mobile ad hoc networks. In *Proceedings of the 2001 ACM International Symposium on Mobile ad hoc networking and computing*, pp. 56–66. ACM Press, New York, 2001.

[15] Sally Floyd and Van Jacobson. Traffic phase effects in packet-switched gateways. *SIGCOMM Computer Communications Review*, 21: 26–42, 1991.

[16] Sally Floyd and Van Jacobson. Random early detection gateways for congestion avoidance. *IEEE/ACM Transactions on Networking*, 1: 397–413, 1993.

[17] Cheng Peng Fu and Soung C. Liew. ATCP: TCP for mobile ad hoc Networks. *IEEE Journal on Selected Areas in Communications*, 21: 216–228, 2003.

[18] Zhenghua Fu, Petros Zerfos, Haiyun Luo, Songwu Lu, Lixia Zhang, and Mario Gerla. The impact of multihop wireless channel on TCP throughput and loss. In *Proceedings of the 22nd Annual Joint Conference of the IEEE Computer and Communications Societies (INFOCOM 2003)*, vol. 3, pp. 1744–1753, March 2003.

[19] Tom Goff, Nael Abu-Ghazaleh, Dhananjay Phatak, and Ridvan Kahvecioglu. Preemptive routing in ad hoc networks. *Journal of Parallel and Distributed Computing*, 63: 123–140, 2003.

[20] The Bluetooth Special Interest Group. http://www.bluetooth.com.

[21] Pawan Kumar Gupta and Joy Kuri. Reliable ELN to enhance throughput of TCP over wireless links via TCP header checksum. In *Proceedings of the Global Telecommunications Conference, 2002. (GLOBECOM '02)*, vol. 1, pp. 1996–2000, November 2002.

[22] Gavin Holland and Nitin Vaidya. Analysis of TCP performance over mobile ad hoc networks. In *Proceedings of the 5th Annual ACM/IEEE International Conference on Mobile Computing and Networking*, pp. 219–230. ACM Press, New York, 1999.

[23] IEEE Standards Association. IEEE P802.11. *The Working Group for Wireless LANs*. http://grouper.ieee.org/groups/802/11/index.html.

[24] University of South California Information Sciences Institute. *Transmission Control Protocol*. Internet Draft, http://www.ietf.org/rfc/rfc793.txt, September 1981. Request for Comments.

[25] V. Jacobson, R. Braden, and D. Borman. *TCP Extensions for High Performance*. Internet Draft, http://www.ietf.org/rfc/rfc1323.txt, May 1992. Request for Comments.

[26] Per Johansson, Tony Larsson, Nicklas Hedman, Bartosz Mielczarek, and Mikael Degermark. Scenario-based performance analysis of routing protocols for mobile ad-hoc networks. In *Proceedings of the 5th Annual ACM/IEEE International Conference on Mobile Computing and Networking*, pp. 195–206. ACM Press, New York, 1999.

[27] David B. Johnson, David A. Maltz, and Yih-Chun Hu. *The Dynamic Source Routing Protocol for Mobile Ad Hoc Networks (DSR)*. Internet Draft, draft-ietf-manet-dsr-10.txt, July 2004. Work in Progress.

[28] Christine E. Jones, Krishna M. Sivalingam, Prathima Agrawal, and Jyh Cheng Chen. A survey of energy efficient network protocols for wireless networks. *Wireless Networks*, 7: 343–358, 2001.

[29] Eun-Sun Jung and Nitin H. Vaidya. A power control MAC protocol for ad hoc networks. In *Proceedings of the 8th Annual International Conference on Mobile Computing and Networking*, pp. 36–47. ACM Press, New York, 2002.

[30] Ad Kamerman and Leo Monteban. WaveLAN II: a high-performance wireless LAN for unlicensed band. *Bell Labs Technical Journal*, 2: 118–133, 1997.

[31] Dongkyun Kim, C.-K. Toh, and Yanghee Choi. TCP-BuS: improving TCP performance in wireless ad hoc networks. *Journal of Communications and Networks*, 3: 175–186, 2001.

[32] R.V. Boppana and S.P. Konduru. An adaptive distance vector routing algorithm for mobile, ad hoc networks. In *Proceedings of the 20th Annual Joint Conference of the IEEE Computer and Communications Societies (INFOCOM 2001)*, vol. 3, pp. 1753–1762, April 2001.

[33] S. Kopparty, S.V. Krishnamurthy, M. Faloutsos, and S.K. Tripathi. Split TCP for mobile ad hoc networks. In *Proceedings of the Global Telecommunications Conference, 2002 (GLOBECOM '02)*, vol. 1, pp. 138–142. IEEE Computer Society Press, Washington, November 2002.

[34] Qun Li, Javed Aslam, and Daniela Rus. Online power-aware routing in wireless ad-hoc networks. In *Proceedings of the 7th Annual International Conference on Mobile Computing and Networking*, pp. 97–107. ACM Press, New York, 2001.

[35] Jian Liu and Suresh Singh. ATCP: TCP for mobile ad hoc networks. *IEEE Journal on Selected Areas in Communications*, 19: 1300–1315, 2001.

[36] Reiner Ludwig, Almudena Konrad, and Anthony D. Joseph. Optimizing the end-to-end performance of reliable flows over wireless links. In *Proceedings of the 5th Annual ACM/IEEE International Conference on Mobile Computing and Networking*, pp. 113–119. ACM Press, New York, 1999.

[37] Tadeusz Wysocki Mehran Abolhasan and Eryk Dutkiewicz. A review of routing protocols for mobile ad hoc networks. *Ad Hoc Networks*, 2: 1–22, 2004.

[38] Stylianos Papanastasiou, Lewis M. MacKenzie, and Mohamed Ould-Khaoua. Reducing the degrading effect of hidden terminal interference in MANETs. In *Proceedings of the 7th ACM International Symposium On Modeling, Analysis and Simulation of Wireless and Mobile Systems*, pp. 311–314. ACM Press, New York, 2004.

[39] Stylianos Papanastasiou and Mohamed Ould-Khaoua. TCP Congestion window evolution and spatial reuse in MANETs. *Journal of Wireless Communications and Mobile Computing*, 4: 669–682, 2004.

[40] V. Paxson and M. Allman. *Computing TCP's Retransmission Timer*. Internet Draft, http://www.ietf.org/rfc/rfc2988.txt, November 2000. Request for Comments.

[41] Charles E. Perkins, Elizabeth M. Belding-Royer, and Samir R. Das. *Ad Hoc On-Demand Distance Vector (AODV) Routing*. http://www.ietf.org/rfc/rfc3561.txt, July 2003, Request for Comments. Experimental RFC.

[42] Charles E. Perkins and Pravin Bhagwat. Highly dynamic destination-sequenced distance-vector routing (DSDV) for mobile computers. In *Proceedings of the Conference on Communications Architectures, Protocols and Applications*, pp. 234–244. ACM Press, New York, 1994.

[43] K.G. Seah and A.L. Ananda. TCP HACK: a mechanism to improve performance over lossy links. *Computer Networks*, 39: 347–361, 2002.

[44] W. Richard Stevens. *TCP/IP Illustrated*, vol. 1. Addison-Wesley, Reading, MA, 1994.

[45] Feng Wang and Yongguang Zhang. Improving TCP performance over mobile ad-hoc networks with out-of-order detection and response. In *Proceedings of the 3rd ACM International Symposium on Mobile Ad Hoc Networking and Computing*, pp. 217–225. ACM Press, New York, 2002.

[46] Kaixin Xu, Sang Bae, Sungwook Lee, and Mario Gerla. TCP behavior across multihop wireless networks and the wired internet. In *Proceedings of the 5th ACM International Workshop on Wireless Mobile Multimedia*, pp. 41–48. ACM Press, New York, 2002.

[47] Kaixin Xu, M. Gerla, and Sang Bae. How effective is the IEEE 802.11 RTS/CTS handshake in ad hoc networks? In *Global Telecommunications Conference, 2002 (GLOBECOM '02)*, vol. 1, pp. 72–76, November 2002.

[48] Kaixin Xu, Mario Gerla, Lantao Qi, and Yantai Shu. Enhancing TCP fairness in ad hoc wireless networks using neighborhood RED. In *Proceedings of the 9th Annual International Conference on Mobile Computing and Networking*, pp. 16–28. ACM Press, New York, 2003.

[49] Shugong Xu and Tarek Saadawi. Performance evaluation of TCP algorithms in multi-hop wireless packet networks. *Wireless Communications and Mobile Computing*, 2 :85–100, 2002.

[50] Shugong Xu and Tarek Saadawi. Revealing the problems with 802.11 medium access control protocol in multi-hop wireless ad hoc networks. *Computer Networks*, 38: 531–548, 2002.

[51] Luqing Yang, Winston K.G. Seah, and Qinghe Yin. Improving fairness among TCP flows crossing wireless ad hoc and wired networks. In *Proceedings of the 4th ACM International Symposium on Mobile Ad Hoc Networking and Computing*, pp. 57–63. ACM Press, New York, 2003.

VII

Algorithms and Protocols for Bluetooth Wireless PAN

31

Intra-Piconet Polling Algorithms in Bluetooth

31.1 Introduction: On Bluetooth Networks and
     Bluetooth Communications **31**-726
31.2 Intra-Piconet Polling Schemes **31**-727
     Traditional Polling Schemes • Dynamic Reordering of Slaves
     • Adaptive Bandwidth Allocation
31.3 BNEP and Segmentation/Reassembly Policies.........**31**-730
31.4 Open Problems and Directions for
     Further Research **31**-731
References **31**-732

Jelena Mišić
Vojislav B. Mišić

Bluetooth is an emerging standard for wireless personal area networks (WPANs): short range, ad hoc wireless networks [1]. Originally, Bluetooth was envisaged as a wireless cable replacement technique, which is why the basic RF range of Bluetooth devices is only about 10 m [2]. Over time, the number of possible uses of Bluetooth have increased to include different networking tasks between computers and computer-controlled devices such as PDAs, mobile phones, smart peripherals, and others. Consequently, performance analysis of Bluetooth networks is an important research topic. Yet Bluetooth networks operate in a rather different manner from other wireless networks and approaches applicable to those networks cannot readily be applied in the Bluetooth environment. Furthermore, many important aspects of communication using Bluetooth networks are not defined by the current Bluetooth specification [2], including the manner in which the master polls its slaves, the scheduling algorithms for Bluetooth scatternets, the preferred scatternet topology (or topologies), support for quality of service, and others. All of these aspects play a crucial role in determining the performance of Bluetooth networks, which is ultimately one of the main criteria for their wider acceptance in the marketplace. In this chapter, we will review and roughly classify existing algorithms for intra-piconet polling. We will also describe some of the topics that deserve future research attention.

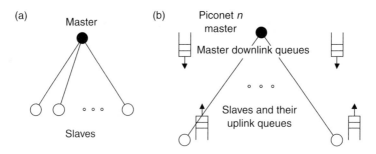

FIGURE 31.1   Bluetooth piconet topology: (a) topology; (b) queueing model of a single piconet.

31.1   Introduction: On Bluetooth Networks and Bluetooth Communications

As mentioned earlier, Bluetooth is a communication technology for short range, ad hoc wireless networks formed by fixed or mobile devices. Bluetooth devices must form networks before the actual communication can start [3]. The simplest network is called piconet: a small, centralized network with up to eight active nodes or devices. One of the nodes is designated as the master, while the others are slaves. At most seven slaves can be active at any given time, and up to 255 others can be *parked* but still listening to the communications in the piconet. This topology is shown schematically in Figure 31.1(a).

Bluetooth uses a set of RF frequencies (79 or 23, in some countries) in the ISM band at about 2.4 GHz. Frequency hopping spread spectrum (FHSS) technique is utilized in order to combat interference. Each piconet hops through the available RF frequencies in a pseudo-random manner. The hopping sequence, which is determined from the Bluetooth device address of the piconet master, is known as the channel [3]. Each channel is divided into time slots of $T = 625\ \mu\text{sec}$, which are synchronized to the clock of the piconet master. In each time slot, a different frequency is used.

All communications in the piconet take place under the control of the piconet master. All slaves listen to downlink transmissions from the master. The slave may reply with an uplink transmission if and only if addressed explicitly by the master, and only immediately after being addressed by the master. Data is transmitted in packets, which take 1, 3, or 5 slots; link management packets also take one slot each. The RF frequency does not change during the transmission of the packet. However, once the packet is sent, the transmission in the next time slot uses the next frequency from the original hopping sequence (i.e., the two or four frequencies from the original sequence are simply skipped). By default, all master transmissions start in even-numbered slots, whilst all slave transmissions start in odd-numbered slots. A downlink packet and the subsequent uplink packet are commonly referred to as a frame. Therefore, the master and the addressed slave use the same communication channel, albeit not at the same time. This communication mechanism, known as time division duplex (TDD) is schematically shown in Figure 31.2.

Because of the TDD communication mechanism, all communications in the piconet must be routed through the master. Each slave will maintain (operate) a queue where the packets to be sent out are stored. The master, on the other hand, operates several such queues, one for each active slave in the piconet. The corresponding queueing model is shown in Figure 31.1(b). We note that these queues may not physically exist, for example, all downlink packets might be stored in a single queue; but the queueing model provides a convenient modeling framework, which facilitates the performance analysis of Bluetooth networks.

The master polls the slave by sending the data packet from the head of the corresponding downlink queue. The slave responds by sending the data packet from the head of its uplink queue. When there is no data packet to be sent, single-slot packets with zero payload are sent — POLL packets in the downlink, and NULL packets in the uplink direction [3]. As the process of polling the slaves is actually embedded in the data transmission mechanism, we will use the term "polling" for every downlink transmission from the master to a slave.

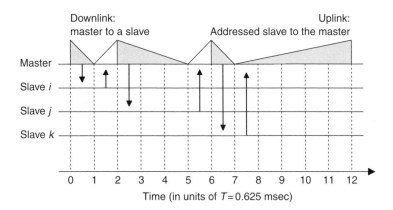

FIGURE 31.2   TDD master–slave communication in Bluetooth. Gray triangles denote data packets, white triangles denote empty (POLL and NULL) packets.

Since packets must wait at the slave and at the master before they can be delivered to their destinations, the delays they experience are mainly queueing delays. Therefore, the performance of the data traffic will be mostly dependent on the choice of the polling scheme used by the master to poll the active slaves in the piconet.

31.2   Intra-Piconet Polling Schemes

The polling scheme is obviously the main determinant of performance of Bluetooth piconets, and one of the main determinants of performance of Bluetooth scatternets. As usual, the main performance indicator is the end-to-end packet delay, with lower delays being considered as better performance. There are, however, at least two other requirements to satisfy. First, the piconet master should try to maintain fairness among the slaves, so that all slaves in the piconet receive equal attention in some shorter or longer time frame. (Of course, their traffic load should be taken into account.) Second, Bluetooth devices are, by default, low power devices, and the polling scheme should be sufficiently simple in terms of computational and memory requirements.

As noted earlier, the current Bluetooth specification does not specifically require or prescribe any specific polling scheme [2]. This may not seem to be too big a problem, since optimal polling schemes for a number of similar single-server, multiple-input queueing system are well known [4,5]. However, the communication mechanisms used in Bluetooth are rather specific, because:

- All communications are bidirectional (i.e., there cannot exist a downlink packet without an uplink packet, or vice versa).
- The master polls the slaves using regular packets, possibly without data payload (i.e., all polls and responses thereto take at least one slot each).
- All slave–slave communications have to be routed through the master (i.e., there can be no direct slave-to-slave communication).
- The master does not know the status of queues at the slaves, because there are no provisions for exchange of such information in the Bluetooth packet structure.

As the consequence, the existing results cannot be applied, and the performance of different polling schemes has to be re-assessed, taking the aforementioned characteristics of the Bluetooth communication mechanisms. It should come as no surprise, then, that a number of polling schemes have been proposed and analyzed [6–9]. Many of the proposed schemes are simply variations of the well-known limited and exhaustive service scheduling [10], but several improved adaptive schemes have been described as well [8,11].

In the discussion that follows, we will present a rough classification of those polling schemes, using the following criteria. First, the polling scheme determines the number of frames exchanged during a single visit to the slave. This number may be set beforehand to a fixed value, or it may be dynamically adjusted on the basis of current and historical traffic information.

Second, different slaves may receive different portions of the bandwidth; again, the allocation may be done beforehand, or it may be dynamically adapted to varying traffic conditions. The latter approach is probably preferable in Bluetooth piconets, which are ad hoc networks formed by mobile users, and the traffic may exhibit considerable variability. In fact, due to users' mobility, even the topology of the piconet may change on short notice. At the same time, the fairness of polling may be more difficult to maintain under dynamic bandwidth allocation.

Finally, the sequence in which slaves are visited may be set beforehand, or it may change from one piconet cycle to another, depending on the traffic information. Slaves that had more traffic in the previous cycle(s) may receive a larger portion of the available bandwidth. Slave that had no traffic may receive less bandwidth, or they may even be ignored for one or more piconet cycles. Again, the main difficulty with such schemes is to ensure that the fairness is maintained.

31.2.1 Traditional Polling Schemes

The simplest polling schemes use a fixed ordering of the slaves and fixed bandwidth allocation per slave. The only variable parameter, then, is the duration of master's visit to each slave.

Under 1-*limited service* polling, the master visits each slave for exactly one frame, and then moves on to the next slave [10]. Data packets are sent if there are any, otherwise empty packets (POLL or NULL) are sent. The scheme is sometimes referred to as (Pure) Round Robin [6] or simply limited service.

Under *exhaustive service* polling, the master stays with the slave as long as there are packets to exchange in either downlink or uplink direction [10]. The absence of packets is detected by a POLL–NULL frame.

Under the *E-limited service* polling, the master stays with a slave until there are no more packets to exchange, or for a fixed number M of frames ($M > 1$), whichever comes first [10]. Packets that arrive during the visit are allowed to enter the uplink queue at the slave and may be serviced — provided the limit of M frames is not exceeded [10]. This scheme is also referred to as Limited Round Robin [6,11].

In fact, 1-limited and exhaustive service polling may be considered as special cases of E-limited service, where the limit M equals 1 and ∞, respectively. In all three cases, the sequence of slaves is fixed and does not change.

In traditional polling systems, exhaustive service performs better than either 1-limited or E-limited service [4]. As Bluetooth piconets are not traditional polling systems (for reasons outlined earlier), hence this result does not hold. Several authors have found that 1-limited performs better than exhaustive service under high load [6,12]. Furthermore, E-limited service has been found to offer better performance than either limited or exhaustive service, and the value of M may be chosen to achieve minimum delays for given traffic burstiness [13].

31.2.2 Dynamic Reordering of Slaves

In fact, even better results may be obtained through the so-called stochastically largest queue (SLQ) policy. Under this policy, the server always services the queue with the highest number of packets [5]. In other words, the master should always poll the slave for which the sum of lengths of uplink and downlink queues is the highest. However, this is not possible in Bluetooth, as there is no way for the master to know the current status of *all* of its slaves' uplink queues. What is possible, though, is to dynamically reorder the slave polling sequence according to the length of the corresponding downlink queues at the master. This reordering may be done once in each piconet cycle, or the master may always poll the slave with the longest downlink queue. In either case, polling may be performed according to limited, exhaustive, or even E-limited scheme.

TABLE 31.1   An Overview of Traditional Polling Algorithms

Polling algorithm	Performance aspect			
	Access delay	End-to-end delay	Adaptivity	Fairness
1-limited	Average	Good (high loads)	None	Inherent
Exhaustive	Best	Good (low and medium loads)	None	Can be unfair
E-limited	Good	Best	Average	Inherent (longer time scale)
EPM [6]	Average	Good (high loads)	Downlink	Can be unfair

One scheme that uses dynamic reordering is the *exhaustive pseudo-cyclic master* (EPM) queue length scheme, proposed in Reference 6, where each slave is visited exactly once per cycle. At the beginning of each cycle, however, the slaves are reordered according to the decreasing length of downlink queues.

One problem with dynamic reordering is that fairness among the slaves cannot be guaranteed when if reordering is done for every poll. For example, two slaves that talk to each other might easily monopolize the network and starve all other slaves. Fairness is easier to achieve when reordering is done on a per cycle basis. Still, some polling schemes (such as the exhaustive scheme) present more of a challenge than others.

Table 31.1 summarizes the characteristics of traditional polling schemes.

31.2.3   Adaptive Bandwidth Allocation

The duration of the visit to each slave may be adjusted according to the current or historical traffic information. This can be done in two ways: by rewarding slaves that have more traffic, or by penalizing slaves that have less traffic. In the former case, the master is allowed to stay longer, and thus exchange more frames, with the slave that has some data to send and receive. In the latter case, the slave that had less traffic or no traffic at all, or the slave which is expected to have no traffic, will simply be ignored for a certain number of piconet cycles.

The former approach is exploited in the *limited and weighted round robin* (LWRR) [6], which tries to increase efficiency by reducing the rate of visits to inactive slaves. Initially, each slave is assigned a weight equal to the so-called maximum priority, or MP. Each slave is polled in E-limited fashion with up to M frames. Whenever there is a data exchange between the slave and the master, the weight of the slave is increased to the value of MP. On the other hand, when a POLL–NULL sequence occurs, the weight for that particular slave is reduced by one. If the slave weight drops to one (which is the lowest value), the slave has to wait a maximum of MP − 1 cycles to be polled again.

A variation of this scheme, labeled *pseudo-random cyclic limited slot-weighted round robin* [11], uses both slave reordering and poll rate reduction. The sequence in which slaves will be polled is determined in a pseudo-random fashion at the beginning of every cycle, and less active slaves are not polled for a certain number of slots (not cycles). In addition, the maximum number of frames that may be exchanged during a single visit to any slave is limited.

Poll rate reduction is also utilized in the *fair exhaustive polling* (FEP) scheme [8], where a pool of "active" slaves is maintained by the master. Slaves are polled with one frame per visit, as in 1-limited service. When a POLL–NULL frame occurs, that slave will be dropped from the pool and will not be polled for some time. The "inactive" slave may be restored to the pool when the master downlink queue that corresponds to that slave contains a packet, or when the entire pool is reset to its original state. The pool is reset when the last slave in the pool is to be dropped, or after a predefined timeout. In this manner, the slaves that have more traffic will receive proportionally larger share of the bandwidth as long as they have traffic to send or receive.

A slightly more sophisticated approach to poll rate reduction is employed in the scheme known as *adaptive E-limited polling* [14]. In this case, each slave is serviced for up to M frames during a single visit. However, the limits are assigned to each slave separately, and they are made variable between predefined bounds. Initially, all slaves are assigned the value of M equal to the upper bound of M^+. When the slave is polled and there is a data packet sent in either direction, the current value of M for that slave is decreased

TABLE 31.2   An Overview of Adaptive Polling Algorithms

Polling algorithm	Performance aspect			
	Access delay	End-to-end delay	Adaptivity	Fairness
LWRR [6]	Average	Good (high loads)	Good (penalizes inactivity)	Inherent
FEP [8]	Good	Bad (under bursty traffic)	Good (penalizes inactivity)	Can be unfair
ACLS [15]	Good	Good	Excellent	Inherent
FPQ [16]	Good	Good	Good (but computationally intensive)	Adjustable

by one, until the lower bound M^- is reached. When a POLL–NULL frame is encountered, the value of M for that slave is reset to the maximum value, but the slave will skip a certain number of cycles. In this manner, the slave that has been idle for some time can send more data immediately after becoming active again. In case there is continuously backlogged traffic, the service gradually decreases to allocate a fair share of available bandwidth to each slave.

The *adaptive cycle-limited service* (ACLS) scheme [15] strives to keep the duration of the piconet cycle as close to a predefined value as possible. Bandwidth is allocated dynamically, partly on the basis of historical data (i.e., the amount of traffic in the previous piconet cycle), and partly on the basis of current traffic (i.e., whether they have some data to exchange or not). However, each of the m slaves is guaranteed a fair share of the available bandwidth over the period of $m - 1$ piconet cycles, plus a certain minimum bandwidth in each piconet cycle. This scheme appears well-suited for piconets in which some of the slaves have tight bandwidth and latency constraints, as is often the case with multimedia traffic.

We also mention the efficient double-cycle polling [17], an adaptive algorithm, which tries to optimize performance for uplink and downlink traffic separately; and fair predictive polling with QoS support (FPQ), a predictive algorithm with rudimentary QoS support [16].

Table 31.2 summarizes the characteristics of adaptive polling schemes.

31.3   BNEP and Segmentation/Reassembly Policies

All of the polling schemes described above focus on the optimization of performance of Bluetooth baseband traffic. As the traffic to be transported originates from the applications running on Bluetooth and other devices, we should also look into the details of packet segmentation and reassembly policies. As most traffic nowadays is based on the transmission control protocol (TCP) family of protocols, it is necessary to examine the ways in which such traffic can be transported over Bluetooth. Fortunately, the bluetooth network encapsulation protocol (BNEP), provides a ready solution to these problems [18].

The BNEP protocol is designed to encapsulate and forward Ethernet frames through Bluetooth networks. Multi-hop traffic, including the slave to slave traffic, may be handled by Bluetooth masters and bridges acting as store-and-forward switches. In other words, the entire TCP PDU, which consists of a number of Bluetooth PDUs, has to be stored in the device before being repackaged (if necessary) and forwarded to the next stop along the route. Routing in this case is done in the IP layer, transparently to Bluetooth. Figure 31.3(a) shows the protocol stack when TCP/IP packets are encapsulated using BNEP. The headers and their typical length are shown in Figure 31.3(b). Note that each TCP message generated will require a total of 59 bytes in appropriate headers throughout the protocol stack.

As Bluetooth baseband packets can have either 1, 3, or 5 slots each, with varying payload of up to 339 bytes, as shown in Table 31.3, L2CAP packets obviously have to be segmented into a number of baseband packets. Upon reception, the payload has to be extracted from the baseband packets, and the L2CAP packet has to be reassembled. Again, the Bluetooth specification offers little guidance in that respects, and several policies for packet segmentation and reassembly have been described [8,9].

We note that the noise and interference levels may play a critical role in choosing the segmentation policy. The Bluetooth frequency hopping sequence has been shown to be fairly efficient, and a large number

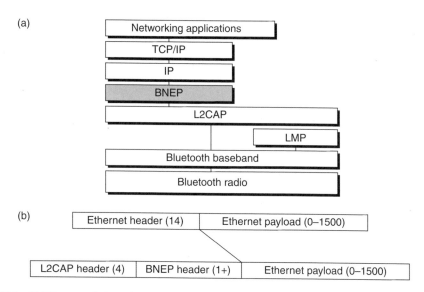

FIGURE 31.3 BNEP: transmission of TCP/IP traffic over Bluetooth: (a) BNEP protocol stack; (b) BNEP with an ethernet packet payload sent using L2CAP (all field sizes are expressed in bytes). (Adapted from: Bluetooth SIG, Technical Report, revision 0.95a, June 2001.)

TABLE 31.3 Packet Types for Communication over an ACL Link

Type	Slot(s)	Payload (bytes)	FEC	Asymmetric data rate (kbps total)
DM1	1	17	2/3	217.6
DH1	1	27	None	341.6
DM3	3	121	2/3	516.2
DH3	3	183	None	692.0
DM5	5	224	2/3	514.1
DH5	5	339	None	780.8

of Bluetooth piconets may coexist in the same radio range [19]. Still, packets can be damaged by noise and interference, in which case retransmission will be needed. We note that in a noisy environment:

- DM-type packets, which include 2/3 forward error correction (FEC), might be preferred over DH-type packets, which have no such provisions;
- Shorter (DH3 and DM3) packets might be preferred over longer ones, despite their smaller payload, because they are less susceptible to damage.

There are many issues related to the transmission of TCP/IP (and other types of) traffic over Bluetooth that ought to be investigated in more detail, in particular reliability, fairness, admission control, and congestion control. An overview of the performance of different polling algorithms in the presence of TCP/IP traffic is given in Reference 14.

31.4 Open Problems and Directions for Further Research

This chapter presents a brief overview and classification of intra-piconet polling schemes for Bluetooth piconets. Some additional issues, in particular the transmission of TCP/IP traffic over Bluetooth, and segmentation and reassembly policies, are discussed as well.

The aspects of Bluetooth networking not covered here include piconet admission control and QoS support. Also, issues related to scatternet formation, operation, and maintenance, are not mentioned here. We have also not addressed the issues related to the Bluetooth radio layer, where problems such as noise, interference, and packet loss, may have a significant impact on performance.

Future work might also analyze the performance of Bluetooth networks under different traffic models, preferably those that try to mimic real life traffic patterns for devices, such as PDAs, that are most likely candidates for communication using Bluetooth technology. Issues related to stability, fairness, reliable data transfer, and others, could also be addressed. Finally, the practical implications of performance analyses on scatternet formation and restructuring, should be examined.

We end this discussion by citing a few among the papers that contain more detailed comparative analyses of different polling schemes. In particular, we note an early report by Johansson et al. [8], a detailed and influential paper by Capone et al. [6], and a more recent overview paper by Chan et al. [14]. It may also be interesting to note that the majority of authors have relied on discrete event simulation to obtain their results. Analytical techniques have been used only recently [12,20], possibly because the well-known results in the area of polling systems [5,10] cannot be directly applied in Bluetooth.

References

[1] "Wireless PAN Medium Access Control MAC and Physical Layer PHY Specification," IEEE standard 802.15, IEEE, New York, 2002.

[2] Bluetooth SIG, *Specification of the Bluetooth System — Architecture and Terminology Overview*, vol. 1, Nov. 2003.

[3] Bluetooth SIG, *Specification of the Bluetooth System*, Feb. 2001.

[4] H. Levy, M. Sidi, and O.J. Boxma, "Dominance relations in polling systems," *Queueing Systems Theory and Applications*, vol. 6, pp. 155–171, 1990.

[5] Z. Liu, P. Nain, and D. Towsley, "On optimal polling policies," *Queueing Systems Theory and Applications*, vol. 11, pp. 59–83, 1992.

[6] A. Capone, R. Kapoor, and M. Gerla, "Efficient polling schemes for Bluetooth picocells," in *Proceedings of the IEEE International Conference on Communications ICC 2001*, vol. 7, Helsinki, Finland, pp. 1990–1994, June 2001.

[7] A. Das, A. Ghose, A. Razdan, H. Saran, and R. Shorey, "Enhancing performance of asynchronous data traffic over the Bluetooth wireless ad-hoc network," in *Proceedings of the 20th Annual Joint Conference of the IEEE Computer and Communications Societies IEEE INFOCOM 2001*, vol. 1, Anchorage, AK, pp. 591–600, Apr. 2001.

[8] N. Johansson, U. Körner, and P. Johansson, "Performance evaluation of scheduling algorithms for Bluetooth," in *Proceedings of the BC'99 IFIP TC 6 5th International Conference on Broadband Communications*, Hong Kong, pp. 139–150, Nov. 1999.

[9] M. Kalia, D. Bansal, and R. Shorey, "MAC scheduling and SAR policies for Bluetooth: a master driven TDD pico-cellular wireless system," in *Proceedings of the 6th IEEE International Workshop on Mobile Multimedia Communications MOMUC'99*, San Diego, CA, pp. 384–388, Nov. 1999.

[10] H. Takagi, *Queueing Analysis*, vol. 1: Vacation and Priority Systems, Amsterdam, The Netherlands: North-Holland, 1991.

[11] Y.-Z. Lee, R. Kapoor, and M. Gerla, "An efficient and fair polling scheme for Bluetooth," in *Proceedings of the MILCOM 2002*, vol. 2, pp. 1062–1068, 2002.

[12] J. Mišić and V.B. Mišić, "Modeling Bluetooth piconet performance," *IEEE Communication Letters*, vol. 7, pp. 18–20, 2003.

[13] J. Mišić, K.L. Chan, and V.B. Mišić, "Performance of Bluetooth piconets under E-limited scheduling," Technical Report TR 03/03, Department of Computer Science, University of Manitoba, Winnipeg, Manitoba, Canada, May 2003.

[14] K.L. Chan, V.B. Mišić, and J. Mišić, "Efficient polling schemes for bluetooth picocells revisited," in *Preceedings of the HICSS-37 Minitrack on Wireless Personal Area Networks*, Big Island, Hawaii, Jan. 2004.

[15] V.B. Mišić, E.W.S. Ko, and J. Mišić, "Adaptive cycle-limited scheduling scheme for Bluetooth piconets," in *Proceedings of the 14th IEEE International Symposium on Personal, Indoor and Mobile Radio Communications PIMRC'2003*, vol. 2, Beijing, China, pp. 1064–1068, Sept. 2003.

[16] J.-B. Lapeyrie and T. Turletti, "FPQ: a fair and efficient polling algorithm with QoS support for Bluetooth piconet," in *Proceedings of the 22nd Annual Joint Conference of the IEEE Computer and Communications Societies IEEE INFOCOM 2003*, vol. 2, New York, NY, pp. 1322–1332, Apr. 2003.

[17] R. Bruno, M. Conti, and E. Gregori, "Wireless access to internet via Bluetooth: performance evaluation of the EDC scheduling algorithm," in *Proceedings of the 1st workshop on Wireless Mobile Internet*, Rome, Italy, pp. 43–49, July 2001.

[18] Bluetooth SIG, "Bluetooth Network Encapsulation Protocol (BNEP) Specification," Technical Report, Revision 0.95a, June 2001.

[19] S. Zürbes, "Considerations on link and system throughput of Bluetooth networks," in *Proceedings of the 11th IEEE International Symposium on Personal, Indoor and Mobile Radio Communications PIMRC 2000*, vol. 2, London, UK, pp. 1315–1319, Sept. 2000.

[20] D. Miorandi, C. Caimi, and A. Zanella, "Performance characterization of a Bluetooth piconet with multi-slot packets," in *Proceedings of the WiOpt'03*, Sophia-Antipolis, France, Mar. 2003.

32

A Survey of Scatternet Formation Algorithms for Bluetooth Wireless Personal Area Networks

32.1 Introduction..**32**-735
32.2 Bluetooth Background **32**-737
    Asymmetric Connection Management • Symmetric
    Connection Management • Connection State
32.3 Scatternet Formation Algorithms......................**32**-739
    Bluetooth Topology Construction Protocol • Bluetrees •
    Bluenet • Scatternet Formation Algorithm Proposed
    by Law et al. • Tree Scatternet Formation Algorithm • Loop
    Scatternet Formation Algorithm • Bluestar • BFS, MST,
    DFS, MDST, and E-MDST • Optimal Topology for Proactive
    Scatternets • Scatternet Route Structure • Two-Phase
    Scatternet Formation Algorithm and its Variant TPSF+
32.4 Performance Comparison **32**-746
    Comparison between TPSF+ and BTCP • Comparison
    between TPSF+ and Bluenet • Comparison between TPSF
    and TPSF+
32.5 Conclusions **32**-751
Acknowledgment...**32**-752
References **32**-752

Vincent W.S. Wong
Chu Zhang
Victor C.M. Leung

32.1 Introduction

Bluetooth is a short-range radio technology that uses the frequency hopping spread spectrum (FHSS) technique with time-division duplex (TDD) in the license-free 2.4 GHz industrial, scientific, and medical (ISM) band. Due to its low cost and low power, Bluetooth is considered as a promising technology for wireless personal area networks (WPANs). Various consumer electronics devices (e.g., personal digital assistants, personal computers, keyboard, and mouse) have begun to provide Bluetooth wireless communication capability. The complete Bluetooth specification is standardized within the Bluetooth special interest group (SIG). The current specification is version 1.2 [1]. To facilitate a wider adoption of this technology, part of the Bluetooth specification (i.e., physical and data link layers) is also standardized within the IEEE 802.15.1a WPAN working group [2].

Bluetooth devices are required to form a *piconet* before exchanging data. Each piconet has a *master* unit that controls the channel access and frequency hopping sequence. Other nodes in the piconet are referred to as the *slave* units. In a Bluetooth piconet, the master node can control up to seven active slaves simultaneously. Several piconets can be interconnected via *bridge* nodes to create a scatternet. Bridge nodes are capable of time-sharing between multiple piconets, receiving packets from one piconet and forwarding them to another. Since a bridge node is connected to multiple piconets, it can be a master in one piconet and act as slave in other piconets. This is called a *master/slave bridge* (*M/S bridge*). Alternatively, a bridge node can act as a slave in all the piconets it is connected to. This is called a *slave/slave bridge* (*S/S bridge*).

If a set of Bluetooth devices are within the transmission range of each other, then a piconet formation algorithm is necessary to create a connected topology for the devices and to assign the role of each device (i.e., master or slave). The current Bluetooth specification [1] has standardized the procedures for piconet formation. This involves a sequence of steps including the inquiry, inquiry scan, page, and page scan phases.

Within the piconet, all nodes share the same channel (i.e., a frequency hopping pattern), which is derived from the node ID and clock information of the master unit. In order to reduce interferences, each data packet transmission uses one of the 79 different frequencies. The medium access control is a slotted TDD mechanism. Two slave nodes cannot transmit packets to each other directly; they must go via the master node. The master node uses a polling-style scheduling mechanism to allocate the time-slots for its slaves. When a master node sends a data packet or control POLL packet to a slave node, it will then reply by either sending a data packet or control NULL packet to its master.

Similarly, given a set of Bluetooth devices, a scatternet formation algorithm is necessary to create a connected topology for the devices, and to assign the role of each device (i.e., master slave or bridge). Although the current Bluetooth specification [1] defines what a scatternet is, the scatternet formation issue is not fully resolved. There are still many issues on how to organize the nodes into a self-configuring operational ad hoc network. Some of the unresolved issues are as follows:

1. How does each node choose its role (i.e., master, slave, or bridge) in the scatternet?
2. How many piconets should there be in a scatternet?
3. How many piconets should a bridge node belong to?
4. How many bridge nodes should a piconet have?
5. How many slave nodes should the master control in one piconet?
6. Which mode (i.e., active, sniff, park, or hold) should the slave use in the scatternet?
7. How should the addition and deletion of nodes be supported as users come and go?
8. How should topology changes due to node mobility be supported?

We now describe how the questions stated above can affect the topology or the performance of the scatternet. Question (1) pertains to the information required to decide the role of each node. Since Bluetooth is intended to interconnect different types of devices, it is desirable for devices, which have higher processing power and battery energy to be the master or bridge nodes. Question (2) raises the packet collision problem in the scatternet. Since all the piconets use the same 79 frequencies with different sequences, packet collisions may occasionally occur for the piconets that are within the same radio range. Packet collisions will degrade the network performance. Although it is possible to reduce the number of piconets within the scatternet, some of the master nodes and bridge nodes may become the bottleneck. This may result in a relatively large average packet transfer delay for some communication sessions.

Consider questions (3) and (4). In the Bluetooth scatternet, different piconets in general are not synchronized with each other. Due to the loss of timing in the switching among different piconets, the bridge node may waste up to two time slots for each switching operation. Thus, the node degree of the bridge node (i.e., the number of piconets the bridge node is connected to) and the number of the bridge nodes in a single piconet should not be large.

For questions (5) and (6). In the Bluetooth specification [1], the slave can be in either active, sniff, hold, or park mode. The selection of the mode of a slave node (including pure slave node, master/slave bridge node, and slave/slave bridge node) in the scatternet is an important issue. If all the pure slaves are always in active mode and most of them do not participate in the exchange of data, network utilization will be reduced due to the polling of these active slaves by their master.

The last two questions pertain to topology maintenance. This includes two scenarios: a node may join or leave the network, and a node may randomly move within the scatternet due to user's mobility. Thus the scatternet formation algorithm also needs to assign some nodes to detect the presence of new nodes and to re-configure the topology of the scatternet due to mobility in a timely manner.

Over the past few years, the active research areas in Bluetooth ad hoc networks include (a) scatternet formation algorithms; (b) intra- and inter-piconet scheduling algorithms; (c) interference mitigation between Bluetooth and IEEE 802.11 systems; and (d) simple and energy efficient receiver design. This paper focuses on the scatternet formation problems. The objectives of this paper are to (1) identify the issues in scatternet formation; (2) describe various scatternet formation algorithms proposed in the literature; and (3) identify some of the open issues for further work.

The rest of this paper is organized as follows. In Section 32.2, we describe both the asymmetric and symmetric link establishment procedures for piconet formation. In Section 32.3, we describe different scatternet formation algorithms proposed in the literature. Section 32.4, presents the performance comparisons of several scatternet formation algorithms. We conclude by stating a number of unresolved problems that need to be addressed for the deployment of the Bluetooth scatternet.

32.2   Bluetooth Background

In this section, we provide some background information on device discovery and link establishment in the Bluetooth system. We begin by describing an asymmetric connection protocol between two Bluetooth devices. It is followed by a description of the symmetric connection management protocol. For an overview of Bluetooth, readers can refer to References 3–6.

32.2.1   Asymmetric Connection Management

Before two Bluetooth devices that are within the same radio range can transmit data packets, they need to discover each other and establish a link connection. This is accomplished by two procedures: *inquiry* and *paging*. The inquiry procedure is used to discover other devices within the same radio range. The paging procedure is used to synchronize and set-up a Bluetooth link connection between the two devices.

32.2.1.1   Inquiry Procedure (Device Discovery)

During an inquiry procedure, the device that wants to collect the information of other Bluetooth devices is called a *sender* and the device that wants itself to be discovered is called a *receiver*. The sender transmits inquiry ID packets asking the neighboring receivers to identity themselves and to provide the synchronization information required for the link establishment procedure later. Each Bluetooth device during the inquiry procedure is in one of the two states: *inquiry* or *inquiry scan*. The sender is in the *inquiry state* and the receiver is in *the inquiry scan* state.

The details of the inquiry procedure are as follows: first, the sender broadcasts an inquiry ID packet to the receivers around it and listens for the response. The receiver scans the ID packets sent by the sender. Both sender and receiver use the same 32 inquiry hop frequencies generated by the general inquiry access code (GIAC). The phase in the sequence is determined by the native clock of the Bluetooth device. The sender and receiver hop at different speeds. The receiver hops at a lower rate over the common frequency pattern (one frequency every 1.28 sec), scanning the ID packets on each hop. The sender transmits at a higher rate and listens in between transmissions for a response packet [1].

Upon receiving the inquiry ID packet, the receiver backs off for a random period. This avoids the contention problem of multiple receivers listening on the same hop frequency and transmitting the response packets simultaneously. After a random period, the receiver wakes up and starts to scan again. Finally, when the receiver receives the inquiry ID packet again, it sends the frequency hop synchronization (FHS) packet to the sender. The FHS packet contains the receiver's address and clock value. At this time, the sender obtains the Bluetooth device address (BD_ADDR) and the clock value of the receiver.

32.2.1.2 Paging Procedure (Link Establishment)

The paging procedure is used to establish the link connection between two devices. The device that initiates the paging procedure will become the master after the link establishment. The device that enters the *page scan* state will later become the slave.

During the paging procedure, the master sends an ID packet on the frequency to which the receiver is listening to. The frequency is determined by the BD_ADDR and the clock information of the receiver discovered during the inquiry procedure. Then, the slave replies with an ID packet. Upon receiving the ID packet from the slave, the master sends an FHS packet, which contains the master's BD_ADDR and clock value. In this way the slave obtains the BD_ADDR and clock value of its master node and thus can synchronize and hop with the master node. After the paging procedure, the master creates a point-to-point connection with the slave on the baseband layer.

32.2.2 Symmetric Connection Management

The inquiry procedure described in the previous section is *asymmetric*. That is, the sender knows the identity of the receiver (via the BD_ADDR) but not vice versa. This is due to the fact that the inquiry ID packet transmitted by the sender does not include any information of the sender itself, while the FHS packet transmitted by the receiver includes the BD_ADDR of the receiver. An asymmetric connection management is adequate as long as the role of the two devices (i.e., master or slave) is predetermined.

To allow for mutual information exchange between two Bluetooth devices during the inquiry procedure, Salonidis et al. proposed that each device should be allowed to alternate between the inquiry and inquiry scan states. The residence time in each state is chosen randomly within a predetermined time period [7]. A number of scatternet formation algorithms also use this symmetric inquiry procedure to allow each device to collect the information of its neighboring devices. Given the local neighboring information exchange, different scatternet formation algorithms will use different schemes (described in the next section) to select the role of each device.

Note that the Bluetooth specification allows the switching of roles between devices. Consider two devices *x* and *y* where device *x* is in the *page* state and device *y* is in the *page scan* state. It is possible that device *y* is already a master of another piconet and therefore, it can host device *x* as one of its slaves. When a piconet is created between devices *x* and *y* with *x* as the master node, the slave node *y* can request a role switch operation. This operation is implemented via the exchange of a specific link manager protocol (LMP) packet that facilitates the two devices to change the frequency hopping sequence of the new master.

32.2.3 Connection State

After successfully performing the inquiry and paging procedures, the two devices are in the connection states, one as the master and the other as the slave. In the connection state, the slave node has four modes, namely active, hold, sniff, or park mode. The hold, sniff, and park modes are also called the low-power modes.

Active mode: A node in active mode listens to the channel all the time. The master can control the transmission to (or from) different slaves by using different scheduling schemes (e.g., pure round robin, deficit round robin). A slave in active mode is assigned by an active member address (AM_ADDR). Each active slave listens to each master-to-slave slots and decides whether the packet received should be accepted depending on the AM_ADDR in the received packet.

Sniff mode: A slave wakes up and listens to the channel during each interval T_{sniff} for $N_{sniff-attempt}$ slots [1]. $N_{sniff-attempt}$ is the number of slots that the slave can listen to during the interval T_{sniff}. If the slave receives a data packet, it will continue listening for subsequent $N_{sniff-timeout}$ time slots. Otherwise, the slave will go into sleep state. $N_{sniff-timeout}$ determines the number of additional time slots that the slave must listen to when it receives a packet.

Hold mode: In this mode, both master and slave agree on a time interval during which the slave needs not to participate in the channels temporarily. During the hold period, the slave can go to some other states such as *inquiry* or *inquiry scan*. A bridge node can use the hold mode to switch between different piconets.

Park mode: In this mode, the slave wakes up only at an interval for synchronization and sleeps at all other times. This interval is agreed between the master node and the slave node before the slave node goes into the park mode. The slave will give up its AM_ADDR and use its parked member address (PM_ADDR), or BD_ADDR after it goes into the park mode. By using the park mode, the total number of slaves in a piconet can be more than seven.

32.3 Scatternet Formation Algorithms

The scatternet formation algorithms can be broadly classified into two categories: proactive and reactive (or on-demand). For proactive scatternet formation algorithms (e.g., [7–18]), the scatternet is formed as soon as the devices are powered on. The same scatternet is used to transport both control and data packets. The control packets carry the messages for nodes joining or leaving, connection set-up, and topology maintenance due to node power off or failure. The data packets carry the application messages between two devices. For reactive scatternet formation algorithms (e.g., [19–22]), there is a decoupling between the control scatternet used to transport control packets and the on-demand scatternet used to transport data packets. The main advantage of this decoupling is the increase of throughput and reduction of average transfer delay for data packets. In this section, we summarize various scatternet formation algorithms proposed in the literature.

32.3.1 Bluetooth Topology Construction Protocol

Salonidis et al. [17] proposed an asynchronous distributed scatternet formation algorithm called the Bluetooth topology construction protocol (BTCP) [7]. The scatternet constructed by BTCP is fully connected and has the following additional features: (a) a bridge node is connected to only two piconets (termed as bridge degree constraint); (b) given a fixed number of nodes, the scatternet consists of the minimum number of piconets; and (c) there exists a bridge node between each pair of piconets (termed as piconet overlap constraint).

The procedure for BTCP scatternet formation is shown in Figure 32.1 and is described as follows: (a) Initially all the Bluetooth devices do not know the existence of other neighboring devices. During a fixed time period (called ALT_TIMEOUT), all the devices collect the information of the other nodes by alternating between the *inquiry* and *inquiry scan* states. (b) The node with the highest Bluetooth device address is elected as the leader among all the nodes. The leader has the information (i.e., Bluetooth address and clock information) of all the other nodes. The leader then determines a connectivity list, which includes the role of each node in the subsequent scatternet. (c) The leader connects the designated masters by paging them and transmitting the connectivity list. (d) The designated masters page their slaves according to the connectivity list. The bridge nodes wait to be paged by two masters. (e) Finally, the BTCP scatternet is formed.

The BTCP causes a small connection establishment delay (about 2 sec) if all the nodes start at the same instant. If the devices cannot start the alternative *inquiry/inquiry scan* at the same time, the connection delay will become longer and the scatternet cannot be guaranteed to be fully connected.

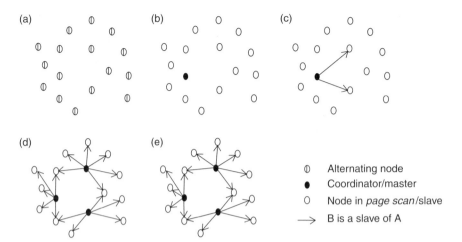

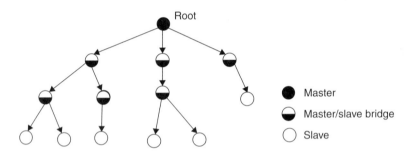

FIGURE 32.1   The BTCP establishment procedure.

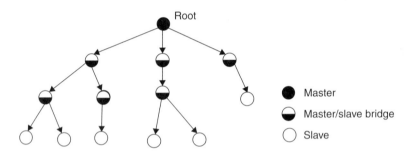

FIGURE 32.2   A Bluetree scatternet. The direction of the arrow is used to indicate the master–slave relationship among the Bluetooth devices.

The BTCP assumes that any two Bluetooth devices can reach each other so that the coverage area of the scatternet is only within one radio range. The resulting scatternet topology has the property that any two piconets are directly connected by a bridge node between them. The number of the nodes in the BTCP scatternet is limited to 36, corresponding to a maximum of 8 piconets.

32.3.2 Bluetrees

Zaruba et al. [8] proposed two scatternet formation algorithms. The topologies created by both algorithms are called Bluetrees. In the first algorithm, the Bluetooth devices begin by the acquiring the Bluetooth address of their neighbors via the boot procedure. The next step is the creation of the rooted spanning tree. An arbitrary node is chosen to be the *blueroot* and acts as the master node. All the intermediate neighboring nodes of blueroot become the slave nodes. These slave nodes will then try to page their neighboring devices, which are not the direct neighbors of the blueroot. If those devices exist, the slave nodes which perform the paging will switch their roles to be the master/slave bridge nodes and establish link connections with those devices. This procedure continues recursively until the leaves of the tree are reached (devices that are pure slave nodes). An example of a rooted Bluetree scatternet topology is shown in Figure 32.2.

Since the topology has a tree structure, there is a unique single path between each source and destination pair. The routing is simplified by having the upstream nodes identify the intermediate downlink links to reach their children.

The second scatternet formation algorithm in [8] aims to reduce the set-up time and avoid the preselection of the blueroot. This is accomplished by choosing multiple nodes as blueroots and then merging the trees created by each blueroot to form a single spanning tree. In this algorithm, devices that have the largest ID among all their neighbors are chosen as the blueroots. By limiting the number of roles on each device and the number of slaves that a master can control in a piconet, the switching cost is reduced in terms of switching delay, scheduling, and re-synchronization.

The shortest path algorithm of network topology graph is used in the performance analysis of Bluetrees. The average route error is calculated by the difference between the actual path of Bluetree and the shortest path. Simulation results show that the average route error is significantly influenced by the density of the network.

As shown in Figure 32.2, a Bluetree scatternet has a hierarchical structure in which all the nodes except the root node and leaf nodes are bridge nodes. Each bridge node acts as the slave of its upstream piconets and master of its downstream piconets. The routing is simple but it may not be efficient since all the inter-piconet traffic has to go through the upstream piconets and may cause a bottleneck if the traffic load is high.

32.3.3 Bluenet

Wang et al. [9] recognized the limitation of the Bluetree scatternet formation and proposed a scatternet formation algorithm called the Bluenet. The algorithm begins with the device discovery phase in which each node learns the information of its neighbor. After that, the master nodes are selected randomly. Each master node invites up to n ($n \leq 5$) neighboring nodes to form a single piconet by paging. Once a node has become a slave in a piconet, it will stop replying to paging messages sent by other nodes. At the end of this phase, the network consists of multiple piconets and some isolated nodes.

In the second phase, the isolated Bluetooth nodes are connected to the piconets. It is possible that these isolated nodes can be connected to more than one piconet. In this case, they become the bridge nodes. In the final phase, the piconets are interconnected with each other to form a connected Bluenet scatternet. Unfortunately, a detailed description is not given in [9] for the second and third phases. An example of a Bluenet scatternet topology is shown in Figure 32.3.

32.3.4 Scatternet Formation Algorithm Proposed by Law et al.

The design goals of the scatternet formation algorithm proposed in [10] are (a) to minimize the number of piconets; (b) to minimize the node degree of the bridge nodes; (c) to minimize the time to create a scatternet (i.e., time complexity); and (d) to reduce the number of signaling messages sent between devices (i.e., message complexity).

The proposed algorithm assumes that all the nodes are within the same transmission range (i.e., they are all one-hop neighbors). When the algorithm is executed, all the Bluetooth devices are partitioned

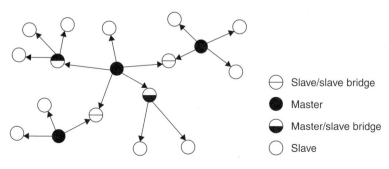

FIGURE 32.3   A Bluenet scatternet.

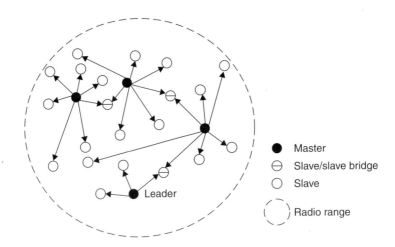

FIGURE 32.4   A scatternet created by the algorithm proposed by Law et al.

into *components*. Each component is a set of interconnected devices. A component can be an isolated node, a piconet, or a scatternet. Each component has a *leader*. The scatternet is formed step-by-step by interconnecting the network components. When two components are interconnected, only one leader remains in the scatternet and the other leader is retired. The pseudo codes for different procedures (e.g., CONNECTED, MERGE, MIGRATE, MOVE) are stated in [10]. Figure 32.4 shows a scatternet formed by this algorithm.

Given the number of nodes n and the degree constraint k, the algorithm has the following properties [10]: (a) the maximum degree of bridge nodes is limited to two; (b) the scatternet contains at least $\lceil (n - 1)/k \rceil$ piconets and at most $\lfloor (n - 2)/(k - 1) \rfloor + 1$ piconets; and (c) the time complexity is of $O(\log n)$ and the message complexity is of $O(n)$.

Once the scatternet has been formed, the leader continues to invoke another procedure periodically (called MAIN in [10]) to detect the presence of new incoming nodes. Thus, new Bluetooth devices can join the scatternet dynamically.

32.3.5   Tree Scatternet Formation Algorithm

Similar to Bluetrees, the tree scatternet formation (TSF) algorithm proposed in [11] also creates a scatternet with a tree structure. At any given time, the topology created by TSF is a collection of one or more rooted spanning trees (called *tree components* in [11]). Each tree component has one *root node* and all the other nodes are called the *tree nodes*. The isolated single nodes (termed as *free nodes*) are not connected to any tree components. The basic operations of TSF are (a) to merge different tree components into a single tree component, (b) to connect a free node to a tree component.

In each tree component, a subnet of nodes (called *coordinators*) is used to detect the presence of other tree components or free nodes. Once a new tree component is detected, a control message will be sent from the coordinator to its root node. The merging of two trees is performed by one tree root node paging the other tree root node. An example of a scatternet created by TSF algorithm is shown in Figure 32.5.

TSF is self-healing in that a Bluetooth device can join or leave the scatternet at any time. When a parent node detects the loss of a child node, the parent node only needs to decide whether it has become a free node. The parent node will then update its node type accordingly. On the other hand, when a child node detects the loss of a parent node, the child node will become either a free node (if it was a leaf node) or a root node in the new tree component.

The TSF algorithm has one limitation. When the coordinators from two different tree components detect the presence of each other, these two tree components can be merged into a single tree component only if those two root nodes are within the same radio range of each other. In that case, the scatternet

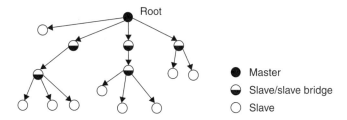

FIGURE 32.5   A scatternet created by TSF.

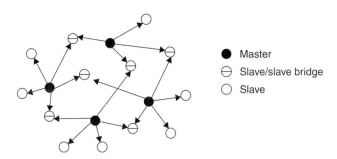

FIGURE 32.6   A scatternet created by the LSF algorithm.

cannot always guarantee to be fully connected. Tan et al. also proposed a distributed scheduling algorithm specifically for the scatternet created by the TSF algorithm. Interested readers can refer to [23] for details.

32.3.6 Loop Scatternet Formation Algorithm

The design goals of the loop scatternet formation (LSF) algorithm proposed in [12] are: (a) to minimize the number of piconets; (b) to reduce the maximum number of hops between any pair of Bluetooth devices (called network diameter); and (c) to reduce the maximum of Bluetooth device pairs that may use the node as a relay device (called maximum node contention).

The LSF algorithm assumes that all the nodes are within the same transmission range (i.e., they are all one-hop neighbors). The algorithm consists of two phases. In the first phase (formation of a ring scatternet), Bluetooth devices are connected to form piconets with at most $k - 2$ slaves, and the piconets are interconnected to form a ring. All bridge nodes are assumed to be slave/slave bridges. Each bridge node is connected to two piconets. The second phase aims to reduce the network diameter and the maximum node contention. This is accomplished by creating more bridge nodes between the piconets. During this phase, a slave/slave bridge is created between piconet i and piconet $i \pm \lceil \sqrt{P} \rceil$ where P is the number of piconets. An example of a scatternet created by LSF algorithm is shown in Figure 32.6 In this example the value of P is equal to 4.

Given the number of nodes n and the degree constraint k, the LSF algorithm has the following properties [12]: (a) the maximum degree of bridge nodes is limited to two; (b) the scatternet contains at least $\lceil n/(k - 1) \rceil$ piconets and at most $\lfloor (n - 2)/(k - 2) \rfloor + 1$ piconets. Under the assumption that $\sqrt{P}$ is an integer, the network diameter is equal to $2\sqrt{P}$.

32.3.7 Bluestar

The scatternet formation algorithm proposed by Petrioli et al. [13] has three phases (Figure 32.7). The first phase is the topology discovery phase (i.e., discovery of neighboring Bluetooth devices). During this phase, each Bluetooth device obtains the ID, weight, and synchronization information of all of its

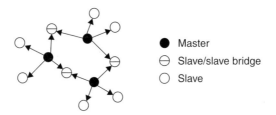

FIGURE 32.7   A scatternet created by interconnecting the BlueStars.

neighboring devices. Note that this discovery phase is symmetric. That is, if a Bluetooth device x has discovered a neighboring device y and has received all of its relevant information (e.g., ID, weight), node y will also have obtained node x's information. This phase is accomplished by alternating between the *inquiry* and *inquiry scan* states, and by forming *temporary piconets*. A temporary piconet is created between two devices for information exchange (e.g., ID, weight). Once the information has been obtained, the temporary piconet is released. A timer is used to limit the duration of the topology discovery phase.

The second phase (called BlueStars formation phase in [13]) aims to create different piconets (called BlueStars). The weight value for each Bluetooth device is used to determine its role. Bluetooth devices which have the highest weight value in their neighborhood become the master nodes. Other devices will become the slave nodes. At the end of phase two, several isolated piconets are created. Each piconet has one master and several slaves. If a piconet has more than seven slaves, some slaves are switched to the park mode. Note that each slave is connected to only one master.

The third phase (called BlueConstellation in [13]) aims to connect different piconets to form a scatternet. Each master node chooses some of its slave nodes to be the bridge nodes. The bridge node selection is performed locally by each master. The selection criterion also depends on the weight and ID information of the masters and slaves in the neighboring piconets.

Petrioli et al. extended their work and renamed the scatternet formation algorithm as Bluemesh in [14]. Simulations results in [14] show that for a given topology, the devices have no more than 2.4 roles on average and the path length for any source and destination pair is not significantly longer than the corresponding shortest path.

32.3.8 BFS, MST, DFS, MDST, and E-MDST

Guerin et al. proposed and analyzed various scatternet formation algorithms in [15]. Given a set of nodes and a set of physical links (a physical link exists between two nodes if they are within the communication range of each other), a physical-topology graph can be constructed. Given a set of nodes and a set of logical Bluetooth links (a logical Bluetooth link exists between two nodes if the Bluetooth topology has a communication link between them), a logical-topology graph can be constructed. According to the Bluetooth specification, Guerin et al. assumed that the node degree of both master nodes and bridge nodes to be less than seven. Given the physical-topology graph, a logical-topology graph exists if the node degree constraints are satisfied. They noticed that a logical-topology graph exists if the physical topology graph has a spanning tree, which satisfies the degree constraints. However, a spanning tree with degree seven or less exists if and only if the maximum degree of a spanning tree, in a graph is upper bounded by seven, and determining this condition is an NP-hard problem. In spite of this fact, polynomial time algorithms are available in a number of practical scenarios.

Guerin et al. began by using three centralized spanning tree algorithms to construct the scatternet. Those spanning tree algorithms include the depth first search (DFS), breadth first search (BFS), and the minimum weight spanning tree (MST). Simulation results showed that the scatternet constructed by using DFS provides a lower average node degree and a lower maximum node degree for the bridge nodes than the scatternet constructed by using either BFS or MST. However, those three algorithms (i.e., BFS, MST, and MST) cannot selectively control the node degrees for the master and bridge nodes.

Guerin et al. then proposed a heuristic called the MDST algorithm. The basic idea of this centralized scheme is to begin with a spanning tree, and then replace edges from nodes of high node degree with those nodes of low degree. They also proposed another heuristic called the extended MDST (E-MDST) algorithm. This centralized scheme provides a mechanism to reduce the node degree of the bridge nodes while maintaining the node degree of the master nodes to be less than seven.

A description for the distributed implementation of the scatternet formation algorithm based on MST is also provided. As mentioned in [15], further work is required for other distributed spanning tree based algorithms.

32.3.9   Optimal Topology for Proactive Scatternets

Marsan et al. [18] proposed a centralized scheme to determine the optimal topology for a proactive Bluetooth scatternet. Given the traffic requirements and some Bluetooth constraints (e.g., a master node can only connect to at most seven active slaves), the objective is to create a scatternet such that the traffic load of the most congested node is minimized. The problem is formulated as an integer linear program. Although the proposed algorithm is centralized and may not be practical for implementation, this algorithm can be used as a benchmark for performance comparisons with other distributed schemes.

32.3.10   Scatternet Route Structure

An on-demand scatternet route formation algorithm was proposed in [19]. A scatternet is only created when there is a need to establish a connection between a source and destination pair. The proposed scatternet formation algorithm combines with on-demand routing together.

When there is no traffic request presents, the Bluetooth devices are in the low-power "STANDBY" state. There is no logical link establishment between these devices. When a source node wants to send packets to a destination node, it first initiates the route discovery process. The control packet called route-discovery-packet (RDP) is sent from the source node to all of its neighboring nodes. Flooding is then used to broadcast this packet to the whole network. Eventually, the RDP packet will reach the destination. The destination will then send a route-reply-packet (RRP) back to the source along the newly discovered route.

The RDP carries various information including the source and destination Bluetooth device addresses (BD_ADDRs), extended ID types, and packet sequence number. Each piece of RDP information is transmitted via two inquiry messages. On the other hand, the RRP is used to carry information such as the master/slave role, address, and clock information. The on-demand scatternet is formed via paging from the destination node to the source node. Each intermediate Bluetooth device in the new scatternet route only needs to page its upstream neighbor to create a physical channel in between and deliver the packet. An example of a scatternet route structure is shown in Figure 32.8.

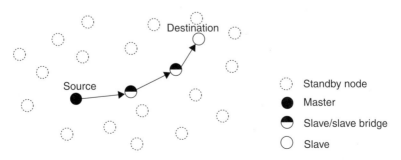

FIGURE 32.8   An example of a scatternet route.

Note that flooding is used during the route discovery phase. Each node has to take part in the route discovery by inquiry even if the source node and destination node are neighbors. The flooding method may cause a large number of unnecessary packet broadcasts in the network. Therefore, the use of flooding may waste the limited bandwidth and limited power of the Bluetooth devices for processing.

32.3.11 Two-Phase Scatternet Formation Algorithm and its Variant TPSF+

In the current Bluetooth specification, all packet transmissions need to be routed via the master or bridge nodes. When the traffic load is high, these master or bridge nodes may become the network bottleneck. Furthermore, in order to support dynamic joining or leaving of mobile devices within a scatternet, a number of time slots in the master nodes may need to be allocated for the dynamic topology configuration updates. This further reduces the time slots available at the master nodes for data transmissions.

Kawamoto et al. [20] proposed a two-phase scatternet formation (TPSF) protocol with the aim of supporting dynamic topology changes while maintaining a high aggregate throughput. In the first phase, a *control scatternet* is constructed for control purpose (i.e., to support dynamic join/leave, route discovery, etc.). The second phase is invoked whenever a node needs to initiate data communications with another node. A dedicated piconet/scatternet is constructed on-demand between the communicating nodes. Since the *on-demand scatternet* can dedicate all the time slots to a single communication session, it has the capability to provide a high throughput and a small end-to-end data transfer delay. The on-demand scatternet is torn down when the data transmissions are finished.

The control scatternet formation consists of three steps. In the first step, each node performs the device discovery by exchanging control information with its neighbor. The second step consists of role determination. The node with the highest number of neighbors among all its neighbors is selected as the master node. The last step is the creation of the control scatternet. The topology of the control scatternet has the following features. Each bridge node belongs to at most two piconets and acts as a slave/slave bridge. Both the master and bridge nodes are in active mode while all the pure slave nodes are in the park mode.

The on-demand scatternet formation consists of two steps. The formation is initiated by the source node that wants to transmit data to the destination node. First, the dynamic source routing (DSR) [26] based protocol is applied to the piconet route discovery in the control scatternet. The piconet route discovery is achieved via the exchange of route request (RREQ) and route reply (RREP) messages. Then, all the master nodes along the piconet route select the participating nodes for the on-demand scatternet. This part is achieved via the exchange of path request (PPEQ) and path reply (PREP) messages. Finally, the destination node initiates the connection set up via the paging procedure. An example of various steps of the TPSF algorithm is shown in Figure 32.9.

In References 21 and 22, Zhang et al. studied the mobility of Bluetooth devices within a limited range and proposed TPSF+, which is a modification of the second phase (i.e., on-demand scatternet formation) of the original TPSF. TPSF+ has the advantage of involving a small number of nodes participating in the on-demand scatternet route discovery procedure so as to avoid unnecessary route discoveries. The on-demand scatternet route discovery is limited to several piconets instead of all the piconets within the control scatternet. Results show that the route discovery procedure for TPSF+ is more efficient than the original TPSF [22].

32.4 Performance Comparison

Apart from the constraints imposed by the Bluetooth system (e.g., a master node can only connect to at most seven active slaves), an efficient scatternet formation algorithm should possess some of the following features:

1. The average connection establishment time (i.e., the average time required to create a scatternet) is low.

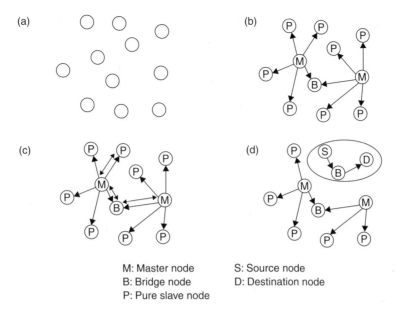

M: Master node   S: Source node
B: Bridge node   D: Destination node
P: Pure slave node

FIGURE 32.9 Two-phase scatternet formation (TPSF) protocol. (a) Distributed nodes without connections. (b) The control scatternet is constructed. (c) RREQ/RREP/PREQ/PREP are exchanged. (d) On-demand scatternet is constructed.

2. The average node degree for the bridge nodes is small.
3. The scatternet is fully connected.
4. The algorithm has a low message complexity and time complexity.
5. The scatternet consists of a small number of piconets.
6. Given a set of source and destination pairs, the scatternet has a high aggregate throughput and a small average packet transfer delay.
7. The delay for topology maintenance is small.

The performance of scatternet formation algorithm can be evaluated either analytically or via simulations. Salonidis et al. [7] used an analytical model to determine the state alternation time period used in the symmetric device discovery phase. Misic et al. [24] used the queueing model to determine the average packet transfer delay of two piconets interconnected by a bridge node. A number of network simulators (e.g., Bluescat [27], BlueHoc [28], Blueware [29]) have been developed for the evaluation of the scatternet performance.

Although there are many scatternet formation algorithms proposed over the past few years, there are only a few performance comparisons reported in the literature.

The Bluenet and Bluetree scatternet formation algorithms are compared in [9] via network simulations. The *average shortest path* and the *maximum traffic flow* are used as the metrics of the performance evaluation. The average shortest path is defined as the average shortest path length (in terms of hop count) between all the 2-node pairs. The maximum traffic flow is used to evaluate the traffic capacity, which can be carried by determining all the possible maximum multi-commodity flows in the network. The simulation results in [9] show that the average shortest path ratio of Bluenet decreases by 0.23 when compared with Bluetree algorithm and the maximum traffic flow in Bluenet is higher than that of Bluetree.

In [25], Basagni et al. compared three scatternet formation algorithms, namely BlueStars [13], Bluetree [8], and the "Yao protocol" [16]. Their results showed that the average device discovery delay is the dominant factor in calculating the average connection establishment time. In addition, the performance

BlueStars is better than Bluetree and the "Yao protocol" in terms of a lower number of piconets, average path length, and the average number of roles per node.

In the remainder of this section, we summarize the results for the performance comparison between TPSF+, BTCP, and Bluenet as reported in [22]. These results are obtained via the use of the Bluescat simulator version 0.6 [27]. As an open source Bluetooth network simulator, Bluescat is the extension of BlueHoc network simulator [28], which is developed based on ns-2 [30] by IBM. The model is extended by including the packet collisions due to simultaneous transmissions of packets from two or more different sources (from different on-demand piconets) within the same radio range. All the master nodes use the deficit round robin scheduling scheme [31]. The random waypoint mobility model [37] is also used for modeling the nodes' movement.

32.4.1 Comparison between TPSF+ and BTCP

In this section, two scatternet formation algorithms TPSF+ [22] and BTCP [7] are being compared. In the simulation model, there are 36 nodes randomly placed within the coverage area of 10 m × 10 m. Since all the nodes are in the same radio range, packet collisions may occur. Two nodes are randomly selected as the source and destination pairs. For each data point, the simulation was run 100 times and each run time was 120 sec. The nonpersistent transmission control protocol (TCP) on/off traffic is used. During the "on" periods, packets are generated at a constant burst rate of 1440 kbps. During the "off" periods, no traffic is generated. Burst times and idle times follow the exponential distributions. The average "on" time is 0.5 sec and the average "off" time is 0.5 sec. The packet size is 1000 bytes.

The performance metrics include the *aggregate throughput* and the *average end-to-end delay*. The aggregate throughput is defined as the total throughput obtained by all the communication sessions. The end-to-end delay is determined from the time when the packet is created to the time when the packet is received.

The performance comparisons between TPSF+ and BTCP are shown in Figure 30.10 and Figure 32.11 The results show that TPSF+ achieves higher aggregate throughput and lower end-to-end delay, compared with BTCP. The comparison between TPSF+ and BTCP for user datagram protocol (UDP) performance can be found in [22].

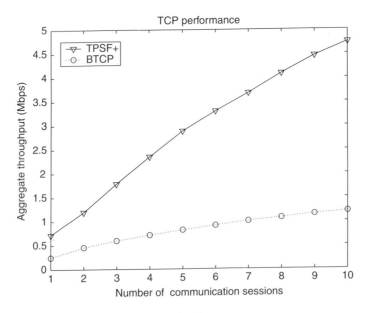

FIGURE 32.10 Aggregate throughput versus number of sessions.

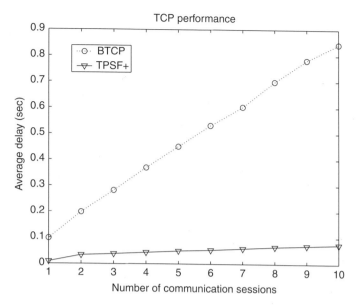

FIGURE 32.11   End-to-end delay versus number of sessions.

32.4.2   Comparison between TPSF+ and Bluenet

In this section, the scatternet formation algorithms Bluenet [9] and TPSF+ [22] are being compared. In the simulation model, there are 40 nodes in total and these nodes are placed in an area of 20 m × 20 m. Note that the scatternet formation algorithm BTCP cannot be included in this experiment due to the fact that BTCP assumed all the nodes are within the same transmission range.

The sample scatternet topologies based on TPSF + and Bluenet scatternet are shown in Figure 32.12 and Figure 32.13, respectively. The arrow denotes the master–slave relationship. The dotted and solid arrows denote that slaves are in park mode and in active mode, respectively. In Figure 32.12, the on-demand scatternet has three hops and it is dedicated to data packet transmission.

Figures 32.14 and Figure 32.15 show the comparisons in terms of the aggregate throughput and end-to-end delay between TPSF+ and Bluenet. TPSF+ achieves a higher aggregate throughput and lower end-to-end delay than Bluenet in multi-hop scenario. The performance bottleneck in Bluenet is due to the traffic load at the master and bridge nodes. TPSF+ avoids this bottleneck problem by setting up dedicated on-demand scatternet for each communication session.

32.4.3   Comparison between TPSF and TPSF+

In this experiment, TPSF+ [22] is compared with the original TPSF [20]. The random waypoint mobility model is used [37]. Each slave is moving within its master's range with variant maximum speeds of 2, 4, 6, 8, and 10 m/sec. For each speed, the simulation was run 2000 times. The performance metric is the *successful on-demand scatternet connection ratio*, which is defined as the ratio of the number of successful connected paths to the total number of communication paths requested.

Figure 32.16 shows that when the node's moving speed increases, the successful path connection ratio is slightly reduced by using TPSF+, whereas it is reduced greatly by using TPSF. In TPSF, the route information is obtained when the nodes access the network. The route information may not always be up-to-date due to the movement of the nodes. In TPSF+, the route information is obtained when a source node needs to send data packets to a destination node. The more updated route information causes the on-demand scatternet be created with a higher successful ratio.

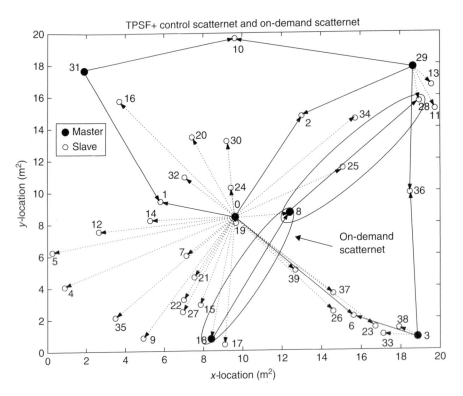

FIGURE 32.12   Sample TPSF+ scatternet topology.

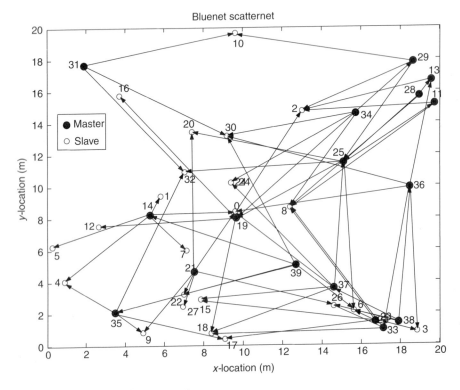

FIGURE 32.13   Sample Bluenet scatternet topology.

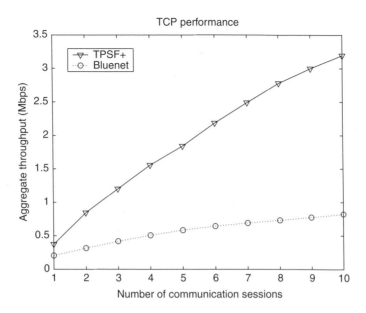

FIGURE 32.14   Aggregate throughput versus number of sessions.

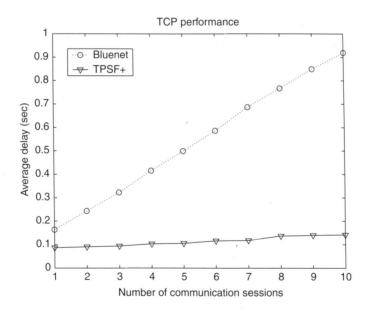

FIGURE 32.15   End-to-end delay versus number of sessions.

32.5   Conclusions

In this paper, we described a number scatternet formation algorithms proposed recently. We conclude this paper by stating a number of unresolved issues that need to be resolved for the deployment of the Bluetooth scatternet:

Implementation and prototyping: Although various scatternet formation algorithms have been proposed in the literature, their performance is mainly obtained via computer simulations. There is very few work

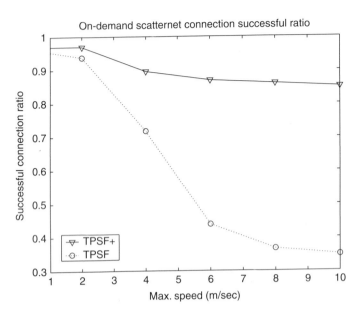

FIGURE 32.16   Successful path connection ratio.

on prototyping of scatternet formation algorithms [32]. A detailed comparison study via a testbed of Bluetooth devices is necessary.

Cross-layer optimization between topology maintenance and scheduling: Over the past few years, there are different inter-piconet scheduling algorithms proposed in the literature (e.g., [33–36]). Although the performance evaluation via network simulations have included different Bluetooth link types (e.g., synchronous connection-oriented [SCO], asynchronous connectionless [ACL]) and traffic models (e.g., constant bit rate traffic, TCP traffic, UDP traffic), the impact of topology maintenance and scheduling have not been investigated. If several time slots in each master node needs to be allocated for topology maintenance, there may be a decrease in the aggregate throughput for Bluetooth ad hoc networks, which use proactive scatternet formation algorithms. Further work on a study of cross-layer optimization between topology maintenance and scheduling is recommended.

Acknowledgment

This work was supported by the Natural Sciences and Engineering Research Council of Canada (NSERC) under grant number STPGP 257684-02. The authors would like to thank Yoji Kawamoto and Xiuying Nie for discussions on different scatternet formation algorithms.

References

[1] *Specification of the Bluetooth System*,   version   1.2,   November   2003,   available   at http://www.bluetooth.org.

[2] The IEEE 802.15 WPAN Task Group 1a, http://www.ieee802.org/15/pub/TG1a.html.

[3] J.C. Haartsen, "The bluetooth radio system," *IEEE Personal Communications*, vol. 7, pp. 28–36, 2000.

[4] P. Johansson, M. Kazantzidis, R. Kapoor, and M. Gerla, "Bluetooth: an enabler for personal area networking," *IEEE Network*, vol. 15, pp. 28–37, 2001.

[5] B. Chatschik, "An overview of the bluetooth wireless technology," *IEEE Communications Magazine*, vol. 39, pp. 86–94, 2001.

[6] K.V. Sairam, N. Gunasekaran, and S.R. Redd, "Bluetooth in wireless communication," *IEEE Communications Magazine*, vol. 40, pp. 90–96, 2002.

[7] T. Salonidis, P. Bhagwat, L. Tassiulas, and R. LaMaire, "Distributed topology construction of bluetooth personal area networks," in *Proceedings of the IEEE Infocom'01*, Anchorage, AK, pp. 1577–1586, April 2001.

[8] G.V. Zaruba, S. Basagni, and I. Chlamtac, "Bluetrees — scatternet formation to enable bluetooth-based ad hoc networks," in *Proceedings of the IEEE ICC'01*, Helsinki, Finland, pp. 273–277, June 2001.

[9] Z. Wang, R.J. Thomas, and Z. Haas, "Bluenet — a new scatternet formation scheme," in *Proceedings of the 35th Hawaii International Conference on System Sciences (HICSS-35)*, Big Island, HI, January 2002.

[10] C. Law, A.K. Mehta, and K.Y. Siu, "A bluetooth scatternet formation algorithm," *ACM Mobile Networks and Applications Journal*, vol. 8, pp. 485–498, 2003.

[11] G. Tan, A. Miu, J. Guttag, and H. Balakrishnan, "An efficient scatternet formation algorithm for dynamic environments," in *Proceedings of the IASTED International Conference on Communications and Computer Networks (CCN)*, Cambridge, MA, November 2002.

[12] H. Zhang, J.C. Hou, and L. Sha, "A bluetooth loop scatternet formation algorithm," in *Proceedings of the IEEE ICC'03*, Anchorage, AK, pp. 1174–1180, May 2003.

[13] C. Petrioli, S. Basagni, and I. Chlamtac, "Configuring BlueStars: multihop scatternet formation for Bluetooth networks," *IEEE Transactions on Computers*, vol. 52, pp. 779–790, 2003.

[14] C. Petrioli, S. Basagni, and I. Chlamtac, "BlueMesh: degree-constrainted multi-hop scatternet formation for bluetooth networks," *ACM/Kluwer Mobile Networks and Applications*, vol. 9, pp. 33–47, 2004.

[15] R. Guerin, J. Rank, S. Sarkar, and E. Vergetis, "Forming connected topologies in bluetooth adhoc networks", in *Proceedings of the International Teletraffic Congress (ITC 18)*, Berlin, Germany, September 2003.

[16] X. Li and I. Stojmenovic, "Partial delaunay triangulation and degree-limited localized bluetooth scatternet formation," in *Proceedings of the Ad-hoc Networks and Wireless (ADHOC-NOW)*, Fields Institute, Toronto, Canada, September 2002.

[17] F. Cuomo, G. Di Bacco, and T. Melodia, "SHAPER: a self-healing algorithm producing multi-hop bluetooth scatternets," in *Proceedings of the IEEE Globecom'03*, San Francisco, CA, pp. 236–240, December 2003.

[18] M.A. Marsan, C.F. Chiasserini, A. Nucci, G. Carello, and L. De Giovanni, "Optimizing the topology of bluetooth wireless personal area networks," in *Proceedings of the IEEE Infocom'02*, New York, pp. 572–579, June 2002.

[19] Y. Liu, M.J. Lee, and T.N. Saadawi, "A bluetooth scatternet-route structure for multihop ad-hoc networks," *IEEE Journal on Selected Areas in Communications*, vol. 21, pp. 229–239, 2003.

[20] Y. Kawamoto, V. Wong, and V. Leung, "A two-phase scatternet formation protocol for bluetooth wireless personal area networks," in *Proceedings of the IEEE Wireless Communications and Networking Conference (WCNC)*, New Orleans, LA, pp. 1453–1458, March 2003.

[21] C. Zhang, V.W.S. Wong, and V.C.M. Leung, "TPSF+: a new two-phase scatternet formation algorithm for bluetooth ad hoc networks," in *Proceedings of the IEEE Globecom'04*, Dallas, TX, pp. 3599–3603, December 2004.

[22] C. Zhang, "A new two-phase scatternet formation algorithm for bluetooth wireless personal area networks," *M.A.Sc.'s Thesis*, The University of British Columbia, Canada, December 2003.

[23] G. Tan and J. Guttag, "A locally coordinated scatternet scheduling algorithm," in *Proceedings of the IEEE Conference on Local Computer Networks (LCN)*, Tampa, FL, November 2002.

[24] J. Misic and V.B. Misic, "Bridges of Bluetooth county: topologies, scheduling, and performance," *IEEE Journal on Selected Areas in Communications*, vol. 21, pp. 240–258, 2003.

[25] S. Basagni, R. Bruno, and C. Petrioli, "A performance comparison of scatternet formation protocols for networks of bluetooth devices," in *Proceedings of the IEEE International Conference on Pervasive Computing and Communications (PerCom'03)*, Dallas — Fort Worth, TX, pp. 341–350, March 2003.

[26] D.B. Johnson, D.A. Maltz, and Y.C. Hu, "The dynamic source routing protocol for mobile ad hoc networks (DSR)," *Internet Engineering Task Force (IETF) Internet-Draft, draft-ietf-manet-dsr-10.txt,* July 2004.

[27] *The Bluescat Simulator* version 0.6. Available at http://www-124.ibm.com/developerworks/oss/cvs/
· bluehoc/bluescat0.6.

[28] *The BlueHoc Simulator* version 1.0, Available at http://www-124.ibm.com/developerworks/ open-source/bluehoc/.

[29] *The Blueware Simulator* version 1.0, Available at http://nms.lcs.mit.edu/projects/blueware.

[30] *The Network Simulator* ns-2.1b7a. Available at http://www.isi.edu/nsnam/ns/.

[31] M. Shreedhar and G. Varghese, "Efficient fair queuing using deficit round robin," *IEEE/ACM Transactions on Networking,* vol. 4, pp. 375–385, 1996.

[32] S. Asthana and D.N. Kalofonos, "Enabling secure ad-hoc group collaboration over bluetooth scatternets," in *Proceedings of the 4th Workshop of Applications and Services of Wireless Networks (ASWN'04),* Boston, MA, August 2004.

[33] P. Johansson, R. Kapoor, M. Kazantzidis, and M. Gerla, "Rendezvous scheduling in bluetooth scatternets," in *Proceedings of the IEEE ICC'02,* New York, April/May 2002.

[34] S. Baatz, M. Frank, C. Kuhl, P. Martini, and C. Scholz, "Bluetooth scatternets: an enhanced adaptive scheduling scheme," in *Proceedings of the IEEE Infocom'02,* New York, pp. 782–790, 2002.

[35] L. Tassiulas and S. Sarkar, "Maxmin fair scheduling in wireless networks," in *Proceedings of the IEEE Infocom'02,* New York, pp. 763–772, June 2002.

[36] R. Kapoor, A. Zanella, and M. Gerla, "A fair and traffic dependent scheduling algorithm for bluetooth scatternets," *ACM/Kluwer Mobile Networks and Applications,* vol. 9, pp. 9–20, 2004.

[37] J. Broch, D. Maltz, D. Johnson, Y. Hu, and J. Jetcheva, "A performance comparison of multi-hop wireless ad-hoc network routing protocols," in *Proceedings of the ACM MobiCom'98,* Dallas, TX, pp. 85–97, October 1998.

33

Bluetooth Voice Access Networks

33.1 Introduction...**33**-755
33.2 Bluetooth Overview **33**-756
33.3 Voice over SCO Links..................................**33**-759
33.4 Voice over ACL Links **33**-764
　　　Bluetooth FH Collision Loss • Extended SCO (eSCO) Links
33.5 Summary **33**-771
　　　Future Work
References **33**-771

Yun Wu
Terence D. Todd

33.1 Introduction

Bluetooth is currently being embedded into a large variety of mobile and portable devices, such as cellphones, Personnel Digital Assistants (PDAs), and laptops. In many devices, real-time voice is a very commonly supported application. Future access networks can use this built-in connectivity to provide real-time voice services in public places such as airports, convention centers, and shopping malls.

Native support for voice is included in the Bluetooth standard through the use of synchronous connection oriented (SCO) links, which provide a circuit-switched type of connection using 64 kbps PCM. In this chapter, we consider the issue of real-time voice support over Bluetooth for future voice access networks. Recent work in this area is summarized and discussed. The native voice capacity of a single Bluetooth piconet is very low, and in many cases multiple Bluetooth access points would be needed to obtain the desired capacity and coverage in access network applications. In order to ensure contiguous service areas many deployments will use heavily overlapped radio coverage using a number of networked single-radio Bluetooth base stations or multichip designs [1]. In such a system the Bluetooth base stations (BBSs) would typically be interconnected via a high-speed wired LAN, which also provides an attachment to the PSTN/Internet. A design referred to as the Bluetooth distributed voice access protocol (DVAP) is introduced. In DVAP, Bluetooth mobiles collect connectivity information, which is provided to the base station infrastructure. Mobile node migration is then used to reduce blocking when a BBS is carrying a full load of active SCO links. A simple heuristic referred to as dynamic base station selection is used for making the base station assignment. To improve the quality of service, the approach referred to as voice over ACL is also discussed in detail.

The remainder of the chapter is organized as follows. Section 33.2 overviews voice transmission in Bluetooth. In Section 33.3, several designs based on SCO links are discussed and summarized. Following this, in Section 33.4 we consider the network designed based on ACL links. Finally, in Section 33.5 some concluding remarks are given.

33.2 Bluetooth Overview

Bluetooth has emerged as a key commercial wireless standard, providing reliable, secure, and low cost local communications. Current projections for Bluetooth are optimistic and suggest that Bluetooth enabled devices will continue to proliferate at a very high rate. Bluetooth is operated in the 2.4-GHz license-free Industrial, Scientific, and Medical (ISM) radio band, and uses a slow frequency hopping air interface at a rate of 1600 hops/sec, with each time slot occupying 625 μsec. The 2.4 GHz ISM band spans about 84 MHz, from 2.4 to 2.4835 GHz; Bluetooth divides this into 79 1-MHz channels, which are used for frequency hopping. Bluetooth has also defined three different radio power classes. Class 1 operates at 100 mW (20 dBm), Class 2 at 2.5 mW (4 dBm), and Class 3 at 1 mW (0 dBm). In a Class 3 system, the typical coverage range is about 10 m, which is targeted at simple cable replacement applications. Class 1 provides a coverage range, which is comparable to that of conventional wireless LANs (WLANs) such as IEEE 802.11.

Interference immunity is an important issue in wireless networks. Since the ISM band is license free, any radio transmitter conforming to public regulations may use it. For this reason the interference characteristics are situation dependent, difficult to generalize, and both interference suppression and interference avoidance can be used to obtain interference immunity. Direct-sequence spectrum spreading (DSSS) is widely used in public cellular systems [2,3]. In an ad hoc environment such as Bluetooth, however, the dynamic range of interfering and intended signals can be huge and near–far ratios in excess of 50 dB are typical [4]. Interference avoidance using frequency-hopping spread spectrum (FHSS) is a more efficient way to achieve interference immunity in this type of situation, and this is the approach that has been taken in Bluetooth. IEEE 802.11b/g devices also operate in the 2.4-GHz ISM band and their Bluetooth interference effects have been the subjects of several recent studies [5]. Other in-band interferers such as microwave ovens may also lead to Bluetooth interference problems [4]. Bluetooth's frequency hopping design is such that multiple overlapping Bluetooth networks are also mutually interfering, that is, interpiconet interference caused by co-located Bluetooth networks provides potential interference, which may be important in certain situations. This issue will be discussed at length later in this chapter.

Bluetooth has a master/slave link layer protocol architecture, and typical nodes have both master and slave functionality. A master can organize up to seven active slaves into a communication unit referred to as a piconet. An example of a piconet is shown in Figure 33.1. All direct communications within a piconet must occur between the master and its slaves, that is, there is no direct slave–slave interaction and thus communication between slaves must occur by relaying packets through the master. When traffic flows call for significant slave-to-slave communication, a reduction in usable capacity will obviously occur.

A Bluetooth master strictly controls transmission activities in its piconet, and the frequency hopping sequence used is uniquely defined and derived from the master's identity. This sequence defines the piconet and therefore there can never be more than one master per piconet. Bluetooth uses slotted time division duplexing (TDD) for communications, using 625 μsec time slots. In a Bluetooth device, its clock is derived from a free running native clock, which is never adjusted, is never turned off, and whose resolution is half that of the basic time slot, that is, 312.5 μsec.

Two basic types of physical links have been defined in the Bluetooth specification, that is, the SCO link and the Asynchronous Connection-Less (ACL) link. SCO links are intended for carrying voice connections, whereas data traffic is sent over ACL links. An SCO link is a symmetric, point-to-point link between a master and a slave, where time slots are reserved for voice packets using time division multiplexing. An SCO link can be viewed as a circuit-switched connection between the master and its slave. An ACL link can be operated either symmetrically or asymmetrically, and time slots are not reserved. Transmissions on ACL links may be interrupted by other activities, such as inquiry, paging, and scanning. Different kinds of packets have been defined in Bluetooth, and some of the relevant packet types will now be discussed.

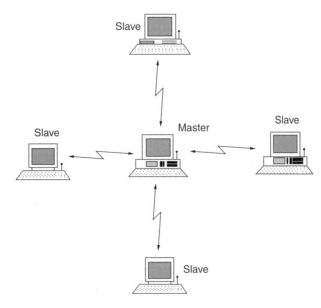

FIGURE 33.1 Bluetooth piconet.

TABLE 33.1 SCO Packets

Type	Payload header (bytes)	User payload (bytes)	FEC	CRC	Symmetric max. rate (kb/sec)
HV1	n/a	10	1/3	No	64.0
HV2	n/a	20	2/3	No	64.0
HV3	n/a	30	No	No	64.0

The DM1 packet is a special type of Bluetooth packet. DM1 cannot only carry regular user data but also can be used to support control messages over any link type, that is, both ACL and SCO. When a DM1 packet is used as a control packet, it can be recognized on the SCO link and can therefore interrupt the synchronous transmission of voice in order to send control information.

The SCO packets are carried over SCO voice links (Table 33.1). Since there is no cyclic redundancy code (CRC) checking in SCO packets, they are never retransmitted even if they have not been correctly received. This means that when real-time voice is carried over SCO links, the SCO packet loss rate must always be such that it satisfies the minimum loss requirements of the voice connection. Three SCO packets, that is, HV1, HV2, and HV3 have been defined. Additionally, the DV packet, which can carry mixed voice/data is defined. Table 33.2 shows the characteristics of the Bluetooth SCO packets. HV stands for high-quality voice. All three HV packets, that is, HV1, HV2, and HV3 packets are transmitted in a single time slot, and deliver a 64 kbps user voice rate. The number following "HV," is the maximum number of voice links that an individual Bluetooth device can support simultaneously. The HV1 and the HV2 packets are protected with a rate 1/3 forward error correction (FEC) and 2/3 FEC, respectively. The HV3 packet is not protected with FEC. Therefore, HV1, HV2, and HV3 packets carry 10, 20, and 30 information bytes respectively. An HV1 packet can carry 1.25 msec of speech at a 64 kbps rate, that is, $10 \, \text{bytes}/(2 \times 625 \, \mu\text{sec}) = 64 \, \text{kbps}$. Every two time slots, an HV1 packet will be transmitted. Therefore the intervoice packet transmission interval of HV1 packets in one direction is two time slots (i.e., $T_{SCO} = 2$). Using the HV1 packet, one 64 kbps voice link occupies the entire capacity of a Bluetooth device, that is, only a single HV1 SCO link can be support by a Bluetooth piconet. The transmission interval using HV2 packets in the same directional link is four time slots (i.e., $T_{SCO} = 4$). One HV2 packet can carry up to 2.5 msec of 64 kb/sec speech. Two HV2 SCO links would be supported per Bluetooth device. An HV3 packet contains 3.75 msec

TABLE 33.2   ACL Packets

Type	Payload header (bytes)	User payload (bytes)	FEC	CRC	Symmetric max. rate (kb/sec)	Asymmetric max. rate (kb/sec) Forward	Reverse
DM1	1	0–17	2/3	Yes	108.8	108.8	108.8
DH1	1	0–27	No	Yes	172.8	172.8	172.8
DM3	2	0–121	2/3	Yes	258.1	387.2	54.4
DH3	2	0–183	No	Yes	390.4	585.6	86.4
DM5	2	0–224	2/3	Yes	286.7	477.8	36.3
DH5	2	0–339	No	Yes	433.9	723.2	57.6
AUX1	1	0–29	No	No	185.6	185.6	185.6

of 64 kbps speech and is transmitted every six time slots (i.e., $T_{SCO} = 6$). Up to three HV3 SCO links can be supported simultaneously. Since error resilience is reduced in going from HV1 to HV3 packets, voice capacity is increased from a single voice link to three voice links.

Conventional data can be also be carried over SCO links using DV packets. A DV packet includes both a voice and data field. A one-byte header in the DV packet is used as a data field. In terms of user payload, 10 bytes belong to voice without FEC protection and up to nine bytes are used by data with a 2/3 rate FEC. There is a two-byte CRC in the DV packet, which is only related to the data. For this reason a DV packet can be retransmitted only to recover the data field. Every two time slots (1.25 msec) a DV packet is sent, therefore each DV packet would carry 1.25 msec of speech at a rate of 64 kbps. For the voice part, the DV packet is the similar to the HV1 packet except that no FEC protection is used.

ACL packets are used on the ACL links. Three DM packets, three DH packets, and one AUX1 packet have been defined for this purpose. Since a CRC check code exists in DM and DH packets, these packets can be retransmitted if no acknowledgment of proper reception is received. However, the AUX1 packet has no CRC and is not retransmitted. DM stands for data-medium rate. There are three types of DM packets, DM1, DM3, and DM5. The number following "DM," represents the number of time slots across which the packet transmission extends. For example, a DM5 packet is continuously transmitted over 5 time slots on the same frequency hop. The information plus CRC bits in DM packets are coded with a rate 2/3 FEC, which adds 5 parity bits to every 10-bit segment. If necessary, extra zeros are appended after the CRC bits to get the total number of bits to a multiple of 10 [6]. The payload header in the DM1, DM3, and DM5 packets is 1, 2, and 3 bytes long, respectively. The user payload of the DM1, DM3, and DM5 packet contains up to 17, 121, and 224 information bytes, respectively, as shown in Table 33.2. The maximum symmetric rate for DM3 packets is 258.1 kbps and the maximum asymmetric rate is 387.2 and 54.4 kbps for forward and reverse links, respectively. When the system is operated in a symmetric manner, both directions use DM3. The maximum rate is 121 bytes/$(6 \times 625 \, \mu sec) = 387.2$ kb/sec. When the system is operated in an asymmetric manner, the forward link uses DM3 packets and reverse link uses DM1 packets. The maximum forward rate is 121 bytes/$(4 \times 625 \, \mu sec) = 387.2$ kb/sec and the maximum reverse rate is 17 bytes/$(4 \times 625 \, \mu sec) = 54.4$ kb/sec.

Table 33.2 also shows the characteristics of DH packets. DH means "data high" rate. There are also three types of DH packets, DH1, DH3, and DH5. As in the DM packets, the number following the word, "DH," represents the number of time slots that a packet uses, that is, a DH3 packet is transmitted across 3 time slots. The payload header in the DH1, DH3, and DH5 packets is 1, 2, and 3 bytes long, respectively. The information bits in DH packets are not FEC encoded. As a result, the user payload of the DH1, DH3, and DH5 packets contains up to 27, 183, and 339 information bytes, respectively. Comparing DM and DH packets for the same packet length, we note that DH packets give higher user capacity than DM packets because there is no FEC overhead. Comparing single time slot and multiple time slot packets in DM or DH packets, single time slot packets result in lower user capacity than multislots packets because there is no loss due to the guard band that separates single slot packets.

Connection establishment in Bluetooth is complicated by the need to synchronize the frequency hopping between nodes. Three mechanisms have been defined to support connection establishment, namely, scanning, paging, and inquiry. Scanning is the process whereby Bluetooth nodes listen for other nodes. When scanning is performed, the frequency hopping rate used is very low, so that the scanning node persists on the same frequency hop for relatively long periods of time. Conversely, a node that performs an inquiry or a page, actively transmits query or paging packets at a high frequency-hopping rate, and thus eventually the two nodes will coincide in time and frequency.

When two devices have never been connected, inquiry and inquiry-scan can be used by the devices to find one another. The inquiring device uses a known general inquiry access code (GIAC) to determine the frequency hopping sequence, and the hopping rate used is twice the nominal slot rate, that is, 3200 hops/sec. Two inquiry packets are sent in a single time slot. The inquiry-scanning device changes its listening frequency every 1.28 sec. When this device receives an inquiry packet, it does not respond immediately but performs a random backoff to avoid inquiry response collisions. After reentering inquiry-scan and receiving a second inquiry packet, it will respond. The inquirer can then proceed to page the device. The paging process is similar except that once a paging device knows the other node's address and clock phase, it can use this information to page the node directly and efficiently. As in inquiry, the paging hop rate is 3200 hops/sec, with two pages per 625 μsec time slot. In the page-scanning device, the listening frequency is changed slowly, every 1.28 sec. Once paging is successful, frequency hop synchronization has been achieved and the slave node can join the piconet. Different data scheduling policies have been studied for Bluetooth [7,8] so that data throughput and delay can be optimized for a single piconet.

One of the key features of Bluetooth is its power saving capabilities. Three power saving modes, that is, sniff, hold, and park, have been defined. In sniff mode, a slave does not listen continuously. Instead, a sniff interval is defined between the slave and its master, and the slave is free to assume a low power sleep/doze state between sniff intervals. In hold mode, the activity of the slave can be further reduced. Before entering hold mode, the master and the slave agree upon the time duration of the hold. When this time expires, the slave will awaken, synchronize to the traffic on the channel, and will wait for communication from the master. Park mode is the most aggressive of the power saving modes. In this case the slave gives up its active member address (AM_ADDR) and does not actively participate in the piconet, however it still remains synchronized to the master. A slave in park mode is given an 8-bit parked member address (PM_ADDR) and 8-bit Access Request Address (AR_ADDR). The PM_ADDR is used in a master-initiated unpark procedure. The AR_ADDR is used by the parked slave in a slave-initiated unpark procedure. A parked slave can be distinguished from the other parked slaves using PM_ADDR. In addition to the PM_ADDR, a parked slave can be unparked by its 48-bit BD_ADDR. The all-zero PM_ADDR is a reserved address; if a parked device has the all-zero PM_ADDR it can only be unparked by the BD_ADDR. In that case, the PM_ADDR has no meaning. By varying the sniff interval (T_{sniff}) or serving time ($N_{sniff-attempt}$), power consumption can be further reduced [9].

Reference 7 proposed an Unparked Queue-based policy (UQP) and Parked-Unparked Queue-based policy (PUQP), which extends PP in Reference 10 to implement a master and parked slave pair. These two policies will reduce park time for slaves that have high backlogs in the master–slave queues. Otherwise, the park time increases if slaves have a small backlog in the master–slave queues. UQP is implemented when the master-initiated unpark procedure is processed. Since the master cannot know the exact queue sizes in slaves, the average queue size information has to use an estimation. PUQP is implemented when the slave-initiated unpark procedure is used. The queue sizes of slaves can be passed along with the packets to the master during the unpark access window. Compared with the constant unpark interval, UQP and PUQP have higher throughput, lower delay, and lower power consumption.

33.3 Voice over SCO Links

The Bluetooth baseband protocol provides both circuit and packet switched modes of operation, that is, time slots can be reserved for packets carrying voice over SCO links (VoSCO), or can be dynamically allocated for data over ACL links. In this section, the characteristics of VoSCO are discussed.

The Bluetooth standard provides a reference profile for simple cordless telephony applications [6]. SCO links are the native mechanism for carrying two-way symmetric voice transmissions based on 64-kbps PCM. For each SCO link, a master transmits a voice packet to the slave in question at regular time intervals, given by the parameter T_{SCO} [11]. The slave then responds with its voice packet in the following slot. There are three ways that a standard SCO link can be packetized, namely, HV1, HV2, and HV3. These correspond to three different values of T_{SCO}, that is, 2, 4, and 6 slots, respectively. In the lower rate schemes, the additional overhead is used for increased FEC coding [6]. Using HV1 packets, one voice link will occupy the entire capacity of the Bluetooth device. Using HV2 packets, the Bluetooth device can provide up to two voice links simultaneously. One HV2 voice link will utilize half of the entire link capacity. A data link using DH1 or DM1 packets in both directions can be supported at the same time when the Bluetooth device has one HV2 voice link. Using HV3 packets, up to three voice links can be provided by the Bluetooth device. With one HV3 voice link, two thirds of the capacity can be used to carry data. If DH1 or DM1 packets are used in both directions, two data links at the rate of 108.8 or 172.8 kbps would be supported simultaneously. In the asymmetric case, one DM3 or DH3 data link, that is, the forward link using DM3 or DH3 packets at the rate of 258.1 or 390.4 kbps, respectively, and the backward link using DM1 or DH1 packets at the rate of 54.4 or 86.4 kbps, respectively, can coexist with an HV3 voice link. Note that the voice cannot be carried simultaneously with a DM5 or DH5 data link. This is also found in References 12 and 13.

Reference 14 considers a single piconet and evaluates the throughput performance of VoSCO. Since the master can support only seven active slaves, at least one slave has to be parked in order to access a new slave when seven are active. A First-Active-First Park (FAFP) scheduling policy is used for the parking or unparking of slaves. In FAFP, the slaves are parked in the order of the time that the slaves actively participate the piconet, that is, the oldest active slave is forced to enter park mode whenever a new slave joins the piconet and the number of active slaves is seven. The parked slave with expired parking time is unparked and becomes active. Up to ten slaves can generate either voice or data traffic. The master polls slaves in a round-robin fashion for ACL connections. Only DH1 or DM1 packets are assumed to be transmitted if at least one SCO connection is present. Data traffic is generated as a Markov-modulated Poisson process (MMPP). Each data source is modeled by a two state Markov chain, where on and off periods are exponentially distributed with a mean of α^{-1} and β^{-1} sec, respectively.

Throughput performance is evaluated when one SCO connection is present (DH-1HV3, DH-1HV2, DM-1HV3, and DM-1HV2). Since we have an SCO connection, 33% of the total available slots must be reserved for the voice (SCO) traffic. When one ACL and one SCO connection are established, the rate of the ACL connection is 114 kbps (DH-1HV3), which can be compared with the rate of 370 kbps with two ACL connections (DH5), that is, a 69.2% decrease. Throughput performance worsens as the number of ACL connections increases. When six ACL connections and one SCO connection are supported, the rate shows 19 and 12 kbps for DH-1HV3 and DM-1HV3, respectively. When nine ACL connections and one SCO connection are supported using park mode, the rate drops to 12 and 8 kbps for DH-1HV3 and DM-1HV3, respectively. The effect of degradation in throughput for an SCO connection is shown to become greater as the number of ACL connections increases, for example, 47 kbps for 10 ACL connections (DM5) versus 8 kbps for 9 ACL and one SCO connections (DM-1HV3), which is 83.1% decrease in throughput.

Performance is also considered when two SCO connections are supported (DH-2HV3 and DM-2HV3). In this case, the two connections occupy more time slots (67% of total available slots) than in the previous case, and thus a lower data rate is available. When four ACL connections with DM packets and two SCO connections with HV3 packets are in operation, the residual throughput is only 8 kbps. This can degrade even further to 3 kbps if more than eight ACL and two SCO connections communicate using park mode (DM-2HV3). In this case, 47 kbps for 10 ACL connections (DM5) reduces to 3 kbps for 8 ACL and two SCO connections (DM-2HV3), which is a 93.6% performance degradation. These low data rates may not be suitable for certain applications.

Data traffic performance in the presence of SCO links was also considered in Reference 8. In this study, up to two SCO links were considered with different values for T_{SCO}, while the piconet was subjected to persistent TCP data transfers. As in Reference 14, similar data capacity reduction effects were observed with increasing numbers of voice links.

Many systems use compressed real-time voice transmission, which can significantly reduce the required rate per call. Bandwidth can be further saved by removing speech silence periods, that is, using voice activity detection (VAD). If T_{SCO} is made larger than that prescribed in the Bluetooth standard, more capacity will be available for ACL traffic [8]. Using this idea, Reference 15 presents an approach called Adaptive T_{SCO}. Using VAD and variable rate voice coding, the SCO interpacket transmission interval can be adaptively determined. In this study, a two-state Markov chain model was used to simulate talkspurt and silence periods for voice. For Adaptive T_{SCO}, the values of T_{SCO} during activity and silence were chosen to be 6 and 12 slots, respectively. With one voice connection, the data throughput is 238.93 kbps and mean packet delay is 21.72 msec using VoSCO. For Adaptive T_{SCO}, the numbers are 257.18 kbps and 18.09 msec, respectively. The performance of Adaptive T_{SCO} is clearly better than that of conventional VoSCO. Data throughput is 119.4 kbps and packet delay is 36.47 msec using VoSCO. For Adaptive T_{SCO}, these numbers are 172.2 kbps and 31.05 msec, respectively. Increasing the number of SCO connections, similar capacity reduction effects were obtained as in References 8 and 14. Note that the Adaptive T_{SCO} scheme is not compatible with the Bluetooth specification and would require changes to the standard.

In some applications, multiple Bluetooth access points would be needed to obtain reasonable coverage. In order to ensure contiguous service areas these networks may use overlapped radio coverage using a number of networked single-radio BBSs. In such a system, the BBSs would be interconnected via a high-speed wired LAN, which also provides an attachment to the PSTN/Internet. This type of system is shown in Figure 33.2. A potential application for this type of system would be like an access network in an airport lounge where phone calls could be made and received using Bluetooth enabled handsets. In the figure, four LAN-interconnected BBSs are shown supporting voice communications for a potentially large number of Bluetooth mobiles (BMs). Locally generated voice calls are originated at the BMs and connected through Bluetooth SCO links onto the PSTN/Internet via the LAN/gateway. It is also assumed that the BMs can receive incoming calls that originate in the PSTN/Internet. A requirement for this is that the mobile nodes register and remain in contact with a base station when they are powered-on and within the coverage area of the system.

In Reference 16 several access point designs were studied based upon this type of architecture. It was assumed that there is strong overlapping of radio coverage between all base stations. The different designs considered the value of base station coordination among the Bluetooth piconets that are supporting SCO voice transmissions. The first design proposed is referred to as Bluetooth Base Station Original (BBSO),

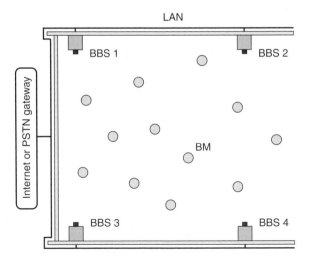

FIGURE 33.2 Bluetooth voice access network.

and corresponds to using the Bluetooth telephony profile directly. Each BBS operates independently and therefore this corresponds to the case where there is no coordination among the base stations. A BM associates with a BBS using the Bluetooth `INQUIRY`/`PAGE` procedures [6]. Once a base station is contacted via `INQUIRY`, `PAGING` is used to establish an ACL link [4]. The mobile node is then parked by the BBS to await incoming and outgoing calls. Once a BM is parked, it must be periodically awakened to determine if a voice call needs to be set up. However, the wakeup procedure is effective only when the BBS has no more than two SCO links in progress [11,16].

In Bluetooth Base Station with Migration (BBSM), "node migration" is used to reduce the effects of base station blocking. In BBSM, when the BBS is about to add a third SCO link, it first unparks and disconnects all of its parked BMs. Each mobile then reenters the `INQUIRY` state so that it can reassociate with a new BBS. In Bluetooth Base Station with Standby Unit (BBSS), one of the BBSs is assigned the role of a `STANDBY_BBS`, dedicated to node discovery and voice call setup functions. The other BBSs are `VOICE_BBS`'s. The main role of the `STANDBY_BBS` is to perform `INQUIRY SCAN`, `PAGE SCAN`, and wakeup functions for BMs that are parked within the coverage area. The `STANDBY_BBS` however, does not provide SCO connections, rather it dynamically moves BMs to an available `VOICE_BBS` when SCO links are required. All nodes are initially parked at the `STANDBY_BBS`, and this avoids the restrictions discussed above, however the SCO link capacity of the `STANDBY_BBS` is wasted. An advantage of this scheme is that BM migration occurs via paging and only when a call is being set up or terminated. A disadvantage with this approach however, is that the system must be carefully designed so that BBS coverage is heavily overlapped. In Reference 16 it was shown that call blocking in BBSS is lower than that of the other designs. Unfortunately, the best design in Reference 16 from a call blocking standpoint (i.e., BBSS) relies on the base stations having full overlap of their coverage areas.

Figure 33.3 shows a comparison of call blocking probability for the different designs proposed in Reference 16 when there are 5 overlapping BBS. The difference in the three plotted curves shows the potential for Bluetooth link sharing. At a 1% blocking rate for the BBSO case, the BBSM protocol achieves a level of blocking, which is over an order of magnitude smaller. The BBSM design can operate with over 60% higher traffic loading than BBSO at the same 1% blocking rate. The improvements obtained by BBSS are even more significant. It can be seen that BBSS can handle about twice the voice traffic loading as BBSO at the 1% call blocking point.

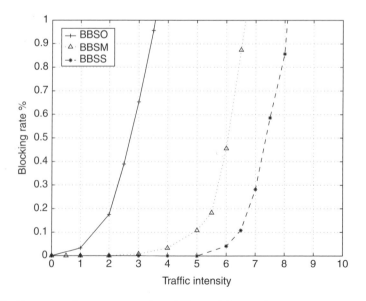

FIGURE 33.3   Blocking rate performance comparison, 5 Bluetooth base stations.

A design referred to as the Bluetooth DVAP was proposed in Reference 1. DVAP is similar to BBSS in that BMs are moved between BBSs to reduce the potential for call blocking. However, in BBSS the standby BBS must be within the range of all the BBSs that it is serving. In DVAP, the role of the standby BBS is performed by all the BBSs in a distributed manner. As in the other designs, an activated mobile associates with a BBS using the standard Bluetooth INQUIRY/PAGE procedures [6]. When a BM is activated within the BBS coverage area, it enters INQUIRY mode immediately to find a base station. Once a base station is contacted via INQUIRY, PAGING is used to establish an ACL link [4]. The mobile node is then parked by the BBS to await incoming and outgoing calls. Periodically, each BM will test the quality of the channel between it and each BBS. In Bluetooth, the most efficient way to do this is via PAGING, that is, to have either the BM or the BBS in question PAGE the other unit. Using LAN communications between the BBSs, Bluetooth clock/phase timing information is exchanged, which can make the PAGING operation very efficient [16]. When call requests arrive, the BBSs intercommunicate so that their state is known to each other. DVAP uses BM migration to reduce call blocking. When a call arrives whose assignment will result in a 3rd SCO link at a target BBS, BMs are moved via Bluetooth PAGING to other BBSs whenever possible [16].

Dynamic base station selection (DBS) is used in DVAP [1]. Multiple base station selection has been considered in the context of cellular systems with overlapping coverage, and the reader is referred to Reference 17 for a comprehensive overview of these techniques. These schemes are based on extensions to directed retry (DR) and directed handoff (DH) [17]. It can be shown that in the general case, base station assignment that minimizes the voice call blocking rate is a difficult discrete optimal control problem, and in DVAP a heuristic is used when a call request arrives. In the following description, we assume that a call request arrives for mobile node i, and this node is denoted by BM_i. BBS_j is defined to be base station j, for $j = 1, \ldots, N_{BBS}$, where N_{BBS} is the number of base stations. L_i is defined to be the current number of SCO links available to service the request from BM_i, and S_i is the set of BBSs, which are within the range of BM_i. M_j is defined to be the set of all mobiles that are within the range of BBS_j. The algorithm is very simple and is based on computing simple worse-case loading factors, which would result from each feasible selection of the base stations available to BM_i.

The loading fraction for BM_i is denoted by LM_i and is defined by $LM_i = 1/L_i$. The load factor for BBS_j is denoted by $BBSM_j$, and is obtained by summing up the station loading fraction values for each mobile within its coverage range, that is,

$$BBSM_i = \sum_{j \in M_i} LM_j. \tag{33.1}$$

When a call arrives for BM_i, the following procedures are used to select a BBS. The algorithm is shown in Figure 33.4. First the minimum worse-case load factor is initialized to ∞. Then all of the potential options

```
1    LF_min  =  ∞
2    for each BBS ∈  S_i
3         begin
4         Assign BM_i to BBS
5         Compute LM_i and BBSM_j for all  i and  j.
6         Find BBSM_max  =   max(BBSM_j)
7         if BBSM_max  <  LF_min then
8              begin
9              LF_min  =   BBSM_max
10             Selected_BBS = BBS
11             end
12        end
```

FIGURE 33.4   Dynamic BBS selection (DBS) algorithm.

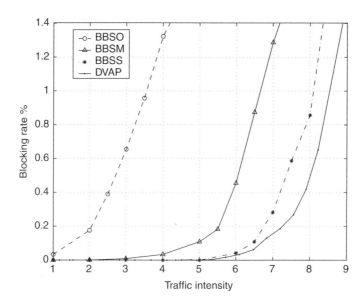

FIGURE 33.5 DVAP blocking rate performance comparison.

for BM_i are tested. This corresponds to the for loop beginning on line 3 in Figure 33.4, which will be executed for each of the BBS options for BM_i. For each option, LM_i and $BBSM_j$ are recomputed. Then the maximum value for $BBSM_j$ is found and denoted by $BBSM_{max}$, in line 6. If this value is found to be the smallest load factor to this point, then LF_{min} is updated and Selected_BBS is set to the current base station. It is clear that once the for loop beginning on line 3 has completed, the option chosen is the one that results in the lowest worse-case load factor for all BBSs. Note that if a BM has no SCO links available, its loading fraction is set to the number of BMs in the system.

Figure 33.5 gives an example of how DVAP compares with the systems studied in Reference 16. As discussed in Reference 1 the designs in Reference 16 are only applicable to deployments which heavily overlap the coverage areas of the base stations. Figure 33.5 shows the simulated voice call blocking rate as a function of traffic intensity for a system with 5 BBSs and 48 BMs. The figure shows that the DVAP system moderately improves upon its call blocking performance. Under full coverage overlap, DVAP is essentially a distributed version of BBSS, and the reason for the improvement is that in BBSS one of the five BBSs is designated as the standby BBS. This leaves only four BBSs to support actual SCO voice transmissions. Additional results in Reference 1 show the value of DVAP under various partial coverage overlapping situations.

33.4 Voice over ACL Links

A number of studies have investigated the use of Bluetooth ACL links for real-time voice transmission [14,15,18]. Using voice compression, the voice vocoding rate can be reduced far below that of conventional 64-kbps PCM, and ACL links are a flexible mechanism for carrying this type of voice traffic. The study in Reference 18 considered the performance of voice over ACL (VoACL). The system studied consists of a full piconet with 7 slaves. TCP data traffic and voice connections were combined in the same piconet with each connection having a slave as a source and another slave as its destination. The total number of connections in the piconet varied from 5 to 15, with a unity ratio of TCP and voice connections. The TCP connections carried large file transfers, with 500-byte packets. The voice-coding rate was 8 kbps and the packetization period was 20 msec, which gives a voice payload size of 20 bytes. DH1 packets were used when carrying VoACL links. This study showed that although voice connection delay is increased slightly using ACL links compared with SCO links, these delays are still quite acceptable for VoIP applications. On the other hand, when voice is carried over ACL, TCP connections achieve large performance improvements compared to operating the same number of calls over SCO links.

Reference 13 introduced a Bluetooth design where time is divided into superframes, which can carry VoACL at a rate of 64 kbps. Bluetooth typically uses 64 kbps log PCM voice coding with either A-law or μ-law companding and transmits 30 bytes every 3.75 msec when HV3 packets are used. VoACL obtains the required 64 kbps by extending the 3.75 msec packetization time to 20 msec and sending 160 voice bytes. A 20-msec frame is a common frame size found in other wireless voice systems. In Bluetooth, 20 msec corresponds to 32 time slots, and this was the superframe size used. The proposed scheme conveys voice information at the start of every superframe. The remainder of the superframe is used to transmit user data or to retransmit corrupted voice packets if necessary. This allows simultaneous data services greater access to the piconet, which increases data throughput. Furthermore, voice information sent in this fashion is protected by a CRC and can be retransmitted in the event of channel error. A motivation of this study was to enable reliable voice connections under marginal link situations, where ARQ retransmission would be beneficial. It should be noted that employing VoACL requires buffering at the receiver that will add to the overall delay of each connection.

The VoACL algorithm constructs an ACL packet from raw voice bytes at the start of every superframe. These packets are referred as VoACL-voice packets and each carries 160 voice bytes. DM5 or DH3 packets can be used to carry voice information in the superframe. These are referred to as VoACL-DH3 and VoACL-DM5, respectively. A packet cycle is defined as the minimum number of slots used to convey both an ACL packet and its acknowledgment. This represents the minimum time that must transpire between two consecutive ACL packet transmissions. The packet cycle for 3- and 5-slot packets is then 4 and 6 slots, respectively. Note that only one direction of the voice link is supported by the superframe, but it is easily extended to the bidirectional case. A superframe always begins with the transmission of a VoACL-voice packet. In the VoACL-DH3 case there will be 28 (32 − 4) slots remaining in the superframe, while the VoACL-DM5 case leaves 26 (32 − 6) slots. The VoACL algorithm uses these residual slots to provide simultaneous data services.

The VoACL algorithm can be operated in either voice priority mode or data priority mode. When operating in voice priority mode, a lost VoACL-voice packet will be retransmitted until received correctly, or until the superframe ends. This scheduling scheme places a higher priority on delivering voice without error rather than on providing simultaneous data throughput and will exhaust an entire superframe if necessary to deliver a single VoACL-voice packet. This may be a favorable approach when transferring important voice information under poor channel conditions, as it affords at least eight transmission attempts in the case of VoACL-DH3 and five in the case of VoACL-DM5. The data priority scheme places the emphasis on supporting simultaneous data applications. In this scheme, VoACL-voice packets are not retransmitted and all of the residual superframe slots are devoted to the transfer of data. This ensures that voice applications do not adversely interfere with simultaneous data applications, and guarantees a minimum level of access to the piconet for data traffic. By using ACL links, VoACL offers a greater degree of scheduling freedom for larger sized packets, which dramatically improves the bandwidth available to simultaneous data applications. Flexible priority assignments between voice and data can be implemented easily in VoACL by allowing varying numbers of retransmissions within a superframe.

33.4.1 Bluetooth FH Collision Loss

In References 1 and 16, several access point designs were introduced for Bluetooth voice access networks. The different designs considered the value of interbase station coordination among the Bluetooth piconets that are supporting SCO voice transmissions. Unfortunately, with increasing overlap of radio coverage, SCO link voice packet loss increases. In Reference 16, lower rate ACL voice links were considered in the design of a voice access network, but the issue of packet loss was not accurately evaluated. To obtain good statistical performance, a large number of potential Bluetooth voice links are provided throughout the desired coverage area. Unfortunately, when coverage is such that significant Bluetooth link sharing is possible, SCO packet loss rates can be very poor. However, when VoACL is used, ARQ retransmission can recover from this loss. It is important in this type of application that a proper statistical measure of voice loss rate is used, which characterizes the link performance over finite time windows. A finite time-window loss model was introduced in this study for this purpose. Analytic and simulation models show that the VoACL design has acceptable voice packet loss performance compared with SCO voice designs [6] and that the call blocking rate is about the same between the ACL and SCO systems.

When there are simultaneous overlapping Bluetooth links, packets can sometimes be lost due to collisions caused by the frequency hopping air interface. This occurs when two links choose the same hopping frequency (or perhaps an adjacent frequency) for transmission. This is a common occurrence considering the fact that there are only 79 hopping frequencies available. The issue of packet loss in Bluetooth voice access networks was first considered in Reference 19. Under worse-case situations the loss rates can be high enough to become a problem, which may affect real-time voice quality, even with a small number of overlapping voice links. The model proposed in Reference 20 assumes that loss occurs whenever packets are overlapped on the same frequency hop. This is clearly a worse-case assumption since it does not account for path-loss effects, which would help to protect a link when the sender and receiver are in close proximity relative to its interferers. In many applications, Bluetooth will benefit from this type of clustering. This type of analysis was given in Reference 21.

Consider the worse-case analysis presented in Reference 19. Assume that there is a set of n_{SCO} SCO links currently active. Note that $n_{SCO} \leq 3N_{vBBS}$, where N_{vBBS} is the number of available VOICE_BBSs. Under worse-case loss assumptions it was shown that the lowest worse-case packet loss rate is given by

$$P_L^{min} = 1 - \gamma^{\lceil n_{sco}/3 \rceil - 1}, \tag{33.2}$$

where $\gamma = 78/79$ for Bluetooth [20]. Similarly the highest worse-case loss rate is given by

$$P_L^{max} = \begin{cases} 1 - \gamma^{n_{sco}-1} & \text{for } n_{sco} \leq N_{vBBS} \\ 1 - \gamma^{N_{vBBS}-1} & \text{otherwise.} \end{cases} \tag{33.3}$$

The derivation of these equations assumes that frequency-hops are generated independently and randomly across the 79 Bluetooth channels [20]. Moreover, the time slots are assumed to be overlapped exactly. When this is not the case, loss rates can be even higher. These results give simplistic minimum/maximum worse-case loss rates, and not mean loss probabilities.

Equations (33.2) and (33.3) are plotted in Figure 33.6 for 3, 5, and 9 BBSs. The lower staircase curve is that for P_L^{min} and the upper curves correspond to P_L^{max}. For a given number of active SCO links, these minimum/maximum worse-case loss rates can be read from the figure. Note that since each base station can support only 3 SCO links, the curves for 3, 5, and 9 BBSs end at 9, 15, and 27 SCO links, respectively. It can be seen that once the number of SCO links operating in a given coverage area equals 4, the minimum loss rate that can be achieved is about 1.2%. Subjective listening tests have shown that for a given loss rate, a random dropping of packets is less objectionable than if losses are bursty. In the Bluetooth case the losses are caused by frequency-hop collisions, which will be highly random. Here, acceptable performance may be obtained up to about 5% packet loss, but good performance requires that a packet loss concealment algorithm is used [22,23]. The results from Figure 33.6 for the 3 BBS case show that the maximum packet loss rate is about 2.5%. As seen in the figure for P_L^{max}, this loss value can be achieved with as few as 3 SCO links, since each of those links could be assigned to the same time-slot pair in three different base stations. On the other hand, if the 3 SCO links were handled by the same BBS, a loss rate of 0 is possible (see P_L^{min}, for 3 SCO links). In the latter case, the single base station will have assigned the SCO links to disjoint time slot pairs. Similarly, in 5 BBS, a packet loss rate of just below 5% is the maximum possible. This loss rate can clearly be obtained with as few as 5 SCO links and yet the curves show that worse-case loss rates less than 4% are possible with as many as 12 SCO links. In the latter case, the 12 SCO links involve only 4 of the 5 BBSs. For 9 BBS in this system we are clearly in an infeasible packet loss region once the number of SCO links exceeds 15. This worse-case analysis suggests that FH packet loss can be a very serious issue when

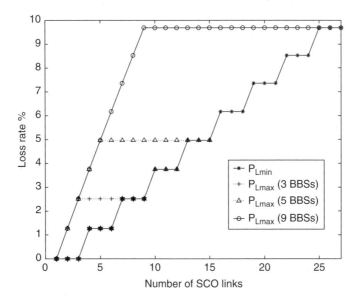

FIGURE 33.6   Packet loss rate, 3, 5, and 9 BBSs.

designing a high capacity voice access network. Very high loss rates are possible even when the number of overlapping SCO links is very low.

Reference 24 included a much more accurate packet loss model and showed that voice over ACL is necessary in order to avoid unacceptable packet loss in high capacity voice access networks. Unlike the scheme proposed in Reference 13, no superframe is used; instead, voice is transmitted directly using VoIP with limited ARQ retransmission. At each station, voice is given priority over data. Table 33.2 shows the characteristics of the Bluetooth ACL links. It can be seen that DM1 and DH1 have a one byte of payload header. The payload in DM1 or DH1 is less than that in HV3. HV3 has the highest payload for a single time slot packet and therefore it has the highest capacity. Using DM1 or DH1 to carry 64 kbps voice calls cannot provide three simultaneous voice links as does HV3. For this reason multislot packets were used. First consider DM5 and DH5. The user payload in a DM5 and DH5 packet is up to 224 and 339 bytes, respectively. The maximum symmetric rate of a DM5 packet is 286.7 kbps. 286.7 kbps $\div$ 64 kbps $\simeq$ 4.5. Using DM5 ACL packets, a Bluetooth enabled device can provide up to four 64-kbps voice links. 433.9 kbps $\div$ 64 kbps $\simeq$ 6.8, up to six 64-kbps voice links can be provided if DH5 ACL packets are used. On the other hand, for a DM3 packet the maximum user data rate is 258.1 kbps and thus up to four voice links can be provided. For a DH3 packet, the maximum rate is 390.4 kbps and up to six voice links can be provided. In the designs considered it was assumed that DM3 packets are used. In order to provide capacity for ARQ retransmissions, a maximum of 3 active voice connections was allowed per base station. Using DM3 packets to carry voice calls, and under this constraint this results in the same raw voice capacity as HV3 SCO links. If a DM3 packet is used for 64-kbps voice, it can carry 15 msec of speech. In that case, a DM3 packet has to be sent every 24 time slots. Every two DM3 packets form a symmetric bidirectional ACL-voice link. When this is done we set an upper limit on the number of ARQ retransmissions allowed, this maximum number is denoted by `Max_Retransmission`. The VoACL scheme over DM3 is very simple. A maximum of 3 voice connections is permitted per Bluetooth base station, and a fixed maximum ACL packet retransmission count is set.

Reference 24 included extensive simulation experiments, which characterize the packet loss rate using a VoACL design. Four different base station selection algorithms were used to test the VoACL (DM3) algorithm mentioned above. The summary of these algorithms is presented in Table 33.3. DBS was presented in Reference 1. DBS uses base station inter-communication to try to balance potential traffic load of each base station. In random base station selection (RBS) [1], BMs randomly choose a BBS that

TABLE 33.3   Summary of Base Station Selection Algorithms

DBS	Interbase station communication is used to compute BBSM loading factors for each potential base station assignment. The algorithm is shown in Figure 33.4 and described in detail in Section 33.3
RBS	An incoming call is randomly assigned to one of available BBSs
QoS	An incoming call is assigned to the BBS, which has the best link to the BM
STATIC OPT	This is a static assignment where each BM is preassigned to a BBS such that the number of BMs are as uniformly distributed as possible

has an available voice link. STATIC OPT corresponds to the optimum *a priori* static assignment of BMs to BBSs [1]. In addition, a scheme referred to as link-QoS was used where an arriving call selects the best link quality among BBSs, which have available voice links. Several different network topologies were considered, which model public access network situations, and were selected so that the level of voice link sharing is different.

The short range propagation model used was originally presented in Reference 4 and given as follows. Distance dependent attenuation, shadowing, and multipath fading are included. The link gain $g = P_r/P_t$, being the ratio between received and transmitted power, is given by

$$g = g_0(d) \cdot s(d) \cdot r^2(d),$$

where $g_0(d)$ is the distance dependent attenuation factor, $s(d)$ is the shadowing factor, and $r^2(d)$ is the multipath factor.

The attenuation factor $g_0(d)$ is modeled as free space propagation

$$g_0(d) = 10^{-4} \cdot d^2.$$

Log-normal shadowing is also included. The shadowing factor $s(d)$ is a random variable with a log-normal distribution, $10 \log s = N(0, \sigma^2)$ is a normally distributed variable in dB with zero mean. The variance σ^2 is dependent on the distance d and is given by

$$\frac{\sigma}{dB} = \min\left\{\frac{d}{2\,m}, 6\,m\right\}.$$

The multipath fading factor $r(d)$ is a Rician random variable with mean power $E[r^2] = 1$. The parameter K in dB is as well modeled as a function of the distance d,

$$\frac{K}{dB} = 12 - \frac{2d}{m}.$$

In Reference 24, a new packet loss figure of merit was used to characterize this type of system. Since the number of calls is a random variable, the loss rate for a connection will typically vary over the lifetime of the connection, as other calls start and end. A *windowed loss rate*, or WLR(t_w, PL_threshold) was used. The window duration parameter t_w is used as a sliding window for every active voice connection. If the packet loss rate across that window ever exceeds PL_threshold, then that voice connection is marked as unacceptable. Then, depending upon the model unacceptable calls can be terminated or allowed to continue. The value of WLR(t_w, PL_threshold) gives the fraction of accepted calls that have been marked unacceptable. WLR defined in this manner gives a much more realistic assessment of the subjective impact of cochannel interference when the interferers are intermittent as is the case when Bluetooth links are in close proximity.

In the results that were presented, each BM is either active, that is, with a call connection pending, in progress, or inactive. An inactive node sleeps for an exponentially distributed time interval with a mean of `Call_Idle_Time` seconds before a call request is generated. The call can originate from either the BM itself or from the PSTN/Internet. When the call has been blocked, the node reenters the call idle state. When the call is set up, a mobile is in the call connection state. The call lasts for a hold time, which is exponentially distributed with a mean of `Call_Hold_Time` seconds. In Reference 24, a default value of 10 sec was used for `Call_Hold_Time` and `Call_Idle_Time` was varied to obtain different normalized traffic intensities. This rather small value was used for the sake of simulation speed, as in Reference 4. In the simulation results the call blocking rate was considered for the various designs. Acceptable values for blocking are generally considered to be less than about 1% [25].

Three different kinds of loss rate were considered, that is, packet loss rate, average call loss rate, and call loss rate using WLR. When the Bit Error Rate (BER) of a packet is below 10^{-3}, the packet is assumed to be unsuccessful. The packet loss rate is the ratio of the total number of lost packets to the total number of packets. When the packet loss rate of a call is over 1%, the call is counted as lost in the average case. The call loss rate in the average case is the ratio of the total number of loss calls in the average case and the total number of calls. Whenever the packet loss rate within the `Check_Window` of a call is over 1%, the call is counted as lost in the WLR case. The call loss rate in the worse case is the ratio of the total number of lost calls in the worse case to the total number of calls. The `Check_Window` is set to 1 sec. The parameter of `Max_Retransmission` is set to 2. The distance attenuation, shadowing factor (log_normal distribution), and multipath fading (Rician distribution) [4] are included in the simulation to determine the BER of a packet.

An example of the results for a moderately overlapped radio case is shown in Figure 33.7. Figure 33.7(a) indicates about a 50% improvement in capacity performance between the DBS algorithm and the STATIC

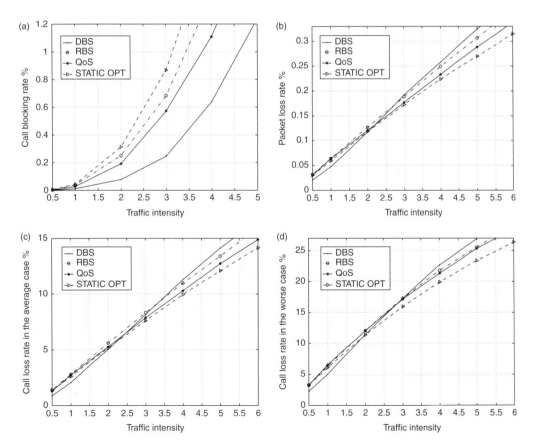

FIGURE 33.7 SCO link call loss and blocking: (a) mean call blocking rate; (b) mean packet loss rate; (c) average call loss rate, and; (d) WLR call loss rate.

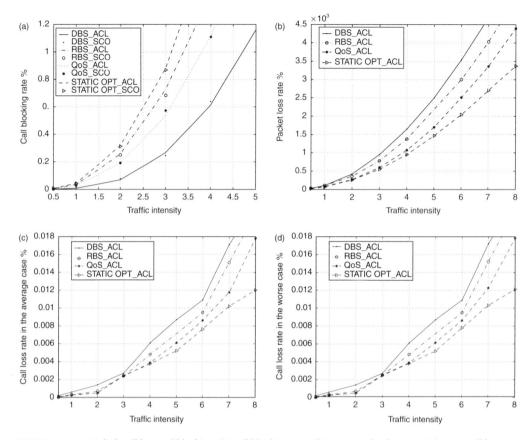

FIGURE 33.8   ACL link call loss and blocking: (a) call blocking rate; (b) mean packet loss rate; (c) mean call loss rate; and (d) WLR call loss rate.

OPT algorithm when the blocking rate is roughly 1% for both algorithms. The blocking rate can be decreased and the system capacity increased by using the DBS algorithm. If we consider only raw packet loss rate shown in Figure 33.7(b), the system performance would be deemed acceptable, that is, the packet loss rate is below 0.35% with the call blocking rate below 1%. However, Figure 33.7(c) and 33.7(d) show that average call loss and WLR performance is unacceptable. This is true even for the link-QoS algorithm. Since SCO packets cannot be retransmitted the call quality is unacceptable. In this case the link sharing gain is not useful because the call loss rate does not meet the minimum threshold. These curves point to a fundamental problem with Bluetooth SCO links, which will prohibit their use in high capacity access network applications.

Call quality over ACL links is shown in Figure 33.8. Using ACL packets to carry voice, the call blocking rate can be maintained but the loss rate can be dramatically reduced. The call blocking performance over SCO link and ACL links is shown in Figure 33.8(a). Figure 33.8(a) shows that the blocking rate for VoACL and SCO links is about the same. This is true for all four different BSS algorithms. However, VoACL can obtain much better loss rate performance as shown in Figure 33.8(b) and 33.8(d) compared with the SCO link case shown in Figure 33.7(b) and 33.7(d). The packet loss rate is below 10^{-3}% at 1% blocking rate for the DBS algorithm. Loss rates remain acceptable even under much higher traffic intensities. When traffic intensity is 10, the packet loss rate is around 8.5×10^{-3}% as shown in Figure 33.8(b). The call loss rate performance shown in Figure 33.8(c) and 33.8(d) is also acceptable. The results are almost same using the two different call loss rate measurements. WLR is less than 0.01% when the blocking rate is below 1% using the DBS algorithm. This is very good and means that the large capacity improvement due to link sharing is feasible when ACL links are used.

33.4.2 Extended SCO (eSCO) Links

Extended SCO links were introduced in the November 2003, Bluetooth 1.2 specification [26]. As in conventional SCO links, slots in eSCO links are reserved and therefore an eSCO link can be used for a circuit-switched connection between the master and the slave. eSCO packets are carried over eSCO links and three types of eSCO packets, namely, EV3, EV4, and EV5 packets, have been defined [26]. Unlike SCO links, ARQ retransmission of voice packets is possible over eSCO links. If an eSCO packet retransmission is required, the retransmission will happen in the slots that follow the reserved slots. Since some capacity may be used for retransmission purposes, a Bluetooth device can only provide up to two symmetric voice connections simultaneously when the maximum number retransmissions is allowed. Compared with the design proposed in Reference 24, eSCO link has less capacity for the same quality of service level.

33.5 Summary

In this chapter we have given an overview of recent work in the area of real-time voice over Bluetooth. Both SCO and ACL links have been considered for voice applications. Future access networks can use Bluetooth connectivity to provide real-time voice services in public places such as airports, convention centers, and shopping malls. We have also considered the design of this type of high capacity public access voice network. To obtain good statistical performance a large number of potential Bluetooth voice links are provided throughout the desired coverage area. Results show that conventional Bluetooth SCO voice links are likely to be inadequate for this purpose. Unfortunately, when coverage is such that significant Bluetooth link sharing is possible, SCO packet loss rates are unacceptable. The alternative is to operate such a network using Bluetooth (VoACL) with ARQ retransmission or to wait for eSCO link support, but in the latter case voice capacity is very limited. Proper statistical measures of voice loss rate are very important in this type of system, which characterizes link performance over finite time windows. Analytic and simulation models show that VoACL designs have acceptable voice packet loss performance compared with SCO link designs [6] and that the call blocking rate is about the same between the ACL and SCO systems.

33.5.1 Future Work

Using voice coding rates lower than 64 kbps, far more than 7 voice connections would be possible per Bluetooth access device. However, a single Bluetooth radio can only support 7 voice connections concurrently due to the limitation of 7 active slave members. Although applying park/unpark procedures can theoretically circumvent this limitation, using this method is not an efficient way to handle this problem. An alternate design is to handle this issue using multihop bridging between multiple piconets. More work is needed to study the practical effectiveness of this approach when carrying real-time services, its affect on delay and jitter, and its negative affects on relay node performance.

References

[1] Y. Wu, T. D. Todd, and S. Shirani. SCO link sharing in Bluetooth voice access networks. *Journal of Parallel and Distributed Computing*, 63: 45–57, 2003.

[2] J. Korhonen. *Introduction to 3G Mobile Communications*. Boston: Artech House, 2001.

[3] W. Webb. *The Future of Wireless Communications*. Boston: Artech House, 2001.

[4] J. Haartsen. The Bluetooth Radio System. *IEEE Personal Communications*, 7, 2000.

[5] A. Kamerman. Coexistence between Bluetooth and IEEE 802.11 CCK: solutions to avoid mutual interference. In *IEEE P802.11 Working Group Contribution IEEE P802.11-00/162r0*, July 2000.

[6] Bluetooth Special Interest Group. *Specification of the Bluetooth System 1.0b*. Bluetooth Special Interest Group, 1999.

[7] M. Kalia, Sumit Garg, and R. Shorey. Efficient policies for increasing capacity in Bluetooth: an indoor pico-cellular wireless system. In *Proceedings of the IEEE 51st Vehicular Technology Conference, 2000. VTC 2000-Spring Tokyo. 2000*, vol. 2, pp. 907–911, 2000.

[8] A. Das, A. Ghose, A. Razdan, A. Saran, and R. Shorey. Enhancing performance of asynchronous data traffic over the Bluetooth wireless ad-hoc network. In *Proceedings of INFOCOM 2001, vol. 1, IEEE*, pp. 591–600, 2001.

[9] S. Garg, M. Kalia, and R. Shorey. MAC scheduling policies for power optimization in Bluetooth: a master driven TDD wireless system. In *Proceedings of the IEEE 51st Vehicular Technology Conference, 2000. VTC 2000 — Spring Tokyo. 2000*, vol. 1, pp. 196–200, 2000.

[10] M. Kalia, D. Bansal, and R. Shorey. Data scheduling and SAR for Bluetooth MAC. In *Proceedings of the IEEE 51st Vehicular Technology Conference, 2000. VTC 2000 — Spring Tokyo. 2000*, vol. 2, pp. 716–720, vol. 2, 2000.

[11] J. Bray and C. F. Sturman. *Bluetooth: Connect Without Cables*. Prentice Hall, 2001.

[12] P. Johansson, N. Johansson, U. Korner, J. Elg, and G. Svennarp. Short range radio based ad-hoc networking: performance and properties. In *Proceedings of the 1999 IEEE International Conference on Communications, 1999. ICC '99*, vol. 3, pp. 1414–1420, 1999.

[13] D. Famolari and F. Anjurn. Improving simultaneous voice and data performance in bluetooth systems. In *Proceedings of the Global Telecommunications Conference, 2002. GLOBECOM '02. IEEE*, vol. 2, pp. 1810–1814, 2002.

[14] T. Lee, K. Jang, H. Kang, and J. Park. Model and performance evaluation of a piconet for point-to-multipoint communications in Bluetooth. In *Proceedings of the Vehicular Technology Conference, 2001. VTC 2001 Spring. IEEE VTS 53rd*, vol. 2, pp. 1144–1148, 2001.

[15] S. Chawla, H. Saran, and M. Singh. QoS based scheduling for incorporating variable rate coded voice in Bluetooth. In *Proceedings of the IEEE International Conference on Communications, 2001. ICC 2001*, vol. 4, pp. 1232–1237, 2001.

[16] J. Xue and T. D. Todd. Basestation collaboration in bluetooth voice networks. *Proceedings of the IEEE Workshop on Wireless Local Networks*, November 2001.

[17] I. Katzela and M. Naghshineh. Channel assignment schemes for cellular mobile telecommunication systems: a comprehensive survey. *Proceedings of IEEE Personal Communications*, 3: 10–31, 1996.

[18] R. Kapoor, Ling-Jyh Chen, Yeng-Zhong Lee, and M. Gerla. Bluetooth:carrying voice over acl links. In *Proceedings of the 4th International Workshop on Mobile and Wireless Communications Network, 2002. On 9–11 Sept.*, pp. 370–383, 2002.

[19] J. Xue and T. D. Todd. Basestation collaboration in Bluetooth voice networks. *Computer Networks*, 41: 289–301, 2003.

[20] A. El-Hoiyhi. Interference between bluetooth networks-upper bound on the packet error rate. *IEEE Communications Letters*, 5: 245 –247, June 2001.

[21] S. Zurbes, W. Stahl, K. Matheus, and J. Haartsen. Radio network performance of bluetooth. In *Proceedings of the IEEE International Conference on Communications, 2000. ICC 2000*, vol. 3, pp. 1563–1567, 2000.

[22] Nortel Inc. Voice Over Packet: An Assessment of Voice Performance on Packet Networks. Nortel White Paper.

[23] A. D. Clark. Modeling the Effects of Burst Packet Loss and Recency on Subjective Voice Quality. IP-Telephony Workshop, Columbia University, 2001.

[24] Y. Wu and T. D. Todd. High Capacity Bluetooth Voice Access Networks Using Voice Over ACL Links (VoACL). Submitted for publication.

[25] J. Bellamy. *Digital Telephony*. Wiley Press, 2000.

[26] Bluetooth Special Interest Group. *Bluetooth Core Specification v1.2*. Bluetooth Special Interest Group, 2003.

34

NTP VoIP Test Bed: A SIP-Based Wireless VoIP Platform

	34.1	Introduction...**34**-773
	34.2	NTP VoIP Test Bed: Architecture and Protocol........**34**-774
		SIP Signal Protocol • Basic SIP Call Flow • NTP VoIP Call Server
	34.3	Wireless VoIP Service **34**-783
		VoIP Mobile Switching Center Approach • WLAN-based GPRS Support Node
	34.4	Conclusions **34**-787
		Acknowledgment...**34**-787
		References **34**-787

Quincy Wu
Whai-En Chen
Ai-Chun Pang
Yi-Bing Lin
Imrich Chlamtac

34.1 Introduction

Over a century, telephony industry was mainly established based on circuit-switched technologies. In the recent years, *Internet protocol* (IP) technology has been utilized to transport voice over packet-switched network [15]. This approach consumes lower bandwidth, and therefore reduces communications costs. For example, a typical coding algorithm used by circuit-switched telephony is based on *International Telecommunications Union* (ITU) Recommendation G.711, which requires 64 Kbps bandwidth in each direction for a telephone call [14]. On the other hand, by adopting silence suppression technique and advanced coding algorithms (such as those described in ITU G.729), voice traffic can be transported over IP network at 8 Kbps [7], with approximately the same voice quality as circuit-switched voice. Many Internet telephony service providers follow this approach to offer cost-effective voice services. Since IP is the de facto standard for data transactions, Voice over IP (VoIP) is advantageous in integration of voice and data [6]. One of the most popular protocols for VoIP signaling and call control is *session initiation protocol* (SIP) that supports functions to facilitate integration of user presence, instant messaging, and multimedia communications [23]. Specifically, SIP is chosen as the signaling protocol in IP multimedia subsystem in 3GPP specifications [17].

Under the *National Telecommunication Development Program* (NTP), we have established a VoIP test bed that allows deployment of SIP-based wireless VoIP applications. This chapter presents the operations and the numbering plan of the NTP VoIP test bed, and the wireless VoIP applications developed in this test bed. We first introduce SIP. Then we discuss the design philosophy of call servers in the NTP VoIP test bed. We propose an approach to support real-time VoIP through *general packet radio service* (GPRS) that interworks with the NTP VoIP test bed. We also describe a *WLAN-based GPRS support node* approach for accessing VoIP services.

34.2   NTP VoIP Test Bed: Architecture and Protocol

Figure 34.1 illustrates the NTP VoIP test bed architecture. Components in this VoIP system interact with each other through the SIP protocol. These components are described as follows:

Call server (Figure 34.1[a]) provides primary capabilities for call-session control in the NTP VoIP test bed. A call server processes SIP requests and responses as a SIP proxy server. It also functions as a registrar that stores the contact information of each SIP user. Details of proxy and registrar servers will be given in Section 34.2.1. At the current stage, the NTP call server is implemented on Window 2000 server running on an industrial PC. Figure 34.2 illustrates the operations, administration, and maintenance (OAM) system of the call server. This figure shows the SIP phone management page of the call server.

PSTN gateway (Figure 34.1[b]) supports interworking between the NTP VoIP test bed and the *public switched telephone network* (PSTN), which allows IP phone users to reach other PSTN users directly or indirectly through *private branch exchange* (PBX; see Figure 34.1[c]). Two types of PSTN gateways have been deployed in the NTP VoIP test bed, including the gateway developed by Industrial Technology Research Institute (ITRI); (see Figure 34.3[a]) and Cisco 2600 (Figure 34.3[b]).

SIP user agent (UA) is a hardware- or a software-based SIP phone client (Figure 34.1[d]) that provides basic call functions such as dial, answer, reject, hold/unhold, and call transfer. In the NTP VoIP test bed, we have installed the SIP UA in terminals including desktop computers, notebooks (with or without WLAN access), and PDAs (with WLAN access only). The GUI for softphone (software-based SIP phone) is shown in Figure 34.4. We also include hardware-based SIP phones (or hardphones) manufactured by Cisco, Leadtek, Pingtel, and Snom (see Figure 34.5).

At the time of writing, the NTP VoIP test bed has been deployed in four sites: National Taiwan University (NTU), National Tsing Hua University (NTHU), National Chiao Tung University (NCTU), and ITRI. NTU is in Taipei City while the other three organizations are in Hsinchu City. Figure 34.1 illustrates the NCTU and NTU sites of the NTP VoIP test bed. A call server and a PSTN gateway are deployed in each site, accompanying several SIP hardphones and softphones.

Depending on the locations of the calling party and the called party, and whether they are SIP phones or PBX/PSTN phones, there are four call setup scenarios in the NTP VoIP test bed.

Scenario 1: If a SIP phone UA1 in NCTU attempts to call another SIP phone UA2 on the same campus, the call setup signaling messages are delivered between UA1 and UA2 indirectly through the NCTU call server. The voice connection is established directly between UA1 and UA2 without involving the NCTU call server.

Scenario 2: If a SIP phone UA1 in NCTU attempts to call another SIP phone UA2 in NTU, UA1 first signals to the NCTU call server. From the destination address, the NCTU call server determines that UA2 is located on the NTU campus, and forwards the call setup request to the NTU call server. The NTU call server retrieves the registry information of the called party and set up the call to UA2. Again, the voice connection is established directly between UA1 and UA2 without involving the NCTU call server.

Scenario 3: If a SIP phone UA1 in NCTU makes a call to a PBX phone P1 in NCTU or a traditional phone in the PSTN, the call setup procedure is similar to that of Scenario 1 with the following exceptions. The NCTU call server determines that the called party is not a SIP phone and routes the call to the NCTU PSTN gateway. The PSTN gateway then sets up the call to P1 directly (if P1 is a PBX phone) or to the PSTN central office for further processing.

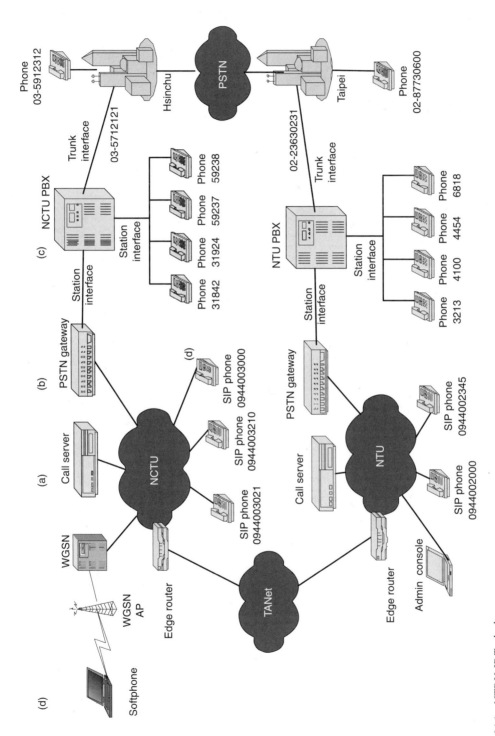

FIGURE 34.1 NTP VoIP Testbed.

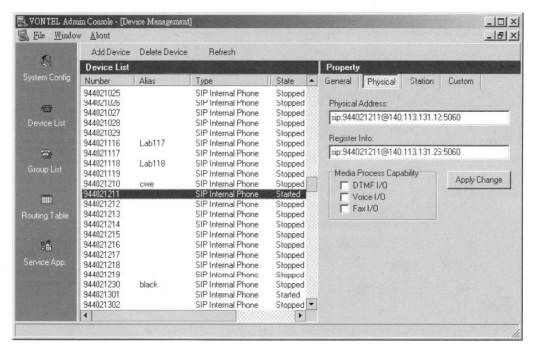

FIGURE 34.2   OAM of the NTP call server.

FIGURE 34.3   NTP PSTN gateways: (a) ITRI PSTN gateway; (b) CISCO 2600.

Scenario 4: If a SIP phone UA1 in NCTU attempts to call a PBX phone P1 in NTU, then the NCTU call server will forward this request to the NTU call server. Similar to Scenario 3, the NTU call server will route this call to the NTU PSTN gateway for further processing.

In order to reach a SIP phone from the PSTN, this SIP phone must be assigned a telephone number. Following Taiwan's numbering plan (based on E.164 recommendation [5]), the telephone number for a SIP phone in the NTP VoIP test bed is of the format 0944-nnn-xxx. The first four digits 0944 are

(a)

(b)

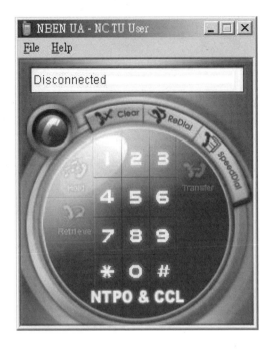

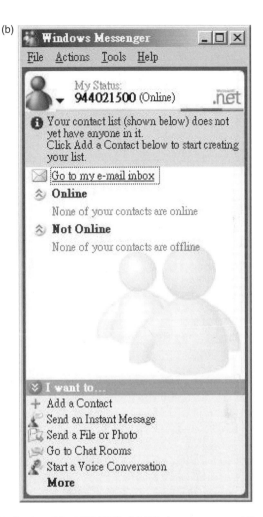

FIGURE 34.4 GUI of NTP VoIP Softphone: (a) SIP UA developed by NTP/ITRI; (b) Windows messenger 4.7 (including a SIP UA) on Windows XP.

the service code for NTP VoIP test bed approved by DGT of Taiwan for experimental usage. In Japan *Electronic Number* (ENUM) trial, the prefix 050 is used for VoIP. In Taiwan, the code 070 will be reserved for VoIP services. In the future, the 0944 code will be replaced by 070. The next three digits nnn refers to the site number. In the current configuration, code 001 represents the NTHU cite, code 002 represents the NTU cite, and code 003 represents the NCTU cite. The remaining digits xxx represent a customer line number automatically generated by the NTP VoIP registration system. Some numbers are preserved for emergency call or special services. To simplify our discussion, in the remainder of this chapter, the IP phones will be identified by SIP URI (to be elaborated) instead of telephone number mentioned above.

In this section, we describe the SIP protocol and call flows. Then we elaborate on the call server developed in the NTP VoIP test bed.

34.2.1 SIP Signal Protocol

Session initiation protocol [21] is a signaling protocol for creating IP multimedia sessions (e.g., voice and video streaming). Following a text-based HTTP-like format, SIP conjuncts with protocols such

(a)   (b)

(c)   (d)

FIGURE 34.5   NTP User terminals: (a) Cisco hardphone; (b) Leadtek hardphone; (c) Pingtel hardphone; (d) Snom hardphone.

as *session description protocol* (SDP) [12] and *real-time transport protocol* (RTP) [22]. SDP describes multimedia information of a session, including media type, IP address, transport port, and codec. RTP transports real-time multimedia data such as voice and video data. SIP supports the following features:

User location determines the end terminal location for communication.
User availability determines the willingness of the called party to engage in communications.
User capability determines the media and media parameters to be used.
Session setup establishes session parameters at both called and calling parties.
Session management transfers and terminates sessions, modifies session parameters, and invokes services.

To handle setup, modification, and tear down of a multimedia session, SIP messages are exchanged between the call parties, including the SIP *requests* from the calling party to the called party, and SIP *responses* from the called party to the calling party. IETF RFC 3261 [21] defines six types of requests that are utilized in the NTP VoIP test bed:

1. REGISTER is sent from a SIP UA to the call server for registering contact information.
2. INVITE is sent from a calling UA to a called UA to initiate a session, for example, audio, video, or a game.
3. ACK is sent from the calling UA to the called UA to confirm that the final response has been received.
4. CANCEL is sent from the calling UA to the called UA to terminate a pending request.
5. BYE is issued by either the calling or the called UAs to terminate sessions.
6. OPTIONS is sent from a SIP UA to query the capabilities of another UA.

All SIP requests are acknowledged by SIP responses. A SIP response consists of a numeric status code and the associated textual phrase. Examples of the codes are 100 (Trying), 180 (Ringing), 200 (OK), 302 (Moved Temporarily), and 487 (Request Terminated).

UA1   UA2

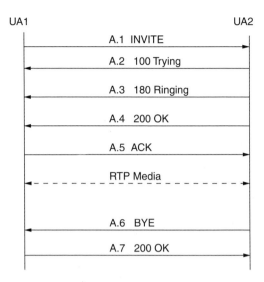

FIGURE 34.6 Basic SIP call setup procedure.

Every SIP entity is identified by a unique identification called SIP *uniform resource identifier* (URI). The SIP URI is of the format `sip:username@hostname:port`. In this format, the prefix `sip:` indicates that the whole string is a SIP URI. The `hostname` is the name of the host for the SIP phone and `username` is a local identifier representing the SIP phone on the host. The `port` is the transport port to receive the SIP message on `hostname`, which has a typical value "5060". A SIP URI example is `sip:george@work.com:5060`.

34.2.2 Basic SIP Call Flow

As its name implies, the primary function of SIP is session initiation (call setup). The basic SIP call setup between two UAs, that is, UA1 (the calling party) and UA2 (the called party), is illustrated in Figure 34.6. The procedure is described as follows:

Step A.1. UA1 sends the `INVITE` request to UA2. The `INVITE` message includes the SIP URI of UA2 and SDP message that describes the RTP information of UA1 (including the IP address and port number). This RTP information will be utilized by UA2 to send VoIP packets to UA1.

Step A.2. Upon receipt of the `INVITE` request, UA2 replies the `100 Trying` response. This message is a *provisional response* to inform UA1 that the `INVITE` request is in progress at UA2.

Step A.3. UA2 plays an audio ringing tone to alarm the called user that an incoming call arrives. UA2 sends the `180 RINGING` response to UA1. This message is a provisional response that asks UA1 to play an audio ringback tone to the calling user.

Step A.4. Once the called user picks up the handset, UA2 sends a `200 OK` response to UA1, which is a final response. The `200 OK` response includes the SDP message that describes the RTP information of UA2 (i.e., IP address and the transport port).

Step A.5. Upon receipt of the `200 OK` response, UA1 sends the `ACK` request to UA2 as an acknow-ledgment. At this point, the session is established and the conversation begins; that is, UA2 starts sending the RTP packets to UA1 based on the SDP parameters obtained at Step A.1. Similarly, UA1 sends the RTP packets to UA2 according to the SDP parameters obtained in Step A.4.

Step A.6. Either one of the call parties can terminate the call. Assume that UA2 terminates this session by sending the `BYE` request to UA1.

Step A.7. Upon receipt of the `BYE` request, UA1 replies the `200 OK` response to confirm that the call has been terminated.

An example of the INVITE request sent in Step A.1 is shown below:

```
INVITE sip:mary@pc3.home.net SIP/2.0
Via: SIP/2.0/UDP station5.work.com
From: George <sip:george@station5.work.com>
To: Mary <sip:mary@pc3.home.net>
Call-ID: 123456@station5.work.com
CSeq: 1 INVITE
Content-Length: 421
Content-Type: application/sdp

v=0
c=IN IP4 140.113.131.23
m=audio 9000 RTP/AVP 0
a=rtpmap:0 PCMU/8000
```

In this example, the From, To, and Call-ID header fields in a SIP message uniquely identify the session. These header fields are never modified during the session, except that some tags may be added to the To/From header fields. The From header field indicates that this call is initiated from sip:george@station5.work.com. SIP allows a display name to be used with the SIP URI. When the called party is alerted, his or her SIP terminal will display the name George instead of the SIP URI sip:george@station5.work.com.

The Content-Type header field indicates that the message body contained in this SIP request is described by SDP. The SDP message provides necessary information to establish a media session.

34.2.3 NTP VoIP Call Server

In addition to user agent (UA), SIP defines three types of servers: proxy server, redirect server, and registrar.

A SIP *proxy server* accepts the SIP requests from a UA, and forwards them to other servers or UAs, perhaps after performing some translations. For example, to resolve the SIP URI in the INVITE request, the proxy server consults the *location service* database to obtain the current IP address and transport port of the called UA. After location information retrieval, the proxy server forwards the INVITE message to the called UA.

A *redirect server* accepts a request from a calling UA, and returns a new contact address to the calling UA. Similar to the proxy server, the redirect server may query the location service database to obtain the called party's contact information. Unlike the proxy server, the redirect server does not forward the INVITE message. Instead, it returns the contact information to the calling UA.

The basic call setup procedure described in Section 34.2.2 assumes that the calling party knows the physical location of the called party. In the real world, a SIP user may change the location, and it may not be possible for the calling party to know the exact location of the called party. To solve this problem, SIP provides a registration network entity called *registrar* to locate the moving users. A SIP registrar only accepts the REGISTER messages. A UA can periodically register its SIP URI and contact information (specifically, the IP address and the transport port) to the registrar. The registrar stores the contact information in the location service database. Therefore, when a UA moves around different networks, the registrar always maintains the actual contact information of the UA. SIP REGISTER request is similar to the registration request (RRQ) message sent from a terminal to a gatekeeper in H.323 [13]. In most implementations, the location service database, the SIP registrar and the proxy server are integrated together. In NTP VoIP test bed, these server functions are implemented in a call server.

Figure 34.7 shows the message flow for SIP registration in the NTP VoIP test bed. When a UA logs in at host station5.work.com, a REGISTER request is sent to the call server (the registrar function). In this message, the Via header field contains the network nodes visited by the request so far. The From header field indicates the address of the individual who initiates the registration request. The To header

FIGURE 34.7   SIP registration procedure.

field indicates the "target user" being registered. The `Contact` header field contains the contact address of the user at which the user can be reached. The call server (the proxy function) retrieves the contact address from the location server database by using the `To` header field as the searching key. Generally, the `From` and `To` fields are identical when a UA registers for itself. In some exceptions, these two fields may have different values. For example, a secretary Mary may use her SIP UA to register the location of the UA for her boss George. In this case, the `From` header field is the SIP URI of Mary's UA, while the `To` header field is the SIP URI of George's UA. Mary may set the `Contact` header field to the SIP phone on her desk, so that she can answer the calls when people call George. She may also set the `Contact` address to a voice mail server. In this case, all SIP calls to George are immediately forwarded to the voice mail system.

If a UA has registered to the call server, then other UAs do not need to know the exact location of the UA when they attempt to make calls to this UA. All they need to do is sending the `INVITE` request to the call server (the proxy function). The call server will look up the location service database to find out the exact contact address of the called party. By introducing the registration mechanism, SIP supports personal mobility where a UA only needs to maintain one externally visible identifier, regardless of its network location. The call flow involving the call server is illustrated in Figure 34.8 with the following steps:

Step B.1. UA1 attempts to make a call to UA2. UA1 first sends the `INVITE` request to the NTP VoIP call server (the proxy function). The call server may query the location service database to obtain the contact information of UA2. Then it forwards the `INVITE` message to UA2.

Step B.2. If the incoming call is accepted, UA2 replies the `200 OK` message to UA1 through the call server (for simplicity, the provisional responses illustrated in Figure 34.6 are not shown here).

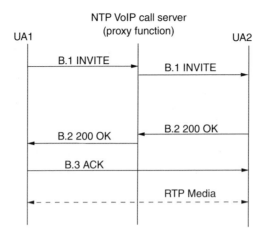

FIGURE 34.8   Call setup via the NTP VoIP call server.

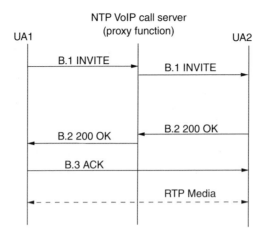

FIGURE 34.9   SIP forking: registration.

Step B.3. Upon receipt of the 200 OK response, UA1 sends the ACK message directly to UA2 without involving the call server. The conversation starts.

In the registration example illustrated in Figure 34.7, the 200 OK response contains more than one Contact header fields. This SIP feature implies that a user is allowed to register for more than one UAs. If multiple contacts are given in registration, then during call setup, the NTP VoIP call server will send the INVITE message to a number of locations at the same time. This operation is known as *forking*.

The forking mechanism provides an interesting feature in NTP VoIP test bed. Suppose that George has two SIP phones sharing a public address sip:george@phone.com, one at home and one at his office. Suppose that the two SIP phones UA1 and UA2 of George share the above public address with the contact addresses sip:george@home.com and sip:george@work.com, respectively. When someone dials George's phone number (his public address), both phones ring. When George picks up the nearest phone, the other phone stops ringing. Figure 34.9 illustrates how this feature is enabled in the NTP VoIP test bed:

Step C.1 When UA1 is turned on, it sends the REGISTER request to the NTP VoIP call server (the registrar function).

Step C.2 The call server binds the contact address of UA1 sip:george@home.com with George's public address sip:george@phone.com.

Step C.3 UA2 is turned on and sends the REGISTER request to the call server.

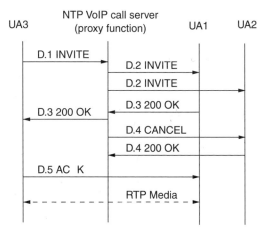

FIGURE 34.10 SIP forking: call setup.

Step C.4 Similarly, the contact address of UA2 `sip:george@work.com` is bound with George's public address `sip:george@phone.com`.

At this point, both contact addresses `sip:george@home.com` and `sip:george@work.com` are associated with George's public address in the call server. When Mary makes a VoIP call to George, the following steps are executed (see Figure 34.10).

Step D.1 Mary's IP phone UA3 sends the `INVITE` request to the NTP VoIP call server (the proxy function). This message specifies George's public address `sip:george@phone.com`. The call server queries the location service database to retrieve the contact addresses of George; that is, `sip:george@home.com` and `sip:george@work.com`.

Step D.2 The call server forwards the `INVITE` request to UA1 and UA2 simultaneously. Both phones ring at the same time.

Step D.3 Suppose that George answers UA1. Then this SIP phone will send the `200 OK` message to the call server. The call server forwards this response to UA3.

Step D.4 The call server sends the `CANCEL` request to UA2. UA2 stops ringing and sends the `200 OK` response to the call server.

Step D.5 UA3 replies `ACK` to UA1 directly, and the RTP session is established between UA3 and UA1.

Forking is useful in combining various services. For example, this feature has been exploited to build a voice mail service in the NTP VoIP test bed.

34.3 Wireless VoIP Service

Several approaches have been proposed to integrate VoIP services with wireless technologies [19,20]. This section describes wireless VoIP services that can be implemented in NTP VoIP test bed. We consider two types of wireless networks that have been interworked with NTP VoIP test bed: GPRS network [16] and *WLAN-based GPRS support Node* (WGSN) network [11].

34.3.1 VoIP Mobile Switching Center Approach

Figure 34.11 shows an integrated system for VoIP services over GSM/GPRS. In this figure, the GPRS network consists of *serving GPRS support node* (SGSN) and *gateway GPRS support node* (GGSN) [2]. The GGSN provides interworking with external data networks (e.g., the NTP VoIP test bed), and is connected

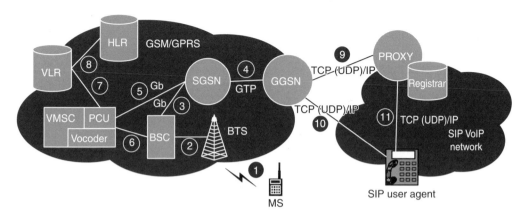

FIGURE 34.11   GSM/GPRS architecture for VoIP.

with SGSNs via an IP-based GPRS backbone network. The SGSN connects the GPRS core network to GSM radio network. Both GGSN and SGSN communicate with the mobility databases including the *home location register* (HLR) and the *visitor location register* (VLR), to track the locations of mobile subscribers.

The GPRS reuses the existing GSM infrastructure to provide end-to-end packet-switched services. In the GSM radio network, a *mobile station* (MS) communicates with a *base station subsystem* (BSS). A BSS consists of several *base transceiver stations* (BTSs) and a *base station controller* (BSC). A BTS contains transmitters, receivers, and signaling equipments specific to the radio interface, which provides the MSs radio access to the GSM/GPRS network. The BSC supports radio channel allocation/release and the handoff management. In the circuit-switched service domain, the *mobile switching center* (MSC) connects to the BSS through one or more E1/T1 lines, which performs the switching functions for mobile applications. In the packet-switched service domain, a *packet control unit* (PCU) is implemented in the BSC to connect to an SGSN. The BSC forwards circuit-switched calls to the MSC, and packet-switched data (through the PCU) to the SGSN.

By introducing a new core network node called *VoIP MSC* (VMSC), we propose real-time VoIP services over GPRS without modifying the existing GSM/GPRS network nodes. A VMSC is a router-based softswitch, which communicates with BSCs, VLR, HLR, and MSCs (either GSM MSCs or VMSCs) just like a standard MSC. The VMSC also interfaces with the PSTN via the SS7 *ISDN User Part* (ISUP) protocol [16]. Unlike MSC, a VMSC communicates with an SGSN through the GPRS Gb interface (based on frame relay). In other words, while an MSC delivers voice traffic through circuit-switched trunks, a VMSC delivers VoIP packets through the IP-based GPRS core network. In Figure 34.11, the data path of a GPRS MS is $(1) \leftrightarrow (2) \leftrightarrow (3) \leftrightarrow (4)$. The voice path in the VMSC approach is $(1) \leftrightarrow (2) \leftrightarrow (6) \leftrightarrow (5) \leftrightarrow (4)$. In this voice path, $(1) \leftrightarrow (2) \leftrightarrow (6)$ is circuit-switched as in GSM, and $(5) \leftrightarrow (4)$ is packet-switched in GPRS. By using the circuit-switched based GSM air interface, the VMSC approach provides better QoS for real-time communications. Both standard GSM MSs and GPRS MSs can setup calls in the VMSC-based network as if they are in the standard GSM/GPRS network. The VMSC is anticipated to be cheaper than an MSC in terms of the implementation costs.

At a VMSC, the voice information is translated into GPRS (i.e., SIP) packets through vocoder and PCU. Then the packets are delivered to the GPRS network through the SGSN (see path $(5) \leftrightarrow (4)$). An IP address is associated with every MS attached to the VMSC. In GPRS, the IP address can be either statically or dynamically created for a GPRS MS. Creation of the IP address is performed by the VMSC through the standard GPRS *packet data protocol* (PDP) context activation procedures [9]. The VMSC maintains an MS table. The table stores the MS *mobility management* (MM) and PDP contexts such as *temporary mobile subscriber identity* (TMSI), *international mobile subscriber identity* (IMSI), and the QoS profile requested. These contexts are the same as those stored in a GPRS MS [9]. In the VMSC approach, SIP is used to support voice applications, where the VMSC serves as the SIP UA for any MS engaged in a VoIP session.

In Figure 34.11 the call server in NTP VoIP test bed performs standard functions for SIP proxy server and registrar (such as request forwarding and user location recording). In a connection between a SIP UA and a GSM MS, the TCP(UDP)/IP protocols are exercised in links (9), (10), and (11). The GPRS tunneling protocol [10] is exercised in link (4), and the GPRS Gb protocol [8] is exercised in links (3) and (5). Standard GSM protocols are exercised in links (1), (2), (6), (7), and (8). The SIP protocol is implemented on top of TCP(UDP)/IP, and is exercised between the VMSC and the nodes in the NTP VoIP test bed. The SIP packets are encapsulated and delivered in the GPRS network through the GPRS tunneling protocol.

In GSM/GPRS, a registration procedure is performed when an MS is turned on or when the MS moves to a new location (referred to as a routing area in GPRS [16,18]). In Figure 34.11, the GSM registration messages are delivered through the path (1) ↔ (2) ↔ (6) ↔ (7) ↔ (8). Then the PDP context for this MS is activated, and the SIP messages (i.e., REGISTER and 200 OK) are delivered through the path (1) ↔ (2) ↔ (6) ↔ (5) ↔ (4) ↔ (9). Details of the SIP registration is described in Figure 34.7 of Section 34.2.3. When outgoing/incoming calls arrive at the MS, the call origination/termination procedures are performed. The signaling path for GSM call setup and SIP INVITE is similar to that for the registration procedure. After the call is established, the voice path is (1) ↔ (2) ↔ (6) ↔ (5) ↔ (4) ↔ (10) in Figure 34.11. Details of SIP call setup procedure is given in Figure 34.8 of Section 34.2.3.

34.3.2 WLAN-based GPRS Support Node

This subsection describes the architecture and features of the *WLAN-based GPRS support node* (WGSN) jointly developed by NCTU and *Computer and Communications Laboratories* (CCL) of ITRI [11]. WGSN provides the functions of SIP *application level gateway* (ALG) and *authentication/authorization/accounting* (AAA) in the NTP VoIP test bed.

Figure 34.12 illustrates the interconnection between the NTP VoIP test bed and the WGSN. The WLAN radio network includes 802.11-based *Access Points* (APs) that provide radio access for the MSs. The WGSN acts as a gateway between the NTP VoIP test bed and the WLAN APs, which obtains the IP address for a MS from a *Dynamic Host Configuration Protocol* (DHCP) server. WGSN routes the packets between the MS and the external data networks (i.e., the NTP VoIP test bed). The WGSN node communicates with the HLR to support GPRS mobility management following 3GPP Technical Specification 23.060 [2]. Therefore, the WLAN authentication and network access procedures are exactly the same as that for GPRS described in Section 34.3.1.

Based on the seven interworking aspects listed in 3GPP Technical Specification 22.934 [3], we describe the following features implemented in WGSN [11].

Service aspects: WGSN provides general Internet access and VoIP services based on SIP. A *network address translator* (NAT) is built in the WGSN node for IP address translation of IP and UDP/TCP headers. However, the NAT does not translate the IP address in SIP/SDP messages, and the VoIP voice packets delivered by the RTP connection cannot pass through the WGSN node. This problem is solved by implementing a SIP ALG in the WGSN node, which interprets SIP messages and modifies the IP address contained in these SIP messages.

Access control aspects: WGSN utilizes the standard GPRS access control for users to access WLAN services. Our mechanism reuses the existing *subscriber identity module* (SIM) card and the subscriber data records in the HLR. Therefore, the WGSN customers do not need a separate WLAN access procedure, and the maintenance for customer information is simplified. User profiles for both GPRS and WLAN are combined in the same database (i.e., the HLR).

Security aspects: WGSN utilizes the existing GPRS authentication mechanism [17]. That is, the WLAN authentication is performed through the interaction between an MS (using SIM card) and the authentication center (residing in the HLR in Figure 34.12). Therefore, WGSN is as secured as existing GPRS networks. We do not attempt to address the WLAN encryption issue [16]. It is

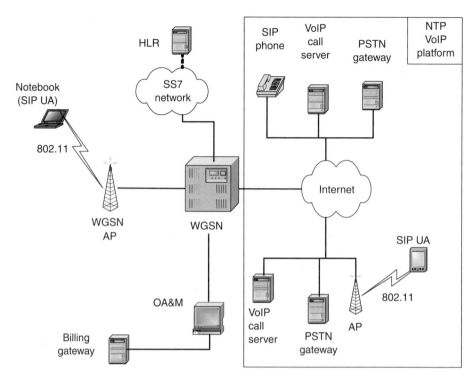

FIGURE 34.12   WGSN architecture.

well-known that WLAN based on IEEE 802.11b is not secured. For a determined attack, *wired equivalent privacy* (WEP) is not safe, which only makes a WLAN network more difficult for an attacker to intrude. The IEEE 802.11 Task Group I is investigating the current 802.11 MAC security. For authentication, 802.1x and its *extensible authentication protocol* (EAP) is the recommended approach. WGSN will follow the resulting solution.

Roaming aspects: WGSN provides roaming between GPRS and WLAN. We utilize the standard GPRS mobility management mechanism without introducing any new roaming procedures.

Terminal aspects: A WGSN MS is installed with a *universal IC card* (UICC) reader (a smart card reader implemented as a standard device on the Microsoft Windows platform). The UICC reader interacts with the SIM card to obtain authentication information for WGSN attach procedure. The SIP UA is implemented at the application layer of a WGSN MS.

Naming and addressing aspects: The WGSN user identification follows the *network access identification* (NAI) format [4] specified in the 3GPP recommendation [3]. Specifically, the IMSI is used as WGSN user identification.

Charging and billing aspects: The WGSN acts as a router, which can monitor and control all traffics for the MSs. The WGSN node provides both offline and online charging (for prepaid services) based on the *call detail records* (CDRs) delivered to the billing gateway.

In addition to the seven aspects listed above, WGSN also provides automatic WLAN network configuration recovery. A WGSN MS can be a notebook, which is used at home or office with different network configurations. The network configuration information includes IP address, subnet mask, default gateway, WLAN *service set identifier* (SSID), and so forth. When the MS enters the WGSN service area, its network configuration is automatically reset to the WGSN WLAN configuration if the MS is successfully authenticated. The original network configuration is automatically recovered when the MS detaches from the WGSN. This WGSN functionality is especially useful for those users who are unfamiliar with network

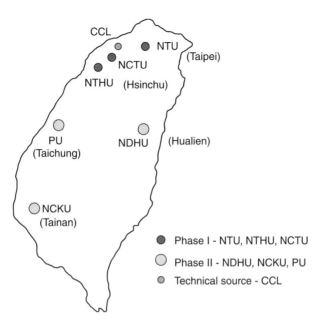

FIGURE 34.13   Connectivity of NTP VoIP sites.

configuration setup. The SIP registration and call setup procedures for WGSN are given in Section 34.2.3. Details of WGSN implementation can be found in Reference 11.

34.4   Conclusions

Under the NTP, we have deployed a SIP-based wireless VoIP test bed. Users participated in this test bed can freely download the SIP-based softphone client (UA) and install it on their personal computers with Windows 2000/XP operating systems. The UA provides digit–number dialing, SIP URI dialing, and E.164 dialing. It also supports functions such as call hold/retrieval and call transfer. At the time of writing, the PDA version of UA program has not been released due to intellectual property issue.

As described in Section 34.3, NTP VoIP test bed has been integrated with GPRS and WLAN to support wireless VoIP services. In the current stage, we are integrating UMTS IP multimedia subsystem [17] with the NTP VoIP test bed, and will develop a wireless VoIP service network based on *open service access* (OSA) specification [1]. Three extra sites are being included in the NTP VoIP test bed. They are located in National Cheng Kung University (NCKU) in Tainan, Providence University (PU) in Taichung, and Dong Hwa University (HDHU) in Hualien. The connectivity of the seven NTP VoIP sites is illustrated in Figure 34.13.

Acknowledgment

This work was sponsored in part by, National Information and Communication Initiative Committee under contract R-0300, National Science Council under contracts NSC 92-2219-E-009-032 and NSC 92-2213-E-002-092, Microsoft Research Asia, Intel, FarEastone, IIS/Academia Sinica, CCL/ITRI, and NCTU Joint Research Center, Chair professorship of Providence University, NTP.

References

[1] 3rd Generation Partnership Project (3GPP); Technical Specification Core Network; Open Service Access (OSA); Application Programming Interface (API); Part 1: Overview; (Release 4). Technical Specification 3G TS 29.198 V4.1.0 (2001–06), 2001.

[2] 3rd Generation Partnership Project (3GPP); Technical Specification Group Services and Systems Aspects; General Packet Radio Service (GPRS); Service Description; Stage 2. Technical Specification 3G TS 23.060 version 4.1.0 (2001–06), 2001.

[3] 3rd Generation Partnership Project (3GPP); Feasibility Study on 3GPP System to Wireless Local Area Network (WLAN) Interworking. Technical Specification 3G TS 22.934 Version 6.0.0, 2002.

[4] Aboba, B. and Beadles, M. The network access identifier. IETF RFC 2486, January 1999.

[5] CCITT. Numbering Plan for the ISDN Era. Technical Report Recommendation E.164 (COM II-45-E), ITU-T, 1991.

[6] Chou, S.L. and Lin, Y.-B. Computer telephony integration and its applications. *IEEE Communications Surveys*, 4, 2000.

[7] Collins, D. *Carrier Grade Voice Over IP*. McGraw-Hill, Singapore, 2001, pp. 46–56.

[8] ETSI/TC. European Digital Cellular Telecommunications System (Phase 2+); GPRS Base Station System (BSS) — Serving GPRS Support Node (SGSN) Interface; Gb Interface Layer 1. Technical Report Recommendation GSM 08.14 Version 6.0.0, ETSI, 1997.

[9] ETSI/TC. GPRS Service Description Stage 2. Technical Report Recommendation GSM GSM 03.60 Version 7.0.0 (Phase 2+), ETSI, 1998.

[10] ETSI/TC. GPRS Tunnelling Protocol (GTP) Across the Gn and Gp Interface. Technical Report Recommendation GSM GSM 09.60 Version 7.0.0 (Phase 2+), ETSI, 1998.

[11] Feng, V.W.-S., Wu, L.-Y., Lin, Y.-B., and Chen, W.E. WGSN: WLAN-based GPRS environment support node with push mechanism. *The Computer Journal*, 47(4): 405–417, 2004.

[12] Handley, M. and Jacobson, V. SDP: session description protocol. IETF RFC 2327, April 1998.

[13] ITU. Packet-Based Multimedia Communications Systems. Technical Report ITU-T H.323, Version 3, International Telecommunication Union, 1999.

[14] ITU-T. Pulse Code Modulation (PCM) of Voice Frequencies, 1988. Recommendation G.711.

[15] Li, B. et al. QoS-enabled voice support in the next-generation internet: issues, existing approaches and challenges. *IEEE Communications Magazine*, 38, 2000.

[16] Lin, Y.-B. and Chlamtac, I. *Wireless and Mobile Network Architectures*. John Wiley & Sons, New York, 2001, pp. 82–88.

[17] Lin, Y.-B., Hanug, Y.-R., Pang, A.-C., and Chlamtac, I. All-IP approach for UMTS third generation mobile networks. *IEEE Network*, 16: 8–19, 2002.

[18] Lin, Y.-B., Huang, Y.-R., Chen, Y.-K., and Chlamtac, I. Mobility management: from GPRS to UMTS. *Wireless Communications and Mobile Computing*, 1: 339–360, 2001.

[19] Pang, A.-C., Lin, P., and Lin, Y.-B. Modeling mis-routing calls due to user mobility in wireless VoIP. *IEEE Communications Letters*, 4: 394–397, 2001.

[20] Rao, H.C.-H., Lin, Y.-B., and Chou, S.-L. iGSM: VoIP service for mobile networks. *IEEE Communications Magazine*, 4: 62–69, 2000.

[21] Rosenberg, J. et al. SIP: session initiation protocol. IETF RFC 3261, June 2002.

[22] Schulzrinne, H. et al. RTP: a transport protocol for real-time applications. IETF RFC 1889, January 1996.

[23] Sinnreich, H. and Johnston, A.B. *Internet Communications Using SIP*. John Wiley & Sons, New York, 2001.

Cross-Layer Algorithms for Video Transmission over Wireless Networks

35.1 Introduction..**35**-789

35.2 Video Transmission over Mobile Wireless Networks..**35**-790
  Application-Based Approaches • Network-Based Approaches
  • Cross-Layer Approaches

35.3 Cross-Layer Error Control Algorithm..................**35**-793
  Error Control Algorithm • Algorithm Analysis •
  Simulation Results • Implementation Considerations

35.4 Cross-Layer Rate Control Algorithm **35**-802
  Rate Control Algorithm • Algorithm Analysis • Simulation
  Results • Implementation Issues

35.5 Summary and Open Problems..........................**35**-808

Acknowledgment..**35**-809

References **35**-809

G. Ding
Bharat Bhargava
X. Wu

35.1 Introduction

The third generation (3G) mobile networks and the rapidly deployed wireless local area networks (WLANs), such as IEEE 802.11 WLAN, have motivated various novel applications for the next generation mobile wireless networks. Multimedia applications, such as video streaming and video telephony, are regarded as "killer applications" in the emerging wireless networks. Video applications usually involve a large volume of data transmitted in a time sensitive fashion. However, the underlying wireless networks only provide time-varying and limited bandwidth, high data error rate, packet delay, and jitter. Extensive research has been done on either video data coding algorithms or wireless network protocols. But the traditional layered network model limits the video transmission over wireless networks because it tries to separate information and functions between different layers. To enable more efficient real-time video transmission over dynamic wireless environments, the video applications and underlying wireless networks should cooperate in order to share information and optimize the transmission process dynamically. This chapter reviews the state-of-the-art research efforts on video coding, error control, and rate control algorithms. A new cross-layer algorithm is presented, which coordinates the algorithms at different layers in order to get better performance than using them separately. The cross-layer rate control algorithm matches the application's future bandwidth requirement to the available bandwidth in the network so that an optimal data transmission rate can be selected. In the cross-layer error control algorithm, lower layers are responsible for error detection and fast retransmission, and application layer conducts an adaptive error correction algorithm with the help of lower layers.

The chapter is organized as follows. Section 35.2 is a review of current research results. Sections 35.3 and 35.4 introduce the cross-layer error control and rate control algorithms, respectively. The theoretical analysis, simulation results, and implementation considerations are presented in detail. Section 35.5 summarizes the chapter and points out some important open problems for future investigation.

35.2   Video Transmission over Mobile Wireless Networks

In this section, we will review current research results on video transmission over wireless networks, including application-based, network-based, and cross-layer approaches.

35.2.1   Application-Based Approaches

At application layer, there are two families of standards to compress the raw video data. The International Telecommunications Union-Telecommunications (ITU-T) H.261 is the first video compression standard gaining widespread acceptance. It was designed for videoconferencing. After that, ITU-T H.263 and its enhanced version H263+ were standardized in 1997 [1], which offered a solution for very low bit-rate (<64 Kb/sec) teleconferencing applications. The Moving Pictures Expert Group (MPEG) series are standardized by International Standard Organization (ISO). MPEG-1 achieves VHS quality digital video and audio at about 1.5 Mb/sec. MPEG-2 is developed for digital television at higher bit rate. In contrast, the recently adopted MEPG-4 standard [2] is more robust and efficient in error-prone environments at variable bit rates, which is achieved by inserting resynchronization markers into the bitstream, partitioning macro blocks within each video packet syntactically, using header extension code to optionally repeat important header information, and using reversible variable-length coding such that data can be decoded in both forward and reverse directions. A completely new algorithm, originally referred to as H.26L, is currently being finalized by both ITU and ISO, known as H.264 or MPEG-4 part 10 Advanced Video Coding (AVC).

Most video applications are characterized as being time sensitive: it will be annoying if the video data does not arrive to the receiver in time. In order to transmit video over error-prone wireless environments, various error-resilient video coding algorithms have been proposed [3]. The scalable coding is one of the most important approaches. The scalable or layered video involves a base layer and at least one enhancement layer. The base layer itself is enough to provide usable result, but the enhancement data can further improve the quality. Scalability can be achieved in many forms, including signal-to-noise-ratio (SNR), temporal, and spatial scalability. A new scalable coding mechanism, called fine granularity scalability (FGS), was recently proposed to MEPG-4 [4]. An FGS encoder uses bitplane coding to represent the enhancement bitstream. Bitplane coding enables continuous sending rate by truncating the enhancement layer bitstream anywhere. This advantage of FGS makes it more flexible than other scalable coding algorithms. An important property of scalable coding is that the base layer has the highest priority and must be transported correctly. In contrast, another coding approach, named multiple description coding (MDC) [5], encodes the signal into multiple bitstreams, or descriptions, of roughly equal importance. So any single description can provide acceptable result, while other descriptions complement to each other to produce better quality. Multiple descriptions can be transmitted simultaneously through diverse paths in order to increase the probability of receiving at least one description. MDC achieves these advantages at the expense of coding efficiency.

Rate control and error control are regarded as application-layer quality of service (QoS) techniques, which maximize the received video quality in the presence of underlying error-prone networks [3].

Rate control determines the data sending rate based on the estimated available bandwidth. A lot of so called transmission control protocol "TCP-friendly" rate control approaches have been proposed for best-effort Internet in order to avoid network congestion [6]. Error control is employed to reduce the effect of transmission error on applications. Two basic approaches are forward error correction (FEC) and automatic repeat request (ARQ). FEC adds parity data to the transmitted packets and this redundant information is used by the receiver to detect and correct errors. FEC maintains constant throughput and has bounded time delay. ARQ is based on packet retransmissions when errors are detected by the receiver. ARQ is simple but the delay is variable. Many alternatives to FEC and ARQ have been introduced in Reference 7. One of the most well-known error coding techniques is Reed–Solomon coding [8] which can deal with burst errors. The field of R–S coding is of the form $GF(2^M)$, where M is any positive integer. Each bit block of 2^M bits is called a symbol. If the original packet length is K symbols, then after adding redundant parity data, the codeword will be of length $N > K$. The original packet can be completely recovered when there are no more than $(N - K)/2$ error symbols during transmission. In order to protect source data with different importance, joint source/channel coding has been proposed [9]. For example, different frames in MPEG-4 coding or different layers in scalable source coding can combine with unequal length of parity data. Error concealment [10] is a technique employed by receivers to minimize the effect of packet errors on the quality of video.

35.2.2 Network-Based Approaches

In addition to the research on video applications, a large body of research has been conducted on improving underlying networks for the benefits of upper applications. We will review network-based approaches from right below application layer down to physical layer, according to the open system interconnect (OSI) reference model.

Between application layer and transport layer, there are several standardized protocols designed for supporting real-time applications, such as real-time transport protocol (RTP), real-time control protocol (RTCP), real-time streaming protocol (RTSP), and session initiation protocol (SIP) [3]. RTP provides extra information to application layer in the form of sequence numbers, time-stamping, payload type, and delivery monitoring. But RTP itself does not ensure timely delivery or other QoS guarantees. RTCP is a control protocol for monitoring RTP packet delivery. RTSP and SIP are designed to initiate and direct delivery of real-time data.

At transport layer, TCP provides reliable transmission of data by flow control, congestion control, and retransmission. However, for most real-time communications, applications can tolerate data errors to some extent, but they have strict time constraint. So another simpler transport protocol, UDP, is widely used for real-time data transmission. UDP only uses cyclic redundancy check (CRC), or checksum, to verify the integrity of received packet. Since UDP does not perform any error correction, it may sacrifice the whole packet only for some minor data errors, which can yield unpredictable degradation and poor application quality. In order to solve this problem, a modified version, called UDP Lite, is introduced [11]. UDP Lite allows partial checksum on packet data by enabling application layer to specify how many bytes of the packet are sensitive and must be protected by checksum. If bit errors occur in the sensitive region, the receiver drops the packet; otherwise it is passed up to the application layer. This approach allows the application to receive partially corrupted packets, which may still generate acceptable video quality.

Most differences between wireless and wired networks exist below transport layers. Wireless networks involve different kinds of radio access networks, such as mobile cellular network and WLAN. Under the umbrella of IMT-2000, 3G mobile network standards provide high date rate up to 384 kbps for mobile users and 2 Mb/sec for users at pedestrian speed. One of the most promising 3G networks is Wideband Code Division Mutilple Access (WCDMA), also known as Universal Mobile Telecommunications System (UMTS). UMTS is standardized by 3GPP [12]. A UMTS network consists of Universal Terrestrial Radio Access Network (UTRAN) and core networks (CN). An IP packet coming from an Internet host is first directed to a 3G gateway general packet radio service (GPRS) support node (3G-GGSN) through Gi interface, then tunneled from 3G-GGSN to UTRAN via 3G Serving GPRS Support Node (3G-SGSN) by GPRS Tunneling Protocol for data transfer (GTP-U). We are most interested in data transmission over the wireless link, which is the air interface (Uu) between UTRAN and mobile stations (called UE in UMTS), because the data transmission rate at wireless link is usually the bottleneck of throughput. Radio Link Control (RLC) [13], located at the upper half of data link layer, segments packets into radio blocks and provides three data transfer modes to upper layers: Transparent Mode without any extra information; Unacknowledged Mode which can only detect erroneous data; Acknowledged Mode, which provides more reliable data delivery by limited number of retransmissions. Media Access Control (MAC) is responsible for mapping logical channels into transport channels provided by physical layer (PHY), as well as reporting of measurements to the Radio Resource Control (RRC). RRC manages and controls the use of resources and therefore has interactions with RLC, MAC, and PHY. In addition to wide area wireless networks, WLANs have been rapidly accepted in enterprise environments, mainly due to the standardization by IEEE 802.11 work group and the low cost of deploying a WLAN. The IEEE 802.11a [14] and the newly approved 802.11g can provide high data rate up to 54 Mb/sec. IEEE 802.11x standards only define the physical layer and MAC layer in order to seamlessly connect WLAN to existing wired LANs. For the channel access control, in contrast to the centralized control in cellular networks, IEEE 802.11 employs carrier sense multiple access with collision avoidance (CSMA/CA) in a highly distributed fashion. The mandatory function implemented in 802.11 stations is distributed coordination function (DCF), by which mobile stations contend for the radio channel and there is only one user at one time. But the DCF does not fit well to the time-sensitive applications. Instead, another optional point coordination function (PCF) should work better because there is a point coordinator (PC), which polls every user in the contention free period (CFP). The contention period (CP) controlled by DCF and the CFP controlled by PCF alternate to accommodate both functions. At the physical layer, IEEE 802.11x provides different transmission rates. For example, 802.11a provides eight data rates from 6 to 54 Mb/sec by changing channel coding parameters [14].

Research activities on video transmission over wireless networks can be found in References 15–31. Liu and Zarki [15] discussed rate control of video source coding for wireless networks. Wu et al. [16] proposed a general adaptive architecture for video transport over wireless networks. Vass et al. [17] proposed a novel joint source/channel coding for wireless links, based on the concept of application-level framing. Wang et al. [18] compared MDC with scalable coding in wireless networks when multiple paths are available. Majumdar et al. [19] investigated unicast and multicast video streaming over WLAN where the wireless network is modeled as a packet erasure model at network layer. Miu et al. [20] presented experimental results to show the advantage of transmitting H.264/MPEG-10 AVC encoded video over multiple access points in IEEE 802.11b WLAN. The above approaches are used at application layer and the underlying wireless network is considered as a high packet loss rate environment. Other research is focused on improving the performance of wireless networks. Fitzek and Reisslein [21] proposed a prefetching protocol for continuous media streaming between a base station and mobile stations in a cell. The protocol used transmission rate adaptation techniques to dynamically allocate transmission capacity for different streams. Singh et al. [22] evaluated the performance of using UDP Lite and transparent mode RLC in global system for mobile (GSM). Wang [23] developed a new MAC protocol for frequency division duplex (FDD) WCDMA in 3G cellular networks. A scheduling scheme, assisted by minimum power allocation and effective connection admission control, was used to fairly queue multimedia traffic with different QoS constraints. Zhang et al. [24] considered scalable video transmission over precoded orthogonal frequency division multiplexing (OFDM) system, enhanced by adaptive vector channel allocation. Zou et al. [25] proposed a MPEG-4 frame drop policy in order to save bandwidth of IEEE 802.11 WLAN in the DCF mode.

35.2.3 Cross-Layer Approaches

According to the above review, most current research activities are focusing on solving part of the problem: there have been a huge number of results based on applications and networks separately, but there is still not much research on the interaction between different layers in order to take full advantage of each other. Girod et al. [26] introduced several advances in channel-adaptive video streaming, which involved several system components jointly. But their integration to wireless networks was not investigated. Zhang et al. [27] integrated their work at different layers into a cross-layer framework supporting multimedia delivery over wireless internet. However, they did not address the cooperation of techniques at different layers. Zheng and Boyce [28] introduced an improved UDP protocol. The protocol captures the frame error information from link layer and uses it for application layer packet error correction. Shan and Zakhor [29] proposed a cross-layer error control scheme in which the application layer implements FEC and requests link layer to retransmit lost packet when necessary. Ding et al. [30] proposed a cross-layer error control algorithm, which can dynamically adjust FEC at application layer according to the link layer retransmission information. Krishnamachari et al. [31] investigated cross-layer data protection for IEEE 802.11a WLAN in the PCF mode. The application layer scalable coding and FEC were adopted, along with MAC layer retransmission and adaptive packetization. The above results investigated cross-layer error control methods in order to deal with high error rate in wireless networks. How to use limited and variant bandwidth in wireless networks is another challenging issue that should be addressed. In the following sections, we will introduce the cross-layer error control and rate control algorithms for video transmission over mobile wireless networks.

35.3 Cross-Layer Error Control Algorithm

Due to channel fading, user mobility and interference from other users, wireless links are less reliable than wired connections. Data link layer and physical layer in wireless networks provide some error control mechanisms, such as error detection and packet retransmission. Yet the lower layers do not know to which extent their error control mechanisms should be used, which is actually better known by the upper layer applications. For example, when using retransmission for error correction in a video application, the lower layers only know whether the data is received correctly, based on which the data will be re-sent or not. Whether the delay is within the time constraint is judged by the application layer. Error control techniques can also be embedded into applications. However, application-layer retransmission is usually less efficient than link layer retransmission [27], and FEC at application layer cannot adapt to network conditions to reduce the redundant error control information. In this section, a novel cross-layer error control algorithm is described. We will go through the algorithm in details, followed by the performance analysis, simulation results and implementation considerations.

35.3.1 Error Control Algorithm

The algorithm with inter-layer signaling is illustrated in the protocol stack in Figure 35.1.

The details for the steps of the algorithm at a sender are as follows:

S1. The compressed real-time data is first processed by error control layer. Data is segmented into packets of K bytes. For each packet, R bytes of parity data are added by R–S coding in the field $GF(2^8)$, and a small header (including sequence number n and parity data length R) of H bytes is also added. The data is then saved into a buffer of $N * (K + R + H)$ bytes row by row and is read column by column. This is for data interleaving. The packet size of the output data is of N bytes. Parameter R is adjusted by

$$R = \max(R - R_{\text{step}}, R_{\text{min}}), \tag{35.1}$$

where R_{min} and R_{step} are the minimal value and decreasing/increasing step of R, respectively.

S2. At data link layer, packet n is fragmented into small radio frames with sequence numbers $n_1, n_2, \ldots,$ and n_m. CRC is added in each radio frames for error detection.

S3. A radio frame n_k is sent and a timer is set.

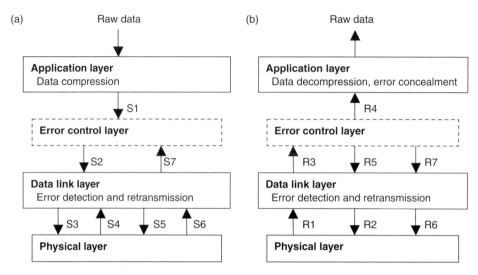

FIGURE 35.1   Protocol architecture for cross-layer error control: (a) sender; (b) receiver.

S4. If a nonacknowledgment (NACK), including a type LOW and a frame sequence number n_k, is received from the receiver, or the timer is timeout,

S5. Then frame n_k is retransmitted. If the sender keeps getting NACKs for the same frame after retx$_{max}$ times of retransmissions, this frame will be discarded and its timer is cleared.

S6. If a NACK, including a type HIGH and a packet sequence number n, is received, it is forwarded to error control layer.

S7. At error control layer, upon receiving a type HIGH NACK with a packet sequence number n, R is updated by

$$R = \min(R_0 + 2^k R_{step}, R_{max}), \qquad (35.2)$$

where R_0 and R_{max} are the initial value and the maximal value of the length R of the parity data, respectively. k is increased by 1 if the packet sequence number in NACK immediately follows the packet sequence number in last NACK. Otherwise k is set to 0.

The details for the steps of the algorithm at a receiver are as follows:

R1. When a frame n_k is received, errors are detected by CRC at the link layer. If any error is detected or the received frame is out of sequence.

R2. Then a NACK, including a type LOW and the frame sequence number n_k, is sent back to the sender. If no error is detected, all frames belonging to packet n are assembled. If the packet n is completed, it is sent up to error control layer.

R3. At error control layer, the received packets are interleaved in order to get the original application packets. R–S decoding algorithm is run to correct errors if there is any. On receiving a duplicated packet, the old packet is replaced with the new one.

R4. The packet is buffered until it is retrieved by the application layer. The application layer uses its own error concealment techniques if an error cannot be recovered by the lower layers.

R5. At error control layer, if a packet cannot be corrected, a NACK including the type HIGH and the sequence number n of the erroneous packet is sent down to data link layer.

R6. At data link layer, upon receiving a NACK from error control layer, the NACK is sent to the sender.

R7. If a packet n is still not available when it is requested by the upper applications, the error control layer sends a request to the data link layer for this packet. The data link layer will immediately send the packet n to upper layers and go to Step 3, even if the requested packet is not completed yet.

There are several remarks on the above algorithm.

1. For the adaptive error control algorithm, the larger the length R of parity data, the more errors can be corrected. On the other hand, this redundancy reduces the bandwidth efficiency. An appropriate R has to be chosen. In the proposed adaptive algorithm, R is dynamically adjusted based on the network information. When there is no HIGH NACK received, sender periodically decreases R from its initial value R_0 by a small step R_{step}. When there is a type HIGH NACK received, which indicates that a higher level error protection is requested, the sender sets R back to R_0. If another HIGH NACK is received immediately following the last HIGH NACK, R is exponentially increased by Equation (35.2). This is used to enhance the error protection capacity in response to the error burst in wireless channels. The initial value of the parity data length R_0 is calculated by

$$R_0 = aR_0 + 2(1 - a)b(K + R)C_{error}/C_{total},$$ (35.3)

where C_{error} and C_{total} are the number of received HIGH NACKs and the total number of sent packets, respectively, and C_{error}/C_{total} is the estimation of packet loss rate. b deals with the variance of the estimation and $b > 1$. a is used to smooth the estimation and $0 < a < 1$. This algorithm works when packet loss rates do not vary very fast.

2. Since each packet can be protected by parity data using different values of R, the proposed algorithm can provide multiple levels of error protection. In an application when data has different importance, the most important part can be assigned a larger R_0 by setting a larger b in Equation (35.3), while less important part can be protected by a smaller R_0.

3. In Figure 35.1, for the sake of illustration and generality, an error control layer is shown between the application layer and the data link layer. It can be actually implemented in application layer, data link layer, or any appropriate layers in between. This virtual error control layer serves as a controller, which coordinates the error control techniques in different layers. Its operations include:

- When the upper layer FEC cannot recover a packet, it requests lower layers to send a type HIGH NACK to the sender in order to increase the parity data length.
- The existing lower layer protocols do not send a packet to the upper layers until every fragment of the packet has been correctly received. This may incur unnecessary packet delay, which is not deserved by real-time services. In the proposed error control algorithm, the virtual control layer can request the lower layer to send an incomplete packet when the upper application requires. The upper layer will try to recover the corrupted packet by its own FEC. In this way, the error control techniques in both the upper and the lower layer are fully utilized.
- Retransmissions in the link layer can be executed up to $retx_{max}$ times or until the packet is requested by upper layers. Since retransmission can be stopped by the upper applications, it does not incur any extra packet delay. However, it increases the probability of obtaining a correct packet before it is requested.

4. Figure 35.1 does not show the intermediate network layers between the application layer and the data link layer. If IP is used for the network layer, IP header is protected by CRC. Packets with corrupted IP header will be discarded. At the transport layer, UDP Lite should be employed in order to pass data payload up as far as there is no error in the header.

35.3.2 Algorithm Analysis

In order to analyze the above algorithm, we adopt the well-known two-state Gilbert Elliot model to represent the variation of wireless channel. The wireless channel can stay at two states: "good" state and "bad" state. The bit error rate (BER) at "good" state, defined as e_g, is far less than the BER at "bad" state, defined as e_b. The state transition matrix is

$$\begin{bmatrix} \mu_{gg} & 1 - \mu_{gg} \\ 1 - \mu_{bb} & \mu_{bb} \end{bmatrix}, \tag{35.4}$$

where μ_{gg} and μ_{bb} are the transition probabilities of staying in "good" and "bad" states, respectively. The steady-state BER can be calculated by:

$$p_b = \frac{1 - \mu_{gg}}{2 - \mu_{gg} - \mu_{bb}} e_g + \frac{1 - \mu_{bb}}{2 - \mu_{gg} - \mu_{bb}} e_b. \tag{35.5}$$

The radio frame error probability at data link layer is:

$$p_L = 1 - (1 - p_b)^{M_0 + H_0} \approx (M_0 + H_0)p_b, \tag{35.6}$$

where M_0 and H_0 are the length of data payload and header of the radio frames, respectively. In order to reflect the effect of retransmission on the system, we define a transmission efficiency parameter as the ratio of transmission time without loss and the transmission time with loss-and-retransmission.

$$\gamma_L = \frac{(M_0 + H_0)}{\sum_{n=1}^{\text{retx}_{\max}-1} (nM_0 + (2n-1)H_0)P_{n0} + (\text{retx}_{\max}M_0 + (2\text{retx}_{\max} - 1)H_0)(1 - \sum_{k=1}^{\text{retx}_{\max}-1} p_{n0})}, \tag{35.7}$$

where $P_{n0} = p_L^{n-1}(1 - p_L)$ represents the probability of $n-1$ radio frames getting errors before a successful (re)transmission. We assume that the transmission rate is constant during retransmissions and the size of NACK packets is dominated by the header length.

At error control layer, due to interleaving, every lower layer frame loss corresponds to one byte error in the application packet. An application packet loss occurs when there is an error in the header of the packet, or there are more than $R/2$ corrupted bytes in the payload. The probability of a packet loss can then be calculated as:

$$p_H = (1 - (1 - p_L)^H)\frac{H}{M + H} + \sum_{i=R/2+1}^{M} \binom{M}{i} p_L^i (1 - p_L)^{M-i} \frac{M}{M + H} \tag{35.8}$$

where $M = K + R$ is the length of data payload. A discrete Markov Chain can be used to model the state of R. Specifically, we let state 0 represent initial value R_0 and state i represent $R = R_0 + i * R_{\text{step}}$, where $i \in \Gamma = [N_L, 2^{N_H}]$. N_L is the negative lowest bound and 2^{N_H} is the upper bound. Letting $p_H(i)$ denote p_H when $R = R_0 + i * R_{\text{step}}$, the nonzero elements in the transition matrix $\{p_{i,j}\}_{i,j \in \Gamma}$ for the Markov Chain is:

$$p_{N_L,N_L} = 1 - p_H(N_L), \quad \text{and} \quad p_{N_L,0} = p_H(N_L),$$

$$p_{i,i-1} = 1 - p_H(i), \quad \text{and} \quad p_{i,0} = p_H(i), \quad \text{for } N_L < i < 0,$$

$$p_{0,-1} = 1 - p_H(0), \quad \text{and} \quad p_{0,1} = p_H(0),$$

$$p_{i,i-1} = 1 - p_H(i), \quad \text{and} \quad p_{i,2^h} = p_H(i), \quad \text{for } 2^{h-1} \le i < 2^h, \quad h = 1, 2, K, N_H,$$

$$p_{2^{N_H},2^{N_H}-1} = 1 - p_H(2^{N_H}), \quad \text{and} \quad p_{2^{N_H},2^{N_H}} = p_H(2^{N_H}).$$

The state transition graph for $N_L = -3$ and $N_H = 3$ is depicted in Figure 35.2.

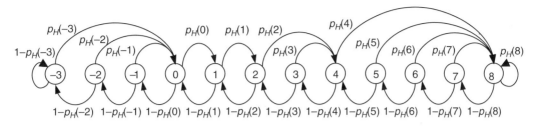

FIGURE 35.2 Transition graph of Markov Chain model ($N_L = -3$ and $N_H = 3$).

The stationary distribution π can be found as

$$\pi(i) = x(i)\pi(0), \quad i \in \Gamma,$$

where

$$\pi(0) = \frac{1}{\sum\limits_{i \in \Gamma} x(i)},$$

$$x(N_L) = \frac{1}{p_H(N_L)} \prod_{j=N_L+1}^{0} (1 - p_H(j)),$$

$$x(i) = \prod_{j=i+1}^{0} (1 - p_H(j)), \quad \text{for } N_L < i < 0,$$

$$x(0) = 1, \quad x(1) = \frac{p_H(0)}{(1 - p_H(1))}, \quad x(2) = \frac{p_H(0)p_H(1)}{(1 - p_H(1))(1 - p_H(2))},$$

$$x(2^{h+1}) = \frac{1}{\prod_{j=2^h+1}^{2^{h+1}} (1 - p_H(j))} \left(\left(1 - \sum_{k=2^{h-1}+1}^{2^h-1} p_H(k) \prod_{j=k+1}^{2^h} (1 - p_H(j)) \right) x(2^h) - p_H(2^h - 1)x(2^h - 1) \right),$$

$$\text{for } h = 1, K, (N_H - 1)$$

$$\text{and } x(i) = \prod_{j=i+1}^{2^h} (1 - p_H(j))x(2^h), \quad \text{for } 2^{h-1} < i < 2^h, \quad h = 2, K, N_H.$$

Figure 35.3 illustrates the packet loss rate (a) and stationary distribution (b) when $N_L = -15, N_H = 5$, and $R_0 = 50$ bytes. The distribution of R is dynamically changing according to the network conditions. For example, when BER is as high as 1%, the parity data length R is more likely to have values larger that R_0. But when BER is equal to 0.5%, R tends to be smaller than R_0. This demonstrates the efficiency of the proposed adaptation algorithm.

35.3.3 Simulation Results

The above error control algorithm is tested by simulations. The video source is a one-hour video trace of movie *Star War IV* [36], which is generated specifically for the research on video transmission over wired or wireless networks [37]. Figure 35.4 shows the data loss with and without cross-layer error control. The parameters for error model (35.4) are $\mu_{gg} = 0.995$, $\mu_{bb} = 0.96$, and $e_g = 10^{-6}$. e_b is chosen so that the steady state BER is 0.02. The parity data length R is fixed at 50 bytes. 1080 video frames are displayed.

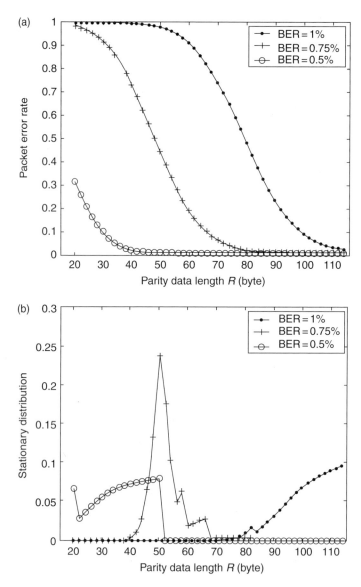

FIGURE 35.3   Packet error rate and distribution of parity data length: (a) packet error rate; (b) stationary distribution of parity data length R.

Using the proposed error control algorithm, most data loss can be avoided. The average loss is 0.256 bytes/frame, in contrast to 6.30 bytes/frame when error control is not used.

In Figure 35.5, the error control scheme using adaptive R is compared with the scheme using fixed R. The BER is varying from 0.02 to 0.1. The adaptive scheme makes a good trade-off between error protection and network traffic so that a nearly constant PSNR is obtained at different BERs.

As stated in Section 35.3.1 (ii), the proposed error control algorithm can provide multiple levels of error protection. For data of higher priority, a larger value is chosen for parameter b in Equation (35.3). For MPEG-4 video streaming, I, P, and B frames can be protected by choosing b as 3, 2, and 1, respectively. As shown in Figure 35.6, the mean parity data length for I frame is the largest, so it has the highest PSNR.

The adaptive error control algorithm is further applied to videos encoded by MPEG-4 at different quality levels, as shown in Table 35.1. The PSNRs after using error control algorithm significantly outperform those without error control.

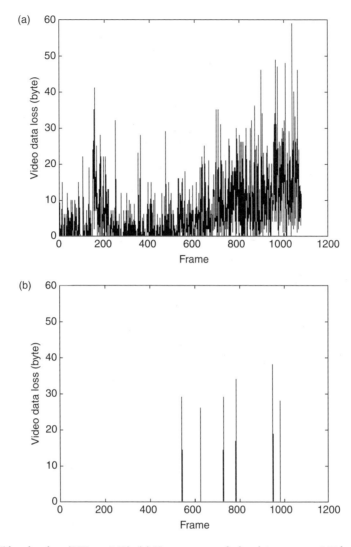

FIGURE 35.4 Video data loss (BER = 0.02). (a) No error control, data loss mean = 6.30 bytes/frame (b) Error control with R = 50 bytes, data loss mean = 0.256 bytes/frame.

35.3.4 Implementation Considerations

The cross-layer algorithms involve at least two layers. This fact may increase the difficulty of implementing them. We here discuss the implementation of the proposed cross-layer algorithm in two most popular wireless networks: 3G UMTS and IEEE 802.11 WLAN.

In UMTS, two implementations are recommended, which use RLC and high speed downlink packet access (HSDPA [32]) for link layer error control, respectively. The modified protocol architectures are shown in Figure 35.7. Only steps in control plane are shown. Other steps belong to data plane and can be treated as regular data transmission in UMTS. The major procedures of implementing cross-layer error control algorithm using RLC include the following:

- The upper layer FEC and interleaving can be integrated into the application layer where developers have more freedom.
- The RLC layer of UMTS can be modified to enable sending NACK frames so that it can handle data link layer ARQ for the cross-layer control algorithm. The related control steps for a sender and a receiver are steps S4, S6, R2, and R6 in Figure 35.1.

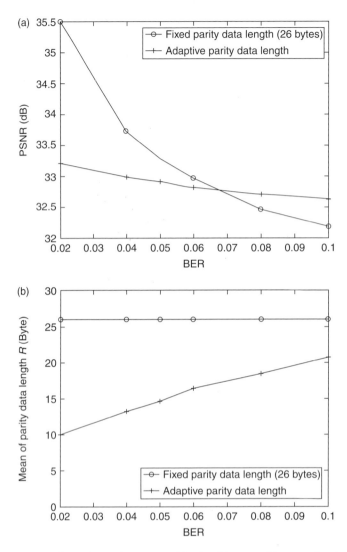

FIGURE 35.5   Simulation results on using adaptive R and fixed R: (a) PSNR; (b) mean of parity data length R.

- The RRC is located at layer 3 (network layer) of UTRAN and has the control interfaces to RLC, MAC, and PHY. RRC can also provide services to upper layers. Therefore, RRC can function as relaying cross-layer control signals. The related control steps are R5, R7, and S7.

The HSDPA is designed to enhance downlink data transmission capacity in UMTS. A new transport channel between MAC and PHY, named shared HSDPA data channel (HS-DSCH), is introduced. HS-DSCH is only used for high speed downlink data transmission from a base station to UE. It is controlled by a new entity in MAC, called MAC-hs, which employs hybrid ARQ (HARQ) to make retransmission decision. Since MAC-hs is closer to physical layer than RLC is, applying error control in MAC-hs is expected to be more efficient than that in RLC. The corresponding control plane protocol stack is shown in Figure 35.7(b).

In IEEE 802.11 WLANs, all the radio access work is conducted in MAC and PHY layers in order to make the underlying networks transparent to the upper applications. This makes it difficult to enable cross-layer information exchange. To minimize cross-layer signaling, the protocol architectures in Figure 35.8 are recommended. Following aspects are discussed: the type HIGH NACK can be an application packet so

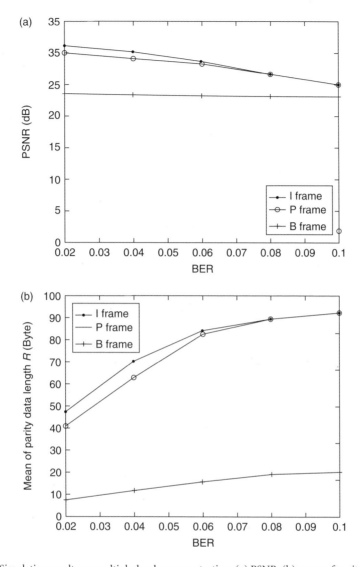

FIGURE 35.6 Simulation results on multiple-level error protection: (a) PSNR; (b) mean of parity data length.

TABLE 35.1 Adaptive Error Control of MPEG-4 Videos of Different Qualities

Quality	Low	Medium	High	64 Kbps	128 Kbps	256 Kbps
PNSR (dB) without error control	27.0317	28.1804	30.1252	28.5134	29.7964	30.2827
PNSR (dB) with error control	27.8962	30.3623	32.9462	30.2560	31.8715	32.8701
Mean of R (bytes)	30.058	29.665	14.598	29.151	29.202	29.240

that step R5 and S7 do not incur any inter-layer communications. HIGH NACK is used to notify the sender of an FEC failure. Yet it does not request any retransmissions at application layer. Since the ACK-based retransmission is mandatory in IEEE 802.11 MAC layer, the MAC layer can be modified by adding the type LOW NACK. This can be implemented by adding an extra field of frame sequence number in the ACK MAC frame. This may also be implemented as a new control MAC frame using one of the reserved subtype numbers between 0001 and 1001 in the field of "frame control" [14].

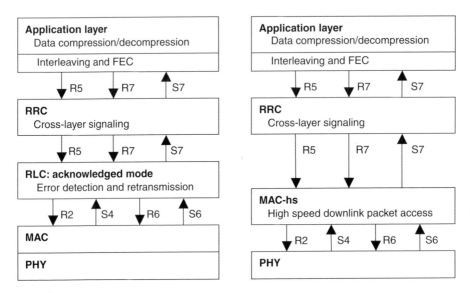

FIGURE 35.7 Implementation of cross-layer error control in UMTS (control plane): (a) error control using RLC; (b) error control using HSDPA.

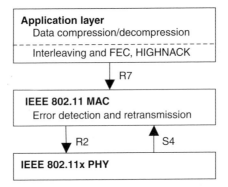

FIGURE 35.8 Implementation of cross-layer error control in IEEE 802.11 WLANs.

35.4 Cross-Layer Rate Control Algorithm

At data link layer, the link-adaptation techniques in wireless networks [33,34] can be employed to adjust transmission bandwidth. At application layer, the data rates of video applications can also be adjusted according to different QoS requirements (e.g., layered application). In this section, we introduce a cross-layer transmission rate control algorithm involving both the application layer and the radio access network. The future data rate requirement of the application is used to determine the network transmission rate.

35.4.1 Rate Control Algorithm

In the data link layer of wireless networks, transmitting rate is determined according to the size of the buffered data at the transmitting site. When the buffer size is too large, a higher transmission rate is needed. When the buffer size is too small, the transmission rate will be reduced. Compared to the cross-layer rate control algorithm, the scheme on monitoring buffer size is lack of reality since it only reflects the past bandwidth requirements. The comparison is illustrated in Figure 35.9. When the application data rate increases during the time interval $(t, t + T)$, the lower layer buffer is rapidly overflowed at point A.

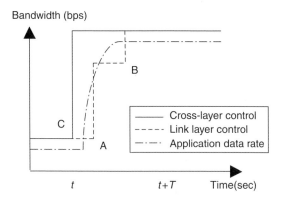

FIGURE 35.9   Illustration of using cross-layer rate control.

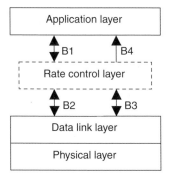

FIGURE 35.10   Protocol architecture for cross-layer rate control.

The bandwidth adjustment is triggered so that, at point A, the transmitting rate is high enough for the application. However, the bandwidth needed for application continues to increase after point A, and another bandwidth adjustment might be needed at point B. Using cross-layer rate control, the bandwidth requirement for time $(t, t + T)$ can be obtained from application layer at point C, and the accurate transmitting bandwidth can be assigned in one step.

The cross-layer rate control algorithm is suitable for layered applications. To address this, we assume that a video application has n layers of data streams, and each extra layer enhances the quality of previous layers but requires more bandwidth. We also assume that the available transmitting bandwidth set in the network is Φ. The protocol architecture is shown in Figure 35.10. Note that the rate control layer is a virtual layer, which can actually be implemented at any appropriate layer for a particular wireless network. Control steps in Figure 35.10 are explained as follows:

B1. Request bandwidth requirement for the coming time interval from the application layer:

$$\mathrm{BW}_H(t, t + T) = [\mathrm{BW}_{H1}, \mathrm{BW}_{H2}, \ldots, \mathrm{BW}_{Hn}].$$

BW_{Hi} is the bandwidth requirement for the application with i layers.

B2. Request current network bandwidth from lower network layers: $\mathrm{BW}_L \in \Phi$.

B3. Call procedure BWSchedule($\mathrm{BW}_H, \mathrm{BW}_L$) with return values $(k, \mathrm{BW}_L^{\mathrm{new}})$.

B4. Inform application layer of the maximal layer, k, that can be sent, as well as the current network bandwidth $\mathrm{BW}_L^{\mathrm{new}}$.

Procedure BWSchedule (BW_H, BW_L)

```
 1:  if  ∫ₜ t  +  T  BW_{Hₙ}(τ)dτ − BW_L · T < Buf− then
 2:              BW_L^{new} = min{BW ∈ Φ| BW ≤ BW_L, and BW ≥ 1/T(∫ₜ t + TBW_{Hₙ}(τ)dτ − Buf+)}
 3:              request BW_L^{new} from lower layers
 4:              return (n, BW_L^{new})
 5:  endif
 6:  if ∫ₜ t + T BW_{Hₙ}(τ)dτ − BW_L · T > Buf+ then
 7:              if BW_L^{new} = min{BW ∈ Φ| BW ≥ 1/T(∫ₜ t + TBW_{Hₙ}(τ)dτ − Buf+)} exists then
 8:                    request BW_L^{new} from lower layers
 9:                    return (n, BW_L^{new})
10:              else
11:                    BW_L^{new} = max Φ
12:                    requestBW_L^{new} from lower layers
13:                    if k = max{i ∈ [1, K, n]| ∫ₜ t + TBW_{Hᵢ}(τ)dτ ≤ T · BW_L^{new} + Buf+)} exists then
14:                          return(k, BW_L^{new})
15:                    else
16:                          return (1, BW_L^{new})
17:                    endif
18:              endif
19:  else
20:         return (n, BW_L)
21:  endif
22:  endprocedure
```

FIGURE 35.11 Procedure BWSchedule.

In this rate control algorithm, the available bandwidth BW_L and the future bandwidth requirement $BW_H(t, t + T)$ are checked periodically. If the future bandwidth requirement reaches the upper limit Buf^* of the buffer, a larger available bandwidth BW_L^{new} that can satisfy the requirement is found. If the maximum bandwidth in Φ cannot satisfy the requirement for all of n layers of the application, a maximal number k is found so that k layers of the application can be accommodated using the available maximum bandwidth. Transmitting bandwidth will be reduced when the future application bandwidth requirement is so small that the negative buffer limit, Buf^-, is reached. The procedure *BWSchedule* is given in Figure 35.11. It runs at every time $t = jT$, where T is the time period and j is a non-negative integer.

35.4.2 Algorithm Analysis

According to the above cross-layer rate control algorithm, since it always chooses the minimal transmission bandwidth that can satisfy the future application requirements, it will neither waste unnecessary bandwidth, nor fail to provide enough bandwidth to upper applications as far as there is some available bandwidth that can satisfy the requirement. This proves the efficiency of the proposed rate control algorithm. In addition, if the elements in Φ have been preordered, the complexity of the algorithm is $O(\log(\max\{n, m\}))$, which is computationally tractable even for large n and m.

One practical issue with this algorithm is that, when it requests a bandwidth $BW_L^{new} \in \Phi$ from lower layers, it may not guarantee to actually obtain this bandwidth at any time. For instance, other applications might have used up all the channels providing the requested transmission capacity, the interference from other users might degrade the quality of requested channel, it might take too much time to communicate with radio access networks to set up the requested radio link, and so on. To theoretically analyze this issue, we assume that there are some other elements, in the preordered set Φ, that are between current bandwidth BW_L and the requested bandwidth BW_L^{new}. At every intermediate bandwidth, it has probability $0 < p < 1$ to reach its neighbor bandwidth closer to BW_L^{new} and probability $1 - p$ to remain the current

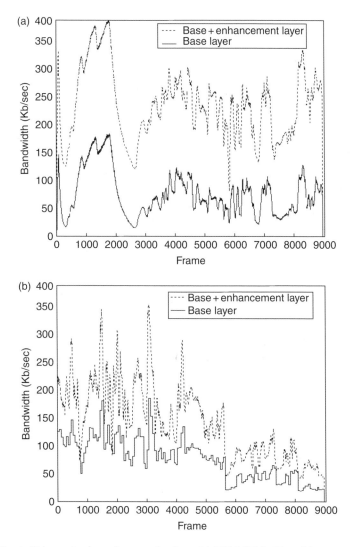

FIGURE 35.12   Bandwidth requirements from applications: (a) FGS MPEG video; (b) temporal scalable MPEG-4 video.

bandwidth. Under these assumptions, we have the following statement:

> *The requested bandwidth* $\mathrm{BW}_L^{\mathrm{new}}$ *can finally be granted with probability 1, but the cost is $1/p$ times of the case when every bandwidth request can be immediately granted.*

This statement can be theoretically proved by discrete markov chain and Wald's equation. We omit the details due to space limit.

35.4.3   Simulation Results

The proposed cross-layer rate control algorithm is used to stream two-layer FGS MPEG video, as well as temporal scalable MPEG-4 video [36]. The required bandwidths for the two encoding methods are shown in Figure 35.12. For the clarity of presentation, only five-minute video results are displayed. Assuming that the available bandwidth from lower layers is in the set $\Phi = [150, 170, 190, 210, 230, 250]$ measured in Kb/sec, the actual transmission bandwidth and the percentage of transmitted data at enhancement layer

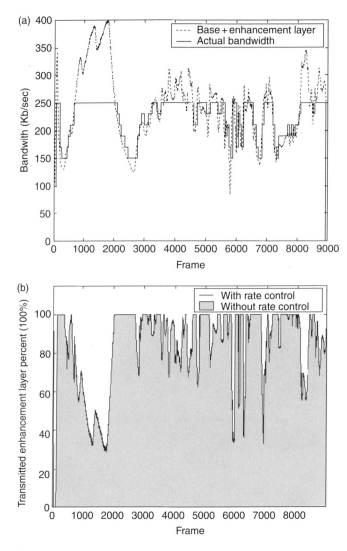

FIGURE 35.13　Simulation results for FGS encoded video: (a) actual transmissions bandwidth; (b) percentage of transmitted data.

for FGS and scalable encoding are depicted in Figure 35.13 and Figure 35.14, respectively. The link layer rate control can only adjust the transmission bandwidth well for FGS, while the cross-layer rate control works well for both cases. The data transmission percentages for cross-layer rate control and link layer rate control are given in Table 35.2.

35.4.4　Implementation Issues

Figure 35.15 shows the implementation of cross-layer rate control algorithm in UMTS and IEEE 802.11 WLANs. Since RRC has the control interfaces to RLC, MAC, and PHY, the proposed rate control algorithm can be implemented in RRC. Along with UTRAN, RRC can control the radio channel configuration in different lower layers in order to satisfy different bandwidth requirements [33]. Examples of the channel configuration for increasing bandwidths are as follows:

1. For time-bounded applications, if a common channel is initially used, it can be reconfigured at physical layer as a dedicated channel.

2. An appropriate transport format (TF) can be chosen from the defined transport format combination (TFC) for the dedicated channel.
3. The radio link can be further reconfigured at MAC layer.

The detailed radio channel configuration in RRC and UTRAN is out of the scope of this chapter.

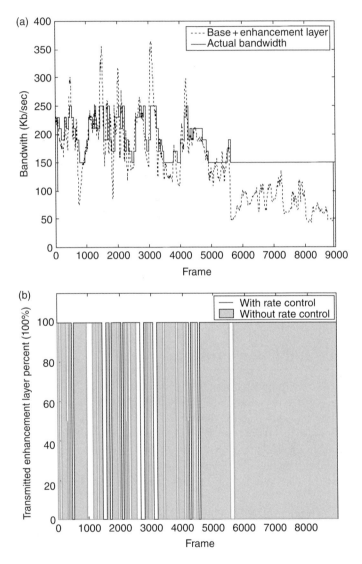

FIGURE 35.14 Simulation results for scalable encoded video: (a) actual transmission bandwidth; (b) percentage of transmitted data.

TABLE 35.2   Percentage of Transmitted Data at Enhancement Layer

	FGS MPEG(%)	Scalable MPEG-4(%)
Cross-layer rate control	91.9	92.9
Link layer rate control	91.8	77.8

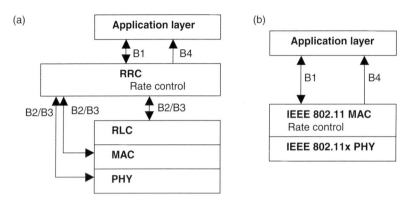

FIGURE 35.15   Implementation of cross-layer rate control: (a) in UMTS (Control plane); (b) in IEEE 802.11 WLANS.

In IEEE 802.11 WLANs, the cross-layer rate control algorithm can be implemented at MAC layer. The cross-layer communications include getting future bandwidth requirement from the application layer (by Step B1) and sending rate control results to the application layer (by Step B4). The physical layer of IEEE 802.11x provides multiple bandwidth choices. Through link adaptation techniques [34], different transmission rates can be chosen according to the application requirements. A great deal of research is undergoing to further improve the QoS capacity in IEEE 802.11 MAC layer. As stated in Section 35.2, although the PCF mode is more suitable for video applications than DCF is, it does not work in the contention period. A new standard, IEEE 802.11e [35], is currently under development. In IEEE 802.11e, DCF is replaced by enhanced DCF (EDCF) and PCF is replaced by hybrid coordination function (HCF). The EDCF provides QoS support by assigning shorter inter-frame space to higher traffic categories, while the HCF enables poll-based data transmission even in contention period.

35.5   Summary and Open Problems

This chapter reviews the current research results on video transmission over mobile wireless networks and presents new cross-layer error control and rate control algorithms. Inter-layer communication is employed to improve the efficiency of error control and the accuracy of transmission rate assignment. The proposed algorithms are theoretically analyzed. Simulation results demonstrate that, for six videos encoded at different quality levels, the proposed error control algorithm improves the video quality from 0.86 to 2.8 dB, and the proposed rate control algorithm improves the percentage of transmitted enhancement layer data by 15.1 and 0.1% for scalable encoded MPEG video and FGS MPEG video, respectively.

Although many research results have been presented in the past several years to enable smooth video transmission over mobile wireless networks, there are still some open problems in need of future study.

- The cross-layer algorithms and protocols for wireless networks are still at their early stage. Some future research topics include cross-layer parameter optimization and integrating cross-layer algorithms to the wireless network specifications.
- The multimedia applications often include different kinds of data, such as image, audio, and video. It will be challenging to successfully deliver and synchronize the combination of such real-time data over wireless networks.
- This chapter considers video transmission over wireless networks. Some video applications, however, involve communications between servers in the wired network and clients in the wireless network. Cheung et al. [38] recently introduced an intermediate agent or proxy, located at the junction of backbone wired network and wireless link, to dynamically feedback the network condition

in order to help sender make adaptive QoS control. More efforts are needed to investigate the rate control and error control algorithms in such hybrid wired and wireless networks.

- Peer-to-peer communication in mobile ad hoc networks introduces some new and challenging problems, such as how to take advantage of multiple paths between a sender and a receiver, how to control real-time data transmission among heterogeneous devices, and how to deal with the mobility of devices, and so on.

Acknowledgment

This work is supported by NSF grants ANI 0219110 and CCR 0001788.

References

[1] "Video coding for low bit rate communication," ITU-T Rec. H.263, ITU, version 2, 1997.

[2] ISO/IEC 14496, "Coding of audio–visual objects," ISO, 1999.

[3] D. Wu, Y.T. Hou, W. Zhu, Y. Zhang, and J.M. Peha, "Streaming video over the internet: approaches and directions," *IEEE Transactions on Circuits and Systems for Video Technology*, 11: 282–300, 2001.

[4] W. Li, "Overview of fine granularity scalability in MPEG-4 video standard," *IEEE Transaction on Circuits and Systems for Video Technology*, 11: 301–317, 2001.

[5] V.K. Goyal, "Multiple description coding: compression meets the network," *IEEE Signal Processing Magazine*, 74–93, 2001.

[6] W. Tan and A. Zakhor, "Real-time internet video using error resilient scalable compression and TCP-Friendly transport protocol," *IEEE Transaction & on Multimedia*, 1: 172–186, 1999.

[7] H. Liu et al., "Error control schemes for networks: an overview," *Mobile Networks and Applications*, 2: 167–182, 1997.

[8] S.B. Wicker, *Error Control Systems for Digital Communication and Storage*, Prentice Hall, New york, 1995.

[9] L.P. Kondi, F. Ishtiaq, and A.K. Katsaggelos, "Joint source-channel coding for scalable video," in *Proceedings of the Conference on Image and Video Communications and Processing*, 2000.

[10] Y. Wang, J. Ostermann, and Y. Zhang, *Video Processing and Communications*, Prentice Hall, New York, 2002.

[11] L Larzon et al., "The UDP lite protocol," internet draft, Jan. 2002.

[12] http://www.3gpp.org.

[13] TS 25.322 V5.3.0, "Radio Link Control (RLC) protocol specification, Release 5", 3GPP, Dec. 2002.

[14] IEEE Std 802.11a-1999: Part 11: wireless LAN Medium Access Control (MAC) and Physical Layer (PHY) specifications: high-speed physical layer in the 5 GHz band, 1999.

[15] H. Liu and M. Zarki, "Adaptive source rate control for real-time wireless video transmission," *Mobile Networks and Applications*, 3: 49–60, 1998.

[16] D. Wu, Y.T. Hou, and Y. Zhang, "Scalable video coding and transport over broadband wireless networks," in *Proceedings of IEEE*, 89: 6–20, 2001.

[17] J. Vass, S. Zhuang, and X. Zhuang, "Scalable, error-resilient, and high-performance video communications in mobile wireless environments," *IEEE Transactions on Circuits and Systems for Video Technology*, 11: 833–847, 2001.

[18] Y. Wang, S. Panwar, S. Lin, and S. Ma, "Wireless video transport using path diversity: multiple description vs. layered coding," in *Proceedings of the International Conference* Image Processing, 1: 21–24, 2002.

[19] A. Majumdar et al., "Multicast and unicast real-time video streaming over wireless LANs," *IEEE Transactions on Circuits and Systems for Video Technology*, 12: 524–534, 2002.

[20] A. Miu, J.G. Apostolopoulos, W. Tan, and M. Trott, "Low-latency wireless video over 802.11 networks using path diversity," in *Proceedings of the International Conference on Multimedia and Expo*, 2003.

[21] F.H.P. Fitzek and M. Reisslein, "A prefetching protocol for continuous media streaming in wireless environments," *IEEE Journal on Selected Areas in Communication*, 19: 2015–2028, 2001.

[22] A. Singh et al., "Performance evaluation of UDP Lite for cellular video," in *Proceedings of the* NOSSDAV'01, 2001.

[23] X. Wang, "An FDD wideband CDMA MAC protocol for wireless multimedia networks," in *Proceedings of the Infocom*, 2003.

[24] H. Zhang, X. Xia, Q. Zhang, and W. Zhu, "Precoded OFDM with adaptive vector channel allocation for scalable video transmission over frequency-selective fading channels," *IEEE Transactions on Mobile Computing*, 1: 132–142, 2002.

[25] S. Zou, H. Wu, and S. Cheng, "A new mechanism of transmitting MPEG-4 video in IEEE 802.11 wireless LAN with DCF," in *Proceedings of the International Conference on* Technology, 2: 1226–1229, 2003.

[26] B. Girod, M. Kalman, Y. Liang, and R. Zhang, "Advances in channel-adaptive video streaming," *Wireless Communications and Mobile Computing*, 2: 549–552, 2002.

[27] Q. Zhang, W. Zhu, and Y-Q. Zhang, "A cross-layer QoS-supporting framework for multimedia delivery over wireless internet," in *Proceedings of Packet Video*, 2002.

[28] H. Zheng and J. Boyce. "An improved UDP protocol for video transmission over internet-to-wireless networks," *IEEE Transactions on Multimedia*, 3: 356–365, 2001.

[29] Y. Shan and A. Zakhor, "Cross-layer techniques for adaptive video streaming over wireless networks," in *Proceedings of the International Conference Multimedia and Expo*, 1: 277–280, 2002.

[30] G. Ding, H. Ghafoor, and B. Bhargava, "Error resilient video transmission over wireless networks," in *Proceedings of the 6th International Symposium Object-Oriented Real-Time Distributed Computing*, 2003.

[31] S. Krishnamachari, M. VanderSchaar, S. Choi, and X. Xu, "Video Streaming over Wireless LANs: a cross-layer Approach," in *Proceedings of Packet Video*, 2003.

[32] TS 25.308 V5.4.0, "High Speed Downlink Packet Access (HSDPA), Overall description, Stage 2, Release 5," 3GPP, Mar. 2003.

[33] TR 25.922 V5.0.0, "Radio resource management strategies, Release 5," 3GPP, Mar. 2002.

[34] D. Qiao, S. Choi, and K.G. Shin, "Goodput analysis and link adaptation for IEEE 802.11a wireless LANs," *IEEE Transactions on Mobile Computing*, 1: 278–292, 2002.

[35] S. Mangold et al., "IEEE 802.11e Wireless LAN for Quality of Service," in *Proceedings of the European Wireless*, 1: 32–39, 2002.

[36] Video Traces for Network Performance Evaluation, http://trace.eas.asu.edu.

[37] F. Fitzek and M. Reisslein, "MPEG-4 and H.263 video traces for network performance evaluation," *IEEE Network*, 15: 40–54, 2001.

[38] G. Cheung, W. Tan, and T. Yoshimura, "Double feedback streaming agent for real-time delivery of media over 3G wireless networks," *IEEE Transactions on Multimedia*, 6: 304–314, 2004.

VIII

Wireless Sensor
Networks

36

Wireless Sensor
Network Protocols

36.1 Introduction to Wireless Sensor Networks.............**36**-813
    Taxonomy of Sensor Networks • Unique Features of Sensor
    Networks • Performance Metrics • Chapter Organization

36.2 Medium Access Control Protocols.....................**36**-816
    Sensor-MAC • Time-Out-MAC • DMAC •
    Traffic-Adaptive Medium Access • Sparse Topology and
    Energy Management

36.3 Network Protocols **36**-819
    Resource-Aware Routing • Data-Centric Routing Protocols
    • Geographic Routing • Clustering for Data Aggregation •
    Querying a Distributed Database • Topology Control

36.4 Protocols for QoS Management **36**-829
    Transport Layer • Providing Coverage of an Environment

36.5 Time Synchronization and Localization Protocols **36**-833
    Time Synchronization • Sensor Localization

36.6 Open Issues..**36**-837

References **36**-839

Mark A. Perillo
Wendi B. Heinzelman

36.1 Introduction to Wireless Sensor Networks

Efficient design and implementation of wireless sensor networks has become a hot area of research in recent years, due to the vast potential of sensor networks to enable applications that connect the physical world to the virtual world. By networking large numbers of tiny sensor nodes, it is possible to obtain data about physical phenomena that was difficult or impossible to obtain in more conventional ways. In the coming years, as advances in micro-fabrication technology allow the cost of manufacturing sensor nodes to continue to drop, increasing deployments of wireless sensor networks are expected, with the networks eventually growing to large numbers of nodes (e.g., thousands). Potential applications for such large-scale wireless sensor networks exist in a variety of fields, including medical monitoring [1,2,3], environmental monitoring [4,5], surveillance, home security, military operations, and industrial machine monitoring. To understand the variety of applications that can be supported by wireless sensor networks, consider the following two examples.

Surveillance. Suppose multiple networked sensors (e.g., acoustic, seismic, and video) are distributed throughout an area such as a battlefield. A surveillance application can be designed on top of this sensor network to provide information to an end user about the environment. In such a sensor network, traffic patterns are many-to-one, where the traffic can range from raw sensor data to a high level description of what is occurring in the environment, if data processing is done locally. The application will have some quality of service (QoS) requirements from the sensor network, such as requiring a minimum percentage sensor coverage in an area where a phenomenon is expected to occur, or requiring a maximum probability of missed detection of an event. At the same time, the network is expected to provide this QoS for a long time (months or even years) using the limited resources of the network (e.g., sensor energy and channel bandwidth) while requiring little to no outside intervention. Meeting these goals requires careful design of both the sensor hardware and the network protocols.

Medical Monitoring. A different application domain that can make use of wireless sensor network technology can be found in the area of medical monitoring. This field ranges from monitoring patients in the hospital using wireless sensors to remove the constraints of tethering patients to big, bulky, wired monitoring devices, to monitoring them in mass casualty situations [6], to monitoring people in their everyday lives to provide early detection and intervention for various types of disease [7]. In these scenarios, the sensors vary from miniature, body-worn sensors to external sensors such as video cameras or positioning devices. This is a challenging environment in which dependable, flexible, applications must be designed using sensor data as input. Consider a personal health monitor application running on a Personnel Digital Assistant (PDA) that receives and analyzes data from a number of sensors (e.g., ECG, EMG, blood pressure, blood flow, and pulse oxymeter). The monitor reacts to potential health risks and records health information in a local database. Considering that most sensors used by the personal health monitor will be battery-operated and use wireless communication, it is clear that this application requires networking protocols that are efficient, reliable, scalable, and secure.

To understand better why traditional network protocols are not suitable for these types of sensor network applications, in the remainder of this section we will categorize the unique features of sensor networks and the performance metrics with which protocols for sensor networks should be evaluated.

36.1.1 Taxonomy of Sensor Networks

As research in sensor networks has grown, so too has the range of applications proposed to make use of this rich source of data. Such diversity of sensor network applications translates to differing requirements from the underlying sensor network. To address these varying needs, many different network models have been proposed, around which protocols for different layers of the network stack have been designed. While there are many ways to classify different sensor network architectures, the following list highlights some fundamental differences in sensor networks that affect protocol design:

- *Data sinks.* One of the most important aspects of a sensor network is the nature of the data sinks. In some situations, end users may be embedded within the sensor network (e.g., actuator that correct abnormalities in environmental conditions, access points that network with the outside world), or they may be less accessible mobile access points that collect data once in a while (e.g., data collectors in the DATA Mules project [8] and in a sensor reachback scenario [9]). This distinction may be important, as efficient distributed data storage techniques may be effective in the latter scenario.
- *Sensor mobility.* Another classification of sensor networks may be made based on the nature of the sensors being deployed. Typically, it can be assumed that sensors are immobile; however, some recent sensor networks projects such as the ZebraNet project [10] have used mobile sensor nodes. In addition, in military operations, additional sensors may be mounted on soldiers or Unattended Air Vehicles (UAVs) to interact with a deployed sensor network. The mobility of sensors can influence protocols at the networking layer as well as those for localization services.
- *Sensor resources.* Sensor nodes may vary greatly in the computing resources available. It is obvious that memory and processing constraints should influence protocol design at nearly every level.
- *Traffic patterns.* Another important aspect to consider is the traffic generated on the network. In many event-driven applications, sensors may operate in a sentry state for the majority of time,

only generating data traffic when an event of interest is detected. In other applications such as environmental monitoring, data should be continuously generated.

As can be seen by the above discussion, there are many features of the sensors, the network and the application that should influence protocol design. Accordingly, much research has gone into designing protocols for these different scenarios.

36.1.2 Unique Features of Sensor Networks

It should be noted that sensor networks do share some commonalities with general ad hoc networks. Thus, protocol design for sensor networks must account for the properties of ad hoc networks, including the following:

- Lifetime constraints imposed by the limited energy supplies of the nodes in the network.
- Unreliable communication due to the wireless medium.
- Need for self-configuration, requiring little or no human intervention.

However, several unique features exist in wireless sensor networks that do not exist in general ad hoc networks. These features present new challenges and require modification of designs for traditional ad hoc networks:

- While traditional ad hoc networks consist of network sizes on the order of 10s, sensor networks are expected to scale to sizes of 1000s.
- Sensor nodes are typically immobile, meaning that the mechanisms used in traditional ad hoc network protocols to deal with mobility may be unnecessary and overweight.
- Since nodes may be deployed in harsh environmental conditions, unexpected node failure may be common.
- Sensor nodes may be much smaller than nodes in traditional ad hoc networks (e.g., PDAs, laptop computers), with smaller batteries leading to shorter lifetimes, less computational power, and less memory.
- Additional services, such as location information, may be required in wireless sensor networks.
- While nodes in traditional ad hoc networks compete for resources such as bandwidth, nodes in a sensor network can be expected to behave more cooperatively, since they are trying to accomplish a similar universal goal, typically related to maintaining an application-level QoS, or fidelity.
- Communication is typically data-centric rather than address-centric, meaning that routed data may be aggregated/compressed/prioritized/dropped depending on the description of the data.
- Communication in sensor networks typically takes place in the form of very short packets, meaning that the relative overhead imposed at the different network layers becomes much more important.
- Sensor networks often have a many-to-one traffic pattern, which leads to a "hot spot" problem.

Incorporating these unique features of sensor networks into protocol design is important in order to efficiently utilize the limited resources of the network. At the same time, to keep the protocols as light-weight as possible, many designs focus on particular subsets of these criteria for different types of applications. This has led to quite a number of different protocols from the data-link layer up to the transport layer, each with the goal of allowing the network to operate autonomously for as long as possible while maintaining data channels and network processing to provide the application's required QoS.

36.1.3 Performance Metrics

Because sensor networks posses these unique properties, some existing performance metrics for wireless network protocols are not suitable for evaluating sensor network protocols. For example, since sensor networks are much more cooperative in nature than traditional ad hoc networks, fairness becomes much less important. In addition, since data sinks are interested in a general description of the environment rather than in receiving all raw data collected by individual nodes, throughput is less meaningful. Depending on the application, delay may be either much more or much less important in sensor networks.

Much more important to sensor network operation is energy-efficiency, which dictates network lifetime, and the high level QoS, or fidelity, that is met over the course of the network lifetime. This QoS is application-specific and can be measured in a number of different ways. For example, in a typical surveillance application, it may be required that one sensor remains active within every subregion of the network, so that any intruder may be detected with high probability. In this case, QoS may be defined by the percentage of the environment that is actually covered by active sensors. In a typical tracking application, this QoS may be the expected accuracy of the target location estimation provided by the network.

36.1.4 Chapter Organization

The rest of this chapter will describe protocols and algorithms that are used to provide a variety of services in wireless sensor networks. Sections 36.2 and 36.3 provide examples of MAC and network protocols, respectively, for use in sensor networks. Section 36.4 presents some high-level protocols for energy-efficient management of sensor networks at the transport layer. Section 36.5 presents time synchronization and localization protocols that are often essential in sensor network applications. Section 36.6 presents a discussion of open research issues in the design of sensor networks.

36.2 Medium Access Control Protocols

Medium access control (MAC) protocols that have been designed for typical ad hoc networks have primarily focused on optimizing fairness and throughput efficiency, with less emphasis on energy conservation. However, the energy constraint is typically considered paramount for wireless sensor networks, and so many MAC protocols have recently been designed that tailor themselves specifically to the characteristics of sensor networks. Protocols such as MACAW [11] and IEEE 802.11 [12] eliminate the energy waste caused by colliding packets in wireless networks. Further, enhancements have been made to these protocols (e.g., PAMAS [13]) to avoid unnecessary reception of packets by nodes that are not the intended destination. However, it has been shown that idle power consumption can be of the same order as the transmit and receive power consumption, and if so, can greatly affect overall power consumption, especially in networks with relatively low traffic rates. Thus, the focus of most MAC protocols for sensor networks is to reduce this idle power consumption by setting the sensor radios into a sleep state as often as possible.

36.2.1 Sensor-MAC

Sensor-MAC (S-MAC) was one of the first MAC protocols to be designed for sensor networks [14]. The basic idea behind SMAC is very simple — nodes create a sleep schedule for themselves that determines at what times to activate their receivers (typically 1 to 10% of a frame) and when to set themselves into a sleep mode. Neighboring nodes are not necessarily required to synchronize sleep schedules, although this will help to reduce overhead (see Figure 36.1[b]). However, nodes must at least share their sleep schedule information with their neighbors through the transmission of periodic SYNC packets. When a source node wishes to send a packet to a destination node, it waits until the destination's wakeup period and sends the packet using CSMA with collision avoidance. S-MAC also incorporates a message passing mechanism, in which long packets are broken into fragments, which are sent and acknowledged successively following the initial request-to-send and clear-to-send (RTS–CTS) exchange. In addition to avoiding lengthy retransmissions, fragmentation helps address the hidden node problem, as fragmented data packets and ACKs can serve the purposes of the RTS and CTS packets for nodes that wakeup in the middle of a transmission, having missed the original RTS–CTS exchange.

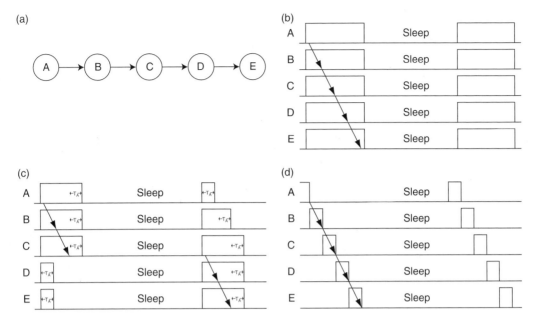

FIGURE 36.1 Chain network scenario (a), Sleep schedule for S-MAC [14] (b), T-MAC [15] (c), and DMAC [16] (d). In S-MAC, nodes synchronize their sleep schedules and remain awake for a predetermined period of time, even if no traffic is being sent, leading to wasted energy due to overhearing. In T-MAC, nodes D and E time-out after not hearing any channel activity for a duration of T_A. This leads to improved energy efficiency but can cause lengthy delays, as the nodes must wait until the next awake phase to complete the last 2 hops. In DMAC, the sleep schedules are staggered in a way so that it offers good energy efficiency and low delay for networks consisting of aggregating trees.

36.2.2 Time-Out-MAC

Several protocols have been developed based on S-MAC that offer solutions for various deficiencies and limitations of the original S-MAC protocol. Time-out-MAC (T-MAC) seeks to eliminate idle energy further by adaptively setting the length of the active portion of the frames [15]. Rather than allowing messages to be sent throughout a predetermined active period, as in S-MAC, messages are transmitted in bursts at the beginning of the frame. If no "activation events" have occurred after a certain length of time, the nodes set their radios into sleep mode until the next scheduled active frame. "Activation events" include the firing of the frame timer or any radio activity, including received or transmitted data, the sensing of radio communication, or the knowledge of neighboring sensors' data exchanges, implied through overheard RTS and CTS packets. An example of how T-MAC works is shown for a transmission from node A to node E in Figure 36.1(c). Since nodes D and E cannot hear node A's transmissions, they time-out after a delay of T_A. The end-to-end transmission from A to E is resumed during the next active period. The gains achieved by T-MAC are due to the fact that S-MAC may require its active period to be longer than necessary to accommodate traffic on the network with a given latency bound. While the duty cycle can always be tuned down, this will not account for bursts of data that can often occur in sensor networks (e.g., following the detection of an event by many surrounding neighboring sensors).

36.2.3 DMAC

As many wireless sensor networks consist of data gathering trees rooted at a single data sink, the direction of packets arriving at a node, if not the arrival times, are fairly stable and predictable. DMAC takes advantage of this by staggering the wakeup times for nodes based on their distance from the data sink [16]. By staggering the wakeup times in this way, DMAC reduces the large delays that can be observed in packets that are forwarded for more than a few hops when synchronizing schedules as in S-MAC and T-MAC. The wakeup scheme consists of a receiving period and a send period, each of length μ (set to accommodate a single transmission), followed by a long sleep period. Nodes on the data gathering tree begin their receiving period after an offset of $d * \mu$, where d represents the node's depth on the tree. In this way, a node's receiving period lines up with its upstream neighbor's send period and a node simply sends during downstream neighbors' receive periods, as shown in Figure 36.1(d). Contention within a sending period is resolved through a simple random backoff scheme, after which a node sends its packet without a preceding RTS–CTS exchange.

36.2.4   Traffic-Adaptive Medium Access

While the above mentioned protocols attempt to minimize power consumption by reducing the time that the radio remains in the idle state, traffic-adaptive medium access (TRAMA) attempts to reduce wasted energy consumption caused by packet collisions [17]. Nodes initially exchange neighborhood information with each other during a contention period via a neighbor protocol (NP) so that each node has knowledge of all 2-hop neighbors. These random access periods are followed by scheduled access periods, where nodes transmit schedule information via the schedule exchange protocol (SEP) as well as actual data packets. Using the neighbor information acquired using NP and the traffic schedule information acquired using SEP, nodes determine their radio state using the adaptive election algorithm (AEA). In AEA, each node calculates a priority for itself and all 2-hop neighbors for the current slot using a hashing function. If a node has the highest priority for that slot and has data to send, it wins that slot and sends its data. If one of its neighbors has the highest priority and the node determines that it should be the intended receiver through information acquired during SEP, it sets itself to the receive mode. Otherwise, it is able to sleep and conserve energy. Since two nodes within the 2-hop neighborhood of a node may consider themselves slot winners if they are hidden from each other, nodes must keep track of an Alternate Winner, as well as the Absolute Winner for a given time slot, so that messages are not lost. For example, consider a node N who determines that the Absolute Winner for a time slot is one of its 2-hop neighbors N_{2-hop}. If a 1-hop neighbor N_{1-hop} who does not know of N_{2-hop} believes that it has won the slot, and wishes to send to N, N must stay awake even though it does not consider N_{1-hop} to have won the slot. Since a node may win more slots than necessary to empty its transmission buffer, some slots may remain unused that could have been used by nodes who won too few slots. To accommodate for this, the AEA assigns priorities for the unused slots to the nodes needing extra slots.

36.2.5   Sparse Topology and Energy Management

In the case of many sensor network applications, it is expected that nodes will continuously sense the environment, but transmit data to a base station very infrequently or only when an event of interest has occurred. In Sparse Topology and Energy Management (STEM), all sensors are left in a sleep state while monitoring the environment but not sending data and are only activated when traffic is generated [18]. In other words, transceivers are activated reactively rather than proactively, as with the other MAC protocols described in this section. When data packets are generated, the sensor generating the traffic uses a paging channel (separate from the data channel) to awaken its downstream neighbors. Two versions of STEM have been proposed — STEM-T, which uses a tone on a separate channel to wake neighboring nodes, and STEM-B, in which the traffic generating node sends beacons on a paging channel and sleeping nodes turn on their radios with a low duty cycle to receive the messages (the paging channel simply consists of synchronized time slots within the main communication channel). While STEM-T guarantees that minimal delay will be met (since receivers are turned on nearly instantaneously after data is generated), it requires more overhead than STEM-B since the receivers on the channel where the tones are sent must be idle, listening all of the time. Moreover, STEM-T may require extra hardware as a separate radio is needed for this channel.

36.3 Network Protocols

When designing network protocols for wireless sensor networks, several factors should be considered. First and foremost, because of the scarce energy resources, routing decisions should be guided by some awareness of the energy resources in the network. Furthermore, sensor networks are unique from general ad hoc networks in that communication channels often exist between events and sinks, rather than between individual source nodes and sinks. The sink nodes are typically more interested in an overall description of the environment, rather than explicit readings from the individual sensor devices. Thus, communication in sensor networks is typically referred to as data-centric, rather than address-centric, and data may be aggregated locally rather than having all raw data sent to the sinks [19]. These unique features of sensor networks have implications in the network layer and thus require a rethinking of protocols for data routing. In addition, sensors often have knowledge of their own location in order to meaningfully assess their data. This location information can be utilized in the network layer for routing purposes. Finally, if a sensor network is well connected (i.e., better than is required to provide communication paths), topology control services should be used in conjunction with the normal routing protocols. This section describes some of the work that has been done to address these sensor network-specific issues in the routing layer.

36.3.1 Resource-Aware Routing

As resources are extremely limited in wireless sensor networks, it is important to consider how to most efficiently use them at all levels of the protocol stack. Many different approaches have been developed that consider the sensors' resources when making routing decisions. Initially, protocols were developed that considered only the sensors' energy resources. Later work considered not only individual sensors' energy but also the sensors' sensing resources.

36.3.1.1 Energy-Aware Routing

Because of the scarce energy supplies available in sensor networks, a great deal of effort has been put forth in creating energy-aware routing protocols that consider the energy resources available at each sensor and that try to balance the power consumption such that certain nodes do not die prematurely. Singh et al. [20] were among the first to develop energy-aware routing metrics. They proposed that the lifetime of the network could be extended by minimizing the cumulative cost c_j of a packet j being sent from node n_1 to node n_k through intermediate nodes n_2, n_3, etc., where

$$c_j = \sum_{i=1}^{k-1} f_i(z_i) \tag{36.1}$$

$$f_i(z_i) = \frac{1}{1 - g(z_i)} \tag{36.2}$$

and $g(z_i)$ represents the normalized remaining lifetime corresponding to node n_i's battery level z_i. Further work by Chang and Tassiulas [21] solved the problem of maximizing network lifetime by finding an optimal energy-aware routing cost. In their work, the routing cost of sending a packet was the sum of the routing costs of the individual links. The cost c_{ij} of a link between node i and node j was set to

$$c_{ij} = e_{ij}^{x_1} \underline{E}_i^{-x_2} E_i^{x_3} \tag{36.3}$$

where e_{ij} represents the energy necessary to transmit from node i to node j, $\underline{E}_i$ represents the residual energy of node i, and E_i represents the initial energy of node i. Brute force simulation methods were used to find the optimal values of x_1, x_2, and x_3.

From the intuition that can be taken from this initial work, several energy-aware routing protocols have been developed for sensor networks, including the one proposed by Shah and Rabaey [22]. In this protocol, query interests are sent from a querying agent by way of controlled flooding toward the source nodes. Each node N_i has a cost $\text{Cost}(N_i)$ associated with it that indicates its reluctance to forward messages. Each upstream neighbor N_j of node N_i calculates a link cost C_{N_j,N_i} associated with N_i that depends on $\text{Cost}(N_i)$ as well as the energy e_{ij} required to transmit over this link and the normalized residual energy R_i at node N_i.

$$C_{N_j,N_i} = \text{Cost}(N_i) + e_{ij}^{\alpha} R_i^{\beta} \tag{36.4}$$

α and β are tunable parameters. Each node N_j builds a forwarding table FT_j consisting of its lowest cost downstream neighbors and the link cost C_{N_j,N_i} associated with those neighbors. Node N_j assigns a probability P_{N_j,N_i} to each neighbor as

$$P_{N_j,N_i} = \frac{1/C_{N_j,N_i}}{\sum_{k\in\text{FT}_j} 1/C_{N_j,N_k}} \tag{36.5}$$

such that received messages will be forwarded over each link with this probability. Before forwarding its message, N_j must determine its own value of $\text{Cost}(N_j)$, which is simply the weighted average of the costs in its forwarding table FT_j

$$\text{Cost}(N_j) = \sum_{i\in\text{FT}_j} P_{N_j,N_i} C_{N_j,N_i} \tag{36.6}$$

36.3.1.2 Fidelity-Aware Routing

Distributed activation based on predetermined routes (DAPR) is similar to these energy-aware routing protocols but was designed specifically for maintaining high-level QoS requirements (e.g., coverage) over long periods of time [23]. Rather than assigning cost to individual nodes based on the residual energy at those nodes, DAPR considers the importance of a node to the sensing application. Since sensors in a coverage application typically cover redundant areas and redundancy can vary throughout the network, some nodes might be considered more important than others. In DAPR, a node first finds the subregion within its region of coverage that is the most poorly covered. The cost assigned to the node is related to the combined energy of all nodes capable of redundantly covering this poorly covered region. Large gains in network lifetime can be seen when considering the importance of a node to the overall sensing task when making routing decisions if the sensor deployment is such that there is a high variation in node density in different subregions of the environment. However, there is an added overhead associated with this approach, as it requires nodes to acquire additional information from neighboring nodes.

36.3.2 Data-Centric Routing Protocols

Sensor networks are fundamentally different from ad hoc networks in the data they carry. While in ad hoc networks individual data items are important, in sensor networks it is the aggregate data or the information carried in the data rather than the actual data itself that is important. This has led to a new paradigm for networking these types of devices — data-centric routing. In data-centric routing, the end nodes, the sensors themselves, are less important than the data itself. Thus, queries are posed for specific data rather than for data from a particular sensor, and routing is performed using knowledge that it is the aggregate data rather than any individual data item that is important.

36.3.2.1 Sensor Protocol for Information via Negotiation

Sensor Protocol for Information via Negotiation (SPIN) is a protocol that was designed to enable data-centric information dissemination in sensor networks [24]. Rather than blindly broadcasting sensor data throughout the network, nodes receiving or generating data first advertise this data through short ADVertisement (ADV) messages. The ADV messages simply consist of an application-specific meta-data

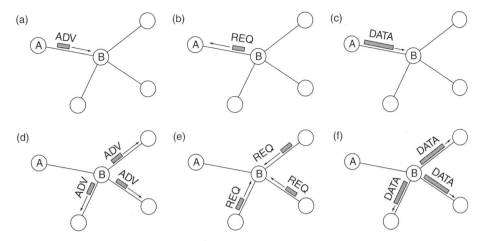

FIGURE 36.2 Illustration of message exchange in the SPIN protocol [24]. Nodes advertise their data with ADV messages (a). Any node interested in receiving the data replies with a REQ message (b), to which the source node replies with the transmission of the actual data (c). The receiving node then advertises this new data (d) and the processes continues (e,f).

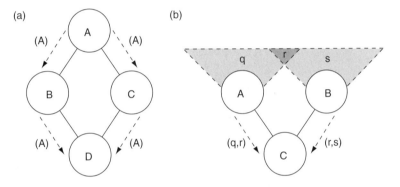

FIGURE 36.3 Problems with blind flooding of sensor data. (a) Implosion occurs in a highly connected network where nodes receive duplicate copies of data, wasting energy and bandwidth resources. As seen in this figure, node D receives two copies of node A's data. (b) Overlap occurs due to the redundant nature of sensor data. This figure shows that C receives data about region *r* from nodes A and B, again wasting valuable sensor resources.

description of the data itself. This meta-data can describe aspects such as the type of data and the location of its origin. Nodes that are interested in this data request the data from the ADV sender through REQuest (REQ) messages. Finally, the data is disseminated to the interested nodes through DATA messages that contain the data. This procedure is illustrated in Figure 36.2.

The advantage of SPIN over blind flooding or gossiping data dissemination methods is that it avoids three costly problems: implosion, overlap, and resource blindness. Implosion occurs in highly connected networks that employ flooding and thus each sensor receives many redundant copies of the data (see Figure 36.3[a]). For large data messages, this wastes considerable energy. In SPIN, on the other hand, short ADV messages will suffer from the implosion problem, but the costly transfer of data messages is greatly reduced. Overlap occurs due to the redundant nature of sensor data. Thus two sensors with some common data will both send their data, causing redundancy in data transmission resulting in energy waste (see Figure 36.3[b]). SPIN is able to solve this problem by naming data so that sensors only request the data or parts of data they are interested in receiving. Finally, in SPIN, there are mechanisms whereby a sensor that is running low on energy will not advertise its data in order to save its dwindling energy resources. Therefore, SPIN solves the resource blindness problem by having sensors make decisions based on their current level of available resources.

36.3.2.2 Directed Diffusion

Directed diffusion is a communication paradigm that has been designed to enable data-centric communication in wireless sensor networks [25]. To perform a sensing task, a querying node creates an interest, which is named according to the attributes of the data or events to be sensed. When an interest is created, it is injected into the network by the sink node by broadcasting an interest message containing the interest type, duration, and an initial reporting rate to all neighbors. For example, one interest might be to count the number of people in a given area every second for the next 10 minutes. Local interest caches at each node contain entries for each interest of which the node is aware that has been created on the network. An entry in the caches contains information about the interest's type, duration, and gradient (a combination of the event rate and direction toward the data sink). Nodes receiving the interest messages find (or create) the relevant interest entry in their caches and update the gradient field toward the node from which the message was received to the rate defined in the interest message. Each gradient also has expiration time information, which must be updated upon the reception of the interest messages.

Interests are diffused throughout the network toward the sink node using one of the several forwarding techniques. For example, Figure 36.4 shows a network in which the interest was sent to the region of interest via controlled flooding. Once the interest reaches the desired region, sensor nodes within the region process the query and begin producing data at the specified rate (if more than one entry for the same interest type exist, data is produced at the maximum rate of these entries). Data pertaining to these interests are then forwarded to each node for which a gradient exists at the rate specified for each individual gradient. After receiving low rate events from the source (recall that the initial reporting rate is set low), the data sink may reinforce higher quality paths, which might be chosen, for example, as those that experience low latency or those in which the confidence in the received data is deemed to be high by some application-specific measure (Figure 36.4). Reinforcement messages simply consist of the original interest messages set to higher reporting rates. These reinforced routes are established more conservatively than the original low rate interest messages so that only a single or a few paths from the event to the sink are used.

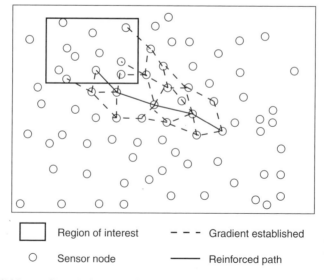

FIGURE 36.4   Establishing gradients in directed diffusion [25]. As the query is routed toward the region of interest, gradients for that interest are established in the reverse direction of the query dissemination. After data begin to arrive at the querying node, the path of highest quality is reinforced.

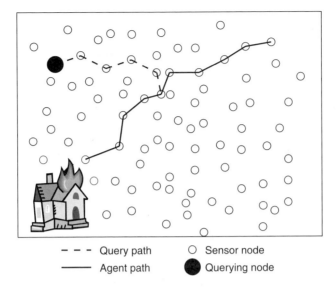

FIGURE 36.5 Query handling in rumor routing [26]. After an event is detected, an agent is initiated and sent on a random path through the network, establishing state at each node on the path. A query packet is similarly sent in a random direction and hopefully crosses paths with the agent, allowing the query to be answered and returned to the querying node.

36.3.2.3 Rumor Routing

While long-lived queries or data flows justify the overhead involved in establishing cost fields in a network, it may not be worth this effort when executing short-lived and one-shot queries. Rumor routing was designed for these types of queries [26]. When an event is detected by a sensor, it probabilistically creates an agent in the form of a data packet, and the agent is forwarded throughout the network in a random manner (solid line in Figure 36.5). Nodes through which the agent is forwarded maintain local state information about the direction and distance to the event. Should an agent traverse a node with knowledge of a path to other events, it adds this information so that subsequent nodes that the agent flows through will maintain state information regarding these events as well. When a node wishes to perform a query related to a given event, it simply forwards a query packet in a random direction so that the query traverses a random walk throughout the network (dashed line in Figure 36.5). Because of the fact that two lines drawn through a given area are likely to cross, there is a high likelihood that the query will eventually reach a node with a path to the specified event, especially if multiple agents carrying that event are sent through the network. If multiple queries do not reach the event, the querying node may resort to flooding queries over the entire network.

36.3.3 Geographic Routing

Wireless sensor networks often require a query packet to be forwarded to a particular region of interest in the network. A natural approach to perform this forwarding is to utilize geographic forwarding. Geographic forwarding reduces the amount of routing overhead, which is largely due to route discovery, and requires little memory utilization for route caching compared to typical address-centric ad hoc routing protocols. Furthermore, geographic routing protocols can enable geographically distributed data storage techniques such as Geographic Hash Tables (GHT) [27].

36.3.3.1 Greedy Perimeter Stateless Routing

Greedy perimeter stateless routing (GPSR) is a geographic routing protocol in which nodes make local packet forwarding decisions according to a greedy algorithm [28]. Under normal circumstances, a packet

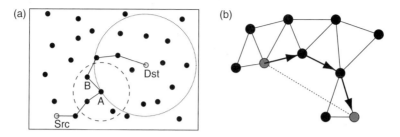

FIGURE 36.6   GPSR [28] greedy forwarding policy (a) and perimeter routing algorithm (b).

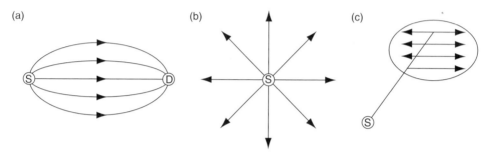

FIGURE 36.7   Possible trajectories to use in TBF [30] for robust multipath routing (a), spoke broadcasting (b), and broadcast within a remote region (c).

that is destined for some node D is forwarded to the node's neighbor that enables the maximum progress toward D (such a greedy forwarding scheme was originally proposed in the work of Takagi and Kleinrock [29]). However, obstacles or a lack of adequate sensor density can cause voids in the network topology so that packets reach a hole, from which the packet cannot be progressed any further without first being sent backward. GPSR accounts for this by incorporating a perimeter routing mechanism. These voids can be detected by the nodes surrounding them, and routes which circumnavigate the voids can be established heuristically. When a packet reaches these voids, these routes can be used (routing by the right-hand rule) until normal greedy routing can be used again. This process is illustrated in Figure 36.6(a). While this approach works well, another more robust perimeter routing algorithm is also proposed. In this algorithm, the graph that can be drawn from the complete network topology is first reduced to a planar graph in which no edges cross. Once a packet reaches a void, the forwarding node N finds the face of the planar graph that is intersected by the line connecting N and the destination (see Figure 36.6[b]). N then forwards the packet to the node along the edge that borders this face. This procedure continues with each forwarding node finding the face that the line connecting N and the destination intersects and routing along an edge bordering the face until the void has been cleared.

36.3.3.2   Trajectory Based Forwarding

Trajectory based forwarding (TBF) is a useful paradigm for geographic routing in wireless sensor networks [30]. Rather than sending a packet along a straight path toward its destination (as methods such as GPSR would do under ideal scenarios with dense deployment and no obstructions), TBF allows packets to follow a source-specified trajectory, increasing the flexibility of an overall forwarding strategy. For example, multipath routing can be achieved by sending multiple copies of a single packet along separate geographic trajectories, increasing resilience to localized failures or congestion in certain parts of the network. Also, TBF can increase the efficiency of many different forwarding techniques, including multipath forwarding (Figure 36.7[a]), spoke broadcasting (Figure 36.7[b]), and broadcast to a remote subregion (Figure 36.7[c]).

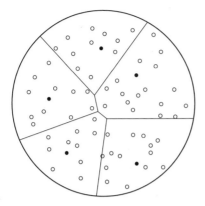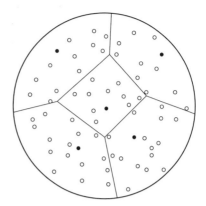

FIGURE 36.8   Adaptive clustering of the network.

36.3.4   Clustering for Data Aggregation

As sensor networks are expected to scale to large numbers of nodes, protocol scalability is an important design criteria. If the sensors are managed directly by the base station, communication overhead, management delay, and management complexity become limiting factors in network performance. Clustering has been proposed by researchers to group a number of sensors, usually within a geographic neighborhood, to form a cluster that is managed by a cluster head. A fixed or adaptive approach may be used for cluster maintenance. In a fixed maintenance scheme, cluster membership does not change over time, whereas in an adaptive clustering scheme, sensors may change their associations with different clusters over time (see, e.g., Figure 36.8).

Clustering provides a framework for resource management. It can support many important network features within a cluster, such as channel access for cluster members and power control, as well as between clusters, such as routing and code separation to avoid intercluster interference. Moreover, clustering distributes the management responsibility from the base station to the cluster heads, and provides a convenient framework for data fusion, local decision making and local control, and energy savings [31,32,33].

36.3.4.1   Low Energy Adaptive Clustering Hierarchy

In-network processing can greatly reduce the overall power consumption of a sensor network when large amounts of redundancy exist between nearby nodes. Rather than requiring all sensors' data to be forwarded to a base station that is monitoring the environment, nodes within a region can collaborate and send only a single summarization packet for the region. This use of clustering was first introduced in the low energy adaptive clustering hierarchy (LEACH) protocol [32]. In LEACH, nodes are divided into clusters, each containing a cluster head whose role is considerably more energy-intensive than the rest of the nodes; for this reason, nodes rotate roles between cluster head and ordinary sensor throughout the lifetime of the network.

At the beginning of each round, each sensor node makes an independent decision through a randomized algorithm about whether or not to assume a cluster head role. Nodes that choose to be cluster heads announce their status to the rest of the network. Based on the received signal strength of these announcements, sensors join the cluster that requires the least power to communicate with the cluster head (assuming transmission power control is available). During the round, the ordinary sensors in each cluster send data to their respective cluster heads according to a time-division multiple access (TDMA) schedule. Intercluster interference is reduced using different spreading codes in neighboring clusters. The cluster head aggregates data from all the cluster members and sends the aggregate data to the base station. The length of each round is chosen such that each node is expected to be able to perform a cluster head role once during its lifetime.

Because there is no interaction between nodes when deciding roles, the cluster heads may be chosen such that there is no uniformity throughout the network and certain sensors are forced to join clusters located at large distances from them. To mitigate this problem, a centralized version of LEACH called LEACH C has been developed. LEACH-C uses simulating annealing to choose the cluster heads for a given round so that the average transmission power between sensors and their cluster heads is minimized.

36.3.4.2 Hybrid Energy-Efficient Distributed Clustering

Nodes in LEACH independently decide to become cluster heads. While this approach requires no communication overhead, it has the drawback of not guaranteeing that the cluster head nodes are well distributed throughout the network. While the LEACH-C protocol solves this problem, it is a centralized approach that cannot scale to very large numbers of sensors.

Many papers have proposed clustering algorithms that create more uniform clusters at the expense of overhead in cluster formation. One approach that uses a distributed algorithm that can converge quickly and has been shown to have a low overhead is called Hybrid energy-efficient distributed clustering (HEED) [34]. HEED uses an iterative cluster formation algorithm, where sensors assign themselves a "cluster head probability" that is a function of their residual energy and a "communication cost" that is a function of neighbor proximity. Using the cluster head probability, sensors decide whether or not to advertise themselves as a candidate cluster head for this iteration. Based on these advertisement messages, each sensor selects the candidate cluster head with the lowest "communication cost" (which could be the sensor itself) as its tentative cluster head. This procedure iterates, with each sensor increasing its cluster head probability at each iteration until the cluster head probability is one and the sensor declares itself a "final cluster head" for this round. The advantages of HEED are that nodes only require local (neighborhood) information to form the clusters, the algorithm terminates in $O(1)$ iterations, the algorithm guarantees that every sensor is part of just one cluster, and the cluster heads are well distributed.

36.3.5 Querying a Distributed Database

Since sensor networks can be thought of as a distributed database system, several architectures (e.g., Cougar [35], SINA [36], TinyDB [37]) propose to interface the application to the sensor network through an SQL-like querying language. However, since sensor networks are so massively distributed, careful consideration should be put into the efficient organization of data and the execution of queries.

36.3.5.1 Tiny AGgregation (TAG) Service

Tiny AGgregation (TAG) is a generic aggregation service for wireless sensor networks that minimizes the amount of messages transmitted during the execution of a query [37]. In contrast to standard database query execution techniques, in which all data is gathered by a central processor where the query is executed, TAG allows the query to be executed in a distributed fashion, greatly reducing the overall amount of traffic transmitted on the network. The standard SQL query types (COUNT, AVERAGE, SUM, MIN, MAX), as well as more sophisticated query types, are included in the service, although certain query types allow more energy savings than others. Time is divided into epochs for queries requiring values to be returned at multiple times. When a query is sent by some node (initially the root), the receiving nodes set their parents to be the sending node and establish an interval within the epoch (intervals may be set to a length of EPOCH_DURATION/d, where d represents the maximum depth of the aggregating tree) during which their eventual children should send their aggregates (this interval should be immediately prior to their sending interval).

36.3.5.2 TinyDB/ACQP

TinyDB is a processing engine that runs acquisitional query processing (ACQP) [38], providing an easy-to-use generic interface to the network through an enhanced SQL-like interface and enabling the execution of queries to be optimized at several levels. ACQP allows storage points containing windows of sensor data to be created so that queries over the data streams can be executed more easily. Such storage points may be beneficial, for example, in sliding window type queries (e.g., find the average temperature in a room over the previous hour once per minute). ACQP also supports queries that should be performed upon the occurrence of specific events as well as queries that allow sensor settings such as the sensing rate to be adapted to meet a certain required lifetime.

Perhaps most importantly, ACQP provides optimization of the scheduling of sensing tasks as well as at the network layer. Since the energy consumption involved in the sensing of certain types of data is not negligible compared to the transmission costs of sending such packets, the scheduling of complex queries should be optimized in order to avoid unnecessary sensing tasks. ACQP optimizes this scheduling based on sensing costs and the expected selectivity of the query so as to minimize the expected power consumption during a query. Significant power savings can also be achieved by the ACQP's batching of event-based queries in some cases.

The topology of an aggregating tree can also be optimized by considering the query in its formation. TinyDB uses semantic routing trees (SRTs). Rather than requiring children to choose a parent node solely based on link quality, the choice of a parent node during the construction of an SRT also depends on the predicates of the query for which the tree is being built (i.e., the conditions that should be met for inclusion in the query). Specifically, children nodes choose a parent either to minimize the difference between their attributes of the predicate in the query or to minimize the spread of the attributes of the children of all potential parents. When a query is processed, a parent knows the attributes of all children and can choose not to forward the message if it determines that none of its children can contribute to the query (based on the query predicate and the attributes of its children).

36.3.5.3 Geographic Hash Table

Geographic Hash Tables provide a convenient, data-centric means to store event-based data in wireless sensor networks [27]. Storing data in a distributed manner provides an energy-efficient alternative in large-scale sensor networks, where the number of messages involved in the querying of the network becomes very large, and in networks where many more events are detected than are queried, where the hot spot around the querying node seen in external storage techniques can be avoided. When an event is sensed, the location at which the data related to the event should be stored is found by hashing its key to a location within the network. This location has no node associated with it when it is hashed, but the data will eventually find a home node closest to the hashed location. Once the location is determined, a data packet is sent using GPSR [28], although with no destination node explicitly included in the routing packet. Eventually the packet will arrive at the closest node to the intended storage location, and GPSR will enter into perimeter mode, routing the packet in a loop around the intended location and eventually sending it back to the node originally initiating the perimeter routing. The node beginning and ending this loop and those on the perimeter path are called the home node and the home perimeter, respectively. To account for dynamic network topologies, a perimeter refresh protocol (PRP) is used, in which the home node periodically sends the packet in a loop on the home perimeter and the home perimeter nodes assume the role of home node if they do not hear these refresh packets after a certain time-out interval.

36.3.6 Topology Control

Research groups have shown that because of the low duty cycles of sensor nodes' radios, the dominant aspect of power consumption is often idle listening. Unless communication is tightly synchronized, even intelligent MAC protocols such as those described in Section 36.2 cannot completely eliminate this wasted power consumption. However, since sensor networks are expected to be characterized by dense sensor deployment, it is not necessary for all sensors' radios to remain on at all times in order for the network to remain fully connected. While traditional topology control protocols attempt to maintain a predetermined number of neighbor nodes through transmission power control (e.g., [39,40,41]) so that congestion is reduced, several topology control protocols designed for ad hoc and sensor networks achieve energy efficiency by assigning the role of router to only enough nodes to keep the network well connected. In other words, the goal of these protocols is to maintain a fully connected dominating set. While some of

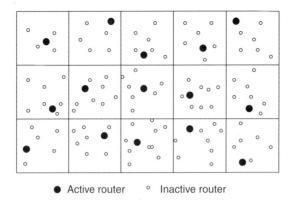

● Active router   ○ Inactive router

FIGURE 36.9   Example of a GAF virtual grid [42]. Only one node per cell is activated as a router.

these protocols were originally designed for use in general ad hoc networks, most are suitable for sensor networks as well.

36.3.6.1 Geographic Adaptive Fidelity

Geographic adaptive fidelity (GAF) is a topology control protocol that was originally designed for use in general ad hoc networks [42]. GAF divides the network into a virtual grid and selects only a single node from each virtual grid cell to remain active as a designated router at a given time, as illustrated in Figure 36.9. As long as the cell dimensions are chosen small enough (transmission_range/$\sqrt{5}$), most nodes in the network, except those near the border of the network, will retain neighbors in all four directions and the network will remain fully connected. Nodes initially enter the discovery state and listen for messages from other nodes within their cell. If another node within the cell is determined to be the designated router for the cell, the node will enter a sleep state and conserve energy. From the sleep state, a node will periodically enter the discovery state. If a node determines that it should be the designated active router for its cell, it will enter the active state and participate in data routing, eventually falling back into the discovery state. As the density of a network implementing GAF increases, the number of activated nodes per grid cell remains constant while the number of nodes per cell increases proportionally. Thus, GAF can allow a network to live for an amount of time approximately proportional to a network's density.

36.3.6.2 Span

Span is a topology control protocol that allows nodes that are not involved in a routing backbone to sleep for extended periods of time [43]. In Span, certain nodes assign themselves the position of "coordinator." These coordinator nodes are chosen to form a backbone of the network, so that the capacity of the backbone approaches the potential capacity of the complete network. Periodically, nodes that have not assigned themselves the coordinator role initiate a procedure to decide if they should become a coordinator. The criteria for this transition is if the minimum distance between any two of the node's neighbors exceeds 3 hops. To avoid the situation where many nodes simultaneously decide to become a coordinator, backoff delays are added to nodes' coordinator announcement messages. The backoff delays are chosen such that nodes with higher remaining energy and those potentially providing more connectivity in their neighborhood are more likely to become a coordinator. To ensure a balance in energy consumption among the nodes in the network, coordinator nodes may fall back from their coordinator role if neighboring nodes can make up for the lost connectivity in the region.

36.3.6.3 Adaptive Self-Configuring Sensor Networks Topologies

Adaptive self-configuring sensor networks topologies (ASCENT) is similar to Span in that certain nodes are chosen to remain active as routers while others are allowed to conserve energy in a sleep state [44]. In ASCENT, the decision to become an active router is based not only on neighborhood connectivity, but also on observed data loss rates, providing the network with the ability to trade energy consumption for communication reliability. Nodes running the ASCENT protocol initially enter a test state where they actively participate in data routing, probe the channel to discover neighboring sensors and learn about data loss rates, and send their own "Neighborhood Announcement" messages. If, based on the current number of neighbors and current data loss rates, the sensor decides that its activation would be beneficial to the network, it becomes active and remains so permanently. If the sensor decides not to become active, it falls into a passive state, where it gathers the same information as it does in the test state (as well as any "Help" messages from neighboring sensors experiencing poor communication links), but it does not actively participate in data routing. From this state, the node may reenter the test state if the information gathered indicates poor neighborhood communication quality, or enter the sleep state, turning its radio off and saving energy. The node periodically leaves the sleep state to listen to the channel from the passive state.

36.3.6.4 Energy-Aware Data Centric Routing

Energy-aware data centric routing (EAD) is an algorithm for constructing a minimum connected dominating set among the sensors in the network, prioritizing nodes so that those with the highest residual energy are most likely to be chosen as nonleaf nodes [45]. To establish a broadcast tree, control messages containing transmitting nodes' type (undefined, leaf node, or nonleaf node), level (in the broadcast tree), parent, and residual energy are flooded throughout the network, starting with the data sink. During the establishment of the tree, undefined nodes listen for control messages. If an undefined node receives a message from a nonleaf node, it becomes a leaf node and prepares to send a message announcing its leaf status after sensing the channel to be idle for some back off time T_2^v. Alternatively, if an undefined node receives a message from a leaf node, it becomes a nonleaf node after sensing the channel idle for some backoff time T_1^v and sending a control message indicating its nonleaf status. However, if a message is received from a nonleaf node during its backoff interval, the node behaves as it would when receiving such a message during its original undecided state. To ensure that nodes with greater residual energy are more likely to assume the more energy-intensive nonleaf roles, T_1^v and T_2^v should be monotonically decreasing functions of the residual energy. Morever, the minimum possible value of T_1^v should be larger than the maximum possible value of T_2^v so that the resulting set of nonleaf nodes is of minimal size. If at any point, a leaf node receives a message from a neighboring nonleaf node indicating that it is the neighbor's parent, it immediately becomes a nonleaf node and broadcasts a message indicating so. Eventually, all connected nodes in the network will assume the role of a leaf or a nonleaf and the resulting nonleaf nodes will comprise an approximation of a minimum connected dominating set with a high priority attached to nodes with the highest remaining energy supplies.

36.4   Protocols for QoS Management

Perhaps one of the most differentiating features of wireless sensor networks is the way in which QoS is redefined. While delay and throughput are typically considered the most important aspects of QoS in general ad hoc wireless and wired networks, new application-specific measures as well as network lifetime are more suitable performance metrics for wireless sensor networks. Because of the redundancy and the application-level importance associated with the data generated by the network, QoS should be determined by the content as well as the amount of data being delivered. In other words, it may be true that the application will be more satisfied with a few pieces of important, unique data than with a large volume of less important, redundant data. Therefore, while it is important to use congestion control in some cases so that the reliability of the sensor network is not reduced due to dropped packets [46], this congestion control can be enhanced by intelligently selecting which nodes should throttle their rates down or stop sending data. Furthermore, the congestion aspect aside, it is important to reduce the amount of traffic generated on the network whenever possible to extend the lifetime of the network because of the tight energy constraints imposed on sensor nodes. This general strategy is often referred to as sensor management, or fidelity control, and is summarized in [47].

36.4.1 Transport Layer

Transport layer protocols are used in many wired and wireless networks as a means to provide services such as loss recovery, congestion control, and packet fragmentation and ordering. While popular transport layer protocols such as TCP may be overweight and many typical transport layer services may not be necessary for most wireless sensor network applications, some level of transport services can be beneficial. This section describes some transport level protocols that are suitable for message delivery in wireless sensor networks.

36.4.1.1 Pump Slowly Fetch Quickly

The pump slowly fetch quickly (PSFQ) protocol was designed to enable reliable distribution of retasking and reprogramming code from a sensor network base station to the sensors in the network [48]. PSFQ provides reliability on a hop-by-hop basis, unlike many end-to-end transport protocols. When new code needs to be distributed, it is fragmented and sent via a pump mechanism that slowly injects packets into the network (with interpacket spacing of at least T_{min}). At a relaying node, a time-to-live (TTL) field is decremented and the message is rebroadcast after a delay chosen in the interval $[T_{min}, T_{max}]$ as long as the local data cache does not indicate a packet loop. To avoid excessive broadcast overlap in dense networks, the packet is removed from the transmit buffer if four copies of the packet have been received by the node. The significant delays are introduced so that normal operation (sensor-to-sink traffic) is not interfered with and to allow the quick recovery of packet losses without requiring large amounts of buffer space at the forwarding nodes. Once a node detects that a packet is received out of sequence, it begins the fetch operation, aggressively trying to quickly recover the lost fragments. PSFQ assumes that most packet loss in sensor networks is caused by poor link quality rather than congestion. Thus, the aggressive recovery approach is not expected to further compound any congestion problem. When packet losses are detected, a node sends a NACK packet indicating the lost packets after a very short delay. If a reply is not received after T_r ($T_r \ll T_{min}$), the NACK is resent up to a threshold number of retires, after which the node gives up. NACKs are withheld if similar NACKs are overheard from neighboring sensors. PSFQ also contains a proactive fetch operation that can be used to detect losses at the end of a sequence (since no subsequent packets will allow the receiving node to know that packets have been lost).

36.4.1.2 Event-to-Sink Reliable Transport

The event-to-sink reliable transport (ESRT) protocol [49] was designed as a solution to the problem posed in Tilak et al.'s work [46]. The protocol achieves energy efficiency by requiring the sensors to send only enough traffic to meet the application's reliability requirements, and it contains mechanisms for detecting and alleviating congestion. From an observed plot of reliability as a function of the sensors' reporting rate (see Figure 36.10), it can be seen that the network can operate in one of five regions:

- No congestion, low reliability (NC, LR)
- The optimal operating region (OOR)
- No congestion, high reliability (NC, HR)
- Congestion, high reliability (C, HR)
- Congestion, low reliability (C, LR)

In ESRT, the data sink in the sensor network periodically broadcasts a revised reporting rate to the sensors, attempting to choose the frequency that will move the network into the OOR. If η_i represents the reliability observed at the sink during interval i, the frequency f_{i+1} for interval $i + 1$ is set to the value as indicated in Table 36.1 and broadcast to the sensors in the network.

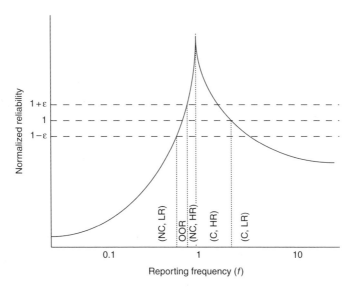

FIGURE 36.10 Example plot of reliability versus sensor reporting rate. ESRT [49] adapts traffic rates according to the region of this plot that the network is operating in.

TABLE 36.1 Rate Adaption in ESRT [49]

Operating region	Updated reporting rate
(NC, LR)	$f_{i+1} = \dfrac{f_i}{\eta_i}$
(OOR)	$f_{i+1} = f_i$
(NC, HR)	$f_{i+1} = \dfrac{f_i}{2}\left(1 + \dfrac{1}{\eta_i}\right)$
(C, HR)	$f_{i+1} = \dfrac{f_i}{\eta_i}$
(C, LR)	$f_{i+1} = f_i^{\eta_i/k}$

Note: After each round, the data sink requires that the new reporting rate is set to f_{i+1}, based on the current operating region, the current reporting rate f_i, the current normalized reliability η_i, and an arbitrary constant k.

While reliability can easily be observed at the sink by counting the number of received packets, congestion is detected by requiring routers to explicitly notify the sink in the event of a buffer overflow.

36.4.2 Providing Coverage of an Environment

Traditional rate control protocols such as ESRT define a network's reliability as the number of total received packets at a base station during a given time interval. However, in many applications, such a definition is only a very coarse approximation of the fidelity of the data that has been aggregated. To get a true measure of fidelity, it is often required to look at the origin and contents of the received packets.

A common application for sensor networks is for the sensors within some region to sense the environment or a subregion in the environment so that it is completely covered. In general, these applications require K-coverage, meaning that each location in the region to be monitored should have K active

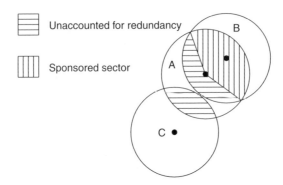

FIGURE 36.11 A sponsored sector, as defined by [51]. Sensor A admits the redundant coverage of sensor B in the vertically shaded regions. The additional redundancy of sensors B and C shown in the horizontally shaded regions is not accounted for.

sensors located within their sensing ranges, with all other sensors turning off in order to save energy (many applications simply require 1-coverage).

36.4.2.1 Probing Environment and Adaptive Sleeping

Probing environment and adaptive sleeping (PEAS) is a protocol that was developed to provide consistent environmental coverage and robustness to unexpected node failures [50]. Nodes begin in a sleeping state, from which they periodically enter a probing state. In the probing state, a sensor transmits a probe packet, to which its neighbors will reply after a random backoff time if they are within the desired probing range. If no replies are received by the probing node, the probing sensor will become active; otherwise, it will return to the sleep state. The probing range is chosen to meet the more stringent of the density requirements imposed by the sensing radius and the transmission radius. The probing rate of PEAS is adaptive and is adjusted to meet a balance between energy savings and robustness. Specifically, a low probing rate may incur long delays before the network recovers following an unexpected node failure. On the other hand, a high probing rate may lead to expensive energy waste. Basically, the probing rate of individual nodes should increase as more node failures arise, so that a consistent expected recovery time is maintained.

36.4.2.2 Node Self-Scheduling Scheme

A node self-scheduling scheme for sensor networks is presented in [51]. In this scheme, a node measures its neighborhood redundancy as the union of the sectors or central angles covered by neighboring sensors within the node's sensing range. At decision time, if the union of a node's "sponsored" sectors covers the full $360°$ (see Figure 36.11), the node will decide to power off. It should be noted that additional redundancy may exist between sensors and that the redundancy model is simplified at a cost of not being able to exploit this redundancy. At the beginning of each round, there is a short self-scheduling phase where nodes first exchange location information and then decide whether or not to turn off after some backoff time. Scenarios of unattended areas due to the simultaneous deactivation of nodes are avoided by requiring nodes to double check their eligibility to turn off after making the decision.

36.4.2.3 Coverage Configuration Protocol

In coverage configuration protocol (CCP), an eligibility rule is proposed to maintain K-coverage [52]. First, each node finds all intersection points between the borders of its neighbors' sensing radii and any edges in the desired coverage area. The CCP rule assigns a node as eligible for deactivation if each of these intersection points is K-covered, where K is the desired sensing degree. The CCP scheme assumes a Span-like protocol and state machine that can use the Span rule for network connectivity or the proposed CCP rule for K-coverage, depending on the application requirements and the relative values of the communication radius and sensing radius. An example of how the CCP rule is applied is given in Figure 36.12. In Figure 36.12(a), node S4, whose sensing range is represented by the bold circle, must decide whether it

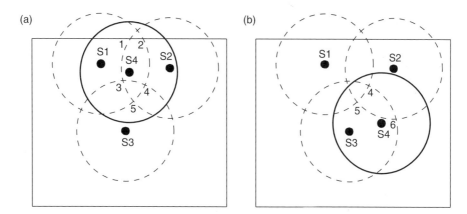

FIGURE 36.12 Illustration of the CCP [52] activation rule for K-coverage, $K = 1$. Node S4 decides whether or not to activate in situations (a) and (b) knowing that neighbors S1, S2, and S3, are already active. In (a), Node S4 may remain inactive since all of its intersection points are K-covered. However, in situation (b), S4 must become active since intersection point 6 is not covered by any of its neighbors.

should become active in order to meet a coverage constraint of $K = 1$. It is assumed S4 knows that S1, S2, and S3, whose sensing ranges are represented by the dashed circles, are currently active. The intersection points within S4's sensing range are found and enumerated 1–5 in the figure. Since S2 covers points 1 and 3, S3 covers points 2 and 4, and S1 covers point 5, S4 determines that the coverage requirements have already been met and remains inactive. In the case illustrated in Figure 36.12(b), there is an intersection point (labeled 6 in the figure) that is not covered by any of S4's neighbors. Thus, S4 must become active and sense the environment.

36.4.2.4 Connected Sensor Cover

The Connected Sensor Cover algorithm provides a joint topology control and sensing mode selection solution [53]. The problem addressed in this work is to find a minimum set of sensors and additional routing nodes necessary in order to efficiently process a query over a given geographical region. In the centralized version of the algorithm, an initial sensor within the query region is randomly chosen, following which additional sensors are added by means of a greedy algorithm. At each step in this algorithm, all sensors that redundantly cover some area that is already covered by the current active subset are considered candidate sensors and calculate the shortest path to one of the sensors already included in the current active subset. For each of these candidate sensors, a heuristic is calculated based on the number of unique sections in the query region that the sensor and its routers would potentially add and the number of sensors on its calculated path. The sensor with the most desirable heuristic value and those along its path are selected for inclusion in the sensor set. This process continues until the query region is entirely covered. The algorithm has been extended to account for node weighting, so that low-energy nodes can be avoided, and to be implemented through distributed means, with little loss in solution optimality compared with the centralized version.

36.5 Time Synchronization and Localization Protocols

One of the main benefits of wireless microsensor networks is the spatial diversity that they provide, enabling applications such as target tracking in which a target's location and speed can be measured as it moves throughout the field where the sensors are deployed. However, such applications require two critical services — localization and time synchronization. These services could potentially be provided by installing GPS radios on the devices; however, in order to deploy microsensors on a mass scale, they should be very inexpensive devices. Furthermore, absolute position and time information is not necessary for many sensor network applications, as relative information can often suffice. If absolute information is necessary, a single or a few high resource nodes can be deployed in the network as references. Thus, there is a need for low-energy distributed algorithms that allows sensors to resolve relative location and time information. There has been a modest amount of research in these areas as wireless sensor networks have grown in popularity over the last several years.

36.5.1 Time Synchronization

To enable applications such as target tracking, sensor networks require time synchronization on a much finer scale than classic synchronization methods such as the network time protocol (NTP) [54]. However, the energy constraints on sensor nodes require that the necessary improvement in synchronization be achieved while at the same time limiting message overhead. Several time synchronization algorithms are described here that try to meet these goals simultaneously.

36.5.1.1   Römer's Algorithm

Römer was among the first to address the time synchronization issue for wireless ad hoc and sensor networks [55]. In the proposed algorithm, nodes do not regularly synchronize clocks; rather, when an event is sensed and a packet needs to be sent to the sinks within the network, the elapsed time since the event was originally sensed is updated within the packet along the path as the packet is routed toward the destination. The forwarding of messages is made somewhat complicated by the uncertainty in time estimation due to clock drift and nondeterministic delays involved in message transfer. Specifically, when transforming some computer clock time delay ΔC from node 1 to node 2, the delay must be estimated by node 2 as an interval $[\Delta C(1-\rho_2)/(1+\rho_1), \Delta C(1+\rho_2)/(1-\rho_1)]$, where ρ_i represents the maximum clock drift of node i. When estimating the elapsed time since the event occurred, the receiving node must make an estimation of the transmission delay between when the packet was sent and when the acknowledgment was received by the previous node. While this estimation is simple to perform at the sending node, it is less obvious at the receiving node. Referring to Figure 36.13, however, it can be seen that this estimation can actually be accomplished without requiring the sending node to send an extra packet explicitly indicating this delay. The round-trip time between a sender and receiver can be estimated at the receiver by the interval $[0, (t_5 - t_4) - (t_2 - t_1)]$ (or $[0, (t_5 - t_4) - (t_2 - t_1)(1 - \rho_r)/(1 + \rho_s)]$) when accounting for clock skew). The time difference $(t_5 - t_4)$ is referred to as rtt$_s$ and may be measured directly by the receiving node while the time difference $(t_2 - t_1)$ is referred to as idle$_s$ and may be piggybacked onto the message packet. It should be noted that this method of delay estimation makes use of two consecutive packet transmissions and the uncertainty in the delay increases with the interpacket delay. Thus, if this delay is too large, it may be necessary to send dummy packets once in a while in order to make these estimations.

If s_i and r_i represent the local time at which node i sends and receives a message, respectively, timestamp estimation can be described as follows. The time at which the event occurs can be estimated by node 2 (first hop) as the time that the packet was received by node 2 (r_2) minus the time that the packet

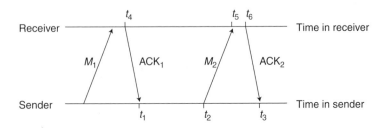

FIGURE 36.13   Timing diagram for message delay estimation needed in the algorithm proposed by Römer [55]. The sender may estimate message delay as $(t_3 - t_2) - (t_6 - t_5)$. The receiver may estimate the message delay as $(t_5 - t_4) - (t_2 - t_1)$.

was waiting at node 1 ($s_1 - r_1$). However, there is uncertainty in the transmission delays, which are lower bounded by 0 and upper bounded by ($\text{rtt}_1 - \text{idle}_1$). Accounting for potential clock skews, the event's time-stamp may be estimated at the second node as

$$\left[r_2 - (s_1 - r_1)\frac{1 + \rho_2}{1 - \rho_1} - \left(\text{rtt}_1 - \text{idle}_1 \frac{1 - \rho_2}{1 + \rho_1} \right), r_2 - (s_1 - r_1)\frac{1 - \rho_2}{1 + \rho_1} \right] \tag{36.7}$$

This estimation process is repeated iteratively so that at the Nth node, the local estimate of the time of the event is

$$\left[r_N - (1 + \rho_N) \sum_{i=1}^{N-1} \frac{s_i - r_i + \text{rtt}_{i-1}}{1 - \rho_i} - \text{rtt}_{N-1} + (1 - \rho_N) \sum_{i=1}^{N-1} \frac{\text{idle}_i}{1 + \rho_i}, r_n - (1 - \rho_N) \sum_{i=1}^{N-1} \frac{s_i - r_i}{1 + \rho_i} \right] \tag{36.8}$$

To implement this algorithm, the three summations in this interval are tracked and updated within the message packet from the source to the destination.

36.5.1.2 Reference-Broadcast Synchronization

While Römer's time synchronization method enables fairly lightweight on-demand event synchronization for sensor networks, finer grained synchronization may be required in certain applications, such as target trajectory estimation. Reference-broadcast synchronization (RBS) allows nodes to synchronize their clocks to the resolution necessary for such sensor network applications [56]. Rather than broadcasting a time-stamp in a synchronization packet as in protocols such as NTP [54], RBS allows the nodes receiving the synchronization packets to use the packet's time of arrival as a reference point for clock synchronization. Because most of the nondeterministic propagation time involved in transmitting a packet over a wireless channel lies between construction of the packet and the sender's transceiver (e.g., sender's queue delay, MAC contention delay, etc.), RBS removes most delay uncertainty involved in typical time synchronization protocols. For single-hop networks, the RBS algorithm is very simple. First, a transmitter broadcasts some number m reference broadcasts. Each receiver that receives these broadcasts exchanges the time that each reference broadcast was received locally with its neighbors. Nodes then calculate phase shifts relative to each other as the average of the difference of the time-stamps of the nodes' local clocks for the m reference broadcasts. In multihop networks, time synchronization can be performed hop by hop between two nodes as long as the nodes on each link along the path have a common node whose reference broadcasts they can synchronize to.

36.5.2 Sensor Localization

Many localization algorithms for sensor networks require nodes to discover relative positioning information (e.g., distance estimations or directional estimations) of neighboring nodes. The ability to attain these estimates is provided in the radio module and is the basis of localization algorithms. Once local distance information is known, simple geometric relations can be used to calculate the local topology, which can then be subsequently disseminated throughout the network, providing globally coordinated localization [57]. This trilateration method is also used in the global positioning system (GPS) [58].

A received signal strength indicator (RSSI) is one method that can be used to infer distances between nodes. However, this method is highly susceptible to errors, especially in environments prone to multipath propagation and shadowing effects. Time of arrival (ToA), which is used in GPS, is another method that can be used; however, clocks on sensor devices may not be able to resolve propagation delays well enough to be able to acquire distance estimation with the required resolution. Time difference of arrival (TDoA), proposed for use in the Cricket platform [59], is a more practical method for estimating distances.

In a TDoA system, an RF signal is transmitted simultaneously with an ultrasonic signal. The difference of times at which the signals are received can be easily translated to distance by multiplying by the difference in speeds. Finally, on sensors in which antenna arrays are used, angle of arrival (AoA) may be used to estimate directional information. In such networks, triangulation is used to find location estimations.

Locations can be estimated by forwarding the local constraints of all nodes in the network to a central server, which can then solve a large program to find location estimates [60,61]. However, there has also been an effort to develop distributed localization protocols for sensor networks, as they scale better and may be more practical for large-scale networks.

36.5.2.1 Reference Point Centroid Scheme

Among the first distributed localization schemes for wireless sensor networks was a simple scheme proposed by Bulusu et al. [62] in which sensors listen for beacons that are broadcast from a few reference points in the network. Sensors hearing these beacons compute their locations to be the centroid of the locations of the reference points whose reference beacons they can hear.

36.5.2.2 Ad-Hoc Localization System

In networks where beacon deployment is sparse enough that location estimations cannot be made directly by each sensor in the network, ad hoc localization system (AHLoS) allows nodes to iteratively resolve their locations through indirect means [63]. In this system, nodes that can acquire position information from and approximate distance to three or more neighboring beacons use a simple atomic multilateration technique. In this technique, nodes estimate their location so as to minimize the mean square error between the estimated location's distances to the beacons and the measured distances to the beacons. Nodes with location estimates in turn become beacon nodes and can be used by their neighbors for atomic multilateration. If a point in this iterative process is reached where no sensor with an unresolved location is within a range of three or more beacons, a cooperative multilateration technique may be used in some cases. In the cooperative multilateration technique, nodes collaboratively solve an overconstrained problem with constraints based on approximated distances to each other and to beacon nodes. For example, in Figure 36.14(b), nodes 5 and 6 can receive messages from beacons 1 and 2, and beacons 3 and 4, respectively, as well as each other. It can be seen that based on the constraints imposed by the measured distances between each pair of communicating nodes, the positions of nodes 5 and 6 can be uniquely determined.

36.5.2.3 DV-Hop

In networks with extremely sparse beacon deployment, the DV-Hop algorithm may be used for node positioning [64]. DV-Hop relies on several landmark nodes located within the network that know their position information through GPS or manual programming. These landmarks first broadcast messages to each other, with the forwarding nodes keeping track of the number of hops these messages are forwarded. Each landmark then calculates a correction factor, which is the average distance to other landmarks (which it can calculate from its own known position and the positions advertised in the broadcast packets) that

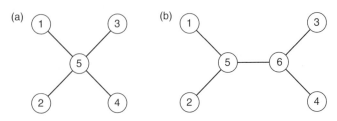

FIGURE 36.14 Scenarios when the AHLoS system [63] can use atomic multilateration (a) and collaborative multilateration (b).

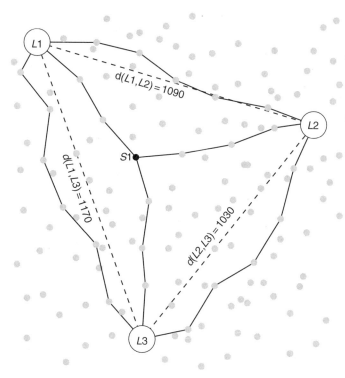

FIGURE 36.15 Position estimation in DV-Hop [64]. The landmark nodes $L1$, $L2$, and $L3$ first forward beacons to each other to find a correction factor in terms of distance per hop. Then node $S1$ can use this correction factor along with its known distance (in hops) to $L1$, $L2$, and $L3$ to find its own location.

it is aware of divided by the average number of hops to these landmarks. The correction factors are then sent to nodes surrounding the landmarks by means of controlled flooding. Once the nodes in the network know this correction factor and the distance (in hops) to at least a few landmark nodes in the network, they can use triangulation techniques in order to calculate a position estimation for themselves.

An example of this procedure is shown in Figure 36.15. Here, landmark nodes $L1$, $L2$, and $L3$ broadcast beacons to each other to calculate the correction factor. If $L2$ and $L3$ are the only landmarks that $L1$ is aware of, it will set the correction factor to

$$C = \frac{d(L1, L2) + d(L1, L3)}{\text{num\_hops}(L1, L2) + \text{num\_hops}(L1, L3)} = \frac{1090\,\text{m} + 1170\,\text{m}}{6 + 7} = 174\,\text{m/hop}$$

The node $S1$ trying to find its position can now perform triangulation using the positions of $L1$, $L2$, and $L3$ and the estimated distance to these landmarks, which are $d(S1, L1) = 174\,\text{m} \times 3 = 522\,\text{m}$, $d(S1, L2) = 174\,\text{m} \times 4 = 696\,\text{m}$, and $d(S1, L3) = 174\,\text{m} \times 4 = 696\,\text{m}$. While DV-Hop provides reasonably accurate location estimation in networks with sparsely deployed landmark nodes, it has been shown that location estimations acquired through methods similar to DV-Hop can be further improved through a refinement stage [65].

36.6 Open Issues

As can be seen by the numerous protocols discussed in this chapter, sensor networks provide many challenges not faced in conventional wireless networks and thus require a rethinking of all layers of the protocol stack. While the current body of work on sensor networks has enabled these networks to produce high quality results for longer periods of time, many open research issues still remain:

- *Appropriate QoS model.* Due to the data-centric nature of sensor networks, describing QoS remains a challenge. In traditional networks, parameters like delay, packet delivery ratio, and jitter can be used to specify application QoS requirements. In sensor networks, on the other hand, these parameters are replaced with ones like probability of missed detection of an event, signal-to-noise ratio, and network sensing coverage. It is much more difficult to translate these data-specific QoS parameters into meaningful protocol parameters.

- *Cross-layer architectures.* To make best use of the limited resources of the sensors, the entire protocol stack should be tailored to the specific needs of the sensor network application. Furthermore, the protocols should be integrated with the hardware, such that any hardware parameters are set to meet the sensor network goals and any protocols are adapted to the specific features of the hardware. While this integrated approach can provide long network lifetime, it trades-off generality and ease of network design to achieve these lifetime increases.

- *Reliability.* In sensor networks, links and sensors themselves may fail, either temporarily or permanently. Designing protocols that provide reliable service in the presence of such failures is an important yet challenging problem.

- *Heterogeneous applications.* The sensor nodes may be shared by multiple applications with differing goals. Sensor network protocols that can efficiently serve multiple applications simultaneously will be very important as the use of sensor networks increases.

- *Heterogeneous sensors.* Much existing work assumes the network is composed of homogeneous nodes. Making best use of the resources in heterogeneous sensor networks remains a challenging problem.

- *Security.* Some initial work has focused on different aspects of security such as ensuring privacy and preventing denial-of-service attacks, but many open questions remain. How much and what type of security is really needed? How can data be authenticated? How can misbehaving nodes be prevented from providing false data? Can energy and security be traded-off such that the level of network security can be easily adapted? These and many other security-related topics must be researched to find low-energy approaches to securing sensor networks.

- *Actuation.* Eventually sensor networks will "close the loop" by providing not only sensing capabilities but also the ability to automatically control the environment based on sensing results. In this case, data do not need to reach any sort of base station or sink points, and thus current models for sensor networks may not be valid. Research is needed to find good protocols for this new sensor network model.

- *Distributed and collaborative data processing.* While much work has been done on architectures to support distributed and collaborative data processing, this is by no means a solved problem. One open question is how to best process heterogeneous data. Furthermore, how much data and what type of data should be processed to meet application QoS goals while minimizing energy drain? These and other questions remain to be solved.

- *Integration with other networks.* Sensor networks may indeed interface with other networks, such as a WiFi network, a cellular network, or the Internet. What is the best way to interface these networks? Should the sensor network protocols support (or at least not compete with) the protocols of the other networks? Or should the sensors have dual network interface capabilities? For some sensor network applications, these questions will be crucial and research is needed to find good solutions.

- *Sensor deployment.* Given that sensor networks suffer from the "hot spot" problem due to the many-to-one traffic patterns, if it is possible to place the sensors at particular locations (or at least certain areas), how should the sensors be deployed so that both sensing and communication goals can be satisfied?

This list highlights just a few of the open research questions. Given the numerous current and envisioned applications for sensor networks, it is likely that research on ways to make these networks better will continue for years to come.

References

[1] C. Kidd et al. The aware home: A living laboratory for ubiquitous computing research. In *Proceedings of the Second International Workshop on Cooperative Buildings (CoBuild)*, 1999.

[2] S. Intille. Designing a home of the future. *IEEE Pervasive Computing*, 1: 76–82, 2002.

[3] L. Schwiebert, S. Gupta, and J. Weinmann. Research challenges in wireless networks of biomedical sensors. In *Proceedings of the Seventh Annual International Conference on Mobile Computing and Networking (MobiCom)*, 2001.

[4] A. Mainwaring, J. Polastre, R. Szewczyk, D. Culler, and J. Anderson. Wireless sensor networks for habitat monitoring. In *Proceedings of the ACM International Workshop on Wireless Sensor Networks and Applications (WSNA)*, 2002.

[5] D. Steere, A. Baptista, D. McNamee, C. Pu, and J. Walpole. Research challenges in environmental observation and forecasting systems. In *Proceedings of the Sixth Annual International Conference on Mobile Computing and Networking (MobiCom)*, 2000.

[6] D. Malan, T. Fulford-Jones, M. Welsh, and S. Moulton. Codeblue: An ad hoc sensor network infrastructure for emergency medical care. In *Proceedings of the International Workshop on Wearable and Implantable Body Sensor Networks*, 2004.

[7] University of Rochester, Center for Future Health.

[8] R. Shah, S. Roy, S. Jain, and W. Brunette. Data MULEs: Modeling a three-tier architecture for sparse sensor networks. In *Proceedings of the First IEEE Workshop on Sensor Network Protocols and Applications (SNPA)*, 2003.

[9] J. Barros and S. Servetto. On the capacity of the reachback channel in wireless sensor networks. In *Proceedings of the IEEE Workshop on Multimedia Signal Processing*, 2002.

[10] P. Juang, H. Oki, Y. Wang, M. Martonosi, L. Peh, and D. Rubenstein. Energy-efficient computing for wildlife tracking: Design tradeoffs and early experiences with zebranet. In *Proceedings of the Tenth International Conference on Architectural Support for Programming Languages and Operating Systems (ASLOS)*, 2002.

[11] V. Bharghavan, A. Demers, S. Shenker, and L. Zhang. MACAW: A media access protocol for wireless LANs. In *Proceedings of ACM SIGCOMM*, 1994.

[12] IEEE Computer Society LAN MAN Standards Committee. Wireless LAN medium access control (MAC) and physical layer (PHY) specifications. In *IEEE Std. 802.11*, 1999.

[13] S. Singh and C. Raghavendra. PAMAS: Power aware multi-access protocol with signalling for ad hoc networks. *ACM SIGCOMM Computer Communication Review*, 28: 5–26, 1998.

[14] Y. Wei, J. Heidemann, and D. Estrin. An energy-efficient MAC protocol for wireless sensor networks. In *Proceedings of the Twenty-First International Annual Joint Conference of the IEEE Computer and Communications Societies (INFOCOM)*, 2002.

[15] T. van Dam and K. Langendoen. An adaptive energy-efficient mac protocol for wireless sensor networks. In *Proceedings of the First ACM Conference on Embedded Networked Sensor Systems (SenSys)*, 2003.

[16] G. Lu, B. Krishnamachari, and C. Raghavendra. An adaptive energy-efficient and low-latency MAC for data gathering in sensor networks. In *Proceedings of the Fourth International Workshop on Algorithms for Wireless, Mobile, Ad Hoc and Sensor Networks (WMAN)*, 2004.

[17] V. Rajendran, K. Obraczka, and J. Garcia-Luna-Aceves. Energy-efficient, collision-free medium access control for wireless sensor networks. In *Proceedings of the First ACM Conference on Embedded Networked Sensor Systems (SenSys)*, 2003.

[18] C. Schurgers, V. Tsiatsis, S. Ganeriwal, and M. Srivastava. Optimizing sensor networks in the energy-latency-density design space. *IEEE Transactions on Mobile Computing*, 1: 70–80, 2002.

[19] S. Pattem, B. Krishnamachari, and R. Govindan. The impact of spatial correlation on routing with compression in wireless sensor networks. In *Proceedings of the Third International Symposium on Information Processing in Sensor Networks (IPSN)*, 2004.

[20] S. Singh, M. Woo, and C. Raghavendra. Power-aware routing in mobile ad hoc networks. In *Proceedings of the Fourth Annual ACM/IEEE International Conference on Mobile Computing and Networking (MobiCom)*, 1998.

[21] J. Chang and L. Tassiulas. Energy conserving routing in wireless ad hoc networks. In *Proceedings of the Nineteenth International Annual Joint Conference of the IEEE Computer and Communications Societies (INFOCOM)*, 2000.

[22] R. Shah and J. Rabaey. Energy aware routing for low energy ad hoc sensor networks. In *Proceedings of the IEEE Wireless Communications and Networking Conference (WCNC)*, 2002.

[23] M. Perillo and W. Heinzelman. DAPR: A protocol for wireless sensor networks utilizing an application-based routing cost. In *Proceedings of the IEEE Wireless Communications and Networking Conference (WCNC)*, 2004.

[24] W. Heinzelman, J. Kulik, and H. Balakrishnan. Adaptive protocols for information dissemination in wireless sensor networks. In *Proceedings of the Fifth Annual ACM/IEEE International Conference on Mobile Computing and Networking (MobiCom)*, 1999.

[25] C. Intanagonwiwat, R. Govindan, and D. Estrin. Directed diffusion: A scalable and robust communication paradigm for sensor networks. In *Proceedings of the Sixth Annual International Conference on Mobile Computing and Networks (MobiCom)*, 2000.

[26] D. Braginsky and D. Estrin. Rumor routing algorithm for sensor networks. In *Proceedings of the First ACM International Workshop on Wireless Sensor Networks and Applications (WSNA)*, 2002.

[27] S. Ratnasamy and B. Karp. GHT: A geographic hash table for data-centric storage. In *Proceedings of the First ACM International Workshop on Wireless Sensor Networks and Applications (WSNA)*, 2002.

[28] B. Karp and H. Kung. GPSR: Greedy perimeter stateless routing for wireless networks. In *Proceedings of the Sixth Annual International Conference on Mobile Computing and Networking (MobiCom)*, 2000.

[29] H. Takagi and L. Kleinrock. Optimal transmission ranges for randomly distributed packet radio terminals. *IEEE Transactions on Communications*, 32: 246–257, 1984.

[30] D. Niculescu and B. Nath. Trajectory based forwarding and its applications. In *Proceedings of the Ninth Annual International Conference on Mobile Computing and Networking (MobiCom)*, 2003.

[31] P. Varshney. *Distributed Detection and Data Fusion*. Springer, New York, 1997.

[32] W. Heinzelman, A. Chandrakasan, and H. Balakrishnan. An application-specific protocol architecture for wireless microsensor networks. *IEEE Transactions on Wireless Communications*, 1: 660–670, 2002.

[33] J. Deng, Y. Han, W. Heinzelmaan, and P. Varshney. Balanced-energy sleep scheduling scheme for high density cluster-based sensor networks. In *Proceedings of the Fourth Workshop on Applications and Services in Wireless Networks (ASWN)*, 2004.

[34] O. Younis and S. Fahmy. Distributed clustering in ad-hoc sensor networks: A hybrid, energy-efficient approach. In *Proceedings of the Twenty-Third Annual Joint Conference of the IEEE Computer and Communications Societies (INFOCOM)*, 2004.

[35] P. Bonnet, J. Gehrke, and P. Seshadri. Querying the physical world. *IEEE Personal Communications*, 7: 10–15, 2000.

[36] C. Shen, C. Srisathapornphat, and C. Jaikaeo. Sensor information networking architecture and applications. *IEEE Personal Communications*, 8: 52–59, 2001.

[37] S. Madden, M. Franklin, J. Hellerstein, and W. Hong. TAG: A tiny aggregation service for ad-hoc sensor networks. In *Proceedings of the ACM Symposium on Operating System Design and Implementation (OSDI)*, 2002.

[38] S. Madden, M. Franklin, J. Hellerstein, and W. Hong. The design of an acquisitional query processor for sensor networks. In *Proceedings of the ACM SIGMOD International Conference on Management of Data*, 2003.

[39] Y. Tseng, Y. Chang, and P. Tseng. Energy-efficient topology control for wireless ad hoc sensor networks. In *Proceedings of the International Computer Symposium*, 2002.

[40] R. Ramanathan and R. Hain. Topology control of multihop wireless networks using transmit power adjustment. In *Proceedings of the Nineteenth International Annual Joint Conference of the IEEE Computer and Communications Societies (INFOCOM)*, 2000.

[41] V. Rodoplu and T. Meng. Minimum energy mobile wireless networks. In *Proceedings of the IEEE International Conference on Communications (ICC)*, 1998.

[42] Y. Xu, J. Heidemann, and D. Estrin. Geography-informed energy conservation for ad hoc routing. In *Proceedings of the Seventh Annual International Conference on Mobile Computing and Networking (MobiCom)*, 2001.

[43] B. Chen, K. Jamieson, H. Balakrishnan, and R. Morris. Span: An energy-efficient coordination algorithm for topology maintenance in ad hoc wireless networks. *ACM Wireless Networks*, 8: 481–494, 2002.

[44] A. Cerpa and D. Estrin. ASCENT: Adaptive self-configuring sensor network topologies. In *Proceedings of the Twenty-First International Annual Joint Conference of the IEEE Computer and Communications Societies (INFOCOM)*, 2002.

[45] X. Cheng, A. Boukerche, and J. Linus. Energy-aware data-centric routing in microsensor networks. In *Proceedings of the Sixth ACM/IEEE International Symposium on Modeling, Analysis, and Simulation of Wireless and Mobile Systems (MSWiM)*, 2003.

[46] S. Tilak, N. Abu-Ghazaleh, and W. Heinzelman. Infrastructure tradeoffs for sensor networks. In *Proceedings of the First ACM International Workshop on Wireless Sensor Networks and Applications (WSNA)*, 2002.

[47] M. Perillo and W. Heinzelman. Sensor management. In C. Raghavendra, K. Sivalingam, and T. Znati, editors, *Wireless Sensor Networks*, pp. 351–372. Kluwer Academic Publishers, Dordrecht, 2004.

[48] C. Wan, A. Campbell, and L. Krishnamurthy. PSFQ: A reliable transport protocol for wireless sensor networks. In *Proceedings of the First ACM International Workshop on Wireless Sensor Networks and Applications (WSNA)*, 2002.

[49] Y. Sankarasubramaniam, O. Akan, and I. Akyildiz. ESRT: Event-to-sink reliable transport in wireless sensor networks. In *Proceedings of the Fourth ACM International Symposium on Mobile Ad Hoc Networking and Computing (MobiHoc)*, 2003.

[50] F. Ye, G. Zhong, J. Cheng, S. Lu, and L. Zhang. PEAS: A robust energy conserving protocol for long-lived sensor networks. In *Proceedings of the Twenty-Third International Conference on Distributed Computing Systems (ICDCS)*, 2003.

[51] D. Tian and N. Georganas. A node scheduling scheme for energy conservation in large wireless sensor networks. *Wireless Communications and Mobile Computing Journal*, 3: 271–290, 2003.

[52] X. Wang, G. Xing, Y. Zhang, C. Lu, R. Pless, and C. Gill. Integrated coverage and connectivity configuration in wireless sensor networks. In *Proceedings of the First ACM Conference on Embedded Networked Sensor Systems (SenSys)*, 2003.

[53] H. Gupta, S. Das, and Q. Gu. Connected sensor cover: Self-organization of sensor networks for efficient query execution. In *Proceedings of the Fourth ACM International Symposium on Mobile Ad Hoc Networking and Computing (MobiHoc)*, 2003.

[54] D. Mills. Internet time synchronization: the network time protocol. *IEEE Transactions on Communications*, 39: 1482–1493, 1991.

[55] K. Römer. Time synchronization in ad hoc networks. In *Proceedings of the Second ACM International Symposium on Mobile Ad Hoc Networking and Computing (MobiHoc)*, 2001.

[56] J. Elson, L. Girod, and D. Estrin. Fine-grained network time synchronization using reference broadcasts. In *Proceedings of the Fifth Symposium on Operating Systems Design and Implementation (OSDI)*, 2002.

[57] S. Capkun, M. Hamdi, and J. Hubaux. GPS-free positioning in mobile ad-hoc networks. In *Proceedings Thirty-Fourth Annual Hawaii International Conference on System Sciences (HICSS-34)*, 2001.

[58] I. Getting. The global positioning system. *IEEE Spectrum*, 30: 36–47, 1993.

[59] N. Priyantha, A. Chakraborty, and H. Balakrishnan. The cricket location-support system. In *Proceedings of the Sixth Annual International Conference on Mobile Computing and Networks (MobiCom)*, 2000.

[60] L. Doherty, K. Pister, and L. Ghaoui. Convex position estimation in wireless sensor networks. In *Proceedings of the Twentieth Annual Joint Conference of the IEEE Computer and Communications Societies (INFOCOM)*, 2001.

[61] Y. Shang, W. Ruml, Y. Zhang, and M. Fromherz. Localization from mere connectivity. In *Proceedings of the Fourth ACM International Symposium on Mobile Ad Hoc Networking and Computing (MobiHoc)*, 2003.

[62] N. Bulusu, J. Heidemannm, and D. Estrin. GPS-less low-cost outdoor localization for very small devices. *IEEE Personal Communications*, 7: 28–34, 2000.

[63] A. Savvides, C. Han, and M. Srivastava. Dynamic fine-grained localization in ad-hoc sensor networks. In *Proceedings of the Seventh Annual International Conference on Mobile Computing and Networking (MobiCom)*, 2001.

[64] D. Niculescu and B. Nath. Ad hoc positioning system (APS). In *Proceedings of the Global Telecommunications Conference (GLOBECOM)*, 2001.

[65] C. Savarese, J. Rabaey, and J. Beutel. Robust positioning algorithms for distributed ad-hoc wireless sensor networks. In *Proceedings of the USENIX Annual Technical Conference*, 2002.

37

Information Fusion Algorithms for Wireless Sensor Networks

37.1	Why Information Fusion?	**37**-843
37.2	Chapter Outline	**37**-844
37.3	Abstract Sensor Algorithms	**37**-845
	Fault-Tolerant Averaging Algorithm • The Fault-Tolerant Interval Function • Other Representative Abstract Sensor Algorithms	
37.4	Estimation Algorithms	**37**-849
	Moving Average Filter Algorithm • Kalman Filter Algorithm • Other Representative Estimation Algorithms	
37.5	Feature Map Algorithms	**37**-854
	Network Scan Algorithms • Other Representative Feature Map Algorithms	
37.6	Inference Methods	**37**-856
	Bayesian Inference • Dempster–Shafer Inference • Other Representative Inference Methods	
37.7	Bibliographic Notes	**37**-861
	Acknowledgment	**37**-862
	References	**37**-862

Eduardo F. Nakamura
Carlos M.S. Figueiredo
Antonio A.F. Loureiro

37.1 Why Information Fusion?

In environments where sensors can be exposed to conditions that might interfere with the measurements provided, wireless sensor networks (WSNs) are used. Such conditions include strong variations in temperature and pressure, electromagnetic noise, and radiation. Therefore, sensor measurements may be imprecise or even useless. Even when environmental conditions are ideal, sensors may not provide the perfect measurements. Essentially, a sensor is a measurement device and an imprecise value is usually associated with its observation. Such imprecisions represent imperfections in the technology and methods used to measure a physical phenomenon or property. In this case, information fusion can be used to process the data provided by multiple sensors and to filter noise measurements providing more accurate information about the monitored entity.

Besides data gathering, it is necessary to understand what the sensor measurements mean, so that events can be detected and undesirable situations avoided. In this case, information fusion can be used to make inference about a monitored entity, for example, given the sensor data and a world model, inference algorithms can be used to provide an interpretation of what is actually happening in the environment.

WSNs are often composed of a large number of sensor nodes posing a new scalability challenge caused by potential collisions and transmissions of redundant data. Regarding energy restrictions, communication should be reduced to increase the lifetime of the sensor nodes. When information fusion is performed during the routing process, that is, sensor data is fused and only the result is forwarded, the number of messages is reduced, collisions are avoided, and energy is saved.

Many definitions of data fusion were provided over the years, and most of them have their origins in the military and remote sensing fields. In 1991, the data fusion work group of the Joint Directors of Laboratories organized an effort to define a lexicon [1] with some terms of reference for data fusion. They define data fusion as a "multi-level, multifaceted process handling the automatic detection, association, correlation, estimation, and combination of data and information from several sources." Klein [2] generalizes this definition stating that data can be provided either by a single source or by multiple sources. Both definitions are general and can be applied in different fields including remote sensing.

Hall and Llinas [3] define data fusion as "the combination of data from multiple sensors, and related information provided by associated databases, to achieve improved accuracy and specific inferences that could not be achieved using a single sensor." Here, data fusion is clearly performed with an objective namely accuracy improvement. However, this definition is restricted to data provided only by sensors and does not foresee the use of data from a single source.

Claiming that all the previous definitions are focused on methods, means, and sensors, Wald [4] changes the focus to the framework used to fuse data. The author states that "data fusion is a formal framework in which are expressed means and tools for the alliance of data originating from different sources. It aims at obtaining information of greater quality; the exact definition of 'greater quality' will depend upon the application." In addition, Wald considers data taken from the same source in different instants as different sources. Note that "quality" is a loose term intentionally adapted to denote that the fused data is somehow more appropriate to the application than the original data. In particular, for WSNs, data can be fused with at least two objectives: improve accuracy and reduce cost of communication.

In Reference 5, Dasarathy suggests the term "information fusion" stating that "in the context of its usage in the society, it encompasses the theory, techniques and tools created and applied to exploit the synergy in the information acquired from multiple sources (sensor, databases, information gathered by human, etc.) such that the resulting decision or action is in some sense better (qualitatively or quantitatively, in terms of accuracy, robustness, etc.) than would be possible if any of these sources were used individually without such synergy exploitation." Possibly, this is the broadest definition embracing any type of source, knowledge, and resource used to fuse different pieces of information. Recently, the term "data aggregation" has been used by the WSN community to refer to information fusion [6,7].

37.2 Chapter Outline

The rest of this chapter will present some information fusion algorithms. Although there are several algorithms that can be classified as information fusion, we discuss only those that we believe to be the most important ones. These algorithms are classified into four different groups, and for each one of these groups, we present two representative methods in detail and discuss some other algorithms. The groups we present are the following:

- *Algorithms for abstract sensors.* These algorithms are suitable for sensor measurements provided in the form of intervals of values.

- *Estimation algorithms.* Estimation algorithms are used to increase data quality by filtering or to make predictions about a physical property.
- *Feature map algorithms.* Features can be associated with geographical positions to provide a map describing how such features are distributed within a given area.
- *Inference algorithms.* These algorithms are used to make inferences about what is the actual meaning of the measurements provided by the sensors.

The remainder of the chapter is organized as follows: Section 37.3 discusses algorithms designed for the acquisition of reliable abstract sensors. Estimation algorithms are presented in Section 37.4. Section 37.5 discusses feature maps. Section 37.6 presents information fusion algorithms used to reason about the data gathered by sensors. We conclude the chapter by presenting some bibliographic notes in Section 37.7.

37.3 Abstract Sensor Algorithms

The algorithms presented in this section were proposed to deal with reliable abstract sensors. The concept of reliable abstract sensor was introduced by Marzullo [8] in the early 1990s. This term defines one of the three sensor types identified in References 8 to 10: concrete, abstract, and reliable abstract sensors. A concrete sensor is a device that perceives the environment by sampling a physical state variable of interest. An abstract sensor is a set of possible values that the physical state variable can assume. Finally, a reliable abstract sensor is the interval (or a set of intervals) that always contains the real value of the physical state variable. A reliable abstract sensor is computed based on several abstract sensors.

37.3.1 Fault-Tolerant Averaging Algorithm

The fault-tolerant averaging algorithm [8] for computing a reliable abstract sensor assumes that at most f of n abstract sensors are faulty (i.e., incorrect), where f is a parameter. Let $\mathcal{I}$ be the set of intervals provided by n abstract sensors that refer to samples of the same physical state variable taken at the same instant. Considering that at most f out of n sensors are faulty, the fault-tolerant averaging computes the Marzullo function $\mathcal{M}_n^f(\mathcal{I})$ that is the smallest interval that surely contains the correct physical value. The implementation is $\mathcal{M}_n^f(\mathcal{I}) = [low, high]$, such that low is the smallest value in at least $n - f$ intervals in $\mathcal{I}$, and $high$ is the largest value in at least $n - f$ intervals in $\mathcal{I}$. In Reference 8, Marzullo shows that the algorithm can be computed in $O(n \log n)$ by sorting the intervals.

As the algorithm computes an intersection of intervals, depending on the intervals in $\mathcal{I}$, the result of $\mathcal{M}_n^f(\mathcal{I})$ can be more accurate than any sensor in $\mathcal{I}$, that is, the resultant interval sometimes can be tighter than the original ones. However, $\mathcal{M}_n^f(\mathcal{I})$ cannot be more accurate than the most accurate sensor in $\mathcal{I}$ when $n = 2f + 1$.

It is true that the result of $\mathcal{M}$ certainly contains the correct value when the number of faulty sensors is at most f. However, it presents an unstable behavior in the sense that minor changes in the input may produce a quite different output.

The pseudo-code of the Marzullo algorithm is depicted in Algorithm 37.1. Lines 1 to 4 specify the input parameters, where:

- n is the number of intervals (sensors' data)
- f is the number of faulty sensors
- *lowData* is an n-vector that stores the low values of the input intervals
- *highData* is an n-vector that stores the high values of the input intervals

For all $1 \leq i \leq n$, the data provided by sensor i is the interval $s_i = [l_i, h_i]$, where $l_i = lowData[i]$ and $h_i = highData[i]$. The local variables in lines 5 to 8 are:

- *low* to store the low limit of the resultant interval
- *high* to store the high limit of the resultant interval

- *lowInterval* is an *n*-vector that stores the number of intervals containing each low value
- *highInterval* is an *n*-vector that stores the number of intervals containing each high value

For all $1 \leq i \leq n$, data from sensor i is $s_i = [l_i, h_i]$, and *lowInterval*$[i]$ is the number of intervals containing l_i, and *highInterval*$[i]$ is the number of intervals containing h_i.

A simple way to compute the vectors *lowInterval* and *highInterval* is to compare all intervals with each other. This is performed in lines 11 to 22. Line 15 evaluates whether the low value of sensor i is contained by the other intervals, and line 18 does the same for the high value of sensor i. Then, the

Algorithm 37.1 Marzullo function, a simple implementation

▷ **Input:**
```
 1:    n; {number of intervals ( sensors)}
 2:    f; {number of faulty intervals ( sensors)}
 3:    lowData[1..n]; {minimum values of the intervals}
 4:    highData[1..n]; {maximum values of the intervals}
```

▷ **Variables:**
```
 5:    low; {stores the resultant low value}
 6:    high; {stores the resultant high value}
 7:    lowInterval[1..n]; {stores the number of intervals containing each low value}
 8:    highInterval[1..n]; {stores number of intervals containing each high value}

 9:    begin
10:        {Intervals computation}
11:        for i ← 1 to n do
12:            lowInterval[i] ← 0;
13:            highInterval[i] ← 0;
14:            for j ← 1 to n do
15:                if lowData[i]∈[lowData[j], highData[j]] then
16:                    lowInterval[i]←lowInterval[i] + 1;
17:                end if
18:                if highData[i] in [lowData[j], highData[j]] then
19:                    highInterval[i]←highInterval[i] + 1;
20:                end if
21:            end for
22:        end for

23:    {Finding the low and high values}
24:    low ← ∞;
25:    high ← −∞;
26:    for i ← 1 to n do
27:        if (lowData[i] < low) ∧ (lowInterval[i] ≥ n − f) then
28:            low ← lowData[i];
29:        end if
30:        if (highData[i] > high) ∧ (highInterval[i] ≥ n − f) then
31:            high ← highData[i];
32:        end if
33:    end for
34:    return [low, high];
35: end
```

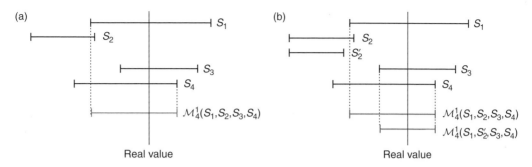

FIGURE 37.1 Example of the $\mathcal{M}$ function.

number of intervals containing the low and high values of each interval s_i is stored in *lowInterval*[i] and *highInterval*[i], respectively. Observe that this part of the algorithm runs in $O(n^2)$.

Once *lowInterval* and *highInterval* are computed we can search for the final *low* and *high* values of the Marzullo function. Then, we search all the elements in *lowData* and *highData* (line 26), determine the smallest value in at least $n - f$ intervals as *low* (line 28), and the largest value in at least $n - f$ intervals as *high* (line 31). The final result is [*low*, *high*]. Since this step runs in $O(n)$, the complexity of this algorithm is $O(n^2) + O(n) = O(n^2)$.

Let us now present an illustrative example to compute the Marzullo function. Figure 37.1(a) shows a scenario with four sensors $\{S_1, S_2, S_3, S_4\}$, where just one of them is a faulty sensor. Note that S_2 and S_3 are the only intervals that do not have any intersection, consequently, one of them is the faulty sensor. Since it is not possible to discover which one provides the correct interval, both must be covered to securely include the *real value*. Thus, $\mathcal{M}_4^1(S_1, S_2, S_3, S_4)$ returns the interval [*low*, *high*], where *low* is the smallest value in at least $n - f = 4 - 1 = 3$ intervals (which is the left edge of S_1), and *high* is the largest value in at least $n - f = 4 - 1 = 3$ intervals (which is the right edge of S_4).

It is possible to see in Figure 37.1(b) that $\mathcal{M}$ presents an unstable behavior. Note that if the right edge of S_2 moves to the left, as given by S_2', then the left edge of the result becomes the left edge of S_3. Thus, a small change in S_2, but large enough to avoid an intersection with S_1, causes a greater variation in the final result.

37.3.2 The Fault-Tolerant Interval Function

Another important representation of the fusion algorithm for abstract sensors is the fault-tolerant interval (FTI) function, or simply the $\mathcal{F}$ function, which was proposed by Schmid and Schossmaier [11].

The algorithm also assumes that at most f of n abstract sensors are faulty, where f is a parameter. Let $\mathcal{I} = \{I_1, \ldots, I_n\}$ be the set of intervals $I_i = [x_i, y_i]$ provided by n abstract sensors. The FTI intersection function is given by $\mathcal{F}_n^f(\mathcal{I}) = [low, high]$, where *low* is the $(f + 1)$th largest of the left edges $\{x_1, \ldots, x_n\}$, and *high* is the $(f + 1)$th smallest of the right edges $\{y_1, \ldots, y_n\}$.

Differently from the Marzullo function, the $\mathcal{F}$ function assures that minor changes in the input intervals will result in minor changes in the integrated result. However, the $\mathcal{F}$ function sacrifices accuracy generating larger output intervals [12].

A possible implementation of the $\mathcal{F}_n^f$ function is depicted in Algorithm 37.2. The input parameters are declared in lines 1 to 4 where:

- n is the number of intervals (sensors' data)
- f is the number of faulty sensors
- *lowData* is an n-vector that stores the low values of the input intervals
- *highData* is an n-vector that stores the high values of the input intervals

Algorithm 37.2 FTI Function

▷ `Input:`
1: n; {*number of intervals (sensors)*}
2: f; {*number of faulty intervals (sensors)*}
3: *lowData*[1..*n*]; {*minimum values of the intervals*}
4: *highData*[1..*n*]; {*maximum values of the intervals*}

▷ `Variables:`
5: *low*; {*stores the resultant low value*}
6: *high*; {*stores the resultant high value*}

7: **begin**
8:    {*sort the input vectors*}
9:    *sort*(*lowData*, *n*);
10:   *sort*(*highData*, *n*);

11:     {*get the resultant values*}
12:     *low* ← *lowData* [*n* − *f*];
13:     *high* ← *highData* [*f* + 1];
14:     **return** [*low*, *high*];
15: **end**

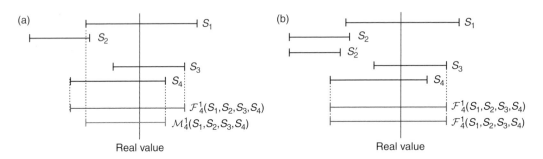

FIGURE 37.2   Example of the $\mathcal{F}$ function.

Both *lowData* and *highData* vectors are sorted in lines 9 and 10, respectively, such as $lowData[i] \leq lowData[j]$ and $highData[i] \leq highData[j]$ for all $1 \leq (i < j) \leq n$. This step can be executed in $O(n \log n)$ using well-known sorting algorithms. Then, the $(f + 1)$th largest element of *lowData* is chosen as *low* (line 12), while the $(f + 1)$th smallest element of *highData* is chosen as *high* (line 13). The final result is [*low*,*high*]. The algorithm runs in $O(n \log n)$.

To exemplify the behavior of $\mathcal{F}$, we consider in Figure 37.2 the same example provided before. The resulting interval is slightly larger than the intervals returned by $\mathcal{M}$ (Figure 37.2[a]). However, the resulting interval does not change when S'_2 is used instead of S_2 (Figure 37.2[b]). As a general result, $\mathcal{M}$ tends to achieve tighter intervals than $\mathcal{F}$. However, $\mathcal{F}$ is less vulnerable to small changes in the input intervals.

37.3.3 Other Representative Abstract Sensor Algorithms

Chew and Marzullo [9] extend the original fault-tolerant averaging algorithm (single dimensional) to fuse data from multidimensional sensors. In Reference 10, Jayasimha extends Marzullo's original work (for the single-dimensional case) to detect all sets (combinations) of possibly-faulty sensors.

After the $\mathcal{M}_n^f$ function was introduced, other fusion methods were proposed for abstract sensors. Besides the $\mathcal{F}_n^f$ function, an overlap function $\mathcal{O}$ was proposed in Reference 13. Compared to $\mathcal{M}_n^f$ and $\mathcal{F}_n^f$, the overlap function provides tighter intervals when the total number of sensors and tamely faulty sensors are large [13]. This technique is generalized in the abstract framework for fault-tolerant integration proposed by Iyengar and Prasad [14]. In References 15 and 16, a multi-resolution decomposition is applied to analyze the overlap function by resolving it at several successively finer scales to isolate the region containing the correct value.

37.4   Estimation Algorithms

Many estimation algorithms for data fusion were inherited from digital signal processing (DSP) and control theory. Algorithms of control theory usually apply the probability laws to calculate a process state vector from a measurement vector or a sequence of measurement vectors [17]. Further information about estimation methods can be found in Kailath et al. [18].

37.4.1   Moving Average Filter Algorithm

The moving average [19] algorithm is widely adopted in DSP solutions because it is very simple to understand and use. Furthermore, this filter is optimal for reducing random noise while retaining a sharp step response. This is the reason that makes the moving average the main filter for encoded signals in the time domain. As the name suggests, this filter computes the arithmetic mean of a number of input measures to produce each point of the output signal. This can be translated in the following equation:

$$\text{output}[i] = \frac{1}{m} \sum_{k=0}^{m-1} \text{input}[i - k] \qquad (37.1)$$

where m is the filter's window, that is, the number of input observations to be fused. Observe that m is also the number of steps the filter takes to detect the change in the signal level. The lower the value of m, the sharper the step edge. On the other hand, the greater the value of m, the cleaner the signal.

The mathematical definition of the moving average filter (37.1) can be translated into Algorithm 37.3. The input parameters declared in lines 1 to 3 are:

- The filter's window m
- The number of input samples n
- The n-vector *input*, with the digital signal to be filtered

The variables used include a temporary accumulator *acc* (line 4) to store the sum of the input samples, and an n-vector *output* (line 5) to store the filtered signal. Initially, the accumulator *acc* is set with the sum of the first m input points (line 10), and the first $m - 1$ output points are set to 0 (line 11). These are undefined points because the moving average takes m input points to generate one filtered point. Then, the filter computes the first filtered point *output*[m] (line 14). After that, to compute the remaining points, the filter adds the next input point, subtracts the first point used to compute the prior point (line 16), and computes the average (line 17).

To exemplify the moving average filter suppose that we have the data traffic received by a sink node of a flat WSN during 1500 sec, as depicted in Figure 37.3(a), and we want to analyze it. In this network, the traffic is kept constant at 7.5 packets/sec (pps) during the first 500 sec, then the traffic level decreases to 7.35 pps, and after 1000 sec, the traffic level decreases again reaching 7.3 pps.

Figure 37.3 shows an example of how the moving average filter works. Figure 37.3(a) and Figure 37.3(b) depict the real traffic and the measured traffic, respectively. The measured traffic presents some noise, due to queue delays, collisions, and clock-drifts. Figure 37.3(c) to Figure 37.3(e), depict the filtered traffic for different values of m. The moving average filter not only smoothes the noise amplitude, but also reduces the sharpness of the edges. Furthermore, the lower the value of m, the sharper the step edge. On the other

Algorithm 37.3　Moving Average Filter

▷ **Input:**
　1: m; {*filter's window*}
　2: n; {*number of input samples*}
　3: $input[1..n]$; {*input samples vector*}

▷ **Variables:**
　4: acc; {*temporary accumulator*}
　5: $output[1..n]$; {*stores the filtered samples*}

　6: **begin**
　7:   {*Initialize the filter with m samples*}
　8: $acc \leftarrow 0$;
　9:   **for** $i \leftarrow 1$ **to** $m-1$ **do**
10:      $acc \leftarrow acc + input[i]$;
11:      $output[i] \leftarrow 0$;
12:    **end for**

13:    {*compute the filtered samples*}
14:    $output[m] \leftarrow acc/m$;
15:    **for** $i \leftarrow m+1$ **to** n **do**
16:      $acc \leftarrow acc + input[i] - input[i-m]$;
17:      $output[m] \leftarrow acc/m$;
18:    **end for**
19:    **return** $output$;
20: **end**

---

hand, the greater the value of m, the cleaner the signal. The amount of noise reduction is equal to the square root of the filter window m, which is also the number of steps the filter takes to detect the change in the signal level.

37.4.2　Kalman Filter Algorithm

The Kalman filter is the most popular fusion method. It was originally proposed in 1960 by Kalman [20] and it has been extensively studied since then [21–23]. For a general introduction, see Maybeck [24] and Bishop and Welsh [23].

The Kalman filter tries to estimate the state x of a discrete-time controlled process that is ruled by the state-space model

$$x(k+1) = \Phi(k)x(k) + \mathbf{G}(k)u(k) + w(k) \tag{37.2}$$

with measurements z represented by

$$z(k) = \mathbf{H}(k)x(k) + v(k) \tag{37.3}$$

where w and v are random variables that represent the Gaussian noise with covariances $\mathbf{Q}(k)$ and $\mathbf{R}(k)$, respectively.

The notations we adopt in the next expressions are defined as follows. For any variable A:

- $A(k \mid k)$ is the current estimate of A
- $A(k+1 \mid k)$ is the prediction of A given the current sample k
- $A(k \mid k-1)$ is the last prediction of A

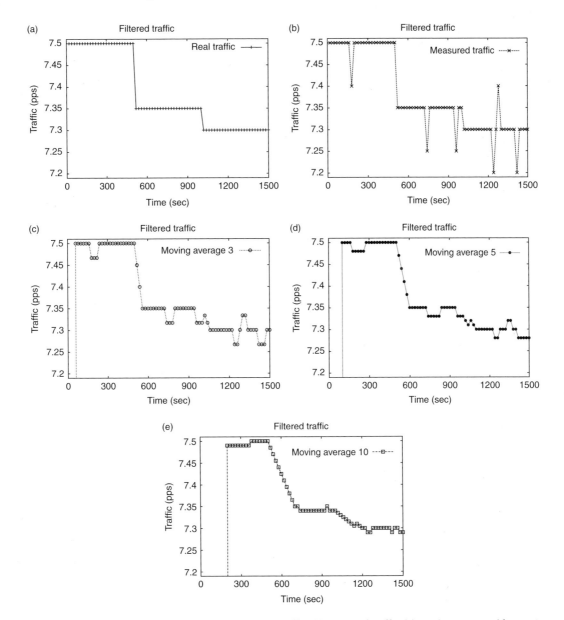

FIGURE 37.3 Example of the moving average: (a) real traffic; (b) measured traffic; (c) moving average with $m = 3$; (d) moving average with $m = 5$; and (e) moving average with $m = 10$.

Thus, based on the measurements z, the estimate of $x(k)$, represented by $\hat{x}(k \mid k)$, and the prediction of $x(k + 1)$, represented by $\hat{x}(k + 1 \mid k)$, have the form

$$\hat{x}(k \mid k) = \hat{x}(k \mid k - 1) + \mathbf{K}(k)[z(k) - \mathbf{H}(k)\hat{x}(k \mid k - 1)], \tag{37.4}$$

$$\hat{x}(k + 1 \mid k) = \Phi(k)\hat{x}(t \mid t) + \mathbf{G}(k)u(k) \tag{37.5}$$

where $\mathbf{K}(k)$ is the Kalman filter gain determined by

$$\mathbf{K}(k) = \mathbf{P}(k \mid k - 1)\mathbf{H}^{\mathrm{T}}(k)[\mathbf{H}(k)\mathbf{P}(k \mid k - 1)\mathbf{H}^{\mathrm{T}}(k) + \mathbf{R}(k)]^{-1} \tag{37.6}$$

where $\mathbf{R}(k)$ is the covariance matrix of the measurement error and $\mathbf{P}(k \mid k-1)$ is the covariance matrix of the prediction error that is determined by

$$\mathbf{P}(k+1 \mid k) = \Phi(k)\mathbf{P}(k \mid k)\Phi^{\mathrm{T}}(k) + \mathbf{Q}(k) \qquad (37.7)$$

and

$$\mathbf{P}(k \mid k) = \mathbf{P}(k \mid k-1) - \mathbf{K}(k)\mathbf{H}(k)\mathbf{P}(k \mid k-1) \qquad (37.8)$$

The Kalman filter is used to fuse low-level redundant data. If the system can be described by a linear model and the error can be modeled as a Gaussian noise, the Kalman filter recursively retrieves statistically optimal estimates [21]. However, other methods should be adopted to deal with nonlinear dynamics and a nonlinear measurement model. According to Jazwinski [25], the variation named Extended Kalman Filter [23] is the most popular approach to implement recursive nonlinear filters. The standard Kalman filter presented above is a simple centralized version that can be extended to improve its performance (e.g., the state prediction [26]) or to be the basis of decentralized implementations [27–30].

Although the provided matrix representation may seem a little complex, an actual implementation of the Kalman filter is very simple as we illustrate in the following. Let us assume that the noisy measurement is the state itself, that is, $\mathbf{H}(k) = 1$. Thus, (37.6) is reduced to

$$\mathbf{K}(k) = \mathbf{P}(k \mid k-1)[\mathbf{P}(k \mid k-1) + \mathbf{R}(k)]^{-1} \qquad (37.9)$$

where $\mathbf{P}(k \mid k-1)$ is the variance of the last estimation of $\hat{x}$, and $\mathbf{R}(k)$ is the variance of the current measurement. Consequently, (37.4) is reduced to

$$\hat{x}(k \mid k) = \hat{x}(k \mid k-1) + \mathbf{K}(k)[z(k) - \hat{x}(k \mid k-1)] \qquad (37.10)$$

Thus, (37.9) and (37.10) can be used to implement a Kalman filter that estimates the real value of n independent and redundant measurements, as described in Algorithm 37.4. The input parameters declared in lines 1 to 3 are:

- The number of input measurements n
- The n-vector $input$, with the actual measurements
- The n-vector R, with the variance (uncertainty) of each measurement

The variables declared are K (line 4), which is used to store the Kalman filter gain, P (line 5), which is used to store the covariance prediction error, and x (line 6), which stores the result of the fusion. Then, expressions (37.9) and (37.10) are recursively applied to each element of the input data, see lines 13 and 14. Also, the covariance prediction error is updated (line 15). Once all data is processed, the algorithm returns the fused value and the resultant variance (line 17).

To exemplify the Kalman filter, suppose that we are monitoring the temperature of a controlled laboratory with four independent sensors $\{S_1, S_2, S_3, S_4\}$, and we have the sensor readings given in Table 37.1. Applying Algorithm 37.4 to $\{S_1, S_2, S_3, S_4\}$, the result will be 19.4 ± 0.5. Note that resultant error, 0.5, is smaller than the error associated with the input data, which means that the result is more precise. The execution of Algorithm 37.4 is shown below:

Step 1 Getting the S_1 (lines 8 and 9):

$$x = 20;$$

$$P = 2.5;$$

Algorithm 37.4 Kalman Filter

▷ **Input:**
  1:   n; {*number of input samples*}
  2:   *input*[1..*n*]; {*input samples (measurements) vector*}
  3:   $R[1..n]$; {*vector of measurements error*}

▷ **Variables:**
  4:   K; {*stores the filter's gain*}
  5:   P; {*stores the estimation error*}
  6:   x; {*stores the current estimation*}

  7: **begin:**
  8:      {*initialize the filter*}
  9:      $x = input[1]$;
 10:      $P = R[1]$;

 11:      {*compute the current estimation and its error*}
 12:      **for** $i \leftarrow 2$ **to** n **do**
 13:         $K \leftarrow P/(P + R[i])$;
 14:         $x \leftarrow x + K(input[i] - x)$;
 15:         $P \leftarrow (P \times R[i])/(P + R[i])$;
 16:      **end for**
 17:      **return** $[x, P]$;
 18: **end**

TABLE 37.1   Applying
the Kalman Filter

Sensor	Measurement (°C)
S_1	20.0 ± 2.5
S_2	19.0 ± 1.7
S_3	18.5 ± 1.5
S_4	21.0 ± 2.3

Step 2   Fuse of the sensor data (lines 11 to 15):
  Step 2.1   Fuse of x and S_2:

$$K = \frac{2.5}{2.5 + 1.7} = 0.6;$$

$$x = 20.0 + 0.6(19.0 - 20.0) = 19.4;$$

$$P = \frac{2.5 \times 1.7}{2.5 + 1.7} = 1.0;$$

  Step 2.2   Fuse of x and S_3:

$$K = \frac{1.0}{1.0 + 1.5} = 0.4;$$

$$x = 19.4 + 0.4(18.5 - 19.4) = 19.0;$$

$$P = \frac{1.0 \times 1.5}{1.0 + 1.5} = 0.6;$$

Step 2.3 Fuse of x and S_4:

$$K = \frac{0.6}{0.6 + 2.3} = 0.2;$$

$$x = 19.0 + 0.2(21.0 - 19.0) = 19.4;$$

$$P = \frac{0.6 \times 2.3}{0.6 + 2.3} = 0.5;$$

37.4.3 Other Representative Estimation Algorithms

The Maximum *a posteriori* Probability [31] is a fusion method based on the Bayesian theory. It is used when parameter x to be discovered is a random variable with a previously known *probability density function* (pdf). Given a measurement sequence, the maximum *a posteriori* probability tries to find the most likely value of x by finding the posterior pdf of x and searching for the value of x that minimizes this pdf.

The least squares [31] is a method used when the parameter to be estimated is not considered a random variable. In contrast to the maximum *a posteriori* probability, this method does not assume any prior probability. The least squares handles the measurements as a deterministic function of the state, and it searches for the value of x that minimizes the sum of the squared errors between the real observations and the predicted ones.

The particle filters are recursive implementations of statistical signal processing [32] known as sequential Monte Carlo methods [33,34]. Although the Kalman filter is the classical approach for state estimation, particle filters represent an alternative for applications with non-Gaussian noise, specially when computational power is rather cheap and the sampling rate is low [35]. The particle filters attempt to build the posterior pdf based on a large number of random samples called particles. The particles are propagated over time sequentially combining sampling and re-sampling steps. At each time step, the re-sampling is used to discard some particles increasing the relevance of regions with high posterior probability.

37.5 Feature Map Algorithms

For some applications, such as resource management, it might be infeasible to directly use individual sensory data. In such cases, features representing some aspects of the environment can be extracted and used by the application. Feature maps are produced by fusion processes to represent how one or more features are distributed along the monitored area.

37.5.1 Network Scan Algorithms

(Sensor) Network Scans are defined by Zhao et al. [36] as a sort of resource/activity map for WSNs. Analogous to a weather map, the network scan depicts the geographical distribution of a resource or activity of a WSN. Considering a resource of interest, instead of providing detailed information about each sensor node in the network, these scans offer a summarized view of the resource distribution.

Regarding a monitored resource or feature. a *Scan* ξ is a tuple

$$\xi(region, min, max) \tag{37.11}$$

where *region* is the region represented by the ξ, *min* is the minimum value of a sensor contained in *region*, and *max* is the maximum value of a sensor inside *region*.

The algorithm is quite simple. First, an aggregation (fusion) tree is formed to determine how the nodes will communicate. Second, each sensor computes its local *Scan* and whenever the energy level drops significantly since the last report, the node sends its *Scan* toward the user. The *Scans* are fused when a node receives two or more topologically adjacent *Scans* that have the same or similar level. The fused *Scan* is a polygon corresponding to a region and the summarized value of the nodes within that region.

Two *Scans* are similar if the relative error is within a given threshold, called tolerance. Although this error can be computed in different ways, Zhao et al. [36] suggest the expression

$$\frac{max(scanA.max, scanB.max) - min(scanA.min, scanB.min)}{avg(scanA.min, scanA.max, scanB.min, scanB.max)} \tag{37.12}$$

to compute the relative error of fusing *scanA* with *scanB*.

Similarly, two *Scans* are adjacent if the distance between the regions of the *Scans* are within a given threshold, called resolution. Again, this distance can be computed in different ways. One possibility is to compute the smallest distance between the regions of the considered *Scans*.

The process of fusing two *Scans* is depicted in Algorithm 37.5. Lines 1 to 4 specify the input parameters where:

- *scanA* is the first scan to be fused.
- *scanB* is the second scan to be fused.
- *T* is the tolerance threshold. Two scans cannot be fused if the resultant error is greater than *T*.
- *R* is the resolution parameter that is used to decide if two regions are adjacent.

Algorithm 37.5 Network Scan

```
▷ Input:
 1:  scanA; {input scan A}
 2:  scanB; {input scan B}
 3:  T; {tolerance parameter}
 4:  B; {resolution parameter}

▷ Variables:
 5:  scanR; {stores the resultant scan}

 6: begin
 7:     {initialize the result}
 8:     scanR ← nil;

 9:  {compute the relative error}
10:  error ← max(scanA.max, scanB.max) − min(scanA.min, scanB.min);
11:  error ← error/avg(scanA.min, scanA.max, scanB.min, scanB.max);

12:  {compute the distance between region of scanA and region of scanB}
13:  dist ← distance(scanA.region, scanB.region);

14:  {If the error is tolerable and the regions are adjacent}
15:  if error ≤ T and dist < R then
16:     {then scanA and scanB are fused}
17:     result.min ← min(scanA.min, scanB.min);
18:     result.max ← max(scanA.max, scanB.max);
19:     result.region ← merge(scanA.region, scanB.region);
20:  endif

21:  return scanR;
22: end
```

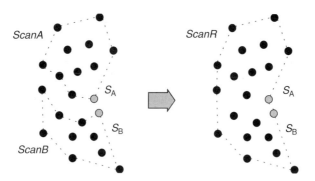

FIGURE 37.4　Example of the network scan.

The sole variable is *scanR* (line 5) and it is used to store the resultant network scan. First, the error is computed in lines 9 to 11 using (37.12). Second, the distance between *scanA* and *scanB* is computed in lines 12 and 13. Then, the algorithm verifies if the two scans are correlated by checking the tolerance and resolution parameters (line 15). If *scanA* and *scanB* are correlated, then they are fused in *scanR* (lines 16 to 19).

To exemplify this algorithm consider two network scans, *scanA* and *scanB*, depicted in Figure 37.4 such that the monitored resource is the residual energy and the input parameters are:

- $scanA.min = 10\%$
- $scanA.max = 15\%$

- $scanB.min = 9\%$
- $scanB.max = 14\%$
- $distance(scanA, scanB) = distance(S_A, S_B) = 1.7u$ ($u = 1$ unit)
- $R = 2.0m$
- $T = 0.9$

Applying (37.12), we obtain an error of 0.5 that is smaller than the tolerance $T = 0.9$. Also, the distance between the two network scans is smaller than the resolution. Thus, the fused result is given by *scanR.min* = 9%, *scanR.max* = 15%, and *scanR.region* is depicted in Figure 37.4.

37.5.2 Other Representative Feature Map Algorithms

Occupancy grids, also called occupancy maps or certainty grids, define a multidimensional (2D or 3D) representation of the environment describing which areas are occupied by an object or which areas are free spaces. The original proposal was provided by Moravec and Elfes [37–39] to build an internal model of static environments based on ultrasonic data. According to Elfes [40], an occupancy grid is "a multidimensional random field that maintains stochastic estimates of the occupancy state of the cells," that is, the observed space is divided into square or cubic cells and each cell contains a value indicating its probability of being occupied. Usually, such probability is computed — based on information provided by several sensors — using different methods, such as Bayesian theory, Dempster–Shafer reasoning, and fuzzy set theory [41].

37.6 Inference Methods

Inference methods are often applied in decision fusion when a decision is taken based on the knowledge of the perceived situation. In this context, inference refers to the transition from one likely true proposition to another the true value of which is believed to result from the previous one. Classical inference methods are based on the Bayesian inference [42] and Dempster–Shafer belief accumulation theory [43,44].

Algorithm 37.6 Bayesian inference

▷ `Input:`
 1: n; {*number of evidences*}
 2: *probXY* [1..n]; {*a priori probabilities of X regarding each hypothesis*}
 3: *probY*[1..n]; {*a priori probability of each hypothesis*}
 4: *probX*; {*a priori probability of X*}

▷ `Variables:`
 5: *probYX* [1..n]; {*stores the a posteriori probability of each hypothesis*}
 6: *mostlikely*; {*stores the most likely hypothesis*}

 7: `begin`
 8:     {*apply the Bayes′s rule*}
 9:     `for` $i \leftarrow 1$ `to` n `do`
10:         *probYX*[i]$\leftarrow$*probXY*[i]$\times$*probY*[i]$/$*probX* ;
11:     `endfor`

12:     {*find the most likely hypothesis*}
13:     *mostlikely* $= 1$;
14:     `for`i $\leftarrow$ 2 `to` n `do`
15:       `if`*probYX*[i] $>$ *probYX* [*mostlikely*] `then`
16:           *mostlikely* $\leftarrow i$;
17:       `end if`
18:     `end for`
19:     `return` *mostlikely* ;
20: `end`

37.6.1 Bayesian Inference

Bayesian inference offers a formalism for information fusion where information is combined according to the rules of probability theory. The uncertainty is represented in terms of conditional probabilities representing the belief, and it can assume values in the [0, 1] interval, where 0 is the absolute disbelief and 1 is the absolute belief. Bayesian inference is based on the rather old Bayes's rule [42], which states that

$$\mathbf{P}(Y \mid X) = \frac{\mathbf{P}(X \mid Y)\mathbf{P}(Y)}{\mathbf{P}(X)} \tag{37.13}$$

where the *a posteriori* probability $\mathbf{P}(Y \mid X)$ represents the belief of hypothesis Y given the information X. This probability is obtained by multiplying $\mathbf{P}(Y)$, the *a priori* probability of hypothesis Y, by $\mathbf{P}(X \mid Y)$, the probability of receiving X given that Y is true. The probability $\mathbf{P}(X)$ is considered as a normalizing constant. The main issue regarding Bayesian inference is that it requires *a priori* knowledge of the probabilities $\mathbf{P}(X)$ and $\mathbf{P}(X \mid Y)$, which may be unavailable. For this reason, subjective probabilities [45] are frequently used.

Bayesian inference can be used in different contexts and different algorithms to search for the most likely hypothesis. However, we can still think about it as an algorithm as we see below. This algorithm searches for the most likely hypothesis given the information X.

The input parameters of Algorithm 37.6 (lines 1 to 4) are:

- the number of hypotheses n
- the vector *probXY* discriminating *a priori* probability of X regarding the occurrence of each hypothesis

- *a priori* probability *probY* of each hypothesis
- the *a priori* probability *probX* of X

The variables used are the vector *probYX* (line 5) to store the computed *a posteriori* probability of each hypothesis and *mostlikely* (line 6) to store the most likely hypothesis given information X. In lines 9 to 11, the Bayes's rule (37.13) is applied to compute the probability of each hypothesis regarding X. Then, the algorithm searches for the most likely hypothesis in lines 13 to 18.

Since Bayesian inference is more like a methodology, each application may apply the Bayes's rule differently, which makes it difficult to present a qualitative example like those we provided for the previous algorithms. However, in Reference 46, Sichitiu and Ramadurai provide an interesting example using Bayesian inference to process information from a mobile beacon so each node can determine its geographical location. In this approach, only a beacon node is aware of its location (e.g., equipped with GPS). After the sensor node has been deployed, the unknown nodes determine their positions by applying the Bayesian inference to some radio frequency (RF) data, and the location information provided by the mobile beacon.

37.6.2　Dempster–Shafer Inference

The Dempster–Shafer inference is based on Dempster–Shafer Belief Accumulation (also referred to as Theory of Evidence or Dempster–Shafer Evidential Reasoning), which is a mathematical theory introduced by Dempster [43] and Shafer [44] that generalizes the Bayesian theory. It deals with beliefs or mass functions just as Bayes's rule does with probabilities. The Dempster–Shafer theory provides a formalism that can be used to represent incomplete knowledge, update beliefs, and combine evidences [47].

According to Reference 21, the use of the Dempster–Shafer theory to fuse sensory data was introduced in 1981 by Garvey et al. [48]. In contrast to Bayesian inference, it allows each source to contribute information in different levels of detail, for example, one sensor can provide information to distinguish individual entities whereas other sensors can provide information to distinguish classes of entities [21]. Furthermore, it does not assign *a priori* probabilities to unknown propositions. Instead, probabilities are assigned only when the supporting information is available.

Choosing between Bayesian inference and the theory of evidence is not a trivial task, which must evaluate the tradeoff between the Bayesian accuracy and the Dempster–Shafer flexibility [17]. Comparisons between these two inference methods can be found in References 49 to 51.

A fundamental concept in a Dempster–Shafer reasoning system is the *frame of discernment*, which is a set of possible states of the system $\Theta = \{\theta_1, \theta_2, \ldots, \theta_N\}$. The frame of discernment must be exhaustive and mutually exclusive in the sense that the system is certainly in one and only one state $\theta_k \in \Theta$.

The elements of the powerset 2^Θ are called hypotheses. Based on an evidence E, a probability is assigned to every hypothesis $H \in 2^\Theta$ according to a *basic probability assignment* (bpa), or mass function, $m: 2^\Theta \rightarrow [0, 1]$ that satisfies

$$m(\emptyset) = 0 \tag{37.14}$$

$$m(H) \geq 0, \quad \forall H \in 2^\Theta \tag{37.15}$$

$$\sum_{H \in 2^\Theta} m(H) = 1 \tag{37.16}$$

To express the overall belief in a hypothesis H, the Dempster–Shafer defines the belief function *bel*: $2^\Theta \rightarrow [0, 1]$ over Θ as

$$bel(H) = \sum_{A \subseteq H} m(A) \tag{37.17}$$

where $bel(\emptyset) = 0$ and $bel(\Theta) = 1$.

If the belief function $bel: 2^{\Theta} \to [0, 1]$ is used to define the belief $bel(H)$ in the hypothesis $H \in \Theta$, then the degree of doubt in H is formulated as

$$dou(H) = bel(\overline{H}) = \sum_{A \cap H = \emptyset} m(A) \qquad (37.18)$$

To express the plausibility of each hypothesis, the function $pl : 2^{\Theta} \to [0, 1]$ over Θ is defined as

$$pl(H) = 1 - dou(H) = 1 - \sum_{A \cap H = \emptyset} m(A) \qquad (37.19)$$

The plausibility intuitively states that the lesser the doubt in hypothesis H, the more plausible the hypothesis. In this context, the confidence interval $[bel(H), pl(H)]$ defines the true belief of the hypothesis H. To combine the effects of two bpa's m_1 and m_2, the Dempster–Shafer theory defines a combination rule, $m_1 \oplus m_2$, that is given by

$$m_1 \oplus m_2(H) = \frac{\sum_{X \cap Y = H} m_1(X) m_2(Y)}{1 - \sum_{X \cap Y = \emptyset} m_1(X) m_2(Y)} \qquad (37.20)$$

where $m_1 \oplus m_2(\emptyset) = 0$.

Although, Dempster–Shafer is a mathematical theory and not an algorithm, we can still write algorithms for the belief function (37.17), the plausibility function (37.19), and the combination rule (37.20).

The belief function (37.17) is implemented in Algorithm 37.7. The input parameters declared in lines 1 and 2 are the hypothesis $H \in 2^{\Theta}$, for which the belief is being computed, and the basic probability

Algorithm 37.7 Dempster–Shafer: Belief

```
▷ Input:
 1: H ∈ 2^Θ; {input hypothesis}
 2: m: 2^Θ → [0, 1]; {associated bpa}

▷ Variables:
 3: belief; {stores the overall belief in H}

 4: begin
 5:     {trivial values}
 6:     if H = ∅ then
 7:         return 0;
 8:     end if
 9:     if H = Θ then
10:         return 1;
11:     end if

12:     {compute the belief in H}
13:     belief ←0;
14:     for each A ∈ 2^Θ do
15:         if A ⊆ H then
16:             belief ←belief + m(A);
17:         end if
18:     end for
19:     return belief;
20: end
```

Algorithm 37.8 Dempster–Shafer: Plausibility

▷ **Input:**
  1: $H \in 2^\Theta$; {*input hypothesis*}
  2: $m : 2^\Theta \rightarrow [0,1]$; {*associated bpa*}

▷ **Variables:**
  3: *doubt*; {*stores the doubt degree*}

  4: **begin**
  5:     {*compute the overall doubt in* H}
  6:     $doubt \leftarrow 0$;
  7:     **for each** $A \in 2^\Theta$ **do**
  8:       **if** $A \cap H = \emptyset$ **then**
  9:         $doubt \leftarrow doubt + m(A)$;
 10:       **end if**
 11:     **end for**
 12:     **return** $1 --- doubt$;
 13: **end**

---

function $m : 2^\Theta \rightarrow [0,1]$. The variable *belief* in line 3 is used to store the belief in H. The belief in $\emptyset$ is always 0 (line 7) while the belief in Θ is always 1 (line 10). Then, for each subset $A \in 2^\Theta$ (line 14), if A is contained in H (line 15), then its bpa $m(A)$ is added to the overall belief in H (line 16).

The plausibility function (37.19) is implemented in Algorithm 37.8. The input parameters declared in lines 1 and 2 are the hypothesis $H \in 2^\Theta$, for which the plausibility is being computed, and the basic probability function $m : 2^\Theta \rightarrow [0,1]$. The variable *doubt* in line 3 is used to store the doubt in H. For each subset $A \in 2^\Theta$ (line 7), if A and H are disjunct (line 8), then we add its bpa $m(A)$ to the overall doubt in H (line 9). The plausibility of H is obtained by subtracting doubt in H from 1 (line 12).

The combination rule (37.20) is implemented in Algorithm 37.9. The input parameters are the hypothesis $H \in \Theta$ (line 1), for which the bpas are being computed, and the basic probability functions $m_1 : 2^\Theta \rightarrow [0,1]$ (line 2) and $m_1 : 2^\Theta \rightarrow [0,1]$ (line 3) are being combined. The variable *comb* in line 4 stores the numerator of (37.20) while variable *empty* (line 5) stores the denominator of (37.20). For each subset $X, Y \in 2^\Theta$ (lines 14 and 15), if $X \cap Y = H$ (line 16), then we update *comb* with the product of $m_1(X) \times m_2(X)$ (line 17). If X and Y are disjunct (line 19), then update *empty* with the product of $m_1(X) \times m_2(X)$ (line 20). Then, the result $m_1 \oplus m_2(H)$ is obtained dividing *comb* per *empty* (line 24).

Similar to the Bayesian inference, it is difficult to provide a quantitative example of Dempster–Shafer inference. Nonetheless, an example is thel denial of service (DoS) detection engine based on Dempster–Shafer proposed by Siaterlis and Maglaris in Reference 52. In this case, Dempster–Shafer's evidential reasoning is used as the mathematical foundation for the development of a DoS detection engine, which combines multiple evidences from simple heuristics and attempts to detect flooding attacks. Dempster–Shafer was chosen because it provides support to express beliefs in some hypotheses, uncertainty, and ignorance in the system and the quantitative measurement of the belief and plausibility in the detection results. This approach is efficient to increase the DoS detection rate and decrease the false alarm rate [52].

37.6.3 Other Representative Inference Methods

Neural networks were originated in the early 1960s [53,54] and they represent an alternative method for inference. They offer structures to provide a supervised learning mechanism that uses a training set to learn patterns from examples. Neural networks are commonly used in conjunction with fuzzy logic. This partnership occurs because fuzzy logic enables the inclusion of qualitative knowledge about a problem into reasoning systems. However, it does not supply adaptation and learning mechanisms. On the other

Algorithm 37.9 Dempster-Shafer: Combination Rule

▷ `Input:`
  1:   $H \in 2^{\hat{}}\Theta$; {*input hypothesis*}
  2:   $m_1 : 2^{\hat{}}\Theta \rightarrow [0, 1]$; {*first bpa*}
  3:   $m_2 : 2^{\hat{}}\Theta \rightarrow [0, 1]$; {*second bpa*}

▷ `Variables:`
  4:   *comb*; {*stores the combination of m_1 and m_2*}
  5:   *empty*; {*stores the empty combination of m_1 and m_2*}

  6: `begin`
  7:       {*trivial values*}
  8:       `if` $H = \emptyset$ `then`
  9:           `return` 0;
 10:       `end if`

 11:   {*combine m_1 and m_2*}
 12:   *comb* $\leftarrow$ 0;
 13:   *empty* $\leftarrow$ 1;
 14:   `for each` $X \in 2^{\hat{}}\Theta$ `do`
 15:       `for each` $Y \in 2^{\hat{}}\Theta$ `do`
 16:           `if` $X \cap Y = H$ `then`
 17:               *comb* $\leftarrow$ *comb* $+ (m_1(X) \times m_2(Y))$;
 18:           `end if`
 19:           `if` $X \cap Y = \emptyset$ `then`
 20:               *empty* $\leftarrow$ *empty* $- (m_1(X) \times m_2(Y))$;
 21:           `end if`
 22:       `end for`
 23:   `end for`
 24:   `return` *comb/empty*;
 25: `end`

---

hand, a key feature of neural networks is the ability to learn from examples of input/output pairs in a supervised fashion. Thus, neural networks are used in learning systems whereas fuzzy logic is used to control its learning rate [55].

37.7 Bibliographic Notes

Although information fusion is essentially applied to sensory data, its applicability extrapolates to other application domains regarding distributed systems for different purposes. For example, methods for reliable abstract sensors have not only been applied to replicated sensors [8] but also to clock synchronization in distributed systems [56].

Regarding estimation methods, Kleine-Ostmann and Bell [57] solve the location problem for cellular network fusing time of arrival (ToA) and time difference of arrival (TDoA) estimates with the least square method and fusing location estimates with a Bayesian estimator. Information fusion is also used by Savvides et al. [58], who use the Kalman filter to refine the initial location estimates computed by the sensor nodes.

Typical applications of feature maps include sensor network management [59,60], position estimation [61,62], robot perception [63,64], and navigation [65,66]. However, there are also applications in computer graphics, such as the simulation of graphical creatures' behavior [67] and detection of volumetric objects collisions [68].

In Reference 69, Chen and Varshney present a Bayesian sampling approach for decision fusion such that Bayesian inference is applied through a hierarchical model to reformulate the distributed detection problem. Other examples of the use of Bayesian inference include map building [38,70] for teams of robots and automatic classification tasks [71–73]. Murphy [74] discusses the use of Dempster–Shafer theory for information fusion (symbol fusion) for autonomous mobile robots exploiting the two particular components: the weight of *conflict* metric and the *enlargement* of the frame of discernment. Boston [75] proposes a fusion algorithm based on Dempster–Shafer theory for signal detection that combines evidence provided by multiple waveform features making explicit the uncertainty in the detection decision.

Acknowledgment

This work has been partially supported by CNPq — Brazilian Research Council under process 55.2111/02-3.

References

[1] U.S. Department of Defence. Data fusion lexicon. Published by Data Fusion Subpanel of the Joint Directors of Laboratories. Technical Panel for C3 (F.E. White, Code 4202, NOSC, San Diego, CA), 1991.

[2] Lawrence A. Klein. *Sensor and Data Fusion Concepts and Applications*, Vol. TT14. SPIE Optical Engineering Press, Bellingham, WA, 1993.

[3] David L. Hall and James Llinas. An introduction to multi-sensor data fusion. In *Proceedings of the IEEE*, Vol., pp. 85: 6–23, 1997.

[4] Lucien Wald. Some terms of reference in data fusion. *IEEE Transactions on Geoscience and Remote Sensing*, 13: 1190–1193, 1999.

[5] B.V. Dasarathy. What, where, why, when, and how? *Information Fusion*, 2: 75–76, 2001 (Editorial).

[6] Konstantinos Kalpakis, Koustuv Dasgupta, and Parag Namjoshi. Efficient algorithms for maximum lifetime data gathering and aggregation in wireless sensor networks. *Computer Networks*, 42: 697–716, 2003.

[7] Robbert van Renesse. The importance of aggregation. In A. Schiper, A. A. Shvartsman, H. Weatherspoon, and B. Y. Zhao, editors, *Future Directions in Distributed Computing: Research and Position Papers*, Lecture Notes in Computer Science, Vol. 2584. Springer-Verlag, Heidelberg, pp. 87–92. January 2003.

[8] Keith Marzullo. Tolerating failures of continuous-valued sensors. *ACM Transactions on Computer Systems*, 8: 284–304, 1990.

[9] P. Chew and K. Marzullo. Masking failures of multidimentional sensors. In *Proceedings of the 10th Symposium on Reliable Distributed Systems*. Pisa, Italy, September/October, IEEE Computer Society Press, Los Alamitos, CA, pp. 32–41, 1991.

[10] D. N. Jayasimha. Fault tolerance in a multisensor environment. In *Proceedings of the 13th Symposium on Reliable Distributed Systems*, Dana Point, CA, IEEE Computer Society Press, Los Alamitos, CA, pp. 2–11, October 1994.

[11] Ulrich Schmid and Klaus Schossmaier. How to reconcile fault-tolerant interval intersection with the lipschitz condition. *Distributed Computing*, 14: 101–111, 2001.

[12] Hairong Qi, S. Sitharama Iyengar, and Krishnendu Chakrabarty. Distributed sensor networks — a review of recent research. *Journal of the Franklin Institute*, 338: 655–668, 2001.

[13] Lakshman Prasad, S. Sitharama Iyengar, R. L. Kashyap, and Rabinder N. Madan. Functional characterization of fault tolerant integration in distributed sensor networks. *IEEE Transactions on Systems, Man and Cybernetics*, 21: 1082–1087, 1991.

[14] S. S. Iyengar and L. Prasad. A general computational framework for distributed sensing and fault-tolerant sensor integration. *IEEE Transactions on Systems, Man and Cybernetics*, 25: 643–650, 1995.

[15] L. Prasad, S. S. Iyengar, R. Rao, and R. L. Kashyap. Fault-tolerant integration of abstract sensor signal estimates using multiresolution decomposition. In *Proceedings of the IEEE International Conference on Systems, Man, and Cybernetics*, Le Touquet, France, pp. 171–176, October 1993.

[16] L. Prasad, S. S. Iyengar, R. L. Rao, and R. L. Kashyap. Fault-tolerant sensor integration using multiresolution decomposition. *Physical Review E*, 49: 3452–3461, 1994.

[17] Boris R. Bracio, Wolfgan Horn, and Dietmar P. F. Möller. Sensor fusion in biomedical systems. In *Proceedings of the 19th Annual International Conference of the IEEE/EMBS*, pp. 1387–1390, Chicago, IL, October/November 1997.

[18] Thomas Kailath, Ali H. Sayed, and Babak Hassibi. *Linear Estimation*. Prentice Hall, New York, 2000.

[19] Steven W. Smith. *The Scientist and Engineer's Guide to Digital Signal Processing*, 2nd ed, California Technical Publishing, San Diego, CA, 1999.

[20] R. E. Kalman. A new approach to linear filtering and prediction problems. *Transaction of the ASME Journal of Basic Engineering*, 82: 35–45, 1960.

[21] R. C. Luo and M. G. Kay. Data fusion and sensor integration: State-of-the-art 1990s. In Mongi A. Abidi and Rafael C. Gonzalez, editors, *Data Fusion in Robotics and Machine Intelligence*. Academic Press, Inc., San Diego, CA, Chapter 3, pp. 7–135, 1992.

[22] O. L. R. Jacobs. *Introduction to Control Theory*, 2nd ed. Oxford University Press, Oxford, 1993.

[23] Boris R. Gary Bishop and Greg Welch. An introduction to the kalman filter. In *SIGGRAPH 2001 Conference Proceedings*, Los Angeles, CA, August 2001, Course 8.

[24] Peter S. Maybeck. *Stochastic Models, Estimation, and Control*, Vol. 1: *Mathematics in Science and Engineering*. Academic Press, New York, 1979.

[25] A. H. Jazwinski. *Stochastic Processes and Filtering Theory*. Academic Press, New York, 1970.

[26] J. B. Gao and C. J. Harris. Some remarks on kalman filters for the multisensor fusion. *Information Fusion*, 3: 191–201, 2002.

[27] H. F. Durrant-Whyte, B. S. Y. Rao, and H. Hu. Toward a fully decentralized architecture for multi-sensor data fusion. In *Proceedings of IEEE International Conference on Robotics and Automation*, Cincinnati, OH, pp. 1331–1336, May 1990.

[28] B. S. Rao and H. F. Durrant-Whyte. Fully decentralised algorithm for multisensor kalman filtering. *IEE Proceedings D Control Theory and Applications*, 138: 413–420, 1991.

[29] Timothy M. Berg and Hugh F. Durrant-Whyte. General decentralised kalman filters. In *Proceedings of the American Control Conference*, Vol. 2, Baltimore, MD, pp. 2273–2274, June 1994.

[30] S. Grime and H. F. Durrant-Whyte. Data fusion in decentralized sensor networks. *Control Engineering Practice*, 2: 849–863, 1994.

[31] C. Brown, H. Durrant-Whyte, J. Leonard, B. Rao, and B. Steer. Distributed data fusion using kalman filtering: A robotics application. In Mongi A. Abidi and Rafael C. Gonzalez, editors, *Data Fusion in Robotics and Machine Intelligence*. Academic Press, Inc., San Diego, CA, Chapter 7, pp. 267–309, 1992.

[32] W. R. Gilks, S. Richardson, and D. J. Spie, editors. *Markov Chain Monte Carlo in Practice*. Chapman & Hall, New York, 1996.

[33] Arnaud Doucet, Nando de Freitas, and Neil Gordon, editors. *Sequential Monte Carlo Methods in Practice*. Springer-Verlag, New York, 2001.

[34] Dan Crisan and Arnaud Doucet. A survey of convergence results on particle filtering methods for practitioners. *IEEE Transactions on Signal Processing*, 50: 736–746, 2002.

[35] Per-Johan Nordlund, Fredrik Gunnarsson, and Fredrik Gustafsson. Particle filters for positioning in wireless networks. In *Proceedings of the European Signal Processing Conference (EURSIPCO'02)*, Toulouse, France, September 2002.

[36] Jerry Zhao, Ramesh Govindan, and Deborah Estrin. Residual energy scans for monitoring wireless sensor networks. In *Proceedings of the IEEE Wireless Communications and Networking Conference (WCNC'02)*, Orlando, FL, March 2002.

[37] H. P. Movarec and A. Elfes. High resolution maps from angle sonar. In *Proceedings of the IEEE International Conference on Robotics and Automation*, St. Louis, MD, pp. 116–121, March 1985.

[38] A. Elfes. A sonar-based mapping and navigation system. In *Proceedings of the IEEE International Conference on Robotics and Automation*, California, 1986.

[39] Alberto Elfes. Sonar-based real-world mapping and navigation. *IEEE Journal of Robotics and Automation*, RA-3: 249–265, 1987.

[40] Alberto Elfes. Using occupancy grids for mobile robot perception and navigation. *IEEE Computer*, 22: 46–57, 1989.

[41] Miguel Ribo and Axel Pinz. A comparison of three uncertainty calculi for building sonar-based occupancy grids. *Robotics and Autonomous Systems*, 35: 201–209, 2001.

[42] Thomas R. Bayes. An essay towards solving a problem in the doctrine of chances. *Philosophical Transactions of the Royal Society*, 53: 370–418, 1763.

[43] A. P. Dempster. A generalization of Bayesian inference. *Journal of the Royal Statistical Society, Series B*, 30: 205–247, 1968.

[44] Glenn Shafer. *A Mathematical Theory of Evidence*. Princeton University Press, Princeton, NJ, 1976.

[45] David Lee Hall. *Mathematical Techniques in Multisensor Data Fusion*. Artech House, Norwood, MA, April 1992.

[46] Mihail L. Sichitiu and Vaidyanathan Ramadurai. Localization of wireless sensor networks with a mobile beacon. Technical Report TR-03/06, Center for Advances Computing and Communications(CACC), Raleigh, NC, July 2003.

[47] G. M. Provan. The validity of Dempster–Shafer belief functions. *International Journal of Approximate Reasoning*, 6: 389–399, 1992.

[48] T. D. Garvey, J. D. Lawrence, and M. A. Fishler. An inference technique for integrating knowledge from disparate sources. In *Proceedings of the 7th International Joint Conference on Artificial Intelligence*, Vancouver, BC, pp. 319–325, 1986.

[49] S. J. Henkind and M. C. Harrison. An analysis of four uncertainty calculi. *IEEE Transactions on Systems, Man and Cybernetics*, 18: 700–714, 1988.

[50] D. M. Buede. Shafer–Dempster and Bayesian reasoning: A response to "Shafer–Dempster reasoning with applications to multisensor target identification systems." *IEEE Transactions on Systems, Man and Cybernetics*, 18: 1009–1011, 1988.

[51] Y. Cheng and R. L. Kashyap. Comparison of Bayesian and Dempster's rules in evidence combination. In G. J. Erickson and C. R. Smith, editors, *Maximum-Entrophy and Bayesian Methods in Science and Engineering*. Kluwer, Dordrecht, The Netherlands, pp. 427–433, 1988.

[52] Christos Siaterlis and Basil Maglaris. Towards multisensor data dusion for DoS detection. In *Proceedings of the 2004 ACM Symposium on Applied Computing (SAC)*, Nicosia, Cyprus, pp. 439–446, 2004.

[53] F. Rosenblatt. Two theorems of statistical separability in the perceptron. In *Mechanization of Thought Processes*, London, UK, pp. 421–456, 1959.

[54] B. Widrow and M. E. Hoff. Adaptive switching circuits. In *IRE Western Electric Show and Convention Record*, Vol. 4, pp. 96–104, 1960.

[55] Piero P. Bonissone. Soft computing: the convergence of emerging reasoning technologies. *Soft Computing*, 1: 6–18, 1997.

[56] Ulrich Schmid. Orthogonal accuracy clock synchronization. *Chicago Journal of Theoretical Computer Science*, 2000: 3–77, 2000.

[57] Thomas Kleine-Ostmann and Amy E. Bell. A data fusion architecture for enhanced position estimation in wireless networks. *IEEE Communications Letters*, 5: 343–345, 2001.

[58] Andreas Savvides, Heemin Park, and Mani B. Srivastava. The bits and flops of the n-hop multilateration primitive for node localization problems. In *Proceedings of the 1st ACM International Workshop on Wireless Sensor Networks and Applications*. Atlanta, GA, ACM Press, New York, pp. 112–121, 2002.

[59] Jerry Zhao, Ramesh Govindan, and Deborah Estrin. Computing aggregates for monitoring wireless sensor networks. In *Proceedings of the 1st IEEE International Workshop on Sensor Network Protocols and Applications (SNPA 2003)*, Anchorage, AK, pp. 139–148, May 2003.

[60] Linnyer Beatrys Ruiz, Jose Marcos Nogueira, and Antonio A. F. Loureiro. MANNA: A management architecture for wireless sensor networks. *IEEE Communications Magazine*, 14: 116–125, 2003.

[61] Clark F. Olson. Probabilistic self-localization for mobile robots. *IEEE Transactions on Robotics and Automation*, 16: 55–66, 2000.

[62] ChindaWongngamnit and Dana Angluin. Robot localization in a grid. *Information Processing Letters*, 77: 261–267, 2001.

[63] Adam Hoover and Bent David Olsen. A real-time occupancy map from multiple video streams. In *Proceedings of the IEEE International Conference Robotics and Automation*, Detroit, MI, USA, pp. 2261–2266, 1999.

[64] Adam Hoover and Bent David Olsen. Sensor network perception for mobile robotics. In *Proceedings of the IEEE International Conference Robotics and Automation*, Vol. 1, San Francisco, CA, pp. 342–347, 2000.

[65] E. A. Puente, L. Moreno, M. A. Salichs, and D. Gachet. Analysis of data fusion methods in certainty grids application to collision danger monitoring. In *Proceedings of the 1991 International Conference on Industrial Electronics, Control and Instrumentation (IECON'91)*, Vol. 2, pp. 1133–1137, October/November 1991.

[66] Daniel Pagac, Eduardo M. Nebot, and Hugh Durrant-Whyte. An evidential approach to map-building for autonomous vehicles. *IEEE Transactions on Robotics and Automation*, 14: 623–629, 1998.

[67] Damian A. Isla and Bruce M. Blumberg. Object persistence for synthetic creatures. In *Proceedings of the 1st International Joint Conference on Autonomous Agents and Multiagent Systems*, pp. 1356–1363, Bologna, Italy, ACM Press, New York, July 2002.

[68] Nikhil Gagvani and Deborah Silver. Shape-based volumetric collision detection. In *Proceedings of the 2000 IEEE Symposium on Volume Visualization*, Salt Lake City, Utah, USA, ACM Press, New York, pp. 57–61, 2000.

[69] Biao Chen and Pramod K. Varshney. A Bayesian sampling approach to decision fusion using hierarchical models. *IEEE Transactions on Signal Processing*, 50: 1809–1818, 2002.

[70] Behzad Moshiri, Mohammad Reza Asharif, and Reza HoseinNezhad. Pseudo information measure: a new concept for extension of Bayesian fusion in robotic map building. *Information Fusion*, 3: 51–68, 2002.

[71] P. Cheeseman and J. Stutz. Bayesian classification (autoclass): theory and results. In U. M. Fayyad, G. Piatetsky-Shapiro, P. Smyth, and R. Uthurusamy, editors, *Advances in Knowledge Discovery and Data Mining*. AAAI Press/MIT Press, Cambridge, MA, 1996.

[72] Robert S. Lynch Jr., and Peter K. Willett. Use of Bayesian data reduction for the fusion of legacy classifiers. *Information Fusion*, 4: 23–34, 2003.

[73] Alexey Tsymbal, Seppo Puuronen, and David W. Patterson. Ensemble feature selection with the simple Bayesian classification. *Information Fusion*, 4: 87–100, 2003.

[74] Robin R. Murphy. Dempster–Shafer theory for sensor fusion in autonomous mobile robots. *IEEE Transactions on Robotics and Automation*, 14: 197–206, 1998.

[75] J. R. Boston. A signal detection system based on Dempster–Shafer theory and comparison to fuzzy detection. *IEEE Transactions on Systems, Man and Cybernetics — Part C: Applications and Reviews*, 30: 45–51, 2000.

38

Node Activity Scheduling Algorithms in Wireless Sensor Networks

38.1 Introduction..**38**-867
38.2 Wireless Sensor Networks **38**-868
38.3 Activity Scheduling Algorithms for Wireless
Sensor Networks **38**-871
  Centralized Algorithms for Sensing Scheduling Problems •
  Distributed Algorithms for Sensing Scheduling Problems •
  Algorithms for Communication Scheduling Problems •
  Integration of Communication Scheduling and Sensing
  Scheduling
38.4 Open Questions and Future Research Direction **38**-880
  Activity Scheduling Schemes Based on the General Sensing
  Model • Activity Scheduling Schemes for 3D Space •
  Activity Scheduling Schemes for Heterogeneous Networks •
  Activity Scheduling Schemes with Mobility Support •
  Integrated Activity Scheduling in Wireless Sensor Networks •
  Impact of Node Density and Network Topology on Coverage
  Capability and Activity Scheduling • Node Deployment
  Strategy to Enhance Coverage Capability
References **38**-881

Di Tian
Nicolas D. Georganas

38.1 Introduction

Recently, the concept of wireless sensor networks has attracted a great deal of research attention. That is due to a wide-range of potential applications that will be enabled by such networks, such as battlefield surveillance, machine failure diagnosis, biological detection, home security, smart spaces, inventory tracking, etc. [1–4]. A wireless sensor network consists of tiny sensing devices, deployed in a region of interest. Each device has processing and wireless communication capabilities, which enable it to gather information about the environment and to generate and deliver report messages to the remote base station (remote user). The base station aggregates and analyzes the report messages received and decides whether there is an unusual or concerned event occurrence in the monitored area.

In wireless sensor networks, the energy source provided for sensors is usually battery power, which has not yet reached the stage for sensors to operate for a long time without recharging. Moreover, since sensors are often intended to work in remote or hostile environments, such as a battlefield or desert, it is undesirable or impossible to recharge or replace the battery power of all the sensors. However, a long system lifetime is always expected by many monitoring applications. The system lifetime, which is measured by the time till all sensors have been drained out of their battery power or the network no longer provides an acceptable event detection ratio, directly affects network usefulness. Therefore, conserving energy resource and prolonging system lifetime is an important issue in design of large-scale wireless sensor networks.

In wireless sensor networks, all sensors cooperate for a common sensing task. This implies that not all sensors are required to continuously monitor environment during the system life, if their subset is sufficient to assure the quality of monitoring service. Furthermore, due to low cost of sensors, it is well-accepted that wireless sensor networks can be deployed with a very high node density (even up to 20 nodes/m^3 [5]). In such high-density networks with energy efficient design requirement, it is desirable to have the nodes sensing the environment alternatively. If all the sensors were vigilant at the same time, a single event would trigger many sensors to generate report messages. When these sensors intend to send the highly redundant report messages simultaneously, the network will suffer excessive energy consumption and packet collision. Recent research advances have shown that energy saving can be achieved by scheduling a subset of nodes for monitoring service at any instant in a highly dense wireless sensor network. In the literature, a few algorithms have been proposed for dynamically configuring sensing status to minimize the number of active nodes while maintaining the original system sensing coverage and the quality of monitoring service without degradation.

Monitoring is just one duty assumed by sensors. Due to limited communication capability, sensors have to act as routers to help forward data packets to remote base stations. Similarly, redundancy also exists in the communication domain in highly dense sensor networks, where adaptively turning off radio of some nodes has no influence on network connectivity and communication performance. In such cases, excessive amount of energy would be consumed unnecessarily if all nodes continuously listen to the media. In the literature, dynamic management of node activity in the communication domain has been broadly studied.

In high-density wireless sensor networks, energy waste always exists if only one kind of redundancy is identified and removed from the network. Minimizing active nodes has to maintain both network connectivity and sensing coverage. Without sufficient coverage, the network cannot guarantee the quality of monitoring service. Without network connectivity, active nodes may not be able to send data back to remote base stations. In the literature, there are a few papers addressing how to integrate the scheduling of sensing and communication activity, how to maintain coverage and connectivity in a single activity scheduling, and how to maximize the number of off-duty nodes in both domains.

The organization of this chapter is as follows: Section 38.2 gives a brief introduction of wireless sensor networks; Section 38.3 is the survey of recent research advances on activity scheduling algorithms in wireless sensor networks; Section 38.4 lists the open questions and future research direction on activity scheduling in wireless sensor networks.

38.2 Wireless Sensor Networks

Recent advances of micro electro-mechanical systems (MEMS) technology, low power and highly integrated digital electronics have fostered the emergence of inexpensive and low-power tiny devices, called micro-sensors. These devices are capable of measuring a wide variety of ambient conditions (such as temperature, pressure, humidity, noise, and lighting) and detecting the abnormal or concerned event occurrence in the vicinity. Although the sensing capability of an individual micro-sensor may be restricted

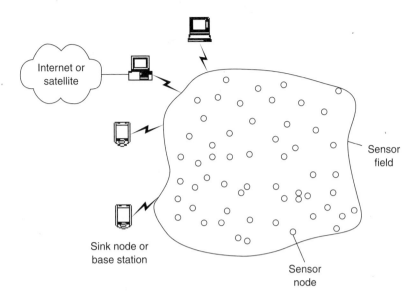

FIGURE 38.1   Typical scenario of wireless sensor network architecture [4].

in range and accuracy, aggregating the detections from multiple micro-sensors will generate a reliable "view" of the environment. Such coordination is enabled by the wireless communication capability possessed by micro-sensors.

A wireless sensor network is envisioned as consisting of hundreds or thousands of tiny sensor devices randomly deployed in a terrain of interest to cooperatively perform a big sensing task. Compared with a traditional system composed of large, expensive, and powerful macro-sensors wired together, wireless sensor networks cost less money in node deployment and network organization. Moreover, small size, light weight, and low cost of micro-sensors allow us to "carelessly" scatter these devices in the monitored area (such as throwing them out from an aircraft), and then making them self-organized in an ad hoc manner. Although a single micro-sensor is limited in sensing capability, the sheer number of nodes in wireless sensor networks guarantees enough redundancy in generated data and therefore the sensing accuracy would not be degraded. Furthermore, wireless sensor networks are more fault-tolerant than traditional macro-sensor networks because failure of a single micro-sensor would not render the failure of the entire system.

Figure 38.1 illustrates a typical scenario of wireless sensor networks [4]. Sensor nodes are scattered in a terrain of interest, called sensor field. Each of them has the capability of collecting data about environment and reporting to sink node (or base station). Exterior users may access data in wireless sensor networks from the sink nodes via Internet or satellite. There can be multiple sink nodes in wireless sensor networks, but the number of them is much smaller than that of sensor nodes. Some researchers suggest deploying another kind of nodes, called gateways, into wireless sensor networks. These gateway nodes are more powerful than sensor nodes and therefore can perform local network management functions such as data aggregation, node organization, status assignment, etc. Sink nodes and gateway nodes can be mobile, but most of recent research results on wireless sensor networks assume that sensor nodes are stationary. That is because the cost and size of current sensor devices would be increased significantly when a mobility component is added.

Communications among sensor nodes, gateways, and sink nodes are all in a wireless manner. Since the communication range of a sensor node is short, due to its constrained energy supply, multiple-hop relaying is necessary for long-distance transmission. The communication protocol stack shared by all nodes, as illustrated in Figure 38.2 is proposed in Reference 4. The protocol stack consists of the application layer, transport layer, network layer, data link layer, physical layer, as well as three planes. The functions of

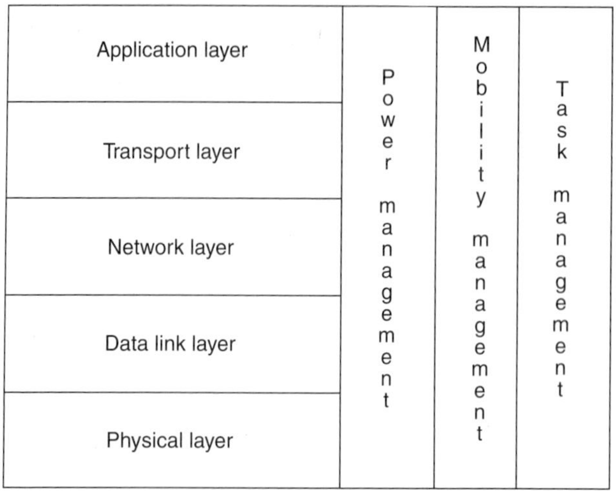

Application layer	P o w e r m a n a g e m e n t	M o b i l i t y m a n a g e m e n t	T a s k m a n a g e m e n t
Transport layer			
Network layer			
Data link layer			
Physical layer			

FIGURE 38.2   Sensor network communication protocol stack [4].

all the layers conform to the specification of the ISO Seven Layer model. Note that although a layered architecture is shown in the figure, it does not mean that different layers are implemented in a completely independent way and there is no interaction among layers. Actually, due to the application-related nature of wireless sensor networks, a cross-layer design (sharing information between layers) can achieve a better performance than an independent-layer design [23]. For instance, SPIN [25] uses data naming and negotiation to suppress the transmission of redundant data in order to reduce unnecessary communication overhead. LEACH [23] and directed diffusion [24] aggregate multiple data packets upon routing to reduce the amount of data transmitted over long distance. In this stack, power, mobility, and task management planes are responsible for providing power, mobility, and task-related information to all the layers, so that these layers can optimize their operations. For instance, the mobility management plane is responsible for detecting movement of neighboring nodes. It can tell the network layer any change of neighbor information, so that the network layer can update routing tables correspondingly. The power management plane is responsible for monitoring power usage of sensors. If the remaining power of a sensor node is not sufficient, its future activity should be carefully controlled. The task management plane is used to determine when a sensor node participates into a sensing task. The sensor nodes that are not joining the sensing task will not generate data reports. The task management plane can also be used to control nodes' communication activity. Not all sensors are necessarily in active status to relay packets for others all the time. Some nodes may enter into sleep mode to save energy and periodically wake up for sending and receiving packets.

The concept of wireless sensor networks has promised many civil and military applications. In the battle-field, timely, detailed, and valuable information about critical terrains and opposing forces can be quickly obtained by throwing micro-sensors from aircrafts. Destruction of some micro-sensors by hostile actions would not influence a military operation as much as that of traditional powerful macro-sensors [23]. In environmental aspects, sensors can be used for detecting fire and floods, tracking the movements of birds, insects, or other animals. Other applications of wireless sensor networks include drug administration in hospitals, tele-monitoring of human physiological data, smart environment, temperature control of an office building, interactive museum, monitoring car thefts, etc. [23].

From random deployment and self-organization's perspective, wireless sensor networks are similar to wireless ad hoc networks. However, there are still a lot of differences between them:

- The number of sensor nodes in a wireless sensor network can be much higher than the nodes in an ad hoc network. It is quite likely that sensors are deployed in a high density, even up to 20 nodes/m^3 [4]. Algorithm and protocol design for wireless sensor networks must scale up to large population. High density in node distribution leads to high correlation in the data generated by multiple sensor nodes for the same event. Such redundancy can be exploited to optimize power and bandwidth utilization.

- Wireless ad hoc networks are based on one-to-one communication. While in wireless sensor networks, data flows are mainly queries diffusing from a single or a few sink nodes to all the sensor nodes, or data collected in the reverse direction. Therefore the design of network layer for wireless sensor networks focuses on efficient broadcasting or reverse-broadcasting protocols.
- Sensor nodes are limited in processing capability, storage space, and transmission range. They are more severely constrained in energy supply than other wireless devices (such as PDA, cellular phones, etc.). Energy efficiency is the primary challenge of wireless sensor networks.
- Due to low-cost manufacturing process, careless deployment and hostile running environment, sensors are prone to failure. Fault-tolerance becomes one of the important requirements in design of wireless sensor networks.

Due to such differences, many protocols have proposed for wireless sensor networks, they are not applicable or usable in wireless sensor networks without any modification. Alternative approaches need to be explored in this new context.

38.3 Activity Scheduling Algorithms for Wireless Sensor Networks

Prolonging system lifetime is one of the main challenges in wireless sensor networks. Configuring node status and selecting a subset of sensor nodes for monitoring service is one of potential solutions to the energy restriction problem in wireless sensor networks. The primary design goal of activity scheduling algorithms is to minimize the number of active nodes and at the same time maintain the quality of monitoring service.

Any sensing activity scheduling algorithm is built on a physical sensing model. In literature, there are two kinds of sensing models that are broadly adapted.

The first model captures the common property of sensing devices with varying complexity and features. It assumes that sensing ability diminishes as distance increases. In the model summarized in Reference 22, the sensing intensity of a node v at a point P is given as:

$$S(v, P) = \begin{cases} \dfrac{a}{d(v, P)^{\beta}}, & A \leq d(v, P) < B \\ 0, & \text{otherwise} \end{cases}$$

where α is the energy emitted by the event occurring at point P, $d(v, P)$ is the Euclidean distance between sensor v and point P, A and B are range-related parameters of sensor v and β is the decaying factor, ranging from 2.0 to 5.0. Whether an event is detectable is determined by an integral function of N sensors with the nearest distance to point P: $I_P = \sum_{i=1}^{N} S(s_i, P)$. When $I_P \geq \theta$, the event at point P can be observed.

The second model is a special case when $N = 1$. We call it Boolean or 0/1 sensing model. This assumes that each sensor can do a 360° observation. Sensing area $S(v)$ of node v is the maximal circular area centered at sensor v that can be well-observed by sensor v. Note that although the maximal observation region of a sensor may be irregular, an irregular region can always contain a circle centered at that sensor. The radius of $S(v)$ is called sensor v's sensing range $r(v)$.

The second model is simple, but the first one is general. Currently, almost all the research results on activity scheduling and sensing coverage are based on the second one.

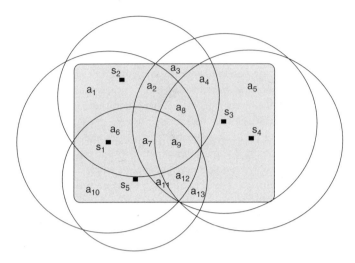

FIGURE 38.3   Sensors and covered areas [17].

38.3.1 Centralized Algorithms for Sensing Scheduling Problems

In Reference 17, Meguerdichian and Potkonjak presented how to use Integer Linear Programming (ILP) techniques to solve optimal 0/1 node-scheduling problems in wireless sensor networks. They considered the shaded area in Figure 38.3 as a monitoring region covered by a sensor network. The sensors are labeled as s_1 to s_5. Based on the 0/1 sensing model, the monitoring region is divided into several areas, denoted as a_j. Different area is covered by a different subset of the sensors.

   s1: {a1, a2, a6, a7, a8, a9, a10, a11, a12, a13}
   s2: {a1, a2, a3, a4, a6, a7, a8, a9}
   s3: {a2, a3, a4, a5, a7, a8, a9, a11, a12, a13}
   s4: {a4, a5, a8, a9, a12, a13}
   s5: {a6, a7, a9, a10, a11, a12, a13}

In brief, ILP is a method used to determine the values of some decision variables in order to maximize or minimize some linear functions of these variables under certain constraints. Based on the above model, the first optimal problem solved by using ILP is minimum 0/1 cover problem. That is to find the smallest set of sensors, which covers the same regions as the complete set of sensors. The formal description of the minimum 0/1 cover problem is as follows [17]:

Given: A sensor set $S = \{s_1, s_2, \ldots, s_{|S|}\}$; a sensor field A, where $\{a_1, a_2, \ldots, a_{|A|}\}$ is a partitioning of A; an area coverage matrix $C_{|A|\times|S|}$, where each element $C_{i,j}$ is 1 if the sensor s_j covers area a_i, and 0 otherwise.

ILP problem:

   Minimize $1_{1\times|S|} \cdot X_{|S|\times 1}$

   Constraint $C_{|A|\times|S|} \cdot X_{|S|\times 1} \geq 1_{|A|\times 1}$, here $X_{|S|\times 1}$ is a vector where X_i is 1 if sensor s_i is in the
   active set, and 0 otherwise.

Another optimal problem is called balanced operation scheduling problem. The formal description of it is as follows [17]:

   Minimize k

   Constraint $C_{|A|\times|S|} \cdot X_{|S|\times T} \geq 1_{|A|\times T}$

   $1_{1\times|S|} \cdot X_{|S|\times T} \leq 1_{1\times T} \cdot k$

where k is the maximum number of time slices that any sensor is active. where $X_{i,j}$ is 1 if sensor s_i is active in the time slice j and 0 otherwise.

The goal of these constraints is to minimize the number of time slices each node is active as well as to minimize the number of active sensor nodes in each time slice.

Besides the 0/1 sensing model, Meguerdichian and Potkonjak [17] gave a ILP formulation for the general sensing model. They defined average sensor field intensity $I(s, a)$ for a region a due to sensor s as:

$$I(s, a) = \left(\frac{1}{|a|}\right) \sum_{i=1}^{|a|} S(s, p_i)$$

Given a sensor field intensity coverage matrix $E_{|A| \times |S|}$, where $E_{i,j}$ represents the average sensor field intensity $I(s_j, a_i)$ in regions a_i due to sensor s_j, the ILP problem is to minimize $1_{1 \times |S|} \cdot X_{|S| \times 1}$ under the constraint of $E_{|A| \times |S|} \cdot X_{|S| \times 1} \geq 1_{|A| \times 1} \cdot E_{\min}$. $E_{\min}$ is a configurable parameter.

In Reference 28, the same approach is proposed to solve other coverage and deployment problems:

i. Given a grid-based region and types of a sensor set, determine where to deploy these sensors to minimize the overall sensor cost under a certain coverage constraint (e.g., each grid is covered by at least k nodes).

ii. How to deploy sensor nodes to ensure that each grid can be covered by a unique node subset, so that target position can be determined by identifying the set of observers.

Besides ILP, other techniques were proposed for centralized solutions to optimal or suboptimal sensing scheduling problems. In Reference 18 Slijepcevic and Potkonjak studied the problem about how to find k-sets of nodes so that each node set can cover the whole monitored area, A, and the cardinality, K, of the node sets is maximal. Since it is NP-complete problem [18], Slijepcevic and Potkonjak designed a heuristic approach to solve it. The algorithm is based on the 0/1 sensing model. Similar to Reference 17, it starts with dividing the monitored area into a set of multiple fields with each field covered by a unique subset of sensor nodes. Then a critical element is identified. The critical element is defined as a member of area set A covered by the minimal number of nodes in node set C. The critical element determines an upper bound on K. For each node subset from C that covers the current critical element, an objective function is calculated. The node subset with the highest result is selected as a cover. All the members of this cover are removed from C. The coverage of A is updated correspondingly. Then a new critical element is determined from A for the new iteration. The design of the objective function is to select a node subset that will cover the largest number of sparsely covered elements. Detailed description of the heuristic approach and objective function can be referred to Reference 18.

In Reference 27, Ko and Rubenstein explained how the graph coloring approach [26] can be used in the context of sensing scheduling in wireless sensor networks. Their algorithm splits time into cycles of K time slots. Each time slot is represented by a kind of color. The algorithm aims to divide nodes into several subsets so that each subset owns one color (representing a time slot) and that the sensing coverages among all the time slots are balanced. The algorithm does it by initially assigning each node a color at random, then gradually increasing the distance of the nodes owning the same color until the algorithm converges. In the algorithm, each time a node changes its color, it propagates its change to all the other nodes in the network in order for each node to update its knowledge of the nearest node owning the same or any other different color. Such update leads to an expensive communication overhead. Furthermore, the sensing coverage is not intended for complete preservation.

In Reference 29, Vieira et al. proposed to use the well-known Voronoi diagram for sensing-scheduling. Voronoi diagram is the partitioning of a plane with n points into n convex polygons such that each polygon contains exactly one point and every point in a given polygon is closer to its central point than to any other. In the algorithm, nodes are first sorted by the size of the Voronoi cell, and then the node with the smallest Voronoi cell is "removed" from the network. Each time a node is "removed," the Voronoi diagram is updated and a new node is identified until a certain threshold reaches.

In Reference 30, Liu et al. used physical constraints (e.g., object shape and mobility) to determine the sensor collaboration region. They developed a centralized approach to determine sensors that are the most sensitive to a detected object. Their approach leverages a dual-space transformation to map the edges of the object (simplified to a half-plane shadow) to a single point in the dual space, and to map the locations of nodes to a set of lines. After such transformation, seeking of the relevant nodes is transformed to tracking of the cell that contains a point (corresponding to the edge of the detected object) in the dual space. Those nodes, which are contributed to the construction of the edges of that cell in the dual space, are identified as the relevant ones. Although not mentioned specifically, their approach is applicable to sensing scheduling, that is, only "frontier" nodes are assigned an active sensing status. However, before the algorithm performs, the shape of detected object should be known and its mobility should be predictable.

38.3.2 Distributed Algorithms for Sensing Scheduling Problems

The algorithms presented in above are all centralized, therefore, suffering the nontrivial overhead of collecting and synchronizing global information. In the literature, some alternative, suboptimal, distributed algorithms were proposed to solve the same or similar sensing scheduling problems.

In Reference 8, a subset of sensor nodes is selected initially and operates in active mode until they run out of their energy resources or are destroyed. Other nodes fall asleep and wake up occasionally to probe their local neighborhood. If there is no working node within their probing range, the sleeping nodes take over the monitoring services, otherwise they reenter the sleeping status. In the algorithm, the relationship between the sensing redundancy (i.e., each point is monitored by at least K working nodes) and the probing range r is derived as $r = \sqrt{(c\pi/KS)}r_m$, where, S is the monitored region; r_m is the sensing range of the 0/1 sensing model; c is a constant whose rough estimation is 0.77. The above equation shows that the required probing range is inversely proportional to the square root of K. For any desired sensing redundancy, there is a corresponding value of the probing range. Therefore, the node density can be controlled through their algorithm. The algorithm is fully distributed and each node does not need to maintain neighbor list and state information. However, the energy consumption is not balanced among the sensor nodes. A subset of the sensors quickly runs out of their energy resources, then another subset of the sensors does the same, and so on. Furthermore, the algorithm does not support the heterogeneous network where sensors have different sensing ranges.

In Reference 6, Tian and Georganas proposed a distributed node-scheduling algorithm that aims to remove as many as possible redundant sensor nodes, while preserving the original sensing coverage. To achieve this goal, they used an intuitive approach that is to calculate the sensing area of each node and then compare it with those of its neighbors. If the neighboring nodes can cover the current node's sensing area, the current node goes to sleep. The off-duty eligibility rule is based on the 0/1 sensing model. As illustrated in Figure 38.4, the part of node As sensing area that can also be covered by node B is the shaded region, denoted as $S_{A \cap B}$. That turning off node A will not reduce the overall sensing coverage, can be expressed

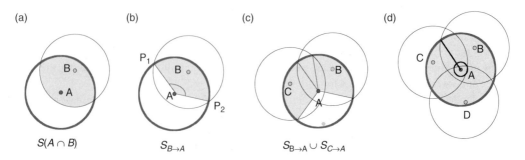

FIGURE 38.4   Off-duty eligibility rule of Reference 6.

as $Y_{I\in N(A)}\, S_{A\cap I} \supseteq S_A$, where, $N(A)$ is the set of sensors whose sensing area is overlapped with that of node A. They observed that the crescent-shaped intersection area $S_{A\cap B}$ always contains a sector $S_{B\to A}$ (the shaded area in Figure 38.4[c]), if the distance between nodes A and B is not more than the sensing range r. Since the area of a sector can be accurately represented by its central angle and the union of two sectors can be transformed to merging two central angles, as illustrated in Figure 38.4(d), to facilitate quick calculation, they changed the evaluation rule of the off-duty eligibility as:

$$\text{if}\quad \underset{I\in N(A)\text{ and }d(A,B)\le r}{Y}\, S_{I\to A} \supseteq S_A, \qquad \text{then}\quad \underset{I\in N(A)}{Y}\, S_{A\cap I} \supseteq S_A$$

Their algorithm theoretically guarantees the preservation of the sensing coverage. Also, the algorithm determines each node's off-duty eligibility through simple arithmetic calculations. However, they conservatively underestimated the overlapping area of sensing areas. Therefore, the off-duty node set is larger than the optimal one.

In Reference 7, Yan et al. proposed another node-scheduling algorithm in the sensing domain. Their algorithm splits time into rounds T. Initially, each sensor node selects a random reference point ref within $[0, T)$, and then exchanges its reference point and its location information with neighbors. Scheduling starts from each sensor node dividing its sensing area into small grids. For each grid point, the node finds all the neighbors that can cover this point and sorts their reference points in an ascending order. Suppose the reference point sequence has n items with the index from 1 to n. node As reference point is located at the ith position, denoted as ref(i). Then node A sets its start work time T_{front} as:

$$[T + \text{ref}(1) - \text{ref}(n)]/2, \qquad \text{the reference point of node A is in the first position}$$

$$[\text{ref}(i) - \text{ref}(i-1)]/2; \qquad \text{otherwise}$$

Node A sets the end work time T_{end} as:

$$[T + \text{ref}(1) - \text{ref}(n)]/2, \qquad \text{the reference point of node } A \text{ is in the last position}$$

$$[\text{ref}(i) - \text{ref}(i-1)]/2; \qquad \text{otherwise}$$

An example is illustrated in Figure 38.5.

For each node and each of its covered grid points, a pair of T_{front} and T_{end} is calculated by the same procedure. The integrated T_{front} and T_{end} of each node is the largest one among all T_{front} and T_{end} for all grid points it can cover, respectively. After the integrated T_{front} and T_{end} are determined, a node is

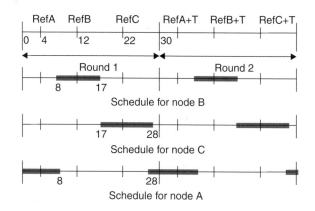

FIGURE 38.5 Scheduling for a single grid point [7].

scheduled to wake up at $T \times i + \text{ref} - T_{\text{front}}$ and to fall asleep at $T \times i + \text{ref} + T_{\text{end}}$. Here, i is a positive integer. The algorithm can support differentiated sensing by simply extending the work time proportionally to the required redundancy along the time line to adjacent working spaces, that is, a node wakes up at $T \times i + \text{ref} - T_{\text{front}} \times \alpha$ and falls asleep at $T \times i + \text{ref} + T_{\text{end}} \times \alpha$. The algorithm in Reference 7 is based on the 0/1 sensing model as well. It has a relatively high computation load. Because sorting has a lowest computation complexity as $O(n \log n)$, the computation complexity of each node is at least $O(N\_g * N\_n * \log(N\_n))$. Here, N_n is the average size of neighbor sets, and N_g is the granularity of grid points (the number of grid points in a sensing area). Furthermore, to reduce energy consumption in exchanging reference points, the algorithm uses a fixed reference point strategy that is performed only once at the first beginning of system operation. So, the algorithm is designed specially for the static networks.

In References 14 and 15 two different distributed protocols were proposed to identify redundant nodes in the sensing domain. In Reference 14, Wang et al. proved that: for 0/1 sensing model, a convex area is completely covered by a set of nodes if (1) all the intersection points between any sensors are covered and (2) the intersection points between the boundary and any sensor are covered. Therefore, a sensor is eligible for off-duty if all the intersection points inside its sensing area are at least k-degree covered by the active nodes. If there is no intersection point inside the sensing area of a sensor, that sensor keeps active when there is less than k active sensors located at the same location. The sensing scheduling protocol in Reference 15 is more complicated than that in Reference 14, but based on the same theorem.

In Reference 31, Tian and Georganas studied several distributed node-scheduling algorithms that do not rely on any location information. Based on the 0/1 sensing model, the relationship between the distance of two neighboring nodes and the size of their overlapped sensing area is given as:

$$|S_{A \cap B}| = \begin{cases} \pi r^2 \cdot \left[\left(2 \arcos \dfrac{d}{2r} - \dfrac{d}{r} \sqrt{1 - \dfrac{d^2}{4r^2}} \right) \cdot \dfrac{1}{\pi} \right], & \text{for all } d \leq 2r \\[4mm] 0, & \text{otherwise} \end{cases}$$

where, r is sensing range and d is the distance of nodes A and B. According to such a relationship, if nodes are aware of the relative distance to their neighbors and the maximal percentage loss of sensing area that can be tolerated after a node is "removed" from the network, is selected, the maximal distance D between two nodes can be determined to guarantee this threshold $\text{AUA}_{\max}$. During scheduling, a node examines if the distance $d_{\min}$ to its nearest active neighbor is less than or equal to the value of D. If affirmative, the node sets its status as off-duty. Besides the distance to the nearest neighbor, the neighbor number can also be used to evaluate the off-duty eligibility of each node. The relationship between the neighbor number and the average size of the overlapped area can be obtained through simulation and numerical analysis methods. Based on the relationship, the minimal neighbor number K can be determined to guarantee the threshold $\text{AUA}_{\max}$. During the scheduling, each node examines if its active neighbor number is equal to or greater than the value of K. If affirmative, the node sets its status as off-duty. In Reference 31, Tian and Georganas approximates the off-duty node percentage ρ as a function of D (or K). Such approximation leads to probability-based sensing scheduling algorithm, which does not rely on any neighbor information.

38.3.3 Algorithms for Communication Scheduling Problems

Besides sensing scheduling, another active research field is to schedule node activity in the communication domain. In this section, the basic ideas of some communication scheduling algorithms will be introduced.

In References 9 and 10, Wu et al. proposed an activity scheduling algorithm based on connected dominating set. A connected dominating set is defined as a connected subset of the vertices of a graph, if every vertex not in the subset is adjacent to at least one vertex in the subset. To reduce system overall energy consumption, only nodes in a dominating set keep active and can relay messages for others. Nondominating set nodes go to a sleep mode and periodically wake up for sending and receiving packets from the associated dominating set node. In Reference 9, Wu and Li proposed to mark a node as in the dominating set if two of its neighbors are not directly connected. To reduce the size of the dominating set, they introduced two rules. The first rule is as follows: if every neighbor of node v is also a neighbor of node u and $id(v) < id(u)$, then node v is not a dominating node. The second rule is: if every neighbor of node v is also a neighbor of node u or a neighbor of node w and $id(v) < min\{id(u), id(w)\}$, then node v is not a dominating node. In Reference 10, the other two rules were proposed to replace the above ones based on node *ID*. The first one is to set the node with higher degree as dominating node whenever there is a tie in order to decrease the size of the dominating set. The formal description of the extended rules is: marking node v as nondominating set node if

---

rule a : 1. $N(v) \subseteq N(u)$ *and* $nd(v) < nd(u)$
         2. $N(v) \subseteq N(u)$ *and* $id(v) < id(u)$ *and* $nd(v) = nd(u)$
rule b : 1. $N(v) \subseteq N(u)YN(w)$ *and* $N(u)! \subseteq N(v)YN(w)$ *and* $N(w)! \subseteq N(u)YN(v)$
         2. $N(v) \subseteq N(u)YN(w)$ *and* $N(u) \subseteq N(v)YN(w)$ *and* $N(w)! \subseteq N(u)YN(v)$ *and*
            {(a) $nd(v) < nd(u)$ *or*
            (b) $nd(v) = nd(u)$ *and* $id(v) < id(u)$}
         3. $N(v) \subseteq N(u)YN(w)$ *and* $N(u) \subseteq N(v)YN(w)$ *and* $N(w) \subseteq N(u)YN(v)$ *and*
            {(a) $nd(v) < nd(u)$ *and* $nd(v) < nd(w)$ *or*
            (b) $nd(v) = nd(u) < nd(w)$ *and* $id(v) < id(u)$
            (c) $nd(v) = nd(u) = nd(w)$ *and* $id(v) < min\{id(u), id(w)\}$

Here, $N(v)$ is the neighbor set of node v. $nd(v)$ is the size of $N(v)$.

The second is to consider energy level owned by competitors in order to balance the energy resources among them. The algorithm always manages to mark those nodes with a higher energy level *el* as dominating set. The formal description of the extended rules based on energy level is the same as the above ones based on node degree, except that *nd* is replaced by *el*.

In the literature, there are other dominating set formation algorithms, although the authors did not explicitly mention the concept of dominating set. In Reference 11, Chen et al. presented a distributed algorithm in which nodes coordinate to make local decisions on whether to stay awake as a coordinator and to participate in the forwarding backbone topology. To preserve network connectivity, a node decides to serve as a coordinator, if it discovers that two of its neighbors cannot communicate with each other directly or through an existing coordinator.

In Reference 12, an algorithm, called geographical adaptive fidelity (GAF), was proposed, which uses geographic information to divide area into fixed square grids. Any node within a square can directly communicate with any node in the adjacent square. Therefore, within each grid, only one node needs to stay awake to forward packets.

In Ascent [32], each node assesses its contribution to the network connectivity based on the measured operating environment (i.e., local node density and packet loss rate), other than on the positional information and local connectivity. If the node is relevant to network connectivity, it serves as a relay station for multi-hop routing.

Besides restricting the number of relay stations to a subset of the nodes in a network graph, other approaches to reduce the overall energy consumption by scheduling communication activity includes letting each node adaptively self-configure its radio status without considering the effect of global connectivity. For instance, in Reference 12, Xu et al. proposed a scheme in which energy is conserved by letting nodes turn off their radio when they are not involved in data communication. They presented how to leverage node density to increase the time when the radio is powered off. Unlike SPAN, GAF, and Ascent, each node independently configures its own communication activity by only using their own application-level information or simple neighbor size.

The third approach is to construct a sparse, connected subgraph from the original unit graph. The unit graph is one in which an edge exists if the distance between its endpoints is at most the maximal transmission range. Such a subgraph is used in routing to optimize network spatial reuse, mitigate MAC-level interference, and reduce energy usage [37].

The sparse subgraph is formed after each node minimizes its transmission power or restricts its directly communicated neighbors, while maintaining the network connectivity.

For instance, the relative neighborhood graph (RNG) is a geometric and graph theoretic concept proposed by Toussaint [38]. Its definition is as follows: edge (u, v) exits if for all $w \neq u, v$: $d(u, v) \leq \max(d(u, w), d(v, w))$. Toussaint proved RNG is a connected and planar graph. Each node has on average about $2.5°$ independent of graph density. The main advantage of RNG is that it can be constructed in a total local fashion. In Reference 37, Li et al. proposed a localized algorithm to construct a local minimal spanning tree (LMST), which contains a minimal spanning tree (MST) as a subset. Suppose there is a weighted graph G. A spanning tree for G is a tree, which is a subgraph of G and contains all the vertices of G. A minimum spanning tree (MST) is one with the least total weight among all the spanning trees. The algorithm in Reference 37 starts from each node building its local MST that consists of itself and its one-hop neighbors. Then an edge (u, v) is kept in LMST if it exists in both node u and node v's local MST. In Reference 39, Ovalle-Martinez et al. proposed to use the longest edge in LMST as a minimal transmission radius shared by all the nodes in network. They designed a scheme to convert LMST to MST by using a loop breakage procedure. The procedure makes upwards messages traversing from leaves in LMST to loops along dangling edges, and then breaks loops by eliminating their longest edges until it ends at a single node.

Rodoplu and Meng [33] and Li and Halpern [34] proposed a distributed algorithm to build a minimal energy topology based on the concept of relay region and enclosure. Suppose node u wishes to transmit to node v. Node v is said to lie in the relay region of another node w, if less power is consumed when node w forwards packets from node u to node v as a relay station than when node u directly transmits packets to node v. The formal definition is $R_{u->w} = \{v \mid P_{u->w->v} < P_{u->v}\}$, where P is the amount of energy consumed in communication. The enclosure of node u, denoted as ε_u, is defined as $\varepsilon_u = I_{k \in N(u)} R^c_{u->k} ID_N$. Here D_N denotes the deployment region for the node set $N \cdot R^c$ is the complement of node set R. $N(u)$ is the original neighbor set of node u. In the constructed sparse subgraph, the neighbor set of node u is redefined as $N(u) = \{v \in N \mid v \in \varepsilon_u, v \neq u\}$. Since such sparse graph induced from enclosures is strongly connected and always contains a minimal power path between any pair of nodes, it is called minimal energy topology. In Reference 35, Wattenhofer et al. came up with another approach to form a minimal energy subgraph by using directional information other than positional information. In the algorithm, each node u sends beacons with a growing power p. If the node finds a new neighboring node v, node u puts node v into its neighbor set $N(u)$. Node u continues increasing its transmission power until every cone with angle α contains at least one of its discovered neighbors. To further decrease the number of directly communicated neighbors, another phase is introduced as: if $v \in N(u)$ and $w \in N(u)$ and $w \in N(v)$ and $p(u, v) + p(v, w) \leq p(u, w)$, then node w is removed from $N(u)$. In Reference 35, Wattenhofer et al. proved that $\alpha \leq 2\pi/3$ is the sufficient condition to make sure that the constructed subgraph is connected. For the smaller angle α, good minimum power routes are guaranteed.

Although the above algorithms are proposed for ad hoc networks, they are adoptable to wireless sensor networks, because highly dense wireless sensor networks have communication redundancy as well.

38.3.4 Integration of Communication Scheduling and Sensing Scheduling

In highly dense wireless sensor networks, it is often a case that two kinds of redundancy exist simultaneously. However, the algorithms introduced in the previous sections only manage to remove one kind of redundancy without addressing the other.

References 14 and 15 were the earliest papers discussing how to integrate activity scheduling in both domains: communication and sensing. Both of them proved that: For 0/1 sensing model, if each point of a convex region is covered by at least one node in a node set, the communication graph consisting of these nodes is connected when $R \geq 2r$. In Reference 14, Wang et al. extended the above conclusion for 1- to k-degree by proving that: if each point of a convex region is covered by at least k nodes in a node set,

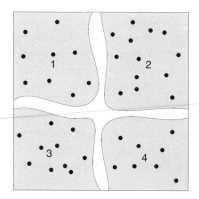

FIGURE 38.6   Illustration of difference between the proof in References 14 and 15 and that in Reference 39.

the communication graph consisting of these nodes is k-degree connected when $R \geq 2r$. Here R and r are the communication range and sensing range shared by all sensor nodes, respectively.

In Reference 39, Tian and Georganas provided the proof for another conclusion: For 0/1 sensing model, if an original network is connected and "removing" some of nodes from the network does not degrade the original sensing coverage, then the subgraph induced by the active nodes are connected when $R \geq 2r$. They also extended the conclusion for 1- to k-degree by proving: $R \geq 2r$ is the sufficient condition to ensure that k-degree coverage preservation implies k-degree connectivity among active nodes if the original network is k-degree connected.

The difference between the conclusion in References 14 and 15 and that in Reference 39 can be better understood through a simple scenario, as illustrated in Figure 38.6. In this scenario, the sensing field can be divided into four subfields 1–4, with each subfield completely covered by a subset of nodes. We call the node subset as subset 1–4 correspondingly. There are two narrow gaps in the middle of the sensing field that cannot be monitored by any node. Furthermore, the communication graph consisting of all the nodes is connected. After node scheduling, redundant nodes are removed from each subset without changing the coverage graph. According to the conclusion in References 14 and 15, we only know that the active nodes within each subset are connected with each other, but cannot determine if the active nodes among different subsets are connected. However, the latter is confirmed affirmative, according to the proof in Reference 39.

Based on the theoretical result, Refernce 14 presented how to integrate the scheduling in the sensing and communication domain in a safe manner. For $R \geq 2r$ cases, if a sensor node is eligible for off-duty in the sensing domain, then the node is assigned an off-duty status in both the sensing and communication domain. For $R < 2r$ cases, if sensors determine their working status in the sensing and communication domain by using independent algorithms separately. Although these algorithms may share the same messages for coordination or status notification, there is no relationship between *sensing_inactive* nodes and *communication_inactive* nodes.

Another approach for $R \geq 2r$ cases can be scheduling the sensing and communication activity in a tightly integrated mode. The scheduling starts from evaluating each node's off-duty eligibility in the sensing domain by using any coverage-preserving sensing scheduling algorithm in the literature. If eligible, the node is assigned inactive status in both the sensing and communication activity. The first phase removes part of the redundancy in the communication domain, while maintaining the connectivity in the subgraph G_{active} induced by the active nodes. To further reduce the communication redundancy, a dominating set formation algorithm in the literature is performed on the constructed subgraph G_{active}. The advantage of this integrated mode is that it always generates the maximal number of *both_inactive* (*sensing inactive and communication inactive*) nodes, which is up bounded by the number of the *sensing inactive* ones.

In Refernce 40, Sanli et al. designed a different two-phase protocol to schedule the sensing and communication activity in an integrated manner. The protocol assumes that there are three kinds of nodes in wireless sensor networks: tiny sensor nodes, three or more powerful reference nodes, and base stations. The reference nodes are equipped with GPS receivers. They can be deployed in a regular pattern, such as in equilateral triangle. The data always are transferred to the reference nodes first, and then forwarded by the reference nodes to base stations. The first phase of the protocol is to build multiple multicast trees rooted at each reference node respectively in a combination way, by considering intermediate nodes' residual energy and the structure of other multicast trees rooted at other reference nodes. Those nodes that are not located at a branch (nonleaf) of any multicast tree are not serving as relay stations. Therefore they can turn off their radio to save energy. After the communication backbone has been established, each node determines whether to turn off its sensing hardware, based on its immediate lower-level neighbor number and its residual energy. The threshold for minimal immediate lower-level neighbor number is derived according to the theoretical analysis on the probability that neighbors can cover the current node's sensing area as a function of the neighbor number. This protocol conserves energy in two aspects: turning off radio of a subset of nodes and turning off sensing units of another subset of nodes. These two kinds of subset may be overlapping.

38.4   Open Questions and Future Research Direction

The following is the list of open questions on activity scheduling in wireless sensor networks that can be future research directions.

38.4.1   Activity Scheduling Schemes Based on the General Sensing Model

Almost all the existing sensing scheduling algorithms are based on the 0/1 sensing model (whether an event can be detected is determined by its distance to sensors). The general sensing model defines that the sensing capability of a single sensor is exponentially proportional to its distance to the event, and that the final sensing result is the integral function of k sensors. The 0/1 sensing model is simple. The general sensing model is realistic. Although the 0/1 sensing model is a special case of the general model when $k = 1$, the existing sensing scheduling algorithms designed for the 0/1 sensing model cannot be adapted to the general sensing model without great changes. New algorithms need to be investigated for the general model.

38.4.2   Activity Scheduling Schemes for 3D Space

All the existing algorithms on coverage, detectability, and activity scheduling were designed and evaluated in two-dimensional space. It is quite difficult to directly transform the existing algorithms to 3D space, although some of them are claimed to have such a feasibility. Activity scheduling in a 3D sensing model is an open question in the literature.

38.4.3   Activity Scheduling Schemes for Heterogeneous Networks

In heterogeneous wireless sensor networks, sensors may be of different capabilities. In other words, different sensors have different maximal communication and sensing ranges. More than one type of sensors is deployed in the same network to cooperate for a common sensing task. Furthermore, there are cases when the sensing and communication capability owned by one sensor is determined by its energy level. As the energy consumption at different nodes are not balanced over time, different sensors may have the different communication and sensing ranges later even if their capabilities are initially identical. Activity scheduling for heterogeneous wireless sensor networks can reduce the overall energy consumption in communication and sensing. It will be much sophisticated than that for homogeneous networks.

38.4.4 Activity Scheduling Schemes with Mobility Support

Mobility support is rarely discussed in activity scheduling algorithms for wireless sensor networks. That may be because most of researchers envision wireless sensor networks consisting of static sensor nodes. However, in some applications such as monitoring marine animals, sensor nodes may move individually or in groups. Scheduling node activity, while maintaining network connectivity and quality of monitoring service, requires nodes to have a capability to predict the future positions of itself and its neighbors, and then to adjust its status to adapt to the dynamic change of network topology.

38.4.5 Integrated Activity Scheduling in Wireless Sensor Networks

Although there have been discussions in the literature on how to integrate activity scheduling in both the sensing and communication domain, further investigation is needed to study what is the optimal integration manner to maximize the energy saving in communication and sensing.

38.4.6 Impact of Node Density and Network Topology on Coverage Capability and Activity Scheduling

The previous work mainly focuses on the algorithm design. More work is needed for theoretical study on node density and network topology on sensing coverage and activity scheduling. For instance, what is the lowest node density to guarantee the desired coverage capability in different network topology, such as grid or random network topology? What is the average redundancy level in a random network topology with a given node density? The answers of such questions can be used to guide node deployment, network planning, and selection of activity scheduling algorithms.

38.4.7 Node Deployment Strategy to Enhance Coverage Capability

In most of wireless sensor network scenarios, sensors are supposed to work in a remote or inhospitable environment. In such an environment, it is impossible or inconvenient or expensive to manually deploy all nodes according to the prior plan. Sensors may be thrown from a plane in such a way that the node distribution is random and unpredictable. Consequently, sensor nodes may not cover the entire monitored region, leaving some holes blinded. In addition, even if the initial deployment can be effectively controlled, more and more sensing holes or blind points will show up due to energy depletion or node destruction. To guarantee quality of monitoring, it is necessary to discover the sensing holes or to alarm users to add new nodes into "empty" areas. Thus, the following problems arise: how to find sensing holes efficiently and quickly; where to put new nodes to enhance the sensing coverage to a desired level while minimizing the number of added nodes. Besides, to eliminate sensing holes or blind points, node deployment schemes have to consider those areas that are weekly covered (i.e., covered by those nodes whose energy resource may be used up before the scheduling for the next node deployment).

References

[1] D. Estrin, R. Govindan, J. Heidemann, and S. Kumar. Next Century Challenges: Scalable Coordination in Sensor Networks. In *Proceedings of the ACM International Conference on Mobile Computing and Networking (MOBICOM'99)*, Washington, DC, 1999; pp. 263–270.

[2] J. Kahn, R. Katz, and K. Pister. Next Century Challenges: Mobile Networking for Smart Dust. In *Proceedings of the ACM International Conference on Mobile Computing and Networking (MOBICOM'99)*, Washington, DC, 1999; pp. 271–278.

[3] A. Cerpa, J. Elson, D. Estrin, L. Girod, M. Hamilton, and J. Zhao. Habitat Monitoring: Application Driver for Wireless Communications Technology. In *Proceedings of the ACM SIGCOMM Workshop on Data Communications in Latin America and the Caribbean*, Costa Rica, 2001; pp. 3–5.

[4] I.F. Akyildiz, W. Su, Y. Sankarasubramaniam, and E. Cayirci. Wireless Sensor Networks: A Survey. *Computer Networks*, March 2002; 393–422.

[5] E. Shih, S. Cho, N. Ickes, R. Min, A. Sinha, A. Wang, and A. Chandrakasan. Physical Layer Driven Protocol and Algorithm Design for Energy-Efficient Wireless Sensor Networks. In *Proceedings of the ACM Special Interest Group on Mobility of Systems Users, Data, and Computing (SIGMOBILE'01)*, Italy, 2001; pp. 272–286.

[6] D. Tian and N.D. Georganas. A Coverage-Preserving Node Scheduling Scheme for Large Wireless Sensor Networks. In *Proceedings of the Processing of ACM Wireless Sensor Network and Application Workshop 2002*, September 2002.

[7] T. Yan, T. He, and J.A. Stankovic. Differentiated Surveillance for Sensor Networks. In *Proceedings of the 1st ACM Conference on Embedded Networked Sensor Systems (SenSys 2003)*, Los Angeles, November 2003.

[8] F. Ye, G. Zhong, S. Lu, and L. Zhang. Energy Efficient Robust Sensing Coverage in Large Sensor Networks. Technical Report, http://www.cs.ucla.edu/~yefan/coverage-tech-report.ps. Page accessed in October 2002.

[9] J. Wu and H. Li. On Calculating Connected Dominating Set for Efficient Routing in Ad Hoc Wireless Networks. In *Proceedings of the 3rd International Workshop on Discrete Algorithms and Methods for Mobile Computing and Communications*, 1999; pp. 7–14.

[10] J. Wu, M. Gao, and I. Stojmenovic. On Calculating Power-aware Connected Dominating Sets for Efficient Routing in Ad Hoc Wireless Networks. In *Proceedings of the International Conference on Parallel Processing*, 2001; pp. 346–356.

[11] B. Chen, K. Jamieson, H. Balakrishnana, and R. Morris. Span: An Energy-Efficient Coordination Algorithm for Topology Maintenance in Ad Hoc Wireless Networks. In *Proceedings of the ACM International Conference on Mobile Computing and Networking (MOBICOM'01)*, Italy, 2001; pp. 85–96.

[12] Y. Xu, J. Heidemann, and D. Estrin. Geography-Informed Energy Conservation for Ad Hoc Routing. In *Proceedings of the ACM International Conference on Mobile Computing and Networking (MOBICOM'01)*, Italy, 2001; pp. 70–84.

[13] A. Cerpa and D. Estrin. ASCENT: Adaptive Self-Configuring Sensor Networks Topologies. In *Proceedings of the 21st International Annual Joint Conference of the IEEE Computer and Communications Societies (INFOCOM 2002)*, New York, NY, USA, June 23–27, 2002.

[14] X. Wang, G. Xing, Y. Zhang, C. Lu, R. Pless, and C.D. Gill. Integrated Coverage and Connectivity Configuration in Wireless Sensor Networks. In *Proceedings of the 1st ACM Conference on Embedded Networked Sensor Systems (SenSys 2003)*, Los Angeles, November 2003.

[15] H. Zhang and J.C. Hou. Maintaining Sensing Coverage and Connectivity in Large Sensor Networks. Technical Report UIUCDCS-R-2003-2351, June 2003.

[16] Y. Xu, J. Heidemann, and D. Estrin. Adaptive Energy-Conserving Routing for Multihop Ad hoc Networks. Technical Report 527, USC/ISI, October 2000. http://www.isi.edu/~johnh/PAPERS/Xu00a.html. Page accessed on October 2002.

[17] S. Meguerdichian and M. Potkonjak. Low Power 0/1 Coverage and Scheduling Techniques in Sensor Networks. UCLA Technical Reports 030001, January 2003.

[18] S. Slijepcevic and M. Potkonjak. Power Efficient Organization of Wireless Sensor Networks. In *Proceedings of the IEEE International Conference on Communications (ICC'01)* Helsinki, Finland, June 2001.

[19] S. Meguerdichian, F. Loushanfar, G. Qu, and M. Potkonjak. Exposure in Wireless Ad-Hoc Sensor Networks. In *Proceedings of the ACM International Conference on Mobile Computing and Networking (MOBICOM'01)*, July 2001; pp. 139–150.

[20] S. Meguerdichian, F. Koushanfar, M. Potkonjak, and M. Srivastava. Coverage Problems in Wireless Ad-Hoc Sensor Networks. In *Proceedings of the IEEE INFOCOM*, vol. 3, April 2001; pp. 1380–1387.

[21] X.Y. Li, P-J. Wan, and O. Frieder. Coverage in Wireless Ad Hoc Sensor Networks. *IEEE Transactions on Computers*, 52: 753–763, 2003.

[22] B. Liu and D. Towsley. A Study of the Coverage of Large-Scale Sensor Networks. *First IEEE Mobile Ad Hoc and Sensor Conference*, October 2004.

[23] W.R. Heizelman, A. Chandrakasan, and H. Balakrishnan. Energy-Efficient Communication Protocol for Wireless Micro Sensor Networks. In *IEEE Proceedings of the Hawaii International Conference on System Sciences*, 2000; pp. 1–10.

[24] C. Intanagonwiwat, R. Govindan, and D. Estrin. Directed Diffusion: A Scalable and Robust Communication Paradigm for Sensor Networks. In *Proceeding of ACM International Conference on Mobile Computing and Networking (MOBICOM'00)*, 2000; pp. 56–57.

[25] J. Kulik, W. Rabiner, and H. Balakrishnana. Adaptive Protocols for Information Dissemination in Wireless Sensor Networks. In *Proceedings of the ACM International Conference on Mobile Computing and Networking (MOBICOM)*, 1999.

[26] B. Ko, K. Ross, and D. Rubenstein. Conserving Energy in Dense Sensor Networks via Distributed Scheduling. http://www-sop.inria.fr/mistral/personnel/K.Avrachenkov/WiOpt/PDFfiles/ko44.pdf

[27] B. Ko and D. Rubenstein. A Greedy Approach to Replicated Content Placement Using Graph Coloring. In *Proceedings of the SPIE ITCom Conference on Scalability and Traffic Control in IP Networks II*, July 2002.

[28] K. Chakrabarty, S.S. Iyengar, H. Qi, and E. Cho. Grid coverage for surveillance and target location in distributed sensor networks. *IEEE Transactions on Computers*, 51: 1448–1453, 2002.

[29] L.F. Vieira, M.A. Vieira, L.B. Ruiz, A.A. Loureiro, and A.O. Fernandes. Applying Voronoi Diagram to Schedule Nodes in Wireless Sensor Networks. Submitted to the Special Issue of MONET on Algorithmic Solutions for Wireless, Mobile, Ad Hoc and Sensor Networks.

[30] J. Liu, P. Cheung, L. Guibas, and F. Zhao. A Dual-Space Approach to Tracking and Sensor Management in Wireless Sensor Networks. In *Proceedings of the First ACM Wireless Sensor Network and Application Workshop*, pp. 131–139, Atlanta, October 2002.

[31] D. Tian and N.D. Georganas. Location and Calculation Free Node-Scheduling Schemes in Large Wireless Sensor Networks. *Ad Hoc Networks Journal*, 2: 65–85, 2004.

[32] A. Cerpa and D. Estrin. ASCENT: Adaptive Self-Configuring Sensor Networks Topologies. In *Proceedings of the 21st International Annual Joint Conference of the IEEE Computer and Communications Societies (INFOCOM 2002)*, New York, USA, June, 23–27, 2002.

[33] V. Rodoplu and T.H. Meng. Minimum Energy Mobile Wireless Networks. *IEEE Journal of Selected Areas in Communications*, 17: 1333–1344, 1999.

[34] L. Li and J.Y. Halpern. Minimum-Energy Mobile Wireless Networks Revisited. In *Proceedings of the IEEE International Conference on Communications (ICC'01)*, Helsinki, Finland, June 2001.

[35] R. Wattenhofer, L. Li, P. Bahl, and Y.-M. Wang. Distributed Topology Control for Power Efficient Operation in Multihop Wireless Ad Hoc Networks. In *Proceedings of the 20th Annual Joint Conference of the IEEE Computer and Communications Societies (INFOCOM)*, April 2001.

[36] F.J. Ovalle-Martinez, I. Stojmenovic, F. Garcia-Nocetti, and J. Solano-Gonzalez, Finding Minimum Transmission Radii and Constructing Minimal Spanning Trees in Ad Hoc and Sensor Networks. In *Proceedings of the 3rd Workshop on Efficient and Experimental Algorithms*, Angra dos Reis, Rio de Janeiro, Brazil, May 25–28, 2004, to appear in LNCS.

[37] N. Li, J.C. Hou, and L. Sha. Design and Analysis of an MST-Based Topology Control Algorithms. In *Proceedings of the INFOCOM 2003*, 2003.

[38] Toussaint, The Relative Neighborhood Graph of a Finite Planar Set, *Pattern Recognition* (1980), no. 4, 261–268.

[39] D. Tian and N.D. Georganas. Connectivity Maintenance on Coverage Preservation in Wireless Sensor Networks. In *Proceedings of the IEEE Canadian Conference on Electrical and Computer Engineering (CCECE 2004)*, Niagara Falls, Canada, May 2–5, 2004.

[40] Hidayet Ozgur Sanli, Hasan Çam, and Xiuzhen Cheng, EQoS: An Energy Efficient QoS Protocol for Wireless Sensor Networks, In *Proceedings of the 2004 Western Simulation MultiConference (WMC'04)*, January 18–21, San Diego, CA, USA.

39

Distributed Collaborative Computation in Wireless Sensor Systems

39.1 Introduction...**39**-885

39.2 Various Approaches.....................................**39**-886
Point Solutions • Application-Specific Network Design •
Model-Based Algorithm Design

39.3 Modeling Wireless Sensor Systems **39**-888

39.4 Models of Computation for Sensor Systems...........**39**-889
Localized Model • Wireless Sensor Network (WSN) Model
• Clustered HeterOgeneouS MOdel for Sensor Networks

39.5 Illustrative Application: Topographic Querying.......**39**-892
Preliminaries • Problem to Be Solved • Algorithm Design
and Analysis • Summarizing the Results

39.6 Concluding Remarks **39**-899

Acknowledgments...**39**-900

References **39**-900

Mitali Singh
Viktor K. Prasanna

39.1 Introduction

Wireless sensor systems consist of hundreds or thousands of small-sized, battery operated, embedded devices (sensor nodes) connected through wireless links. Such systems are being deployed for a host of applications that involve monitoring and responding to physical environments. Examples of such applications include target tracking, ecological micro-sensing, building surveillance, health monitoring, and fault diagnosis systems [4,8,14,23]. While much ongoing research has focused on addressing the hardware design and networking issues in these systems, efficient realization of these systems also involves solving many nontrivial computation problems. Designing algorithms for these systems is significantly challenging owing to the large number of sensor nodes involved, severe power, computation and memory constraints, limited communication bandwidth, and frequent node and link failures in the system.

Consider a sensor system deployed over a two-dimensional terrain consisting of sensor nodes equipped with temperature sensors. The end user desires to extract a variety of topographic information about the temperature characteristics of the monitored terrain. He is interested in enumeration or description of geographically contiguous regions in the terrain that contain sensor nodes with readings in a specific range. Moreover, he wants to periodically compute the area or estimate the boundary of the hottest region in the terrain. The end user may also want to gather similar information about node properties such as residual battery power for resource management, dynamic retasking, and preventive maintenance of these systems. The above problems require information exchange between a large set of nodes in the system and cannot be solved using localized computation at individual nodes.

Centralized solutions have been devised for solving several computation problems in sensor systems, where the sensor system merely acts as a *data collection network* (as opposed to a *distributed computing system*). The sensor nodes collect data from the environment, perform local signal processing to filter out erroneous samples, and wirelessly route the data to a centralized node (base station). All processing takes place at the centralized node. No in-network computation is performed. This approach is relatively easy to implement, as it requires little functionality within the system, but has many drawbacks. In state-of-the-art sensor systems, the energy dissipation and latency involved in communicating a bit of data between sensor nodes is several orders of magnitude higher than that of executing a computation instruction [16] at the local processor. For problems that require input data sets from a large number of sensor nodes, the centralized approach results in high latency, large energy dissipation, and communication bottlenecks in the system. The system is required to have a more powerful node that has sufficient computation and memory resources to gather and process data from all the nodes in the system in a centralized manner. Moreover, the processing node forms a single point of failure.

The above drawbacks can be avoided by designing algorithms that perform distributed data processing and information storage to minimize communication in the system. There are several challenges involved in designing distributed algorithms for these systems. The algorithms must scale well to the network size. This implies that as the number of collaborating nodes increases, the algorithm is capable of distributing computation tasks in the system such that the computation and storage requirements do not surpass the resource availability at the individual nodes. The energy depletion in the system is uniform so that the life longevity of the system is not undermined by early die-outs of a few overloaded nodes. The algorithms must be robust to node failures and to unreliable communication in the system.

There have been many different approaches used for designing distributed algorithms for sensor systems; some of these are discussed in the following section. Section 39.3 discusses some key issues involved in modeling of wireless sensor systems. Section 39.4 presents the key attributes of some of the existing models of computation for these systems. To illustrate the pros and cons of the various design techniques, Section 39.5 presents various algorithms for extracting topographic information in sensor systems. Section 39.6 concludes the chapter.

39.2 Various Approaches

39.2.1 Point Solutions

Several signal processing algorithms involve collaboration only between small numbers of spatially nearby nodes. Efficient implementations of beamforming algorithms onto a spatial cluster of three to seven sensor nodes have been discussed in References 7 and 22. Since the number of collaborating nodes is small, task allocation between the sensor nodes is hand optimized. Architectural models of the sensor hardware are used to estimate costs for computation and broadcasts. The algorithms are validated through experiments. This approach is analogous to manual optimization of assembly code for performance optimization. While suitable for small-size clusters, this approach is not scalable for designing algorithms involving hundreds or thousands of sensor nodes.

39.2.2 Application-Specific Network Design

This approach exploits application-specific customization of the network protocols for performing collaborative computation in the network. The routing layer not only facilitates sending of data to from data sources to sink, but also performs data aggregation at the intermediate sensor nodes. For example, in Reference 26, the authors discuss implementation of a routing protocol that implements computation primitives such as summing and finding max or min. This approach can be compared with the application specific integrated circuit (ASIC) design methodology in traditional systems. The resulting system is highly optimized for a specific application, but must be redesigned from scratch if the hardware is changed or the network is required to support an additional application over time. It requires the developer to understand both the application semantics as well as the intricacies of the underlying network and hardware for designing such systems.

39.2.3 Model-Based Algorithm Design

As sensor systems evolve, realization of these systems will involve solving more and more challenging computation problems. As opposed to a single application, future sensor systems may support multiple applications simultaneously. Application-specific design techniques will not scale well to such systems. Ease-of-design, design reusability, and rapid design space exploration will be important concerns along with efficiency in algorithm design for advanced sensor systems. The above challenges motivate use of a model-based approach in designing these systems, which exploit high-level modeling to decouple algorithm design from system implementation. By hiding low-level details, it simplifies design and facilitates rapid, high-level analysis of distributed algorithms. Figure 39.1 illustrates this approach and is described as follows.

Algorithm design begins with identification of a *realistic* model of computation that provides a system level abstraction of the underlying system to the algorithm designer, while hiding away the unnecessary implementation details. Some of the system properties that are abstracted by the model include node topology and density, medium of communication, and the energy and time costs for performing various operations in the system. The model also identifies the variable system parameters or knobs that can be tuned by the designer to enhance the performance of the algorithms. For example, if the sensor radios support variable transmission range, it is abstracted as a system knob. The designer can evaluate various settings of this knob to reduce energy dissipation and interference in the system [3]. Acceptable

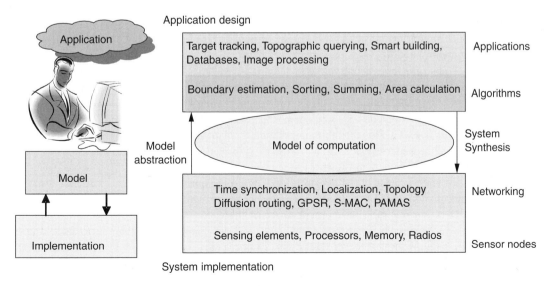

FIGURE 39.1 Model-based algorithm design.

ranges of values are defined for the system knobs to ensure realistic evaluation of the algorithms. The model forms the basis for design and performance analysis of distributed algorithms for these systems. Using the model, the designer is able to formally verify whether an algorithm of the desired functionality can be realized on the sensor system. The model is exploited to formulate and evaluate various algorithm designs. A coarse (theoretical) performance analysis is performed using the cost functions defined by the model for identifying the most suitable algorithms. The selected algorithms are then subjected to low-level simulations and prototyping for analysis that is more accurate and hardware-specific optimizations.

The model forms the basis for formal design and analysis of algorithms. It facilitates algorithm design and performance evaluation at a conceptual level without requiring real implementation. The design is thus portable across several sensor platforms, and is stable with respect to small changes in the hardware or the networking protocols. Performance evaluation is performed using analytical techniques as opposed to time-consuming simulations. Thus, in a relatively short time, a larger design space can be explored.

39.3   Modeling Wireless Sensor Systems

As algorithm designers, we must contemplate as to what is a good model of computation for sensor systems. Can existing computation models be used for abstracting these systems? What features of the sensor system must the model abstract? How to model the networking functionality provided by the lower level protocols? What are the costs of executing various operations in the system? Note that the many low-level details must be ignored to facilitate simplicity.

Even though a large number of computation models have been defined by the theoretical community for design of parallel and distributed systems [13,20], we observe, that they must be redefined in the context of wireless sensor systems. For example, parallel computation research is mainly concerned with exploiting concurrency. In order to address the needs of distributed computing in wireless ad hoc sensor systems, one has to address how key goals, such as power minimization, low latency, scalability, and robustness are affected by the algorithms used. Moreover, most theoretical models assume a well-defined network topology. The topology in a sensor system depends on a large number of factors such as the distribution and density of the sensor nodes, the terrain on which the network is deployed, and the probability of node and link failure in the system. However, some commonalities are also present between conventional distributed systems and sensor systems such as the high relative cost of communication to computation, issues concerning load balancing, and so forth.

Since a model is merely an abstraction, it can be defined at various levels of details. The portability, complexity, and computation power of a model of computation depends on the features it abstracts. A simpler model abstracts fewer details of the underlying system, and consequently, is relevant to a larger set of systems. However, since it is defined at a higher level of abstraction it may be computationally less powerful. For example, consider an algorithm designed using an anonymous system model. The algorithm can be implemented on a sensor system where sensor nodes have no IDs and also on a system where sensor nodes have IDs. However, the computation of the model is limited to symmetric functions, and algorithms such as leader-election cannot be designed using this model.

A less-detailed model may not only restrict the complexity of the algorithms that can be solved using it, but can also affect the efficiency of the algorithms. For example, consider a model that does not abstract the variable transmission range of the sensor nodes. Algorithms designed using this model cannot exploit techniques such as dynamic range assignment to sensor nodes for energy and interference reduction. A delicate balance is required between model complexity and its capabilities such that the model is simple to use and yet capable of implementing a desired set of operations. The definition of a model is largely influenced by the metrics a designer wishes to optimize. Until now, time and memory have been the key optimization metrics for designing an algorithm. Sensor systems, on the other hand, are also desired to be energy efficient, robust, and scalable. Some of the key attributes of sensor systems that must be abstracted by a model are as follows:

1. *Node density and distribution:* Network connectivity and density are important considerations while designing algorithms for sensor systems. Strong connectivity is essential to solve computation problems such as finding the global maximum value. While dense networks are better connected than sparse networks, signal interference is a major concern. The node distribution in a sensor system is often defined by the geometry of the physical phenomenon. Uniformly deployed networks have larger bandwidth and fault tolerance as opposed to systems with nonuniform node deployments.

2. *Network composition:* Sensor nodes constituting a system may be homogeneous or heterogeneous. Node heterogeneity may be in terms of different communication and computation capabilities, or different sensing modalities. Sensor nodes may be mobile or stationary. These must be defined as part of the computation model.

3. *Global information:* Algorithm design (as well as what can be realized in the system) can benefit significantly from globally shared information in the network such as a global coordinate system, a global clock or support for unique identification of the sensor nodes. Such information must be captured by the model of computation.

4. *Wireless technology:* Sensor nodes communicate among themselves wirelessly. Various types of radios have been proposed for communication in these systems such as radio frequency (RF), directional RF, laser, and optical radios. The model must abstract wireless features of sensor nodes such as the radio transmission range and angle, the channel access mechanism, and the contention resolution strategy.

5. *Time and energy costs:* The model must define the time and energy costs for the various operations performed by the sensor nodes such as sensing, local computation and data transmission and reception. If the sensor node supports power management, the model must define each power state as well as the overheads and mechanisms for transition from one state to another.

6. *Robustness:* Robustness must be defined in terms of the high-level operations supported by the model as opposed to the probability of individual node or link failure in the network. While an individual link or node failure is highly probable, lower level protocols may provide robust implementation of the high-level operations. For example, consider a communication primitive used for transfer of information from a data source to a data sink in the network. A robust routing protocol ensures successful execution of the operation, provided a path exists between the two nodes. The failure probability of this operation must be calculated in terms of the failure of all possible paths between the two nodes as opposed to a single link or node failure.

Experience from sensor system design tells us that each of the above listed features could vary from one system to another. Based on his intuition and knowledge, a computer scientist may choose to omit certain features to simplify the model. Alternatively, additional features may be included to enhance the accuracy of the model. Consequently, a large number of models can be defined for sensor systems, which is neither surprising nor dismissive of this approach. It merely reflects upon the diverse nature of sensor systems that cannot be captured by a single model of computation.

39.4 Models of Computation for Sensor Systems

In recent years, several models of computation have been defined and used for designing algorithms for sensor systems. We give a brief overview of some of the existing models.

39.4.1 Localized Model

This model abstracts the network as a collection of loosely coupled nodes. Each sensor node performs computation only based on its local states and the data received from other nodes in its vicinity. No global states or synchronization is assumed in the network. Since the operations at any node in the network are independent of other nodes, algorithms designed using this model are extremely robust and scalable. These properties make this model attractive for designing algorithms for sensor systems with unreliable communication links and frequent node failures. In such systems, the overheads for maintaining and exchanging global information are exorbitant. Several algorithms have been designed for sensor systems using this model [6,10,11]. The limitations of this model are in terms of the types of computation problems that can be solved. Problems involving global coordination or consensus cannot be solved using this model. Lack of global states makes it challenging to detect the termination of the algorithms. It is difficult to analyze the global behavior and performance of the system using this model.

39.4.2 Wireless Sensor Network (WSN) Model

The Wireless Sensor Network (WSN) model [5] abstracts the entire network as a single distributed system. The key assumptions of this model are described as follows:

1. *Network topology:* The system consists of n homogeneous sensor nodes uniformly distributed over a two-dimensional terrain. The terrain is divided into equi-sized cells, each containing a sensor node. The cell size is chosen such that the sensor nodes in the (north, south, east, west) adjacent cells are able to communicate with each other. The sensor nodes are thus organized in a mesh-like topology in the network (see Figure 39.2).
2. *Node characteristics:* The sensor nodes are uniquely identifiable by their location in the grid. Each sensor node is equipped with a local processor, a small-sized memory, a RF radio with fixed transmission range r, and a clock. The clocks of the sensor nodes are globally time synchronized.
3. *Wireless Communication:* The sensors communicate over a shared wireless channel. The area within distance r of a sensor node is called the *intensity zone* of the sensor node and the area between distance r and $2r$ is called the *fading zone*. A collision is said to occur if two sensor nodes that are within the transmission range of each other transmit concurrently. Collisions result in loss of information. A sensor node is guaranteed to receive a transmission from a sensor node in its intensity zone provided no collision occurs. It does not receive any transmission from sensor nodes outside the fading zone.
4. *Communication costs:* The model assumes that a sensory node can transmit or receive one unit of data in a single time slot. The overall communication time is measured as the total number of time slots required for execution of the algorithms. Energy dissipation of a sensor node is measured in terms of the number of time slots in which a sensor node sends or receives data.

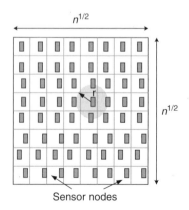

FIGURE 39.2   WSN model.

The WSN model greatly simplifies the algorithm design for sensor systems by making assumption about a regular mesh-like topology, global time synchronization, and localization in the system. The model has been used for design and analysis of algorithms for solving nontrivial computation problems such as sorting in sensor networks [5]. It allows a designer to define system-wide communication schedules that minimize latency and interference in the system. The model has large similarity to several well-known parallel and distributed models (such as the main circulating circuit (MCC) model). Thus, algorithms design for these models can be easily adapted for execution in the WSN model. The main drawback of the model is that it is applicable to a very small class of sensor systems. Global time synchronization is not achievable in large-sized sensor systems without incurring large costs (e.g., providing each node with a GPS). Moreover, it may be difficult to maintain a regular grid topology in the system in the presence of frequent node or link failures.

39.4.3 Clustered HeterOgeneouS MOdel for Sensor Networks

COSMOS abstracts the class of dense, uniformly deployed, hierarchical sensor systems consisting of heteregeneous sensor nodes communicating over the locally shared wireless channels. Unlike the WSN model, COSMOS does not assume global time synchronization or globally unique IDs for all sensor nodes. It abstracts spatial clustering, which is a key feature of several sensor systems. It also models power management schemes supported by the state-of-the-art sensor hardware platforms that can be exploited for reducing energy dissipation in the system. The key assumptions of this model are discussed as follows:

1. *Hierarchical network architecture*: Two types of nodes are present in the network — a large number of sensor nodes with limited computation, storage, and communication capability, and fewer clusterheads that are more powerful than the sensor nodes in terms of computation, communication, and memory resources. The sensor nodes are primarily responsible for data collection in the network, while the clusterheads are used for routing, power management, and implementation of sophisticated computation and communication primitives.

2. *Network topology*: The sensor nodes are uniformly distributed over a two-dimensional terrain. The clusterheads organize themselves into a grid array as illustrated in Figure 39.3. The sensor nodes in each cluster are locally time synchronized with the clusterhead. There is no global time synchronization among the clusterheads. The clusterheads have globally unique IDs, whereas the sensor nodes are uniquely identifiable only within a cluster.

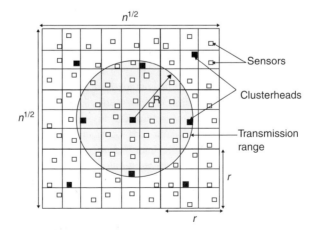

FIGURE 39.3 COSMOS.

3. *Communication*: Nodes communicate over one or more wireless channels using omnidirectional antennas. Within a cluster, communication is single-hop, and all nodes are time synchronized with the clusterhead. Between clusterheads, communication is multihop and asynchronous. Collision occurs if a recipient is within the receiving range of two or more concurrent transmissions over the same channel.

4. *Energy and time costs*: For a sensor node, transmission, reception, or local computation of one unit of data consumes one unit of energy and requires one time step. A clusterhead with range r consumes r^2 units of energy for transmission and one unit for reception, and one unit for computation per unit of data.

5. *Power management*: A sensor node can be either in the *active* state or *asleep* state. Computation and communication can be performed only in the active state. Negligible energy is dissipated in the asleep state. Internal timers or an external paging channel can be used to wake up asleep sensor nodes. Energy dissipation in maintaining the internal timer is negligible. For the paging channel, energy dissipation is proportional to the total number of messages transmitted.

6. *Complexity Analysis*: As is customary [12,13], only for the sake of analysis, the model assumes that all sensor nodes operate at the same clock frequency and there exists a logical clock in the system [12]. The time complexity of an algorithm is defined as the total number of *logical* time steps required for executing the algorithm. The overall energy dissipation in the system is the sum of the computation, communication, and paging energy.

Similar to the WSN model, COSMOS abstracts the network topology of the network as a grid. However, unlike WSN, it assumes clustering in the network and the grid topology is only maintained by the clusterheads. In case a clusterhead fails, an alternate node in the cluster is elected to be the clusterhead, without affecting the grid topology. Thus, the model is robust to a significant number of node failures in the network. References 17 and 19 discuss design of several distributed algorithms using COSMOS.

The following section describes topographic querying, which is one of the most fundamental applications of wireless sensor systems. For the purpose of illustrating the various approaches towards algorithm design in sensor systems, the following section describes and compares three algorithms for identifying and labeling topographic regions in the network.

39.5   Illustrative Application: Topographic Querying

Topographic querying [15,18] is the process of extracting data from a sensor system for understanding the (graphic) delineation of features ((meta)data) of interest in a particular terrain. The end user might be interested in visualizing gradients of sensor readings across the entire field; that is, in constructing and periodically updating a topographic map of the terrain showing the exact boundaries of different feature regions and the average (or min or max) value of readings in each region. Other useful queries, which are not of much scope could include enumeration and description of feature regions with a specific (range of) sensor reading. Application areas in which such topographic querying is particularly useful include contaminant monitoring, water temperature estimation of river beds, oceanography studies related to temperature, direction, and velocity of flow of ocean currents, and so forth [21,23–25].

One approach to supporting topographic queries is query-initiated explicit self-organization, followed by data aggregation and exfiltration [11]. Query responses can be routed to the querying node using flooding or preexisting data gradients [9]. This approach is attractive because it is simple, robust, and requires minimal infrastructure. Unfortunately, the energy dissipation depends on the size, shape, and number of regions. Moreover, since each query route is established independently in a greedy manner, it may lead to congestion in some parts of the network. An alternate approach involves periodic identification and labeling of homogeneous regions (components) in the network. Once this information is gathered and stored in the network, many other queries can be rapidly answered. Processing and responding to queries could be, in most cases, decoupled from the actual data-gathering and boundary-estimation process, which can occur independently and at a low frequency (depending on the application). The number of participating nodes and the communication required for responding to a query can be minimized. In this chapter, we discuss various algorithms for solving the problem of identifying and labeling homogeneous regions in the system.

39.5.1 Preliminaries

A *feature* is a user-defined predicate on sensor data. Consider a sensor system deployed for measuring the temperature of the environment. An example of a feature for this system is: *"Is temperature greater than forty?"*. A sensor node is said to satisfy a feature, if the result of applying the corresponding predicate on the sensor data is true. The sensor data used for determining whether a given feature is satisfied or not is called the *feature value* of the node, and a sensor node that satisfies the feature is called a *feature node*. All sensor nodes with temperature readings greater than 40 are the feature nodes for the above example feature. The temperature readings stored at the sensor nodes are the feature values. We assume that the total number of features defined in the network is a constant. Each feature partitions the network into a set of feature nodes and nonfeature nodes. A sensor node can locally determine whether or not it satisfies a given feature.

A *feature region* refers to a geographically contiguous area in the network consisting of sensor nodes that satisfy the same feature. Given a feature f, the feature region is defined as follows. Consider a partition of the network into a grid consisting of cells of size $\delta \times \delta$ as illustrated in Figure 39.4. Two sensor nodes are said to be *g*-connected (geographically contiguous) if they belong to the same or neighboring (north, south, west, east) cells. Let $G_f = (V, E_f)$ denote the *feature graph* for f. The vertices $v \in V$ represent the sensor nodes, and an edge $e(u, v) \in E_f$ exists between two vertices u and v iff they are *g*-connected and they both satisfy feature f. Each connected component in G_f represents a feature region. Here, δ defines the resolution factor for region identification. Our assumption about dense deployment ensures that $\delta \leq r$, implying that nodes in neighboring cells are also wireless neighbors.

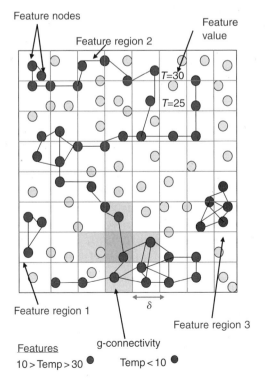

FIGURE 39.4 Feature regions.

The *diameter* of a feature region is defined as the number of feature nodes on the longest feature chain in the region. A feature chain represents the shortest sequence of g-connected feature nodes belonging to a region that connect any two feature nodes in the same region.

39.5.2　Problem to Be Solved

Consider a network of n sensor nodes uniformly deployed over a two-dimensional terrain, where the sensor nodes store features of interest to the end users. Construct the topographic map of the network based on the features stored at the sensor nodes in a time and energy-efficient manner. This involves the following:

1. Given a list of (user-specified) features, identify and label the feature regions in the network.
2. Assignment of region labels to the sensor nodes in a globally consistent manner, which implies the following. All sensor nodes are informed of the termination of the algorithm, and the label of the regions to which they belong. All the sensor nodes belonging to a feature region have the same region label and the sensor nodes in distinct regions have different labels.

39.5.3　Algorithm Design and Analysis

In this section, we describe two distributed algorithms that have been used for solving the above problem. The centralized algorithm is used as the baseline for comparing the performance of the above two algorithms. The *Forest-of-trees* algorithm is a localized algorithm, whereas the *recursive* algorithm is a global algorithm designed using COSMOS.

While the algorithms do not make any assumptions about the total number of feature regions; for the sake of simplicity, the analysis is performed only for a single feature region. We evaluate the performance of the algorithms as a function of the following parameters: n, the number of sensor nodes in the network, s, the number of nodes in a single-hop cluster (see Section 39.4.3), f, the number of feature nodes in the region, and d, the diameter of the region. We focus only on analyzing the communication costs for the above algorithms. The computation costs are assumed as negligible in comparison to communication costs (see problem exercises at the end). For short-range radio communication, the transmission costs are equal to reception costs. We assume that a sensor node requires T time steps and dissipates E units of energy for transmission or reception of one unit of data over the wireless channel:

1. Centralized algorithm:　This approach involves routing of information (such as feature and ID) from all feature nodes to a centralized node (or base station). All processing takes place at the centralized node. Once, the feature regions have been identified and labeled, the labels are transmitted by the centralized node to all feature nodes in the network. Each of the feature nodes transmits one unit of data to the centralized node. Consider the scenario when feature nodes are at the boundary of the network. Here, each data unit must traverse at least $(\sqrt{n}/2\sqrt{s})$ hops to the centralized node. Total number of transmissions in the network is $(f \cdot \sqrt{n}/2\sqrt{s})$. If we assume perfect scheduling (no delay incurred in the network due to congestion), at least $(\sqrt{n}/2\sqrt{s}) \cdot T$ time steps are required for communication of data from the feature nodes to the centralized node. A sensor node can receive only one data unit at a time. At least $f \cdot T$ time steps are required by the centralized node to receive f units of data. Thus, overall time complexity of the algorithm is $O((f + (\sqrt{n}/\sqrt{s})) \cdot T)$. An efficient paging scheme would switch off redundant sensor nodes to ensure that each transmission is received by exactly one sensor node. In such a scenario, the overall energy dissipation in the network is twice the transmission energy, which is $O((f \cdot (\sqrt{n}/\sqrt{s})) \cdot E)$.

2. Forest-of-trees algorithm:　This approach has been discussed in Reference 11. Only feature nodes participate in the algorithm. Every node maintains the variables m_{id} and m_h that represent the ID of the lowest ID sensor node in the network and the minimum number of hops to reach this node. At the start of the algorithm, each sensor node initializes m_{id} to its own label and m_h to infinity. All feature nodes broadcast their IDs with hop initialized to one. When a feature node receives a broadcast, it compares the received value to the stored m_{id}. If the value is smaller, the node updates the variable m_{id}, increments and stores the hop count, and broadcasts a message containing the m_{id} and the hop count incremented by one. The node sets as parent the node from which it received the broadcast. The above step is also repeated if the m_{id} received is equal to the stored value but the hop count is smaller. The system stabilizes when all feature nodes have the minimum value of m_{id} and the shortest path to the feature node with ID $= m_{id}$. This results in a spanning tree with the minimum ID node acting as the root. The ID of the root can be used to label the feature region. All feature nodes transmit the feature information up the tree to the root node. The root node collects data from all the feature nodes. The topographic meta data and boundary information about the feature region are maintained at the root node.

We analyze the energy and time complexity of the algorithm. Consider a feature region that consists of a single sequence of feature nodes (chain) of size l. The sequence is monotonically increasing in terms of the node IDs. For $1 \le k \le l$, let f_k denote the sensor node in the kth position in the sequence. In the worst scenario (when the channel access scheme gives priority to higher ID node), f_k receives $k - 1$ labels and transmits k labels. $O(l)$ time steps are required to route the label of the min ID node to all feature nodes, and l^2 transmissions take place. In the best scenario f_k node receives and transmits $(k - 1)/\sqrt{s}$ labels. This reduces the time to $O(l/\sqrt{s} \cdot T)$ time steps and energy to $O(l \cdot l/\sqrt{s} \cdot E) = O(l^2/\sqrt{s} \cdot E)$ energy units. Next, we consider a region with diameter d and f feature nodes such that all feature nodes lie on monotonically increasing chains of length d. As discussed earlier, above routing of the min ID label on a chain requires $O(d/\sqrt{s} \cdot T)$ time steps and $O(d/\sqrt{s})$ transmissions. There are f/d chains in the network, and therefore the total number of data transmissions is $O(f \cdot d/\sqrt{s})$. Each transmission is received by $O(s)$ neighboring nodes, resulting in energy dissipation of $O(f \cdot d \cdot \sqrt{s} \cdot E)$ energy units. Thus, the overall time complexity of this algorithm is $O(d/\sqrt{s} \cdot T)$ and energy complexity is $O(f \cdot d \cdot \sqrt{s} \cdot E)$ energy units.

A major drawback of this algorithm is that sensor nodes cannot detect the termination of the labeling phase (stabilization of paths to the root) in a consistent manner. A sensor node on a shorter route to the root might stabilize much before the most distant sensor node. An additional termination detection phase must be included in the algorithm after the labeling phase. One heuristic could involve introducing a wait state after the labeling phase. Once a sensor node detects local stabilization it waits for time $O(d)$ to ensure global stabilization in the system. Since the value of d is not known and $d = n$ in the worst case, termination detection will incur an additional overhead of $O(n \cdot T)$ time steps.

3. Recursive algorithm: The problem of identification and labeling of regions in the network can be abstracted as the widely studied component labeling problem in classical image processing [1,2]. The algorithm is described in Reference 15, and it proceeds in the following steps:

Step I: Intracluster identification and labeling of feature regions. Every feature node in the network that is not a clusterhead, transmits its location and reading to its clusterhead. After the transmission, all nodes that are not clusterheads switch themselves off for a prespecified duration depending upon the duty cycle of sampling. Only the clusterheads are active at this point.

Based on the information sent to it by feature nodes within its cluster, each clusterhead constructs a local *feature graph*. Vertices of this graph correspond to feature nodes, and an edge exists between two vertices if (i) the corresponding feature nodes are g-connected and (ii) the sensor readings denote the same feature. The clusterhead runs a depth-first-search algorithm to identify connected components in its local feature graph. Each connected component represents a feature region within the cluster. Boundary estimation and labeling of regions that lie entirely within the cluster can be accomplished at the clusterhead. For regions that cross cluster boundaries, communication with other clusterheads is required to identify their extent.

Each local region is represented as a single vertex and assigned a unique label $v = \langle I, i \rangle$, where I is the ID of the clusterhead, and i is a locally unique region ID. For each vertex v corresponding to a region with feature x, the clusterhead maintains l_v, a list of the feature nodes that lie on the cluster boundary. Each boundary crossing is represented as an outgoing edge denoted by $\langle v, x, l_v, d \rangle$. Here $d \in \{N, S, E, W\}$ identifies the adjacent cluster with which the boundary is shared.

At the end of Step I, only clusterheads are in the active state and all other nodes are switched off. Each clusterhead has a (possibly null) feature graph that represents the feature regions within its cluster.

Step II: Network-wide distributed graph formation. Each clusterhead inspects its feature graph and identifies the regions whose feature nodes lie on one or more boundaries of the cluster. The boundary information is transmitted to the corresponding neighbor. This, effectively results in an exchange of information between neighboring clusterheads that share a feature region.

For each outgoing edge $\langle v, x, l_v, d \rangle$, the clusterhead identifies $v' = \langle I', i' \rangle$ the corresponding label in the neighboring cluster. If there exists more than one outgoing edge to the same vertex in the neighboring cluster from one or more vertices of the local feature graph, these edges and vertices are reduced to a single vertex since they belong to the same feature region.

At the end of Step II, a distributed graph is constructed in the network. Vertices in the graph that share an edge belong to the same region. Each connected component of the distributed graph represents a region in the network. At this stage, vertices belonging to the same region are not guaranteed to have the same label.

Step III: Recursive graph reduction. Recursive reduction of the distributed graph is performed using a quad-tree approach that can be informally described as follows. In the first step of recursion, the network is conceptually divided into blocks of 2×2 clusterheads each. Within each block, one clusterhead is the leader and the other three clusterheads send information about their edges to the leader, which now computes the connected components within the subgraph, which correspond to the feature regions in its area of oversight (which corresponds to four clusters). Each connected component is reduced to a single vertex, and the minimum of all the vertex labels in the connected component is assigned to this new vertex. The set of outgoing edges of the connected component is used to define the edges of the new vertex. If this edge set is empty, it means the vertex represents a feature region. Now, the block leaders organize themselves into 2×2 blocks and one of them is elected as a leader for the next level of recursion, and the process is repeated until all components have been identified and labeled. In each stage, *information from exactly four clusterheads is merged, which leads to the quad-tree structure.* More formally, the above process is accomplished in a distributed manner at each clusterhead as follows.

Each clusterhead maintains the current level of recursion k, initialized to 1. It is also aware of its own ID I, which denotes its position in the mesh. The following routines are also implemented on each clusterhead:

1. REDUCE($g1, g2, g3, g4, G$): This routine takes four sub-graphs ($g1, g2, g3, g4$) as input, and produces a reduced graph G as the output. The semantics of what the subgraph representations mean and how the reduction is accomplished were described previously.
2. LEADER(*id, rec-Level*): This routine takes as input a clusterhead ID and a level of recursion. Note that the clusterhead ID implies a unique block at that level of recursion. This function returns the ID of the leader of that block, based on the leader selection scheme being used. The scheme adopted in our algorithm is shown in Figure 39.5, and is intended to always move the data to the center of the mesh. Due to space constraints, we do not discuss the exact labeling scheme (a variation of shuffled row major labeling) and the expressions used at each node to compute the next function. In the figure, L1 and L2 denotes leaders of 2×2 and 4×4 blocks, respectively. The arrows show the sources and destinations of data transfers in the 2×2 blocks.
3. SEND(*data, id*): This routine sends the data item to the clusterhead with the given ID.

Every clusterhead performs the following sequence of actions. First, it uses the LEADER function with its own identifier as input, and determines the leader id for its block. If the leader ID is not the same as its own ID (which means it is not a leader), it uses the SEND function to send its subgraph to the leader; else, it waits till subgraphs from the three nonleader clusterheads are received, and uses the REDUCE function to reduce the four subgraphs (including its own) into a new subgraph G. If G contains one or more vertices that correspond to regions entirely within the geographic oversight of that block leader, they are not part of the subgraph that is transmitted up the quad-tree. Only the information about cross-boundary regions is communicated at the higher level of recursion. Note that if all regions are within the block boundary,

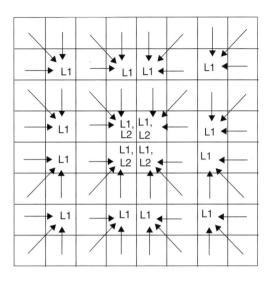

FIGURE 39.5   Leader selection scheme.

the subgraph that is sent up the quad-tree could be empty. Even in this case, the clusterhead transmits an empty subgraph to the leader of the next higher level to facilitate termination detection of the algorithm.

Next, k is incremented by one and the process is repeated till $k = (\log(\sqrt{n}/\sqrt{s}) - 1)$.[1] When this happens, the clusterhead sends its subgraph to a uniquely labeled *root* node. The bottom-right leader at the final recursion level is selected to be the *root*. Since every clusterhead knows the values of n and s, the ID of the *root* can be determined.

At the end of Step III, all regions in the network have been assigned a unique label, which is stored at the leader of the smallest block that contains the entire region.

Step IV: Label propagation through backtracking. The region labels stored at different clusterheads are propagated down the quad-tree, which is accomplished by backtracking. This process is basically an inversion of Step III earlier. When a clusterhead receives new labels for the vertices from the block leader, it updates the locally stored graph. Vertices belonging to different local components are merged if they are assigned the same label. The above steps are repeated till the labels are routed to every clusterhead in the mesh. The clusterhead now activates the feature nodes in turn by using the paging mechanism, and transmits to each feature node the labels of the region to which it belongs.

At the end of Step IV, each feature node knows the label of the regions it belongs to. Each clusterhead stores detailed information about the feature nodes in its cluster. When a sensor node receives a label, it is informed of the termination of the algorithm.

We analyze the communication time and energy of the algorithm. In Step I, each of the f feature nodes contends for the wireless channel and on gaining access, transmits one unit of data to the clusterhead. This involves f one-to-one transmissions resulting in an overall energy dissipation of $(2f \cdot E)$ units. Any clusterhead in the network has $5s$ neighbors: s within its cluster and $4s$ in the four (north, south, west, east) adjacent clusters. In the worst case, all the neighbors are feature nodes and $(5s \cdot T)$ time steps are required to schedule the data communication between the feature nodes and the clusterheads.

Step II of the algorithm requires each clusterhead to exchange boundary information with its neighbors. In the worst-case scenario, all feature nodes are on cluster boundaries and thus f units of data are exchanged between the clusterheads. This results in f transmissions, dissipating $(2f \cdot E)$ units of energy in the system. A clusterhead transmits and receives no more than $\min(f, 4\sqrt{s})$ units of data. There are at most nine

---

[1] All logs are to the base 2 unless specified otherwise.

clusterheads in a common transmission area that may contend for the shared channel at any given time. The above communication can be scheduled in fewer than $(\min(9f, 36\sqrt{s}) \cdot T)$ time steps.

Step III consists of at most $(\log(\sqrt{n}/\sqrt{s}) - 1)$ recursive steps. We consider communication in the kth recursive step. Consider a block of size $2^k \times 2^k$. It consists of four smaller blocks (we call them subblocks) of size $2^{k-1} \times 2^{k-1}$. The leader of one of these subblocks is assigned the role of the block leader. The leaders of the other three subblocks communicate information about their subgraph to the block leader (of the larger block). A subblock leader only transmits information about the vertices that have edges crossing the subblock boundaries. There are at most $(4 \cdot 2^{k-1} \cdot \sqrt{s})$ such edges (and vertices). Note the subblock leader of one of the blocks is the block leader itself. Each of the other three subblock leaders is less than 2^k hops away from the block leader. Thus, there are at most $(12 \cdot 2^{k-1} \cdot \sqrt{s})$ data units in a block that must be communicated over 2^k hops. There are $(n/s \cdot 1/2^{2k})$ such blocks in the network. The communication can take place in parallel in the various blocks. Efficient in-block routing can avoid contention over shared channels (we do not discuss these in this paper). Thus, the kth recursive step can be completed in $(12 \cdot 2^{k-1} \cdot \sqrt{s} \cdot T)$ time steps. In the worst case (when the region spans over the entire network), $((n/s \cdot 1/2^{2k}) \cdot (2 \cdot 12 \cdot 2^{k-1} \cdot \sqrt{s}) \cdot 2^k \cdot E = 12n/\sqrt{s} \cdot E)$ units of energy is dissipated in the kth recursive step. Since there are $\log(\sqrt{n}/\sqrt{s}) - 1$ recursive steps, Step IV requires $(6\sqrt{n} \cdot T)$ time steps and dissipates $(12n/\sqrt{s} \cdot (\log(\sqrt{n}/\sqrt{s}) - 1) \cdot E)$ units of energy in the worst case.

The above analysis assumes the worst case scenario (region consisting of n^2 feature nodes). Consider a region with diameter d consisting of $f < n$ feature nodes. Step III may be completed in fewer recursive steps if the region is contained in a smaller-sized block $(d/\sqrt{s} \cdot T$ time steps in the best case, when the region lies in a block of size $d \times d$). When the region is located in the center of the network, $\log(\sqrt{n}/\sqrt{s}) - 1$ recursive steps are required as in the worst case. However, energy dissipation is lesser for a smaller region. In a region with f feature nodes, in the worst case, all information is routed over $\sqrt{n}/\sqrt{s}$ hops in the network since the manner we elect leaders ensures that information flow is toward the center. The overall energy dissipation would be $O(f \cdot \sqrt{n}/\sqrt{s} \cdot E)$ in the worst scenario. Thus, the overall energy dissipation in this step is $\min(O((n/\sqrt{s} \cdot \log \sqrt{n}/\sqrt{s}) \cdot E), O(f \cdot \sqrt{n}/\sqrt{s} \cdot E))$.

In Step IV, the labels are transmitted to the sensor nodes by a backtracking procedure, and we do not analyze it in detail. However, it is easy since the time and energy of this step is not larger than Steps I–III combined.

Thus, overall energy dissipation for the algorithm is $O((f + \min(O(n/\sqrt{s} \cdot \log \sqrt{n}/\sqrt{s}), O(f \cdot \sqrt{n}/\sqrt{s}))) \cdot E)$ units and it requires $O((\sqrt{n} + s) \cdot T)$ time steps.

39.5.4 Summarizing the Results

The time and energy costs for the three algorithms are summarized in Tables 39.1 and 39.2. We observe that for small size regions ($f = O(\sqrt{n})$ and $d = O(\sqrt{n})$), all the three algorithms have similar time and energy performance. As the region size grows, the centralized algorithm performs the worst, whereas the recursive algorithm has the best time and energy performance.

Note that the above analysis compares the algorithm for the worst-case performance. Moreover it only provides asymptotic bounds on the performance of these algorithms. For the purpose of obtaining a more realistic performance evaluation, simulations were performed to compare the time and energy performance of the two decentralized algorithms. The results are illustrated in Figure 39.6. The y-axis represents the energy and time costs for the two algorithms, the x-axis denotes the diameter of the

TABLE 39.1 Energy Dissipation

Algorithm	$f = O(n) d = O(n)$	$f = O(n) d = O(\sqrt{n})$	$f = O(\sqrt{n}) d = O(\sqrt{n})$
Centralized	$O(n^{3/2} \cdot E)$	$O(n^{3/2} \cdot E)$	$O(n \cdot E)$
Decentralized	$O(n^2 \cdot E)$	$O(n^{3/2} \cdot E)$	$O(n \cdot E)$
Recursive	$O(n \cdot \log n \cdot E)$	$O(n \cdot \log n \cdot E)$	$O(n \cdot E)$

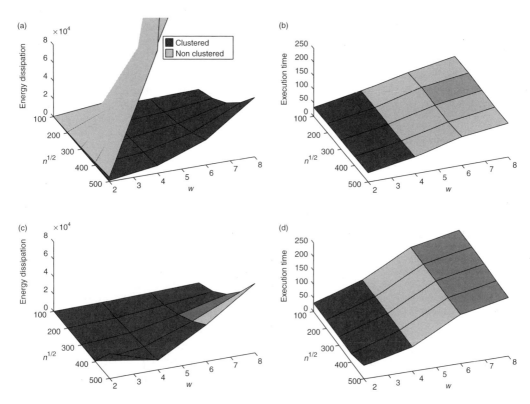

FIGURE 39.6 Simulation results. (a) Forest-of-trees: energy, (b) Forest-of-trees: latency, (c) Recursive: energy, and (d) Recursive: latency.

network, and the z-axis shows the average diameter of the feature region (the regions were assumed to be circular). We observed that the time performance of the two algorithms was almost the same. The energy dissipation of the recursive algorithms was significantly lower than the (nonclustered) forest-of-trees approach. The former algorithm conserves significant energy by allowing all nonclusterhead nodes to sleep for most of the time. The forest-of-trees algorithm was adapted to support clustering. The sensor nodes transmitted the feature information to the clusterheads and the forest-of-trees algorithm was executed only on the clusterheads. This (clustered) approach significanfly reduced the overall energy dissipation of the algorithm.

39.6 Concluding Remarks

In this chapter, we discussed various approaches to algorithm design in wireless sensor systems. We believe high-level modeling and analysis will play an important role in meeting the challenges of the next generation sensor systems (such as efficiency, ease-of-design, design reusability, scalability, and portability).

TABLE 39.2 Time Performance

Algorithm	$f = O(n)d = O(n)$	$f = O(n)d = O(\sqrt{n})$	$f = O(\sqrt{n})d = O(\sqrt{n})$
Centralized	$O(n \cdot T)$	$O(n \cdot T)$	$O(\sqrt{n} \cdot T)$
Forest-of-trees	$O(n \cdot T)$	$O(\sqrt{n} \cdot T)$	$O(\sqrt{n} \cdot T)$
Recursive	$O(\sqrt{n} \cdot T)$	$O(\sqrt{n} \cdot T)$	$O(\sqrt{n} \cdot T)$

However, significant efforts will be required in defining and standardizing models of computation for the various application domains.

In the current scenario, most research efforts have focused on designing localized algorithms for reasons of robustness and scalability. However, it is important to realize that several sensor applications (such as topographic querying) can benefit by maintaining some global state in the system. Therefore it is useful to define models that abstract the network as a single distributed system facilitating global state maintenance and system level optimizations. Moreover, the system will also benefit from low-level services that provide support for global state maintenance with low overheads.

Acknowledgments

We would like to thank Amol Bakshi for his constructive comments and help during the preparation of this work. This work is supported by the DARPA Power Aware Computing and Communication Program under contract no. F33615-02-2-4005 and the NSF under grant number IIS-0330445.

References

[1] H.M. Alnuweiri and V.K. Prasanna. Fast image labelling using local operators on mesh connected computers. *IEEE Transactions on Pattern Analysis and Machine Intelligence*, 13: 202–207, 1991.

[2] H.M. Alnuweiri and V.K. Prasanna. Parallel architectures and algorithms for image component labeling. *IEEE Transactions on Pattern Analysis and Machine Intelligence*, 14: 1014–1034, 1992.

[3] T. Antoniou, I. Chatzigiannakis, G. Mylonas, S. Nikoletseas, and A. Boukerche. A new energy efficient and fault-tolerant protocol for data propagation in smart dust networks using varying transmission range. In *Proceedings of the IEEE/SCS Annual Simulation Symposium*, April 2004.

[4] P. Bonnet, J.E. Gehrke, and P. Seshadri. Towards sensor database systems. In *Proceedings of the International Conference on Mobile Data Management (MDM)*, January 2001.

[5] J.L. Bordim, K. Nakano, and H. Shen. Sorting on single-channel wireless sensor networks. In *Proceedings of the International Symposium on Parallel Architectures, Algorithms, and Networks (ISPAN)*, May 2002.

[6] A. Boukerche, R.W.N. Pazzi, and R. Araujo. A novel fault tolerant and energy-aware based algorithm for wireless sensor networks. In *Proceedings of the International Workshop on Algorithmic Aspects of Wireless Sensor Networks (ALGOSENSORS)*, July 2004.

[7] J.C. Chen, K. Yao, and R.E. Hudson. Source localization and beamforming. *IEEE Signal Processing Magazine*, 30–39, March 2002.

[8] D. Estrin, L. Girod, G. Pottie, and M. Srivastava. Instrumenting the world with wireless sensor networks. In *Proceedings of the International Conference on Acoustics, Speech, and Signal Processing (ICASSP)*, May 2001.

[9] C. Intanagonwiwat, R. Govindan, and D. Estrin. Directed diffusion: scalable and robust communication paradigm for sensor networks. In *Proceedings of the International Conference on Mobile Computing and Networking*, August 2000.

[10] D. Kempe, A. Dobra, and J. Gehrke. Gossip-based computation of aggregate information. In *Proceedings of the IEEE Symposium on Foundations of Computer Science*, October 2003.

[11] B. Krishnamachari and S.S. Iyengar. Efficient and fault-tolerant feature extraction in sensor networks. In *Proceedings of the International Workshop on Information Processing in Sensor Networks (IPSN)*, April 2003.

[12] L. Lamport and N. Lynch. Distributed computing: models and methods. *Source Handbook of Theoretical Computer Science*, (vol. B), MIT Press Cambridge, MA, USA, pp. 1157–1199, 1990.

[13] C. Leopold. *Parallel and Distributed Computing: A Survey of Models, Paradigms and Approaches*. John and Sons, Wiley New York, November 2000.

[14] D. Li, K. Wong, Y. Hu, and A. Sayeed. Detection, classification, tracking of targets in micro-sensor networks. *IEEE Signal Processing Magazine*, pp. 17–29, March 1992.

[15] M. Singh, A. Bakshi, and V.K. Prasanna. Constructing topographic maps in networked sensor systems. In *Proceedings of the International Workshop on Algorithms for Wireless and Mobile Networks*, August 2004.

[16] M. Singh and V.K. Prasanna. System level energy trade-offs for collaborative computation in wireless networks. In *Proceedings of the International Conference on Communications (ICC), Workshop on Integrated Management of Power Aware Communications, Computing and networking (IMPACCT)*, May 2002.

[17] M. Singh and V.K. Prasanna. A hierarchical model for distributed collaborative computation in wireless sensor networks. In *Proceedings of the International Parallel and Distributed Processing Symposium, Workshop on Advances in Parallel and Distributed Computational Models*, April 2003.

[18] M. Singh and V.K. Prasanna. Supporting topographic queries in a class of networked sensor systems. In *Proceedings of the International Workshop on Sensor Networks and Systems for Pervasive Computing*, March 2005.

[19] M. Singh, V.K. Prasanna, J. Rolim, and C.S. Raghavendra. Collaborative and distributed computation in mesh-like wireless sensor arrays. In *Personal Wireless Communications*, September 2003.

[20] D.B. Skillicorn and D. Talia. Models and languages for parallel computation. *ACM Computing Surveys*, 30(2): 123–169, June 1998.

[21] Global soil moisture data bank. http://climate.envsci.rutgers.edu/soil-moisture.

[22] A. Wang and A. Chandrakasan. Energy efficient system partitioning for distributed wireless sensor networks. In *Proceedings of the International Conference on Acoustics, Speech, and Signal Processing (ICASSP)*, May 2001.

[23] Contaminant transport monitoring. http://cens.ucla.edu/Research/Applications/ctm.htm.

[24] The klamath river flow studies, http://www.krisweb.com/krisweb-kt/klamflow.htm.

[25] Physical Oceanography Distributed Active Archive Center (PODAAC). http://podaac.jpl.nasa.gov.

[26] J. Zhao, R. Govindan, and D. Estrin. Computing aggregates for monitoring wireless sensor networks. In *Proceedings of the International Conference on Communications (ICC), Workshop on Sensor Network Protocols and Applications*, May 2003.

IX

Security Issues in
Wireless Networks

40

Security Architectures in Wireless LANs

40.1 Introduction..40-905
IEEE 802.11b Security Mechanisms • WEP Security Issues •
WEP Improvements

40.2 Wireless Security Threats and Risks 40-909
Types of Security Threats • Physical Security of Wireless
Devices • WLAN Attacks

40.3 Risk Mitigation and Countermeasures.................40-911
Basic Countermeasures • AAA Infrastructure Solutions •
Additional Enhancements

40.4 AAA Infrastructure 40-914
RADIUS

40.5 IEEE 802.1X Standard 40-916
802.1X Architecture • EAP • 802.1X over 802.11

40.6 VPN...40-920
VPN Techniques • Layer 2 VPNs — PPTP and L2TP •
Layer 3 VPN — IPSec • Protocol Comparison

40.7 Performance Issues with Security Architectures.......40-924
Security Configuration Levels • 802.1X Model • VPN
Model • 802.1X Model Implementation • VPN Model
Implementation

40.8 Performance Test Results 40-929
HTTP and FTP Testing Scenarios • 802.1X Model Test
Results • VPN Model Test Results • Overall Performance

40.9 Summary 40-933

References 40-934

Ray Hunt
Jenne Wong

40.1 Introduction

Wireless Local Area Networks (WLANs) have gained increasing market popularity in locations such as airports, cafés, universities, and businesses, but WLAN security remains an ongoing concern. The rapid deployment of WLANs has further emphasized the security vulnerabilities of the Institute of Electrical and Electronics Engineers 802.11 standard that in its initial deployment, suffered from a number of security limitations. In this chapter, we investigate existing and proposed WLAN security technologies designed to improve the 802.11 standard including some of the current proposals coming from the IEEE 802.11i

TABLE 40.1　802.11 Authentication and Privacy Methods

Authentication Method	Open System and Shared Key
Privacy Method	WEP (optional; to be used with Shared Key)

security working group. This chapter also addresses a variety of security vulnerabilities specific to WLANs and discusses techniques for mitigation as well as reporting on a range of relevant papers that address these topics. Two authentication architectures (IEEE 802.1X and VPNs [Virtual Private Networks]) are discussed and finally some performance evaluation studies of these two security mechanisms are described.

40.1.1 IEEE 802.11b Security Mechanisms

In this section, we examine the security mechanisms provided by the 802.11 standard. Vulnerabilities such as weak key derivation and key management in the wired equivalent privacy (WEP) protocol are identified and improvements in key management and authentication to fix these limitations are described.

The 802.11 standard provides two types of authentication (open system and shared key), and a privacy method (WEP), as shown in Table 40.1. These mechanisms mainly deal with the security provisions in the link and physical layers of the *open systems interconnection* (OSI) model. They do not support either end-to-end or user-authentication. The standard only aims to make wireless networks as secure as its wired counterparts.

40.1.1.1 Authentication

Open system authentication: This is the default authentication service. It is in fact "null" authentication (i.e., no authentication).

Shared key authentication: The same secret (thus global) key is shared between an Access Point (AP) and client stations to authenticate stations joining a network. The key resides in each station's *management information base* (MIB) in write-only form and is available only to the Medium Access Control layer. This method requires the use of the WEP mechanism.

The process of the shared key authentication operates in four steps:
- The requesting station sends an authentication frame to the AP.
- The AP receives the authentication frame and replies with a random challenge text generated by the WEP encryption engine, using the *pseudorandom number generator* (PRNG).
- The requesting station copies the challenge text in an authentication frame and then encrypts it with the shared secret key. The encrypted frame is sent back to the AP.
- The receiving AP decrypts the text with the same shared key, and compares it to the challenge text sent earlier. Confirmation is sent if a match occurs; else a negative authentication is generated.

Clearly the main limitation, and risk of this system results from the shared key being copied by an attacker. Thus the distribution and overall key management mechanism is the main risk of using WEP rather than the strength of the cryptographic system itself.

40.1.1.2 Encryption and the WEP Protocol

The 802.11 standard includes the WEP as an optional protocol (ANSI/IEEE. Std 802.11, 1999). WEP ensures the confidentiality and integrity of data transmitted between users when using shared key authentication but does not specify key distribution and management systems.

WEP encryption uses the same 60-bit shared secret key to encrypt and decrypt data. In WEP, it is necessary to generate a different *Rivest Cipher 4* (RC4) key for each packet from a shared key. WEP was not designed for high security, but rather to be at least as secure as its wired counterpart. Two processes work inside the algorithm; one encrypts the plain text while the other protects the data's integrity. The main components of this algorithm use an *initialization vector* (IV) of 24 bits, PRNG, and an *integrity*

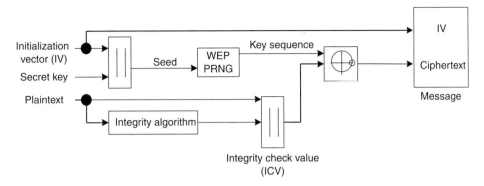

FIGURE 40.1   WEP encipherment block diagram.

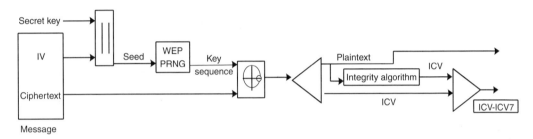

FIGURE 40.2   WEP decipherment block diagram.

check value (ICV) of 32 bits, as shown in Figure 40.1. The secret key is concatenated with an IV and the resulting *seed* becomes the input to a PRNG.

A frame contains the encrypted data (ciphertext block) together with the ICV and the cleartext IV (Figures 40.1 and Figure 40.2). Data integrity is provided by the ICV. The output message is generated by concatenating IV and the secret key (the key remains constant), while the IV changes periodically when connections between devices are made (Weatherspoon, 2000). The IV can be reused after a limited time — which represents one of the major weaknesses of WEP (see Section 40.1.2). Furthermore, static key management of WEP requires manual configurations on wireless devices. Thus, as the number of clients grow, administration overheads increase.

Vendors have provided support for a stronger WEP key length of 128-bits in their WLAN solutions. However, this method is still vulnerable to attacks and the key management issues described above remain.

40.1.2 WEP Security Issues

In spite of the overall concerns about wireless security and the limitations of the early solutions provided by 802.11 via WEP, in many reported incidents, attacks on WLANs showed that WEP was not even activated (Held, 2001; WECA, 2001a). Vendors provide a deactivated WEP as a default installation, and a survey conducted by Gartner found that about "60% of Wi-Fi (802.11b) installations did not even turn the WEP on" (McDonough, 2002).

The WEP in the 802.11 standard provides only limited support in safeguarding the network. The limitations in WEP's cryptography and authentication can lead to security problems such as modification of data, redirection attacks, rogue APs, and replay attacks. Borisov et al. (2001) discovered that WEP has fundamental flaws in the cryptographic design of the implementation of the RC4 encryption algorithm and checksum methods. In particular, this relates to key reuse and weak message authentication. They claimed that WEP's goals of securing confidentiality, access control, and data integrity had all failed. The classical 40-bit and extended versions of the 128-bit key size provided by WEP did not deter hackers, indicating that key size was largely irrelevant. The 24-bit IV is prone to reuse frequency because the vector size is limited. IV collisions produce identical WEP keys when the same IV is used with the same shared secret key for more than one data frame, and this is the weakness that attackers exploit. This claim is supported by an early study by Walker (2000), which showed that regardless of key size, the vulnerability in RC4 prevented WEP from providing "a meaningful notion of privacy to users." In addition, weak message authentication made it possible to inject traffic into the network. Although long key-length versions of WEP were released to the market, the limitations in WEP were not caused by a shorter key.

An experiment on the security mechanisms mentioned previously was conducted by Arbaugh et al. (2001); they discovered weaknesses in the 802.11 access control mechanism even when it was used with the WEP authentication method. Thus they stated, "all of these (even vendors' proprietary security) mechanisms are completely ineffective." This weakness increases the risk of rogue AP networks and station redirection attacks (see Section 40.2).

Considering these three papers, the security mechanisms defined in 802.11 are compromised. Until this point, attacks on the WEP were based on the design of the system, and users assumed the underlying cryptography, Rivest Shamir Adleman's RC4 algorithm, was sound. Fluhrer et al. (2001) presented the final blow to WEP security when they found "weaknesses in the key scheduling algorithm" of RC4. These flaws made the RC4 keys fundamentally weak, and the authors designed an attack that would allow a passive user to recover the secret WEP key simply by collecting a sufficient number of frames encrypted with weak keys. An implementation of the attack was carried out by Stubblefield et al. (2001). Tools to plan an attack (such as AirSnort and WEPCrack) exist, and key recovery with AirSnort takes only a few seconds when a sufficient number of weakly encrypted frames are gathered. These studies showed that the WEP security mechanism is ineffective in the absence of good key management. The vulnerabilities in 802.11 can be generalized as follows:

- No dynamically generated session keys
- Static WEP keys
- No mutual authentication

40.1.3   WEP Improvements

The link layer security provisions in the 802.11 standards are all vulnerable to attacks. Therefore, systems should deploy "additional higher-level security mechanisms such as access control, end-to-end encryption, password protection, authentication, virtual private networks, or firewalls" (WECA, 2001a) and assume WEP as a very basic layer of security only.

The IEEE 802.11 committee set up task group 802.11i (Task Group i (of IEEE), 2002) to enhance the security and authentication mechanism of the current 802.11 MAC. Their work has resulted in the development of:

- Replacement of the 802.11 authentication standard with 802.1X authentication and key management.
- Improvement of the existing WEP with *temporal key integrity protocol* (TKIP), also known as WPA (WiFi Protected Access). WPA was a temporary solution based upon existing hardware only.
- Deployment of a *robust security network* (RSN) solution with a stronger encryption algorithm — the Advanced Encryption Standard (AES). This solution requires replacement of existing wireless NICs (Network Interface Cards) and APs and is now referred to as WPA2 and is consistent with the IEEE 802.11i standard that was ratified in 2004.

40.1.3.1   Replacement of 802.11 Authentication and Key Management with IEEE 802.1X

The 802.1X standard has been introduced to provide a centralized authentication and dynamic key distribution for 802.11 architecture utilizing the 802.1X standard with RADIUS (Roshan, 2001; Task Group i, 2002). 802.1X is an authentication standard for 802-based LANs using port-based network access control. It is used for communication between wireless clients and APs, while RADIUS operates between an AP and an authentication server (see Section 40.5).

40.1.3.2 Improvement of WEP with TKIP

The TKIP solution deploys a hashing technique that generates a temporal key (a unique RC4 key) to derive a per data packet key. This strengthens the RC4 key-scheduling algorithm by pre-processing the key and the IV by passing them through a hashing process. The solution consists of:

- An encryptor and decryptor that share an RC4 104-bit or 128-bit secret key. This key is called the *temporal key* (TK).
- An encryptor and decryptor that use the RC4 stream cipher.
- An IV value that is not used more than once with each TK. Implementations must ensure that the TK is updated before the full 16-bit IV (Housley and Whiting, 2001) or 48-bit IV (Housley et al., 2002) space is exhausted. The 48-bit IV solution provides a longer key life span than the 16-bit IV.
- This solution has been given the name WiFi protected access (WPA) but was a stepping stone en route to the full IEEE 802.11i security working group solution now referred to as WPA2.

40.1.3.3 Deployment of an RSN Solution (IEEE 802.11i)

The 802.11i standard addresses the user authentication and encryption weaknesses of WEP-based wireless security. The components of 802.11i include the 802.1X port-based mutual authentication framework, a Message Integrity Check, and the TKIP, all of which are included in WPA2.

The AES encryption algorithm (TGi, 2002) (which replaces RC4 encryption and WEP keys) provides key hierarchy and management features, and cipher and authentication negotiation. In addition, 802.11i adds the Wireless Robust Authentication Protocol, which consists of a sequence counter and a cipher block chaining message authentication code. Thus HMAC[1]-SHA1-128 can be used as the hashing function to support message authentication with AES. With these features, 802.11i addresses the security of the earlier WLAN standards.

The AES encryption algorithm has a stronger cryptographic strength than the RC4 algorithm used by WEP and WPA. However, the use of AES requires the replacement of older APs because of the need for higher performing processors. As a result, 802.11i has been targeted at next-generation APs and clients.

40.2 Wireless Security Threats and Risks

Wired and wireless LANs share some common security risks: physical security, insider attacks, unauthorized access, and eavesdropping. A study published by the Wi-Fi Alliance in October 2001 found that "security has been, and remains, the overriding concern regarding wireless networking deployment" (WECA, 2001b), among 72% of wireless intenders and 50% of wireless adopters.

40.2.1 Types of Security Threats

A taxonomy of security threats is depicted in Figure 40.3. These attacks can come from internal or external sources.

40.2.1.1 Passive Attacks

An unauthorized party gains access to a network and does not modify the resources on the network. Types of passive attack include:

- *Eavesdropping*: An attacker simply monitors message content. For example, an unauthorized person drives through the city and listens to various WLAN transmissions within different organizations (i.e., war driving, see Section 40.2.3 for more detail).
- *Traffic analysis*: An attacker monitors the traffic for communication pattern analysis. The statistics collected can be used to perform a dictionary attack.

---

[1] *Hashed message authentication code* (HMAC) with either *message digest algorithm* (MD5) or *secure hashing algorithm* (SHA1).

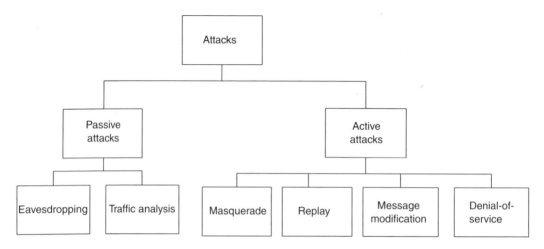

FIGURE 40.3   Taxonomy of security threats.

40.2.1.2   Active Attacks

An unauthorized party gains access to a network and modifies the resources on the network. An attacker must first spoof either the MAC address and/or Internet Protocol address of a user/device to gain network access, and then change the content of the network resources. Types of attack include:

- *Masquerading*:   An attacker impersonates an authorized user and thereby gains certain unauthorized privileges. Masquerading includes the use of spoofing, rogue APs, and redirection attacks. An attacker can fool users to log in to the rogue AP by placing a rogue AP in the same area as a valid AP, sending the same Service Set Identifier but with a stronger signal than the valid AP. The attacker is able to decipher the shared key from the traffic collected. The rogue AP can be used to redirect users' transmissions to an invalid destination, or to insert deauthentication packets.
- *Replay*:   An attacker monitors the traffic (passive attack) and then retransmits the message as the legitimate user.
- *Message modification*:   An attacker alters the legitimate message by deleting, adding, changing, or recording it. Furthermore, an attacker may wish to alter the configuration of a device, using, for example, *simple network management protocol* (SNMP) to configure APs.
- *Denial-of-service* (DoS):   An attacker prevents or renders the normal use or management of network systems useless by issuing malicious commands or injecting a large amount of traffic that fills up the radio frequency. This type of attack can be further extended to *distributed DoS* (DDoS) attacks.

40.2.2   Physical Security of Wireless Devices

Physical security is the most fundamental step in ensuring that only authorized users have access to wireless devices such as laptops, handhelds, and APs. APs (or base stations) must be difficult to access to prevent security breaches, and the AP must be placed in an appropriate place for an adequate coverage area. For example, the improper placement of an AP might allow an attacker to bypass other security measures, such as modifying the AP configuration by direct connection. Physical locks, cipher locks, and biometric systems can be used to counter thefts.

40.2.3   WLAN Attacks

Wireless networks without proper security implementation can be penetrated easily. The physical freedom of a WLAN is also its vulnerability; traffic is no longer confined to a wire. Privacy concerns over data transmission increase because data on a WLAN is "broadcast for all to hear" (O'Hara and Petrick, 1999); eavesdropping becomes easy. In addition, RF-based networks are open to packet interception by any receiver within range of a data transmitter. 802.11 beacon frames, used to broadcast network parameters, are sent unencrypted. Thus, by monitoring beacon frames, wandering users with an 802.11 receiver can discover wireless networks in the area.

In addition, the deployment of wireless networks opens a "back door" (Arbaugh et al., 2001) into the internal network that permits an attacker access beyond the physical security perimeter of the organization. The lack of a physical boundary allows attacks such as sniffing, resource stealing, traffic redirection, DoS, SSID, and MAC address masquerading to occur.

War driving is similar to *war dialling* (dialling every number looking for a modem backdoor into a network), and the Wall Street Journal reported an incident in which two people armed with wireless tools were able to drive around Silicon Valley and intercept traffic such as emails over unprotected WLANs (Bansal, 2001). Another example was an audit carried out of four major airports in the United States (Brewin and Verton, 2002). This study found that WLANs in applications such as passenger check-in and baggage transfers were operating without even some of the most basic forms of security protection. Using NetStumbler, an AP-detection tool, the authors discovered that only 32 of the 112 WLAN APs had the WEP protocol turned on, and most of the APs were broadcasting plaintext SSIDs.

Other new types of attacks include *warspammers* and *warchalking*. Warspamming takes advantage of unprotected WLANs to bombard email users with unsolicited and unwelcome messages. Warchalking takes place when hackers draw a chalk symbol on a wall or piece of pavement to indicate the presence of a wireless networking node. Warchalking was originally developed to alert system administrators to their wireless network security lapses (Wearden, 2002).

40.3   Risk Mitigation and Countermeasures

WLAN risks can be mitigated by applying both basic and authentication, authorization, accounting (AAA) infrastructure countermeasures to address specific attacks and threats. Basic countermeasures involve altering the existing security functions provided within wireless equipment. These countermeasures provide only limited defense against casual attacks; for determined adversaries, organizations should consider AAA integrated solutions. The AAA infrastructure countermeasures provide integrated solutions using existing AAA infrastructure components such as the RADIUS protocol and public key infrastructure (PKI), with network solutions such as VPN and the 802.1X standards.

40.3.1   Basic Countermeasures

Every wireless device comes with different default settings; such built-in configurations can be prone to security vulnerabilities. This section discusses basic security protection to prevent casual attacks.

40.3.1.1   Change Default Password

Default passwords such as "public" or blank passwords are not sufficient protection. Administrators should deploy strong passwords of at least eight characters that are consistent with the organization's security policies. When combining this with the AAA infrastructure solutions, two-factor authentication can be implemented.

40.3.1.2   Change SSID

The factory default SSID for an AP may provide network names, which can easily be detected. The SSID should not provide information on the function or location of an AP. The naming, convention of SSIDs should be formed according to an organization's security policy; an example is shown in Figure 40.4.

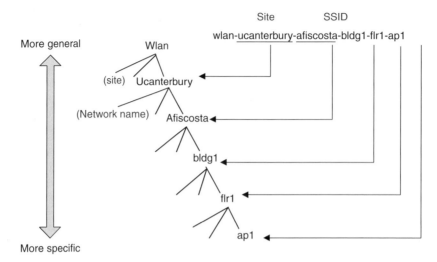

FIGURE 40.4   An Example of naming conventions for wireless devices. (Adapted from Gast, M. [2000] *Wireless Networks: The Definitive Guide.* O'Reilly, April.)

40.3.1.3   Enable MAC Authentication

A MAC address is a hardware address that uniquely identifies each device on a network (e.g., a wireless network adapter), identifying devices, not users. Vendors[2] provide MAC-address filtering capability to regulate communication among different computers on a network. During the initial connection procedures, wireless APs can check the MAC address of connecting stations to ensure the station is on the list of known good MAC addresses, called a MAC *access control list* (ACL). However, a MAC ACL does not provide strong authentication against determined attacks. MAC addresses are transmitted in cleartext from a wireless NIC to an AP and eavesdropping software can easily capture MAC addresses. Furthermore, users with sufficient operating system privileges can alter addresses to masquerade as an allowed wireless-network user.

40.3.1.4   Protect the AP Placement

Improper placement of APs (Section 40.2.2) can lead to security breaches. If malicious users can gain physical access to an AP, they would be able to change the configuration of the AP. Thus organizations need physical security on their wireless devices that should include regular security audits to ensure risk mitigation.

40.3.1.5   Enable WEP Authentication and Encryption

Built-in security configurations do not enable the WEP security function by default. To provide at least some basic defense against attacks, shared authentication should be used in preference to open system authentication. Deployment of the strongest encryption option available is also desirable whenever possible. Note that enabling the WEP mechanism can only prevent basic attacks (Section 40.2.3), and for determined adversaries, the WEP key length is unsafe at any size (Section 40.1.2).

40.3.1.6   Change SNMP Parameters

SNMP agents can be configured in APs to allow network management software tools to monitor the status of an AP and its clients. The default SNMP community string commonly used by SNMP agents is the word "public" with assigned "read" or "read and write" privileges. To prevent unauthorized users from

---

[2]For larger organizations with multiple APs, consider storing the MAC list at a centralized location using a RADIUS server.

writing data to an AP (an integrity breach), SNMP parameter settings must be changed or disabled (if SNMP is not required in the organization).

Several vendors use SNMP as an AP management mechanism and thereby increase vulnerabilities. For example, one vendor uses SNMPv1 for AP management, thus all management traffic traverses the network unencrypted. Another vendor allows SNMP read access to WEP keys, even though WEP keys must remain secret. Some vendors use cleartext telnet for remote command-line interfaces. Web-based interfaces are nearly all simple Hypertext Transfer Protocol and do not use *secure socket layer* (SSL) or *transport layer security* (TLS) for protection.

Improvements for network resource management may utilze SNMPv3 or *policy based network management* (PBNM; Wong, 2001) with *common open policy service* (COPS) protocol for better security.

40.3.1.7 Change the Default Channel

To prevent DoS attacks and radio interferences between two or more APs in close location, the channel setting of an AP should be changed to operate in a different frequency band, particularly avoiding the default channels of 1, 6, and 11.

40.3.1.8 Dynamic Host Control Protocol (DHCP) Usage

Automatic network connections involve the use of a DHCP server. The DHCP automatically assigns IP addresses to clients that associate with an AP. Using DHCP allows users the advantages of roaming or establishing ad hoc networks. However, the threat with DHCP is that a malicious user can gain unauthorized access using a mobile device, because DHCP may not know which wireless access devices have access permission, and might automatically assign the device a valid IP address. Depending on the size of the network, disabling DHCP and using static IP addresses may be feasible.

A solution to overcome DHCP threats might involve placing the DHCP behind the wired network's firewall, which grants access to a wireless network located outside the firewall. If user authentication and access control are moved to the link layer, then threats to the DHCP are limited to insider attacks. An AAA integrated solution such as 802.1X authentication can assist in militating against the DHCP server.

40.3.2 AAA Infrastructure Solutions

AAA infrastructure provides centralized network management and several AAA protocol architectures exist — RADIUS (Section 40.4) being the most common and it is used in conjunction with both 802.1X and VPNs. Enhancing WLAN security requires integration with an AAA infrastructure to overcome security vulnerabilities. Two security solutions have been recommended in various studies (Borisov et al., 2001; Caballero and Malmkvist, 2002; Convery and Miller, 2001; Karygiannis and Owens, 2002; TGi, 2002):

- 802.1X Solution
- VPN Solution

40.3.2.1 802.1X Solution

This solution utilizes the existing 802.1X standard, incorporating port-based network access control for 802.11 infrastructures, and using AAA protocols to support wireless LAN security. Authentication methods are based on an *extensible authentication protocol* (EAP) to provide integration capability (with future authentication methods,[3] see Section 40.5).

40.3.2.2 VPN Solution

This solution applies the existing wired network solution to the wireless counterparts, using security mechanisms such as *IP Security* (IPSec) and tunneling (see Section 40.6).

---

[3]This group includes *secure remote password* (SRP) and *tunneled transport layer security* (TTLS) authentication methods, but they have not been widely adopted yet, but may become future mainstream authentication choices.

40.3.3 Additional Enhancements

Additional enhancements can be applied regardless of whether or not an integrated network security solution exists in the organization.

40.3.3.1 Firewalls

Placing a firewall between the trusted wired network and untrusted wireless networks provides an extra layer of access control (Harris, 1998). Interoperability between vendors' products must be considered when a firewall facilitates the traffic flow between wired and wireless networks.

Implementing personal firewall software on client computers can provide some protection against attacks, especially for clients accessing public WLANs. Organizations can set up these personal firewalls to be centrally or individually managed.

40.3.3.2 Intrusion Detection Systems

An *intrusion detection system* (IDS) provides effective security against unauthorized attempts to alter network resources, whether the network has been compromised or accessed. It minimizes the risk of an intruder breaking into authentication servers and compromising databases. Some IDS systems are specially designed for the WLAN environment, such as Air Defense IDS. A range of new products are emerging entitled "IPSs" (Intrusion Prevention Systems). Such systems combine firewall and IDS capabilities into a single device.

40.3.3.3 Anti-Virus Software

In common with wired networks, anti-virus systems provide another level of security against attacks, such as virus, worms, and Trojans. Organizations should deploy anti-virus software on both authentication servers and wireless clients to ensure system integrity.

40.3.3.4 Software Upgrades and Patches

Regular updates on software patches and upgrades, such as AP management software, will assist in ensuring that as many security vulnerabilities have been identified and corrected as possible.

40.3.3.5 Application Layer Security

Additional access control can be provided using applications with strong built-in cryptographic systems. In particular, web-based systems can be secured with SSL or TLS and host logins can be secured with *secure shell* (SSH). Other environments may have already deployed a framework such as Kerberos for application layer security, which may be able to be configured in support of wireless networks (e.g., Windows 2000-based networks).

40.4 AAA Infrastructure

AAA (Mitton et al., 2001) is important in both wired and wireless networks to ensure network security as well as providing a mechanism for billing. AAA protocols currently include RADIUS, *terminal access controller access control system* (TACACS; Finseth, 1993), and DIAMETER. RADIUS and TACACS were developed for remote user dial-in services and thus were not specifically intended to cater for wireless and mobile networks. DIAMETER (Calhoun et al., 2002) in contrast, takes wireless network access into consideration and supports applications such as IP mobility and roaming. RADIUS is the most widely deployed protocol for services such as dialup *point-to-point protocol* (PPP) and terminal server access. Infrastructure networks often deploy AAA architecture to ensure network access control (examples include VPN solutions). RADIUS is the most common AAA protocol used in the deployment of WLAN security solutions.

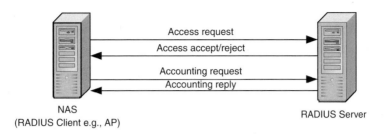

FIGURE 40.5 RADIUS communication exchange.

40.4.1 RADIUS

RADIUS is defined in Request for Comment 2865 (Rigney et al., 2000b) and RFC 2869 (Rigney et al., 2000a), and is used in a client/server environment (Figure 40.5), and has been widely used by businesses and *Internet Service Providers* (ISPs) to control remote access.

A RADIUS client is a type of *network access server* (NAS), and sends authentication and accounting requests to the RADIUS server in order to gain network access. RADIUS servers are responsible for authenticating users' requests from a NAS and to grant or reject access based on user credentials. RADIUS accounting is used to collect accounting data on users for billing, auditing, or trend analysis purposes. Interactions between the client and RADIUS server are authenticated through a shared secret that is never transmitted over the network. User Datagram Protocol (rather than Transmission Control Protocol) is used as the transport protocol for RADIUS (port 1812 for authentication and port 1813 for accounting). RADIUS supports a variety of authentication methods, including EAP. AAA servers based on the RADIUS protocol include Microsoft's Internet Authentication Server (IAS), Cisco's Access Control Server (ACS), and FreeRADIUS. Caballero and Malmkvist (2002) investigated and designed a NAS using the RADIUS protocol for public WLAN access networks, while Lee (2002) evaluated different interactions of AAA architecture with mobile IP for WLANs.

40.4.1.1 Authentication Protocols

An AAA server can support different types of user authentication methods:

1. *Password authentication protocol* (PAP) uses cleartext passwords. It provides low security protection against unauthorized access.
2. *Challenge handshake authentication protocol* (CHAP) is a challenge-response authentication protocol using an industry-standard *message hashing5* (MD5) one-way encryption scheme to encrypt the response.[4]
3. Microsoft-CHAP (MS-CHAP) is a proprietary protocol that supports one-way, encrypted password authentication. If the AAA server supports MS-CHAP, data encryption can be carried out using *Microsoft point-to-point encryption* (MPPE), which is based on the RC4 algorithm. MS-CHAP2 is an improved version and offers mutual authentication.
4. *Shiva password authentication protocol* (SPAP) is a two-way reversible encryption mechanism employed by Shiva.
5. The EAP protocol defined in RFC 2284 (Blunk and Vollbrecht, 1998), is a general protocol for PPP authentication that supports multiple authentication mechanisms. EAP does not select a specific authentication mechanism at the link control phase, but rather postpones this until the authentication phase. The client and AAA server negotiate the exact authentication method to be used (see Section 40.5.2 for more detail). EAP supports the following authentication types:
   - MD5-CHAP encrypts user names and passwords with an MD5 algorithm. This is equivalent to CHAP and called Wireless CHAP.
   - TLS uses digital certificates or smartcard devices. Authentication requires a user certificate and private key.

---

[4]In a Microsoft environment, users' passwords need to be stored in reversible encrypted format.

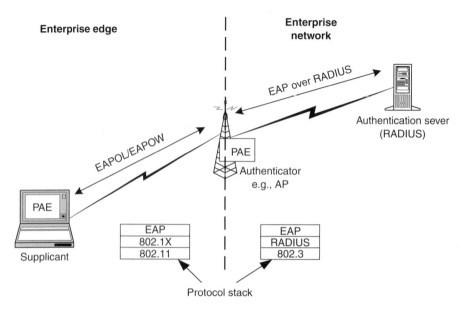

FIGURE 40.6   802.1X over 802.11 topology. (Adapted from ORiNOCO [2002] Technical Bulletin 048/B, Lucent, April.)

- Additional support for third-party authentication such as *tunneled TLS* (TTLS) from Funk.
- PEAP (Protected EAP) that is very similar to TTLS. Although proposed by Microsoft and CISCO it is largely an open standard.

40.5   IEEE 802.1X Standard

The 802.11 working group adopted the IEEE 802.1X standard, providing *port-based network access control*, a mechanism that uses the physical access characteristics of the IEEE 802.LAN infrastructure. The 802.11 TGi has taken the IEEE Std. 802.1X (IEEE Std. 802.1X, 2001) as its base to control network access on point-to-point connections, and adds several features for 802.11 LANs (WLANs). These features include dynamic key management and user-based authentication (whereas WEP only provides device-based authentication).

Using open standards such as EAP (Blunk and Vollbrecht, 1998), and RADIUS, the 802.1X standard (Figure 40.6) enables interoperable user identification and centralized management. Mutual authentication (e.g., TLS) can be carried out to ensure that the derived keys arrive at the right entity, avoiding attacks from rogue APs. Two communication protocols are used to facilitate the 802.1X exchange. Together they form the underlying EAP framework (see Section 40.5.1 for more details):

- EAP over LAN (EAPOL) is used between the authenticator's *port access entity* (PAE) and supplicant's (client's) PAE.
- EAP over Wireless (EAPOW) is another EAPOL packet type defined for use in transporting global keys (EAPOW-keys).
- EAP over RADIUS (de-facto), the authenticator PAE communicates with the *authentication server* (AS) using the EAP protocol carried in a higher layer protocol, such as RADIUS.

Key management provides the ability for an AP to distribute or obtain global[5] key information to/from attached stations, through the use of the EAPOL-key messages, after successful authentication. EAPOL

---

[5] Global key are the same shared keys for every client; they are also referred to as multicast keys.

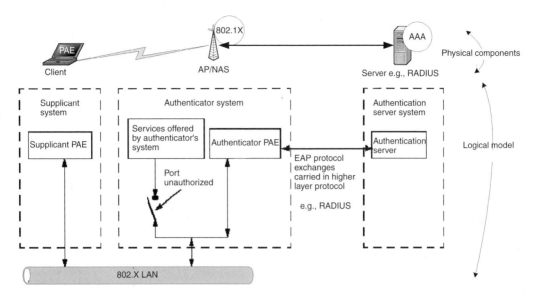

FIGURE 40.7   802.1X architecture. (Adapted from IEEE Std. 802.1X [2001] New York, IEEE Inc., ISBN 0-7381-2627-5, 25 October.)

defines the encapsulation techniques used in order to carry EAP packets between the client and an AP. Note that 802.1X authentication occurs after 802.11 associations. Keys are derived on the client and the AAA server (in this case the RADIUS server).

If negotiation of a session key during authentication is required, the use of EAP-TLS protocol is recommended. Thus, 802.1X can derive keys that provide per session key authentication, integrity, and confidentiality. However, it does not provide these per se, but only assists in deriving keys for the session; cryptographic support from WEP, *triple data encryption standard* (3DES), or AES is required to provide encryption. 802.1X requires an authentication protocol and EAP-TLS (RFC 2246; Dierks and Allen, 1999) is commonly used to support the key derivation algorithm.

40.5.1   802.1X Architecture

The standard defines the following components, shown in Figure 40.7. The architecture can be used with any 802 networks, such as 802.3 and 802.11. At the top of the diagram, we illustrate the physical components above the logical model that may be used for a WLAN environment.

- *Port*: This is a single point of attachment to the LAN infrastructure. It may be a physical network interface or a virtual MAC such as the *logical* ports managed by an AP in an 802.11 LAN.
- *PAE*: The protocol entity associated with a port. It can support the protocol functionality associated with either the authenticator, the supplicant, or both.
- *Supplicant*: A component at one end of a point-to-point LAN segment, it is authenticated by an authenticator attached to the other end of that link. Supplicants can also be called client stations.
- *Authenticator*: The authenticator is at the other end of the LAN segment and facilitates authentication of the component attached to the other end of that link. It is important to note that the authenticator's functionality is independent of the actual authentication method. It simply acts as a pass-through for the authentication server.
- *AS*: This is a component that provides an authentication service to an authenticator. It determines whether a supplicant is authorized to access the services by the authenticator based on the supplicant's credentials.

It is possible to authenticate using just the AP only, although in practice all authentication should be carried out by the AS. It is more desirable to implement an AAA server (in this case a RADIUS server) to centrally manage access control. In such situations, the authenticator is expected to act as an AAA client.

40.5.2　EAP

An EAP authentication protocol allows greater flexibility and scalability for future authentication support without the need to change the AP or the NIC. Authentication and key exchange can be upgraded without hardware modifications, avoiding limitations commonly associated with WEP.

Authentication processes can be performed using a username/password combination or digital certificate. A certificate is issued by a trusted authority such as a PKI to provide a strong one-to-one relationship. Furthermore, EAP authentication methods can support encryption key generation to protect the information transferred between the supplicant and the AS. Major EAP types include:

- *EAP-SRP* (secure remote password) assumes that both the client and the AS are authenticated using a password supplied by the client (ORiNOCO, 2002; Wu, 2000). Clients are not required to store or manage any long-term keys, which eliminate the reusable password problem. Encryption keys are generated during the exchange.
- *EAP-MD5* is equivalent to CHAP (Simpson, 1996). The client is authenticated by the AS using the password supplied by the client. No mutual authentication can be performed, as the client cannot authenticate the AS. There are no encryption keys generated during the authentication process.
- *EAP-TLS* (Aboba and Simon, 1999) provides mutual authentication of the client and the AS, and is carried out using digital certificates. The AS requires access to the certificate repository, and encryption keys are generated during the exchange. Certificates may be replaced by smartcards; Windows 2000 platform supports such an authentication process (Aboba and Simon, 1999).
- *EAP-TTLS* is a Funk proprietary (Funk and Blake-Wilson, 2002; ORiNOCO, 2002) authentication mechanism. It combines EAP-TLS and traditional password-based methods (ORiNOCO, 2002) such as CHAP, and *one time passwords* (OTP). The client does not need a digital certificate, but can be authenticated using a password. The client authenticates the server via an X.509 certificate, while the server authenticates the client by its encrypted password. Encryption keys are generated during the exchange.
- *PEAP* is a more recent open standard (proposed by Microsoft and CISCO) very similar to the (proprietary) TTLS. PEAP securely transports authentication data, including passwords by using tunneling between PEAP clients and the authentication server. PEAP makes it possible to authenticate wireless LAN clients without requiring them to have certificates. PEAP provides an improvement over EAP-MD5 (Wireless CHAP) that was vulnerable to a dictionary attack on the encrypted handshake.

EAP-TLS and EAP-MD5 protocols are the most common authentication methods associated with 802.1X deployments. The performance studies described in Section 40.7 evaluate these two EAP methods.

40.5.3　802.1X over 802.11

Applying the 802.1X structure to the 802.11 network architecture (Figure 40.8) provides a controlled wireless network with user identification, centralized authentication, and key management. Dynamic key management in an 802.1X framework rectifies the limitations in the WEP security mechanism by deploying per-user session keys.

802.1X authentication is carried out after an 802.11 association. Prior to authentication, the AP filters all non-802.1X traffic to and from the client. Only after the AS has successfully authenticated the client, can it be allowed to access the network. The Internet Engineering Task Force Network Working Group presented a draft on 802.1X RADIUS usage (Congdon et al., 2002), supporting a RADIUS server as the backend AS. Keys derived between the client and the RADIUS server are transferred from the RADIUS

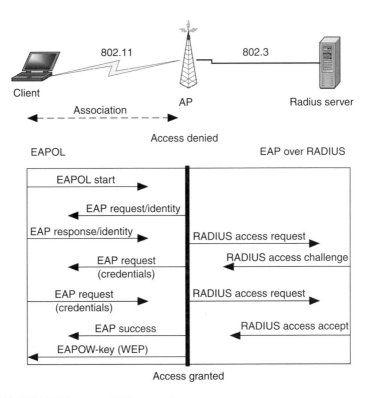

FIGURE 40.8   802.1X/RADIUS over an 802.11 network.

server to the AP after successful authentication. The secret shared between the RADIUS server and its client will be used to encrypt the attribute containing the keys, in order to avoid eavesdropping.

One session key can be derived for each user per session. However, if global keys (multicast/WEP keys) are used, then the session key (unicast key) sent from the AS to the AP is only used to encrypt the global key, providing per-packet authentication and integrity. An EAPOW-key packet is used for the global keys.

The 802.1X standard also supports per-station session keys, but most practical implementations only support global keys. This is because if global keys are supported, the session key is only used to encrypt the global key. The problem associated with global keys (that secrets shared among many people can lead to compromise of the secret), is solved by deploying the 802.1X per-session user keys.

Windows XP (Microsoft, 2002) is an example of a supplicant that has integrated support for the 802.1X protocol. Its authentication types support EAP-MD5 and EAP-TLS.

40.5.3.1   EAP-MD5

The EAP-MD5 authentication method provides one-way password-based authentication of the client performed by the AS. It is popular due to its easy deployment of passwords and usernames. However, it provides limited security, and no encryption keys are generated during the exchange. EAP-MD5 can be useful in public areas if encryption is provided at the application layer. The process is illustrated in Figure 40.9.

40.5.3.2   EAP-TLS

EAP-TLS is the most commonly implemented EAP type for WLANs. Its authentication is based on PKI support using X.509 certificates. Both the AS and clients must possess certificates validated by a trusted authority, which ensures explicit mutual authentication. After the authentication exchange, a shared session key is generated between the AS and the client. After the AS supplies the secret key to the AP via a secured link (using EAP over RADIUS), the AS and the client can use it to bootstrap their per-packet

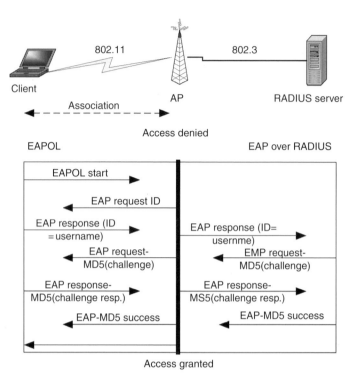

FIGURE 40.9   EAP-MD5 authentication process. (Adapted from ORiNOCO [2002] ORiNOCO Technical Bulletin 048/B, Lucent, April.)

authenticated and secured communication. Deploying EAP-TLS is complicated; thus, depending on the scale of an organization's network, administrative burdens might outweigh the security advantages. Figure 40.10 illustrates the process.

40.6   VPN

VPN technology is a rapidly growing security solution that provides secure data transmission over shared or public networks such as the Internet. Data is transmitted over the network by creating an encrypted, virtual, point-to-point connection between the client and a gateway VPN server (which may also be a firewall) that resides in a private network (Figure 40.11). VPNs provide organizations with a range of security tools to protect their internal infrastructures from external compromises (Hansen, 2001; Maier, 2000). Electronic-commerce (e-commerce) deployment (Maier, 2000), for example, requires a VPN or some other extranet security protection.

VPNs can be implemented on wireless LANs in a similar manner. Several options are associated with WLAN placement:

- Placing the WLAN outside an organization's firewall and employing a VPN to "obviate" the need for link-layer security and provide extra access control. This method might be derived from the same method used for Internet access (Maier, 2000) — treating wireless deployment as an extranet access. This approach may be expensive depending on an organization's needs.
- Other literature (Convery and Miller, 2001; InterLink Networks, 2001; Seng, 2002) suggests enhancing the VPN with the IPSec protocol by overlaying it on clear text 802.11 wireless traffic.
- Alternatively, an organization can set up a *wireless demilitarized zone* (WDMZ) (Enterasys Network, 2002) inside the organization's network to filter out unauthorized access. This alternative ensures end-to-end security and prevents threats such as replay and traffic analysis attacks.

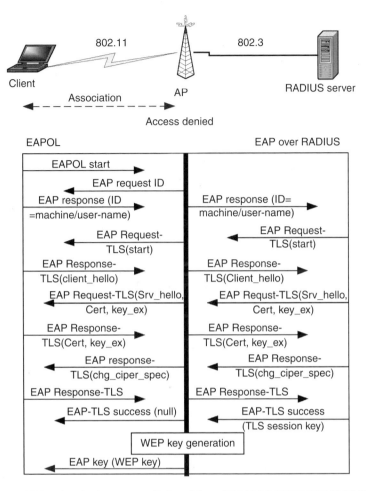

FIGURE 40.10   EAP-TLS authentication process. (Adapted from ORiNOCO [2002] ORiNOCO Technical Bulletin 048/B, Lucent, April.)

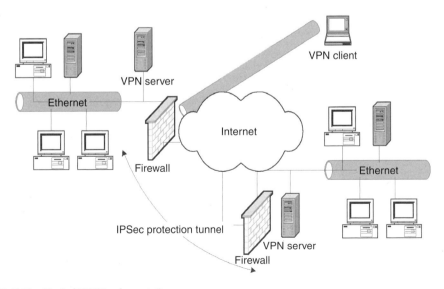

FIGURE 40.11   Typical VPN implementation.

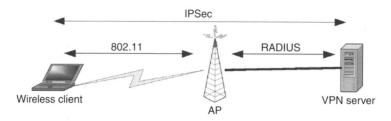

FIGURE 40.12   WLAN VPN structure.

Rincon (2002) provided a secured WLAN for home and *small-to-medium enterprise* (SME) users with an IPSec-based VPN. As a result of this, he proposed a prototype to automate an IPSec configuration and provide nomadic mobility for users roaming between multiple WLANs. All the WLAN clients are secured from a single workstation acting as a gateway. Caballero and Malmkvist (2002) also suggested using IPSec combined with application level access control for public WLANs.

The first two points earlier are illustrated in the VPN structure shown in Figure 40.11 and were central to much of this study.

40.6.1   VPN Techniques

VPNs employ a variety of security mechanisms, including cryptographic techniques and device- or user-based authentication. Tunneling offers encapsulation (of one protocol packet inside another) for encrypted VPN traffic, so that third parties cannot view the contents of packets transported over the network.

From a technology perspective, two categories of VPN can be identified based upon whether they operate across Layer 2 or Layer 3 network in the OSI Model. The common implementations of these two layers include the *point-to-point tunneling protocol* (PPTP) and *layer 2 tunneling protocol* (L2TP) at Layer 2 and the IPSec at Layer 3 (Halpern et al., 2001). Depending on the scale of an organization's network, and how much it values its data, both VPN security technologies can be applied. Most VPNs today make use of the IPSec protocol suite, as it ensures stronger protection against attacks. Figure 40.12 demonstrates how IPSec interacts with a WLAN.

40.6.2   Layer 2 VPNs — PPTP and L2TP

As indicated above, there are two types of Layer 2 VPN technology: PPTP and L2TP.

40.6.2.1   PPTP

PPTP RFC 2637 (Davies, 2001; Hamzeh et al., 1999) uses a username and password to provide authenticated and encrypted communications between a client and a gateway or between two gateways, without PKI support. PPTP uses a TCP connection for tunnel maintenance and *generic routing encapsulation* (GRE) encapsulated PPP frames for tunneled data. The payloads of the encapsulated PPP frames can be encrypted and/or compressed using Microsoft's proprietary encryption mechanism, MPPE based on RC4. Authentication methods include CHAP, PAP, and MS-CHAPv2. The use of PPP provides the ability to negotiate authentication, encryption, IP address assignment, and a variety of other operational characteristics for various protocols.

40.6.2.2   L2TP

L2TP (RFC 2661; Townsley et al., 1999) encapsulates PPP frames to be sent over a wide range of communication types, such as IP, frame relay, or ATM networks. When configured to use IP as its transport, L2TP can be used as a VPN tunneling protocol over the Internet. It uses UDP to send L2TP control messages for tunnel maintenance and L2TP-encapsulated PPP frames as the tunneled data on UDP port 1701. The encapsulated PPP frames can be encrypted or compressed.

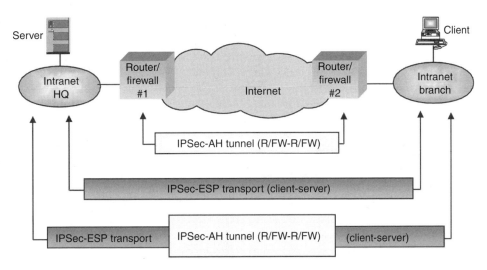

FIGURE 40.13   IPSec transport and tunnel modes in operation.

Through its use of PPP, L2TP gains multi-protocol support and provides a wide range of user authentication options, including CHAP, MS-CHAP, MS-CHAPv2, and EAP.

40.6.3 Layer 3 VPN — IPSec

IPSec, defined in RFC 2401 and 2411 (Kent and Atkinson, 1998; Rodgers, 2001; Thayer et al., 1998), provides integrity protection, authentication, and (optional) privacy and replay protection services for IP traffic. IPSec is currently the most popular protocol due to perceived security and its ability to use a single technology for both remote and Intranet/Extranet applications. IPSec sets up *security associations* (SAs), to negotiate security services between two points during the session. These SAs can be nested, allowing different IPSec relationships to be active on the same link, such as *quick mode* and *main mode*.[6] In order to establish an SA, IPSec relies on the *Internet security association and key management protocol* (ISAKMP, RFC 2408; Maughan et al., 1998) and *internet key exchange* (IKE, RFC 2409; Harkins and Carrel, 1998; Maughan et al., 1998), which define protocol formats and procedures for security negotiations, such as authentication type, encryption method, key lifetime, etc.

The IPSec protocol suite specifies cryptographic support of the *data encryption standard* (DES) and *triple DES* (3DES). Hashing functions can be selected from: HMAC-MD5 or HMAC-SHA1; HMAC-SHA1 is computationally more expensive than HMAC-MD5.

IPSec supports two main architectures (and corresponding packet types):

- *Encapsulating security payload* (ESP) header, which provides privacy, authenticity, and integrity.
- *Authentication header* (AH), which provides only integrity and authenticity for packets, but not privacy.

Two operational modes are provided in IPSec (see Figure 40.13):

- *Transport mode* secures an existing IP packet from source to destination. This mode allows end-to-end points to communicate over a secured tunnel.
- *Tunnel mode* puts an existing IP packet inside a new IP packet that is sent to a tunnel end point in the IPSec format, typically between a pair of firewalls/security gateways over an untrusted network.

---

[6]Microsoft Windows XP provides an IPSec monitor to view these SAs.

Both transport and tunnel modes can be encapsulated in ESP or AH headers. IPSec transport mode was designed to provide end-to-end security for IP traffic between two communicating systems (e.g., to secure a TCP connection or a UDP datagram). IPSec tunnel mode was designed primarily for network edge nodes, such as routers, or gateways, to secure other IP traffic inside an IPSec tunnel that connects one private IP network to another private IP network over a public or untrusted IP network. In both cases, the IKE negotiates secret keys and secured communication parameters to be used between two parties.

40.6.3.1　Device Authentication

This is also known as the machine-level authentication, which authenticates the device, not the user. The authentication methods include preshared secret key, and PKI certificates for mutual authentication. Kerberos may be used if the software provides the functionality, for example, Microsoft Windows 2000-based networks.

The preshared secret key method does not scale well as the size of a network grows. PKI provides scalability but can be complicated to deploy. However PKI is a good method for device authentication as it is supported with EAP-TLS and is included in the performance tests described in Section 40.7.

40.6.3.2　L2TP/IPSec Interworking

L2TP/IPSec have been defined to interwork in RFC 3193 (Patel et al., 2001), and the IPSec protocol suite is used to protect L2TP traffic over IP networks. By placing L2TP (see Section 40.6.2.2) as the payload within an IPSec packet, communications benefit from the standards-based encryption and authenticity of IPSec, as well as the interoperability to accomplish user authentication, tunnel address assignment, multi-protocol support, and multicast support using PPP. The combination, L2TP/IPSec, offers an IPSec solution to interoperable client-to-gateway VPN scenarios.

Due to incompatibilities between the IKE protocol and *network address translation* (NAT), it is currently not possible to support L2TP/IPsec or IPSec transport mode through a NAT device such as a firewall. Thus the deployment of IPSec transport mode should consider the requirements for NAT and scalability of access control (PKI is desirable in large scale networks).

40.6.4　Protocol Comparison

A comparison of the tunneling technologies is presented in Table 40.2.

40.7　Performance Issues with Security Architectures

Implementation of any security system necessarily has some impact on performance. In this section we test the basic security architectures discussed in the previous sections, thus quantifying the effect on throughput and response time for typical applications. These results are useful as they assist network engineers in quantifying the effect of implementing a selected security architecture that might be specified by the security policy for the operation of the WLAN. These tests are carried out on lightly loaded networks and do not address conditions of saturation and overload (Baghaei and Hunt, 2004). Further, only basic applications such as transaction-based HTTP systems and file transfer were evaluated. The range of security configurations for both 802.1X and VPNs are detailed in the following section.

40.7.1　Security Configuration Levels

Table 40.3 shows the different 802.1X and VPN security options that were evaluated. They range from level one (lowest level of security) to level ten (highest level of security) and include a combination of authentication, authorization, and encryption mechanisms.

TABLE 40.2   Network Security Protocol Differences (Microsoft, 1999a)

Feature	Description	PPTP/ PPP	L2TP/ PPP	L2TP/ IPSec	IPSec Tunnel
User authentication	Can authenticate the user that is initiating the communications	Yes	Yes	Yes	Work in Progress (WIP)
Machine authentication	Authenticates the machines involved in the communications	Yes[a]	Yes	Yes	Yes
NAT capable	Can pass through Network Address Translators to hide one or both end-points of the communications	Yes	Yes	No	No
Multi-protocol support	Defines a standard method for carrying IP and non-IP traffic	Yes	Yes	Yes	WIP
Dynamic tunnel IP address assignment	Defines a standard way to negotiate an IP address for the tunneled part of the communications. Important so that returned packets are routed back through the same session rather than through a non-tunneled and unsecured path and to eliminate static, manual end-system configuration	Yes	Yes	Yes	WIP
Encryption	Can encrypt traffic it carries	Yes	Yes	Yes	Yes
Uses PKI	Can use PKI to implement encryption and/or authentication	Yes	Yes	Yes	Yes
Packet authenticity	Provides an authenticity method to ensure packet content is not changed in transit	No	No	Yes	Yes
Multicast support	Can carry IP multicast traffic in addition to IP unicast traffic	Yes	Yes	Yes	Yes

[a] When used as a client VPN connection, machine-based authentication authenticates the user, not the computer. When used as a gateway-to-gateway connection, the computer is assigned a user ID and is authenticated.

TABLE 40.3   Security Levels

Level	802.1X Model	VPN Model (using firewall)
1	No Security (Default Installation)	
2	MAC authentication	PPTP Tunneling & CHAP
3	WEP authentication	IPSec Tunneling & CHAP
4	WEP (40 bit) authentication & encryption	PPTP & CHAP
5	WEP (128 bit) authentication & encryption	IPSec & CHAP
6	IEEE 802.1X EAP-MD5	IPSec & EAP-TLS
7	IEEE 802.1X EAP-TLS	IPSec & CHAP & DES
8	IEEE 802.1X EAP-MD5 & WEP encryption	IPSec & EAP-TLS & DES
9	IEEE 802.1X EAP-TLS & WEP encryption	IPSec & CHAP & 3DES
10	—	IPSec & EAP-TLS & 3DES

40.7.2   802.1X Model

There are *nine* security levels available from both 802.11 and 802.1X standards which are:

- *Level 1 No security*: This is the default security setting provided by vendors. There is no security mechanism activated with default configuration.

- *Level 2 MAC address authentication*: This level provides MAC address authentication carried out at the AP.
- *Level 3 WEP authentication*: The shared key authentication method specified in the 802.11 standard is used.
- *Level 4 WEP authentication with 40-bit WEP encryption*: This level combines the encryption algorithm to provide data privacy.
- *Level 5 WEP authentication with 128-bit WEP encryption*: The 128-bit shared key used is proprietary-based (in the case of Lucent).
- *Level 6 EAP-MD5 authentication*: This is one of the 802.1X standard's authentication methods, using password/username.
- *Level 7 EAP-TLS authentication*: This is the PKI-based authentication method supported by 802.1X.
- *Level 8 EAP-MD5 with 128-bit WEP encryption*: The combined effect of these tools provides strong data protection.
- *Level 9 EAP-TLS with 128-bit WEP encryption*: The combined effect of these tools provides the strongest level of encryption and authentication using per-session keys.

The Security Levels 2 to 5 of the 802.1X model are consistent with the 802.11 standard. Security Levels 6 to 9 are provided by the 802.1X standard.

40.7.3 VPN Model

The *VPN Model* is based upon the IPSec tunneling protocol suite. Tunneling is achieved by operating L2TP/IPSec (transport mode option of IPSec) and we refer to this as the IPSec tunneling technology. This is the end-to-end IPSec solution provided by Microsoft. There are two types of authentication methods deployed in the experiments:

- Device authentication using PKI with X.509 certificates
- User authentication based on open-standards — CHAP and EAP-TLS

These two methods provide direct comparison with the authentication methods deployed in the 802.1X model. PPTP has been selected to provide a performance comparison with tunneling techniques. Rincòn (2002) observed in his research that L2TP/IPSec tunneling produced greater performance overheads than PPTP.

The *ten* security levels available from the VPN/IPSec model are:

- *Level 1 No security*: This is the default security setting. Both the 802.lX and VPN models have this in common.
- *Level 2 PPTP tunneling with CHAP*: Authenticated tunnel provided using PPTP tunneling and CHAP authentication.
- *Level 3 IPSec tunneling with CHAP*: Authenticated tunnel using IPSec tunnel and CHAP authentication.
- *Level 4 Firewall with PPTP and CHAP*: Introducing a firewall into the architecture to filter the network traffic.
- *Level 5 Firewall with IPSec and CHAP*: A firewall is introduced into an IPSec-based network. From this level onward, all the security levels will be based on IPSec design.
- *Level 6 Firewall with IPSec and EAP-TLS*: Applying user-based PKI with device-based certificate authentication.
- *Level 7 IPSec with CHAP and DES*: Provides DES encryption to IPSec with CHAP user authentication.
- *Level 8 IPSec with EAP-TLS and DES*: Applies DES encryption to EAP-TLS user authentication.
- *Level 9 IPSec with CHAP and 3DES*: Provides strongest encryption (3DES) with CHAP.

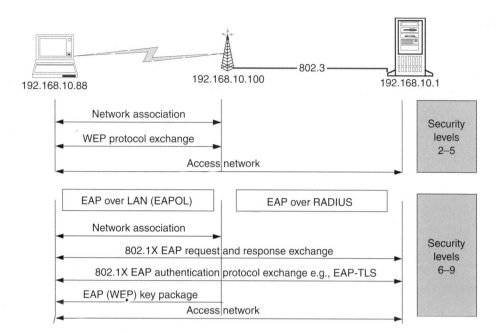

FIGURE 40.14   802.1X model logical flow.

- *Level 10 IPSec with EAP-TLS and 3DES*: Encrypts data traffic with the strongest encryption and user authentication methods.

The VPN model can be grouped into two parts; Security Levels 2 to 4 require authentication and tunneling using either PPTP or L2TP/IPSec before and after the firewall. Security Levels 5 to 10 require the IPSec protocol suite with a firewall to carry out authentication and encryption.

40.7.4   802.1X Model Implementation

The 802.1X model consists of the 802.11 access mechanism using open and shared key authentication, WEP encryption, and the 802.1X port-based authentication. By combining 802.1X with 802.11 protocols (as security Levels 6 to 9 in Table 40.3), the model provides a controlled wireless network with user identification, centralized authentication, and dynamic key management.

The 802.11 access mechanism was tested for security Levels 2 to 5. Static key management and basic network access was facilitated by the access point. For security Levels 6 to 9, the integration of 802.1X and 802.11 provides a dynamic key management and centralized authentication by the RADIUS server (Figure 40.14). Authentication methods chosen for the experiment were the EAP-MD5 and EAP-TLS; other proprietary authentication methods such as EAP-TTLS were not considered. This model did not support end-to-end security, because privacy and confidentiality were only ensured on the wireless link by WEP, but not enforced on the wired counterparts.

Wireless users were treated as if they existed in one subnetwork in an organization's intranet. A specific IP address was assigned to the wireless user, AP, and different components of the server.

RADIUS server and certificate authorities were added to the basic network structure to provide the 802.1X authentication support (Figure 40.15). The RADIUS server supports wireless user sign-on, and a certificate authority was used to issue certificates to users for EAP-TLS authentication.

40.7.4.1   Remote Access Policies

In the 802.1X model, the RADIUS server and its client, AP-2000, performed the policy check and authentication. Policies are effective after the activation of the RADIUS server. The 802.1X standard specifies the use

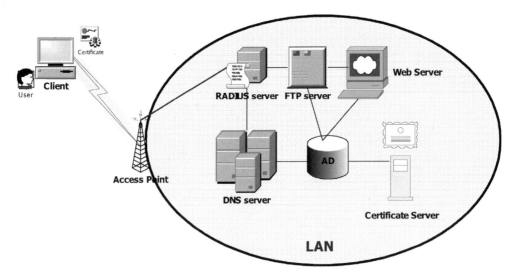

FIGURE 40.15   802.1X model implementation.

of a RADIUS server for enhanced user authentication. In the 802.1X model, a RADIUS server was activated after security Level 5 (see Section 40.7.3). The following remote access policies are based on this point:

1. General wireless access to the Intranet.
2. Wireless user groups can access the organization's intranet via IEEE 802.11 wireless transmission based upon the following two criteria:
   - Wireless access control using EAP-MD5:
     - Access is granted to an organization's intranet if the wireless user possesses the right username and password.
   - Wireless access control using EAP-TLS:
     - Access is granted to an organization's intranet if the wireless user possesses the correct X.509 digital certificate.

40.7.5   VPN Model Implementation

The VPN model deployed the IPSec mechanism to support an end-to-end secured communication from wireless to wired links. Tunneling protocols used were IPSec and PPTP. Most of the testing was based on the IPSec mechanism. However, as PPTP is a popular VPN tunneling choice used by organizations, we tested the effect of using PPTP tunneling against IPSec at the early stage of the model. PKI was selected over preshared keys to identify a user, and X.509 certificates were used for distributing and authenticating the keys. User authentication alternatives elected were CHAP and EAP-TLS. Section 40.4.1.1 explained that CHAP authentication is equivalent to EAP-MD5 authentication.

User authentication and tunneling options were tested before and after the firewall installation in security Levels 2 to 5. From there onwards, IPSec was the sole security protocol used with different user authentication alternatives. Security Level 3 and Levels 5 to 10 provided both user and device authentication. Encryption mechanisms such as DES and 3DES were used to ensure end-to-end data protection. The network was divided into subnetworks and treated the wireless subnetwork as an extranet (Figure 40.16).

40.7.5.1   Additional Components

In the VPN model, additional components were introduced into the network: RADIUS server, VPN server, certificate authority, and a firewall (Figure 40.17). The RADIUS server and certificate authority

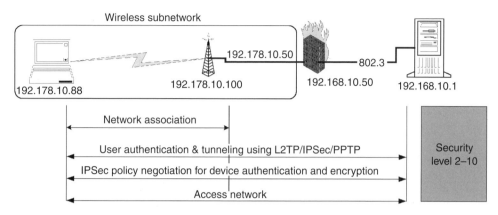

FIGURE 40.16 VPN model logical structure.

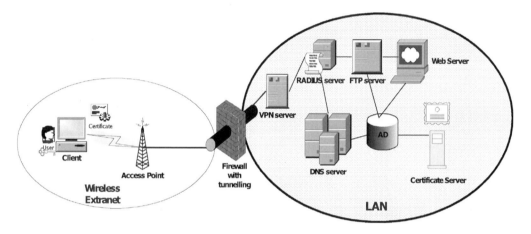

FIGURE 40.17 VPN model implementation.

were installed during the 802.1X model experiment; thus, only a VPN server and a firewall were newly implemented.

We configured the VPN server to be the RADIUS client and a firewall was set up between the access point and the VPN server.

40.8   Performance Test Results (Hunt et al., 2003)

40.8.1   HTTP and FTP Testing Scenarios

At the higher security levels (Table 40.3), the general trend is increased response times and decreased throughputs. The graphs shown in Figure 40.18 to Figure 40.21 present overviews of the security mechanisms' impacts on performance. The 802.1X model provides better response times and throughputs than the VPN model; the IPSec-based VPN model provides end-to-end security that results in higher performance overheads. We found that File Transfer Protocol performed better than HTTP because the later requires more interaction between the server and the client. However, some security mechanisms impacted FTP transactions more than they did HTTP-based applications.[7] Providing the same data file sizes for both traffic types represents a better measurement for HTTP.

---

[7]For example, EAP-MD5 authentication has less impact on HTTP performance than does WEP.

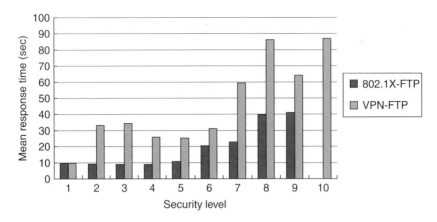

FIGURE 40.18   FTP mean response times for 802.1X and VPN models.

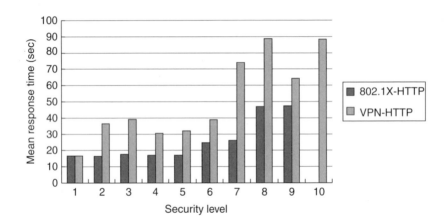

FIGURE 40.19   HTTP mean response times for 802.1X and VPN models.

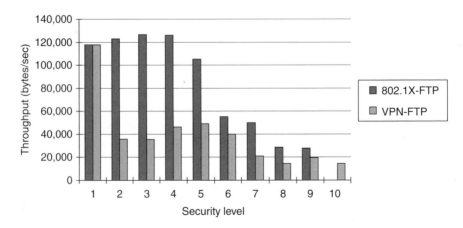

FIGURE 40.20   FTP mean throughputs for 802.1X and VPN models.

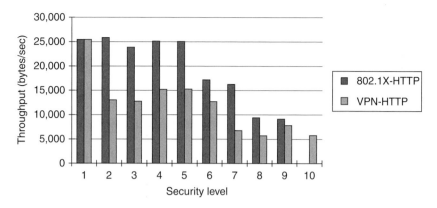

FIGURE 40.21 HTTP mean throughputs for 802.1X and VPN models.

An inverse relationship was found in both the 802.1X and VPN models between response time and throughput: as response time increased throughput decreased. See Figure 40.18 and Figure 40.20 for FTP traffic and Figure 40.19 and Figure 40.21 for HTTP traffic.

40.8.2   802.1X Model Test Results

Deploying the 802.1X infrastructure causes increased performance degradation compared to the 802.11 standard.

40.8.2.1   Authentication

MAC address authentication produces no performance overheads when compared to the default security setting, and thus should be used at all times. The 802.11 standard's WEP (shared key) authentication creates a minor positive effect on FTP throughput but decreases HTTP throughput by 7.5%.[8] Since the effect is small, WEP authentication should also be deployed.

The result of deploying AAA architecture for authentication imposes significant performance degradation. Using 802.1X authentication methods degraded network performance significantly compared to WEP authentication. Furthermore, EAP-TLS produced greater performance impacts than EAP-MD5, as it provides mutual authentication and key management.

40.8.2.2   Encryption

WEP encryption improved the network performance slightly (approximately 5% for HTTP, although again this may be as a result of vendors' implementation). However, longer key lengths of 128 bits impacted FTP performance by less than 20%. Thus depending on the nature of traffic transmission, the strongest key length should be deployed.

Use of WEP encryption in conjunction with 802.1X authentication (EAP-MD5) resulted in significant degradation — 100% increase in response times and 50% decreases in throughputs for both traffic types. Certificate-based authentication with WEP encryption provided similar results.

40.8.2.3   Interaction Between Authentication and Encryption

When comparing the combined effects of authentication and encryption, the overall degradation was severe. For FTP, the response time increased by more than 268% and throughput reduced by 73%. HTTP experienced similar results at a smaller rate. These results indicate that encrypting traffic can pose a substantial burden on performance, but it depends upon the type of authentication deployed. Only a minimal performance impact difference of 3% was found between EAP-MD5 and EAP-TLS.

---

[8]Vendors' implementation of these security architectures does create minor differences in performance.

40.8.3   VPN Model Test Results

Implementing a VPN model provides stronger protection as a result of its complex cryptographic methods, as well as providing improved key management.

40.8.3.1   Tunneling

An authenticated tunnel creates a dramatic response time delay of more than 245% for FTP and 113% for HTTP. Throughputs were reduced by more than 50%. The latency was greater than Rodgers' finding of 85%, but his study did not observe any impact on throughput. This may be due to the transaction size or the transmission medium differences. Different tunneling technologies impacted only the response time of the traffic, with HTTP traffic affected slightly. Certificate-based authentication (EAP-TLS) generates more than 20% delay and reduces throughput by 17% for all traffic types.

40.8.3.2   Encryption

Encryption poses a substantially greater burden than merely tunneling the traffic. Certain cryptographic methods produced definite performance degradation on FTP traffic, but HTTP performance improved. Deploying DES encryption created more than a 130% delay and 50% reduction in throughput for CHAP and EAP-TLS authentication methods. For stronger encryption only FTP traffic was degraded regardless of which authentication methods were used. Mixed results were found on traffic response time, with 8.1% increase for FTP and 13.4% improvement for HTTP accompanying CHAP authentication. FTP throughput decreased by 7.9%, and HTTP throughput improved by 16%. The type of user authentication chosen also played a role when interacting with the encryption algorithm; when using digital certificates, there was no difference between encryption methods chosen.

40.8.3.3   Interaction of Authentication and Encryption

Combing different authentication and encryption methods produced significant impacts on performance. Keeping the encryption algorithm constant, the choice of user authentication methods created substantial performance overheads and longer response times regardless of the encryption algorithm used. For DES, FTP is affected more than double the HTTP effect resulting in a 35.8% delay and 31.3% drop in throughput. But for 3DES, using CHAP or EAP-TLS authentication produced similar effects.

40.8.4   Overall Performance

These results are limited to a single cell WLAN. Integrating these models with a backbone network would provide further information on the effect of different traffic types and loads transferred over the network.

The two measured indicators of performance, throughput and response time were affected by the security mechanisms deployed. We can generalize our findings as follows (for more details, see Table 40.4):

1. Deploying MAC and WEP authentication created no overheads.
2. Different authentication methods created different levels of performance overhead; EAP-TLS generated the longest delay and decreased throughput. A comparison of the authentication mechanisms can be summarized as follows:
    - EAP-TLS > EAP-MD5/CHAP > WEP > MAC
3. Tunneling produced large overheads; IPSec overheads > PPTP overheads.
4. WEP encryption impact on performance varied; key length affected only response times. However, when WEP encryption was used in conjunction with 802.1X-based authentication, network performance was dramatically degraded.
5. Deploying DES cryptographic methods introduced large overheads, however, there was not much difference between 3DES and DES, especially when used with a certificate-based authentication.
6. The interaction of authentication and encryption generated different results from those resulting from adding encryption to the same authentication methods for FTP and HTTP traffic. EAP-TLS produced the most adverse impact.

TABLE 40.4 Summary of Security Impact on Performance

Security Impact on Performance (significant % change)	802.1X Model				VPN Model			
	Res. Time		Throughput		Res. Time		Throughput	
	FTP	HTTP	FTP	HTTP	FTP	HTTP	FTP	HTTP
Authentication								
MAC	0	0	0	0				
WEP	0	<10	0	<10				
EAP-MD5 > WEP	>100	<50	50–100	<50				
EAP-TLS > MD5/CHAP	<50	<10	<10	<10	<50	<50	<50	<50
Tunneling								
PPTP					>100	>100	50–100	<50
IPSec > PPTP					<10	<10	0	0
Encryption								
WEP (40-bit)	0	Imp.	0	Imp.				
WEP 40-bit versus 128-bit	<50	0	<50	0				
EAP-MD5 with WEP	<100	<100	<50	<50				
EAP-TLS with WEP	<100	<100	<50	<50				
CHAP with DES					>100	>100	50–100	100–100
TLS with DES					>100	>100	50–100	50–100
3DES > DES with CHAP					<10	Imp	<10	Imp
3DES > DES with TLS					0	0	0	0
Integrated Authentication & Encryption								
WEP versus MD5 with WEP	>100	>100	50–100	50–100				
TLS-WEP versus MD5 WEP	<10	0	<10	<10				
TLS > CHAP with DES					<50	<50	<50	<50
TLS > CHAP with 3DES					<50	<50	<50	<50

Imp. = improves the network performance.

TABLE 40.5 OSI Model and Security Mechanisms

OSI Layer	Name	Security Mechanism
7	Application	PKI, RADIUS, Kerberos, IPSec (IKE)
6	Presentation	
5	Session	SSL/TLS/SSH
4	Transport	
3	Network	IPSec (AH & ESP), PPTP
2	Link	802.1X, L2TP, CHAP
1	Physical	WEP

Source: Adopted from Caballoro, J., and D. Malmknist (2002) Master Thesis, Department of microelectronics and informations Technology, Royal Institute of Technology, RTH, Stockholm, January.

40.9 Summary

WLAN systems are beginning to incorporate a variety of new security architectures. It is important that an appropriate security mechanism is selected in order to comply with an organization's business policy. The security mechanisms discussed in this section are summarized showing their respective layers in the OSI model (Table 40.5).

Polices and resources developed for remote dial-up users may be helpful because of the similarity between a wireless and a dial-up client. Both are unknown users that must be authenticated before network access is granted, and the use of an untrusted network means that strong encryption is required.

Our experimental evaluation investigated the performance impact incurred with various security mechanisms, and found that the more secured a network became, the higher the performance impact. The VPN model incurred greater performance degradation than the 802.1X model as we expected; the VPN model provides end-to-end security with double authentication (device and user), a stronger encryption method, as well as better key management and tunneling technology. The general pattern for various application protocols carried over a network also followed our expectation that request-reply (HTTP) applications created more performance overheads than the FTP, as the interaction between two endpoints increased due to security negotiation and management. Our evaluation of different security levels showed that the higher the security level the greater the performance impact. There were exceptions however, and adding MAC address authentication, WEP authentication and encryption (40-bit), and 3DES encryption in the VPN model actually created minimal impact and yet provided at least a basic level of security.

Further research would be required to investigate different traffic characteristics for different cryptographic methods, the effects of multiple clients with resulting contention, use of 802.11g as well as 802.11b interfaces, new developments from the 802.11i working group (WPA2, AES, RSN etc.), as well as the performance resulting from secure handoff between WLAN (802.11) and Wireless Wide Area Network (cdma) networks.

References

Aboba, B. and D. Simon (1999) *PPP EAP TLS Authentication Protocol*. RFC 2716, IETF, October.

ANSI/IEEE. Std 802.11 (1999) *Part 11: Wireless LAN Medium Access Control (MAC) and Physical Layer (PHY) Specifications*, 1st ed. New York: IEEE, Inc., ISBN 0-7381-1658-0, 20 August.

Arbaugh, W. A., N. Shankar, and Y.C. Wan (2001) *Your 802.11 Wireless Network has No Clothes*. University of Maryland, Maryland, 30 March.

Baghaei, N. and R. Hunt (2004) Security performance of loaded IEEE 802.11b wireless networks. *Computer Communications*, Elsevier, U.K., Vol. 27, No. 17, 2004, pp. 1746–1756.

Bansal, R. (2001) Wireless networks: an electronic battlefield? *IEEE Microwave Magazine*, pp. 32–34, December.

Blunk, L. and J. Vollbrecht (1998) *PPP Extensible Authentication Protocol (EAP)*, RFC 2284, IETF.

Borisov, B., I. Goldberg, and D. Wagner (2001) Intercepting mobile communications: the insecurity of 802.11. In *Proceedings of the 7th Annual International Conference on Mobile Computing and Networking*, ACM, 16–21 July.

Brewin, B. and D. Verton (2002) Airport WLANs lack safeguards. *Computerworld*, 16 September.

Caballero, J. and D. Malmkvist (2002) Experimental study of a network access server for a public WLAN access network. Master Thesis, Department of Microelectronics and Information Technology, Royal Institute of Technology, KTH, Stockholm, January.

Calhoun, P., J. Arkko, G. Zorn, and J. Loughney (2002) *DIAMETER Base Protocol*. Internet Draft, IETF, June.

Congdon, P., B. Aboba, T. Moore, A. Smith, G. Zorn, and J. Roese (2002) *IEEE 802.1X RADIUS Usage Guidelines*. Internet-Draft, IETF, 17 June.

Convery, S. and D. Miller (2001) *SAFE: Wireless LAN Security in Depth*. Cisco Systems, Inc., December, www.cisco.com/warp/public/cc/so/cuso/epso/sqfr/safw1_wp.pdf

Davies, J. (2001) Virtual Private Networking with Windows 2000: Deploying Remote Access VPNs, Microsoft, August.

Dierks, T. and C. Allen (1999) *The TLS Protocol Version 1.0*, RFC 2246. IETF.

Enterasys Network (2002) Wireless Demilitarised Zone (WDMZ) — Enterasys Network's Best Practices Approach to an Interoperable WLAN Security Solution. Enterasys Network. www.bitpipe.com/data/detail?id=1015306519_611&type=RES&x=187349215

Finseth, C. (1993) An Access Control Protocol, Sometimes Called TACAS, RFC1492, IETF.

Fluhrer, S., I. Mantin, and A. Shamir (2001) Weaknesses in the key scheduling algorithm of RC4. In *Proceedings of the 8th Annual Workshop on Selected Areas in Cryptography*, August.

Funk, P. and S. Blake-Wilson (2002) *EAP Tunnelled TLS Authentication Protocol (EAP-TTLS)*. Internet Draft, IETF, November.

Gast, M. (2002) 802.11 Network Deployment, In *802.11 Wireless Networks: The Definitive Guide*, O'Reilly, ISBN 0-596-00183-5, April.

Halpern, J., S. Convery, and R. Saville (2001) *SAFE VPN: IPSec Virtual Private Networks in Depth.* Cisco System Inc., www.mnemonic.no/linker/pdf/IPSec_VPN_in_Depth.pdf

Hamzeh, K., G. Pall, W. Verthein, J. Taarud, W. Little, and G. Zorn (1999) *Point-to-Point Tunnelling Protocol (PPTP)*. RFC2637, IETF, July.

Hansen, J. V. (2001) Internet commerce security: issues and models for control checking. *Journal of the Operational Research Society*, 52: 1159–1164.

Harkins, D. and D. Carrel (1998) *The Internet Key Exchange (IKE)*. RFC 2409, IETF, November.

Harris, B. A. (1998) Firewall and virtual private networks. Master Thesis, Department of Computer Science, University of Canterbury, Christchurch.

Held, G. (2001) The ABCs of IEEE 802.11. *IT Professional*, 3: 49–52.

Housley, R. and D. Whiting (2001) IEEE P802.11 Wireless LANs: temporal key hash. Document Number: IEEE 802.11-01/550r3, IEEE Task Group I, 20 December, http://grouper.ieee.org/groups/802/11/Documents/DocumentHolder/1-550.zip

Housley, R., D. Whiting, and N. Ferguson (2002) *IEEE P802.11 Wireless LANs: Alternate Temporal Key Hash.* Document Number: IEEE 802.11-02/282r2, IEEE Task Group I, 23 April, http://grouper.ieee.org/groups/802/11/Documents/DocumentHolder/2-282.zip

Hunt, R., J. Vargo, and J. Wong (2003) Security architectures in wireless and mobile network performance. In *Proceedings of the 5th IEEE International Conference on Mobile and Wireless Communications Networks (MWCN, 2003)*, Singapore, October.

IEEE Std. 802.1X (2001) *Port-Based Network Access Control.* New York: IEEE, Inc. ISBN 0-7381-2627-5, 25 October.

InterLink Networks (2001) *Wireless LAN Access Control and Authentication.* InterLink Networks, www.interlinknetworks.com/references/WLAN_Access_Control.html

Karygiannis, T. and L. Owens (2002) *Draft: Wireless Network Security — 802.11, Bluetooth and Handheld Devices.* USA, National Institute of Standards and Technology.

Kent, S. and R. Atkinson (1998) *Security Architecture for the Internet Protocol.* RFC 2401, IETF, November.

Lee, Y.-J. (2002) Mobile IP and AAA Architecture for Wireless LAN. Master's Thesis, Department of Electrical Engineering, National Taiwan University, Taipei, June.

Maier, P. (2000) *Ensuring Extranet Security and Performance.* Information Systems Management, 17: 33-40.

Maughan, D., M. Schertler, M. Schneider, and J. Turner (1998) *Internet Security Association and Key Management Protocol (ISAKMP)*. RFC2408, IETF.

McDonough, J. (2002) *Wireless LANs That Tell All.* Wireless NewsFactor, 13 March, www.wirelessnewsfactor.com/perl/story/16748.html

Microsoft (1999) Microsoft Privacy Protected Network Access: Virtual Private Networking and Intranet Security, Microsoft, 13 May. www.microsoft.com/windows2000/docs/nwpriv.doc

Microsoft (2002) *Wireless 802.11 Security with Windows XP*. Microsoft, www.microsoft.com//windowsxp/pro/techinfo/administration/wirelesssecurity/XP80211Security.doc

Mitton, D., M. St. Johns, S. Barkley, D. Nelson, B. Patil, M. Stevens, and B. Wolff (2001) *Authentication, Authorisation, and Accounting: Protocol Evaluation*. RFC3127, IETF, June.

O'Hara, B. and A. Petrick (1999) *The IEEE 802.11 Handbook: A Designer's Companion.* New York: The IEEE, Inc. ISBN 0-7381-1855-9.

ORiNOCO (2002) *Principles of 802.1X Security.* ORiNOCO Technical Bulletin 048/B, Lucent, April.

Patel, B., B. Aboba, W. Dixon, G. Zorn, and S. Booth (2001) *Securing L2TP Using IPSec.* RFC3193, IETF, November.

Rigney, C., W. Willats, and P. Calhoun (2000a) *RADIUS Extensions*, RFC 2869: IETF. June.

Rigney, C., W. Willats, A. Rubens, and W. Simpson (2000b). *Remote Authentication Dial in User Service (RADIUS)*. RFC2865, IETF, June.

Rincón, R. B. (2002) Secure WLAN Operation and deployment in home and small to medium size office environment. Master's Thesis, Department of Engineering and Computer Science, Technische Universitat Berlin, Berlin, 6 March.

Rodgers, C. (2001) Virtual private networks: strong security at what cost? Department of Computer Science, University of Canterbury, Christchurch, November.

Roshan, P. (2001) *802.1X Authenticates 802.11 Wireless.* Network World Fusion, 24 September.

Seng, N. T. (2002) *Secured Public Access WLAN.* In *Proceedings of the Asia Pacific Regional Internet Conference on Operational Technologies (APRICOT)*, Bangkok, 27 February – 27 March.

Simpson, W. (1996) PPP Challenge Handshake Authentication Protocol (CHAP). RFC 1334, IETF, August.

Stubblefield, A., J. Ioannidis, and A. D. Rubin (2001) Using the Fluhrer, Mantin, and Shamir *Attack to Break WEP*: AT&T Lab Technical Report TD-4ZCPZZ.

Task Group i. (2002) *TGi Security Overview*, IEEE, Inc. Document Number IEEE 802.11-02/114r1.

TGi. (2002) *Task Group 802.11i*: IEEE, Inc. www.ieee802.org/11.

Thayer, R., N. Doraswamy, and R. Glenn (1998) IP Security Document Roadmap. RFC2411, IETF, November.

Townsley, W., A. Valencia, G. Pall, G. Zorn, and B. Palter (1999) *Layer Two Tunnelling Protocol "L2TP"*. RFC 2661, IETF, August.

Walker, J.R. (2000) *Unsafe at Any Key Size; An Analysis of the WEP Encapsulation.* Technical Report 03628E, IEEE. 802.11 Committee, March. http://grouper.ieee.org/groups/802/11/Documents/DocumentHolder/0-362.zip

Wearden, G. (2002) *Heard of Drive-by Hacking?* Meet Drive-by Spamming, ZDNetUK, 5 September.

Weatherspoon, S. (2000) Overview of IEEE 802.11b Security. *Intel Technology Journal*, Q2.

WECA (2001a) *802.11b Wired Equivalent Privacy (WEP) Security*: WECA, 19 February, www.wi-fi.org/WI-FiWEPSecurity.pdf

WECA (2001b) *Wireless LAN Research Study. Survey*, WECA, http://wifi.org/pptfiles/Wireless_LANResearch_ExecutiveSummary.ppt

Wong, J. (2001) Policy based network management. Department of Accountancy, Finance, and Information Systems, University of Canterbury, Christchurch, October.

Wu, T. (2000) The SRP Authentication and Key Exchange System. RFC 2945, IETF, September.

41

Security in Wireless Ad Hoc Networks

41.1 Why Is Security an Issue?...............................**41**-938
      Wireless Ad Hoc Networking Environment • Applications of
      Ad Hoc Networks • Vulnerabilities of Ad Hoc Networks
41.2 Security Challenges **41**-942
      Security Goals • Security Requirements
41.3 Security Technologies..................................**41**-944
      Symmetric Encryption • Public Key Encryption • Hashing
      Function • Digital Signature and Certificate •
      Identity-Based Cryptography • Threshold Secret Sharing
      Scheme • Intrusion Detection
41.4 Current Countermeasures.............................**41**-948
      Secure Routing • Trust and Key Management • Service
      Availability Protection
41.5 Conclusions **41**-954
      Acknowledgments.......................................**41**-954
      References **41**-954

Hongmei Deng
Dharma P. Agrawal

The last few years have seen a tremendous growth in the telecommunications world, aiming to provide "anywhere, anytime" communication services to people in every location on the surface of the earth. With the advent of wireless transmission, people are no longer bounded by the harness of wired networks and are able to access and share information on a global scale. Due to the increasing availability of many lightweight, compact, portable computing devices, like laptops, personal digital assistants (PDAs), cellular phones, and electronic organizers, various wireless transmission paradigms have been developed, some of which are already in daily use for business as well as personal applications.

Wireless local area networks (WLANs) have gained in usefulness and acceptability in providing a wider range of coverage and increased transfer rates. The most well-known representatives of WLANs are based on standards like IEEE 802.11, HiperLAN, and their variants. Recently, a new type of WLAN called the wireless ad hoc networks has emerged. A wireless ad hoc network is formed without the support of any fixed infrastructure, with functions based on the collaborative efforts of multiple self-organizing mobile nodes. The underlying infrastructureless property makes it different from traditional WLANs and also provides an extremely flexible method for establishing communications in situations where geographical or terrestrial constraints demand a totally distributed network system. A set of applications, such as military tracking, hazardous environment exploration, reconnaissance surveillance, and instant conference setup, is the key motivation for the recent research efforts in this area.

While the inherent characteristics of an ad hoc network make it useful for many applications, they also bring in a lot of research challenges. One such important issue is the security; the problem becomes more serious in these applications where security is a critical factor. Recent research have shown that wireless ad hoc networks are more vulnerable than their traditional counterparts, and the conventional security approaches adopted for traditional networks are not directly applicable to wireless ad hoc networks. Designing an efficient security mechanism faces several new challenges. A distributed and light-weight security scheme is necessary due to its infrastructure-less nature and low resource availability. Recently, a lot of research efforts have been made with special focus on enhancing the security of wireless ad hoc networks.

This chapter provides a general understanding of the security issues in wireless ad hoc networks. Beginning with a brief introduction on an ad hoc networking environment and its wide applications, we highlight the associated security threats and challenges that are now facing wireless ad hoc networks. We also describe some basic network security concepts and strategies, and then present a detailed overview of the existing security schemes.

41.1   Why Is Security an Issue?

This section introduces the basic characteristics of a wireless ad hoc network environment, describes its various existing and further applications, and highlights its security vulnerabilities and threats to address why security is an important issue in wireless ad hoc networks.

41.1.1   Wireless Ad Hoc Networking Environment

Wireless local area networks support two types of architectures by offering two modes of operation: client–server mode and ad hoc mode. The client–server mode (Figure 41.1[a]) is chosen in architectures where individual network devices connect to a wired network via a dedicated infrastructure (known as access point) that serves as a bridge between the wireless mobile devices and the wired network. This type of connection can be compared with the centralized local area network (LAN) architecture that has servers offering services and clients accessing them. A larger area can be covered by installing several access points, with overlapping access areas as in cellular structure. In the ad hoc (also known as peer-to-peer) mode (Figure 41.1[b]), the connections between two or more devices are established in an instantaneous manner without the support of a central controller. These two architectures are commonly referred to as the infrastructure-based and the infrastructure-less (or ad hoc) networks.

The idea behind wireless ad hoc networking is to mitigate the requirement of a wireless system with a fixed permanent infrastructure as its backbone by allowing out-of-range communication through a multi-hop fashion, thereby forming a network when needed, even in specific situations where building a fixed infrastructure is difficult due to geographical, terrestrial, time, or feasibility constraints. Wireless

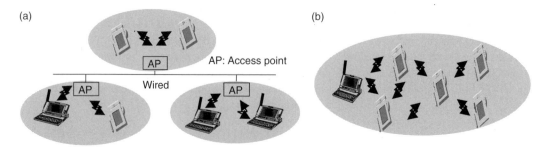

FIGURE 41.1   WLAN architectures. (a) Infrastructure-based wireless network and (b) infrastructure-less wireless network (ad-hoc network).

ad hoc network is considered to be a collection of wireless mobile nodes, forming a multi-hop packet radio network without the support of any established infrastructure or centralized administration [1]. The mobile nodes in ad hoc networks are not bound to any centralized control like base stations or mobile switching centers. Each node acts not only as a host, but also as a router, forwarding packets for those nodes that are not in direct transmission range to another. This feature offers unrestricted mobility and dynamically changing connectivity to the nodes of an ad hoc network, while requiring the nodes to cooperate with each other in maintaining links and routing information to each other. By nature, ad hoc networks are highly dynamic and self-organizing with few channels. They generally have the following characteristics:

1. *"By air" transmission:* There is no physical link between the nodes of an ad hoc network and, therefore, the users transmit information through the "air." Using wireless transmission brings with it unlimited freedom and "anytime, anywhere" services to its mobile users.
2. *Absence of a fixed infrastructure:* In contrast to the well-known infrastructure, wireless ad hoc networks do not have any *a priori* fixed infrastructure. Not relying on any such support, all the mobile nodes have equal functionality and perform the network operations (packet forwarding and routing) in a distributed way.
3. *Dynamic network topology:* Wireless nodes roam independently and are capable of moving freely at an arbitrary speed and direction. Thus, the topology of the network may change in a frequently random fashion.
4. *Constrained network bandwidth:* The use of wireless communication typically implies a lower bandwidth than that of traditional wired networks. Moreover, achievable throughput of wireless communications take into account the effects of multiple access, fading, noise, interference conditions, and so on, and is often much less than the maximum transmission rate of a radio.
5. *Energy constrained nodes:* To support mobility, almost all the mobile nodes in ad hoc networks have batteries or other exhaustible means as their energy source; hence they are considered as energy constrained mobile hosts. Moreover, they are also resource-constraint in terms of storage memory, computational capability, weight, and size.

41.1.2 Applications of Ad Hoc Networks

The infrastructure-less and self-configurable properties make ad hoc networks a natural option for many problems that require a totally distributed system. When properly combined with satellite-based information delivery, some new safety/rescue scenarios of ad hoc networks can be developed. As more advanced features of wireless mobile systems are being developed, including support for multimedia data rates, global roaming capabilities, and integration among different network structures, various new applications are expected to emerge. Here, we summarize some possible applications, although many are still in the early phase of exploration.

Military tactical networks: is the most critical application of ad hoc networks. Since it is impractical to have a prebuilt infrastructure in a battlefield, ad hoc networking can instantaneously provide anywhere and anytime communications between military units. As an inherent benefit of decentralized network, the loss of any one unit does not disrupt the network operation, as long as there are other units that can still provide the network services. Examples of military applications are the Tactical Internet [2] and the Saab NetDefense concept [3].

Personal area network: is a simple application of ad hoc networks that works by interconnecting different devices of an individual, for example, desktop computers, notebooks, PDAs, cellular phones, and the like. This enables a wireless network of mobile and static devices to communicate with each other without any centralized administration.

Instant network: allows multiple mobile users to dynamically set up a network to share data and documents without the support of an external LAN or Internet connection. This may be the most intuitive application of ad hoc networks because the mobile users may not have a prior relationship and these devices may also have different capabilities in terms of power source, CPU performance, and memory size.

Disaster area network: is another potential application of ad hoc networks that can be used to allow communication among rescue workers and other personnels, thereby supporting relief efforts in areas where a fixed infrastructure may not be present or may have been destroyed by natural disasters like wildfire, earthquake, flooding, oil spill, gas leak, toxic material spill, volcanic eruption, and so on. For example, telemedicine is an application by which a victim of a traffic accident in a remote location can be assisted either by providing emergency video conference assistance from a surgeon or by accessing the medical records in a hospital.

Sensor network: is a special class of ad hoc networks with very little or no mobility. It may consist of hundreds or thousands of communication-enabled sensor nodes, with each node containing one or more sensors to measure movement, texture, sound, gas, light, contaminants, or temperature. Such a network differs significantly from the other types of ad hoc networks in terms of extremely limited processing power and resources of the sensor nodes, and can be used in both military and nonmilitary domains, such as surveillance and environmental monitoring as illustrated in Figure 41.2 and Figure 41.3.

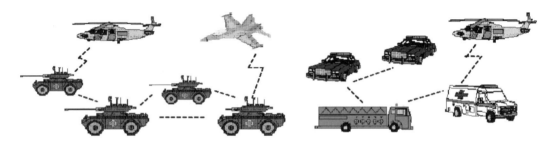

FIGURE 41.2   Defense and emergency applications of ad hoc networks.

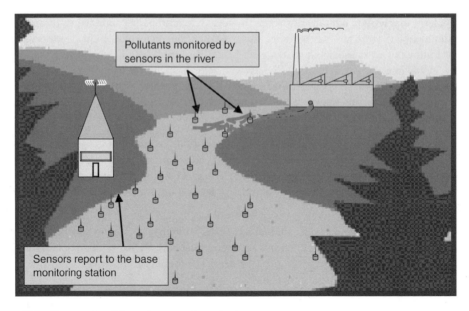

FIGURE 41.3   Sensor network in environmental monitoring.

41.1.3 Vulnerabilities of Ad Hoc Networks

While ad hoc networks provide great flexibility in establishing on-the-fly communication links, they also bring a lot of research challenges at the same time. As mentioned earlier, one important issue is security.

The truth is that the underlying characteristics of wireless ad hoc networks make them highly vulnerable. First, the use of wireless communication and the freedom of mobility place the network nodes at a greater risk of being captured, compromised or hijacked physically than their wired counterparts. Second, lack of a predefined infrastructure means that there is no centralized control for the network services. The network functions in a distributed fashion by cooperative participation of all nodes. The decentralized decision making is prone to attacks that are designed to break the cooperative algorithms. Third, dynamically changing topology aids the attackers to update routing information maliciously by pretending to be legitimate topological changes. Fourth, a constrained network bandwidth may limit the number and size of the message transmitted during a protocol execution. Fifth, energy- and resource-constrained mobile nodes are sensitive to the use of computationally complex algorithms. Finally, node selfishness is a security issue specific to ad hoc network. Since routing and network management are carried by all the available nodes in ad hoc networks, some nodes may selfishly deny the routing request from other nodes to save their own resources (e.g., battery power, memory, CPU).

Based on the fundamental vulnerabilities of wireless hoc networks, various potential attacks have been widely exploited [4–7] and can be classified into *passive attacks* or *active attacks*, depending on the nature of the attack. Figure 41.4 shows a categorization of the attacks.

Passive attack: is a kind of attack that listens to the routing traffic, aiming to gather valuable information, such as network connectivity, location, traffic distribution, and so on. The main goal of a passive attack is to create a threat against network privacy, rather than disrupt the operation of the routing protocol. Since passive attackers do not involve any alteration of transmitted message or disruption of the network activity, they are very difficult to prevent and detect.

Active attack: is an attempt to improperly modify routing message, inject erroneous messages, or impersonate a node in order to confuse the routing procedure and degrade the network performance. Depending up on the purpose of the attackers, active attacks can be further categorized into *routing disruption attacks* and *resource consumption attacks.* In the former the attacker attempts to route legitimate packets in a malicious way, whereas in the later the attacker aims to consume valuable network resources, for example, bandwidth, limited memory space of other nodes, and computational power.

The following are several common attack subcategories, depending on the actions taken by an attacker.

Eavesdropping: This is a type of a passive attack that can be performed easily in a wireless environment. An attacker can obtain direct knowledge of the network by intercepting the transmitted data packets. Passive eavesdropping can be prevented by various encryption schemes to protect the privacy of data transmission.

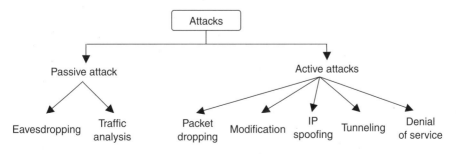

FIGURE 41.4 Classification of attacks.

Traffic analysis: An attack may not only collect information by passive eavesdropping, but it may also extract valuable information from the characteristics of the transmission like identity of the source and destination nodes, the number of data packets transmitted, and frequency of data transmission, and the like. The extracted information may allow the attacker to do a further analysis and deduce some sensitive knowledge. This type of attack is especially useful when encryption schemes are used in data transmission. Passive traffic analysis can be considered as one of the most difficult attacks to deal with.

Packet dropping: An attacker can hide itself from other communicating nodes by simply destroying or discarding all the packets it receives. This type of attack, also called *ignorance attack*, has limited impact on network performance for as long as there are multiple routes between the source and the destination nodes. Furthermore, an attacker can launch a *black hole attack* by sending forged routing packets, so that it can route all the packets to some destinations to itself, and then discard them. As a special case, an attacker can choose to selectively drop packets instead of all the packets, thus creating a *gray hole attack*. This type of attack can be easily deployed by impersonating the destination nodes. Moreover, impersonation may also happen to the source node or any other nodes in the network.

Modification: This attack maliciously modifies the message during transmission between the communicating nodes, implying that the communicating nodes do not share the same view of the transmitted packet. An example could be that of an attacker trying to *shorten* or *lengthen* the node list of the routing packet transmitted in the source routing protocol.

IP spoofing: An attacker can send a packet containing its own message authentification code (MAC) address and a victim's IP address, thereby usurping the IP-to-MAC address binding of the victim from the other neighbor's address resolution protocol (ARP) cache. This causes the attacker to receive the packets intended for the victim.

Tunneling attack: An attacker can also create another type of routing disruption, named *tunneling attack* [6], using a pair of malicious nodes linked together via a private network connection. Every packet that node A receives from the ad hoc network it forwards to node B using a private network connection, and then node B forwards this packet normally; similarly, node B may forward all the packets received to node A. Such an attack potentially disrupts routing by short-circuiting the normal flow of routing packets.

Denial of service: It can be caused by any malfunction of the packet forwarding service. An attacker can mount a *replay attack* by sending some old messages to a node, aiming to overload the network and deplete the node's resources. More seriously, an attack can create a *rushing attack* by sending many routing request packets with high frequency, in an attempt to keep other nodes busy with the route discovery process, so the network service cannot be achieved by other legitimate nodes. *Replay attack* and *rushing attack* are also considered as resource consumption attacks.

We note that all the attacks discussed above can be generated either by an outside intruder or by an inside intruder. The attacks caused by the nodes that do not belong to the network are considered as *external attacks*, while the attacks created by the compromised internal nodes are called *internal attacks*. Internal attacks usually have a more serious effect in degrading network performance, since the compromised node would have access to all the cryptographic keys of the network and it may also cooperate with other adversaries or compromised nodes.

Therefore, we have emphasized again that ad hoc network presents a large security problem and securing ad hoc networks is critical in the development of any real application of wireless ad hoc networks.

41.2   Security Challenges

This section describes the challenges encountered in designing an efficient security protocol in wireless ad hoc networks.

41.2.1 Security Goals

Similar to traditional networks, the goals of securing an ad hoc network can be characterized by the following attributes [8]: availability, confidentiality, integrity, authentication, and non-repudiation:

1. *Availability* ensures that the intended network services are available to the intended parties whenever needed.
2. *Confidentiality* ensures that the transmitted information can be accessed only by the intended receivers and is never disclosed to unauthorized entities. It protects the information from threats based on eavesdropping.
3. *Authentication* allows a node to validate the identity of the peer node it is communicating with. It guards messages against impersonation, substitution, or spoofing. Without authentication, an adversary can masquerade as a node, thus gaining unauthorized access to resource and sensitive information and interfering with the operation of other nodes.
4. *Integrity* guarantees that information is never corrupted during transmission. Only the authorized parties are able to modify the information. It ensures that any modification of the messages will be detected. Message modification includes writing, changing, deleting, creating, and delaying, or replaying of a transmitted message.
5. *Non-repudiation* ensures that an entity can prove the transmission or reception of information by another entity, that is, a sender/receiver cannot falsely deny having received or sent certain data.

41.2.2 Security Requirements

Besides these basic security goals, designing an efficient security scheme to protect ad hoc networks faces several new challenges.

First, the critical issue concerning the design of security service in ad hoc networks is not to rely on any centralized entity, and the security management should be implemented in a distributed fashion. An ad hoc network is a distributed network in which each node has equal functionality. There are no dedicated service nodes that can work as a trusted authority to generate and distribute the network keys or provide certificates to the nodes, as the certificate authority (CA) does in the traditional public key infrastructure (PKI) [9,10] supported approach. Even if a service node can be defined, ensuring the availability of the service node to all the nodes in such a dynamic network is not an easy task. Moreover, with limited physical protection, the service node is prone to a single point of failure, that is, by just damaging the service node, it is possible to paralyze the whole network . Thus, the two commonly used architectures in regular wired and wireless networks — centralized and hierarchical architectures — are vulnerable to the loss of the central controlling node and are not suitable for ad hoc networks. Instead, distributed strategies are often preferred and have shown significance in recent research.

Second, the low resource availability necessitates its efficient utilization and prevents the use of complex authentication and encryption algorithms. Light-weight authentication and encryption schemes with resource awareness are required. Traditional public key cryptography-based authentication and encryption mechanisms are fully developed in securing networks. Unfortunately, generation and verification of digital signatures are relatively expensive, which limits its wide application to ad hoc networks. Symmetric cryptography is more efficient than public key-based asymmetric primitives as it requires both the sender and the receiver to share a secret. In ad hoc networks, the problem is how to distribute the shared secret safely so that only the two parties (correct sender and receiver) receive it and not anyone else. It is thus challenging to define some new efficient cryptography mechanisms for designing a light-weight authentication and encryption scheme.

Third, an integrated security scheme that combines intrusion prevention and intrusion detection mechanism is necessary. Intrusion prevention implies developing secured protocols or modifying the logic of existing protocols to make them secure. Most of the key-based security protocols belong to this type. Intrusion prevention approach, usually as a first line of defense, can provide a certain level of security to wireless ad hoc networks, but it is far from being sufficient. Developing a system that is absolutely secure is generally impossible. There are always some weak links by which attackers can get access to the network and damage the information system. However, while encryption and authentication mechanisms can efficiently prevent intrusion, they cannot totally eliminate it, especially those initiated from inside. For example, any compromised node can easily launch an *ignorance attack* or create a *rushing attack*, since the encryption and authentication mechanisms become meaningless for them as they owe the keys for encryption and authentication. To mitigate the problem, intrusion detection is designed to protect ad hoc networks as a complementary mechanism. The two approaches are not independent of each other and should work together to provide security services. Intrusion prevention approaches can efficiently deal with the attacks from outsiders by constraining the network access control, and intrusion detection can efficiently detect some active attacks from inside the network. Therefore, a security scheme combining these two mechanisms is better suited to secure ad hoc networks.

Fourth, various ad hoc network applications exhibit numerous inconsistencies, and designing an efficient security scheme for a generalized ad hoc network makes the problem complex and more difficult to address. The level of security should depend on the type of network application under consideration. For example, all the devices in a personal area network can be considered to be a trusted entity and the network is usually operated in a friendly environment; thus, there is no need to spend time and consume limited network bandwidth on authentication. Encryption may be needed to protect secrecy. To secure the application type of instant conference, a totally distributed key management mechanism and a suitable encryption/decryption mechanism are required. There is a tradeoff between the network capabilities and security, as a very secure system may become too complex, slow, and difficult to administer. If the information transmitted is not critical and the network is used in a friendly environment, a lower level of security can be achieved using some simple security mechanisms. While the above two applications can be considered as the two extremes, military network stands in between. The military network, in general, is a planned network in which nodes usually have the same origin, for example, coming from a same military unit or control center; thus, it can perhaps be assumed that the nodes are supplied with some initial data structures such as certificates, passwords, user names, and the like. Moreover, the requirements of military networks are generally high from both safety and efficiency point of view.

Therefore, an efficient security scheme for ad hoc networks should be an integration of intrusion prevention and intrusion detection mechanisms, which must be implemented in a distributed fashion with resource awareness and a reasonable degree of security.

41.3 Security Technologies

To discuss the security strategies in ad hoc networks, we first need to familiarize ourselves with basic security technologies. This section covers some of the security concepts and strategies employed in ad hoc networks. Cryptography is one such strategy by which a sender can disguise data so that a receiver can recover the data from the disguised message, but an intruder cannot gain any information from intercepted data; digital signature and certificate provides authentication and nonrepudiation; message digest and hashing function ensures message integrity; threshold secret sharing presents an efficient method for group communication; and intrusion detection provides one more line of defense.

41.3.1 Symmetric Encryption

In its simplest form, symmetric encryption requires that the sender and receiver of a message use one value (key) to encrypt and decrypt the traffic, respectively. The process of symmetric key encryption is illustrated in Figure 41.5. A plain text message m is encrypted using a shared key k, resulting in the cipher text $c = E_k(m)$. To recover the plain text message, the cipher text is decrypted using the same key. Since symmetric encryption involves a single key to encrypt and decrypt, the shared key must be distributed over a secured communication channel.

The Data Encryption Standard (DES) [11] is an example of symmetric cryptography algorithms. DES was established in conjunction with IBM and published in 1977 as the first standardized cryptography

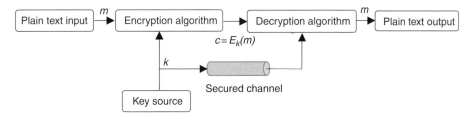

FIGURE 41.5 Symmetric key encryption.

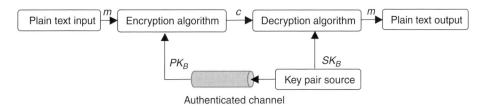

FIGURE 41.6 Public key encryption.

algorithm for the industry. DES utilizes a key of 64 bits, of which 56 are used directly for encryption and decryption and 8 are used for parity checking. The encryption and decryption algorithms are identical, and each operates on a block of 64 bits. DES performs 16 rounds of confusion and diffusion iteratively on each group of 8 plain text letters. DES has been in use for many years and is widely available in both hardware and software versions. However, the 56-bit key in DES is not considered to be long enough for many applications. To thwart the exhaustive search attack on the key space, triple DES increases the key length to 112 bits and performs three times encryptions with the DES using two different keys. That is, data is encrypted via DES three times, the first time by the first key, the second time by the second key, and the third time by the first key again.

There are also some new symmetric key encryption algorithms that have been developed, such as the fast encryption algorithm (FEAL) [12], International Data Encryption Algorithm (IDEA) [13], RC6 [14], Rijndael [15], Serpent [16], and so on. Among these, RC6, Rijndael, and Serpent were submitted as candidates for the Advanced Encryption Standard (AES) competition in 1998. Finally, the Rijndael algorithm won and is currently proposed to be the new encryption standard designed to protect sensitive information in the next century. AES was adopted by the federal government in 2001 and will replace the aging DES to encrypt information.

41.3.2 Public Key Encryption

Instead of using a shared key between the communicating parties, public key encryption relies on the use of two different but mathematically related keys, called public/private key pair. The public key is related to the private key, but the private key cannot be derived from it without additional information. Public key is used for encryption and private key for decryption. The process of public key encryption is illustrated in Figure 41.6.

If an entity A wishes to transmit an encrypted message to another entity B, A first needs to obtain B's public key PK_B. This public key must be authenticated so that A can be assured that the public key it believes really belongs to B. Once A has B's authentic public key, it encrypts the message using it. When B receives the encrypted message, B decrypts it using its private key SK_B. The security of public key encryption relies on the fact that it is infeasible to determine the private key, even if the public key is known, and also that it is infeasible to determine the plain text without the private key, even when the cipher text is given.

The framework of public key cryptography was proposed by Diffie and Hellman in 1976 [9]. The Rivest, Shamir, and Adleman (RSA) [10] and ElGamal [17] cryptosystems are two of the most popular public key designs. The RSA system applies the integer factorization problem to construct a public key cryptosystem, relying on the fact that factoring a big integer is intractable with the current factoring algorithm. The ElGamal system uses the difficulty of a discrete logarithm problem to design a public key cryptosystem. The discrete logarithm problem for a fixed prime number p can be stated as given an integer g between 0 and $p - 1$ and an exponential relation $y = g^x(\mod p)$ for some x, determine the integer x. Currently there is no known efficient algorithm to solve the discrete logarithm problem. The ElGamal cryptosystem has been used as the U.S. government's digital signature algorithm (DSA).

In 1985, Koblitz and Miller, mathematicians, proposed a public key system called the elliptic curve cryptosystem (ECC) that used groups of points on an elliptic curve [18,19]. The work utilized a discrete logarithm problem over the points on an elliptic curve with an advantage that the cryptographic keys used by the elliptic curve system are shorter. Later in 1991, Koyama et al. [20] presented an implementation of RSA using elliptic curves and Menezes and Vanstone [21] also implemented the elliptic variant of the ElGamal cryptosystem in 1993. These systems are relatively new, but there is a common belief that the cryptosystems based on elliptic curves provide stronger security since the elliptic curve discrete logarithm problem is believed to be harder than both the integer factorization problem and the discrete logarithm problem. In offering similar levels of security, the 160-bit ECC is comparable with the 1024-bit RSA and DSA.

Compared with symmetric encryption, public key encryption only requires an authenticated channel as opposed to a secure channel for the distribution of keys. However, public key encryption requires much more computational resources than symmetric key encryption and is therefore relatively slower in terms of performance. Public key encryption is typically only used to encrypt small amounts of data, for example, symmetric encryption keys.

41.3.3 Hashing Function

Public key encryption is generally much slower when compared to symmetric key encryption methods. Hashing function was thus invented to produce a relatively short fingerprint of a much longer message in order to make symmetric key signing and authentication more efficient. The fingerprint is also called a message digest. Hashing function can be divided into keyed and nonkeyed functions. Nonkeyed hashing function is a process by which messages of arbitrary lengths are compressed into digests (known as message integrity code, MIC) of fixed length without using any keys as input, while keyed hashing function combines a message and a secret key to generate a message digest, also known as MAC.

Secure Hash Algorithm (SHA-1) is the hashing standard currently recommended by the U.S. National Institute of Standards and Technology (NIST). SHA-1 is considered secure because it is computationally infeasible to recover a message corresponding to a given message digest or to find two different messages that hash to the same message digest value.

41.3.4 Digital Signature and Certificate

In public key cryptography, a digital signature is used to validate the identity of the sender. Figure 41.7 illustrates how a digital signature is used. An entity A wishes to send a message to another entity B. A wants the message transmitted without any modification and B wants to make sure that the message really comes from A. What A does is generate its signature by hashing digest of the message and encrypting it with its private key SK_A. A sends both the message and its signature to B. B can verify the signature by computing the hash digest of the message it received and compares it with the digest it gets when decrypting the signature using A's public key PK_A.

As we discussed earlier, in public key encryption, the sender needs to obtain an authentic public key of the receiver B to encrypt the message, otherwise if the public key belongs to an attacker, the message intended for B can be decrypted and read by the attacker. Digital certificate is used to provide this service.

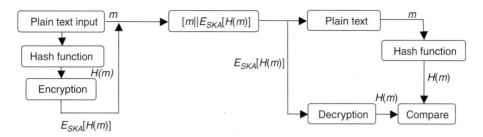

FIGURE 41.7 Example of a digital signature.

The digital certificate is a statement issued by some trusted party, named CA, indicating that it verifies that the public key PK_A in fact belongs to A. The CA digitally signs this statement and therefore anyone with the public key of the CA can decrypt the signature and verify the public key.

41.3.5 Identity-Based Cryptography

Identity-based cryptosystem was proposed by Shamir [22] with the original motivation of simplifying certificate management in e-mail system, thus avoiding the high cost of the public key management and signature authentication in PKI-supported cryptosystem. The basic idea is to find an approach in which each entity's public key can be defined by an arbitrary string. In other words, users may use some well-known information, such as an e-mail address, IP address, or identity as their public key; thus, there is no need to propagate this common information through the network. The idea remained a theoretical concept until the first practical identity-based encryption scheme was proposed by Boneh and Franklin [23]. Since then several other identity-based cryptography schemes [24–28] have been proposed.

For such a system to work, there is a *private key generator* (PKG), which generates the system parameters, and the master public/private key pair, with the master public key known to everyone, but the master private key kept only to itself. PKG is also responsible for generating private keys corresponding to the defined public keys for the network user by employing the master private key and the user's public key. There are four algorithms in the identity-based encryption scheme: Setup, Extract, Encrypt, and Decrypt:

- *Setup*: Generates the master public/private key pair and the common system parameters.
- *Extract*: Given an arbitrary string as a public key, this algorithm generates the corresponding private key.
- *Encrypt*: Takes a message and encrypts it using the public key.
- *Decrypt*: Decrypts a message using the corresponding private key.

41.3.6 Threshold Secret Sharing Scheme

The secret sharing scheme allows a group of users (also called shareholders) to share a secret in such a way that no single user can deduce the secret from his share alone. To construct the secret, one needs to combine a sufficient number of shares. A (k, n) threshold secret sharing represents a secret distributed to n shareholders, and any k out of the n shareholders can reconstruct the secret, but any collection of less than k partial shares cannot get any information about the secret. Here, k is the threshold parameter such that $1 \leq k \leq n$.

A classical (k, n) secret sharing algorithm was proposed by Adi Shamir [29] in 1979, which is based on polynomial interpolation. To distribute a secret x among n users, a trust authority chooses a large prime q, and randomly selects a polynomial $f(z)$ over Z_q of degree $k - 1$, such that $f(0) = x$. The trust authority computes each user's share using $S_i = f(i) \bmod q$ and securely sends the share S_i to user i. Then any k shareholders can reconstruct the secret using the Lagrange interpolation by $f(z) = \sum_{i=1}^{k} S_i l_i(z) (\mathrm{mod}\ q)$, where $l_i(z) = \prod_{j=1, j \neq i}^{k} ((z - j)/(i - j)) (\mathrm{mod}\ q)$ is the Lagrange coefficient. Then, the secret reconstructed is $f(z)|_{z=0}$. Shamir's secret sharing scheme suffers from a lack of trust authority and share verification, for which many developments have been proposed. Also, work has been done on the issues related to verifiable secret sharing [30] and verifiable secret redistribution [31].

41.3.7 Intrusion Detection

The idea of intrusion detection is to set user characteristics as normal behavior within the network in terms of a set of relevant system features. Once the set of system features is selected, the classification model is built to detect the anomalies from its normal behavior. Intrusion detection has been an active area of research since its proposal by Denning in 1987 [32], and many intrusion detection systems have been developed to protect conventional networks, employing methods from various branches of science such as mathematics, statistics, and machine learning. A survey of these techniques can be found in [33]. Here we categorize different intrusion detection techniques.

41.3.7.1 Misuse and Anomaly Detection

From the perspective of analytical techniques, intrusion detection can also be defined either as misuse or as anomaly detection. Misuse detection techniques [34–36], or signature recognition techniques as they are often called, use the signature of known attacks (i.e., the patterns of an attacker's behavior) to identify the matched activity as an attack. Although they can accurately and efficiently detect instances of known attacks, they lack the ability to detect new types of attacks since it is impossible to anticipate all the different attacks that could occur. Anomaly detection techniques [37,38] detect intrusions by finding deviations from established user profiles, based on the assumptions that certain functionality performed by each user within the network is observable and does not change substantially over time. The main advantage of anomaly detection is that it can detect both novel and known attacks if they demonstrate large differences from the normal profile. Since anomaly detection techniques models all anomalies as intrusions, higher false alarm rates are expected when anomalies are caused by behavioral irregularity, instead of intrusions.

41.3.7.2 Supervised and Unsupervised Intrusion Detection

From the perspective of the classifier that is used to actually detect intrusions, intrusion detection can be classified as supervised or unsupervised. The supervised method of detecting intrusion relies on the presence of a set of examples corresponding to normal usage and malicious activity. Labeling to generate the training data is time consuming and is prone to errors. Moreover, the volume of data corresponding to intrusions is much smaller when compared to data corresponding to normal usage, which leads to an imbalanced dataset wherein normal behavior is very well characterized while intrusions are characterized by a rather limited set. Unsupervised intrusion detection (also known as anomaly detection over noisy data) is better capable of discovering buried detections [38,39]. While unsupervised intrusion detection has some preferred practical properties as compared to supervised intrusion detection, a higher rate of false alarm may limit its applicability. It is conceivable that an intrusion detection system, in which a combination of supervised and unsupervised learning is used, could provide improved detection accuracy.

41.4 Current Countermeasures

In recent research on security of wireless ad hoc networks, several good security approaches have been proposed, which generally fall into three categories: secure routing, trust and key management, and service availability protection.

41.4.1 Secure Routing

Establishing the correct route between communicating nodes in an ad hoc network is a prerequisite for guaranteeing that messages are delivered in a timely manner. If routing is misdirected, the entire network can be paralyzed. As the function of route discovery is performed by routing protocols, securing routing protocols has warranted more attention. The routing protocols designed for ad hoc networks [40,41] assume that all the nodes within the network behave properly according to the routing protocols and no malicious nodes exist in the network. Obviously this assumption is too strong to be practical. Current research efforts toward the design of secure routing protocols (SRPs) include symmetric key-based schemes, asymmetric schemes, and intrusion detection [42–49].

41.4.1.1 Symmetric Schemes

Researchers have proposed the use of *symmetric* key cryptography for authenticating ad hoc routing protocols, based on the assumption that a security association (a shared key K_{ST}) between the source node S and the destination node T exists.

41.4.1.1.1 Secure Routing Protocol

The SRP [42] is proposed as an extension of a variety of existing reactive routing protocols. SRP makes efficient use of the security association (K_{ST}) between the two communicating nodes S and T and can combat attacks that disrupt the route discovery process and guarantees the acquisition of correct connectivity information.

In SRP, source node S not only maintains a monotonically increased *Query sequence number* Q_{seq} for each destination, but it also generates a random *Query identifier* Q_{ID} that is used by intermediate nodes as a means to identify the request. The random *Query identifier* is unpredictable by an adversary with limited computational power; thus, it can combat the attack where malicious nodes simply broadcast fabricated requests only to cause subsequent legitimate queries to be dropped.

The security association K_{ST} is used to generate a MAC in routing request message by a keyed hash algorithm with the input of an entire IP header, the basic protocol RREQ packet, and the shared key K_{ST}. SRP uses the computed MAC to authenticate the communicating nodes. Upon receipt of a routing request (RREQ), the destination node verifies the integrity and authenticity of the RREQ by calculating keyed hash of the request fields and comparing them with the MAC contained in the SRP header. The source node performs a similar operation when receiving route reply (RREP) message to verify the destination node.

41.4.1.1.2 Secure Efficient Ad Hoc Distance Vector Routing Protocol

The secure efficient ad hoc distance (SEAD) vector routing protocol [43] is presented as a proactive SRP based on the design of the distance-sequenced distance vector (DSDV) protocol. The essential idea of SEAD is to authenticate the sequence number and metric in a routing update entry using one-way hash chain elements, thus preventing any node from advertising a route to some destination claiming a greater sequence number than that destination's own current sequence number. SEAD assumes some mechanism for a node to distribute an authentic element of the hash chain that can be used to authenticate all the other elements of the chain.

To create a one-way hash chain, a node chooses a random initial value $x \in \{0, 1\}^{\rho}$, where ρ is the length in bits of the output of the hash function, and computes the list of values $h_0, h_1, h_2, \ldots, h_n$, where $h_0 = x$, and $h_i = Hash(h_{i-1})$ for $0 < i < n$. The node at initialization generates the elements of its hash chain from "left to right" and then uses certain (authentic) elements of the chain to secure its routing updates. Based on its authentic element, it is possible to verify the next element in the sequence and within the chain. For an example, given an authenticated h_i value, a node can authenticate h_{i-3} by recursively computing the hash function three times on the value of h_{i-3}.

In SEAD, two mechanisms of TELSA [44] and MAC are also presented to authenticate the source address of each routing update message so that an attacker creating routing loops through the impersonation attack can be prevented.

41.4.1.1.3 ARIADNE

Another approach called the ARIADNE [45] was proposed by the same authors as SEAD to secure DSR [40] routing protocol. ARIADNE can provide security against one compromised node and arbitrary active attackers relying on a highly efficient symmetric cryptography. Like the SRP protocol, ARIADNE assumes some mechanism to bootstrap the authentic keys required by the protocol. In other words, a shared secret key (K_{ST}) between the source node S and the destination node T, an authentic TESLA [44] key for each node in the network, and an authentic element from the route discovery chain of every node initiating route discoveries are required.

In the design of ARIADNE, the target node of a route discovery process authenticates the initiator; the initiator can then authenticate each intermediate node on the path in RREP message and no immediate node can be removed from the node list of the RREQ or RREP during the route discovery process. For point-to-point authentication of a message, MAC (e.g., HMAC) and a shared key between the two parties are used. For authentication of a broadcast packet such as RREQ, a TESLA broadcast authentication protocol is used. Moreover, one-way hashing is used to verify that no intermediate node was omitted from the path in RREQ.

41.4.1.2 Asymmetric Schemes

Symmetric encryption is suitable for ad hoc networks due to its comparatively lower resource consumption. The problem is how to generate and distribute the symmetric keys in the first place, such that only the correct sender and receiver keep them. To mitigate this problem, several asymmetric-based schemes have also been proposed. Authenticated routing for ad hoc network (ARAN) [46] and Secure ad hoc on-demand distance vector (SAODV) [47] are considered as *asymmetric*-based security approaches.

41.4.1.2.1 Authenticated Routing for Ad Hoc Network

Authenticated routing for ad hoc network (ARAN) [46] is used to detect and protect against malicious actions carried out by third parties and peer nodes in the ad hoc network environment. ARAN employs cryptographic certificates to provide authentication, message integrity, and non-repudiation as part of a minimal security policy. A trusted certificate server is required in ARAN to distribute the certificates to all the nodes in the network.

The basic idea of ARAN is that every node forwarding a route request and route reply message must sign it. For example, suppose source node S would like to send a message to destination node T. If nodes A and B are the intermediate nodes along the path of route discovery packet, that is, $S \rightarrow A \rightarrow B \rightarrow T$. S first initializes the route discovery process by broadcasting a route discovery packet named PDP, and the packets exchanged between the nodes are shown in Figure 41.8.

The routing request packet includes a packet type identifier ("RDP"), the IP address of the destination node (IP_T), certificate of source node S ($cert_S$), a random value (called nonce) N_S, the current time t, and the signature of S generated using its private key ($SK - S$). The neighbor node of the source node signs the contents to the message, adds its certificate, and forwards the message to its neighbors. Upon receiving the broadcast, the neighbor node validates the signature with the given certificate, removes the previous node's signature, signs the message again using its private key and rebroadcasts the RDP.

$$S \rightarrow *: [RDP, IP_T, cert_S, N_S, t]_{SK\text{-}S}$$
$$A \rightarrow *: [[RDP, IP_T, cert_S, N_S, t]_{SK\text{-}S}]_{SK\text{-}A}, cert_A$$
$$B \rightarrow *: [[RDP, IP_T, cert_S, N_S, t]_{SK\text{-}S}]_{SK\text{-}B}, cert_B$$

$$T \rightarrow B: [REP, IP_S, cert_T, N_S, t]_{SK\text{-}T}$$
$$B \rightarrow A: [[REP, IP_S, cert_T, N_S, t]_{SK\text{-}T}]_{SK\text{-}B}, cert_B$$
$$A \rightarrow S: [[REP, IP_S, cert_T, N_S, t]_{SK\text{-}T}]_{SK\text{-}A}, cert_A$$

FIGURE 41.8   Packets exchanged between nodes in ARAN.

Eventually, the message is received by the destination node T, who replies to the first RDP that it receives for a source and a given nonce. The destination node T verifies the RDP packet and then unicasts a Reply (REP) packet back along the reverse path to the source in the same way. When the source node receives the REP, it verifies whether the correct nonce was returned by the destination node along with its signature.

The ARAN protocol is capable of providing strong security against modification, fabrication, and impersonation, but performing a digital signature on every routing packet could become the performance bottleneck both on the bandwidth requirement and the amount of computation.

41.4.1.2.2 Secure Ad Hoc On-Demand Distance Vector Routing Protocol

The SAODV routing protocol [47] is designed to be a secure extension of the existing ad hoc on-demand distance vector (AODV) routing protocol [41]. SAODV mainly addresses authentication and integrity by using digital signature and hash chains. The basic idea of SAODV is to first separate the routing messages transmitted to mutable and nonmutable types, and then use digital signatures to authenticate the nonmutable fields of the messages, and hash chains to secure the mutable fields.

In the AODV RREQ and RREP messages, the only mutable information is hop count. SAODV protects the *Hop_Count* information by applying one-way hash functions recursively in each node to ensure that the *Hop_Count* has not been decremented by an attacker. Every time a node originates an RREQ or an RREP message, it performs the following operations:

- Generate a random number (*seed*).
- Set the *Max_Hop_Count* field: *Max_Hop_Count=TimeToLive*.
- Set the *Hash* field: *Hash=seed*.
- Calculate the *Top_Hash*: *Top_Hash*$=h^{Max\_Hop\_Count}(seed)$, where h is a hash function, $h^i(x)$ is the result of applying the function h to xi times.

Every time a node receives an RREQ or an RREP message, it performs the following operations in order to verify the hop count:

- Verify the *Top_Hash* field: *Top_Hash*$==h^{Max\_Hop\_Count-Hop\_Count}(Hash)$.
- Apply the hash function to the *Hash* value in the Signature Extension: *Hash*$= h(Hash)$.
- Rebroadcast the RREQ or forward the RREP.

In SAODV, using digital signatures provides end-to-end authentication and applying one-way hash chains obtains a certain type of protection to mutable information with less expensive computation. However, the effectiveness of SAODV is sensitive to tunneling attacks. A novel distributed routing protocol for ad hoc networks called SDAR has been introduced in [50] that guarantees security, anonymity, and high reliability of the established route. This is done by letting all trustworthy intermediate nodes construct the path while maintaining their anonymity. Moreover, malicious nodes are identified and are avoided in forming reliable paths.

41.4.1.3 Intrusion Detection

All previously introduced symmetric and asymmetric approaches belong to intrusion prevention technique. Several intrusion detection schemes are designed to enhance the security of ad hoc networks.

Zhang and Lee [48] present a distributed intrusion detection and response architecture that provides an excellent guide on designing intrusion detection system in wireless ad hoc networks. Figure 41.9(a) shows the proposed completely distributed intrusion detection model, in which every node in the network participates in the process of intrusion detection. Each node is responsible for detecting intrusion locally and independently based on the data collected. When a malicious node is found by the local detector, it is broadcasted to the entire network. Each node also makes a final decision based on the detection reports from other nodes. In [49], a distributed hierarchical detection model (Figure 41.9[b]) is proposed and the intrusion detection can be performed in a supervised or unsupervised way depending on the availability of attack data. The distributed hierarchical detection model consists of two layers, cluster member layer and cluster head layer. The whole network is logically divided into several clusters, each of which consists

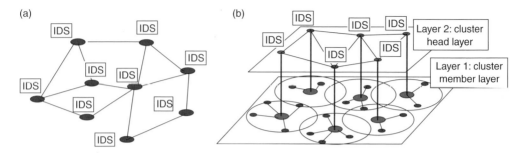

FIGURE 41.9   Two intrusion detection models. (a) A completely distributed detection model and (b) a distributed heirarchical detection model.

of one special node as the cluster head and several normal nodes as the cluster members. The functions of local data collection and local intrusion detection are preformed on the cluster head layer. The cluster head also collects detection results from other cluster heads. The final decision is based on a cooperative rule for detecting reports at the cluster heads.

Marti et al. [7] introduced the *Watchdog* and *Pathrater* techniques that improve throughput in an ad hoc network by identifying misbehaving nodes that agree to forward the packets but never do so. The *Watchdog* can be considered as a simpler version of the intrusion detection agent to identify misbehaving nodes, and the *Pathrater* works as the response agent to help routing protocols avoid these nodes. However, the *Watchdog* can only detect the nodes that do not forward the packets, and the method works only on the source routing protocol since 2-hop routing information is needed.

The main problems of the intrusion detection approach are: first, not all malicious behaviors are detectable, in particular, the dynamically changing topology in ad hoc networks makes detection more difficult; second, even if some attacks can be detected, a high false alarm rate is still expected to be present. Therefore, intrusion detection usually works as a complementary approach to provide a second line of defense to the network.

41.4.2   Trust and Key Management

Most of the intrusion prevention approaches discussed above make an assumption that efficient key distribution and management has been implemented by some kind of a key distribution center, or by a CA to sign certificates binding public keys to nodes. The CA has the power to keep connecting to the network and cannot be compromised; however, to maintain the server safely and keep it available when needed is another major issue that cannot be easily solved. To mitigate this problem, the concept of threshold secret sharing is introduced and there are three proposed approaches.

41.4.2.1   Partially Distributed Certificate Authority

Zhou and Hass [51] present a partially distributed CA scheme by adopting the PKI and threshold secret sharing. In this approach, the certificate generation service, as a whole, has a public/private key pair. The service is distributed to a group of special nodes in a (k, n) threshold secret sharing fashion such that the service private key is shared by n special nodes. Any special service node can generate a partial certificate using its share of the service private key. All nodes in the system know the public key of the service and trust any partial certificates generated by the special service node. By combining any k partial certificates, the complete certificate can be recovered.

This scheme is one of the first to introduce the threshold scheme into security protocols in ad hoc networks and provides an excellent guide to the following work. However, the approach requires an administrative infrastructure to be available to distribute the shares to the special nodes and issue public/private key pairs to all the nodes. How to keep the n special nodes available when needed and how the normal nodes know how to locate the server nodes makes system maintenance difficult.

41.4.2.2 Fully Distributed Certificate Authority

In [52], Kong et al. proposed another threshold cryptography-based scheme in which certification services like issuing, renewal, and revocation of certificates are distributed among all the nodes in the network, and a single node holds just a share of the complete certificate signing key. Any coalition of k nodes can serve as the CA and jointly provide certification services for a requesting mobile user. The authors also localized the trust model to make it scalable to large ad hoc wireless networks. A mobile user is trusted if any k neighboring nodes of the user claim so and a locally trusted user is globally accepted over the network. The fully localized and widely available features provide ubiquitous availability for mobile users since any mobile user can easily obtain the CA service by contacting k neighbors. We consider it as having a fully distributed CA, in which the capabilities of CA are distributed to all the nodes and any operations requiring the CA's private key can only be performed by a coalition of k or more nodes.

This solution has more advantage than the partial distributed CA scheme in the sense that it is easier for a node to locate k neighbor nodes and request the CA service since all nodes are part of the CA service. However, it requires a set of complex maintenance protocols. Also, the scheme requires a bootstrapping phase where a trusted dealer has to distribute the certificates to and share secrets with each mobile user.

41.4.2.3 Identity-Based Key Management

The traditional PKI-based key management is well known for its expensive computation. In order to achieve better efficiency, an identity-based key management mechanism is proposed [53]. In this approach, the PKG service is controlled by a pair of master public/private key and is distributed to all the nodes in the network in such a way that the master public key is well known to all the nodes in the network, and the master private key is shared by all of them in a (k, n) threshold fashion. Each user in the network uses the identity as its public key, and needs to obtain its personal private key corresponding to its identity before utilizing any network service. The private key is obtained by contacting at least k PKG service nodes and each PKG service node replies by sending the sub-share of the private key of the requesting node.

One advantage of this approach is that it does not need the support of a trusted third party to safely compute a master key, separate it into multiple pieces (sharing), and then distribute the shares to shareholders. Instead, the master key pair is computed collaboratively by the initial network nodes without constructing the master private key at any single node. Moreover, since the personal public key is derived from the node's identity, which is well-known information, there is no need for certificate generation, propagation, and storage. Therefore, the approach can obtain similar security with the conventional PKI-based approaches, but with reduced computational consumption and communicational overhead.

41.4.3 Service Availability Protection

As there is no infrastructure in wireless ad hoc networks, the mobile nodes have to cooperate in order to communicate. Intentional nonoperation can be caused by malicious nodes that are intent on attacking the network, and it can also be caused by a few selfish nodes that are keen on saving computational resources (power, CPU, memory, etc.). These selfish nodes tend to use services provided by other nodes, but not to provide services to the community. Selfishness can cause serious damage in terms of global network throughput and delay, which is shown by the simulation study on the DSR routing protocol [54]. The problem of selfishness has only recently been addressed and few mechanisms are proposed to protect the network from the problem of service unavailability due to the existence of selfish nodes. Generally, there are two types of solutions: reputation-based solutions and monitoring solutions. The reputation-based approach uses some forms of reputation like digital cash, credit, or rating as an incentive for cooperation, and the monitoring approach employs neighborhood watching based on the principle that misbehaving nodes will be detected through the shared observation information of a majority of legitimate nodes.

41.4.3.1 CONFIDANT

Buchegger and Boudec presented a technique called cooperation of nodes, fairness in dynamic ad hoc networks (CONFIDANT) [55] that aimed at detecting and isolating misbehaving nodes, thus making it unattractive to deny cooperation. Trust relationships and routing decisions are based on experienced, observed, or reported routing and forwarding behavior of other nodes. CONFIDANT is designed as an extension to the DSR routing protocol, and consists of four components: the monitor, the reputation system, the path manager, and the trust manager. The monitor is equivalent to a neighborhood watch, where nodes locally look for deviating nodes by either listening to the transmission of the next node (so-called passive acknowledgement) or by observing route protocol behavior. The reputation records for first-hand or trusted second-hand observations about routing and forwarding behavior of other nodes. The trust manager controls trust given to ALARM messages and the path manager adapts the behavior of the local node according to its reputation and takes actions against malicious nodes.

41.4.3.2 Collaborative Reputation Mechanism

Another approach called the collaborative reputation (CORE) mechanism [56] was proposed, in which node cooperation is stimulated by a collaborative monitoring and a reputation mechanism. Each network user keeps track of other users' collaboration using a technique called *reputation*. The reputation metric is calculated based on data monitored by the local users and some information provided by other nodes involved in each operation. Based on the collected observations, each node computes a reputation value for every neighbor using an aggregate reputation mechanism instead of simply using subjective reputation (observations), indirect reputation (positive reports by others), and functional reputation (task-specific behavior). The reputation mechanism allows the nodes of networks to gradually isolate the selfish nodes. For example, when the reputation of a node decreases below a predefined threshold, service provision to the misbehaving node will be avoided. In CORE, since there is no incentive for a node to maliciously spread negative information about other nodes, simple denial of service attacks using collaborative technique itself is prevented.

41.5 Conclusions

We have discussed in this chapter the various security issues faced by wireless ad hoc technology. By analyzing various security threats, we find that wireless ad hoc networks are vulnerable due to their underlying characteristics. We have described several existing countermeasures to secure wireless ad hoc networks. However, we have seen that many security solutions have been proposed to secure wireless ad hoc networks, but none can claim that it can address all or even most, of the security problems. Many security problems such as tunneling attack, packet dropping, and node selfishness are expected to be solved. In essence, secure wireless ad hoc networks would be a long-term ongoing research topic.

With increasing maturity of wireless ad hoc networks, more comprehensive security solutions are expected to appear that could take care of all the prevalent issues such as tunneling attack handling strategies, combination of intrusion prevention and intrusion detection, integrated network security, more efficient key management and cooperation enforcement mechanisms, and the like.

Acknowledgments

This work is supported in part by the Ohio Board of Regents Doctoral Enhancements Funds and the National Science Foundation under grant CLR-0113361.

References

[1] M. Corson and A. Ephremides, "A distributed routing algorithm for mobile radio networks," in *Proceedings of the MILCOM*, 1989.

[2] Tactical Internet, Available on-line http://www.fas.org/man/dod-101/sys/land/internet-t.htm.

[3] Saab NetDefence, Available on-line, http://www.saab.se/.

[4] D.P. Agrawal and Q.-A. Zeng, *Introduction to Wireless and Mobile Systems*, Brooks/Cole Publisher, New York, 2002.

[5] M. Jakobsson and S. Wetzel, "Stealth attacks on ad-hoc wireless networks," in *Proceedings of the IEEE Vehicular Technology Conference*, 2003.

[6] A. Perrig, Y.C. Hu, and D.B. Johnson, "Wormhole protection in wireless ad hoc networks," Technical Report *TR01-384*, Department of Computer Science, Rice University, December 2001.

[7] S. Marti, T. Giuli, K. Lai, and M. Baker, "Mitigating routing misbehavior in mobile ad hoc networks," in *Proceedings of the 6th International Conference on Mobile Computing and Networking (MOBICOM'00)*, pp. 255–265, August 2000.

[8] W. Stallings, *Cryptography and Network Security: Principles and Practice*, Prentice-Hall, New York, 1999.

[9] W. Diffie and M.E. Hellman, "New directions in cryptography," *IEEE Transactions on Information Theory*, 22, 644–654, 1976.

[10] R.L Rivest, A. Shamir, and L.M. Adleman, "A method for obtaining digital signatures and public-key cryptosystems," *Communications of the ACM*, 21; 120–126, 1978.

[11] National Soviet Bureau of Standards, "Announcing the data encryption standard," Technical Report FIPS Publication 46, US Commence Department, National Bureau of Standards, January 1977.

[12] A. Shimizu and S. Miyaguchi, "Fast data encipherment Algorithm FEAL," in D. Chaum and W.L. Price, eds., *Advances in Cryptography (EUROCRYPT'87)*, Lecture notes in Computer Science, No. 304, pp. 267–280, Springer, Berlin, Heidelberg, New York, 1988.

[13] X. Lai, J. Massey, and S. Murphy, "A proposal for a new block encryption standard," in I.B. Damgard ed., *Advances in Cryptography (EUROCRYPT'87)*, Lecture notes in Computer Science, No. 473, pp. 389–404, Springer, Berlin, Heidelberg, New York, 1990.

[14] R. Rivest, M. Robshaw, R. Sidney, et al., RC6 Block Cipher, http://www.rsasecurity.com/rsalabs/rc6/.

[15] J. Daemen and V. Rijmen, *The Design of Rijndael*, Springer, Berlin, Heidelberg, New York, 2002. www.nist.gov/aes.

[16] E. Biham, R. Anderson, and L. Knudsen, "Serpent: a new block cipher proposal," in S. Vaudenay, ed., *Proceedings of the 5th International Workshop on Fast Software Encryption*, Lecture notes in Computer Science, No. 1372, pp. 222–238, Springer, Berlin, Heidelberg, New York, 1998.

[17] T. ElGamal, "A public key cryptosystem and a signature scheme based on discrete logarithms," *IEEE Transactions on Information Theory*, IT-31, 469–472, 1985.

[18] N. Koblitz, "Elliptic curve cryptosystems," *Mathematics of Computation*, 48, 203–209, 1987.

[19] V.S. Miller, "Use of elliptic curves in cryptography," In H.C. Williams, ed., *Advances in Cryptography (CRYPTO'90)*, Lecture Notes in Computer Science, No. 537, pp. 627–638, Springer, Berlin, Heidelberg, New York, 1991.

[20] K. Koyama, U.M. Maurer, et al., "New public-key schemes based on elliptic curves over the ring Z_n," in C. Pomerance, ed., *Advances in Cryptography (CRYPTO'91)*, Lecture notes in Computer Science, No. 576, pp. 252–266, Springer, Berlin, Heidelberg, New York, 1992.

[21] A. Menezes and S. Vanstone, "Elliptic curve cryptosystems and their implementation," *Journal of Cryptology*, 6, 209–224, 1993.

[22] A. Shamir, "Identity based cryptosystems and signatures schemes," in *Proceedings of the Advances in Cryptology*, 1984.

[23] D. Bonh and M. Franklin, "Identity-based encryption from weil pairing," *Advances in Cryptology, CRYPTO 2001*, Lecture Notes in Computer Science, Vol. 2139, pp. 213–229, Springer-Verlag, Berlin, 2001.

[24] C. Cocks, "An identity based encryption scheme based on quadratic residues," *Cryptography and Coding*, 2001.

[25] F. Hess, "Efficient identity based signature schemes based on pairings," in *Proceedings of the 9th Workshop on Selective Areas on Cryptography (SAC 2002)*, Lecture Notes in Computer Science, Springer-Verlag, Berlin, 2002.

[26] J.C. Cha and J.H. Cheon, "An identity-based signature from gap Diffie-Hellman groups," *Cryptology ePrint Archive*, Report 2002/018, http://eprint.iaccr.org/.

[27] K. G. Paterson, "ID-based signatures from pairing on elliptic curves," *Cryptology ePrint Archive*, Report 2002/004, http://eprint.iacr.org/.

[28] R. Sakai, K. Ohgishi, and M. Kasahara, "Cryptosystems based on Pairing," *SCIS*, 2000.

[29] A. Shamir, "How to share a secret," *Communications of the ACM*, 22, 612–613, 1979.

[30] T.P. Pederson, "Non-interactive and information-theoretic secure verifiable secret sharing," Lecture Notes in Computer Science, pp. 129–140, 1992.

[31] Y. Desmedt and S. Jajodia, "Redistribution secret shares to new access structures and its applications," Technical Report ISSE TR-97-01, George Mason University, Fairfax, VA, July 1997.

[32] D. Denning, "An intrusion detection model," *IEEE Transactions on Software Engineering*, 12, 222–232, 1987.

[33] C. Warrender, S. Forrest, and B. Pearlmutter, "Detecting intrusions using system calls: alternative data models," in *Proceedings of the 1999 IEEE Symposium on Security and Privacy*, pp. 133–145, 1999.

[34] L. Zirkle, "What is host-based intrusion detection?," Virginia Tech CNS, *SANS Institute Resources, Intrusion Detection FAQ*, 2000.

[35] S. Kumar and E.H. Spafford, "A software architecture to support misuse intrusion detection," in *Proceedings of the 18th National Information Security Conference*, pp. 194–204, 1995.

[36] K. Ilgun, R.A. Kemmerer, and P.R. Porras, "State transition analysis: a rule-based intrusion detection approach," *IEEE Transactions on Software Engineering*, 21, 181–199, 1995.

[37] W. Lee and S. Stolfo, "A framework for constructing features and models for intrusion detection systems," *ACM Transactions on Information and System Security*, 3, 227–261, 2000.

[38] E. Eskin, "Anomaly detection over noisy data using learned probability distributions," in *Proceedings of the International Conference on Machine Learning*, pp. 255–262, 2000.

[39] E. Eskin, A. Arnold, and M. Prerau, "A geometric framework for unsupervised anomaly detection: detecting intrusions in unlabeled data," *Data Mining for Security Applications*, 2002.

[40] D.B. Johnson and D.A. Maltz, "Dynamic source routing in ad hoc networks," *Ad Hoc Networking*, pp. 139–172, 2001.

[41] C.E. Perkins and E.M. Royer. "Ad hoc on-demand distance vector routing," in *Proceedings of the 2nd IEEE Workshop on Mobile Computing Systems and Applications*, New Orleans, LA, February, pp. 90–100, 1999.

[42] P. Papadimitratos and Z. Haas, "Secure routing for mobile ad hoc networks," in *Proceedings of the SCS Communication Networks and Distributed Systems Modeling and Simulation Conference (CNDS'02)*, January 2002.

[43] Y.C. Hu and D.B. Johnson, and A. Perrig, "SEAD: secure efficient distance vector routing in mobile wireless ad-hoc networks," in *Proceedings of the 4th IEEE Workshop on Mobile Computing Systems and Applications (WMCSA '02)*, pp. 3–13, 2002.

[44] A. Perrig, R. Canetti, and B. Whillock, "TELSA: multicast source authentication transform specification," *draft-ietf-msec-tesla-spec-00*, October 2002.

[45] Y.C. Hu, A. Perrig, and D.B. Johnson, "Ariadne: a secure on-demand routing protocol for ad hoc networks," in *Proceedings of the 8th ACM International Conference on Mobile Computing and Networking*, September 2002.

[46] B. Dahill, B.N. Levine, E. Royer, and C. Shields, "A secure routing protocol for ad hoc networks," in *Proceedings of the 10th Conference on Network Protocols (ICNP)*, November 2002.

[47] M.G. Zapata, "Secure ad hoc on-demand distance vector routing," *ACM SIGMOBILE Mobile Computing and Communications Review*, 6, 106–107, 2002.

[48] Y. Zhang and W. Lee, "Intrusion detection in wireless ad-hoc networks," in *Proceedings of the 6th International Conference on Mobile Computing and Networking (MobiCom'2000)*, August 2000.

[49] H. Deng, Q.-A. Zeng, and D.P. Agrawal, "SVM-based intrusion detection system for wireless ad hoc networks," *IEEE Vehicular Technology Conference*, Orlando, October 6–9, Fall 2003.

[50] A. Boukerche, K. El-Khatib, X. Li, and L. Korba, "SDAR: a secure distributed anonymous routing protocol for wireless and mobile ad hoc networks," in *Proceedings of the 29th IEEE Conference on Local Computer Networks*, pp. 618–624, 2004.

[51] L. Zhou and Z. J. Hass, "Securing ad hoc networks," *IEEE Networks Special Issue on Network Security*, November/December 1999.

[52] J. Kong, P. Zerfos, H. Luo, S. Lu, and L. Zhang, "Providing robust and ubiquitous security support for mobile ad-hoc networks," in *Proceedings of the IEEE 9th International Conference on Network Protocols (ICNP'01)*, 2001.

[53] H. Deng, A. Mukherjee, and D.P. Agrawal, "Threshold and identity-based key management and authentication for wireless ad hoc networks," in *Proceedings of the IEEE International Conferences on Information Technology (ITCC'04)*, April 5–7, 2004.

[54] P. Michiardi and R. Molva, "Simulation-based analysis of security exposures in mobile ad hoc networks," in *Proceedings of European Wireless Conference*, 2002.

[55] S. Buchegger and J. L. Boudec, "Performance analysis of the CONFIDANT protocol (Cooperation Of Nodes: Fairness In Dynamic Adhoc NeTworks)", in *Proceedings of MobiHoc*, 2002.

[56] P. Michiardi and R. Molva, "CORE: A COllaborative REputation mechanism to enforce node cooperation in mobile ad hoc networks," in *Proceedings of the Conference on Communication and Multimedia Security*, 2002.

42

Field-Based Motion Coordination in Pervasive Computing Scenarios

42.1 Introduction...**42**-959

42.2 Adaptive Motion Coordination: Applications, Issues, and Requirements.....................................**42**-961
A Case Study Scenario • Inadequacy of Traditional Approaches

42.3 Field-Based Approach to Motion Coordination.......**42**-963
Key Concepts in the Field-Based Approach • The State of the Art in Field-Based Coordination

42.4 Modeling Field-Based Coordination **42**-967
Flock Field: Moving while Maintaining a Formation • Person Presence Field: Surrounding a Prey • Room Field and Crowd Field: Load Balancing • Room Field and Crowd Field: Meetings

42.5 Programming Motion Coordination Protocols........**42**-976
Architecture and Implementation of a Field-Based System • Specifying Tuples and Agents

Marco Mamei
Franco Zambonelli

42.6 Open Issues and Research Directions **42**-980

References **42**-981

42.1 Introduction

As computing is becoming pervasive, autonomous computer-based systems are going to be embedded in all everyday objects and in the physical environment. In such a scenario, mobility too, in different forms, will be pervasive. Mobile users, mobile devices, computer-enabled vehicles, as well as mobile software components, define an open and dynamic networked world that will be the stage of near-future software applications.

Traditionally, software applications have always been built by adopting programming paradigms rooted on not-autonomous components coupled at design time by fixed interaction patterns. Although simple in principle, this approach seems inadequate to deal with the dynamism of the above sketched application scenario and it is likely to lead to even-more brittle and fragile applications.

Only in recent years, the research on software agents has fostered new programming paradigms based on autonomous components (i.e., components with a separated thread of execution and control) interacting to realize an application [3, 6, 25]. In this chapter, the term agent can refer not only to software components, but also to any autonomous real-world entity with computing and networking capability (e.g., a user carrying on a Wi-Fi PDA, a robot, or a modern car). Autonomous interacting components are intrinsically robust, in that, for example, if a component breaks down or goes out of the wireless network, the others can autonomously and dynamically reorganize their interaction patterns to account for such a problem. The key element leading to such robust, scalable, and flexible behaviors is coordination: the agents' fundamental capability to dynamically decide on their own actions in the context of their dynamic operational environment. Autonomous components, in fact, must be able to coordinate their activities' patterns to achieve goals, that possibly exceed their capabilities as singles, despite — and possibly taking advantage — of environment dynamics and unexpected situations [24, 31].

Unfortunately, general and widely applicable approaches to program and manage these autonomous interacting systems are unknown. The main conceptual difficulty is that it is possible to apply a direct engineered control only on the single agents' activities, while the application task is often expressed at the global-ensemble scale [4, 24]. Bridging the gap between single and global activities is not nearly easy, but it is although possible: distributed algorithms for autonomous sensor networks have been proposed and successfully verified [9], routing protocols in mobile ad hoc network (MANET) (in which devices coordinate to let packets flow from sources to destinations) have been already widely used [26]. The problem is still that the above successful approaches are ad hoc to a specific application domain and it is very difficult to generalize them to other scenarios. There is a great need for general, widely applicable engineering methodologies, middleware, and APIs to embed, support, and control coordination in multi-agent systems [1, 17, 27, 32].

A significant application scenario related to the above general problem is distributed motion coordination. Here the task is of coordinating the movements of a large set of autonomous agents in a distributed environment. The goals of agents' coordinated movements can be various: letting them meet somewhere, distribute themselves accordingly to specific spatial patterns, or simply move in the environment without interfering with each other and avoiding the emergence of traffic jams.

With this regard, protocols based on the concept of force fields (field-based coordination) [15, 16, 18] suit well the needs of distributed motion coordination. These approaches are inspired by the physical world, that is, from the way masses and particles in the universe move and globally self-organize driven by gravitational and electromagnetic fields.

In these protocols, the agents' operation environment is abstracted and described by means of fields capable of expressing forces driving agents' activities. Agents can locally perceive these fields and decide where to go on the basis of which fields they sense and their magnitudes.

Specifically, on the one hand, it is possible to assume that each agent can generate application-specific fields, conveying some information about the local environment and about itself. Fields will be propagated in the environment according to field-specific laws and to be possibly composed with other fields. On the other hand, agents can perceive these fields and move accordingly (or, if the agent is a program running on a mobile device, it can suggest its user how to move), for example, by following a field gradient downhill or uphill. Since agents' movements are simply driven by these abstract force fields, without any central controller, the global coordination emerges in a self-organized way from the interrelated effects of agents following the fields' shape and of dynamic fields re-shaping due to agents' movements.

Engineering a coordination protocol within a field-based approach is basically a bottom-up approach and consists in specifying how agents generate fields, how these fields are propagated, and how agents subscribe and react to the fields. The result is a robust specification that automatically adapts to changing environmental conditions that require limited computational efforts by agents, and only strictly local communication efforts. The analogy of field-based protocols and physical laws carries on a further advantage: field-based coordination activities can be effectively modeled in dynamical systems terms, that is, via differential equations that can be easily numerically integrated for the sake of rapid verification.

Further, this chapter is structured as follows: Section 42.2 describes the problem of motion coordination along with its possible applications and its main requirements. Section 42.3 presents field-based solutions, surveying systems that exploited this approach. Section 42.4 shows how field-based coordination can be modeled, in terms of a dynamical system formalism. Section 42.5 deals with programming field-based approaches, presenting a reference middleware and programming model. Finally, Section 42.6 outlines open issues and research directions.

42.2 Adaptive Motion Coordination: Applications, Issues, and Requirements

Enabling a set of agents to coordinate their movements in an environment and engineering a coordination protocol accordingly is not a problem if (i) agents know each other; (ii) they have a global knowledge about the environment in which they situate; (iii) they are not resource constrained. In these cases, it is possible to define and optimize any needed protocol to orchestrate agents' movements, and it is possible for agents to communicate with each other and to exploit environment knowledge as needed to implement such protocol. Unfortunately, the above assumptions cannot be made in the design and development of the software support for a mobile and wireless computing scenario, where (i) the number and the identities of agents, together with their operational environment are likely to vary in time, and the software systems must be engineered so as to adaptively deal with these situations; (ii) the need for scalability forbids a global perception of the environment and instead calls for a strictly local view, enabling the system to scale linearly with its size; (iii) agents are likely to be executed on resource-limited devices. This implies the need for cost-effective solutions that should not require the execution of complex algorithmic task or of complex communication protocols.

The challenge of identifying adaptive and effective protocols for the engineering of motion coordination activities in mobile computing scenarios is strictly tied to the problems of having agents be "context-aware." Coordination, by its own nature, requires some sort of context awareness: an agent can coordinate with other agents only if it is aware of "what is around." In mobile computing scenarios, however, such contextual information must be dynamically obtained by agents, so as to enable adaptivity; simple and cheap to be obtained and expressive enough to facilitate the definition of coordination activities and their execution by agents.

It is worth noting that the focus is not on how contextual information can be produced (e.g., it is possible to assume that localization mechanisms are available [11]), but how and in which form it can be made available to agents.

The following of this section introduces a simple case study and show that current coordination models are, form a software engineering point of view, inadequate to effectively address motion coordination problems.

42.2.1 A Case Study Scenario

The presented case study involves the problem of providing support for tourists visiting a museum. Specifically, it focuses on how tourists can be supported in planning their movements across a possibly large and unfamiliar museum and in coordinating such movements with other, possible unknown, tourists. It is possible to assume that each tourist is provided with a wireless-enabled computer assistant (e.g., a PDA) providing support for motion coordination (i.e., suggests the user on where to go). The kind of coordination activities involved may include:

- Scheduling attendance at specific exhibitions occurring at specific times. Here the application goal is simply to route tourists across the museum to let them reach the right place on time.
- Having a group of students split in the museum according to teacher-specific laws. Also in this case, the application goal is guide the students to let them visit relevant rooms accordingly to a specified schedule.

- Letting a group of agents (e.g., museum security guards) to move maintaining a formation. Here the PDAs could instruct guards to maintain a specified distance form each other so as to improve museum coverage.
- Helping a tourist to avoid crowd or queues while visiting the museum. Here the PDA can route the tourist to avoid a crowded corridor, or can change the visit schedule so as to postpone rooms that are overcrowded at the moment.
- Support a group of user to meet in the museum: meet in a specific room, join a specific person, meet in the most convenient room, accordingly to the participants actual location. Here the PDA would lead the user to the meeting location.
- In case of emergency, the system can enforce a evacuation plan, guiding tourists to nearby emergency exits, while avoiding crowds.

Accordingly with the assumptions made before, it is possible to assume that (i) the museum environment is unknown and dynamic: the museum floor-plan may be unknown, agents can enter and leave the museum at will and they are free to roam across the building. Agents are enabled to interact one another either directly or via an embedded infrastructure, possibly installed in the building. Since agents are not provided with any a priori information on the museum floor-plan or crowd condition, they have to acquire relevant information dynamically in-place, by interacting with other agents or with available services. (ii) They are able to perceive only a local view on the museum. This is both to support the scalability of the system, and (iii) not to overload the resource-limited PDAs with large amount of information.

It is worth noting that the above scenario and the associated motion coordination problems are of a very general nature, being isomorphic to scenarios such as, traffic management and forklifts activity in a warehouse, where navigators' equipped vehicles hint their pilots on where to go, or software agents exploring the Web, where mobile software agents can coordinate distributed researches by moving on various Web-sites. Therefore, also all the considerations presented in this chapter are of a more general validity, besides the addressed case study.

42.2.2 Inadequacy of Traditional Approaches

Most coordination models and middleware used so far in the development of distributed applications appear inadequate in supporting coordination activities in mobile and wireless computing scenarios.

In the following paragraphs will be presented various coordination models and middleware to illustrate their inadequacies form a software engineering perspective. The analysis will be mainly focused on evaluating how those middleware provide agents with contextual information and whether if the information provided is suitable for the motion coordination task.

Models based on *direct communication* promote designing a distributed application by means of a group of components that are in charge to communicate with each other in a direct and explicit way. Systems like Jini [13] and FIPA-based agent systems [3] are examples of middleware infrastructures rooted on a direct communication model. The problem of this approach is that agents are placed in a "void" space: the model, per se, does not provide any contextual information. Agents can only perceive and interact with other components/agents, and the middleware support is mostly reduced to helping in finding communication partners. Thus, each agent has to "manually" become context aware by discovering and explicitly interacting with the other entities in the environment. For instance, in the case study, an agent would have to explicitly retrieve and communicate with some services to gain the museum map, it would have to discover which other tourists are currently populating the museum, and explicitly negotiate with them to agree on a specific motion coordination policy (e.g., to move avoiding the formation of queues or to meet each other at a suitable place). From a software engineering perspective, this imposes a notable burden in the agent code and typically ends up with ad hoc, not scalable, and not adaptable solutions.

Models based on *shared data-spaces* support inter-agent interactions through shared localized data structures. These data structures can be hosted in some data-space (e.g., tuple space), as in EventHeap [14], JavaSpaces [10], or they can be carried on by agents themselves and dynamically merged to enable interactions, as in Lime [25] or XMiddle [21]. In these cases, agents are no longer placed in a totally void space, but live in an environment that can be modeled and described in terms of the information stored in the data spaces. Such information, typically referring to local conditions, can provide some sort of context-awareness to agents without forcing them to directly communicate with each other. For instance, in the case study, one can assume to have, at each room and corridor, a local data-space storing both local map information and messages left by the other agents about their presence and possibly about their intended next position. Still, to enforce a specific motion coordination pattern, agents may have to access several data-spaces to access all the required information (e.g., which agents are in which rooms), to build an internal representation of the current situation (e.g., where crowds and queues are), and then decide the next movements by (again) negotiating with other agents or accordingly to a predefined, nonadaptable, set of rules. The result is again that lots of code and computational effort is required to "understand" and exploit the contextual information for the achievement of a specific motion coordination task. This makes also this approach ineffective from a software engineering perspective.

Event-based models relying on *publish/subscribe* mechanisms make agents interact with each other by generating events and by reacting to events of interest, without having them to interact explicitly with each other. Typical infrastructures rooted on this model are: Jedi [8] and Siena [7]. Without doubt, event-based model promotes stronger context-awareness, in that components can be considered as embedded in an active environment able of notifying them about what is happening around, freeing agents from the need of explicitly querying other agents or the environment (as in direct-communication and data-space models) to discover what's happening. For instance, in the case study, a possible use of this approach would be to let each agent notify its movements across the museum, update (accordingly to other agents' notified movements) its internal representation of the crowd distribution in the museum (i.e., context-awareness) and then move properly by continuously adapting its plans accordingly to the newly received events. However, from a software engineering perspective, the information conveyed by events tends to be too low-level to be effectively used by agents, forcing them to catch and analyze a possibly large number of inter-related events to achieve a specific motion coordination pattern. This process is of course a big burden in the agent code, making also such kind of middleware unsuitable from a software engineering point of view.

42.3 Field-Based Approach to Motion Coordination

The common general problem, of the above interaction models, is that they are typically used to gather/communicate only a general purpose, not expressive, description of the context. This general purpose representation tends to be strongly separated from its usage by the agents, typically forcing them to execute complex algorithms to elaborate, interpret, and decide what to do with that information. In the above mentioned meeting application, for example, agents could be provided with the museum floor-plan and the position of other agents in the meeting group. Still, even if that information is a *complete* representation of their operational environment, agents will have to execute complex algorithms to decide and negotiate where to meet and how to go there.

On the contrary, if the context would have been represented expressively to facilitate agents in the achievement of a specific task, agents would trivially use that information to decide what to do. For example, in the above meeting application, if the agents would be able to perceive in their environment something like a "red carpet" leading to the meeting room, it would be trivial for them to exploit the information: just walk on the red carpet.

So the point is: how is it possible to create the "red carpet"? How to effectively represent context for the sake of motion coordination?

An intriguing possibility is to take inspiration from the physical world, and in particular from the way masses and physical particles move and globally self-organize their movements accordingly to that local contextual information that is represented by gravitational and electro-magnetic fields. These fields are sorts of "red carpets": particles achieve their tasks simply by following the fields.

This is the basis of the field-based approach. Following this approach, agents achieve their goals not because of their capabilities as single individuals, but because they are part of an (auto)organized system that leads them to the goals achievement. Such characteristics also imply that agents' activities are automatically adapted to the environmental dynamic, which is reflected in a changing view of the environment, without forcing agents to re-adapt themselves.

42.3.1 Key Concepts in the Field-Based Approach

Field-based approach is mainly driven by the above analysis and aims at providing agents with abstract — simple yet effective — representations of the context. To this end, the field-based approach delegates to a middleware infrastructure or to the agents themselves (connected in a peer-to-peer mobile ad hoc network) the task of constructing and automatically updating an essential distributed "view" of the system situation — possibly tailored to application-specific motion coordination problems — that "tells" agents what to do (i.e., how to move to implement a specific motion coordination patterns). Agents are simply let with the decision of whether to follow such a suggestion or not. Operatively, the key points of the field-approach can be schematized in the following four points:

1. The environment is represented and abstracted by "computational fields," spread by agents or by the infrastructure. These fields convey useful information for the agents' coordination tasks and provide agents with strong coordination-task-tailored context awareness.
2. The coordination protocol is realized by letting agents move following the "waveform" of these fields.
3. Environmental dynamics and agents' movements induce changes in the fields' surface, composing a feedback cycle that influences agents' movement (point 2).
4. This feedback cycle let the system (agents, environment, and infrastructure) auto-organize, so that the coordination task is eventually achieved.

A computational field can be considered a sort of distributed data structure characterized by a unique identifier, a location-dependent numeric value. Fields are locally accessible by agents depending on their location, providing them and a local perspective of the global situation of the system. For instance, with reference to the case study, the guide of the museum can spread in the environment a computational field whose value monotonically increase as it gets farther from his/her (see Figure 42.1), and thus implicitly enabling an agent, from wherever, of "sensing" how distant the guide is and where (in which direction) (s)he is.

Fields can be generated by the agents or by the environment, and their existence must be supported by a proper infrastructure. From the agent point of view, it does not matter if the infrastructure is implemented in some external servers accessed by the agent, or is part of the agent itself. What that matters is that it provide an up-to-date, local, field-based representation of the environment.

In field-based protocols, the simple principle to enforce motion coordination is by having agents move following the local shape of specific fields. For instance, a tourist looking for a guide can simply follow downhill the corresponding computational field.

Dynamic changes in the environment and agents' movements induce changes in the fields' surface, producing a feedback cycle that consequently influences agents' movement. For instance, should the museum guide be moving around in the museum, the field-supporting infrastructure would automatically update the corresponding computational field and would, consequently, have any agent looking for a guide re-adapt its movement accordingly. Should there be multiple guides in the museum, they could decide to sense each other's fields so as to stay as far as possible from each other to improve their reachability by tourists, and possibly dynamically re-shaping their formation when a guide, for contingent problems, has to move or go away.

Further examples follow later on. Here, it is worth noting that a field-based system — following its physical inspiration — can be considered as a simple dynamical system: agents are seen as balls rolling

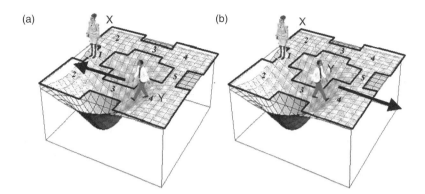

FIGURE 42.1   The tourist Y senses the field generated by the tourist guide X (increasing with the distance from the source, as the numeric values in each room indicates). (a) Y agent follows the field downhill to approach X; (b) Y follows the field uphill to get farther from X. In this figure, agent Y uses the field generated by X directly as an application-specific *coordination field*.

upon a surface, and complex movements are achieved not because of the agents will, but because of dynamic re-shaping of this surface.

In a given environment, several different types of fields can exist and be propagated by agents and by the infrastructure, accordingly to field-specific laws. Of course, the agents and the infrastructure can decide to propagate any type of application-specific fields, accordingly to propagation rules that can facilitate the achievement of application-specific problems.

The achievement of an application-specific coordination task, however, is rarely relying on the evaluation, as it is, of an existing computational field (as in the case of a tourist looking for a guide and simply following the specific field of that guide). Rather, in most cases, an application-specific task relies on the evaluation of an application-specific *coordination field*, as a combination (e.g., linear) of some of the locally perceived fields. The *coordination field* is a new field in itself, and it is built with the goal of encoding in its shape the agent's coordination task. Once a proper *coordination field* is computed, agents can achieve their coordination task by simply following (deterministically or with some probabilistic rule) the shape of their *coordination field* uphill, downhill, or along its equipotential lines (depending on the specific problem) as if they were walking upon the *coordination field* associated surface (see Figure 42.1). For instance, in the case study, for guides to stay as far as possible from each other, they have to follow uphill a *coordination field* resulting from the combination of all the computational fields of each guide.

In the following, the *coordination field* will be referred as the only field (eventually computed by the agent) driving the agent movements.

42.3.2   The State of the Art in Field-Based Coordination

Several proposals addressing the problem of distributed motion coordination with field-based protocols have been proposed in the last few years. The basic idea to implement the concept of a field is to rely on distributed data structure actually spread in the environment. Such distributed data structures must be supported by a distributed infrastructure. Specifically, a field would be injected in the network from a specific node. Then it will be propagated across the infrastructure hop-by-hop. During propagation the field's value would be changed so as to let the field assume the intended shape.

In the case study, it is possible to assume the presence, in the museum, of a densely distributed network of computer-based devices, associated with rooms, corridors, art pieces, alarm systems, climate conditioning systems, etc. All these devices will be capable of communicating with each other and with the mobile devices located in their proximity via the use of a short-range wireless link. Moreover, it is important to make the following two assumptions (i) devices are provided with localization mechanisms enabling them to know neighbors' coordinates in a private local coordinate frame [11]. (ii) The rough topology of the ad hoc network being formed by computer-based devices resemble the museum's topology (i.e., floor-plan). This means in particular that there are not network links between physical barriers (like walls). Several mechanism may be employed to achieve this property, it can be assumed that either the devices are able to detect and drop those network links crossing physical barriers (e.g., relying on signal strength attenuation or some other sensor installed on the device) or that the museum building is pre-installed with a network backbone — reflecting its floor-plan topology — and that all the nodes can connect only to the backbone. These two properties are needed to assure that the fields are propagated coherently with the museum floor-plan (i.e., agents can follow fields without stumbling into walls).

Co-Fields [18] is the model developed by the authors of this chapter and is the one that closely targets the application scenario described so far. This model has been implemented upon the Tuples On The Air (TOTA) [20] middleware infrastructure. The TOTA middleware would be installed on a large number of the above devices, that will connect in an ad hoc network suitable to support distributed data structures implementing the field abstraction. Agents would connect to this infrastructure, inject field-like data structures that propagate hop-by-hop across the infrastructure, sense locally the stored data structures and move accordingly.

Similar to Co-Fields is the multi-layered multi-agent situated system (MMASS) formal model for multi-agent coordination, described in Reference 2. This model represents the environment as a multi-layered graph in which agents can spread abstract fields representing different kinds of stimuli through the nodes of this graph. The agents' behavior is then influenced by the stimuli they perceive in their location. In fact agents can associate reactions to these stimuli, like in an event-based model, with the add-on of the location-dependency that is associated to events and reactions. The main application scenarios of this model is quite different from the one discussed in this chapter since it focuses on simulation of artificial societies and social phenomena. For this reason MMASS lacks of an implemented infrastructure supporting the model and it is only realized via simulation based on cellular automata.

An area in which the problem of achieving effective context-awareness and adaptive coordination has been addressed via a field-based approach is Amorphous Computing [5,23]. The particles constituting an amorphous computer have the basic capabilities of propagating sorts of abstract computational fields in the network, and to sense and react to such fields. In particular, particles can transfer an activity state towards directions described by fields' gradients, so as to make coordinated patterns of activities emerge in the system independently of the specific structure of the network. Such mechanism can be used, among other possibilities, to drive particles' movements and let the amorphous computer self-assemble in a specific shape.

A very similar approach to self-assembly has been proposed in the area of modular robots [28]. A modular robot is a collection of simple autonomous actuators with few degrees of freedom connected with each other. A distributed control algorithm is executed by all the actuators that coordinate to let the robot assume a global coherent shape or a global coherent motion pattern (i.e., gait). Currently proposed approaches [28] adopts the biologically inspired idea of hormones to control such a robot. Hormone signals are similar to fields in that: they propagate through the network without specific destinations, their content can be modified during propagation and they may trigger different actions for different receivers.

Shifting from physical to virtual movements, the popular videogame "The Sims" [29] exploits sorts of computational fields, called "happiness landscapes" and spread in the virtual city in which characters live, to drive the movements of nonplayer characters. In particular, nonplayer characters autonomously move in the virtual Sims city with the goal of increasing their happiness by climbing the gradients of specific computational fields. For instance, if a character is hungry, it perceives and follows a happiness landscape whose peaks correspond to places where food can be found, that is, a fridge. After having eaten, a new landscape will be followed by the character depending on its needs.

It is interesting to report that there are also systems that adopt a field-based approach, without requiring the presence of any needed distributed infrastructure. The field-based representation, in fact, is computed internally by the agents by exploiting data coming from sensors (typically a camera) installed on the agents themselves. For example, an agent seeing an obstacle through its camera would represent the obstacle as a repeating field. The main advantage of this approach is that it does not require any communication between the agents. The drawback is the need of complex sensors to be mounted on agents (e.g., the camera) and the need for complex algorithms mapping images into fields. Examples of this kind of approach can be found in robotics.

The idea of fields driving robots movement is not new [16]. For instance, one of the most recent re-issue of this idea, the electric field approach (EFA) [15], has been exploited in the control of a team of Sony Aibo legged robots in the RoboCup domain. Following this approach, each Aibo robot builds a field-based representation of the environment from the images captured by its head mounted cameras (stereo-vision), and decides its movements by examining the fields' gradients of this representation.

Apart from the presented differences, the modeling of the field-based approach in both the earlier two kinds of models is rather similar and it basically consists in representing the environment by means of the fields and having each agent to follow the gradient of a *coordination field* computed by combining fields relevant for its application task.

42.4 Modeling Field-Based Coordination

The physical inspiration of the field-based approach invites modeling field-based protocols in terms of a dynamical system. Specifically, this implies modeling fields as continuous functions and writing the differential equations governing the motion of a point (the agent) driven by the gradient of a specific *coordination field* (the field driving agents movements, evaluated by the agent, by locally combining other fields in the environment). Once the analytical shape of the *coordination field* is defined, writing and solving numerically the differential equations of the system is rather simple. This provides an effective way to simulate the system, and it can be regarded as a tool to make experiments, to quickly verify that a *coordination field* correctly expresses a coordination task, and to tune coefficients.

More in detail, it is rather easy to see that considering the agent i, denoting its coordinates (for sake of notation simplicity restricted to the two-dimensional case) as $(x_i(t), y_i(t))$ and its *coordination field* as $CF_i(x, y, t)$, the differential equations governing i behavior are in the form:

$$\frac{dx_i}{dt} = v\frac{\partial CF_i(x, y, t)}{\partial x}(x, y, t)$$

$$\frac{dy_i}{dt} = v\frac{\partial CF_i(x, y, t)}{\partial y}(x, y, t)$$

If i follows uphill the *coordination field*. They are

$$\frac{dx_i}{dt} = -v\frac{\partial CF_i(x, y, t)}{\partial x}(x, y, t)$$

$$\frac{dy_i}{dt} = -v\frac{\partial CF_i(x, y, t)}{\partial y}(x, y, t)$$

If i follows downhill the *coordination field*.

This is because the direction of the gradient of the agent's *coordination field*, evaluated toward the spatial coordinates, points to the direction in which the *coordination field* increases. So the agent i will follow this gradient or will go in the opposite direction depending on if it wants to follow its *coordination field* uphill or downhill. v is a term that can model the speed of an agent.

The case in which the agent follows an equipotential line of the coordination space, is slightly more complicated, because if the space dimension is greater than 2, it is not possible to talk about an equipotential line, but only about an equipotential hyper-surface. This hyper surface will be the one orthogonal to the gradient of the *coordination field* evaluated only toward the space coordinates, which will be thus a function of the time:

$$\nabla CF_i(x_1, x_2, \ldots, x_n)(t)$$

So the only dynamical equation it is possible to write in this case, is the one that specifies that the agent's movements belong to this hyper surface:

$$\left(\frac{dx_1^i}{dt}, \frac{dx_2^i}{dt}, \ldots, \frac{dx_n^i}{dt}\right) \cdot \nabla CF_i(x_1^i, x_2^i, \ldots, x_n^i)(t) = 0$$

where the *dot* stands for the classical scalar product.

In the following of this section field-based protocols and the dynamical system formalism will be applied to a number of motion coordination problems. Before proceeding, it is worth noting that, considering those situations in which agent movements do not occur in an open space but are constrained by some environmental conditions, the dynamical system description become more complex. The definition of "constrained" equations would involve either some artificial force fields to constrain agents to stay within the building plan or a domain not based on $\Re^n$, but on a more general and complex manifold. Due to the complexities involved in the description of such equations, this chapter considers and discusses field based protocols only in open, unconstrained spaces, an approach which is anyway valuable to evaluate the correctness of a coordination protocol.

However, to overcome this limitation, a multi-agent simulation of the museum application has been used to complement the dynamical system description. Such simulation has been developed using the Swarm toolkit [30] and enables to model: any required spatial environment (e.g., a specific museum map); the presence in such environment of any needed number of fields; and the presence of any needed number of agents each its own goals (e.g., number of tourists each with a specific plan of visit to the museum).

42.4.1 Flock Field: Moving while Maintaining a Formation

Consider the case study of having agents distribute according to specific geometrical patterns (flocks), and to let them preserve such patterns while moving. More specifically, agents can represent security guards (or security robots) in charge of cooperatively monitor a building by spreading in the building so as to stay at specified distances from each other [12]. To this end, it is possible to take inspiration from the work done in the swarm intelligence research [4]: flocks of birds stay together, coordinate turns, and avoid each other, by following a very simple swarm algorithm. Their coordinated behavior can be explained by assuming that each bird tries to maintain a specified separation from the nearest birds and to match nearby birds' velocity. To implement such a coordinated behavior with a field-based protocol and apply it to the case study, each agent can generate a field *(flock-field)* whose magnitude assumes the minimal value at specific distance from the source, distance expressing the intended spatial separation between agents. The final shape of this field approaches the function depicted in Figure 42.2(a). Fields are always updated to reflect peers' movements. To coordinate movements, peers have simply to locally perceive the generated fields, and to follow downhill the gradient of the fields. The result is a globally coordinated movement in which peers maintain an almost regular grid formation (see Figure 42.2[b]).

Analytically, the *(flock-field)* generated by an agent i located at (X_P^i, Y_P^i) can be simply described as follows:

$$d = \sqrt{(x - X_P^i)^2 + (y - Y_P^i)^2}$$

$$FLOCK_i(x, y, t) = d^4 - 2a^2d^2$$

where a is again the distance at which agents must stay away from each other. Starting from these simple equations, one can write, for each of the agents in the set, the differential equations ruling the dynamic

(a)

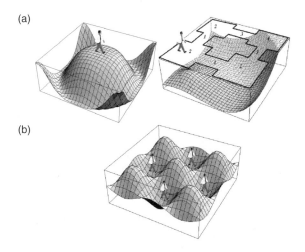

(b)

FIGURE 42.2   (a) Ideal shape of the flock field. (b) When all the agents follow other agents' fields they collapse in a regular grid formation.

behavior of the system, that is, expressing that agents follow downhill the minimum of the *(flock-field)*. These have the form:

$$\frac{dx_i}{dt} = -v\frac{\partial CF_i(x, y, t)}{\partial x}(x_i, y_i) \qquad\qquad i = 1, 2\ldots, n$$

$$\frac{dy_i}{dt} = -v\frac{\partial CF_i(x, y, t)}{\partial y}(x_i, y_i) \qquad\qquad i = 1, 2\ldots, n$$

$$\frac{dx_i}{dt} = -v\frac{\partial \min(FLOCK_1, FLOCK_2, \ldots, FLOCK_n)}{\partial x}(x_i, y_i) \qquad i = 1, 2\ldots, n$$

$$\frac{dy_i}{dt} = -v\frac{\partial \min(FLOCK_1, FLOCK_2, \ldots, FLOCK_n)}{\partial y}(x_i, y_i) \qquad i = 1, 2\ldots, n$$

Such equations can be numerically integrated by making use of any suitable mathematical software. In this study, the Mathematica package [22] has been used. Figure 42.3(a) shows the results obtained by integrating the above equations, for different initial conditions, and for a system composed by four agents. The figure shows an x–y plane with the trajectories of the agents of the system (i.e., the solutions of $(x_i(t), y_i(t))$ evaluated for a certain time interval) while moving in a open space. It is rather easy to see that the four agents maintain a formation, trying to maintain specified distances form each other. Figure 42.3(b) shows the result of the motion coordination in the simulated environment. Specifically, a simple museum map has been drawn, with three agents (the white dots) moving in it. The formation consists in having the agents remain in adjacent rooms from each other.

42.4.2   Person Presence Field: Surrounding a Prey

The aim of this coordination task is to allow a group of agents ("predators") moving in the building to surround and eventually capture another agent ("prey"). As an example, one may, think at a group of security guards in charge of searching and catching a child that got lost in a big building and that is frightened by those big military men. This task of surrounding a prey relies on a field that must be generated by every agent in the building. This field will be referred as the *person presence* field to stress the fact it represent the presence of a person in the building. The *person presence* field simply has value that increases monotonically as the distance from the person increases. The analytical description of this field is straightforward. If the agent i is at the coordinates (X_P^i, Y_P^i), then it generates a *person presence* field

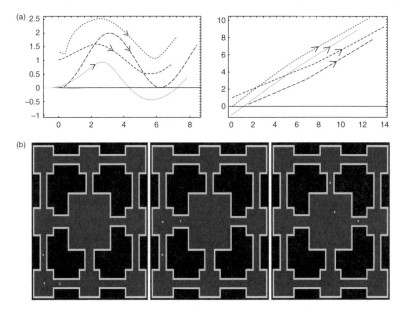

FIGURE 42.3 (a) Solutions of the flock fields' differential equations, for different initial conditions. (b) Flocking: from left to right, different stages in the movement of a group of agents through the building and coordinating with each other so as to stay always in neighbor rooms.

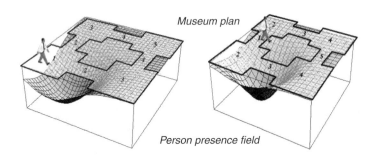

FIGURE 42.4 Person presence fields in a building.

whose equation is (see Figure 42.4):

$$\text{PRES}_i(x, y, t) = 1 + k_p - k_p \, e^{-h_p((x-X_p^i)^2 + (y-Y_p^i)^2)}$$

$$k_p, h_p > 0; \quad 1 + k_p - k_p \, e^{-2h_p} \approx 1$$

Now, considering the prey's *coordination field* consisting in the linear combination of all the predators' fields, by following the decrease of the resulting field, the prey runs away from the predators.

$$\text{CF}^{\text{prey}}(x, y, t) = \sum_{i=1}^{n} -\text{PRES}_i^{\text{pred}}(x, y, t)$$

$\text{PRES}_i^{\text{pred}}$ is the *person presence* field of the predator agent *i*. Similarly, considering each predator's *coordination field* consisting of the linear combination between the prey's field and all the other predators' fields a predator is directed toward the prey, but avoids other predators.

$$\text{CF}_i^{\text{pred}}(x, y, t) = \text{PRES}^{\text{prey}}(x, y, t) + \sum_{j=1, j \neq i}^{n} -\text{PRES}_i^{\text{pred}}(x, y, t)$$

PRES^{prey} is the *person presence* field of the prey agent. Obtaining the differential equations governing the system is now just a matter of substituting the fields' analytical description together with the *coordination field*'s description in the differential equations describing the field-based protocol. The substitution will be reported just for this first example.

$$\frac{\mathrm{d}x^{\text{prey}}}{\mathrm{d}t} = -v^{\text{prey}} \frac{\partial \text{CF}^{\text{prey}}(x, y, t)}{\partial x} (x^{\text{prey}}, y^{\text{prey}})$$

$$\frac{\mathrm{d}y^{\text{prey}}}{\mathrm{d}t} = -v^{\text{prey}} \frac{\partial \text{CF}^{\text{prey}}(x, y, t)}{\partial y} (x^{\text{prey}}, y^{\text{prey}})$$

$$\frac{\mathrm{d}x_i^{\text{pred}}}{\mathrm{d}t} = -v^{\text{pred}} \frac{\partial \text{CF}_i^{\text{pred}}(x, y, t)}{\partial x} (x_i^{\text{pred}}, y_i^{\text{pred}}) \qquad i = 1, 2, \ldots, n$$

$$\frac{\mathrm{d}y_i^{\text{pred}}}{\mathrm{d}t} = -v^{\text{pred}} \frac{\partial \text{CF}_i^{\text{pred}}(x, y, t)}{\partial y} (x_i^{\text{pred}}, y_i^{\text{pred}}) \qquad i = 1, 2, \ldots, n$$

The results of the numerical integration of these equations in the case of three predators and one prey are displayed in Figure 42.5(a), which shows a common x–y plane with the trajectories of the elements of the system (i.e., the solutions of $(x_i(t), y_i(t))$ evaluated for a certain time interval). Here it is possible to notice (Figure 42.5[a]-left) that if the predators do not repeal one another they are not able to surround the prey and all reach the prey from the same direction. On the contrary (Figure 42.5(a)-right), if they repeal each other, they reach the prey from different directions, surrounding it. Figure 42.5(b) shows the

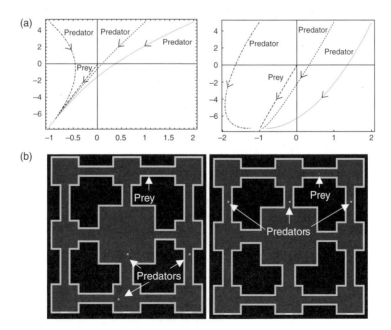

FIGURE 42.5 (a) Surrounding a prey: predators that simply follow the prey without enacting any coordination strategy are not able to surround the prey (top); predators being repulsed by other predators' fields are able to surround the prey (bottom). (b) surrounding in the simulator.

result of the motion coordination in the simulated environment. Maintaining a certain distance from each other, predator agents surround the prey without leaving any escape path.

42.4.3 Room Field and Crowd Field: Load Balancing

The aim of this coordination task is to allow the users to avoid queues while visiting the museum. For this reason their agents will drive them to visit the rooms in their schedule trying to avoid crowded areas. At the global scale, this realize a load balancing policy between users and museum's rooms. To this end other two fields need to be introduced: the *room field* and the *crowd field*. The *room field* is generated by every building's room, it has a value that increases with the distance from the source. The analytical description of this field is straightforward: considering the room i to be located — the center of the room is taken as a reference — at the coordinates (X_R^i, Y_R^i), then it is possible to describe its *room field* by the following equation (see Figure 42.6-top):

$$\text{RF}_i(x, y, t) = 1 + k_r - k_r \, e^{-h_r((x-X_R^i)^2 + (y-Y_R^i)^2)}$$

$$k_r, h_r > 0; \qquad 1 + k_r - k_r \, e^{-2h_r} \approx 1$$

The *crowd field* measures the amount of crowd in a room. It is evaluated by the infrastructure manipulating the *person presence* fields of the people in the building. The analytical description of this field is a bit more complicated than the previous ones, because it tries to abstract an entity that is strictly coupled with the discrete nature of the space considered by means of a continuous field.

The crowd field, in fact, should in principle have a constant value in the area of a room and drop to zero (or to another value) outside that area. However, in order to avoid discontinuities, the crowd field, generated by the room i at the coordinates (X_R^i, Y_R^i), can be modeled by means of the following function (see Figure 42.6-bottom):

$$\text{CWF}_i(x, y, t) = k_c^i \, e^{-h_c((x-X_R^i)^2 + (y-Y_R^i)^2)}$$

where h_c is chosen so as the magnitude of the field outside the room is almost 0, while $k_c^i > 0$ is a coefficient that represents how much crowded the room is. It is important to understand, that because of

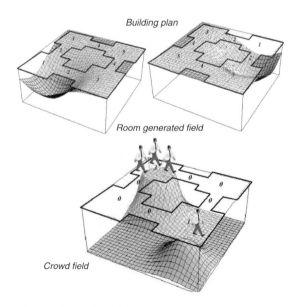

FIGURE 42.6   Room Fields (top) and Crowd Field (bottom) in a building.

its role, k_c^i cannot be a fixed coefficient, but it must vary over time to reflect the different crowd conditions, so in general $k_c^i = k_c^i(t) \cdot k_c^i(t)$ is simply given by the number of people present in room i at time t (whose coordinates are within the room's perimeter) normalized to the dimension of the room. It is clear that the $k_c^i(t)$ defined in this way is an effective measure of how much a room is crowded. Finally the global *crowd field* will be defined by means of the sum of the rooms' *crowd fields*:

$$\mathrm{CWF}(x, y, t) = \sum_{i \in \mathrm{rooms}} \mathrm{CWF}_i(x, y, t)$$

Let us start considering the load balancing coordination protocol, whose aim is to help a user to avoid crowds or lines while roaming through the building. Basically, each user specifies his/her path through the building by selecting the rooms he/she wants to visit. This can be accomplished by letting the users' agents evaluate their *coordination field* as a minimum combination (fields are combined by taking in each point the minimum one) of the *room fields* (RF) to which their users are interested.

$$\mathrm{CF}(x, y, t) = \min(\mathrm{RF}_i(x, y, t) : i = 1, 2, \ldots, n)$$

The *coordination field* produced models a surface having minimum points in correspondence of the rooms the user wants to visit. The user follows greedily the decrease of the *coordination field* and thus visiting the associated rooms. In order not to get trapped in a minimum, when the user completed the visit of a room, the corresponding field is removed from the combination and so it does not represent a minimum anymore. To take into consideration the load balancing service, the *crowd field* (CWF) has to be taken into account and for this purpose the combination computed by the agent can become:

$$\mathrm{CF}(x, y, t) = \min(\mathrm{RF}_i(x, y, t) : i = 1, 2, \ldots, n) + \mu \mathrm{CWF}(x, y, t)$$

The term $\mu \mathrm{CWF}(x, y, t)$ with $\mu \geq 0$ is a field that has its maximum points where the crowd is particularly intense, as depicted in Figure 42.6-bottom. This term manages the load balancing policy: in fact when this term is added to the minimum combination it changes the steepness of the *coordination field* in the crowded zones. In particular a crowded zone tends to be a peak in the *coordination field* and thus it tends to repulse the income of other agents. It is easy in fact to understand that the agent will follow the "greed path" — the one indicated by $\min(\mathrm{RF}_i(x, y, t) : i = 1, 2, \ldots, n)$ — only if CF decreases toward the same direction indicated by $\min(\mathrm{RF}_i(x, y, t) : i = 1, 2, \ldots, n)$. Indicating this direction with v, this condition can be restated as:

$$\nabla(\min(\mathrm{RF}_i(x, y, t) : i = 1, 2, \ldots, n) + \mu \mathrm{CWF})(v) \leq 0 \Rightarrow$$
$$\mu \leq -\frac{\nabla \min(\mathrm{RF}_i(x, y, t) : i = 1, 2, \ldots, n)(v)}{\nabla \mathrm{CWF}(v)}$$

Observe that $\nabla \min(\mathrm{RF}_i(x, y, t) : i = 1, 2, \ldots, n)(v) \leq 0$ and thus the condition $\mu \geq 0$ can be satisfied. For this reason μ can be regarded as a term specifying the relevance of the *crowd field*. If μ is too high the agent will suggest the user to follow alternative (possibly longer) uncrowned paths toward the destination whenever the "greed" one will be a bit crowded. If μ is too low the agent will accept always to remain in the "greed" path disregarding the traffic conditions. The result of the numerical integration of the differential equations is depicted in Figure 42.7(a). Here an agent is willing to proceed from $(0, 0)$ to $(10, 0)$. If it does not consider crowd (Figure 42.7[a]-left) it follows a straight line form source to destination (eventually queuing in the middle). Otherwise, it is able to avoid the crowded area in the middle, by walking around it (Figure 42.7[a]-right). Figure 42.7(b) shows the result of the motion coordination in two different simulated environments. On the left the crowd is not considered in the agents' *coordination field* and thus crowds and queues arise. On the right the crowd term is considered and thus crowds are avoided.

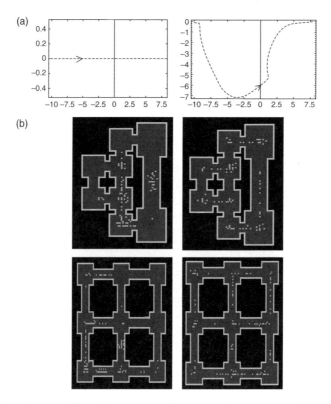

FIGURE 42.7 (a) The trajectory followed by an agent when no crowd is encountered (left). Following downhill *coordination field* the agent avoids a crowd between itself and its target (right). (b) Load balancing in two different simulated museums. When agents do not exploit Co-Fields, crowded zones appear (left). These crowded zones are mostly avoided when agents follow the proper *coordination field* (right).

42.4.4 Room Field and Crowd Field: Meetings

Let us turn the attention to a "meeting" protocol whose aim is to help a group of users to dynamically find and move toward the most suitable room for a meeting.

The definition of this coordination protocol can rely on the previously introduced field by changing the *coordination field* perceived by agents. Also, this coordination protocol can be built upon the previous coordination protocol so as to retain the load-balanced movements also when agents are moving to meet each other. Several different protocols can be though related to how a group of agents should meet:

1. The group of users wants to meet in a particular room x. This is the simplest case and each of the user has to compute the following *coordination field*:

$$CF(x, y, t) = RF_x(x, y, t) + \mu CWF(x, y, t)$$

   In this way every agent is directed to the minimum of that leads to the meeting room. The crowd field term enforces the load balancing policy also in this case.

2. The group of users wants to meet in the room where person x is located. This is very simple as well: each of the user has to compute the following *coordination field*:

$$CF(x, y, t) = PRES_x(x, y, t) + \mu CWF(x, y, t)$$

   where $PRES_x$ is the field generated by person x. In this way every agent is directed to the minimum of $PRES_x$ that leads to the meeting room (where person x is located). It is interesting to notice that this approach works even if person x moves after the meeting has been scheduled. The meeting will be automatically rescheduled in the new minimum of PRES_x.

3. The group of users wants to meet in the room that is between them (their "barycenter"). To this purpose each user i can compose its *coordination field* by combining the fields of all the other users:

$$\text{CF}(x, y, t) = \sum_{i \neq x} \text{PRES}_x(x, y, t) + \mu\text{CWF}(x, y, t)$$

In this way all the users "fall" toward each other, and they meet in the room that is in the middle. It is interesting to notice, that this "middle room" is evaluated dynamically and the process takes into consideration the crowd. So if a room is overcrowded it will not be chosen, even if it is the room, which is physically in the middle of the users. For the same reason if some users encounter a crowd in their path to the meeting room, the meeting room is automatically changed to one closer to these unlucky users. The strength of this approach is that it is fully integrated with the field concept and that the meeting room is chosen to encounter the difficulties found by the user in real time.

By considering the third of the above three possibilities, the dynamical system description for this problem is straightforward and the integration of the differential equations is depicted in Figure 42.8(a), showing how agents effectively converge to each other in the barycenter in an open space. Specifically, it depicts the x–y plane where agents live, with the trajectories of the agents of the system (i.e. the solutions of $(x_i(t), y_i(t))$ evaluated for a certain time interval). Figure 42.8(b) shows the various stages in the meeting process in two different simulated environments.

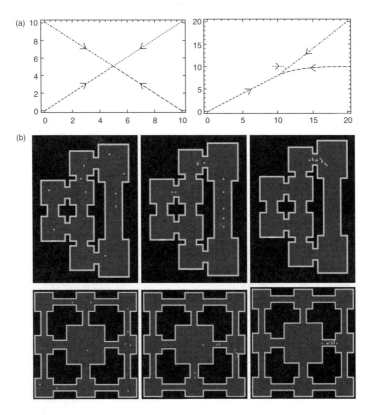

FIGURE 42.8   (a) Meeting: solutions of the system's differential equations, for different initial conditions. (b) From left to right different stages in the meeting process, showing that agents converge toward each other.

42.5   Programming Motion Coordination Protocols

Turning the discussion more concrete: what software architecture is required to support field-based protocols? How can be implemented? How can be programmed?

The following of this section will focus on field-based protocols implemented through distributed data structures actually spread in the environment.

42.5.1   Architecture and Implementation of a Field-Based System

The middleware considered in this section, to support field-based coordination is Tuples On The Air (TOTA) [19]. TOTA proposes relying on distributed tuples to represent fields. Tuples are not associated to a specific node (or to a specific data space). Instead, tuples are injected in the network and can autonomously propagate and diffuse across the network accordingly to a specified pattern. Thus, TOTA tuples form spatially distributed data structures suitable to represent fields. To support this idea, TOTA is composed by a peer-to-peer network of possibly mobile nodes, each running a local version of the TOTA middleware. Upon the distributed space identified by the dynamic network of TOTA nodes, each component is capable of locally storing tuples and letting them diffuse through the network. Tuples are injected in the system from a particular node, and spread hop-by-hop accordingly to their propagation rule. In fact, a TOTA tuple is defined in terms of a "content" and a "propagation rule."

$$T = (C, P)$$

The content C is an ordered set of typed fields representing the information carried on by the tuple. The propagation rule P determines how the tuple should be distributed and propagated across the network. This includes determining the "scope" of the tuple (i.e., the distance at which such tuple should be propagated and possibly the spatial direction of propagation) and how such propagation can be affected by the presence or the absence of other tuples in the system. In addition, the propagation rule can determine how the tuple's content should change while it is propagated.

The spatial structures created by tuples propagation must be maintained coherent despite network dynamism. To this end, the TOTA middleware supports tuples propagation actively and adaptively: by constantly monitoring the network local topology and the income of new tuples, the middleware automatically re-propagates tuples as soon as appropriate conditions occur. For instance, when new nodes get in touch with a network, TOTA automatically checks the propagation rules of the already stored tuples and eventually propagates the tuples to the new nodes. Similarly, when the topology changes due to nodes' movements, the distributed tuples automatically change to reflect the new topology (see Figure 42.9).

From the architecture point of view, each TOTA middleware is constituted by three main parts (see Figure 42.10): (i) the TOTA API, is the main interface to access the middleware. It provides functionalities to let the application to inject new tuples in the system (*inject* method), to read tuples both from the local tuple space and from the node's one-hop neighborhood, either via pattern matching or via key-access to a tuple's unique id (*read, readOneHop, keyrd, keyrdOneHop* methods), to delete tuples from the local middleware (*delete* method), or to place and remove subscriptions in the event interface (*subscribe, unsubscribe* methods). Moreover, two methods allow tuples to be actually stored in he local tuple space or to be propagated to neighboring nodes (*store, move* methods). (ii) The EVENT INTERFACE is the component in charge of asynchronously notifying the application about subscribed events. Basically upon a matching event, the middleware invokes on the subscribed components a *react* method to handle the event. (iii) The TOTA ENGINE is the core of TOTA: it is in charge of maintaining the TOTA network by storing the references to neighboring nodes and to manage tuples' propagation by executing their propagation methods and by sending and receiving tuples. Within this component is located the tuple space in which to store TOTA tuples.

From an implementation point of view, a first prototype of TOTA running on Laptops and on Compaq IPAQs equipped with 802.11b, Familiar LINUX and J2ME-CDC (Personal Profile) has been developed.

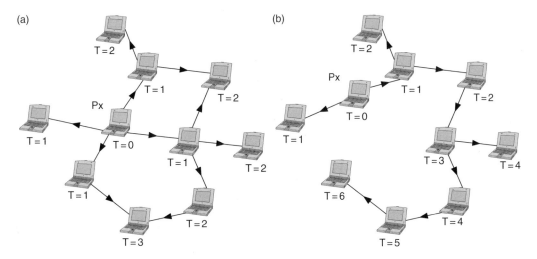

FIGURE 42.9 (a) *Px* propagates a tuple that increases its value by one at every hop. (b) When the tuple source *Px* moves, all tuples are updated to take into account the new topology.

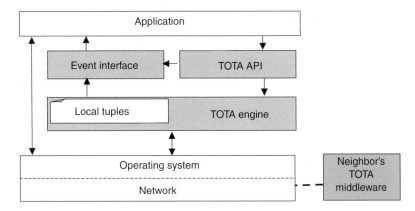

FIGURE 42.10 The TOTA middleware architecture.

IPAQs connect locally in the MANET mode (i.e., without requiring access points) creating the skeleton of the TOTA network. Tuples are being propagated through multicast sockets to all the nodes in the one-hop neighborhood. The use of multicast sockets has been chosen to improve the communication speed by avoiding 802.11b unicast handshake. By considering the way in which tuples are propagated, TOTA is very well suited for this kind of broadcast communication. This is a very important feature, because it could allow implementing TOTA also on really simple devices (e.g., micro sensors) that cannot be provided with sophisticate (unicast enabled) communication mechanisms. Other than this communication mechanism, at the core of the TOTA middleware there is a simple event-based engine, that monitors network reconfigurations and the income of new tuples and react either by triggering propagation of already stored tuples or by generating an event directed to the event interface.

42.5.2 Specifying Tuples and Agents

Relying on an object oriented methodology, TOTA tuples are actually objects: the object state models the tuple content, while the tuple's propagation has been encoded by means of a specific *propagate* method.

When a tuple is injected in the network, it receives a reference to the local instance of the TOTA middleware, then its code is actually executed (the middleware invokes the tuple's *propagate* method).

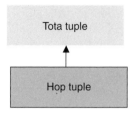

FIGURE 42.11   Tuples' class hierarchy.

If during the tuple's execution, the middleware *move* method is called, the tuple is actually sent to all the one-hop neighbors, where it will be executed recursively. During migration, the object state (i.e., tuple content) is properly serialized to be preserved and rebuilt upon the arrival in the new host.

Following this schema, it is possible to define an abstract class TotaTuple, that provides a general framework for tuples programming. The code of this class is in the following:

```
abstract class TotaTuple {
 protected TotaInterface tota;
 /* the state is the tuple content */
 /* this method inits the tuple, by giving a reference
     to the current TOTA middleware */
 public void init(TotaInterface tota) {
  this.tota = tota;
 }
 /* this method codes the tuple actual actions */
 public abstract void propagate();
 /* this method enables the tuple to react
 to happening events... see later in the article */
 public void react(String reaction, String event)
{}}
```

It is worth noting that in TOTA, a tuple does not own a thread, but it is actually executed by the middleware that runs the tuple's *init* and *propagate* methods. Tuples, however, must remain active even after the middleware has run their code. This is fundamental because their self-maintenance algorithm, for example, must be executed whenever the right condition appears (e.g., when a new node connects to the network, the tuples must propagate to this newly arrived node). To this end, tuples can place subscriptions as provided by the standard TOTA API. These subscriptions let the tuples remain "alive," being able to execute upon triggering conditions.

A programmer can obtain new tuples by subclassing the *TotaTuple* class. Moreover, to facilitate the creation of tuples, a tuples' class hierarch from which the programmer can inherit has been developed. The class hierarchy enable a programmer to create custom tuples without worrying about the intricacies of dealing with tuple propagation and maintenance (a piece of the hierarchy is depicted in Figure 42.11). More detailed information about the complete class hierarchy can be found in Reference 20.

The class *HopTuple* inherits from *TotaTuple*. This class is a template to create self-maintained distributed data structures over the network. Specifically, it implements the superclass method *propagate*, as shown in the following:

```
public final void propagate() {
 if(decideEnter()) {
  boolean prop = decidePropagate();
  changeTupleContent();
  this.makeSubscriptions();
  tota.store(this);
  if(prop)
   tota.move(this);
}}
```

In particular, this propagation method is realized by executing a sequence of other specific methods: *decideEnter*, *decidePropagate*, *changeTupleContent*, and *makeSubscriptions* so as to realize a breadth first, expanding ring propagation. The result is simply a tuple that floods the network without changing its content:

1. When a tuple arrives in a node (either because it has been injected or it has been sent from a neighbor node) the middleware executes the *decideEnter* method that returns true if the tuple can enter the middleware and actually execute there, false otherwise. The standard implementation returns true if the middleware does not already contain that tuple.
2. If the tuple is allowed to enter the method *decidePropagate* is run. It returns true if the tuple has to be further propagated, false otherwise. The standard implementation of this method returns always true, realizing a tuple's that floods the network being recursively propagated to all the peers.
3. The method *changeTupleContent* changes the content of the tuple. The standard implementation of this method does not change the tuple content.
4. The method *makeSubscriptions* allows the tuple to place subscriptions in the TOTA middleware. As stated before, in this way the tuple can react to events even when they happen after the tuple completes its execution. The standard implementation does not subscribe to anything.
5. After that, the tuple is inserted in the TOTA tuple space by executing *tota.store(this)*.
6. Then, if the *decidePropagate* method returned true, the tuple is propagated to all the neighbors via the command *tota.move(this)*. The tuple will eventually reach neighboring nodes, where it will be executed again. It is worth noting that the tuple will arrive in the neighboring nodes with the content changed by the last run of the *changeTupleContent* method.

Programming a TOTA tuple to create a distributed data structure basically reduces at inheriting from the above class and overloading the above four methods to customize the tuple behavior. Here, in the following, two examples (*Gradient* and *Flock* tuples) are presented to clarify the TOTA framework.

A *Gradient* tuple creates a tuple that floods the network in a breadth-first way and have an integer hop-counter that is incremented by one at every hop. *Flock* tuple are slightly more complicated in that they overload the *changeTupleContent* method to create the flocking shape. The actual code of these tuples is in the following:

```
public class Gradient extends HopTuple {
 public String name;
 public int val = 0;

 protected void changeTupleContent() {
 /* The correct hop value is maintained in the superclass HopBasedTuple */
 super.changeTupleContent();
 val = hop;
}}

public class FlockTuple extends HopTuple {
 public String peerName;
 public int val = 2;
 /* always propagate */
 public HopBasedTuple changeTupleContent() {
 super.changeTupleContent();
 /* The correct hop value is maintained in the superclass HopBasedTuple */
 if(hop <= 2) val--;
  else if(hop >2) val++;
 return this;
}}
```

It is rather easy now to program the agents required in the case study. Let us focus the attention on the meeting coordination protocol described in Section 42.4.4, in the case in which agents are going to meet in their barycenter room. With this regard, the algorithm followed by meeting agents is very simple: agents have to determine the furthest agent, and then move by following downhill that agent's presence field (implemented by means of a *Gradient* tuple). In this way agents will meet in their barycenter. More in detail, at start-up each agent will propagate its presence field, by injecting the associated tuple. Then it will read its local tuple space to determine the farthest agent (the one whose presence field is bigger). Then it will inspect its one-hop neighborhood to find the local shape of the presence fields of the selected agent. Finally, it will move by following the tuple's gradient downhill. More information on this and on the problems that arise can be found in Reference 20. The actual code is in the following:

```
public class MeetingAgent extends Thread implements AgentInterface
{
  private TotaMiddleware tota;

  public void run() {
  /* inject the meeting tuple to participate the meeting */
  Gradient mt = new Gradient ();
  mt.setContent(peer.toString());
  tota.inject(mt);

while(true) {
 /* read other agents' meeting tuples */
 Gradient presence = new Gradient();
 Vector v = tota.read(presence);
 /* evaluate the gradients and select the peer to
     which the gradient goes downhill */
 Gradient farther = getMaximum(v); Vector w = tota.
 KeyrdOneHop(farther);
 GenPoint destination = getDestination(w);
 /* move downhill following the meeting tuple */
 peer.move(destination);
}}}
```

42.6   Open Issues and Research Directions

This chapter presented the field-based approach to coordinate the movements of a large number of autonomous agents in a wireless and mobile computing scenario. At the moment the main weakness of this approach is the lack of a general methodology to help identifying, given a specific motion pattern to be enforced, which fields have to be defined, how they should be propagated, and how they should be combined in a *coordination field*. Nevertheless, the immediate applicability of field-based protocols is guaranteed by the possibility of getting inspiration from (and of reverse engineering) a wide variety of motion patterns present in nature. Phenomena such as diffusion, birds' flocking, ants' foraging, bee dances [4], to mention a few examples, can all be easily modeled with fields (i.e., in terms of agents climbing/descending a *coordination field* obtained as the composition of some computational fields), and all have practical application in mobile computing scenarios. However, it is likely that a complete general methodology could be found in the future and it would allow to develop sound engineering procedures to develop field-based protocols. In pursuing this long-term goal, deployment of applications will definitely help identifying current shortcomings and directions of improvement. With this regard, particularly significant results can possibly be obtained by understanding how the field approach could encode coordination tasks that do not deal with physical movements, but with other kind of actions. In these cases, a *coordination field* can be thought as spread in an abstract coordination space to encode any kind of coordination actions. Agents would be driven by the fields, not by moving from one place to another, but by making other kind of activities. In this direction, in the next future it will be interesting to apply the field-based model, in the development of new applications for sensor networks with a particular interest in those algorithms exploiting ideas taken from manifold geometry [31].

References

[1] H. Abelson et al., "Amorphous Computing," *Communications of the ACM*, 43: 74–82, May 2000.

[2] S. Bandini, S. Manzoni, and C. Simone, "Heterogeneous agents situated in heterogeneous spaces," in *Proceedings of the 3rd International Symposium From Agent Theories to Agent Implementations*, Vienna (A), April 2002.

[3] F. Bellifemine, A. Poggi, and G. Rimassa, "JADE — A FIPA2000 compliant agent development environment," in *Proceedings of the 5th International Conference on Autonomous Agents (Agents 2001)*, Montreal, CA, May 2001.

[4] E. Bonabeau, M. Dorigo, and G. Theraulaz, *Swarm Intelligence*, Oxford University Press, Oxford, 1999.

[5] W. Butera, "Programming a Paintable Computer," PhD thesis, MIT Media Lab, February 2002.

[6] G. Cabri, L. Leonardi, and F. Zambonelli, "Engineering mobile agent applications via context-dependent coordination," *IEEE Transaction on Software Engineering*, 28: 1040–1056, 2002.

[7] A. Carzaniga, D. Rosenblum, and A. Wolf, "Design and evaluation of a wide-area event notification service," *ACM Transactions on Computer Systems*, 19: 332–383, 2001.

[8] G. Cugola, A. Fuggetta, and E. De Nitto, "The JEDI event-based infrastructure and its application to the development of the OPSS WFMS," *IEEE Transactions on Software Engineering*, 27: 827–850, 2001.

[9] D. Estrin, D. Culler, K. Pister, and G. Sukjatme, "Connecting the physical world with pervasive networks," *IEEE Pervasive Computing*, 1: 59–69, 2002.

[10] E. Freeman, S. Hupfer, and K. Arnold, *Javaspaces Principles, Patterns, and Practice*, Addison-Wesley, Reading, MA, 1999.

[11] J. Hightower and G. Borriello, "Location systems for ubiquitous computing," *IEEE Computer*, 34: 57–66, 2001.

[12] A. Howard, M. Mataric, and G. Sukhatme, "An incremental self-deployment algorithm for mobile sensor networks," *Autonomous Robots* (Special issue on intelligent embedded systems), 2002, pp. 113–126.

[13] S. Li, R. Ashri, M. Buurmeijer, E. Hol, B. Flenner, J. Scheuring, A. Schneider, "Professional Jini," Wrox Press Ltd., 2000.

[14] B. Johanson and A. Fox, "The event heap: a coordination infrastructure for interactive workspaces," in *Proceedings of the 4th IEEE Workshop on Mobile Computer Systems and Applications (WMCSA-2002)*, Callicoon, New York, USA, June 2002.

[15] S. Johansson and A. Saffiotti, "Using the electric field approach in the robocup domain," in *Proceedings of the International RoboCup Symposium*, Seattle, WA, 2001.

[16] O. Khatib, "Real-time obstacle avoidance for manipulators and mobile robots," *The International Journal of Robotics Research*, 5: 90–98, 1986.

[17] J. Kephart and D.M. Chess, "The vision of autonomic computing," *IEEE Computer*, 36: 41–50, 2003.

[18] M. Mamei, F. Zambonelli, and L. Leonardi, "Co-fields: a physically inspired approach to distributed motion coordination," *IEEE Pervasive Computing*, 3: 52–61, 2004.

[19] M. Mamei, F. Zambonelli, and L. Leonardi, "Tuples on the air: a middleware for context-aware computing in dynamic networks," in *Proceedings of the Workshops at the 23rd International Conference on Distributed Computing Systems*, IEEE CS Press, Providence, RI, pp. 342–347, May 2003.

[20] M. Mamei and F. Zambonelli, "Self-maintained distributed data structure over mobile ad-hoc network," Technical Report No. DISMI-2003-23, University of Modena and Reggio Emilia, August 2003.

[21] C. Mascolo, L. Capra, and W. Emmerich, "An XML based middleware for peer-to-peer computing," in *Proceedings of the IEEE International Conference of Peer-to-Peer Computing (P2P2001)*, Linkoping, Sweden, August 2001.

[22] Mathematica, http://www.wolfram.com.

[23] R. Nagpal, A. Kondacs, and C. Chang, "Programming methodology for biologically-inspired self-assembling systems," in *Proceedings of the AAAI Spring Symposium on Computational Synthesis: From Basic Building Blocks to High Level Functionality*, March 2003.

[24] H.V. Parunak, S. Brueckner, and J. Sauter, "ERIM's approach to fine-grained agents," in *Proceedings of the NASA Workshop on Radical Agent Concepts*, Greenbelt, MD, USA, January 2002.

[25] G.P. Picco, A.L. Murphy, and G.C. Roman, "LIME: a middleware for logical and physical mobility," in *Proceedings of the 21st International Conference on Distributed Computing Systems*, IEEE CS Press, Washington, DC, pp. 524–536, July 2001.

[26] Poor, "Embedded networks: pervasive, low-power, wireless connectivity," PhD thesis, MIT, Cambridge 2001.

[27] D. Servat and A. Drogoul, *Combining Amorphous Computing and Reactive Agent-Based Systems: A Paradigm for Pervasive Intelligence?*, AAMAS, Bologna, Italy, July 2002.

[28] W. Shen, B. Salemi, and P. Will, "Hormone-inspired adaptive communication and distributed control for CONRO self-reconfigurable robots," *IEEE Transactions on Robotics and Automation*, 18: 1–12, 2002.

[29] The Sims, http://thesims.ea.com.

[30] The Swarm Simulation Toolkit, http://www.swarm.org.

[31] F. Zambonelli and M. Mamei, "The Cloak of Invisibility: Challenges and Applications," *IEEE Pervasive Computing*, 1: 62–70, 2002.

[32] F. Zambonelli and V. Parunak, "From design to intentions: sign of a revolution," in *Proceedings of the International Conference on Autonomous Agents and Multi-agent Systems*, Bologna, Italy, July 2002.

43

Technical Solutions for Mobile Commerce

43.1 Introduction **43**-983
43.2 Mobile Commerce Applications and Environment ... **43**-984
43.3 Infrastructure Requirements for
Mobile Commerce **43**-985
    Infrastructure Requirements for m-Commerce •
    A Dependable, Interoperable, and Secure Wireless Solution for
    m-Commerce
43.4 Location Management Support for
Mobile Commerce **43**-987
43.5 Support for Mobile Commerce Transactions **43**-988
    Unicast versus Multicast m-Commerce Transactions •
    Reliable m-Commerce Transactions • Protocols for
    Group-Oriented m-Commerce Transactions
43.6 Modeling Mobile Commerce Applications **43**-992
    Modeling a Generic m-Commerce Application • Steps of the
    Simulation Model
43.7 Open Research Problems in Mobile Commerce **43**-993
43.8 Conclusions and Future of Mobile Commerce **43**-994
References **43**-994

Upkar Varshney

43.1 Introduction

In the last few years, mobile commerce has attracted significant attention among users, service providers, vendors, content developers, businesses, and researchers [1]. However, there are several challenges that must be overcome before m-commerce could become a major success. These challenges include identifying suitable applications and services, designing and utilizing infrastructure support, providing support for transactions, addressing security and privacy concerns, providing support for mobile payments, and implementing new business models. These important issues are not necessarily independent, for example, security and privacy relate to m-payments, while infrastructure limitations directly affect applications and services. There has been some work on addressing these challenges including introduction and m-commerce framework [1], wireless networking requirements [2,3], services and discovery [4–7], security [8], group-oriented m-commerce services [9,10], location-based services [11], mobile payments [12], database [13], and usability issues [14].

In this chapter, we address several technical challenges in mobile commerce including infrastructure issues, location management, transactions support, and also present several interesting research problems. Some of the proposed technical solutions are described in terms of algorithms and protocols. The chapter is organized as follows: mobile commerce applications in Section 43.2, a wireless architecture and infrastructure requirements in Section 43.3, location management in Section 43.4, transaction support in Section 43.5, m-commerce modeling in Section 43.6, open research problems in Section 43.7, and some conclusions in Section 43.8.

43.2 Mobile Commerce Applications and Environment

Several m-commerce applications have been proposed [1–7], however, only few of these including mobile financial applications, mobile advertising, and location-based services have been initiated by wireless service providers.

Mobile Financial Applications consist of mobile banking and brokerage services, mobile money transfer, and mobile payments. These could lead to the evolution of mobile devices into highly personalized assistants, replacing bank, ATM, and credit cards by allowing financial transactions using mobile money. In the near future, the number of users making mobile payments will reach to several millions in Western Europe, Asia, and North America. Many banks in Europe are supporting basic mobile financial applications to reach to a large base of mobile and wireless users.

Location and user-sensitive advertising is another m-commerce application in progress. By combining user's purchasing habits, specified preferences, and its current location, a highly targeted advertising can be performed. In one possible scenario, he/she could be informed about various ongoing specials in his/her close vicinity or a selected area of interest. These messages can be sent to all users who are currently in a certain area (identified by advertisers or even by users) or to certain users in all locations. Depending on interests and personality types of individual users, advertisers could decide whether "push" or "pull" form of advertising is more suitable. It should be noted that there are issues of privacy and sharing of user information with other providers. It is likely that an "opt-in" approach would be implemented where explicit user permission is obtained before "pushing" any advertising contents.

Location-based services utilize location information to provide specialized contents to m-commerce users. These contents could include information on desired restaurants, devices, users, and products. A user could be interested in knowing availability and waiting time at one or more restaurants close to his/her current location (pull). Another user would like to be alerted when one of his/her friends is in the same general area (push). Location information of fixed entities can be kept in separate databases for each area, while location tracking of mobile and portable entities could be performed as and when needed (on-demand). When a user enters a designated area, user information from previous networks and locations will be accessed. This will also allow a determination of location-aware services the user has subscribed to or is authorized to access. Currently, there are a few examples of location-based services, not necessarily personalized or user-specific. These include mapping, routing, and list of places in users' vicinity [5]. Although not m-commerce offerings have become instant successes, there is considerable interest among users and businesses.

In addition to the basic versions of these applications, more sophisticated applications involving increased user personalization and context awareness must be offered. Mobile games, personalized contents, entertainment services, mobile auction and trading, and product recommendation systems could be widely deployed and used. The mobile and multiparty games are likely to become major drivers of m-commerce, especially if group connectivity for wireless users can be maintained. Entertainment applications will also attract some users especially if the contents can be tailored to different user groups and interests. Applications such as mobile office, mobile distance education, and wireless data center (application where a large amount of stored data to be made available to mobile users for making "intelligent" decisions) could add value to m-commerce services [3].

43.3 Infrastructure Requirements for Mobile Commerce

m-Commerce can be visualized as a multi-layer system with a set of functions for multiple players including users, wireless providers, and content providers [1]. One problem with layering of functions is the amount of communications that occur between neighboring layers. Another way to visualize m-commerce is to show functions and roles of different players in m-commerce on the same level. Three major entities in m-commerce are:

1. Users and user infrastructure such as devices and applications
2. Network infrastructure and multiple service providers
3. Content providers and third party service providers

As shown in Figure 43.1, the infrastructure for m-commerce includes devices and user interface, wireless networks, mobile middleware, servers, and databases [11]. Here a typical m-commerce transaction could involve several components as the request is forwarded to a transaction server, which checks with an authentication server. Then the preference and pricing databases are accessed. Also, to locate other users and servers, a location database is contacted. If the transaction requires a group session, then a multicast server is contacted [10].

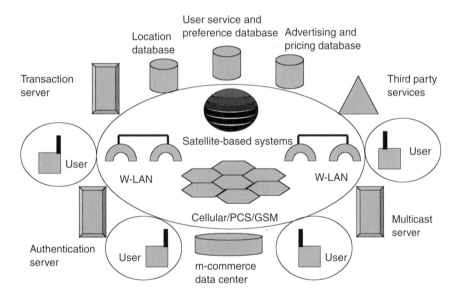

W-LAN: Wireless local area network; PCS: Personal communications systems; GSM: Global systems for mobile

1. Identify APP_REQ
2. If TRANS_SUPPORT = YES
    Locate TRANS_SERVER
3. IF PERSONALIZATION_LEVEL = HIGH
    Download USER_PREFERENCE
4. IF LOC_SUPPORT = YES, Locate LOC_SERVER
5. IF NUM_USER > 2, Locate MCAST_SERVER
6. IF OTHER_SERVICES = YES, Locate 3rd PARTY_SERVER

FIGURE 43.1 An architecture and basic protocol for mobile commerce.

43.3.1 Infrastructure Requirements for m-Commerce

One major challenge is how to support diverse requirements of m-commerce applications by using the functionalities and capabilities of wireless infrastructure. The first step is to derive specific infrastructure requirements of m-commerce applications, and then match these to specific capabilities of current and emerging mobile and wireless networks. From a detailed analysis of several m-commerce applications, some important m-commerce requirements have been derived [3]. In this chapter, we discuss location management, wireless dependability, multicast, and interoperability and scalability requirements. These requirements are shown in Table 43.1 and possible wireless networks are included and discussed [15].

Many m-commerce applications require location information of users, devices, servers, products, and services. Additionally, these applications have widely different location precision, response time, and scalability requirements. In addition to supporting these requirements, more work is necessary for location coordination among multiple wireless networks, location negotiation protocols for m-commerce, and, prioritization of location requests using applications requirements, context (emergency, anxiety etc), and processing delays.

M-commerce applications would be significantly affected by the dependability problems of existing wireless networks. The situation will be even more complex in the emerging "3rd generation and beyond" wireless networks as increased heterogeneity and inter-carrier roaming will increase the vulnerability of wireless infrastructure and a higher impact of increased fault propagation. Additionally, these emerging wireless networks will support group-oriented m-commerce applications; thus impact of dependability problems could propagate to multiple locations serving wireless users. One way to address the dependability of wireless networks for m-commerce transactions is to deploy fault-tolerant architectures to isolate and bypass failures. The cost of deployment for fault-tolerant architectures could be reduced by a more "selective" fault-tolerant design.

Some m-commerce applications such as mobile auction and interactive mobile games would require continued group connectivity. This could be achieved by using wireless multicast connecting a group of users where information sharing is supported by efficient routing in wireless networks. The challenge in wireless multicast for m-commerce is in maintaining group communications with user mobility and under variable quality of wireless links. This combined with brief-disconnectivity or intermittent connectivity could lead to a loss of user input, thus affecting the outcome of applications such as mobile auction [3].

There are several existing and emerging wireless networks that can be used for m-commerce including global system for mobile communications (GSM), DoCoMo's iMode service, and enhanced data rates for GSM evolution (EDGE), and general packet radio service (GPRS). Also, proprietary wireless WANs, wireless LANs, satellite systems, and wireless local loops are access choices for m-commerce. Since these networks differ considerably, interoperability becomes a major challenge.

TABLE 43.1   Specific Wireless Infrastructure Requirements

Requirements	Attributes	Candidate networks
Location management	Location accuracy and response time	Satellite-based network (higher accuracy, slow response time). Cellular and PCS networks (location support not available everywhere). Wireless LANs (low accuracy but better response time)
Wireless dependability	Availability and survivability	Current wireless networks support availability between 90 and 98%. Increased dependability is possible using fault-tolerance and backup components in wireless networks
Multicast	Group membership and connectivity	Satellite-based networks can support multicast using broadcast technologies. Cellular and PCS systems do not support network or physical-layer multicast. M-commerce application could support multicast at application layer
Interoperability	Ability to use multiple networks	Difficult to achieve as most wireless networks differ considerably in protocols, frequencies, power and interface requirements. Multi-network access possible by using several interface cards

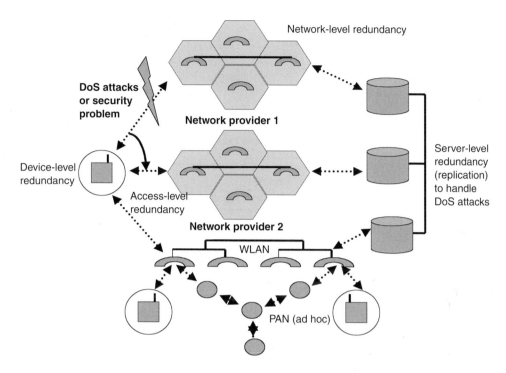

1. Generate APP_REQS & USER_REQS
2. Check for all Networks
   IF (APP_REQ & USER_REQ)<= NETWORK[I]_OFFER
       Select NETWORK[I] using LOAD DIVISION
   ELSE RENEGOTIATE
3. IF (APP=ON & NET_PROB=YES)
   SWAP-ACCESS (NETWORK[I], NETWORK[J])

FIGURE 43.2 A dependable, secure, and scalable solution for m-commerce.

43.3.2 A Dependable, Interoperable, and Secure Wireless Solution for m-Commerce

Secure and dependable wireless solutions must be designed and implemented before mission-critical data can be put on wireless infrastructure. A possible network architecture for m-commerce is shown in Figure 43.2. An increased dependability is achieved by fault-tolerance or "added" redundancy in wireless infrastructure. The fault-tolerance allows m-commerce transactions to be executed even when there are one or more failures. This network and access redundancy would also allow overcoming or at least alleviating denial of service attacks. If an attack occurs at device, access, network, or server levels, the redundancy would allow switching to another network, server, or device interface. The architecture is also scalable in terms of number of users, transactions, and network size [15].

43.4 Location Management Support for Mobile Commerce

Depending on the network architecture, all location requests from m-commerce applications go to a location server. The middleware negotiates or requests specific requirements (precision, response time, etc.) from multiple networks or schemes (Figure 43.3). The middleware can translate application's

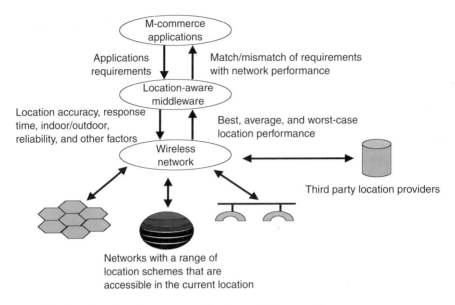

FIGURE 43.3   Selecting a location scheme or network using applications requirements.

TABLE 43.2   The Middleware Algorithm in the Location Management Architecture

Step	Actions	Comments
Step 1: Application sending the request	Receive and put App_Req in Queue_Priority	Priority is based on context and application requirements

Step 2: Middleware processing the request	If Object_Tracked = Fixed,
	    Query Loc_database
	Else
	    Generate Net_Req(Min_Loc_Acc,
	Max_Res_Time, Min_Depend, ...)
	
	  While Match=0
	    Send Net_Req(Desired) to WNet_I
	    Receive Net_Res(Levels)
	    Match=Net_Res(Levels) > Net_Req(Desired)
		The application requirements are translated and matched with the capabilities of underlying wireless networks

Step 3: Middleware responding to application	Send App_Resp	

requirements into specific location attributes (location precision and response time) for the underlying wireless infrastructure. It can also translate network performance attributes into specific requirement support. The protocol and steps used by the middleware are shown in Table 43.2. Also, a tradeoff between the number of users needing location tracking and possible response time (or accuracy) would have to be made in case of limited wireless resources. Middleware could also negotiate support for location management even when the user is roaming in another network with different location accuracy, response time, location scheme, and format of location information. The location information requested by the roaming user is more likely to be needed and is certainly more critical.

43.5   Support for Mobile Commerce Transactions

In general, a transaction can be defined as a series of steps that should be executed in a certain sequence. m-Commerce is likely to be transaction-oriented due to applications such as mobile auction, mobile entertainment services, wireless data center, mobile financial services, mobile advertising and shopping, and location-based services. Also, the transactions are likely to exhibit a high level of diversity in terms of the number of entities involved, resource requirements, and the response time. The transactions could also be local or end-to-end, symmetric or asymmetric, requiring different levels of reliability, and could last from less than a second to minutes and even longer. The requirements for some mobile transactions could also include atomicity, so if a step fails due to user mobility or brief disconnectivity, the whole transaction must be aborted after a certain timeout, this would release some resources temporarily. However, additional resources would be necessary to support the transaction when re-attempted at a later instant. An m-commerce transaction can involve multiple heterogeneous wireless networks. The steps and resource requirements would vary from application to application and could also depend on "pull" or "push" transactions. Some of the steps will require little resources, however one or more steps could present significant resource requirements and quick response time. Although it is not clear what would be the precise characteristics of these transactions, we could attempt to derive several general characteristics in terms of the following attributes.

43.5.1 Unicast versus Multicast m-Commerce Transactions

Many of the m-commerce transactions would be unicast requiring involvement of single or two mobile users. Such transactions could still involve several other fixed entities. Unicast transactions are much easier to support, as group management and coordination of user inputs are not required. The multicast transactions would be much harder to support as managing interactions among mobile users who could be experiencing network connectivity problems (such as brief disconnectivity and intermittent connectivity) becomes a complex task.

43.5.2 Reliable m-Commerce Transactions

Some of the mobile commerce transactions would require reliability and even atomicity. For example, mobile commerce transactions for financial services would require that either all or none of the steps are executed. Such atomicity would require that an aborted transaction leads to the same state as was there in the beginning and then the whole transaction is retried. On the other hand, many transactions do not require strict reliability from the underlying infrastructure. The example of such transaction would be mobile advertising, where reaching to all the customers is desirable but not critical for the operation of the transaction.

43.5.3 Protocols for Group-Oriented m-Commerce Transactions

The group-oriented mobile services could include mobile auction, mobile games, mobile financial services, mobile and locational advertising, mobile entertainment services, mobile distance education, proactive service management, product recommendation systems, mobile inventory management, and product location and search [1]. The focus here is to derive the requirements of these mobile services by classifying services with similar requirements in a group. This classification would reduce the complexity of requirement analysis and will also help in modeling and performance evaluation. The criteria used were the type of communications, number of entities, response time, and connectivity requirements (Table 43.3). It should be noted here that this is not an exhaustive list of group-oriented mobile services, however any current and emerging mobile service can be added to this classification with little additional effort. We believe that by deriving general requirements of group-oriented mobile services, better protocols can be designed and deployed [10].

From Table 43.3, it is clear that group-oriented mobile services present diverse set of requirements where some services need real-time multicast while others could be supported well even if the delays were higher. Although the requirements of group-oriented mobile services could vary considerably, for a general case we assume a three-stage sequence: preprocessing, inputs from users, and postprocessing. The preprocessing stage could involve service advertisement and discovery, joining of members and setting

TABLE 43.3   Group-Oriented Mobile Services and Requirements

Group-oriented mobile services	Description	Type of communication and number of entities	Multicast requirements and response time
Mobile Auction, interactive games, financial services	Users to buy or sell certain items, play multiparty games, conduct multi-party financial transactions	Real-time multicast with active participation from multiple users	Very low delays. Continued connectivity required
Mobile and locational advertising	Applications turning the wireless devices into a powerful marketing medium	Asymmetric nonreal-time multicast involving hundreds or more devices	Higher delays can be tolerated Intermittent connectivity or brief disconnectivity may be tolerated
Mobile entertainment services/mobile distance education	Entertainment or distance education services to users on per event or subscription basis	Asymmetric real-time Multicast involving multiple users	Large bandwidth and low delays. Intermittent connectivity or brief disconnectivity significantly affects the overall experience
Proactive service management	Providing users the information they will need in very-near-future	Asymmetric nonreal-time multicast involving a few entities	Higher delays can be tolerated. Intermittent connectivity or brief disconnectivity can be tolerated
Product recommendation systems	To receive recommendation on products & services from a third party or customers	Asymmetric nonreal-time multicast involving large number of entities	Higher delays can be tolerated. Intermittent connectivity or brief disconnectivity can be tolerated
Mobile inventory management/ product location	Reducing the inventory by managing in-house and inventory-on-move/to find the location of product and services	Unicast/multicast involving few entities	Response time of few seconds. Intermittent connectivity or brief disconnectivity increases the delay

up of rules and regulations. The impact of disconnectivity would differ considerably among these stages of group-oriented services. For example, brief disconnectivity and intermittent connectivity could be tolerated during postprocessing stage while such events would affect the application and its performance significantly during Stage 2.

The user mobility and wireless link characteristics could make the continuous communications among group users difficult. This would affect the performance and outcome of group-oriented mobile services such as mobile auction, games, and financial services where user participation is important. No response from a certain user for some time could lead an ambiguity about the user leaving the group or experiencing a brief disconnectivity. If there is a reason to believe that user is experiencing connectivity problems, then the state of the application could be maintained until a time-out occurs or the user sends some information again. On the other hand, if the user response could not have affected the outcome of the group application, then it would be wasteful for everyone in the group to wait for this user. Such determination could be difficult for all possible states of group-oriented mobile services.

In supporting group-oriented transactions, one major issue is the determination of the reasonable time to wait for response from mobile users. As time-out is a possible technique for group management (remove a user from the group application if do not hear anything for a given time) in such applications, the value of time-outs could affect the performance of group-oriented mobile services. If a smaller value is chosen, it will lead to frequent removal of mobile users from the group when experiencing a brief disconnectivity. Later on, these users could attempt to connect to the group but since the state of application will likely to have changed significantly (such as bidding in a mobile auction), these users would have lost the chance to affect the outcome of the group application. On the other side, if a higher value of time-out is chosen, then all the connected users are forced to wait for these users who are either experiencing significant

TABLE 43.4 Major Steps of Three Protocols

Protocol	Description	Comments

1: All users equal and UW	Derive Num_TranStages and Users_TranStages
	
	While Done_Stages < Num_TranStages
	  For Every User
	    While (User_Input=False) and (User_LV=False)
	      Wait
	    Process Trans_Stage
	    Increment Done_Stages
		Simplest to implement and works well for transactions with tolerance for higher delays. Also, will have the least affect of brief disconnectivity or intermittent connectivity (highest probability of transaction completion)

2: All users equal and LW	Derive Num_TranStages, Req_Users_TranStages, and, TIMEOUT
	
	While Done_Stages < Num_TranStages
	  For Every ReqUser
	    While (User_Input=False)
	      If Wait-time>=TIMEOUT
	        Trans_Satge=TERMINATED
	      Else Wait
	    Process Trans_Stage
	    Increment Done_Stages
		Simple to implement and works well for transactions with lower delay requirement. Also, will have the effect of brief disconnectivity or intermittent connectivity (reasonable probability of transaction completion)

3: SBW	Derive Num_TranStages, Req_Users_TranStages, Members_Status, and, TIMEOUT_Status
	
	While Done_Stages < Num_TranStages
	  For Every ReqUser
	    While (User_Input=False)
	      If Wait-time>=TIMEOUT_Status
	        Trans_Stage=TERMINATED
	      Else Wait
	    Process Trans_Stage
	    Increment Done_Stages
		More complex to implement, but works well for transactions with lower delay requirement. It can lead to a more optimal combination of response time and transaction completion probability under intermittent connectivity

connectivity problem or have been permanently disconnected. This will also reduce the performance (speed) of mobile services for most users. To avoid these extremes, finding an optimal or near optimal time-out is necessary for achieving a high level of performance from mobile applications for most users. The value of time-outs could affect the transaction completion probability and transaction response time.

Three protocols for supporting transactions in group-oriented mobile services (Table 43.4) are named unlimited wait (UW) limited wait (LW), and status-based wait (SBW) using their basic operation technique [10]. The mobile services that could tolerate larger response time would work better with UW due to its simplicity and less overhead, although it has no upper bound on delays. The mobile services requiring lower delays will benefit from LW or SBW. These protocols result in different levels of transaction completion probabilities and response times.

For UW, the group is forced to wait until the user sends an input. LW involves waiting only for some time as determined by previous values and using information on brief disconnectivity. In SBW, the group waits for different time based on the type of user: passive listener, active listener, and active participant. For the first type, the user can be removed from the group and can join later without affecting the status of the group. For the second type, the group can wait for some time, however for the third type of users, the group should wait longer as the input from such users can affect the outcome of the group application.

The group-oriented mobile services can be modeled using multiple stages and participants (Figure 43.4). The number of stages and interdependency among these differ substantially among mobile applications. The lack of input from a certain user in a stage could affect one or more inputs in the next stage, thus resulting in a different outcome. Both the number and actual "critical" users whose response is needed for the correctness of application can be identified and waited differently. The number of stages can be determined by the total number of users, minimum and maximum number of users per stage, delay tolerance per stage, processing delay per stage, and the required transaction delay.

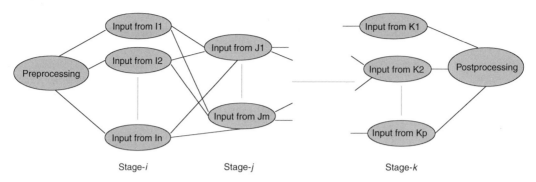

FIGURE 43.4 Stages and participants in group applications.

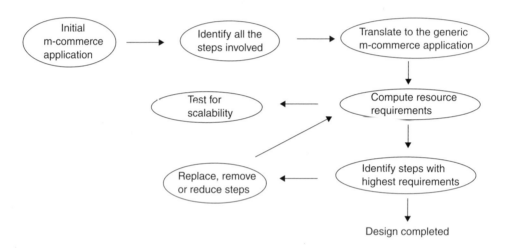

FIGURE 43.5 A possible applications design approach.

43.6 Modeling Mobile Commerce Applications

A possible simulation-oriented approach for designing and improving m-commerce applications is shown in Figure 43.5. It starts with a basic application framework and identifies all the steps involved in the application. Then, these steps are translated into the generic m-commerce application to compute the resource requirements. Then, certain steps with highest amount of resource requirements are identified and efforts are made to replace, remove, or reduce these steps. The proposed approach would allow to prototype new applications, test the user requirements, and compute the processing and communication requirements. The approach can also be used for testing the scalability of a new m-commerce application as the number of users or number of transactions per user increases. So this environment would allow to build, test, and implement current and emerging mobile commerce applications in an accelerated way.

43.6.1 Modeling a Generic m-Commerce Application

In modeling m-commerce applications, several steps have been identified (Figure 43.6). These steps are dependent on whether "push" or "pull" version is employed. In "push" version, the network or server initiates a transaction based on an event related to the current time (network pushes some information every hour), user's current location (whenever the user comes within a mile of a store, network sends an advertising message), or a predetermined threshold.

(a)

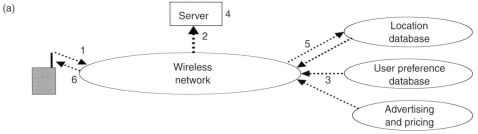

1. TRANS_REQ (App, Users, Desired)
2. SEVER _LOCATE (REQUIREMENTS_TRANS) and SERVER_FOUND(ADD_SERVER)
3. USER_ATTRIBUTES_DOWNLOAD, USER_PREFERENCE_DOWNLOAD, and USER-LOCATION_DOWNLOAD
4. TRANS_PROCESSING
5. USER_ATTRIBUTES_UPDATE, USER_PREFERENCE_UPDATE, and, USER_LOCATION_UPDATE
6. TRANS_RESP(Results)

(b)

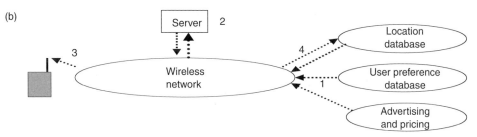

1. USER_ATTRIBUTES_DOWNLOAD, USER_PREFERENCE_DOWNLOAD, and USER-LOCATION_DOWNLOAD
2. TRANS_PROCESSING
3. TRANS(Results)
4. USER_ATTRIBUTES_UPDATE, USER_PREFERENCE_UPDATE, and, USER_LOCATION_UPDATE

FIGURE 43.6   (a) Pull model of m-commerce applications and generic steps and (b) push model of m-commerce applications and generic steps.

43.6.2 Steps of the Simulation Model

1. Receive the number of steps involved in the application OR derive the minimum number of steps needed to perform the requested transaction.
2. Receive the number of users involved, application version (push or pull), maximum allowable wireless traffic, minimum capability of devices, number of databases involved, and their processing and storage capabilities.
3. Compute the amount of traffic generated for each step for the given number of users and compare this against the constraints provided in Step 2, if not possible, go to Step 5, else Step 6.
4. Compute the amount of processing needed in each step for the given number of users and compare this against the constraints provided in Step 2, if not possible, go to Step 5, else Step 6.
5. Re-compute the requirements with reduced number of users and reduced number (or size) of messages per step.
6. Output the results and the processing and communications requirements of the given application.

43.7 Open Research Problems in Mobile Commerce

There are several research problems in m-commerce and we are currently working on few of these problems:

1. *Design and development of m-commerce applications*: Certainly more work is needed in designing new m-commerce applications that are both suitable to users and can be supported by the current

and emerging wireless infrastructure. Using our proposed approach (or other approaches), processing and storage requirements of different applications can be estimated. This can also help in determining if an application can be scaled to an increased number of users and transactions.

2. *Interoperability and integrated access for m-commerce*: One major issue in m-commerce is the focus on personalization and not so much on the type of network access used. To ensure un-interrupted access to m-commerce applications, wireless networks should interoperate and provide an integrated access to users by supporting handoffs and information exchange among networks.

3. *Design of algorithms and protocols*: One important area of research is to design algorithms and protocols for supporting reliability, multicast and quality of service for m-commerce transactions. It should be noted that most m-commerce transactions could not be supported by traditional connection-level resource allocation due to varying nature of both arrival and duration of transactions. Some work is also necessary in designing protocols for user and location-specific service discovery. To support highly valued m-commerce transactions, more work is needed in either adapting existing security mechanisms or design new m-commerce specific security protocols.

4. *Software and database support*: In addition to networks and applications, the role of software and database is becoming important. The software for m-commerce applications must be able to deal with disconnected operations, while databases should be able to store and update a large amount of m-commerce data in a reliable way.

5. *Performance evaluation of architectures, systems or networks for m-commerce*: As most of the current architectures and networks are not designed with m-commerce requirements, more work is necessary in evaluating the effectiveness for m-commerce environment.

43.8 Conclusions and Future of Mobile Commerce

m-Commerce has the potential of utilizing instant access of devices, flexibility, and convenience of wireless networks, and personalization and location-awareness of mobile applications. These could lead to a new level of user-empowerment not seen or observed before. For fully realizing the potential of m-commerce, we believe one of the major challenges is the infrastructure support for location tracking, transaction-level QoS, and group transactions. To evaluate the requirements of m-commerce applications, the performance modeling is required. These can be facilitated by algorithms and protocols designed specifically for m-commerce. In this chapter, we discussed requirements of m-commerce applications and possible ways to support these by networking infrastructure. We believe that much more work is necessary in designing algorithms and protocols for m-commerce and also in the modeling for performance evaluation. Some efforts must also be spent toward devices and middleware, user interface design, and introduction of context-awareness in m-commerce applications.

References

[1] U. Varshney, R. Vetter, and R. Kalakota, "M-commerce: a new frontier," *IEEE Computer*, vol. 33, no. 10, October 2000, pp. 32–38.

[2] U. Varshney and R. Vetter, "Framework, applications, and networking support for M-commerce," *ACM/Kluwer Journal on Mobile Network and Applications*, vol. 7, no. 3, June 2002, pp. 185–198.

[3] M-commerce tutorial by U. Varshney, IEEE WCNC 2003 (slides available from the author via e-mail: uvarshney@gsu.edu).

[4] Session on mobile commerce, *in Proceedings of the 2nd ACM International Workshop on Mobile Commerce*, September 2002.

[5] O. Rastimor, V. Korolev, A. Joshi, and T. Finin, "Agents2Go: an infrastructure for location-dependent service discover in the mobile electronic commerce environment," *in Proceedings of the 1st ACM International Workshop on Mobile Commerce*, July 2001.

[6] J. Sairamesh, I. Stanoi, C.-S. Li, and B. Topol, "Wireless trading in B2B markets: concepts, architecture, and experiences," *in Proceedings of the 1st International Workshop on Mobile Commerce,* July 2001, pp. 7–13.

[7] O. Ratsimor, D. Chakraborty, A. Joshi, and T. Finin, "Allia: alliance-based service discovery for ad-hoc environments," *in Proceedings of the 2nd ACM International Workshop on Mobile Commerce,* September 2002, pp. 1–9.

[8] X. Geng, Y. Huang, and A. Whinston, "Defending wireless infrastructure against the challenge of DDoS attacks," *ACM/Kluwer Journal on Mobile Networks and Applications,* vol. 7, 2002, pp. 213–223.

[9] U. Varshney, "Multicast over wireless networks," *Communications of the ACM,* vol. 45, 2002, pp. 31–37.

[10] U. Varshney, "Performance Evaluation of Group-oriented Mobile Services," *ACM/Kluwer Journal on Mobile Networks and Applications,* vol. 10, no. 4, pp. 465–474, June 2005.

[11] U. Varshney, "Location management for mobile commerce applications in wireless internet," *ACM Transactions on Internet Technologies,* vol. 3, 2003, pp. 236–255.

[12] U. Varshney, "Mobile payments," *IEEE Computer,* vol. 35, no. 12, 2002, pp. 120–121.

[13] K. Kuramitsu and K. Sakamura, "Towards ubiquitous database in mobile commerce," *In Proceedings of the 2nd ACM International Workshop on Data Engineering for Wireless and Mobile Access,* 2001, pp. 84–89.

[14] S. AlShaali and U. Varshney, "On the usability of mobile commerce," International *Journal of Mobile Communications,* vol. 3, no. 1, 2005, pp. 29–37.

[15] U. Varshney, "Mobile commerce: current status, challenges and future," *IEEE Vehicular Technology Society News,* vol. 51, 2004, pp. 4–9.

Index

A

Absolute guarantee (AG) algorithm, 425
Abstract sensor algorithms, 845–849
Access category (AC), 627–628
Access control aspect, 785
Access control servers (ACS), 915
Access point (AP), *see* Base station (BS)
Accumulation, 856
ACE, 268
Acknowledgement packet (ACK), 102–103, 109, 112,
    113, 202, 242, 268, 359, 550, 563, 587, 588–589,
    592–593, 595, 599–600, 605, 615, 616, 622, 702,
    704, 706, 710, 717, 718, 719, 816
  duplication sequence number (ADSN), 709
  feedback, 692
  pacing, 683, 689
ACL links, 755–756, 760, 761, 764–765, 766, 767, 770,
    771
Acquisitional query processing (ACQP), 826–827
Active member address (AM_ADDR), 738–739, 759
Active mobile probability, 450
Activity scheduling, 260
Activity scheduling algorithms for wireless sensor
    networks, 871–880
  mobility support, 880
  impact of node density, 881
  for 3D space, 880
Actual borrowable bandwidth (ABB), 656, 658, 664
Actuation, 838
Ad hoc multicast routing protocols (AMRoute), 207
  utilizing increasing id-numberS (AMRIS), 209
Ad hoc networks, 17–18, 19, 45, 50, 72, 107, 114, 116,
    118, 135–136, 138–139, 147, 165–166, 181, 218,
    234, 244, 252, 271, 325–326, 330, 506–507, 526,
    582, 613, 640, 641, 816, 828, 938
  see also peer-to-peer (P2P) network
  communication infrastructure, 167–173, 181
  clustering, in, 527–530
  contention-based protocols, 79–84, 95
  destination location, 349
  geographic routing protocols, 348, 350–351, 357,
      358, 823
  Hop spanners, 171–173
  localization systems, 343, 344
  mobile devices, 536–537
  modeling, 166, 174
  multichannel support, 106–107
  power management, 561–567, 604
  resource discovery, 357
  support of quality of service (QoS), 100, 637

security applications, 939–940
self-organizing approach, 521
test for processing, 169–171
  Gabriel test, 169–170
  Morelia test, 170–171
throughput analysis, 62–69, 70
  geometric modeling, 62–63
vulnerabilities, 941–942
Ad hoc on-demand distance vector routing (AODV), 17,
    185, 187–188, 201, 208, 220–221, 240, 271, 319,
    521, 524, 640, 951
  see also Transmission control protocols (TCP),
      Unicast routing protocol
  quality of service (QoS), 642
Ad hoc routing protocols, 183, 700
Ad hoc TCP (ATCP), 712, 713, 714–715
Ad hoc traffic indication messages (ATIM), 118, 563, 605
Ad hoc wireless networks, 628, 725, 726
ADAPT protocols, 36
Adaptation, concept, 579
Adaptive backbone-based multicast protocols, 210–211
Adaptive bandwidth allocation, 729–730
Adaptive bandwidth reservation schemes, 651
Adaptive contention-based access, 602
Adaptive CWL setting, 707, 711
Adaptive cycle-limited service (ACLS), 730
Adaptive distance vector (ADV), 705, 708
Adaptive dynamic backbone (ADB), 210–211
Adaptive election algorithm (AEA), 818
Adaptive e-limited polling, 729
Adaptive enhanced distributed coordination function
    (AEDCF), 38
Adaptive hierarchical routing protocols, 218
Adaptive motion coordination, applications, issues and
    requirements, 961–963
  case study scenario, 961–962
  inadequacy of traditional approaches, 962–963
Adaptive pacing, 706, 715–716
Adaptive protocols and cross layering, 579
Adaptive rectangular forwarding zone (ARFZ), 352
Adaptive self-configuring sensor network topologies,
    828–829
Adaptive-step power control (ASPC), 432
Additive increase multiple decrease (AIMD) congestion
    control mechanism, 699
Additive white Gaussian noise channel, 277
Address-centric systems, 267, 268, 271
ADE, 268
Ad-hoc localization systems (AHLoS), 836
Ad-hoc wireless networks, 259

ADisjoinMessage, 223
Adjacent-constraints (ACC), 461, 465
Adjustable power energy model, 239
Admission control algorithms for downlink packet
        transmission in WCDMA, 427–428
Admission control strategy, 668
Admission control, 648–649
Advanced configuration and power interface
        (ACPI), 276
Advanced Encryption Standards (AES), 12, 908, 909, 945
Advanced mobile phone service (AMPS), 27, 416
Advanced video coding (AVC), 790
Advertisement message, 528–530, 820–821
AFECA, 564, 602
Aggregation factor, 272
AGREED, 469, 470, 472
AIRMAIL, 688, 689, 690, 691
ALARM message, 954
Algorithmic challenges
    cross-layer approach, 579, 700, 712, 714, 716, 792–793
    cluster-based energy aware, 527–530
    location awareness, 166, 181, 344
    location information, 183, 198, 200, 320–325,
            331–334, 336–337, 343, 344, 345, 350, 358,
            395, 396, 403–409, 501, 502, 858, 876, 984,
            986, 988
    location management, 2, 4, 358, 365–366, 368–370,
            385, 478–479, 480–483, 496, 498, 984, 987–988
    in mobile environments, 128–129
    scatternet formation in Bluetooth, 736, 738, 740–741,
            746, 747, 751–752
    scheduling for contention-free protocols, 84–94,
            95, 582
    self-organization, 519–520, 521, 536
Algorithms
    for high mobility network, 135–154
    for leader election, 129–135
    notations and basics, 86
    centralized and decentralized, 504
ALOHA protocol, 31, 32, 46, 78, 586–587
ALT_TIMEOUT, 739
Amorphous computing, 966
Analog amplitude, 10
Anchored Geodesic Packet Forwarding, 348
Angle of arrival (AoA), 502, 836
Anode-electrolyte interface, 539
Anomaly detection techniques, 948
Anonymity, 406
Anonymous on-demand routing (ANODR), 401–402
Ant behavior, 333–334
Ant colony algorithm (ACA), 367–368, 379–380,
        386, 389
    escape operator, 383
    framework, 380–385
    local heuristic, 382
    pheromone update, 382–383
    transition function, 382
    Type-1 rank variant, 383–384
Antennas, 100, 594
    Omnidirectional, 40, 45
    pattern choice, 460
    smart antennas, 40–41
Anti-virus software, 914
Application layer security, 914

Application level gateway (ALG), 785
Application specific coordination, 965
Application-specific integrated circuits (ASIC), 887
Application-specific network design, 887
Approximation algorithms, 241, 273, 276
Approximation ratio, 245–247, 248, 314
Arbitration interframe space (AIFS), 618, 627, 628
Arborescence, 245
Area-based decision, 258–259
ARI, 358
ARIADNE, 950
Arpanet, 5
Artificial intelligence, 375
Association/disassociation, 614
Associativity-based routing (ABR), 185, 190–191,
        506, 713
Asymmetric schemes, 950–951
Asymmetry in network, 692
Asymptotic optimality, 495
Asymptotically optimal backoff (AOB), 562, 598
Asynchronous connectionless (ACL), 752, 756, 769
ATIM windows, 83–84
ATM, 636, 922, 984
ATMEL ATmega, 276
Atomic constraints, 49–50
Audio on demand (AoD), 12
Authenticated routing for ad hoc networks (ARAN), 950
Authentication, 614, 915–916, 918, 923, 926–928, 931,
        943, 934
    and encryption, interaction, 932, 944
Authentication headers (AH), 923, 924
Authentication/authorization/accounting (AAA)
        infrastructure, 785, 911, 913, 914–915, 918, 924
Auto regressive moving average (ARMA) model, 454
Automatic network connections, 913
Automatic repeat request (ARQ), 351, 637, 699, 791, 799
    retransmission, 765, 766–767, 771
Automation, 393, 396
Automotive telematics, framework for security and
        privacy, 398–399
Autonomous components, 960
Autonomous sensor networks, 960
Average shortest path, 747

B

Backbone communication and spectrum, 468
Backbone connectivity, 278
Backbone network, 415, 416
Backbone routers, 636
Backoff protocol, strategies, 114, 591–592, 595–598, 602,
        619, 622–623, 625, 628, 703
    for collision resolution, 34
    binary exponential, 34
    polynomial, 34
    for rendezvous region (RR), 359
Backtracking, 897
Bandwidth, 260, 305–307, 310, 439, 440, 441, 633–634,
        648, 649, 651, 652, 654–655, 657, 669, 728,
        761, 789
    see also Ek-Kadi's rate-based bandwidth borrowing
        scheme, (EBS)
    aggregation, 685

allocation, 648, 654–655, 656
asymmetry, 693
congestion, 41
control and management, 11
delay, 706, 710–711, 712
management, 498
protocol, 642
requirement, 803–804, 806, 808
reservation, 452, 469, 651–652
routine, 640
utilization, 454, 648, 649, 660, 665, 667, 671–673
waste, 702
Bandwidth estimate (BWE), 683–684
Bandwidth sharing system (BWSS), 617, 685
Bandwidth units (BUs), 450–451
Base stations (BSs) (*also known as* access points), 78, 84,
    85, 86, 88, 92, 287, 365, 415–416, 418–419, 420,
    421, 422, 423, 424, 425, 427, 428–429, 430, 431,
    433, 440, 450, 451, 455, 467, 468, 470, 472, 478,
    481, 549–551, 556, 557, 558, 562, 579, 582, 583,
    604, 605, 613, 650–652, 653, 679, 681, 686, 689,
    690, 695, 785, 906, 909–913, 919; selection, 460;
    see also cellular networks
Base station controllers (BSC), 460, 478, 784
Base station subsystems (BSS), 784
Base transceiver stations (BTSs), 403, 784
Basic access mode, 515–516
Basic borrowing algorithm with reassignment, 466
Basic multiple access techniques, 24–26
Basic service set (BSS), 559, 582, 604
Basic service set identifier (BSSID), 407
Battery energy limitations, 578, 643
Battery technology, 536, 538–540, 600
Bayesian inference, 857, 860
Bayesian sampling approach, 862
Bayesian theory, 854, 856
Beacon, beaconing process, 531, 549–552, 562–563,
    604–605, 606
Beamforming algorithms, 886
Bellman–Ford routing algorithm, 185
Best effort service, 633, 634, 640, 642, 647
Binary exponential backoff (BEB), 591, 592, 595–596,
    598, 605, 614, 615–616, 620, 625
Binomial distribution, 448
Biocomputing, 463
Biologically inspired techniques, 367, 370
Bit error rate (BER), 235, 419, 421, 430, 432, 433, 441,
    442, 592, 601, 679, 700, 701, 707, 714, 715, 769,
    795–797
Bitplane coding, 790
Black burst (BB), 617
    Medium access control (MAC) protocols, 619, 621
Blackout detection, 691
BLOCK, 470
Blocker tag, 408–409
Bandwidth_loss tolerance (BLT), 660, 673
Blue constellation, 744
Bluenet, 741, 747, 748
Bluestars, 743–744 , 748
Bluetooth networks and communication, 4, 11, 13,
    14–15, 19, 395, 603, 699, 725, 726, 730, 731,
    735–736, 742–745, 746, 751, 756–759, 760, 761,
    764–765, 766–767, 770, 771

Bluetooth base stations (BBSs), 755, 761, 763–764,
    767–768, 770
    with migration (BBSM), 763–764
    original (BBSO), 761
    with standby unit (BBSS), 762
Bluetooth device addresses (BD_ADDRs), 738,
    745, 759
Bluetooth DVAP, 763
Bluetooth FH collision loss, 765–771
Bluetooth INQUIRY/PAGE, 762–763
Bluetooth mobiles (BMs), 761, 762, 763–764, 767–769
Bluetooth network encapsulation protocol (BNEP)
    and segmentation/reassembly policies,
    730–731
Bluetooth telephony, 762
Bluetooth topology construction protocol (BTCP),
    739–743, 748
Bluetrees, 740–741, 747, 748
BMSs, 228
BMT, 220, 223, 229
Body area networks (BAN), 559–560
Boolean or 0/1 sensing model, 871
Bootstrapping multicast routing, 358 `
Border nodes, *see* gateways
Bordercast resolution protocol (BRP), 192
    see also Unicast routing protocol
Borrowing-based dynamic channel allocation
    algorithms, 460
Boukerche–Hong–Jacob Algorithm, 470
Bounded slowdown (BSD), 552–553
Branch reconstruction (BR) process, 209
Breadth first search (BFS), 744–745
Broadcast communication, 23, 261, 274–275, 886
Broadcast ID (broadcast_id), 187, 188
Broadcast query (BQ), 190
Broadcast routing protocols, 183, 184, 185, 201–205, 213
Broadcast scheduling problem (BSP), 48–50
Broadcast state (BT), 60, 70
Broadcasting, 248, 257, 354, 356, 871
Broadcasting and multicasting protocols, 240–242
    centralized methods, 244–247
    localized methods, 248–259
    performance measurement, 243
    reliability, 242
    theoretical analysis, 245–247
Broadcasting average incremental power (BAIP), 245
Broadcasting incremental power (BIP), 243, 244–245,
    246–247, 252, 353
Broken links, 185, 189, 190, 501, 502, 506, 512
Buffer auto-tuning, 692
Buffer zone, notion of, 509, 512, 515
Burst errors, 637
Busy tone multiple access (BTMA), 34, 116, 590
Byte (versus ACK) counting, 692
Byte counting, 695

C

Caching and prefetching techniques, 491, 547
Call adaptability, 657
Call admission for asymmetric traffic, 452–454

Call admission control (CAC), algorithm, 418, 442, 455, 652, 653, 654, 667
  dynamic approaches, 447–455
  looking around algorithm, 425–426
  quality of service (QoS), 446, 447, 448, 451, 452, 455
  for resource management, 423–430
  classification, 424–425
  in power controlled mobile systems, 427–428
  static approaches, 446–447
  using shadow cluster, 449–451
Call blocking (system capacity measurement), 424
Call blocking probability (CBP), 290–291, 295, 297–301, 305, 306, 309, 312–313, 314, 315, 408, 423, 441, 445, 446–447, 448, 451, 454, 648, 649, 651, 654–655, 660, 664, 667, 673, 762, 763
Call blocking rate, 764, 766, 770
Call dropping probability (CDP), 295–301, 309, 313, 315, 422, 424, 433–434, 441, 443–449, 451–452, 454, 648, 649, 651, 652, 653, 654–655, 660, 664, 665, 667, 668, 669, 671–674
Call duration, 429
CALL-hold-time, 769
CALL-ID, 780
CALL-Idle-Time, 769
Call server, 774
Call-session control, 774
Call setup, 404
Cambridge Bat, 394
Cao's Qos enhancement channel allocation scheme, 471
Cao-Singhal's
  DCA algorithm, 468, 469–470
  fault tolerant enhanced channel allocations, 472
Capture effect, 584–585, 590, 703, 714, 716, 717
Carrier sense multiple access (CSMA), 31–32, 36, 37, 39, 46, 62, 68–69, 80, 118, 589–590, 593, 617, 688, 696, 816
  MAC, 587–588, 589
  multi-channel, 81
  use in wireless, 102–104
Carrier sense multiple access with collision avoidance (CSMA/CA), 11–12, 16, 33–34, 40, 46, 62, 68–69, 79, 103, 242, 561, 582, 590, 594, 595, 601–602, 605, 614, 615, 620, 628, 792
  based access procedure, 626
Carrier sense multiple access with collision detection (CSMA/CD), 32–33, 589
Carrier sensing (CS) mechanism, 101, 102, 103, 105, 113, 117, 584, 589, 590–591, 593, 596, 597, 601, 603, 618
Carrier sensing range (R_{cs}), 701, 702
Carrier-to-noise ratio (CNR), 235
CATA protocols, 36
Cathode-electrolyte interface, 539
CCI, 420, 422
CDPD, 679
Cell and connection parameters, 656–657
Cell load, 427
Cell planning, 481
Cells, 260, 287–289, 294, 304, 314, 440, 651
Cellular automata, 966
Cellular mobile communications, 439
Cellular networks/systems and standards, 1, 2, 6–11, 23, 24, 26–28, 29, 41, 46, 218, 277, 287, 289, 299, 314, 322, 345, 366, 385, 403, 416, 418, 422, 424, 446,

447, 479, 481–482, 483, 491, 498, 535, 613, 650–652, 653, 654, 763, 838, 861
  see also Mobile telephony
  base stations (BSs), 27, 29, 287
  channel assignment strategies, 442
  hierarchical architecture, 478
  land based, 654
  location area based, 481
  mobile stations (MSs), 27
  mobile switching center (MSC), 27, 46, 416, 420, 422, 478, 481, 650, 784
  mobility, 291–298, 301, 315
  power management, 420–423, 434
  quality of service (QoS) provisioning schemes, 654–667, 671
  radio coverage, 287
  radio systems, limitations, 417–418
Central processing unit (CPU), 276–277, 537, 687, 700, 940
  computation time, 270
  power consumption, 540–541, 543, 548, 704
  scheduling, 547
Centralized algorithms for sensing scheduling problems, 872–874
Centralized clustering, 247
Certificate management, 947
Certification services, 953
Challenge handshake authentication protocol (CHAP), 915, 918, 922, 926, 928, 932
Channel access, 266, 278, 521, 580
Channel access scheduling, 55–61
  modeling of network and contention, 55–56
  research challenges, 72–74
Channel allocation algorithms, protocols, 315–316, 442, 460, 461–462, 651, 673
  biologically inspired, 463–466
  borrowing based, 466
  classification, 462–463
  for fault-tolerant enhancement, 472
  for quality of service (QoS) enhancement, 471–472
  versus mutual exclusion, 468
Channel assignment, 418
Channel asymmetry, 681, 691
Channel bandwidth, 78, 100, 104, 113, 578
Channel holding time, 289, 291, 293–294, 314
  call duration and dwell time, 293, 310–313, 314, 315, 316
Channel hopping, 114
Channel Hoping Access with Trains (CHAT), 116
Channel interference, 183, 580
Channel losses, 681
Channel reservation scheme, 100, 103, 298–303, 315, 316, 446–447, 589, 654
  see also cellular systems, mobile telephony
Channel reuse, quality of service (QoS), 460
Channel selection, 107–114, 117–118
Channel-sharing schemes, 654
Channel utilization, 218, 446, 449, 587, 589, 596, 598, 606
Channel waste, 595
Charging and billing aspects, 786
Check_Window, 769
Choy–Singh DCA algorithm, 468, 469

Chromosomes, 370–372, 374–375, 386
    crossover, 372–373
    elitism, 372
    fitness, 371, 372
    mutation, 372, 374
    selection process, 272
Chvatal's algorithm, 248
Circuit switched services, technologies, 305, 755,
        771, 773
CISCO, 915, 916, 918
    Aironnet 350, 552
Class topology, 716
Classical image processing, 895
Clear Channel Assessment (CCA), 101, 102, 108, 109,
        242, 603
Clear-to-send (CTS) control packet, 33, 40, 69, 79, 81,
        82, 83, 103, 109, 110, 112, 113, 116, 117, 242,
        590–591, 592–596, 603, 604, 605, 615–617, 622,
        626, 701–702, 707, 816, 817, 818
    see also Request-to-send (RTS) control packet
Clear-to-Send scheme (RTS/CTS),
Clint–server mode, 938
Cloaking, temporal and spatial, 408
Clock-synchronization mechanisms, 603, 605–606
Closed loop power control (CLPC), 430, 431, 432
CLR packet, 190
Cluster networks, 241
Cluster-based energy conservation (CEC) algorithm,
        527–530
Cluster-based infrastructures, 602
Cluster-based algorithms, protocols, 236, 528–530
    for MANETs, 603
Clustered heterogeneous model for sensor networks
        (COSMOS), 891–892, 894
    hierarchical network architecture, 891
Clusterhead gateway source routing (CGSR), 218–220,
        221, 228, 229
Clustering, Clusterheads, 218–225, 229, 236, 249, 348,
        353, 567, 891, 895, 896–899
    for data aggression, 825–826
    with geometry property, 248–250
    combining with low weight, 256–357
CMMBCR, 565
CMOS technology, 543
Co-channel constraints (CCC), 461, 465
Co-channel interference, 651, 468, 470
Co-channel reuse distance (R_U), 461
Co-channel reuse ratio, 461
Co-cluster constraints (CCC), 461
Code assignment, 56–57, 59, 60, 70
Code division (CD), 461, 586–587, 589
Code division multiple access (CDMA), 6–7, 8–9, 10, 15,
        25, 26, 29, 47, 55, 81, 106, 218, 409, 416, 417, 425,
        427, 428–429, 441, 461
    see also tree-walking singulation protocol
    cdmaOne, 8
    CDMA2000, 8, 9, 10, 28
    power control, 430, 434
Code separation, 218
Coding process, techniques, 271, 492, 583, 601, 777
Coexisting and overlapping technologies, 6
Co-field, 966
Co-group interference, 470
Collaborative reputation mechanism (CORE), 954

Collision avoidance and resolution multiple access
        (CARMA-MC), 81–82
    ready-to-receive (RTR), 81–82
Collision avoidance, 589, 591
    schemes, analysis, 593–594
Collision detection (CD) techniques, 102, 585–586, 593
    in wireless systems, 588–589
Collision domains, capture effect, 584–585
Collision resolution protocols, 594–595
Combinatorial optimization problems, 464
Combinatorics (graph theory), 127
Common open policy service (COPS), 913
Communication, 23, 124, 183, 184, 892
Communication algorithms, 521–524
    for high mobility networks, 135–154
        experimental evaluation, 149–154
        motion of the hosts–adversaries, 138
        motion space model, 137–138
        restricted motion adversary, 142–145
        RUNNERS algorithm, 140–141, 152–154
        SNAKE algorithm, 139–140, 141–148, 150–154
Communication and information dissemination,
        166–167
Communication channels, reuse, 461–462
Communication graph, 250, 879
Collision avoidance (CA) access scheme, 595
Communication infrastructure, 166, 167–173, 521
Communication links, 130
Communication needs, 165
Communication-based power management, 556–557
Communication protocols, 961
Communication range and sensing range, 879
Communication scheduling and sensing scheduling,
        integration, 878–880
Communication scheduling problems, algorithms,
        876–878
Communication time, 145–148
Communication topology, 167, 172, 173
Compaq PAQs, 976–977
Compass routing, 345, 347
Competition, 468, 470
Complete partitioning (CP), 446
Complete sharing (CS), 446
Complimentary code keying (CCK), 106
Component labeling problem, 895
COMPOW protocol, 604
Computation, Computational, 900, 963
    complexity, 876
    cost, 268, 886, 888
    geometry techniques, 240
    intensive functions, 274, 636
    learning theory, 36
    problems, 889–890
Computer and communications laboratories (CCL), 785
Computer simulations, 751
Confidentiality, 943
Congestion avoidance (CA), 680–681, 683–685,
        688–689, 692–694, 700, 706, 708, 709–710, 712
Congestion variations, 602
Congestion window (cwnd), 680, 684, 712
    size, 706
    update mechanism, 691

Connected dominating set (CDS), 172, 202, 239, 244,
        247–252, 260, 501–505, 507–514, 516, 827
    see also Minimum connected dominating set (MCDS)
    based broadcast algorithms, 513
    determining for efficient routing, 502
    distributed, 248–249
    distributed weighted, 250–252
    formation process, 504, 507, 511, 513, 516
    localized, 248, 512, 514
    protocol, 503, 507
Connected sensor cover, 833
Connection admission control (CAC), 433–434,
        440–441
Connection dropping rate (CDR), 471
Connection state, 738
Connectivity, 199, 206, 209, 218, 269, 329, 478, 537, 594,
        602, 636, 939, 986, 989–991
Connectivity information, 949
Connectivity ratio, 514–515
Connector selection, 251
Conservative CTS-Reply (CCR), 594
Consistency issues, 503, 515
Constant bit rate (CBR), 656, 657, 658
Constrained equations, 968
Consumer electric devices, 735
CONTACT, 782
Contention and collision avoidance in multi-hop
        scenarios, 598–600
Contention-based access operations, 625
Contention-based collision avoidance, 591
Contention-based path selection (COPAS), 717
Contention-based protocols, 79–84, 582
Contention context, 45, 52, 54
Contention control protocols, 602
Contention frame, 594
Contention free period (CFP), 792
Contention free period generator (CFPD), 627
Contention-free polling periods, 11
Contention-free protocols, 84–94
Contention period (CP), 626, 792
Contention pressure, 592
Contention slots, 79
Contention windows, 319, 580, 615, 618–619
Context awareness, 963, 964, 966, 984, 994
Contiguous Allocation algorithm for Multiple Priorities
        (CAMP), 85, 91–93, 94
Contiguous Allocation algorithm for Single Priority
        (CASP), 85, 86–87
Contiguous Sorted Sequential Allocation (CSSA), 85, 89
Control coordination, 960
Control scatternet, 746
Convergence, 2, 5, 6, 18, 633
Cooperation of nodes, fairness in dynamic ad hoc
        networks (CONFIDENT), 954
Coordination, Coordinators, 266, 270, 564
Copying algorithm, 592, 690
Copying overhead, 695
Core based tree (CBT), 207
Core broadcast, 210
Core extraction based distributed ad-hoc routing
        (CEDAR), 194, 210, 640–641
    see also Unicast routing protocol
Core extraction, 210
Core networks (CN), 791

Core routers, 636
Core zone (CZ), 425–427
CORE_FWD message, 211
Core-assisted mesh protocols (CAMP), 209
Correlation factor, 272
Co-site constraints (CSC), 465
Cost-efficiency, 643
Counter-based scheme, 258
Coverage capability, impact of node density, 881
Coverage configuration protocol, 832–833
Credit cards, 984
Cricket location-support system, 394, 406–407, 408
Cross-border feedback, 717
Cross-layer approach, 579, 700, 712, 714, 716, 792–793
Cross-layer error control algorithm, 793–802
    algorithm analysis, 795–797
    implementation considerations, 799–801
    simulation results, 797–799
Cross-layer feedback, 707
Cross-layer hybrid solutions, 602
Cross-layer optimization, 752
    for energy efficiency, 266, 269–271, 276–279, 280, 752
        application layer, 278
        hardware layer, 276–277
        MAC layer, 276, 277–278, 279
        physical layer, 277
        routing layer, 278
        between topology maintenance and scheduling, 752
Cross layer power management, 566, 568–569, 602
Cross-layer rate control algorithm, 790, 802–808
    algorithm analysis, 804–805
    implementation issues, 806–808
    simulation results, 805–806
Cryptographic
    coprocessors, 399
    keys, 403, 404
    trapdoors, 402
Cryptography, 908, 944, 946, 953
    based authentication and encryption
            mechanisms, 943
    identity based, 947
Cumulative acknowledgements (ACKs), 690, 699
Cumulative distribution function (CDF), 310
CW scaling factor, 625
CW_size-copying algorithm, 592
CWL, 707, 711, 712
CW-size, 596–597
Cyclic redundancy code (CRC), 701, 758, 765, 791

D

Data access, 344
Data aggregation, 399, 567, 268, 269, 272, 274, 825–826
Data-centric computation and communication
        paradigm, 266
Data-centric information dissemination in sensor
        networks, 820–822
Data-centric routing protocols, 267, 268, 271–272,
        820–823
Data collection networks, 886
Data communication network, 521, 897
Data compression, 491, 498
Data encryption standard (DES), 926, 932, 944–945

Data error rate, 789
DATA frame, 550, 593–594, 702, 717
Data fusion, 844, 849
Data-gathering and boundary-estimation process, 892
Data gathering tree, 818
Data high, 758
Data link layers, 802
Data loss, 637
Data-medium rate, 758
Data protection managers (DPM), 398, 399
Data sensing, 270
Data sinks, 814
Data slots, 79–80
Data spaces, 962–963
Data structures, 962
Data transfer, 402
Data transmission rate, 277, 586
Database, 983
    retrieval, 442
Datagram traffic, 647
DBASE, 626
Dead reckoning method (DRM), 334–335, 337
Dead-end recovery, 349
Decryption, 945, 947
Default passwords, 911
Degree limitation, 39
Delay, 441, 834
Delayed congestion response (DCR), 472
Delayed duplicate ACKs (DDA), 688, 689–690
Delivery TIM (DTIM), 604
Demand access multiple access (DAMA), 29
Dempster–Shafer belief accumulation, 858
Dempster–Shafer inference, 858–860, 862
Demster–Shafer reasoning, 856
Denial-of-service (DoS), 910, 911, 913, 942
    detection rate, 860
Density control algorithms, 524–527, 531
Density of network, 250
Dependability, 986, 987
Depth first search (DFS), 178, 320, 744–745
Destination address (dst_addr), 187, 188
Destination location, 345, 349–350, 351
Destination sequential distance vector (DSDV) protocol,
        185–186, 187, 188, 191, 194, 642, 949
    transmission control protocol (TCP)
            performance, 705
    see also Unicast routing protocol
Destination to source (DtoS_stream), 599
Destination unreachable, 714
Detection area, 584, 589
Deterministic distributed approach, 470–471
Device authentication, 924
DIAMETER, 914
Differential destination multicast (DDM), 211–212
Differential quadrature phase shift keying (DQPSK)
        modulation, 235
Differentiated services (DiffServ), 635, 636, 637
    assured, 635
    premium, 635
Digital communication systems, 235, 395
Digital encoding, 7, 10
Digital enhanced cordless telecommunication (DECT),
        7, 11, 13
Digital search tree, 487

Digital signal processors (DSPs), 276
Digital signature algorithm (DSA), 946
Digital signature and certificate, 946–947
Digital subscriber lines (DSL), 16
Digital video broadcast (DVB), 11
Dijkstra's algorithm, 245
DIMENSIONS, 358
Direct communication, 962
Direct sequence spread spectrum (DSSS), 12, 55, 56,
        581, 756
Directed acyclic graphs (DAGs), 126, 130, 131–132, 133,
        134, 136, 188, 200
Directed diffusion, 271, 567, 822–823
Directed handoff (DH), 763, 767
Directed retry, 763
Directional network allocation vector (DNAV), 40
Directional virtual carrier sensing (DVCS)
        mechanism, 40
Directionally request-to-send (DRTS), 40
Disaster area networks, 940
Dissemination algorithms, 567
Distance-based schemes, 258
    location update, 323
Distance dependent attenuation, 768, 769
Distance effect, 335
Distance information, 407
Distance routing effect algorithm for mobility
        (DREAM), 194–195, 320, 330, 345, 346, 350
Distance vector multicast routing protocol
        (DVMRP), 207
Distance vector protocols, 348
Distributed activation based on predetermined routes
        (DAPR), 820
Distributed algorithms, 241
    for sensing scheduling problems, 874–876
Distributed and collaborative date processing, 838
Distributed call admission control (DCA) algorithms,
        447–448, 454
    based on graph coloring, 470–71
    based on borrowing from richest, first available,
            466–467
    based on dynamic load balancing strategy, 467
Distributed computing system, 126–128, 886, 888
    leader election problem, 126, 127, 129–135
    motion coordination, 137
Distributed contention-based medium access control
        (MAC) protocols, evolutionary perspective,
        586–600
Distributed control algorithms, 600
Distributed coordination function (DCF), 11, 12, 33, 38,
        103, 550, 578, 582, 595, 596, 603, 604, 605, 614,
        617–618, 627, 628, 792, 808
    collision avoidance and contention control, 583
    quality of service, 621
Distributed data structure, 965
Distributed database, querying, 826–827
Distributed denial of service (DDoS), 910
Distributed dynamic resource channel allocation
        algorithm (DDRA), 468, 470
Distributed fair scheduling (DFS), 614, 623–625
Distributed foundation wireless medium access control
        (DFWMAC), 582–583
Distributed IEEE 802.11 MAC, 648
Distributed in-network processing algorithms, 274

Distributed interframe space (DIFS), 103, 235, 358, 582, 595, 596, 615, 617–618, 619, 621 627
Distributed priority access, 621–623
Distributed priority scheduling (DPS), 621–623
Distributed reservation access, 614, 625–627
Distributed source coding, 275–276
Distributed voice access protocol (DVAP), 755, 763, 764
    see also Bluetooth
Distributed weighted fair queuing (DWFQ), 624, 625
Distributed weighted fair scheduling (DWFS), 629
Distribution contention control, (DCC), 598
Distribution system (DS), 582
Divide-and-conquer algorithm, 275
DM packets, 758, 767
D-Medium access control (D-MAC), 40, 278, 817–818
    quality of service (QoS) leverages, 617–619
DoCoMo's iMode service, 986
Dominating set, 877
Dominators, 248, 250, 502
Double-covered broadcast (DCB), 201, 202–203
Doubling circles, 328–329
Downlinks (forward links), 85, 86, 277, 416, 429–430, 433, 434, 442, 452–453, 726, 727, 729, 730; power control for, 431, 433
DREAM location service (DLS), 330–331, 337
DS/SSMA, 100
Dual busy tone multiple access (DBTMA), 34, 116
Dual mode advanced mobile phone service (D-AMPS), 7, 27
Duplicate ACK (DUPAK), 680–681, 683, 684, 685, 687–691, 695, 700, 714
    see also transmission control protocol (TCP)
DV-Hop algorithm, 836
Dynamic base station selection (DBS), 763, 767, 769–770
Dynamic channel allocation (DCA) algorithms, protocols, 117, 442, 463–463, 464, 465
    based on borrowing for, 466–467
    genetic-based, 465–466
    based on mutual exclusion paradigm, 468–70
    neural network-based, 464–465
    simulated annealing (SA)-based, 463–464
Dynamic delayed ACK, 707, 710–711
Dynamic frequency scaling (DFS), 543
Dynamic host configuration protocol (DHCP), 785, 913
Dynamic multiple-threshold bandwidth reservation (DMTBR), 451–452
Dynamic power management (DPM), 542
Dynamic priority access, 614
Dynamic source routing (DSR) protocol, 17, 185, 186, 187, 188, 201, 212, 220–221, 240, 271, 319, 334, 336, 704, 709, 746, 950, 953, 954
    Transmission control protocol (TCP), 705–706, 707, 708, 712, 714
Dynamic voltage scaling (DVS), 543

E

E-AODV, 642
Earlier deadline first (EDF) paradigm, 621–622
Eavesdropping, 909, 911, 919, 941
EBS, see Ek-Kadi's rate-based bandwidth borrowing scheme
Edge routers, 636

Effective load for the target cell (ELT), 426–427
Eight-phase-shift keying (8PSK), 8–9
Ek-Kadi's rate-based bandwidth borrowing scheme, (EBS), 656–661, 664, 668, 671, 673
    fairness, 657
    handoff management, 658–659
    simulation results, 659–661, 665, 667
Elasticity, 244
Electric field approach (EFA), 967
Electronic commerce (e-commerce), 920
Electronic Industries Association, 27
Electronic number (ENUM), 777
ElGamal cryptosystem system, 946
Eliminate_then_Select (ETS), 329
E-limited scheme, 728
Elitism, 380, 383
Elliptic curve cryptosystem (EEC), 946
Encapsulating security payload (ESP), 923, 924
Encapsulation techniques, 917
Encoding, 649
Encryption, 7, 399, 401, 403, 405, 924–927, 931, 932, 934, 941, 943, 944–947
    see also security in ad WEP protocol, 906–907
    and authentication integration, 931
End-to-end data transfer delay, 746
End-to-end encryption, 908
End-to-end mechanisms, 691, 700
End-to-end modifications, 707–712, 719
Energy-aware data (EAD) centric routing, 819–820, 829
Energy conservation, 240, 272
Energy-conserving medium access control (EC-MAC), 79, 84, 85
Energy dissipation and interference, 887, 890, 897
Energy efficiency, 260, 266, 276–279, 691–692, 871
Energy-efficient information routing and coding, 271–276
Energy management, 818–819
Energy-usage perspective, 600
Energy utilization, 600
Enhanced data rates, 14
Enhanced data rates for GSM evolution (EDGE), 7, 8–9, 28, 986
Enhanced distributed coordination function (EDCF), 38, 627–628, 808
Enterasys Networks RoamAbout interfaces, 549
Entropy-aware routing, 271, 272
Entropy coding, 479, 497
Environmental dynamics and agents' movements, 964
Erlang-B formula, 289–91, 315
Error coding techniques is Reed-Solomon coding, 791
Error-congestion loss distinction, 709–10
Error connection techniques, 637
Error control algorithm, 793–794, 798, 809
Error detection mechanisms, 701
Error rate, 648
Error-resilient video coding algorithms, 790
Estimation algorithms, 849–853, 854
Estimation error, 350
Ethernet, 11, 577, 578, 586, 589, 730
Euclidean distance, 871
European Telecommunication Standards Institute (ETSI), 13, 27–28
Event-based models, 963
EVENT INTERFACE, 976

Event-to-sink reliable transport, 830–831
Evolution loop, 374
Exhaustive pseudo-cycle master (EPM), 729
Exhaustive token-controlled networks, 236
Expanding ring algorithm, 529–530
Expected number of transmissions (ETX), 278
Expected drift rate, 454
Explicit link failure notification (ELFN), 707, 712, 717, 719
Explicit bad state notification (EBSN), 684, 685, 690
Explicit congestion notification (ECN), 692, 694, 714, 715
Explicit loss notification (ELN) strategies, 685, 690
Explicit route disconnection notification (ERDN), 713
Explicit route successful notification (ERSN), 713
Exponential backoff mechanisms, 706
Exposed terminal problem, *see* Hidden and exposed terminals
Extended MDST (E-MDST), 745
Extended synchronous connection-oriented (ESCO) links, 771
Extensible authentication protocols (EAP), 786, 915–917, 918
    MD5, 919, 926, 928, 932
    PEAP, 918
    Secure routing protocol (SRP), 918, 948, 949–952
    Transport layer security (TLS), 913–916, 918–920, 926–928, 932
    Tunneled transport layer security (TTLS), 916, 918

F

FACE, 325
Face routing, 351, 356
FACK, 681, 694
Fair exhaustive polling (FEP), 729
Fairness problem, 706, 954
Fast encryption algorithm (FEAL), 945
Fast recovery, 681, 684
Fast retransmit algorithms, 680
Fault diagnosis systems, 885
Fault propagation, 986
Fault-tolerant architectures, 986, 987
Fault-tolerant averaging algorithm, 845–847
Fault-tolerant integration, 849
Fault-tolerant interval (FTI) function, 847
Feature graph, 895
Feature map algorithms, 854–856
Feature node, 893
Feedback control messages, 640
Feedback-control-based approach, 349, 625
Feedback cycle, 964
Feedback mechanism, 707, 711
Fidelity-aware routing, 820
Fidelity control, 830
Field based coordination systems, 963–967
    architecture and implementation, 960–961, 976–977
    key concepts, 964–965
    modeling, 967–975
    programming, 976–977
    state of art, 965–967
File transfer protocol (FTP), 703, 931, 932, 934
Fine granularity scalability (FGS), 790, 805–806, 808

Fingerprint, 946
Finite state machines (FSM), 543, 545, 547
Finite-state predictability, 492
FIPA-based agent systems, 962
Firewalls, 914, 926, 928, 929
1G systems, 10
First-active-first park (FAFP), 760
First-come-first-serve (FIFO) discipline, 54, 57, 60, 130, 423, 427, 444, 446, 716
Fitness calculation, 385–388
Five-Phase Reservation Protocol (FPRP), 47–48
    collision report (CR), 48
    packing and elimination (P/E), 48
    reservation acknowledgement (RA), 48
    reservation confirm (RC), 48
    reservation request (RR), 48
Fixed channel allocation (FCA), 442, 446, 462, 463
Fixed rectangular forwarding zone (FRFZ), 352
Fixed-retransmission time-out (RTO), 707, 708, 717
Fixed transmitter – tunable receiver (FT–TR), 78
Fixed wire network traffic, 368
Flexible quality of service (QoS) model for MANET (FQMM), 637
Flock field, moving while maintaining a formation, 968–969
Flooding, 197, 198, 201, 217, 240–241, 252, 257–259, 324, 337, 344, 345, 346, 348, 349, 352, 353, 355, 356, 357, 567, 714, 746, 820
Floor acquisition multiple access (FAMA), 33, 593, 595, 688, 696
Floor-plan topology, 966
Flow control, 689
Flows pressures, 592
Forest-of-trees algorithm, 894, 899
Forward control channel, 478
Forward error correction (FEC), 637, 687, 688, 730, 757–758, 760, 791, 793, 799
    coding, 760
Forward link, *see* downlink
Forward node set selection process (FNSSP) algorithm, 202
Forwarding backbone topology, 564, 877
Forwarding equivalence class (FEC), 636
Forwarding group (FG), 205–206
Forwarding group multicast protocols (FGMP), 205–206
    see also multicast routing protocols
4G cellular mobile networks, 3, 10–11
Four-way handshaking, 616
Fractional guard channel (FGC), 443–444
Frame control, 801
Frame synchronization message (FSM), 84
Free RADIUS, 915
Frequency division (FD), 461
Frequency division duplexing (FDD), 24, 25, 26, 27, 28, 586, 792
Frequency division multiple access (FDMA), 10, 12, 25, 26, 27, 29, 41, 47, 81, 106, 416, 417, 441, 461
    power control, 430
Frequency hop collisions, 766
Frequency hop synchronization, (FHS), 738, 759
Frequency hopped multiple access (FHMA), 26
Frequency-hopping rate, 759
Frequency hopping spread spectrum (FHSS), 12, 14, 103, 114, 116, 581, 596, 726, 735, 756

Frequency modulation (FM), 10
Frequency reuse, 101
Full-duplex communication, 24, 26
Full-duplex link, 586
Functionality, 267, 270
Fuzzy set theory, 856

G

G zone routing protocol (GZRP), 194, 200, 201, 320
Gabriel graph (GG), 168–169, 197, 348
Gafni and Bertsekas (GB) algorithm, 130, 131
    Partial reversal algorithm, 131
    Full reversal algorithm, 131
Galois fields, 35, 50
Gamma distribution, 314
Gateway GPRS support node (GGSN), 783–784
Gateway node selection, 260
Gateway nodes, 869
Gateway selection algorithm, 225–227
Gateway source routing infrastructure creation
        protocol, 219
Gateways, 219, 249, 567
Gaussian approximation, 449
Gaussian distribution, 448
Gaussian minimum-shift keying (GMSK), 8–9
Gaussian noise, 850, 852, 854
General inquiry access code (GIAC), 759
General packet radio service (GPRS), 4, 7, 8, 28, 535,
        679, 774, 783–785, 787, 791–792, 986
    Tunneling Protocol for data transfer (GTP-U), 792
    wireless local area network (WLAN)-based support
        node, 785–786
Genetic algorithms (GA), 244, 367–368, 370–374, 383,
        384, 386, 389, 465–466, 652
    gene, 370–373
    chromosome, 370–373, 374
    framework, 374–375
    population, 370–372, 374
    rank-and tournament selection variant, 367
    three operators, 372–74
Genetic-based dynamic channel allocation protocols,
        466–467
Geo temporally ordered routing algorithm (GeoTORA),
        194, 200, 353
Geocasting protocols, 3, 343, 344, 352–357, 361
Geographic distance routing (GEDIR), 194, 198–199
Geographic forwarding routing protocols, 320, 321, 356
    greedy, 345–346, 348, 349, 351, 352, 355
    non-deterministic, 349
Geographic hash table (GHT), 357–358, 827
Geographic protocols, 343–344
Geographic region summary service, 334
Geographic rendezvous mechanisms, 344
Geographic routing protocols, 272, 343, 344–352, 354,
        358–360, 823–824
    effect of link loss, 351–352
    location in accuracy, 350–351
        effect on face routing, 351
Geographical adaptive fidelity (GAF), 278, 349, 358,
        525–526, 531, 564, 566–567, 602, 828, 877
    pseudo code, 526
Geographical energy aware routing (GEAR), 348

Geographical forwarding protocol, 324, 335
Geographical information, 496
Geographical Routing Algorithm (GRA), 194, 198,
        320, 321
Geographic-based rendezvous mechanism, 351, 357–360
Geographic-forwarding geocast (GFG), 354–355, 356
Geographic-forwarding-perimeter-geocast (GFPG),
        354, 355–357
Geometric graph, 166
Geometric spanners, 167–69, 171
    gabriel graph (GG), 168–169, 197, 252, 348
    nearest neighbor graph (NNG), 168, 169
    relative neighbor graph (RNG), 168, 197, 252,
        253–254, 255–257, 348, 504–505
Geometry properties, 249–250
Georganas approximates, 876
Geostationary Earth Orbit (GEO) satellite links, 693
Geosynchronous-orbit satellite (GEO), 18, 29
Gigabit Ethernet (1 Gbps), 6
Gilbert Elliot model, 795
Global coordination problems, 890
Global information, 889
Global positioning system (GPS), 29, 128, 166, 183,
        200–201, 205, 240, 258, 320, 343, 350–351, 394,
        407, 506, 531, 564, 602, 643, 835, 836, 858, 880
    Ant-Like routing algorithm (GPSAL), 333–334
    based route discovery optimization (GDSR), 194,
        200–201
Global state information, 241
Global state maintenance, 900
Global system for mobile communications (GSM), 6–7,
        8, 13, 27–28, 305, 399, 403, 417, 423, 481, 483,
        535, 783–784, 785, 792, 986
    variant, 9
Global time synchronization, 891
GOAFR, 348
Gossip and probabilistic schemes, 258
Gossiping, 567
Graph coloring approach, 873
graph coloring, 470–471
Graph theoretic approaches, 48–50
Gravitational and electromagnetic fields, 960
GRDY, 250–251
Greedy-Face-Greedy (GFG), 320
Greedy perimeter stateless routing (GPSR) protocols,
        194, 197–198, 320–321, 337, 348, 358, 531,
        823–824, 827
Greedy set cover algorithm, 257
GRID, 194, 199, 349, 873, 876
Grid location service (GLS), 195, 325–328, 334, 337,
        350, 357
Grid network topology, 114, 716
Group allocation multihop multiple access (GAMMA),
        79–80
Group oriented mobile services, 989–992
GROUPREQUEST, 212
Group-transmission period, 626
GRSS, 337
Guard channels, 442–443, 446, 448
GUI, 774

H

Half-duplex radio systems, 30, 55, 586, 688
Handoff algorithms, 200, 418, 419, 420–423, 433, 440,
        446, 449, 453, 454, 649, 650, 651, 652, 653–654,
        655–656, 660, 664, 668–669, 671, 673, 684, 685
    see also call blocking probability (CBP), call dropping
        probability (CPP)
    arrival rate, 452
    conventional, 421–422
    desirable features, 420
    distance-based, 421, 422
    fractional guard channel (FGC), 443–444
    guard channel, 422, 442–443
    increased frequency, 441
    minimum power algorithm, 422
    and mobility management, 418–423, 471, 478, 695,
        653, 658–659, 665, 667–668
    prioritization, 422–423, 433, 441
    queuing, 422–423, 444–446, 653, 654, 673
    types, 419–420
Happiness landscapes, 966
Hard-disk management, 537
Hard handoff, (HHO), 419, 420
Hardware components, 540
    general scheme for power management, 541–543
    quality of service (QoS), 541–542
Hardware specific optimization, 888
Harris theory, 143
Hash-table-like mapping scheme, 350, 358
Hashed message authentication code (HMAC), 909, 923
    see also Secure hashing algorithm (SHA 1)
Hashing function, 52, 73, 909, 923, 946
Head of line (HOL), 617, 621, 622, 624
Header checksum option (HACK) for TCP, 710
Header compression, 692
Hello message, 201, 211, 223, 225, 235, 242, 321, 328,
        331–332, 502, 503, 509, 510–512, 513
Heterogeneous wireless sensor networks, 838
    activity scheduling schemes, 880
Hidden and exposed terminals, 47, 78, 102, 103, 105,
        589–91, 593, 594, 604, 616, 701–702
Hidden node problem, 241, 242
Hierarchical legend-based service, 332
Hierarchical matching tree, 273
Hierarchical MRSVP (HMRSVP), 638–639
Hierarchical state routing (HSR), 192, 193, 194, 324
    see also Unicast routing protocol
High error rate links, 692
High speed circuit switched data (HSCSD), 7, 28
High-speed circuit-switched (HSCS) services, 305
HiperLAN, 11, 13, 937
Home location register (HLR), 403–405, 784, 785
Home region location services, 323–324
HomeRF, 16, 19
Homezone, 195–196
Hop reservation multiple access (HRMA), 116
Hopping sequence, 726
Hops to reach (hop_cnt), 188
HopsToCore, 211
Hop-stretch factor, 167, 171, 172
Host connectivity, 601, 604
Host mobility, 578
HSDPA, 800

HS-DSCH, 800
HTTP, 777, 924
    and FTP testing scenarios, 929–934
Hybrid activation multiple access (HAMA) protocols,
        46, 60–61, 67–68, 69–72
    throughput analysis, 62, 63–65
Hybrid ARQ (HARQ), 800
Hybrid channel allocation (HCA), 461, 463
Hybrid coordination function (HCF), 808
Hybrid energy-efficient distributed (HEED)
        clustering, 826
Hybrid MAC protocols, 36, 39
Hybrid routing tree algorithm, 272–274
Hyper-exponential distribution, 310, 313, 314

I

IamDominator, 249
iAmNoLongerYourCH flag, 223
ICMP QoS-LOST message, 642
ID-based flat routing protocols, 184, 185–192
ID-based hierarchical protocols, 184, 192–194
IDLE channels, 81
IEEE 801.11, 46, 699
IEEE 802.1, 79
IEEE 802.11, 3, 4, 11–13, 15, 33–34, 38, 40, 47, 78, 82, 83,
        100, 103–104, 106, 107, 108, 112, 113, 259, 336,
        521, 578, 592, 596, 598, 606, 613–614, 700, 706,
        756, 786, 789, 792, 916, 937
    authentication, 906
    contention control in, 595–598
    distributed coordination function (DCF) MAC
        protocol, 578, 579, 586, 591, 594, 595,
        597–599, 602–603, 606, 614, 615–617, 624,
        628, 700
    enhanced DCF (EDCF), 617–618, 627–628
    link layer format, 701
    medium access control (MAC), 578, 579, 586, 606,
        624, 711, 716
    mobile area networks (MANETs), 598
    point coordination function (PCF), 603
    quality of service (QoS), 648, 808
    replacement and key management with 802.1X, 908,
        918–919
    security architecture, 905–906, 925–926, 927
    sleep-mode, 601
    (Wi-Fi) power saving mode (PSM), 549–552, 556,
        562–564, 580, 581–583, 594, 604–606
    wireless local area network (WLAN), 560–561, 562,
        595, 792, 800, 806, 808
IEEE 802.15, 14, 15, 78, 560
    wireless personal area networks (WPANs), 736
IEEE 802.1X standards, 911, 913, 916–917, 924,
        925–926, 929–930, 934
    architecture, 917–918
    model implementation, 927–928
    over 802.11, 918–919
    test results, 931
IEEE 802.3, 589, 701
IEEE802.16, see Wireless Metropolitan Area Networks
        (WiMAX)
IKE, 924
IMRG, 243, 255–257

IMT-2000 standards, 8, 9
Inactive times (or user think times), 556
In-band signaling, 639, 640
Independent basic service set (IBSS), 562–563, 582, 604
Indirect transmission control protocol (TCP) (I-TCP), 556, 557, 686, 689
Industrial Technology Research Institute (ITRI), 774
Industrial, scientific and medical (ISM) frequency bands, 13, 107, 613, 726, 735, 756
Inference methods, 396, 409, 856, 860–861
Information and communication technologies (ICT), 287
Information coding perspective, 269, 274
Information dissemination, 166–167, 173–181, 184
    routing in directed planar graphs, 174–178
        directed link model, 174
        Eulerian networks, 175–77
        outerplanar networks, 177–178
    routing in undirected planar graphs, 173–174
        compass routing algorithm, 173, 174
        face routing algorithm, 174
    traversal of nonplanar graphs, 178–179
        quasi-planar graph travesal algorithm, 179–181
Information exchange, 275
Information frames (IFs), 47
Information fusion, 844, 861–862
Information processing and information routing, 266, 267–269, 271–272, 274
    collaborative, 270–272
Information replication, 195
Information retrieval, 6
Information sharing, 266, 270
Information slots (ISs), 47
Information-theoretic, 479
Infrared data association (IrDA), 16, 19
Infrared signals, 103
Infrastructure-based networks, 19, 535, 537, 569, 582, 601–602, 613, 938
    power management, 548–559, 563
    wireless technology, 6
In-network data processing, 268, 567
Inquiry scanning device, 739, 759
Insignia, 639–640
Integer factorization problem, 946
Integer linear programming (ILP), 872–873
Integrated service (IntServ), 634–635
    access, 636, 637, 638
    control load service, 635
    over Differentiated services (DiffServ), 635–636
    guaranteed service, 634
Integrated services over specific link layer (ISSLL), 635, 638
Integrity check value (ICV), 906–907
Integrity protection, 923, 943
Inter frame space (IFS), 582, 615, 621
Interference, 428, 637, 726
Interference range (IR), 701, 702
Interim Standard 54 (IS-54), 7, 27, 28; IS-136, 27, 28
Intermediate node initiated rerouting (INIR), 643
Intermediate receivers (IR), 599
Internal node concept (CDS), 256–257
International data encryption algorithm (IDEA), 945
International mobile subscriber identity (IMSI), 784, 786

International Standard Organization (ISO), 790
    Seven layer model, 870
International Telecommunications Union (ITU), 9, 10, 773, 790
International Telecommunications Union-Telecommunications (ITU-T), 790
Internet, 5, 6, 29, 166, 183, 266, 331, 345, 352, 416, 478, 535, 536, 559, 578, 755, 761, 838, 869, 920, 939
Internet control message protocols (ICMP), 714
Internet Engineering Task Force (IETF), 634, 635, 636, 692–694, 779
    Network Working Group, 918
Internet protocols (IP), 11, 638, 699, 773–774, 786, 910, 922, 924, 950
    see also virtual private network (VPN)
    mobility, 914
    quality of service (QoS) models, 634–637
    related issues, 636–637
    routing, 636
    security (IPSec), 691, 913, 922, 923–924, 925–927, 928, 929, 932
    spoofing, 942
Internet security association and key management protocols (ISAKMP), 923
Internet service providers (ISPs), 915
Internet telephony, 635
Interoperability, 13, 986, 987
Inter-piconet scheduling algorithms, 752
Inter-piconet traffic, 741
Interpolation technique, 546
Interrupted poisson process (IPP), 314
Interzone routing protocols (IERP), 192
    see also Unicast routing protocol
    intra-cluster identification and labeling of feature regions, 895
    intra-handoff mechanism, 469–70
    intra-piconet polling schemes for bluetooth, 725, 727–730
Intrusion detection system (IDS), 914, 943, 944, 948, 951–952
INVITE request, 779–780, 782–783
Iridium satellite system, 668
ISDN, 478, 784
ISDN user part (ISUP), 784
Iterative maximum-branch minimization (IMBM) algorithm, 245
ItryDominator, 251

J

Japanese Digital Cellular (JDC), see Pacific Digital Cellular (PDC)
Java, 655
Jitter, 633, 634, 643, 649, 771, 789
    and random assessment delay (RAD), 242
Join reply message, 206
Join request message, 206, 207, 208, 209, 211
Join table message, 206, 207
JOIN_ACK, 209, 210, 211
Joint data compression, 268–269, 271, 273, 274
Joint information routing and coding, 274–275

K

Kalman filter algorithm, 850–853, 854
    extended Kerberos, 914
Key management, 908, 916, 918, 927, 944, 947, 948,
        952–953
    identity-based, 953
Knowledge representation, 483
Kruskal's algorithm, 245

L

label distribution protocol (LDP), 636
label propagation through backtracking, 897
lack-off procedure, 617–619
Lampel-2iv Symbol wise model, 492, 495–96
Laptops, 11, 538, 613, 704, 937, 976
Last-hop wireless access, 613
Later 2 tunneling protocol (L2TP), 922, 924
    internet protocol security (IPSec), 926, 927
Latin squares, 50–51
Layer 3 virtual private networks (VPN)–IPSec, 923–924
Layer scheduling, 270
Leader election algorithm, 129–135
    analytic considerations, 133–135
    pseudo-code description, 132–133, 158
    for reconfiguration of robot chains, 154–160
LEADER, 896
Learning, 483, 492, 498
Least clusterhead change (LCC), 219
LEDA, 149
Legend exchange and augmentation protocol (LEAP),
        331–334, 337
Lezi-update algorithm, 490–496
    incremental parse tree, 492–493
    profile-based paging, 495–496
    sampling, 491
    tradeoff between update and paging, 496–497
    update cost and message complexity, 493–494
Limited and weighted round robin (LWRR), 729
Limited wait (LW), 991
Line of sight (LOS), 422
Link access probability, 39
Link activation, 2, 46, 48, 57–60, 62, 66, 67–68, 73
Link activation multiple access (LAMA) protocols, 46,
        57–58, 59, 67–68, 69–72
    throughput analysis, 62, 67
Link-adaptation techniques, 802
Link availability, 506–509
Link capacity, 685
Link fading, 637
Link failure, 131, 136, 187, 210, 472, 578, 642–643, 707,
        708, 717, 885
    detection, 513
Link latency, 681
Link layer, 686, 688, 706, 716, 793
    error control, 799
    information, 706
    rate control, 806
    retransmissions, 695
    security provisions, 908, 920
    and TCP interaction, 689
    transmission delays, 593

Link layer control (LLC), 578, , 581, 584, 588, 592
    see also MAC
Link management, 726
Link manager protocol (LMP), 738
Link quality, 421, 434
Link-quality of service algorithm, 770
Link random early detection (LRED), 706, 712
    and adaptive pacing, 715–716
Link reversal routing (LRR), 130, 131, 136, 188, 200
Link selection strategies, 351
Link state information, 640
Link state propagation, 194, 210, 641
LINUX, 976
Listen before talk, see Carrier sense multiple access with
        collision avoidance (CSMA/CA)
Listen while transmitting, concept, 586
Lithium-ion cells, 538
Load balancing, 221, 228, 972–974
Local acknowledgement (LACKs), 718
Local area networks (LANs), 3, 19, 33, 46, 78, 79, 84,
        99–100, 103–104, 107, 118, 276, 345, 394, 407,
        478, 577–578, 579, 589, 590, 761, 763, 916,
        920, 939
    MAC protocols, 586
Local coordinate system (LCS), 531
Local estimation algorithm, 449
Local maximum node, 197, 198
Local network management, 6, 869
Local predictive resource reservation, 454–455
Local recovery mechanisms, 687–688, 689, 691, 695
Localization mechanisms, protocols, 15, 320, 343,
        350–351, 352, 359, 531, 833–837, 891, 961, 965
Localized low weight structures, 252–256
    negative results, 256
Localized minimum spanning tree (LMST), 243,
        252–253, 256, 509, 878
    combining relative neighborhood graph (RNG),
        255–257
    structure based, 254–255
Localized query (LQ), 190–191
Location-aided routing (LAR), 194, 195–196, 241, 320,
        333, 336, 345, 346
Location areas (LA), 366, 480, 482, 483, 496
    cell planning, 484
    partitioning, 481
Location awareness, 166, 181, 344, 393, 994
Location based broadcasting algorithms, 256
Location-based routing protocols, 194–201, 258
    see also Unicast routing protocol
Location-based services in mobile commerce, 983, 984,
        985, 989
Location data protection, 403
Location dissemination system, 322
Location-guided k-ary (LGK) tree, 212
Location-guided steiner (LGS) tree, 212
Location-guided tree (LGT) construction
        algorithms, 212
Location information, 183, 198, 200, 241, 320–325,
        331–334, 336–337, 343, 344, 345, 350, 358, 395,
        396, 397, 403–409, 501, 502, 858, 876, 984,
        986, 988
Location management, 2, 4, 358, 365–366, 368–370,
        478–479, 482–483, 496, 498, 653
    cost, 368–370, 385

Location management (*contd.*)
  always-update strategy, 366
  never-update strategy, 366
  reporting cells, 366
  support for mobile commerce, 984, 987–988
  taxonomy, 480–483
Location packet (LP), 330–331
Location privacy, 337, 402
  attacks on, 395–396
  first-hand communication, 395, 401, 409
  inference, 396, 409
  second-hand communication, 395, 409
  solutions, 396–409
Location query, failures, 328
Location query request, 327
Location request, 336
Location server system, 326–327
Location server update, 327–328
Location services, 320, 321–322
Location trace aided routing (LOTAR), 336
Location tracking systems, 406–407, 498, 988
Location uncertainty, 478, 479, 481, 482, 497–498
  and entropy, 488–90
Location update, 329, 479, 480
Location and user-sensitive advertising, 984
Logic topology, 511
Loop breakage procedure, 878
Loop scatternet formation (LSF) algorithm, 743
Loss detection and dependence on timers, 690
Lossy quantization algorithms, 496
Low-earth-orbit (LEO) satellite networks, 18, 29, 648,
      650, 653–654, 673, 693
  quality of service (QoS) provisioning, 667–671
Low-energy adaptive clustering hierarchy (LEACH)
      protocol, 272, 527–529, 567, 825–826, 870
Low-power sleep mode, 601–602

M

M/G/I queuing system, 53–54, 289
Main circulating circuit (MCC), 891
Management information base (MIB), 906
Mapping, 984
Marking process, 504, 507–508, 509, 511
Markov chain, 299, 304, 306–309, 314, 315, 454, 467,
      486–88, 546–547, 760, 796–797, 805
Markov modulated poisson process (MMPP), 314
Marzullo function, 845–847, 849
Masquerading, 910
Master/slave bridge (M/S bridge), 736, 740
Master/slave link layer protocol architecture, 756, 759
Mauve classification, 322, 329, 331, 336
Maximum effective load for adjacent cells (MELA), 427
Maximum independent set (MIS), 48, 247, 250–251
Maximum segment size, 680
Maximum traffic flow, 747
Maximum transmission units (MTUs), 692–693
Maximum-degree vertex first approach, 50
Max-min fair bandwidth borrowing scheme (MBS),
      661–667, 671, 673
  fairness, 661–664
  simulation results, 665–667
MAX-RETRANSMISSION, 767
MDST, 744–745

Measurement based prioritization scheme (MBPS), 446
Measurement vector, 849
Medium access control (MAC) protocols, 2, 3, 15, 24, 30,
      37, 78, 95, 99–100, 113, 114, 116, 117–118, 217,
      221, 228, 229, 232, 241, 242, 243, 258, 270, 348,
      395, 401, 407, 514, 566, 569, 578, 579, 586, 613,
      615, 626, 688, 701, 702, 706, 711, 716, 736, 786,
      792, 800–1, 807–808, 818, 827, 828, 911, 926,
      932, 949
  access control list (MAC ACL), 912
  address, 910
  ALOHA protocol, 31, 32, 46, 78
  Carrier sense multiple access (CSMA), 31–32, 37
  Channel access scheduling, 55–61
  contention and collision avoidance, 595, 598–599
  contention under the power saving viewpoint,
      602–603
  distributed, contention-based, 586–588
  IEEE 802.11, 100, 103–104, 106, 107, 108, 112, 113,
      114, 521
  layer policies, 549–553, 556, 561, 59
  layer sleep scheduling, 266
  Mobile ad hoc networks (MANETs), 31–37,
      101–102, 107
  Multichannel (MMAC), 82–84, 104–106, 107–118
  power saving solutions, 600, 601–602, 603, 605, 606
  quality of service (QoS), 614, 617, 628–629
  radio sleep scheduling, 266, 276, 277–278, 279
  research directions, 38–41
  scheduling policies, 269
  wireless assumption, 583–586
Medium earth orbit (MEO), 693
Medium earth-orbit satellite (MEO), 18
Memory effect, 596
Merge and disjoin algorithms, 223–225
Merger-accept message, 222, 225
Merger-request message, 222–223
Mesh-based multicast routing protocols, 210
Message contents, 242
Message digest algorithm (MD5), 909, 915
Message integrity check, 909
Message integrity code (MIC), 946
Message modification, 910
Metamorphic systems, 137, 155
Metropolitan area networks (MANs), 11, 16, 19
Micro electro-mechanical systems (MEMS), 37, 265, 868
Micro-rocket engines, 538
Microsoft CHAP (MS-CHAP), 915
Microsoft point-point encryption (MPPE), 915, 922
Microsoft Windows 2000, 924
Microsoft, 916, 918
Middleware infrastructure, 962
Military tactical networks, 939
Miniaturization techniques, 265
Minimal energy topology, 878
Minimal spanning tree, 276
Minimal steiner tree, 266, 268, 272, 273–274
Minimum connected dominating set (MCDS), 202, 239,
      244, 247, 248, 250
Minimum dominating set (MDS), 247, 248
Minimum expected (MEX) bandwidth, 657
Minimum spanning tree (MST), 203, 204, 241, 243,
      244–247, 251, 255–256, 508, 509, 512, 878;
      distributed, 252; link-based, 246–247

Minimum weight spanning tree (MST), 744–745
Minimum weighted connected dominating set
    (MWCDS), 250
Minimum-energy broadcast/multicast problem, 244,
    245–246
MIPS, 276
Mix zones, 405–406
MIXes in mobile communication systems, 403–405
MIX-Nets, 401
Mobile ad hoc networks (MANETs), 2, 3, 4, 6, 17, 23, 24,
    30–38, 40, 41, 47, 48, 99–100, 107, 135–136, 150,
    195, 204–207, 212, 239, 240, 241, 266, 348, 501,
    502, 504, 506, 509, 510, 514, 515, 516, 566, 578,
    579–581, 594, 595, 597–600, 634, 637, 639–640,
    641, 642, 643–644, 648, 699, 809, 960, 977
    adapting transmission control protocol (TCP),
        707–720
    algorithmic problems, 128–129
    algorithms, 131–132
    allocation MAC protocol, 35–36
    characteristics, 700–705
    contention based MAC protocols, 31–34, 580–581
    location information, 319–320
    packet drop, 700
    performance, 705–707
    power management, 563, 600–603, 606
    quality of service (QoS), 580
    real time applications, 639–640, 643
    routing failure, 705
    Wireless channel characteristics, 701–704
Mobile algorithms, 124–126
Mobile and wireless computing, coordination
    activities, 962
Mobile base station (MBS), 468–69
Mobile cellular communication network, 460–462
    architect for, 460–461
    see also channel allocation
Mobile commerce, applications and environment, 984
    infrastructure requirements, 985–987
    modeling, 992
    open research problem, 993–994
    simulation model, 993; technical solutions, 983–994
Mobile communication capability, 1
Mobile computing and communication devices, 4, 393,
    406, 407, 460, 477, 535–536, 959
    ad hoc, 124
    algorithmic problems, 128–129
    basic problems, 126–127
    global model, 124, 125, 128
    hybrid, 124
    local, 124, 125
    mathematical tools for, 127
    power consumption, 555
    power management, 536–537, 540–549
    protocols, behavior, 127–128
    routing, 126, 128
    synchronization, 562–563
    throughput, 127
Mobile end transport protocol (METP), 686, 689
Mobile financial applications, 984
Mobile hosts (MHs), 460–461, 469, 579, 582–583, 587,
    588, 589, 590, 591, 592, 594–595, 599, 600, 602,
    604, 650–652, 653–654, 660, 668, 673, 685
    failure, 472

Mobile information systems, 124
Mobile IP, 323
    registration protocol, 639
Mobile networks (MN), 637, 638–639, 640
Mobile nodes (MNs), 496–497, 700, 704, 939, 976
Mobile payments, 983
Mobile and pervasive computing
    power management, 568–569
    power needs, 538
    research trends, 536–537
    three classes, 538
    see also battery technology
Mobile radio communication networks, 460, 567
Mobile RSVP (MRSVP), 638–639
Mobile sensor nodes, 814
Mobile stations (MS), 403, 405, 784–785
Mobile subscriber number (MSISDN), 403, 404
Mobile switching center (MSC), 27, 416, 420, 422, 460,
    478, 481, 650, 784
Mobile system architecture, 536
Mobile transmission control protocol (M-TCP), 686,
    689, 691
Mobile telephony, 287–288
    see also call blocking probability (CBP); calls
        dropping probability (CDP)
    arrival process, 292–293, 306, 310–311, 314
    basic model for performance evaluation and design of
        cells, 288, 289–291, 315
    dimensioning problem, 301–302
    failure probability, 295, 297
    finite population case, 307–310, 314
    mobility model, 291–298, 301, 303, 304–305,
        307–308, 315
    model extensions, 303–310
    model for generally distributed dwell time and call
        duration, 293, 310–313, 314, 315, 316
    performance metrics, 295–298
    quality of service, 288, 291, 298, 301, 315
    services with different bandwidth, 305–307
    steady-state probabilities, 299, 306, 309, 310, 311
Mobile terminals (MTs), 218, 220–225, 227, 229–232,
    234, 297, 365–366, 368, 415, 418–419, 420, 421,
    422, 423, 433, 449–451, 478, 495, 535, 613, 614
Mobile update schemes, 481–482
    distance-based, 481–482
    movement-based, 482, 484, 490–491, 498
    time-based, 482, 484, 490, 498
Mobile versus distributed computing, 124–129
Mobile wireless networks, 440, 808
    video transmission, 790–793
Mobility, 602, 637, 638, 649, 653–654, 684, 705, 719, 809,
    870, 939, 959
Mobility databases, 784
Mobility management (MM), mobility control, 2, 350,
    416, 418–423, 478–479, 501, 502, 503, 505–506,
    513, 515, 784, 785, 870
Mobility track framework, 479, 483
Mobility and traffic control, 668–669
Model-based algorithm design, 887–888
Modeling wireless sensor systems, 888–889
Modified offered load techniques, 314
Modified predictive call ad con (MPCAC)
    algorithm, 429
Modulation scaling, 277

Modulation schemes, 566

Module advertising, 989

Monitoring and control systems, 15, 274, 868, 874, 885–886

Most forward with fixed radius (MFR), 321, 345, 346

Motion coordination, 4, 155, 960–961, 969, 973
protocols programming, 976–980

Motion graph, 137–140, 145, 150–151

Motion pattern, 125

Motion planning, 154, 155

Motorola, 686

Movement across cells, 681

Moving average filter algorithm, 849

Moving pictures expert group (MPEG), 790–791, 792, 798, 805, 808

Multi hop communication, 606

Multi-agent system, 960

Multicast core-extraction distributed ad hoc routing (MCEDAR), 210

Multicast operation of AODV routing (MAODV) protocol, 208
see also multicast routing protocol

Multicast routing protocols, 183, 184, 205–213, 239, 261

Multi-channel hidden terminal problem, 83

Multi-channel medium access control (MMAC), 82–84, 104–106
with cooperative channel selection (MMAC-CC), 111–112, 114
with dynamic channel selection, 107–114, 117–118;
using clearest channel at receiver (MMAC-CR), 110–111, 112, 114
using clearest channel at transmitter (MMAC-CT), 109–110, 112, 114
network model, 113
optimization, 113–114
performance, 112–114
related work on, 114–118
throughput performance, 114

Multi-channel support in ad hoc standards, 106–107

Multi-channel wireless networks, 81–82

MULTI-FIT, 86

Multi-hop 802.11 ad hoc network, 710

Multi-hop communication links, 277, 278, 579, 581, 582, 599

Multi-hop wireless communication networks, 55–56, 240, 343, 565–566, 568, 601–602, 603, 642, 656, 705, 710, 719

Multi-layered multi-agent situated system (MMASS), 966

Multimedia applications, 633, 654, 789

Multimedia connections, 649–650

Multimedia sensors, 848

Multi-hop scenario, 730, 938–939
contention and collision avoidance, 598–600
power management in, 563–565

Multipath fading, 701, 769

Multiple access collision avoidance (MACA), 33, 278, 591–593, 595, 603, 620

Multiple access collision avoidance by invitation (MACA BI), 591

Multiple access collision avoidance or wireless (MACAW), 33, 592–593, 688, 689, 691, 696
see also transmission control protocol (TCP)

Multiple access technologies, 10, 416

Multiple data channels with channel selection, 117–118

multiple description coding (MDC), 790

Multiple service types, 441–442

Multiple traffic classes, admission control, 446

Multiple-input queuing systems, 727

Multiplicative increase and linear decrease (MILD), 292

Multipoint relaying (MPR), 504

Multi-protocol label switching (MPLS), 636–637

Mutual exclusion-based dynamic channel allocation algorithms, 460

N

NAK, 690

Naming and addressing aspects, 786

Narrowband (NB), 446

National Telecommunication Development Program (NTP), 773–774

Near-to-optimal network-interface power consumption, 603

Negative acknowledgements (NAKs), 683

Neighbor-aware contention resolution (NCR), 45–46, 51–54, 56, 70–71
contention resolution, 595
dynamic resource allocation, 52–53
multiplication, 53
pseudo-identities, 52–53
root operation, 53
performance, 53–54
system delay, 53–54

Neighbor connectivity, 828

Neighbor coverage-based decision, 259

Neighbor discovery process, 211, 605–606

Neighbor elimination, 259

Neighbor knowledge information, 241, 248–249, 870

Neighbor selection schemes, 352

Neighborhood announcement messages, 829

Neighborhood congestion notification, 716

Neighborhood RED (NRED), 716, 719

Neighborhood reservation (NR), 643

Neighborhood zone, (NZ), 425–427

Network access identification (NAI), 786

Network access server (NAS), 915

Network address translation (NAT), 785, 924

Network allocation vector (NAVs), 33–34, 113, 603–604, 616

Network-based approaches, 791–792

Network composition, 889

Network connectivity, 6, 243, 277, 343, 504–505, 506, 514, 563–564, 868, 877, 889

Network file system (NFS), 551

Network information flow, 269

Network interface (NI), 589, 600, 601, 602, 603

Network interface cards (NICs), 13, 908, 918

Network layer and transmission control protocol (TCP), interactions, 689

Network management, 16, 419, 579–580, 601

Network nodes, 784

Network organization, 269

Network partitioning, 704, 705

Network protocols, 537, 888

Network resources, 258

Network scan algorithms, 854–856

Network simulations, 747
Network structure, 368–370
  call arrival weight, 369
  reporting ells, 368–369
  vicinity value, 369–370
Network time protocol (NTP), 834, 835
Network topology, 17, 18, 47–48, 49, 50, 62, 70, 99, 113,
    114, 129, 185, 188, 197, 205, 209, 220, 228, 252,
    266, 268, 269, 273, 279, 332, 350, 358–359, 477,
    483–84, 564, 566, 578, 601, 641, 741, 824, 939
  impact of node density, 881
Network utilization, 592, 679
Networked sensor system (NSSs), 265, 266, 267,
    269–272, 274, 277, 278, 280
Network-wide distributed graph formation, 896
Networking subsystem, 541
Neural network-based dynamic channel allocation
    protocols, 464–465
New call admission protocol, 658
New Session message, 209
Newreno, 681, 706, 711
  see also time control protocols (TCP)
Node activation multiple access (NAMA) protocols, 46,
    56–57, 59, 61, 67–68, 69–72; throughput analysis,
    62, 63
Node activation, 2, 46, 48, 56–57, 59, 60, 61, 67–68, 73
Node-centric paradigm, 266
Node characteristics, 890
Node density and distribution, 260, 874, 881, 887, 889
Node deployment strategy, 869
  to Enhance Coverage Capability, 881
Node distribution, 871
Node link, 59
Node location, 350, 519
Node mobility, 183, 190, 195, 202, 213, 244, 260, 350,
    477, 614, 642, 704
Node scheduling algorithms, 876
Node self-scheduling scheme, 832
Node synchronization, 503
Node topology, 887, 888, 890, 891
Nodes transmission range, 501, 502–503, 506, 509
Non acknowledgement (NACK), 794–795, 796, 799,
    801, 830
Nonadjustable power energy model, 239, 247
Non-contigous allocation algorithm for multiple
    priorities (NCAMP), 85, 93–95
Non-contigous allocation algorithm for single priority
    (NCASP), 85, 87–90, 91, 93
Non-contigous round robin allocation (NCRRA), 85,
    89–91
Non-contigous sequential round robin allocation
    (NCSRRA), 85, 91
Nongeographic ad hoc routing protocols, 344, 345
Non-line of sight (NLOS), 422
Non-planar physical topology, 347
Non-preemptive dynamic priority discipline, 446
Non-real-time traffic, 451, 548
Non-repudiation, 943, 944, 950
Non-work conserving scheduling algorithm, 716–716
Normal time applications, 652
Normalized link failure frequency (NLFF), 211
NP-complete, 48, 86, 243, 244, 247, 366, 482, 502, 873
NP-hard, 246, 268, 276
NP-problem, 642

National Telecommunication Development Program
    (NTP), 4, 787
  voice over IP (VoIP) call server, 780–783
  voice over IP (VoIP) test bed, architecture and
      protocol, 774–783, 785
NULL packets, 190, 726, 727, 728, 729–730, 736
NumDpart, 467

O

On demand protocols, 236
On-demand ad hoc routing algorithms, 714
On-demand multicast routing protocol (ODMRP),
    206–207
On-demand scatternet formation, 746
One time password (OTP), 918
Onion routing, 401, 409
Online algorithm, 482, 491
On-the-fly communication links, 941
Open mobile alliance (OMA), 686
Open shortest path first (OSPF), 330–331
Open systems interconnections (OSI), 566, 579, 791,
    906, 922, 933
  layering structure, 579
Open-loop power control (OLPC), 430–431
Operations, administration, and maintenance (OAM)
    system, 774
Optimal channel reservation rule, 314
Optimal topology for proactive scatternets, 745
Optimality criterion, 433
Optimization problems, 482–483, 498, 545, 547
Optimization techniques, 244, 252, 266, 267,
    276–280, 424
Optimized link state routing protocol (OLSR), 17, 504
Orthogonal arrays (OA), 35
Orthogonal channels, 117
Orthogonal frequency division modulation (OFDM),
    10, 12, 78, 106, 792
Orthogonal matrices, 51
Orthogonal transmissions, 100
Out-of-order (OO) detection mechanism, 708–709
Overhearing, 40

P

Pacific digital cellular (PDC), 28
Packet collision, 101, 102, 103, 104–105, 868
Packet control unit (PCU), 784
Packet data protocol (PDP), 202, 784
Packet delay, 633, 634, 648, 649, 789
Packet dropping, 700, 942
Packet forwarding module, 198, 199, 210, 564, 639, 823,
    939, 942
Packet interception, 911
Packet loss, 706, 796
Packet loss rate, 633, 769, 770
Packet reception rate (PRR), 351
Packet replication, 643
Packet reservation multiple access (PRMA), 46
Packet scheduling algorithms, 623
Packet switched network (PSN), 773
Packet switched wireless networks (PSWN), 78

Packet transmission, power management, 561–562
Paging systems, 2, 23, 24, 26–28, 41, 365, 478–479, 480,
        482–483, 496–497, 498, 647–648, 745, 756, 759
    cluster paging, 480
    profile-based, 495
    selective paging, 480
Pairwise communication, 136
Pairwise link activation multiple access (PAMA)
        protocols, 46, 59, 69
    throughput analysis, 62, 66–67
Park mode, 739
Parked member address (PM_ADDR), 739, 759
Parked-unpark procedures, 771
Parked-unparked queue-based policy (PUQP), 759
PARO, 565
Password authentication protocol (PAP), 915, 922
Path asymmetry, effect, 690
Path discovery process, 187, 191, 521
Path loss characteristics, 100, 240, 766
Path maintenance, 187
PATH messages, 638
Path redundancy scheme, 641
Path reply (PREP), 746
Pathrater, 952
PBX, 774, 776
PCM, 755, 760, 765
PCS (1900 MHz), 7
PCS networks, 480, 496, 498
PEAP (Protected EAP), 916, 918
Peer-to-peer (P2P) based communication, 267, 271
Peer-to-peer (P2P) mobile ad hoc network, 964
Peer-to-peer (P2P) network management, 15, 560, 579,
        582, 614, 938
Peleg, D., 171
Perfect uplink PC, 428
Performance comparison, 385–389
    results, 388–389
    stopping conditions, 385–388
Performance enhancing proxies (PEPs), 692
Performance implications of link characteristics
        (PILC), 692
Performance optimization, 886
Per-hop behaviors (PHBs), 635
Perimeter refresh protocols (PRP), 827
Personal area networks (PAN), 19, 559–560, 939
Personal communication service, 483
Personal digital cellular (PDC), 7
Personal location privacy policies, 399–400
Personnel digital assistants (PDAs), 6, 11, 14, 394, 535,
        537, 538, 559, 613, 704, 725, 755, 814, 937, 960,
        961, 962
    power consumption, 540–541
Pervasive computing, 566
    see also Mobile and pervasive computing
    field-based motion coordination, 959–960
    power management, 559–568
    research trends, 536–537
Phase-shift keying (PSK), 106
Pheromone trail, 381
Physical layer (PHY), 792, 800, 806
PI-broadcast algorithm, 256
PIC, 407
Picocells, 307

Piconets, 4, 14, 726, 727–730, 736, 741, 743, 756, 765, 771
    admission control, 731
    temporary, 744
Ping-pong process, 191
Place bar, 407
PlaceLab, 407
Planar connectivity graphs, 347
Planar graphs, 347–348, 354, 356
Planarization, 348, 351, 355
Plausibility functions, 859–860
PL-Threshold, 768
Plugging in, 14
POCA, 56, 59
Point coordination function (PCF), 11, 12, 578,
        582–583, 603, 604, 792, 808
Point coordination function inter-frame space (PIFS),
        582, 615, 617, 621
Point to point modification, 719
Point-to-point link, 756
Point-to-point modifications, 712–718
Point-to-point transmission, 700, 702
Point-to-point tunneling protocol (PPTP), 922, 926,
        928, 932
Poisson distribution, 66, 113, 289, 292, 307, 314, 587
Policy based network management (PBNM), 913
Policy custodian directory, 399
POLL packets, 726, 727, 728, 736
Pollaczek–Kinchin formula, 54
Poll-base data transmission, 808
Poll-Null frame, 729–730
Polling analysis, exhaustive, 229–232
    partially gated, 233–234
Polling-authorization mechanism, 581
Polling process, 726
    traditional schemes, 728
Polling-style scheduling mechanism, 736
Polling systems, 728–729
Poly-logarithmic approximation routing scheme, 274
Polynomial interpolation, 947
Population density and mobile telephony, 307–310, 314
Portable computing and communication devices, 613
Portable electronic devices, 393
Port-based network access control, 916
Positioning-based routing protocols, 320
Positive feedback, 432
Power analysis, 234–236
Power-attenuation model, 240, 253
Power aware collision avoidance and contention
        control, 602
Power aware multiple access (PAMAS) protocols, 40,
        117, 278, 603
Power aware routing, 602
Power aware virtual base station (PA-VBS)
        infrastructure creation protocol, 221
Power breakdown of typical mobile devices, 540–541
Power conservation, 37, 117, 704
Power consumption, 259
Power control, 277, 433, 434, 578, 692
Power control algorithms, 418, 431–433
    for 3G WCDMA system, 431, 432
Power controlled multiple access (PCMA) protocols,
        40, 604
Power-controlled networks, 128

Power management, 3, 39–40, 536–537, 614, 685,
  870, 892
  application-dependent adaptive, 557–558
  application driven, 553–554
  in cellular systems, 430–433
  in infrastructure-based networks, 548–559
  MAC-layer Policies, 549–553, 556, 561
  for mobile devices, 540–549
  in multi-hop ad hoc networks, 563–565
  and pervasive computing, 559–568
  in single-hop ad hoc networks, 561–563
  quality of service (QoS), 554, 557, 559
  in sensor network, 565–568
  strategies, overview, 542–543
  upper-layer policies, 553–559
Power manager (PM), 541–542, 546–548
Power-manageable component (PMC), 541, 548
Power-save distributed contention control (PS-DCC),
  602–603
Power saving mode (PSM) protocols, 549–552, 581,
  600–606
  adaptive, 552
  in IEEE 802.11 ad hoc networks, 562–563
  solutions at MAC layer, 601–602
Power wastage, 561
PPP, 922
Prakash-Shivratri-Singhal's
  DCA algorithm, 468–469
  fault tolerant enhanced channel allocation
    scheme, 472
PRAS, 668, 673
  bandwidth reservation strategy, 669–670
  call admission strategy, 670–671
  simulation results, 671
Predetermined static differentiation, 620–621
Prediction by partial match (PPM), 495
Predictive call admission control (PCAC) algorithm, 428
Predictive distance-based location update scheme, 323
Predictive policies of power management, 544–545
Preemptive ad hoc on-demand distance vector routing
  (PAODV), 185, 191–192
  see also Unicast routing protocol
Preferable channel list (PCL), 83–84, 118
Preliminary active mobile probabilities, 450
Price fields, concept, 960
Prim's algorithm, 245
Privacy awareness system (paws) for ubiquitous
    computing environment, 396–398
Privacy enabled resource manager (PERM), 399
Privacy preferences project (P3P), 396
Private key generator (PKG), 947, 953
Proactive (table-driven) routing protocol, 184, 185, 198
Proactive location database systems, 3, 320, 322–330, 337
Proactive location dissemination systems, 320,
    330–335, 337
Proactive location services, 322
Proactive route selection and maintenance, 714
Proactive routing algorithms, 565
Probabilistic metric approximation, 274
Probability based broadcast protocols, 241
Probability density function (PDF), 64, 310
Probability index scheme, 652
Probing environment and adaptive sleeping (PEAS), 832
Probing packet transmission, 682, 704, 713

Procedure 3 (ModelHandover balance), 298, 300
Programming motion coordination protocols, 976–980
Progressively closer nodes (PCN), 352
Propagation delay, 593
Propagation models, 583
Protection against wrapped sequence space (PAWS), 693
Protocol-adaptation solutions, 578
Protocol independent multicast-dense mode
    (PIM-DM), 207
Protocol independent multicast-SM (PIM-SM), 358
Protocols for group oriented m-commerce transactions,
    989–992
Protocols for quality of service management, 829–833
Prototype sensor nodes, 538
Proxy server, 780
Pruning rules, 504
Pseudo codes, 742
Pseudo link, 219
Pseudonyms, 406
Pseudorandom number generator (PRNG), 906, 907
PSNR, 798
PS-Poll, 550
PTAS, 247
pTCP, 684, 691, 696
PTX, 562
Public key cryptosystem, encryption, 945–946
Public key infrastructure (PKI), 911, 924, 943, 953
Public switched telephone network (PSTN), 16, 416,
    478, 650, 776, 784
  gateway, 774, 776
  Internet, 755, 761, 769, 774
Publish/subscribe mechanisms, 963
Pulsed mode, 540
Pump slowly tech quickly, 830

Q

Quality of service (QoS), 3, 11, 12, 13, 16, 194, 236, 269,
    288, 291, 298, 301, 315, 365, 416, 417, 418,
    423–424, 427, 433–434, 439–441, 446, 447, 448,
    451, 452, 455, 460, 471–472, 498, 541–542, 554,
    557, 559, 578, 580, 600, 604, 614–615, 617,
    628–629, 636, 654–667, 671, 673–674, 730, 731,
    768, 784, 790, 802, 809, 838

  In ad hoc routing protocols, 194
  in cellular networks, 288, 291, 298, 301, 315,
      654–667, 671
  differentiation, 619
  enhancement proposals, 619–627
  leverage of distributed MAC protocols, 617–619, 627
  in medium access control (MAC) protocols for
      wireless networks, 24, 35, 38–39, 41, 614–615,
      617, 628–629
  models in Ad Hoc Networks, 637
  models in IP networks, 634–637
  of multimedia traffic, 3
  in multichannel MAC protocols for mobile ad hoc
      networks, 100, 106
  related issues, 636–637
  provisioning in low-earth-orbit (LEO) satellite
      networks, 667–671
  in pervasive computing, 541–542, 554, 557, 559

Quasi-local broadcasting, 241
routing protocols, 640–643
in sensor networks, 820
signaling, 637–640
standards strategies, 650–654
support in mobile ad hoc networks, 633–634, 636
in wireless cellular networks, 416, 417, 418, 423–424, 427, 433–434, 460, 471–472
in wireless and mobile networks, 236, 439–441, 446, 447, 448, 451, 452, 455, 580, 600, 604, 647, 649–650, 802, 809, 838
in wireless sensor networks, 814
Quad-tree algorithms, 408
Quality of service (QoS) access point (QAP), 628
Query (QRY) packet, 189
Query sequence number (Q_{SEQ}), 949
Quorum-based location services, 195, 324

R

Radiation energy, 70
Radio access network, 802
Radio channels, availability, 289
Radio frequency (RF), 14, 15, 16, 276, 407, 416, 726, 836, 858, 911
Radio frequency identification (RFID), 394, 408–409
Radio link control (RLC), 419, 792, 799, 806
Radio propagation and interference models, 11, 166
Radio resource management (RRM), 3, 418, 433, 434
Radio resources, 314, 418
Radio transmission radius, 270
Radio transmission range, 101
Radio transmission speed, 270
Radio transmission technology (RTT), 10
RADIUS, 908, 911, 913, 914, 915, 916–918, 919, 927, 928–929
RA-LeZi, 497
Random access memory (RAM), 407
Random assessment delay (RAD), 242, 258–259
Random backoffs, 103, 105–106, 113, 235, 759
Random base station selection (RBS), 767
Random early detection (RED), 694, 714, 715–716, 719
Random walks (stochastic), 127, 141–142
Randomized distributed approach, 471
Randomized-destination sequential distance vector (R-DSDV), 185, 191
  see also Unicast routing protocol
Rapid recovery phase, 684
Rate adaptation, 277
Rate-based bandwidth borrowing scheme, see Ek-Kadi's rate-based bandwidth borrowing scheme, (EBS)
Rate-based transmission control, 683
Rate capacity effect, 539
RBOP, 252, 353
RC4 encryption algorithm, 907–908, 915
Reactive (on-demand) routing protocols, 184, 185, 192
Reactive location systems (RLSs), 320, 321–322, 336–337
Reactive policies of power management, 542–547
  timeout-based policies, 544–545
  see also taxonomy
Ready to receive (RTR), 591
Real-time applications, 567
Real-time control protocol (RTCP), 791

Real-time multimedia traffic, 647
Real-time packets, 626
Real-time streaming protocol (RTSP), 791
Real-time traffic, 451
Real-time transparent protocol (RTP), 778, 779, 780, 785, 791
Real-time voice quality, 766
Real-time voice services, 752
Real-time voice transmission, 761
REB, 429
Received power call admission control (RPCAC) algorithm, 427, 429–430
Received signal strength (RSS), 418, 419, 420, 421, 422, 423, 467
Received signal strength indicator, 835
Receiver advertising (RA), 205–206
Receiver-based congestion control, 685
Receiver-based protocols, 692
Receiver initiated busy tone multiple access (RI-BTMA), 116
Receiver initiated collision avoidance (RICH), 116
  with dual polling (RICH-DP), 116
Receiver initiated multiple access (RIMA) protocol, 33, 591
Reception control protocols (RCP), 685
Recursive algorithm, 895
REDUCE, 896
Redundancy, 273, 868, 874, 876, 878–879, 987
REF, 922
Reference-broadcast synchronization (RBS), 835
Reference point centroid scheme, 836
REFMACSIM simulator, 112
REFUSE, 469
Region gap problem, 355
REGISTER message, 780
Registration areas (RAs), 496–497
Registration request (RRQ), 780, 782
REJECT, 470
Relative neighborhood graph (RNG), 197, 243, 252, 256, 348, 504–505, 878
  structures based on, 253–254
Relaxation effect, 538, 540
Relay region and enclosure, 878
Reliability, 124, 838
Reliable link, 588
Remote activated switch (RAS), 604
Remote procedure calls (RPC), 551
Remote sensing policies, 927–928
Rendezvous Regions (RR), 350, 358–360, 361
  bootstrap, 360
  failures, 360
  insertion, 359, 360
  lookup, 359
  region detection, 359
  replication, 359
  server election, 359
Reno TCP congestion control, 680–681, 692, 705, 706, 709, 712, 717
REPLY message, 468, 470, 472, 951
Reporting cell problem, 377, 380–381, 384, 389, 482
Reputation, 954
Request for data (REQ), 685
REQUEST message, 468, 821
Request-response transactions, 551

Request-to-send (RTS) control packet, 33, 40, 69, 79, 81, 82, 83, 103, 109, 110, 112, 113, 116–117, 242, 590–591, 592–596, 600, 603, 604, 605, 615–617, 626, 701–702, 707, 816, 817, 818
   RTS–CTS–DATA exchange, 592
   RTS–CTS–DATA–ACK, 592–593, 599, 616, 622
Reservation-based access schemes, 601
Reservation-based algorithms, 626
Reservation-based distributed quality of service (QoS) MAC protocols, 626
Reservation frame (RF), 627
Reservation slots (RSs), 47
Resource allocation, 2, 6, 429
Resource auction multiple access (RAMA), 46
Resource aware routing, 819–820
Resource discovery, 343, 344, 357
Resource management, 460, 482, 886
Reservation and polling, 229–234
Resource provisioning, 635
Resource reservation protocol (RSVP), 48, 454–455, 580–581, 635–636, 638–639, 640, 644, 654–655
Resource stealing, 911
Resource utilization, 606, 640, 644, 654
Responsiveness, 712
Restricted Delaunay graph (RDG), 348
RESV messages, 638
Re-synchronization, 741
Retransmission approaches and ACKs, 689
Retransmission time-out (RTO), 700, 704, 705, 706, 707, 708, 713
   frozen, 707
Reverse channels, 478–479
Reverse link, *see* uplink
Reverse path, 188
   congestion, 683
Reversed-multicast tree, 267
RFC, 693–694, 915, 923
Risk mitigation and countermeasures, 911
Rivest, Shamir, and Adleman (RSA) system, 946
RoboCup, 967
Robotic devices, 124, 127
   an algorithm for reconfiguration, 127, 128, 154–160
Robust security networks (RSN), 908, 909
Robustness, 149, 203, 210, 343, 346, 359, 360–361, 367, 888, 889, 900
ROCA, 56
ROM, 407
Romer's algorithm, 834–835 (Diacritic)
Room field and crowd field
   load balancing, 972–974
   meeting, 974–975
Round-robin fashion, 760
Round trip time measurements (RTTM), 693
Round trip times (RTTs), 553–554, 550–551, 681, 682–683, 685, 689, 690, 691, 695, 696, 705, 707, 712
Round-trip hop count (RTHC), 706
Route computation, 194, 210, 640, 641
Route discovery process, 186, 187, 191, 199, 236, 345, 348, 402, 641, 705, 714, 746, 949
Route-discovery-packet (RDP), 745, 951
Route-driven gossip (RDG), 212–213
Route error (RERR), 187, 707
Route failure, 708

Route failure notification (RFN), 712–713
Route maintenance, 186, 187, 190, 199, 402
Route pseudonyms, 401–402
Route query, 217, 236
Route reconstruction (RRC) process, 190–191, 792, 800, 806–807
Route reestablishment notification (RRN), 713
Route reply packet (RREP), 186, 188, 190, 208, 642, 717, 745, 746, 949, 950, 951
Route request (RREQ), 186–187, 188, 197, 199, 201, 208, 402, 521, 642, 714, 717, 746, 949, 950, 951
Route selection, hop-by-hop and explicit routing, 636
Routing algorithms, routing issues, 266, 270, 566–567, 704–705
Routing disruption attacks, 941
Routing failures, 704, 705, 714
Routing information protocol (RIP), 331
Routing protocols, 166, 167, 269, 700, 706, 713, 714, 949, 960, 984
Routing tree algorithm 524
Routing tree-based structure, 277
RRTS, 592
Rumor routing, 823
RUNNERS algorithm, 140–141; and SNAKE, experimental comparison, 152–154

S

Sampling mechanism, 498
Satellite networks/systems, 2, 18, 23, 28–30, 41, 654, 986
Scalability, 18, 252, 270, 320, 343, 348, 400, 639, 685, 825, 844, 889–890, 961,962, 986
Scalable broadcast algorithm (SBA), 201–202, 259
Scalable location update-based routing protocol (SLURP), 320, 321
Scale free aggression tree, 273
Scaling laws for traffic, energy and delay, 275, 279–280
Scan, 854–856
Scatternet formation algorithms 736, 738, 740–741, 746, 747, 751–752
Schedule exchange protocols (SEP), 818
Scheduling active and sleep periods, 259–260
Scheduling communication activity, 877
Scheduling policies, 759
Screen blanking, 537
2G cellular mobile networks, 6, 7–8, 10, 27, 417
   FDMA/TDMA, 441
2.5G cellular networks, 6, 8–9, 28
Selective acknowledgements (SACK), 681, 682, 683, 689, 692, 693–694, 696, 706, 710
   see also Transmission control protocol (TCP)
Secret sharing algorithm, 947–948
Secure ad hoc on-demand distance vector routing (SAODV), 950, 951
Secure distributed ad-hoc routing (SDAR), 951
Secure efficient ad hoc distance (SEAD) vector routing protocol, 949, 950
Secure hashing algorithm (SHA 1), 909, 923, 946
Secure remote password, 918
Secure routing protocols (SRP), 948, 949–952
Secure shell (SSH), 914
Secure socket layers (SSL), 913, 914

Security aspect, security issues, 520, 785, 838, 983, 987
    challenges wireless ad ho networks, 942–944
    performance issues, 924–925
    surveillance, 15
    techniques, 944–948
    in wireless local area network (WLAN), 905
Security associations (SAs), 923
Select_then_Eliminate (STE), 329–330
Selecting forwarding neighbors, 257–258
Self-clocked fair queuing (SCFQ), 619, 624, 625
Self-organizing medium access control for sensor
        networks (SMACS), 521
Self-positioning algorithm (SPA), 531
Self tuning power management, 554–556
Semantic routing trees (SRTs), 827
SEND, 896
Sender advertising (SA), 205–206
Sensing and communication scheduling, 868
Sensing scheduling problems, 872–874
    distributed algorithms 874–876
Sensing_inactive nodes and communication_inactive
        nodes, 879
Sensor collaboration region, 874
Sensor deployment, 838
Sensor field intensity, 873
Sensor localization, 835–836
Sensor-medium access control (S-MAC), 37, 278 ,
        816–817, 818
Sensor mobility, 814
Sensor networks, systems, 1, 2, 4, 6, 17–18, 23, 24, 37–38,
        41, 519–520, 527, 861, 888–891, 893, 899,
        940, 980
    clustered heterogeneous model, 891–892
    clustering, in, 527–530
    data-centric storage, 358
    destination location, 349
    fairness, 39, 41
    geographic protocols, 343, 344, 348, 350, 360
    power management, 39–40, 565–68
    quality of service (QoS), 38–39
    self-organizing approach, 521
    smart antennas, 40–41
    traffic patterns, 814–815
Sensor nodes, 266, 268, 270, 272–278, 280, 519, 538,
        540–541, 565–566, 567–568, 813, 815, 830, 844,
        869, 871, 874, 879, 885–886, 875, 879, 886,
        891–892, 893, 894–896, 899
Sensor protocol for information via negotiation (SPIN),
        820–822
Sensor resources, 814
Sequence ID (SID), 209, 626
Sequence number of destination (dst_seq_no), 188
Service availability protection, 948, 953
Service location, 343, 344
Service options, 419
Service provider (SP), 541–543, 545–546
    reactive policies, 542–547
Service requester (SR), 541–542, 543, 545–546, 547
Service set identifier (SSID), 786, 911
Serving GPRS support node (SGSN), 783–784
Session description protocol (SDP), 778

Session initiation protocol (SIP), 773–774, 776, 777–779,
        787, 791
    ALG, 785
    basic call flow, 779–780
    proxy server, 785; uniform resource identifier (URI),
        779, 781
    user agent (UA), 774, 776, 779, 781, 782–783,
        784–785, 787
Session management transfers, 779
Session-state information, 498
Set cover problem, 245
Settling time table, 185–186
78 LZ compression algorithm, 498
Shadow cluster concept, 449–451, 651–652
Shadow fading, 421
Shallow light tree (STL), 272–273
Shared data spaces, 962
Shared radio spectrum, 277
Shared wireless access protocol (SWAP), 16
Shiva password authentication protocol (SPAP), 915
Short inter-frame space (SIFS), 103, 113, 582, 595, 615
Short message service (SMS), 7
Short range propagation, 768
Shortest path algorithm, 335, 741
Shortest path tree (SPT), 243, 244–245, 266, 272,
        273–274, 275
Short-hop communication (Spectrumbs), 468
Short-range wireless communication, 13–16, 19
Signal attenuation, 701
Signal fading, 681
Signal power measurements, 117
Signal processing algorithms, 269, 270, 278, 886
Signal propagation, 70, 100, 430, 701
Signal stability-based adaptive routing (SSA), 506, 705
Signal strength hole, 419, 421
Signal strength measurements, 714
Signal to interference ratio (SIR), 419, 420, 421–422,
        424, 427, 428, 431, 432, 433
Signal-to-interference-plus-noise ratio (SINR), 101, 102,
        106, 110, 601
Signal-to-noise quality, 277
Signal-to-noise ratio (SNR), 203, 601, 637, 701, 790
Simple location service (SLS), 331, 337
Simple network management protocols (SNMP), 910,
        912–913
Sims, 966
Simulated annealing (SA)-based dynamic channel
        allocation protocols, 463–464
Simulation models, 569, 655–656
Simulation results, 112, 514–515
Simulations, 69–72, 137, 149, 245, 260, 355, 356, 454,
        596, 718
Single celled organisms, 370
Single-hop ad hoc networks
    communication links, 579, 581
    distributed fair scheduling (DFS), 623–625
    hotspot networks, 535
    power management, 561–563
    wireless communication, 537, 565, 568
Single-radio Bluetooth, 755
Single layer approach, 700
Single spanning tree, 741
Single-value approximation, 448
Single wireless communication channel, 585

Sink nodes, 869
Slave to slave traffic, 730
Slave/slave bridge (S/S bridge), 736
Slaves, dynamic recording, 728–729
Slave-slave communication, 756
Sleep scheduling, 266, 270, 276, 277–278, 279
Sleep synchronization, 604
Sleep time management, 601
Sleeping nodes, 874, 876
Slepian-wolf coding, 269, 275–276
Slot utilization 598
Slotted access schemes, 591
Slotted-Aloha MAC protocol, 587
Slotted CSMA MAC protocol, 588, 589
Slow congestion avoidance (SCA), 706, 707, 711–712
Small-to-medium enterprise (SME), 922
SMART, 689, 690
    retransmissions, 682
Smart neighbor discovery, 513–514
SmartNode, 604
SNAKE algorithm, 139–140
    analysis, 141–148
    validation of performance, 150–152
    and RUNNER, experimental comparison, 152–154
Sniff mode, 739
Snoop protocols, 687, 688, 689, 690, 691
Soft channel reservation, 81, 109, 117
Soft handoff (SHO), 419
Soft handoff zone (SHZ), 425–427
Soft-state mode, 638
Software architecture, systems, 540, 548, 960, 961, 976
Software upgrades and patches, 914
Solar cells, 538
Source address (src_addr), 187, 188
Source destination, 191
Source initiated rerouting (SIRR), 643
Source sequence number (src_seq_no), 187
Source to destination (StoD_stream), 599, 641
Space division multiple access (SDMA), 409
Spaghetti-design principle, 579
SPAN, 260, 278, 504, 564, 566–567, 602, 828, 877
Spanning tree, 247, 248
Sparse topology and energy management (STEM), 278, 567, 818–819
Spatial contention and reuse, 702–703, 712
Spatial scalability, 790
SPEED, 349
Split-channel reservation multiple access (SRMA) protocol, 116, 590
Split connection approaches, 691
Split connection mechanisms, schemes, 680, 685–686, 690, 692, 695
Split TCP, 717–718
Spotbeams, 653–654
Spread spectrum cellular technology, 7
Spread spectrum multiple access, 25
Spread spectrum transmission, 417
Stable dynamic call admission control (SDCA), 448–449
Standard deviation rate, 454
STANDBY BBS, 762
Standby routing, 642, 707
Star topology, 15
Static differentiation, 614
STATIC OPT algorithm, 768, 769–770

Status-based wait (SBW), 991
Steady-state location probability, 480
Steiner systems, 35
Steiner tree algorithm, 247
Stochastic policies of power management, 545–547
Stochastic variability, 597
Stochastically largest queue (SLQ), 728
Streaming, 441
Stream-oriented fairness, 592
Strength and SIR combined power controlled algorithm, 432
String topology, 716
Subscriber identity module (SIM) smart card, 7
Support management subprotocol, 142, 149
support message authentication, 909
Switching delay, 741
Symmetric connection management, 738
SYNC, 816
Synchronization, 139, 140–141, 470, 592, 604, 614, 874
Synchronization protocol, 503, 510–511
Synchronized sleep periods, 600
Synchronous connection-oriented (SCO), 752, 755, 756–758, 763, 765, 766–767, 770, 771
System integration, 579
System utilization, 441, 455

T

Tabu search (TS), 367–368, 375–377, 384, 386, 389
    framework, 377–379
    reporting cell problem, 377
    search diversification, 376–379
    search intensification, 376–377
    tabu list, 376–377, 378, 384, 389
Tahoe, 705, 706
    see also Transmission control protocols (TCP)
Talk-to-live (TTL), 336
Target user, 781
Task management planes, 870
Taxonomy and reactive policies, 543
TDM schemes, 50
TDP, 202
Telecommunication Industry Association, 27
Telematic applications, 408
Telephony, Telephone network, 638, 647–748, 773
TELSA, 949, 950
Temporal key integrity protocols (TKIP), 12, 907–908, 909
Temporally ordered routing algorithm (TORA), 130, 131–132, 185, 188–190, 200, 353
    see also Unicast routing protocol
Temporary mobile subscriber identity (TMSI), 403, 786
Terminal access controller access control system (TACACS), 914
Terminode remote routing (TRR), 320
Terminode routing, 348, 351
Text compression schemes, 495
3G cellular mobile network, 3, 6, 8, 9–10, 28, 41, 305, 307, 417, 480, 789, 791–792, 986
    3G Partnership Project (3GPP), 28, 785, 786, 791 code division multiple access (CDMA), 441
        Universal mobile telecommunications system (UMTS), 799

Wide code division multiple access (WCDMA), 432,
    791–792
3G Gateway GPRS support node (3G-GGSN), 791
3G Serving GPRS support node (3G-SGSN), 791
3D Space, 880
    activity scheduling schemes, 880
Threshold flag, 221–225, 227, 228
Threshold secret sharing scheme, 947–948
Ticket-based routing, 641–642
    see also
Time and energy costs, 889
Time of arrival (ToA), 835, 861
Time based location update scheme, 323
Time difference of arrival (TDoA), 835–836, 861
Time division (TD), 461
Time-division channel access, 73
Time division duplexing (TDD), 24, 25, 26, 85, 586, 726,
    735, 736, 756
Time division multiple access (TDMA) protocols, 2,
    6–7, 8, 9, 10, 13, 16, 25, 29, 35, 36, 39, 41, 45, 46,
    47, 51, 73, 80, 106, 277–278, 409, 416, 417, 441,
    461, 521, 825
    see also Code division multiple access (CDMA)
    dynamic (D-TDMA), 46
    dynamic reservations, 47–48
    power control, 430
    tree-walking singulation protocol
Time-frequency plane, 416–417
Time-indexed semi-Markov decision processes
    (TISMDP), 547
Time-out-MAC (T-MAC), 817–818
Time slots, 49, 51, 55, 60, 70, 73
Time spread multiple access (TSMA), 47, 50
Time synchronization, 55, 642, 834–835
    and localization protocols, 833–837
Timing synchronization function (TSF), 603, 604–605
Tiny aggregation (TAG) service, 827
TinyDB/acquisitional query processing (ACQOP),
    827–828
T-Medium access control (T-MAC), 278
Topographic querying, 892–899
    algorithm design and analysis, 894–898
    preliminaries, 893–894
Topology, 277, 346, 636, 700, 703, 711, 726, 736, 742,
    744, 752
    connectivity, 136, 244
    change, 49, 129, 130, 136, 187, 189, 191, 207, 209, 242,
        344, 358, 521, 638, 712, 713, 746, 939
    control, 3, 270, 501, 502–03, 504–05, 506, 509, 510,
        512, 514, 516, 827–828
    information, 45, 641
    management, 566–567
        algorithmic aspect, 564
Topology-aware route-driven gossip (TA-RDG), 213
Topology dissemination based on reverse-path
        forwarding (TBRPF), 17
Topology-transparent protocols, 35, 38, 39, 46, 47
    scheduling, 50–51
Traffic-adaptive medium access (TRAMA), 818
Traffic adaptive resource allocation, 73
Traffic analysis, characteristics, 626, 648, 668–669, 942
Traffic demand matrix, 86
Traffic handling techniques, 637
Traffic indication map (TIM), 550, 604

Traffic routing, 99
Traffic scheduling algorithm, 85
Trajectory based forwarding (TBF), 824
TRANSFER, 468, 472
Transistor technology, 539
Transmission (bit), rate, 418
Transmission area, 584
Transmission bandwidth, 802
Transmission control protocols (TCP), 73, 545, 550–551,
    556, 567,599, 694, 699, 705–706, 707, 708, 712,
    713, 714, 730, 748, 752, 760, 764–765, 785, 830,
    915, 922, 924
    ACKS, 551, 680, 682, 683, 684, 685, 686, 689, 712
    aware, 689, 691, 710
    with Buffering capability and sequence information
        (TCP-BuS), 712–713
    congestion, 680–681, 683–685, 688, 689, 692–694,
        700, 708
    detection of out-of-order and response (DOOR), 707,
        708–709
    with delayed congestion response (TCP-DCR), 688
    fairness, 712, 716
    Feedback (TCP-F), 712–713
    header checksum option (HACK), 707, 710
    IP protocols, 3–4, 8, 550, 686, 730
    and link layer interaction, 689
    interaction with MAC layer (MACAW), 688
    in Manet, 699–720
    and the network layer, interaction, 689
    packet sequence number (TPSN), 709
    Peach, 684, 690, 691, 696
    probing, 689, 691
    Real, 685, 692
    Reno, 680–681, 692, 705, 706, 709, 712, 717
    Santa Cruz, 683, 689, 690
    over satellites, 693–694
    segment size (SS), 683
    SYN, 551
    Tahoe, 691
    throughput, 705–706, 712, 716
    Vegas, 681, 683, 706, 709
    Veno, 709
    westwood, 683–684, 689, 690, 691, 692
    window scaling option, 693
    in wireless networks, 679–696
Transmission power, 564, 601
Transmission range (R_{TX}), 240, 701, 702
Transmission scheduling, 278
Transmitter-oriented channel access (TOCA), 56, 59, 60
Transmitter power management, 460
Transport format (TF), 807
Transport format combination (TFC), 807
Transport layer security (TLS), 913–916, 918–920,
    926–928, 932
Transport layer, 830
Transport mode, 923
Transport unaware link improvement protocol
        (TULIP), 688, 689, 690, 691, 695
Trapdoors, 402
Tree-based collision resolution mechanisms, 594–595
Tree initialization, 209
Tree scatternet formation (TSF) algorithm, 742–743
Tree-walking singulation protocol, 409
Triangulation, 408

Trigger-based distributed routing (TDR), 642–643
Trilaterationmethod, 835
Trojans, 914
Trust and key management, 948, 952–953
    fully distributed certificate authority, 953
    partially distributed certificate authority, 952
TRYING response, 779
Tunable compression, routing with, 274
Tunable transmitter–fixed receiver (TT–FR), 78
Tunable transmitter–tunable receiver (TT–TR), 78–79
Tunnel mode, 923
Tunneled TLS (TTLS), 916, 918
Tunneling technologies, 926–928, 932
    attack, 942
    comparison, 924
Tuples on the air (TOTA), 966, 976–980
Two-phase scatternet formation (TPSF) algorithm and
        its variant TPSF+, 746
Two-phase scatternet formation +(TPSF+), 748
    and BTCP, comparison between, 748
    path request (PREQ), 746
Two-tier data dissemination (TDD), 349
Type of service (TOS) field, 635

U

Ubiquitous networks, 1
Ultra wideband (UWB), 15, 78
Unicast routing protocols, 183, 184, 184–201, 205, 213,
        239, 240, 320, 336, 344, 353, 354
    ID-based flat, 184, 185–192
    location-based, 184, 185
Unicast state (Ut and DT), 60, 63–65, 70
Unicast versus multicast m-commerce transactions, 989
Unified T/O/C division multiple access (UxDMA), 46,
        47, 49–50, 69–72
Uniform quorum system (UQS), 324–325, 337
Unit disk graph (UDG), 166, 167, 239, 240, 247, 248,
        249, 250, 254, 256, 258, 347, 502, 504
United States Digital Cellular (USDC), 27
    Dual mode advanced mobile phone service
        (D-AMPS), 7, 27
Universal IS card (UICC), 786
Universal mobile telecommunications system (UMTS),
        8–9, 10, 28, 399, 442, 535, 791–792, 799–800, 806
    see also Universal terrestrial radio access networks
        (UTRAN)
    IMT-2000, 417
    traffic asymmetry, 442
    traffic classes, 441
Universal terrestrial radio access networks (UTRAN),
        791–792, 800, 807
Unlimited wait (UW), 991
Unparked queue-based policy (UQP), 759
UPD, 189
Uplink (reverse link), 86, 416, 429–430, 434, 442,
        452–453, 726, 727, 728, 730
    non-distributed, 427
    power control for, 431–433
Upper-layer policies of power management, 553–559
Usability issues, 983
User capability, 779
Users classes, 304–305, 316

User datagram protocols (UDP), 748, 791, 793, 915, 922;
        traffic, 752
User identification, 927
User location, 778
User mobility, 310, 315, 316, 418, 451, 454, 455, 477, 478,
        479, 483–488, 536, 579, 643, 989, 990
    and call duration, 304
    finite-context model, 486–488
    complexity, 483
    optimality, 483, 491
    movement history, 484–485
    mobility model, 485–487
User personalization, 984
User privacy policies, 399

V

Vanadium and molybdenum oxide batteries, 538
Variable bit rate (VBR), 626
Variable-share update algorithm, 558
VHS, 790
Video conferencing, 635, 647–748
Video on demand (VoD), 12
Video streaming, 647, 652, 789
Video surveillance, 274
Video telephony, 789
Video transmission over mobile wireless networks,
        790–793, 808
View consistency, 509–514
    conservative local view, 511–514
    consistent local view, 510–511
VirtG graph, 249, 251
VirtG minimum spanning tree (VMST), 251–252
Virtual backbone (VB), 324–328
Virtual base station (VBS), 219, 221, 228–229, 232, 235
    disadvantages, 221
    infrastructure creation protocol, 220–221
    routing in, 220–221
Virtual base station on-demand (VBS-O) routing
        protocol, 227–229, 234
Virtual bottleneck cell, 652
Virtual carrier sensing (VCS), 33, 594, 600, 603
Virtual grid, 278
Virtual home region (VHR), 350
Virtual private networks (VPNs), 635, 906, 908, 911,
        913, 920–922, 924, 926, 927, 934
    Internet protocols, security (IPSec), 922, 926, 928, 929
    model implementation, 928–929
    Point-to-point tunneling protocol (PPTP), 926,
        928, 932
    test results, 932
Visitor location register (VLR), 784
VOCAL, 767, 770, 771
Voice, 77, 439, 441, 649
Voice activity detection (VAD), 761
VOICE-base station subsystems, 766
Voice calls, 306
Voice connection delay, 765
Voice over ACL links (VoACL), 764–765, 766
Voice over IP (VoIP), 4, 12, 19, 773
Voice over Synchronous connection-oriented (VoSCO)
        links, 759–764
Voice service to multimedia service, 441, 442

Voice transmission, 10
Voice over IP (VoIP) Mobile switching center (VMSC), 784–785
Voice over IP (VoIP) test bed, 787
Voltage scaling, 270
Voronoi diagram, 873
Voronoi partitions, 353

W

WAKE_UP message, 556
Wald's equation, 805
warchalking, 911
Warning energy aware clusterhead (WEAC) infrastructure creation protocol, 222–227, 228
WarningThreshold flag, 223
warspammers, 911
Watchdog, 952
WatchPad, 496
Waveform (channel) selection, 94, 460
web-based interfaces, 913
Web browsing, 442, 553, 556, 557, 647–648
Web server, 558
Weighted fair queuing (WFQ), 624
Weighted sum scheme, 652
Weiner process, 454
Wide area networks (WAN), 19, 693, 986
Wideband (WB), 446
Wideband-code division multiple access (W-CDMA), 8, 9, 10, 28, 417, 427, 431
  power control, 432
Wi-Fi networks, 6, 12, 13, 553, 838, 960
  access points, 535–536, 537
Wi-fi protected area (WPA) see Temporal key integrity protocol (TKIP)
Window-based systems, 546, 692
Window loss rate (WLR), 768–769
Window synchronization, 605
Windows XP (Microsoft 2002), 919
Wired communication infrastructure, networks, 102, 165, 417, 425, 440, 478, 600, 623, 679–680, 700, 701, 704
Wired equivalent privacy (WEP), 786, 906–907, 913, 917–918, 926, 927, 932, 934
  authentication and encryption, 908, 912, 931, 943
  improvements, 908
Wired local area networks, 755
Wired networks, 633–634, 647, 648, 686, 691, 705, 712, 719, 808, 937
Wired telephone systems, 298
Wireless access protocols (WAPs), 407
Wireless ad hoc networks, 1, 118, 172, 239, 240–241, 243, 247, 259, 260, 360, 589, 870–871, 937
  security issue/challenges, 938–944
    current countermeasures, 948–954
  security technologies, 944–954
  vulnerability, 941–942
Wireless Application Environment (WAE), 686
Wireless application protocols (WAP), 686–687, 693
Wireless cellular networks (WCNs), 3, 415–418, 425
Wireless channel characteristics, 701–704, 720
Wireless communication capabilities, 867, 869

Wireless communication networks, 3, 4, 5–6, 16, 37, 49, 78, 129–130, 166, 171, 181, 252, 261, 269, 277, 319, 402, 439–441, 447, 477, 498, 535, 537, 600, 606, 633–634, 647, 679, 700, 793, 890, 960
  see also Transmission control protocol (TCP)
  connectivity, 347, 578
  geographic routing protocols, 343, 344, 348
  medium access problem, 100–106
  packet drop, 700
  power consumption, 540
  quality of service (QoS), 634, 637, 647, 648, 649
  resource allocation, 455
  resource management, 460
Wireless data protocol (WDP), 687
Wireless demilitarized zone (WDMZ), 920
Wireless devices, 39, 41, 165
  physical security, 910
Wireless Ethernet, see Carrier sense multiple access with collision avoidance (CSMA/CA)
Wireless infrastructure, 986–988
Wireless interface, 551, 554, 556, 558
  power consumption, 551
Wireless link, 685, 686, 691, 695, 885
  effect on round trip times (RTT), 690
Wireless local area networks (WLANs), 4, 6, 11–13, 19, 23, 77, 395, 537, 543, 547, 548, 559, 561, 578, 579–581, 613, 614, 617, 625, 679, 688, 756, 774, 785, 787, 789, 791, 792, 806, 986
  see also Transmission control protocol (TCP)
  based GPRS support node (WGSN), 785–786, 787
  energy consumption, 562
  IEEE 802.11 standard, 581–582, 756
  logical (bi-directional) links, 586
  MAC level perspective, 580–581
  power saving solutions, 600, 601–602, 603, 606
  quality of service (QoS), 498, 614–615, 617, 621, 623, 627, 628–629
  security aspect, 905, 907, 910, 914, 916, 917, 919, 920–921, 924, 932, 933, 938–939
    active attacks, 910, 941
    passive attacks, 909, 941
    threats and risks, 909–911, 941
  two types of architecture, 938–939
Wireless local loop (WLL), 287
Wireless media characteristics, 681
Wireless Metropolitan Area Networks (WiMAX), 11, 16–17
Wireless Multicast Tree with Hitch-Hiking (WMH), 201, 203–205
Wireless Multimedia Services (WMSs), 3, 449
Wireless nodes, 614
  transmission range, 240, 701
Wireless personal area networks (WPANs), 4, 13, 14, 77, 725, 735–736
  see also Bluetooth
Wireless protocols design, role, 578–581
  quality of service (QoS), 578
wireless robust authentication protocol, 909
Wireless sensor networks (WSNs), 17–18, 519–520, 523, 524–525, 527, 565–566, 567, 813, 843–844, 890–891
  fault-tolerance, 871
  integrated activity scheduling, 880
  modeling, 888–889

node activity scheduling algorithms, 867–881
performance metrics, 815–816
quality of service, 814
network protocols, 819–830
quality of service (QoS), 816, 820
taxonomy, 814–815
unique features, 815
Wireless session control (WSP), 686
Wireless signals, 583–584
Wireless transmission control protocol (TCP)
        mechanisms, 681–688, 718, 719
comparison, 688–692
suitability to different network environments, 691
wireless technology, 889
        Transmission control protocol (TCP), 679–696
Wireless transaction protocol (WTP), 687
management entity, 687
Wireless transport layer security (WTLS), 687
Wireless VoIP service, 783–787
Wireless wide area networks (WWANs), 679, 689, 691
Witness, 347
WML, 686
Work-conserving scheduling schemes, 717
Workload dependent, 545

Workload modifications, 547–548
mixed policies, 547–548
proactive policies, 542–543, 547
Worst case approximation ratio, 249
WTCP99, 689, 691–692, 696

X

XML, 396

Y

Yao graph, 252
Yao protocol, 747–748

Z

Zebra net project, 814
Zigbee and IEEE 802.15, 14, 15, 19
Zone routing protocols (ZRP), 192–193, 200, 201,
        241, 319
        see also Unicast routing protocol
Zone_ mobile terminals (MTs), 220, 221, 222, 227, 229
Zone-based hierarchical link state (ZHLS), 194, 199